SEVENTH EDITION

SPECTROMETRIC IDENTIFICATION OF ORGANIC COMPOUNDS

ROBERT M. SILVERSTEIN

FRANCIS X. WEBSTER

DAVID J. KIEMLE

State University of New York
College of Environmental Science & Forestry

JOHN WILEY & SONS, INC.

Acquisitions Editor *Debbie Brennan*
Project Editor *Jennifer Yee*
Production Manager *Pamela Kennedy*
Production Editor *Sarah Wolfman-Robichaud*
Marketing Manager *Amanda Wygal*
Senior Designer *Madelyn Lesure*
Senior Illustration Editor *Sandra Rigby*
Project Management Services *Penny Warner/Progressive Information Technologies*

This book was set in 10/12 Times Ten by Progressive Information Technologies and printed and bound by Courier Westford. The cover was printed by Lehigh Press.

This book is printed on acid free paper. ∞

To order books or for customer service please, call 1-800-CALL WILEY (225-5945).

ISBN 0-471-39362-2
WIE ISBN 0-471-42913-9

Printed in the United States of America

10 9 8 7 6 5 4 3 2 1

PREFACE

The first edition of this problem-solving textbook was published in 1963 to teach organic chemists how to identify organic compounds from the synergistic information afforded by the combination of mass (MS), infrared (IR), nuclear magnetic resonance (MNR), and ultraviolet (UV) spectra. Essentially, the molecule is perturbed by these energy probes, and the responses are recorded as spectra. UV has other uses, but is now rarely used for the identification of organic compounds. Because of its limitations, we discarded UV in the sixth edition with our explanation.

The remarkable development of NMR now demands four chapters. Identification of difficult compounds now depends heavily on 2-D NMR spectra, as demonstrated in Chapters 5, 6, 7, and 8.

Maintaining a balance between theory and practice is difficult. We have avoided the arcane areas of electrons and quantum mechanics, but the alternative black-box approach is not acceptable. We avoided these extremes with a pictorial, non-mathematical approach presented in some detail. Diagrams abound and excellent spectra are presented at every opportunity since interpretations remain the goal.

Even this modest level of expertise will permit solution of a gratifying number of identification problems. Of course, in practice other information is usually available: the sample source, details of isolation, a synthesis sequence, or information on analogous material. Often, complex molecules can be identified because partial structures are known, and specific questions can be formulated; the process is more confirmation than identification. In practice, however, difficulties arise in physical handling of minute amounts of compound: trapping, elution from adsorbents, solvent removal, prevention of contamination, and decomposition of unstable compounds. Water, air, stopcock greases, solvent impurities, and plasticizers have frustrated many investigations. For pedagogical reasons, we deal only with pure organic compounds. "Pure" in this context is a relative term, and all we can say is the purer, the better. In many cases, identification can be made on a fraction of a milligram, or even on several micrograms of sample. Identification on the milligram scale is routine. Of course, not all molecules yield so easily. Chemical manipulations may be necessary, but the information obtained from the spectra will permit intelligent selection of chemical treatments.

To make all this happen, the book presents relevant material. Charts and tables throughout the text are extensive and are designed for convenient access. There are numerous sets of Student Exercises at the ends of the chapters. Chapter 7 consists of six compounds with relevant spectra, which are discussed in appropriate detail. Chapter 8 consists of Student Exercises that are presented (more or less) in order of increasing difficulty.

The authors welcome this opportunity to include new material, discard the old, and improve the presentation. Major changes in each chapter are summarized below.

Mass Spectrometry (Chapter 1)

The strength of this chapter has been its coverage of fragmentation in EI spectra and remains so as a central theme. The coverage of instrumentation has been rewritten and greatly expanded, focusing on methods of ionization and of ion separation. All of the spectra in the chapter have been redone; there are also spectra of new compounds. Fragmentation patterns (structures) have been redone and corrected. Discussion of EI fragmentation has been partially rewritten. Student Exercises at the end of the chapter are new and greatly expanded.

The Table of Formula Masses (four decimal places) is convenient for selecting tentative, molecular formulas, and fragments on the basis of unit-mass peaks. Note that in the first paragraph of the Introduction to Chapter 7, there is the statement: *"Go for the molecular formula."*

Infrared Spectrometry (Chapter 2)

It is still necessary that an organic chemist understands a reasonable amount of theory and instrumentation in IR spectrometry. We believe that our coverage of "characteristic group absorptions" is useful, together with group-absorption charts, characteristic spectra, references, and Student Exercises. This chapter remains essentially the same except the Student Exercises at the end of the chapter. Most of the spectra have been redone.

Proton NMR Spectrometry (Chapter 3)

In this chapter, we lay the background for nuclear magnetic resonance in general and proceed to develop proton NMR. The objective is the interpretation of proton spectra. From the beginning, the basics of NMR spectrometry evolved with the proton, which still accounts for most of the NMR produced.

Rather than describe the 17 Sections in this chapter, we simply state that the chapter has been greatly expanded and thoroughly revised. More emphasis is placed on FT NMR, especially some of its theory. Most of the figures have been updated, and there are many new figures including many 600 MHz spectra. The number of Student Exercises has been increased to cover the material discussed. The frequent expansion of proton multiplets will be noted as students master the concept of "first-order multiplets." This important concept is discussed in detail.

One further observation concerns the separation of ^{1}H and ^{13}C spectrometry into Chapters 3 and 4. We are convinced that this approach, as developed in earlier editions, is sound, and we proceed to Chapter 4.

Carbon-13 NMR Spectrometry (Chapter 4)

This chapter has also been thoroughly revised. All of the Figures are new and were obtained either at 75.5 MHz (equivalent to 300 MHz for protons) or 150.9 MHz (equivalent to 600 MHz for protons). Many of the tables of ^{13}C chemical shifts have been expanded.

Much emphasis is placed on the DEPT spectrum. In fact, it is used in all of the Student Exercises in place of the obsolete decoupled ^{13}C spectrum. The DEPT spectrum provides the distribution of carbon atoms with the number of hydrogen atoms attached to each carbon.

Correlation NMR Spectrometry; 2-D NMR (Chapter 5)

Chapter 5 still covers 2-D correlation but has been reorganized, expanded, and updated, which reflects the ever increasing importance of 2-D NMR. The reorganization places all of the spectra together for a given compound and treats each example separately: ipsenol, caryophyllene oxide, lactose, and a tetrapeptide. Pulse sequences for most of the experiments are given. The expanded treatment also includes many new 2-D experiments such as ROESY and hybrid experiments such as HMQC-TOCSY. There are many new Student Exercises.

NMR Spectrometry of Other Important Nuclei Spin 1/2 Nuclei (Chapter 6)

Chapter 6 has been expanded with more examples, comprehensive tables, and improved presentation of spectra. The treatment is intended to emphasize chemical correlations and include several 2-D spectra. The nuclei presented are:

^{15}N, ^{19}F, ^{29}Si, and ^{31}P

Solved Problems (Chapter 7)

Chapter 7 consists of an introduction followed by six solved "Exercises." Our suggested approaches have been expanded and should be helpful to students. We have refrained from being overly prescriptive. Students are urged to develop their own approaches, but our suggestions are offered and caveats posted. The six exercises are arranged in increasing order of difficulty. Two Student Exercises have been added to this chapter, structures are provided, and the student is asked to make assignments and verify the structures. Additional Student Exercises of this type are added to the end of Chapter 8.

Assigned Problems (Chapter 8)

Chapter 8 has been completely redone. The spectra are categorized by structural difficulty, and 2-D spectra are emphasized. For some of the more difficult examples, the structure is given and the student is asked to verify the structure and to make all assignments in the spectra.

Answers to Student Exercises are available in PDF format to teachers and other professionals, who can receive the answers from the publisher by letterhead request. Additional Student Exercises can be found at http://www.wiley.com/college/silverstein.

Final Thoughts

Most spectrometric techniques are now routinely accessible to organic chemists in walk-up laboratories. The generation of high quality NMR, IR, and MS data is no longer the rate-limiting step in identifying a chemical structure. Rather, the analysis of the data has become the primary hurdle for the chemist as it has been for the skilled spectroscopist for many years. Software tools are now available for the estimation and prediction of NMR, MS, and IR spectra based on a structural input and the dream solution of automated structural elucidation based on spectral input is also becoming increasingly available. Such tools offer both the skilled and non-skilled experimentalist much-needed assistance in interpreting the data. There are a number of tools available today for predicting spectra, (see http://www.acdlabs.com for more explicit details), which differ in both complexity and capability.

In summary, this textbook is designed for upper-division undergraduates and for graduate students. It will also serve practicing organic chemists. As we have reiterated throughout the text, the goal is to interpret spectra by utilizing the synergistic information. Thus, we have made every effort to present the requisite spectra in the most "legible" form. This is especially true of the NMR spectra. Students soon realize the value of first-order multiplets produced by the 300 and 600 MHz spectrometers, and they will appreciate the numerous expanded insets. As will the instructors.

ACKNOWLEDGMENTS

We thank Anthony Williams, Vice President and Chief Science Officer of Advanced Chemistry Development (ACD), for donating software for IR/MS processing, which was used in four of the eight chapters; it allowed us to present the data easily and in high quality. We also thank Paul Cope from Bruker BioSpin Corporation for donating NMR processing software. Without these software packages, the presentation of this book would not have been possible.

We thank Jennifer Yee, Sarah Wolfman-Robichaud, and other staff of John Wiley and Sons for being highly cooperative in transforming the various parts of a complex manuscript into a handsome Seventh Edition.

The following reviewers offered encouragement and many useful suggestions. We thank them for the considerable time expended: John Montgomery, Wayne State University; Cynthia McGowan, Merrimack College; William Feld, Wright State University; James S. Nowick, University of California, Irvine; and Mary Chisholm, Penn State Erie, Behrend College.

Finally, we acknowledge Dr. Arthur Stipanovic Director of Analytical and Technical services for allowing us the use of the Analytical facilities at SUNY ESF, Syracuse.

Our wives (Olive, Kathryn, and Sandra) offered constant patience and support. There is no adequate way to express our appreciation.

Robert M. Silverstein
Francis X. Webster
David J. Kiemle

From left to right: Robert M. Silverstein, Francis X. Webster, and David J. Kiemle.

PREFACE TO FIRST EDITION

During the past several years, we have been engaged in isolating small amounts of organic compounds from complex mixtures and identifying these compounds spectrometrically.

At the suggestion of Dr. A. J. Castro of San Jose State College, we developed a one unit course entitled "Spectrometric Identification of Organic Compounds," and presented it to a class of graduate students and industrial chemists during the 1962 spring semester. This book has evolved largely from the material gathered for the course and bears the same title as the course.*

We should first like to acknowledge the financial support we received from two sources: The Perkin-Elmer Corporation and Stanford Research Institute.

A large debt of gratitude is owed to our colleagues at Stanford Research Institute. We have taken advantage of the generosity of too many of them to list them individually, but we should like to thank Dr. S. A. Fuqua, in particular, for many helpful discussions of NMR spectrometry. We wish to acknowledge also the cooperation at the management level, Dr. C. M. Himel, chairman of the Organic Research Department, and Dr. D. M. Coulson, chairman of the Analytical Research Department.

Varian Associates contributed the time and talents of its NMR Applications Laboratory. We are indebted to Mr. N. S. Bhacca, Mr. L. F. Johnson, and Dr. J. N. Shoolery for the NMR spectra and for their generous help with points of interpretation.

The invitation to teach at San Jose State College was extended to Dr. Bert M. Morris, head of the Department of Chemistry, who kindly arranged the administrative details.

The bulk of the manuscript was read by Dr. R. H. Eastman of the Stanford University whose comments were most helpful and are deeply appreciated.

Finally, we want to thank our wives. As a test of a wife's patience, there are few things to compare with an author in the throes of composition. Our wives not only endured, they also encouraged, assisted, and inspired.

R. M. Silverstein
G. C. Bassler

Menlo Park, California
April 1963

* A brief description of the methodology had been published: R. M. Silverstein and G. C. Bassler, *J. Chem. Educ.* **39,** 546 (1962).

CONTENTS

CHAPTER 5 *CORRELATION NMR SPECTROMETRY; 2-D NMR* 245

CHAPTER 6 *NMR SPECTROMETRY OF OTHER IMPORTANT SPIN 1/2 NUCLEI* 316

CHAPTER 7 *SOLVED PROBLEMS* 341

CHAPTER 8 *ASSIGNED PROBLEMS* 381

CHAPTER 1

MASS SPECTROMETRY

1.1 INTRODUCTION

The concept of mass spectrometry is relatively simple: A compound is ionized (ionization method), the ions are separated on the basis of their mass/charge ratio (ion separation method), and the number of ions representing each mass/charge "unit" is recorded as a spectrum. For instance, in the commonly used electron-impact (EI) mode, the mass spectrometer bombards molecules in the vapor phase with a high-energy electron beam and records the result as a spectrum of positive ions, which have been separated on the basis of mass/charge (m/z).*

To illustrate, the EI mass spectrum of benzamide is given in Figure 1.1 showing a plot of abundance (vertical peak intensity) versus m/z. The positive ion peak at m/z 121 represents the intact molecule (M) less one electron, which was removed by the impacting electron beam; it is designated the molecular ion, $M^{\cdot+}$. The energetic molecular ion produces a series of fragment ions, some of which are rationalized in Figure 1.1.

It is routine to couple a mass spectrometer to some form of chromatographic instrument, such as a gas chromatograph (GC-MS) or a liquid chromatograph (LC-MS). The mass spectrometer finds widespread use in the analysis of compounds whose mass spectrum is known and in the analysis of completely unknown compounds. In the case of known compounds, a computer search is conducted comparing the mass spectrum of the compound in question with a library of mass spectra. Congruence of mass spectra is convincing evidence for identification and is often even admissible in court. In the case of an unknown compound, the molecular ion, the fragmentation pattern, and evidence from other forms of spectrometry (e.g., IR and NMR) can lead to the identification of a new compound. Our focus and goal in this chapter is to develop skill in the latter use. For other applications or for more detail,

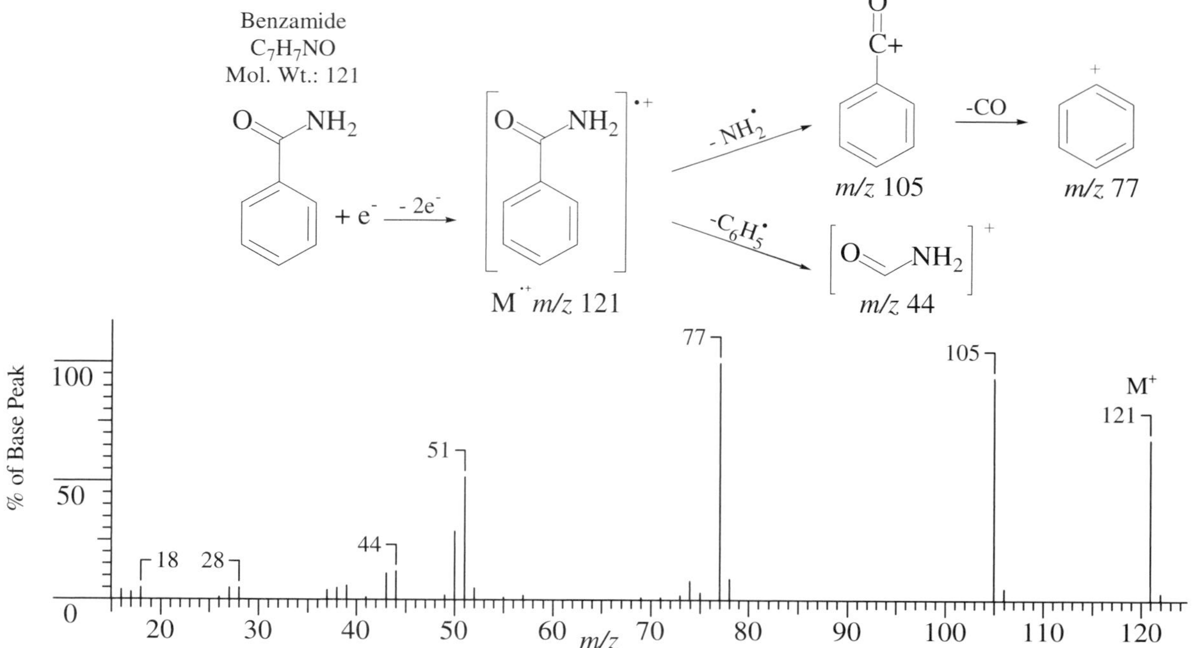

FIGURE 1.1 The EI mass spectrum of benzamide above which is a fragmentation pathway to explain some of the important ions.

* The unit of mass is the Dalton (Da), defined as 1/12 of the mass of an atom of the isotope ^{12}C, which is arbitrarily 12.0000 . . . mass units.

mass spectrometry texts and spectral compilations are listed at the end of this chapter.

1.2 INSTRUMENTATION

This past decade has been a time of rapid growth and change in instrumentation for mass spectrometry. Instead of discussing individual instruments, the type of instrument will be broken down into (1) ionization methods and (2) ion separation methods. In general, the method of ionization is independent of the method of ion separation and vice versa, although there are exceptions. Some of the ionization methods depend on a specific chromatographic front end (e.g., LC-MS), while still others are precluded from using chromatography for introduction of sample (e.g., FAB and MALDI). Before delving further into instrumentation, let us make a distinction between two types of mass spectrometers based on resolution.

The minimum requirement for the organic chemist is the ability to record the molecular weight of the compound under examination to the nearest whole number. Thus, the spectrum should show a peak at, say, mass 400, which is distinguishable from a peak at mass 399 or at mass 401. In order to select possible molecular formulas by measuring isotope peak intensities (see Section 1.5.2.1), adjacent peaks must be cleanly separated. Arbitrarily, the valley between two such peaks should not be more than 10% of the height of the larger peak. This degree of resolution is termed "unit" resolution and can be obtained up to a mass of approximately 3000 Da on readily available "unit resolution" instruments.

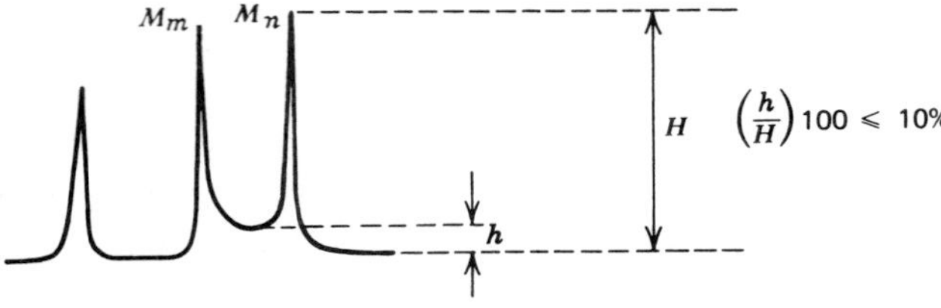

To determine the resolution of an instrument, consider two adjacent peaks of approximately equal intensity. These peaks should be chosen so that the height of the valley between the peaks is less than 10% of the intensity of the peaks. The resolution (R) is $R = M_n/(M_n - M_m)$, where M_n is the higher mass number of the two adjacent peaks, and M_m is the lower mass number.

There are two important categories of mass spectrometers: low (unit) resolution and high resolution. Low-resolution instruments can be defined arbitrarily as the instruments that separate unit masses up to m/z 3000 [$R = 3000/(3000 - 2999) = 3000$]. A high-resolution instrument (e.g., $R = 20{,}000$) can distinguish between $C_{16}H_{26}O_2$ and $C_{15}H_{24}NO_2$ [$R = 250.1933/(250.1933 - 250.1807) = 19857$]. This important class of mass spectrometers, which can have R as large as 100,000, can measure the mass of an ion with sufficient accuracy to determine its atomic composition (molecular formula).

All mass spectrometers share common features. (See Figure 1.2) Some sort of chromatography usually accomplishes introduction of the sample into the mass spectrometer, although many instruments also allow for direct insertion of the sample into the ionization chamber. All mass spectrometers have methods for ionizing the sample and for separating the ions on the basis of m/z. These methods are discussed in detail below. Once separated, the ions must be detected and quantified. A typical ion collector consists of collimating slits that direct only one set of ions at a time into the collector, where they are detected and amplified by an electron multiplier. The method of ion detection is dependent to some extent on the method of ion separation.

Nearly all mass spectrometers today are interfaced with a computer. Typically, the computer controls the operation of the instrument including any chromatography, collects and stores the data, and provides either graphical output (essentially a bar graph) or tabular lists of the spectra.

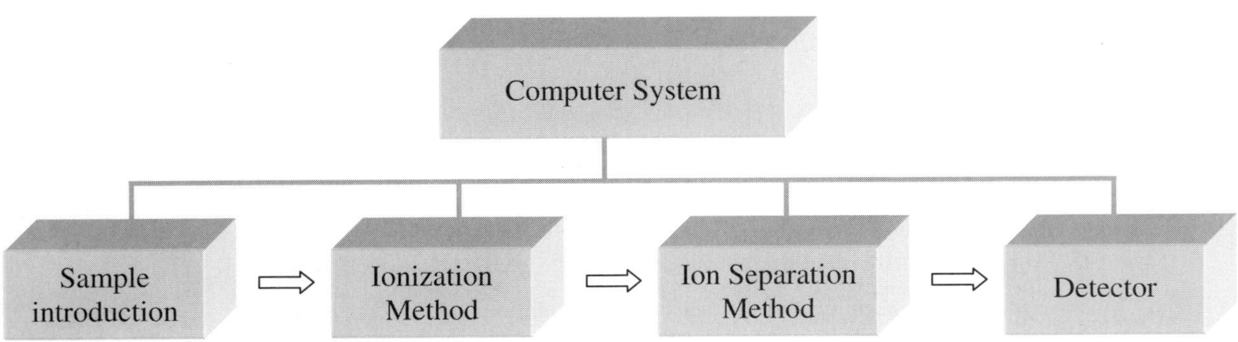

FIGURE 1.2 Block diagram of features of a typical mass spectrometer.

1.3 IONIZATION METHODS

The large number of ionization methods, some of which are highly specialized, precludes complete coverage. The most common ones in the three general areas of gas-phase, desorption, and evaporative ionization are described below.

1.3.1 Gas-Phase Ionization Methods

Gas-phase methods for generating ions for mass spectrometry are the oldest and most popular methods. They are applicable to compounds that have a minimum vapor pressure of ca. 10^{-6} Torr at a temperature at which the compound is stable; this criterion applies to a large number of nonionic organic molecules with MW < 1000.

1.3.1.1 Electron Impact Ionization. Electron impact (EI) is the most widely used method for generating ions for mass spectrometry. Vapor phase sample molecules are bombarded with high-energy electrons (generally 70 eV), which eject an electron from a sample molecule to produce a radical cation, known as the molecular ion. Because the ionization potential of typical organic compounds is generally less than 15 eV, the bombarding electrons impart 50 eV (or more) of excess energy to the newly created molecular ion, which is dissipated in part by the breaking of covalent bonds, which have bond strengths between 3 and 10 eV.

Bond breaking is usually extensive and critically, highly reproducible, and characteristic of the compound. Furthermore, this fragmentation process is also "predictable" and is the source of the powerful structure elucidation potential of mass spectrometry. Often, the excess energy imparted to the molecular ion is too great, which leads to a mass spectrum with no discernible molecular ion. Reduction of the ionization voltage is a commonly used strategy to obtain a molecular ion; the strategy is often successful because there is greatly reduced fragmentation. The disadvantage of this strategy is that the spectrum changes and cannot be compared to "standard" literature spectra.

To many, mass spectrometry is synonymous with EI mass spectrometry. This view is understandable for two reasons. First, historically, EI was universally available before other ionization methods were developed. Much of the early work was EI mass spectrometry. Second, the major libraries and databases of mass spectral data, which are relied upon so heavily and cited so often, are of EI mass spectra. Some of the readily accesible databases contain EI mass spectra of over 390,000 compounds and they are easily searched by efficient computer algorithms. The uniqueness of the EI mass spectrum for a given organic compound, even for stereoisomers, is an almost certainty. This uniqueness, coupled with the great sensitivity of the method, is what makes GC-MS such a powerful and popular analytical tool.

1.3.1.2 Chemical Ionization. Electron impact ionization often leads to such extensive fragmentation that no molecular ion is observed. One way to avoid this problem is to use "soft ionization" techniques, of which chemical ionization (CI) is the most important. In CI, sample molecules (in the vapor phase) are not subjected to bombardment by high energy electrons. Reagent gas (usually methane, isobutane, ammonia, but others are used) is introduced into the source, and ionized. Sample molecules collide with ionized reagent gas molecules (CH_5^+, $C_4H_9^+$, etc) in the relatively high-pressure CI source, and undergo secondary ionization by proton transfer producing an $[M + 1]^+$ ion, by electrophilic addition producing $[M + 15]^+$, $[M + 24]^+$, $[M + 43]^+$, or $[M + 18]^+$ (with NH_4^+) ions, or by charge exchange (rare) producing a $[M]^+$ ion. Chemical ionization spectra sometimes have prominent $[M - 1]^+$ ions because of hydride abstraction. The ions thus produced are even electron species. The excess energy transfered to the sample molecules during the ionization phase is small, generally less than 5 eV, so much less fragmentation takes place. There are several important consequences, the most valuable of which are an abundance of molecular ions and greater sensitity because the total ion current is concentrated into a few ions. There is however, less information on structure. The quasimolecular ions are usually quite stable and they are readily detected. Oftentimes there are only one or two fragment ions produced and sometimes there are none.

For example, the EI mass spectrum of 3, 4-dimethoxyacetophenone (Figure 1.3) shows, in addition to the molecular ion at m/z 180, numerous fragment peaks in the range of m/z 15–167; these include the base peak at m/z 165 and prominent peaks at m/z 137 and m/z 77. The CI mass spectrum (methane, CH_4, as reagent gas) shows the quasimolecular ion ($[M + 1]^+$, m/z 181) as the base peak (100%), and virtually the only other peaks, each of just a few percent intensity, are the molecular ion peak, m/z 180, m/z 209 ($[M + 29]^+$ or $M + C_2H_5^+$), and m/z 221 ($[M + 41]^+$ or $M + C_3H_5^+$). These last two peaks are a result of electrophilic addition of carbocations and are very useful in indentifing the molecular ion. The excess methane carrier gas is ionized by electron impact to the primary ions CH_4^+ and CH_3^+. These react with the excess methane to give secondary ions.

$$CH_3^+ + CH_4 \longrightarrow C_2H_5^+ \text{ and } H_2$$

$$CH_4 + C_2H_5^+ \longrightarrow C_3H_5^+ \text{ and } 2H_2$$

The energy content of the various secondary ions (from, respectively, methane, isobutane, and ammonia) decrease in the order: $CH_5^+ > t\text{-}C_4H_9^+ > NH_4^+$. Thus,

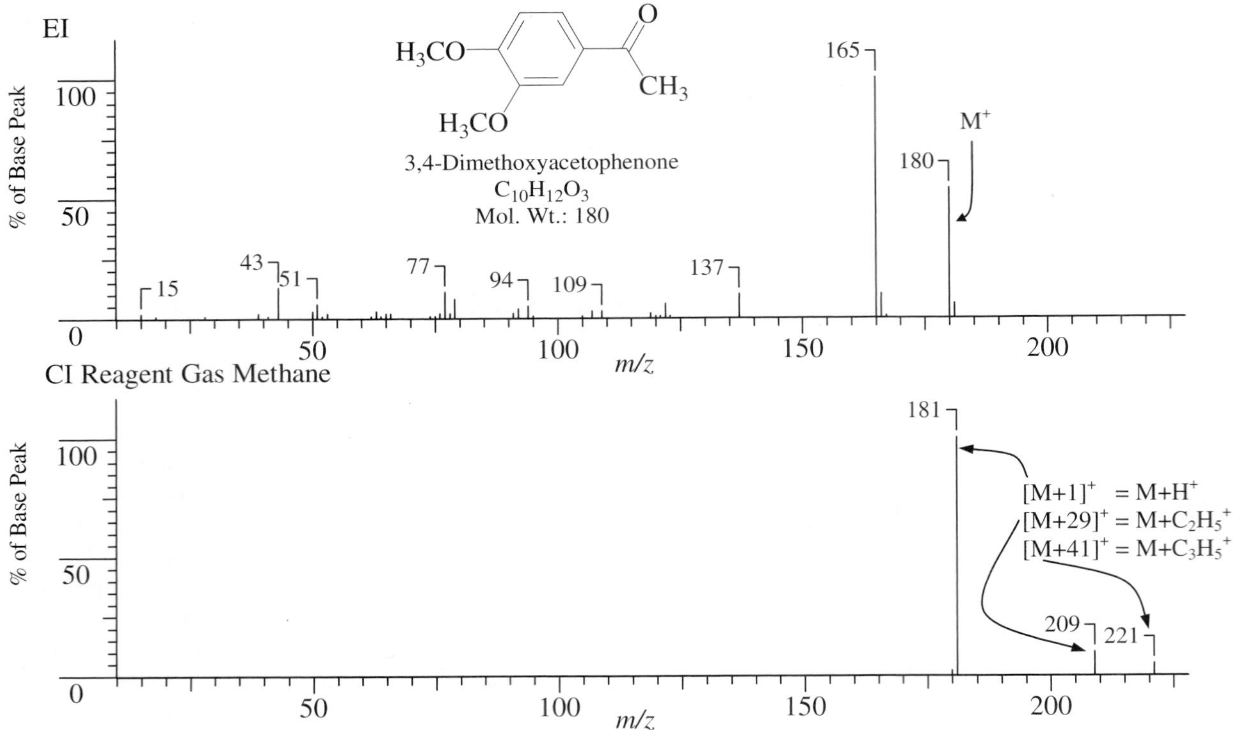

FIGURE 1.3 The EI and CI mass spectra of 3,4-dimethoxyacetophenone.

by choice of reagent gas, we can control the tendency of the CI produced $[M + 1]^+$ ion to fragment. For example, when methane is the reagent gas, dioctyl phthalate shows its $[M + 1]^+$ peak (*m/z* 391) as the base peak; more importantly, the fragment peaks (e.g., *m/z* 113 and 149) are 30–60% of the intensity of the base beak. When isobutane is used, the $[M + 1]^+$ peak is still large, while the fragment peaks are only roughly 5% as intense as the $[M + 1]^+$ peak.

Chemical ionization mass spectrometry is not useful for peak matching (either manually or by computer) nor is it particularly useful for structure elucidation; its main use is for the detection of molecular ions and hence molecular weights.

1.3.2 Desorption Ionization Methods

Desorption ionization methods are those techniques in which sample molecules are emitted directly from a condensed phase into the vapor phase as ions. The primary use is for large, nonvolatile, or ionic compounds. There can be significant disadvantages. Desorption methods generally do not use available sample efficiently. Oftentimes, the information content is limited. For unknown compounds, the methods are used primarily to provide molecular weight, and in some cases to obtain an exact mass. However, even for this purpose, it should be used with caution because the molecular ion or the quasimolecular ion may not be evident. The resulting spectra are often complicated by abundant matrix ions.

1.3.2.1 Field Desorption Ionization. In the field desorption (FD) method, the sample is applied to a metal emitter on the surface of which is found carbon microneedles. The microneedles activate the surface, which is maintained at the accelerating voltage and functions as the anode. Very high voltage gradients at the tips of the needles remove an electron from the sample, and the resulting cation is repelled away from the emitter. The ions generated have little excess energy so there is minimal fragmentation, i.e., the molecular ion is usually the only significant ion seen. For example with cholesten-5-ene-3,16,22,26-tetrol the EI and CI do not see a molecular ion for this steroid. However, the FD mass spectrum (Figure 1.4) shows predominately the molecular ion with virtually no fragmentation.

Field desorption was eclipsed by the advent of FAB (next section). Despite the fact that the method is often more useful than FAB for nonpolar compounds and does not suffer from the high level of background ions that are found in matrix-assisted desorption methods, it has not become as popular as FAB probably because the commercial manufacturers have strongly supported FAB.

1.3.2.2 Fast Atom Bombardment Ionization. Fast atom bombardment (FAB) uses high-energy xenon or argon atoms (6–10 keV) to bombard samples dissolved in a liquid of low vapor pressure (e.g., glycerol). The matrix protects the sample from excessive radiation damage. A related method, liquid secondary

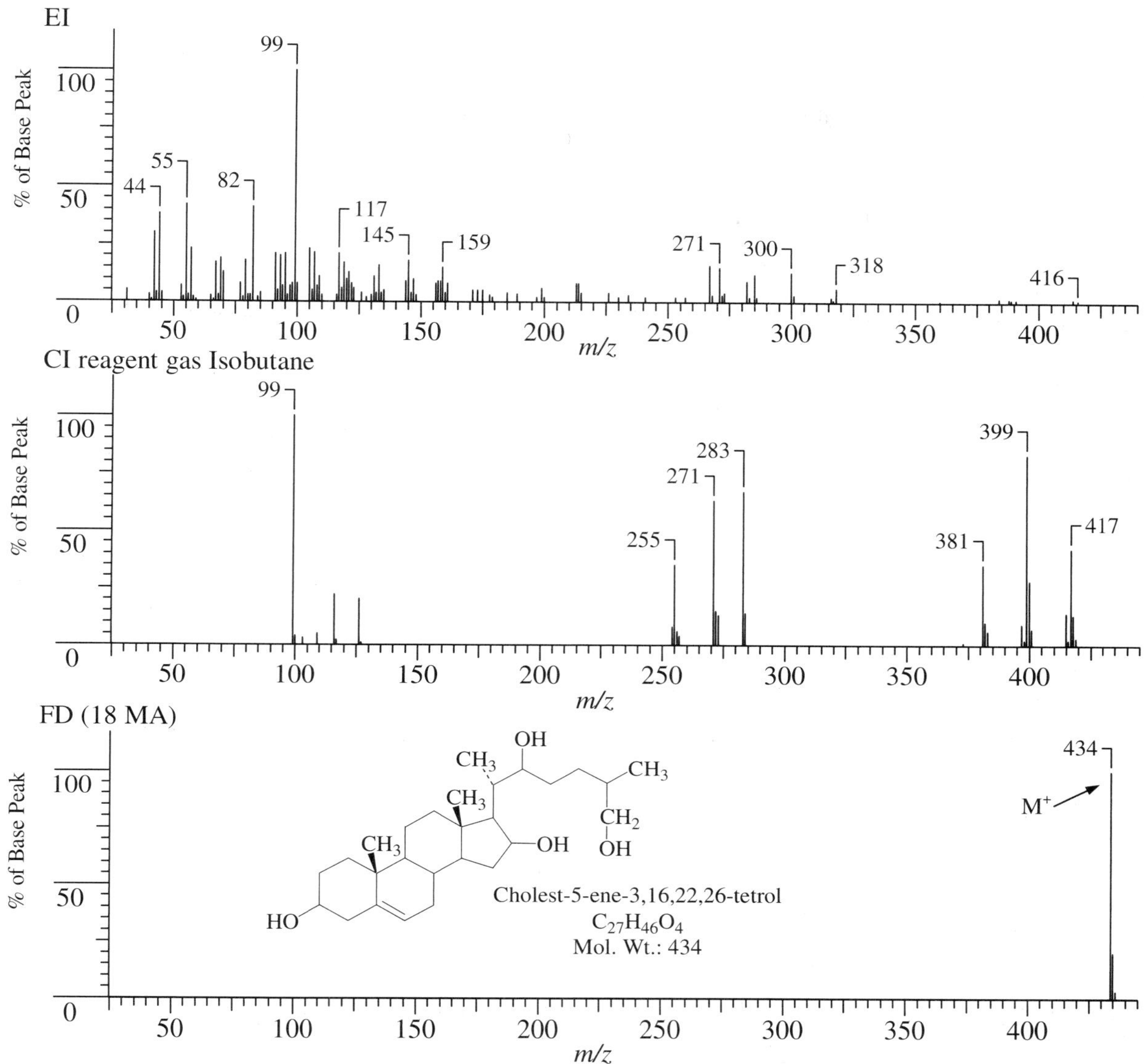

FIGURE 1.4 The electron impact (EI), chemical ionization (CI), and field desorption (FD) mass spectra of cholest-5-ene-3, 16, 22, 26-tetrol.

ionization mass spectrometry, LSIMS, is similar except that it uses somewhat more energetic cesium ions (10–30 keV).

In both methods, positive ions (by cation attachment ($[M + 1]^+$ or $[M + 23, Na]^+$) and negative ions (by deprotonation $[M - 1]^+$) are formed; both types of ions are usually singly charged and, depending on the instrument, FAB can be used in high-resolution mode. FAB is used primarily with large nonvolatile molecules, particularly to determine molecular weight. For most classes of compounds, the rest of the spectrum is less useful, partially because the lower mass ranges may be composed of ions produced by the matrix itself. However, for certain classes of compounds that are composed of "building blocks," such as polysaccharides and peptides, some structural information may be obtained because fragmentation usually occurs at the glycosidic and peptide bonds, respectively, thereby affording a method of sequencing these classes of compounds.

The upper mass limit for FAB (and LSIMS) ionization is between 10 and 20 kDa, and FAB is really most useful up to about 6 kDa. FAB is seen most often with double focusing magnetic sector instruments where it has a resolution of about 0.3 *m/z* over the entire mass range; FAB can, however, be used with most types of mass analyzers. The biggest drawback to using FAB is that the spectrum always shows a high level of matrix generated ions, which limit sensitivity and which may obscure important fragment ions.

1.3.2.3 Plasma Desorption Ionization.

Plasma desorption ionization is a highly specialized technique used almost exclusively with a time of flight mass analyzer (Section 1.4.4). The fission products from Californium 252 (^{252}Cf), with energies in the range of 80–100 MeV, are used to bombard and ionize the sample. Each time a ^{252}Cf splits, two particles are produced moving in opposite directions. One of the particles hits a triggering detector and signals a start time. The other particle strikes the sample matrix ejecting some sample ions into a time of flight mass spectrometer (TOF-MS). The sample ions are most often released as singly, doubly, or triply protonated moieties. These ions are of fairly low energy so that structurally useful fragmentation is rarely observed and, for polysaccharides and polypeptides, sequencing information is not available. The mass accuracy of the method is limited by the time of flight mass spectrometer. The technique is useful on compounds with molecular weights up to at least 45 kDa.

1.3.2.4 Laser Desorption Ionization. A pulsed laser beam can be used to ionize samples for mass spectrometry. Because this method of ionization is pulsed, it must be used with either a time of flight or a Fourier transform mass spectrometer (Section 1.4.5). Two types of lasers have found widespread use: A CO_2 laser, which emits radiation in the far infrared region, and a frequency-quadrupled neodymium/yttriumaluminum-garnet (Nd/YAG) laser, which emits radiation in the UV region at 266 nm. Without matrix assistance, the method is limited to low molecular weight molecules (<2 kDa).

The power of the method is greatly enhanced by using matrix assistance (matrix assisted laser desorption ionization, or MALDI). Two matrix materials, nicotinic acid and sinapinic acid, which have absorption bands coinciding with the laser employed, have found widespread use and sample molecular weights of up to two to three hundred thousand Da have been successfully analyzed. A few picomoles of sample are mixed with the matrix compound followed by pulsed irradiation, which causes sample ions (usually singly charged monomers but occasionally multiply charged ions and dimers have been observed) to be ejected from the matrix into the mass spectrometer.

The ions have little excess energy and show little propensity to fragment. For this reason, the method is fairly useful for mixtures. The mass accuracy is low when used with a TOF-MS, but very high resolution can be obtained with a FT-MS. As with other matrix-assisted methods, MALDI suffers from background interference from the matrix material, which is further exacerbated by matrix adduction. Thus, the assignment of a molecular ion of an unknown compound can be uncertain.

1.3.3 Evaporative Ionization Methods

There are two important methods in which ions or, less often, neutral compounds in solution (often containing formic acid) have their solvent molecules stripped by evaporation, with simultaneous ionization leaving behind the ions for mass analysis. Coupled with liquid chromatography instrumentation, these methods have become immensely popular.

1.3.3.1 Thermospray Mass Spectrometry. In the thermospray method, a solution of the sample is introduced into the mass spectrometer by means of a heated capillary tube. The tube nebulizes and partially vaporizes the solvent forming a stream of fine droplets, which enter the ion source. When the solvent completely evaporates, the sample ions can be mass analyzed. This method can handle high flow rates and buffers; it was an early solution to interfacing mass spectrometers with aqueous liquid chromatography. The method has largely been supplanted by electrospray.

1.3.3.2 Electrospray Mass Spectrometry. The electrospray (ES) ion source (Figure 1.5) is operated at or near atmospheric pressure and, thus is also called atmospheric pressure ionization or API. The

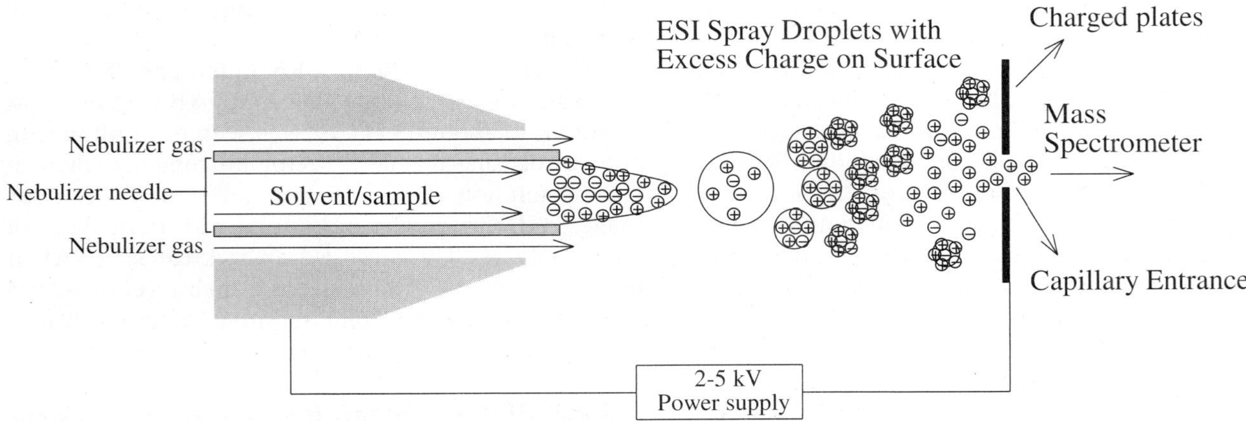

FIGURE 1.5 A diagram showing the evaporation of solvent leading to individual ions in an electrospray instrument.

sample in solution (usually a polar, volatile solvent) enters the ion source through a stainless steel capillary, which is surrounded by a co-axial flow of nitrogen called the nebulizing gas. The tip of the capillary is maintained at a high potential with respect to a counter-electrode. The potential difference produces a field gradient of up to 5 kV/cm. As the solution exits the capillary, an aerosol of charged droplets forms. The flow of nebulizing gas directs the effluent toward the mass spectrometer.

Droplets in the aerosol shrink as the solvent evaporates, thereby concentrating the charged sample ions. When the electrostatic repulsion among the charged sample ions reaches a critical point, the droplet undergoes a so-called "Coulombic explosion," which releases the sample ions into the vapor phase. The vapor phase ions are focused with a number of sampling orifices into the mass analyzer.

Electrospray MS has undergone an explosion of activity since about 1990, mainly for compounds that have multiple charge bearing sites. With proteins, for example, ions with multiple charges are formed. Since the mass spectrometer measures mass to charge ratio (m/z) rather than mass directly, these multiply charged ions are recorded at apparent mass values of 1/2, 1/3, . . . $1/n$ of their actual masses, where n is the number of charges (z). Large proteins can have 40 or more charges so that molecules of up to 100 kDa can be detected in the range of conventional quadrupole, ion trap, or magnetic sector mass spectrometers. The appearance of the spectrum is a series of peaks increasing in mass, which correspond to pseudo molecular ions possessing sequentially one less proton and therefore one less charge.

Determination of the actual mass of the ion requires that the charge of the ion be known. If two peaks, which differ by a single charge, can be identified, the calculation is reduced to simple algebra. Recall that each ion of the sample molecule (M_s) has the general form $(M_s + zH)^{z+}$ where H is the mass of a proton (1.0079 Da). For two ions differing by one charge, $m_1 = [M_s + (z + 1)H]/(z + 1)$ and $m_2 = [(M_s + zH)/z]$. Solving the two simultaneous equations for the charge z, yields $z = (m_1 - H)/(m_2 - m_1)$. A simple computer program automates this calculation for every peak in the spectrum and calculates the mass directly.

Many manufacturers have introduced inexpensive mass spectrometers dedicated to electrospray for two reasons. First, the method has been very successful while remaining a fairly simple method to employ. Second, the analysis of proteins and smaller peptides has grown in importance, and they are probably analyzed best by the electrospray method.

Figure 1.6 compares the EI mass spectrum (lower portion of the figure) of lactose to its ES mass spectrum (upper portion of figure). Lactose is considered in more detail in Chapter 5. The EI mass spectrum is completely

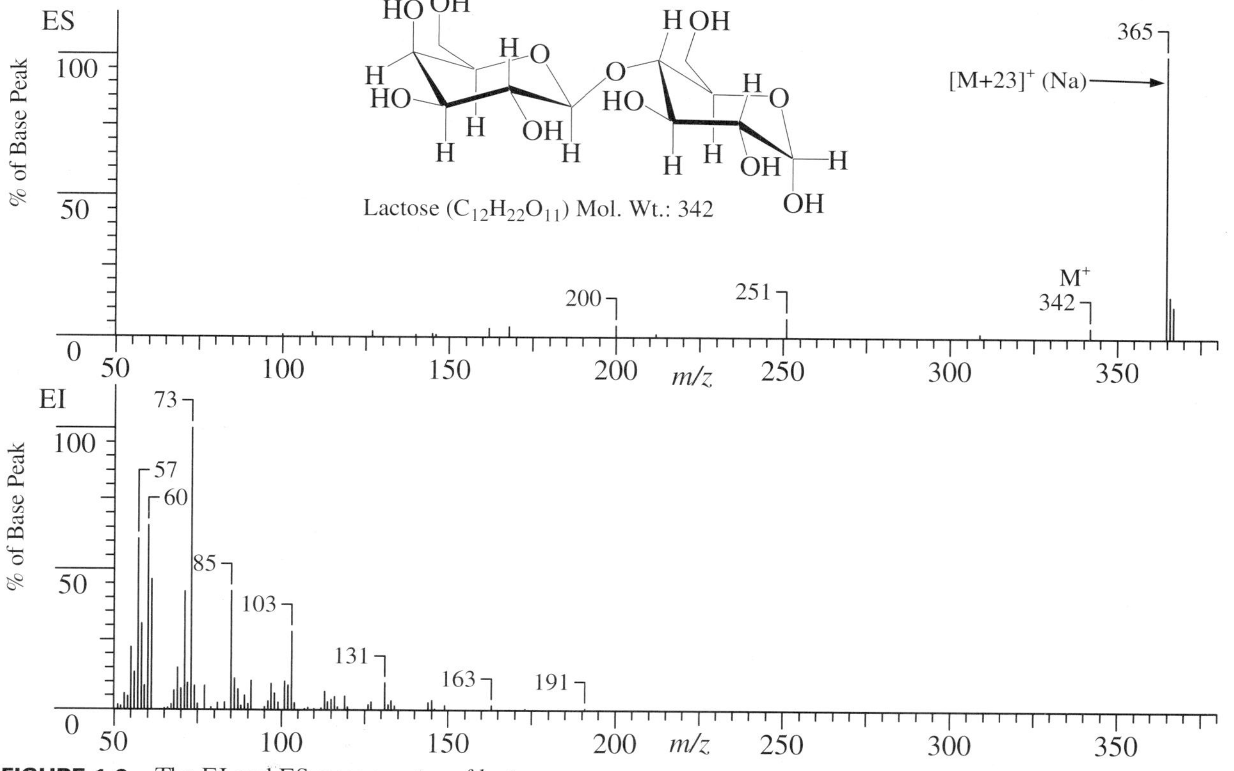

FIGURE 1.6 The EI and ES mass spectra of lactose.

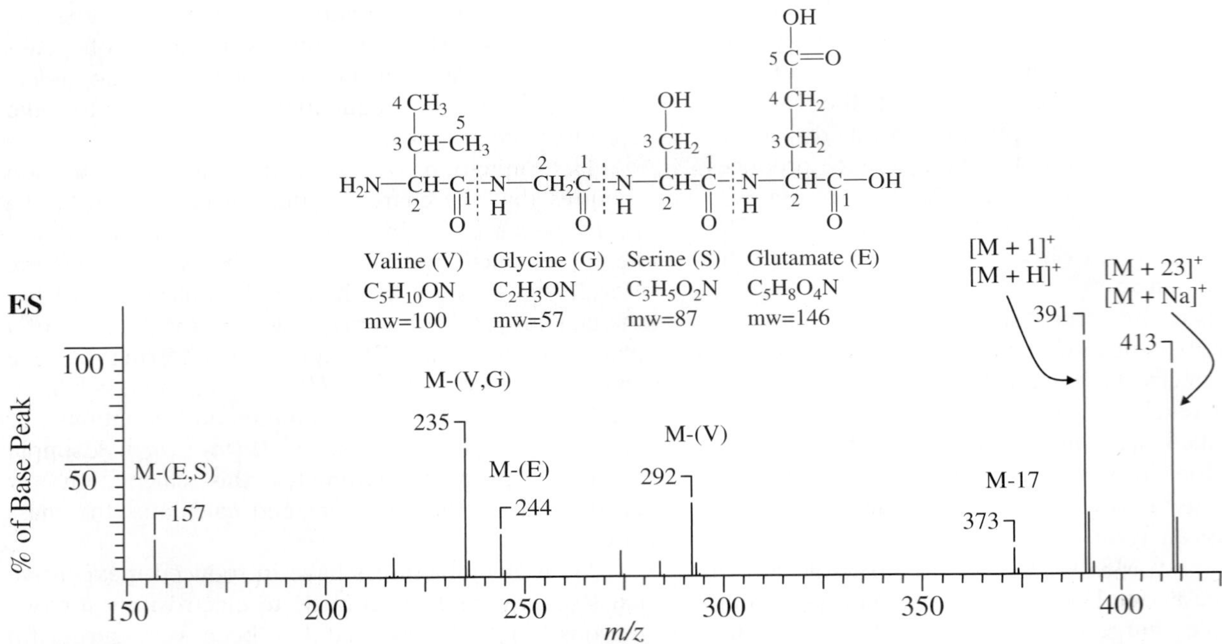

FIGURE 1.7 The electrospray (ES) mass spectrum for the tetra-peptide whose structure is given in the figure. See text for explanation.

useless because lactose has low vapor pressure, it is thermally labile, and the spectrum shows no characteristic peaks. The ES mass spectrum shows a weak molecular ion peak at *m/z* 342 and a characteristic $[M + 23]^+$, the molecular ion peak plus sodium. Because sodium ions are ubiquitous in aqueous solution, these sodium adducts are very common.

The ES mass spectrum of a tetra-peptide comprised of valine, glycine, serine, and glutamic acid (VGSE) is given in Figure 1.7. VGSE is also an example compound in Chapter 5. The base beak is the $[M + 1]^+$ ion at *m/z* 391 and the sodium adduct, $[M + 23]^+$, is nearly 90% of the base peak. In addition, there is some useful fragmentation information characteristic of each of the amino acids. For small peptides, it is not uncommon to find some helpful fragmentation, but for proteins it is less likely.

Methods of ionization are summarized in Table 1.1.

1.4 MASS ANALYZERS

The mass analyzer, which separates the mixture of ions that are generated during the ionization step by *m/z* in order to obtain a spectrum, is the heart of each mass spectrometer, and there are several different types with

TABLE 1.1 Summary of Ionization Methods.

Ionization Method	Ions Formed	Sensitivity	Advantage	Disadvantage
Electron impact	M^+	ng–pg	Data base searchable Structural information	M^+ occasionally absent
Chemical ionization	$M + 1, M + 18$, etc	ng–pg	M+ usually present	Little structural information
Field desorption	M^+	μg–ng	Non volatile compounds	Specialized equipment
Fast atom bombardment	$M + 1, M +$ cation $M +$ matrix	μg–ng	Non volatile compounds Sequencing information	Matrix interference Difficult to interpret
Plasma desorption	M+	μg–ng	Non volatile compounds	Matrix interference
Laser desorption	$M + 1, M +$ matrix	μg–ng	Non volatile compounds Burst of ions	Matrix interference
Thermospray	M^+	μg–ng	Non volatile compounds	Outdated
Electrospray	M^+, M^{++}, M^{+++}, etc.	ng–pg	Non volatile compounds interfaces w/ LC Forms multiply charged ions	Limited classes of compounds Little structural information

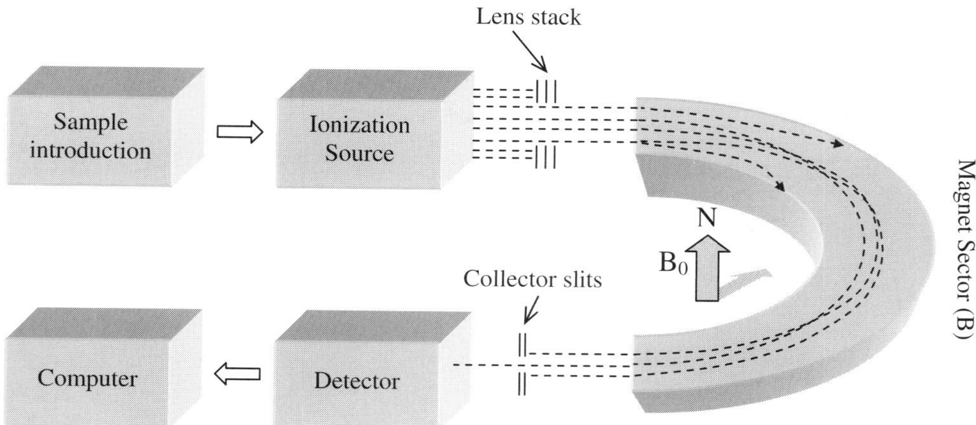

FIGURE 1.8 Schematic diagram of a single focusing, 180° sector mass analyzer. The magnetic field is perpendicular to the page. The radius of curvature varies from one instrument to another.

different characteristics. Each of the major types of mass analyzers is described below. This section concludes with a brief discussion of tandem MS and related processes.

1.4.1 Magnetic Sector Mass Spectrometers

The magnetic sector mass spectrometer (MS-MS) uses a magnetic field to deflect moving ions around a curved path (see Figure 1.8). Magnetic sector mass spectrometers were the first commercially available instruments, and they remain an important choice. Separation of ions occurs based on the mass/charge ratio with lighter ions deflected to a greater extent than are the heavier ions. Resolution depends on each ion entering the magnetic field (from the source) with the same kinetic energy, accomplished by accelerating the ions (which have a charge z) with a voltage V. Each ion acquires kinetic energy $E = zV = mv^2/2$. When an accelerated ion enters the magnetic field (B), it experiences a deflecting force (Bzv), which bends the path of the ion orthogonal to its original direction. The ion is now traveling in a circular path of radius r, given by $Bzv = mv^2/r$. The two equations can be combined to give the familiar magnetic sector equation: $m/z = B^2r^2/2V$. Because the radius of the instrument is fixed, the magnetic field is scanned to bring the ions sequentially into focus. As these equations show, a magnetic sector instrument separates ions on the basis of momentum, which is the product of mass and velocity, rather than mass alone; therefore, ions of the same mass but different energies will come into focus at different points.

An electrostatic analyzer (ESA) can greatly reduce the energy distribution of an ion beam by forcing ions of the same charge (z) and kinetic energy (regardless of mass) to follow the same path. A slit at the exit of the ESA further focuses the ion beam before it enters the detector. The combination of an

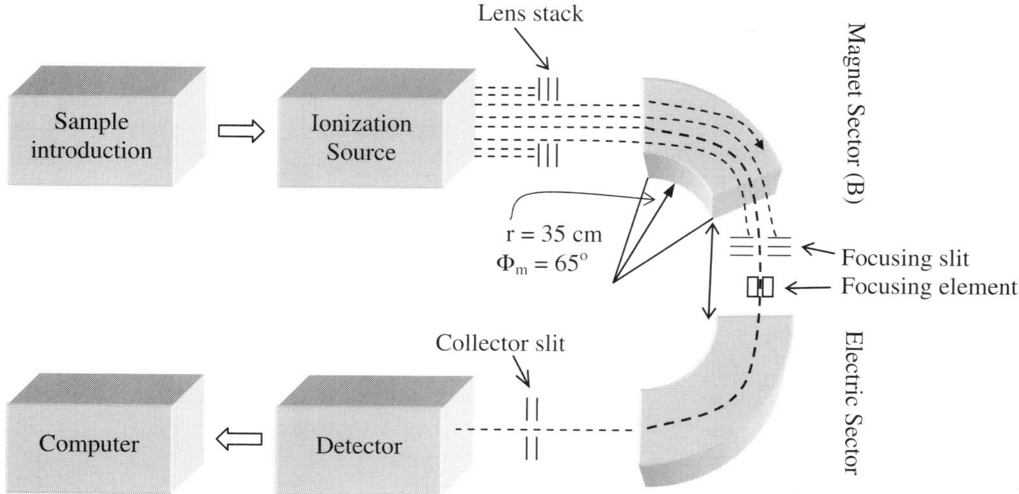

FIGURE 1.9 Schematic of double-focusing mass spectrometer.

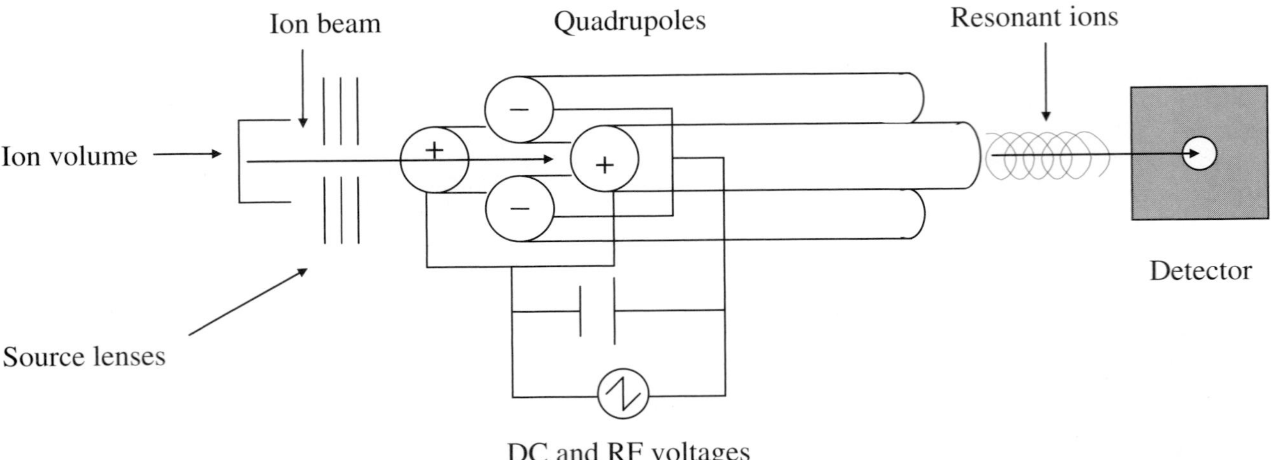

FIGURE 1.10 Schematic representation of a quadrupole "mass filter" or ion separator.

ESA and a magnetic sector is known as double focusing, because the two fields counteract the dispersive effects each has on direction and velocity.

The resolution of a double focusing magnetic sector instrument (Figure 1.9) can be as high as 100,000 through the use of extremely small slit widths. This very high resolution allows the measurement of "exact masses," which unequivocally provide molecular formulas, and is enormously useful. Such high-resolution instruments sacrifice a great deal of sensitivity. By comparison, slits allowing an energy distribution for about 5000 resolution give at least 0.5 m/z accuracy across the entire mass range, i.e., the "unit resolution" that is used in a standard mass spec. The upper mass limit for commercial magnetic sector instruments is about m/z 15,000. Raising this upper limit is theoretically possible but impractical.

1.4.2 Quadrupole Mass Spectrometers

The quadrupole mass analyzer is much smaller and cheaper than a magnetic sector instrument. A quadrupole setup (seen schematically in Figure 1.10) consists of four cylindrical (or of hyperbolic cross-section) rods (100–200 mm long) mounted parallel to each other, at the corners of a square. A complete mathematical analysis of the quadrupole mass analyzer is complex but we can discuss how it works in a simplified form. A constant DC voltage modified by a radio frequency voltage is applied to the rods. Ions are introduced to the "tunnel" formed by the four rods of the quadrupole in the center of the square at one end to the rods, and travel down the axis.

For any given combination of DC voltage and modified voltage applied at the appropriate frequency, only ions with a certain m/z value possess a stable trajectory and therefore are able to pass all the way to the end of the quadrupole to the detector. All ions with different m/z values travel unstable or erratic paths and collide with one of the rods or pass outside the quadrupole. An easy way to look at the quadrupole mass analyzer is as a tunable mass filter. In other words, as the ions enter at one end, only one m/z ion will pass through. In practice, the filtering can be carried out at a very fast rate so that the entire mass range can be scanned in considerably less than 1 second.

With respect to resolution and mass range, the quadrupole is generally inferior to the magnetic sector. For instance, the current upper mass range is generally less than 5000 m/z. On the other hand, sensitivity is generally high because there is no need for resolving slits, which would remove a portion of the ions. An important advantage of quadrupoles is that they operate most efficiently on ions of low velocity, which means that their ion sources can operate close to ground potential (i.e., low voltage). Since the entering ions generally have energies of less than 100 eV, the quadrupole mass spectrometer is ideal for interfacing to LC systems and for atmospheric pressure ionization (API) techniques such as electrospray (see Section 1.3.3.2). These techniques work best on ions of low energy so that fewer high-energy collisions will occur before they enter the quadrupole.

1.4.3 Ion Trap Mass Spectrometer

The ion trap is sometimes considered as a variant of the quadrupole, since the appearance and operation of the two are related. However, the ion trap is potentially much more versatile and clearly has greater potential for development. At one time the ion trap had a bad reputation because the earliest versions gave inferior results compared to quadrupoles. The results were oftentimes "concentration dependent"; relatively large sample sizes usually gave many peaks with the mass of the [ion + 1], which renders the resulting spectra useless in a search with standard EI libraries. These problems have been overcome and

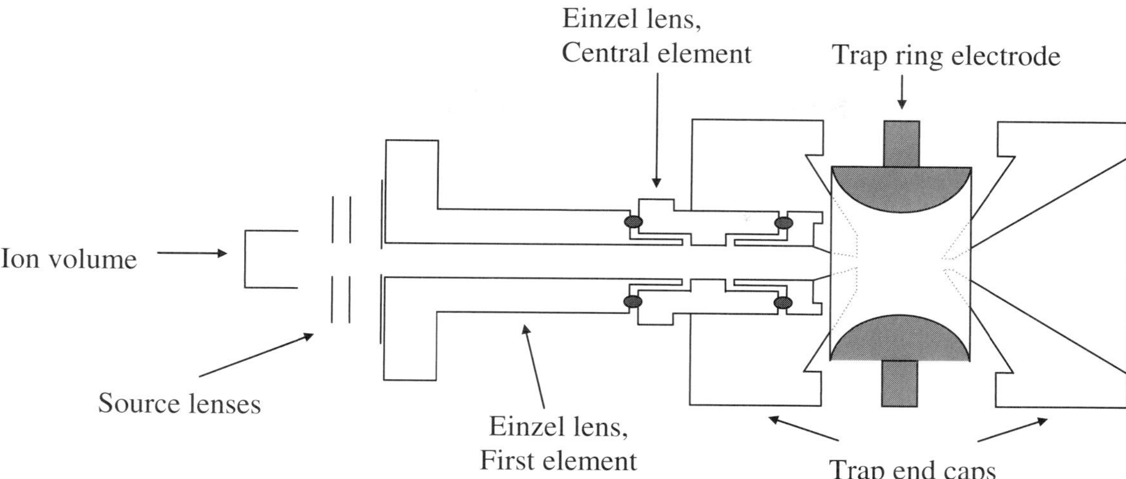

FIGURE 1.11 Cross sectional view of an ion trap.

the EI spectra obtained with an ion trap are now fully searchable with commercial databases. Furthermore, the ion trap is more sensitive than the quadrupole arrangement, and the ion trap is routinely configured to carry out tandem experiments with no extra hardware needed.

In one sense, an ion trap is aptly named because, unlike the quadrupole, which merely acts as a mass filter, it can "trap" ions for relatively long periods of time, with important consequences. The simplest use of the trapped ions is to sequentially eject them to a detector, producing a conventional mass spectrum. Before other uses of trapped ions are briefly described, a closer look at the ion trap itself will be helpful.

The ion trap generally consists of three electrodes, one ring electrode with a hyperbolic inner surface and two hyperbolic endcap electrodes at either end (a cross section of an ion trap is found in Figure 1.11). The ring electrode is operated with a sinusoidal radio frequency field while the endcap electrodes are operated in one of three modes. The endcap may be operated at ground potential, or with either a DC or an AC voltage.

The mathematics that describes the motion of ions within the ion trap is given by the Mathieu equation. Details and discussions of three-dimensional ion stability diagrams can be found in either March and Hughes (1989) or Nourse and Cooks (1990). The beauty of the ion trap is that by controlling the three parameters of RF voltage, AC voltage, and DC voltage, a wide variety of experiments can be run quite easily (for details see March and Hughes 1989).

There are three basic modes in which the ion trap can be operated. First, when the ion trap is operated with a fixed RF voltage and no DC bias between the endcap and ring electrodes, all ions above a certain cutoff m/z ratio will be trapped. As the RF voltage is raised, the cutoff m/z is increased in a controlled manner and the ions are sequentially ejected and detected. The result is the standard mass spectrum and this procedure is called the "mass-selective instability" mode of operation. The maximum RF potential that can be applied between the electrodes limits the upper mass range in this mode. Ions of mass contained beyond the upper limit are removed after the RF potential is brought back to zero.

The second mode of operation uses a DC potential across the endcaps; the general result is that there is now both a low and high-end cutoff (m/z) of ions. The possibilities of experiments in this mode of operation are tremendous, and most operations with the ion trap use this mode. As few as one ion mass can be selected. Selective ion monitoring is an important use of this mode of operation. There is no practical limit on the number of ions masses that can be selected.

The third mode of operation is similar to the second, with the addition of an auxiliary oscillatory field between the endcap electrodes, which results in adding kinetic energy selectively to a particular ion. With a small amplitude auxiliary field, selected ions gain kinetic energy slowly, during which time they usually undergo a fragmenting collision; the result can be a nearly 100% MS-MS efficiency. If the inherent sensitivity of the ion trap is considered along with the nearly 100% tandem efficiency, the use of the ion trap for tandem MS experiment greatly outshines the so called "triple quad" (see below).

Another way to use this kinetic energy addition mode is to selectively reject unwanted ions from the ion trap. These could be ions derived from solvent or from the matrix in FAB or LSIMS experiments. A constant frequency field at high voltage during the ionization period will selectively reject a single ion. Multiple ions can also be selected in this mode.

1.4.4 Time-of-Flight Mass Spectrometer

The concept of time-of-flight (TOF) mass spectrometers is simple. Ions are accelerated through a potential (V) and are then allowed to "drift" down a tube to a detector. If the assumption is made that all of the ions arriving at the beginning of the drift tube have the same energy given by $zeV = mv^2/2$, then ions of different mass will have different velocities: $v = (2zeV/m)^{1/2}$. If a spectrometer possesses a drift tube of length L, the time of flight for an ion is given by: $t = (L^2m/2zeV)^{1/2}$, from which the mass for a given ion can be easily calculated.

The critical aspect of this otherwise simple instrument is the need to produce the ions at an accurately known start time and position. These constraints generally limit TOF spectrometers to use pulsed ionization techniques, which include plasma and laser desorption (e.g., MALDI, matrix assisted laser desorption ionization).

The resolution of TOF instruments is usually less than 20,000 because some variation in ion energy is unavoidable. Also, since the difference in arrival times at the detector can be less than 10^{-7} s, fast electronics are necessary for adequate resolution. On the positive side, the mass range of these instruments is unlimited, and, like quadrupoles, they have excellent sensitivity due to lack of resolving slits. Thus, the technique is most useful for large biomolecules.

1.4.5 Fourier Transform Mass Spectrometer

Fourier transform (FT) mass spectrometers are not very common now because of their expense; in time, they may become more widespread as advances are made in the manufacture of superconducting magnets. In a Fourier transform mass spectrometer, ions are held in a cell with an electric trapping potential within a strong magnetic field. Within the cell, each ion orbits in a direction perpendicular to the magnetic field, with a frequency proportional to the ion's m/z. A radiofrequency pulse applied to the cell brings all of the cycloidal frequencies into resonance simultaneously to yield an interferogram, conceptually similar to the free induction decay (FID) signal in NMR or the interferogram generated in FTIR experiments. The interferogram, which is a time domain spectrum, is Fourier transformed into a frequency domain spectrum, which then yields the conventional m/z spectrum. Pulsed Fourier transform spectrometry applied to nuclear magnetic resonance spectrometry is discussed in Chapters 3, 4, and 5.

Because the instrument is operated at fixed magnetic field strength, extremely high field superconducting magnets can be used. Also, because mass range is directly proportional to magnetic field strength, very high mass detection is possible. Finally, since all of the ions from a single ionization event can be trapped and analyzed, the method is very sensitive and works well with pulsed ionization methods. The most compelling aspect of the method is its high resolution, making FT mass spectrometers an attractive alternative to other mass analyzers. The FT mass spectrometer can be coupled to chromatographic instrumentation and various ionization methods, which means that it can be easily used with small molecules. Further information on FT mass spectrometers can be found in the book by Gross (1990).

1.4.6 Tandem Mass Spectrometry

Tandem mass spectrometry or MS-MS ("MS squared") is useful in studies with both known and unknown compounds; with certain ion traps, MS to the nth ($MS^{(n)}$) is possible where $n = 2$ to 9. In practice, n rarely exceeds 2 or 3. With MS-MS, a "parent" ion from the initial fragmentation (the initial fragmentation gives rise to the conventional mass spectrum) is selected and allowed or induced to fragment further thus giving rise to "daughter" ions. In complex mixtures, these daughter ions provide unequivocal evidence for the presence of a known compound. For unknown or new compounds, these daughter ions provide potential for further structural information.

One popular use of MS-MS involves ionizing a crude sample, selectively "fishing out" an ion characteristic for the compound under study, and obtaining the diagnostic spectrum of the daughter ions produced from that ion. In this way, a compound can be unequivocally detected in a crude sample, with no prior chromatographic (or other separation steps) being required. Thus, MS-MS can be a very powerful screening tool. This type of analysis alleviates the need for complex separations of mixtures for many routine analyses. For instance, the analysis of urine samples from humans (or from other animals such as race horses) for the presence of drugs or drug metabolites can be carried out routinely on whole urine (i.e., no purification or separation) by MS-MS. For unknown compounds, these daughter ions can provide structural information as well.

One way to carry out MS-MS is to link two or more mass analyzers in series to produce an instrument capable of selecting a single ion, and examining how that ion (either a parent or daughter ion) fragments. For instance, three quadrupoles can be linked (a so called "triple quad") to produce a tandem mass spectrometer. In this arrangement, the first quadrupole selects a specific ion for further analysis, the second

TABLE 1.2 Summary of Mass Analyzers.

Mass Analyzer	Mass Range	Resolution	Sensitivity	Advantage	Disadvantage
Magnetic Sector	1–15,000 m/z	0.0001	Low	High res.	Low sensitivity Very expensive High technical expertise
Quadrupole	1–5000 m/z	unit	High	Easy to use Inexpensive High sensitivity	Low res. Low mass range
Ion trap	1–5000 m/z	unit	High	Easy to use Inexpensive High sensitivity Tandem MS (MS^n)	Low res. Low mass range
Time of flight	Unlimited	0.0001	High	High mass range Simple design	Very high res.
Fourier transform	up to 70 kDa	0.0001	High	Very High res. and mass range	Very expensive High technical expertise

quadrupole functions as a collision cell (collision induced decomposition, CID) and is operated with radiofrequency only, and the third quadrupole separates the product ions, to produce a spectrum of daughter ions. The field of tandem mass spectrometry is already rather mature with good books available (Benninghoven et al. 1987 and Wilson et al. 1989).

In order for an instrument to carry out MS-MS, it must be able to do the three operations outlined above. As we have seen however, ion-trap systems capable of MS-MS and $MS^{(n)}$ do not use a tandem arrangement of mass analyzers at all, but rather use a single ion trap for all three operations simultaneously. As has already been stated, these ion-trap tandem mass spectrometer experiments are very sensitive and are now user friendly. The ion trap brings the capability for carrying out MS-MS experiments to the benchtop at relatively low cost.

A summary of mass analyzers and ionization methods is displayed in Table 1.2.

1.5 INTERPRETATION OF EI MASS SPECTRA

Our discussion of interpreting mass spectra is limited to EI mass spectrometry. Fragmentation in EI mass spectra is rich with structural information; mastery of EI mass spectra is especially useful for the organic chemist.

EI mass spectra are routinely obtained at an electron beam energy of 70 eV. The simplest event that occurs is the removal of a single electron from the molecule in the gas phase by an electron of the electron beam to form the molecular ion, which is a radical cation. For example, methanol forms a molecular ion in which the single dot represents the remaining odd electron as seen in Scheme 1.1. When the charge can be localized on one particular atom, the charge is shown on that atom:

$$CH_3\overset{\cdot +}{\underset{\cdot\cdot}{O}}H$$

$$CH_3OH + e^- \longrightarrow CH_3OH^{\cdot +} (m/z\ 32) + 2e^-$$

(Sch 1.1)

Many of these molecular ions disintegrate in 10^{-10}–10^{-3} s to give, in the simplest case, a positively charged fragment and a radical. Many fragment ions are thus formed, and each of these can cleave to yield smaller fragments; examples of possible cleavages for methanol are given in Scheme 1.2.

$$CH_3OH^{\cdot +} \longrightarrow CH_2OH^+ (m/z\ 31) + H^{\cdot}$$
$$CH_3OH^{\cdot +} \longrightarrow CH_3^+ (m/z\ 15) + {}^{\cdot}OH$$
$$CH_2OH^+ \longrightarrow CHO^+ (m/z\ 29) + H_2$$

(Sch 1.2)

If some of the molecular ions remain intact long enough to reach the detector, we see a molecular ion peak. It is important to recognize the molecular ion peak because this gives the molecular weight of the compound. With unit resolution, this weight is the molecular weight to the nearest whole number.

A mass spectrum is a presentation of the masses of the positively charged fragments (including the molecular ion) versus their relative concentrations. The most intense peak in the spectrum, called the base peak, is assigned a value of 100%, and the intensities (height × sensitivity factor) of the other peaks, including the molecular ion peak, are reported as percentages of the base peak. Of course, the molecular ion peak may sometimes be the base peak. In Figure 1.1, the molecular ion peak is m/z 121, and the base peak is m/z 77.

A tabular or graphic presentation of a spectrum may be used. A graph has the advantage of presenting patterns that, with experience, can be quickly recognized. However, a graph must be drawn so that there is no difficulty in distinguishing mass units. Mistaking a peak at, say, m/z 79 for m/z 80 can result in total confusion. The molecular ion peak is usually the peak of highest mass number except for the isotope peaks.

1.5.1 Recognition of the Molecular Ion Peak

Quite often, under electron impact (EI), recognition of the molecular ion peak $(M)^+$ poses a problem. The peak may be very weak or it may not appear at all; how can we be sure that it is the molecular ion peak and not a fragment peak or an impurity? Often the best solution is to obtain a chemical ionization spectrum (see Section 1.3.1.2). The usual result is an intense peak at $[M + 1]^+$ and little fragmentation.

Many peaks can be ruled out as possible molecular ions simply on grounds of reasonable structure requirements. The "nitrogen rule" is often helpful. It states that a molecule of even-numbered molecular weight must contain either no nitrogen or an even number of nitrogen atoms; an odd-numbered molecular weight requires an odd number of nitrogen atoms.* This rule holds for all compounds containing carbon, hydrogen, oxygen, nitrogen, sulfur, and the halogens, as well as many of the less usual atoms such as phosphorus, boron, silicon, arsenic, and the alkaline earths.

A useful corollary states that fragmentation at a single bond gives an odd-numbered ion fragment from an even-numbered molecular ion, and an even-numbered ion fragment from an odd-numbered molecular ion. For this corollary to hold, the ion fragment must contain all of the nitrogen (if any) of the molecular ion.

Consideration of the breakdown pattern coupled with other information will also assist in identifying molecular ions. It should be kept in mind that Appendix A contains fragment formulas as well as molecular formulas. Some of the formulas may be discarded as trivial in attempts to solve a particular problem.

The intensity of the molecular ion peak depends on the stability of the molecular ion. The most stable molecular ions are those of purely aromatic systems. If substituents that have favorable modes of cleavage are present, the molecular ion peak will be less intense, and the fragment peaks relatively more intense. In general, the following group of compounds will, in order of decreasing ability, give prominent molecular ion peaks: aromatic compounds > conjugated alkenes > cyclic compounds > organic sulfides > short, normal alkanes > mercaptans. Recognizable molecular ions are usually produced for these compounds in order of decreasing ability: ketones > amines > esters > ethers > carboxylic acids ~ aldehydes ~ amides ~ halides. The molecular ion is frequently not detectable in aliphatic alcohols, nitrites, nitrates, nitro compounds, nitriles, and in highly branched compounds.

The presence of an M − 15 peak (loss of CH_3), or an M − 18 peak (loss of H_2O), or an M − 31 peak (loss of OCH_3 from methyl esters), and so on, is taken as confirmation of a molecular ion peak. An M − 1 peak is common, and occasionally an M − 2 peak (loss of H_2 by either fragmentation or thermolysis), or even a rare M − 3 peak (from alcohols) is reasonable. Peaks in the range of M − 3 to M − 14, however, indicate that contaminants may be present or that the presumed molecular ion peak is actually a fragment ion peak. Losses of fragments of masses 19–25 are also unlikely (except for loss of F = 19 or HF = 20 from fluorinated compounds). Loss of 16 (O), 17 (OH), or 18 (H_2O) are likely only if an oxygen atom is in the molecule.

* For the nitrogen rule to hold, only unit atomic masses (i.e., integers) are used in calculating the formula masses.

1.5.2 Determination of a Molecular Formula

1.5.2.1 Unit-Mass Molecular Ion and Isotope Peaks. So far, we have discussed the mass spectrum in terms of unit resolutions: The unit mass of the molecular ion of C_7H_7NO (Figure 1.1) is m/z 121—that is, the sum of the unit masses of the most abundant isotopes: (7 × 12 [for ^{12}C]) + (7 × 1 [for ^{1}H]) + (1 × 14 [for ^{14}N] + (1 × 16 [for ^{16}O]) = 121.

In addition, molecular species exist that contain the less abundant isotopes, and these give rise to the "isotope peaks" at M + 1, M + 2, etc. In Figure 1.1, the M + 1 peak is approximately 8% of the intensity of the molecular ion peak, which for this purpose, is assigned an intensity of 100%. Contributing to the M + 1 peak are the isotopes, ^{13}C, ^{2}H, ^{15}N, and ^{17}O. Table 1.3 gives the abundances of these isotopes relative to those of the most abundant isotopes. The only contributor to the M + 2 peak of C_7H_7NO is ^{18}O, whose relative abundance is very low: thus the M + 2 peak is undetected. If only C, H, N, O, F, P, and I are present, the approximate expected percentage (M + 1) and percentage (M + 2) intensities can be calculated by use of the following equations for a compound of formula $C_nH_mN_xO_y$ (note: F, P, and I are monoisotopic and do not contribute and can be ignored for the calculation):

$$\%\,(M + 1) \approx (1.1 \cdot n) + (0.36 \cdot x) \text{ and } \%\,(M + 2) \approx (1.1 \cdot n)^2/200 + (0.2 \cdot y)$$

TABLE 1.3 Relative Isotope Abundances of Common Elements.

Elements	Isotope	Relative Abundance	Isotope	Relative Abundance	Isotope	Relative Abundance
Carbon	^{12}C	100	^{13}C	1.11		
Hydrogen	^{1}H	100	^{2}H	0.016		
Nitrogen	^{14}N	100	^{15}N	0.38		
Oxygen	^{16}O	100	^{17}O	0.04	^{18}O	0.2
Fluorine	^{19}F	100				
Silicon	^{28}Si	100	^{29}Si	5.1	^{30}Si	3.35
Phosphorus	^{31}P	100				
Sulfur	^{32}S	100	^{33}S	0.78	^{34}S	4.4
Chlorine	^{35}Cl	100			^{37}Cl	32.5
Bromine	^{79}Br	100			^{81}Br	98
Iodine	^{127}I	100				

If these isotope peaks are intense enough to be measured accurately, the above calculations may be useful in determining the molecular formula.*

If sulfur or silicon is present, the M + 2 will be more intense. In the case of a single sulfur atom, ^{34}S contributes approximately 4.40% to the M + 2 peak; for a single silicon in the molecule, ^{30}Si contributes about 3.35% to the M + 2 peak (see Section 1.6.15). The effect of several bromine and chlorine atoms is described in Section 1.6.16. Note the appearance of additional isotope peaks in the case of multiple bromine and chlorine atoms. Obviously the mass spectrum should be routinely scanned for the relative intensities of the M + 2, M + 4, and higher isotope peaks, and the relative intensities should be carefully measured. Since F, P, and I are monoisotopic, they can be difficult to spot.

For most of the Problems in this text, the unit-resolution molecular ion, used in conjunction with IR and NMR, will suffice for determining the molecular formula by browsing in Appendix A. For several more difficult Problems, the high-resolution formula masses—for use with Appendix A (see Section 1.5.2.2)—have been supplied.

Table 1.3 lists the principal stable isotopes of the common elements and their relative abundance calculated on the basis of 100 molecules containing the most common isotope. Note that this presentation differs from many isotope abundance tables, in which the sum of all the isotopes of an element adds up to 100%.

1.5.2.2 High-Resolution Molecular Ion. A unique molecular formula (or fragment formula) can often be derived from a sufficiently accurate mass measurement alone (high-resolution mass spectrometry). This is possible because the nuclide masses are not integers (see Table 1.4). For example, we can distinguish at a unit mass of 28 among CO, N_2, CH_2N, and C_2H_4. The exact mass of CO is: 12.0000 (for ^{12}C) + 15.9949 (for ^{16}O) = 27.9949; the exact mass of N_2 is: 2×14.0031 (for ^{14}N) = 28.0062. Similar calculations give an exact mass of 28.0187 for CH_2N and 28.0312 for C_2H_4.

Thus, the mass observed for the molecular ion of CO, for example, is the sum of the exact formula masses of the most abundant isotope of carbon and of oxygen. This differs from a molecular weight of CO based on atomic weights that are the average of

TABLE 1.4 Exact Masses of Isotopes.

Element	Atomic Weight	Nuclide	Mass
Hydrogen	1.00794	^{1}H	1.00783
		D(^{2}H)	2.01410
Carbon	12.01115	^{12}C	12.00000 (std)
		^{13}C	13.00336
Nitrogen	14.0067	^{14}N	14.0031
		^{15}N	15.0001
Oxygen	15.9994	^{16}O	15.9949
		^{17}O	16.9991
		^{18}O	17.9992
Fluorine	18.9984	^{19}F	18.9984
Silicon	28.0855	^{28}Si	27.9769
		^{29}Si	28.9765
		^{30}Si	29.9738
Phosphorus	30.9738	^{31}P	30.9738
Sulfur	32.0660	^{32}S	31.9721
		^{33}S	32.9715
		^{34}S	33.9679
Chlorine	35.4527	^{35}Cl	34.9689
		^{37}Cl	36.9659
Bromine	79.9094	^{79}Br	78.9183
		^{81}Br	80.9163
Iodine	126.9045	^{127}I	126.9045

* There are limitations beyond the difficulty of measuring small peaks: The $^{13}C/^{12}C$ ratio differs with the source of the compound—synthetic compared with a natural source. A natural product from different organisms or regions may show differences. Furthermore, isotope peaks may be more intense than the calculated value because of ion–molecule interactions that vary with the sample concentration or with the class of compound involved.

weights of all natural isotopes of an element (e.g., C = 12.01, O = 15.999).

Table 1.4 gives the masses to four or five decimal places for the common nuclides; it also gives the familiar atomic weights (average weights for the elements).

Appendix A lists molecular and fragment formulas in order of the unit masses. Under each unit mass, the formulas are listed in the standard *Chemical Abstract* system. The formula mass (FM) to four decimal places is given for each formula. Appendix A is designed for browsing, on the assumption that the student has a unit molecular mass from a unit-resolution mass spectrometer and clues from other spectra. Note that the table includes only C, H, N, and O.

1.5.3 Use of the Molecular Formula. Index of Hydrogen Deficiency

If organic chemists had to choose a single item of information above all others that are usually available from spectra or from chemical manipulations, they would certainly choose the molecular formula.

In addition to the kinds and numbers of atoms, the molecular formula gives the index of hydrogen deficiency. The index of hydrogen deficiency is the number of *pairs* of hydrogen atoms that must be removed from the corresponding "saturated" formula to produce the molecular formula of the compound of interest. The index of hydrogen deficiency is also called the number of "sites (or degrees) of unsaturation"; this description is incomplete since hydrogen deficiency can result from cyclic structures as well as from multiple bonds. The index is thus the sum of the number of rings, the number of double bonds, and twice the number of triple bonds.

The index of hydrogen deficiency can be calculated for compounds containing carbon, hydrogen, nitrogen, halogen, oxygen, and sulfur having the generalized molecular formula, $C_nH_mX_xN_yO_z$, from the equation

$$\text{Index} = (n) - (m/2) - (x/2) + (y/2) + 1$$

Thus, the compound C_7H_7NO has an index of $7 - 3.5 + 0.5 + 1 = 5$. Note that divalent atoms (oxygen and sulfur) are not counted in the formula.

For the generalized molecular formula $\alpha_{I}\beta_{II}\gamma_{III}\delta_{IV}$, the index = (IV) − (I/2) + (III/2) + 1, where α is H, D, or halogen (i.e., any monovalent atom), β is O, S, or any other bivalent atom, γ is N, P, or any other trivalent atom, and δ is C, Si, or any other tetravalent atom. The numerals I–IV designate the numbers of the mono-, di-, tri-, and tetravalent atoms, respectively.

For simple molecular formulas, we can arrive at the index by comparison of the formula of interest with the molecular formula of the corresponding saturated compound. Compare C_6H_6 and C_6H_{14}; the index is 4 for the former and 0 for the latter.

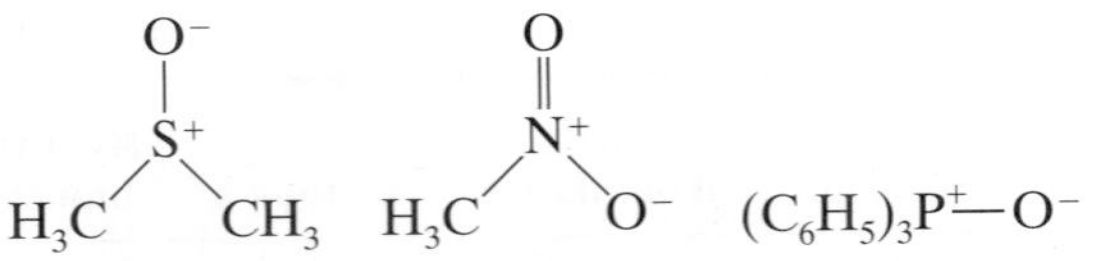

$H_3C-S^+(O^-)-CH_3$ Dimethyl sulfoxide; $H_3C-N^+(=O)-O^-$ nitromethane; $(C_6H_5)_3P^{\pm}-O^-$ triphenylphosphine oxide

FIGURE 1.12 "Polar" Lewis structures of dimethyl sulfoxide, nitromethane, and triphenylphosphine oxide that correctly account for the index of hydrogen deficiency.

The index for C_7H_7NO is 5, and a possible structure is benzamide (see Figure 1.1). Of course, other isomers (i.e., compounds with the same molecular formula) are possible, such as

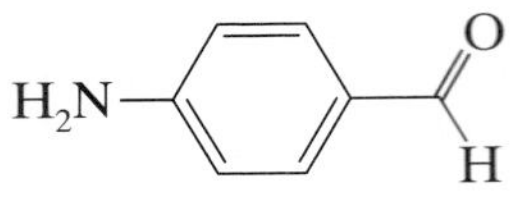

$H_2N-C_6H_4-CHO$

Note that the benzene ring itself accounts for four "sites of unsaturation": three for the double bonds and one for the ring.

Polar structures must be used for compounds containing an atom in a higher valence state, such as sulfur or phosphorus. Thus, if we treat sulfur in dimethyl sulfoxide (DMSO) formally as a divalent atom, the calculated index, 0, is compatible with the structure in Figure 1.12. We must use only formulas with filled valence shells; that is, the Lewis octet rule must be obeyed.

Similarly, if we treat the nitrogen in nitromethane as a trivalent atom, the index is 1, which is compatible with Figure 1.12. If we treat phosphorus in triphenylphosphine oxide as trivalent, the index is 12, which fits the Lewis structure in Figure 1.12. As an example, let us consider the molecular formula $C_{13}H_9N_2O_4BrS$. The index of hydrogen deficiency would be $13 - 10/2 + 2/2 + 1 = 10$ and a consistent structure would be

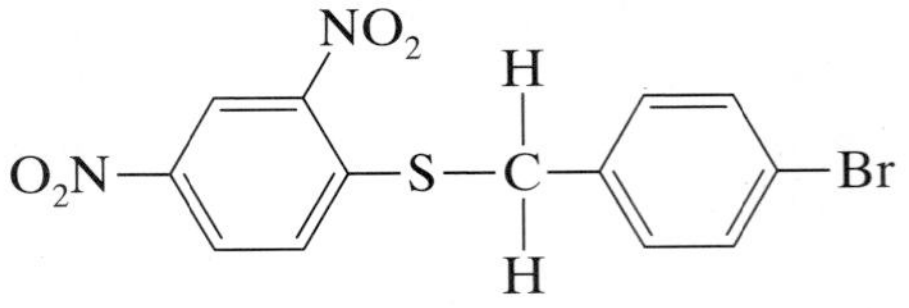

$O_2N-C_6H_3(NO_2)-S-CH_2-C_6H_4-Br$

(Index of hydrogen deficiency = 4 per benzene ring and 1 per NO_2 group.)

The formula above for the index can be applied to fragment ions as well as to the molecular ion. When it is applied to even-electron (all electrons paired) ions, the result is always an odd multiple of 0.5. As an example, consider $C_7H_5O^+$ with an index of 5.5. A reasonable structure is

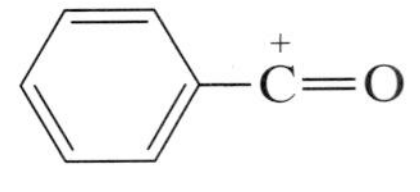

$C_6H_5-\overset{+}{C}=O$

since 5 1/2 pairs of hydrogen atoms would be necessary to obtain the corresponding saturated formula $C_7H_{16}O$ ($C_nH_{2n+2}O$). Odd-electron fragment ions will always give integer values of the index.

Terpenes often present a choice between a double bond and a ring structure. This question can readily be resolved on a microgram scale by catalytically hydrogenating the compound and rerunning the mass spectrum. If no other easily reducible groups are present, the increase in the mass of the molecular ion peak is a measure of the number of double bonds and other "unsaturated sites" must be rings.

Such simple considerations give the chemist very ready information about structure. As another example, a compound containing a single oxygen atom might quickly be determined to be an ether or a carbonyl compound simply by counting "unsaturated sites."

1.5.4 Fragmentation

As a first impression, fragmenting a molecule with a huge excess of energy would seem a brute-force approach to molecular structure. The rationalizations used to correlate spectral patterns with structure, however, can only be described as elegant, though sometimes arbitrary. The insight of such pioneers as McLafferty, Beynon, Stenhagen, Ryhage, and Meyerson led to a number of rational mechanisms for fragmentation. These were masterfully summarized and elaborated by Biemann (1962), Budzikiewicz (1967), and others.

Generally, the tendency is to represent the molecular ion with a localized charge. Budzikiewicz et al. (1967) approach is to localize the positive charge on either a π bond (except in conjugated systems), or on a heteroatom. Whether or not this concept is totally rigorous, it is, at the least, a pedagogic *tour de force.* We shall use such locally charged molecular ions in this book.

Structures **A, B,** and **C** in Figure 1.13, for example, represent the molecular ion of cyclohexadiene. Compound **A** is a delocalized structure with one less electron than the original uncharged diene; both the electron and the positive charge are delocalized over the π system. Since the electron removed to form the molecular ion is a π electron, other structures, such as **B** or **C** (valence bond structures) can be used. Structures such as **B** and **C** localize the electron and the positive charge and thus are useful for describing fragmentation processes.

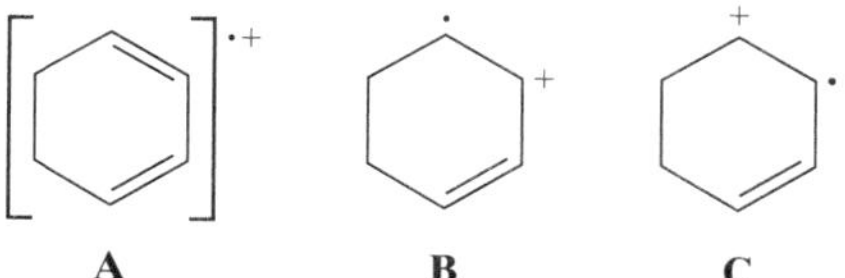

FIGURE 1.13 Different representations of the radical cation of cyclohexadiene.

Fragmentation is initiated by electron impact. Only a small part of the driving force for fragmentation is energy transferred as the result of the impact. The major driving force is the cation-radical character that is imposed upon the structure.

Fragmentation of the odd-electron molecular ion (radical-cation, $M^{\cdot+}$) may occur by homolytic or heterolytic cleavage of a single bond. In homolytic cleavage (Scheme 1.3, ***I***) each electron "moves" independently as shown by a (single-barbed) fishhook: the fragments are an even-electron cation and a free radical (odd electron). To prevent clutter, only one of each pair of fishhooks need be shown (Scheme 1.3, ***II***). In heterolytic cleavage, a pair of electrons "move" together toward the charged site as shown by the conventional curved arrow; the fragments are again an even-electron cation and a radical, but here the final charge site is on the alkyl product. (Scheme 1.3, ***III***)

$$I\quad CH_3{-}CH_2{-}\overset{+\cdot}{O}{-}R \longrightarrow \dot{C}H_3 + H_2C{=}\overset{+}{O}{-}R$$

$$II\quad CH_3{-}CH_2{-}\overset{+\cdot}{O}{-}R \longrightarrow \dot{C}H_3 + H_2C{=}\overset{+}{O}{-}R$$

$$III\quad CH_3{-}CH_2{-}CH_2{-}\overset{\cdot+}{Br} \longrightarrow CH_3{-}CH_2{-}CH_2^+ + Br^{\cdot}$$

$$IV\quad CH_3{-}CH_2{-}CH_2^+ \longrightarrow CH_3^+ + H_2C{=}CH$$

(Sch 1.3)

In the absence of rings (whose fragmentation requires cleavage of two or more bonds), most of the prominent fragments in a mass spectrum are even-electron cations formed as above by a single cleavage. Further fragmentation of an even-electron cation usually results in another even-electron cation and an-even-electron neutral molecule or fragment (Scheme 1.3, ***IV***).

Simultaneous or consecutive cleavage of several bonds may occur when energy benefits accrue from formation of a highly stabilized cation and/or a stable radical, or a neutral molecule, often through a well-defined low-energy pathway. These are treated in Section 1.5.5 (rearrangements) and in Section 1.6 under individual chemical classes.

The probability of cleavage of a particular bond is related to the bond strength, to the possibility of low energy transitions, and to the stability of the fragments, both charged and uncharged, formed in the fragmentation process. Our knowledge of pyrolytic cleavages can be used, to some extent, to predict likely modes of cleavage of the molecular ion. Because of the extremely low pressure in the mass spectrometer, there are very few fragment collisions; we are dealing largely with unimolecular decompositions. This assumption, backed by a large collection of reference spectra, is the basis for the vast amount of information available from the fragmentation pattern of a molecule. Whereas conventional organic chemistry deals with reactions initiated by chemical reagents, by thermal energy, or by light, mass spectrometry is concerned with the consequences suffered by an organic molecule at a vapor pressure of about 10^{-6} mm Hg struck by an ionizing electron beam.

A number of general rules for predicting prominent peaks in EI spectra can be written and rationalized by using standard concepts of physical organic chemistry.

1. The relative height of the molecular ion peak is greatest for the straight-chain compound and decreases as the degree of branching increases (see rule 3).
2. The relative height of the molecular ion peak usually decreases with increasing molecular weight in a homologous series. Fatty esters appear to be an exception.
3. Cleavage is favored at alkyl-substituted carbon atoms: the more substituted, the more likely is cleavage. This is a consequence of the increased stability of a tertiary carbocation over a secondary, which in turn is more stable than a primary.

Cation stability order:

$CH_3^+ < R_2CH_2^+ < R_3CH^+ < R_3C^+$

Generally, the largest substituent at a branch is eliminated most readily as a radical, presumably because a long-chain radical can achieve some stability by delocalization of the lone electron.

4. Double bonds, cyclic structures, and especially aromatic (or heteroaromatic) rings stabilize the molecular ion and thus increase the probability of its appearance.
5. Double bonds favor allylic cleavage and give the resonance-stabilized allylic carbocation. This rule does not hold for simple alkenes because of the ready migration of the double bond, but it does hold for cycloalkenes.
6. Saturated rings tend to lose alkyl side chains at the α bond. This is merely a special case of branching (rule 3). The positive charge tends to stay with the ring fragment. See Scheme 1.4.

R ·+ −R· +

(Sch 1.4)

Unsaturated rings can undergo a *retro*-Diels-Alder reaction Scheme 1.5:

·+ ·+ + CH_2 ‖ CH_2

(Sch 1.5)

7. In alkyl-substituted aromatic compounds, cleavage is very probable at the bond β to the ring, giving the resonance-stabilized benzyl ion or, more likely, the tropylium ion (see Scheme 1.6).

R · + −R· CH_2 + $\overset{+}{C}H_2$ H H H H H H H H H^- shift H H H + H H H H

(Sch 1.6)

8. The C—C bonds next to a heteroatom are frequently cleaved, leaving the charge on the fragment containing the heteroatom whose nonbonding electrons provide resonance stabilization.
9. Cleavage is often associated with elimination of small, stable, neutral molecules, such as carbon monoxide, olefins, water, ammonia, hydrogen sulfide, hydrogen cyanide, mercaptans, ketene, or alcohols, often with rearrangement (Section 1.5.5).

It should be kept in mind that the fragmentation rules above apply to EI mass spectrometry. Since other ionizing (CI, etc.) techniques often produce molecular ions with much lower energy or quasimolecular ions with very different fragmentation patterns, different rules govern the fragmentation of these molecular ions.

1.5.5 Rearrangements

Rearrangement ions are fragments whose origin cannot be described by simple cleavage of bonds in the molecular ion but are a result of intramolecular atomic rearrangement during fragmentation. Rearrangements involving migration of hydrogen atoms in molecules that contain a heteroatom are especially common. One important example is the so-called McLafferty rearrangement; it is illustrated in Scheme 1.7 for the general case.

where Y = H, R, OH, OR, NH_2

$-RHC{=}CH_2$

(Sch 1.7)

To undergo a McLafferty rearrangement, a molecule must possess an appropriately located heteroatom (e.g., O), a π system (usually a double bond), and an abstractable hydrogen atom γ to the C=O system.

Such rearrangements often account for prominent characteristic peaks and are consequently very useful for our purpose. They can frequently be rationalized on the basis of low-energy transitions and increased stability of the products. Rearrangements resulting in elimination of a stable neutral molecule are common (e.g., the alkene product in the McLafferty rearrangement) and will be encountered in the discussion of mass spectra of chemical classes.

Rearrangement peaks can be recognized by considering the mass (m/z) number for fragment ions and for their corresponding molecular ions. A simple (no rearrangement) cleavage of an even-numbered molecular ion gives an odd-numbered fragment ion and simple cleavage of an odd-numbered molecular ion gives an even-numbered fragment. Observation of a fragment ion mass different by 1 unit from that expected for a fragment resulting from simple cleavage (e.g., an even-numbered fragment mass from an even-numbered molecular ion mass) indicates rearrangement of hydrogen has accompanied fragmentation. Rearrangement peaks may be recognized by considering the corollary to the "nitrogen rule" (Section 1.5.1). Thus, an even-numbered peak derived from an even-numbered molecular ion is a result of two cleavages, which may involve a rearrangement.

"Random" rearrangements of hydrocarbons were noted by the early mass spectrometrists in the petroleum industry. For example, the rearrangement of the *neo*-pentyl radical-cation to the ethyl cation, shown in Scheme 1.8, defies a straightforward explanation.

$$\left[H_3C{-}C(CH_3)_2{-}CH_3 \right]^{\cdot+} \longrightarrow [C_2H_5]^+$$

(Sch 1.8)

1.6 MASS SPECTRA OF SOME CHEMICAL CLASSES

Mass spectra of a number of chemical classes are briefly described in this section in terms of the most useful generalizations for identification. For more details, the references cited should be consulted (in particular, the thorough treatment by Budzikiewicz, Djerassi, and Williams, 1967). Databases are available both from publishers and as part of instrument capabilities. The references are selective rather than comprehensive. A table of frequently encountered fragment ions is given in Appendix B. A table of fragments (uncharged) that are commonly eliminated and some structural inferences are presented in Appendix C. More exhaustive listings of common fragment ions have been compiled (see References).

1.6.1 Hydrocarbons

1.6.1.1 Saturated Hydrocarbons. Most of the early work in mass spectrometry was done on hydrocarbons of interest to the petroleum industry. Rules 1–3, (Section 1.5.4) apply quite generally; rearrangement peaks, though common, are not usually intense (random rearrangements), and numerous reference spectra are available.

The molecular ion peak of a straight-chain, saturated hydrocarbon is always present, though of low intensity for long-chain compounds. The fragmentation pattern is characterized by clusters of peaks, and the corresponding peaks of each cluster are 14 mass units (CH_2) apart. The largest peak in each cluster represents a C_nH_{2n+1} fragment and thus occurs at $m/z = 14n + 1$; this is accompanied by C_nH_{2n} and C_nH_{2n-1} fragments. The most abundant fragments are at C_3 and C_4, and the fragment abundances decrease in a smooth curve down to $[M - C_2H_5]^+$; the $[M - CH_3]^+$ peak is characteristically very weak or missing. Compounds containing more than eight carbon atoms show fairly similar spectra; identification then depends on the molecular ion peak.

Spectra of branched saturated hydrocarbons are grossly similar to those of straight-chain compounds, but the smooth curve of decreasing intensities is broken by preferred fragmentation at each branch. The smooth curve for the *n*-alkane in Figure 1.14 (top) is in contrast to the discontinuity at C_{12} for the branched alkane (Figure 1.14, bottom). This discontinuity indicates that the longest branch of 5-methylpentadecane has 10 carbon atoms.

In the bottom spectrum of Figure 1.14, the peaks at m/z 169 and 85 represent cleavage on either side of the branch with charge retention on the substituted carbon atom. Subtraction of the molecular weight from the sum of these fragments accounts for the fragment —CH—CH_3. Again, we appreciate the absence of a C_{11} unit, which cannot form by a single cleavage. Finally, the presence of a distinct M − 15 peak also indicates a methyl branch. The fragment resulting from cleavage at a branch tends to lose a single hydrogen atom so that the resulting C_nH_{2n} peak is prominent and sometimes more intense than the corresponding C_nH_{2n+1} peak.

A saturated ring in a hydrocarbon increases the relative intensity of the molecular ion peak and favors cleavage at the bond connecting the ring to the rest of the molecule (rule 6, Section 1.5.4). Fragmentation of the ring is usually characterized by loss of two carbon atoms as C_2H_4 (28) and C_2H_5 (29). This tendency to lose even-numbered fragments, such as C_2H_4 gives a spectrum that contains a greater proportion of even-numbered mass ions than the spectrum of an acyclic hydrocarbon. As in branched hydrocarbons, C—C cleavage is accompanied by loss of a hydrogen atom. The characteristic peaks are therefore in the C_nH_{2n-1} and C_nH_{2n-2} series.

The mass spectrum of cyclohexane (Figure 1.15) shows a much more intense molecular ion than those of acyclic compounds, since fragmentation requires the cleavage of two carbon-carbon bonds. This spectrum has its base peak at m/z 56 (because of loss of C_2H_4) and a large peak at m/z 41, which is a fragment in the C_nH_{2n-1} series with $n = 3$.

1.6.1.2 Alkenes (Olefins) The molecular ion peak of alkenes, especially polyalkenes, is usually distinct. Location of the double bond in acyclic alkenes is difficult because of its facile migration in the fragments. In cyclic (especially polycyclic) alkenes, location of the double bond is frequently evident as a result of a strong tendency for allylic cleavage without much double-bond migration (rule 5, Section 1.5.4). Conjugation with a carbonyl group also fixes the position of the double bond. As with saturated hydrocarbons, acyclic alkenes are characterized by clusters of peaks at intervals of 14 units. In these clusters the C_nH_{2n-1} and C_nH_{2n} peaks are more intense than the C_nH_{2n+1} peaks.

The mass spectrum of β-myrcene, a terpene, is shown in Figure 1.16. The peaks at m/z 41, 55, and 69 correspond to the formula C_nH_{2n-1} with $n = 3, 4$, and 5, respectively. Formation of the m/z 41 peak must

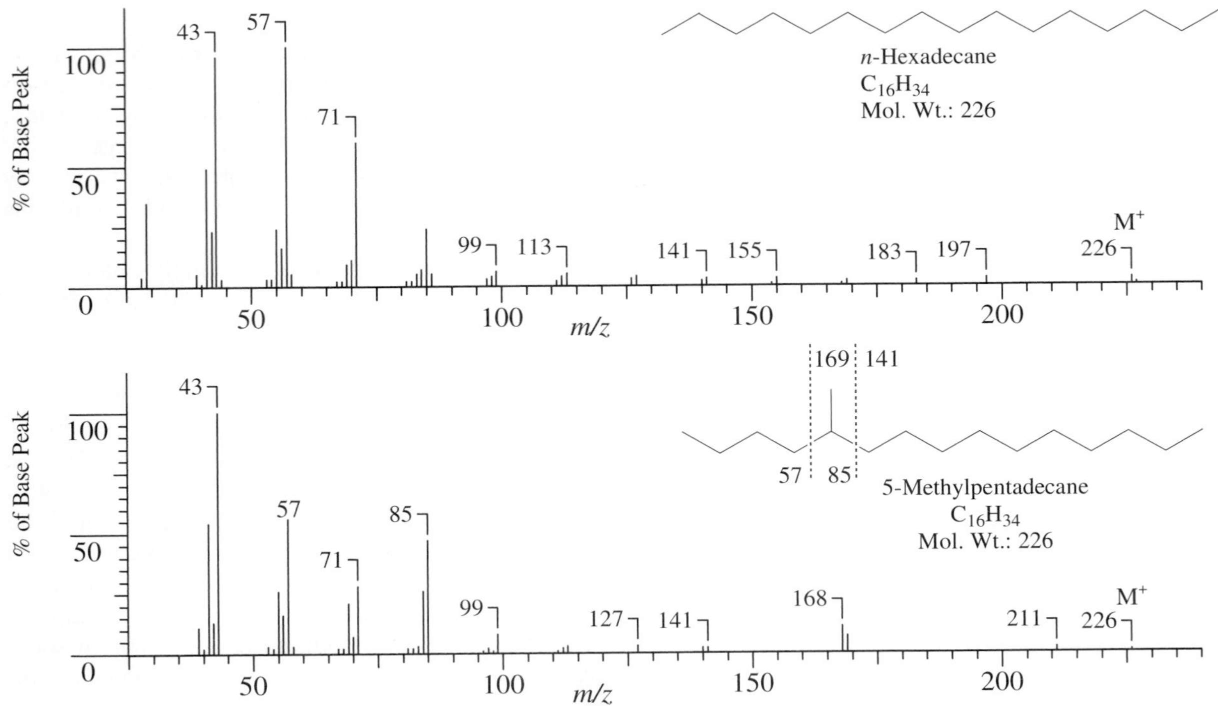

FIGURE 1.14 EI mass spectra of isomeric C_{16} hydrocarbons.

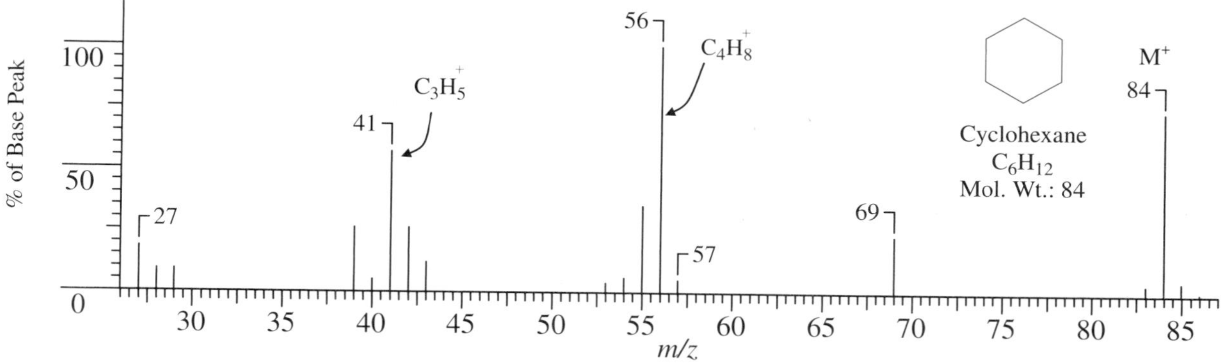

FIGURE 1.15 EI mass spectrum of cyclohexane.

involve isomerization. The peaks at m/z 67 and 69 are the fragments from cleavage of a bi-allylic bond, which is shown in Scheme 1.9.

m/z 69 or m/z 67

(Sch 1.9)

The peak at m/z 93 is rationalized in Scheme 1.10 as a structure of formula $C_7H_9^+$ formed by double bond isomerization (resulting in increased conjugation), followed by allylic cleavage. The ion at m/z 93 has at least two important resonance forms that contribute to its stability. As an exercise, the student is encouraged to draw them.

m/z 93

(Sch 1.10)

Cyclic alkenes usually show a distinct molecular ion peak. A unique mode of cleavage is the *retro*-Diels-Alder reaction. This reaction is illustrated with limonene in Scheme 1.11. A retro-Diels-Alder reaction in this example gives two isoprene molecules. Since the reaction is an example of a rearrangement, one of the isoprene moieties is a neutral molecule.

(Sch 1.11)

1.6.1.3 Aromatic and Aralkyl Hydrocarbons. An aromatic ring in a molecule stabilizes the molecular ion peak (rule 4, Section 1.5.4), which is usually sufficiently large that accurate intensity measurements can be made on the M + 1 and M + 2 peaks.

Figure 1.17 is the mass spectrum of naphthalene. The molecular ion peak is also the base peak, and the largest fragment peak, m/z 51, is only 12.5% as intense as the molecular ion peak.

An alkyl-substituted benzene ring frequently gives a prominent peak (often the base peak) at m/z 91

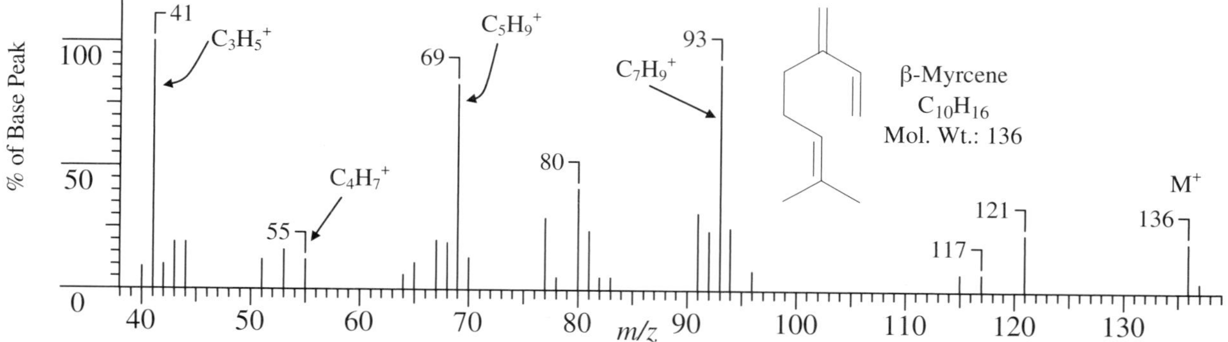

FIGURE 1.16 EI mass spectrum of β-myrcene.

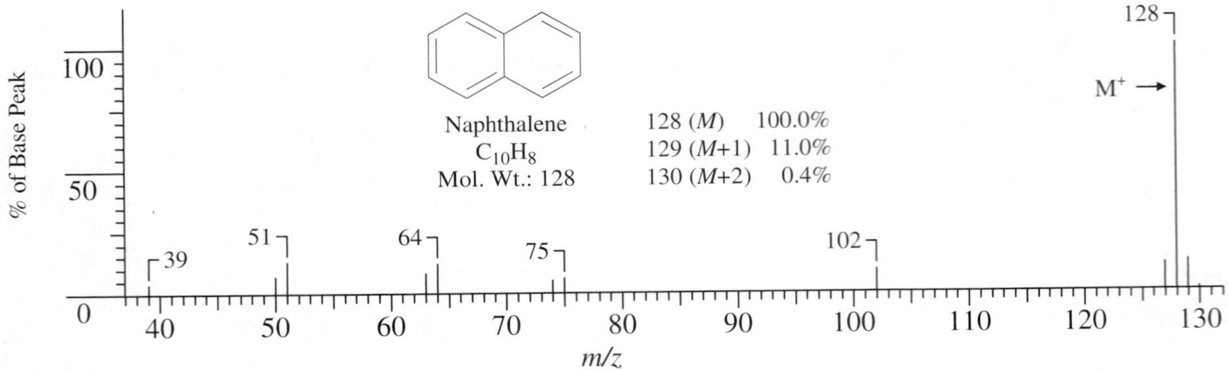

FIGURE 1.17 EI mass spectrum of naphthalene.

($C_6H_5CH_2^+$). Branching at the α-carbon leads to masses higher than 91, by increments of 14, the largest substituent being eliminated most readily (rule 3, Section 1.5.4). The mere presence of a peak at mass 91, however, does not preclude branching at the α-carbon because this highly stabilized fragment may result from rearrangements. A distinct and sometimes prominent M − 1 peak results from similar benzylic cleavage of a C—H bond.

It has been proposed that, in most cases, the ion of mass 91 is a tropylium rather than a benzylic cation. This explains the ready loss of a methyl group from xylenes (Scheme 1.12), although toluene does not easily lose a methyl group. The incipient molecular radical ion of xylene rearranges to the methylcyloheptatriene radical ion, which then cleaves to the tropylium ion ($C_7H_7^+$). The frequently observed peak at m/z 65 results from elimination of a neutral acetylene molecule from the tropylium ion.

$-CH_3\cdot$ m/z 91

(Sch 1.12)

Hydrogen migration with elimination of a neutral alkene molecule accounts for the peak at m/z 92 observed when the alkyl group is longer than C_2. Scheme 1.13 illustrates with a general example. Note again that this is an example of a rearrangement.

$-H_2C{=}CHR$ m/z 92

(Sch 1.13)

A characteristic cluster of ions resulting from an α cleavage and hydrogen migration in monoalkylbenzenes appears at m/z 77 ($C_6H_5^+$), 78 ($C_6H_6^+$), and 79 ($C_6H_7^+$).

Alkylated polyphenyls and alkylated polycyclic aromatic hydrocarbons exhibit the same β cleavage as alkylbenzene compounds.

1.6.2 Hydroxy Compounds

1.6.2.1 Alcohols The molecular ion peak of a primary or secondary alcohol is usually quite small and for a tertiary alcohol is often undetectable. The molecular ion of 1-pentanol is extremely weak compared with its near homologs. Expedients such as CI, or derivatization, may be used to obtain the molecular weight.

Cleavage of the C—C bond next to the oxygen atom is of general occurrence (rule 8, Section 1.5.4). Thus, primary alcohols show a prominent peak resulting from $^+CH_2$—OH (m/z 31). Secondary and tertiary alcohols cleave analogously to give a prominent peak resulting from $^+$CHR—OH (m/z 45, 59, 73, etc.), and $^+$CRR′—OH (m/z 59, 73, 87, etc.), respectively. The largest substituent is expelled most readily (rule 3). Occasionally, the C—H bond next to the oxygen atom is cleaved; this less (or least) favored pathway gives rise to an M − 1 peak.

Primary alcohols, in addition to the principal C—C cleavage next to the oxygen atom, show a homologous series of peaks of progressively decreasing intensity resulting from cleavage at C—C bonds successively removed from the oxygen atom. In long-chain ($>C_6$) alcohols, the fragmentation becomes dominated by the hydrocarbon pattern; in fact, the spectrum resembles that of the corresponding alkene. The spectrum in the vicinity of a very weak or missing molecular ion peak of a primary alcohol is sometimes complicated by weak M − 2 and M − 3 peaks.

A distinct and sometimes prominent peak can usually be found at M − 18 from loss of water. This peak is most noticeable in spectra of primary alcohols. This elimination by electron impact has been rationalized and a mechanism in which a δ-hydrogen is lost as shown in Scheme 1.14 ***I***. A similar mechanism can be drawn in which a γ-hydrogen is lost. The M − 18 peak is frequently exaggerated by thermal decomposition of higher alcohols on hot inlet surfaces. Elimination of water, together with elimination of an alkene from primary alcohols (see Scheme 1.14 ***II***), accounts for the presence of a peak at M − (alkene + H_2O), that is, a peak at M − 46, M − 74, M − 102, . . .

I $\xrightarrow{}$ $\xrightarrow{-H_2O}$

II $\xrightarrow[-H_2O]{-H_2C=CH_2}$ $[H_2C{=}CHR]^{\cdot+}$

Sch 1.14

Alcohols containing branched methyl groups (e.g., terpene alcohols) frequently show a fairly strong peak at M − 33 resulting from loss of CH_3 and H_2O.

Cyclic alcohols undergo fragmentation by complicated pathways; for example, cyclohexanol (M = *m/z* 100) forms $C_6H_{11}O^+$ by simple loss of the α-hydrogen, loses H_2O to form $C_6H_{10}^{+}$ (which appears to have more than one possible bridged bicyclic structure), and forms $C_3H_5O^+$ (*m/z* 57) by a complex ring cleavage pathway.

A peak at *m/z* 31 (see above) is quite diagnostic for a primary alcohol provided it is more intense than peaks at *m/z* 45, 59, 73 . . . . However, the first-formed ion of a secondary alcohol can decompose further to give a moderately intense *m/z* 31 ion.

Figure 1.18 gives the characteristic spectra of isomeric primary, secondary, and tertiary C_5 alcohols.

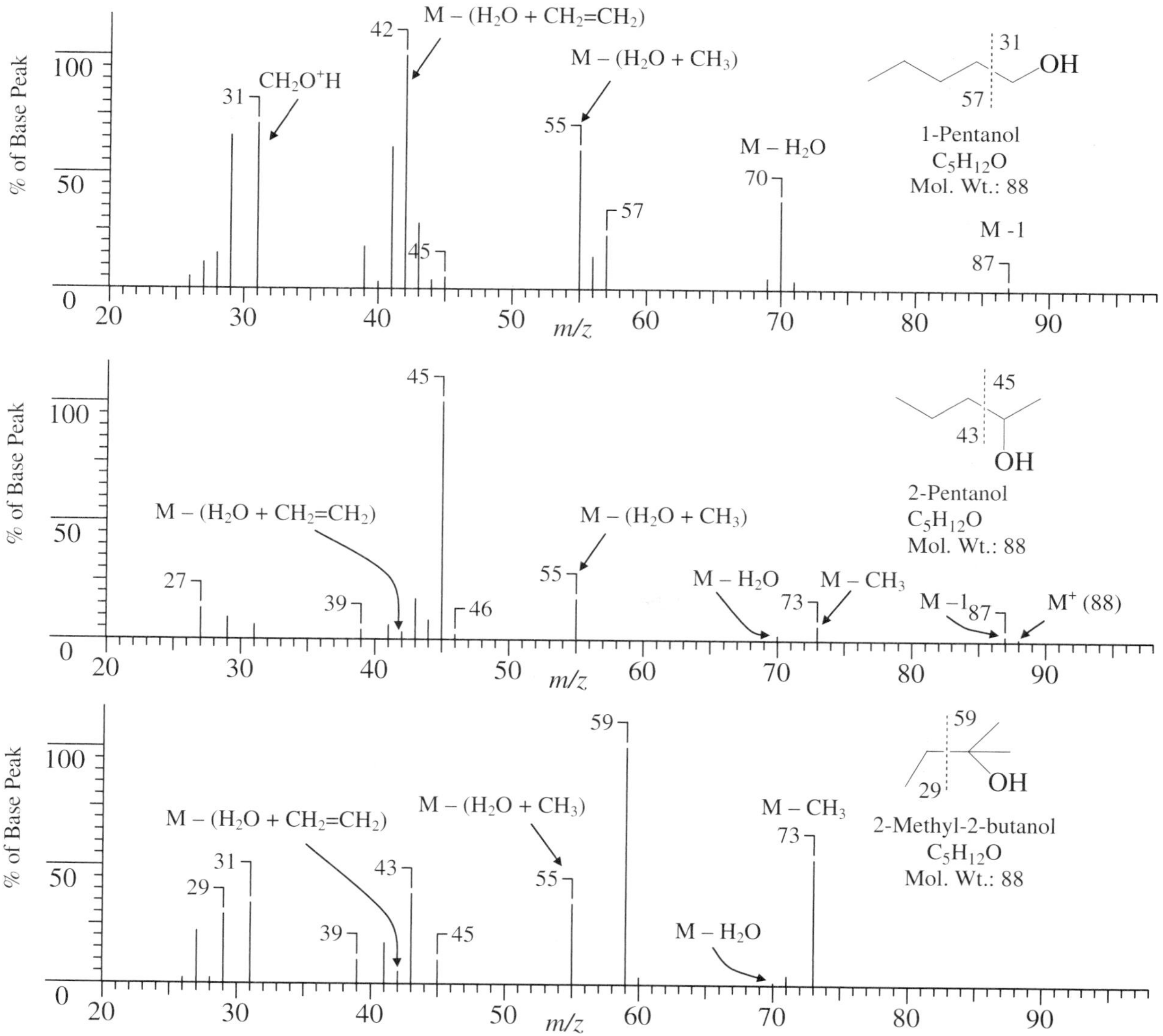

FIGURE 1.18 EI mass spectra of isomeric pentanols.

Benzyl alcohols and their substituted homologs and analogs constitute a distinct class. Generally, the parent peak is strong. A moderate benzylic peak (M − OH) may be present as expected from cleavage β to the ring. A complicated sequence leads to prominent M − 1, M − 2, and M − 3 peaks. Benzyl alcohol itself fragments to give sequentially the M − 1 ion, the $C_6H_7^+$ ion by loss of CO, and the $C_6H_5^+$ ion by loss of H_2 (see Scheme 1.15).

(Sch 1.15)

Loss of H_2O to give a distinct M − 18 peak is a common feature, especially pronounced and mechanistically straightforward in some *ortho*-substituted benzyl alcohols. The loss of water shown in Scheme 1.16 works equally well with an oxygen atom at the *ortho*-position (a phenol). The aromatic cluster at m/z 77, 78, and 79 resulting from complex degradation is prominent here also.

(Sch 1.16)

1.6.2.2 Phenols. A conspicuous molecular ion peak facilitates identification of phenols. In phenol itself, the molecular ion peak is the base peak, and the M − 1 peak is small. In cresols, the M − 1 peak is larger than the molecular ion as a result of a facile benzylic C—H cleavage. A rearrangement peak at m/z 77 and peaks resulting from loss of CO (M − 28) and CHO (M − 29) are usually found in phenols.

The mass spectrum of ethyl phenol, a typical phenol, is shown in Figure 1.19. This spectrum shows that a methyl group is lost much more readily than an α-hydrogen.

1.6.3 Ethers

1.6.3.1 Aliphatic Ethers (and Acetals). The molecular ion peak (two mass units larger than that of an analogous hydrocarbon) is small, but larger sample size usually will make the molecular ion peak or the M + 1 peak obvious ($H^\cdot$ transfer during ion-molecule collision).

The presence of an oxygen atom can be deduced from strong peaks at m/z 31, 45, 59, 73, ... These peaks represent the RO^+ and $ROCH_2^+$ fragments. Fragmentation occurs in two principal ways

1. Cleavage of the C—C bond next to the oxygen atom (α, β bond, rule 8, Section 1.5.4). One or the other of these oxygen-containing ions may account for the base peak. In the case shown in Figure 1.20, the first cleavage (i.e., at the branch position to lose the larger fragment) is preferred. However, the first-formed fragment decomposes further by loss of ethylene to give the base peak; this decomposition is important when the α-carbon is substituted (see McLafferty rearrangement, Section 1.5.5)
2. C—O bond cleavage with the charge remaining on the alkyl fragment. The spectrum of long-chain ethers becomes dominated by the hydrocarbon pattern.

Acetals are a special class of ethers. Their mass spectra are characterized by an extremely weak molecular ion peak, by the prominent peaks at M − R and M − OR (and/or M − OR′), and a weak peak at M − H. Each of these cleavages is mediated by an oxygen atom and thus facile. As usual, elimination of the largest group is preferred. As with aliphatic ethers, the first-formed oxygen-containing fragments can decompose further with hydrogen migration and alkene elimination. Ketals behave similarly.

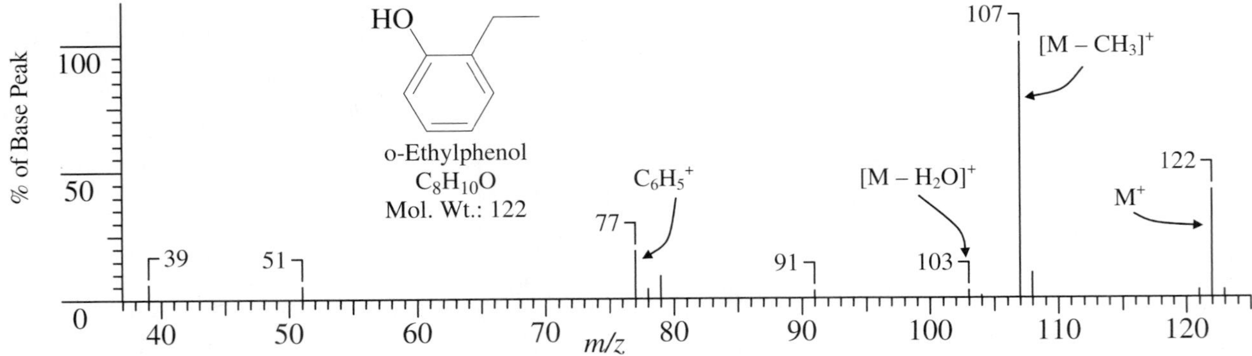

FIGURE 1.19 EI mass spectrum of *o*-ethylphenol.

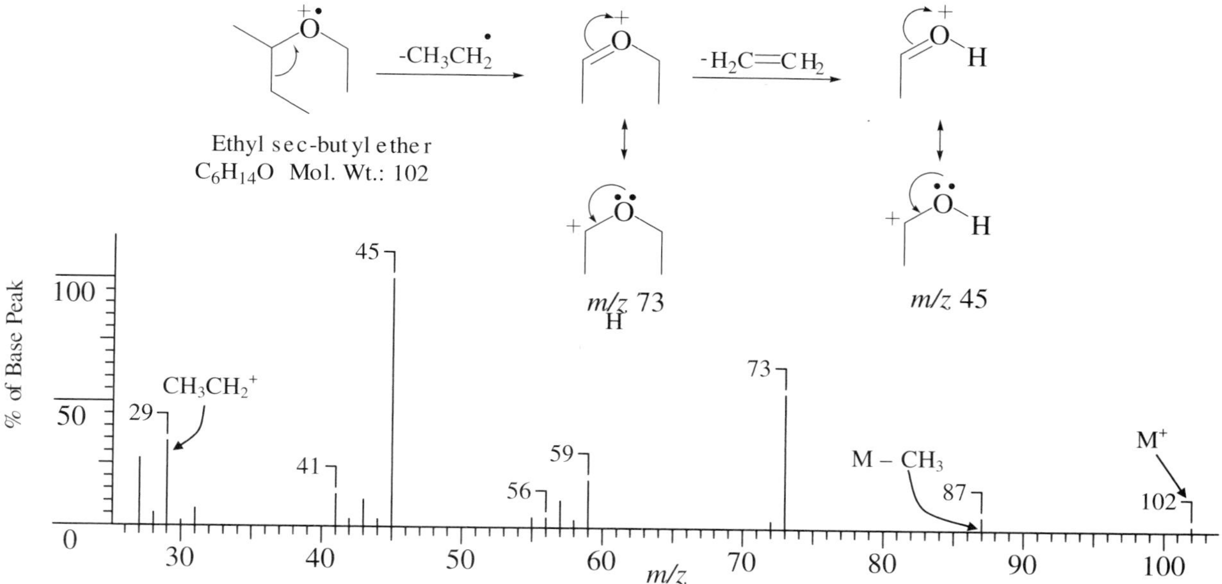

FIGURE 1.20 EI mass spectrum of ethyl *sec*-butyl ether.

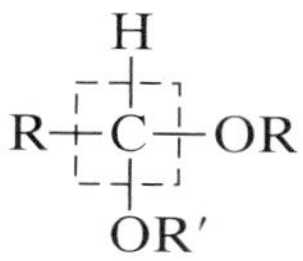

R—C(H)(OR′)—OR

1.6.3.2 Aromatic Ethers.

The molecular ion peak of aromatic ethers is prominent. Primary cleavage occurs at the bond β to the ring, and the first-formed ion can decompose further. Thus, anisole (Figure 1.21, MW 108) gives ions of *m/z* 93 and 65. The characteristic aromatic peaks at *m/z* 78 and 77 may arise from anisole.

When the alkyl portion of an aromatic alkyl ether is C_2 or larger, cleavage β to the ring is accompanied by hydrogen migration (Scheme 1.17) as noted above for alkylbenzenes. Clearly, cleavage is mediated by the ring rather than by the oxygen atom; C—C cleavage next to the oxygen atom is insignificant.

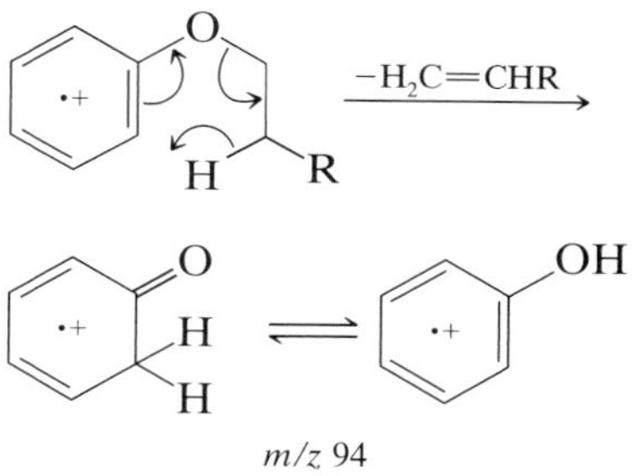

$-H_2C{=}CHR$

m/z 94

(Sch 1.17)

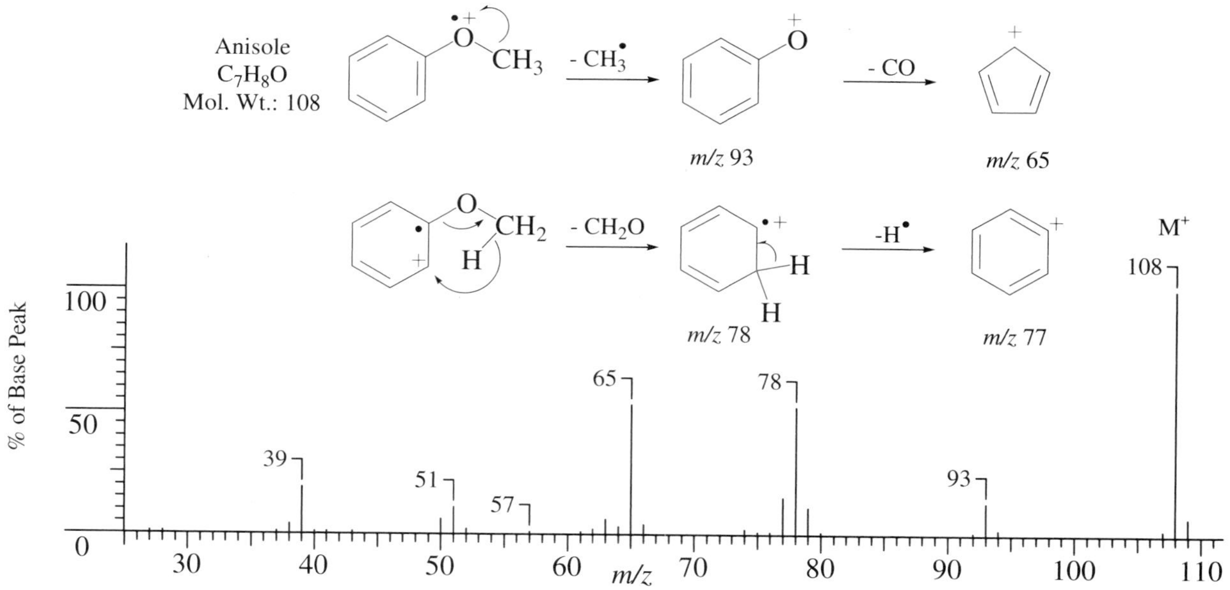

Anisole
C_7H_8O
Mol. Wt.: 108

- $CH_3^\bullet$ → *m/z* 93 → - CO → *m/z* 65

- CH_2O → *m/z* 78 → $-H^\bullet$ → *m/z* 77

% of Base Peak: 100, 50, 0; peaks 39, 51, 57, 65, 78, 93, M$^+$ 108; *m/z* 30–110

FIGURE 1.21 EI mass spectrum of anisole.

Diphenyl ethers show peaks at M − H, M − CO, and M − CHO by complex rearrangements.

1.6.4 Ketones

1.6.4.1 Aliphatic Ketones. The molecular ion peak of ketones is usually quite pronounced. Major fragmentation peaks of aliphatic ketones result from cleavage at one of the C—C bonds adjacent to the oxygen atom, the charge remaining with the resonance-stabilized acylium ion (Scheme 1.18). Thus, as with alcohols and ethers, cleavage is mediated by the oxygen atom. This cleavage gives rise to a peak at m/z 43 or 57 or 71 . . . . The base peak very often results from loss of the larger alkyl group.

$$R_2C{=}\ddot{O}^{+\cdot} \xrightarrow{-R^\cdot} R{-}C{\equiv}\ddot{O}^{+} \longleftrightarrow R{-}\overset{+}{C}{=}\ddot{O}$$

(Sch 1.18)

When one of the alkyl chains attached to the C=O group is C_3 or longer, cleavage of the C—C bond once removed (α, β-bond) from the C=O group occurs with hydrogen migration to give a major peak (McLafferty rearrangement, Scheme 1.19). Simple cleavage of the α, β-bond, which does not occur to any extent, would give an ion of low stability because it would have two adjacent positive centers.

$$\xrightarrow{-RHC{=}CH_2}$$

(Sch 1.19)

Note that in long-chain ketones the hydrocarbon peaks are indistinguishable (without the aid of high-resolution techniques) from the acyl peaks, since the mass of the C=O unit (28) is the same as two methylene units. The multiple cleavage modes in ketones sometimes make difficult the determination of the carbon chain configuration. Reduction of the carbonyl group to a methylene group yields the corresponding hydrocarbon whose fragmentation pattern leads to the carbon skeleton.

1.6.4.2 Cyclic Ketones. The molecular ion peak in cyclic ketones is prominent. As with acyclic ketones, the primary cleavage of cyclic ketones is adjacent to the C=O group, but the ion thus formed must undergo further cleavage in order to produce a fragment. The base peak in the spectrum of cyclopentanone and of cyclohexanone (Figure 1.22) is m/z 55. The mechanisms are similar in both

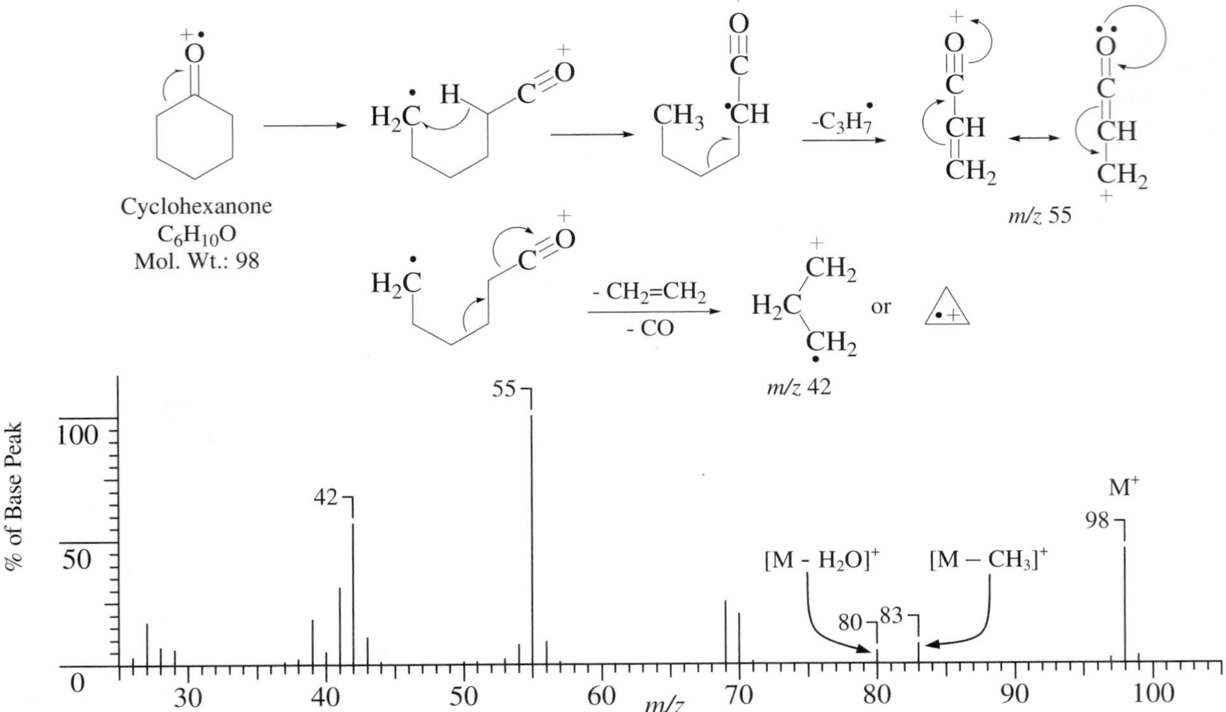

FIGURE 1.22 EI mass spectrum of cyclohexanone.

cases: hydrogen shift to convert a primary radical to a conjugated secondary radical followed by formation of the resonance-stabilized ion, *m/z* 55. The other distinctive peaks at *m/z* 83 and 42 in the spectrum of cyclohexanone have been rationalized as depicted in Figure 1.22.

1.6.4.3 Aromatic Ketones. The molecular ion peak of an aromatic ketone is prominent. Cleavage of aryl alkyl ketones occurs at the bond β to the ring, leaving a characteristic $ArC{\equiv}O^+$ fragment (*m/z* 105 when Ar = phenyl), which usually accounts for the base peak. Loss of CO from this fragment gives the "aryl" ion (*m/z* 77 in the case of acetophenone). Cleavage of the bond adjacent to the ring to form a $RC{\equiv}O^+$ fragment (R = alkyl) is less important though somewhat enhanced by electron-withdrawing groups (and diminished by electron-donating groups) in the *para*-position of the Ar group.

When the alkyl chain is C_3 or longer, cleavage of the C—C bond once removed from the C=O group occurs with hydrogen migration. This is the same cleavage noted for aliphatic ketones that proceeds through a cyclic transition state and results in elimination of an alkene and formation of a stable ion.

The mass spectrum of an unsymmetrical diaryl ketone, *p*-chlorobenzophenone, is displayed in Figure 1.23. The molecular ion peak (*m/z* 216) is prominent and the intensity of the M + 2 peak (33.99%, relative to the molecular ion peak) demonstrates that chlorine is in the structure (see the discussion of Table 1.5 and Figure 1.29 in Section 1.6.16).

Since the intensity of the *m/z* 141 peak is about 1/3 the intensity of the *m/z* 139 peak, these peaks, which contain chlorine, correspond to the same fragment. The same can be said about the fragments producing the *m/z* 111 and 113 peaks.

The fragmentation leading to the major peaks is sketched in Figure 1.23. The $Cl{-}ArC{\equiv}O^+$ peak is larger than the $Cl{-}Ar^+$ peak, and the $ArC{\equiv}O^+$ peak is larger than the Ar^+ peak (β cleavage favored). If the [fragment + 2] peaks for the Cl-substituted moieties are taken into account however, there is little difference in abundance between $Cl{-}ArCO^+$ and $ArCO^+$, or between $Cl{-}Ar^+$ and Ar^+; the inductive (electron withdrawing) and resonance (electron releasing) affects of the *para*-substituted Cl are roughly balanced out as they are in electrophilic aromatic substitution reactions.

1.6.5 Aldehydes

1.6.5.1 Aliphatic Aldehydes. The molecular ion peak of aliphatic aldehydes is usually discernible. Cleavage of the C—H and C—C bonds next to the oxygen atom results in an M − 1 peak and in an M − R peak (*m/z* 29, CHO^+). The M − 1 peak is a good diagnostic peak even for long-chain aldehydes, but the *m/z* 29 peak present in C_4 and higher aldehydes results from the hydrocarbon $C_2H_5^+$ion.

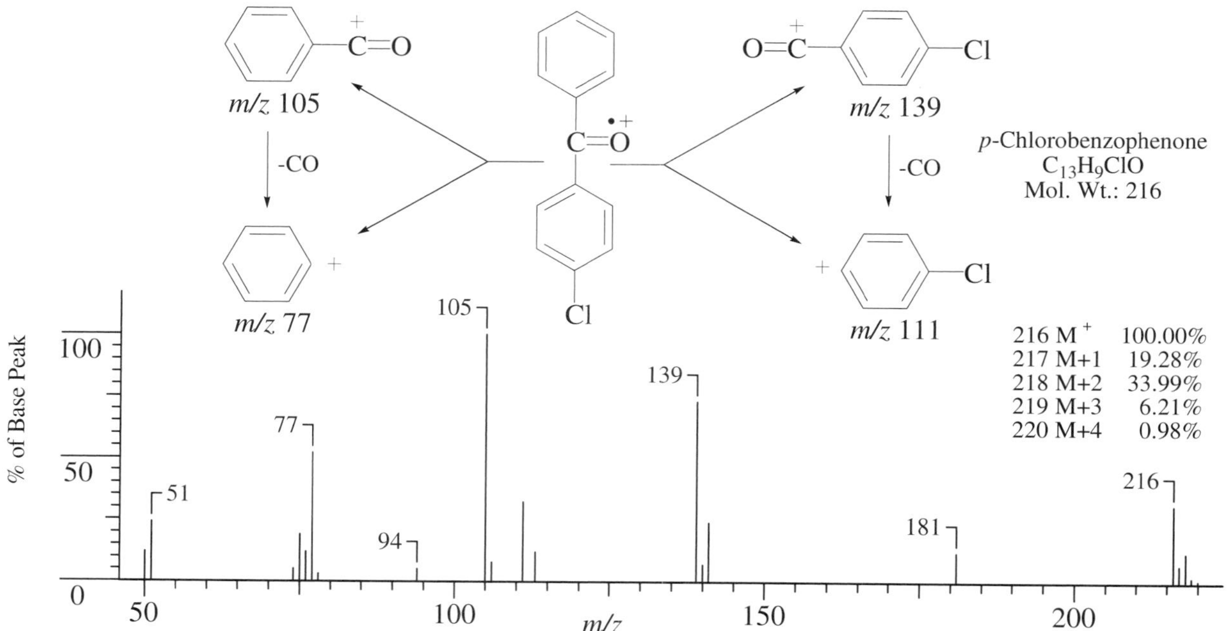

FIGURE 1.23 EI mass spectrum of *p*-chlorobenzophenone.

In the C_4 and higher aldehydes, McLafferty cleavage of the α, β C—C bond occurs to give a major peak at *m/z* 44, 58, or 72, . . . , depending on the α substituents. This is the resonance-stabilized (Scheme 1.20) ion formed through the cyclic transition state as shown in Scheme 1.7, where Y = H.

(Sch 1.20)

In straight-chain aldehydes, the other unique, diagnostic peaks are at M − 18 (loss of water), M − 28 (loss of ethylene), M − 43 (loss of CH_2═CH—O$^{\bullet}$), and M − 44 (loss of CH_2═CH—OH). The rearrangements leading to these peaks have been rationalized (see Budzikiewicz et al., 1967). As the chain lengthens, the hydrocarbon pattern (*m/z* 29, 43, 57, 71, . . .) becomes dominant. These features are evident in the spectrum of nonanal (Figure 1.24).

1.6.5.2 Aromatic Aldehydes. Aromatic aldehydes are characterized by a large molecular ion peak and by an M − 1 peak (Ar—C≡O^+) that is always large and may be larger than the molecular ion peak. The M − 1 ion, C_6H_5—CO^+ when Ar = phenyl, eliminates CO to give the phenyl ion (*m/z* 77), which in turn eliminates acetylene to give the $C_4H_3^+$ ion (*m/z* 51).

1.6.6 Carboxylic Acids

1.6.6.1 Aliphatic Acids. The molecular ion peak of a straight-chain monocarboxylic acid is weak but usually discernible. The most characteristic (sometimes the base) peak is *m/z* 60 resulting from the McLafferty rearrangement (Scheme 1.21). Branching at the α-carbon enhances this cleavage.

(Sch 1.21)

In short-chain acids, peaks at M − OH and M − CO_2H are prominent: these represent cleavage of bonds next to C═O. In long-chain acids, the spectrum consists of two series of peaks resulting from cleavage at each C—C bond with retention of charge either on the oxygen-containing fragment (*m/z* 45, 59, 73, 87, . . .) or on the alkyl fragment (*m/z* 29, 43, 57, 71, 85, . . .). As previously discussed, the hydrocarbon pattern also shows peaks at *m/z* 27, 28; 41, 42; 55, 56; 69, 70; . . . . In summary, besides the McLafferty rearrangement peak, the spectrum of a long-chain acid resembles the series of "hydrocarbon" clusters at intervals of 14 mass units. In each cluster, however, is a prominent peak at $C_nH_{2n-1}O_2$. Decanoic acid, Figure 1.25, nicely illustrates many of the points discussed above.

Dibasic acids usually have low volatility and hence are converted to esters to increase vapor pressure. Trimethylsilyl esters are often successful.

1.6.6.2 Aromatic Acids. The molecular ion peak of aromatic acids is large. The other prominent peaks are formed by loss of OH (M − 17) and of CO_2H (M − 45). Loss of H_2O (M − 18) is prominent if a hydrogen-bearing *ortho* group is available as outlined in Scheme 1.22. This is one example of the general "*ortho* effect" noted when the substituents can be in a six-membered transition state to facilitate loss of a neutral molecule of H_2O, ROH, or NH_3.

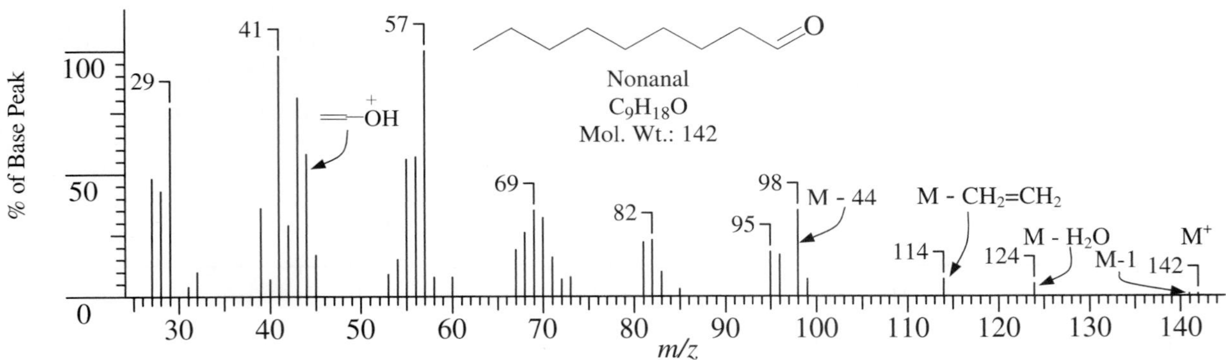

FIGURE 1.24 EI mass spectrum of nonanal.

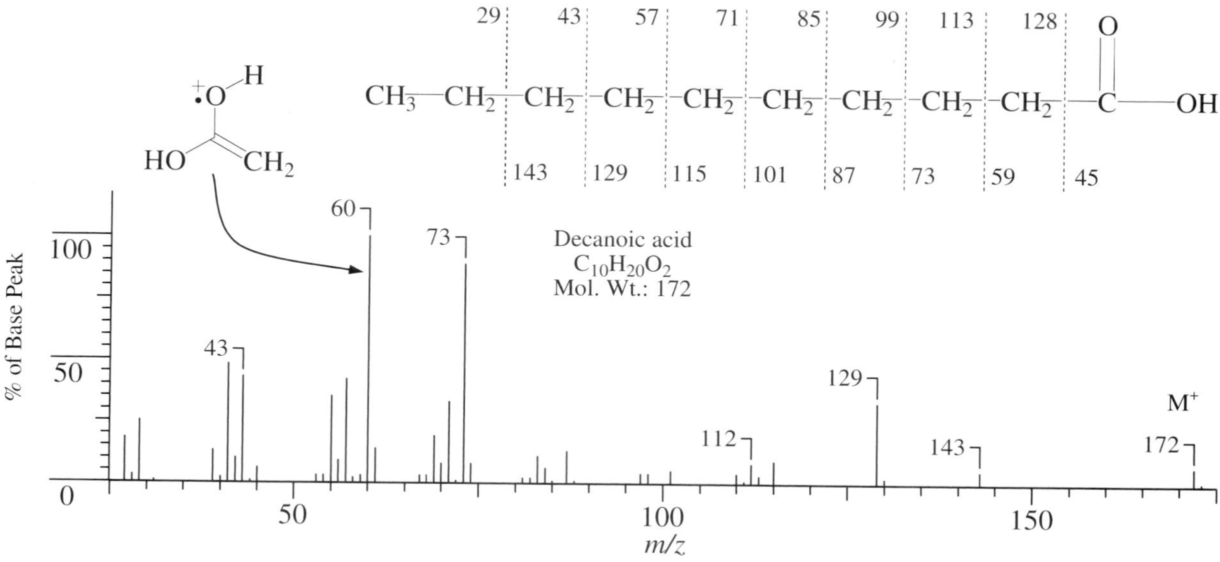

FIGURE 1.25 EI mass spectrum of decanoic acid.

−HZ

Z = OH, OR, NH_2; Y = CH_2, O, NH

(Sch 1.22)

1.6.7 Carboxylic Esters

1.6.7.1 Aliphatic Esters. The molecular ion peak of a methyl ester of a straight-chain aliphatic acid is usually distinct. Even waxes usually show a discernible molecular ion peak. The molecular ion peak is weak in the range m/z 130 to ~200, but becomes somewhat more intense beyond this range. The most characteristic peak results from the familiar McLafferty rearrangement (Scheme 1.23 gives the rearrangement for an ester) and cleavage one bond removed from the C=O group. Thus, a methyl ester of an aliphatic acid unbranched at the α-carbon gives a strong peak at m/z 74, which in fact, is the base peak in straight-chain methyl esters from C_6 to C_{26}. The alcohol moiety and/or the α substituent can often be deduced by the location of the peak resulting from this cleavage.

−RHC=CH_2

(Sch 1.23)

For the general ester, four ions can result from bond cleavage next to C=O.

$$R{-}C({=}O){-}OR'$$

The ion R^+ is prominent in the short-chain esters but diminishes rapidly with increasing chain length and is barely perceptible in methyl hexanoate. The ion $R{-}C{\equiv}O^+$ gives an easily recognizable peak for esters. In methyl esters, it occurs at M − 31. It is the base peak in methyl acetate and is still 4% of the base peak in the C_{26} methyl ester. The ions $[OR']^+$ and $[C({=}O)OR']^+$ are usually of little importance. The latter is discernible when $R' = CH_3$ (see m/z 59 peak of Figure 1.26).

First, consider esters in which the acid portion is the predominant portion of the molecule. The fragmentation pattern for methyl esters of straight-chain acids can be described in the same terms used for the pattern of the free acid. Cleavage at each C—C bond gives an alkyl ion (m/z 29, 43, 57, . . .) and an oxygen-containing ion, $C_nH_{2n-1}O_2^+$ (m/z 59, 73, 87, . . .). Thus, there are hydrocarbon clusters at intervals of 14 mass units; in each cluster is a prominent peak at $C_nH_{2n-1}O_2^+$. The peak (m/z 87) formally represented by the ion $[CH_2CH_2COOCH_3]^+$ is always more intense than its homologs, but the reason is not immediately obvious. However, it seems clear that the $C_nH_{2n-1}O_2^+$ ions do not at all arise from simple cleavage.

The spectrum of methyl octanoate is presented as Figure 1.26. This spectrum illustrates one difficulty in using the M + 1 peak to arrive at a molecular formula (previously mentioned, Section 1.5.2.1). The measured value for the M + 1 peak is 12%. The calculated value is 10.0%. The measured value is high because of an ion-molecule reaction induced by the relatively large sample that was used to see the weak molecular ion peak.

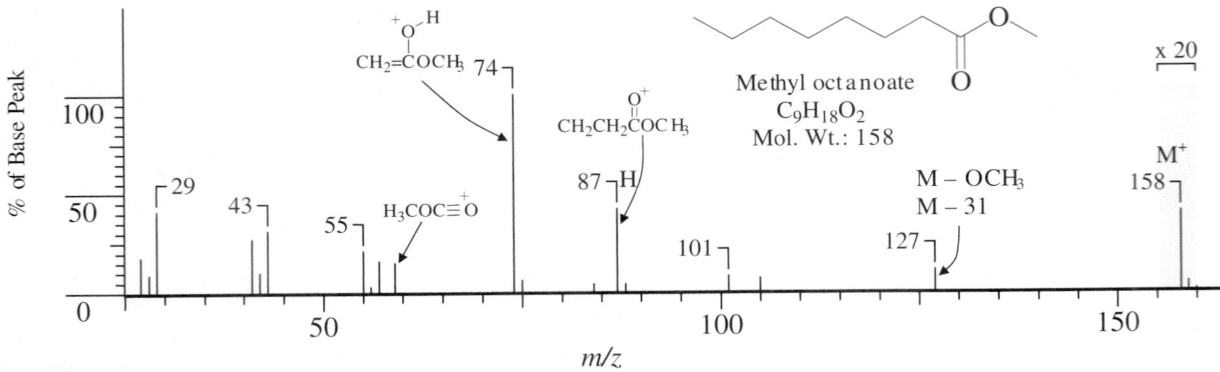

FIGURE 1.26 EI mass spectrum of methyl octanoate.

Now let us consider esters in which the alcohol portion is the predominant portion of the molecule. Esters of fatty alcohols (except methyl esters) eliminate a molecule of acid in the same manner that alcohols eliminate water. A scheme similar to that described earlier for alcohols, involving a single hydrogen transfer to the alcohol oxygen of the ester, can be written. An alternative mechanism involves a hydride transfer to the carbonyl oxygen (McLafferty rearrangement).

The loss of acetic acid by the mechanism described above is so facile in steroidal acetates that they frequently show no detectable molecular ion peak. Steroidal systems also seem unusual in that they often display significant molecular ions as alcohols, even when the corresponding acetates do not.

Esters of long-chain alcohols show a diagnostic peak at *m/z* 61, 75, or 89 . . . . from elimination of the alkyl moiety as an alkene and transfer of *two* hydrogen atoms to the fragment containing the oxygen atoms, which in essence is the protonated carboxylic acid.

Esters of dibasic acids $ROOC(CH_2)_nCOOR$, in general, give recognizable molecular ion peaks. Intense peaks are found at $[ROOC(CH_2)_nC{=}O]^+$ and at $[ROOC(CH_2)_n]^+$.

1.6.7.2 Benzyl and Phenyl Esters. Benzyl acetate (also furfuryl acetate and other similar acetates) and phenyl acetate eliminate the neutral molecule ketene (Scheme 1.24); frequently this gives rise to the base peak.

$$\xrightarrow{-H_2C{=}C{=}O}$$

m/z 108

(Sch 1.24)

Of course, the *m/z* 43 peak $(CH_3C{=}O)^+$ and *m/z* 91 $(C_7H_7)^+$ peaks are prominent for benzyl acetate.

1.6.7.3 Esters of Aromatic Acids. The molecular ion peak of methyl esters of aromatic acids is prominent (ArCOOR, R = CH_3). As the size of the alcohol moiety increases, the intensity of the molecular ion peak decreases rapidly to practically zero at C_5. The base peak results from elimination of ·OR, and elimination of ·COOR accounts for another prominent peak. In methyl esters, these peaks are at M − 31 and M − 59, respectively.

As the alkyl moiety increases in length, three modes of cleavage become important: (1) McLafferty rearrangement, (2) rearrangement of two hydrogen atoms with elimination of an allylic radical, and (3) retention of the positive charge by the alkyl group.

The familiar McLafferty rearrangement pathway gives rise to a peak for the aromatic acid, $(ArCOOH)^+$. The second, similar pathway gives the protonated aromatic acid, $(ArCOOH_2)^+$. The third mode of cleavage gives the alkyl cation, R^+.

Appropriately, *ortho*-substituted benzoates eliminate ROH through the general *"ortho"* effect described above under aromatic acids. Thus, the base peak in the spectrum of methyl salicylate is *m/z* 120; this ion eliminates carbon monoxide to give a strong peak at *m/z* 92.

A strong characteristic peak at mass 149 is found in the spectra of all esters of phthalic acid, starting with the diethyl ester. This peak is not significant in the dimethyl or methyl ethyl ester of phthalic acid, nor in esters of isophthalic or terephthalic acids, all of which give the expected peaks at M − R, M − 2R, M − CO_2R, and M − $2CO_2R$. Since long-chain phthalate esters are widely used as plasticizers, a strong peak at *m/z* 149 may indicate contamination. The *m/z* 149 fragment (essentially a protonated phthalic anhydride) is probably formed by two ester cleavages involving the shift of two hydrogen atoms and then another hydrogen atom, followed by elimination of H_2O.

1.6.8 Lactones

The molecular ion peak of five-membered ring lactones is distinct but is weaker when an alkyl substituent is present at C_4. Facile cleavage of the side chain at C_4 (rules 3 and 8, Section 1.5.4) gives a strong peak at M − alkyl.

The base peak (*m*/*z* 56) of γ-valerolactone (Figure 1.27) and the same strong peak of butyrolactone are rationalized, which shows the elimination of acetaldehyde.

Labeling experiments indicate that some of the *m*/*z* 56 peak in γ-valerolactone arises from the $C_4H_8^+$ ion. The other intense peaks in γ-valerolactone are at *m*/*z* 27 ($C_2H_3^+$), 28 ($C_2H_4^+$), 29 ($C_2H_5^+$), 41 ($C_3H_5^+$), and 43 ($C_3H_7^+$), and 85 ($C_4H_5O_2^+$, loss of the methyl group). In butyrolactone, there are strong peaks at *m*/*z* 27, 28, 29, 41, and 42 ($C_3H_6^+$).

1.6.9 Amines

1.6.9.1 Aliphatic Amines. The molecular ion peak of an aliphatic monoamine is an odd number, but it is usually quite weak and, in long-chain or highly branched amines, undetectable. The base peak frequently results from C—C cleavage next to the nitrogen atom (α, β rule 8, Section 1.5.4); for primary amines unbranched at the α-carbon, this is *m*/*z* 30 ($CH_2NH_2^+$) shown in Scheme 1.25. This cleavage accounts for the base peak in all primary amines and secondary and tertiary amines that are not branched at the α-carbon. Loss of the largest branch ($\cdot R''$ in Scheme 1.25) from the α-C atom is preferred.

$$H_2\dot{N}^{+}\!-\!C(R)(R')(R'') \xrightarrow{-\cdot R''} H_2\overset{+}{N}\!=\!C(R)(R') \longleftrightarrow H_2\ddot{N}\!-\!\overset{+}{C}(R)(R')$$

(Sch 1.25)

When branching at the α-carbon is absent, an M − 1 peak is usually visible. This is the same type of cleavage noted above for alcohols. The effect is more pronounced in amines because of the better resonance stabilization of the ion fragment by the less electronegative N atom compared with the O atom.

Primary straight-chain amines show a homologous series of peaks of progressively decreasing intensity (the cleavage at the ϵ-bond is slightly more important than at the neighboring bonds) at *m*/*z* 30, 44, 58, . . . resulting from cleavage at C—C bonds successively removed from the nitrogen atom with retention of the charge on the *N*-containing fragment. These peaks are accompanied by the hydrocarbon pattern of C_nH_{2n+1}, C_nH_{2n}, and C_nH_{2n-1} ions. Thus, we note characteristic clusters at intervals of 14 mass units, each cluster containing a peak resulting from a $C_nH_{2n+2}N$ ion. Because of the very facile cleavage to form the base peak, the fragmentation pattern in the high mass region becomes extremely weak.

Cyclic fragments apparently occur during the fragmentation of longer chain amines. The fragment shown in Scheme 1.26 gives a six-membered ring; five-membered rings are also commonly formed.

R, $\dot{N}H_2^{+}$ $\xrightarrow{-\cdot R}$ $\overset{+}{N}H_2$

m/*z* 86

(Sch 1.26)

A peak at *m*/*z* 30 is good though not conclusive evidence for a straight-chain primary amine. Further decomposition of the first-formed ion from a secondary or tertiary amine leads to a peak at *m*/*z* 30, 44, 58, 72, ... This is a process similar to that described for aliphatic alcohols and ethers above and, similarly, is enhanced by branching at one of the α-carbon atoms.

Cleavage of amino acid esters occurs at both C—C bonds (dashed lines below) next to the nitrogen

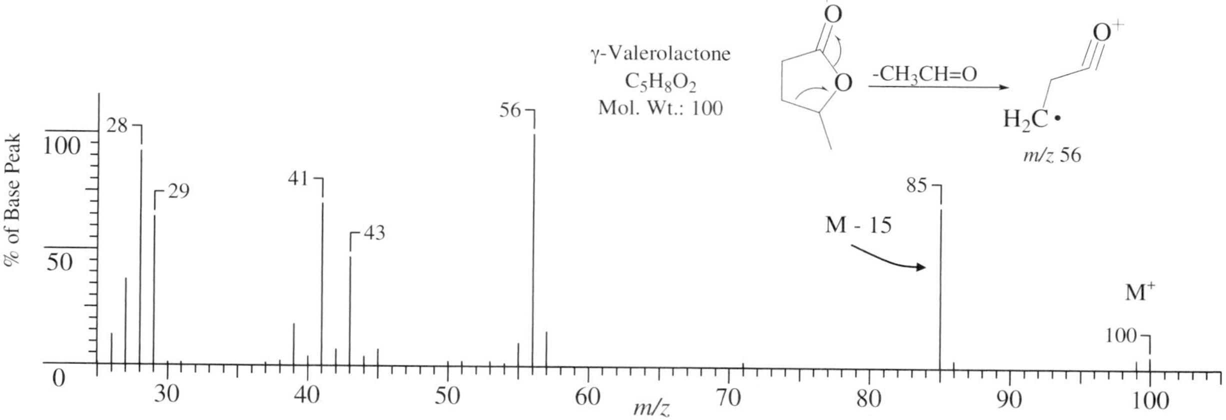

FIGURE 1.27 EI mass spectrum of γ-valerolactone.

atom; loss of the carbalkoxy group (—COOR′) is preferred. The aliphatic amine fragment ($^{+}NH_2=CHCH_2CH_2R$) decomposes further to give a peak at *m/z* 30.

$$R'OOC - \overset{\overset{+\cdot}{NH_2}}{\underset{H}{C}} - CH_2CH_2R$$

1.6.9.2 Cyclic Amines.

In contrast to acyclic amines, the molecular ion peaks of cyclic amines are usually intense unless there is substitution at the α position; for example, the molecular ion peak of pyrrolidine is strong. Primary cleavage at the bonds next to the N atom leads either to loss of an α-hydrogen atom to give a strong M − 1 peak or to opening of the ring; the latter event is followed by elimination of ethylene to give $\cdot CH_2—^{+}NH=CH_2$ (*m/z* 43, base peak), hence by loss of a hydrogen atom to give $CH_2=N^{+}=CH_2$ (*m/z* 42). *N*-methyl pyrrolidine also gives a $C_2H_4N^{+}$ (*m/z* 42) peak, apparently by more than one pathway.

Piperidine likewise shows a strong molecular ion and M − 1 (base) peak. Ring opening followed by several available sequences leads to characteristic peaks at *m/z* 70, 57, 56, 44, 43, 42, 30, 29, and 28. Substituents are cleaved from the ring (rule 6, Section 1.5.4).

1.6.9.3 Aromatic Amines (Anilines).

The molecular ion peak (odd number) of an aromatic monoamine is intense. Loss of one of the amino H atoms of aniline gives a moderately intense M − 1 peak; loss of a neutral molecule of HCN followed by loss of a hydrogen atom gives prominent peaks at *m/z* 66 and 65, respectively.

It was noted above that cleavage of alkyl aryl ethers occurs with rearrangement involving cleavage of the ArO—R bond: that is, cleavage was controlled by the ring rather than by the oxygen atom. In the case of alkyl aryl amines, cleavage of the C—C bond next to the nitrogen atom is dominant (Scheme 1.27): that is, the heteroatom controls cleavage.

$$C_6H_5\overset{+\cdot}{N}H-CH_2-CH_2R \xrightarrow{-\dot{C}H_2R} C_6H_5\overset{+}{N}H=CH_2$$

m/z 106

(Sch 1.27)

1.6.10 Amides

1.6.10.1 Aliphatic Amides.

The molecular ion peak of straight-chain monoamides is usually discernible. The dominant modes of cleavage depend on the length of the acyl moiety and on the lengths and number of the alkyl groups attached to the nitrogen atom.

The base peak (*m/z* 59, $H_2NC(=OH^{+})CH_2^{\cdot}$) in all straight-chain primary amides higher than propionamide results from the familiar McLafferty rearrangement. Branching at the α-carbon (CH_3, etc.) gives a homologous peak at *m/z* 73 or 87, ...

Primary amides give a strong peak at *m/z* 44 from cleavage of the R—$CONH_2$ bond: ($O=C=^{+}NH_2$); this is the base peak in C_1—C_3 primary amides and in isobutyramide. A moderate peak at *m/z* 86 results from γ,δ C—C cleavage, possibly accompanied by cyclization (Scheme 1.28).

$$O=C(\overset{+\cdot}{N}H_2)CH_2CH_2CH_2-R \xrightarrow{-R^{\cdot}} \text{cyclic } O=C\text{–}\overset{+}{N}H_2 \text{ ring}$$

m/z 86

(Sch 1.28)

Secondary and tertiary amides with an available hydrogen on the γ-carbon of the acyl moiety and methyl groups on the N atom show the dominant peak resulting from the McLafferty rearrangement. When the *N*-alkyl groups are C_2 or longer and the acyl moiety is shorter than C_3, another mode of cleavage predominates. This is cleavage of the *N*-alkyl group β to the nitrogen atom, and cleavage of the carbonyl C—N bond with migration of an α-hydrogen atom of the acyl moiety (expelling a neutral ketene molecule) and leaving $^{+}NH_2=CH_2$ (*m/z* 30).

1.6.10.2 Aromatic Amides.

Benzamide (Figures 1.1) is a typical example. Loss of NH_2 from the molecular ion yields a resonance-stabilized benzoyl cation that in turn undergoes cleavage to a phenyl cation. A separate fragmentation pathway gives rise to a modest *m/z* 44 peak.

1.6.11 Aliphatic Nitriles

The molecular ion peaks of aliphatic nitriles (except for acetonitrile and propionitrile) are weak or absent, but the M + 1 peak can usually be located by its behavior on increasing the sample size (Section 1.5.2.1). A weak but diagnostically useful M − 1 peak is formed by loss of an α-hydrogen to form the stable ion: $RCH=C=N^{+}$.

The base peak of straight-chain nitriles between C4 and C9 is m/z 41. This peak is the ion resulting from hydrogen rearrangement in a six-membered transition state, similar to a McLafferty rearrangement giving a peak at m/z 41 $CH_2{=}C{=}N^+{-}H$. However, this peak lacks diagnostic value because of the presence of the C_3H_5 (m/z 41) for all molecules containing a hydrocarbon chain.

A peak at m/z 97 is characteristic and intense (sometimes the base peak) in straight-chain nitriles C_8 and higher. The mechanism depicted in Scheme 1.29 has been proposed to account for this ion.

$-H_2C{=}CHR$

m/z 97

(Sch 1.29)

Simple cleavage at each C—C bond (except the one next to the N atom) gives a characteristic series of homologous peaks of even mass number down the entire length of the chain (m/z 40, 54, 68, 82, . . .) resulting from the $(CH_2)_nC{\equiv}N^+$ ions. Accompanying these peaks are the usual peaks of the hydrocarbon pattern.

1.6.12 Nitro Compounds

1.6.12.1 Aliphatic Nitro Compounds. The molecular ion peak (odd number) of an aliphatic mononitro compound is weak or absent (except in the lower homologs). The main peaks are attributable to the hydrocarbon fragments up to M − NO_2. Presence of a nitro group is indicated by an appreciable peak at m/z 30 (NO^+) and a smaller peak at mass 46 (NO_2^+).

1.6.12.2 Aromatic Nitro Compounds. The molecular ion peak of aromatic nitro compounds (odd number for one N atom) is strong. Prominent peaks result from elimination of an NO_2 radical (M − 46, the base peak in nitrobenzene), and of a neutral NO molecule with rearrangement to form the phenoxy cation (M − 30); both are good diagnostic peaks. Loss of acetylene from the M − 46 ion accounts for a strong peak at M − 72; loss of CO from the M − 30 ion gives a peak at M − 58. A diagnostic peak at m/z 30 results from the NO^+ ion.

The isomeric *o*-, *m*-, and *p*-nitroanilines each give a strong molecular ion (even number). They all give prominent peaks resulting from two sequences. The first pathway entails a loss of an NO_2 group (M − 46) to give an m/z 92; this ion loses HCN to give an m/z 65. The second sequence records a loss of NO (M − 30) to give m/z 108, which loses CO to give m/z 80.

Aside from differences in intensities, the three isomers give very similar spectra. The *meta* and *para* compounds give a small peak at m/z 122 from loss of an O atom, whereas the *ortho* compound eliminates OH as depicted in Scheme 1.30 to give a small peak at m/z 121.

$\cdot OH$

m/z 121

(Sch 1.30)

1.6.13 Aliphatic Nitrites

The molecular ion peak (odd number) of aliphatic nitrites (one N present) is weak or absent. The peak at m/z 30 (NO^+) is always large and is often the base peak. There is a large peak at m/z 60 ($CH_2{=}^+ONO$) in all nitrites unbranched at the α-carbon; this represents cleavage of the C—C bond next to the ONO group. An α branch can be identified by a peak at m/z 74, 88, or 102 . . . . Absence of a large peak at m/z 46 permits differentiation from nitro compounds. Hydrocarbon peaks are prominent, and their distribution and intensities describe the arrangement of the carbon chain.

1.6.14 Aliphatic Nitrates

The molecular ion peak (odd number) of aliphatic nitrates (one nitrogen present) is weak or absent. A prominent (frequently the base) peak is formed by cleavage of the C—C bond next to the ONO_2 group with loss of the heaviest alkyl group attached to the α-carbon. The NO_2^+ peak at m/z 46 is also prominent. As in the case of aliphatic nitrites, the hydrocarbon fragment ions are distinct.

1.6.15 Sulfur Compounds

The contribution (4.4%, see Table 1.3 and Figure 1.28) of the ^{34}S isotope to the M + 2 peak, and often to a (fragment + 2) peak, affords ready recognition of sulfur-containing compounds. A homologous series of sulfur containing fragments is four mass units higher than the hydrocarbon fragment series. The number of sulfur atoms can be determined from the size of the contribution of the ^{34}S isotope to the M + 2 peak. The mass of the sulfur atom(s) present is subtracted from the molecular weight. In diisopentyl disulfide, for example, the molecular weight is 206, and the molecule contains two sulfur atoms. The formula

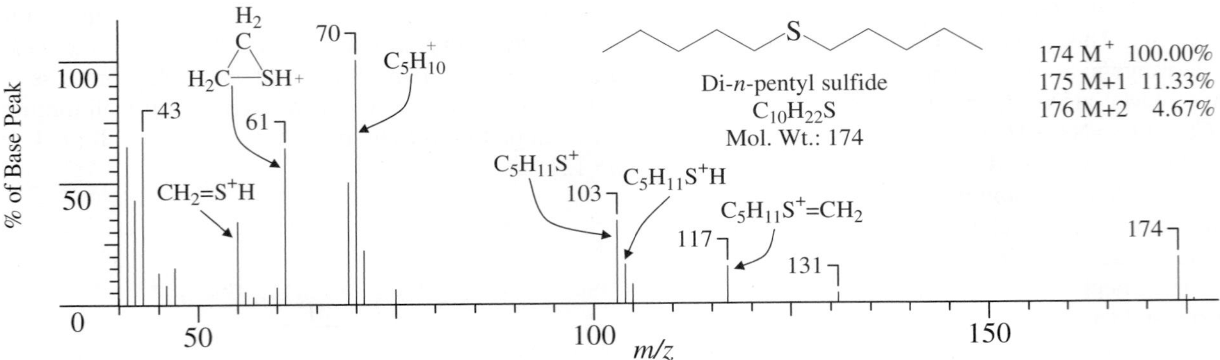

FIGURE 1.28 EI mass spectrum of di-*n*-pentyl sulfide.

for the rest of the molecule is therefore found under mass 142, that is, 206 − (2 × 32).

1.6.15.1 Aliphatic Mercaptans (Thiols). The molecular ion peak of aliphatic mercaptans, except for higher tertiary mercaptans, is usually strong enough so that the M + 2 peak can be accurately measured. In general, the cleavage modes resemble those of alcohols. Cleavage of the C—C bond (α,β-bond) next to the SH group gives the characteristic ion $CH_2=SH^+$ (*m/z* 47). Sulfur is poorer than nitrogen, but better than oxygen, at stabilizing such a fragment. Cleavage at the β,γ bond gives a peak at *m/z* 61 of about one-half the intensity of the *m/z* 47 peak. Cleavage at the γ,δ-bond gives a small peak at *m/z* 75, and cleavage at the δ,ϵ-bond gives a peak at *m/z* 89 that is more intense than the peak at *m/z* 73; presumably the *m/z* 89 ion is stabilized by cyclization:

$\overset{+}{S}$—H

Again analogous to alcohols, primary mercaptans split out H_2S to give a strong M − 34 peak, the resulting ion then eliminating ethylene: thus the homologous series M − H_2S − $(CH_2=CH_2)_n$ arises.

Secondary and tertiary mercaptans cleave at the α-carbon atom with loss of the largest group to give a prominent peak M − CH_3, M − C_2H_5, M − C_3H_7, ... However, a peak at *m/z* 47 may also appear as a rearrangement peak of secondary and tertiary mercaptans. A peak at M − 33 (loss of HS) is usually present for secondary mercaptans.

In long-chain mercaptans, the hydrocarbon pattern is superimposed on the mercaptan pattern. As for alcohols, the alkenyl peaks (i.e., *m/z* 41, 55, 69, . . .) are as large or larger than the alkyl peaks (*m/z* 43, 57 71, . . .).

1.6.15.2 Aliphatic Sulfides. The molecular ion peak of aliphatic sulfides is usually intense enough so that the M + 2 peak can be accurately measured. The cleavage modes generally resemble those of ethers. Cleavage of one or the other of the α, β C—C bonds occurs, with loss of the largest group being favored. These first-formed ions decompose further with hydrogen transfer and elimination of an alkene. The steps for aliphatic ethers also occur for sulfides (Scheme 1.31); the end result is the ion $RCH=SH^+$ (see Figure 1.28 for an example.)

$\xrightarrow{-CH_3^{\cdot}}$ $\xrightarrow{-C_2H_4}$ *m/z* 61

(Sch 1.31)

For a sulfide unbranched at either δ-carbon atom, this ion is $CH_2=SH^+$ (*m/z* 47), and its intensity may lead to confusion with the same ion derived from a mercaptan. However, the absence of M − H_2S or M − SH peaks in sulfide spectra makes the distinction.

A moderate to strong peak at *m/z* 61 is present (see alkyl sulfide cleavage, Figure 1.28) in the spectrum of all except tertiary sulfides. When an α-methyl substituent is present, *m/z* 61 is the ion, $CH_3CH=SH^+$, resulting from the double cleavage. Methyl primary sulfides cleave at the α, β-bond to give the *m/z* 61 ion, $CH_3—S^+=CH_2$

However, a strong *m/z* 61 peak in the spectrum of a straight-chain sulfide calls for a different explanation. Scheme 1.32 offers a plausible explanation.

$\xrightarrow{-H_2C=CHR}$ $\xrightarrow{-R\dot{C}H_2}$ *m/z* 61

(Sch 1.32)

Sulfides give a characteristic ion by cleavage of the C—S bond with retention of charge on sulfur. The resulting RS^+ ion gives a peak at *m/z* 32 + CH_3, 32 + C_2H_5, 32 + C_3H_7, ... The ion of *m/z* 103 seems especially favored possibly because of formation of a rearranged cyclic ion (Scheme 1.33). These features are illustrated by the spectrum of di-*n*-pentyl sulfide (Figure 1.28).

m/z 103

(Sch 1.33)

As with long-chain ethers, the hydrocarbon pattern may dominate the spectrum of long-chain sulfides; the C_nH_{2n} peaks seem especially prominent. In branched chain sulfides, cleavage at the branch may reduce the relative intensity of the characteristic sulfide peaks.

1.6.15.3 Aliphatic Disulfides. The molecular ion peak at least up to C_{10} disulfides, is strong. A major peak found in these spectra results from cleavage of one of the C—S bonds with retention of the charge on the alkyl fragment. Another major peak results from the same cleavage along with a shift of a hydrogen atom to form the RSSH fragment, which retains the charge. Other peaks apparently result from cleavage between the sulfur atoms without rearrangement, and with migration of one or two hydrogen atoms to give, respectively, RS^+, $RS^+ - 1$, and $RS^+ - 2$.

1.6.16 Halogen Compounds

A compound that contains one chlorine atom will have an M + 2 peak approximately one-third the intensity of the molecular ion peak because of the presence of a molecular ion containing the ^{37}Cl isotope (see Table 1.4). A compound that contains one bromine atom will have an M + 2 peak almost equal in intensity to the molecular ion because of the presence of a molecular ion containing the ^{81}Br isotope. A compound that contains two chlorines, or two bromines, or one chlorine and one bromine will show a distinct M + 4 peak, in addition to the M + 2 peak, because of the presence of a molecular ion containing two atoms of the heavy isotope. In general, the number of chlorine and/or bromine atoms in a molecule can be ascertained by the number of alternate peaks beyond the molecular ion peak. Thus, three chlorine atoms in a molecule will give peaks at M + 2, M + 4, and M + 6; in polychloro compounds, the peak of highest mass may be so weak as to escape notice.

The relative abundances of the peaks (molecular ion, M + 2, M + 4, and so on) have been calculated by Beynon et al. (1968) for compounds containing chlorine and bromine (atoms other than chlorine and bromine were ignored). A portion of these results is presented here, somewhat modified, as Table 1.5. We can now tell what combination of chlorine and bromine atoms is present. It should be noted that Table 1.4 presents the isotope contributions in terms of percent of the molecular ion peak. Figure 1.29 provides the corresponding bar graphs.

As required by Table 1.5, the M + 2 peak in the spectrum of *p*-chlorobenzophenone (Figure 1.23) is about one-third the intensity of the molecular ion peak (*m/z* 218). As mentioned earlier, the chlorine containing fragments (*m/z* 141 and 113) show [fragment + 2] peaks of the proper intensity.

Unfortunately, the application of isotope contributions, though generally useful for aromatic halogen compounds, is limited by the weak molecular ion peak of many aliphatic halogen compounds of more than about six carbon atoms for a straight chain, or fewer for a branched chain. However, the halogen-containing fragments are recognizable by the ratio of the (fragment + 2) peaks to fragment peaks in monochlorides or monobromides. In polychloro or polybromo compounds, these (fragment + isotope) peaks form a distinctive series of multiplets (Figure 1.30). Coincidence of a fragment ion with one of the isotope fragments, with another disruption of the characteristic ratios, must always be kept in mind.

Neither fluorine nor iodine has a heavier isotope.

TABLE 1.5 Intensities of Isotope Peaks (Relative to the Molecular Ion) for Combination of Chlorine and Bromine.

Halogen Present	% M+2	% M+4	% M+6	% M+8	% M+10	% M+12
Cl	32.6					
Cl_2	65.3	10.6				
Cl_3	97.8	31.9	3.5			
Cl_4	131.0	63.9	14.0	1.2		
Cl_5	163.0	106.0	34.7	5.7	0.4	
Cl_6	196.0	161.0	69.4	17.0	2.2	0.1
Br	97.9					
Br_2	195.0	95.5				
Br_3	293.0	286.0	93.4			
BrCl	130.0	31.9				
$BrCl_2$	163.0	74.4	10.4			
Br_2Cl	228.0	159.0	31.2			

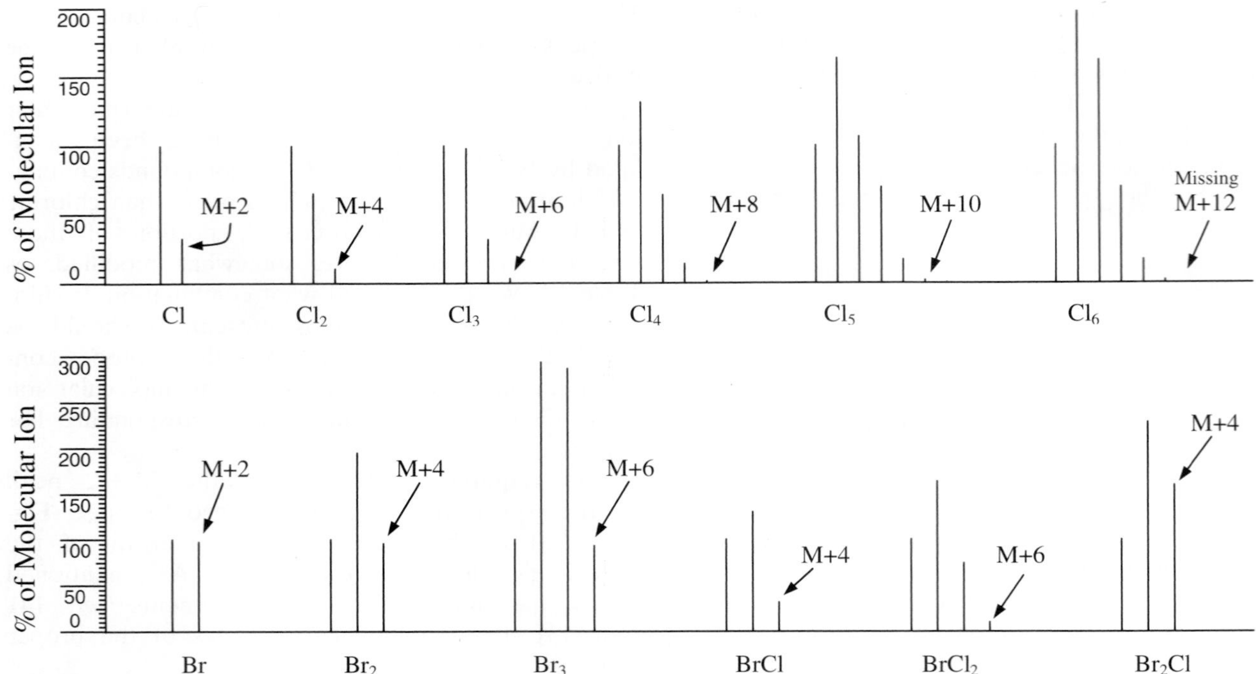

FIGURE 1.29 Predicted patterns of M, M + 2, M + 4, . . . for compounds with various combinations of chlorine and bromine.

1.6.16.1 Aliphatic Chlorides. The molecular ion peak is detectable only in the lower monochlorides. Fragmentation of the molecular ion is mediated by the chlorine atom but to a much lesser degree than is the case in oxygen-, nitrogen-, or sulfur-containing compounds. Thus, cleavage of a straight-chain monochloride at the C—C bond adjacent to the chlorine atom accounts for a small peak at *m/z* 49, $CH_2{=}Cl^+$ (and, of course, the isotope peak at *m/z* 51).

Cleavage of the C—Cl bond leads to a small Cl^+ peak and to a R^+ peak, which is prominent in the lower chlorides but quite small when the chain is longer than about C_5.

Straight-chain chlorides longer than C_6 give $C_3H_6Cl^+$, $C_4H_8Cl^+$, and $C_5H_{10}Cl^+$ ions. Of these, the $C_4H_8Cl^+$ ion forms the most intense (sometimes the base) peak; a five-membered cyclic structure (Scheme 1.34) may explain its stability.

R
Cl $\xrightarrow{-R^{\cdot}}$ Cl^+

(Sch 1.34)

Loss of HCl occurs, possibly by 1,3 elimination, to give a peak (weak or moderate) at M − 36.

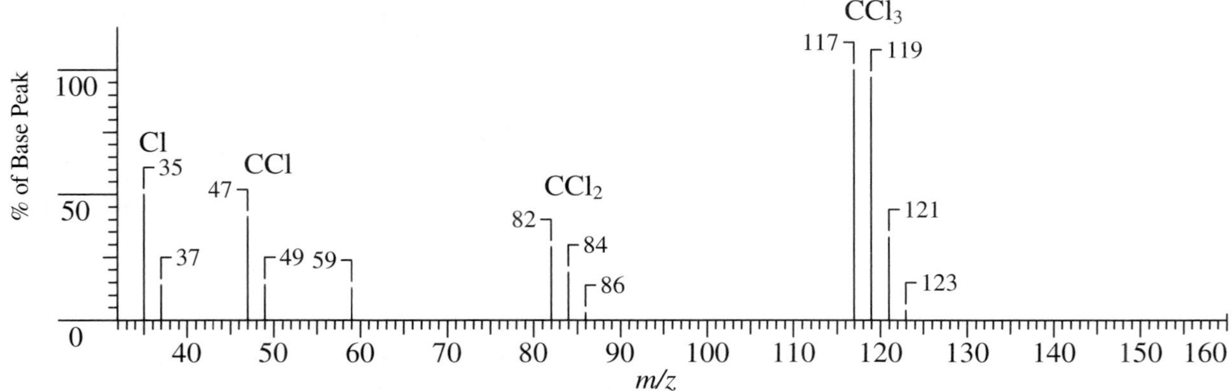

FIGURE 1.30 EI mass spectrum of carbon tetrachloride (CCl_4).

In general, the spectrum of an aliphatic monochloride is dominated by the hydrocarbon pattern to a greater extent than that of a corresponding alcohol, amine, or mercaptan.

1.6.16.2 Aliphatic Bromides. The remarks under aliphatic chlorides apply quite generally to the corresponding bromides.

1.6.16.3 Aliphatic Iodides. Aliphatic iodides give the strongest molecular ion peak of the aliphatic halides. Since iodine is monoisotopic, there is no distinctive isotope peak. The presence of an iodine atom can sometimes be deduced from isotope peaks that are suspiciously low in relation to the molecular ion peaks, and from several distinctive peaks; in polyiodo compounds, the large interval between major peaks is characteristic.

Iodides cleave much as do chlorides and bromides, but the $C_4H_8I^+$ ion is not as evident as the corresponding chloride and bromide ions.

1.6.16.4 Aliphatic Fluorides. Aliphatic fluorides give the weakest molecular ion peak of the aliphatic halides. Fluorine is monoisotopic, and its detection in polyfluoro compounds depends on suspiciously small isotopic peaks relative to the molecular ion, on the intervals between peaks, and on characteristic peaks. Of these, the most characteristic is *m/z* 69 resulting from the ion CF_3^+, which is the base peak in all perfluorocarbons. Prominent peaks are noted at *m/z* 119, 169, 219 . . . ; these are increments of CF_2. The stable ions $C_3F_5^+$ and $C_4F_7^+$ give large peaks at *m/z* 131 and 181. The M − F peak is frequently visible in perfluorinated compounds. In monofluorides, cleavage of the α, β C—C bond is less important than in the other monohalides, but cleavage of a C—H bond on the α-carbon atom is more important. This reversal is a consequence of the high electronegativity of the F atom and is rationalized by placing the positive charge on the α-carbon atom. The secondary carbonium ion thus depicted in Scheme 1.35 by a loss of a hydrogen atom is more stable than the primary carbonium ion resulting from loss of an alkyl radical.

$$\left[\text{R—CH}_2\text{—F}\right]^{\cdot+} \xrightarrow{-\text{H}^\cdot} \text{R—}\overset{+}{\text{C}}\text{H—F}$$

$$\left[\text{R—CH}_2\text{—F}\right]^{\cdot+} \xrightarrow{-\text{R}^\cdot} \overset{+}{\text{C}}\text{H}_2\text{—F}$$

(Sch 1.35)

1.6.16.5 Benzyl Halides. The molecular ion peak of benzyl halides is usually detectable. The benzyl (or tropylium) ion from loss of the halide (rule 8, Section 1.5.4) is favored even over β-bond cleavage of an alkyl substituent. A substituted phenyl ion (α-bond cleavage) is prominent when the ring is polysubstituted.

1.6.16.6 Aromatic Halides. The molecular ion peak of an aryl halide is readily apparent. The M − X peak is large for all compounds in which X is attached directly to the ring.

1.6.17 Heteroaromatic Compounds

The molecular ion peak of heteroaromatics and alkylated heteroaromatics is intense. Cleavage of the bond β to the ring, as in alkylbenzenes, is the general rule; in pyridine, the position of substitution determines the ease of cleavage of the β-bond (see below).

Localizing the charge of the molecular ion on the heteroatom, rather than in the ring π structure, provides a satisfactory rationale for the observed mode of cleavage. The present treatment follows that used by Djerassi (Budzikiewicz et al., 1967).

The five-membered ring heteroaromatics (furan, thiophene, and pyrrole) show very similar ring cleavage patterns. The first step in each case is cleavage of the carbon-heteroatom bond, followed by loss of either a neutral acetylene molecule or by loss of radical fragments. Thus, furan exhibits two principal peaks: $C_3H_3^+$ (*m/z* 39) and HC≡O$^+$ (*m/z* 29). For thiophene, there are three peaks, $C_3H_3^+$ (*m/z* 39), HC≡S$^+$ (*m/z* 45), and $C_2H_2S^+$ (*m/z* 58). And for pyrrole, there are three peaks: $C_3H_3^+$ (*m/z* 39), HC═NH$^+$ (*m/z* 28) and $C_2H_2NH^+$ (*m/z* 41). Pyrrole also eliminates a neutral molecule of HCN to give an intense peak at *m/z* 40. The base peak of 2,5-dimethylfuran is *m/z* 43 (CH_3C≡O$^+$).

Cleavage of the β C—C bond in alkylpyridines (Scheme 1.36) depends on the position of the ring substitution, being more pronounced when the alkyl group is in the 3 position. An alkyl group of more than three carbon atoms in the 2 position can undergo migration of a hydrogen atom to the ring nitrogen.

−RHC═CH$_2$
N H R CH$_2$ *m/z* 93

(Sch 1.36)

A similar cleavage is found in pyrazines since all ring substituents are necessarily ortho to one of the nitrogen atoms.

REFERENCES

General

Beynon, J. H. (1960). *Mass Spectrometry and Its Application to Organic Chemistry.* Amsterdam: Elsevier.

Beynon, J. H., and Brenton, A. G. (1982). *Introduction to Mass Spectrometry.* Cardiff: University of Wales Publications.

Beynon, J. H., Saunders, R. A., and Williams A. E. (1968). *The Mass Spectra of Organic Molecules.* New York: Elsevier.

Biemann. K. (1962). *Mass Spectrometry, Applications to Organic Chemistry.* New York: McGraw-Hill.

Budzikiewicz, H., Djerassi, C., and Williams. D. H. (1967). *Mass Spectrometry of Organic Compounds.* San Francisco: Holden-Day.

Budzikiewicz, H., Djerassi, C., and Williams, D. H. (1964). *Structure Elucidation of Natural Products by Mass Spectrometry,* Vols. I and 11. San Francisco: Holden-Day.

Chapman. J. R. (1993). *Practical Organic Mass Spectrometry,* 2nd ed. New York: Wiley.

Constatin, E., Schnell, A., and Thompson, M. (1990). *Mass Spectrometry.* Englewood Cliffs, NJ: Prentice-Hall.

Davis, R., and Frierson, M. (1987). *Mass Spectrometry.* New York: Wiley. Self-study guide.

Hamming, M., and Foster, N. (1972). *Interpretation of Mass Spectra of Organic Compounds.* New York: Academic Press.

Hoffmann, E.D., and Stroobant, V. (2002). *Mass Spectrometry: Principles and Applications.* John Wiley and Sons, Ltd.

Howe. I., Williams, D. H., and Bowen. R. D. (1981). *Mass Spectrometry—Principles and Application.* New York: McGraw-Hill.

McLafferty, F. W., and Turecek. A. (1993). *Interpretation of Mass Spectra,* 4th ed. Mill Valley, CA: University Scientific Books.

McLafferty, F. W., and Venkataraghavan. R. (1982). *Mass Spectral Correlations.* Washington, DC: American Chemical Society.

McNeal, C. J., Ed. (1986). *Mass Spectrometry in the Analysis of Large Molecules.* New York: Wiley.

Middleditch, B. S.. Ed. (1979). *Practical Mass Spectrometry.* New York: Plenum Press.

Milne, G. W. A. (1971). *Mass Spectrometry: Techniques and Applications.* New York: Wiley-Interscience.

Rose, M., and Johnston, R. A. W. (1982). *Mass Spectrometry for Chemists and Biochemists.* New York: Cambridge University Press.

Shrader, S. R. (1971). *Introduction to Mass Spectrometry.* Boston: Allyn and Bacon.

Smith, R. M. (1998). *Understanding Mass Spectra: A Basic Approach.* Wiley-Interscience.

Watson, J. T. (1985). *Introduction to Mass Spectrometry.* 2nd ed. New York: Raven Press.

Williams, D. H. (June 1968–June 1979). *Mass Spectrometry. A Specialist Periodical Report,* Vols. I–V. London: Chemical Society.

(1980). *Advances in Mass Spectrometry; Applications in Organic and Analytical Chemistry.* New York: Pergamon Press.

Mass Spectrometry Reviews. New York: Wiley, 1982 to date.

Mass Spectrometry Bulletin. Aldermaston, England, 1966 to date.

Data and Spectral Compilations

American Petroleum Institute Research Project 44 and Thermodynamics Research Center (formerly MCA Research Project). (1947 to date). *Catalog of Selected Mass Spectral Data.* College Station, TX; Texas A & M University, Dr. Bruno Zwohnski, Director.

Beynon, J. H., and Williams. A. E. (1963). *Mass and Abundance Tables for Use in Mass Spectrometry.* Amsterdam: Elsevier.

(1992). *Eight Peak Index of Mass Spectra,* 4th ed. Boca Raton, FL: CRC Press.

Heller, S. R., and Milne, G. W. EPA/NIH Mass Spectral Search System (MSSS). A Division of CIS. Washington, DC: U.S. Government Printing Office. An interactive computer searching system containing the spectra of over 32,000 compounds. These can be searched on the basis of peak intensities as well as by Biemann and probability matching techniques.

McLafferty, F. W. (1982). *Mass Spectral Correlations,* 2nd ed. Washington, DC: American Chemical Society.

McLafferty, F. W., and Stauffer, D. B. (1988). *The Wiley/NBS Registry of Mass Spectral Data* (7 volumes). New York: Wiley-Interscience.

McLafferty, F. W., and Stauffer, D. B. (1992). *Registry of Mass Spectral Data,* 5th ed. New York: Wiley. 220,000 spectra.

McLafferty, F. W., and Stauffer, D. B. (1991). *The Important Peak Index of the Registry of Mass Spectral Data,* 3 Vols. New York: Wiley.

Special Monographs

Ardrey, R. E. (2003). *Liquid Chromatography—Mass Spectrometry: An Introduction.* New York: John Wiley & Sons.

Ashcroft, A. E. (1997). *Ionization Methods in Organic Mass Spectrometry.* London: Royal Society of Chemistry.

Cech, N. B., and Enke, C. G. (2002). Practical implications of some recent studies in electrospray ionization fundamentals. *Mass Spectrometry Reviews,* **20,** 362–387.

Cole, R. B. Ed. (1997). *Electrospray Ionization Mass Spectrometry.* New York: Wiley-Interscience.

Harrison, G. (1992). *Chemical Ionization Mass Spectrometry,* 2nd ed. Boca Raton. FL: CRC Press.

Kinter, M., and Sherman, N. E. (2000). *Protein Sequencing and Identification Using Tandem Mass Spectrometry.* New York: Wiley-Interscience.

Linskens, H. F., and Berlin, J. Eds. (1986). *Gas Chromatography-Mass Spectrometry.* New York: Springer-Verlag.

March, R. E., and Hughes. R. J. (1989). *Quadrupole Storage Mass Spectrometry.* New York: Wiley-Interscience.

McLafferty, F. W. (1983). *Tandem Mass Spectrometry,* 2nd ed. New York: Wiley-Interscience.

Message, G. M. (1984). *Practical Aspects of Gas Chromatography-Mass Spectrometry.* New York: Wiley.

Siuzdak, G. (1996). *Mass Spectrometry for Biotechnology.* New York: Academic Press.

Waller: G. R., Ed. (1972). *Biochemical Applications of Mass Spectrometry.* New York: Wiley-Interscience.

Waller, G. R., and Dermer, O. C., Eds. (1980). *Biochemical Applications of Mass Spectrometry,* First Suppl. Vol. New York: Wiley-Interscience.

Exercise 1.1

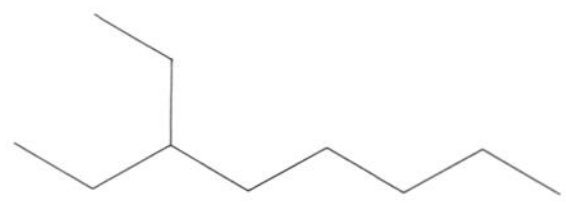

a 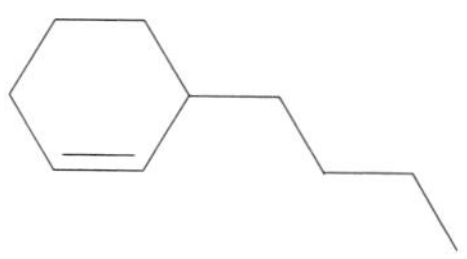 b 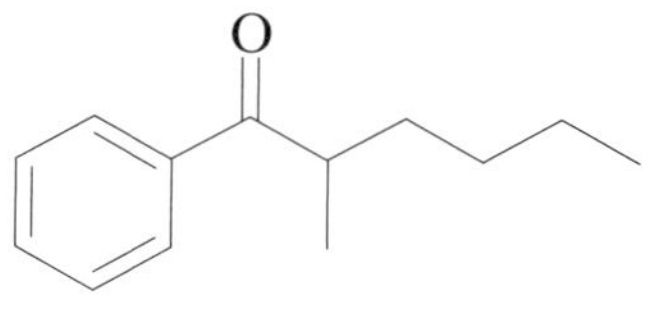 c

d 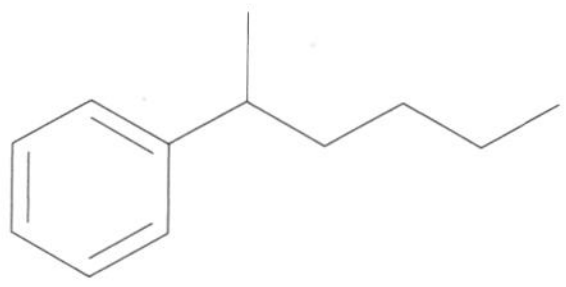 e 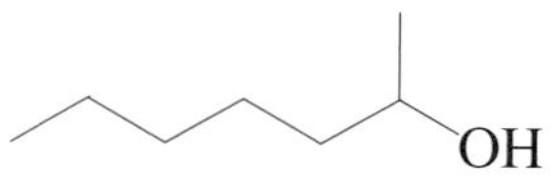 f

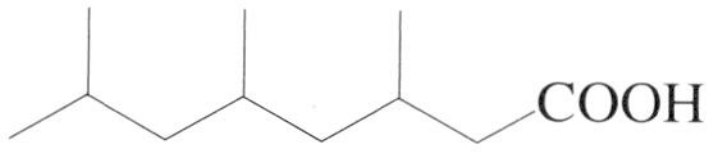

COOH OH OH

g  h 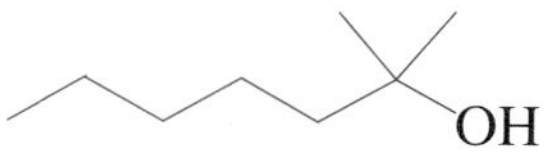 i

S OH O

j 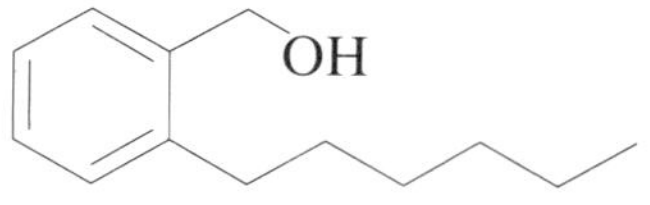 k 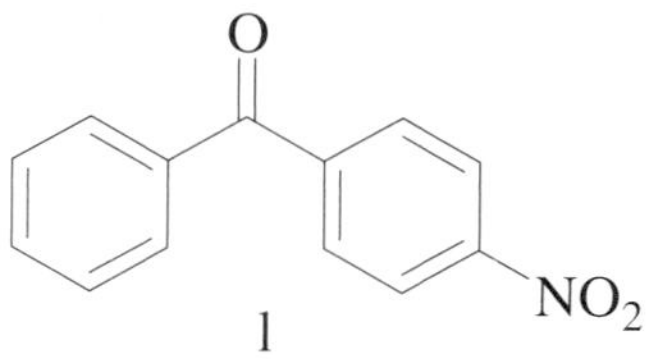 l NO$_2$

O O OAc

H

Br

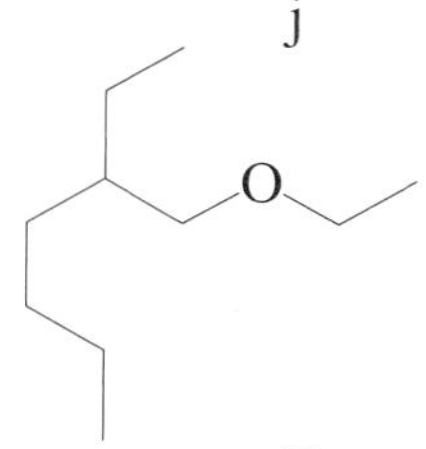

m 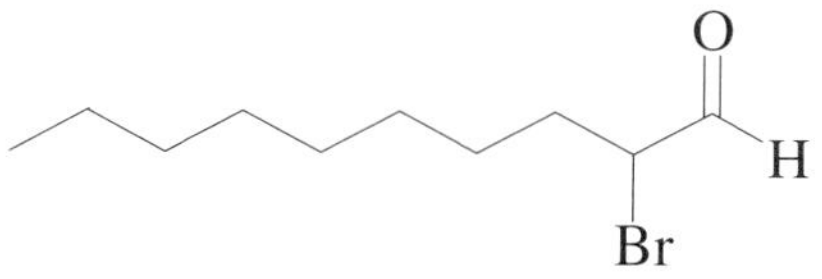 n 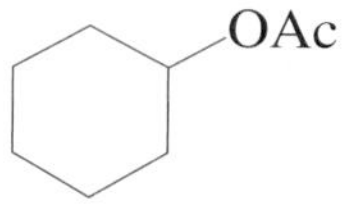 o

STUDENT EXERCISES

1.1 Using Table 1.4, calculate the exact mass for the above compounds (**a**–**o**).

1.2 Determine the index of hydrogen deficiency for the above compounds.

1.3 Write the structure for the molecular ion for each compound (**a**–**o**) showing, when possible, the location of the radical cation.

1.4 Predict three major fragmentation/rearrangement pathways for the above compounds. For each pathway, cite the rule from Section 1.5.4 that supports your prediction.

1.5 For each fragmentation/rearrangement pathway from exercise 1.4, show a detailed mechanism using either single barbed or double barbed arrows as appropriate.

1.6 Match each of the exact masses to the following mass spectra. (**A**–**W**) Note that two compounds have the same exact mass, and you will need to consider the CI mass spectrum when given: (a) 56.0264, (b) 73.0896, (c) 74.0363, (d) 89.0479, (e) 94.0535, (f) 96.0572, (g) 98.0736, (h) 100.0893, (i) 102.0678, (j) 113.0845, (k) 114.1043, (l) 116.0841, (m) 116.1206, (n) 122.0733, (o) 122.0733, (p)126.1041, (q) 138.0687, (r) 150.0041, (s) 152.0476, (t) 156.9934, (u) 161.9637, (v) 169.9735, (w) 208.0094.

1.7 For each of the mass spectra (**A**–**W**), determine if there are any of the following heteroatoms in the compound: S, Cl, Br.

1.8 For each exact mass corresponding to mass spectra **A**–**W**, determine the molecular formula. Remember to look at the heteroatoms that were determined in exercise 1.7.

1.9 Determine the index of hydrogen deficiency for each of the formulas in exercise 1.6.

1.10 List the base peak and molecular ion peak for each of the EI mass spectra (**A**–**W**).

1.11 Choose three ions (besides the molecular ion) from each EI mass spectrum (**A**–**W,** except for H), and determine the molecular formula for each fragment ion, and give the molecular formula for the portion that is lost from the molecular ion. Indicate which ions result from a rearrangement.

Exercise 1.6 (A-C)

A

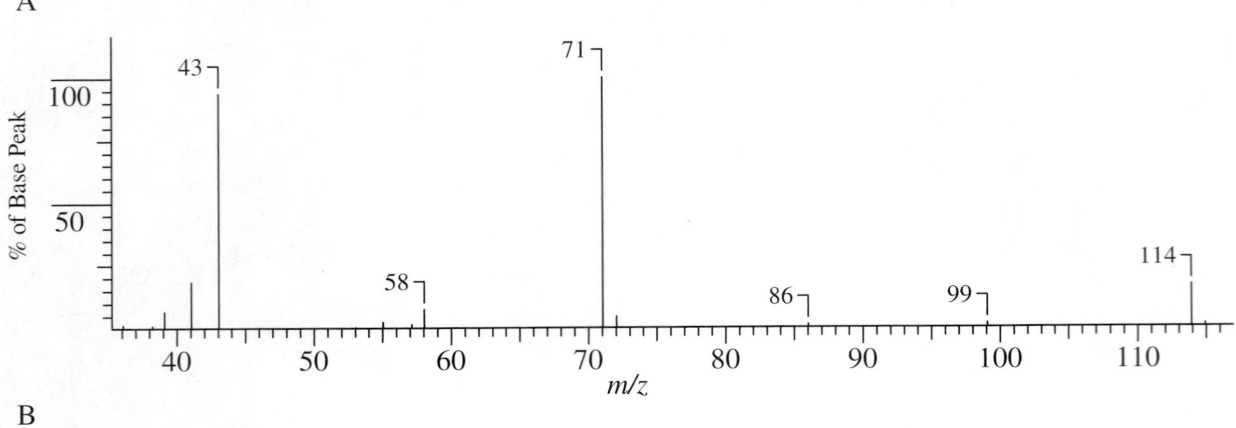

B

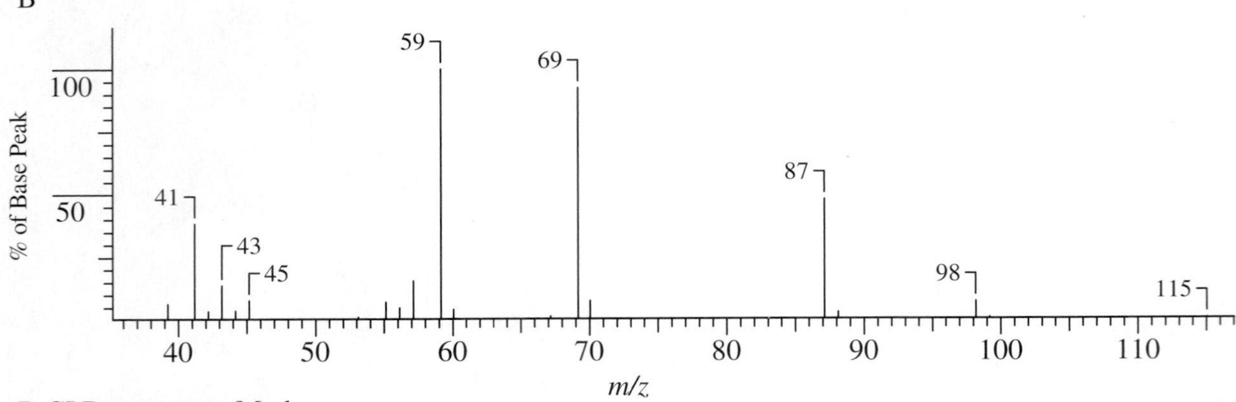

B CI Reagent gas Methane

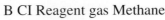

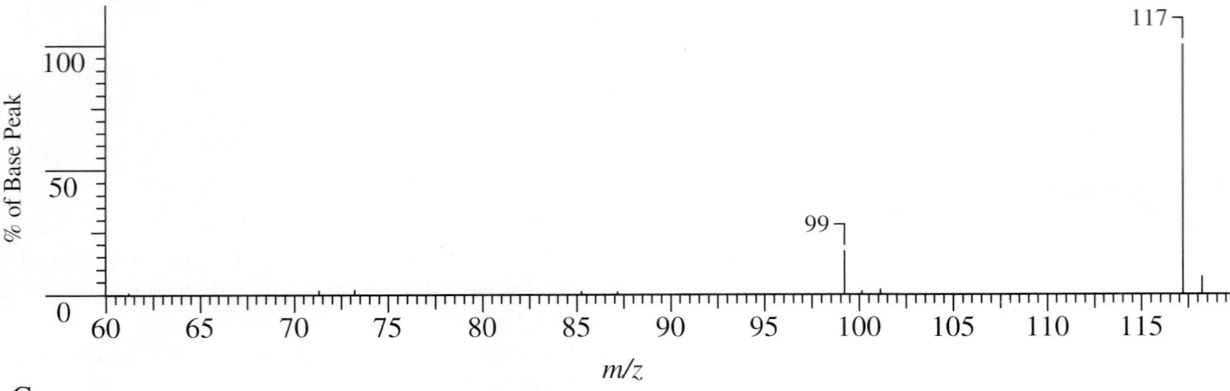

C

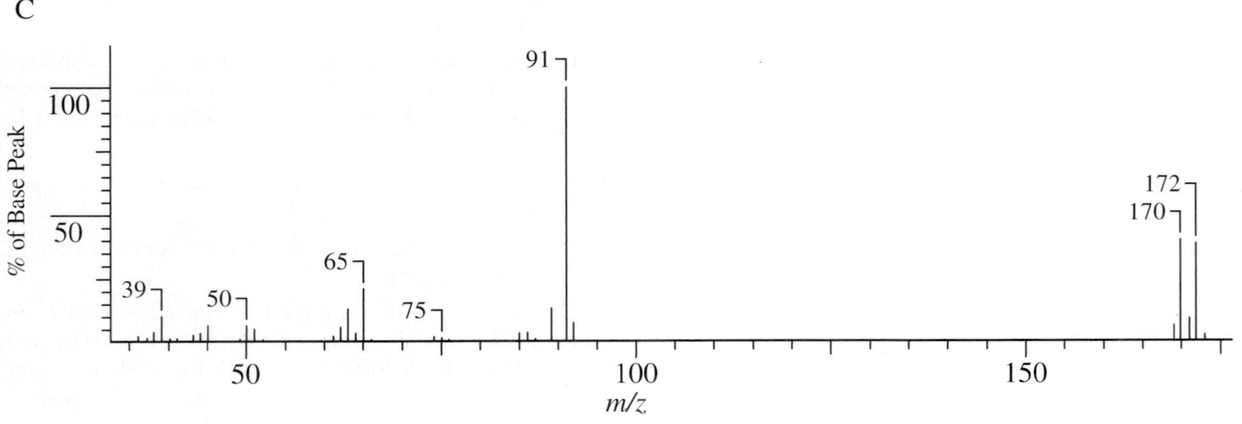

Exercise 1.6 (D-G)

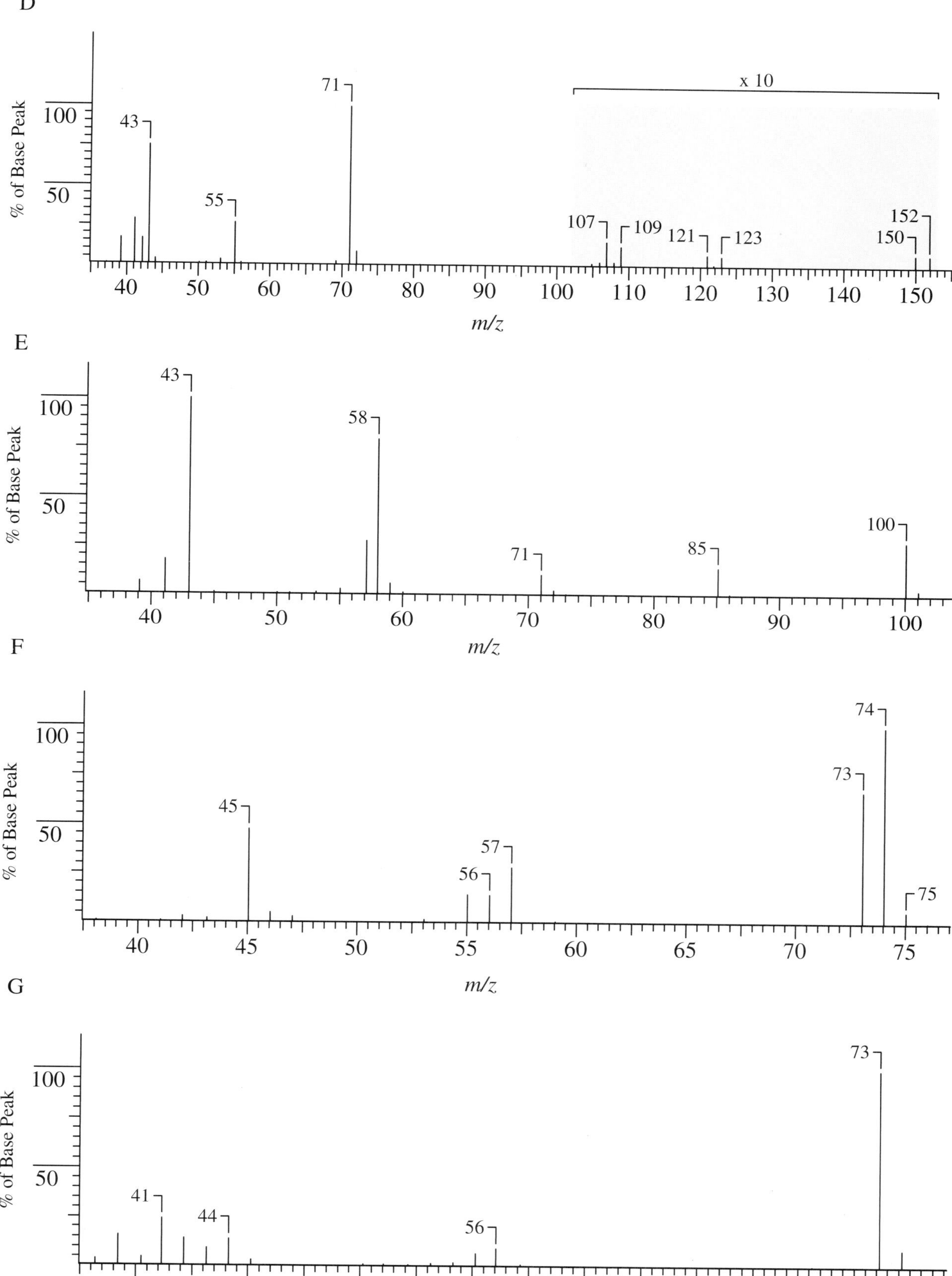

Exercise 1-6 (H-I)

H

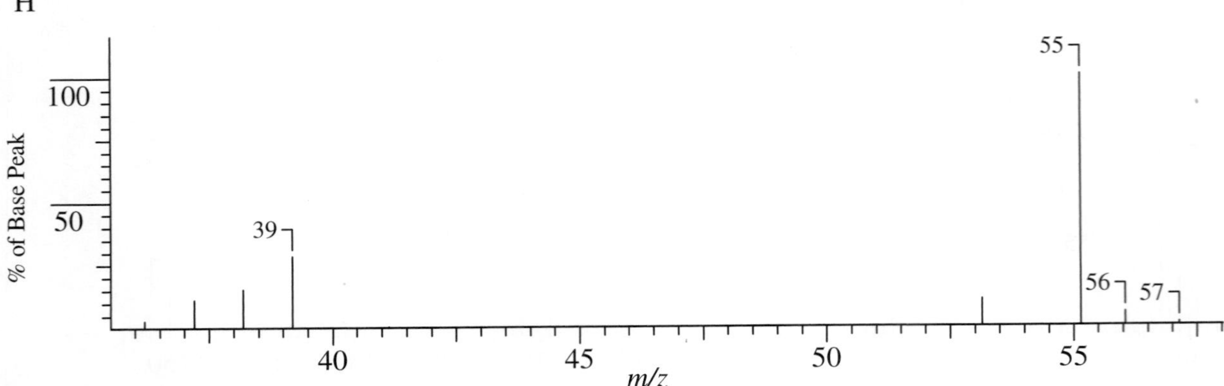

H CI Reagent gas Methane

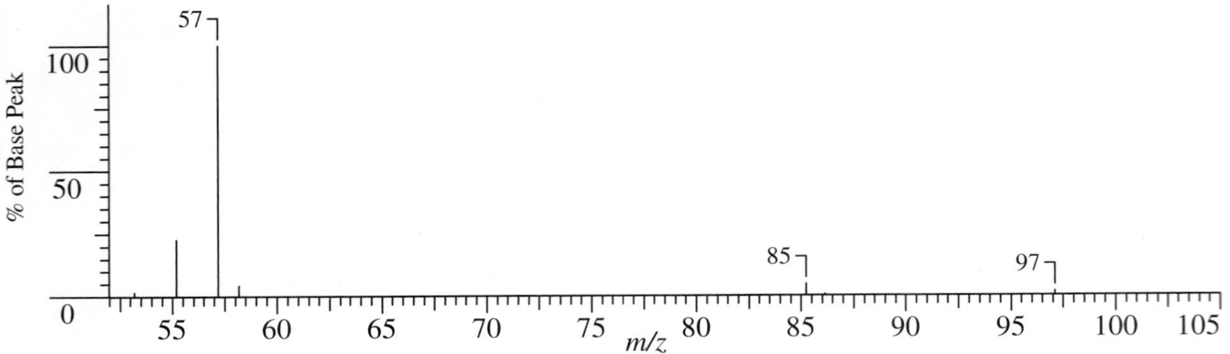

I

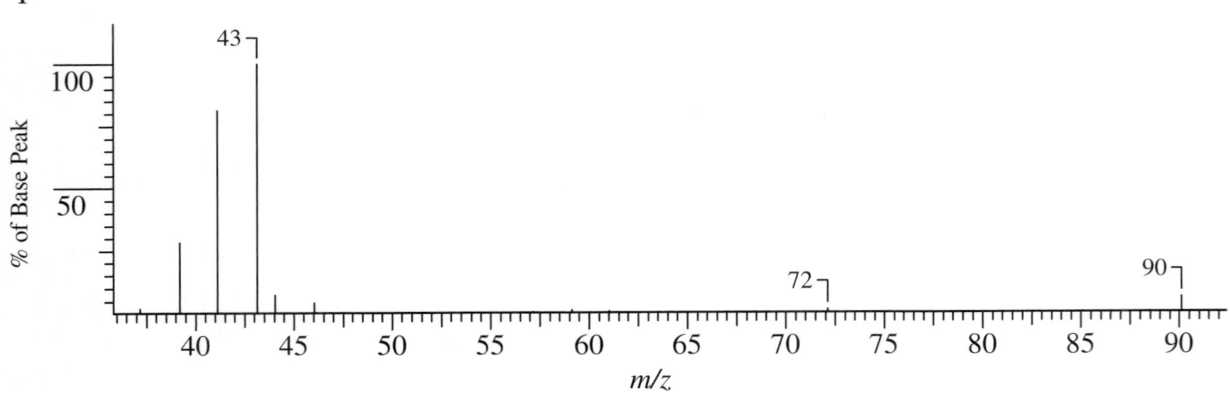

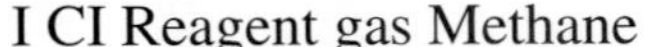

I CI Reagent gas Methane

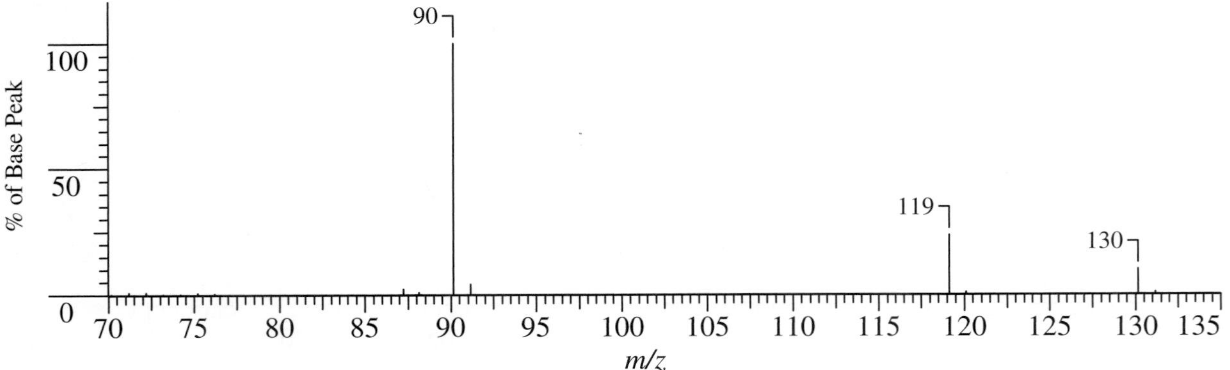

Exercise 1.6 (J-M)

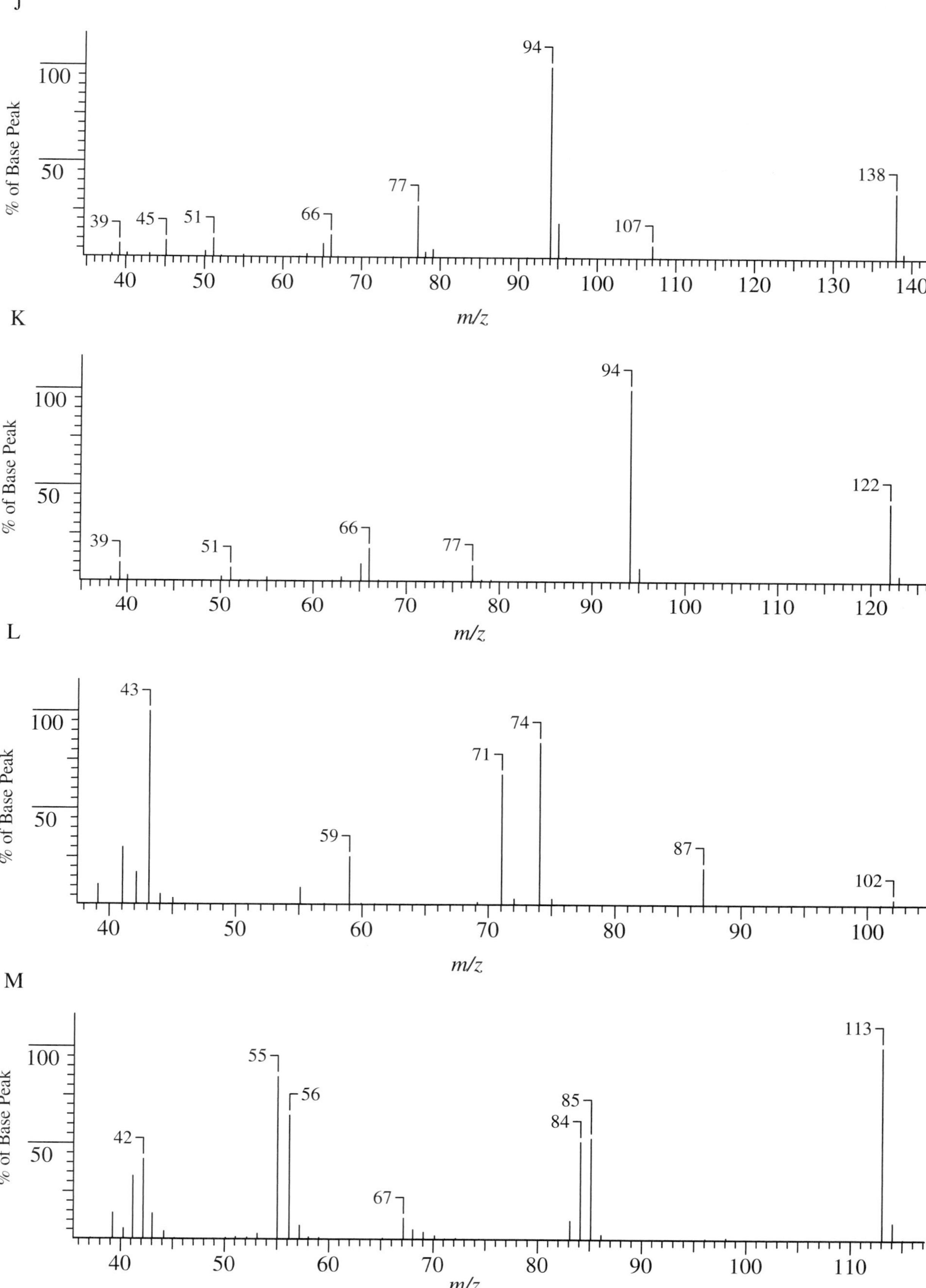

Exercise 1.6 (N-Q)

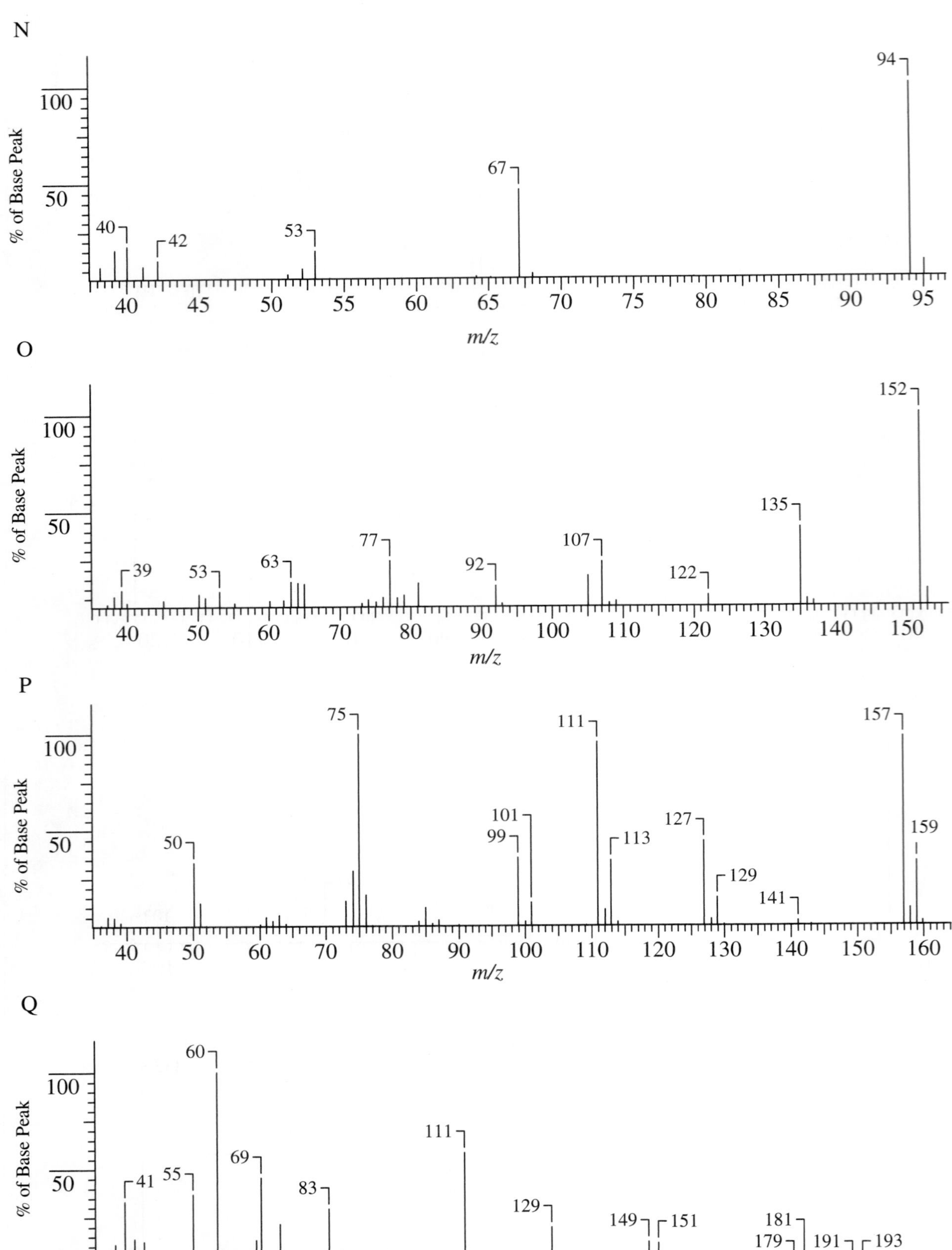

Exercise 1.6 (Q-T)

Q CI Reagent gas Methane

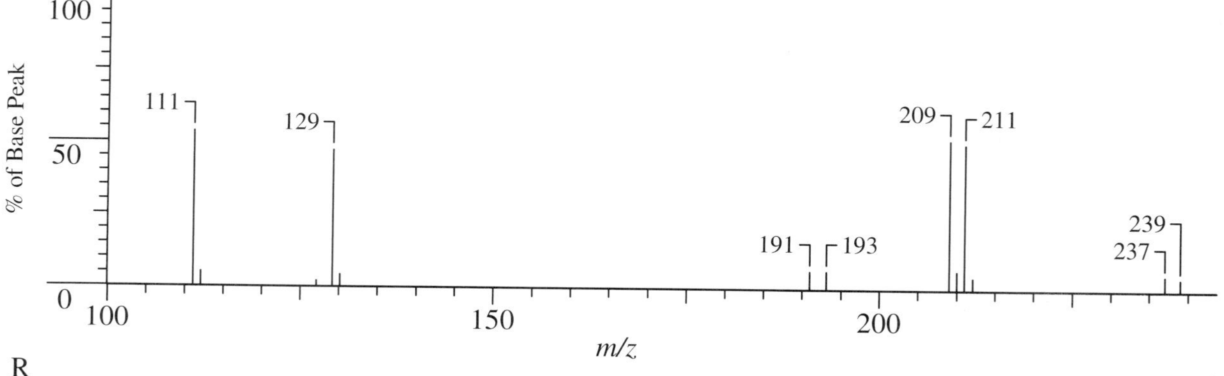

R

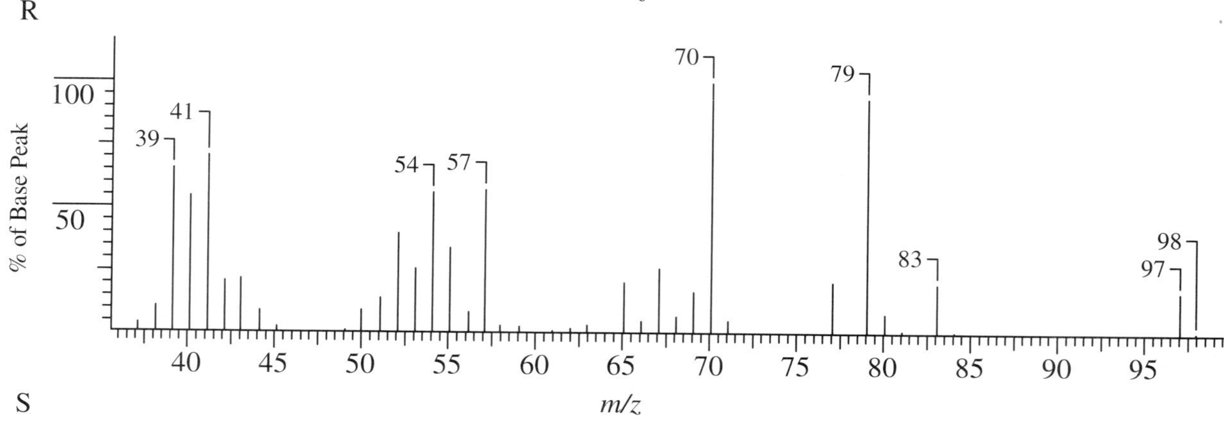

S

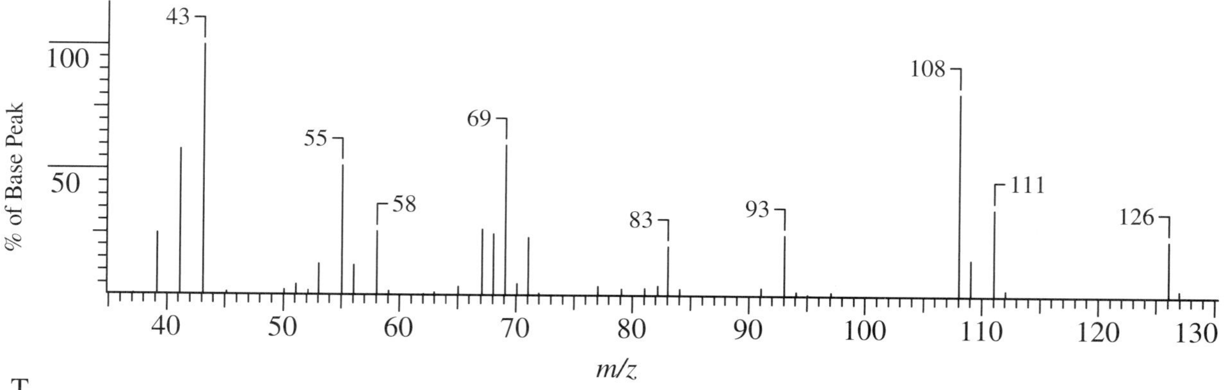

T

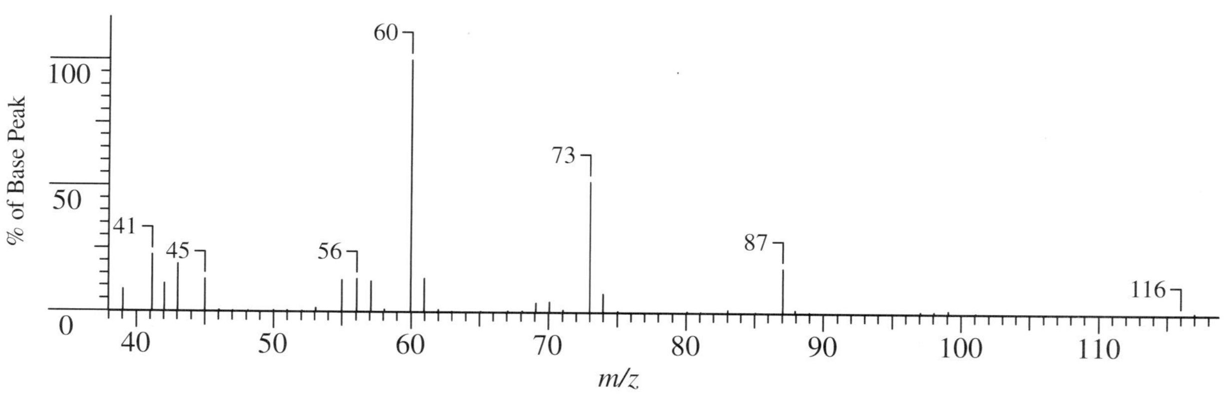

Exercise 1.6 (T-W)

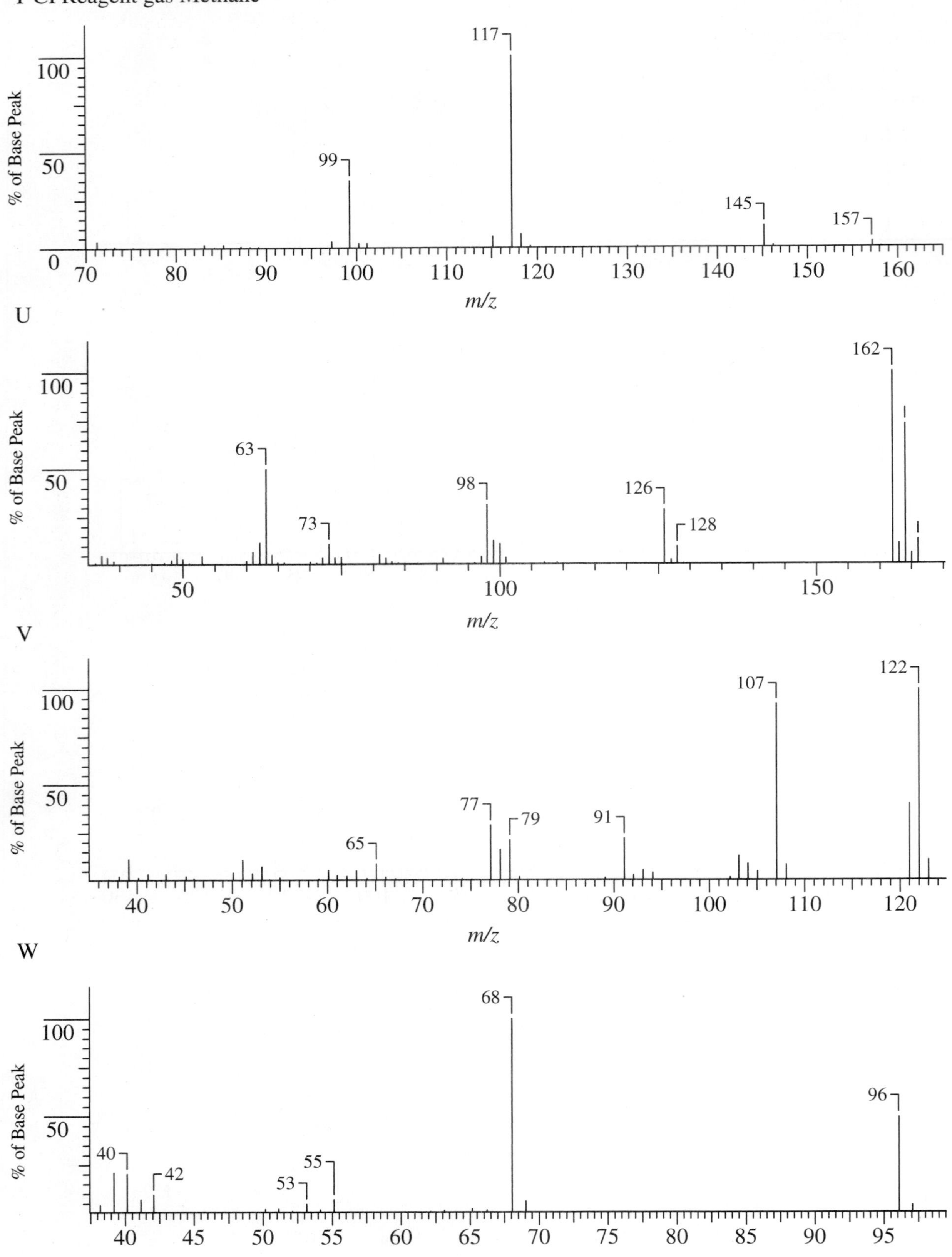

APPENDIX A FORMULA MASSES (FM) FOR VARIOUS COMBINATIONS OF CARBON, HYDROGEN, NITROGEN, AND OXYGEN[a]

	FM
12	
C	12.0000
13	
CH	13.0078
14	
N	14.0031
CH_2	14.0157
15	
HN	15.0109
CH_3	15.0235
16	
O	15.9949
H_2N	16.0187
CH_4	16.0313
17	
HO	17.0027
H_3N	17.0266
18	
H_2O	18.0106
24	
C_2	24.0000
26	
CN	26.0031
C_2H_2	26.0157
27	
CHN	27.0109
C_2H_3	27.0235
28	
N_2	28.0062
CO	27.9949
CH_2N	28.0187
C_2H_4	28.0313
29	
HN_2	29.0140
CHO	29.0027
CH_3N	29.0266
C_2H_5	29.0391
30	
NO	29.9980
H_2N_2	30.0218
CH_2O	30.0106
CH_4N	30.0344
C_2H_6	30.0470
31	
HNO	31.0058
H_3N_2	31.0297
CH_3O	31.0184
CH_5N	31.0422
32	
O_2	31.9898
H_2NO	32.0136
H_4N_2	32.0375
CH_4O	32.0262
33	
HO_2	32.9976
H_3NO	33.0215
34	
H_2O_2	34.0054
38	
C_3H_2	38.0157
39	
C_2HN	39.0109
C_3H_3	39.0235
40	
C_2H_2N	40.0187
C_3H_4	40.0313
41	
CHN_2	41.0140
C_2H_3N	41.0266
C_3H_5	41.0391
42	
N_3	42.0093
CNO	41.9980
CH_2N_2	42.0218
C_2H_2O	42.0106
C_2H_4N	42.0344
C_3H_6	42.0470
43	
HN_3	43.0170
$CHNO$	43.0058
CH_3N_2	43.0297
C_2H_3O	43.0184
C_2H_5N	43.0422
C_3H_7	43.0548
44	
N_2O	44.0011
CO_2	43.9898
CH_2NO	44.0136
CH_4N_2	44.0375
C_2H_4O	44.0262
C_2H_6N	44.0501
C_3H_8	44.0626
45	
CH_3NO	45.0215
CH_5N_2	45.0453
C_2H_5O	45.0340
C_2H_7N	45.0579
46	
NO_2	45.9929
CH_2O_2	46.0054
CH_4NO	46.0293
CH_6N_2	46.0532
C_2H_6O	46.0419
47	
HNO_2	47.0007
CH_3O_2	47.0133
CH_5NO	47.0371
48	
O_3	47.9847
H_2NO_2	48.0085
H_4N_2O	48.0324
CH_4O_2	48.0211
49	
H_3NO_2	49.0164
52	
C_4H_4	52.0313
53	
C_3H_3N	53.0266
C_4H_5	53.0391
54	
$C_2H_2N_2$	54.0218
C_3H_2O	54.0106
C_3H_4N	54.0344
C_4H_6	54.0470
55	
$C_2H_3N_2$	55.0297
C_3H_3O	55.0184
C_3H_5N	55.0422
C_4H_7	55.0548
56	
C_2O_2	55.9898
C_2H_2NO	56.0136
$C_2H_4N_2$	56.0375
C_3H_4O	56.0262
C_3H_6N	56.0501
C_4H_8	56.0626
57	
C_2H_3NO	57.0215
$C_2H_5N_2$	57.0453
C_3H_5O	57.0340
C_3H_7N	57.0579
C_4H_9	57.0705
58	
CH_2N_2O	58.0167
CH_4N_3	58.0406
$C_2H_2O_2$	58.0054
C_2H_4NO	58.0293
$C_2H_6N_2$	58.0532
C_3H_6O	58.0419
C_3H_8N	58.0657
C_4H_{10}	58.0783
59	
$CHNO_2$	59.0007
CH_3N_2O	59.0246
CH_5N_3	59.0484
$C_2H_3O_2$	59.0133
C_2H_5NO	59.0371
$C_2H_7N_2$	59.0610
C_3H_7O	59.0497
C_3H_9N	59.0736
60	
CH_2NO_2	60.0085
CH_4N_2O	60.0324
CH_6N_3	60.0563
$C_2H_4O_2$	60.0211
C_2H_6NO	60.0450
$C_2H_8N_2$	60.0688
C_3H_8O	60.0575
C_5	60.0000
61	
CH_3NO_2	61.0164
CH_5N_2O	61.0402
CH_7N_3	61.0641
$C_2H_5O_2$	61.0289
C_2H_7NO	61.0528
62	
CH_2O_3	62.0003
CH_4NO_2	62.0242
CH_6N_2O	62.0480
$C_2H_6O_2$	62.0368
63	
HNO_3	62.9956
CH_5NO_2	63.0320
64	
C_5H_4	64.0313
65	
C_4H_3N	65.0266
C_5H_5	65.0391
66	
C_4H_4N	66.0344
C_5H_6	66.0470
67	
$C_3H_3N_2$	67.0297
C_4H_3O	67.0184
C_4H_5N	67.0422
C_5H_7	67.0548
68	
$C_3H_4N_2$	68.0375
C_4H_4O	68.0262
C_4H_6N	68.0501
C_5H_8	68.0626
69	
C_3H_3NO	69.0215
$C_3H_5N_2$	69.0453

[a] With permission from J.H. Beynon, *Mass Spectrometry and its Application to Organic Chemistry,* Amsterdam, 1960. The columns headed FM contain the *formula masses* based on the exact mass of the most abundant isotope of each element; these masses are based on the most abundant isotope of carbon having a mass of 12.0000. Note that the table includes only C, H, N, and O.

APPENDIX A *(Continued)*

	FM
C_4H_5O	69.0340
C_4H_7N	69.0579
C_5H_9	69.0705
70	
$C_2H_4N_3$	70.0406
$C_3H_2O_2$	70.0054
C_3H_4NO	70.0293
$C_3H_6N_2$	70.0532
C_4H_6O	70.0419
C_4H_8N	70.0657
C_5H_{10}	70.0783
71	
$C_2H_3N_2O$	71.0246
$C_2H_5N_3$	71.0484
$C_3H_3O_2$	71.0133
C_3H_5NO	71.0371
$C_3H_7N_2$	71.0610
C_4H_7O	71.0497
C_4H_9N	71.0736
C_5H_{11}	71.0861
72	
$C_2H_2NO_2$	72.0085
$C_2H_4N_2O$	72.0324
$C_2H_6N_3$	72.0563
$C_3H_4O_2$	72.0211
C_3H_6NO	72.0449
$C_3H_8N_2$	72.0688
C_4H_8O	72.0575
$C_4H_{10}N$	72.0814
C_5H_{12}	72.0939
73	
$C_2H_3NO_2$	73.0164
$C_2H_5N_2O$	73.0402
$C_2H_7N_3$	73.0641
$C_3H_5O_2$	73.0289
C_3H_7NO	73.0528
$C_3H_9N_2$	73.0767
C_4H_9O	73.0653
$C_4H_{11}N$	73.0892
74	
$C_2H_2O_3$	74.0003
$C_2H_4NO_2$	74.0242
$C_2H_6N_2O$	74.0480
$C_2H_8N_3$	74.0719
$C_3H_6O_2$	74.0368
C_3H_8NO	74.0606
$C_3H_{10}N_2$	74.0845
$C_4H_{10}O$	74.0732
75	
$C_2H_3O_3$	75.0082
$C_2H_5NO_2$	75.0320
$C_2H_7N_2O$	75.0559
$C_2H_9N_3$	75.0798
$C_3H_7O_2$	75.0446
C_3H_9NO	75.0684
76	
$C_2H_4O_3$	76.0160
$C_2H_6NO_2$	76.0399
$C_2H_8N_2O$	76.0637
$C_3H_8O_2$	76.0524
C_5H_2N	76.0187
C_6H_4	76.0313
77	
CH_3NO_3	77.0113
$C_2H_5O_3$	77.0238
$C_2H_7NO_2$	77.0477
C_6H_5	77.0391
78	
$C_2H_6O_3$	78.0317
C_5H_4N	78.0344
C_6H_6	78.0470
79	
C_5H_5N	79.0422
C_6H_7	79.0548
80	
$C_3H_2N_3$	80.0249
$C_4H_4N_2$	80.0375
C_5H_4O	80.0262
C_5H_6N	80.0501
C_6H_8	80.0626
81	
$C_3H_3N_3$	81.0328
$C_4H_5N_2$	81.0453
C_5H_5O	81.0340
C_5H_7N	81.0579
C_6H_9	81.0705
82	
$C_3H_4N_3$	82.0406
C_4H_4NO	82.0293
$C_4H_6N_2$	82.0532
C_5H_6O	82.0419
C_5H_8N	82.0657
C_6H_{10}	82.0783
83	
$C_3H_5N_3$	83.0484
$C_4H_3O_2$	83.0133
C_4H_5NO	83.0371
$C_4H_7N_2$	83.0610
C_5H_7O	83.0497
C_5H_9N	83.0736
C_6H_{11}	83.0861
84	
$C_3H_6N_3$	84.0563
$C_4H_4O_2$	84.0211
C_4H_6NO	84.0449
$C_4H_8N_2$	84.0688
C_5H_8O	84.0575
$C_5H_{10}N$	84.0814
C_6H_{12}	84.0939
85	
$C_3H_5N_2O$	85.0402
$C_3H_7N_3$	85.0641
$C_4H_5O_2$	85.0289
C_4H_7NO	85.0528
$C_4H_9N_2$	85.0767
C_5H_9O	85.0653
$C_5H_{11}N$	85.0892
C_6H_{13}	85.1018
86	
$C_2H_2N_2O_2$	86.0116
$C_2H_4N_3O$	86.0355
$C_2H_6N_4$	86.0594
$C_3H_4NO_2$	86.0242
$C_3H_6N_2O$	86.0480
$C_3H_8N_3$	86.0719
$C_4H_6O_2$	86.0368
C_4H_8NO	86.0606
$C_4H_{10}N_2$	86.0845
$C_5H_{10}O$	86.0732
$C_5H_{12}N$	86.0970
C_6H_{14}	86.1096
87	
$C_2H_7N_4$	87.0672
$C_3H_3O_3$	87.0082
$C_3H_5NO_2$	87.0320
$C_3H_7N_2O$	87.0559
$C_3H_9N_3$	87.0798
$C_4H_7O_2$	87.0446
C_4H_9NO	87.0684
$C_4H_{11}N_2$	87.0923
$C_5H_{11}O$	87.0810
$C_5H_{13}N$	87.1049
88	
$C_2H_4N_2O_2$	88.0273
$C_2H_6N_3O$	88.0511
$C_2H_8N_4$	88.0750
$C_3H_4O_3$	88.0160
$C_3H_6NO_2$	88.0399
$C_3H_8N_2O$	88.0637
$C_3H_{10}N_3$	88.0876
$C_4H_8O_2$	88.0524
$C_4H_{10}NO$	88.0763
$C_4H_{12}N_2$	88.1001
$C_5H_{12}O$	88.0888
89	
$C_2H_5N_2O_2$	89.0351
$C_2H_7N_3O$	89.0590
$C_2H_9N_4$	89.0829
$C_3H_5O_3$	89.0238
$C_3H_7NO_2$	89.0477
$C_3H_9N_2O$	89.0715
$C_3H_{11}N_3$	89.0954
$C_4H_9O_2$	89.0603
$C_4H_{11}NO$	89.0841
C_7H_5	89.0391
90	
$C_2H_4NO_3$	90.0191
$C_2H_6N_2O_2$	90.0429
$C_2H_8N_3O$	90.0668
$C_2H_{10}N_4$	90.0907
$C_3H_6O_3$	90.0317
$C_3H_8NO_2$	90.0555
$C_3H_{10}N_2O$	90.0794
$C_4H_{10}O_2$	90.0681
C_7H_6	90.0470
91	
$C_2H_3O_4$	91.0031
$C_2H_5NO_3$	91.0269
$C_2H_7N_2O_2$	91.0508
$C_2H_9N_3O$	91.0746
$C_3H_7O_3$	91.0395
$C_3H_9NO_2$	91.0634
C_6H_5N	91.0422
C_7H_7	91.0548
92	
$C_2H_4O_4$	92.0109
$C_2H_6NO_3$	92.0348
$C_2H_8N_2O_2$	92.0586
$C_3H_8O_3$	92.0473
C_6H_4O	92.0262
C_6H_6N	92.0501
C_7H_8	92.0626
93	
$C_2H_5O_4$	93.0187
$C_2H_7NO_3$	92.0426
$C_5H_5N_2$	93.0453
C_6H_5O	93.0340
C_6H_7N	93.0579
C_7H_9	93.0705
94	
$C_2H_6O_4$	94.0266
$C_4H_4N_3$	94.0406
C_5H_4NO	94.0293
$C_5H_6N_2$	94.0532
C_6H_6O	94.0419
C_6H_8N	94.0657
C_7H_{10}	94.0783
95	
$C_4H_5N_3$	95.0484
C_5H_5NO	95.0371
$C_5H_7N_2$	95.0610
C_6H_7O	95.0497
C_6H_9N	95.0736
C_7H_{11}	95.0861
96	
$C_4H_6N_3$	96.0563
$C_5H_4O_2$	96.0211
C_5H_6NO	96.0449
$C_5H_8N_2$	96.0688
C_6H_8O	96.0575
$C_6H_{10}N$	96.0814
C_7H_{12}	96.0939
97	
$C_3H_5N_4$	97.0515
$C_4H_5N_2O$	97.0402
$C_5H_5O_2$	97.0289
C_5H_7NO	97.0528
$C_5H_9N_2$	97.0767

APPENDIX A *(Continued)*

	FM
C_6H_9O	97.0653
$C_6H_{11}N$	97.0892
C_7H_{13}	97.1018
98	
$C_3H_4N_3O$	98.0355
$C_3H_6N_4$	98.0594
$C_4H_4NO_2$	98.0242
$C_4H_6N_2O$	98.0480
$C_4H_8N_3$	98.0719
$C_5H_6O_2$	98.0368
C_5H_8NO	98.0606
$C_5H_{10}N_2$	98.0845
$C_6H_{10}O$	98.0732
$C_6H_{12}N$	98.0970
C_7H_{14}	98.1096
99	
$C_3H_5N_3O$	99.0433
$C_3H_7N_4$	99.0672
$C_4H_3O_3$	99.0082
$C_4H_5NO_2$	99.0320
$C_4H_7N_2O$	99.0559
$C_4H_9N_3$	99.0798
$C_5H_7O_2$	99.0446
C_5H_9NO	99.0685
$C_5H_{11}N_2$	99.0923
$C_6H_{11}O$	99.0810
$C_6H_{13}N$	99.1049
C_7H_{15}	99.1174
100	
$C_2H_4N_4O$	100.0386
$C_3H_4N_2O_2$	100.0273
$C_3H_6N_3O$	100.0511
$C_3H_8N_4$	100.0750
$C_4H_4O_3$	100.0160
$C_4H_6NO_2$	100.0399
$C_4H_8N_2O$	100.0637
$C_4H_{10}N_3$	100.0876
$C_5H_8O_2$	100.0524
$C_5H_{10}NO$	100.0763
$C_5H_{12}N_2$	100.1001
$C_6H_{12}O$	100.0888
$C_6H_{14}N$	100.1127
C_7H_{16}	100.1253
101	
$C_3H_3NO_3$	101.0113
$C_3H_5N_2O_2$	101.0351
$C_3H_7N_3O$	101.0590
$C_3H_9N_4$	101.0829
$C_4H_5O_3$	101.0238
$C_4H_7NO_2$	101.0477
$C_4H_9N_2O$	101.0715
$C_4H_{11}N_3$	101.0954
$C_5H_9O_2$	101.0603
$C_5H_{11}NO$	101.0841
$C_5H_{13}N_2$	101.1080
$C_6H_{13}O$	101.0967
$C_6H_{15}N$	101.1205

	FM
102	
$C_2H_6N_4O$	102.0542
$C_3H_4NO_3$	102.0191
$C_3H_6N_2O_2$	102.0429
$C_3H_8N_3O$	102.0668
$C_3H_{10}N_4$	102.0907
$C_4H_6O_3$	102.0317
$C_4H_8NO_2$	102.0555
$C_4H_{10}N_2O$	102.0794
$C_4H_{12}N_3$	102.1032
$C_5H_{10}O_2$	102.0681
$C_5H_{12}NO$	102.0919
$C_5H_{14}N_2$	102.1158
$C_6H_{14}O$	102.1045
C_8H_6	102.0470
103	
$C_2H_5N_3O_2$	103.0382
$C_2H_7N_4O$	103.0621
$C_3H_3O_4$	103.0031
$C_3H_5NO_3$	103.0269
$C_3H_7N_2O_2$	103.0508
$C_3H_9N_3O$	103.0746
$C_3H_{11}N_4$	103.0985
$C_4H_7O_3$	103.0395
$C_4H_9NO_2$	103.0634
$C_4H_{11}N_2O$	103.0872
$C_4H_{13}N_3$	103.1111
$C_5H_{11}O_2$	103.0759
$C_5H_{13}NO$	103.0998
C_7H_5N	103.0422
C_8H_7	103.0548
104	
$C_2H_4N_2O_3$	104.0222
$C_2H_6N_3O_2$	104.0460
$C_2H_8N_4O$	104.0699
$C_3H_4O_4$	104.0109
$C_3H_6NO_3$	104.0348
$C_3H_8N_2O_2$	104.0586
$C_3H_{10}N_3O$	104.0825
$C_3H_{12}N_4$	104.1063
$C_4H_8O_3$	104.0473
$C_4H_{10}NO_2$	104.0712
$C_4H_{12}N_2O$	104.0950
$C_5H_{12}O_2$	104.0837
$C_6H_4N_2$	104.0375
C_7H_4O	104.0262
C_7H_6N	104.0501
C_8H_8	104.0626
105	
$C_2H_5N_2O_3$	105.0300
$C_2H_7N_3O_2$	105.0539
$C_2H_9N_4O$	105.0777
$C_3H_5O_4$	105.0187
$C_3H_7NO_3$	105.0426
$C_3H_9N_2O_2$	105.0664
$C_3H_{11}N_3O$	105.0903
$C_4H_9O_3$	105.0552

	FM
$C_4H_{11}NO_2$	105.0790
$C_6H_5N_2$	105.0453
C_7H_5O	105.0340
C_7H_7N	105.0579
C_8H_9	105.0705
106	
$C_2H_4NO_4$	106.0140
$C_2H_6N_2O_3$	106.0379
$C_2H_8N_3O_2$	106.0617
$C_2H_{10}N_4O$	106.0856
$C_3H_6O_4$	106.0266
$C_3H_8NO_3$	106.0504
$C_3H_{10}N_2O_2$	106.0743
$C_4H_{10}O_3$	106.0630
C_6H_4NO	106.0293
$C_6H_6N_2$	106.0532
C_7H_6O	106.0419
C_7H_8N	106.0657
C_8H_{10}	106.0783
107	
$C_2H_5NO_4$	107.0218
$C_2H_7N_2O_3$	107.0457
$C_2H_9N_3O_2$	107.0695
$C_3H_7O_4$	107.0344
$C_3H_9NO_3$	107.0583
$C_5H_5N_3$	107.0484
C_6H_5NO	107.0371
$C_6H_7N_2$	107.0610
C_7H_7O	107.0497
C_7H_9N	107.0736
C_8H_{11}	107.0861
108	
$C_2H_6NO_4$	108.0297
$C_2H_8N_2O_3$	108.0535
$C_3H_8O_4$	108.0422
$C_4H_4N_4$	108.0437
$C_5H_4N_2O$	108.0324
$C_5H_6N_3$	108.0563
$C_6H_4O_2$	108.0211
C_6H_6NO	108.0449
$C_6H_8N_2$	108.0688
C_7H_8O	108.0575
$C_7H_{10}N$	108.0814
C_8H_{12}	108.0939
109	
$C_2H_7NO_4$	109.0375
$C_4H_5N_4$	109.0515
$C_5H_5N_2O$	109.0402
$C_5H_7N_3$	109.0641
$C_6H_5O_2$	109.0289
C_6H_7NO	109.0528
$C_6H_9N_2$	109.0767
C_7H_9O	109.0653
$C_7H_{11}N$	109.0892
C_8H_{13}	109.1018
110	
$C_4H_4N_3O$	110.0355

	FM
$C_4H_6N_4$	110.0594
$C_5H_6N_2O$	110.0480
$C_5H_8N_3$	110.0719
$C_6H_6O_2$	110.0368
C_6H_8NO	110.0606
$C_6H_{10}N_2$	110.0845
$C_7H_{10}O$	110.0732
$C_7H_{12}N$	110.0970
C_8H_{14}	110.1096
111	
$C_4H_5N_3O$	111.0433
$C_4H_7N_4$	111.0672
$C_5H_5NO_2$	111.0320
$C_5H_7N_2O$	111.0559
$C_5H_9N_3$	111.0789
$C_6H_7O_2$	111.0446
C_6H_9NO	111.0684
$C_6H_{11}N_2$	111.0923
$C_7H_{11}O$	111.0810
$C_7H_{13}N$	111.1049
C_8H_{15}	111.1174
112	
$C_3H_4N_4O$	112.0386
$C_4H_4N_2O_2$	112.0273
$C_4H_6N_3O$	112.0511
$C_4H_8N_4$	112.0750
$C_5H_4O_3$	112.0160
$C_5H_6NO_2$	112.0399
$C_5H_8N_2O$	112.0637
$C_5H_{10}N_3$	112.0876
$C_6H_8O_2$	112.0524
$C_6H_{10}NO$	112.0763
$C_6H_{12}N_2$	112.1001
$C_7H_{12}O$	112.0888
$C_7H_{14}N$	112.1127
C_8H_{16}	112.1253
113	
$C_3H_5N_4O$	113.0464
$C_4H_5N_2O_2$	113.0351
$C_4H_7N_3O$	113.0590
$C_4H_9N_4$	113.0829
$C_5H_5O_3$	113.0238
$C_5H_7NO_2$	113.0477
$C_5H_9N_2O$	113.0715
$C_5H_{11}N_3$	113.0954
$C_6H_9O_2$	113.0603
$C_6H_{11}NO$	113.0841
$C_6H_{13}N_2$	113.1080
$C_7H_{13}O$	113.0967
$C_7H_{15}N$	113.1205
C_8H_{17}	113.1331
114	
$C_3H_6N_4O$	114.0542
$C_4H_4NO_3$	114.0191
$C_4H_6N_2O_2$	114.0429
$C_4H_8N_3O$	114.0668
$C_4H_{10}N_4$	114.0907

APPENDIX A *(Continued)*

	FM
$C_5H_6O_3$	114.0317
$C_5H_8NO_2$	114.0555
$C_5H_{10}N_2O$	114.0794
$C_5H_{12}N_3$	114.1032
$C_6H_{10}O_2$	114.0681
$C_6H_{12}NO$	114.0919
$C_6H_{14}N_2$	114.1158
$C_7H_{14}O$	114.1045
$C_7H_{16}N$	114.1284
C_8H_{18}	114.1409
C_9H_6	114.0470
115	
$C_3H_5N_3O_2$	115.0382
$C_3H_7N_4O$	115.0621
$C_4H_5NO_3$	115.0269
$C_4H_7N_2O_2$	115.0508
$C_4H_9N_3O$	115.0746
$C_4H_{11}N_4$	115.0985
$C_5H_7O_3$	115.0395
$C_5H_9NO_2$	115.0634
$C_5H_{11}N_2O$	115.0872
$C_5H_{13}N_3$	115.1111
$C_6H_{11}O_2$	115.0759
$C_6H_{13}NO$	115.0998
$C_6H_{15}N_2$	115.1236
$C_7H_{15}O$	115.1123
$C_7H_{17}N$	115.1362
C_9H_7	115.0548
116	
$C_2H_4N_4O_2$	116.0335
$C_3H_4N_2O_3$	116.0222
$C_3H_6N_3O_2$	116.0460
$C_3H_8N_4O$	116.0699
$C_4H_4O_4$	116.0109
$C_4H_6NO_3$	116.0348
$C_4H_8N_2O_2$	116.0586
$C_4H_{10}N_3O$	116.0825
$C_4H_{12}N_4$	116.1063
$C_5H_8O_3$	116.0473
$C_5H_{10}NO_2$	116.0712
$C_5H_{12}N_2O$	116.0950
$C_5H_{14}N_3$	116.1189
$C_6H_{12}O_2$	116.0837
$C_6H_{14}NO$	116.1076
$C_6H_{16}N_2$	116.1315
$C_7H_4N_2$	116.0375
$C_7H_{16}O$	116.1202
C_8H_6N	116.0501
C_9H_8	116.0626
117	
$C_2H_5N_4O_2$	117.0413
$C_3H_3NO_4$	117.0062
$C_3H_5N_2O_3$	117.0300
$C_3H_7N_3O_2$	117.0539
$C_3H_9N_4O$	117.0777
$C_4H_5O_4$	117.0187
$C_4H_7NO_3$	117.0426
$C_4H_9N_2O_2$	117.0664
$C_4H_{11}N_3O$	117.0903
$C_4H_{13}N_4$	117.1142
$C_5H_9O_3$	117.0552
$C_5H_{11}NO_2$	117.0790
$C_5H_{13}N_2O$	117.1029
$C_5H_{15}N_3$	117.1267
$C_6H_{13}O_2$	117.0916
$C_6H_{15}NO$	117.1154
C_8H_7N	117.0579
C_9H_9	117.0705
118	
$C_2H_4N_3O_3$	118.0253
$C_2H_6N_4O_2$	118.0491
$C_3H_4NO_4$	118.0140
$C_3H_6N_2O_3$	118.0379
$C_3H_8N_3O_2$	118.0617
$C_3H_{10}N_4O$	118.0856
$C_4H_6O_4$	118.0266
$C_4H_8NO_3$	118.0504
$C_4H_{10}N_2O_2$	118.0743
$C_4H_{12}N_3O$	118.0981
$C_4H_{14}N_4$	118.1220
$C_5H_{10}O_3$	118.0630
$C_5H_{12}NO_2$	118.0868
$C_5H_{14}N_2O$	118.1107
$C_6H_{14}O_2$	118.0994
$C_7H_6N_2$	118.0532
C_8H_6O	118.0419
C_8H_8N	118.0657
C_9H_{10}	118.0783
119	
$C_2H_5N_3O_3$	119.0331
$C_2H_7N_4O_2$	119.0570
$C_3H_5NO_4$	119.0218
$C_3H_7N_2O_3$	119.0457
$C_3H_9N_3O_2$	119.0695
$C_3H_{11}N_4O$	119.0934
$C_4H_7O_4$	119.0344
$C_4H_9NO_3$	119.0583
$C_4H_{11}N_2O_2$	119.0821
$C_4H_{13}N_3O$	119.1060
$C_5H_{11}O_3$	119.0708
$C_5H_{13}NO_2$	119.0947
$C_6H_5N_3$	119.0484
C_7H_5NO	119.0371
$C_7H_7N_2$	119.0610
C_8H_7O	119.0497
C_8H_9N	119.0736
C_9H_{11}	119.0861
120	
$C_2H_6N_3O_3$	120.0410
$C_2H_8N_4O_2$	120.0648
$C_3H_6NO_4$	120.0297
$C_3H_8N_2O_3$	120.0535
$C_3H_{10}N_3O_2$	120.0774
$C_3H_{12}N_4O$	120.1012
$C_4H_8O_4$	120.0422
$C_4H_{10}NO_3$	120.0661
$C_4H_{12}N_2O_2$	120.0899
$C_5H_4N_4$	120.0437
$C_5H_{12}O_3$	120.0786
$C_6H_4N_2O$	120.0324
$C_6H_6N_3$	120.0563
C_7H_6NO	120.0449
$C_7H_8N_2$	120.0688
C_8H_8O	120.0575
$C_8H_{10}N$	120.0814
C_9H_{12}	120.0939
121	
$C_2H_5N_2O_4$	121.0249
$C_2H_7N_3O_3$	121.0488
$C_2H_9N_4O_2$	121.0726
$C_3H_7NO_4$	121.0375
$C_3H_9N_2O_3$	121.0614
$C_3H_{11}N_3O_2$	121.0852
$C_4H_9O_4$	121.0501
$C_4H_{11}NO_3$	121.0739
$C_5H_5N_4$	121.0515
$C_6H_5N_2O$	121.0402
$C_6H_7N_3$	121.0641
$C_7H_5O_2$	121.0289
C_7H_7NO	121.0528
$C_7H_9N_2$	121.0767
C_8H_9O	121.0653
$C_8H_{11}N$	121.0892
C_9H_{13}	121.1018
122	
$C_2H_6N_2O_4$	122.0328
$C_2H_8N_3O_3$	122.0566
$C_2H_{10}N_4O_2$	122.0805
$C_3H_8NO_4$	122.0453
$C_3H_{10}N_2O_3$	122.0692
$C_4H_{10}O_4$	122.0579
$C_5H_6N_4$	122.0594
$C_6H_4NO_2$	122.0242
$C_6H_6N_2O$	122.0480
$C_6H_8N_3$	122.0719
$C_7H_6O_2$	122.0368
C_7H_8NO	122.0606
$C_7H_{10}N_2$	122.0845
$C_8H_{10}O$	122.0732
$C_8H_{12}N$	122.0970
C_9H_{14}	122.1096
123	
$C_2H_7N_2O_4$	123.0406
$C_2H_9N_3O_3$	123.0644
$C_3H_9NO_4$	123.0532
$C_5H_5N_3O$	123.0433
$C_5H_7N_4$	123.0672
$C_6H_5NO_2$	123.0320
$C_6H_7N_2O$	123.0559
$C_6H_9N_3$	123.0798
$C_7H_7O_2$	123.0446
C_7H_9NO	123.0684
$C_7H_{11}N_2$	123.0923
$C_8H_{11}O$	123.0810
$C_8H_{13}N$	123.1049
C_9H_{15}	123.1174
124	
$C_2H_8N_2O_4$	124.0484
$C_4H_4N_4O$	124.0386
$C_5H_4N_2O_2$	124.0273
$C_5H_6N_3O$	124.0511
$C_5H_8N_4$	124.0750
$C_6H_4O_3$	124.0160
$C_6H_6NO_2$	124.0399
$C_6H_8N_2O$	124.0637
$C_6H_{10}N_3$	124.0876
$C_7H_8O_2$	124.0524
$C_7H_{10}NO$	124.0763
$C_7H_{12}N_2$	124.1001
C_8N_2	124.0062
$C_8H_{12}O$	124.0888
$C_8H_{14}N$	124.1127
C_9H_{16}	124.1253
125	
$C_4H_3N_3O_2$	125.0226
$C_4H_5N_4O$	125.0464
$C_5H_5N_2O_2$	125.0351
$C_5H_7N_3O$	125.0590
$C_5H_9N_4$	125.0829
$C_6H_5O_3$	125.0238
$C_6H_7NO_2$	125.0477
$C_6H_9N_2O$	125.0715
$C_6H_{11}N_3$	125.0954
$C_7H_9O_2$	125.0603
$C_7H_{11}NO$	125.0841
$C_7H_{13}N_2$	125.1080
$C_8H_{13}O$	125.0967
$C_8H_{15}N$	125.1205
C_9H_{17}	125.1331
126	
$C_3H_2N_4O_2$	126.0178
$C_4H_4N_3O_2$	126.0304
$C_4H_6N_4O$	126.0542
$C_5H_4NO_3$	126.0191
$C_5H_6N_2O_2$	126.0429
$C_5H_8N_3O$	126.0668
$C_5H_{10}N_4$	126.0907
$C_6H_6O_3$	126.0317
$C_6H_8NO_2$	126.0555
$C_6H_{10}N_2O$	126.0794
$C_6H_{12}N_3$	126.1032
$C_7H_{10}O_2$	126.0681
$C_7H_{12}NO$	126.0919
$C_7H_{14}N_2$	126.1158
$C_8H_{14}O$	126.1045
$C_8H_{16}N$	126.1284
C_9H_{18}	126.1409
127	

APPENDIX A (Continued)

	FM
$C_3H_3N_4O_2$	127.0257
$C_4H_5N_3O_2$	127.0382
$C_4H_7N_4O$	127.0621
$C_5H_5NO_3$	127.0269
$C_5H_7N_2O_2$	127.0508
$C_5H_9N_3O$	127.0746
$C_5H_{11}N_4$	127.0985
$C_6H_7O_3$	127.0395
$C_6H_9NO_2$	127.0634
$C_6H_{11}N_2O$	127.0872
$C_6H_{13}N_3$	127.1111
$C_7H_{11}O_2$	127.0759
$C_7H_{13}NO$	127.0998
$C_7H_{15}N_2$	127.1236
$C_8H_{15}O$	127.1123
$C_8H_{17}N$	127.1362
C_9H_{19}	127.1488
128	
$C_3H_4N_4O_2$	128.0335
$C_4H_4N_2O_3$	128.0222
$C_4H_6N_3O_2$	128.0460
$C_4H_8N_4O$	128.0699
$C_5H_4O_4$	128.0109
$C_5H_6NO_3$	128.0348
$C_5H_8N_2O_2$	128.0586
$C_5H_{10}N_3O$	128.0825
$C_5H_{12}N_4$	128.1063
$C_6H_8O_3$	128.0473
$C_6H_{10}NO_2$	128.0712
$C_6H_{12}N_2O$	128.0950
$C_6H_{14}N_3$	128.1189
$C_7H_{12}O_2$	128.0837
$C_7H_{14}NO$	128.1076
$C_7H_{16}N_2$	128.1315
$C_8H_{16}O$	128.1202
$C_8H_{18}N$	128.1440
C_9H_{20}	128.1566
$C_{10}H_8$	128.0626
129	
$C_3H_3N_3O_3$	129.0175
$C_3H_5N_4O_2$	129.0413
$C_4H_5N_2O_3$	129.0300
$C_4H_7N_3O_2$	129.0539
$C_4H_9N_4O$	129.0777
$C_5H_5O_4$	129.0187
$C_5H_7NO_3$	129.0426
$C_5H_9N_2O_2$	129.0664
$C_5H_{11}N_3O$	129.0903
$C_5H_{13}N_4$	129.1142
$C_6H_9O_3$	129.0552
$C_6H_{11}NO_2$	129.0790
$C_6H_{13}N_2O$	129.1029
$C_6H_{15}N_3$	129.1267
$C_7H_{13}O_2$	129.0916
$C_7H_{15}NO$	129.1154
$C_7H_{17}N_2$	129.1393
$C_8H_{17}O$	129.1280
$C_8H_{19}N$	129.1519
C_9H_7N	129.0579
$C_{10}H_9$	129.0705
130	
$C_3H_4N_3O_3$	130.0253
$C_3H_6N_4O_2$	130.0491
$C_4H_4NO_4$	130.0140
$C_4H_6N_2O_3$	130.0379
$C_4H_8N_3O_2$	130.0617
$C_4H_{10}N_4O$	130.0856
$C_5H_6O_4$	130.0266
$C_5H_8NO_3$	130.0504
$C_5H_{10}N_2O_2$	130.0743
$C_5H_{12}N_3O$	130.0981
$C_5H_{14}N_4$	130.1220
$C_6H_{10}O_3$	130.0630
$C_6H_{12}NO_2$	130.0868
$C_6H_{14}N_2O$	130.1107
$C_6H_{16}N_3$	130.1346
$C_7H_4N_3$	130.0406
$C_7H_{14}O_2$	130.0994
$C_7H_{16}NO$	130.1233
$C_7H_{18}N_2$	130.1471
$C_8H_6N_2$	130.0532
$C_8H_{18}O$	130.1358
C_9H_8N	130.0657
$C_{10}H_{10}$	130.0783
131	
$C_3H_3N_2O_4$	131.0093
$C_3H_5N_3O_3$	131.0331
$C_3H_7N_4O_2$	131.0570
$C_4H_5NO_4$	131.0218
$C_4H_7N_2O_3$	131.0457
$C_4H_9N_3O_2$	131.0695
$C_4H_{11}N_4O$	131.0934
$C_5H_7O_4$	131.0344
$C_5H_9NO_3$	131.0583
$C_5H_{11}N_2O_2$	131.0821
$C_5H_{13}N_3O$	131.1060
$C_5H_{15}N_4$	131.1298
$C_6H_{11}O_3$	131.0708
$C_6H_{13}NO_2$	131.0947
$C_6H_{15}N_2O$	131.1185
$C_6H_{17}N_3$	131.1424
$C_7H_5N_3$	131.0484
$C_7H_{15}O_2$	131.1072
$C_7H_{17}NO$	131.1311
$C_8H_7N_2$	131.0610
C_9H_7O	131.0497
C_9H_9N	131.0736
$C_{10}H_{11}$	131.0861
132	
$C_3H_4N_2O_4$	132.0171
$C_3H_6N_3O_3$	132.0410
$C_3H_8N_4O_2$	132.0648
$C_4H_6NO_4$	132.0297
$C_4H_8N_2O_3$	132.0535
$C_4H_{10}N_3O_2$	132.0774
$C_4H_{12}N_4O$	132.1012
$C_5H_8O_4$	132.0422
$C_5H_{10}NO_3$	132.0661
$C_5H_{12}N_2O_2$	132.0899
$C_5H_{14}N_3O$	132.1138
$C_5H_{16}N_4$	132.1377
$C_6H_4N_4$	132.0437
$C_6H_{12}O_3$	132.0786
$C_6H_{14}NO_2$	132.1025
$C_6H_{16}N_2O$	132.1264
$C_7H_9N_3$	132.0563
$C_7H_{16}O_2$	132.1151
C_8H_6NO	132.0449
$C_8H_8N_2$	132.0688
C_9H_8O	132.0575
$C_9H_{10}N$	132.0814
$C_{10}H_{12}$	132.0939
133	
$C_3H_5N_2O_4$	133.0249
$C_3H_7N_3O_3$	133.0488
$C_3H_9N_4O_2$	133.0726
$C_4H_7NO_4$	133.0375
$C_4H_9N_2O_3$	133.0614
$C_4H_{11}N_3O_2$	133.0852
$C_4H_{13}N_4O$	133.1091
$C_5H_9O_4$	133.0501
$C_5H_{11}NO_3$	133.0739
$C_5H_{13}N_2O_2$	133.0978
$C_5H_{15}N_3O$	133.1216
$C_6H_5N_4$	133.0515
$C_6H_{13}O_3$	133.0865
$C_6H_{15}NO_2$	133.1103
$C_7H_5N_2O$	133.0402
$C_7H_7N_3$	133.0641
C_8H_7NO	133.0528
$C_8H_9N_2$	133.0767
C_9H_9O	133.0653
$C_9H_{11}N$	133.0892
$C_{10}H_{13}$	133.1018
134	
$C_3H_6N_2O_4$	134.0328
$C_3H_8N_3O_3$	134.0566
$C_3H_{10}N_4O_2$	134.0805
$C_4H_8NO_4$	134.0453
$C_4H_{10}N_2O_3$	134.0692
$C_4H_{12}N_3O_2$	134.0930
$C_4H_{14}N_4O$	134.1169
$C_5H_{10}O_4$	134.0579
$C_5H_{12}NO_3$	134.0817
$C_5H_{14}N_2O_2$	134.1056
$C_6H_4N_3O$	134.0355
$C_6H_6N_4$	134.0594
$C_6H_{14}O_3$	134.0943
$C_7H_6N_2O$	134.0480
$C_7H_8N_3$	134.0719
$C_8H_6O_2$	134.0368
C_8H_8NO	134.0606
$C_8H_{10}N_2$	134.0845
$C_9H_{10}O$	134.0732
$C_9H_{12}N$	134.0970
$C_{10}H_{14}$	134.1096
135	
$C_3H_7N_2O_4$	135.0406
$C_3H_9N_3O_3$	135.0644
$C_3H_{11}N_4O_2$	135.0883
$C_4H_9NO_4$	135.0532
$C_4H_{11}N_2O_3$	135.0770
$C_4H_{13}N_3O_2$	135.1009
$C_5H_3N_4O$	135.0308
$C_5H_{11}O_4$	135.0657
$C_5H_{13}NO_3$	135.0896
$C_6H_5N_3O$	135.0433
$C_6H_7N_4$	135.0672
$C_7H_5NO_2$	135.0320
$C_7H_7N_2O$	135.0559
$C_7H_9N_3$	135.0798
$C_8H_7O_2$	135.0446
C_8H_9NO	135.0684
$C_8H_{11}N_2$	135.0923
$C_9H_{11}O$	135.0810
$C_9H_{13}N$	135.1049
$C_{10}H_{15}$	135.1174
136	
$C_3H_8N_2O_4$	136.0484
$C_3H_{10}N_3O_3$	136.0723
$C_3H_{12}N_4O_2$	136.0961
$C_4H_{10}NO_4$	136.0610
$C_4H_{12}N_2O_3$	136.0848
$C_5H_2N_3O_2$	136.0147
$C_5H_4N_4O$	136.0386
$C_5H_{12}O_4$	136.0735
$C_6H_4N_2O_2$	136.0273
$C_6H_6N_3O$	136.0511
$C_6H_8N_4$	136.0750
$C_7H_4O_3$	136.0160
$C_7H_6NO_2$	136.0399
$C_7H_8N_2O$	136.0637
$C_7H_{10}N_3$	136.0876
$C_8H_8O_2$	136.0524
$C_8H_{10}NO$	136.0763
$C_8H_{12}N_2$	136.1001
$C_9H_{12}O$	136.0888
$C_9H_{14}N$	136.1127
$C_{10}H_{16}$	136.1253
137	
$C_3H_9N_2O_4$	137.0563
$C_3H_{11}N_3O_3$	137.0801
$C_4H_{11}NO_4$	137.0688
$C_5H_3N_3O_2$	137.0226
$C_5H_5N_4O$	137.0464
$C_6H_5N_2O_2$	137.0351
$C_6H_7N_3O$	137.0590
$C_6H_9N_4$	137.0829

APPENDIX A *(Continued)*

	FM
$C_7H_5O_3$	137.0238
$C_7H_7NO_2$	137.0477
$C_7H_9N_2O$	137.0715
$C_7H_{11}N_3$	137.0954
$C_8H_9O_2$	137.0603
$C_8H_{11}NO$	137.0841
$C_8H_{13}N_2$	137.1080
$C_9H_{13}O$	137.0967
$C_9H_{15}N$	137.1205
$C_{10}H_{17}$	137.1331
138	
$C_3H_{10}N_2O_4$	138.0641
$C_5H_4N_3O_2$	138.0304
$C_5H_6N_4O$	138.0542
$C_6H_4NO_3$	138.0191
$C_6H_6N_2O_2$	138.0429
$C_6H_8N_3O$	138.0668
$C_6H_{10}N_4$	138.0907
$C_7H_6O_3$	138.0317
$C_7H_8NO_2$	138.0555
$C_7H_{10}N_2O$	138.0794
$C_7H_{12}N_3$	138.1032
$C_8H_{10}O_2$	138.0681
$C_8H_{12}NO$	138.0919
$C_8H_{14}N_2$	138.1158
$C_9H_{14}O$	138.1045
$C_9H_{16}N$	138.1284
$C_{10}H_{18}$	138.1409
139	
$C_4H_3N_4O_2$	139.0257
$C_5H_3N_2O_3$	139.0144
$C_5H_5N_3O_2$	139.0382
$C_5H_7N_4O$	139.0621
$C_6H_5NO_3$	139.0269
$C_6H_7N_2O_2$	139.0508
$C_6H_9N_3O$	139.0747
$C_6H_{11}N_4$	139.0985
$C_7H_7O_3$	139.0395
$C_7H_9NO_2$	139.0634
$C_7H_{11}N_2O$	139.0872
$C_7H_{13}N_3$	139.1111
$C_8H_{11}O_2$	139.0759
$C_8H_{13}NO$	139.0998
$C_8H_{15}N_2$	139.1236
$C_9H_3N_2$	139.0297
$C_9H_{15}O$	139.1123
$C_9H_{17}N$	139.1362
$C_{10}H_{19}$	139.1488
$C_{11}H_7$	139.0548
140	
$C_4H_4N_4O_2$	140.0335
$C_5H_4N_2O_3$	140.0222
$C_5H_6N_3O_2$	140.0460
$C_5H_8N_4O$	140.0699
$C_6H_4O_4$	140.0109
$C_6H_6NO_3$	140.0348

	FM
$C_6H_8N_2O_2$	140.0586
$C_6H_{10}N_3O$	140.0825
$C_6H_{12}N_4$	140.1063
$C_7H_8O_3$	140.0473
$C_7H_{10}NO_2$	140.0712
$C_7H_{12}N_2O$	140.0950
$C_7H_{14}N_3$	140.1189
$C_8H_{12}O_2$	140.0837
$C_8H_{14}NO$	140.1076
$C_8H_{16}N_2$	140.1315
$C_9H_{16}O$	140.1202
$C_9H_{18}N$	140.1440
$C_{10}H_6N$	140.0501
$C_{10}H_{20}$	140.1566
$C_{11}H_8$	140.0626
141	
$C_4H_3N_3O_3$	141.0175
$C_4H_5N_4O_2$	141.0413
$C_5H_3NO_4$	141.0062
$C_5H_5N_2O_3$	141.0300
$C_5H_7N_3O_2$	141.0539
$C_5H_9N_4O$	141.0777
$C_6H_5O_4$	141.0187
$C_6H_7NO_3$	141.0426
$C_6H_9N_2O_2$	141.0664
$C_6H_{11}N_3O$	141.0903
$C_6H_{13}N_4$	141.1142
$C_7H_9O_3$	141.0552
$C_7H_{11}NO_2$	141.0790
$C_7H_{13}N_2O$	141.1029
$C_7H_{15}N_3$	141.1267
$C_8H_{13}O_2$	141.0916
$C_8H_{15}NO$	141.1154
$C_8H_{17}N_2$	141.1393
$C_9H_{17}O$	141.1280
$C_9H_{19}N$	141.1519
$C_{10}H_7N$	141.0579
$C_{10}H_{21}$	141.1644
$C_{11}H_9$	141.0705
142	
$C_4H_4N_3O_3$	142.0253
$C_4H_6N_4O_2$	142.0491
$C_5H_4NO_4$	142.0140
$C_5H_6N_2O_3$	142.0379
$C_5H_8N_3O_2$	142.0617
$C_5H_{10}N_4O$	142.0856
$C_6H_6O_4$	142.0266
$C_6H_8NO_3$	142.0504
$C_6H_{10}N_2O_2$	142.0743
$C_6H_{12}N_3O$	142.0981
$C_6H_{14}N_4$	142.1220
$C_7H_{10}O_3$	142.0630
$C_7H_{12}NO_2$	142.0868
$C_7H_{14}N_2O$	142.1107
$C_7H_{16}N_3$	142.1346
$C_8H_{14}O_2$	142.0994

	FM
$C_8H_{16}NO$	142.1233
$C_8H_{18}N_2$	142.1471
$C_9H_6N_2$	142.0532
$C_9H_{18}O$	142.1358
$C_9H_{20}N$	142.1597
$C_{10}H_8N$	142.0657
$C_{10}H_{22}$	142.1722
$C_{11}H_{10}$	142.0783
143	
$C_4H_3N_2O_4$	143.0093
$C_4H_5N_3O_3$	143.0331
$C_4H_7N_4O_2$	143.0570
$C_5H_5NO_4$	143.0218
$C_5H_7N_2O_3$	143.0457
$C_5H_9N_3O_2$	143.0695
$C_5H_{11}N_4O$	143.0934
$C_6H_7O_4$	143.0344
$C_6H_9NO_3$	143.0583
$C_6H_{11}N_2O_2$	143.0821
$C_6H_{13}N_3O$	143.1060
$C_6H_{15}N_4$	143.1298
$C_7H_{11}O_3$	143.0708
$C_7H_{13}NO_2$	143.0947
$C_7H_{15}N_2O$	143.1185
$C_7H_{17}N_3$	143.1424
$C_8H_{15}O_2$	143.1072
$C_8H_{17}NO$	143.1311
$C_8H_{19}N_2$	143.1549
$C_9H_7N_2$	143.0610
$C_9H_{19}O$	143.1436
$C_9H_{21}N$	143.1675
$C_{10}H_7O$	143.0497
$C_{10}H_9N$	143.0736
$C_{11}H_{11}$	143.0861
144	
$C_4H_4N_2O_4$	144.0171
$C_4H_6N_3O_3$	144.0410
$C_4H_8N_4O_2$	144.0648
$C_5H_6NO_4$	144.0297
$C_5H_8N_2O_3$	144.0535
$C_5H_{10}N_3O_2$	144.0774
$C_5H_{12}N_4O$	144.1012
$C_6H_8O_4$	144.0422
$C_6H_{10}NO_3$	144.0661
$C_6H_{12}N_2O_2$	144.0899
$C_6H_{14}N_3O$	144.1138
$C_6H_{16}N_4$	144.1377
$C_7H_{12}O_3$	144.0786
$C_7H_{14}NO_2$	144.1025
$C_7H_{16}N_2O$	144.1264
$C_7H_{18}N_3$	144.1502
$C_8H_6N_3$	144.0563
$C_8H_{16}O_2$	144.1151
$C_8H_{18}NO$	144.1389
$C_8H_{20}N_2$	144.1628
C_9H_6NO	144.0449

	FM
$C_9H_8N_2$	144.0688
$C_9H_{20}O$	144.1515
$C_{10}H_8O$	144.0575
$C_{10}H_{10}N$	144.0814
$C_{11}H_{12}$	144.0939
145	
$C_4H_5N_2O_4$	145.0249
$C_4H_7N_3O_3$	145.0488
$C_4H_9N_4O_2$	145.0726
$C_5H_7NO_4$	145.0375
$C_5H_9N_2O_3$	145.0614
$C_5H_{11}N_3O_2$	145.0852
$C_5H_{13}N_4O$	145.1091
$C_6H_9O_4$	145.0501
$C_6H_{11}NO_3$	145.0739
$C_6H_{13}N_2O_2$	145.0978
$C_6H_{15}N_3O$	145.1216
$C_6H_{17}N_4$	145.1455
$C_7H_5N_4$	145.0515
$C_7H_{13}O_3$	145.0865
$C_7H_{15}NO_2$	145.1103
$C_7H_{17}N_2O$	145.1342
$C_7H_{19}N_3$	145.1580
$C_8H_5N_2O$	145.0402
$C_8H_7N_3$	145.0641
$C_8H_{17}O_2$	145.1229
$C_8H_{19}NO$	145.1467
C_9H_7NO	145.0528
$C_9H_9N_2$	145.0767
$C_{10}H_9O$	145.0653
$C_{10}H_{11}N$	145.0892
$C_{11}H_{13}$	145.1018
146	
$C_4H_6N_2O_4$	146.0328
$C_4H_8N_3O_3$	146.0566
$C_4H_{10}N_4O_2$	146.0805
$C_5H_8NO_4$	146.0453
$C_5H_{10}N_2O_3$	146.0692
$C_5H_{12}N_3O_2$	146.0930
$C_5H_{14}N_4O$	146.1169
$C_6H_{10}O_4$	146.0579
$C_6H_{12}NO_3$	146.0817
$C_6H_{14}N_2O_2$	146.1056
$C_6H_{16}N_3O$	146.1295
$C_7H_6N_4$	146.0594
$C_7H_{14}O_3$	146.0943
$C_7H_{16}NO_2$	146.1182
$C_7H_{18}N_2O$	146.1420
$C_8H_2O_3$	146.0003
$C_8H_6N_2O$	146.0480
$C_8H_8N_3$	146.0719
$C_8H_{18}O_2$	146.1307
$C_9H_6O_2$	146.0368
C_9H_8NO	146.0606
$C_9H_{10}N_2$	146.0845
$C_{10}H_{10}O$	146.0732

APPENDIX A *(Continued)*

	FM
$C_{10}H_{12}N$	146.0970
$C_{11}H_{14}$	146.1096
147	
$C_4H_7N_2O_4$	147.0406
$C_4H_9N_3O_3$	147.0644
$C_4H_{11}N_4O_2$	147.0883
$C_5H_9NO_4$	147.0532
$C_5H_{11}N_2O_3$	147.0770
$C_5H_{13}N_3O_2$	147.1009
$C_5H_{15}N_4O$	147.1247
$C_6H_{11}O_4$	147.0657
$C_6H_{13}NO_3$	147.0896
$C_6H_{15}N_2O_2$	147.1134
$C_6H_{17}N_3O$	147.1373
$C_7H_5N_3O$	147.0433
$C_7H_7N_4$	147.0672
$C_7H_{15}O_3$	147.1021
$C_7H_{17}NO_2$	147.1260
$C_8H_5NO_2$	147.0320
$C_8H_7N_2O$	147.0559
$C_8H_9N_3$	147.0798
$C_9H_7O_2$	147.0446
C_9H_9NO	147.0684
$C_9H_{11}N_2$	147.0923
$C_{10}H_{11}O$	147.0810
$C_{10}H_{13}N$	147.1049
$C_{11}H_{15}$	147.1174
148	
$C_4H_8N_2O_4$	148.0484
$C_4H_{10}N_3O_3$	148.0723
$C_4H_{12}N_4O_2$	148.0961
$C_5H_{10}NO_4$	148.0610
$C_5H_{12}N_2O_3$	148.0849
$C_5H_{16}N_4O$	148.1325
$C_6H_4N_4O$	148.0386
$C_6H_{12}O_4$	148.0735
$C_6H_{14}NO_3$	148.0974
$C_6H_{16}N_2O_2$	148.1213
$C_7H_6N_3O$	148.0511
$C_7H_8N_4$	148.0750
$C_7H_{16}O_3$	148.1100
$C_8H_6NO_2$	148.0399
$C_8H_8N_2O$	148.0637
$C_8H_{10}N_3$	148.0876
$C_9H_8O_2$	148.0524
$C_9H_{10}NO$	148.0763
$C_9H_{12}N_2$	148.1001
$C_{10}H_{12}O$	148.0888
$C_{10}H_{14}N$	148.1127
$C_{11}H_{16}$	148.1253
149	
$C_4H_9N_2O_4$	149.0563
$C_4H_{11}N_3O_3$	149.0801
$C_4H_{13}N_4O_2$	149.1040
$C_5H_{11}NO_4$	149.0688
$C_5H_{13}N_2O_3$	149.0927
$C_5H_{15}N_3O_2$	149.1165
$C_6H_5N_4O$	149.0464
$C_6H_{13}O_4$	149.0814
$C_6H_{15}NO_3$	149.1052
$C_7H_5N_2O_2$	149.0351
$C_7H_7N_3O$	149.0590
$C_7H_9N_4$	149.0829
$C_8H_5O_3$	149.0238
$C_8H_7NO_2$	149.0477
$C_8H_9N_2O$	149.0715
$C_8H_{11}N_3$	149.0954
$C_9H_9O_2$	149.0603
$C_9H_{11}NO$	149.0841
$C_9H_{13}N_2$	149.1080
$C_{10}H_{13}O$	149.0967
$C_{10}H_{15}N$	149.1205
$C_{11}H_{17}$	149.1331
150	
$C_4H_{10}N_2O_4$	150.0641
$C_4H_{12}N_3O_3$	150.0879
$C_4H_{14}N_4O_2$	150.1118
$C_5H_{12}NO_4$	150.0766
$C_5H_{14}N_2O_3$	150.1005
$C_6H_4N_3O_2$	150.0304
$C_6H_6N_4O$	150.0542
$C_6H_{14}O_4$	150.0892
$C_7H_6N_2O_2$	150.0429
$C_7H_8N_3O$	150.0668
$C_7H_{10}N_4$	150.0907
$C_8H_6O_3$	150.0317
$C_8H_8NO_2$	150.0555
$C_8H_{10}N_2O$	150.0794
$C_8H_{12}N_3$	150.1032
$C_9H_{10}O_2$	150.0681
$C_9H_{12}NO$	150.0919
$C_9H_{14}N_2$	150.1158
$C_{10}H_{14}O$	150.1045
$C_{10}H_{16}N$	150.1284
$C_{11}H_{18}$	150.1409
151	
$C_4H_{11}N_2O_4$	151.0719
$C_4H_{13}N_3O_3$	151.0958
$C_5H_3N_4O_2$	151.0257
$C_5H_{13}NO_4$	151.0845
$C_6H_3N_2O_3$	151.0144
$C_6H_5N_3O_2$	151.0382
$C_6H_7N_4O$	151.0621
$C_7H_5NO_3$	151.0269
$C_7H_7N_2O_2$	151.0508
$C_7H_9N_3O$	151.0746
$C_7H_{11}N_4$	151.0985
$C_8H_7O_3$	151.0395
$C_8H_9NO_2$	151.0634
$C_8H_{11}N_2O$	151.0872
$C_8H_{13}N_3$	151.1111
$C_9H_{11}O_2$	151.0759
$C_9H_{13}NO$	151.0998
$C_9H_{15}N_2$	151.1236
$C_{10}H_{15}O$	151.1123
$C_{10}H_{17}N$	151.1362
$C_{11}H_{19}$	151.1488
152	
$C_4H_{12}N_2O_4$	152.0797
$C_5H_4N_4O_2$	152.0335
$C_6H_4N_2O_3$	152.0222
$C_6H_6N_3O_2$	152.0460
$C_6H_8N_4O$	152.0699
$C_7H_6NO_3$	152.0348
$C_7H_8N_2O_2$	152.0586
$C_7H_{10}N_3O$	152.0825
$C_7H_{12}N_4$	152.1063
$C_8H_8O_3$	152.0473
$C_8H_{10}NO_2$	152.0712
$C_8H_{12}N_2O$	152.0950
$C_8H_{14}N_3$	152.1189
$C_9H_{12}O_2$	152.0837
$C_9H_{14}NO$	152.1076
$C_9H_{16}N_2$	152.1315
$C_{10}H_{16}O$	152.1202
$C_{10}H_{18}N$	152.1440
$C_{11}H_6N$	152.0501
$C_{11}H_{20}$	152.1566
$C_{12}H_8$	152.0626
153	
$C_5H_3N_3O_3$	153.0175
$C_5H_5N_4O_2$	153.0413
$C_6H_5N_2O_3$	153.0300
$C_6H_7N_3O_2$	153.0539
$C_6H_9N_4O$	153.0777
$C_7H_5O_4$	153.0187
$C_7H_7NO_3$	153.0426
$C_7H_9N_2O_2$	153.0664
$C_7H_{11}N_3O$	153.0903
$C_7H_{13}N_4$	153.1142
$C_8H_9O_3$	153.0552
$C_8H_{11}NO_2$	153.0790
$C_8H_{13}N_2O$	153.1029
$C_8H_{15}N_3$	153.1267
$C_9H_{13}O_2$	153.0916
$C_9H_{15}NO$	153.1154
$C_9H_{17}N_2$	153.1393
$C_{10}H_{17}O$	153.1280
$C_{10}H_{19}N$	153.1519
$C_{11}H_7N$	153.0579
$C_{11}H_{21}$	153.1644
$C_{12}H_9$	153.0705
154	
$C_5H_4N_3O_3$	154.0253
$C_5H_6N_4O_2$	154.0491
$C_6H_4NO_4$	154.0140
$C_6H_6N_2O_3$	154.0379
$C_6H_8N_3O_2$	154.0617
$C_6H_{10}N_4O$	154.0856
$C_7H_6O_4$	154.0266
$C_7H_8NO_3$	154.0504
$C_7H_{10}N_2O_2$	154.0743
$C_7H_{12}N_3O$	154.0981
$C_7H_{14}N_4$	154.1220
$C_8H_{10}O_3$	154.0630
$C_8H_{12}NO_2$	154.0868
$C_8H_{14}N_2O$	154.1107
$C_8H_{16}N_3$	154.1346
$C_9H_{14}O_2$	154.0994
$C_9H_{16}NO$	154.1233
$C_9H_{18}N_2$	154.1471
$C_{10}H_{18}O$	154.1358
$C_{10}H_{20}N$	154.1597
$C_{11}H_8N$	154.0657
$C_{11}H_{22}$	154.1722
$C_{12}H_{10}$	154.0783
155	
$C_5H_3N_2O_4$	155.0093
$C_5H_5N_3O_3$	155.0331
$C_5H_7N_4O_2$	155.0570
$C_6H_5NO_4$	155.0218
$C_6H_7N_2O_3$	155.0457
$C_6H_9N_3O_2$	155.0695
$C_6H_{11}N_4O$	155.0934
$C_7H_7O_4$	155.0344
$C_7H_9NO_3$	155.0583
$C_7H_{11}N_2O_2$	155.0821
$C_7H_{13}N_3O$	155.1060
$C_8H_{11}O_3$	155.0708
$C_8H_{13}NO_2$	155.0947
$C_8H_{15}N_2O$	155.1185
$C_8H_{17}N_3$	155.1424
$C_9H_{15}O_2$	155.1072
$C_9H_{17}NO$	155.1311
$C_9H_{19}N_2$	155.1549
$C_{10}H_7N_2$	155.0610
$C_{10}H_{19}O$	155.1436
$C_{10}H_{21}N$	155.1675
$C_{11}H_7O$	155.0497
$C_{11}H_9N$	155.0736
$C_{11}H_{23}$	155.1801
$C_{12}H_{11}$	155.0861
156	
$C_5H_4N_2O_4$	156.0171
$C_5H_6N_3O_3$	156.0410
$C_5H_8N_4O_2$	156.0648
$C_6H_6NO_4$	156.0297
$C_6H_8N_2O_3$	156.0535
$C_6H_{10}N_3O_2$	156.0774
$C_6H_{12}N_4O$	156.1012
$C_7H_8O_4$	156.0422
$C_7H_{10}NO_3$	156.0661
$C_7H_{12}N_2O_2$	156.0899
$C_7H_{14}N_3O$	156.1138

APPENDIX A *(Continued)*

	FM
$C_7H_{16}N_4$	156.1377
$C_8H_{12}O_3$	156.0786
$C_8H_{14}NO_2$	156.1025
$C_8H_{16}N_2O$	156.1264
$C_8H_{18}N_3$	156.1502
$C_9H_6N_3$	156.0563
$C_9H_{16}O_2$	156.1151
$C_9H_{18}NO$	156.1389
$C_9H_{20}N_2$	156.1628
$C_{10}H_6NO$	156.0449
$C_{10}H_8N_2$	156.0688
$C_{10}H_{20}O$	156.1515
$C_{10}H_{22}N$	156.1753
$C_{11}H_8O$	156.0575
$C_{11}H_{10}N$	156.0814
$C_{11}H_{24}$	156.1879
$C_{12}H_{12}$	156.0939
157	
$C_5H_5N_2O_4$	157.0249
$C_5H_7N_3O_3$	157.0488
$C_5H_9N_4O_2$	157.0726
$C_6H_7NO_4$	157.0375
$C_6H_9N_2O_3$	157.0614
$C_6H_{11}N_3O_2$	157.0852
$C_6H_{13}N_4O$	157.1091
$C_7H_9O_4$	157.0501
$C_7H_{11}NO_3$	157.0739
$C_7H_{13}N_2O_2$	157.0978
$C_7H_{15}N_3O$	157.1216
$C_7H_{17}N_4$	157.1455
$C_8H_5N_4$	157.0515
$C_8H_{13}O_3$	157.0865
$C_8H_{15}NO_2$	157.1103
$C_8H_{17}N_2O$	157.1342
$C_8H_{19}N_3$	157.1580
$C_9H_5N_2O$	157.0402
$C_9H_7N_3$	157.0641
$C_9H_{17}O_2$	157.1229
$C_9H_{19}NO$	157.1467
$C_9H_{21}N_2$	157.1706
$C_{10}H_7NO$	157.0528
$C_{10}H_9N_2$	157.0767
$C_{10}H_{21}O$	157.1593
$C_{10}H_{23}N$	157.1832
$C_{11}H_9O$	157.0653
$C_{11}H_{11}N$	157.0892
$C_{12}H_{13}$	157.1018
158	
$C_5H_6N_2O_4$	158.0328
$C_5H_8N_3O_3$	158.0566
$C_5H_{10}N_4O_2$	158.0805
$C_6H_8NO_4$	158.0453
$C_6H_{10}N_2O_3$	158.0692
$C_6H_{12}N_3O_2$	158.0930
$C_6H_{14}N_4O$	158.1169
$C_7H_{10}O_4$	158.0579
$C_7H_{12}NO_3$	158.0817
$C_7H_{14}N_2O_2$	158.1056
$C_7H_{16}N_3O$	158.1295
$C_7H_{18}N_4$	158.1533
$C_8H_6N_4$	158.0594
$C_8H_{14}O_3$	158.0943
$C_8H_{16}NO_2$	158.1182
$C_8H_{18}N_2O$	158.1420
$C_8H_{20}N_3$	158.1659
$C_9H_6N_2O$	158.0480
$C_9H_8N_3$	158.0719
$C_9H_{18}O_2$	158.1307
$C_9H_{20}NO$	158.1546
$C_{10}H_6O_2$	158.0368
$C_{10}H_8NO$	158.0606
$C_{10}H_{10}N_2$	158.0845
$C_{10}H_{22}O$	158.1672
$C_{11}H_{10}O$	158.0732
$C_{11}H_{12}N$	158.0970
$C_{12}H_{14}$	158.1096
159	
$C_5H_7N_2O_4$	159.0406
$C_5H_9N_3O_3$	159.0644
$C_5H_{11}N_4O_2$	159.0883
$C_6H_9NO_4$	159.0532
$C_6H_{11}N_2O_3$	159.0770
$C_6H_{13}N_3O_2$	159.1009
$C_6H_{15}N_4O$	159.1247
$C_7H_{11}O_4$	159.0657
$C_7H_{13}NO_3$	159.0896
$C_7H_{15}N_2O_2$	159.1134
$C_7H_{17}N_3O$	159.1373
$C_8H_5N_3O$	159.0433
$C_8H_7N_4$	159.0672
$C_8H_{15}O_3$	159.1021
$C_8H_{17}NO_2$	159.1260
$C_8H_{19}N_2O$	159.1498
$C_8H_{21}N_3$	159.1737
$C_9H_5NO_2$	159.0320
$C_9H_7N_2O$	159.0559
$C_9H_9N_3$	159.0798
$C_9H_{19}O_2$	159.1385
$C_9H_{21}NO$	159.1624
$C_{10}H_7O_2$	159.0446
$C_{10}H_9NO$	159.0684
$C_{10}H_{11}N_2$	159.0923
$C_{11}H_{11}O$	159.0810
$C_{11}H_{13}N$	159.1049
$C_{12}H_{15}$	159.1174
160	
$C_5H_8N_2O_4$	160.0484
$C_5H_{10}N_3O_3$	160.0723
$C_5H_{12}N_4O_2$	160.0961
$C_6H_{10}NO_4$	160.0610
$C_6H_{12}N_2O_3$	160.0848
$C_6H_{14}N_3O_2$	160.1087
$C_6H_{16}N_4O$	160.1325
$C_7H_{12}O_4$	160.0735
$C_7H_{14}NO_3$	160.0974
$C_7H_{16}N_2O_2$	160.1213
$C_7H_{18}N_3O$	160.1451
$C_7H_{20}N_4$	160.1690
$C_8H_6N_3O$	160.0511
$C_8H_8N_4$	160.0750
$C_8H_{16}O_3$	160.1100
$C_8H_{18}NO_2$	160.1338
$C_8H_{20}N_2O$	160.1577
$C_9H_6NO_2$	160.0399
$C_9H_8N_2O$	160.0637
$C_9H_{10}N_3$	160.0876
$C_9H_{20}O_2$	160.1464
$C_{10}H_8O_2$	160.0524
$C_{10}H_{10}NO$	160.0763
$C_{10}H_{12}N_2$	160.1001
$C_{11}H_{12}O$	160.0888
$C_{11}H_{14}N$	160.1127
$C_{12}H_{16}$	160.1253
161	
$C_5H_9N_2O_4$	161.0563
$C_5H_{11}N_3O_3$	161.0801
$C_5H_{13}N_4O_2$	161.1040
$C_6H_{11}NO_4$	161.0688
$C_6H_{13}N_2O_3$	161.0927
$C_6H_{15}N_3O_2$	161.1165
$C_6H_{17}N_4O$	161.1404
$C_7H_5N_4O$	161.0464
$C_8H_5N_2O_2$	161.0351
$C_8H_7N_3O$	161.0590
$C_8H_9N_4$	161.0829
$C_8H_{17}O_3$	161.1178
$C_8H_{19}NO_2$	161.1416
$C_9H_5O_3$	161.0238
$C_9H_7NO_2$	161.0477
$C_9H_9N_2O$	161.0715
$C_9H_{11}N_3$	161.0954
$C_{10}H_9O_2$	161.0603
$C_{10}H_{11}NO$	161.0841
$C_{10}H_{13}N_2$	161.1080
$C_{11}H_{13}O$	161.0967
$C_{11}H_{15}N$	161.1205
$C_{12}H_{17}$	161.1331
162	
$C_5H_{10}N_2O_4$	162.0641
$C_5H_{12}N_3O_3$	162.0879
$C_5H_{14}N_4O_2$	162.1118
$C_6H_{12}NO_4$	162.0766
$C_6H_{14}N_2O_3$	162.1005
$C_6H_{16}N_3O_2$	162.1244
$C_6H_{18}N_4O$	162.1482
$C_7H_6N_4O$	162.0542
$C_7H_{14}O_4$	162.0892
$C_7H_{16}NO_3$	162.1131
$C_7H_{18}N_2O_2$	162.1369
$C_8H_6N_2O_2$	162.0429
$C_8H_8N_3O$	162.0668
$C_8H_{10}N_4$	162.0907
$C_8H_{18}O_3$	162.1256
$C_9H_6O_3$	162.0317
$C_9H_8NO_2$	162.0555
$C_9H_{10}N_2O$	162.0794
$C_9H_{12}N_3$	162.1032
$C_{10}H_{10}O_2$	162.0681
$C_{10}H_{12}NO$	162.0919
$C_{10}H_{14}N_2$	162.1158
$C_{11}H_{14}O$	162.1045
$C_{11}H_{16}N$	162.1284
$C_{12}H_{18}$	162.1409
163	
$C_5H_{11}N_2O_4$	163.0719
$C_5H_{13}N_3O_3$	163.0958
$C_5H_{15}N_4O_2$	163.1196
$C_6H_{13}NO_4$	163.0845
$C_6H_{15}N_2O_3$	163.1083
$C_6H_{17}N_3O_2$	163.1322
$C_7H_5N_3O_2$	163.0382
$C_7H_7N_4O$	163.0621
$C_7H_{15}O_4$	163.0970
$C_7H_{17}NO_3$	163.1209
$C_8H_5NO_3$	163.0269
$C_8H_7N_2O_2$	163.0508
$C_8H_9N_3O$	163.0746
$C_8H_{11}N_4$	163.0985
$C_9H_7O_3$	163.0395
$C_9H_9NO_2$	163.0634
$C_9H_{11}N_2O$	163.0872
$C_9H_{13}N_3$	163.1111
$C_{10}H_{11}O_2$	163.0759
$C_{10}H_{13}NO$	163.0998
$C_{10}H_{15}N_2$	163.1236
$C_{11}H_{15}O$	163.1123
$C_{11}H_{17}N$	163.1362
$C_{12}H_{19}$	163.1488
164	
$C_5H_{12}N_2O_4$	164.0797
$C_5H_{14}N_3O_3$	164.1036
$C_5H_{16}N_4O_2$	164.1275
$C_6H_4N_4O_2$	164.0335
$C_6H_{14}NO_4$	164.0923
$C_6H_{16}N_2O_3$	164.1162
$C_7H_6N_3O_2$	164.0460
$C_7H_8N_4O$	164.0699
$C_7H_{16}O_4$	164.1049
$C_8H_6NO_3$	164.0348
$C_8H_8N_2O_2$	164.0586
$C_8H_{10}N_3O$	164.0825
$C_8H_{12}N_4$	164.1063
$C_9H_8O_3$	164.0473
$C_9H_{10}NO_2$	164.0712
$C_9H_{12}N_2O$	164.0950
$C_9H_{14}N_3$	164.1189
$C_{10}H_{12}O_2$	164.0837
$C_{10}H_{14}NO$	164.1076

APPENDIX A *(Continued)*

	FM
$C_{10}H_{16}N_2$	164.1315
$C_{11}H_{16}O$	164.1202
$C_{11}H_{18}N$	164.1440
$C_{12}H_{20}$	164.1566
165	
$C_5H_{13}N_2O_4$	165.0876
$C_5H_{15}N_3O_3$	165.1114
$C_6H_5N_4O_2$	165.0413
$C_6H_{15}NO_4$	165.1001
$C_7H_5N_2O_3$	165.0300
$C_7H_7N_3O_2$	165.0539
$C_7H_9N_4O$	165.0777
$C_8H_5O_4$	165.0187
$C_8H_7NO_3$	165.0426
$C_8H_9N_2O_2$	165.0664
$C_8H_{11}N_3O$	165.0903
$C_8H_{13}N_4$	165.1142
$C_9H_9O_3$	165.0552
$C_9H_{11}NO_2$	165.0790
$C_9H_{13}N_2O$	165.1029
$C_9H_{15}N_3$	165.1267
$C_{10}H_{13}O_2$	165.0916
$C_{10}H_{15}NO$	165.1154
$C_{10}H_{17}N_2$	165.1393
$C_{11}H_{17}O$	165.1280
$C_{11}H_{19}N$	165.1519
$C_{12}H_7N$	165.0579
$C_{12}H_{21}$	165.1644
$C_{13}H_9$	165.0705
166	
$C_5H_{14}N_2O_4$	166.0954
$C_6H_4N_3O_3$	166.0253
$C_6H_6N_4O_2$	166.0491
$C_7H_6N_2O_3$	166.0379
$C_7H_8N_3O_2$	166.0617
$C_7H_{10}N_4O$	166.0856
$C_8H_6O_4$	166.0266
$C_8H_8NO_3$	166.0504
$C_8H_{10}N_2O_2$	166.0743
$C_8H_{12}N_3O$	166.0981
$C_8H_{14}N_4$	166.1220
$C_9H_{10}O_3$	166.0630
$C_9H_{12}NO_2$	166.0868
$C_9H_{14}N_2O$	166.1107
$C_9H_{16}N_3$	166.1346
$C_{10}H_{14}O_2$	166.0994
$C_{10}H_{16}NO$	166.1233
$C_{10}H_{18}N_2$	166.1471
$C_{11}H_{18}O$	166.1358
$C_{11}H_{20}N$	166.1597
$C_{12}H_8N$	166.0657
$C_{12}H_{22}$	166.1722
$C_{13}H_{10}$	166.0783
167	
$C_6H_5N_3O_3$	167.0331
$C_6H_7N_4O_2$	167.0570
$C_7H_5NO_4$	167.0218
$C_7H_7N_2O_3$	167.0457
$C_7H_9N_3O_2$	167.0695
$C_7H_{11}N_4O$	167.0934
$C_8H_7O_4$	167.0344
$C_8H_9NO_3$	167.0583
$C_8H_{11}N_2O_2$	167.0821
$C_8H_{13}N_3O$	167.1060
$C_8H_{15}N_4$	167.1298
$C_9H_{11}O_3$	167.0708
$C_9H_{13}NO_2$	167.0947
$C_9H_{15}N_2O$	167.1185
$C_9H_{17}N_3$	167.1424
$C_{10}H_{15}O_2$	167.1072
$C_{10}H_{17}NO$	167.1311
$C_{10}H_{19}N_2$	167.1549
$C_{11}H_7N_2$	167.0610
$C_{11}H_{19}O$	167.1436
$C_{11}H_{21}N$	167.1675
$C_{12}H_9N$	167.0736
$C_{12}H_{23}$	167.1801
$C_{13}H_{11}$	167.0861
168	
$C_6H_4N_2O_4$	168.0171
$C_6H_6N_3O_3$	168.0410
$C_6H_8N_4O_2$	168.0648
$C_7H_6NO_4$	168.0297
$C_7H_8N_2O_3$	168.0535
$C_7H_{10}N_3O_2$	168.0774
$C_7H_{12}N_4O$	168.1012
$C_8H_8O_4$	168.0422
$C_8H_{10}NO_3$	168.0661
$C_8H_{12}N_2O_2$	168.0899
$C_8H_{14}N_3O$	168.1138
$C_8H_{16}N_4$	168.1377
$C_9H_{12}O_3$	168.0786
$C_9H_{14}NO_2$	168.1025
$C_9H_{16}N_2O$	168.1264
$C_9H_{18}N_3$	168.1502
$C_{10}H_{16}O_2$	168.1151
$C_{10}H_{18}NO$	168.1389
$C_{10}H_{20}N_2$	168.1628
$C_{11}H_8N_2$	168.0688
$C_{11}H_{20}O$	168.1515
$C_{11}H_{22}N$	168.1753
$C_{12}H_8O$	168.0575
$C_{12}H_{10}N$	168.0814
$C_{12}H_{24}$	168.1879
$C_{13}H_{12}$	168.0939
169	
$C_6H_5N_2O_4$	169.0249
$C_6H_7N_3O_3$	169.0488
$C_6H_9N_4O_2$	169.0726
$C_7H_7NO_4$	169.0375
$C_7H_9N_2O_3$	169.0614
$C_7H_{11}N_3O_2$	169.0852
$C_7H_{13}N_4O$	169.1091
$C_8H_9O_4$	169.0501
$C_8H_{11}NO_3$	169.0739
$C_8H_{13}N_2O_2$	169.0978
$C_8H_{15}N_3O$	169.1216
$C_8H_{17}N_4$	169.1455
$C_9H_{13}O_3$	169.0865
$C_9H_{15}NO_2$	169.1103
$C_9H_{17}N_2O$	169.1342
$C_9H_{19}N_3$	169.1580
$C_{10}H_7N_3$	169.0641
$C_{10}H_{17}O_2$	169.1229
$C_{10}H_{19}NO$	169.1467
$C_{10}H_{21}N_2$	169.1706
$C_{11}H_7NO$	169.0528
$C_{11}H_9N_2$	169.0767
$C_{11}H_{21}O$	169.1593
$C_{11}H_{23}N$	169.1832
$C_{12}H_9O$	169.0653
$C_{12}H_{11}N$	169.0892
$C_{12}H_{25}$	169.1957
$C_{13}H_{13}$	169.1018
170	
$C_6H_6N_2O_4$	170.0328
$C_6H_8N_3O_3$	170.0566
$C_6H_{10}N_4O_2$	170.0805
$C_7H_8NO_4$	170.0453
$C_7H_{10}N_2O_3$	170.0692
$C_7H_{12}N_3O_2$	170.0930
$C_7H_{14}N_4O$	170.1169
$C_8H_{10}O_4$	170.0579
$C_8H_{12}NO_3$	170.0817
$C_8H_{14}N_2O_2$	170.1056
$C_8H_{16}N_3O$	170.1295
$C_8H_{18}N_4$	170.1533
$C_9H_6N_4$	170.0594
$C_9H_{14}O_3$	170.0943
$C_9H_{16}NO_2$	170.1182
$C_9H_{18}N_2O$	170.1420
$C_9H_{20}N_3$	170.1659
$C_{10}H_6N_2O$	170.0480
$C_{10}H_8N_3$	170.0719
$C_{10}H_{18}O_2$	170.1307
$C_{10}H_{20}NO$	170.1546
$C_{10}H_{22}N_2$	170.1784
$C_{11}H_8NO$	170.0606
$C_{11}H_{10}N_2$	170.0845
$C_{11}H_{22}O$	170.1671
$C_{11}H_{24}N$	170.1910
$C_{12}H_{10}O$	170.0732
$C_{12}H_{12}N$	170.0970
$C_{12}H_{26}$	170.2036
$C_{13}H_{14}$	170.1096
171	
$C_6H_7N_2O_4$	171.0406
$C_6H_9N_3O_3$	171.0644
$C_6H_{11}N_4O_2$	171.0883
$C_7H_9NO_4$	171.0532
$C_7H_{11}N_2O_3$	171.0770
$C_7H_{13}N_3O_2$	171.1009
$C_7H_{15}N_4O$	171.1247
$C_8H_{11}O_4$	171.0657
$C_8H_{13}NO_3$	171.0896
$C_8H_{15}N_2O_2$	171.1134
$C_8H_{17}N_3O$	171.1373
$C_8H_{19}N_4$	171.1611
$C_9H_5N_3O$	171.0433
$C_9H_7N_4$	171.0672
$C_9H_{15}O_3$	171.1021
$C_9H_{17}NO_2$	171.1260
$C_9H_{19}N_2O$	171.1498
$C_9H_{21}N_3$	171.1737
$C_{10}H_7N_2O$	171.0559
$C_{10}H_9N_3$	171.0798
$C_{10}H_{19}O_2$	171.1385
$C_{10}H_{21}NO$	171.1624
$C_{10}H_{23}N_2$	171.1863
$C_{11}H_7O_2$	171.0446
$C_{11}H_9NO$	171.0684
$C_{11}H_{11}N_2$	171.0923
$C_{11}H_{23}O$	171.1750
$C_{11}H_{25}N$	171.1988
$C_{12}H_{11}O$	171.0810
$C_{12}H_{13}N$	171.1049
$C_{13}H_{15}$	171.1174
172	
$C_6H_8N_2O_4$	172.0484
$C_6H_{10}N_3O_3$	172.0723
$C_6H_{12}N_4O_2$	172.0961
$C_7H_{10}NO_4$	172.0610
$C_7H_{12}N_2O_3$	172.0848
$C_7H_{14}N_3O_2$	172.1087
$C_7H_{16}N_4O$	172.1325
$C_8H_{12}O_4$	172.0735
$C_8H_{14}NO_3$	172.0974
$C_8H_{16}N_2O_2$	172.1213
$C_8H_{18}N_3O$	172.1451
$C_8H_{20}N_4$	172.1690
$C_9H_6N_3O$	172.0511
$C_9H_8N_4$	172.0750
$C_9H_{16}O_3$	172.1100
$C_9H_{18}NO_2$	172.1338
$C_9H_{20}N_2O$	172.1577
$C_9H_{22}N_3$	172.1815
$C_{10}H_6NO_2$	172.0399
$C_{10}H_8N_2O$	172.0637
$C_{10}H_{10}N_3$	172.0876
$C_{10}H_{20}O_2$	172.1464
$C_{10}H_{22}NO$	172.1702
$C_{10}H_{24}N_2$	172.1941
$C_{11}H_8O_2$	172.0524
$C_{11}H_{10}NO$	172.0763
$C_{11}H_{12}N_2$	172.1001
$C_{11}H_{24}O$	172.1828
$C_{12}H_{12}O$	172.0888
$C_{12}H_{14}N$	172.1127

APPENDIX A *(Continued)*

	FM
$C_{13}H_{16}$	172.1253
173	
$C_6H_9N_2O_4$	173.0563
$C_6H_{11}N_3O_3$	173.0801
$C_6H_{13}N_4O_2$	173.1040
$C_7H_{11}NO_4$	173.0688
$C_7H_{13}N_2O_3$	173.0927
$C_7H_{15}N_3O_2$	173.1165
$C_7H_{17}N_4O$	173.1404
$C_8H_{13}O_4$	173.0814
$C_8H_{15}NO_3$	173.1052
$C_8H_{17}N_2O_2$	173.1291
$C_8H_{19}N_3O$	173.1529
$C_8H_{21}N_4$	173.1768
$C_9H_7N_3O$	173.0590
$C_9H_9N_4$	173.0829
$C_9H_{17}O_3$	173.1178
$C_9H_{19}NO_2$	173.1416
$C_9H_{21}N_2O$	173.1655
$C_{10}H_5O_3$	173.0238
$C_{10}H_7NO_2$	173.0477
$C_{10}H_9N_2O$	173.0715
$C_{10}H_{11}N_3$	173.0954
$C_{10}H_{21}O_2$	173.1542
$C_{10}H_{23}NO$	173.1781
$C_{11}H_9O_2$	173.0603
$C_{11}H_{11}NO$	173.0841
$C_{11}H_{13}N_2$	173.1080
$C_{12}H_{13}O$	173.0967
$C_{12}H_{15}NO_2$	173.1205
$C_{13}H_{17}$	173.1331
174	
$C_6H_{10}N_2O_4$	174.0641
$C_6H_{12}N_3O_3$	174.0879
$C_6H_{14}N_4O_2$	174.1118
$C_7H_{12}NO_4$	174.0766
$C_7H_{14}N_2O_3$	174.1005
$C_7H_{16}N_3O_2$	174.1244
$C_7H_{18}N_4O$	174.1482
$C_7H_{16}N_4O$	174.1244
$C_8H_6N_4O$	174.0542
$C_8H_{14}O_4$	174.0892
$C_8H_{16}NO_3$	174.1131
$C_8H_{18}N_2O_2$	174.1369
$C_8H_{20}N_3O$	174.1608
$C_8H_{22}N_4$	174.1846
$C_9H_6N_2O_2$	174.0429
$C_9H_{10}N_4$	174.0907
$C_9H_{18}O_3$	174.1256
$C_9H_{20}NO_2$	174.1495
$C_9H_{22}N_2O$	174.1733
$C_{10}H_6O_3$	174.0317
$C_{10}H_8NO_2$	174.0555
$C_{10}H_{10}N_2O$	174.0794
$C_{10}H_{12}N_3$	174.1032
$C_{10}H_{22}O_2$	174.1620
$C_{11}H_{10}O_2$	174.0681
$C_{11}H_{12}NO$	174.0919
$C_{11}H_{14}N_2$	174.1158
$C_{12}H_{14}O$	174.1045
$C_{12}H_{16}N$	174.1284
$C_{13}H_{18}$	174.1409
175	
$C_6H_{11}N_2O_4$	175.0719
$C_6H_{13}N_3O_3$	175.0958
$C_6H_{15}N_4O_2$	175.1196
$C_7H_{13}NO_4$	175.0845
$C_7H_{15}N_2O_3$	175.1083
$C_7H_{17}N_3O_2$	175.1322
$C_7H_{19}N_4O$	175.1560
$C_8H_7N_4O$	175.0621
$C_8H_{15}O_4$	175.0970
$C_8H_{17}NO_3$	175.1209
$C_8H_{19}N_2O_2$	175.1447
$C_8H_{21}N_3O$	175.1686
$C_9H_5NO_3$	175.0269
$C_9H_7N_2O_2$	175.0508
$C_9H_9N_3O$	175.0746
$C_9H_{11}N_4$	175.0985
$C_9H_{19}O_3$	175.1334
$C_9H_{21}NO_2$	175.1573
$C_{10}H_7O_3$	175.0395
$C_{10}H_9NO_2$	175.0634
$C_{10}H_{11}N_2O$	175.0872
$C_{10}H_{13}N_3$	175.1111
$C_{11}H_{11}O_2$	175.0759
$C_{11}H_{13}NO$	175.0998
$C_{11}H_{15}N_2$	175.1236
$C_{12}H_{15}O$	175.1123
$C_{12}H_{17}N$	175.1362
$C_{13}H_3O$	175.0184
$C_{13}H_{19}$	175.1488
176	
$C_6H_{12}N_2O_4$	176.0797
$C_6H_{14}N_3O_3$	176.1036
$C_6H_{16}N_4O_2$	176.1275
$C_7H_{14}NO_4$	176.0923
$C_7H_{16}N_2O_3$	176.1162
$C_7H_{18}N_3O_2$	176.1400
$C_7H_{20}N_4O$	176.1639
$C_8H_6N_3O_2$	176.0460
$C_8H_8N_4O$	176.0699
$C_8H_{16}O_4$	176.1049
$C_8H_{18}NO_3$	176.1287
$C_8H_{20}N_2O_2$	176.1526
$C_9H_6NO_3$	176.0348
$C_9H_8N_2O_2$	176.0586
$C_9H_{10}N_3O$	176.0825
$C_9H_{12}N_4$	176.1063
$C_9H_{20}O_3$	176.1413
$C_{10}H_8O_3$	176.0473
$C_{10}H_{10}NO_2$	176.0712
$C_{10}H_{12}N_2O$	176.0950
$C_{10}H_{14}N_3$	176.1189
$C_{11}H_{12}O_2$	176.0837
$C_{11}H_{14}NO$	176.1076
$C_{11}H_{16}N_2$	176.1315
$C_{12}H_{16}O$	176.1202
$C_{12}H_{18}N$	176.1440
$C_{13}H_{20}$	176.1566
177	
$C_6H_{13}N_2O_4$	177.0876
$C_6H_{15}N_3O_3$	177.1114
$C_6H_{17}N_4O_2$	177.1353
$C_7H_5N_4O_2$	177.0413
$C_7H_{15}NO_4$	177.1001
$C_7H_{17}N_2O_3$	177.1240
$C_7H_{19}N_3O_2$	177.1478
$C_8H_5N_2O_3$	177.0300
$C_8H_7N_3O_2$	177.0539
$C_8H_9N_4O$	177.0777
$C_8H_{17}O_4$	177.1127
$C_8H_{19}NO_3$	177.1365
$C_9H_7NO_3$	177.0426
$C_9H_9N_2O_2$	177.0664
$C_9H_{11}N_3O$	177.0903
$C_9H_{13}N_4$	177.1142
$C_{10}H_9O_3$	177.0552
$C_{10}H_{11}NO_2$	177.0790
$C_{10}H_{13}N_2O$	177.1029
$C_{10}H_{15}N_3$	177.1267
$C_{11}H_{13}O_2$	177.0916
$C_{11}H_{15}NO$	177.1154
$C_{11}H_{17}N_2$	177.1393
$C_{12}H_{17}O$	177.1280
$C_{12}H_{19}N$	177.1519
$C_{13}H_{21}$	177.1644
178	
$C_6H_{14}N_2O_4$	178.0954
$C_6H_{16}N_3O_3$	178.1193
$C_6H_{18}N_4O_2$	178.1431
$C_7H_6N_4O_2$	178.0491
$C_7H_{16}NO_4$	178.1080
$C_7H_{18}N_2O_3$	178.1318
$C_8H_6N_2O_3$	178.0379
$C_8H_8N_3O_2$	178.0617
$C_8H_{10}N_4O$	178.0856
$C_8H_{18}O_4$	178.1205
$C_9H_6O_4$	178.0266
$C_9H_8NO_3$	178.0504
$C_9H_{10}N_2O_2$	178.0743
$C_9H_{12}N_3O$	178.0981
$C_9H_{14}N_4$	178.1220
$C_{10}H_{10}O_3$	178.0630
$C_{10}H_{12}NO_2$	178.0868
$C_{10}H_{14}N_2O$	178.1107
$C_{10}H_{16}N_3$	178.1346
$C_{11}H_{14}O_2$	178.0994
$C_{11}H_{16}NO$	178.1233
$C_{11}H_{18}N_2$	178.1471
$C_{12}H_{18}O$	178.1358
$C_{12}H_{20}N$	178.1597
$C_{13}H_8N$	178.0657
$C_{13}H_{22}$	178.1722
$C_{14}H_{10}$	178.0783
179	
$C_6H_{15}N_2O_4$	179.1032
$C_6H_{17}N_3O_3$	179.1271
$C_7H_5N_3O_3$	179.0331
$C_7H_7N_4O_2$	179.0570
$C_7H_{17}NO_4$	179.1158
$C_8H_5NO_4$	179.0218
$C_8H_7N_2O_3$	179.0457
$C_8H_9N_3O_2$	179.0695
$C_8H_{11}N_4O$	179.0934
$C_9H_7O_4$	179.0344
$C_9H_9NO_3$	179.0583
$C_9H_{11}N_2O_2$	179.0821
$C_9H_{13}N_3O$	179.1060
$C_9H_{15}N_4$	179.1298
$C_{10}H_{11}O_3$	179.0708
$C_{10}H_{13}NO_2$	179.0947
$C_{10}H_{15}N_2O$	179.1185
$C_{10}H_{17}N_3$	179.1424
$C_{11}H_{15}O_2$	179.1072
$C_{11}H_{17}NO$	179.1311
$C_{11}H_{19}N_2$	179.1549
$C_{12}H_{19}O$	179.1436
$C_{12}H_{21}N$	179.1675
$C_{13}H_9N$	179.0736
$C_{13}H_{23}$	179.1801
$C_{14}H_{11}$	179.0861
180	
$C_6H_{16}N_2O_4$	180.1111
$C_7H_6N_3O_3$	180.0410
$C_7H_8N_4O_2$	180.0648
$C_8H_6NO_4$	180.0297
$C_8H_8N_2O_3$	180.0535
$C_8H_{10}N_3O_2$	180.0774
$C_8H_{12}N_4O$	180.1012
$C_9H_8O_4$	180.0422
$C_9H_{10}NO_3$	180.0661
$C_9H_{12}N_2O_2$	180.0899
$C_9H_{14}N_3O$	180.1138
$C_9H_{16}N_4$	180.1377
$C_{10}H_{12}O_3$	180.0786
$C_{10}H_{14}NO_2$	180.1025
$C_{10}H_{16}N_2O$	180.1264
$C_{10}H_{18}N_3$	180.1502
$C_{11}H_{16}O_2$	180.1151
$C_{11}H_{18}NO$	180.1389
$C_{11}H_{20}N_2$	180.1628
$C_{12}H_8N_2$	180.0688
$C_{12}H_{20}O$	180.1515
$C_{12}H_{22}N$	180.1753
$C_{13}H_8O$	180.0575
$C_{13}H_{10}N$	180.0814
$C_{13}H_{24}$	180.1879

APPENDIX A *(Continued)*

	FM
$C_{14}H_{12}$	180.0939
181	
$C_7H_5N_2O_4$	181.0249
$C_7H_7N_3O_3$	181.0488
$C_7H_9N_4O_2$	181.0726
$C_8H_7NO_4$	181.0375
$C_8H_9N_2O_3$	181.0614
$C_8H_{11}N_3O_2$	181.0852
$C_8H_{13}N_4O$	181.1091
$C_9H_9O_4$	181.0501
$C_9H_{11}NO_3$	181.0739
$C_9H_{13}N_2O_2$	181.0978
$C_9H_{15}N_3O$	181.1216
$C_9H_{17}N_4$	181.1455
$C_{10}H_{13}O_3$	181.0865
$C_{10}H_{15}NO_2$	181.1103
$C_{10}H_{17}N_2O$	181.1342
$C_{10}H_{19}N_3$	181.1580
$C_{11}H_7N_3$	181.0641
$C_{11}H_{17}O_2$	181.1229
$C_{11}H_{19}NO$	181.1467
$C_{11}H_{21}N_2$	181.1706
$C_{12}H_7NO$	181.0528
$C_{12}H_9N_2$	181.0767
$C_{12}H_{21}O$	181.1593
$C_{12}H_{23}N$	181.1832
$C_{13}H_9O$	181.0653
$C_{13}H_{11}N$	181.0892
$C_{13}H_{25}$	181.1957
$C_{14}H_{13}$	181.1018
182	
$C_7H_6N_2O_4$	182.0328
$C_7H_8N_3O_3$	182.0566
$C_7H_{10}N_4O_2$	182.0805
$C_8H_8NO_4$	182.0453
$C_8H_{10}N_2O_3$	182.0692
$C_8H_{12}N_3O_2$	182.0930
$C_8H_{14}N_4O$	182.1169
$C_9H_{10}O_4$	182.0579
$C_9H_{12}NO_3$	182.0817
$C_9H_{14}N_2O_2$	182.1056
$C_9H_{16}N_3O$	182.1295
$C_9H_{18}N_4$	182.1533
$C_{10}H_6N_4$	182.0594
$C_{10}H_{14}O_3$	182.0943
$C_{10}H_{16}NO_2$	182.1182
$C_{10}H_{18}N_2O$	182.1420
$C_{10}H_{20}N_3$	182.1659
$C_{11}H_8N_3$	182.0719
$C_{11}H_{18}O_2$	182.1307
$C_{11}H_{20}NO$	182.1546
$C_{11}H_{22}N_2$	182.1784
$C_{12}H_8NO$	182.0606
$C_{12}H_{10}N_2$	182.0845
$C_{12}H_{22}O$	182.1671
$C_{12}H_{24}N$	182.1910
$C_{13}H_{10}O$	182.0732
$C_{13}H_{12}N$	182.0970
$C_{13}H_{26}$	182.2036
$C_{14}H_{14}$	182.1096
183	
$C_7H_7N_2O_4$	183.0406
$C_7H_9N_3O_3$	183.0644
$C_7H_{11}N_4O_2$	183.0883
$C_8H_9NO_4$	183.0532
$C_8H_{11}N_2O_3$	183.0770
$C_8H_{13}N_3O_2$	183.1009
$C_8H_{15}N_4O$	183.1247
$C_9H_{11}O_4$	183.0657
$C_9H_{13}NO_3$	183.0896
$C_9H_{15}N_2O_2$	183.1134
$C_9H_{17}N_3O$	183.1373
$C_9H_{19}N_4$	183.1611
$C_{10}H_7N_4$	183.0672
$C_{10}H_{15}O_3$	183.1021
$C_{10}H_{17}NO_2$	183.1260
$C_{10}H_{19}N_2O$	183.1498
$C_{10}H_{21}N_3$	183.1737
$C_{11}H_7N_2O$	183.0559
$C_{11}H_9N_3$	183.0798
$C_{11}H_{19}O_2$	183.1385
$C_{11}H_{21}NO$	183.1624
$C_{11}H_{23}N_2$	183.1863
$C_{12}H_7O_2$	183.0446
$C_{12}H_9NO$	183.0684
$C_{12}H_{11}N_2$	183.0923
$C_{12}H_{23}O$	183.1750
$C_{12}H_{25}N$	183.1988
$C_{13}H_{11}O$	183.0810
$C_{13}H_{13}N$	183.1049
$C_{13}H_{27}$	183.2114
$C_{14}H_{15}$	183.1174
184	
$C_7H_8N_2O_4$	184.0484
$C_7H_{10}N_3O_3$	184.0723
$C_7H_{12}N_4O_2$	184.0961
$C_8H_{10}NO_4$	184.0610
$C_8H_{12}N_2O_3$	184.0848
$C_8H_{14}N_3O_2$	184.1087
$C_8H_{16}N_4O$	184.1325
$C_9H_{12}O_4$	184.0735
$C_9H_{14}NO_3$	184.0974
$C_9H_{16}N_2O_2$	184.1213
$C_9H_{18}N_3O$	184.1451
$C_9H_{20}N_4$	184.1690
$C_{10}H_6N_3O$	184.0511
$C_{10}H_8N_4$	184.0750
$C_{10}H_{16}O_3$	184.1100
$C_{10}H_{18}NO_2$	184.1338
$C_{10}H_{20}N_2O$	184.1577
$C_{10}H_{22}N_3$	184.1815
$C_{11}H_8N_2O$	184.0637
$C_{11}H_{10}N_3$	184.0876
$C_{11}H_{20}O_2$	184.1464
$C_{11}H_{22}NO$	184.1702
$C_{11}H_{24}N_2$	184.1941
$C_{12}H_8O_2$	184.0524
$C_{12}H_{10}NO$	184.0763
$C_{12}H_{12}N_2$	184.1001
$C_{12}H_{24}O$	184.1828
$C_{12}H_{26}N$	184.2067
$C_{13}H_{12}O$	184.0888
$C_{13}H_{14}N$	184.1127
$C_{13}H_{28}$	184.2192
$C_{14}H_{16}$	184.1253
185	
$C_7H_9N_2O_4$	185.0563
$C_7H_{11}N_3O_3$	185.0801
$C_7H_{13}N_4O_2$	185.1040
$C_8H_{11}NO_4$	185.0688
$C_8H_{13}N_2O_3$	185.0927
$C_8H_{15}N_3O_2$	185.1165
$C_8H_{17}N_4O$	185.1404
$C_9H_{13}O_4$	185.0814
$C_9H_{15}NO_3$	185.1052
$C_9H_{17}N_2O_2$	185.1291
$C_9H_{19}N_3O$	185.1529
$C_9H_{21}N_4$	185.1768
$C_{10}H_7N_3O$	185.0590
$C_{10}H_9N_4$	185.0829
$C_{10}H_{17}O_3$	185.1178
$C_{10}H_{19}NO_2$	185.1416
$C_{10}H_{21}N_2O$	185.1655
$C_{10}H_{23}N_3$	185.1894
$C_{11}H_9N_2O$	185.0715
$C_{11}H_{11}N_3$	185.0954
$C_{11}H_{21}O_2$	185.1542
$C_{11}H_{23}NO$	185.1781
$C_{11}H_{25}N_2$	185.2019
$C_{12}H_9O_2$	185.0603
$C_{12}H_{11}NO$	185.0841
$C_{12}H_{13}N_2$	185.1080
$C_{12}H_{25}O$	185.1906
$C_{12}H_{27}N$	185.2145
$C_{13}H_{13}O$	185.0967
$C_{13}H_{15}N$	185.1205
$C_{14}H_{17}$	185.1331
186	
$C_7H_{10}N_2O_4$	186.0641
$C_7H_{12}N_3O_3$	186.0879
$C_7H_{14}N_4O_2$	186.1118
$C_8H_{12}NO_4$	186.0766
$C_8H_{14}N_2O_3$	186.1005
$C_8H_{16}N_3O_2$	186.1244
$C_8H_{18}N_4O$	186.1482
$C_9H_6N_4O$	186.0542
$C_9H_{14}O_4$	186.0892
$C_9H_{16}NO_3$	186.1131
$C_9H_{18}N_2O_2$	186.1369
$C_9H_{20}N_3O$	186.1608
$C_{10}H_6N_2O_2$	186.0429
$C_{10}H_8N_3O$	186.0668
$C_{10}H_{10}N_4$	186.0907
$C_{10}H_{18}O_3$	186.1256
$C_{10}H_{20}NO_2$	186.1495
$C_{10}H_{22}N_2O$	186.1733
$C_{10}H_{24}N_3$	186.1972
$C_{11}H_8NO_2$	186.0555
$C_{11}H_{10}N_2O$	186.0794
$C_{11}H_{12}N_3$	186.1032
$C_{11}H_{22}O_2$	186.1620
$C_{11}H_{24}NO$	186.1859
$C_{11}H_{26}N_2$	186.2098
$C_{12}H_{10}O_2$	186.0681
$C_{12}H_{12}NO$	186.0919
$C_{12}H_{14}N_2$	186.1158
$C_{12}H_{26}O$	186.1985
$C_{13}H_{14}O$	186.1045
$C_{13}H_{16}N$	186.1284
$C_{14}H_{18}$	186.1409
187	
$C_7H_{11}N_2O_4$	187.0719
$C_7H_{13}N_3O_3$	187.0958
$C_7H_{15}N_4O_2$	187.1196
$C_8H_{13}NO_4$	187.0845
$C_8H_{15}N_2O_3$	187.1083
$C_8H_{17}N_3O_2$	187.1322
$C_8H_{19}N_4O$	187.1560
$C_9H_7N_4O$	187.0621
$C_9H_{15}O_4$	187.0970
$C_9H_{17}NO_3$	187.1209
$C_9H_{19}N_2O_2$	187.1447
$C_9H_{21}N_3O$	187.1686
$C_9H_{23}N_4$	187.1925
$C_{10}H_7N_2O_2$	187.0508
$C_{10}H_9N_3O$	187.0746
$C_{10}H_{11}N_4$	187.0985
$C_{10}H_{19}O_3$	187.1334
$C_{10}H_{21}NO_2$	187.1573
$C_{10}H_{23}N_2O$	187.1811
$C_{10}H_{25}N_3$	187.2050
$C_{11}H_7O_3$	187.0395
$C_{11}H_9NO_2$	187.0634
$C_{11}H_{11}N_2O$	187.0872
$C_{11}H_{13}N_3$	187.1111
$C_{11}H_{23}O_2$	187.1699
$C_{11}H_{25}NO$	187.1937
$C_{12}H_{11}O_2$	187.0759
$C_{12}H_{13}NO$	187.0998
$C_{12}H_{15}N_2$	187.1236
$C_{13}H_{15}O$	187.1123
$C_{13}H_{17}N$	187.1362
$C_{14}H_{19}$	187.1488
188	
$C_7H_{12}N_2O_4$	188.0797
$C_7H_{14}N_3O_3$	188.1036
$C_7H_{16}N_4O_2$	188.1275
$C_8H_{14}NO_4$	188.0923

APPENDIX A *(Continued)*

	FM
$C_8H_{16}N_2O_3$	188.1162
$C_8H_{18}N_3O_2$	188.1400
$C_8H_{20}N_4O$	188.1639
$C_9H_6N_3O_2$	188.0460
$C_9H_8N_4O$	188.0699
$C_9H_{16}O_4$	188.1049
$C_9H_{18}NO_3$	188.1287
$C_9H_{20}N_2O_2$	188.1526
$C_9H_{22}N_3O$	188.1764
$C_9H_{24}N_4$	188.2003
$C_{10}H_8N_2O_2$	188.0586
$C_{10}H_{10}N_3O$	188.0825
$C_{10}H_{12}N_4$	188.1063
$C_{10}H_{20}O_3$	188.1413
$C_{10}H_{22}NO_2$	188.1651
$C_{10}H_{24}N_2O$	188.1890
$C_{11}H_8O_3$	188.0473
$C_{11}H_{10}NO_2$	188.0712
$C_{11}H_{12}N_2O$	188.0950
$C_{11}H_{14}N_3$	188.1189
$C_{11}H_{24}O_2$	188.1777
$C_{12}H_{12}O_2$	188.0837
$C_{12}H_{14}NO$	188.1076
$C_{12}H_{16}N_2$	188.1315
$C_{13}H_{16}O$	188.1202
$C_{13}H_{18}N$	188.1440
$C_{14}H_{20}$	188.1566
189	
$C_7H_{13}N_2O_4$	189.0876
$C_7H_{15}N_3O_3$	189.1114
$C_7H_{17}N_4O_2$	189.1353
$C_8H_{15}NO_4$	189.1001
$C_8H_{17}N_2O_3$	189.1240
$C_8H_{19}N_3O_2$	189.1478
$C_8H_{21}N_4O$	189.1717
$C_9H_7N_3O_2$	189.0539
$C_9H_9N_4O$	189.0777
$C_9H_{17}O_4$	189.1127
$C_9H_{19}NO_3$	189.1365
$C_9H_{21}N_2O_2$	189.1604
$C_9H_{23}N_3O$	189.1842
$C_{10}H_7NO_3$	189.0426
$C_{10}H_9N_2O_2$	189.0664
$C_{10}H_{11}N_3O$	189.0903
$C_{10}H_{13}N_4$	189.1142
$C_{10}H_{21}O_3$	189.1491
$C_{10}H_{23}NO_2$	189.1730
$C_{11}H_9O_3$	189.0552
$C_{11}H_{11}NO_2$	189.0790
$C_{11}H_{13}N_2O$	189.1029
$C_{11}H_{15}N_3$	189.1267
$C_{12}H_{13}O_2$	189.0916
$C_{12}H_{15}NO$	189.1154
$C_{12}H_{17}N_2$	189.1393
$C_{13}H_{17}O$	189.1280
$C_{13}H_{19}N$	189.1519
$C_{14}H_{21}$	189.1644
190	
$C_7H_{14}N_2O_4$	190.0954
$C_7H_{16}N_3O_3$	190.1193
$C_7H_{18}N_4O_2$	190.1431
$C_8H_6N_4O_2$	190.0491
$C_8H_{16}NO_4$	190.1080
$C_8H_{18}N_2O_3$	190.1318
$C_8H_{20}N_3O_2$	190.1557
$C_8H_{22}N_4O$	190.1795
$C_9H_8N_3O_2$	190.0617
$C_9H_{10}N_4O$	190.0856
$C_9H_{18}O_4$	190.1205
$C_9H_{20}NO_3$	190.1444
$C_9H_{22}N_2O_2$	190.1682
$C_{10}H_8NO_3$	190.0504
$C_{10}H_{10}N_2O_2$	190.0743
$C_{10}H_{12}N_3O$	190.0981
$C_{10}H_{14}N_4$	190.1220
$C_{10}H_{22}O_3$	190.1569
$C_{11}H_{10}O_3$	190.0630
$C_{11}H_{12}NO_2$	190.0868
$C_{11}H_{14}N_2O$	190.1107
$C_{11}H_{16}N_3$	190.1346
$C_{12}H_{14}O_2$	190.0994
$C_{12}H_{16}NO$	190.1233
$C_{12}H_{18}N_2$	190.1471
$C_{13}H_{18}O$	190.1358
$C_{13}H_{20}N$	190.1597
$C_{14}H_{22}$	190.1722
$C_{15}H_{10}$	190.0783
191	
$C_7H_{15}N_2O_4$	191.1032
$C_7H_{17}N_3O_3$	191.1271
$C_7H_{19}N_4O_2$	191.1509
$C_8H_7N_4O_2$	191.0570
$C_8H_{17}NO_4$	191.1158
$C_8H_{19}N_2O_3$	191.1396
$C_8H_{21}N_3O_2$	191.1635
$C_9H_7N_2O_3$	191.0457
$C_9H_9N_3O_2$	191.0695
$C_9H_{11}N_4O$	191.0934
$C_9H_{19}O_4$	191.1284
$C_9H_{21}NO_3$	191.1522
$C_{10}H_7O_4$	191.0344
$C_{10}H_9NO_3$	191.0583
$C_{10}H_{11}N_2O_2$	191.0821
$C_{10}H_{13}N_3O$	191.1060
$C_{10}H_{15}N_4$	191.1298
$C_{11}H_{11}O_3$	191.0708
$C_{11}H_{13}NO_2$	191.0947
$C_{11}H_{15}N_2O$	191.1185
$C_{11}H_{17}N_3$	191.1424
$C_{12}H_{15}O_2$	191.1072
$C_{12}H_{17}NO$	191.1311
$C_{12}H_{19}N_2$	191.1549
$C_{13}H_{19}O$	191.1436
$C_{13}H_{21}N$	191.1675
$C_{14}H_9N$	191.0736
$C_{14}H_{23}$	191.1801
$C_{15}H_{11}$	191.0861
192	
$C_7H_{16}N_2O_4$	192.1111
$C_7H_{18}N_3O_3$	192.1349
$C_7H_{20}N_4O_2$	192.1588
$C_8H_6N_3O_3$	192.0410
$C_8H_8N_4O_2$	192.0648
$C_8H_{18}NO_4$	192.1236
$C_8H_{20}N_2O_3$	192.1475
$C_9H_6NO_4$	192.0297
$C_9H_8N_2O_3$	192.0535
$C_9H_{10}N_3O_2$	192.0774
$C_9H_{12}N_4O$	192.1012
$C_9H_{20}O_4$	192.1362
$C_{10}H_8O_4$	192.0422
$C_{10}H_{10}NO_3$	192.0661
$C_{10}H_{12}N_2O_2$	192.0899
$C_{10}H_{14}N_3O$	192.1138
$C_{10}H_{16}N_4$	192.1377
$C_{11}H_{12}O_3$	192.0786
$C_{11}H_{14}NO_2$	192.1025
$C_{11}H_{16}N_2O$	192.1264
$C_{11}H_{18}N_3$	192.1502
$C_{12}H_{16}O_2$	192.1151
$C_{12}H_{18}NO$	192.1389
$C_{12}H_{20}N_2$	192.1628
$C_{13}H_8N_2$	192.0688
$C_{13}H_{20}O$	192.1515
$C_{13}H_{22}N$	192.1753
$C_{14}H_{10}N$	192.0814
$C_{14}H_{24}$	192.1879
$C_{15}H_{12}$	192.0939
193	
$C_7H_{17}N_2O_4$	193.1189
$C_7H_{19}N_3O_3$	193.1427
$C_8H_7N_3O_3$	193.0488
$C_8H_9N_4O_2$	193.0726
$C_8H_{19}NO_4$	193.1315
$C_9H_7NO_4$	193.0375
$C_9H_9N_2O_3$	193.0614
$C_9H_{11}N_3O_2$	193.0852
$C_9H_{13}N_4O$	193.1091
$C_{10}H_9O_4$	193.0501
$C_{10}H_{11}NO_3$	193.0739
$C_{10}H_{13}N_2O_2$	193.0978
$C_{10}H_{15}N_3O$	193.1216
$C_{10}H_{17}N_4$	193.1455
$C_{11}H_{13}O_3$	193.0865
$C_{11}H_{15}NO_2$	193.1103
$C_{11}H_{17}N_2O$	193.1342
$C_{11}H_{19}N_3$	193.1580
$C_{12}H_{17}O_2$	193.1229
$C_{12}H_{19}NO$	193.1467
$C_{12}H_{21}N_2$	193.1706
$C_{13}H_9N_2$	193.0767
$C_{13}H_{21}O$	193.1593
$C_{13}H_{23}N$	193.1832
$C_{14}H_9O$	193.0653
$C_{14}H_{11}N$	193.0892
$C_{14}H_{25}$	193.1957
$C_{15}H_{13}$	193.1018
194	
$C_7H_{18}N_2O_4$	194.1267
$C_8H_6N_2O_4$	194.0328
$C_8H_8N_3O_3$	194.0566
$C_8H_{10}N_4O_2$	194.0805
$C_9H_8NO_4$	194.0453
$C_9H_{10}N_2O_3$	194.0692
$C_9H_{12}N_3O_2$	194.0930
$C_9H_{14}N_4O$	194.1169
$C_{10}H_{10}O_4$	194.0579
$C_{10}H_{12}NO_3$	194.0817
$C_{10}H_{14}N_2O_2$	194.1056
$C_{10}H_{16}N_3O$	194.1295
$C_{10}H_{18}N_4$	194.1533
$C_{11}H_{14}O_3$	194.0943
$C_{11}H_{16}NO_2$	194.1182
$C_{11}H_{18}N_2O$	194.1420
$C_{11}H_{20}N_3$	194.1659
$C_{12}H_8N_3$	194.0719
$C_{12}H_{18}O_2$	194.1307
$C_{12}H_{20}NO$	194.1546
$C_{12}H_{22}N_2$	194.1784
$C_{13}H_8NO$	194.0606
$C_{13}H_{10}N_2$	194.0845
$C_{13}H_{22}O$	194.1671
$C_{13}H_{24}N$	194.1910
$C_{14}H_{10}O$	194.0732
$C_{14}H_{12}N$	194.0970
$C_{14}H_{26}$	194.2036
$C_{15}H_{14}$	194.1096
195	
$C_8H_7N_2O_4$	195.0406
$C_8H_9N_3O_3$	195.0644
$C_8H_{11}N_4O_2$	195.0883
$C_9H_9NO_4$	195.0532
$C_9H_{11}N_2O_3$	195.0770
$C_9H_{13}N_3O_2$	195.1009
$C_9H_{15}N_4O$	195.1247
$C_{10}H_{11}O_4$	195.0657
$C_{10}H_{13}NO_3$	195.0896
$C_{10}H_{15}N_2O_2$	195.1134
$C_{10}H_{17}N_3O$	195.1373
$C_{10}H_{19}N_4$	195.1611
$C_{11}H_7N_4$	195.0672
$C_{11}H_{15}O_3$	195.1021
$C_{11}H_{17}NO_2$	195.1260
$C_{11}H_{19}N_2O$	195.1498
$C_{11}H_{21}N_3$	195.1737
$C_{12}H_7N_2O$	195.0559
$C_{12}H_9N_3$	195.0798
$C_{12}H_{19}O_2$	195.1385

APPENDIX A *(Continued)*

	FM
$C_{12}H_{21}NO$	195.1624
$C_{12}H_{23}N_2$	195.1863
$C_{13}H_9NO$	195.0684
$C_{13}H_{11}N_2$	195.0923
$C_{13}H_{23}O$	195.1750
$C_{13}H_{25}N$	195.1988
$C_{14}H_{11}O$	195.0810
$C_{14}H_{13}N$	195.1049
$C_{14}H_{27}$	195.2114
$C_{15}H_{15}$	195.1174
196	
$C_8H_8N_2O_4$	196.0484
$C_8H_{10}N_3O_3$	196.0723
$C_8H_{12}N_4O_2$	196.0961
$C_9H_{10}NO_4$	196.0610
$C_9H_{12}N_2O_3$	196.0848
$C_9H_{14}N_3O_2$	196.1087
$C_9H_{16}N_4O$	196.1325
$C_{10}H_{12}O_4$	196.0735
$C_{10}H_{14}NO_3$	196.0974
$C_{10}H_{16}N_2O_2$	196.1213
$C_{10}H_{18}N_3O$	196.1451
$C_{10}H_{20}N_4$	196.1690
$C_{11}H_8N_4$	196.0750
$C_{11}H_{16}O_3$	196.1100
$C_{11}H_{18}NO_2$	196.1338
$C_{11}H_{20}N_2O$	196.1577
$C_{11}H_{22}N_3$	196.1815
$C_{12}H_8N_2O$	196.0637
$C_{12}H_{10}N_3$	196.0876
$C_{12}H_{20}O_2$	196.1464
$C_{12}H_{22}NO$	196.1702
$C_{12}H_{24}N_2$	196.1941
$C_{13}H_8O_2$	196.0524
$C_{13}H_{10}NO$	196.0763
$C_{13}H_{12}N_2$	196.1001
$C_{13}H_{24}O$	196.1828
$C_{13}H_{26}N$	196.2067
$C_{14}H_{12}O$	196.0888
$C_{14}H_{14}N$	196.1127
$C_{14}H_{28}$	196.2192
$C_{15}H_{16}$	196.1253
197	
$C_8H_9N_2O_4$	197.0563
$C_8H_{11}N_3O_3$	197.0801
$C_8H_{13}N_4O_2$	197.1040
$C_9H_{11}NO_4$	197.0688
$C_9H_{13}N_2O_3$	197.0927
$C_9H_{15}N_3O_2$	197.1165
$C_9H_{17}N_4O$	197.1404
$C_{10}H_{13}O_4$	197.0814
$C_{10}H_{15}NO_3$	197.1052
$C_{10}H_{17}N_2O_2$	197.1291
$C_{10}H_{19}N_3O$	197.1529
$C_{10}H_{21}N_4$	197.1768
$C_{11}H_7N_3O$	197.0590
$C_{11}H_9N_4$	197.0829
$C_{11}H_{17}O_3$	197.1178
$C_{11}H_{19}NO_2$	197.1416
$C_{11}H_{21}N_2O$	197.1655
$C_{11}H_{23}N_3$	197.1894
$C_{12}H_9N_2O$	197.0715
$C_{12}H_{11}N_3$	197.0954
$C_{12}H_{21}O_2$	197.1542
$C_{12}H_{23}NO$	197.1781
$C_{12}H_{25}N_2$	197.2019
$C_{13}H_9O_2$	197.0603
$C_{13}H_{11}NO$	197.0841
$C_{13}H_{13}N_2$	197.1080
$C_{13}H_{25}O$	197.1906
$C_{13}H_{27}N$	197.2145
$C_{14}H_{13}O$	197.0967
$C_{14}H_{15}N$	197.1205
$C_{14}H_{29}$	197.2270
$C_{15}H_{17}$	197.1331
198	
$C_8H_{10}N_2O_4$	198.0641
$C_8H_{12}N_3O_3$	198.0879
$C_8H_{14}N_4O_2$	198.1118
$C_9H_{12}NO_4$	198.0766
$C_9H_{14}N_2O_3$	198.1005
$C_9H_{16}N_3O_2$	198.1244
$C_9H_{18}N_4O$	198.1482
$C_{10}H_{14}O_4$	198.0892
$C_{10}H_{16}NO_3$	198.1131
$C_{10}H_{18}N_2O_2$	198.1369
$C_{10}H_{20}N_3O$	198.1608
$C_{10}H_{22}N_4$	198.1846
$C_{11}H_8N_3O$	198.0668
$C_{11}H_{10}N_4$	198.0907
$C_{11}H_{18}O_3$	198.1256
$C_{11}H_{20}NO_2$	198.1495
$C_{11}H_{22}N_2O$	198.1733
$C_{11}H_{24}N_3$	198.1972
$C_{12}H_8NO_2$	198.0555
$C_{12}H_{10}N_2O$	198.0794
$C_{12}H_{12}N_3$	198.1032
$C_{12}H_{22}O_2$	198.1620
$C_{12}H_{24}NO$	198.1859
$C_{12}H_{26}N_2$	198.2098
$C_{13}H_{10}O_2$	198.0681
$C_{13}H_{12}NO$	198.0919
$C_{13}H_{14}N_2$	198.1158
$C_{13}H_{26}O$	198.1985
$C_{13}H_{28}N$	198.2223
$C_{14}H_{14}O$	198.1045
$C_{14}H_{16}N$	198.1284
$C_{14}H_{30}$	198.2349
$C_{15}H_{18}$	198.1409
199	
$C_8H_{11}N_2O_4$	199.0719
$C_8H_{13}N_3O_3$	199.0958
$C_8H_{15}N_4O_2$	199.1196
$C_9H_{13}NO_4$	199.0845
$C_9H_{15}N_2O_3$	199.1083
$C_9H_{17}N_3O_2$	199.1322
$C_9H_{19}N_4O$	199.1560
$C_{10}H_7N_4O$	199.0621
$C_{10}H_{15}O_4$	199.0970
$C_{10}H_{17}NO_3$	199.1209
$C_{10}H_{19}N_2O_2$	199.1447
$C_{10}H_{21}N_3O$	199.1686
$C_{10}H_{23}N_4$	199.1925
$C_{11}H_7N_2O_2$	199.0508
$C_{11}H_9N_3O$	199.0746
$C_{11}H_{11}N_4$	199.0985
$C_{11}H_{19}O_3$	199.1334
$C_{11}H_{21}NO_2$	199.1573
$C_{11}H_{23}N_2O$	199.1811
$C_{11}H_{25}N_3$	199.2050
$C_{12}H_9NO_2$	199.0634
$C_{12}H_{11}N_2O$	199.0872
$C_{12}H_{13}N_3$	199.1111
$C_{12}H_{23}O_2$	199.1699
$C_{12}H_{25}NO$	199.1937
$C_{12}H_{27}N_2$	199.2176
$C_{13}H_{11}O_2$	199.0759
$C_{13}H_{13}NO$	199.0998
$C_{13}H_{15}N_2$	199.1236
$C_{13}H_{27}O$	199.2063
$C_{13}H_{29}N$	199.2301
$C_{14}H_{15}O$	199.1123
$C_{14}H_{17}N$	199.1362
$C_{15}H_{19}$	199.1488
200	
$C_8H_{12}N_2O_4$	200.0797
$C_8H_{14}N_3O_3$	200.1036
$C_8H_{16}N_4O_2$	200.1275
$C_9H_{14}NO_4$	200.0923
$C_9H_{16}N_2O_3$	200.1162
$C_9H_{18}N_3O_2$	200.1400
$C_9H_{20}N_4O$	200.1639
$C_{10}H_8N_4O$	200.0699
$C_{10}H_{16}O_4$	200.1049
$C_{10}H_{18}NO_3$	200.1287
$C_{10}H_{20}N_2O_2$	200.1526
$C_{10}H_{22}N_3O$	200.1764
$C_{10}H_{24}N_4$	200.2003
$C_{11}H_8N_2O_2$	200.0586
$C_{11}H_{10}N_3O$	200.0825
$C_{11}H_{12}N_4$	200.1063
$C_{11}H_{20}O_3$	200.1413
$C_{11}H_{22}NO_2$	200.1651
$C_{11}H_{24}N_2O$	200.1890
$C_{11}N_{26}N_3$	200.2129
$C_{12}H_8O_3$	200.0473
$C_{12}H_{10}NO_2$	200.0712
$C_{12}H_{12}N_2O$	200.0950
$C_{12}H_{14}N_3$	200.1189
$C_{12}H_{24}O_2$	200.1777
$C_{12}H_{26}NO$	200.2015
$C_{12}H_{28}N_2$	200.2254
$C_{13}H_{12}O_2$	200.0837
$C_{13}H_{14}NO$	200.1076
$C_{13}H_{16}N_2$	200.1315
$C_{13}H_{28}O$	200.2141
$C_{14}H_{16}O$	200.1202
$C_{14}H_{18}N$	200.1440
$C_{15}H_{20}$	200.1566
201	
$C_8H_{13}N_2O_4$	201.0876
$C_8H_{15}N_3O_3$	201.1114
$C_8H_{17}N_4O_2$	201.1353
$C_9H_{15}NO_4$	201.1001
$C_9H_{17}N_2O_3$	201.1240
$C_9H_{19}N_3O_2$	201.1478
$C_9H_{21}N_4O$	201.1717
$C_{10}H_7N_3O_2$	201.0539
$C_{10}H_9N_4O$	201.0777
$C_{10}H_{17}O_4$	201.1127
$C_{10}H_{19}NO_3$	201.1365
$C_{10}H_{21}N_2O_2$	201.1604
$C_{10}H_{23}N_3O$	201.1842
$C_{10}H_{25}N_4$	201.2081
$C_{11}H_7NO_3$	201.0426
$C_{11}H_9N_2O_2$	201.0664
$C_{11}H_{11}N_3O$	201.0903
$C_{11}H_{13}N_4$	201.1142
$C_{11}H_{21}O_3$	201.1491
$C_{11}H_{23}NO_2$	201.1730
$C_{11}H_{25}N_2O$	201.1968
$C_{11}H_{27}N_3$	201.2207
$C_{12}H_9O_3$	201.0552
$C_{12}H_{11}NO_2$	201.0790
$C_{12}H_{13}N_2O$	201.1029
$C_{12}H_{15}N_3$	201.1267
$C_{12}H_{25}O_2$	201.1855
$C_{12}H_{27}NO$	201.2094
$C_{13}H_{13}O_2$	201.0916
$C_{13}H_{15}NO$	201.1154
$C_{13}H_{17}N_2$	201.1393
$C_{14}H_{17}O$	201.1280
$C_{14}H_{19}N$	201.1519
$C_{15}H_{21}$	201.1644
202	
$C_8H_{14}N_2O_4$	202.0954
$C_8H_{16}N_3O_3$	202.1193
$C_8H_{18}N_4O_2$	202.1431
$C_9H_6N_4O_2$	202.0491
$C_9H_{16}NO_4$	202.1080
$C_9H_{18}N_2O_3$	202.1318
$C_9H_{20}N_3O_2$	202.1557
$C_9H_{22}N_4O$	202.1795
$C_{10}H_8N_3O_2$	202.0617
$C_{10}H_{10}N_4O$	202.0856
$C_{10}H_{18}O_4$	202.1205
$C_{10}H_{20}NO_3$	202.1444
$C_{10}H_{22}N_2O_2$	202.1682

APPENDIX A (Continued)

	FM
$C_{10}H_{24}N_3O$	202.1921
$C_{10}H_{26}N_4$	202.2160
$C_{11}H_8NO_3$	202.0504
$C_{11}H_{10}N_2O_2$	202.0743
$C_{11}H_{12}N_3O$	202.0981
$C_{11}H_{14}N_4$	202.1220
$C_{11}H_{22}O_3$	202.1569
$C_{11}H_{24}NO_2$	202.1808
$C_{11}H_{26}N_2O$	202.2046
$C_{12}H_{10}O_3$	202.0630
$C_{12}H_{12}NO_2$	202.0868
$C_{12}H_{14}N_2O$	202.1107
$C_{12}H_{16}N_3$	202.1346
$C_{12}H_{26}O_2$	202.1934
$C_{13}H_{14}O_2$	202.0994
$C_{13}H_{16}NO$	202.1233
$C_{13}H_{18}N_2$	202.1471
$C_{14}H_{18}O$	202.1358
$C_{14}H_{20}N$	202.1597
$C_{15}H_{22}$	202.1722
203	
$C_8H_{15}N_2O_4$	203.1032
$C_8H_{17}N_3O_3$	203.1271
$C_8H_{19}N_4O_2$	203.1509
$C_9H_7N_4O_2$	203.0570
$C_9H_{17}NO_4$	203.1158
$C_9H_{19}N_2O_3$	203.1396
$C_9H_{21}N_3O_2$	203.1635
$C_9H_{23}N_4O$	203.1873
$C_{10}H_7N_2O_3$	203.0457
$C_{10}H_9N_3O_2$	203.0695
$C_{10}H_{11}N_4O$	203.0934
$C_{10}H_{19}O_4$	203.1284
$C_{10}H_{21}NO_3$	203.1522
$C_{10}H_{23}N_2O_2$	203.1761
$C_{10}H_{25}N_3O$	203.1999
$C_{11}H_7O_4$	203.0344
$C_{11}H_9NO_3$	203.0583
$C_{11}H_{11}N_2O_2$	203.0821
$C_{11}H_{13}N_3O$	203.1060
$C_{11}H_{15}N_4$	203.1298
$C_{11}H_{23}O_3$	203.1648
$C_{11}H_{25}NO_2$	203.1886
$C_{12}H_{11}O_3$	203.0708
$C_{12}H_{13}NO_2$	203.0947
$C_{12}H_{15}N_2O$	203.1185
$C_{12}H_{17}N_3$	203.1424
$C_{13}H_{15}O_2$	203.1072
$C_{13}H_{17}NO$	203.1311
$C_{13}H_{19}N_2$	203.1549
$C_{14}H_{19}O$	203.1436
$C_{14}H_{21}N$	203.1675
$C_{15}H_9N$	203.0736
$C_{15}H_{23}$	203.1801
204	
$C_8H_{16}N_2O_4$	204.1111
$C_8H_{18}N_3O_3$	204.1349
$C_8H_{20}N_4O_2$	204.1588
$C_9H_6N_3O_3$	204.0410
$C_9H_8N_4O_2$	204.0648
$C_9H_{18}NO_4$	204.1236
$C_9H_{20}N_2O_3$	204.1475
$C_9H_{22}N_3O_2$	204.1713
$C_9H_{24}N_4O$	204.1952
$C_{10}H_8N_2O_3$	204.0535
$C_{10}H_{10}N_3O_2$	204.0774
$C_{10}H_{12}N_4O$	204.1012
$C_{10}H_{20}O_4$	204.1362
$C_{10}H_{22}NO_3$	204.1600
$C_{10}H_{24}N_2O_2$	204.1839
$C_{11}H_8O_4$	204.0422
$C_{11}H_{10}NO_3$	204.0661
$C_{11}H_{12}N_2O_2$	204.0899
$C_{11}H_{14}N_3O$	204.1138
$C_{11}H_{16}N_4$	204.1377
$C_{11}H_{24}O_3$	204.1726
$C_{12}H_{12}O_3$	204.0786
$C_{12}H_{14}NO_2$	204.1025
$C_{12}H_{16}N_2O$	204.1264
$C_{12}H_{18}N_3$	204.1502
$C_{13}H_{16}O_2$	204.1151
$C_{13}H_{18}NO$	204.1389
$C_{13}H_{20}N_2$	204.1628
$C_{14}H_{20}O$	204.1515
$C_{14}H_{22}N$	204.1753
$C_{15}H_{10}N$	204.0814
$C_{15}H_{24}$	204.1879
$C_{16}H_{12}$	204.0939
205	
$C_8H_{17}N_2O_4$	205.1189
$C_8H_{19}N_3O_3$	205.1427
$C_8H_{21}N_4O_2$	205.1666
$C_9H_7N_3O_3$	205.0488
$C_9H_9N_4O_2$	205.0726
$C_9H_{19}NO_4$	205.1315
$C_9H_{21}N_2O_3$	205.1553
$C_9H_{23}N_3O_2$	205.1791
$C_{10}H_7NO_4$	205.0375
$C_{10}H_9N_2O_3$	205.0614
$C_{10}H_{11}N_3O_2$	205.0852
$C_{10}H_{13}N_4O$	205.1091
$C_{10}H_{21}O_4$	205.1440
$C_{10}H_{23}NO_3$	205.1679
$C_{11}H_9O_4$	205.0501
$C_{11}H_{11}NO_3$	205.0739
$C_{11}H_{13}N_2O_2$	205.0978
$C_{11}H_{15}N_3O$	205.1216
$C_{11}H_{17}N_4$	205.1455
$C_{12}H_{13}O_3$	205.0865
$C_{12}H_{15}NO_2$	205.1103
$C_{12}H_{17}N_2O$	205.1342
$C_{12}H_{19}N_3$	205.1580
$C_{13}H_{17}O_2$	205.1229
$C_{13}H_{19}NO$	205.1467
$C_{13}H_{21}N_2$	205.1706
$C_{14}H_9N_2$	205.0767
$C_{14}H_{21}O$	205.1593
$C_{14}H_{23}N$	205.1832
$C_{15}H_9O$	205.0653
$C_{15}H_{11}N$	205.0892
$C_{15}H_{25}$	205.1957
$C_{16}H_{13}$	205.1018
206	
$C_8H_{18}N_2O_4$	206.1267
$C_8H_{20}N_3O_3$	206.1506
$C_8H_{22}N_4O_2$	206.1744
$C_9H_6N_2O_4$	206.0328
$C_9H_8N_3O_3$	206.0566
$C_9H_{10}N_4O_2$	206.0805
$C_9H_{20}NO_4$	206.1393
$C_9H_{22}N_2O_3$	206.1631
$C_{10}H_8NO_4$	206.0453
$C_{10}H_{10}N_2O_3$	206.0692
$C_{10}H_{12}N_3O_2$	206.0930
$C_{10}H_{14}N_4O$	206.1169
$C_{10}H_{22}O_4$	206.1518
$C_{11}H_{10}O_4$	206.0579
$C_{11}H_{12}NO_3$	206.0817
$C_{11}H_{14}N_2O_2$	206.1056
$C_{11}H_{16}N_3O$	206.1295
$C_{11}H_{18}N_4$	206.1533
$C_{12}H_{14}O_3$	206.0943
$C_{12}H_{16}NO_2$	206.1182
$C_{12}H_{18}N_2O$	206.1420
$C_{12}H_{20}N_3$	206.1659
$C_{13}H_8N_3$	206.0719
$C_{13}H_{18}O_2$	206.1307
$C_{13}H_{20}NO$	206.1546
$C_{13}H_{22}N_2$	206.1784
$C_{14}H_{10}N_2$	206.0845
$C_{14}H_{22}O$	206.1671
$C_{14}H_{24}N$	206.1910
$C_{15}H_{10}O$	206.0732
$C_{15}H_{12}N$	206.0970
$C_{15}H_{26}$	206.2036
$C_{16}H_{14}$	206.1096
207	
$C_8H_{19}N_2O_4$	207.1345
$C_8H_{21}N_3O_3$	207.1584
$C_9H_7N_2O_4$	207.0406
$C_9H_9N_3O_3$	207.0644
$C_9H_{11}N_4O_2$	207.0883
$C_9H_{21}NO_4$	207.1471
$C_{10}H_9NO_4$	207.0532
$C_{10}H_{11}N_2O_3$	207.0770
$C_{10}H_{13}N_3O_2$	207.1009
$C_{10}H_{15}N_4O$	207.1247
$C_{11}H_{11}O_4$	207.0657
$C_{11}H_{13}NO_3$	207.0896
$C_{11}H_{15}N_2O_2$	207.1134
$C_{11}H_{17}N_3O$	207.1373
$C_{11}H_{19}N_4$	207.1611
$C_{12}H_{15}O_3$	207.1021
$C_{12}H_{17}NO_2$	207.1260
$C_{12}H_{19}N_2O$	207.1498
$C_{12}H_{21}N_3$	207.1737
$C_{13}H_9N_3$	207.0798
$C_{13}H_{19}O_2$	207.1385
$C_{13}H_{21}NO$	207.1624
$C_{13}H_{23}N_2$	207.1863
$C_{14}H_9NO$	207.0684
$C_{14}H_{11}N_2$	207.0923
$C_{14}H_{23}O$	207.1750
$C_{14}H_{25}N$	207.1988
$C_{15}H_{11}O$	207.0810
$C_{15}H_{13}N$	207.1049
$C_{15}H_{27}$	207.2114
$C_{16}H_{15}$	207.1174
208	
$C_8H_{20}N_2O_4$	208.1424
$C_9H_8N_2O_4$	208.0484
$C_9H_{10}N_3O_3$	208.0723
$C_9H_{12}N_4O_2$	208.0961
$C_{10}H_{10}NO_4$	208.0610
$C_{10}H_{12}N_2O_3$	208.0848
$C_{10}H_{14}N_3O_2$	208.1087
$C_{10}H_{16}N_4O$	208.1325
$C_{11}H_{12}O_4$	208.0735
$C_{11}H_{14}NO_3$	208.0974
$C_{11}H_{16}N_2O_2$	208.1213
$C_{11}H_{18}N_3O$	208.1451
$C_{11}H_{20}N_4$	208.1690
$C_{12}H_8N_4$	208.0750
$C_{12}H_{16}O_3$	208.1100
$C_{12}H_{18}NO_2$	208.1338
$C_{12}H_{20}N_2O$	208.1577
$C_{12}H_{22}N_3$	208.1815
$C_{13}H_8N_2O$	208.0637
$C_{13}H_{10}N_3$	208.0876
$C_{13}H_{20}O_2$	208.1464
$C_{13}H_{22}NO$	208.1702
$C_{13}H_{24}N_2$	208.1941
$C_{14}H_{10}NO$	208.0763
$C_{14}H_{12}N_2$	208.1001
$C_{14}H_{24}O$	208.1828
$C_{14}H_{26}N$	208.2067
$C_{15}H_{12}O$	208.0888
$C_{15}H_{14}N$	208.1127
$C_{15}H_{28}$	208.2192
$C_{16}H_{16}$	208.1253
209	
$C_9H_9N_2O_4$	209.0563
$C_9H_{11}N_3O_3$	209.0801
$C_9H_{13}N_4O_2$	209.1040
$C_{10}H_{11}NO_4$	209.0688
$C_{10}H_{13}N_2O_3$	209.0927
$C_{10}H_{15}N_3O_2$	209.1165
$C_{10}H_{17}N_4O$	209.1404

APPENDIX A *(Continued)*

	FM
$C_{11}H_{13}O_4$	209.0814
$C_{11}H_{15}NO_3$	209.1052
$C_{11}H_{17}N_2O_2$	209.1291
$C_{11}H_{19}N_3O$	209.1529
$C_{11}H_{21}N_4$	209.1768
$C_{12}H_9N_4$	209.0829
$C_{12}H_{17}O_3$	209.1178
$C_{12}H_{19}NO_2$	209.1416
$C_{12}H_{21}N_2O$	209.1655
$C_{12}H_{23}N_3$	209.1894
$C_{13}H_9N_2O$	209.0715
$C_{13}H_{11}N_3$	209.0954
$C_{13}H_{21}O_2$	209.1542
$C_{13}H_{23}NO$	209.1781
$C_{13}H_{25}N_2$	209.2019
$C_{14}H_9O_2$	209.0603
$C_{14}H_{11}NO$	209.0841
$C_{14}H_{13}N_2$	209.1080
$C_{14}H_{25}O$	209.1906
$C_{14}H_{27}N$	209.2145
$C_{15}H_{13}O$	209.0967
$C_{15}H_{15}N$	209.1205
$C_{15}H_{29}$	209.2270
$C_{16}H_{17}$	209.1331
210	
$C_9H_{10}N_2O_4$	210.0641
$C_9H_{12}N_3O_3$	210.0879
$C_9H_{14}N_4O_2$	210.1118
$C_{10}H_{12}NO_4$	210.0766
$C_{10}H_{14}N_2O_3$	210.1005
$C_{10}H_{16}N_3O_2$	210.1244
$C_{10}H_{18}N_4O$	210.1482
$C_{11}H_{14}O_4$	210.0892
$C_{11}H_{16}NO_3$	210.1131
$C_{11}H_{18}N_2O_2$	210.1369
$C_{11}H_{20}N_3O$	210.1608
$C_{11}H_{22}N_4$	210.1846
$C_{12}H_8N_3O$	210.0668
$C_{12}H_{10}N_4$	210.0907
$C_{12}H_{18}O_3$	210.1256
$C_{12}H_{20}NO_2$	210.1495
$C_{12}H_{22}N_2O$	210.1733
$C_{12}H_{24}N_3$	210.1972
$C_{13}H_8NO_2$	210.0555
$C_{13}H_{10}N_2O$	210.0794
$C_{13}H_{12}N_3$	210.1032
$C_{13}H_{22}O_2$	210.1620
$C_{13}H_{24}NO$	210.1859
$C_{13}H_{26}N_2$	210.2098
$C_{14}H_{10}O_2$	210.0681
$C_{14}H_{12}NO$	210.0919
$C_{14}H_{14}N_2$	210.1158
$C_{14}H_{26}O$	210.1985
$C_{14}H_{28}N$	210.2223
$C_{15}H_{14}O$	210.1045
$C_{15}H_{16}N$	210.1284
$C_{15}H_{30}$	210.2349
$C_{16}H_{18}$	210.1409
211	
$C_9H_{11}N_2O_4$	211.0719
$C_9H_{13}N_3O_3$	211.0958
$C_9H_{15}N_4O_2$	211.1196
$C_{10}H_{13}NO_4$	211.0845
$C_{10}H_{15}N_2O_3$	211.1083
$C_{10}H_{17}N_3O_2$	211.1322
$C_{10}H_{19}N_4O$	211.1560
$C_{11}H_7N_4O$	211.0621
$C_{11}H_{15}O_4$	211.0970
$C_{11}H_{17}NO_3$	211.1209
$C_{11}H_{19}N_2O_2$	211.1447
$C_{11}H_{21}N_3O$	211.1686
$C_{11}H_{23}N_4$	211.1925
$C_{12}H_9N_3O$	211.0746
$C_{12}H_{11}N_4$	211.0985
$C_{12}H_{19}O_3$	211.1334
$C_{12}H_{21}NO_2$	211.1573
$C_{12}H_{23}N_2O$	211.1811
$C_{12}H_{25}N_3$	211.2050
$C_{13}H_9NO_2$	211.0634
$C_{13}H_{11}N_2O$	211.0872
$C_{13}H_{13}N_3$	211.1111
$C_{13}H_{23}O_2$	211.1699
$C_{13}H_{25}NO$	211.1937
$C_{13}H_{27}N_2$	211.2176
$C_{14}H_{11}O_2$	211.0759
$C_{14}H_{13}NO$	211.0998
$C_{14}H_{15}N_2$	211.1236
$C_{14}H_{27}O$	211.2063
$C_{14}H_{29}N$	211.2301
$C_{15}H_{15}O$	211.1123
$C_{15}H_{17}N$	211.1362
$C_{15}H_{31}$	211.2427
$C_{16}H_{19}$	211.1488
212	
$C_9H_{12}N_2O_4$	212.0797
$C_9H_{14}N_3O_3$	212.1036
$C_9H_{16}N_4O_2$	212.1275
$C_{10}H_{14}NO_4$	212.0923
$C_{10}H_{16}N_2O_3$	212.1162
$C_{10}H_{18}N_3O_2$	212.1400
$C_{10}H_{20}N_4O$	212.1639
$C_{11}H_8N_4O$	212.0699
$C_{11}H_{16}O_4$	212.1049
$C_{11}H_{18}NO_3$	212.1287
$C_{11}H_{20}N_2O_2$	212.1526
$C_{11}H_{22}N_3O$	212.1764
$C_{11}H_{24}N_4$	212.2003
$C_{12}H_8N_2O_2$	212.0586
$C_{12}H_{10}N_3O$	212.0825
$C_{12}H_{12}N_4$	212.1063
$C_{12}H_{20}O_3$	212.1413
$C_{12}H_{22}NO_2$	212.1651
$C_{12}H_{24}N_2O$	212.1890
$C_{12}H_{26}N_3$	212.2129
$C_{13}H_8O_3$	212.0473
$C_{13}H_{10}NO_2$	212.0712
$C_{13}H_{12}N_2O$	212.0950
$C_{13}H_{14}N_3$	212.1189
$C_{13}H_{24}O_2$	212.1777
$C_{13}H_{26}NO$	212.2015
$C_{13}H_{28}N_2$	212.2254
$C_{14}H_{12}O_2$	212.0837
$C_{14}H_{14}NO$	212.1076
$C_{14}H_{16}N_2$	212.1315
$C_{14}H_{28}O$	212.2141
$C_{14}H_{30}N$	212.2380
$C_{15}H_{16}O$	212.1202
$C_{15}H_{18}N$	212.1440
$C_{15}H_{32}$	212.2505
$C_{16}H_{20}$	212.1566
213	
$C_9H_{13}N_2O_4$	213.0876
$C_9H_{15}N_3O_3$	213.1114
$C_9H_{17}N_4O_2$	213.1353
$C_{10}H_{15}NO_4$	213.1001
$C_{10}H_{17}N_2O_3$	213.1240
$C_{10}H_{19}N_3O_2$	213.1478
$C_{10}H_{21}N_4O$	213.1717
$C_{11}H_7N_3O_2$	213.0539
$C_{11}H_9N_4O$	213.0777
$C_{11}H_{17}O_4$	213.1127
$C_{11}H_{19}NO_3$	213.1365
$C_{11}H_{21}N_2O_2$	213.1604
$C_{11}H_{23}N_3O$	213.1842
$C_{11}H_{25}N_4$	213.2081
$C_{12}H_9N_2O_2$	213.0664
$C_{12}H_{11}N_3O$	213.0903
$C_{12}H_{13}N_4$	213.1142
$C_{12}H_{21}O_3$	213.1491
$C_{12}H_{23}NO_2$	213.1730
$C_{12}H_{25}N_2O$	213.1968
$C_{12}H_{27}N_3$	213.2207
$C_{13}H_9O_3$	213.0552
$C_{13}H_{11}NO_2$	213.0790
$C_{13}H_{13}N_2O$	213.1029
$C_{13}H_{15}N_3$	213.1267
$C_{13}H_{25}O_2$	213.1855
$C_{13}H_{27}NO$	213.2094
$C_{13}H_{29}N_2$	213.2332
$C_{14}H_{13}O_2$	213.0916
$C_{14}H_{15}NO$	213.1154
$C_{14}H_{17}N_2$	213.1393
$C_{14}H_{29}O$	213.2219
$C_{15}H_{17}O$	213.1280
$C_{15}H_{19}N$	213.1519
$C_{16}H_{21}$	213.1644
214	
$C_9H_{14}N_2O_4$	214.0954
$C_9H_{16}N_3O_3$	214.1193
$C_9H_{18}N_4O_2$	214.1431
$C_{10}H_{16}NO_4$	214.1080
$C_{10}H_{18}N_2O_3$	214.1318
$C_{10}H_{20}N_3O_2$	214.1557
$C_{10}H_{22}N_4O$	214.1795
$C_{11}H_8N_3O_2$	214.0617
$C_{11}H_{10}N_4O$	214.0856
$C_{11}H_{18}O_4$	214.1205
$C_{11}H_{20}NO_3$	214.1444
$C_{11}H_{22}N_2O_2$	214.1682
$C_{11}H_{24}N_3O$	214.1921
$C_{11}H_{26}N_4$	214.2160
$C_{12}H_8NO_3$	214.0504
$C_{12}H_{10}N_2O_2$	214.0743
$C_{12}H_{12}N_3O$	214.0981
$C_{12}H_{14}N_4$	214.1220
$C_{12}H_{22}O_3$	214.1569
$C_{12}H_{24}NO_2$	214.1808
$C_{12}H_{26}N_2O$	214.2046
$C_{12}H_{28}N_3$	214.2285
$C_{13}H_{10}O_3$	214.0630
$C_{13}H_{12}NO_2$	214.0869
$C_{13}H_{14}N_2O$	214.1107
$C_{13}H_{16}N_3$	214.1346
$C_{13}H_{26}O_2$	214.1934
$C_{13}H_{28}NO$	214.2172
$C_{13}H_{30}N_2$	214.2411
$C_{14}H_{14}O_2$	214.0994
$C_{14}H_{16}NO$	214.1233
$C_{14}H_{18}N_2$	214.1471
$C_{15}H_{18}O$	214.1358
$C_{15}H_{20}N$	214.1597
$C_{16}H_{22}$	214.1722
215	
$C_9H_{15}N_2O_4$	215.1032
$C_9H_{17}N_3O_3$	215.1271
$C_9H_{19}N_4O_2$	215.1509
$C_{10}H_7N_4O_2$	215.0570
$C_{10}H_{17}NO_4$	215.1158
$C_{10}H_{19}N_2O_3$	215.1396
$C_{10}H_{21}N_3O_2$	215.1635
$C_{10}H_{23}N_4O$	215.1873
$C_{11}H_7N_2O_3$	215.0457
$C_{11}H_9N_3O_2$	215.0695
$C_{11}H_{11}N_4O$	215.0934
$C_{11}H_{19}O_4$	215.1284
$C_{11}H_{21}NO_3$	215.1522
$C_{11}H_{23}N_2O_2$	215.1761
$C_{11}H_{25}N_3O$	215.1999
$C_{11}H_{27}N_4$	215.2238
$C_{12}H_9NO_3$	215.0583
$C_{12}H_{11}N_2O_2$	215.0821
$C_{12}H_{13}N_3O$	215.1060
$C_{12}H_{15}N_4$	215.1298
$C_{12}H_{23}O_3$	215.1648
$C_{12}H_{25}NO_2$	215.1886
$C_{12}H_{27}N_2O$	215.2125
$C_{12}H_{29}N_3$	215.2363
$C_{13}H_{11}O_3$	215.0708

APPENDIX A *(Continued)*

	FM
$C_{13}H_{13}NO_2$	215.0947
$C_{13}H_{15}N_2O$	215.1185
$C_{13}H_{17}N_3$	215.1424
$C_{14}H_{15}O_2$	215.1072
$C_{14}H_{17}NO$	215.1311
$C_{14}H_{19}N_2$	215.1549
$C_{15}H_{19}O$	215.1436
$C_{15}H_{21}N$	215.1675
$C_{16}H_{23}$	215.1801
216	
$C_9H_{16}N_2O_4$	216.1111
$C_9H_{18}N_3O_3$	216.1349
$C_9H_{20}N_4O_2$	216.1588
$C_{10}H_8N_4O_2$	216.0648
$C_{10}H_{18}NO_4$	216.1236
$C_{10}H_{20}N_2O_3$	216.1475
$C_{10}H_{22}N_3O_2$	216.1713
$C_{10}H_{24}N_4O$	216.1952
$C_{11}H_8N_2O_3$	216.0535
$C_{11}H_{10}N_3O_2$	216.0774
$C_{11}H_{12}N_4O$	216.1012
$C_{11}H_{20}O_4$	216.1362
$C_{11}H_{22}NO_3$	216.1600
$C_{11}H_{24}N_2O_2$	216.1839
$C_{11}H_{26}N_3O$	216.2077
$C_{11}H_{28}N_4$	216.2316
$C_{12}H_8O_4$	216.0422
$C_{12}H_{10}NO_3$	216.0661
$C_{12}H_{12}N_2O_2$	216.0899
$C_{12}H_{14}N_3O$	216.1138
$C_{12}H_{16}N_4$	216.1377
$C_{12}H_{24}O_3$	216.1726
$C_{12}H_{26}NO_2$	216.1965
$C_{12}H_{28}N_2O$	216.2203
$C_{13}H_{12}O_3$	216.0786
$C_{13}H_{14}NO_2$	216.1025
$C_{13}H_{16}N_2O$	216.1264
$C_{13}H_{18}N_3$	216.1502
$C_{13}H_{28}O_2$	216.2090
$C_{14}H_{16}O_2$	216.1151
$C_{14}H_{18}NO$	216.1389
$C_{14}H_{20}N_2$	216.1628
$C_{15}H_{20}O$	216.1515
$C_{15}H_{22}N$	216.1753
$C_{16}H_{10}N$	216.0814
$C_{16}H_{24}$	216.1879
$C_{17}H_{12}$	216.0939
217	
$C_9H_{17}N_2O_4$	217.1189
$C_9H_{19}N_3O_3$	217.1427
$C_9H_{21}N_4O_2$	217.1666
$C_{10}H_7N_3O_3$	217.0488
$C_{10}H_9N_4O_2$	217.0726
$C_{10}H_{19}NO_4$	217.1315
$C_{10}H_{21}N_2O_3$	217.1553
$C_{10}H_{23}N_3O_2$	217.1791
$C_{10}H_{25}N_4O$	217.2030
$C_{11}H_7NO_4$	217.0375
$C_{11}H_9N_2O_3$	217.0614
$C_{11}H_{11}N_3O_2$	217.0852
$C_{11}H_{13}N_4O$	217.1091
$C_{11}H_{21}O_4$	217.1440
$C_{11}H_{23}NO_3$	217.1679
$C_{11}H_{25}N_2O_2$	217.1917
$C_{11}H_{27}N_3O$	217.2156
$C_{12}H_9O_4$	217.0501
$C_{12}H_{11}NO_3$	217.0739
$C_{12}H_{13}N_2O_2$	217.0978
$C_{12}H_{15}N_3O$	217.1216
$C_{12}H_{17}N_4$	217.1455
$C_{12}H_{25}O_3$	217.1804
$C_{12}H_{27}NO_2$	217.2043
$C_{13}H_{13}O_3$	217.0865
$C_{13}H_{15}NO_2$	217.1103
$C_{13}H_{17}N_2O$	217.1342
$C_{13}H_{19}N_3$	217.1580
$C_{14}H_{17}O_2$	217.1229
$C_{14}H_{19}NO$	217.1467
$C_{14}H_{21}N_2$	217.1706
$C_{15}H_9N_2$	217.0767
$C_{15}H_{21}O$	217.1593
$C_{15}H_{23}N$	217.1832
$C_{16}H_{11}N$	217.0892
$C_{16}H_{25}$	217.1957
$C_{17}H_{13}$	217.1018
218	
$C_9H_{18}N_2O_4$	218.1267
$C_9H_{20}N_3O_3$	218.1506
$C_9H_{22}N_4O_2$	218.1744
$C_{10}H_8N_3O_3$	218.0566
$C_{10}H_{10}N_4O_2$	218.0805
$C_{10}H_{20}NO_4$	218.1393
$C_{10}H_{22}N_2O_3$	218.1631
$C_{10}H_{24}N_3O_2$	218.1870
$C_{10}H_{26}N_4O$	218.2108
$C_{11}H_8NO_4$	218.0453
$C_{11}H_{10}N_2O_3$	218.0692
$C_{11}H_{12}N_3O_2$	218.0930
$C_{11}H_{14}N_4O$	218.1169
$C_{11}H_{22}O_4$	218.1518
$C_{11}H_{24}NO_3$	218.1757
$C_{11}H_{26}N_2O_2$	218.1996
$C_{12}H_{10}O_4$	218.0579
$C_{12}H_{12}NO_3$	218.0817
$C_{12}H_{14}N_2O_2$	218.1056
$C_{12}H_{16}N_3O$	218.1295
$C_{12}H_{18}N_4$	218.1533
$C_{12}H_{26}O_3$	218.1883
$C_{13}H_{14}O_3$	218.0943
$C_{13}H_{16}NO_2$	218.1182
$C_{13}H_{18}N_2O$	218.1420
$C_{13}H_{20}N_3$	218.1659
$C_{14}H_{18}O_2$	218.1307
$C_{14}H_{20}NO$	218.1546
$C_{14}H_{22}N_2$	218.1784
$C_{15}H_{10}N_2$	218.0845
$C_{15}H_{22}O$	218.1671
$C_{15}H_{24}N$	218.1910
$C_{16}H_{10}O$	218.0732
$C_{16}H_{12}N$	218.0970
$C_{16}H_{26}$	218.2036
$C_{17}H_{14}$	218.1096
219	
$C_9H_{19}N_2O_4$	219.1345
$C_9H_{21}N_3O_3$	219.1584
$C_9H_{23}N_4O_2$	219.1822
$C_{10}H_7N_2O_4$	219.0406
$C_{10}H_9N_3O_3$	219.0644
$C_{10}H_{11}N_4O_2$	219.0883
$C_{10}H_{21}NO_4$	219.1471
$C_{10}H_{23}N_2O_3$	219.1710
$C_{10}H_{25}N_3O_2$	219.1948
$C_{11}H_9NO_4$	219.0532
$C_{11}H_{11}N_2O_3$	219.0770
$C_{11}H_{13}N_3O_2$	219.1009
$C_{11}H_{15}N_4O$	219.1247
$C_{11}H_{23}O_4$	219.1597
$C_{11}H_{25}NO_3$	219.1835
$C_{12}H_{11}O_4$	219.0657
$C_{12}H_{13}NO_3$	219.0896
$C_{12}H_{15}N_2O_2$	219.1134
$C_{12}H_{17}N_3O$	219.1373
$C_{12}H_{19}N_4$	219.1611
$C_{13}H_{15}O_3$	219.1021
$C_{13}H_{17}NO_2$	219.1260
$C_{13}H_{19}N_2O$	219.1498
$C_{13}H_{21}N_3$	219.1737
$C_{14}H_9N_3$	219.0798
$C_{14}H_{19}O_2$	219.1385
$C_{14}H_{21}NO$	219.1624
$C_{14}H_{23}N_2$	219.1863
$C_{15}H_9NO$	219.0684
$C_{15}H_{11}N_2$	219.0923
$C_{15}H_{23}O$	219.1750
$C_{15}H_{25}N$	219.1988
$C_{16}H_{11}O$	219.0810
$C_{16}H_{13}N$	219.1049
$C_{16}H_{27}$	219.2114
$C_{17}H_{15}$	219.1174
220	
$C_9H_{20}N_2O_4$	220.1424
$C_9H_{22}N_3O_3$	220.1662
$C_9H_{24}N_4O_2$	220.1901
$C_{10}H_8N_2O_4$	220.0484
$C_{10}H_{10}N_3O_3$	220.0723
$C_{10}H_{12}N_4O_2$	220.0961
$C_{10}H_{22}NO_4$	220.1549
$C_{10}H_{24}N_2O_3$	220.1788
$C_{11}H_{10}NO_4$	220.0610
$C_{11}H_{12}N_2O_3$	220.0848
$C_{11}H_{14}N_3O_2$	220.1087
$C_{11}H_{16}N_4O$	220.1325
$C_{12}H_{12}O_4$	220.0735
$C_{12}H_{14}NO_3$	220.0974
$C_{12}H_{16}N_2O_2$	220.1213
$C_{12}H_{18}N_3O$	220.1451
$C_{12}H_{20}N_4$	220.1690
$C_{13}H_8N_4$	220.0750
$C_{13}H_{16}O_3$	220.1100
$C_{13}H_{18}NO_2$	220.1338
$C_{13}H_{20}N_2O$	220.1577
$C_{13}H_{22}N_3$	220.1815
$C_{14}H_{10}N_3$	220.0876
$C_{14}H_{20}O_2$	220.1464
$C_{14}H_{22}NO$	220.1702
$C_{14}H_{24}N_2$	220.1941
$C_{15}H_{10}NO$	220.0763
$C_{15}H_{12}N_2$	220.1001
$C_{15}H_{24}O$	220.1828
$C_{15}H_{26}N$	220.2067
$C_{16}H_{12}O$	220.0888
$C_{16}H_{14}N$	220.1127
$C_{16}H_{28}$	220.2192
$C_{17}H_{16}$	220.1253
221	
$C_9H_{21}N_2O_4$	221.1502
$C_9H_{23}N_3O_3$	221.1741
$C_{10}H_9N_2O_4$	221.0563
$C_{10}H_{11}N_3O_3$	221.0801
$C_{10}H_{13}N_4O_2$	221.1040
$C_{10}H_{23}NO_4$	221.1628
$C_{11}H_{11}NO_4$	221.0688
$C_{11}H_{13}N_2O_3$	221.0927
$C_{11}H_{15}N_3O_2$	221.1165
$C_{11}H_{17}N_4O$	221.1404
$C_{12}H_{13}O_4$	221.0814
$C_{12}H_{15}NO_3$	221.1052
$C_{12}H_{17}N_2O_2$	221.1291
$C_{12}H_{19}N_3O$	221.1529
$C_{12}H_{21}N_4$	221.1768
$C_{13}H_9N_4$	221.0829
$C_{13}H_{17}O_3$	221.1178
$C_{13}H_{19}NO_2$	221.1416
$C_{13}H_{21}N_2O$	221.1655
$C_{13}H_{23}N_3$	221.1894
$C_{14}H_9N_2O$	221.0715
$C_{14}H_{11}N_3$	221.0954
$C_{14}H_{21}O_2$	221.1542
$C_{14}H_{23}NO$	221.1781
$C_{14}H_{25}N_2$	221.2019
$C_{15}H_9O_2$	221.0603
$C_{15}H_{11}NO$	221.0841
$C_{15}H_{13}N_2$	221.1080
$C_{15}H_{25}O$	221.1906
$C_{15}H_{27}N$	221.2145
$C_{16}H_{13}O$	221.0967
$C_{16}H_{15}N$	221.1205
$C_{16}H_{29}$	221.2270

APPENDIX A *(Continued)*

	FM
$C_{17}H_{17}$	221.1331
222	
$C_{9}H_{22}N_{2}O_{4}$	222.1580
$C_{10}H_{10}N_{2}O_{4}$	222.0641
$C_{10}H_{12}N_{3}O_{3}$	222.0879
$C_{10}H_{14}N_{4}O_{2}$	222.1118
$C_{11}H_{12}NO_{4}$	222.0766
$C_{11}H_{14}N_{2}O_{3}$	222.1005
$C_{11}H_{16}N_{3}O_{2}$	222.1244
$C_{11}H_{18}N_{4}O$	222.1482
$C_{11}N_{3}O_{3}$	221.9940
$C_{12}H_{14}O_{4}$	222.0892
$C_{12}H_{16}NO_{3}$	222.1131
$C_{12}H_{18}N_{2}O_{2}$	222.1369
$C_{12}H_{20}N_{3}O$	222.1608
$C_{12}H_{22}N_{4}$	222.1846
$C_{13}H_{8}N_{3}O$	222.0668
$C_{13}H_{10}N_{4}$	222.0907
$C_{13}H_{18}O_{3}$	222.1256
$C_{13}H_{20}NO_{2}$	222.1495
$C_{13}H_{22}N_{2}O$	222.1733
$C_{13}H_{24}N_{3}$	222.1972
$C_{14}H_{10}N_{2}O$	222.0794
$C_{14}H_{12}N_{3}$	222.1032
$C_{14}H_{22}O_{2}$	222.1620
$C_{14}H_{24}NO$	222.1859
$C_{14}H_{26}N_{2}$	222.2098
$C_{15}H_{10}O_{2}$	222.0681
$C_{15}H_{12}NO$	222.0919
$C_{15}H_{14}N_{2}$	222.1158
$C_{15}H_{26}O$	222.1985
$C_{15}H_{28}N$	222.2223
$C_{16}H_{14}O$	222.1045
$C_{16}H_{16}N$	222.1284
$C_{16}H_{30}$	222.2349
$C_{16}NO$	221.9980
$C_{17}H_{18}$	222.1409
223	
$C_{10}H_{11}N_{2}O_{4}$	223.0719
$C_{10}H_{13}N_{3}O_{3}$	223.0958
$C_{10}H_{15}N_{4}O_{2}$	223.1196
$C_{11}H_{13}NO_{4}$	223.0845
$C_{11}H_{15}N_{2}O_{3}$	223.1083
$C_{11}H_{17}N_{3}O_{2}$	223.1322
$C_{11}H_{19}N_{4}O$	223.1560
$C_{12}H_{7}N_{4}O$	223.0621
$C_{12}H_{15}O_{4}$	223.0970
$C_{12}H_{17}NO_{3}$	223.1209
$C_{12}H_{19}N_{2}O_{2}$	223.1447
$C_{12}H_{21}N_{3}O$	223.1686
$C_{12}H_{23}N_{4}$	223.1925
$C_{13}H_{9}N_{3}O$	223.0746
$C_{13}H_{11}N_{4}$	223.0985
$C_{13}H_{19}O_{3}$	223.1334
$C_{13}H_{21}NO_{2}$	223.1573
$C_{13}H_{23}N_{2}O$	223.1811
$C_{13}H_{25}N_{3}$	223.2050
$C_{14}H_{9}NO_{2}$	223.0634
$C_{14}H_{11}N_{2}O$	223.0872
$C_{14}H_{13}N_{3}$	223.1111
$C_{14}H_{23}O_{2}$	223.1699
$C_{14}H_{25}NO$	223.1937
$C_{14}H_{27}N_{2}$	223.2176
$C_{15}H_{11}O_{2}$	223.0759
$C_{15}H_{13}NO$	223.0998
$C_{15}H_{27}O$	223.2063
$C_{15}H_{29}N$	223.2301
$C_{16}H_{15}O$	223.1123
$C_{16}H_{17}N$	223.1362
$C_{16}H_{31}$	223.2427
$C_{17}H_{19}$	223.1488
224	
$C_{10}H_{12}N_{2}O_{4}$	224.0797
$C_{10}H_{14}N_{3}O_{3}$	224.1036
$C_{10}H_{16}N_{4}O_{2}$	224.1275
$C_{11}H_{14}NO_{4}$	224.0923
$C_{11}H_{16}N_{2}O_{3}$	224.1162
$C_{11}H_{18}N_{3}O_{2}$	224.1400
$C_{11}H_{20}N_{4}O$	224.1639
$C_{12}H_{8}N_{4}O$	224.0699
$C_{12}H_{16}O_{4}$	224.1049
$C_{12}H_{18}NO_{3}$	224.1287
$C_{12}H_{20}N_{2}O_{2}$	224.1526
$C_{12}H_{22}N_{3}O$	224.1764
$C_{12}H_{24}N_{4}$	224.2003
$C_{13}H_{8}N_{2}O_{2}$	224.0586
$C_{13}H_{10}N_{3}O$	224.0825
$C_{13}H_{12}N_{4}$	224.1063
$C_{13}H_{20}O_{3}$	224.1413
$C_{13}H_{22}NO_{2}$	224.1651
$C_{13}H_{24}N_{2}O$	224.1890
$C_{13}H_{26}N_{3}$	224.2129
$C_{14}H_{10}NO_{2}$	224.0712
$C_{14}H_{12}N_{2}O$	224.0950
$C_{14}H_{14}N_{3}$	224.1189
$C_{14}H_{24}O_{2}$	224.1777
$C_{14}H_{26}NO$	224.2015
$C_{14}H_{28}N_{2}$	224.2254
$C_{15}H_{12}O_{2}$	224.0837
$C_{15}H_{14}NO$	224.1076
$C_{15}H_{16}N_{2}$	224.1315
$C_{15}H_{28}O$	224.2141
$C_{15}H_{30}N$	224.2380
$C_{16}H_{16}O$	224.1202
$C_{16}H_{18}N$	224.1440
$C_{16}H_{32}$	224.2505
$C_{17}H_{20}$	224.1566
225	
$C_{10}H_{13}N_{2}O_{4}$	225.0876
$C_{10}H_{15}N_{3}O_{3}$	225.1114
$C_{10}H_{17}N_{4}O_{2}$	225.1353
$C_{11}H_{15}NO_{4}$	225.1001
$C_{11}H_{17}N_{2}O_{3}$	225.1240
$C_{11}H_{19}N_{3}O_{2}$	225.1478
$C_{11}H_{21}N_{4}O$	225.1717
$C_{12}H_{9}N_{4}O$	225.0777
$C_{12}H_{17}O_{4}$	225.1127
$C_{12}H_{19}NO_{3}$	225.1365
$C_{12}H_{21}N_{2}O_{2}$	225.1604
$C_{12}H_{23}N_{3}O$	225.1842
$C_{12}H_{25}N_{4}$	225.2081
$C_{13}H_{9}N_{2}O_{2}$	225.0664
$C_{13}H_{11}N_{3}O$	225.0903
$C_{13}H_{13}N_{4}$	225.1142
$C_{13}H_{21}O_{3}$	225.1491
$C_{13}H_{23}NO_{2}$	225.1730
$C_{13}H_{25}N_{2}O$	225.1968
$C_{13}H_{27}N_{3}$	225.2207
$C_{14}H_{9}O_{3}$	225.0552
$C_{14}H_{11}NO_{2}$	225.0790
$C_{14}H_{13}N_{2}O$	225.1029
$C_{14}H_{15}N_{3}$	225.1267
$C_{14}H_{25}O_{2}$	225.1855
$C_{14}H_{27}NO$	225.2094
$C_{14}H_{29}N_{2}$	225.2332
$C_{15}H_{13}O_{2}$	225.0916
$C_{15}H_{15}NO$	225.1154
$C_{15}H_{17}N_{2}$	225.1393
$C_{15}H_{29}O$	225.2219
$C_{15}H_{31}N$	225.2458
$C_{16}H_{17}O$	225.1280
$C_{16}H_{19}N$	225.1519
$C_{16}H_{33}$	225.2584
$C_{17}H_{21}$	225.1644
226	
$C_{10}H_{14}N_{2}O_{4}$	226.0954
$C_{10}H_{16}N_{3}O_{3}$	226.1193
$C_{10}H_{18}N_{4}O_{2}$	226.1431
$C_{11}H_{16}NO_{4}$	226.1080
$C_{11}H_{18}N_{2}O_{3}$	226.1318
$C_{11}H_{20}N_{3}O_{2}$	226.1557
$C_{11}H_{22}N_{4}O$	226.1795
$C_{12}H_{8}N_{3}O_{2}$	226.0617
$C_{12}H_{10}N_{4}O$	226.0856
$C_{12}H_{18}O_{4}$	226.1205
$C_{12}H_{20}NO_{3}$	226.1444
$C_{12}H_{22}N_{2}O_{2}$	226.1682
$C_{12}H_{24}N_{3}O$	226.1929
$C_{12}H_{26}N_{4}$	226.2160
$C_{13}H_{10}N_{2}O_{2}$	226.0743
$C_{13}H_{12}N_{3}O$	226.0981
$C_{13}H_{14}N_{4}$	226.1220
$C_{13}H_{22}O_{3}$	226.1569
$C_{13}H_{24}NO_{2}$	226.1808
$C_{13}H_{26}N_{2}O$	226.2046
$C_{13}H_{28}N_{3}$	226.2285
$C_{14}H_{10}O_{3}$	226.0630
$C_{14}H_{12}NO_{2}$	226.0868
$C_{14}H_{14}N_{2}O$	226.1107
$C_{14}H_{16}N_{3}$	226.1346
$C_{14}H_{26}O_{2}$	226.1934
$C_{14}H_{28}NO$	226.2172
$C_{14}H_{30}N_{2}$	226.2411
$C_{15}H_{14}O_{2}$	226.0994
$C_{15}H_{16}NO$	226.1233
$C_{15}H_{18}N_{2}$	226.1471
$C_{15}H_{30}O$	226.2298
$C_{15}H_{32}N$	226.2536
$C_{16}H_{18}O$	226.1358
$C_{16}H_{20}N$	226.1597
$C_{16}H_{34}$	226.2662
$C_{17}H_{22}$	226.1722
227	
$C_{10}H_{15}N_{2}O_{4}$	227.1032
$C_{10}H_{17}N_{3}O_{3}$	227.1271
$C_{10}H_{19}N_{4}O_{2}$	227.1509
$C_{11}H_{17}NO_{4}$	227.1158
$C_{11}H_{21}N_{3}O_{2}$	227.1635
$C_{11}H_{23}N_{4}O$	227.1873
$C_{12}H_{7}N_{2}O_{3}$	227.0457
$C_{12}H_{9}N_{3}O_{2}$	227.0695
$C_{12}H_{11}N_{4}O$	227.0934
$C_{12}H_{19}O_{4}$	227.1284
$C_{12}H_{21}NO_{3}$	227.1522
$C_{12}H_{23}N_{2}O_{2}$	227.1761
$C_{12}H_{25}N_{3}O$	227.1999
$C_{12}H_{27}N_{4}$	227.2238
$C_{13}H_{9}NO_{3}$	227.0583
$C_{13}H_{11}N_{2}O_{2}$	227.0821
$C_{13}H_{13}N_{3}O$	227.1060
$C_{13}H_{15}N_{4}$	227.1298
$C_{13}H_{25}NO_{2}$	227.1886
$C_{13}H_{27}N_{2}O$	227.2125
$C_{13}H_{29}N_{3}$	227.2363
$C_{14}H_{11}O_{3}$	227.0708
$C_{14}H_{13}NO_{2}$	227.0947
$C_{14}H_{15}N_{2}O$	227.1185
$C_{14}H_{17}N_{3}$	227.1424
$C_{14}H_{27}O_{2}$	227.2012
$C_{14}H_{29}NO$	227.2250
$C_{15}H_{15}O_{2}$	227.1072
$C_{15}H_{17}NO$	227.1311
$C_{15}H_{19}N_{2}$	227.1549
$C_{15}H_{31}O$	227.2376
$C_{15}H_{33}N$	227.2615
$C_{16}H_{19}O$	227.1436
$C_{16}H_{21}N$	227.1675
$C_{17}H_{23}$	227.1801
228	
$C_{10}H_{16}N_{2}O_{2}$	228.1111
$C_{10}H_{18}N_{3}O_{3}$	228.1349
$C_{10}H_{20}N_{4}O_{2}$	228.1588
$C_{11}H_{8}N_{4}O_{2}$	228.0648
$C_{11}H_{18}NO_{4}$	228.1236
$C_{11}H_{20}N_{2}O_{3}$	228.1475
$C_{11}H_{22}N_{3}O_{2}$	228.1713
$C_{11}H_{24}N_{4}O$	228.1952
$C_{12}H_{8}N_{2}O_{3}$	228.0535

APPENDIX A *(Continued)*

	FM
$C_{12}H_{12}N_4O$	228.1012
$C_{12}H_{20}O_4$	228.1362
$C_{12}H_{22}NO_3$	228.1600
$C_{12}H_{24}N_2O_2$	228.1839
$C_{12}H_{26}N_3O$	228.2077
$C_{12}H_{28}N_4$	228.2316
$C_{13}H_8O_4$	228.0422
$C_{13}H_{10}NO_3$	228.0661
$C_{13}H_{12}N_2O_2$	228.0899
$C_{13}H_{14}N_3O$	228.1138
$C_{13}H_{24}O_3$	228.1726
$C_{13}H_{26}NO_2$	228.1965
$C_{13}H_{28}N_2O$	228.2203
$C_{13}H_{30}N_3$	228.2442
$C_{14}H_{12}O_3$	228.0786
$C_{14}H_{14}NO_2$	228.1025
$C_{14}H_{16}N_2O$	228.1264
$C_{14}H_{18}N_3$	228.1502
$C_{14}H_{28}O_2$	228.2090
$C_{14}H_{30}NO$	228.2329
$C_{14}H_{32}N_2$	228.2567
$C_{15}H_{16}O_2$	228.1151
$C_{15}H_{18}NO$	228.1389
$C_{15}H_{20}N_2$	228.1628
$C_{15}H_{32}O$	228.2454
$C_{16}H_{20}O$	228.1515
$C_{16}H_{22}N$	228.1753
$C_{17}H_{10}N$	228.0814
$C_{17}H_{24}$	228.1879
$C_{18}H_{12}$	228.0939
229	
$C_{10}H_{17}N_2O_4$	229.1189
$C_{10}H_{19}N_3O_3$	229.1427
$C_{10}H_{21}N_4O_2$	229.1666
$C_{11}H_7N_3O_3$	229.0488
$C_{11}H_9N_4O_2$	229.0726
$C_{11}H_{19}NO_4$	229.1315
$C_{11}H_{21}N_2O_3$	229.1553
$C_{11}H_{23}N_3O_2$	229.1791
$C_{11}H_{25}N_4O$	229.2030
$C_{12}H_9N_2O_3$	229.0614
$C_{12}H_{11}N_3O_2$	229.0852
$C_{12}H_{13}N_4O$	229.1091
$C_{12}H_{21}O_4$	229.1440
$C_{12}H_{23}NO_3$	229.1679
$C_{12}H_{25}N_2O_2$	229.1917
$C_{12}H_{27}N_3O$	229.2156
$C_{12}H_{29}N_4$	229.2394
$C_{13}H_9O_4$	229.0501
$C_{13}H_{11}NO_3$	229.0739
$C_{13}H_{13}N_2O_2$	229.0978
$C_{13}H_{15}N_3O$	229.1216
$C_{13}H_{17}N_4$	229.1455
$C_{13}H_{25}O_3$	229.1804
$C_{13}H_{27}NO_2$	229.2043
$C_{13}H_{29}N_2O$	229.2281
$C_{13}H_{31}N_3$	229.2520
$C_{14}H_{13}O_3$	229.0865
$C_{14}H_{15}NO_2$	229.1103
$C_{14}H_{17}N_2O$	229.1342
$C_{14}H_{19}N_3$	229.1580
$C_{14}H_{29}O_2$	229.2168
$C_{14}H_{31}NO$	229.2407
$C_{15}H_{17}O_2$	229.1229
$C_{15}H_{19}NO$	229.1467
$C_{15}H_{21}N_2$	229.1706
$C_{16}H_{21}O$	229.1593
$C_{16}H_{23}N$	229.1832
$C_{17}H_9O$	229.0653
$C_{17}H_{11}N$	229.0892
$C_{18}H_{13}$	229.1018
230	
$C_{10}H_{18}N_2O_4$	230.1267
$C_{10}H_{20}N_3O_3$	230.1506
$C_{10}H_{22}N_4O_2$	230.1744
$C_{11}H_8N_3O_3$	230.0566
$C_{11}H_{10}N_4O_2$	230.0805
$C_{11}H_{20}NO_4$	230.1393
$C_{11}H_{22}N_2O_3$	230.1631
$C_{11}H_{24}N_3O_2$	230.1870
$C_{11}H_{26}N_4O$	230.2108
$C_{12}H_8NO_4$	230.0453
$C_{12}H_{10}N_2O_3$	230.0692
$C_{12}H_{12}N_3O_2$	230.0930
$C_{12}H_{14}N_4O$	230.1169
$C_{12}H_{22}O_4$	230.1518
$C_{12}H_{24}NO_3$	230.1757
$C_{12}H_{26}N_2O_2$	230.1996
$C_{12}H_{28}N_3O$	230.2234
$C_{12}H_{30}N_4$	230.2473
$C_{13}H_{10}O_4$	230.0579
$C_{13}H_{12}NO_3$	230.0817
$C_{13}H_{14}N_2O_2$	230.1056
$C_{13}H_{16}N_3O$	230.1295
$C_{13}H_{18}N_4$	230.1533
$C_{13}H_{26}O_3$	230.1883
$C_{13}H_{28}NO_2$	230.2121
$C_{13}H_{30}N_2O$	230.2360
$C_{14}H_{14}O_3$	230.0943
$C_{14}H_{16}NO_2$	230.1182
$C_{14}H_{18}N_2O$	230.1420
$C_{14}H_{20}N_3$	230.1659
$C_{14}H_{30}O_2$	230.2247
$C_{15}H_{18}O_2$	230.1307
$C_{15}H_{20}NO$	230.1546
$C_{15}H_{22}N_2$	230.1784
$C_{16}H_{10}N_2$	230.0845
$C_{16}H_{22}O$	230.1671
$C_{16}H_{24}N$	230.1910
$C_{17}H_{10}O$	230.0732
$C_{17}H_{12}N$	230.0970
$C_{17}H_{26}$	230.2036
$C_{18}H_{14}$	230.1096
231	
$C_{10}H_{19}N_2O_4$	231.1345
$C_{10}H_{21}N_3O_3$	231.1584
$C_{10}H_{23}N_4O_2$	231.1822
$C_{11}H_7N_2O_4$	231.0406
$C_{11}H_9N_3O_3$	231.0644
$C_{11}H_{11}N_4O_2$	231.0883
$C_{11}H_{21}NO_4$	231.1471
$C_{11}H_{23}N_2O_3$	231.1710
$C_{11}H_{25}N_3O_2$	231.1948
$C_{11}H_{27}N_4O$	231.2187
$C_{12}H_9NO_4$	231.0532
$C_{12}H_{11}N_2O_3$	231.0770
$C_{12}H_{13}N_3O_2$	231.1009
$C_{12}H_{15}N_4O$	231.1247
$C_{12}H_{23}O_4$	231.1597
$C_{12}H_{25}NO_3$	231.1835
$C_{12}H_{27}N_2O_2$	231.2074
$C_{12}H_{29}N_3O$	231.2312
$C_{13}H_{11}O_4$	231.0657
$C_{13}H_{13}NO_3$	231.0896
$C_{13}H_{15}N_2O_2$	231.1134
$C_{13}H_{17}N_3O$	231.1373
$C_{13}H_{19}N_4$	231.1611
$C_{14}H_{15}O_3$	231.1021
$C_{14}H_{17}NO_2$	231.1260
$C_{14}H_{19}N_2O$	231.1498
$C_{14}H_{21}N_3$	231.1737
$C_{15}H_9N_3$	231.0798
$C_{15}H_{19}O_2$	231.1385
$C_{15}H_{21}NO$	231.1624
$C_{15}H_{23}N_2$	231.1863
$C_{16}H_9NO$	231.0684
$C_{16}H_{11}N_2$	231.0923
$C_{16}H_{23}O$	231.1750
$C_{17}H_{11}O$	231.0810
$C_{17}H_{13}N$	231.1049
$C_{17}H_{27}$	231.2114
$C_{18}H_{15}$	231.1174
232	
$C_{10}H_{20}N_2O_4$	232.1424
$C_{10}H_{22}N_3O_3$	232.1662
$C_{10}H_{24}N_4O_2$	232.1901
$C_{11}H_8N_2O_4$	232.0484
$C_{11}H_{10}N_3O_3$	232.0723
$C_{11}H_{12}N_4O_2$	232.0961
$C_{11}H_{22}NO_4$	232.1549
$C_{11}H_{24}N_2O_3$	232.1788
$C_{11}H_{26}N_3O_2$	232.2026
$C_{11}H_{28}N_4O$	232.2265
$C_{12}H_{10}NO_4$	232.0610
$C_{12}H_{12}N_2O_3$	232.0848
$C_{12}H_{14}N_3O_2$	232.1087
$C_{12}H_{16}N_4O$	232.1325
$C_{12}H_{24}O_4$	232.1675
$C_{12}H_{26}NO_3$	232.1914
$C_{12}H_{28}N_2O_2$	232.2152
$C_{13}H_{12}O_4$	232.0735
$C_{13}H_{14}NO_3$	232.0974
$C_{13}H_{16}N_2O_2$	232.1213
$C_{13}H_{18}N_3O$	232.1451
$C_{13}H_{20}N_4$	232.1690
$C_{13}H_{28}O_3$	232.2039
$C_{14}H_{16}O_3$	232.1100
$C_{14}H_{18}NO_2$	232.1338
$C_{14}H_{20}N_2O$	232.1577
$C_{14}H_{22}N_3$	232.1815
$C_{15}H_{10}N_3$	232.0876
$C_{15}H_{20}O_2$	232.1464
$C_{15}H_{22}NO$	232.1702
$C_{15}H_{24}N_2$	232.1941
$C_{16}H_{10}NO$	232.0768
$C_{16}H_{12}N_2$	232.1001
$C_{16}H_{24}O$	232.1828
$C_{16}H_{26}N$	232.2067
$C_{17}H_{12}O$	232.0888
$C_{17}H_{14}N$	232.1127
$C_{17}H_{28}$	232.2192
$C_{18}H_{16}$	232.1253
233	
$C_{10}H_{23}N_3O_3$	233.1741
$C_{10}H_{25}N_4O_2$	233.1979
$C_{11}H_9N_2O_4$	233.0563
$C_{11}H_{11}N_3O_3$	233.0801
$C_{11}H_{23}NO_4$	233.1628
$C_{11}H_{25}N_2O_3$	233.1866
$C_{11}H_{27}N_3O_2$	233.2105
$C_{12}H_{11}NO_4$	233.0688
$C_{12}H_{13}N_2O_3$	233.0927
$C_{12}H_{15}N_3O_2$	233.1165
$C_{12}H_{17}N_4O$	233.1404
$C_{12}H_{25}O_4$	233.1753
$C_{12}H_{27}NO_3$	233.1992
$C_{13}H_{13}O_4$	233.0814
$C_{13}H_{15}NO_3$	233.1052
$C_{13}H_{17}N_2O_2$	233.1291
$C_{13}H_{19}N_3O$	233.1529
$C_{13}H_{21}N_4$	233.1768
$C_{14}H_9N_4$	233.0829
$C_{14}H_{17}O_3$	233.1178
$C_{14}H_{19}NO_2$	233.1416
$C_{14}H_{21}N_2O$	233.1655
$C_{15}H_9N_2O$	233.0715
$C_{15}H_{11}N_3$	233.0954
$C_{15}H_{21}O_2$	233.1542
$C_{15}H_{23}NO$	233.1781
$C_{15}H_{25}N_2$	233.2019
$C_{16}H_9O_2$	233.0603
$C_{16}H_{11}NO$	233.0841
$C_{16}H_{13}N_2$	233.1080
$C_{16}H_{25}O$	233.1906
$C_{16}H_{27}N$	233.2145
$C_{17}H_{13}O$	233.0967
$C_{17}H_{15}N$	233.1205
$C_{17}H_{29}$	233.2270
$C_{18}H_{17}$	233.1331
234	
$C_{10}H_{22}N_2O_4$	234.1580
$C_{10}H_{24}N_3O_3$	234.1819

APPENDIX A *(Continued)*

	FM
$C_{10}H_{26}N_4O_2$	234.2057
$C_{11}H_{10}N_2O_4$	234.0641
$C_{11}H_{12}N_3O_3$	234.0879
$C_{11}H_{14}N_4O_2$	234.1118
$C_{11}H_{24}NO_4$	234.1706
$C_{11}H_{26}N_2O_3$	234.1945
$C_{12}H_{12}NO_4$	234.0766
$C_{12}H_{14}N_2O_3$	234.1005
$C_{12}H_{16}N_3O_2$	234.1244
$C_{12}H_{18}N_4O$	234.1482
$C_{12}H_{26}O_4$	234.1832
$C_{13}H_{14}O_4$	234.0892
$C_{13}H_{16}NO_3$	234.1131
$C_{13}H_{18}N_2O_2$	234.1369
$C_{13}H_{20}N_3O$	234.1608
$C_{13}H_{22}N_4$	234.1846
$C_{14}H_{10}N_4$	234.0907
$C_{14}H_{18}O_3$	234.1256
$C_{14}H_{20}NO_2$	234.1495
$C_{14}H_{22}N_2O$	234.1733
$C_{14}H_{24}N_3$	234.1972
$C_{15}H_{10}N_2O$	234.0794
$C_{15}H_{12}N_3$	234.1032
$C_{15}H_{22}O_2$	234.1620
$C_{15}H_{24}NO$	234.1859
$C_{15}H_{26}N_2$	234.2098
$C_{16}H_{10}O_2$	234.0681
$C_{16}H_{12}NO$	234.0919
$C_{16}H_{14}N_2$	234.1158
$C_{16}H_{26}O$	234.1985
$C_{16}H_{28}N$	234.2223
$C_{17}H_{16}N$	234.1284
$C_{17}H_{30}$	234.2349
$C_{18}H_{18}$	234.1409
235	
$C_{10}H_{23}N_2O_4$	235.1659
$C_{10}H_{25}N_3O_3$	235.1897
$C_{11}H_{11}N_2O_4$	235.0719
$C_{11}H_{13}N_3O_3$	235.0958
$C_{11}H_{15}N_4O_2$	235.1196
$C_{11}H_{25}NO_4$	235.1784
$C_{12}H_{13}NO_4$	235.0845
$C_{12}H_{15}N_2O_3$	235.1083
$C_{12}H_{17}N_3O_2$	235.1322
$C_{12}H_{19}N_4O$	235.1560
$C_{13}H_{15}O_4$	235.0970
$C_{13}H_{17}NO_3$	235.1209
$C_{13}H_{19}N_2O_2$	235.1447
$C_{13}H_{21}N_3O$	235.1686
$C_{13}H_{23}N_4$	235.1925
$C_{14}H_9N_3O$	235.0746
$C_{14}H_{11}N_4$	235.0985
$C_{14}H_{19}O_3$	235.1334
$C_{14}H_{21}NO_2$	235.1573
$C_{14}H_{23}N_2O$	235.1811
$C_{14}H_{25}N_3$	235.2050
$C_{15}H_9NO_2$	235.0634
$C_{15}H_{11}N_2O$	235.0872
$C_{15}H_{13}N_3$	235.1111
$C_{15}H_{23}O_2$	235.1699
$C_{15}H_{25}NO$	235.1937
$C_{15}H_{27}N_2$	235.2176
$C_{16}H_{11}O_2$	235.0759
$C_{16}H_{13}NO$	235.0998
$C_{16}H_{15}N_2$	235.1236
$C_{16}H_{27}O$	235.2063
$C_{16}H_{29}N$	235.2301
$C_{17}H_{15}O$	235.1123
$C_{17}H_{17}N$	235.1362
$C_{17}H_{31}$	235.2427
$C_{18}H_{19}$	235.1488
236	
$C_{10}H_{24}N_2O_4$	236.1737
$C_{11}H_{12}N_2O_4$	236.0797
$C_{11}H_{14}N_3O_3$	236.1036
$C_{11}H_{16}N_4O_2$	236.1275
$C_{12}H_2N_3O_3$	236.0096
$C_{12}H_4N_4O_2$	236.0335
$C_{12}H_{14}NO_4$	236.0923
$C_{12}H_{16}N_2O_3$	236.1162
$C_{12}H_{18}N_3O_2$	236.1400
$C_{12}H_{20}N_4O$	236.1639
$C_{13}H_8N_4O$	236.0699
$C_{13}H_{16}O_4$	236.1049
$C_{13}H_{18}NO_3$	236.1287
$C_{13}H_{20}N_2O_2$	236.1526
$C_{13}H_{22}N_3O$	236.1764
$C_{13}H_{24}N_4$	236.2003
$C_{14}H_{10}N_3O$	236.0825
$C_{14}H_{12}N_4$	236.1063
$C_{14}H_{20}O_3$	236.1413
$C_{14}H_{22}NO_2$	236.1651
$C_{14}H_{24}N_2O$	236.1890
$C_{14}H_{26}N_3$	236.2129
$C_{15}H_{10}NO_2$	236.0712
$C_{15}H_{12}N_2O$	236.0950
$C_{15}H_{14}N_3$	236.1189
$C_{15}H_{24}O_2$	236.1777
$C_{15}H_{26}NO$	236.2015
$C_{15}H_{28}N_2$	236.2254
$C_{16}H_{12}O_2$	236.0837
$C_{16}H_{14}NO$	236.1076
$C_{16}H_{16}N_2$	236.1315
$C_{16}H_{28}O$	236.2141
$C_{16}H_{30}N$	236.2380
$C_{17}H_{16}O$	236.1202
$C_{17}H_{18}N$	236.1440
$C_{17}H_{32}$	236.2505
$C_{18}H_{20}$	236.1566
237	
$C_{11}H_{13}N_2O_4$	237.0876
$C_{11}H_{15}N_3O_3$	237.1114
$C_{11}H_{17}N_4O_2$	237.1353
$C_{12}H_{15}NO_4$	237.1001
$C_{12}H_{17}N_2O_3$	237.1240
$C_{12}H_{19}N_3O_2$	237.1478
$C_{12}H_{21}N_4O$	237.1717
$C_{13}H_9N_4O$	237.0777
$C_{13}H_{17}O_4$	237.1127
$C_{13}H_{19}NO_3$	237.1365
$C_{13}H_{21}N_2O_2$	237.1604
$C_{13}H_{23}N_3O$	237.1842
$C_{13}H_{25}N$	237.2081
$C_{14}H_9N_2O_2$	237.0664
$C_{14}H_{11}N_3O$	237.0903
$C_{14}H_{13}N_4$	237.1142
$C_{14}H_{21}O_3$	237.1491
$C_{14}H_{23}NO_2$	237.1730
$C_{14}H_{25}N_2O$	237.1968
$C_{14}H_{27}N_3$	237.2207
$C_{15}H_9O_3$	237.0552
$C_{15}H_{11}NO_2$	237.0790
$C_{15}H_{13}N_2O$	237.1029
$C_{15}H_{15}N_3$	237.1267
$C_{15}H_{25}O_2$	237.1855
$C_{15}H_{27}NO$	237.2094
$C_{15}H_{29}N_2$	237.2332
$C_{16}H_{13}O_2$	237.0916
$C_{16}H_{15}NO$	237.1154
$C_{16}H_{17}N_2$	237.1393
$C_{16}H_{29}O$	237.2219
$C_{16}H_{31}N$	237.2458
$C_{17}H_{17}O$	237.1280
$C_{17}H_{19}N$	237.1519
$C_{17}H_{33}$	237.2584
$C_{18}H_{21}$	237.1644
238	
$C_{11}H_{14}N_2O_4$	238.0954
$C_{11}H_{16}N_3O_3$	238.1193
$C_{11}H_{18}N_4O_2$	238.1431
$C_{12}H_{16}NO_4$	238.1080
$C_{12}H_{18}N_2O_3$	238.1318
$C_{12}H_{20}N_3O_2$	238.1557
$C_{12}H_{22}N_4O$	238.1795
$C_{13}H_8N_3O_2$	238.0617
$C_{13}H_{10}N_4O$	238.0856
$C_{13}H_{18}O_4$	238.1205
$C_{13}H_{20}NO_3$	238.1444
$C_{13}H_{22}N_2O_2$	238.1682
$C_{13}H_{24}N_3O$	238.1921
$C_{13}H_{26}N_4$	238.2160
$C_{14}H_{10}N_2O_2$	238.0743
$C_{14}H_{12}N_3O$	238.0981
$C_{14}H_{14}N_4$	238.1220
$C_{14}H_{22}O_3$	238.1569
$C_{14}H_{24}NO_2$	238.1808
$C_{14}H_{26}N_2O$	238.2046
$C_{14}H_{28}N_3$	238.2285
$C_{15}H_{10}O_3$	238.0630
$C_{15}H_{12}NO_2$	238.0868
$C_{15}H_{14}N_2O$	238.1107
$C_{15}H_{16}N_3$	238.1346
$C_{15}H_{26}O_2$	238.1934
$C_{15}H_{28}NO$	238.2172
$C_{15}H_{30}N_2$	238.2411
$C_{16}H_{14}O_2$	238.0994
$C_{16}H_{16}NO$	238.1233
$C_{16}H_{18}N_2$	238.1471
$C_{16}H_{30}O$	238.2298
$C_{16}H_{32}N$	238.2536
$C_{17}H_{18}O$	238.1358
$C_{17}H_{20}N$	238.1597
$C_{17}H_{34}$	238.2662
$C_{18}H_{22}$	238.1722
239	
$C_{11}H_{15}N_2O_4$	239.1032
$C_{11}H_{17}N_3O_3$	239.1271
$C_{11}H_{19}N_4O_2$	239.1509
$C_{12}H_{17}NO_4$	239.1158
$C_{12}H_{19}N_2O_3$	239.1396
$C_{12}H_{21}N_3O_2$	239.1635
$C_{12}H_{23}N_4O$	239.1873
$C_{13}H_9N_3O_2$	239.0695
$C_{13}H_{11}N_4O$	239.0934
$C_{13}H_{19}O_4$	239.1284
$C_{13}H_{21}NO_3$	239.1522
$C_{13}H_{23}N_2O_2$	239.1761
$C_{13}H_{25}N_3O$	239.1999
$C_{13}H_{27}N_4$	239.2238
$C_{14}H_9NO_3$	239.0583
$C_{14}H_{11}N_2O_2$	239.0821
$C_{14}H_{13}N_3O$	239.1060
$C_{14}H_{15}N_4$	239.1298
$C_{14}H_{23}O_3$	239.1648
$C_{14}H_{25}NO_2$	239.1886
$C_{14}H_{27}N_2O$	239.2125
$C_{14}H_{29}N_3$	239.2363
$C_{15}H_{11}O_3$	239.0708
$C_{15}H_{13}NO_2$	239.0947
$C_{15}H_{15}N_2O$	239.1185
$C_{15}H_{17}N_3$	239.1424
$C_{15}H_{27}O_2$	239.2012
$C_{15}H_{29}NO$	239.2250
$C_{15}H_{31}N_2$	239.2489
$C_{16}H_{15}O_2$	239.1072
$C_{16}H_{17}NO$	239.1311
$C_{16}H_{19}N_2$	239.1549
$C_{16}H_{31}O$	239.2376
$C_{16}H_{33}N$	239.2615
$C_{17}H_{19}O$	239.1436
$C_{17}H_{21}N$	239.1675
$C_{17}H_{35}$	239.2740
$C_{18}H_{23}$	239.1801
240	
$C_{11}H_{16}N_2O_4$	240.1111
$C_{11}H_{18}N_3O_3$	240.1349
$C_{11}H_{20}N_4O_2$	240.1588
$C_{12}H_8N_4O_2$	240.0648
$C_{12}H_{18}NO_4$	240.1236
$C_{12}H_{20}N_2O_3$	240.1475
$C_{12}H_{22}N_3O_2$	240.1713
$C_{12}H_{24}N_4O$	240.1952

APPENDIX A *(Continued)*

	FM
$C_{13}H_8N_2O_3$	240.0535
$C_{13}H_{10}N_3O_2$	240.0774
$C_{13}H_{12}N_4O$	240.1012
$C_{13}H_{20}O_4$	240.1362
$C_{13}H_{22}NO_3$	240.1600
$C_{13}H_{24}N_2O_2$	240.1839
$C_{13}H_{28}N_4$	240.2316
$C_{14}H_8O_4$	240.0422
$C_{14}H_{10}NO_3$	240.0661
$C_{14}H_{12}N_2O_2$	240.0899
$C_{14}H_{14}N_3O$	240.1138
$C_{14}H_{16}N_4$	240.1377
$C_{14}H_{24}O_3$	240.1726
$C_{14}H_{26}NO_2$	240.1965
$C_{14}H_{28}N_2O$	240.2203
$C_{14}H_{30}N_3$	240.2442
$C_{15}H_{12}O_3$	240.0786
$C_{15}H_{14}NO_2$	240.1025
$C_{15}H_{16}N_2O$	240.1264
$C_{15}H_{18}N_3$	240.1502
$C_{15}H_{28}O_2$	240.2090
$C_{15}H_{30}NO$	240.2329
$C_{15}H_{32}N_2$	240.2567
$C_{16}H_{16}O_2$	240.1151
$C_{16}H_{20}N_2$	240.1628
$C_{16}H_{18}NO$	240.1389
$C_{16}H_{32}O$	240.2454
$C_{16}H_{34}N$	240.2693
$C_{17}H_{20}O$	240.1515
$C_{17}H_{22}N$	240.1753
$C_{17}H_{36}$	240.2819
$C_{18}H_{24}$	240.1879
241	
$C_{11}H_{17}N_2O_4$	241.1189
$C_{11}H_{19}N_3O_3$	241.1427
$C_{11}H_{21}N_4O_2$	241.1666
$C_{12}H_{19}NO_4$	241.1315
$C_{12}H_{21}N_2O_3$	241.1553
$C_{12}H_{23}N_3O_2$	241.1791
$C_{12}H_{25}N_4O$	241.2030
$C_{13}H_{11}N_3O_2$	241.0852
$C_{13}H_{13}N_4O$	241.1091
$C_{13}H_{21}O_4$	241.1440
$C_{13}H_{25}N_2O_2$	241.1679
$C_{13}H_{25}N_2O_2$	241.1917
$C_{13}H_{27}N_3O$	241.2156
$C_{13}H_{29}N_4$	241.2394
$C_{14}H_{11}NO_3$	241.0739
$C_{14}H_{13}N_2O_2$	241.0978
$C_{14}H_{15}N_3O$	241.1216
$C_{14}H_{17}N_4$	241.1445
$C_{14}H_{25}O_3$	241.1804
$C_{14}H_{27}NO_2$	241.2043
$C_{14}H_{29}N_2O$	241.2281
$C_{14}H_{31}N_3$	241.2520
$C_{15}H_{13}O_3$	241.0865
$C_{15}H_{15}NO_2$	241.1103
$C_{15}H_{17}N_2O$	241.1342
$C_{15}H_{19}N_3$	241.1580
$C_{15}H_{29}O_2$	241.2168
$C_{15}H_{31}NO$	241.2407
$C_{15}H_{33}N_2$	241.2646
$C_{16}H_{17}O_2$	241.1229
$C_{16}H_{19}NO$	241.1467
$C_{16}H_{21}N_2$	241.1706
$C_{16}H_{33}O$	241.2533
$C_{16}H_{35}N$	241.2771
$C_{17}H_{21}O$	241.1593
$C_{17}H_{23}N$	241.1832
$C_{18}H_{25}$	241.1957
242	
$C_{11}H_{18}N_2O_4$	242.1267
$C_{11}H_{20}N_3O_3$	242.1506
$C_{11}H_{22}N_4O_2$	242.1744
$C_{12}H_8N_3O_3$	242.0566
$C_{12}H_{10}N_4O_2$	242.0805
$C_{12}H_{20}NO_4$	242.1393
$C_{12}H_{22}N_2O_3$	242.1631
$C_{12}H_{24}N_3O_2$	242.1870
$C_{12}H_{26}N_4O$	242.2108
$C_{13}H_8NO_4$	242.0453
$C_{13}H_{10}N_2O_3$	242.0692
$C_{13}H_{12}N_3O_2$	242.0930
$C_{13}H_{14}N_4O$	242.1169
$C_{13}H_{22}O_4$	242.1518
$C_{13}H_{24}NO_3$	242.1757
$C_{13}H_{26}N_2O_2$	242.1996
$C_{13}H_{28}N_3O$	242.2234
$C_{13}H_{30}N_4$	242.2473
$C_{14}H_{10}O_4$	242.0579
$C_{14}H_{12}NO_3$	242.0817
$C_{14}H_{14}N_2O_2$	242.1056
$C_{14}H_{16}N_3O$	242.1295
$C_{14}H_{18}N_4$	242.1533
$C_{14}H_{26}O_3$	242.1883
$C_{14}H_{28}NO_2$	242.2121
$C_{14}H_{30}N_2O$	242.2360
$C_{14}H_{32}N_3$	242.2598
$C_{15}H_{14}O_3$	242.0943
$C_{15}H_{16}NO_2$	242.1182
$C_{15}H_{18}N_2O$	242.1420
$C_{15}H_{20}N_3$	242.1659
$C_{15}H_{30}O_2$	242.2247
$C_{15}H_{32}NO$	242.2485
$C_{15}H_{34}N_2$	242.2724
$C_{16}H_{18}O_2$	242.1307
$C_{16}H_{20}NO$	242.1546
$C_{16}H_{22}N_2$	242.1784
$C_{16}H_{34}O$	242.2611
$C_{16}H_{18}O_2$	242.1307
$C_{16}H_{20}NO$	242.1546
$C_{16}H_{22}N_2$	242.1784
$C_{16}H_{34}O$	242.2611
$C_{17}H_{22}O$	242.1871
$C_{17}H_{24}N$	242.1910
$C_{18}H_{10}O$	242.0732
$C_{18}H_{12}N$	242.0970
$C_{18}H_{26}$	242.2036
$C_{19}H_{14}$	242.1096
243	
$C_{11}H_{19}N_2O_4$	243.1345
$C_{11}H_{21}N_3O_3$	243.1584
$C_{11}H_{23}N_4O_2$	243.1822
$C_{12}H_7N_2O_4$	243.0406
$C_{12}H_9N_3O_3$	243.0644
$C_{12}H_{11}N_4O_2$	243.0883
$C_{12}H_{21}NO_4$	243.1471
$C_{12}H_{23}N_2O_3$	243.1710
$C_{12}H_{25}N_3O_2$	243.1948
$C_{12}H_{27}N_4O$	243.2187
$C_{13}H_9NO_4$	243.0532
$C_{13}H_{11}N_2O_3$	243.0770
$C_{13}H_{13}N_3O_2$	243.1009
$C_{13}H_{15}N_4O$	243.1247
$C_{13}H_{23}O_4$	243.1597
$C_{13}H_{25}NO_3$	243.1835
$C_{13}H_{27}N_2O_2$	243.2074
$C_{13}H_{29}N_3O$	243.2312
$C_{13}H_{31}N_4$	243.2551
$C_{14}H_{11}O_4$	243.0657
$C_{14}H_{13}NO_3$	243.0896
$C_{14}H_{15}N_2O_2$	243.1134
$C_{14}H_{17}N_3O$	243.1373
$C_{14}H_{19}N_4$	243.1611
$C_{14}H_{27}O_3$	243.1961
$C_{14}H_{29}NO_2$	243.2199
$C_{14}H_{31}N_2O$	243.2438
$C_{14}H_{33}N_3$	243.2677
$C_{15}H_{15}O_3$	243.1021
$C_{15}H_{17}NO_2$	243.1260
$C_{15}H_{19}N_2O$	243.1498
$C_{15}H_{21}N_3$	243.1737
$C_{15}H_{31}O_2$	243.2325
$C_{15}H_{33}NO$	243.2564
$C_{16}H_{19}O_2$	243.1385
$C_{16}H_{21}NO$	243.1624
$C_{16}H_{23}N_2$	243.1863
$C_{17}H_{23}O$	243.1750
$C_{17}H_{25}N$	243.1988
$C_{18}H_{11}O$	243.0810
$C_{18}H_{13}N$	243.1049
$C_{18}H_{27}$	243.2114
$C_{19}H_{15}$	243.1174
244	
$C_{11}H_{20}N_2O_4$	244.1424
$C_{11}H_{22}N_3O_3$	244.1662
$C_{11}H_{24}N_4O_2$	244.1901
$C_{12}H_8N_2O_4$	244.0484
$C_{12}H_{10}N_3O_3$	244.0723
$C_{12}H_{12}N_4O_2$	244.0961
$C_{12}H_{22}NO_4$	244.1549
$C_{12}H_{24}N_2O_3$	244.1788
$C_{12}H_{26}N_3O_2$	244.2026
$C_{12}H_{28}N_4O$	244.2265
$C_{13}H_{10}NO_4$	244.0610
$C_{13}H_{12}N_2O_3$	244.0848
$C_{13}H_{14}N_3O_2$	244.1087
$C_{13}H_{16}N_4O$	244.1325
$C_{13}H_{24}O_4$	244.1675
$C_{13}H_{26}NO_3$	244.1914
$C_{13}H_{28}N_2O_2$	244.2152
$C_{13}H_{30}N_3O$	244.2391
$C_{13}H_{32}N_4$	244.2629
$C_{14}H_{12}O_4$	244.0735
$C_{14}H_{14}NO_3$	244.0974
$C_{14}H_{16}N_2O_2$	244.1213
$C_{14}H_{18}N_3O$	244.1451
$C_{14}H_{20}N_4$	244.1690
$C_{14}H_{28}O_3$	244.2039
$C_{14}H_{30}NO_2$	244.2278
$C_{14}H_{32}N_2O$	244.2516
$C_{15}H_{16}O_3$	244.1100
$C_{15}H_{18}NO_2$	244.1338
$C_{15}H_{20}N_2O$	244.1577
$C_{15}H_{22}N_3$	244.1815
$C_{15}H_{32}O_2$	244.2403
$C_{16}H_{10}N_3$	244.0876
$C_{16}H_{20}O_2$	244.1464
$C_{16}H_{22}NO$	244.1702
$C_{16}H_{24}N_2$	244.1941
$C_{17}H_{10}NO$	244.0763
$C_{17}H_{12}N_2$	244.1001
$C_{17}H_{24}O$	244.1828
$C_{17}H_{26}N$	244.2067
$C_{18}H_{12}O$	244.0888
$C_{18}H_{14}N$	244.1127
$C_{18}H_{28}$	244.2192
$C_{19}H_{16}$	244.1253
245	
$C_{11}H_{21}N_2O_4$	245.1502
$C_{11}H_{23}N_3O_3$	245.1741
$C_{11}H_{25}N_4O_2$	245.1979
$C_{12}H_9N_2O_4$	245.0563
$C_{12}H_{11}N_3O_3$	245.0801
$C_{12}H_{13}N_4O_2$	245.1040
$C_{12}H_{23}NO_4$	245.1628
$C_{12}H_{25}N_2O_3$	245.1866
$C_{12}H_{27}N_3O_2$	245.2105
$C_{12}H_{29}N_4O$	245.2343
$C_{13}H_{11}NO_4$	245.0688
$C_{13}H_{13}N_2O_3$	245.0927
$C_{13}H_{15}N_3O_2$	245.1165
$C_{13}H_{17}N_4O$	245.1404
$C_{13}H_{25}O_4$	245.1753
$C_{13}H_{27}NO_3$	245.1992
$C_{13}H_{29}N_2O_2$	245.2230
$C_{13}H_{31}N_3O$	245.2469
$C_{14}H_{13}O_4$	245.0814

APPENDIX A (Continued)

	FM
$C_{14}H_{15}NO_3$	245.1052
$C_{14}H_{17}N_2O_2$	245.1291
$C_{14}H_{19}N_3O$	245.1529
$C_{14}H_{21}N_4$	245.1768
$C_{14}H_{29}O_3$	245.2117
$C_{14}H_{31}NO_2$	245.2356
$C_{15}H_{17}O_3$	245.1178
$C_{15}H_{19}NO_2$	245.1416
$C_{15}H_{21}N_2O$	245.1655
$C_{15}H_{23}N_3$	245.1894
$C_{16}H_9N_2O$	245.0715
$C_{16}H_{11}N_3$	245.0954
$C_{16}H_{21}O_2$	245.1542
$C_{16}H_{23}NO$	245.1781
$C_{16}H_{25}N_2$	245.2019
$C_{17}H_{11}NO$	245.0841
$C_{17}H_{13}N_2$	245.1080
$C_{17}H_{25}O$	245.1906
$C_{17}H_{27}N$	245.2145
$C_{18}H_{13}O$	245.0967
$C_{18}H_{15}N$	245.1205
$C_{18}H_{29}$	245.2270
$C_{19}H_{17}$	245.1331
246	
$C_{11}H_{22}N_2O_4$	246.1580
$C_{11}H_{24}N_3O_3$	246.1819
$C_{11}H_{26}N_4O_2$	246.2057
$C_{12}H_{10}N_2O_4$	246.0641
$C_{12}H_{12}N_3O_3$	246.0879
$C_{12}H_{14}N_4O_2$	246.1118
$C_{12}H_{24}NO_4$	246.1706
$C_{12}H_{26}N_2O_3$	246.1945
$C_{12}H_{28}N_3O_2$	246.2183
$C_{12}H_{30}N_4O$	246.2422
$C_{13}H_{12}NO_4$	246.0766
$C_{13}H_{14}N_2O_3$	246.1005
$C_{13}H_{16}N_3O_2$	246.1244
$C_{13}H_{18}N_4O$	246.1482
$C_{13}H_{26}O_4$	246.1832
$C_{13}H_{28}NO_3$	246.2070
$C_{13}H_{30}N_2O_2$	246.2309
$C_{14}H_{14}O_4$	246.0892
$C_{14}H_{16}NO_3$	246.1131
$C_{14}H_{18}N_2O_2$	246.1369
$C_{14}H_{20}N_3O$	246.1608
$C_{14}H_{22}N_4$	246.1846
$C_{14}H_{30}O_3$	246.2196
$C_{15}H_{10}N_4$	246.0907
$C_{15}H_{18}O_3$	246.1256
$C_{15}H_{20}NO_2$	246.1495
$C_{15}H_{22}N_2O$	246.1733
$C_{15}H_{24}N_3$	246.1972
$C_{16}H_{10}N_2O$	246.0794
$C_{16}H_{12}N_3$	246.1032
$C_{16}H_{22}O_2$	246.1620
$C_{16}H_{24}NO$	246.1859
$C_{16}H_{26}N_2$	246.2098
$C_{17}H_{10}O_2$	246.0681
$C_{17}H_{12}NO$	246.0919
$C_{17}H_{14}N_2$	246.1158
$C_{17}H_{26}O$	246.1985
$C_{17}H_{28}N$	246.2223
$C_{18}H_{14}O$	246.1045
$C_{18}H_{16}N$	246.1284
$C_{18}H_{30}$	246.2349
$C_{19}H_{18}$	246.1409
247	
$C_{11}H_{23}N_2O_4$	247.1659
$C_{11}H_{25}N_3O_3$	247.1897
$C_{11}H_{27}N_4O_2$	247.2136
$C_{12}H_{11}N_2O_4$	247.0719
$C_{12}H_{13}N_3O_3$	247.0958
$C_{12}H_{15}N_4O_2$	247.1196
$C_{12}H_{25}NO_4$	247.1784
$C_{12}H_{27}N_2O_3$	247.2023
$C_{12}H_{29}N_3O_2$	247.2261
$C_{13}H_{13}NO_4$	247.0845
$C_{13}H_{15}N_2O_3$	247.1083
$C_{13}H_{17}N_3O_2$	247.1322
$C_{13}H_{19}N_4O$	247.1560
$C_{13}H_{27}O_4$	247.1910
$C_{13}H_{29}NO_3$	247.2148
$C_{14}H_{15}O_4$	247.0970
$C_{14}H_{17}NO_3$	247.1209
$C_{14}H_{19}N_2O_2$	247.1448
$C_{14}H_{21}N_3O$	247.1686
$C_{14}H_{23}N_4$	247.1925
$C_{15}H_9N_3O$	247.0746
$C_{15}H_{11}N_4$	247.0985
$C_{15}H_{19}O_3$	247.1334
$C_{15}H_{21}NO_2$	247.1573
$C_{15}H_{23}N_2O$	247.1811
$C_{15}H_{25}N_3$	247.2050
$C_{16}H_{11}N_2O$	247.0872
$C_{16}H_{13}N_3$	247.1111
$C_{16}H_{23}O_2$	247.1699
$C_{16}H_{25}NO$	247.1937
$C_{16}H_{27}N_2$	247.2176
$C_{17}H_{11}O_2$	247.0759
$C_{17}H_{13}NO$	247.0998
$C_{17}H_{15}N_2$	247.1236
$C_{17}H_{27}O$	247.2063
$C_{17}H_{29}N$	247.2301
$C_{18}H_{15}O$	247.1123
$C_{18}H_{17}N$	247.1362
$C_{18}H_{31}$	247.2427
$C_{19}H_{19}$	247.1488
248	
$C_{11}H_{24}N_2O_4$	248.1737
$C_{11}H_{26}N_3O_3$	248.1976
$C_{11}H_{28}N_4O_2$	248.2214
$C_{12}H_{12}N_2O_4$	248.0797
$C_{12}H_{14}N_3O_3$	248.1036
$C_{12}H_{16}N_4O_2$	248.1275
$C_{12}H_{26}NO_4$	248.1863
$C_{12}H_{28}N_2O_3$	248.2101
$C_{13}H_{14}NO_4$	248.0923
$C_{13}H_{16}N_2O_3$	248.1162
$C_{13}H_{18}N_3O_2$	248.1400
$C_{13}H_{20}N_4O$	248.1639
$C_{13}H_{28}O_4$	248.1988
$C_{14}H_{16}O_4$	248.1049
$C_{14}H_{20}N_2O_2$	248.1526
$C_{14}H_{22}N_3O$	248.1764
$C_{14}H_{24}N_4$	248.2003
$C_{15}H_{10}N_3O$	248.0825
$C_{15}H_{12}N_4$	248.1063
$C_{15}H_{20}O_3$	248.1413
$C_{15}H_{22}NO_2$	248.1651
$C_{15}H_{24}N_2O$	248.1890
$C_{15}H_{26}N_3$	248.2129
$C_{16}H_{10}NO_2$	248.0712
$C_{16}H_{12}N_2O$	248.0950
$C_{16}H_{14}N_3$	248.1189
$C_{16}H_{24}O_2$	248.1777
$C_{16}H_{26}NO$	248.2015
$C_{16}H_{28}N_2$	248.2254
$C_{17}H_{12}O_2$	248.0837
$C_{17}H_{14}NO$	248.1076
$C_{17}H_{16}N_2$	248.1315
$C_{17}H_{28}O$	248.2141
$C_{17}H_{30}N$	248.2380
$C_{18}H_{16}O$	248.1202
$C_{18}H_{18}N$	248.1440
$C_{18}H_{32}$	248.2505
$C_{19}H_{20}$	248.1566
249	
$C_{11}H_{25}N_2O_4$	249.1815
$C_{11}H_{27}N_3O_3$	249.2054
$C_{12}H_{13}N_2O_4$	249.0876
$C_{12}H_{15}N_3O_3$	249.1114
$C_{12}H_{17}N_4O_2$	249.1353
$C_{12}H_{27}NO_4$	249.1941
$C_{13}H_{15}NO_4$	249.1001
$C_{13}H_{17}N_2O_3$	249.1240
$C_{13}H_{19}N_3O_2$	249.1478
$C_{13}H_{21}N_4O$	249.1717
$C_{14}H_9N_4O$	249.0777
$C_{14}H_{17}O_4$	249.1127
$C_{14}H_{19}NO_3$	249.1365
$C_{14}H_{21}N_2O_2$	249.1604
$C_{14}H_{23}N_3O$	249.1842
$C_{14}H_{25}N_4$	249.2081
$C_{15}H_9N_2O_2$	249.0664
$C_{15}H_{11}N_3O$	249.0903
$C_{15}H_{13}N_4$	249.1142
$C_{15}H_{21}O_3$	249.1491
$C_{15}H_{23}NO_2$	249.1730
$C_{15}H_{25}N_2O$	249.1968
$C_{15}H_{27}N_3$	249.2207
$C_{16}H_{11}NO_2$	249.0790
$C_{16}H_{13}N_2O$	249.1029
$C_{16}H_{15}N_3$	249.1267
$C_{16}H_{25}O_2$	249.1855
$C_{16}H_{27}NO$	249.2094
$C_{16}H_{29}N_2$	249.2332
$C_{17}H_{13}O_2$	249.0916
$C_{17}H_{15}NO$	249.1154
$C_{17}H_{17}N_2$	249.1393
$C_{17}H_{29}O$	249.2219
$C_{17}H_{31}N$	249.2458
$C_{18}H_{17}O$	249.1280
$C_{18}H_{19}N$	249.1519
$C_{18}H_{33}$	249.2584
$C_{19}H_{21}$	249.1644
250	
$C_{11}H_{26}N_2O_4$	250.1894
$C_{12}H_{14}N_2O_4$	250.0954
$C_{12}H_{16}N_3O_3$	250.1193
$C_{12}H_{18}N_4O_2$	250.1431
$C_{13}H_{16}NO_4$	250.1080
$C_{13}H_{18}N_2O_3$	250.1318
$C_{13}H_{20}N_3O_2$	250.1557
$C_{13}H_{22}N_4O$	250.1795
$C_{14}H_{10}N_4O$	250.0856
$C_{14}H_{20}NO_3$	250.1444
$C_{14}H_{22}N_2O_2$	250.1682
$C_{14}H_{24}N_3O$	250.1921
$C_{14}H_{26}N_4$	250.2160
$C_{15}H_{10}N_2O_2$	250.0743
$C_{15}H_{12}N_3O$	250.0981
$C_{15}H_{14}N_4$	250.1220
$C_{15}H_{22}O_3$	250.1569
$C_{15}H_{24}NO_2$	250.1808
$C_{15}H_{26}N_2O$	250.2046
$C_{15}H_{28}N_3$	250.2285
$C_{16}H_{10}O_3$	250.0630
$C_{16}H_{12}NO_2$	250.0868
$C_{16}H_{14}N_2O$	250.1107
$C_{16}H_{16}N_3$	250.1346
$C_{16}H_{26}O_2$	250.1934
$C_{16}H_{28}NO$	250.2172
$C_{16}H_{30}N_2$	250.2411
$C_{17}H_{14}O_2$	250.0994
$C_{17}H_{16}NO$	250.1233
$C_{17}H_{18}N_2$	250.1471
$C_{17}H_{30}O$	250.2298
$C_{17}H_{32}N$	250.2536
$C_{18}H_{18}O$	250.1358
$C_{18}H_{20}N$	250.1597
$C_{18}H_{34}$	250.2662
$C_{19}H_{22}$	250.1722

APPENDIX B COMMON FRAGMENT IONS

All fragments listed bear +1 charges. To be used in conjunction with Appendix C. Not all members of homologous and isomeric series are given. The list is meant to be suggestive rather than exhaustive. Appendix II of Hamming and Foster (1972). Table A-7 of McLafferty's (1993) interpretative book, and the high-resolution ion data of McLafferty (1982) are recommended as supplements.

m/z	**Ions**[a]
14	CH_2
15	CH_3
16	O
17	OH
18	H_2O, NH_4
19	F, H_3O
26	$C{\equiv}N$, C_2H_2
27	C_2H_3
28	C_2H_4, CO, N_2 (air), CH=NH
29	C_2H_5, CHO
30	CH_2NH_2, NO
31	CH_2OH, OCH_3
32	O_2 (air)
33	SH, CH_2F
34	H_2S
35	^{35}Cl[b]
36	$H^{35}Cl$[b]
39	C_3H_3
40	$CH_2C{\equiv}N$, Ar (air)
41	C_3H_5, $CH_2C{\equiv}N$ + H, C_2H_2NH
42	C_3H_6, C_2H_2O
43	C_3H_7, $CH_3C{=}O$, C_2H_5N
44	$CH_2C({=}O)H$ + H, CH_3CHNH_2, CO_2 (air), $NH_2C{=}O$, $(CH_3)_2N$
45	$CH_3CH(OH)$, CH_2CH_2OH, CH_2OCH_3, $C({=}O)OH$
46	NO_2
47	CH_2SH, CH_3S
48	CH_3S + H
49	$CH_2{}^{35}Cl$[b]
51	CH_2F_2, C_4H_3
53	C_4H_5
54	$CH_2CH_2C{\equiv}N$
55	C_4H_7, $CH_2{=}CHC{=}O$
56	C_4H_8
57	C_4H_9, $C_2H_5C{=}O$
58	$CH_3C({=}O)CH_2$ + H, $C_2H_5CHNH_2$, $(CH_3)_2NCH_2$, $C_2H_5NHCH_2$, C_2H_5S
59	$(CH_3)_2COH$, $CH_2OC_2H_5$, CO_2CH_3, $NH_2C({=}O)CH_2$ + H, CH_3OCHCH_3, CH_3CHCH_2OH, C_2H_5CHOH
60	CH_2CO_2H + H, CH_2ONO
61	CH_3CO_2 + 2H, CH_2CH_2SH, CH_2SCH_3
65	C_5H_5
66	H_2S_2, [cyclopentadiene ring] $\equiv C_5H_6$
67	C_5H_7
68	$CH_2CH_2CH_2C{\equiv}N$
69	C_5H_9, CF_3, $CH_3CH{=}CHC{=}O$, $CH_2{=}C(CH_3)C{=}O$
70	C_5H_{10}
71	C_5H_{11}, $C_3H_7C{=}O$
72	$C_2H_5C({=}O)CH_2$ + H, $C_3H_7CHNH_2$, $(CH_3)_2N{=}C{=}O$, $C_2H_5NHCHCH_3$ and isomers
73	Homologs of 59, $(CH_3)_3Si$
74	$CH_2CO_2CH_3$ + H
75	$CO_2C_2H_5$ + 2H, $C_2H_5CO_2$ + 2H, $CH_2SC_2H_5$, $(CH_3)_2CSH$, $(CH_3O)_2CH$, $(CH_3)_2SiOH$
76	C_6H_4 (C_6H_4XY)
77	C_6H_5 (C_6H_5X)
78	C_6H_5 + H
79	C_6H_5 + 2H, ^{79}Br[b]
80	CH_3SS + H, $H^{79}Br$[b], [pyrrole ring, N–H, with CH_2], [pyrrole ring, N–CH_2]
81	[furan ring]–CH_2, C_6H_9
82	$(CH_2)_4C{\equiv}N$, C_6H_{10}, $C^{35}Cl_2$[b]
83	C_6H_{11}, $CH^{35}Cl_2$[b], [thiophene ring]
85	[dihydropyran ring], [lactone ring], C_6H_{13}, $C_4H_9C{=}O$, $Cl^{35}ClF_2$[b]
86	$C_3H_7C({=}O)CH_2$ + H, $C_4H_9CHNH_2$ and isomers
87	$C_3H_7CO_2$, Homologs of 73, $CH_2CH_2CO_2CH_3$
88	$CH_2CO_2C_2H_5$ + H
89	$CO_2C_3H_7$ + 2H, [benzene ring]–C
90	[benzene ring]–CH, CH_3CHONO_2
91	$(C_6H_5)CH_2$, $(C_6H_5)CH$ + H, $(C_6H_5)C$ + 2H, $(CH_2)_4{}^{35}Cl$[b], $(C_6H_5)N$
92	[pyridine ring]–CH_2, $(C_6H_5)CH_2$ + H

APPENDIX B (Continued)

m/z **Ions**[a]

93 $CH_2{}^{79}Br^b$, C_7H_9, $(C_6H_5)O$, [structure: 2-pyrrolyl with N and C=O]

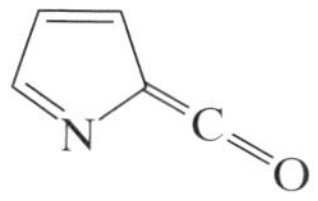

94 $(C_6H_5)O + H$, [structure: pyrrole ring (N–H) bearing C=O]

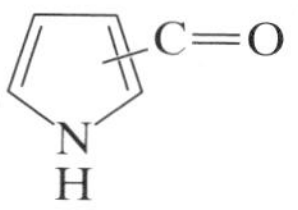

95 [structure: furan ring bearing C=O]

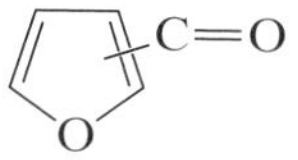

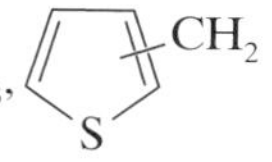

96 $(CH_2)_5C{\equiv}N$

97 C_7H_{13}, [structure: thiophene ring bearing CH_2]

98 [structure: furan ring bearing CH_2O] + H

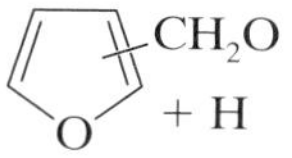

99 C_7H_{15}, $C_6H_{11}O$, [structure: six-membered ring lactone with O and C=O]

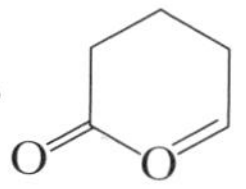

100 $C_4H_9C({=}O)CH_2 + H$, $C_5H_{11}CHNH_2$
101 $CO_2C_4H_9$
102 $CH_2CO_2C_2H_7 + H$
103 $CO_2C_4H_9 + 2H$, $C_5H_{11}S$, $CH(OCH_2CH_3)_2$
104 $C_2H_5CHONO_2$
105 $C_6H_5C{=}O$, $C_6H_5CH_2CH_2$, $C_6H_5CHCH_3$
106 $C_6H_5NHCH_2$
107 $C_6H_5CH_2O$, $HO(C_6H_4)CH_2$, $C_2H_4{}^{79}Br^b$

108 $C_6H_5CH_2O + H$, [structure: N-methylpyrrole ring (N–CH_3) bearing C=O]

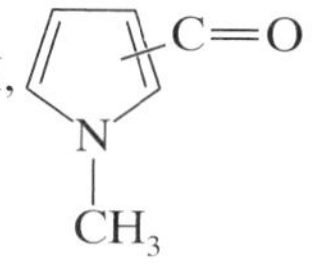

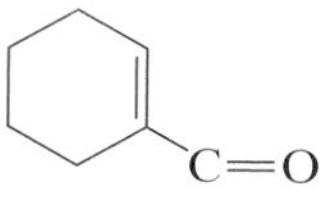

109 [structure: cyclohexene ring bearing C=O]

111 [structure: thiophene ring bearing C=O]

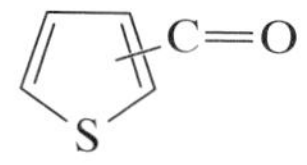

119 CF_3CF_2, $(C_6H_5)C(CH_3)_2$, $CH_3CH(C_6H_4)CH_3$, $CO(C_6H_4)CH_3$

120 [structure: cyclohexadiene ring bearing C=O and =O]

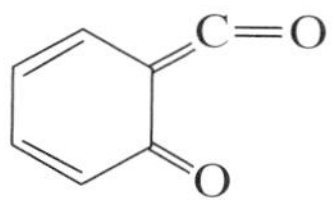

121 C_9H_{13}, [structure: benzene ring bearing C=O and OH], [structure: benzene ring bearing $O{-}CH_3$ and CH_2], [structure: cyclohexadiene ring bearing N=O and =NH]

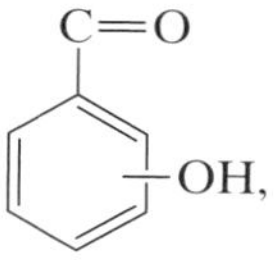

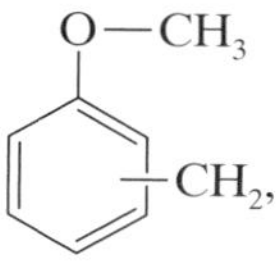

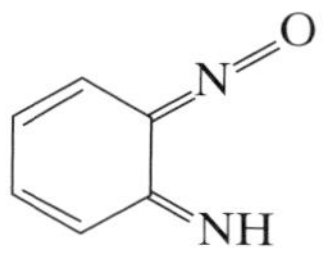

122 $C_6H_5CO_2 + H$
123 $F(C_6H_4)C{=}O$, $C_6H_5CO_2 + 2H$
125 C_6H_5SO
127 I
128 HI

130 [structure: indole ring (N–H) bearing CH_2]

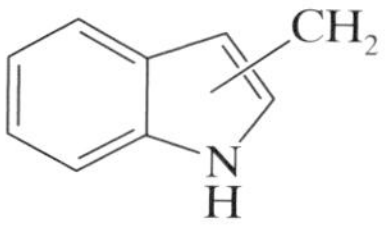

131 C_3F_5, $C_6H_5CH{=}CHC{=}O$
135 $(CH_2)_4{}^{79}B_r{}^b$
138 $CO_2(C_6H_4)OH + H$
139 $^{35}Cl(C_6H_4)C{=}O^b$
141 CH_2I
147 $(CH_3)_2Si{=}O{-}Si(CH_3)_3$

149 [structure: phthalic anhydride] + H

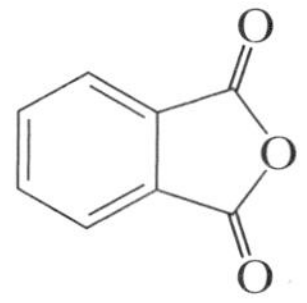

154 $(C_6H_5)_2$

---

[a] Ions indicated as a fragment $+n$H (n + 1,2,3, . . .) are ions that arise via rearrangement involving hydrogen transfer.
[b] Only the more abundant isotope is indicated.

APPENDIX C COMMON FRAGMENTS LOST

This list is suggestive rather than comprehensive. It should be used in conjunction with Appendix B. Table 5-19 of Hamming and Foster (1972) and Table A-5 of McLafferty (1993) are recommended as supplements. All of these fragments are lost as neutral species.

Molecular Ion Minus	Fragment Lost (Inference Structure)
1	$H\cdot$
2	$2H\cdot$
15	$CH_3\cdot$
16	O ($ArNO_2$, amine oxides, sulfoxides); $\cdot NH_2$ (carboxamides, sulfonamides)
17	$HO\cdot$
18	H_2O (alcohols, aldehydes, ketones)
19	$F\cdot$
20	HF
26	$CH{\equiv}CH$, $\cdot CH{\equiv}N$
27	$CH_2{=}CH\cdot$, $HC{\equiv}N$ (aromatic nitrites, nitrogen heterocycles)
28	$CH_2{=}CH_2$, CO, (quinones) (HCN + H)
29	$CH_3CH_2\cdot$, (ethyl ketones, $ArCH_2CH_2CH_3$), $\cdot CHO$
30	$NH_2CH_2\cdot$, CH_2O ($ArOCH_3$), NO ($ArNO_2$), C_2H_6
31	$\cdot OCH_3$(methyl esters), $\cdot CH_2OH$, CH_3NH_2
32	CH_3OH, S
33	$HS\cdot$ (thiols), ($\cdot CH_3$ and H_2O)
34	H_2S (thiols)
35	$Cl\cdot$
36	HCl, $2H_2O$
37	H_2Cl (or HCl + H)
38	C_3H_2, C_2N, F_2
39	C_3H_3, HC_2N
40	$CH_3C{\equiv}CH$
41	$CH_2{=}CHCH_2\cdot$
42	$CH_2{=}CHCH_3$, $CH_2{=}C{=}O$, $H_2C\overset{H_2}{\overset{C}{\triangle}}CH_2$, NCO, $NCNH_2$
43	$C_3H_7\cdot$(propyl ketones, $ArCH_2{-}C_3H_7$), $CH_3\overset{O}{\overset{\|}{C}}\cdot$(methyl ketones, $CH_3\overset{O}{\overset{\|}{C}}G$, where G = various functional groups), $CH_2{=}CH{-}O\cdot$, ($CH_3\cdot$ and $CH_2{=}CH_2$), HCNO
44	$CH_2{=}CHOH$, CO_2 (esters, anhydrides), N_2O, $CONH_2$, $NHCH_2CH_3$
45	CH_3CHOH, $CH_3CH_2O\cdot$(ethyl esters), CO_2H, $CH_3CH_2NH_2$
46	(H_2O and $CH_2{=}CH_2$), CH_3CH_2OH, $\cdot NO_2$ ($ArNO_2$)
47	$CH_3S\cdot$
48	CH_3SH, SO (sulfoxides), O_3
49	$\cdot CH_2Cl$
51	$\cdot CHF_2$
52	C_4H_4, C_2N_2
53	C_4H_5
54	$CH_2{=}CH{-}CH{=}CH_2$
55	$CH_2{=}CHCHCH_3$

APPENDIX C *(Continued)*

Molecular Ion Minus	Fragment Lost (Inference Structure)
56	$CH_2{=}CHCH_2CH_3$, $CH_3CH{=}CHCH_3$, 2CO
57	$C_4H_9\cdot$ (butyl ketones), C_2H_5CO (ethyl ketones, EtC=OG, G = various structural units)
58	·NCS, (NO + CO), CH_3COCH_3, C_4H_{10}
59	$CH_3OC(=O)\cdot$, $CH_3C(=O)NH_2$,  (thiiranyl ring with S(H)·)
60	C_3H_7OH, $CH_2{=}C(OH)_2$ (acetate esters)[a]
61	$CH_3CH_2S\cdot$,  (thiiranyl ring with S(H)·)
62	(H_2S and $CH_2{=}CH_2$)
63	$\cdot CH_2CH_2Cl$
64	C_5H_4, S_2, SO_2
68	$CH_2{=}C(CH_3){-}CH{=}CH_2$
69	$CF_3\cdot$, $C_5H_9\cdot$
71	$C_5H_{11}\cdot$
73	$CH_3CH_2OC(=O)\cdot$
74	C_4H_9OH
75	C_6H_3
76	C_6H_4, CS_2
77	C_6H_5, CS_2H
78	C_6H_6, CS_2H_2, C_5H_4N
79	Br·, C_5H_5N
80	HBr
85	$\cdot CClF_2$
100	$CF_2{=}CF_2$
119	$CF_3{-}CF_2\cdot$
122	C_6H_5COOH
127	I·
128	HI

[a] McLafferty rearrangement.

CHAPTER 2

INFRARED SPECTROMETRY

2.1 INTRODUCTION

Infrared (IR) radiation refers broadly to that part of the electromagnetic spectrum between the visible and microwave regions. Of greatest practical use to the organic chemist is the limited portion between 4000 and 400 cm^{-1}. There has been some interest in the near-IR (14,290–4000 cm^{-1}) and the far-IR regions, 700–200 cm^{-1}.

From the brief theoretical discussion that follows, it is clear that even a very simple molecule can give an extremely complex spectrum. The organic chemist takes advantage of this complexity when matching the spectrum of an unknown compound against that of an authentic sample. A peak-by-peak correlation is excellent evidence for identity. Any two compounds, except enantiomers, are unlikely to give exactly the same IR spectrum.

Although the IR spectrum is characteristic of the entire molecule, it is true that certain groups of atoms give rise to bands at or near the same frequency regardless of the structure of the rest of the molecule. It is the persistence of these characteristic bands that permits the chemist to obtain useful structural information by simple inspection and reference to generalized charts of characteristic group frequencies. We shall rely heavily on these characteristic group frequencies.

Since we are not solely dependent on IR spectra for identification, a detailed analysis of the spectrum will not be required. Following our general plan, we shall present only sufficient theory to accomplish our purpose: utilization of IR spectra in conjunction with other spectral data in order to determine molecular structure.

The importance of IR spectrometry as a tool of the practicing organic chemist is readily apparent from the number of books devoted wholly or in part to discussions of applications of IR spectrometry (see the references at the end of this chapter). There are many compilations of spectra as well as indeces to spectral collections and to the literature. Among the more commonly used compilations are those published by Sadtler (1994) and by Aldrich (1989).

2.2 THEORY

Infrared radiation of frequencies less than about 100 cm^{-1} is absorbed and converted by an organic molecule into energy of molecular rotation. This absorption is quantized; thus a molecular rotation spectrum consists of discrete lines.

Infrared radiation in the range from about 10,000–100 cm^{-1} is absorbed and converted by an organic molecule into energy of molecular vibration. This absorption is also quantized, but vibrational spectra appear as bands rather than as lines because a single vibrational energy change is accompanied by a number of rotational energy changes. It is with these vibrational-rotational bands, particularly those occurring between 4000 and 400 cm^{-1}, that we shall be concerned. The frequency or wavelength of absorption depends on the relative masses of the atoms, the force constants of the bonds, and the geometry of the atoms.

Band positions in IR spectra are presented here as wavenumbers ($\bar{\nu}$) whose unit is the reciprocal centimeter (cm^{-1}); this unit is proportional to the energy of vibration and modern instruments are linear in reciprocal centimeters. Wavelength (λ) was used in the older literature in units of micrometers ($\mu m = 10^{-6}$ m; earlier called microns). Wavenumbers are reciprocally related to wavelength.

$$cm^{-1} = 10^4/\mu m$$

Note that wavenumbers are sometimes called "frequencies." However, this is incorrect since wavenumbers ($\bar{\nu}$ in units of cm^{-1}) are equal to $1 \times 10^4/\lambda$ in units of μm, whereas frequencies (ν in Hz) are equal to c/λ in centimeters, c being the speed of light (3×10^{10} cm/s). The symbol $\bar{\nu}$ is called "nu bar." The IR spectra in this text are generally given as a linear function of cm^{-1} unless otherwise indicated. Note that spectra that are recorded as a function of μm are quite different in appearance from those plotted in cm^{-1} (see Figure 2.6).

Band intensities can be expressed either as transmittance (T) or absorbance (A). Transmittance is the ratio of the radiant power transmitted by a sample to the radiant power incident on the sample. Absorbance is the logarithm, to the base 10, of the reciprocal of the transmittance; $A = \log_{10}(1/T)$. Organic chemists usually report intensity in semiquantitative terms (s = strong, m = medium, w = weak).

There are two types of molecular vibrations: stretching and bending. A stretching vibration is a rhythmical movement along the bond axis such that the interatomic distance is increasing or decreasing. A bending vibration may consist of a change in bond angle between bonds with a common atom or the movement of a group of atoms with respect to the remainder of the molecule without movement of the atoms in the group with respect to one another. For example, twisting, rocking, and torsional vibrations involve a change in bond angles with reference to a set of coordinates arbitrarily set up within the molecule.

Only those vibrations that result in a rhythmical change in the dipole moment of the molecule are observed in the IR. The alternating electric field, produced by the changing charge distribution accompanying a vibration, couples the molecule vibration with the oscillating electric field of the electromagnetic radiation.

A molecule has as many degrees of freedom as the total degrees of freedom of its individual atoms. Each atom has three degrees of freedom corresponding to the Cartesian coordinates (x, y, z) necessary to describe its position relative to other atoms in the molecule. A molecule of n atoms therefore has $3n$ degrees of freedom. For nonlinear molecules, three degrees of freedom describe rotation and three describe translation; the remaining $3n - 6$ degrees of freedom are vibrational degrees of freedom or fundamental vibrations. Linear molecules have $3n - 5$ vibrational degrees of freedom, for only two degrees of freedom are required to describe rotation.

Fundamental vibrations involve no change in the center of gravity of the molecule. The three fundamental vibrations of the nonlinear, triatomic water molecule are depicted in the top portion of Figure 2.1. Note the very close spacing of the interacting or coupled asymmetric and symmetric stretching compared with the far-removed scissoring mode.

The CO_2 molecule is linear and contains three atoms; therefore it has four fundamental vibrations $[(3 \times 3) - 5]$ as shown in the middle section of Figure 2.1. The symmetrical stretching vibration in (1) is inactive in the IR since it produces no change in the dipole moment of the molecule. The bending vibrations in (3) and (4) above are equivalent and are the resolved components of bending motion oriented at any angle to the internuclear axis; they have the same frequency and are said to be doubly degenerate.

The various stretching and bending modes for an AX_2 group appearing as a portion of a molecule, for example, the CH_2 group in a hydrocarbon molecule, are shown in Figure 2.1. The $3n - 6$ rule does not apply since the CH_2 group represents only a portion of a molecule.

The theoretical number of fundamental vibrations (absorption frequencies) will seldom be observed because overtones (multiples of a given frequency) and combination tones (sum of two other vibrations) increase the number of bands, whereas other phenomena reduce the number of bands. The following will reduce the theoretical number of bands.

1. Fundamental frequencies that fall outside of the 4000–400 cm^{-1} region.
2. Fundamental bands that are too weak to be observed.
3. Fundamental vibrations that are so close that they coalesce.
4. The occurrence of a degenerate band from several absorptions of the same frequency in highly symmetrical molecules.
5. The failure of certain fundamental vibrations to appear in the IR because of the lack of change in molecular dipole.

Assignments for stretching frequencies can be approximated by the application of Hooke's law. In the application of the law, two atoms and their connecting bond are treated as a simple harmonic oscillator composed of two masses joined by a spring. The following equation, derived from Hooke's law, states the relationship between frequency of oscillation, atomic masses, and the force constant of the bond.

$$\bar{\nu} = \frac{1}{2\pi c}\sqrt{\frac{f}{(MxMy)/(Mx + My)}}$$

where $\bar{\nu}$ = the vibrational frequency (cm^{-1})
c = velocity of light (cm/s)
f = force constant of bond (dyne/cm)
Mx and My = mass (g) of atom x and atom y, respectively.

The value of f is approximately 5×10^5 dyne/cm for single bonds and approximately two and three times this value for double and triple bonds, respectively (see Table 2.1). The force constant, f, can be thought of as a measure of bond "stiffness." This force constant can be correlated with such properties as bond order and bond strength. Because the frequency is directly related to the square root of the force constant, we know that the frequency of bond vibrations should decrease as bonds decrease in strength.

Application of the formula to C—H stretching using 2.10×10^{-23} and 1.67×10^{-24} g as mass values for

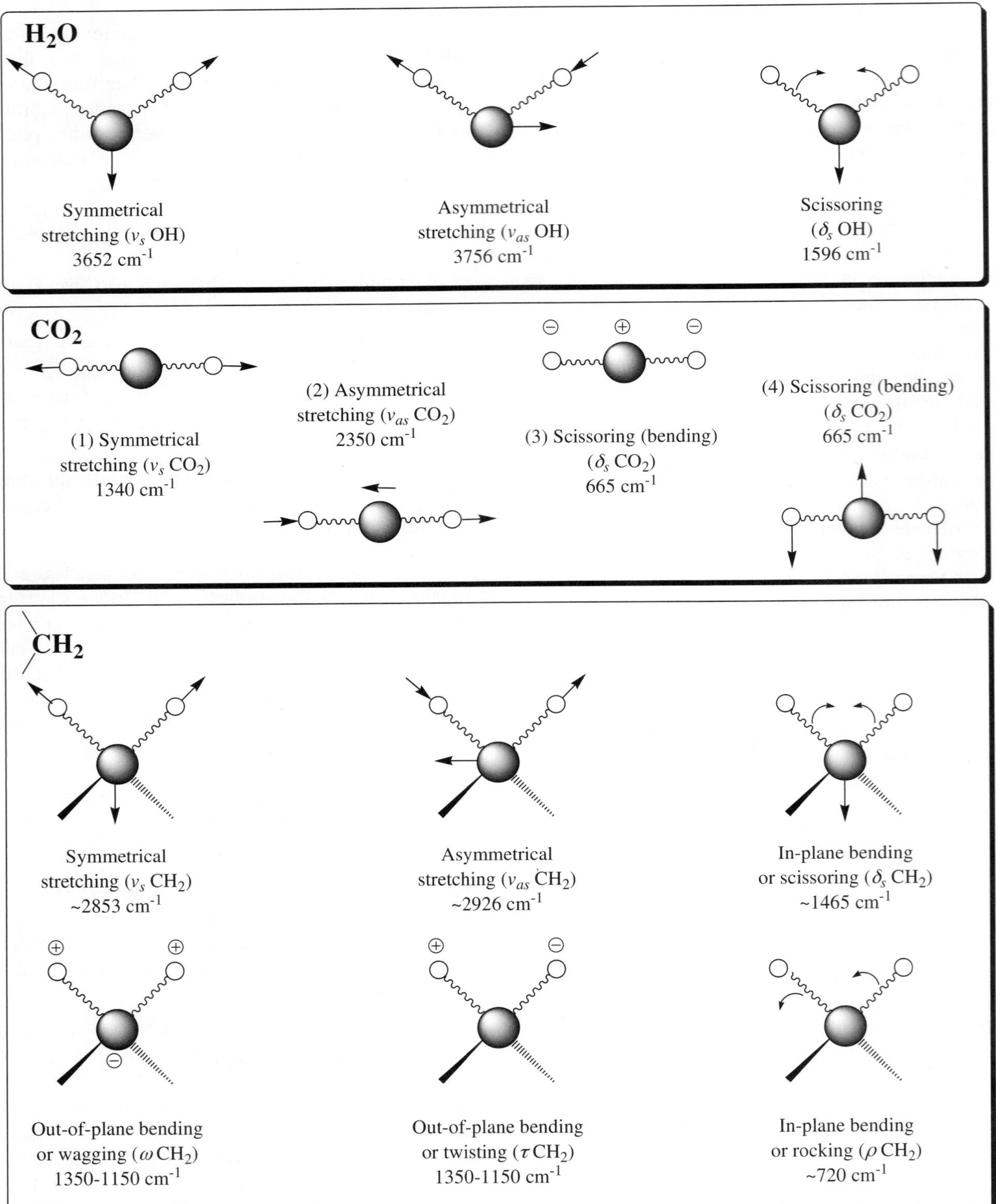

FIGURE 2.1. (Top) Vibrational modes for H_2O. (Middle) Vibrational modes for linear CO_2. (Bottom) Vibrational modes for a CH_2 group (+ and − indicate movement perpendicular to the plane of the page).

TABLE 2.1 IR Absorption Regions Using Hooke's Law

Bond Type	Force Constant f in dyne/cm	Absorption Region (cm^{-1})	
		Calculated	Observed
C—O	5.0×10^5	1113	1300–800
C—C	4.5×10^5	1128	1300–800
C—N	4.9×10^5	1135	1250–1000
C=C	9.7×10^5	1657	1900–1500
C=O	12.1×10^5	1731	1850–1600
C≡C	15.6×10^5	2101	2150–2100
C—D	5.0×10^5	2225	2250–2080
C—H	5.0×10^5	3032	3000–2850
O—H	7.0×10^5	3553	3800–2700

C and H, respectively, places the frequency of the C—H bond vibration at 3032 cm^{-1}. Actually, C—H stretching vibrations, associated with methyl and methylene groups, are generally observed in the region between 2960 and 2850 cm^{-1}. The calculation is not highly accurate because effects arising from the environment of the C—H within a molecule have been ignored. The frequency of IR absorption is commonly used to calculate the force constants of bonds.

The shift in absorption frequency following deuteration is often employed in the assignment of C—H stretching frequencies. The above equation can be used to estimate the change in stretching frequency as the result of deuteration. The term $MxMy/(Mx + My)$ will be equal to $M_C M_H/(M_C + M_H)$ for the C—H compound. Since $M_C >> M_H$, this term is approximately equal to $M_C M_H/M_C$ or to M_H. Thus, for the C—D compound, the term is equal to M_D; the frequency by Hooke's law application is inversely proportional to the square root of the mass of the isotope of hydrogen. The ratio of the C—H to C—D stretching frequencies should therefore, equal $\sqrt{2}$. If the ratio of the frequencies, following deuteration, is much less than $\sqrt{2}$, we can assume that the vibration is not simply a C—H stretching vibration but instead a mixed vibration involving interaction (coupling) with another vibration.

Crude approximations based on Hooke's law can be made by calculating the stretching frequencies for certain bond types as indicated in Table 2.1

To approximate the vibrational frequencies of bond stretching by Hooke's law, the relative contributions of bond strengths and atomic masses must be considered. For example, a superficial comparison of the C—H group with the F—H group, on the basis of atomic masses, might lead to the conclusion that the stretching frequency of the F—H bond should occur at a lower frequency than that for the C—H bond. However, the increase in the force constant from left to right across the first two rows of the periodic table has a greater effect than the mass increase. Thus, the F—H group absorbs at a higher frequency (4138 cm^{-1}) than the C—H group (3040 cm^{-1}).

In general, functional groups that have a strong dipole give rise to strong absorptions in the IR.

2.2.1 Coupled Interactions

When two bond oscillators share a common atom, they seldom behave as individual oscillators unless the individual oscillation frequencies are widely different. This is because there is mechanical coupling interaction between the oscillators. For example, the carbon dioxide molecule (see Figure 2.1), which consists of two C=O bonds with a common carbon atom, has two fundamental stretching vibrations: an asymmetrical and a symmetrical stretching mode. The symmetrical stretching mode consists of an in-phase stretching or contracting of the C=O bonds, and absorption occurs at a wavelength longer than that observed for the carbonyl group in an aliphatic ketone. The symmetrical stretching mode produces no change in the dipole moment (μ) of the molecule and is therefore "inactive" in the IR, but it is easily observed in the Raman spectrum* near 1340 cm^{-1}. In the asymmetrical stretching mode, the two C=O bonds stretch out of phase; one C=O bond stretches as the other contracts. The asymmetrical stretching mode, since it produces a change in the dipole moment, is IR active; the absorption (2350 cm^{-1}) is at a higher frequency (shorter wavelength) than observed for a carbonyl group in aliphatic ketones.

$$\longleftarrow O{=}C{=}O \longrightarrow \qquad \longleftarrow O{=}\overrightarrow{C}{=}O \longleftarrow$$

Symmetrical $\mu = 0$ — Asymmetrical $\mu \neq 0$

This difference in carbonyl absorption frequencies displayed by the carbon dioxide molecule results from strong mechanical coupling or interaction. In contrast, two ketonic carbonyl groups separated by one or more carbon atoms show normal carbonyl absorption near 1715 cm^{-1} because appreciable coupling is prevented by the intervening carbon atom(s).

Coupling accounts for the two N—H stretching bands in the 3497–3077 cm^{-1} region in primary amine and primary amide spectra; for the two C=O stretching bands in the 1818–1720 cm^{-1} region in carboxylic anhydride and imide spectra, and for the two C—H stretching bands in the 3000–2760 cm^{-1} region for both methylene and methyl groups.

Useful characteristic group frequency bands often involve coupled vibrations. The spectra of alcohols have a strong band in the region between 1260 and 1000 cm^{-1}, which is usually designated as the "C—O stretching band." In the spectrum of methanol this band is at 1034 cm^{-1}; in the spectrum of ethanol it occurs at

* Band intensity in Raman spectra depends on bond polarizability rather than molecular dipole changes.

1053 cm^{-1}. Branching and unsaturation produce absorption characteristic of these structures (see alcohols). It is evident that we are dealing not with an isolated C—O stretching vibration but rather a coupled asymmetric vibration involving C—C—O stretching.

Vibrations resulting from bond angle changes frequently couple in a manner similar to stretching vibrations. Thus, the ring C—H out-of-plane bending frequencies of aromatic molecules depend on the number of adjacent hydrogen atoms on the ring; coupling between the hydrogen atoms is affected by the bending of the C—C bond in the ring to which the hydrogen atoms are attached.

Interaction arising from coupling of stretching and bending vibrations is illustrated by the absorption of secondary acyclic amides. Secondary acyclic amides, which exist predominantly in the *trans* conformation, show strong absorption in the 1563–1515 cm^{-1} region; this absorption involves coupling of the N—H bending and C—N stretching vibrations.

The requirements for effective coupling interaction may be summarized as follows:

1. The vibrations must be of the same symmetry species if interaction is to occur.
2. Strong coupling between stretching vibrations requires a common atom between the groups.
3. Interaction is greatest when the coupled groups absorb, individually, near the same frequency.
4. Coupling between bending and stretching vibrations can occur if the stretching bond forms one side of the changing angle.
5. A common bond is required for coupling of bending vibrations.
6. Coupling is negligible when groups are separated by one or more carbon atoms and the vibrations are mutually perpendicular.

As we have seen in our discussion of interaction, coupling of two fundamental vibrational modes will produce two new modes of vibration, with frequencies higher and lower than that observed when interaction is absent. Interaction can also occur between fundamental vibrations and overtones or combination-tone vibrations. Such interaction is known as Fermi resonance. One example of Fermi resonance is afforded by the absorption pattern of carbon dioxide. In our discussion of interaction, we indicated that the symmetrical stretching band of CO_2 appears in the Raman spectrum near 1340 cm^{-1}. Actually two bands are observed: one at 1286 cm^{-1} and one at 1388 cm^{-1}. The splitting results from coupling between the fundamental C=O stretching vibration, near 1340 cm^{-1}, and the first overtone of the bending vibration. The fundamental bending vibration occurs near 666 cm^{-1}, the first overtone near 1334 cm^{-1}.

Fermi resonance is a common phenomenon in IR and Raman spectra. It requires that the vibrational levels be of the same symmetry species and that the interacting groups be located in the molecule so that mechanical coupling is appreciable.

An example of Fermi resonance in an organic structure is the "doublet" appearance of the C=O stretch of certain cyclic ketones under sufficient resolution conditions. Figure 2.2 shows the appearance of the spectrum of cycloheptanone under the usual conditions of resolution; the carbonyl peak at 1709 cm^{-1} is a "singlet." With adequate resolution however, the IR spectra of cyclopentanone's carbonyl region, which are given in Figure 2.3 for four different conditions, show a doublet for the carbonyl group. These doublets are due to Fermi resonance of the carbonyl group with an overtone or combination band of an α-methylene group.

2.2.2 Hydrogen Bonding

Hydrogen bonding can occur in any system containing a proton donor group (X–H) and a proton acceptor ($\ddot{Y}$) if the *s* orbital of the proton can effectively

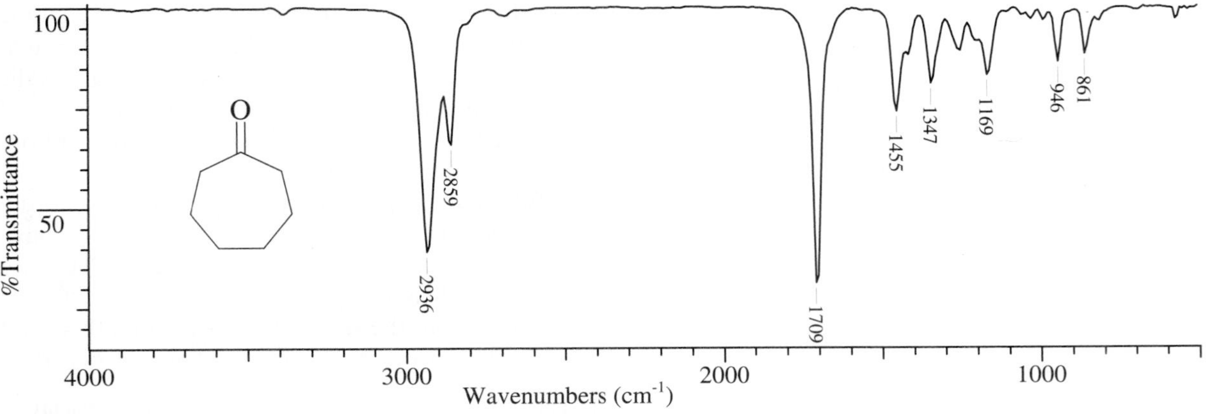

FIGURE 2.2. Cycloheptanone, neat.

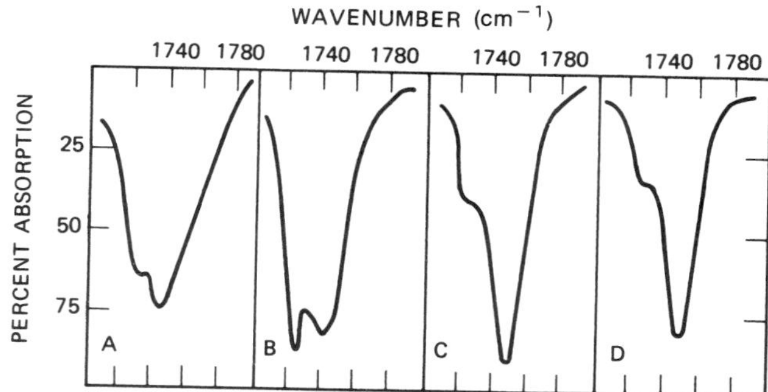

FIGURE 2.3. Infrared spectrum of cyclopentanone in various media. A. Carbon tetrachloride solution (0.15 *M*). B. Carbon disulfide solution (0.023 *M*). C. Chloroform solution (0.025 *M*). D. Liquid state (thin film). (Computed spectral slit width 2 cm^{-1}.)

overlap the *p* or π orbital of the acceptor group. Atoms X and $\ddot{Y}$ are electronegative, with $\ddot{Y}$ possessing lone pair electrons. The common proton donor groups in organic molecules are carboxyl, hydroxyl, amine, or amide groups. Common proton acceptor atoms are oxygen, nitrogen, and the halogens. Unsaturated groups, such as the C=C linkage, can also act as proton acceptors.

The strength of the hydrogen bond is at a maximum when the proton donor group and the axis of the lone pair orbital are collinear. The strength of the bond decreases as the distance between X and Y increases.

Hydrogen bonding alters the force constant of both groups; thus, the frequencies of both stretching and bending vibrations are altered. The X—H stretching bands move to lower frequencies (longer wavelengths) usually with increased intensity and band widening. The stretching frequency of the acceptor group, for example, C=O, is also reduced but to a lesser degree than the proton donor group. The H—X bending vibration usually shifts to a shorter wavelength when bonding occurs; this shift is less pronounced than that of the stretching frequency.

Intermolecular hydrogen bonding involves association of two or more molecules of the same or different compounds. Intermolecular bonding may result in dimer molecules (as observed for carboxylic acids) or in polymeric molecular chains, which exist in neat samples or concentrated solutions of monohydroxy alcohols. Intramolecular hydrogen bonds are formed when the proton donor and acceptor are present in a single molecule under spatial conditions that allow the required overlap of orbitals, for example, the formation of a five- or six-membered ring. The extent of both inter- and intra-molecular bonding is temperature dependent. The effect of concentration on intermolecular and intra-molecular hydrogen bonding is markedly different. The bands that result from intermolecular bonding generally disappear at low concentrations (less than about 0.01 *M* in nonpolar solvents). Intramolecular hydrogen bonding is an internal effect and persists at very low concentrations.

The change in frequency between "free" OH absorption and bonded OH absorption is a measure of the strength of the hydrogen bond. Ring strain, molecular geometry, and the relative acidity and basicity of the proton donor and acceptor groups affect the strength of bonding. Intramolecular bonding involving the same bonding groups is stronger when a six-membered ring is formed than when a smaller ring results from bonding. Hydrogen bonding is strongest when the bonded structure is stabilized by resonance.

The effects of hydrogen bonding on the stretching frequencies of hydroxyl and carbonyl groups are summarized in Table 2.2. Figure 2.14 (spectrum of cyclohexylcarbinol in the O—H stretch region) clearly illustrates this effect.

TABLE 2.2 Stretching Frequencies in Hydrogen Bonding

	Intermolecular Bonding			**Intramolecular Bonding**		
X—H···Y Strength	**Frequency Reduction** ν_{OH}	$\nu_{C=O}$	**Compound Class**	**Frequency Reduction** ν_{OH}	$\nu_{C=O}$	**Compound Class**
Weak	300[a]	15[b]	Alcohols, phenols, and intermolecular hydroxyl to carbonyl bonding	<100[a]	10	1,2-Diols, α- and most β-hydroxy ketones; *o*-chloro and *o*-alkoxy phenols
Medium				100–300[a]	50	1,3-Diols; some β-hydroxy ketones; β-hydroxy amino compounds; nitro compounds
Strong	>500[a]	50[b]	RCO_2H dimers	>300[a]	100	*o*-Hydroxy aryl ketones; *o*-hydroxy aryl acids; β-diketones; tropolones

[a] Frequency shift relative to "free" stretching frequencies.
[b] Carbonyl stretching only where applicable.

An important aspect of hydrogen bonding involves interaction between functional groups of solvent and solute. If the solute is polar, then it is important to note the solvent used and the solute concentration.

2.3 INSTRUMENTATION

2.3.1 Dispersion IR Spectrometer

For many years, an infrared spectrum was obtained by passing an infrared beam though the sample and scanning the spectrum with a dispersion device (the familiar diffraction grating). The spectrum was scanned by rotating the diffraction grating; the absorption areas (peaks) were detected and plotted as frequencies versus intensities.

Figure 2.4 demonstrates a sophisticated double-beam dispersion instrument, operation of which involves splitting the beam and passing one portion through the sample cell and the other portion through the reference cell. The individual beams are then recombined into a single beam of alternating segments by means of the rotating sector mirror, *M*7, and the absorption intensities of the segments are balanced by the attenuator in the reference beam. Thus, the solvent in the reference cell and in the sample cell are balanced out, and the spectrum contains only the absorption peaks of the sample itself.

2.3.2 Fourier Transform Infrared Spectrometer (Interferometer)

Fourier transform infrared (FT IR) spectrometry has been extensively developed over the past decade and provides a number of advantages. Radiation containing all IR wavelengths (e.g., 4000–400 cm^{-1}) is split into two beams (Figure 2.5). One beam is of fixed length, the other of variable length (movable mirror).

The varying distances between two pathlengths result in a sequence of constructive and destructive interferences and hence variations in intensities: an interferogram. Fourier transformation converts this interferogram from the time domain into one spectral point on the more familiar form of the frequency domain. Smooth and continuous variation of the length of the piston adjusts the position of mirror B and varies the length of beam B; Fourier transformation at successive points throughout this variation gives rise to the complete IR spectrum. Passage of this radiation through a sample subjects the compound to a broad band of energies. In principle, the analysis of one broadbanded pass of radiation through the sample will give rise to a complete IR spectrum.

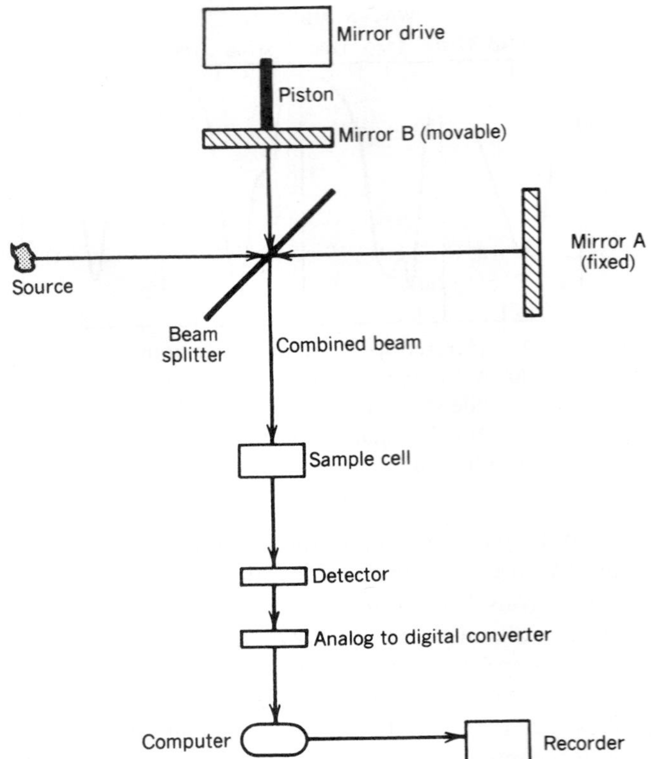

FIGURE 2.5. Schematic of an FT IR spectrometer.

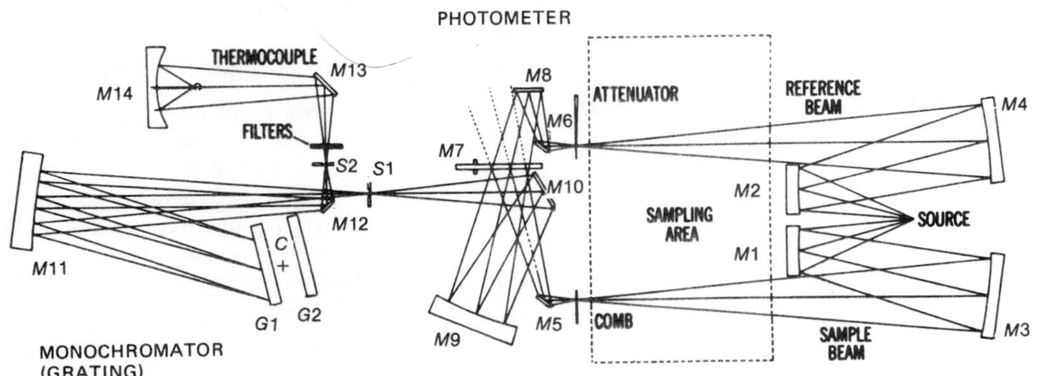

FIGURE 2.4. Optical system of double-beam IR spectrophotometer.

There are a number of advantages to FT IR methods. Since a monochromator is not used, the entire radiation range is passed through the sample simultaneously and much time is saved (Felgett's advantage). FT IR instruments can have very high resolution ($\leq 0.001\ cm^{-1}$). Moreover, since the data undergo analog-to-digital conversion, IR results are easily manipulated; results of several scans are combined to average out random absorption artifacts, and excellent spectra from very small samples can be obtained. An FT IR unit can therefore be used in conjunction with HPLC or GC. As with any computer-aided spectrometer, spectra of pure samples or solvents (stored in the computer) can be subtracted from mixtures. Flexibility in spectral printout is also available; for example, spectra linear in either wavenumber or wavelength can be obtained from the same data set.

Several manufacturers offer GC-FT IR instruments with which a vapor-phase spectrum can be obtained on nanogram amounts of a compound eluting from a capillary GC column. Vapor-phase spectra resemble those obtained at high dilution in a nonpolar solvent: Concentration-dependent peaks are shifted to higher frequency compared with those obtained from concentrated solutions, thin films, or the solid state (see Aldrich, 1989).

2.4 SAMPLE HANDLING

Infrared spectra may be obtained for gases, liquids, or solids. The spectra of gases or low-boiling liquids may be obtained by expansion of the sample into an evacuated cell. Gas cells are available in lengths of a few centimeters to 40 m. The sampling area of a standard IR spectrophotometer will not accommodate cells much longer than 10 cm; long paths are achieved by multiple reflection optics.

Liquids may be examined neat or in solution. Neat liquids are examined between salt plates, usually without a spacer. Pressing a liquid sample between flat plates produces a film 0.01 mm or less in thickness, the plates being held together by capillary action. Samples of 1–10 mg are required. Thick samples of neat liquids usually absorb too strongly to produce a satisfactory spectrum. Volatile liquids are examined in sealed cells with very thin spacers. Silver chloride plates may be used for samples that dissolve sodium chloride plates.

Solutions are handled in cells of 0.1–1 mm thickness. Volumes of 0.1–1 ml of 0.05–10% solutions are required for readily available cells. A compensating cell, containing pure solvent, is placed in the reference beam. The spectrum thus obtained is that of the solute except in those regions in which the solvent absorbs strongly. For example, thick samples of carbon tetrachloride absorb strongly near $800\ cm^{1}$; compensation for this band is ineffective since strong absorption prevents any radiation from reaching the detector.

The solvent selected must be dry and transparent in the region of interest. When the entire spectrum is of interest, several solvents must be used. A common pair of solvents is carbon tetrachloride (CCl_4) and carbon disulfide (CS_2). Carbon tetrachloride is relatively free of absorption at frequencies above $1333\ cm^{-1}$, whereas CS_2 shows little absorption below $1333\ cm^{-1}$. Solvent and solute combinations that react must be avoided. For example, CS_2 cannot be used as a solvent for primary or secondary amines. Amino alcohols react slowly with CS_2 and CCl_4. The absorption patterns of selected solvents and mulling oils are presented in Appendix A.

Solids are usually examined as a mull, as a pressed disk, or as a deposited glassy film. Mulls are prepared by thoroughly grinding 2–5 mg of a solid in a smooth agate mortar. Grinding is continued after the addition of 1 or 2 drops of the mulling oil. The suspended particles must be less than 2 μm to avoid excessive scattering of radiation. The mull is examined as a thin film between flat salt plates. Nujol® (a high-boiling petroleum oil) is commonly used as a mulling agent. When hydrocarbon bands interfere with the spectrum, Fluorolube® (a completely halogenated polymer containing F and Cl) or hexachlorobutadiene may be used. The use of both Nujol and Fluorolube mulls makes possible a scan, essentially free of interfering bands, over the 4000–250 cm^{-1} region.

The pellet (pressed-disk) technique depends on the fact that dry, powdered potassium bromide (or other alkali metal halides) can be compacted under pressure to form transparent disks. The sample (0.5–1.0 mg) is intimately mixed with approximately 100 mg of dry, powdered KBr. Mixing can be effected by thorough grinding in a smooth agate mortar or, more efficiently, with a small vibrating ball mill, or by lyophilization. The mixture is pressed with special dies under a pressure of 10,000–15,000 psi into a transparent disk. The quality of the spectrum depends on the intimacy of mixing and the reduction of the suspended particles to 2 μm or less. Microdisks, 0.5–1.5 mm in diameter, can be used with a beam condenser. The microdisk technique permits examination of samples as small as 1 μg. Bands near 3448 and $1639\ cm^{-1}$, resulting from moisture, frequently appear in spectra obtained by the pressed-disk technique.

The use of KBr disks or pellets has often been avoided because of the demanding task of making good pellets. Such KBr techniques can be less formidable through the Mini-Press, which affords a simple procedure; the KBr-sample mixture is placed in the nut portion of the assembly with one bolt in place. The second bolt is introduced, and pressure is applied by tightening the bolts. Removal of the bolts leaves a pellet in the nut that now serves as a cell.

Deposited films are useful only when the material can be deposited from solution or cooled from a melt as microcrystals or as a glassy film. Crystalline films generally lead to excessive light scattering. Specific crystal orientation may lead to spectra differing from those observed for randomly oriented particles such as exist in a mull or halide disk. The deposited film technique is particularly useful for obtaining spectra of resins and plastics. Care must be taken to free the sample of solvent by vacuum treatment or gentle heating.

A technique known as attenuated total reflection or internal reflection spectroscopy is now available for obtaining qualitative spectra of solids regardless of thickness. The technique depends on the fact that a beam of light that is internally reflected from the surface of a transmitting medium passes a short distance beyond the reflecting boundary and returns to the transmitting medium as a part of the process of reflection. If a material (i.e., the sample) of lower refractive index than the transmitting medium is brought in contact with the reflecting surface, the light passes through the material to the depth of a few micrometers, producing an absorption spectrum. An extension of the technique provides for multiple internal reflections along the surface of the sample. The multiple internal reflection technique results in spectra with intensities comparable to transmission spectra.

In general, a dilute solution in a nonpolar solvent furnishes the best (i.e., least distorted) spectrum. Nonpolar compounds give essentially the same spectra in the condensed phase (i.e., neat liquid, a mull, a KBr disk, or a thin film) as they give in nonpolar solvents. Polar compounds, however, often show hydrogen-bonding effects in the condensed phase. Unfortunately, polar compounds are frequently insoluble in nonpolar solvents, and the spectrum must be obtained either in a condensed phase or in a polar solvent; the latter introduces the possibility of solute–solvent hydrogen bonding.

Reasonable care must be taken in handling salt cells and plates. Moisture-free samples should be used. Fingers should not come in contact with the optical surfaces. Care should be taken to prevent contamination with silicones, which are hard to remove and have strong absorption patterns.

2.5 INTERPRETATION OF SPECTRA

There are no rigid rules for interpreting an IR spectrum. Certain requirements, however, must be met before an attempt is made to interpret a spectrum.

1. The spectrum must be adequately resolved and of adequate intensity.
2. The spectrum should be that of a reasonably pure compound.
3. The spectrophotometer should be calibrated so that the bands are observed at their proper frequencies or wavelengths. Proper calibration can be made with reliable standards, such as polystyrene film.
4. The method of sample handling must be specified. If a solvent is employed, the solvent, concentration, and the cell thickness should be indicated.

A precise treatment of the vibrations of a complex molecule is not feasible; thus, the IR spectrum must be interpreted from empirical comparison of spectra and extrapolation of studies of simpler molecules. Many questions arising in the interpretation of an IR spectrum can be answered by data obtained from the mass (Chapter 2) and NMR spectra (Chapter 3–6).

Infrared absorption of organic molecules is summarized in the chart of characteristic group absorptions in Appendix B. Many of the group absorptions vary over a wide range because the bands arise from complex interacting vibrations within the molecule. Absorption bands may, however, represent predominantly a single vibrational mode. Certain absorption bands, for example, those arising from the C—H, O—H, and C=O stretching modes, remain within fairly narrow regions of the spectrum. Important details of structure may be revealed by the exact position of an absorption band within these narrow regions. Shifts in absorption position and changes in band contours, accompanying changes in molecular environment, may also suggest important structural details.

The two important areas for a preliminary examination of a spectrum are the regions 4000–1300 and 900–650 cm^{-1}. The high-frequency portion of the spectrum is called the functional group region. The characteristic stretching frequencies for important functional groups such as OH, NH, and C=O occur in this portion of the spectrum. The absence of absorption in the assigned ranges for the various functional groups can usually be used as evidence for the absence of such groups in the molecule. Care must be exercised, however, in such interpretations since certain structural characteristics may cause a band to become extremely broad so that it may go unnoticed. For example, intramolecular hydrogen bonding in the enolic form of acetylacetone results in a broad OH band, which may be overlooked. The absence of absorption in the 1850–1540 cm^{-1} region excludes a structure containing a carbonyl group.

Weak bands in the high-frequency region, resulting from the fundamental absorption of functional groups, such as S—H and C=C, are extremely valuable in the determination of structure. Such weak bands would be of little value in the more complicated regions of the spectrum. Overtones and combination tones of lower frequency bands frequently appear in the high-frequency region of the spectrum. Overtone and combination-tone bands are characteristically weak except when Fermi resonance occurs. Strong skeletal bands for aromatics and heteroaromatics fall in the 1600–1300 cm^{-1} region of the spectrum.

The lack of strong absorption bands in the 900–650 cm^{-1} region generally indicates a nonaromatic structure. Aromatic and heteroaromatic compounds display strong out-of-plane C—H bending and ring bending absorption bands in this region that can frequently be correlated with the substitution pattern. Broad, moderately intense absorption in the low-frequency region suggests the presence of carboxylic acid dimers, amines, or amides, all of which show out-of-plane bending in this region. If the region is extended to 1000 cm^{-1}, absorption bands characteristic of alkene structures are included.

The intermediate portion of the spectrum, 1300–900 cm^{-1}, is usually referred to as the "fingerprint" region. The absorption pattern in this region is frequently complex, with the bands originating in interacting vibrational modes. This portion of the spectrum is extremely valuable when examined in reference to the other regions. For example, if alcoholic or phenolic O—H stretching absorption appears in the high-frequency region of the spectrum, the position of the C—C—O absorption band in the 1260–1000 cm^{-1} region frequently makes it possible to assign the O—H absorption to alcohols and phenols with highly specific structures. Absorption in this intermediate region is probably unique for every molecular species.

Any conclusions reached after examination of a particular band should be confirmed where possible by examination of other portions of the spectrum. For example, the assignment of a carbonyl band to an aldehyde should be confirmed by the appearance of a band or a pair of bands in the 2900–2695 cm^{-1} region of the spectrum, arising from C—H stretching vibrations of the aldehyde group. Similarly, the assignment of a carbonyl band to an ester should be confirmed by observation of a strong band in the C—O stretching region, 1300–1100 cm^{-1}.

Similar compounds may give virtually identical spectra under normal conditions, but fingerprint differences can be detected with an expanded vertical scale or with a very large sample (major bands off scale). For

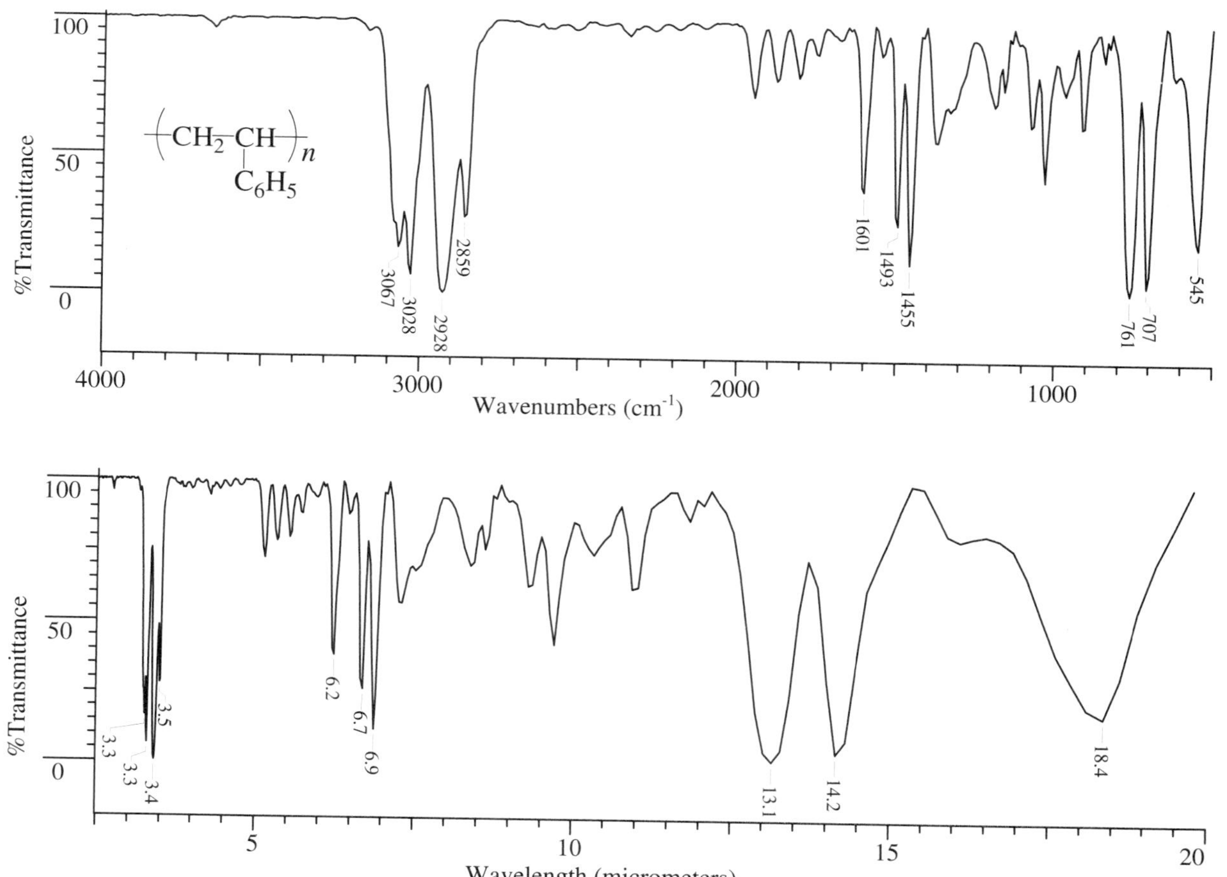

FIGURE 2.6. Polystyrene, same sample for both (a) and (b). Spectrum (a) linear in wavenumber (cm^{-1}); spectrum (b) linear in wavelength (μm).

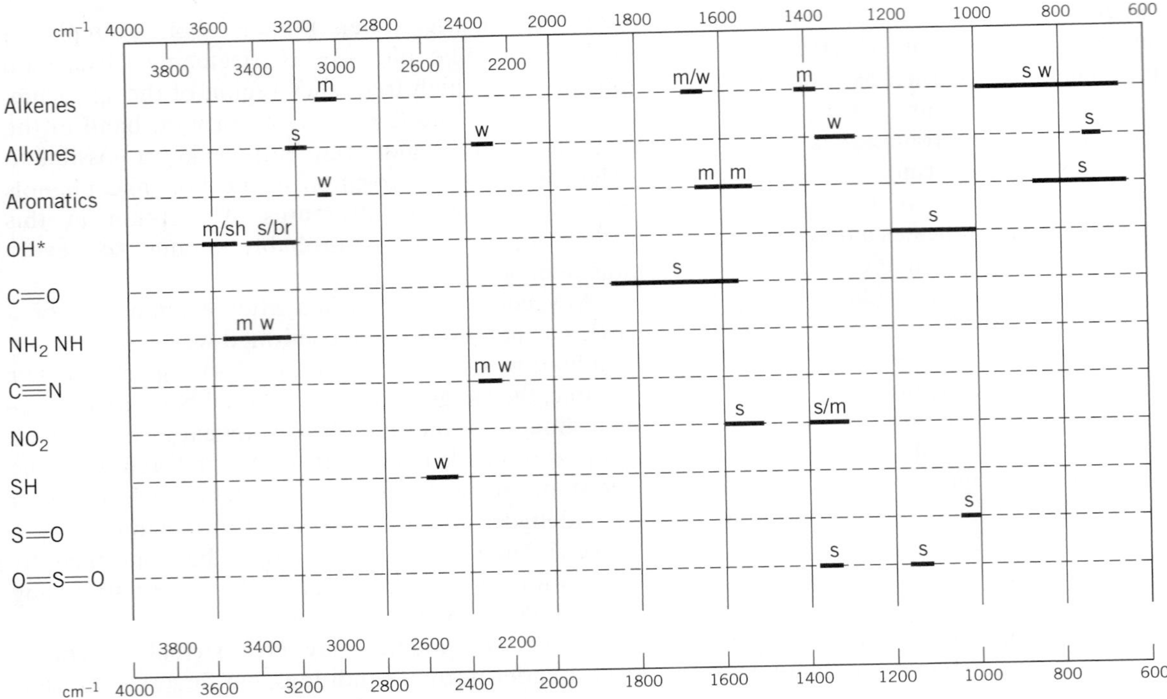

*Free OH, medium and sharp; bonded OH, strong and broad

FIGURE 2.7. Simplified chart of several common functional groups with very characteristic absorptions. s = strong, m = medium, w = weak, sh = sharp, br = broad.

example, pentane and hexane are essentially indistinguishable under normal conditions and can be differentiated only at very high recorder sensitivity.

Finally, in a "fingerprint" comparison of spectra, or any other situation in which the shapes of peaks are important, we should be aware of the substantial differences in the appearance of the spectrum in changing from a spectrum that is linear in wavenumber to one that is linear in wavelength (Figure 2.6).

Admittedly, the full chart of characteristic absorption groups (Appendix B) is intimidating. The following statements and a simplified chart may help (Figure 2.7).

The first bit of advice is negative: Do not attempt a frontal, systematic assault on an infrared spectrum. Rather, look for evidence of the presence or absence of a few common functional groups with very characteristic absorptions. Start with OH, C=O, and NH groups in Figure 2.7 since a "yes/no" answer is usually available. A "yes" answer for any of these groups sharpens the focus considerably. Certainly the answer will contribute to development of a molecular formula from the mass spectrum (Chapter 1) and to an entry point for the NMR spectra (Chapters 3–6). These other spectra, in turn, will suggest further leads in the IR spectrum.

Figure 2.7 lists the common groups that provide distinctive, characteristic absorptions. Section 2.6 furnishes more detailed information-including a number of caveats.

2.6 CHARACTERISTIC GROUP ABSORPTIONS OF ORGANIC MOLECULES

A table of characteristic group absorptions is presented as Appendix B. The ranges presented for group absorptions have been assigned following the examination of many compounds in which the groups occur. Although the ranges are quite well defined, the precise frequency or wavelength at which a specific group absorbs is dependent on its environment within the molecule and on its physical state.

This section is concerned with a comprehensive look at these characteristic group absorptions and their relationship to molecular structure. As a major type or class of molecule or functional group is introduced in the succeeding sections, an example of an IR spectrum with the important peak assignments will be given for most.

2.6.1 Normal Alkanes (Paraffins)

The spectra of normal alkanes (paraffins) can be interpreted in terms of four vibrations, namely, the stretching and bending of C—H and C—C bonds. Detailed analysis of the spectra of the lower members of the alkane series has made detailed assignments of the spectral positions of specific vibrational modes possible.

Not all of the possible absorption frequencies of the paraffin molecule are of equal value in the assignment of structure. The C—C bending vibrations occur at very low frequencies (below 500 cm^{-1}) and therefore do not appear in our spectra. The bands assigned to C—C stretching vibrations are weak and appear in the broad region of 1200–800 cm^{-1}; they are generally of little value for identification.

The most characteristic vibrations are those arising from C—H stretching and bending. Of these vibrations, those arising from methylene twisting and wagging are usually of limited diagnostic value because of their weakness and instability. This instability is a result of strong coupling to the remainder of the molecule.

The vibrational modes of alkanes are common to many organic molecules. Although the positions of C—H stretching and bending frequencies of methyl and methylene groups remain nearly constant in hydrocarbons, the attachment of CH_3 or CH_2 to atoms other than carbon, or to a carbonyl group or aromatic ring, may result in appreciable shifts of the C—H stretching and bending frequencies.

The spectrum of dodecane, Figure 2.8, is that of a typical straight-chain hydrocarbon.

2.6.1.1 C—H Stretching Vibrations Absorption arising from C—H stretching in the alkanes occurs in the general region of 3000–2840 cm^{-1}. The positions of the C—H stretching vibrations are among the most stable in the spectrum.

Methyl Groups An examination of a large number of saturated hydrocarbons containing methyl groups showed, in all cases, two distinct bands occurring at approximately 2962 and 2872 cm^{-1}. The first of these results from the asymmetrical (as) stretching mode in which two C—H bonds of the methyl group are extending while the third one is contracting ($\nu_{as}CH_3$). The second arises from symmetrical (s) stretching (ν_sCH_3) in which all three of the C—H bonds extend and contract in phase. The presence of several methyl groups in a molecule results in strong absorption at these positions.

Methylene Groups The asymmetrical stretching ($\nu_{as}CH_2$) and symmetrical stretching (ν_sCH_2) occur, respectively, near 2926 and 2853 cm^{-1}. The positions of these bands do not vary more than ±10 cm^{-1} in the aliphatic and nonstrained cyclic hydrocarbons. The frequency of methylene stretching is increased when the methylene group is part of a strained ring.

2.6.1.2 C—H Bending Vibrations Methyl Groups Two bending vibrations can occur within a methyl group. The first of these, the symmetrical bending vibration, involves the in-phase bending of the C—H bonds (**I**). The second, the asymmetrical bending vibration, involves out-of-phase bending of the C—H bonds (**II**).

I **II**

In **I**, the C—H bonds are moving like the closing petals of a flower; in **II**, one petal opens as two petals close.

The symmetrical bending vibration (δ_sCH_3) occurs near 1375 cm^{-1}, the asymmetrical bending vibration ($\delta_{as}CH_3$) near 1450 cm^{-1}.

The asymmetrical vibration generally overlaps the scissoring vibration of the methylene groups (see below). Two distinct bands are observed, however, in compounds such as diethyl ketone, in which the methylene scissoring band has been shifted to a lower frequency, 1439–1399 cm^{-1}, and increased in intensity because of its proximity to the carbonyl group.

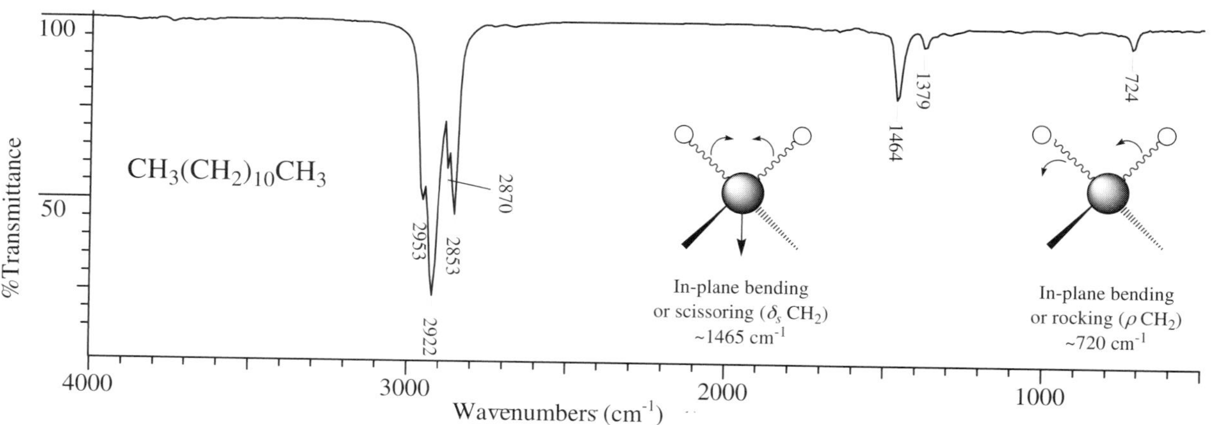

FIGURE 2.8. Dodecane. C—H stretch: 2953 cm^{-1} $\nu_{as}CH_3$, 2870 cm^{-1} ν_sCH_3, 2922 cm^{-1} $\nu_{as}CH_2$, 2853 cm^{-1} ν_sCH_2. C—H bend: 1464 cm^{-1} δ_sCH_2, 1450 cm^{-1} $\delta_{as}CH_3$, 1379 cm^{-1} δ_s CH_3. CH_2 rock: 724 cm^{-1} ρ CH_2.

The absorption band near 1375 cm^{-1}, arising from the symmetrical bending of the methyl C—H bonds, is very stable in position when the methyl group is attached to another carbon atom. The intensity of this band is greater for each methyl group in the compound than that for the asymmetrical methyl bending vibration or the methylene scissoring vibration.

Methylene Groups The bending vibrations of the C—H bonds in the methylene group have been shown schematically in Figure 2.1. The four bending vibrations are referred to as *scissoring, rocking, wagging,* and *twisting.*

The scissoring band ($\delta_s CH_2$) in the spectra of hydrocarbons occurs at a nearly constant position near 1465 cm^{-1} (see Figure 2.8).

The band resulting from the methylene rocking vibration (ρ CH_2), in which all of the methylene groups rock in phase, appears near 720 cm^{-1} for straight-chain alkanes of seven or more carbon atoms. This band may appear as a doublet in the spectra of solid samples. In the lower members of the *n*-alkane series, the band appears at somewhat higher frequencies.

Absorption of hydrocarbons, because of methylene twisting and wagging vibrations, is observed in the 1350–1150 cm^{-1} region. These bands are generally appreciably weaker than those resulting from methylene scissoring. A series of bands in this region, arising from the methylene group, is characteristic of the spectra of solid samples of long-chain acids, amides, and esters.

2.6.2 Branched-Chain Alkanes

In general, the changes brought about in the spectrum of a hydrocarbon by branching result from changes in skeletal stretching vibrations and methyl bending vibrations; these occur below 1500 cm^{-1}. The spectrum of isooctane in Figure 2.9 is that of a typical branched alkane.

2.6.2.1 C—H Stretching Vibrations Tertiary C—H Groups

Absorption resulting from this vibrational mode is very weak and is usually lost in other aliphatic C—H absorption. Absorption in hydrocarbons occurs near 2890 cm^{-1}.

2.6.2.2 C—H Bending Vibrations gem-Dimethyl Groups

Configurations in which two methyl groups are attached to the same carbon atom exhibit distinctive absorption in the C—H bending region. The isopropyl group shows a strong doublet, with peaks of almost equal intensity at 1385–1380 and 1370–1365 cm^{-1}. The *tertiary* butyl group gives rise to two C—H bending bands, one in the 1395–1385 cm^{-1} region and one near 1370 cm^{-1}. In the *t*-butyl doublet, the long wavelength band is more intense. When the *gem*-dimethyl group occurs at an internal position, a doublet is observed in essentially the same region where absorption occurs for the isopropyl and *t*-butyl groups. Doublets are observed for *gem*-dimethyl groups because of interaction between the in-phase and out-of-phase CH_3 bending of the two methyl groups attached to a common carbon atom.

Weak bands result from methyl rocking vibrations in isopropyl and *t*-butyl groups. These vibrations are sensitive to mass and interaction with skeletal stretching modes and are generally less reliable than the C—H bending vibrations. The following assignments have been made: isopropyl group, 922–919 cm^{-1}, and *t*-butyl group, 932–926 cm^{-1}.

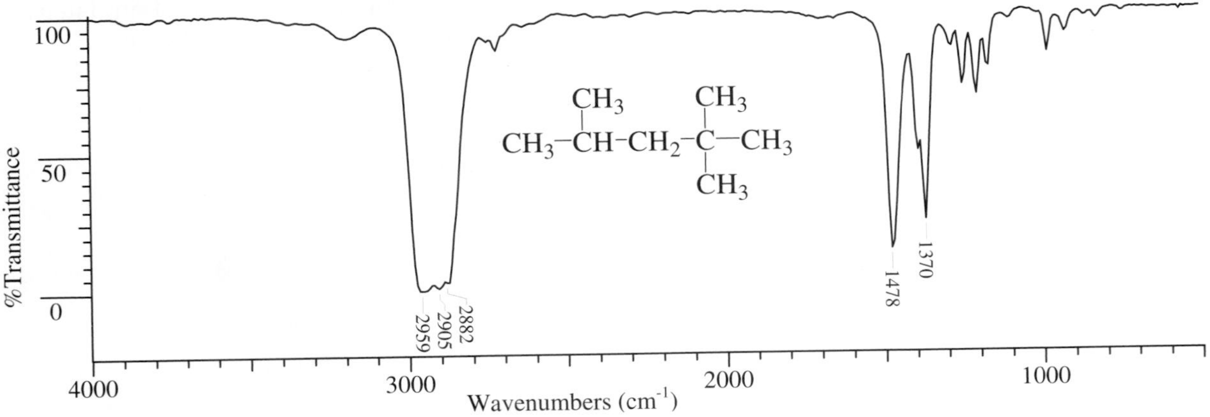

FIGURE 2.9. 2,2,4-Trimethylpentane.* C—H stretch (see Figure 2.8). C—H bend (see Figure 2.8). There are overlapping doublets for the *t*-butyl and the isopropyl groups at 1400–1340 cm^{-1}. Compare the absence of a methylene rocking band(s) 1000–800 cm^{-1} to Figure 2.8.

* Isooctane is the trivial name for 2,2,4-trimethylpentane.

2.6.3 Cyclic Alkanes

2.6.3.1 C—H Stretching Vibrations The methylene stretching vibrations of unstrained cyclic poly(methylene) structures are much the same as those observed for acyclic paraffins. Increasing ring strain moves the C—H stretching bands progressively to high frequencies. The ring CH_2 and CH groups in a monoalkylcyclopropane ring absorb in the region of 3100–2990 cm^{-1}.

2.6.3.2 C—H Bending Vibrations Cyclization decreases the frequency of the CH_2 scissoring vibration. Cyclohexane absorbs at 1452 cm^{-1}, whereas *n*-hexane absorbs at 1468 cm^{-1}. Cyclopentane absorbs at 1455 cm^{-1}, cyclopropane absorbs at 1442 cm^{-1}. This shift frequently makes it possible to observe distinct bands for methylene and methyl absorption in this region.

2.6.4 Alkenes

Alkene (olefinic) structures introduce several new modes of vibration into a hydrocarbon molecule: a C═C stretching vibration, C—H stretching vibrations in which the carbon atom is present in the alkene linkage, and in-plane and out-of-plane bending of the alkene C—H bond. The spectrum of Figure 2.10 is that of a typical terminal alkene.

***2.6.4.1 C═C Stretching Vibrations* Unconjugated Linear Alkenes** The C═C stretching mode of unconjugated alkenes usually shows moderate to weak absorption at 1667–1640 cm^{-1}. Monosubstituted alkenes, that is, vinyl groups, absorb near 1640 cm^{-1}, with moderate intensity. Disubstituted *trans*-alkenes, tri- and tetraalkyl-substituted alkenes absorb at or near 1670 cm^{-1}; disubstituted *cis*-alkenes and vinylidene alkenes absorb near 1650 cm^{-1}.

The absorption of symmetrical disubstituted *trans*-alkenes or tetrasubstituted alkenes may be extremely weak or absent. The *cis*-alkenes, which lack the symmetry of the *trans* structure, absorb more strongly than *trans*-alkenes. Internal double bonds generally absorb more weakly than terminal double bonds because of pseudosymmetry.

Abnormally high absorption frequency is observed for —CH═CF_2 and —CF═CF_2 groups. The former absorbs near 1754 cm^{-1}, the latter near 1786 cm^{-1}. In contrast, the absorption frequency is reduced by the attachment of chlorine, bromine, or iodine.

Cycloalkenes Absorption of the internal double bond in the unstrained cyclohexene system is essentially the same as that of a *cis*-isomer in an acyclic system. The C═C stretch vibration is coupled with the C—C stretching of the adjacent bonds. As the angle α (C–C═C, C⟷C, α) becomes smaller, the interaction becomes less until it is at a minimum at 90° in cyclobutene (1566 cm^{-1}). In the cyclopropene structure, interaction again becomes appreciable, and the absorption frequency increases (1641 cm^{-1}).

The substitution of alkyl groups for a hydrogen atom in strained ring systems serves to increase the frequency of C═C absorption. Cyclobutene absorbs at 1566 cm^{-1}, 1-methylcyclobutene at 1641 cm^{-1}.

The absorption frequency of external exocyclic bonds increases with decreasing ring size. Methylenecyclohexane absorbs at 1650 cm^{-1}, methylenecyclopropane at 1781 cm^{-1}.

Conjugated Systems The alkene bond stretching vibrations in conjugated dienes without a center of symmetry interact to produce two C═C stretching bands. The spectrum of an unsymmetrical conjugated diene, such as 1,3-pentadiene, shows absorption near

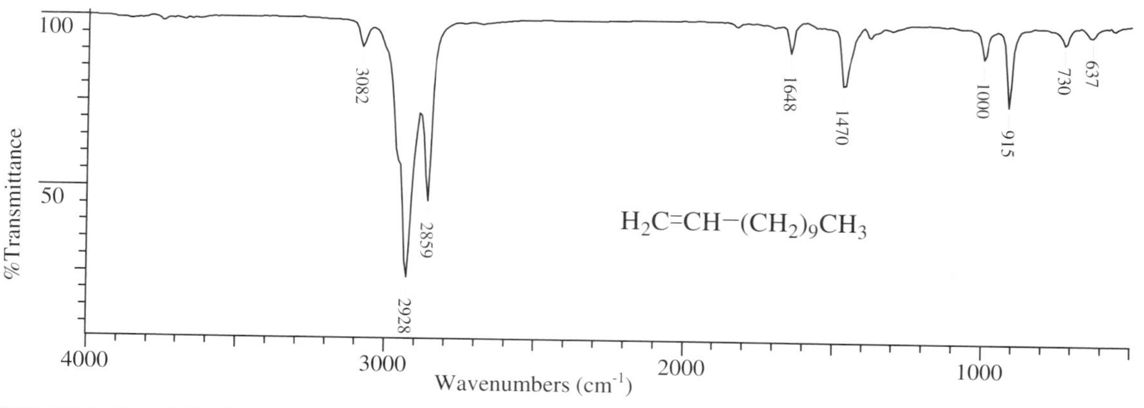

FIGURE 2.10. 1-Dodecene. C—H stretch (see Figure 2.8). Note alkene C—H stretch at 3082 cm^{-1}. C═C stretch, 1648 cm^{-1}, see Appendix Table C-1. Out-of-plane C—H bend: 1000 cm^{-1}, (alkene) 915 cm^{-1}. Methylene rock: 730 cm^{-1}.

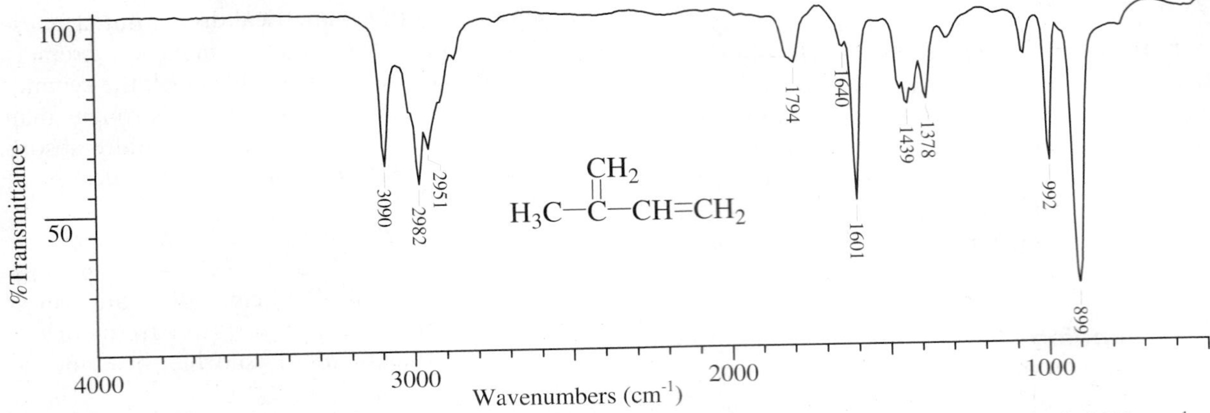

FIGURE 2.11. Isoprene. C—H stretch: =C—H 3090 cm^{-1}. Coupled C=C—C=C stretch: symmetric 1640 cm^{-1} (weak), asymmetric 1601 cm^{-1} (strong). C—H bend (saturated, alkene in-plane). C—H out-of-plane bend: 992 cm^{-1}, 899 cm^{-1} (see vinyl, Appendix Table C-1.)

1650 and 1600 cm^{-1}. The symmetrical molecule 1, 3-butadiene shows only one band near 1600 cm^{-1}, resulting from asymmetric stretching; the symmetrical stretching band is inactive in the IR. The IR spectrum of isoprene (Figure 2.11) illustrates many of these features.

Conjugation of an alkene double bond with an aromatic ring produces enhanced alkene absorption near 1625 cm^{-1}.

The absorption frequency of the alkene bond in conjugation with a carbonyl group is lowered by about 30 cm^{-1}; the intensity of absorption is increased. In *s-cis* structures, the alkene absorption may be as intense as that of the carbonyl group. *s-Trans* structures absorb more weakly than *s-cis* structures.

Cumulated Alkenes A cumulated double-bond system, as occurs in the allenes $\left(\gt C{=}C{=}CH_2\right)$ absorbs near 2000–1900 cm^{-1}. The absorption results from asymmetric C=C=C stretching. The absorption may be considered an extreme case of exocyclic C=C absorption.

2.6.4.2 Alkene C—H Stretching Vibrations

In general, any C—H stretching bands above 3000 cm^{-1} result from aromatic, heteroaromatic, alkyne, or alkene C—H stretching. Also found in the same region are the C—H stretching in small rings, such as cyclopropane, and the C—H in halogenated alkyl groups. The frequency and intensity of alkene C—H stretching absorption are influenced by the pattern of substitution. With proper resolution, multiple bands are observed for structures in which stretching interaction may occur. For example, the vinyl group produces three closely spaced C—H stretching bands. Two of these result from symmetrical and asymmetrical stretching of the terminal C—H groups, and the third from the stretching of the remaining single C—H.

2.6.4.3 Alkene C—H Bending Vibrations

Alkene C—H bonds can undergo bending either in the same plane as the C=C bond or perpendicular to it; the bending vibrations can be either in phase or out of phase with respect to each other.

Assignments have been made for a few of the more prominent and reliable in-plane bending vibrations. The vinyl group absorbs near 1416 cm^{-1} because of a scissoring vibration of the terminal methylene. The C—H rocking vibration of a *cis*-disubstituted alkene occurs in the same general region.

The most characteristic vibrational modes of alkenes are the out-of-plane C—H bending vibrations between 1000 and 650 cm^{-1}. These bands are usually the strongest in the spectra of alkenes. The most reliable bands are those of the vinyl group, the vinylidene group, and the *trans*-disubstituted alkene. Alkene absorption is summarized in Appendix Tables C-1 and C-2.

In allene structures, strong absorption is observed near 850 cm^{-1}, arising from $=CH_2$ wagging. The first overtone of this band may also be seen.

2.6.5 Alkynes

The two stretching vibrations in alkynes (acetylenes) involve C≡C and C—H stretching. Absorption due to C—H bending is characteristic of acetylene and monosubstituted alkynes. The spectrum of Figure 2.12 is that of a typical terminal alkyne.

2.6.5.1 C≡C Stretching Vibrations The weak C≡C stretching band of alkyne molecules occurs in the region of 2260–2100 cm^{-1}. Because of symmetry, no C≡C band is observed in the IR for symmetrically substituted alkynes. In the IR spectra of monosubstituted alkynes, the band appears at 2140–2100 cm^{-1}.

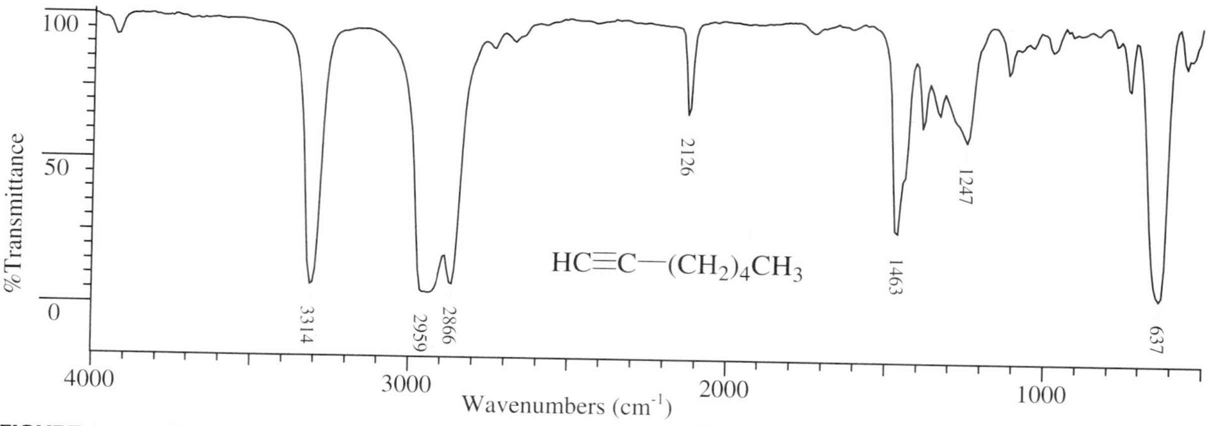

FIGURE 2.12. 1-Heptyne. ≡C—H stretch, 3314 cm^{-1}. Alkyl C—H stretch 1450–1360 cm^{-1} (see Figure 2.8), 2960–2860 cm^{-1}. C≡C stretch, 2126 cm^{-1}. C—H bend: 1463 cm^{-1} $\delta_s CH_2$, 1450 cm^{-1} $\delta_{as} CH_3$. ≡C—H bend overtone, 1247 cm^{-1}. ≡C—H bend fundamental, 637 cm^{-1}.

Disubstituted alkynes, in which the substituents are different, absorb near 2260–2190 cm^{-1}. When the substituents are similar in mass, or produce similar inductive and resonance effects, the band may be so weak as to be unobserved in the IR spectrum. For reasons of symmetry, a terminal C≡C produces a stronger band than an internal C≡C (pseudosymmetry). The intensity of the C≡C stretching band is increased by conjugation with a carbonyl group.

2.6.5.2 C—H Stretching Vibrations The C—H stretching band of monosubstituted alkynes occurs in the general region of 3333–3267 cm^{-1}. This is a strong band and is narrower than the hydrogen-bonded OH and NH bands occurring in the same region.

2.6.5.3 C—H Bending Vibrations The C—H bending vibration of alkynes or monosubstituted alkynes leads to strong, broad absorption in the 700–610 cm^{-1} region. The first overtone of the C—H bending vibration appears as a weak, broad band in the 1370–1220 cm^{-1} region.

2.6.6 Mononuclear Aromatic Hydrocarbons

The most prominent and most informative bands in the spectra of aromatic compounds occur in the low-frequency range between 900 and 675 cm^{-1}. These strong absorption bands result from the out-of-plane ("oop") bending of the ring C—H bonds. In-plane bending bands appear in the 1300–1000 cm^{-1} region. Skeletal vibrations, involving carbon—carbon stretching within the ring, absorb in the 1600–1585 and 1500–1400 cm^{-1} regions. The skeletal bands frequently appear as doublets, depending on the nature of the ring substituents.

Aromatic C—H stretching bands occur between 3100 and 3000 cm^{-1}. Weak combination and overtone bands appear in the 2000–1650 cm^{-1} region. The pattern of the overtone bands is not a reliable guide to the substitution pattern of the ring. Because they are weak, the overtone and combination bands are most readily observed in spectra obtained from thick samples. The spectrum of Figure 2.13 is that of a typical aromatic (benzenoid) compound.

2.6.6.1 Out-of-Plane C—H Bending Vibrations The in-phase, out-of-plane bending of a ring hydrogen atom is strongly coupled to adjacent hydrogen atoms. The position of absorption of the out-of-plane bending bands is therefore characteristic of the number of adjacent hydrogen atoms on the ring. The bands are frequently intense and appear at 900–675 cm^{-1}.

Assignments for C—H out-of-plane bending bands in the spectra of substituted benzenes appear in the chart of characteristic group absorptions (Appendix B). These assignments are usually reliable for alkyl-substituted benzenes, but caution must be observed in the interpretation of spectra when polar groups are attached directly to the ring, for example, in nitrobenzenes, aromatic acids, and esters or amides of aromatic acids.

The absorption band that frequently appears in the spectra of substituted benzenes near 600–420 cm^{-1} is attributed to out-of-plane ring bending.

2.6.7 Polynuclear Aromatic Hydrocarbons

Polynuclear aromatic compounds, like the mononuclear aromatics, show characteristic absorption in three regions of the spectrum.

The aromatic C—H stretching and the skeletal vibrations absorb in the same regions as observed for the mononuclear aromatics. The most characteristic

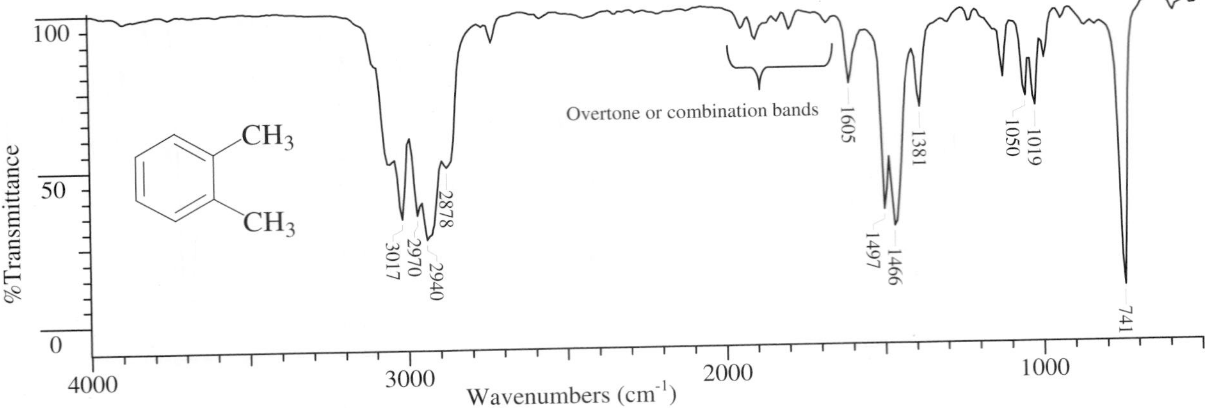

FIGURE 2.13. *o*-Xylene. Aromatic C—H stretch, 3017 cm^{-1}. Methyl bands, C—H stretch 3970, 2940, 2875 cm^{-1}. Overtone or combination bands, 2000–1667 cm^{-1} (see Figure 2.16). C═C ring stretch, 1605, 1497, 1466 cm^{-1}. In-plane C—H bend, 1050, 1019 cm^{-1}. Out-of-plane —C—H bend, 741 cm^{-1}.

absorption of polynuclear aromatics results from C—H out-of-plane bending in the 900–675 cm^{-1} region. These bands can be correlated with the number of adjacent hydrogen atoms on the rings. Most β-substituted naphthalenes, for example, show three absorption bands resulting from out-of-plane C—H bending; these correspond to an isolated hydrogen atom and two adjacent hydrogen atoms on one ring and four adjacent hydrogen atoms on the other ring.

In the spectra of α-substituted naphthalenes, the bands for the isolated hydrogen and the two adjacent hydrogen atoms of β-naphthalenes are replaced by a band for three adjacent hydrogen atoms. This band is near 810–785 cm^{-1}. Additional bands may appear because of ring bending vibrations (see Table 2.3). The position of absorption bands for more highly substituted naphthalenes and other polynuclear aromatics are summarized by Colthup et al. (1990) and by Conley (1972).

2.6.8 Alcohols and Phenols

The characteristic bands observed in the spectra of alcohols and phenols result from O—H stretching and C—O stretching. These vibrations are sensitive to hydrogen bonding. The C—O stretching and O—H bending modes are not independent vibrational modes because they couple with the vibrations of adjacent groups.

TABLE 2.3 C—H Out-of-Plane Bending Vibrations of a β-Substituted Naphthalene

Substitution Pattern	Absorption Range (cm^{-1})
Isolated hydrogen	862–835
Two adjacent hydrogen atoms	835–805
Four adjacent hydrogen atoms	760–735

2.6.8.1 O—H Stretching Vibrations The nonhydrogen-bonded or "free" hydroxyl group of alcohols and phenols absorbs strongly in the 3700–3584 cm^{-1} region. These sharp, "free" hydroxyl bands are observed in the vapor phase, in very dilute solution in nonpolar solvents or for hindered OH groups. Intermolecular hydrogen bonding increases as the concentration of the solution increases, and additional bands start to appear at lower frequencies, 3550–3200 cm^{-1}, at the expense of the free hydroxyl band. This effect is illustrated in Figure 2.14, in which the absorption bands in the O—H stretching region are shown for two

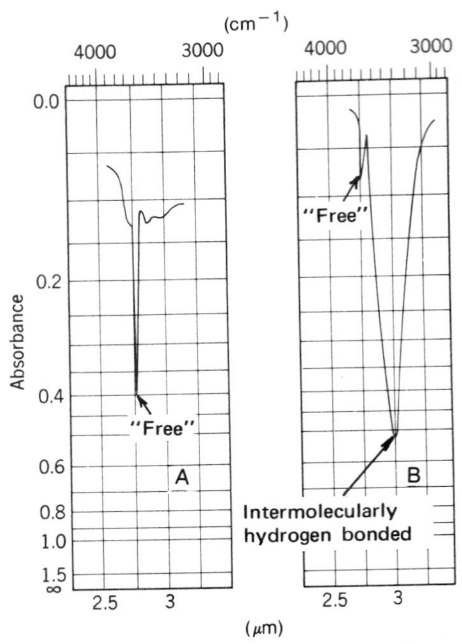

FIGURE 2.14. Infrared spectrum of the O—H stretching region of cyclohexylcarbinol in CCl_4. Peak A at 0.03 *M* (0.406-mm cell); Peak B at 1.00 *M* (0.014-mm cell).

different concentrations of cyclohexylcarbinol in carbon tetrachloride. For comparisons of this type, the path length of the cell must be altered with changing concentration, so that the same number of absorbing molecules will be present in the IR beam at each concentration. The band at 3623 cm^{-1} results from the monomer, whereas the broad absorption near 3333 cm^{-1} arises from "polymeric" structures.

H R R
O O O
R H H etc.

Strong intramolecular hydrogen bonding occurs in *o*-hydroxyacetophenone. The resulting absorption at 3077 cm^{-1} is broad, shallow, and independent of concentration (Figure 2.15).

O H
O
CH_3

In contrast, *p*-hydroxyacetophenone

Ar
C=O---HO— —C(=O---etc.)
H_3C CH_3

shows a sharp free hydroxy peak at 3600 cm^{-1} in dilute CCl_4 solution as well as a broad, strong intermolecular peak at 3100 cm^{-1} in the spectrum of a neat sample. In structures such as 2,6-di-*t*-butylphenol, in which steric hindrance prevents hydrogen bonding, no bonded hydroxyl band is observed, not even in spectra of neat samples.

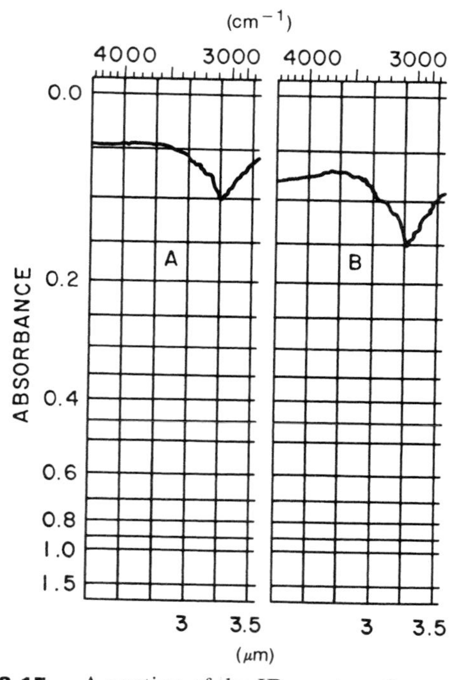

FIGURE 2.15. A portion of the IR spectra of *o*-hydroxyacetophenone. Peak A at 0.03 *M*, cell thickness: 0.41 mm. Peak B at 1.0 *M*, cell thickness: 0.015 mm.

2.6.8.2 C—O Stretching Vibrations The C—O stretching vibrations in alcohols (Figure 2.16 and 2.18) and phenols (Figure 2.17) produce a strong band in the 1260–1000 cm^{-1} region of the spectrum.

The C—O stretching mode is coupled with the adjacent C—C stretching vibration; thus in primary alcohols the vibration might better be described as an asymmetric C—C—O stretching vibration. The vibrational mode is further complicated by branching and

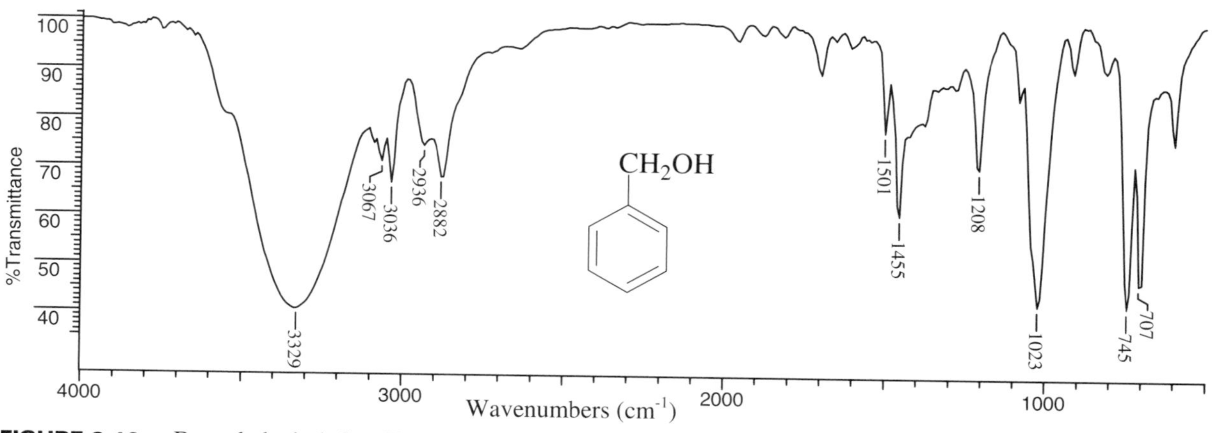

FIGURE 2.16. Benzyl alcohol. O—H stretch: intermolecular hydrogen bonded, 3329 cm^{-1}. C—H stretch: aromatic 3100–3000 cm^{-1}. C—H stretch: methylene, 2940–2860 cm^{-1}. Overtone or combination bands, 2000–1667 cm^{-1}. C⩦C ring stretch, 1501, 1455 cm^{-1}, overlapped by CH_2 scissoring, about 1471 cm^{-1}. O—H bend, possibly augmented by C—H in-plane bend, 1209 cm^{-1}. C—O stretch, primary alcohol (see Table 2.5) 1023 cm^{-1}. Out-of-plane aromatic C—H bend, 745 cm^{-1}. Ring C⩦C bend, 707 cm^{-1}.

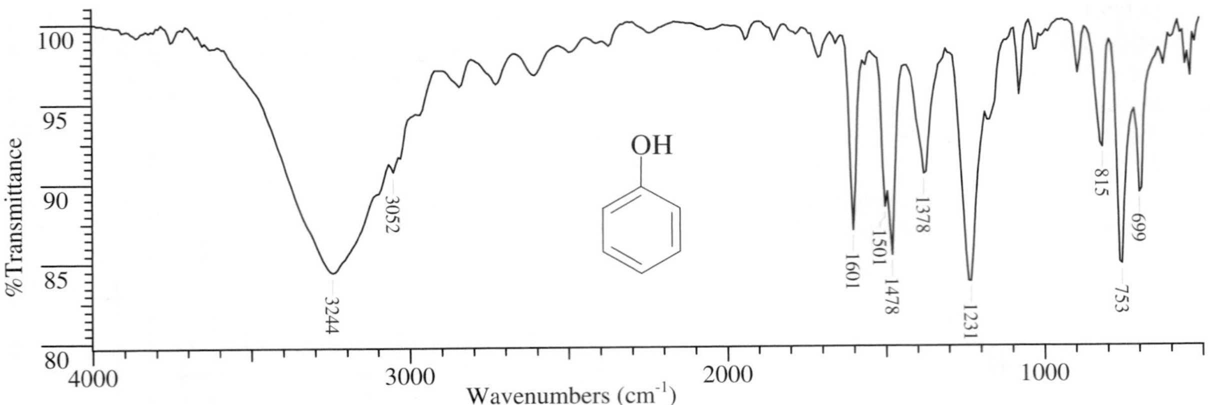

FIGURE 2.17. Phenol (Melt). Broad intermolecular hydrogen bonded, O—H stretch, 3244 cm^{-1}. Aromatic C—H stretch, 3052 cm^{-1}. Overtone or combination bands, 2000–1667 cm^{-1}. C═C ring stretch, 1601, 1501, 1478 cm^{-1}. In-plane O—H bend, 1378 cm^{-1}. C—O stretch, 1231 cm^{-1}. Out-of-plane C—H bend, 815, 753 cm^{-1}. Out-of-plane ring C═C bend, 699 cm^{-1}. (Broad) hydrogen-bonded, out-of-plane O—H bend, about 650 cm^{-1}.

α,βunsaturation. These effects are summarized in Table 2.4 for a series of secondary alcohols (neat samples).

TABLE 2.4 C—C—O Stretching Vibrations of Secondary Alcohols

Secondary Alcohol	Absorption (cm^{-1})
2-Butanol	1105
3-Methyl-2-butanol	1091
1-Phenylethanol	1073
3-Buten-2-ol	1058
Diphenylmethanol	1014

The absorption ranges of the various types of alcohols appear in Table 2.5. These values are for neat samples of the alcohols.

Mulls, pellets, or melts of phenols absorb at 1390–1330 and 1260–1180 cm^{-1}. These bands apparently result from interaction between O—H bending and C—O stretching. The long-wavelength band is the stronger and both bands appear at longer wavelengths in spectra observed in solution. The spectrum of phenol in Figure 2.17 was determined on a melt, to show a high degree of association.

2.6.8.3 O—H Bending Vibrations The O—H in-plane bending vibration occurs in the general region of 1420–1330 cm^{-1}. In primary and secondary alcohols, the O—H in-plane bending couples with the C—H wagging vibrations to produce two bands: the first near 1420 cm^{-1}, the second near 1330 cm^{-1}. These bands are of little diagnostic value. Tertiary alcohols, in which no coupling can occur, show a single band in this region, the position depending on the degree of hydrogen bonding.

The spectra of alcohols and phenols determined in the liquid state, show a broad absorption band in the 769–650 cm^{-1} region because of out-of-plane bending of the bonded O—H group.

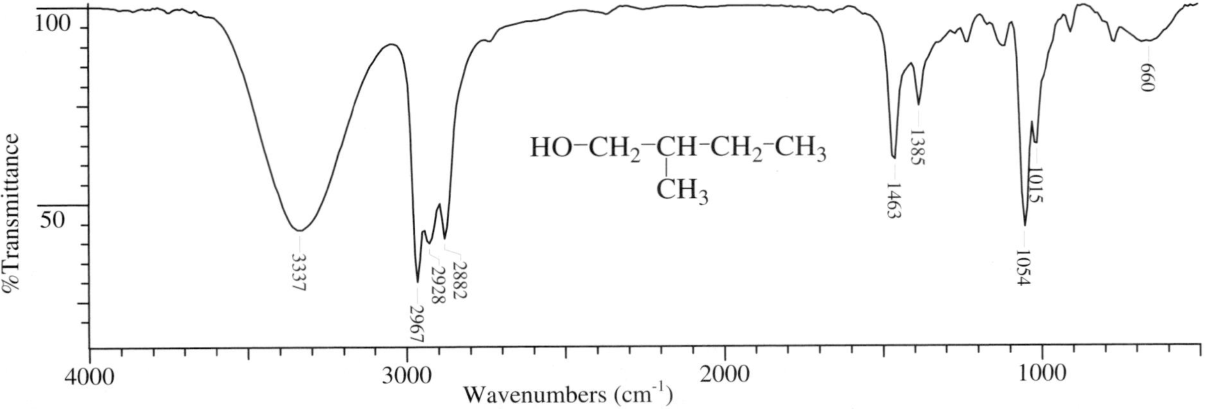

FIGURE 2.18. 2-Methyl-1-butanol. O—H stretch, intermolecular hydrogen bonding 3337 cm^{-1}. C—H stretch (see Figure 2.8, 3000–2800 cm^{-1}). C—H bend (see Figure 2.8). C—O stretch 1054 cm^{-1}.

TABLE 2.5 Alcoholic C—O Stretch Absorptions

Alcohol Type	Absorption Range (cm^{-1})
(1) Saturated tertiary (2) Secondary, highly symmetrical	1205–1124
(1) Saturated secondary (2) α-Unsaturated or cyclic tertiary	1124–1087
(1) Secondary, α-unsaturated (2) Secondary, alicyclic five- or six-membered ring (3) Saturated primary	1085–1050
(1) Tertiary, highly α-unsaturated (2) Secondary, di-α-unsaturated (3) Secondary, α-unsaturated and a-branched (4) Secondary, alicyclic seven- or eight-membered ring (5) Primary, α-unsaturated and/or a-branched	<1050

2.6.9 Ethers, Epoxides, and Peroxides

2.6.9.1 C—O Stretching Vibrations The characteristic response of ethers in the IR is associated with the stretching vibration of the C—O—C system. Since the vibrational characteristics of this system would not be expected to differ greatly from the C—C—C system, it is not surprising to find the response to C—O—C stretching in the same general region. However, since vibrations involving oxygen atoms result in greater dipole moment changes than those involving carbon atoms, more intense IR bands are observed for ethers. The C—O—C stretching bands of ethers, as is the case with the C—O stretching bands of alcohols, involve coupling with other vibrations within the molecule. The spectrum of anisole in Figure 2.19 is that of a typical aryl alkyl ether.

In the spectra of aliphatic ethers, the most characteristic absorption is a strong band in the 1150–1085 cm^{-1} region because of asymmetrical C—O—C stretching; this band usually occurs near 1125 cm^{-1}. The symmetrical stretching band is usually weak and is more readily observed in the Raman spectrum.

The C—O—C group in a six-membered ring absorbs at the same frequency as in an acyclic ether. As the ring becomes smaller, the asymmetrical C—O—C stretching vibration moves progressively to lower wavenumbers (longer wavelengths), whereas the symmetrical C—O—C stretching vibration (ring breathing frequency) moves to higher wavenumbers.

Branching on the carbon atoms adjacent to the oxygen usually leads to splitting of the C—O—C band. Isopropyl ether shows a triplet structure in the 1170–1114 cm^{-1} region, the principal band occurring at 1114 cm^{-1}.

Spectra of aryl alkyl ethers display an asymmetrical C—O—C stretching band at 1275–1200 cm^{-1} with symmetrical stretching near 1075–1020 cm^{-1}. Strong absorption caused by asymmetrical C—O—C stretching in vinyl ethers occurs in the 1225–1200 cm^{-1} region with a strong symmetrical band at 1075–1020 cm^{-1}. Resonance, which results in strengthening of the C—O bond, is responsible for the shift in the asymmetric absorption band of arylalkyl and vinyl ethers.

The C=C stretching band of vinyl ethers occurs in the 1660–1610 cm^{-1} region. This alkene band is characterized by its higher intensity compared with the C=C stretching band in alkenes. This band frequently appears as a doublet resulting from absorption of rotational isomers.

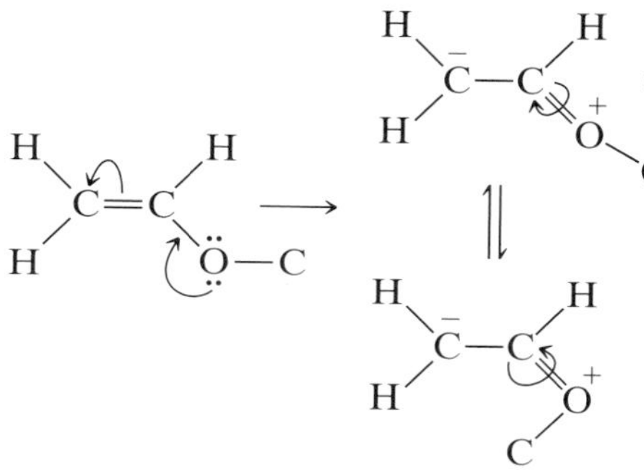

trans ~ 1620 cm^{-1}

cis ~ 1640 cm^{-1}

Coplanarity in the *trans*-isomer allows maximum resonance, thus more effectively reducing the double-bond character of the alkene linkage. Steric hindrance reduces resonance in the *cis*-isomer.

The two bands arising from =C—H wagging in terminal alkenes occur near 1000 and 909 cm^{-1}. In the spectra of vinyl ethers, these bands are shifted to longer wavelengths because of resonance.

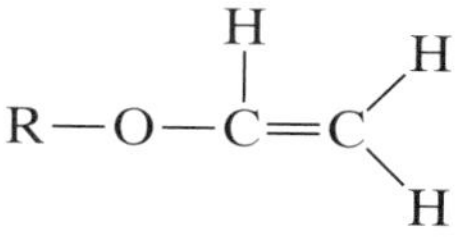

terminal CH_2 wag, 813 cm^{-1}
trans CH wag, 960 cm^{-1}

Alkyl and aryl peroxides display C—C—O absorption in the 1198–1176 cm^{-1} region. Acyl and aroyl peroxides display two carbonyl absorption bands in the 1818–1754 cm^{-1} region. Two bands are observed because of mechanical interaction between the stretching modes of the two carbonyl groups.

The symmetrical stretching, or ring breathing frequency, of the epoxy ring, all ring bonds stretching and

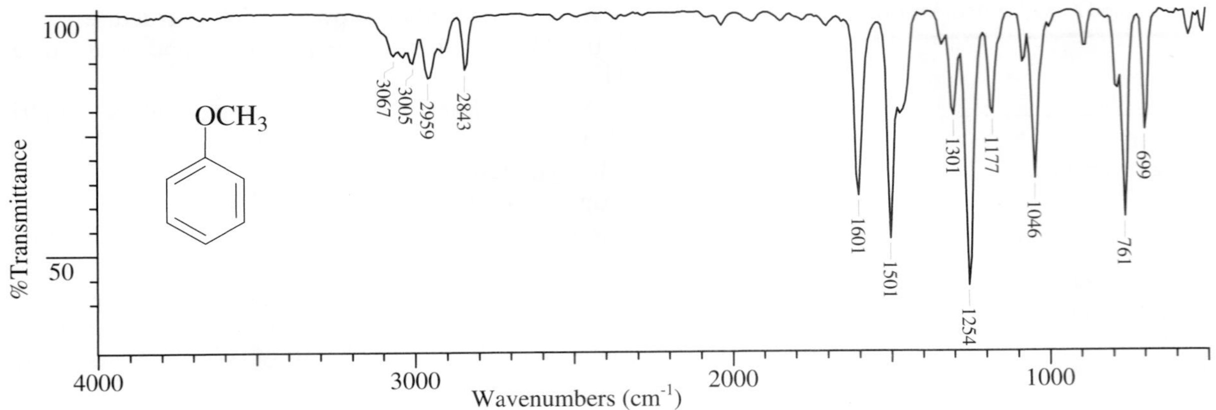

FIGURE 2.19. Anisole. Aromatic C—H stretch, 3067, 3030, 3005 cm^{-1}. Methyl C—H stretch, 2950, 2843 cm^{-1}. Overtone-combination region, 2000–1650 cm^{-1}. C==C ring stretch, 1601, 1501 cm^{-1}. Asymmetric C—O—C stretch, 1254 cm^{-1}. Symmetric C—O—C stretch, 1046 cm^{-1}. Out-of-plane C—H bend, 784, 761 cm^{-1}. Out-of-plane ring C==C bend, 699 cm^{-1}.

contracting in phase, occurs near 1250 cm^{-1}. Another band appears in the 950–810 cm^{-1} region attributed to asymmetrical ring stretching in which the C—C bond is stretching during contraction of the C—O bond. A third band, referred to as the "12 micron band," appears in the 840–750 cm^{-1} region. The C—H stretching vibrations of epoxy rings occur in the 3050–2990 cm^{-1} region of the spectrum.

2.6.10 Ketones

2.6.10.1 C=O Stretching Vibrations Ketones, aldehydes, carboxylic acids, carboxylic esters, lactones, acid halides, anhydrides, amides, and lactams show a strong C=O stretching absorption band in the region of 1870–1540 cm^{-1}. Its relatively constant position, high intensity, and relative freedom from interfering bands make this one of the easiest bands to recognize in IR spectra.

Within its given range, the position of the C=O stretching band is determined by the following factors: (1) the physical state, (2) electronic and mass effects of neighboring substituents, (3) conjugation, (4) hydrogen bonding (intermolecular and intramolecular), and (5) ring strain. Consideration of these factors leads to a considerable amount of information about the environment of the C=O group.

In a discussion of these effects, it is customary to refer to the absorption frequency of a neat sample of a saturated aliphatic ketone, 1715 cm^{-1}, as "normal." For example, acetone and cyclohexanone absorb at 1715 cm^{-1}. Changes in the environment of the carbonyl can either lower or raise the absorption frequency from this "normal" value. A typical ketone spectrum is displayed in Figure 2.20.

The absorption frequency observed for a neat sample is increased when absorption is observed in nonpolar solvents. Polar solvents reduce the frequency of absorption. The overall range of solvent effects does not exceed 25 cm^{-1}.

Replacement of an alkyl group of a saturated aliphatic ketone by a hetero atom (G) shifts the

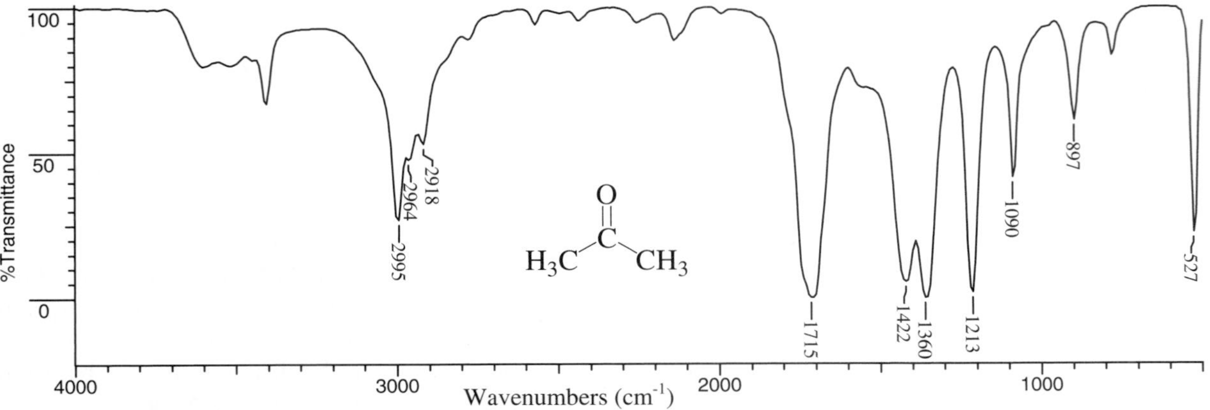

FIGURE 2.20. Acetone. ν_{as}, Methyl, 2995 cm^{-1}. ν_{as}, Methylene, 2964 cm^{-1}. ν_s, Methyl, 2918 cm^{-1}. Normal C=O stretch, 1715 cm^{-1}. δ_{as}, CH_3 1422 cm^{-1}. δ_s, CH_3 1360 cm^{-1}. C—CO—C stretch and bend, 1213 cm^{-1}.

carbonyl absorption. The direction of the shift depends on whether the inductive effect (a) or resonance effect (b) predominates.

(a) (b)

The inductive effect reduces the length of the C=O bond and thus increases its force constant and the frequency of absorption. The resonance effect increases the C=O bond length and reduces the frequency of absorption.

The absorptions of several carbonyl compound classes are summarized in Table 2.6.

Conjugation with a C=C bond results in delocalization of the π electrons of both unsaturated groups. Delocalization of the π electrons of the C=O group reduces the double-bond character of the C—O bond, causing absorption at lower wavenumbers (longer wavelengths). Conjugation with an alkene or phenyl group causes absorption in the 1685–1666 cm^{-1} region. Additional conjugation may cause a slight further reduction in frequency. This effect of conjugation is illustrated in Figure 2.21.

Steric effects that reduce the coplanarity of the conjugated system reduce the effect of conjugation. In the absence of steric hindrance, a conjugated system will tend toward a planar conformation. Thus, α,β-unsaturated ketones may exist in *s-cis* and *s-trans* conformations. When both forms are present, absorption for each of the forms is observed. The absorption of benzalacetone in CS_2 serves as an example; both the *s-cis* and *s-trans* forms are present at room temperature.

TABLE 2.6 The Carbonyl Absorption of Various RC(=O)G Compounds

G	ν C=O (cm^{-1})
G Effect Predominantly Inductive	
Cl	1815–1785
F	~1869
Br	1812
OH (monomer)	1760
OR	1750–1735
G Effect Predominantly Resonance	
NH_2	1695–1650
SR	1720–1690

s-trans 1674 cm^{-1} *s-cis* 1699 cm^{-1}

The absorption of the alkene bond in conjugation with the carbonyl group occurs at a lower frequency than that of an isolated C=C bond; the intensity of the conjugated double-bond absorption, when in an *s-cis* system, is greater than that of an isolated double bond.

Intermolecular hydrogen bonding between a ketone and a hydroxylic solvent such as methanol causes a slight decrease in the absorption frequency of the carbonyl group. For example, a neat sample of ethyl methyl ketone absorbs at 1715 cm^{-1},

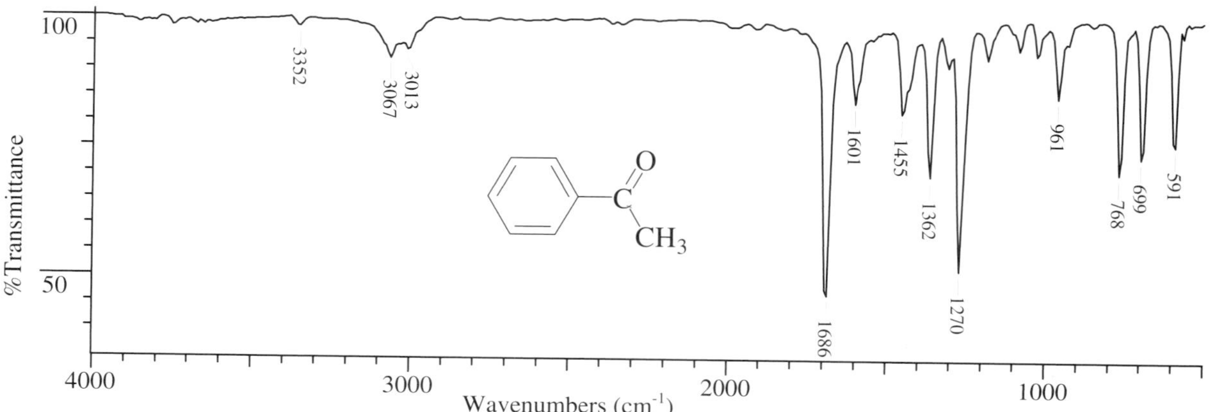

FIGURE 2.21. Acetophenone. Overtone of C=O stretch 3352 cm^{-1}; frequency about twice that of C=O stretch. C=O stretch, 1686 cm^{-1}, lower frequency than observed in Figure 2.20 because of the conjugation with the phenyl group.

whereas a 10% solution of the ketone in methanol absorbs at 1706 $\mathrm{cm^{-1}}$.

β-Diketones usually exist as mixtures of tautomeric keto and enol forms. The enolic form does not show the normal absorption of conjugated ketones. Instead, a broad band appears in the 1640–1580 $\mathrm{cm^{-1}}$ region, many times more intense than normal carbonyl absorption. The intense and displaced absorption results from intramolecular hydrogen bonding, the bonded structure being stabilized by resonance.

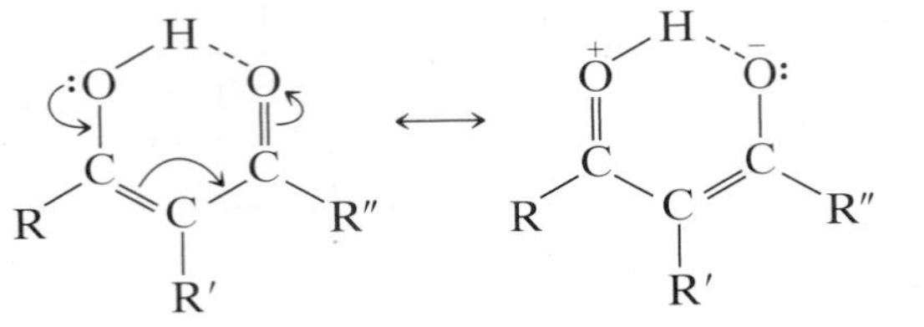

Acetylacetone as a liquid at 40°C exists to the extent of 64% in the enolic form that absorbs at 1613 $\mathrm{cm^{-1}}$. The keto form and a small amount of unbonded enolic form may be responsible for two bands centering near 1725 $\mathrm{cm^{-1}}$. Interaction between the two carbonyl groups in the keto form was suggested as a cause for this doublet. The enolic O—H stretching absorption is seen as a broad shallow band at 3000–2700 $\mathrm{cm^{-1}}$.

α-Diketones, in which carbonyl groups exist in formal conjugation, show a single absorption band near the frequency observed for the corresponding monoketone. Biacetyl absorbs at 1718 $\mathrm{cm^{-1}}$, benzil at 1681 $\mathrm{cm^{-1}}$. Conjugation is ineffective for α-diketones and the C=O groups of these diketones do not couple as do, for example, the corresponding groups in acid anhydrides (see Section 2.6.16).

Quinones, which have both carbonyl groups in the same ring, absorb in the 1690–1655 $\mathrm{cm^{-1}}$ region. With extended conjugation, in which the carbonyl groups appear in different rings, the absorption shifts to the 1655–1635 $\mathrm{cm^{-1}}$ region.

Acyclic α-chloro ketones absorb at two frequencies because of rotational isomerism. When the chlorine atom is near the oxygen, its negative field repels the nonbonding electrons of the oxygen atom, thus increasing the force constant of the C=O bond. This conformation absorbs at a higher frequency (1745 $\mathrm{cm^{-1}}$) than that in which the carbonyl oxygen and chlorine atom are widely separated (1725 $\mathrm{cm^{-1}}$). In rigid molecules such as the monoketo steroids, α-halogenation results in equatorial or axial substitution. In the equatorial orientation, the halogen atom is near the carbonyl group and the "field effect" causes an increase in the C=O stretching frequency. In the isomer in which the halogen atom is axial to the ring, and distant from the C=O, no shift is observed.

In cyclic ketones, the bond angle of the C(C)C=O group influences the absorption frequency of the carbonyl group. The C=O stretching undoubtedly is affected by adjacent C—C stretching. In acyclic ketones and in ketones with a six-membered ring, the angle is near 120°. In strained rings in which the angle is less than 120°, interaction with C—C bond stretching increases the energy required to produce C=O stretching and thus increases the stretching frequency. Cycloheptanone absorbs at 1709 $\mathrm{cm^{-1}}$, cyclohexanone absorbs at 1715 $\mathrm{cm^{-1}}$, cyclopentanone absorbs at 1751 $\mathrm{cm^{-1}}$, and cyclobutanone absorbs at 1775 $\mathrm{cm^{-1}}$.

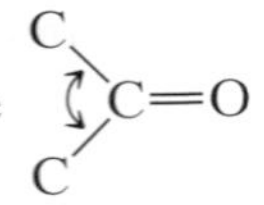

2.6.10.2 C—C(=O)—C Stretching and Bending Vibrations Ketones show moderate absorption in the 1300–1100 $\mathrm{cm^{-1}}$ region as a result of C—C—C stretching and C—C(=O)—C bending in the C—C—C group. The absorption may consist of multiple bands. Aliphatic ketones absorb in the 1230–1100 $\mathrm{cm^{-1}}$ region; aromatic ketones absorb at the higher frequency end of the general absorption region.

2.6.11 Aldehydes

The spectrum of octanal, illustrating typical aldehydic absorption characteristics, is shown in Figure 2.22.

2.6.11.1 C=O Stretching Vibrations The carbonyl groups of aldehydes absorb at slightly higher frequencies than that of the corresponding methyl ketones. Aliphatic aldehydes absorb near 1740–1720 $\mathrm{cm^{-1}}$. Aldehydic carbonyl absorption responds to structural changes in the same manner as ketones. Electronegative substitution on the α-carbon increases the frequency of carbonyl absorption. Acetaldehyde absorbs at 1730 $\mathrm{cm^{-1}}$, trichloroacetaldehyde absorbs at 1768 $\mathrm{cm^{-1}}$. Conjugate unsaturation, as in α,β-unsaturated aldehydes and benzaldehydes, reduces the frequency of carbonyl absorption. α,β-Unsaturated aldehydes and benzaldehydes absorb in the region of 1710–1685 $\mathrm{cm^{-1}}$. Internal hydrogen bonding, such as occurs in salicylaldehyde, shifts the absorption (1666 $\mathrm{cm^{-1}}$ for salicylaldehyde) to lower wavenumbers. Glyoxal, like the α-diketones, shows only one carbonyl absorption peak with no shift from the normal absorption position of monoaldehydic absorption.

2.6.11.2 C—H Stretching Vibrations The majority of aldehydes show aldehydic C—H stretching absorption in 2830–2695 $\mathrm{cm^{-1}}$ region. Two

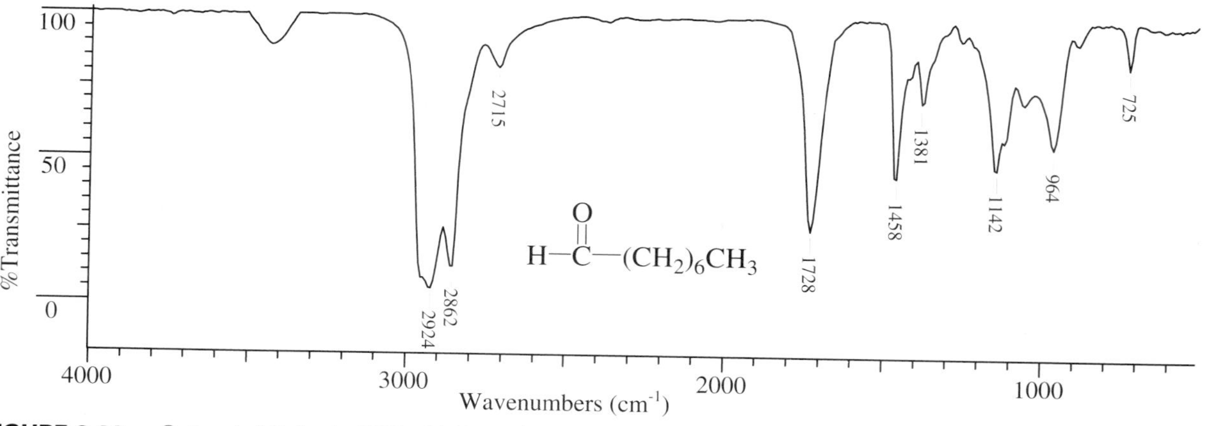

FIGURE 2.22. Octanal. Aliphatic, 2980–2860 cm^{-1} (see Figures 2.8). Aldehydic, C—H stretch, 2715 cm^{-1}. Normal aldehydic C═O stretch, 1728 cm^{-1}. Aldehydic C—H bend, 1381 cm^{-1}.

moderately intense bands are frequently observed in this region. The appearance of two bands is attributed to Fermi resonance between the fundamental aldehydic C—H stretch and the first overtone of the aldehydic C—H bending vibration that usually appears near 1390 cm^{-1}. Only one C—H stretching band is observed for aldehydes, whose C—H bending band has been shifted appreciably from 1390 cm^{-1}.

Some aromatic aldehydes with strongly electronegative groups in the *ortho* position may absorb as high as 2900 cm^{-1}.

An absorption of medium intensity near 2720 cm^{-1}, accompanied by a carbonyl absorption band is good evidence for the presence of an aldehyde group.

2.6.12 Carboxylic Acids

2.6.12.1 O—H Stretching Vibrations In the liquid or solid state, and in CCl_4 solution at concentrations much over 0.01 *M*, carboxylic acids exist as dimers due to strong hydrogen bonding.

R—C(=O⋯H—O)(—O—H⋯O=)C—R ⟷ R—C(—O^{-}⋯H—O^{+}=)(=O^{+}—H⋯O^{-}—)C—R

The exceptional strength of the hydrogen bonding is explained on the basis of the large contribution of the ionic resonance structure. Because of the strong bonding, a free hydroxyl stretching vibration (near 3520 cm^{-1}) is observed only in very dilute solution in nonpolar solvents or in the vapor phase.

Carboxylic acid dimers display very broad, intense O—H stretching absorption in the region of 3300–2500 cm^{-1}. The band usually centers near 3000 cm^{-1}. The weaker C—H stretching bands are generally seen superimposed upon the broad O—H band. Fine structure observed on the long-wavelength side of the broad O—H band represents overtones and combination tones of fundamental bands occurring at longer wavelengths. The spectrum of a typical aliphatic carboxylic acid is displayed in Figure 2.23.

Other structures with strong hydrogen bonding, such as β-diketones, also absorb in the 3300–2500 cm^{-1} region, but the absorption is usually less intense. Also, the C═O stretching vibrations of structures such as β-diketones are shifted to lower frequencies than those observed for carboxylic acids.

Carboxylic acids can bond intermolecularly with ethers, such as dioxane and tetrahydrofuran, or with other solvents that can act as proton acceptors. Spectra determined in such solvents show bonded O—H absorption near 3100 cm^{-1}.

2.6.12.2 C═O Stretching Vibrations The C═O stretching bands of acids are considerably more intense than ketonic C═O stretching bands. The monomers of saturated aliphatic acids absorb near 1760 cm^{-1}.

The carboxylic dimer has a center of symmetry; only the asymmetrical C═O stretching mode absorbs in the IR. Hydrogen bonding and resonance weaken the C═O bond, resulting in absorption at a lower frequency than the monomer. The C═O group in dimerized saturated aliphatic acids absorbs in the region of 1720–1706 cm^{-1}.

Internal hydrogen bonding reduces the frequency of the carbonyl stretching absorption to a greater degree than does intermolecular hydrogen bonding. For example, salicylic acid absorbs at 1665 cm^{-1}, whereas *p*-hydroxybenzoic acid absorbs at 1680 cm^{-1}.

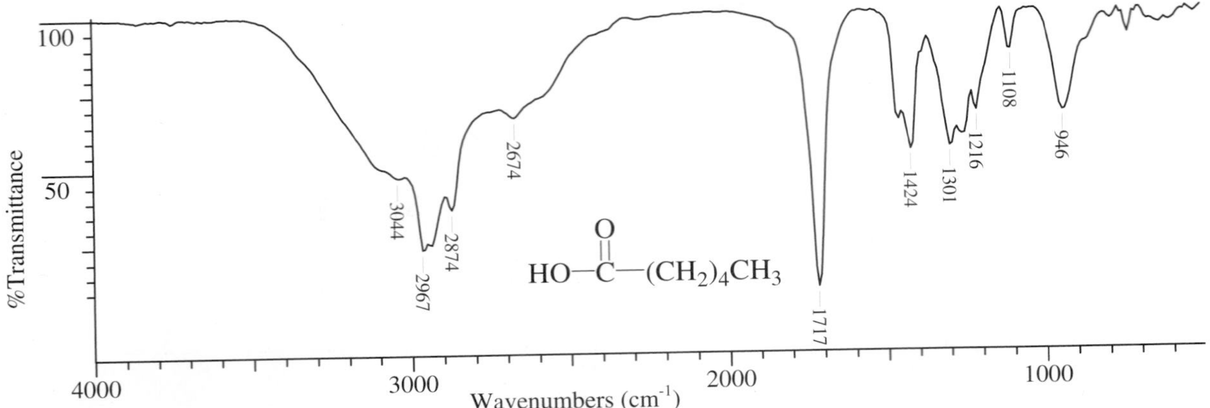

FIGURE 2.23. Hexanoic acid. Broad O—H stretch, 3300–2500 cm^{-1}. C—H stretch (see Figure 2.8), 2967, 2874, 2855 cm^{-1}. Superimposed upon O—H stretch. Normal, dimeric carboxylic C=O stretch, 1717 cm^{-1}. C—O—H in-plane bend, 1424 cm^{-1}. C—O stretch, dimer, 1301 cm^{-1}. F. O—H out-of-plane bend, 946 cm^{-1}.

Unsaturation in conjugation with the carboxylic carbonyl group decreases the frequency (increases the wavelength) of absorption of both the monomer and dimer forms only slightly. In general, α,β-unsaturated and aryl conjugated acids show absorption for the dimer in the 1710–1680 cm^{-1} region. Extension of conjugation beyond the α,β-position results in very little additional shifting of the C=O absorption.

Substitution in the α-position with electronegative groups, such as the halogens, brings about a slight increase in the C=O absorption frequency (10–20 cm^{-1}). The spectra of acids with halogens in the α-position, determined in the liquid state or in solution, show dual carbonyl bands resulting from rotational isomerism (field effect). The higher frequency band corresponds to the conformation in which the halogen is in proximity to the carbonyl group.

2.6.12.3 *C—O Stretching and O—H Bending Vibrations* Two bands arising from C—O stretching and O—H bending appear in the spectra of carboxylic acids near 1320–1210 and 1440–1395 cm^{-1}, respectively. Both of these bands involve some interaction between C—O stretching and in-plane C—O—H bending. The more intense band, near 1315–1280 cm^{-1} for dimers, is generally referred to as the C—O stretching band and usually appears as a doublet in the spectra of long-chain fatty acids. The C—O—H bending band near 1440–1395 cm^{-1} is of moderate intensity and occurs in the same region as the CH_2 scissoring vibration of the CH_2 group adjacent to the carbonyl.

One of the characteristic bands in the spectra of dimeric carboxylic acids results from the out-of-plane bending of the bonded O—H. The band appears near 920 cm^{-1} and is characteristically broad with medium intensity.

2.6.13 Carboxylate Anion

The carboxylate anion has two strongly coupled C⩵O bonds with bond strengths intermediate between C=O and C—O.

—C(⋯O)(⋯O)⁻

The carboxylate ion gives rise to two bands: a strong asymmetrical stretching band near 1650–1550 cm^{-1} and a weaker, symmetrical stretching band near 1400 cm^{-1}.

The conversion of a carboxylic acid to a salt can serve as confirmation of the acid structure. This is conveniently done by the addition of a tertiary aliphatic amine, such as triethylamine, to a solution of the carboxylic acid in chloroform (no reaction occurs in CCl_4). The carboxylate ion thus formed shows the two characteristic carbonyl absorption bands in addition to an "ammonium" band in the 2700–2200 cm^{-1} region. The O—H stretching band, of course, disappears. The spectrum of ammonium benzoate, Figure 2.24, demonstrates most of these features.

2.6.14 Esters and Lactones

Esters and lactones have two characteristically strong absorption bands arising from C=O and C—O stretching. The intense C=O stretching vibration occurs at higher frequencies (shorter wavelength) than that of normal ketones. The force constant of the carbonyl bond is increased by the electron-attracting nature of the adjacent oxygen atom (inductive effect). Overlapping occurs between esters in which the carbonyl frequency is lowered, and ketones in which the normal ketone frequency is raised. A distinguishing feature of esters and lactones, however, is the strong

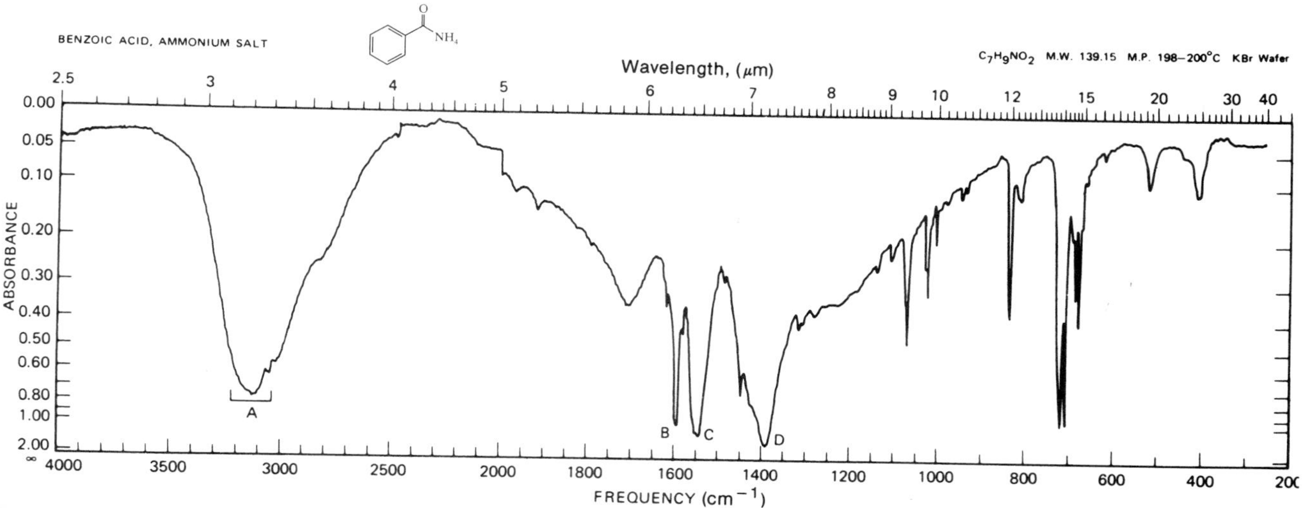

FIGURE 2.24. Benzoic acid, ammonium salt. A. N—H and C—H stretch, 3600–2500 cm^{-1}. B. Ring C$\cdots$O stretch, 1600 cm^{-1}. C. Asymmetric carboxylate anion C($\cdots$O)$_2^-$ stretch, 1550 cm^{-1}. D. Symmetric carboxylate C($\cdots$O)$_2^-$ stretch, 1385 cm^{-1}.

C—O stretching band in the region where a weaker band occurs for ketones. There is overlapping in the C=O frequency of esters or lactones and acids, but the OH stretching and bending vibrations and the possibility of salt formation distinguish the acids.

The frequency of the ester carbonyl responds to environmental changes in the vicinity of the carbonyl group in much the same manner as ketones. The spectrum of phenyl acetate, Figure 2.25, illustrates most of the important absorption characteristics for esters.

2.6.14.1 C=O Stretching Vibrations The C=O absorption band of saturated aliphatic esters (except formates) is in the 1750–1735 cm^{-1} region. The C=O absorption bands of formates, α,β-unsaturated, and benzoate esters are in the region of 1730–1715 cm^{-1}. Further conjugation has little or no additional effect upon the frequency of the carbonyl absorption.

In the spectra of vinyl or phenyl esters, with unsaturation adjacent to the C—O— group, a marked rise in the carbonyl frequency is observed along with a lowering of the C—O frequency. Vinyl acetate has a carbonyl band at 1776 cm^{-1}; phenyl acetate absorbs at 1771 cm^{-1}.

α-Halogen substitution results in a rise in the C=O stretching frequency. Ethyl trichloroacetate absorbs at 1770 cm^{-1}.

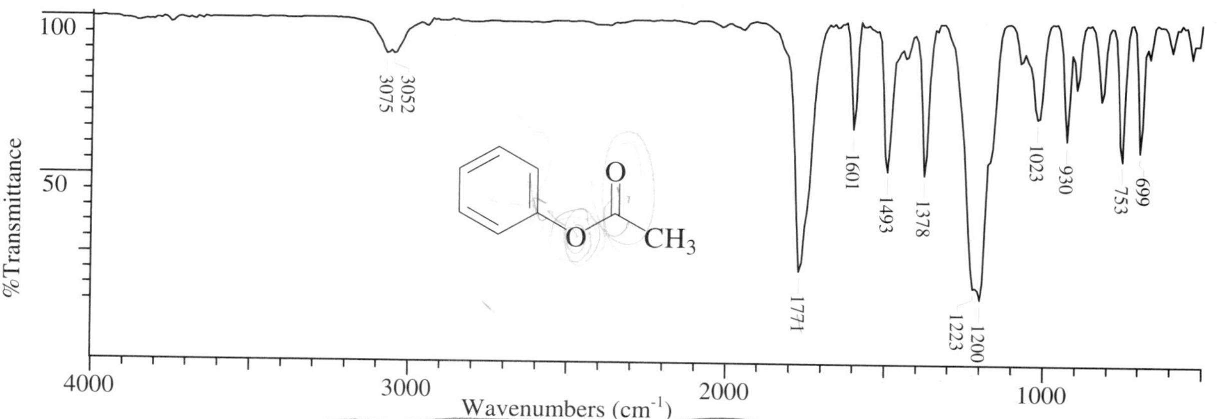

FIGURE 2.25. Phenyl acetate. Aromatic C—H stretch, 3075, 3052 cm^{-1}. C=O stretch, 1771 cm^{-1}: this is higher frequency than that from a normal ester C=O stretch (1740 cm^{-1}: see Table 2.6) because of phenyl conjugation with alcohol oxygen; conjugation of an aryl group or other unsaturation with the carbonyl group causes this C=O stretch to be at lower than normal frequency (e.g., benzoates absorb at about 1724 cm^{-1}). Ring C$\cdots$C stretch, 1601 cm^{-1}. $\delta_{as}CH_3$, 1493 cm^{-1}, δ_sCH_3, 1378 cm^{-1}. Acetate C(=O)—O stretch, 1223 cm^{-1}. O—C$\cdots$C asymmetrical stretch, 1200 cm^{-1}.

In oxalates and α-keto esters, as in α-diketones, there appears to be little or no interaction between the two carbonyl groups so that normal absorption occurs in the region of 1755–1740 cm^{-1}. In the spectra of β-keto esters, however, where enolization can occur, a band is observed near 1650 cm^{-1} that results from hydrogen bonding between the ester C=O and the enolic hydroxyl group.

The carbonyl absorption of saturated δ-lactones (six-membered ring) occurs in the same region as straight-chain, unconjugated esters. Unsaturation α to the C=O

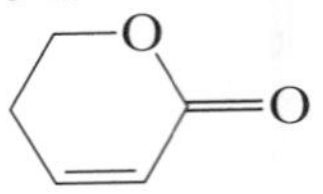

1720 cm^{-1} 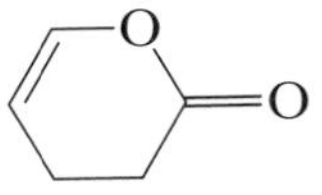 1760 cm^{-1}

reduces the C=O absorption frequency. Unsaturation α to the —O— group increases it.

α-Pyrones frequently display two carbonyl absorption bands in the 1775–1715 cm^{-1} region, probably because of Fermi resonance.

Saturated γ-lactones (five-membered ring) absorb at shorter wavelengths than esters or δ-lactones: 1795–1760 cm^{-1}; δ-valerolactone absorbs at 1770 cm^{-1}. Unsaturation in the γ-lactone molecule affects the carbonyl absorption in the same manner as unsaturation in δ-lactones.

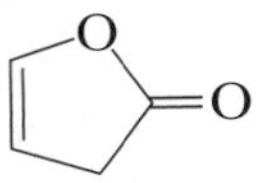

1800 cm^{-1} 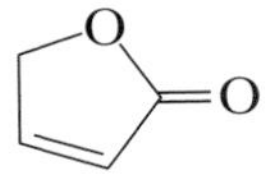 1750 cm^{-1}

In unsaturated lactones, when the double bond is adjacent to the —O—, a strong C=C absorption is observed in the 1685–1660 cm^{-1} region.

2.6.14.2 C—O Stretching Vibrations The "C—O stretching vibrations" of esters actually consist of two asymmetrical coupled vibrations: C—C(=O)—O and O—C—C, the former being more important. These bands occur in the region of 1300–1000 cm^{-1}. The corresponding symmetric vibrations are of little importance. The C—O stretch correlations are less reliable than the C=O stretch correlations.

The C—C(=O)—O band of saturated esters, except for acetates, shows strongly in the 1210–1163 cm^{-1} region. It is often broader and stronger than the C=O stretch absorption. Acetates of saturated alcohols display this band at 1240 cm^{-1}. Vinyl and phenyl acetates absorb at a somewhat lower frequency, 1190–1140 cm^{-1}; for example, see Figure 2.25. The C—C(=O)—O stretch of esters of α,β-unsaturated acids results in multiple bands in the 1300–1160 cm^{-1} region. Esters of aromatic acids absorb strongly in the 1310–1250 cm^{-1} region. The analogous type of stretch in lactones is observed in the 1250–1111 cm^{-1} region.

The O—C—C band of esters ("alcohol" carbon—oxygen stretch) of primary alcohols occurs at about 1164–1031 cm^{-1} and that of esters of secondary alcohols occurs at about 1100 cm^{-1}. Aromatic esters of primary alcohols show this absorption near 1111 cm^{-1}.

Methyl esters of long-chain fatty acids present a three-band pattern with bands near 1250, 1205, and 1175 cm^{-1}. The band near 1175 cm^{-1} is the strongest.

2.6.15 Acid Halides

2.6.15.1 C=O Stretching Vibrations Acid halides show strong absorption in the C=O stretching region. Unconjugated acid chlorides absorb in the 1815–1785 cm^{-1} region. Acetyl fluoride in the gas phase absorbs near 1869 cm^{-1}. Conjugated acid halides absorb at a slightly lower frequency because resonance reduces the force constant of the C=O bond; aromatic acid chlorides absorb strongly at 1800–1770 cm^{-1}. A weak band near 1750–1735 cm^{-1} appearing in the spectra of aroyl chlorides probably results from Fermi resonance between the C=O band and the overtone of a lower wavenumber band near 875 cm^{-1}. The spectrum of a typical aromatic acid chloride is given in Figure 2.26.

2.6.16 Carboxylic Acid Anhydrides

2.6.16.1 C=O Stretching Vibrations Anhydrides display two stretching bands in the carbonyl region. The two bands result from asymmetrical and symmetrical C=O stretching modes. Saturated acyclic anhydrides absorb near 1818 and 1750 cm^{-1}. Conjugated acyclic anhydrides show absorption near 1775 and 1720 cm^{-1}; the decrease in the frequency of absorption is caused by resonance. The higher frequency band is the more intense.

Cyclic anhydrides with five-membered rings show absorption at higher frequencies (lower wavelengths) than acyclic anhydrides because of ring strain; succinic anhydride absorbs at 1865 and 1782 cm^{-1}. The lower frequency (longer wavelength) C=O band is the stronger of the two carbonyl bands in five-membered ring cyclic anhydrides.

2.6.16.2 C—O Stretching Vibrations Other strong bands appear in the spectra of anhydrides as a result of C—C(=O)—O—C(=O)—C stretching vibrations. Unconjugated straight-chain anhydrides absorb near 1047 cm^{-1}. Cyclic anhydrides display bands near

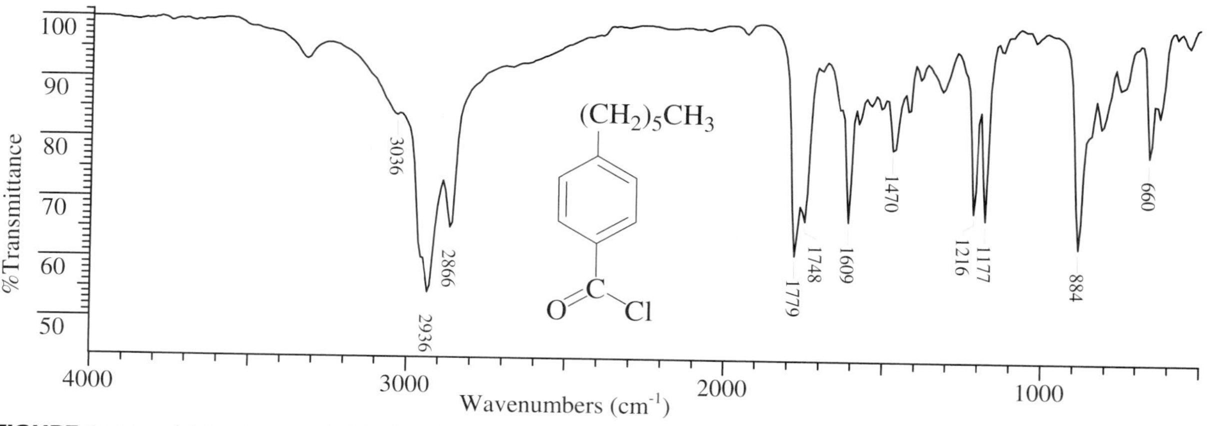

FIGURE 2.26. 4-Hexylbenzoyl chloride. Aromatic C—H stretch, 3036 cm^{-1}. C—H stretch 2936, 2866 cm^{-1} C═O stretch, 1779 cm^{-1} (see Table 2.6). (Acid chloride C═O stretch position shows very small dependence on conjugation; aroyl chlorides identified by band such as at 1748 cm^{-1} Fermi resonance band (of C═O stretch and overtone of 884 cm^{-1} band), 1748 cm^{-1}.

952–909 and 1299–1176 cm^{-1}. The C—O stretching band for acetic anhydride is at 1125 cm^{-1}.

The spectrum of benzoic anhydride in Figure 2.27 is that of a typical aromatic anhydride.

2.6.17 Amides and Lactams

All amides show a carbonyl absorption band known as the amide I band. Its position depends on the degree of hydrogen bonding and, thus, on the physical state of the compound.

Primary amides show two N—H stretching bands resulting from symmetrical and asymmetrical N—H stretching. Secondary amides and lactams show only one N—H stretching band. As in the case of O—H stretching, the frequency of the N—H stretching is reduced by hydrogen bonding, though to a lesser degree. Overlapping occurs in the observed position of N—H and O—H stretching frequencies so that an unequivocal differentiation in structure is sometimes impossible.

Primary amides and secondary amides, and a few lactams, display a band or bands in the region of 1650–1515 cm^{-1} caused primarily by NH_2 or NH bending, the amide II band. This absorption involves coupling between N—H bending and other fundamental vibrations and requires a *trans* geometry.

Out-of-plane N—H wagging is responsible for a broad band of medium intensity in the 800–666 cm^{-1} region.

The spectrum of acrylamide in Figure 2.28 is that of a typical primary amide of an unsaturated acid.

2.6.17.1 N—H Stretching Vibrations In dilute solution in nonpolar solvents, primary amides show two moderately intense N—H stretching

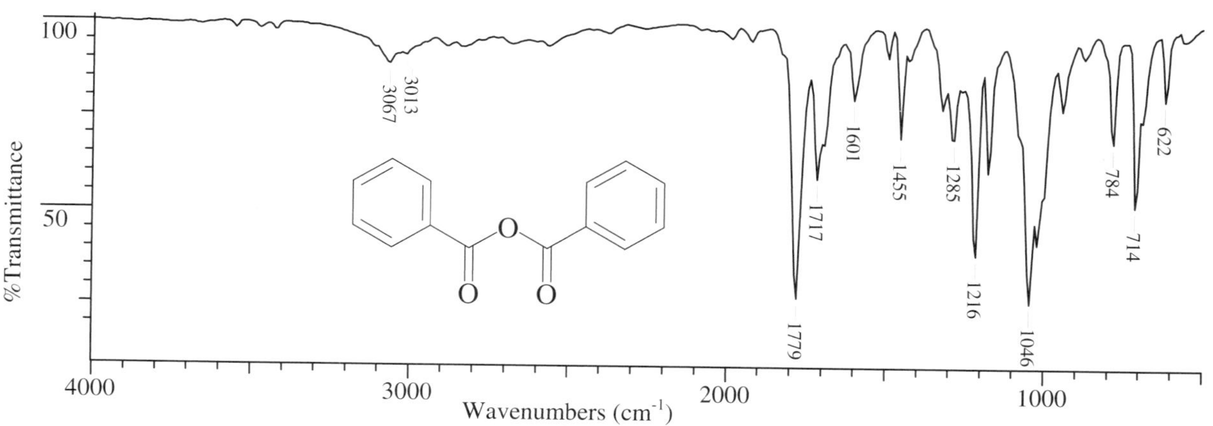

FIGURE 2.27. Benzoic anhydride. Aromatic C—H stretch, 3067, 3013 cm^{-1}. Asymmetric and symmetric C═O coupled stretching, respectively: 1779, 1717 cm^{-1}. See Table 2.6. C—CO—O—CO—C stretch, 1046 cm^{-1}.

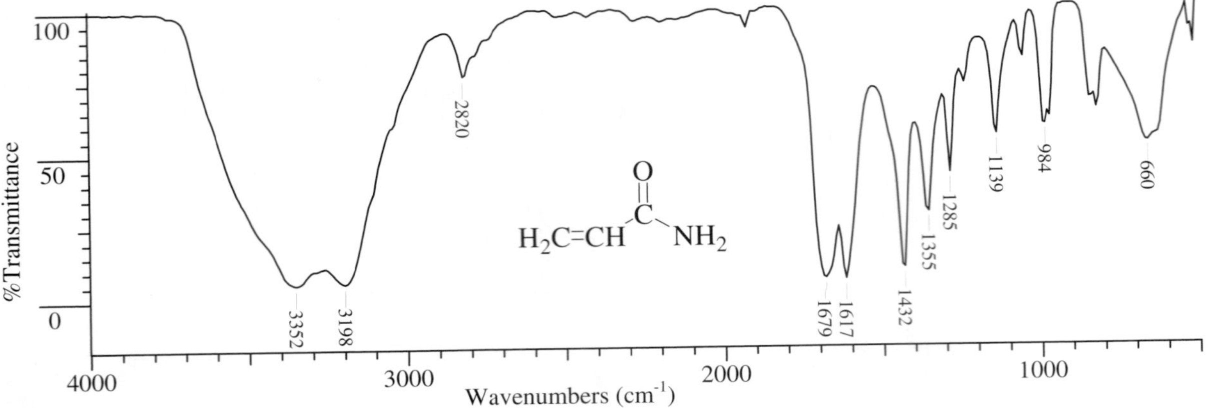

FIGURE 2.28. Acrylamide. N—H stretch, coupled, primary amide, hydrogen bonded; asymmetric, 3352 cm^{-1}; symmetric, 3198 cm^{-1}. Overlap C═O stretch, amide I band, 1679 cm^{-1}; see Table 2.6. N—H bend, amide II band, 1617 cm^{-1}. C—N stretch, 1432 cm^{-1}. Broad N—H out-of-plane bend 700–600 cm^{-1}.

frequencies corresponding to the asymmetrical and symmetrical N—H stretching vibrations. These bands occur near 3520 and 3400 cm^{-1}, respectively. In the spectra of solid samples, these bands are observed near 3350 and 3180 cm^{-1} because of hydrogen bonding.

In IR spectra of secondary amides, which exist mainly in the *trans* conformation, the free N—H stretching vibration observed in dilute solutions occurs near 3500–3400 cm^{-1}. In more concentrated solutions and in solid samples, the free N—H band is replaced by multiple bands in the 3330–3060 cm^{-1} region. Multiple bands are observed since the amide group can bond to produce dimers with an *s-cis* conformation and polymers with an *s-trans* conformation.

s-cis   *s*-trans

2.6.17.2 C═O Stretching Vibrations (Amide I Band)

The C═O absorption of amides occurs at lower frequencies than "normal" carbonyl absorption due to the resonance effect (see Section 2.6.10.1). The position of absorption depends on the same environmental factors as the carbonyl absorption of other compounds.

Primary amides (except acetamide, whose C═O bond absorbs at 1694 cm^{-1}) have a strong amide I band in the region of 1650 cm^{-1} when examined in the solid phase. When the amide is examined in dilute solution, the absorption is observed at a higher frequency, near 1690 cm^{-1}. In more concentrated solutions, the C═O frequency is observed at some intermediate value, depending on the degree of hydrogen bonding.

Simple, open-chain, secondary amides absorb near 1640 cm^{-1} when examined in the solid state. In dilute solution, the frequency of the amide I band may be raised to 1680 cm^{-1} and even to 1700 cm^{-1} in the case of the anilides. In the anilide structure there is competition between the ring and the C═O for the nonbonded electron pair of the nitrogen.

The carbonyl frequency of tertiary amides is independent of the physical state since hydrogen bonding with another tertiary amide group is impossible. The C═O absorption occurs in the range of 1680–1630 cm^{-1}. The absorption range of tertiary amides in solution is influenced by hydrogen bonding with the solvent: *N,N*-Diethylacetamide absorbs at 1647 cm^{-1} in dioxane and at 1615 cm^{-1} in methanol.

Electron-attracting groups attached to the nitrogen increase the frequency of absorption since they effectively compete with the carbonyl oxygen for the electrons of the nitrogen, thus increasing the force constant of the C═O bond.

2.6.17.3 N—H Bending Vibrations (Amide II Band)

All primary amides show a sharp absorption band in dilute solution (amide II band) resulting from N—H, bending at a somewhat lower frequency than the C═O band. This band has an intensity of one-half to one-third of the C═O absorption band. In mulls and pellets the band occurs near 1655–1620 cm^{-1} and is usually under the envelope of the amide I band. In dilute solutions, the band appears at lower frequency,

1620–1590 cm^{-1}, and normally is separated from the amide I band. Multiple bands may appear in the spectra of concentrated solutions, arising from the free and associated states. The nature of the R group R—C(=O)—NH_2 has little effect upon the amide II band.

Secondary acyclic amides in the solid state display an amide II band in the region of 1570–1515 cm^{-1}. In dilute solution, the band occurs in the 1550–1510 cm^{-1} region. This band results from interaction between the N—H bending and the C—N stretching of the C—N—H group. A second, weaker band near 1250 cm^{-1} also results from interaction between the N—H bending and C—N stretching.

2.6.17.4 Other Vibration Bands The C—N stretching band of primary amides occurs near 1400 cm^{-1}. A broad, medium band in the 800–666 cm^{-1} region in the spectra of primary and secondary amides results from out-of-plane N—H wagging.

In lactams of medium ring size, the amide group is forced into the *s-cis* conformation. Solid lactams absorb strongly near 3200 cm^{-1} because of the N—H stretching vibration. This band does not shift appreciably with dilution since the *s-cis* form remains associated at relatively low concentrations.

2.6.17.5 C=O Stretching Vibrations of Lactams The C=O absorption of lactams with six-membered rings or larger is near 1650 cm^{-1}. Five-membered ring (γ) lactams absorb in the 1750–1700 cm^{-1} region. Four-membered ring (β) lactams, unfused, absorb at 1760–1730 cm^{-1}. Fusion of the lactam ring to another ring generally increases the frequency by 20–50 cm^{-1}.

Most lactams do not show a band near 1550 cm^{-1} that is characteristic of *s-trans* noncyclic secondary amides. The N—H out-of-plane wagging in lactams causes broad absorption in the 800–700 cm^{-1} region.

2.6.18 Amines

The spectrum of a typical primary, aliphatic diamine appears in Figure 2.29.

2.6.18.1 N—H Stretching Vibrations Primary amines, examined in dilute solution, display two weak absorption bands: one near 3500 cm^{-1} and the other near 3400 cm^{-1}. These bands represent, respectively, the "free" asymmetrical and symmetrical N—H stretching modes. Secondary amines show a single weak band in the 3350–3310 cm^{-1} region. These bands are shifted to longer wavelengths by hydrogen bonding. The associated N—H bands are weaker and frequently sharper than the corresponding O—H bands. Aliphatic primary amines (neat) absorb at 3400–3300 and 3330–3250 cm^{-1}. Aromatic primary amines absorb at slightly higher frequencies (shorter wavelengths). In the spectra of liquid primary and secondary amines, a shoulder usually appears on the low-frequency side of the N—H stretching band, arising from the overtone of the NH bending band intensified by Fermi resonance. Tertiary amines do not absorb in this region.

2.6.18.2 N—H Bending Vibrations The N—H bending (scissoring) vibration of primary amines is observed in the 1650–1580 cm^{-1} region of the spectrum. The band is medium to strong in intensity and is

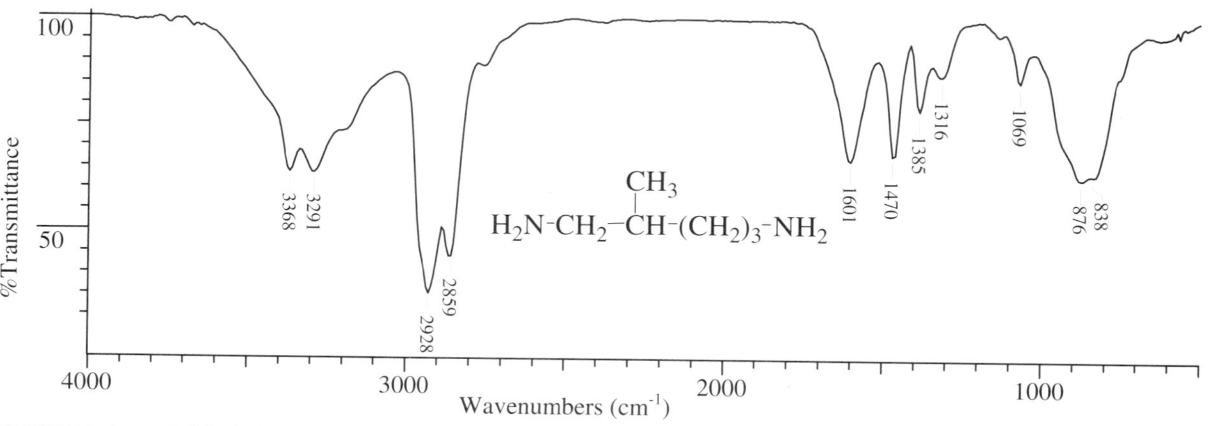

FIGURE 2.29. 2-Methyl-1,5-pentanediamine. N—H stretch, hydrogen-bonded, primary amine coupled doublet: asymmetric, 3368 cm^{-1}. Symmetric, 3291 cm^{-1}. (Shoulder at about 3200 cm^{-1}, Fermi resonance band with overtone of band at 1601 cm^{-1}. Aliphatic C—H stretch, 2928, 2859 cm^{-1}. N—H bend (scissoring) 1601 cm^{-1}. $\delta_s CH_2$ (scissoring), 1470 cm^{-1}. C—N stretch, 1069 cm^{-1}. N—H wag (neat sample), ~900–700 cm^{-1}.

moved to slightly higher frequencies when the compound is associated. The N—H bending band is seldom detectable in the spectra of aliphatic secondary amines, whereas secondary aromatic amines absorb near 1515 cm^{-1}.

Liquid samples of primary and secondary amines display medium to strong broad absorption in the 909–666 cm^{-1} region of the spectrum arising from N—H wagging. The position of this band depends on the degree of hydrogen bonding.

2.6.18.3 C—N Stretching Vibrations Medium to weak absorption bands for the unconjugated C—N linkage in primary, secondary, and tertiary aliphatic amines appear in the region of 1250–1020 cm^{-1}. The vibrations responsible for these bands involve C—N stretching coupled with the stretching of adjacent bonds in the molecule. The position of absorption in this region depends on the class of the amine and the pattern of substitution on the α-carbon.

Aromatic amines display strong C—N stretching absorption in the 1342–1266 cm^{-1} region. The absorption appears at higher frequencies (shorter wavelengths) than the corresponding absorption of aliphatic amines because the force constant of the C—N bond is increased by resonance with the ring.

Characteristic strong C—N stretching bands in the spectra of aromatic amines have been assigned as in Table 2.7.

TABLE 2.7 C—N Stretch of Aromatic Amines

Aromatic Amine	Absorption Region (cm^{-1})
Primary	1340–1250
Secondary	1350–1280
Tertiary	1360–1310

2.6.19 Amine Salts

2.6.19.1 N—H Stretching Vibrations The ammonium ion displays strong, broad absorption in the 3300–3030 cm^{-1} region because of N—H stretching vibrations (see Figure 2.24). There is also a combination band in the 2000–1709 cm^{-1} region.

Salts of primary amines show strong, broad absorption between 3000 and 2800 cm^{-1} arising from asymmetrical and symmetrical stretching in the NH_3^+ group. In addition, multiple combination bands of medium intensity occur in the 2800–2000 cm^{-1} region, the most prominent being the band near 2000 cm^{-1}. Salts of secondary amines absorb strongly in the 3000–2700 cm^{-1} region with multiple bands extending to 2273 cm^{-1}. A medium band near 2000 cm^{-1} may be observed. Tertiary amine salts absorb at longer wavelengths than the salts of primary and secondary amines (2700–2250 cm^{-1}). Quaternary ammonium salts can have no N—H stretching vibrations.

2.6.19.2 N—H Bending Vibrations The ammonium ion displays a strong, broad NH_4^+ bending band near 1429 cm^{-1}. The NH_3^+ group of the salt of a primary amine absorbs near 1600–1575 and 1550–1504 cm^{-1}. These bands originate in asymmetrical and symmetrical NH_3^+ bending, analogous to the corresponding bands of the CH_3 group. Salts of secondary amines absorb near 1620–1560 cm^{-1}. The N—H bending band of the salts of tertiary amines is weak and of no practical value.

2.6.20 Amino Acids and Salts of Amino Acids

Amino acids are encountered in three forms:

1. The free amino acid (zwitterion).*

$$-\overset{|}{\underset{\underset{NH_3^+}{|}}{C}}-CO_2^-$$

2. The hydrochloride (or other salt).

$$-\overset{|}{\underset{\underset{NH_3^+\ Cl^-}{|}}{C}}-CO_2H$$

3. The sodium (or other cation) salt.

$$-\overset{|}{\underset{\underset{NH_2}{|}}{C}}-CO_2^-\ Na^+$$

Free primary amino acids are characterized by the following absorptions (most of the work was done with α-amino acids, but the relative positions of the amino and carboxyl groups seem to have little effect):

1. A broad, strong NH_3^+ stretching band in the 3100–2600 cm^{-1} region. Multiple combination

* Aromatic amino acids are not zwitterions. Thus *p*-aminobenzoic acid is

$$H_2N-C_6H_4-COOH$$

and overtone bands extend the absorption to about 2000 cm^{-1}. This overtone region usually contains a prominent band near 2222–2000 cm^{-1} assigned to a combination of the asymmetrical NH_3^+ bending vibration and the torsional oscillation of the NH_3^+ group. The torsional oscillation occurs near 500 cm^{-1}. The 2000 cm^{-1} band is absent if the nitrogen atom of the amino acid is substituted.

2. A weak asymmetrical NH_3^+ bending band near 1660–1610 cm^{-1}, a fairly strong symmetrical bending band near 1550–1485 cm^{-1}.
3. The carboxylate ion group $-C(\!=\!O)_2^-$ absorbs strongly near 1600–1590 cm^{-1} and more weakly near 1400 cm^{-1}. These bands result, respectively, from asymmetrical and symmetrical $C(\!=\!O)_2$ stretching.

The spectrum of the amino acid leucine, including assignments corresponding to the preceding three categories, is shown in Figure 2.30.

Hydrochlorides of amino acids present the following patterns:

1. Broad, strong absorption in the 3333–2380 cm^{-1} region resulting from superimposed O—H and NH_3^+ stretching bands. Absorption in this region is characterized by multiple fine structure on the low-wavenumber side of the band.
2. A weak, asymmetrical NH_3^+ bending band near 1610–1590 cm^{-1}; a relatively strong, symmetrical NH_3^+ bending band at 1550–1481 cm^{-1}.
3. A strong band at 1220–1190 cm^{-1} arising from C—C(=O)—O stretching.
4. Strong carbonyl absorption at 1755–1730 cm^{-1} for α-amino acid hydrochlorides, and at 1730–1700 cm^{-1} for other amino acid hydrochlorides.

Sodium salts of amino acids show the normal N—H stretching vibrations at 3400–3200 cm^{-1} common to other amines. The characteristic carboxylate ion bands appear near 1600–1590 cm^{-1} and near 1400 cm^{-1}.

2.6.21 Nitriles

The spectra of nitriles (R—C≡N) are characterized by weak to medium absorption in the triple-bond stretching region of the spectrum. Aliphatic nitriles absorb near 2260–2240 cm^{-1}. Electron-attracting atoms, such as oxygen or chlorine, attached to the carbon atom α to the C≡N group reduce the intensity of absorption. Conjugation, such as occurs in aromatic nitriles, reduces the frequency of absorption to 2240–2222 cm^{-1} and enhances the intensity. The spectrum of a typical nitrile is shown in Figure 2.31.

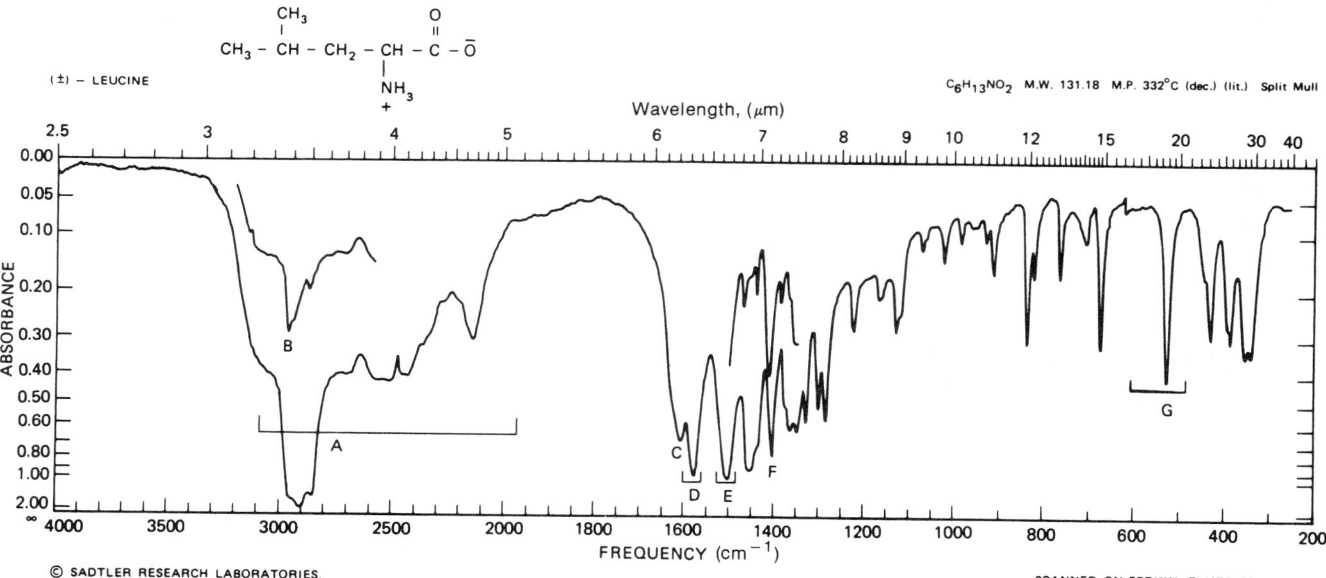

FIGURE 2.30. (±)-Leucine. A. Broad (—NH_3^+) N—H stretch, 3100–2000 cm^{-1}, extended by combination band at 2140 cm^{-1}, and other combination-overtone bands. B. Aliphatic C—H stretch (superimposed on N—H stretch), 2967 cm^{-1}. C. Asymmetric (—NH_3^+) N—H bend, 1610 cm^{-1}. D. Asymmetric carboxylate $(C\!=\!O)_2$ stretch, 1580 cm^{-1}. E. Symmetric (—NH_3^+) N—H bend, 1505 cm^{-1}. F. Symmetric carboxylate $(C\!=\!O)_2$ stretch, 1405 cm^{-1}. G. Torsional (—NH_3^+) N—H oscillation, 525 cm^{-1}.

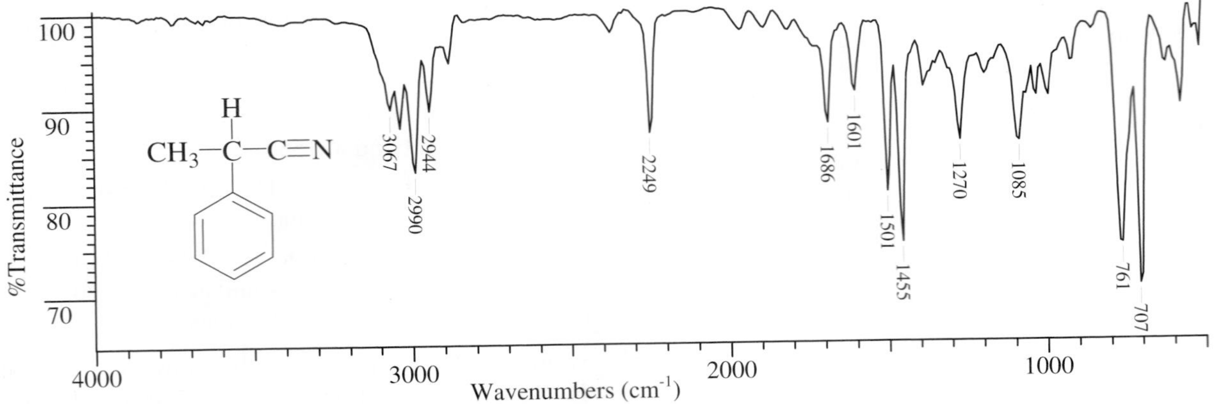

FIGURE 2.31. α-Methylbenzyl cyanide. Aromatic C—H stretch, 3067, 3030 cm^{-1}. Aliphatic C—H stretch, 2990, 2944 cm^{-1}. C≡N stretch, 2249 cm^{-1}. Out-of-plane C—H bend (aromatic ring), 761 cm^{-1}.

2.6.22 Isonitriles ($R—\overset{+}{N}\equiv\overset{-}{C}$), Cyanates (R—O—C≡N), Isocyanates (R—N=C=O), Thiocyanates (R—S—C≡N), and Isothiocyanates (R—N=C=S)

These groups show the triple bond or cumulative bond stretch in the 2280–2000 cm^{-1} region.

2.6.23 Compounds Containing —N=N Group

The N=N stretching vibration of a symmetrical *trans*-azo compound is forbidden in the IR but absorbs in the 1576 cm^{-1} region of the Raman spectrum. Unsymmetrical *para*-substituted azobenzenes in which the substituent is an electron-donating group absorb near 1429 cm^{-1}. The bands are weak because of the nonpolar nature of the bond.

2.6.24 Covalent Compounds Containing Nitrogen–Oxygen Bonds

Nitro compounds, nitrates, and nitramines contain an NO_2 group. Each of these classes shows absorption caused by asymmetrical and symmetrical stretching of the NO_2 group. Asymmetrical absorption results in a strong band in the 1661–1499 cm^{-1} region; symmetrical absorption occurs in the region between 1389–1259 cm^{-1}. The exact position of the bands is dependent on substitution and unsaturation in the vicinity of the NO_2 group.

2.6.24.1 N=O Stretching Vibrations Nitro Compounds In the nitroalkanes, the bands occur near 1550 and 1372 cm^{-1}. Conjugation lowers the frequency of both bands, resulting in absorption near 1550–1500 and 1360–1290 cm^{-1}. Attachment of electronegative groups to the α-carbon of a nitrocompound causes an increase in the frequency of the asymmetrical NO_2 band and a reduction in the frequency of the symmetrical band; chloropicrin, Cl_3CNO_2, absorbs at 1610 and 1307 cm^{-1}.

Aromatic nitro groups absorb near the same frequencies as observed for conjugated aliphatic nitro compounds. Interaction between the NO_2 out-of-plane bending and ring C—H out-of-plane bending frequencies destroys the reliability of the substitution pattern observed for nitroaromatics in the long-wavelength region of the spectrum. Nitroaromatic compounds show a C—N stretching vibration near 870 cm^{-1}. The spectrum of nitrobenzene, with assignments corresponding to the preceding discussion, is shown in Figure 2.32.

Because of strong resonance in aromatic systems containing NO_2 groups and electron-donating groups such as the amino group, *ortho* or *para* to one another, the symmetrical NO_2 vibration is shifted to lower frequencies and increases in intensity. *p*-Nitroaniline absorbs at 1475 and 1310 cm^{-1}.

The positions of asymmetric and symmetric NO_2 stretching bands of nitramines $\rangle N—NO_2$ and the NO stretch of nitrosoamines are given in Appendix B.

Nitrates Organic nitrates show absorption for N—O stretching vibrations of the NO_2 group and for the O—N linkage. Asymmetrical stretching in the NO_2 group results in strong absorption in the 1660–1625 cm^{-1} region; the symmetrical vibration absorbs strongly near 1300–1255 cm^{-1}. Stretching of the π bonds of the N—O linkage produces absorption near 870–833 cm^{-1}. Absorption observed at longer wavelengths, near 763–690 cm^{-1}, likely results from NO_2 bending vibrations.

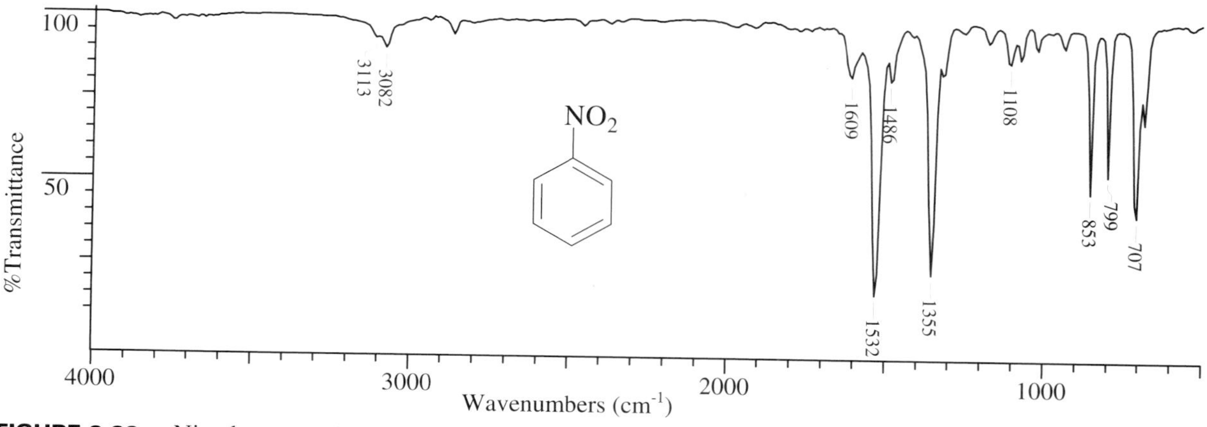

FIGURE 2.32. Nitrobenzene. Aromatic C—H stretch, 3113, 3082 cm^{-1}. Asymmetric ($ArNO_2$) $(N{=}O)_2$ stretch, 1532 cm^{-1}. Symmetric ($ArNO_2$) $(N{=}O)_2$ stretch 1355 cm^{-1}. C—N stretch for $ArNO_2$, 853 cm^{-1}. Low-frequency bands are of little use in determining the nature of ring substitution since these absorption patterns result from interaction of NO_2 and C—H out-of-plane bending frequencies. The inability of the "oop" region to reveal structural information is typical of aromatic compounds with highly polar substituents.

Nitrites Nitrites display two strong N=O stretching bands. The band near 1680–1650 cm^{-1} is attributed to the *trans* isomer; the *cis* isomer absorbs in the 1625–1610 cm^{-1} region. The N—O stretching band appears in the region between 850 and 750 cm^{-1}. The nitrite absorption bands are among the strongest observed in IR spectra.

Nitroso Compounds Primary and secondary aliphatic *C*-nitroso compounds are usually unstable and rearrange to oximes or dimerize. Tertiary and aromatic nitroso compounds are reasonably stable, existing as monomers in the gaseous phase or in dilute solution and as dimers in neat samples. Monomeric, tertiary, aliphatic nitroso compounds show N=O absorption in the 1585–1539 cm^{-1} region; aromatic monomers absorb between 1511 and 1495 cm^{-1}.

The N → O stretching absorption of dimeric nitroso compounds are categorized in Appendix B as to *cis* versus *trans* and aliphatic versus aromatic. Nitrosoamine absorptions are given in Appendix B.

2.6.25 Organic Sulfur Compounds

2.6.25.1 S—H Stretching Vibrations Mercaptans Aliphatic mercaptans and thiophenols, as liquids or in solution, show S—H stretching absorption in the range of 2600–2550 cm^{-1}. The S—H stretching band is characteristically weak and may go undetected in the spectra of dilute solutions or thin films. However, since few other groups show absorption in this region, it is useful in detecting S—H groups. The spectrum of 1,6-hexanedithiol in Figure 2.33 is that of a mercaptan with a detectable S—H stretch band. The band may be obscured by strong carboxyl absorption in the same region. Hydrogen bonding is much weaker for S—H groups than for O—H and N—H groups.

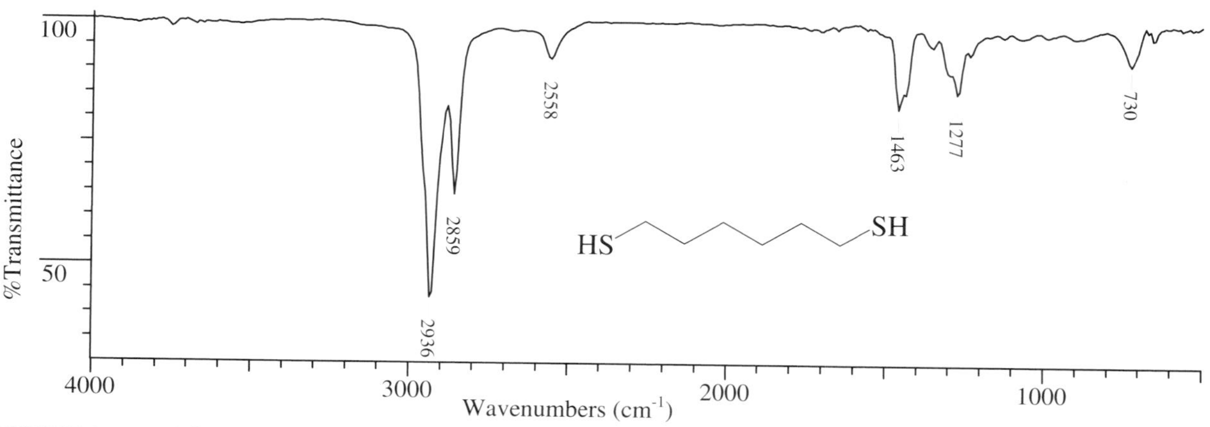

FIGURE 2.33. 1,6-Hexanedithiol Aliphatic C—H stretch, 2936, 259 cm^{-1}. Moderately weak S—H stretch, 2558 cm^{-1}. C—S stretch 730 cm^{-1}.

The S—H group of thiol acids absorbs in the same region as mercaptans and thiophenols.

3.6.25.2 C—S and C=S Stretching Vibrations Sulfides The stretching vibrations assigned to the C—S linkage occur in the region of 700–600 cm^{-1}. The weakness of absorption and variability of position make this band of little value in structural determination.

Disulfides The S—S stretching vibration is very weak and falls between 500 and 400 cm^{-1}.

Thiocarbonyl Compounds Aliphatic thials or thiones exist as trimeric, cyclic sulfides. Aralkyl thiones may exist either as monomers or trimers, whereas diaryl thiones, such as thiobenzophenone, exist only as monomers. The C=S group is less polar than the C=O group and has a considerably weaker bond. In consequence, the band is not intense, and it falls at lower frequencies, where it is much more susceptible to coupling effects. Identification is therefore difficult and uncertain.

Compounds that contain a thiocarbonyl group show absorption in the 1250–1020 cm^{-1} region. Thiobenzophenone and its derivatives absorb moderately in the 1224–1207 cm^{-1} region. Since the absorption occurs in the same general region as C—O and C—N stretching, considerable interaction can occur between these vibrations within a single molecule.

Spectra of compounds in which the C=S group is attached to a nitrogen atom show an absorption band in the general C=S stretching region. In addition, several other bands in the broad region of 1563–700 cm^{-1} can be attributed to vibrations involving interaction between C=S stretching and C—N stretching.

Thioketo compounds that can undergo enolization exist as thioketo-thioenol tautomeric systems; such systems show S—H stretching absorption. The thioenol tautomer of ethyl thiobenzoylacetate,

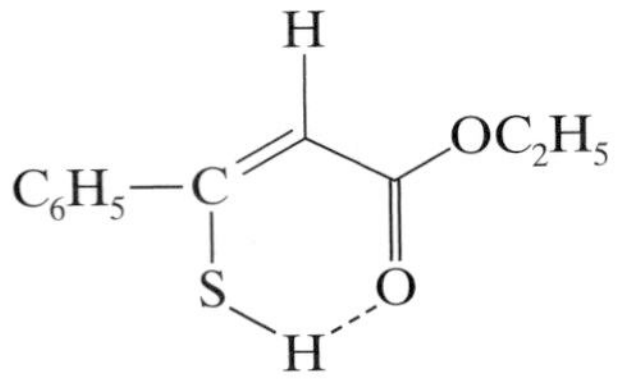

absorbs broadly at 2415 cm^{-1} because of hydrogen bonded S—H stretching absorption.

2.6.26 Compounds Containing Sulfur—Oxygen Bonds

2.6.26.1 S=O Stretching Vibrations Sulfoxides Alkyl and aryl sulfoxides as liquids or in solution show strong absorption in the 1070–1030 cm^{-1} region. This absorption occurs at 1050 cm^{-1} for dimethyl sulfoxide (DMSO, methyl sulfoxide). Conjugation brings about a small change in the observed frequency, in contrast to the marked reduction in frequency of the C=O bond accompanying conjugation. Diallyl sulfoxide absorbs at 1047 cm^{-1}. Phenyl methyl sulfoxide and cyclohexyl methyl sulfoxide absorb at 1055 cm^{-1} in dilute solution in carbon tetrachloride. The sulfoxide group is susceptible to hydrogen bonding, the absorption shifting to slightly lower frequencies from dilute solution to the liquid phase. The frequency of S=O absorption is increased by electronegative substitution.

Sulfones Spectra of sulfones show strong absorption bands at 1350–1300 and 1160–1120 cm^{-1}. These bands arise from asymmetric and symmetric SO_2 stretching, respectively. Hydrogen bonding results in absorption near 1300 and 1125 cm^{-1}. Splitting of the high-frequency band often occurs in CCl_4 solution or in the solid state.

Sulfonyl Chlorides Sulfonyl chlorides absorb strongly in the regions of 1410–1380 and 1204–1177 cm^{-1}. This increase in frequency, compared with the sulfones, results from the electronegativity of the chlorine atom.

Sulfonamides Solutions of sulfonamides absorb strongly at 1370–1335 and 1170–1155 cm^{-1}. In the solid phase, these frequencies are lowered by 10–20 cm^{-1}. In solid samples, the high-frequency band is broadened and several submaxima usually appear.

Primary sulfonamides show strong N—H stretching bands at 3390–3330 and 3300–3247 cm^{-1} in the solid state; secondary sulfonamides absorb near 3265 cm^{-1}.

Sulfonates, Sulfates, and Sulfonic Acids The asymmetric (higher frequency, shorter wavelength) and symmetric S=O stretching frequency ranges for these compounds are as follows:

TABLE 2.8 Stretching Frequency of Sulfonates, Sulfates, Sulfonic acids, and Sulfonate salts

Class	Stretching Frequency (cm^{-1})
Sulfonates (covalent)	1372–1335, 1195–1168
Sulfates (organic)	1415–1380, 1200–1185
Sulfonic acids	1350–1342, 1165–1150
Sulfonate salts	~1175, ~1055

The spectrum of a typical alkyl arenesulfonate is given in Figure 2.34. In virtually all sulfonates, the asymmetric stretch occurs as a doublet. Alkyl and aryl sulfonates show negligible differences; electron-donating groups in the *para* position of arenesulfonates cause higher frequency absorption.

Sulfonic acids are listed in narrow ranges above; these apply only to anhydrous forms. Such acids hydrate readily to give bands that are probably a

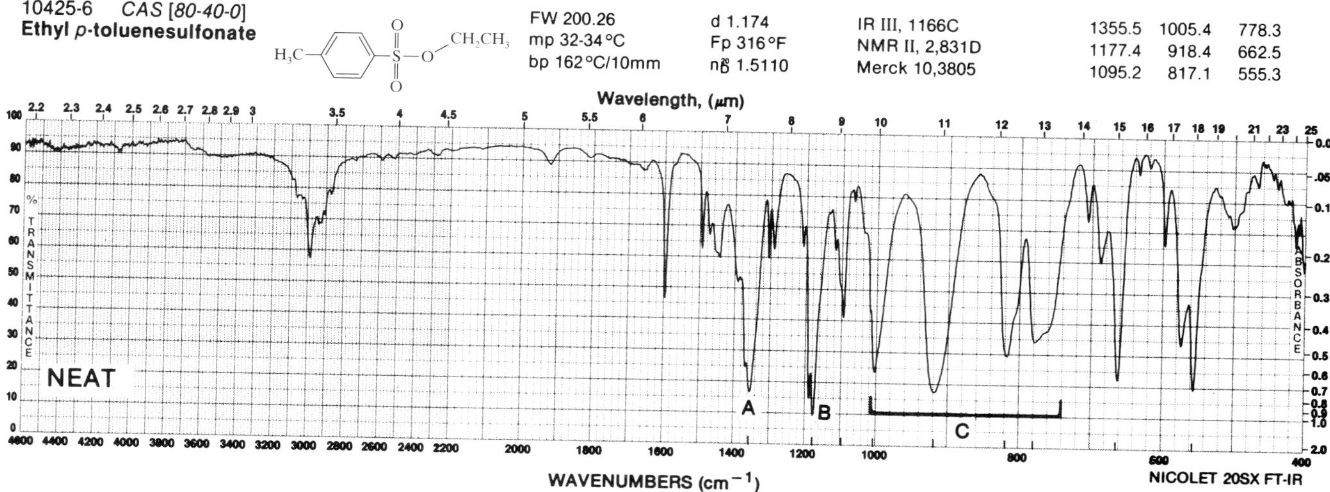

FIGURE 2.34. Ethyl *p*-toluenesulfonate. A. Asymmetric $S(=O)_2$ stretch, 1355 cm^{-1}. B. Symmetric $S(=O)_2$ stretch, 1177 cm^{-1}. C. Various strong S—O—C stretching, 1000–769 cm^{-1}.

result of the formation of hydronium sulfonate salts, in the 1230–1120 cm^{-1} range.

2.6.27 Organic Halogen Compounds

The strong absorption of halogenated hydrocarbons arises from the stretching vibrations of the carbon-halogen bond.

Aliphatic C—Cl absorption is observed in the broad region between 850 and 550 cm^{-1}. When several chlorine atoms are attached to one carbon atom, the band is usually more intense and at the high-frequency end of the assigned limits. Carbon tetrachloride shows an intense band at 797 cm^{-1}. The first overtones of the intense fundamental bands are frequently observed. Brominated compounds absorb in the 690–515 cm^{-1} region, iodo compounds in the 600–500 cm^{-1} region. A strong CH_2 wagging band is observed for the CH_2X (X = Cl, Br, and I) group in the 1300–1150 cm^{-1} region.

Fluorine-containing compounds absorb strongly over a wide range between 1400 and 1000 cm^{-1} because of C—F stretching modes. A monofluoroalkane shows a strong band in the 1100–1000 cm^{-1} region. As the number of fluorine atoms in an aliphatic molecule increases, the band pattern becomes more complex, with multiple strong bands appearing over the broad region of C—F absorption. The CF_3 and CF_2 groups absorb strongly in the 1350–1120 cm^{-1} region.

Chlorobenzenes absorb in the 1096–1089 cm^{-1} region. The position within this region depends on the substitution pattern. Aryl fluorides absorb in the 1250–1100 cm^{-1} region of the spectrum. A monofluorinated benzene ring displays a strong, narrow absorption band near 1230 cm^{-1}.

2.6.28 Silicon Compounds

2.6.28.1 Si—H Vibrations Vibrations for the Si—H bond include the Si—H stretch (~2200 cm^{-1}) and the Si—H bend (800–950 cm^{-1}). The Si—H stretching frequencies are increased by the attachment of an electronegative group to the silicon.

2.6.28.2 SiO—H and Si—O Vibrations The OH stretching vibrations of the SiOH group absorb in the same region as the alcohols, 3700–3200 cm^{-1}, and strong Si—O bands are at 830–1110 cm^{-1}. As in alcohols, the absorption characteristics depend on the degree of hydrogen bonding.

2.6.28.3 Silicon—Halogen Stretching Vibrations Absorption caused by Si—F stretch is in the 800–1000 cm^{-1} region.

Bands resulting from Si—Cl stretching occur at frequencies below 666 cm^{-1}.

2.6.29 Phosphorus Compounds

2.6.29.1 P=O and P—O Stretching Vibrations Such absorptions are listed in Appendix D, Table D-1.

2.6.30 Heteroaromatic Compounds

The spectra of heteroaromatic compounds result primarily from the same vibrational modes as observed for the aromatics.

2.6.30.1 C—H Stretching Vibrations Heteroaromatics, such as pyridines, pyrazines, pyrroles, furans, and thiophenes, show C—H stretching bands in the 3077–3003 cm^{-1} region.

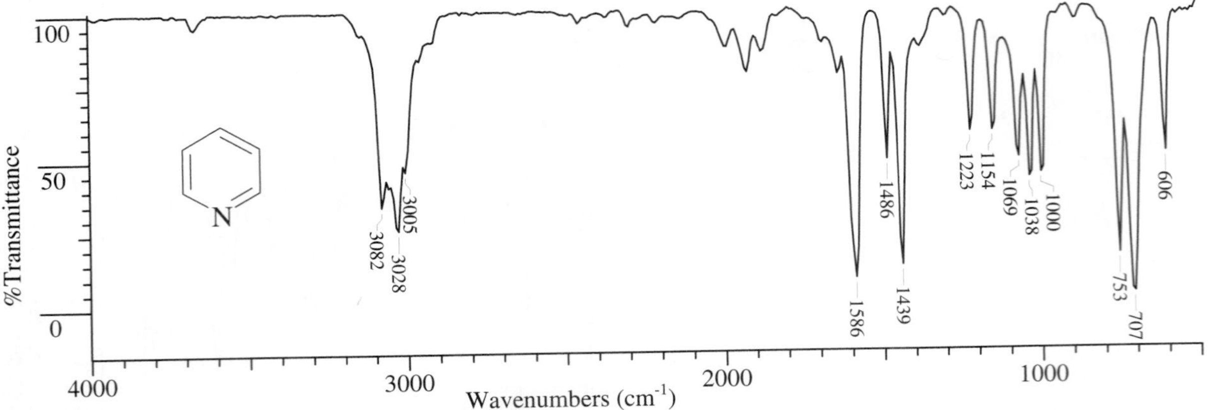

FIGURE 2.35. Pyridine. Aromatic C—H stretch, 3090–3000 cm^{-1}. C=C, C=N ring stretching (skeletal bands), 1600–1430 cm^{-1}. C—H out-of-plane bending, 753, 707 cm^{-1}. See Appendix E, Table E-1 for patterns in region C for substituted pyridines.

2.6.30.2 N—H Stretching Frequencies Heteroaromatics containing an N—H group show N—H stretching absorption in the region of 3500–3220 cm^{-1}. The position of absorption within this general region depends on the degree of hydrogen bonding, and hence upon the physical state of the sample or the polarity of the solvent. Pyrrole and indole in dilute solution in nonpolar solvents show a sharp band near 3495 cm^{-1}; concentrated solutions show a widened band near 3400 cm^{-1}. Both bands may be seen at intermediate concentrations.

2.6.30.3 Ring Stretching Vibrations (Skeletal Bands) Ring stretching vibrations occur in the general region between 1600 and 1300 cm^{-1}. The absorption involves stretching and contraction of all of the bonds in the ring and interaction between these stretching modes. The band pattern and the relative intensities depend on the substitution pattern and the nature of the substituents.

Pyridine (Figure 2.35) shows four bands in this region and, in this respect, closely resembles a monosubstituted benzene. Furans, pyrroles, and thiophenes display two to four bands in this region.

2.6.30.4 C—H Out-of-Plane Bending The C—H out-of-plane bending (γ-CH) absorption pattern of the heteroaromatics is determined by the number of adjacent hydrogen atoms bending in phase. The C—H out-of-plane and ring bending (β ring) absorption of the alkylpyridines are summarized in Appendix E, Table E-1.

Absorption data for the out-of-phase C—H bending (γ-CH) and ring bending (β ring) modes of three common five-membered heteroaromatic rings are presented in Appendix E, Table E-2. The ranges in Table E-2 include polar as well as nonpolar substituents on the ring.

REFERENCES

Introductory and Theoretical

Bellamy, L.J. (1975). *The Infrared Spectra of Complex Molecules*, 3rd ed. London: Chapman and Hall: New York: Halsted-Wiley.

Bellamy, L.J. (1968). *Advances in Infrared Group Frequencies.* London: Methuen.

Coleman, P.A. (1991). *Practical Sampling Techniques for Infrared Analysis.* Boca Raton, FL: CRC Press.

Colthup, N.B., Daly, L.H., and Wiberley, S.E. (1990). *Introduction to Infrared and Raman Spectroscopy,* 3rd ed. New York and London: Academic Press.

Conley, R.T. (1972). *Infrared Spectroscopy,* 2nd ed. Boston: Allyn and Bacon.

Durig, J.R. (1985). *Chemical, Biological, and Industrial Applications of Infrared Spectra.* New York: Wiley.

Ferraro, J.R., and Basile, L.J., Eds. (1985). *Fourier Transform Infrared Spectroscopy,* 4 vols. New York: Academic Press. Applications.

George, W.O., and McIntire, P. (1986). *Infrared Spectroscopy.* New York: Wiley.

Griffiths, P.R., and DeHaseth, J.A. (1986). *FTIR,* New York: Wiley, 1986.

Herres, W. (1987). *HRGC-FTIR, Capillary GC and FTIR Spectroscopy: Theory and Applications.* New York: Huthig.

Mattson, J.S., Ed. (1977). *Infrared, Correlation and Fourier Transform Spectroscopy.* New York: Marcel Dekker.

Nakanishi, K., and Solomon, P.H. (1977). *Infrared Absorption Spectroscopy-Practical,* 2nd ed. San Francisco: Holden-Day.

Roeges, N.P.G. (1984). *Guide to Interpretation of Infrared Spectra of Organic Structures.* New York: Wiley.

Smith, B.C. (1995). *Fundamentals of Fourier Transform Infrared Spectroscopy.* Boca Raton, FL: CRC Press.

Szymanski, H.A. Ed. (1969, 1970). *Raman Spectroscopy: Theory and Practice,* Vols. 1, 2. New York: Plenum Press.

Szymanski, H.A. (1989). *Correlation of Infrared and Raman Spectra of Organic Compounds.* Hertillon Press.

Szymanski, H.A. (1964). IR Theory and Practice of *Infrared Spectroscopy.* New York: Plenum Press.

Szymanski, H.A. (1964–1967). *Interpreted Infrared Spectra,* Vols. I–III. New York: Plenum Press.

Szymanski, H.A. (1963). *Infrared Band Handbook.* New York: Plenum Press.

Compilations of Spectra or Data; Workbooks, Special Topics

Aldrich Library of FTIR Spectra (1989), 3 Vols. Milwaukee, WI: Includes vapor-phase spectrum Vol. 3

ASTM-Wyandotte Index, Molecular Formula List of Compounds, Names and References to Published Infrared Spectra. Philadelphia, PA: ASTM Special Technical Publications 131 (1962) and 131-A (1963). Lists about 57,000 compounds. Covers IR, near-IR, and far-IR spectra.

Bentley, F.F., Smithson, L.D., and Rozek, A.L. (1988). *Infrared Spectra and Characteristic Frequencies—700 to 300 cm^{-1}, A Collection of Spectra, Interpretation and Bibliography.* New York: Wiley-Interscience.

Brown, C.R., Ayton, M.W., Goodwin, T.C., and Derby, T.J. (1954). *Infrared-A Bibliography.* Washington, DC: Library of Congress, Technical Information Division.

Catalog of Infrared Spectrograms, American Petroleum Institute Research Project 44. Pittsburgh, PA: Carnegie Institute of Technology.

Catalog of Infrared Spectrograms. Philadelphia: Sadder Research Laboratories, PA. 19104. Spectra are indexed by name and by the Spec-Finder. The latter is an index that tabulates major bands by wavelength intervals. This allows quick identification of unknown compounds.

Catalog of Infrared Spectral Data, Manufacturing Chemists Association Research Project, Chemical and Petroleum Research Laboratories. Pittsburgh, PA: Carnegie Institute of Technology, to June 30, 1960; College Station, TX: Chemical Thermodynamics Properties Center, Agriculture and Mechanical College of Texas, from July 1, 1960.

Craver, C., Ed. (1982). *Desk Book of Infrared Spectra,* 2nd ed. Kirkwood MO: Coblenz Society. 900 dispersive spectra.

Dobriner, K., Katzenellenbogen, E.R., and Jones, R.N. (1953). *Infrared Absorption Spectra of Steriods-An Atlas,* Vol. 1. New York: Wiley-Interscience.

Documentation of Molecular Spectroscopy (DMS), London: Butterworths Scientific Publications, and Weinheim/Bergstrasse. West Germany: Verlag Chemie GMBH, in cooperation with the Infrared Absorption Data Joint Committee, London, and the Institut für Spectrochemie and Angewandte Spectroskopie, Dortmund. Spectra are presented on coded cards. Coded cards containing abstracts of articles relating to IR spectrometry are also issued.

Dolphin, D., and Wick, A.E. (1977). *Tabulation of Infrared Spectral Data.* New York: Wiley.

Finch, A., Gates, P.N., Radcliffe, K., Dickson, F.N., and Bentley, F.F. (1970). *Chemical Applications of Far Infrared Spectroscopy.* New York: Academic Press.

Flett, M. St. C. (1969). *Characteristic Frequencies of Chemical Groups in the Infrared.* New York: American Elsevier.

Hershenson, H.H. (1959, 1964). *Infrared Absorption Spectra Index.* New York and London: Academic Press. Two volumes cover 1945–1962.

Infrared Band Handbook. Supplements 1 and 2. (1964). New York: Plenum Press, 259 pp.

Infrared Band Handbook. Supplements 3 and 4. (1965). New York: Plenum Press.

Infrared Band Handbook. 2nd rev. ed. (1970). New York: IFI/Plenum Press. Two volumes.

IRDC Cards: Infrared Data Committee of Japan (S. Mizushima). Haruki-cho, Tokyo: Handled by Nankodo Co.

Katritzky, A.R., Ed. (1963). Physical Methods in Heterocyclic Chemistry, Vol. II. New York and London: Academic Press.

Lin-Vien, D., Colthup, N.B. Fately, W.G., and Grasselli, J.G. (1991). *Handbook of Infrared and Raman Characteristic Frequencies of Organic Molecules.* New York: Academic Press.

Loader, E.J. (1970). *Basic Laser Raman Spectroscopy.* London, New York: Heyden-Sadtler.

Maslowsky, E. (1977). *Vibrational Spectra of Organometallic Compounds.* New York: Wiley-Interscience.

Ministry of Aviation Technical Information and Library Services, Ed. (1960). An Index of Published Infra-Red Spectra, Vols. 1 and 2. London: Her Majesty's Stationery Office.

NRC-NBS (Creitz) File of Spectrograms. Issued by National Research Council-National Bureau of Standards Committee on Spectral Absorption Data, National Bureau of Standards, Washington 25, DC. Spectra presented on edge punched cards.

Nyquist, R.A. (1984). *Interpretation of Vapor-Phase Infrared Spectra.* Philadelphia: Sadtler-Heyden.

Porro, T.J., and Pattacini, S.C. (1993). *Spectroscopy* 8(7), 40–47. Sample handling.

Sadtler Standard Infrared Grating Spectra. (1972). Philadelphia: Sadtler Research Laboratories, Inc. 26,000 spectra in 26 volumes.

Sadtler Standard Infrared Prism Spectra. (1972). Philadelphia: Sadtler Research Laboratories, Inc. 43,000 spectra in 43 volumes.

Sadtler Reference Spectra-Commonly Abused Drugs IR & UV Spectra. (1972). Philadelphia: Sadtler Research Laboratories, Inc. 600 IR Spectra, 300 UV.

Sadtler Reference Spectra—Gases & Vapors High Resolution Infrared. (1972). Philadelphia: Sadtler Research Laboratories, Inc. 150 Spectra.

Sadtler Reference Spectra- Inorganics IR Grating. (1967). Philadelphia: Sadtler Research Laboratories, Inc. 1300 spectra.

Sadtler Reference Spectra—Organometallics IR Grating. (1966). Philadelphia: Sadtler Research Laboratories, Inc. 400 spectra.

Sadtler Digital FTIR Libraries. (1994). Philadelphia: Sadtler Research Labs. 100,000 digitized spectra.

Socrates, G. (1994). IR Characteristic Group Frequencies. New York: Wiley.

Stewart, J.E. (1965). Far infrared spectroscopy. In S.K. Freeman (Ed.), *Interpretive Spectroscopy*. New York: Reinhold, p. 131.

Tichy, M. (1964). The determination of intramolecular hydrogen bonding by infrared spectroscopy and its applications in stereochemistry. In R.R. Raphael (Ed.), *Advances in Organic Chemistry: Methods and Results*, Vol. 5. New York: Wiley-Interscience.

STUDENT EXERCISES

2.1 Either benzonitrile or phenylacetonitrile shows a band of medium intensity at 2940 cm^{-1}; the other compound shows nothing in the range 3000–2500 cm^{-1}. Explain.

2.2 Select a compound that best fits each of the following sets of IR bands (in cm^{-1}). Each set corresponds to a list of just a few important bands for each compound.

Benzamide	Diphenyl sulfone
Benzoic acid	Formic acid
Benzonitrile	Isobutylamine
Biphenyl	1-Nitropropane dioxane

a. 3080 (w), nothing 3000–2800, 2230 (s), 1450 (s), 760 (s), 688 (s)

b. 3380 (m), 3300 (m), nothing 3200–3000, 2980 (s), 2870 (m), 1610 (m), ~900–700 (b)

c. 3080 (w), nothing 3000–2800, 1315 (s), 1300 (s), 1155 (s)

d. 2955 (s), 2850 (5), 1120 (s)

e. 2946 (s), 2930 (m), 1550 (s), 1386 (m)

f. 2900 (b, s), 1720 (b, s)

g. 3030 (m), 730 (s), 690 (s)

h. 3200–2400 (5), 1685 (b, s), 705 (s)

i. 3350 (s), 3060 (m), 1635 (s)

s = strong, m = medium, w = weak, b = broad

For Exercises 2.3–2.6, match the name from each list to the proper IR spectrum. Identify the diagnostic bands in each spectrum.

Spectra for Exercises 2.3 to 2.8 can be found on the Wiley website at *http://www.wiley.com/college/silverstein* in PDF format.

2.3 SPECTRA A–D
1,3-Cyclohexadiene
Diphenylacetylene
1-Octene
2-Pentene

2.4 SPECTRA E–I
Butyl acetate
Butyramide
Isobutylamine
Lauric acid
Sodium propionate

2.5 SPECTRA J–M
Allyl phenyl ether
Benzaldehyde
o-Cresol
m-Toluic acid

2.6 SPECTRA N–R
Aniline
Azobenzene
Benzophenone oxime
Benzylamine
Dimethylamine hydrochloride

2.7 Deduce the structure of compound (S) whose formula is C_2H_3NS from the spectrum below.

2.8 Point out evidence for enol formation of 2,4-pentanedione (Compound T). Include explanations of the two bands in the 1700–1750 cm^{-1} range, the 1650 band, and the very broad band with multiple maxima running from 3400–2600 cm^{-1} (only the peaks at 3000–2900 result from C—H stretching).

2.9 For each of the following IR spectra (A–W) list functional groups that a) are present, and b) are absent. The mass spectra of these compounds are in Chapter 1, (Exercise 1.6.)

Problem 2.9 Spectrum A

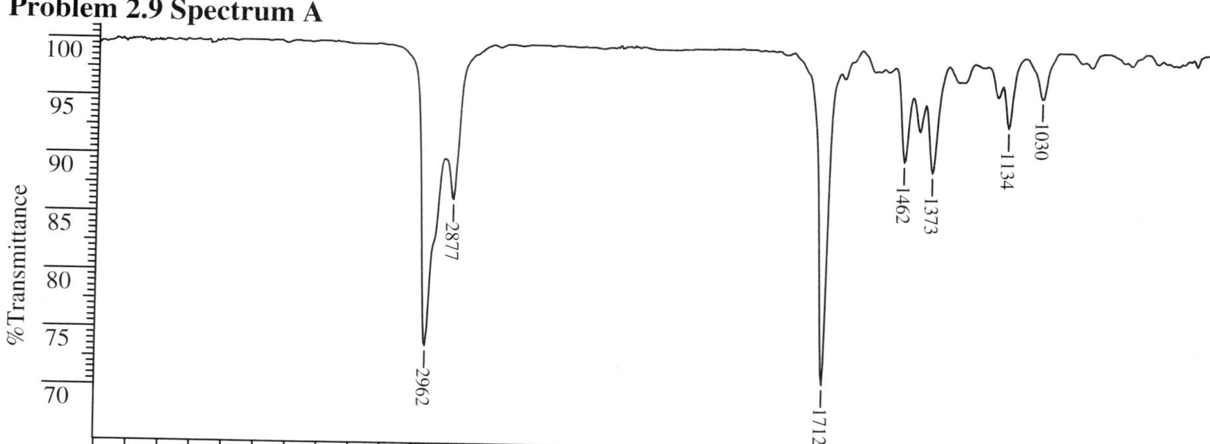

Problem 2.9 Spectrum B

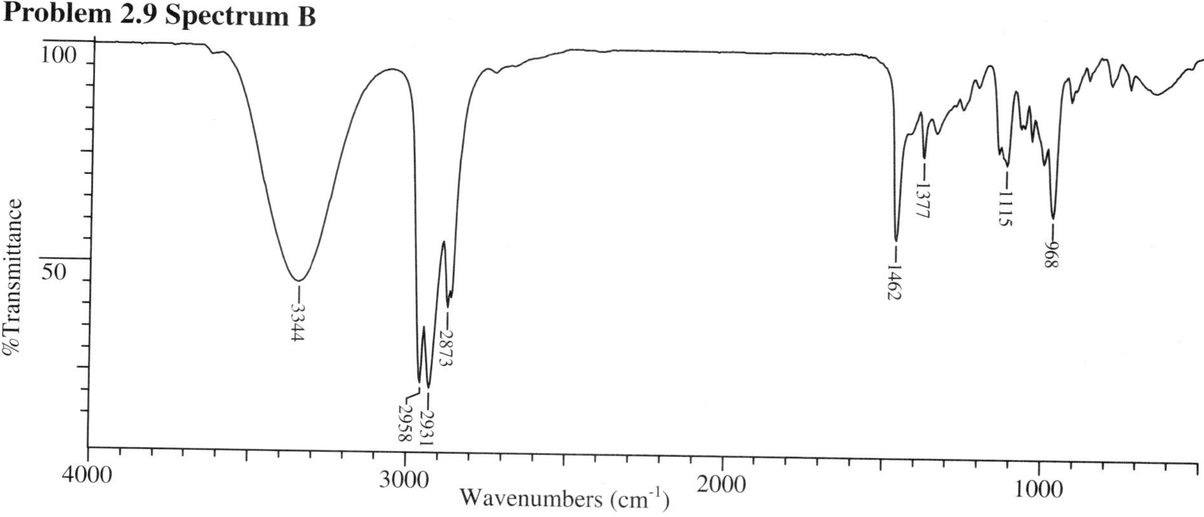

Problem 2.9 Spectrum C

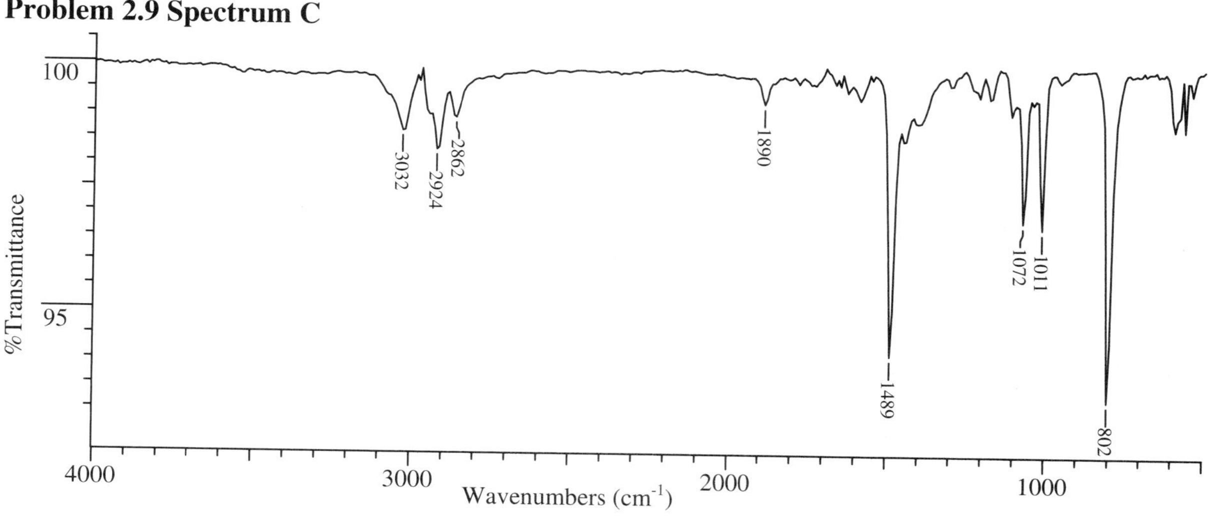

Problem 2.9 Spectrum D

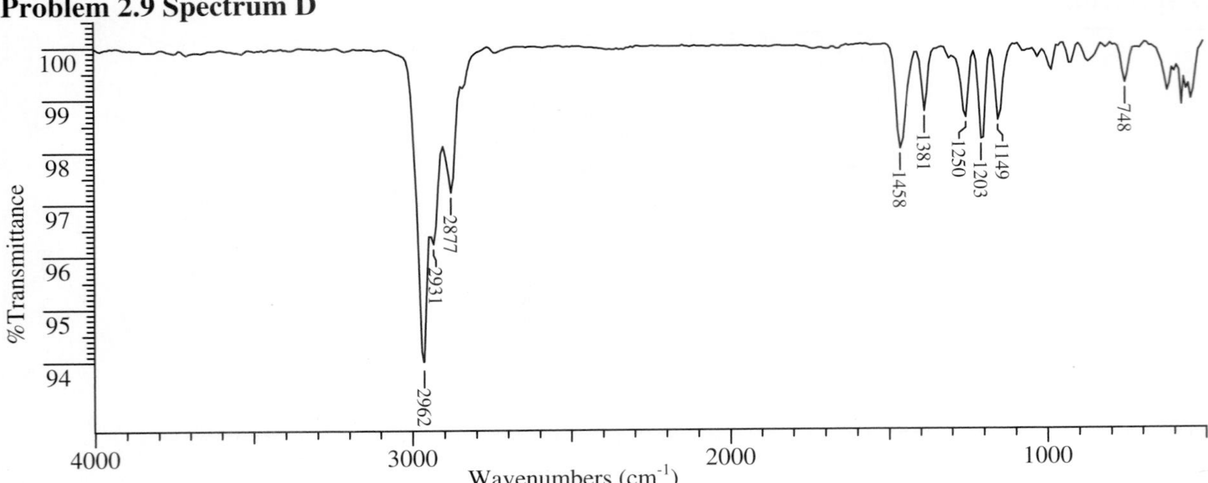

Problem 2.9 Spectrum E

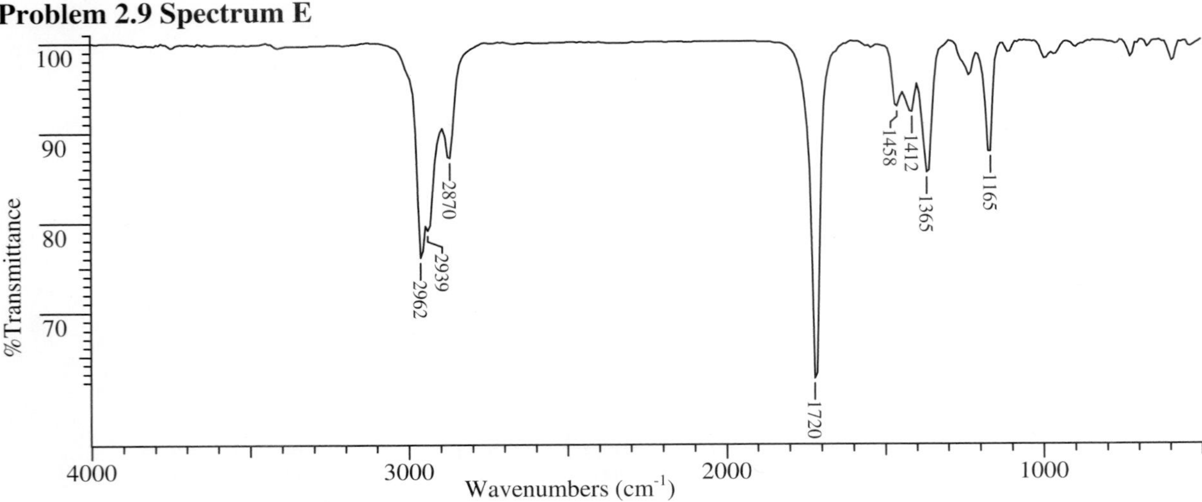

Problem 2.9 Spectrum F

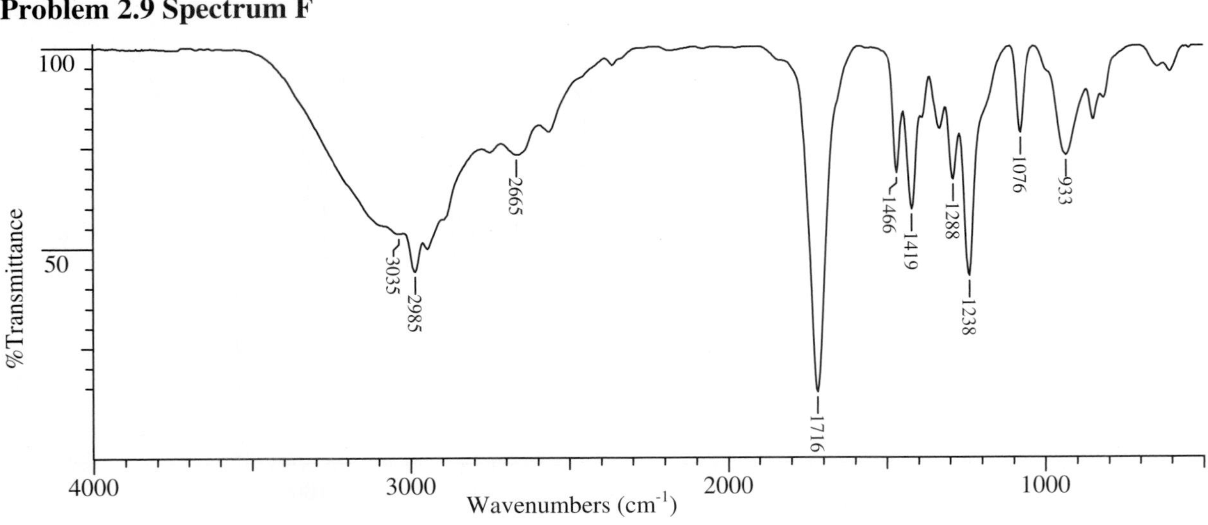

Exercise 2.9

Problem 2.9 Spectrum G

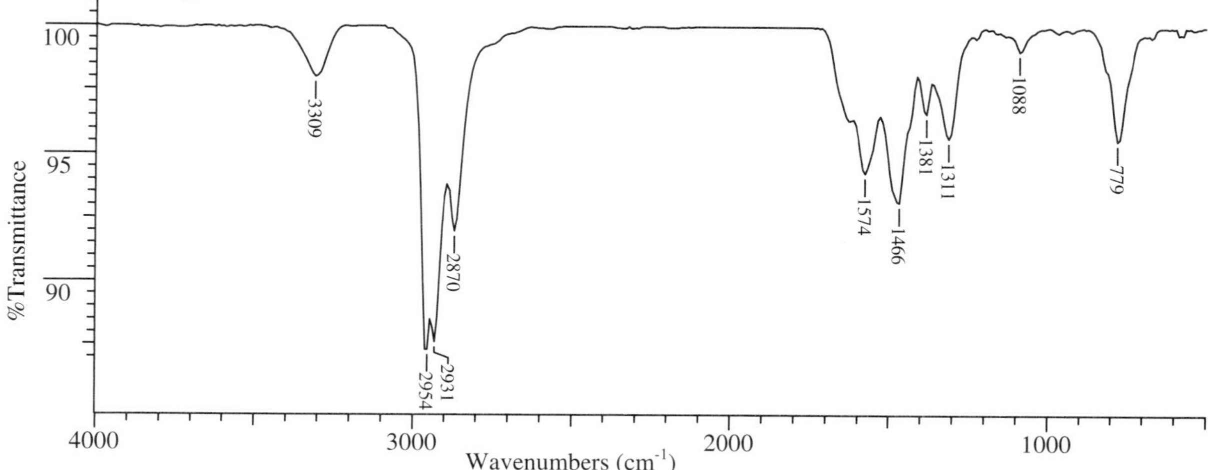

Problem 2.9 Spectrum H

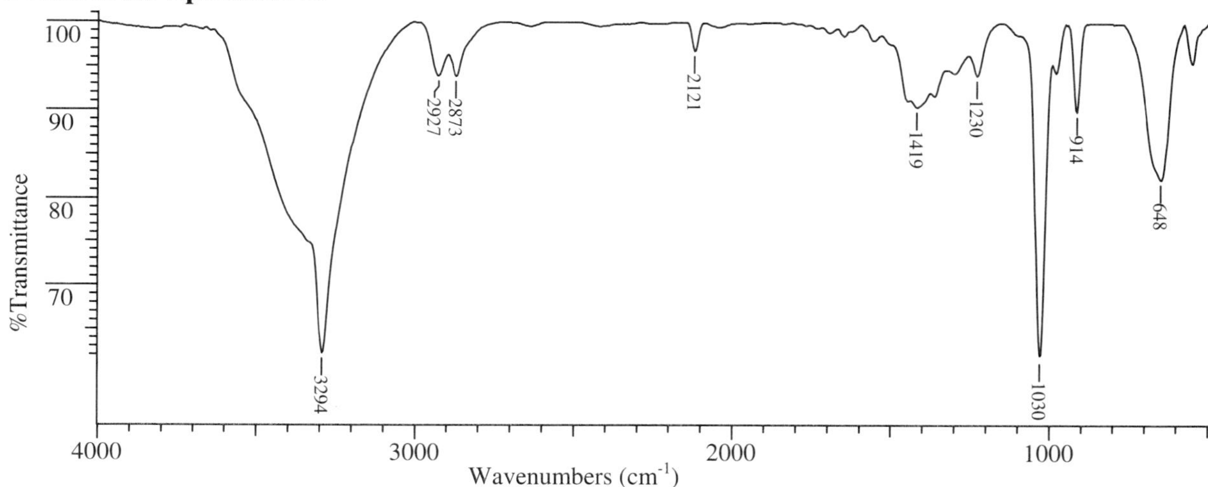

Problem 2.9 Spectrum I

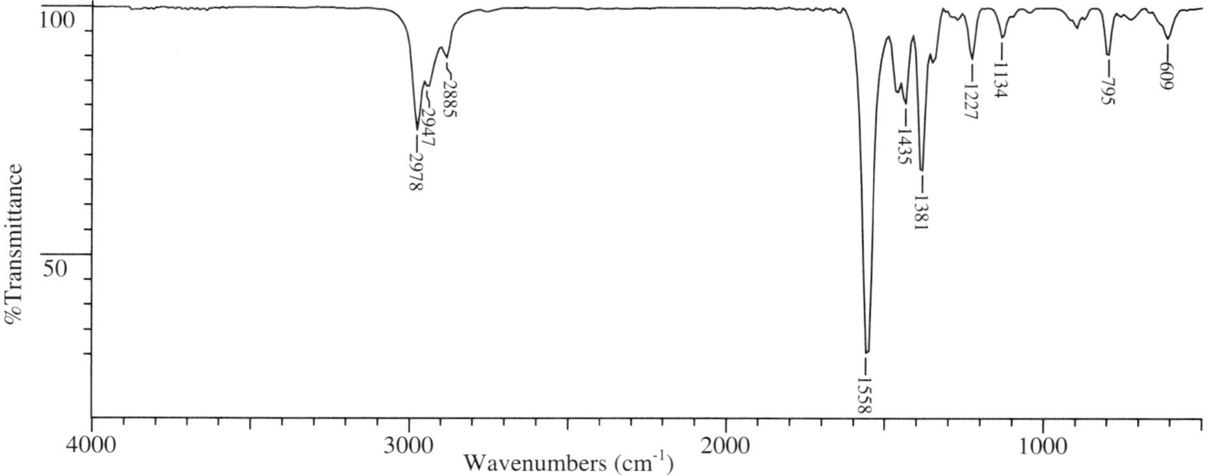

Problem 2.9 Spectrum J

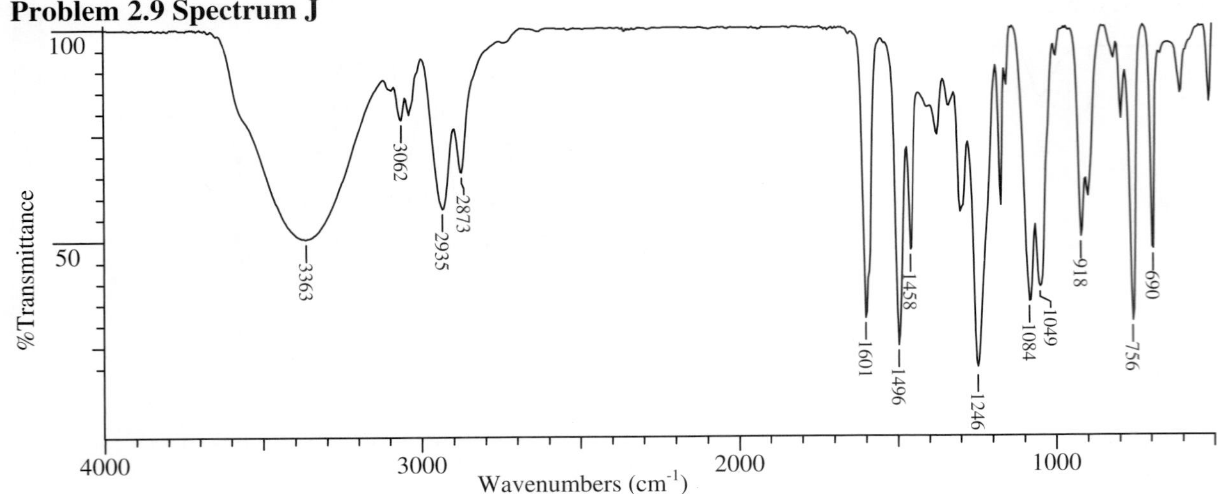

Problem 2.9 Spectrum K

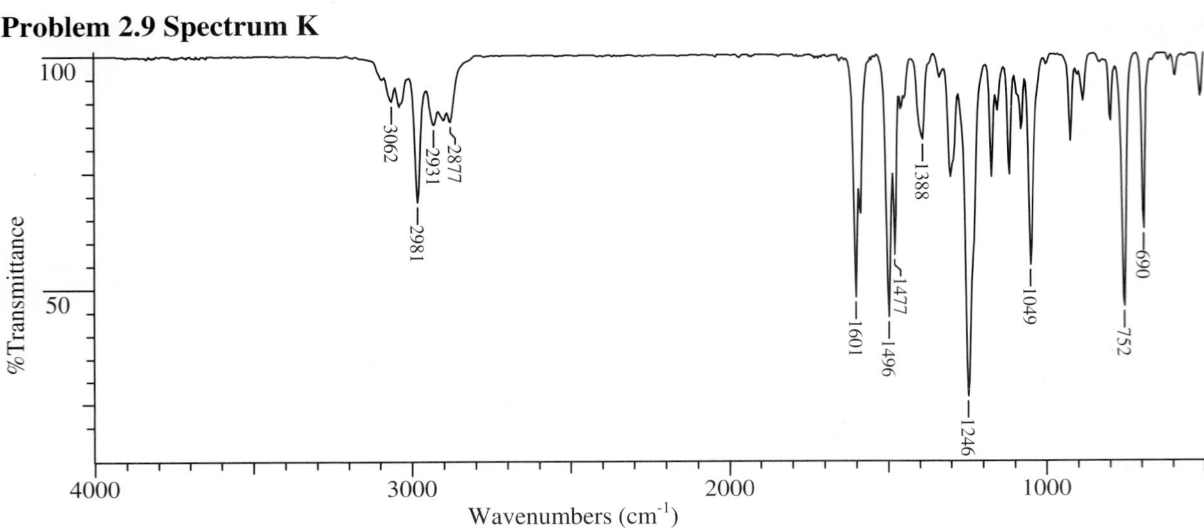

Problem 2.9 Spectrum L

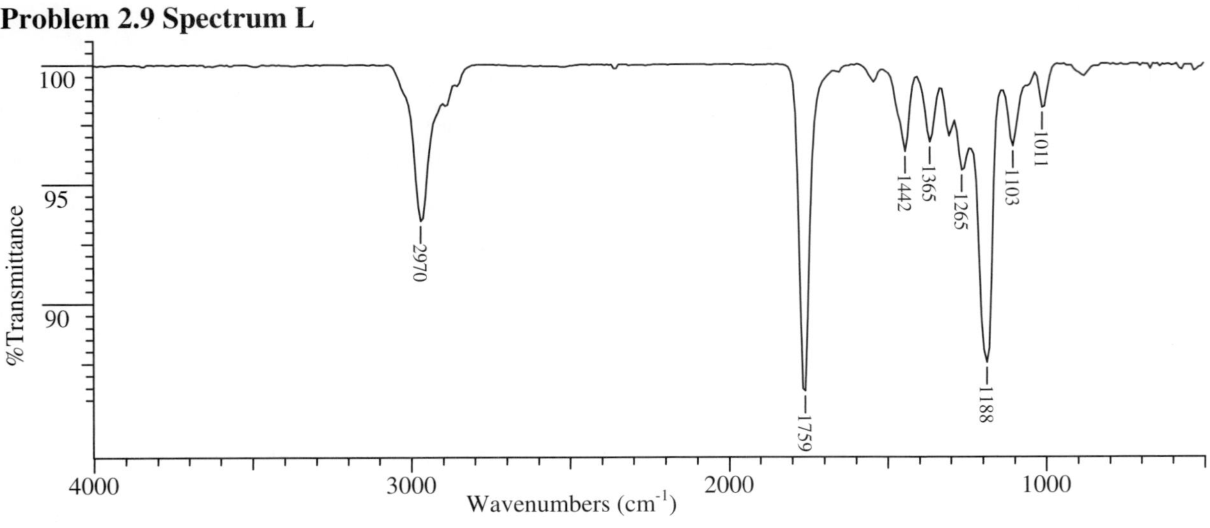

Exercise 2.9

Problem 2.9 Spectrum M

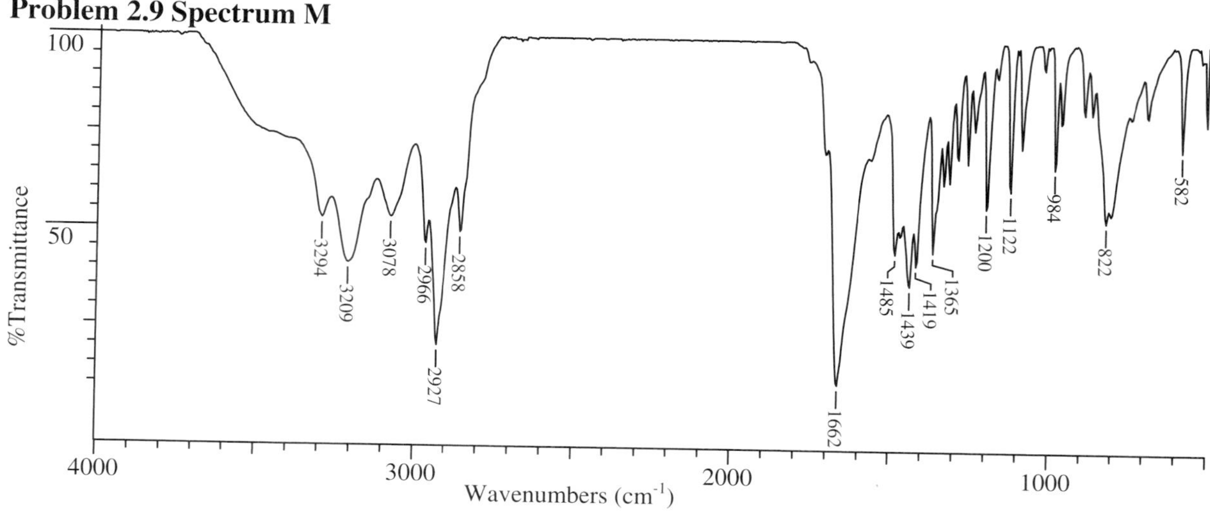

Problem 2.9 Spectrum N

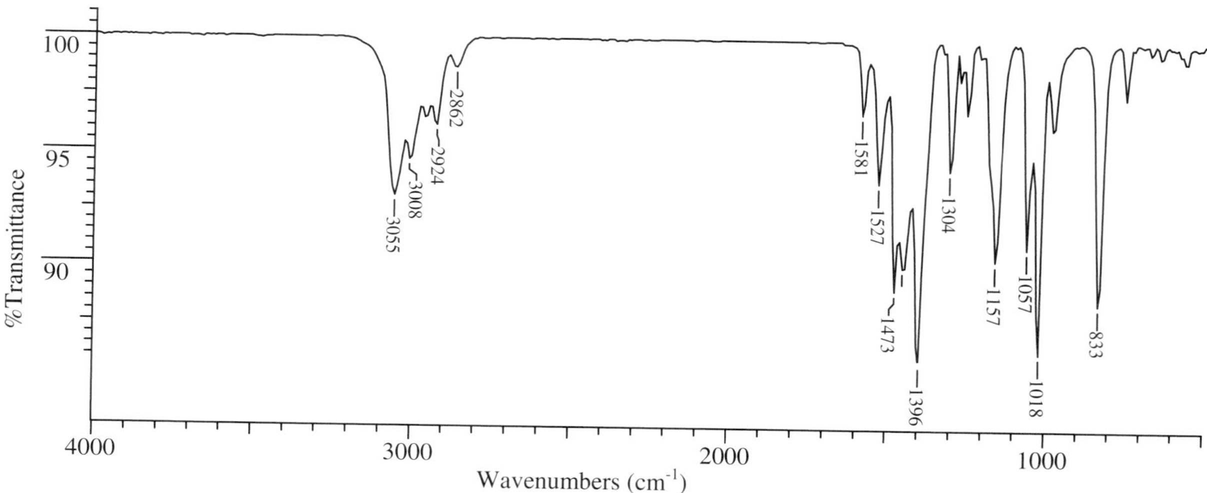

Problem 2.9 Spectrum O

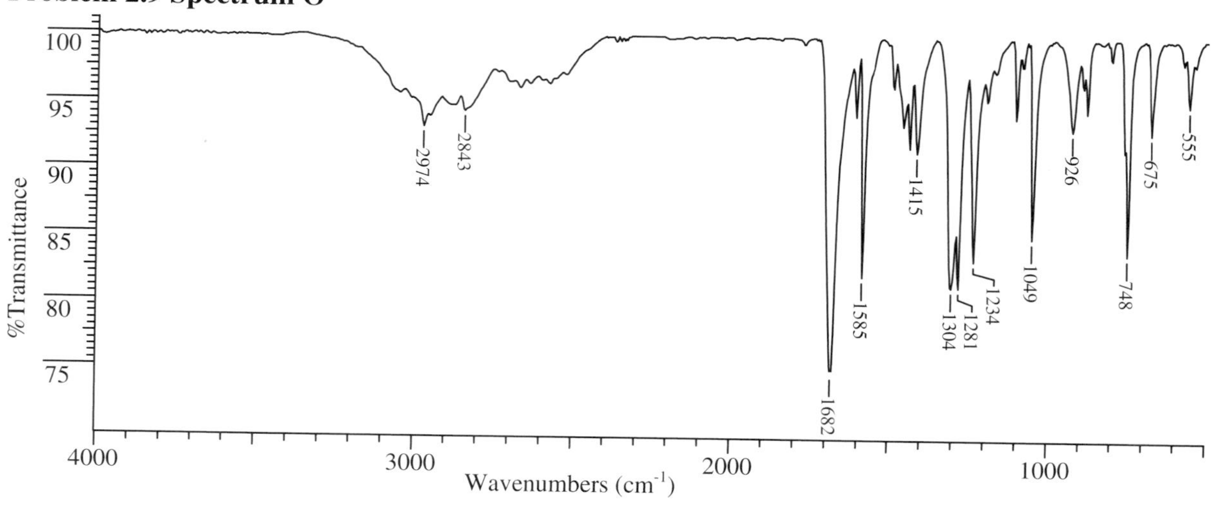

Problem 2.9 Spectrum P

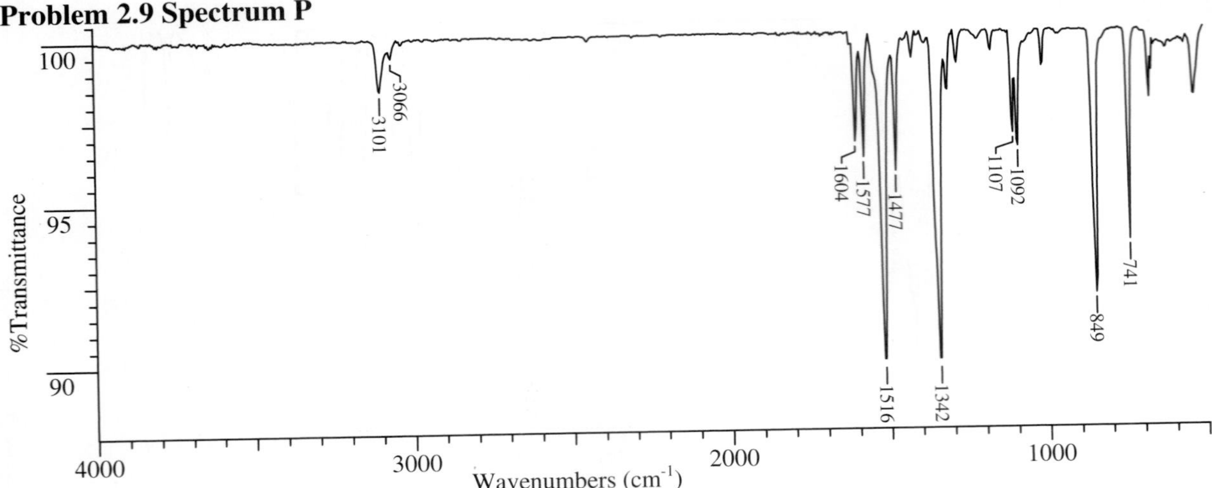

Problem 2.9 Spectrum Q

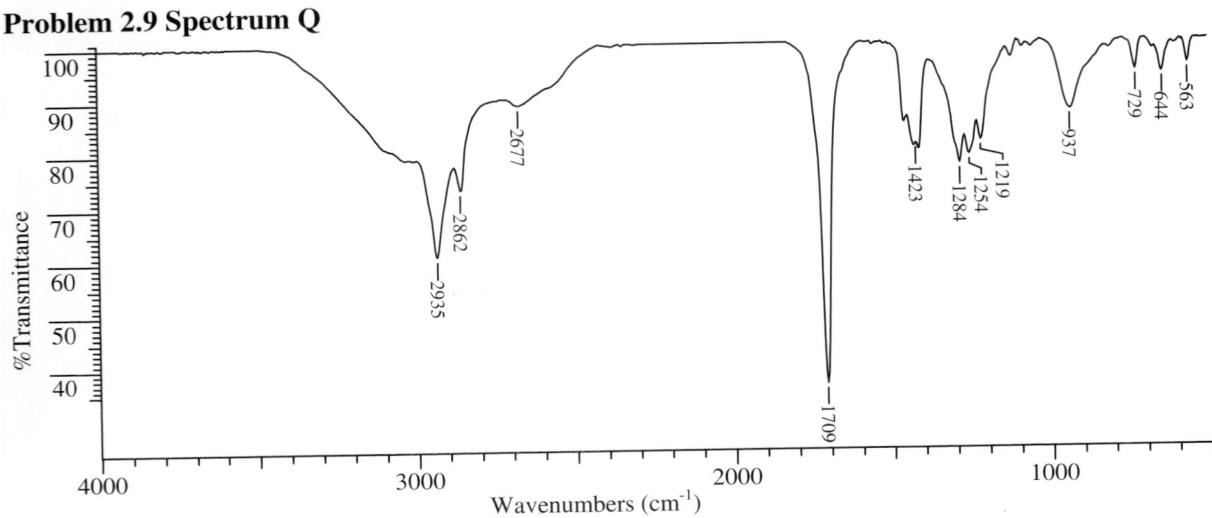

Problem 2.9 Spectrum R

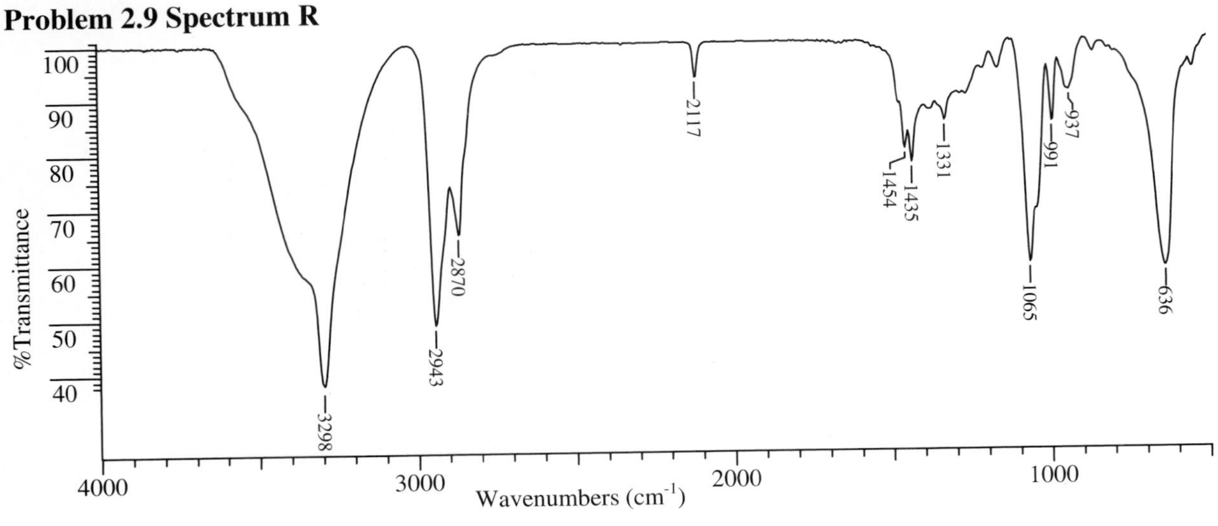

Exercise 2.9

Problem 2.9 Spectrum S

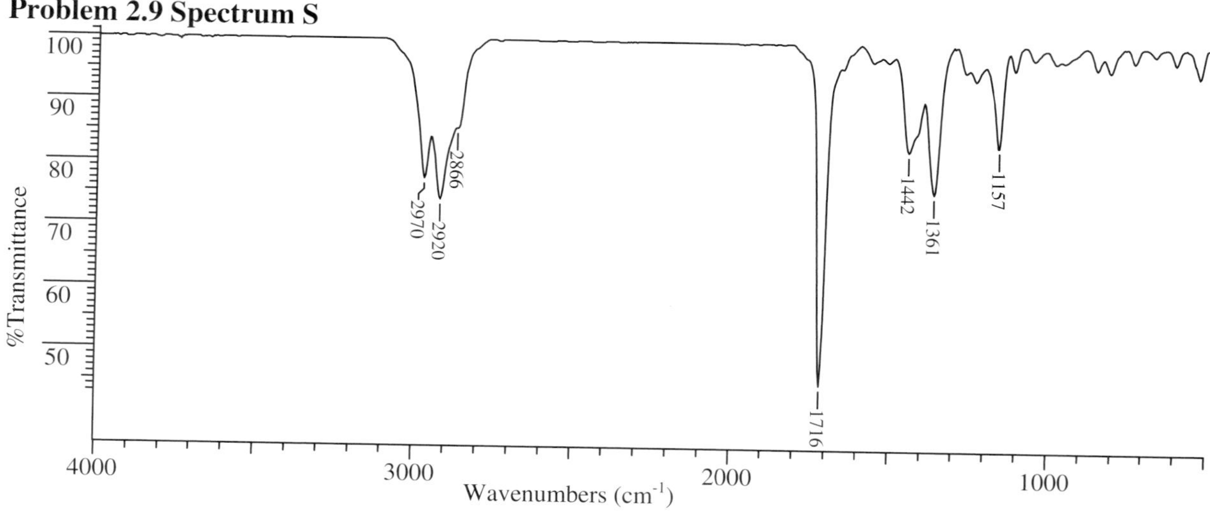

Problem 2.9 Spectrum T

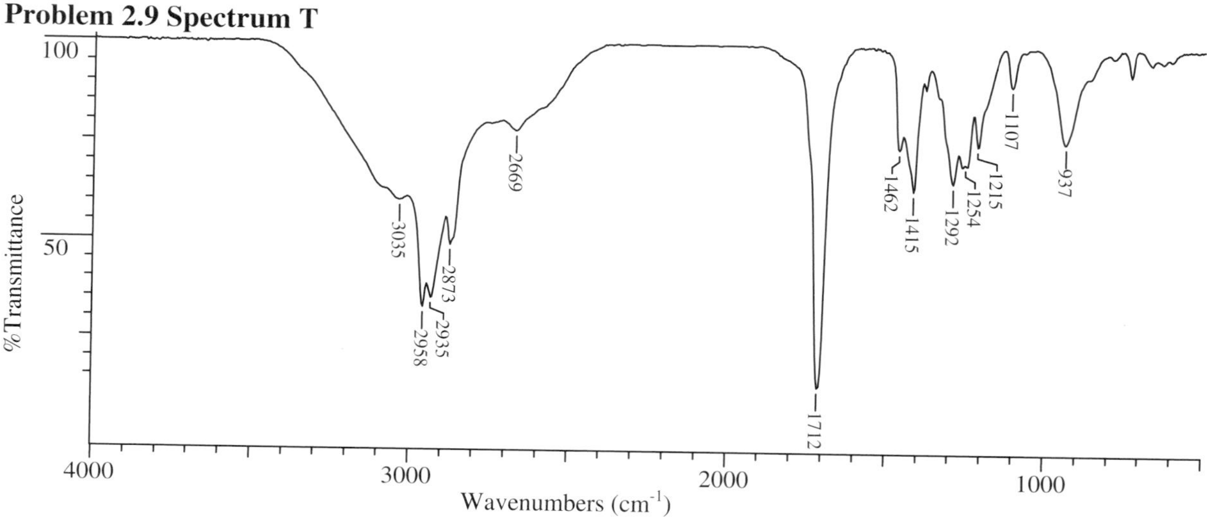

Problem 2.9 Spectrum U

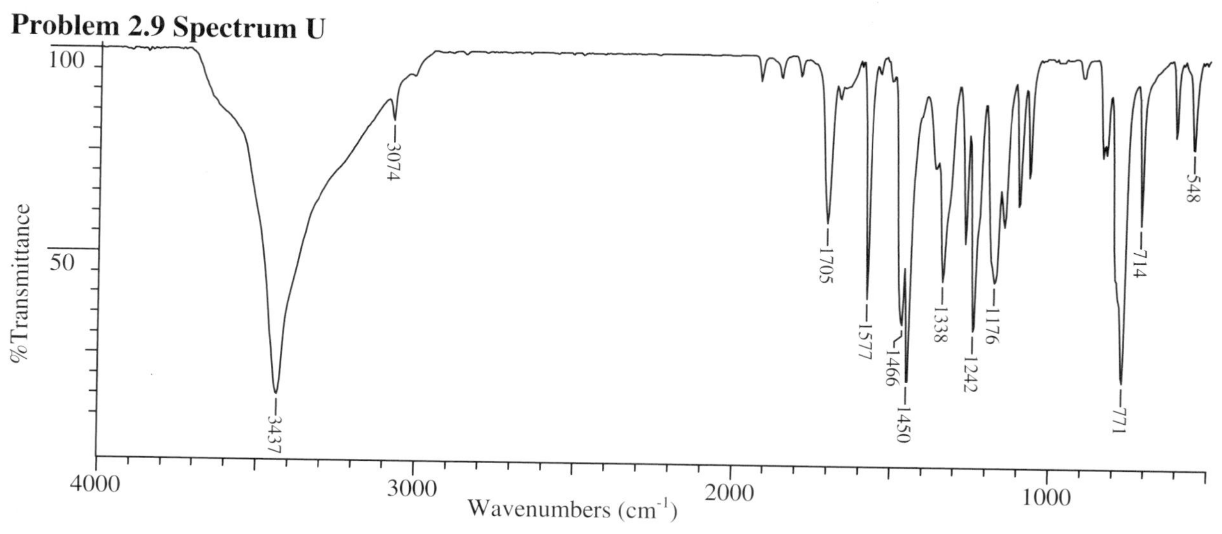

Problem 2.9 Spectrum V

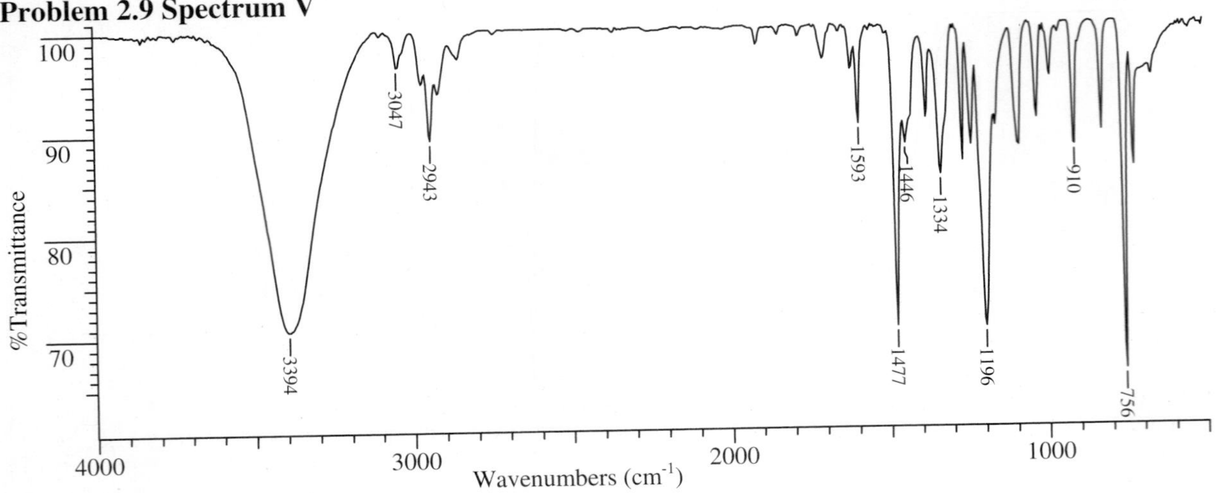

Problem 2.9 Spectrum W

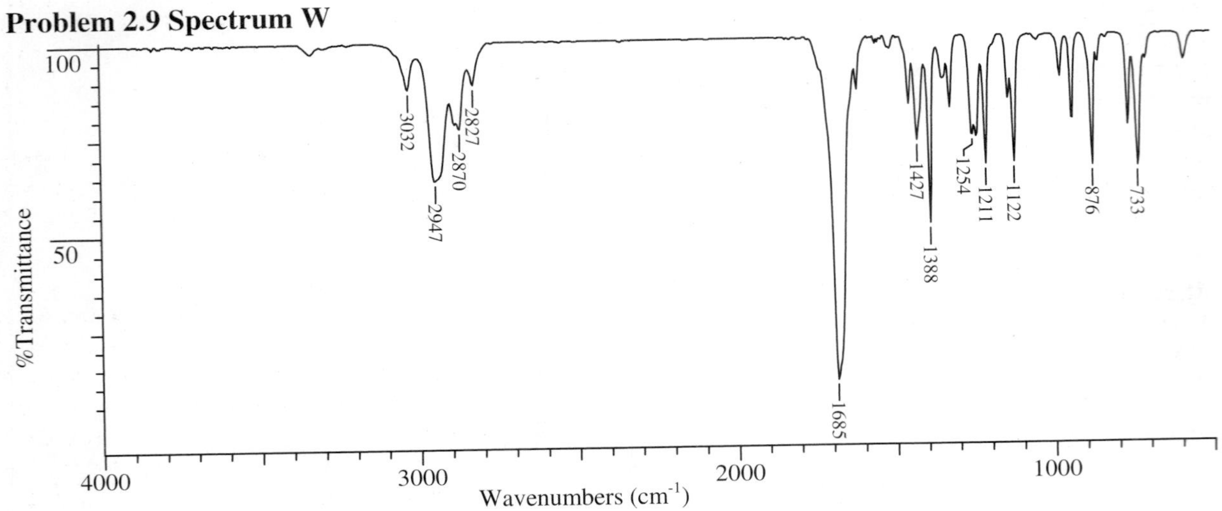

CHART AND SPECTRAL PRESENTATIONS OF ORGANIC SOLVENTS, MULLING OILS, AND OTHER COMMON LABORATORY SUBSTANCES

APPENDIX A TRANSPARENT REGIONS OF SOLVENTS AND MULLING OILS

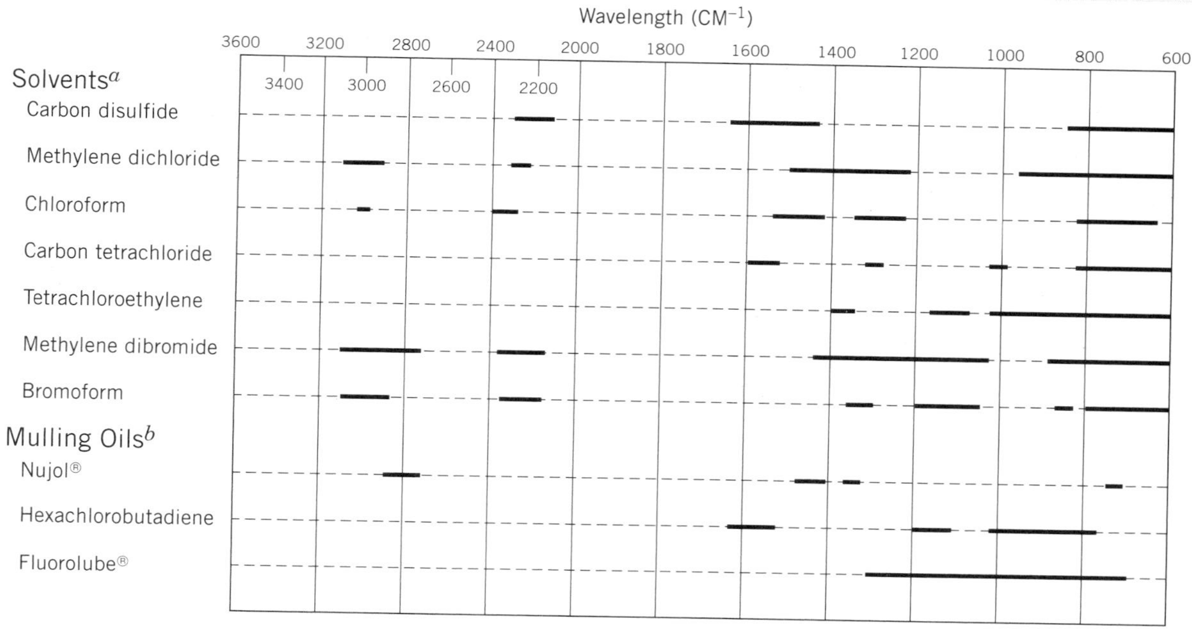

[a] The open regions are those in which the solvent transmits more than 25% of the incident light at 1 mm thickness.
[b] The open regions for mulling oils indicate transparency of thin films.

APPENDIX B CHARACTERISTIC GROUP ABSORPTIONS[a]

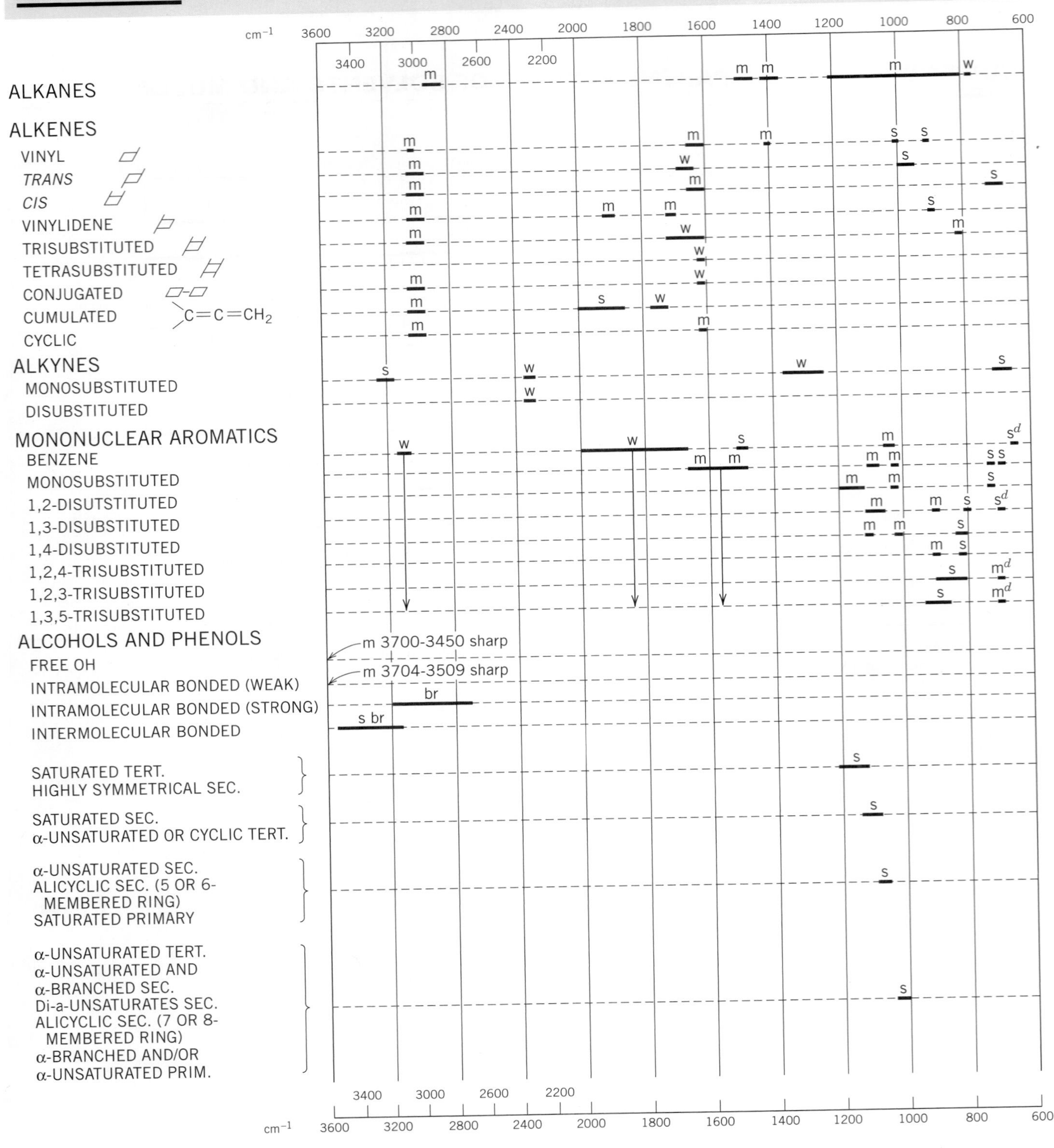

[a] Absorptions are shown by heavy bars. s = strong, m = medium, w = weak, sh = sharp, br = broad. Two intensity designations over a single bar indicate that two peaks may be present.

[b] May be absent.

[c] Frequently a doublet.

[d] Ring bending bands.

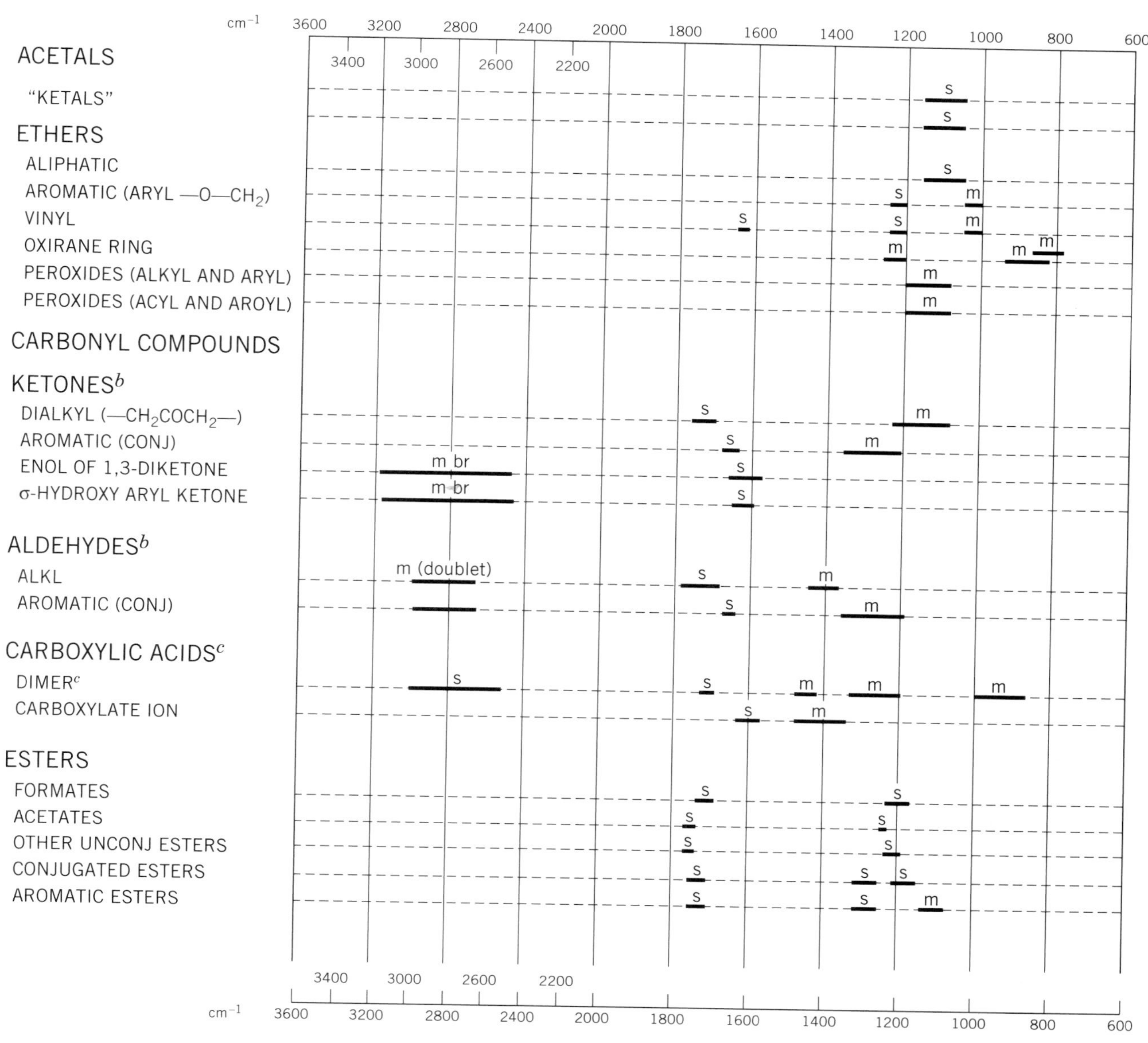

[a] Three bands, sometimes a fourth for ketals, and a fifth band for acetals.
[b] Conjugated aliphatic examples show C=O stretch at virtually the same position as aromatic structures.
[c] Conjugated examples show C=O stretch at lower wavenumbers (1710–1680 cm^{-1}). The O—H stretch (3300–2600 cm^{-1}) is very broad.

APPENDIX B *(Continued)*

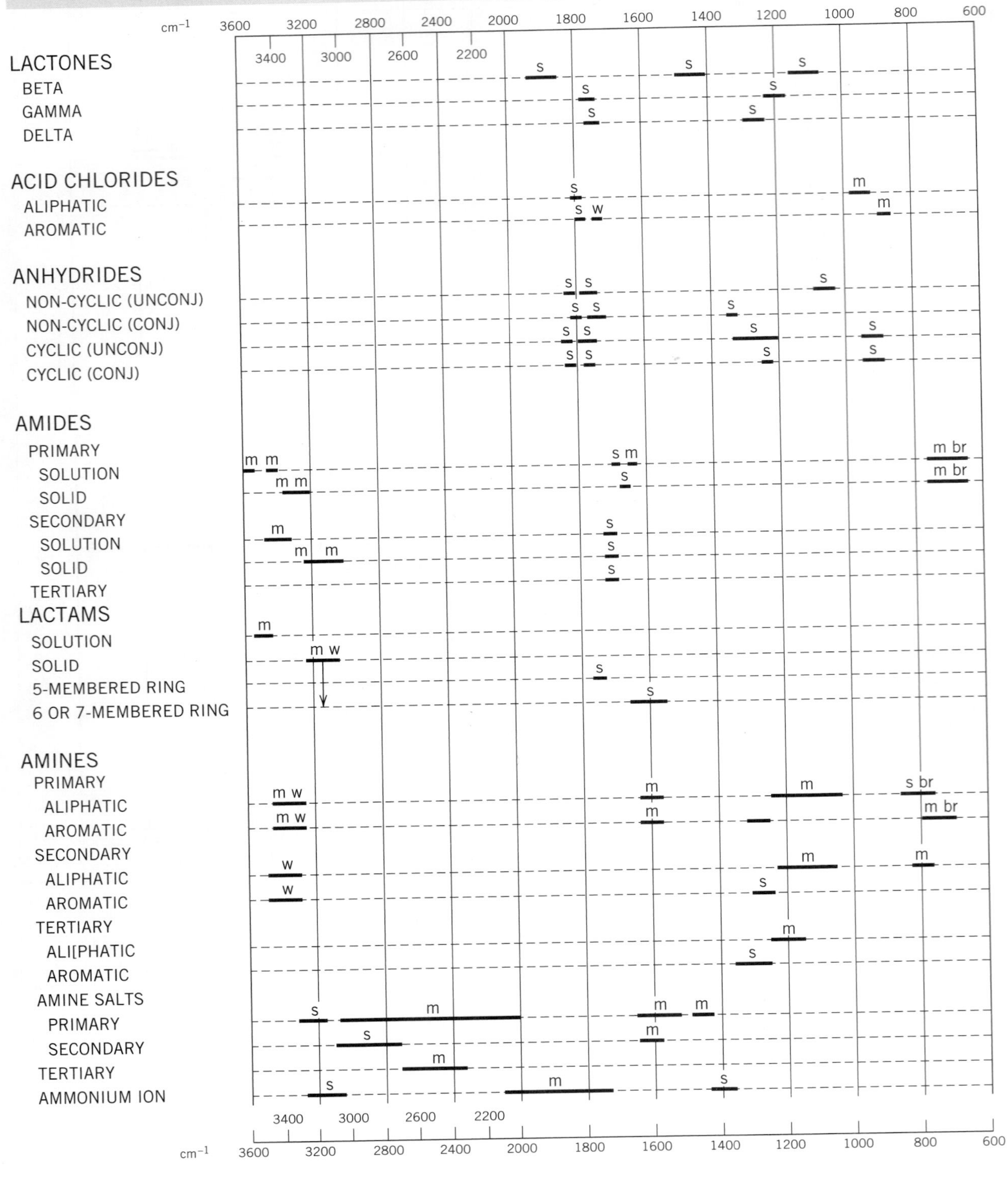

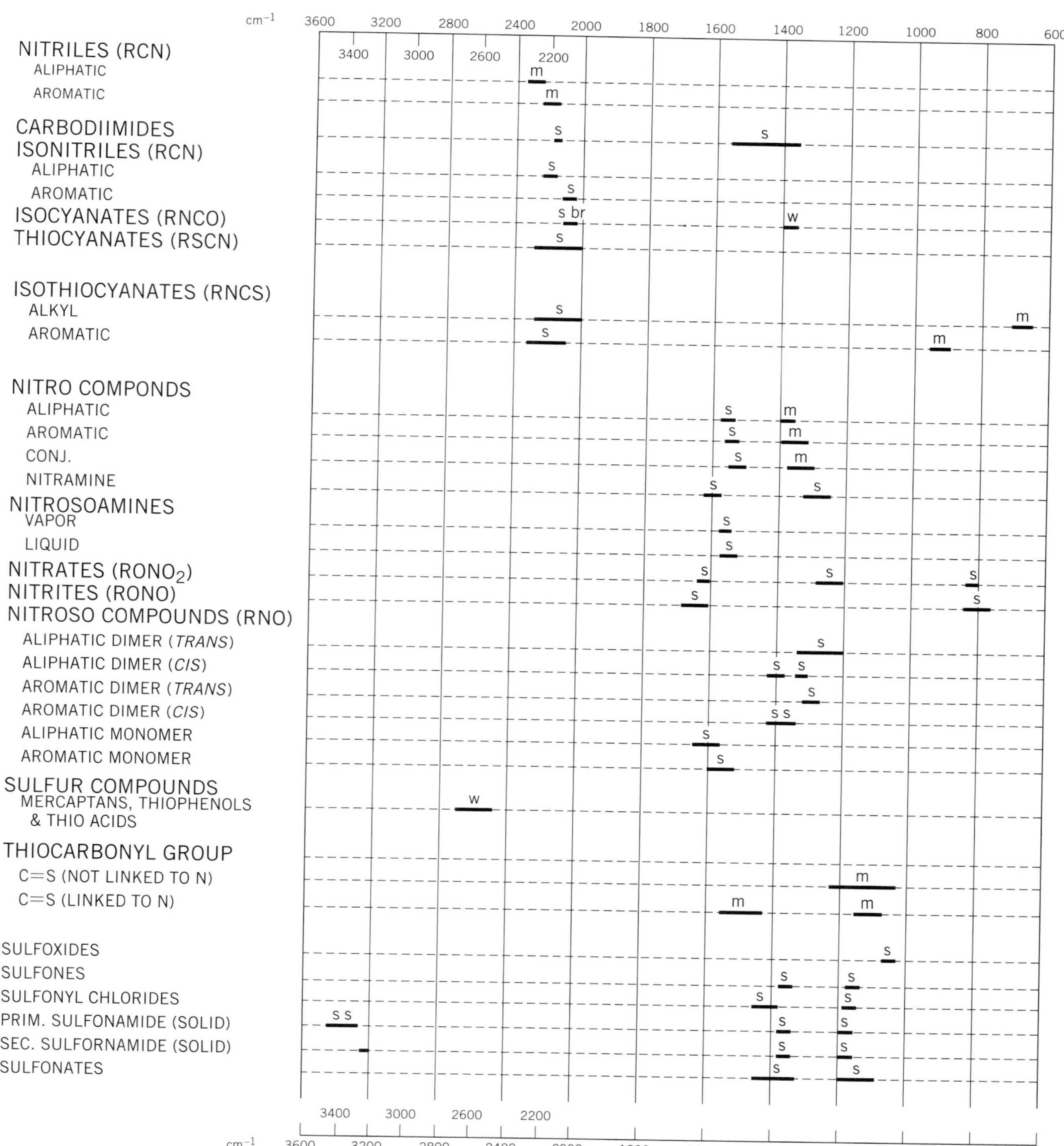
cm⁻¹
3600 3200 2800 2400 2000 1800 1600 1400 1200 1000 800 600
3400 3000 2600 2200
NITRILES (RCN)
ALIPHATIC
AROMATIC
CARBODIIMIDES
ISONITRILES (RCN)
ALIPHATIC
AROMATIC
ISOCYANATES (RNCO)
THIOCYANATES (RSCN)
ISOTHIOCYANATES (RNCS)
ALKYL
AROMATIC
NITRO COMPONDS
ALIPHATIC
AROMATIC
CONJ.
NITRAMINE
NITROSOAMINES
VAPOR
LIQUID
NITRATES (RONO2)
NITRITES (RONO)
NITROSO COMPOUNDS (RNO)
ALIPHATIC DIMER (TRANS)
ALIPHATIC DIMER (CIS)
AROMATIC DIMER (TRANS)
AROMATIC DIMER (CIS)
ALIPHATIC MONOMER
AROMATIC MONOMER
SULFUR COMPOUNDS
MERCAPTANS, THIOPHENOLS
& THIO ACIDS
THIOCARBONYL GROUP
C=S (NOT LINKED TO N)
C=S (LINKED TO N)
SULFOXIDES
SULFONES
SULFONYL CHLORIDES
PRIM. SULFONAMIDE (SOLID)
SEC. SULFORNAMIDE (SOLID)
SULFONATES
3400 3000 2600 2200
cm⁻¹
3600 3200 2800 2400 2000 1800 1600 1400 1200 1000 800 600

APPENDIX B *(Continued)*

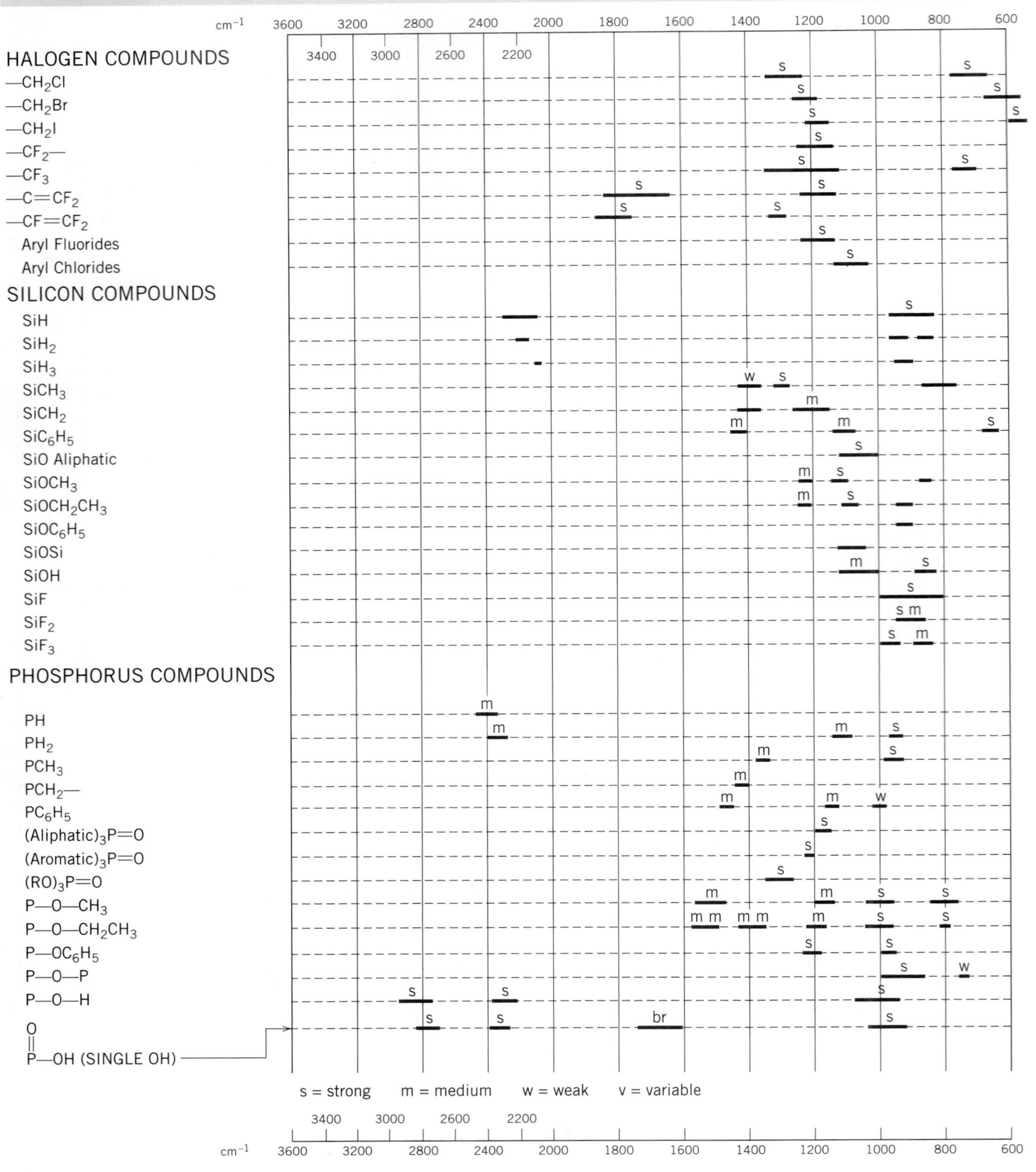

APPENDIX C ABSORPTIONS FOR ALKENES

TABLE C-1 Alkene Absorptions[a]

H H C=C R H Vinyl 1648–1638 cm^{-1} 995–985 cm^{-1} (s)[b] 915–905 cm^{-1} (s)	H H C=C R R *cis* 1662–1626 cm^{-1} (v) 730–665 cm^{-1} (s)	R H C=C H R *trans* 1678–1668 cm^{-1} (v) 980–960 cm^{-1} (s)[c]
R H C=C R H Vinylidine 1658–1648 cm^{-1} (m) 895–885 cm^{-1} (s)	R R C=C H R Trisubstituted 1675–1665 cm^{-1} (w) 840–790 cm^{-1} (m)	R R C=C R R Tetrasubstituted 1675–1665 cm^{-1} very weak or absent.

[a] s = strong, m = medium, w = weak, v = variable.
[b] This band also shows a strong overtone band.
[c] This band occurs near 1000 cm^{-1} in conjugated *trans–trans* systems such as the esters of sorbic acid.

TABLE C-2 C=C Stretching Frequencies in Cyclic and Acyclic Systems (cm^{-1})

Ring[a] or Chain	H H C=C C C	H CH_3 C=C C C	CH_3 CH_3 C=C C C	C C=CH_2 C
Chain *cis*	1661	1681	1672	1661
Chain *trans*	1676			
Three-membered ring	1641		1890	1780
Four-membered ring	1566		1685	1678
Five-membered ring	1611	1658	1686	1657
Six-membered ring	1649	1678	1685	1651
Seven-membered ring	1651	1673		
Eight-membered ring	1653			

[a] All rings have *cis* double bonds.

APPENDIX D ABSORPTIONS FOR PHOSPHORUS COMPOUNDS

TABLE D-1 P=O and P—O Stretching Vibrations

Group	Position cm^{-1} Intensity[a]	ν_{P-O} Bands[a] (cm^{-1})
P=O stretch		
Phosphine oxides		
Aliphatic	~1150	
Aromatic	~1190	
Phosphate esters[b]	1299–1250	
P—OH	1040–910 (s)	
P—O—P	1000–870 (s)	~700 w
P—O—C (aliph)	1050–970 (s)[c]	830–740 (s)[d]
P—O—C (arom)	1260–1160 (s)	994–855 (s)

[a] s = strong; w = weak
[b] The increase in P=O stretching frequency of the ester, relative to the oxides, results from the electronegativity of the attached alkoxy groups.
[c] May be a doublet.
[d] May be absent.

APPENDIX E ABSORPTIONS FOR HETEROAROMATICS

TABLE E-1 γ-CH and Ring Bending (β-Ring) Bands of Pyridines[a]

Substitution	Number Adjacent H Atoms	γ-CH (cm^{-1})	β-Ring
2-	4	781–740	752–746
3-	3	810–789	715–712
4-	2	820–794	775–709

[a] The γ and β notations are explained in the text (Section 3.6.30.4) and in the book by Katritzky (1963).

TABLE E-2 Characteristic γ-CH or β-Ring Bands of Furans, Thiophenes, and Pyrroles

Ring	Position of Substitution	Phase	γ-CH or β-Ring Modes[a]			
			cm^{-1}	cm^{-1}	cm^{-1}	cm^{-1}
Furan	2-	$CHCl_3$	~925	~884	835–780	
	2-	Liquid	960–915	890–875		780–725
	2-	Solid	955–906	887–860	821–793	750–723
	3-	Liquid		885–870	741	
Thiophene	2-	$CHCl_3$	~925	~853	843–803	
	3-	Liquid				755
Pyrrole	2-Acyl	Solid			774–740	~755

[a] The γ and β notations are explained in the text (Section 3.6.30.4) and in the book by Katritzky (1963).

CHAPTER 3

PROTON NMR SPECTROMETRY

3.1 INTRODUCTION

Nuclear magnetic resonance (NMR) spectrometry is basically another form of absorption spectrometry, akin to IR or UV spectrometry. Under appropriate conditions *in a magnetic field,* a sample can absorb electromagnetic radiation in the radio frequency (rf) region at frequencies governed by the characteristics of the sample. Absorption is a function of certain nuclei in the molecule. A plot of the frequencies of the absorption peaks versus peak intensities constitutes an NMR spectrum. This chapter covers proton magnetic resonance (^{1}H NMR) spectrometry.

With some mastery of basic theory, interpretation of NMR spectra merely by inspection is usually feasible in greater detail than is the case for IR or mass spectra. The present account will suffice for the immediate limited objective: identification of organic compounds in conjunction with other spectrometric information. References are given at the end of this chapter.

3.2 THEORY

3.2.1 Magnetic Properties of Nuclei

We begin by describing some magnetic properties of nuclei. All nuclei carry a charge. In some nuclei, this charge "spins" on the nuclear axis, and this circulation of nuclear charge generates a magnetic dipole along the axis (Figure 3.1). The angular momentum of the spinning charge can be described in terms of its quantum spin number I; these numbers have values of 0, 1/2, 1, 3/2, and so on ($I = 0$ denotes no spin). The intrinsic magnitude of the generated dipole is expressed in terms of the nuclear magnetic moment, μ.

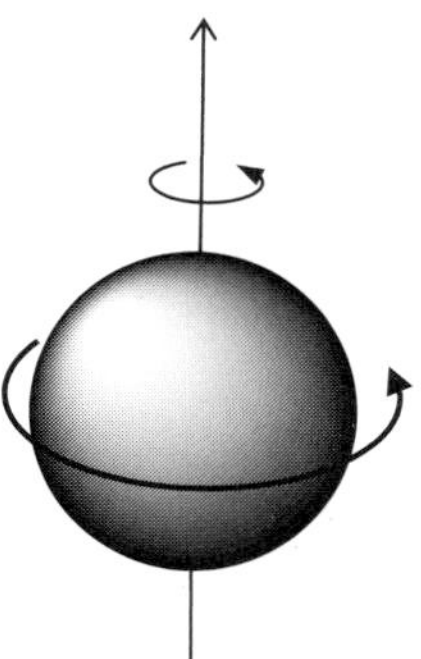

FIGURE 3.1 Spinning charge on proton generates magnetic dipole.

Relevant properties, including the spin number, I, of several nuclei are given in Chapter 6, Appendix A. The spin number I can be determined from the atomic mass and the atomic number as shown in Table 3.1.

TABLE 3.1 Type of nuclear spin number, I, with various combinations of atomic mass and atomic number.

I	Atomic Mass	Atomic Number	Example of Nuclei
Half-integer	Odd	Odd	$^1_1$H($\frac{1}{2}$), $^3_1$H($\frac{1}{2}$), $^{15}_7$N($\frac{1}{2}$), $^{19}_9$F($\frac{1}{2}$), $^{31}_{15}$P($\frac{1}{2}$)
Half-integer	Odd	Even	$^{13}_6$C($\frac{1}{2}$), $^{17}_8$O($\frac{1}{2}$), $^{29}_{14}$Si($\frac{1}{2}$)
Integer	Even	Odd	$^2_1$H(1), $^{14}_7$N(1), $^{10}_5$B(3)
Zero	Even	Even	$^{12}_6$C(0), $^{16}_8$O(0), $^{34}_{16}$S(0)

Spectra of several nuclei can be readily obtained (e.g., $^1_1$H, $^3_1$H, $^{13}_6$C, $^{15}_7$N, $^{19}_9$F, $^{31}_{15}$P) since they have spin numbers I of 1/2 and a uniform spherical charge distribution (Figure 3.1). Of these by far, the most widely used in NMR spectrometry are ^{1}H (this chapter) and ^{13}C (Chapter 4).

Nuclei with a spin number I of one or higher have a nonspherical charge distribution. This asymmetry is described by an electrical quadrupole moment, which, as we shall see later, affects the relaxation time and, consequently, the linewidth of the signal and coupling with neighboring nuclei. In quantum mechanical terms, the spin number I determines the number of orientations a nucleus may assume in an external uniform magnetic field in accordance with the formula $2I + 1$. We are concerned with the proton whose spin number I is 1/2.

3.2.2 Excitation of Spin 1/2 Nuclei

For spin 1/2 nuclei in an external magnetic field (Figure 3.2), there are two energy levels and a slight excess of proton population in the lower energy state ($N_\alpha > N_\beta$) in accordance with the Boltzmann distribution. The states are labeled α and β or 1/2 and $-1/2$; ΔE is given by

$$\Delta E = (h\gamma/2\pi)\boldsymbol{B}_0$$

where h is Planck's constant, which simply states that ΔE is proportional to $\boldsymbol{B}_0$ since h, γ, and π are constants. $\boldsymbol{B}_0$ represents the magnetic field strength.*

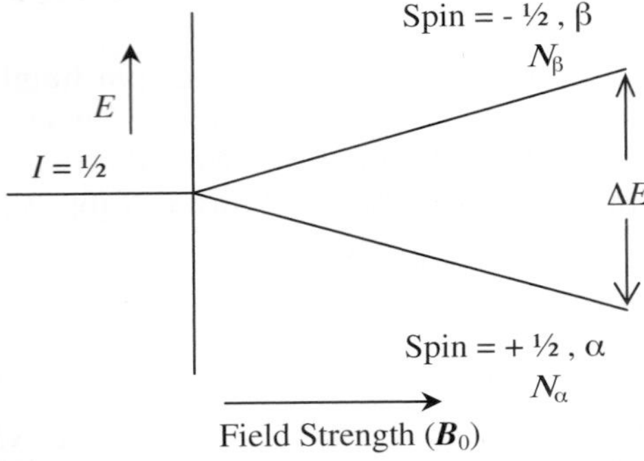

FIGURE 3.2 Two proton energy levels in a magnetic field of magnitude $\boldsymbol{B}_0$. N is population of spins in the upper (N_β) and lower (N_α) energy states. The direction of the magnetic field ($\boldsymbol{B}_0$) is up, parallel to the ordinate, and field strength ($\boldsymbol{B}_0$) increases to the right. Larger ($\boldsymbol{B}_0$) fields increase ΔE.

Once two energy levels for the proton have been established, it is possible to introduce energy in the form of radiofrequency radiation (ν_1) to effect a transition between these energy levels in a stationary magnetic field of given strength $\boldsymbol{B}_0$. *The fundamental NMR equation correlating the applied radiofrequency ν, with the magnetic field strength is*

$$\nu_1 = (\gamma/2\pi)\,\boldsymbol{B}_0$$

since

$$\Delta E = h\nu$$

The introduced radiofrequency ν_1 is given in megahertz (MHz). A frequency of 100 MHz is needed at a magnetic field strength $\boldsymbol{B}_0$ of 2.35 tesla (T) for the proton (or any other desired combination of ν and $\boldsymbol{B}_0$ at the same ratio. See Chapter 6, Appendix A.). At this ratio, the system is in *resonance;* energy is absorbed by the proton, raising it to the higher energy state, and a spectrum can be recorded. Hence, the name *nuclear magnetic resonance spectrometry* is applied. The constant γ is called the magnetogyric ratio, a fundamental nuclear constant; it is the proportionality constant between the magnetic moment, μ, and the spin number, I.

$$\gamma = 2\pi\mu/hI$$

The radiofrequency energy ν_1 can be introduced either by continuous-wave (CW) scanning of the frequency range or by pulsing the entire range of frequencies with a single burst of radiofrequency energy. The two methods result in two distinct "classes" of NMR spectrometers.

In the CW spectrometer, the sample is placed in a magnetic field and irradiated by slowly sweeping the required frequency range. The resulting spectrum is obtained by recording the absorption of energy as a function of frequency. In the pulsed spectrometer, the sample is placed in the magnetic field and irradiated with a pulse of high power radiofrequency energy with a frequency range sufficiently large to cover the entire range. This pulse simultaneously excites all of the nuclei in the sample. Immediately following the pulse, the excited nuclei begin to return to the ground state and radiate the absorbed energy. A detector collects this energy producing a free induction decay (FID), which is the sum of all the nuclei radiating over time. The information in the FID, which is a function of time, is converted (Fourier transform, FT) to a readable spectrum, which is a function of frequency.

The problem is how to apply radiofrequency (rf) electromagnetic energy to protons aligned in a stationary magnetic field and how to measure the energy thus absorbed as the protons are raised to the higher spin state. This can be explained in classical mechanical terms, wherein we visualize the proton as spinning in an external magnetic field. The magnetic axis of the proton precesses about the z axis of the stationary magnetic field $\boldsymbol{B}_0$ in the same manner in which an off-perpendicular spinning top (or a gyroscope) precesses under the influence of gravity (Figure 3.3).

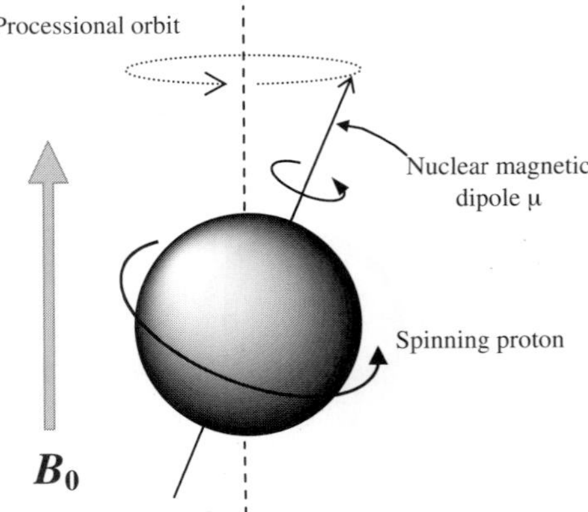

FIGURE 3.3 Classical representation of a proton precessing in a magnetic field of magnitude $\boldsymbol{B}_0$ in analogy with a precessing spinning top.

* The designation $\boldsymbol{B}$ (magnetic induction or flux density) supercedes $\boldsymbol{H}$ (magnetic intensity). The SI term tesla (T), the unit of measurement for $\boldsymbol{B}$, supercedes the term gauss (G); T = 10^4 G. The frequency term hertz (Hz) supercedes cycles per second (cps). MHz is megahertz (10^6 Hz).

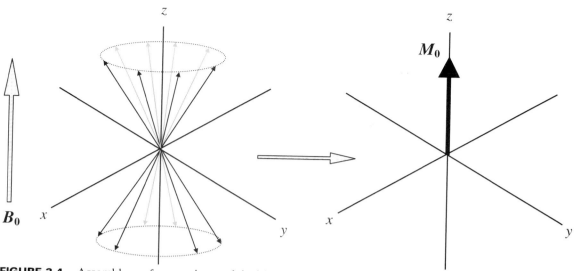

FIGURE 3.4 Assemblage of precessing nuclei with net macroscopic magnetization $\boldsymbol{M}_0$ in the direction of the stationary magnetic field $\boldsymbol{B}_0$.

An assemblage of equivalent protons precessing in random phase around the z axis (i.e., in the direction of the stationary magnetic field $\boldsymbol{B}_0$) has a net macroscopic magnetization $\boldsymbol{M}_0$ along the z axis, but none in the xy plane (Figure 3.4).

When an applied rf (ν_1) is equal to the precessional frequency of the equivalent protons (Larmor frequency ν_L in MHz), the state of nuclear magnetic resonance is attained, and the basic NMR relationship can be written:

$$\nu_L = \nu_1 = (\gamma/2\pi)\boldsymbol{B}_0$$

This equation applies to an isolated proton assemblage (see Sections 3.5).

The aim is to tip the net magnetization $\boldsymbol{M}_0$ toward the xy horizontal plane of the stationary Cartesian frame of reference and measure the resulting component of magnetization in that plane. Rf electromagnetic energy ν_1 is applied so that its magnetic component $\boldsymbol{B}_1$ is at right angles to the main magnetic field $\boldsymbol{B}_0$ and is rotating with the precessing proton assemblage. This is accomplished by an rf oscillator with its axis (conventionally along the x coordinate) perpendicular to the axis of the main magnetic field $\boldsymbol{B}_0$. Such an oscillator will generate a CW, oscillating, magnetic field $\boldsymbol{B}_1$ along the direction of the x axis. An oscillating magnetic field can be resolved into two components rotating in opposite directions (Figure 3.5). One of these components is rotating in the same direction as the precessional orbit of the protons; the oppositely rotating component is ineffective.

When the oscillator frequency ν, is varied (frequency "scan"), the frequency of the rotating magnetic field will come into *resonance* with the precessing Larmor frequencies ν_L of the protons, induce phase coherence, and tip the net magnetiza-

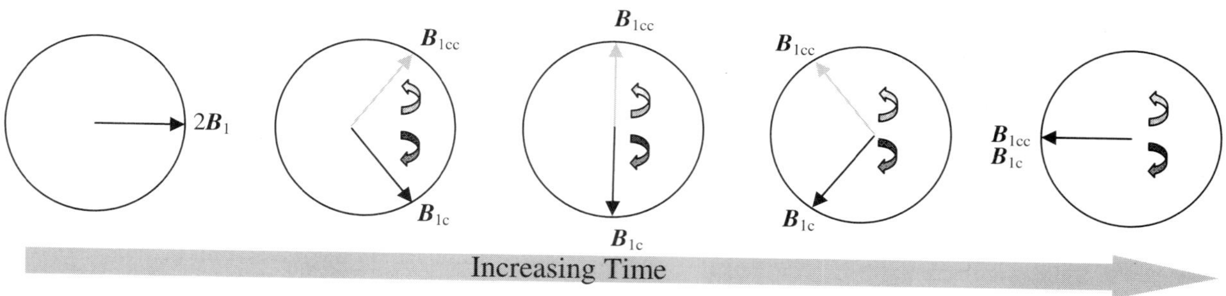

FIGURE 3.5 An oscillating magnetic field can be resolved into two counterrotating components. $\boldsymbol{B}_{1c}$ and $\boldsymbol{B}_{1cc}$ stand for $\boldsymbol{B}_1$ clockwise and $\boldsymbol{B}_1$ counter clockwise, respectively.

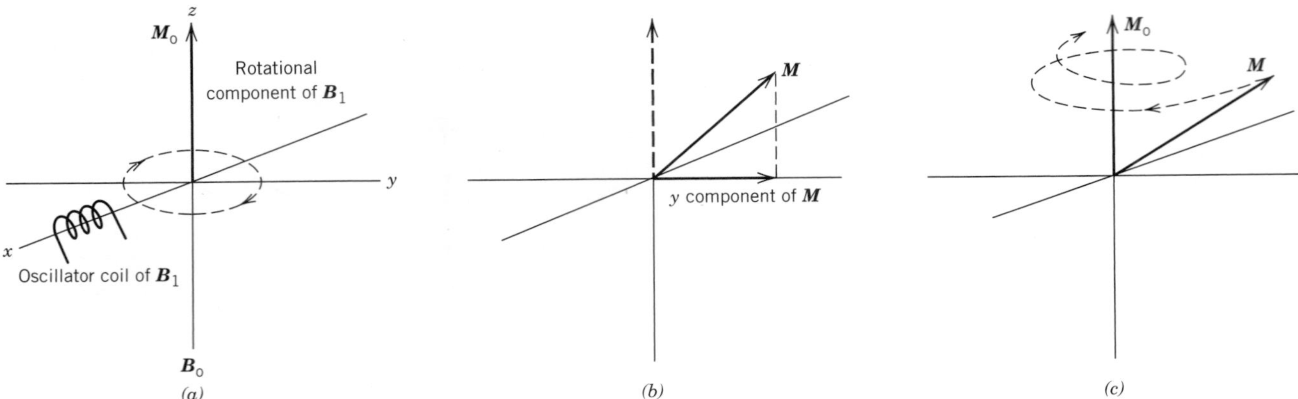

FIGURE 3.6 (a and b) Oscillator generates rotating component of applied magnetic field B_1. The net magnetization M_0 is tipped to M, which precesses about the z axis generating a component of magnetization in the horizontal plane. (c) Longitudinal relaxation of M to M_0 follows a decreasing spiral. Transverse relaxation T_2 (dephasing of M) is omitted. The Cartesian frame is stationary.

tion M_0 toward the horizontal plane (Figure 3.6. a and b). The magnetic component generated in the xy plane can be detected by the receiver coil mounted in the xy plane. Thus, the maximum signal intensity is obtained at the oscillator frequency matching the precession frequency

Small "spinning side bands" (Figure 3.7) are sometimes seen in a CW spectrum symmetrically disposed on both sides of a strong absorption peak; these result from inhomogeneities in the magnetic field and in the spinning tube. They are readily recognized because of their symmetrical appearance and because their separation from the absorption peak is equal to the rate of spinning, typically 10–20 Hz.

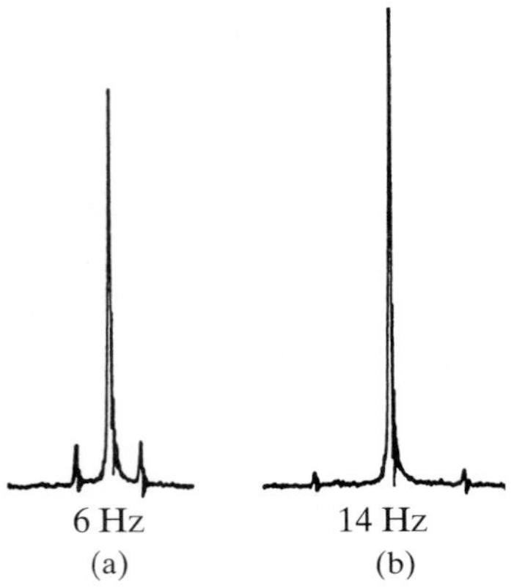

FIGURE 3.7 Signal of neat chloroform with spinning side bands produced by spinning rate of 6 Hz, (a), and 14 Hz (b). (From Bovey, F.A. (1969). *NMR Spectroscopy*. New York: Academic Press. With permission.)

The oscillations, also seen only in scanned (CW) spectra, at the low-frequency end of a strong sharp peak are called "ringing" (Figure 3.8). These are "beat" frequencies resulting from passage through the absorption peak and are common in the older literature.

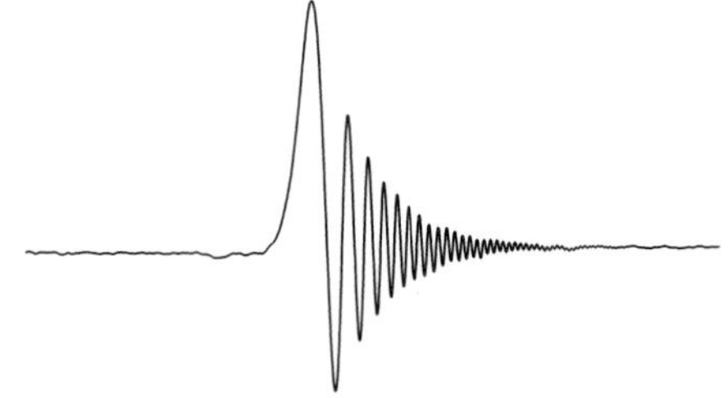

FIGURE 3.8 Ringing (or wiggles) seen after passage through resonance in a scanned spectrum. Direction of scan is from left to right.

The biggest drawback of the CW instrument is that it takes several minutes to scan the entire spectrum (0–14 ppm) and is inherently of low sensitivity with limited sample or with less sensitive nuclei. Repeated scanning does little to improve the signal/noise ratio. Typically this type of instrument is used with neat or very concentrated samples. Other nuclei, such as ^{13}C (approximately 6000 times less sensitive than ^{1}H), are not usually analyzed by CW.

So far, we have described the interaction between the net magnetization M_0 of an assemblage of identical protons in a static, homogeneous magnetic field B_0 and an oscillating rf field ν_1—actually one of two circular components of the rf-generated magnetic field B_1 (Figure 3.5). To obtain a spectrum, the oscillator frequency ν_1 is scanned over the proton frequency range; alternatively, the oscillator frequency may be held constant and the field B_0 scanned. Each different kind of proton must be brought into resonance one by one. This mode is called CW spectrometry and was employed in the early instruments. CW is still used in some of the lower resolution instruments, but CW has been almost completely superceded by pulsed FT. However, since the CW mode is grasped more readily, we have discussed it first, using the familiar, stationary *xyz* Cartesian frame of reference. We now present an introduction to pulsed FT spectrometry and the rotating Cartesian frame of reference.

The pulsed technique was developed largely in response to the need for much higher sensitivity in ^{13}C spectrometry (Chapter 4). This higher sensitivity is achieved by exciting all of the nuclei of interest simultaneously (in this chapter, protons), then collecting all of the signals simultaneously (Figure 3.9a). In a sense, a pulse may be described as an instantaneous "scan." A short (microseconds, μs), powerful rf pulse of center frequency ν_1 applied along the *x* axis generates the entire, desired frequency range and has essentially the same effect as the scanning oscillator: It tips the net magnetization $\boldsymbol{M}_0$ toward the *xy* plane (usually a 90° tip) but does so for all of the protons simultaneously (Figure 3.9b). The magnetization signals are almost immediately detected after the pulse in the *xy* plane and collected by a computer (following analog to digital conversion) over a period of time, which is called the acquisition period (t_2) (Figure 3.9c). During this period, the signals ($\boldsymbol{M}$), which are precessing about the *z* axis, are detected as projections on the *y* axis. For example, instantaneously after the pulse, $\boldsymbol{M}$ is located on the *y* axis and the signal is at a maximum; after a period of time the signals rotate to the *x* axis. The projection on the *y* axis is now 0, continuing to rotate the signals go the $-y$ axis (minimum). This rotation about the *z* axis continues until the magnetization ($\boldsymbol{M}$) relaxes back to equilibrium (Section 3.2.3).

The result is a so-called FID, which may be described as a decaying interferogram (see Section 3.2.3 for examples). The signals collected represent the *difference* between the applied frequency ν_1, and the Larmor frequency ν_L of each proton. The FIDs are then Fourier transformed by a computer algorithm into a conventional NMR spectrum. Since relaxation times for protons are usually on the order of a few seconds or fractions of a second, rapid repetitive pulsing with signal accumulation is possible. Some ^{13}C nuclei—those that have no attached protons to provide T_1 relaxation—require much longer intervals between pulses to allow for relaxation; lack of adequate intervals results in weak signals and inaccurate peak areas (see Section 4.3.1).

3.2.3 Relaxation

Having in classical mechanical terms tipped the net magnetization ($\boldsymbol{M}_0$) toward the *xy* plane, we need to discuss how $\boldsymbol{M}_0$ returns to the *z* axis.

There are two "relaxation" processes. The spin-lattice or longitudinal relaxation process, designated by the time T_1, involves transfer of energy from the "excited" protons to the surrounding protons that are tumbling at the appropriate frequencies. Figure 3.6c shows the loss of the *xy* component by the T_1 process as the net magnetization returns to the *z* axis in a decreasing spiral.

The spin-spin or transverse relaxation, characterized by the time T_2, involves transfer of energy among the precessing protons, which results in dephasing (fanning out), line broadening, and signal loss (Figure 3.10). The designation T_2^* is used to denote the time for all the contributory factors to the transverse signal loss. This term includes both T_2 (the time of the actual spin dynamics) and the effect of magnetic field inhomogeneities, which usually dominates.

Typically, with small molecules, T_1 and T_2 are similar in magnitude; the relationship between T_1 and T_2 is illustrated below:

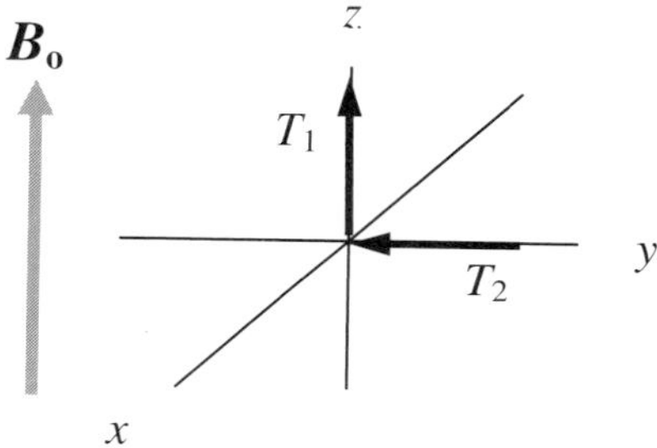

As T_2 relaxation occurs, it decreases the $\boldsymbol{M}$ in the *xy* plane transferring the energy to the *z* axis. With larger molecules (and other nuclei), T_1 relaxation times are typically much longer than T_2.

For protons in the usual nonviscous solutions, the T_2 relaxation times are sufficiently long that sharp peaks are obtained, and T_1 relaxation times are short enough that the intensities of the peaks are proportional to the number of protons involved. Thus, the relative number of different kinds of protons in a spectrum can be determined by measuring the areas under the peaks.

However, for ^{13}C and ^{15}N nuclei, the T_1 relaxation times must be considered since they are longer than those of protons and vary widely. Several preliminary, general statements follow, and further relevant details are found in Chapters 4 and 6.

^{13}C and ^{15}N nuclei undergo T_1 dipole–dipole interactions with attached protons and, to a lesser extent, with other nearby protons. There are further complications with ^{15}N nuclei. In routine spectra of ^{13}C and ^{15}N, large T_1 values result in only partial recovery of the signal so that a delay interval must be inserted between the individual pulses (see Figure 4.2a). Thus, we see that T_1 relaxation is intimately involved with peak intensity.

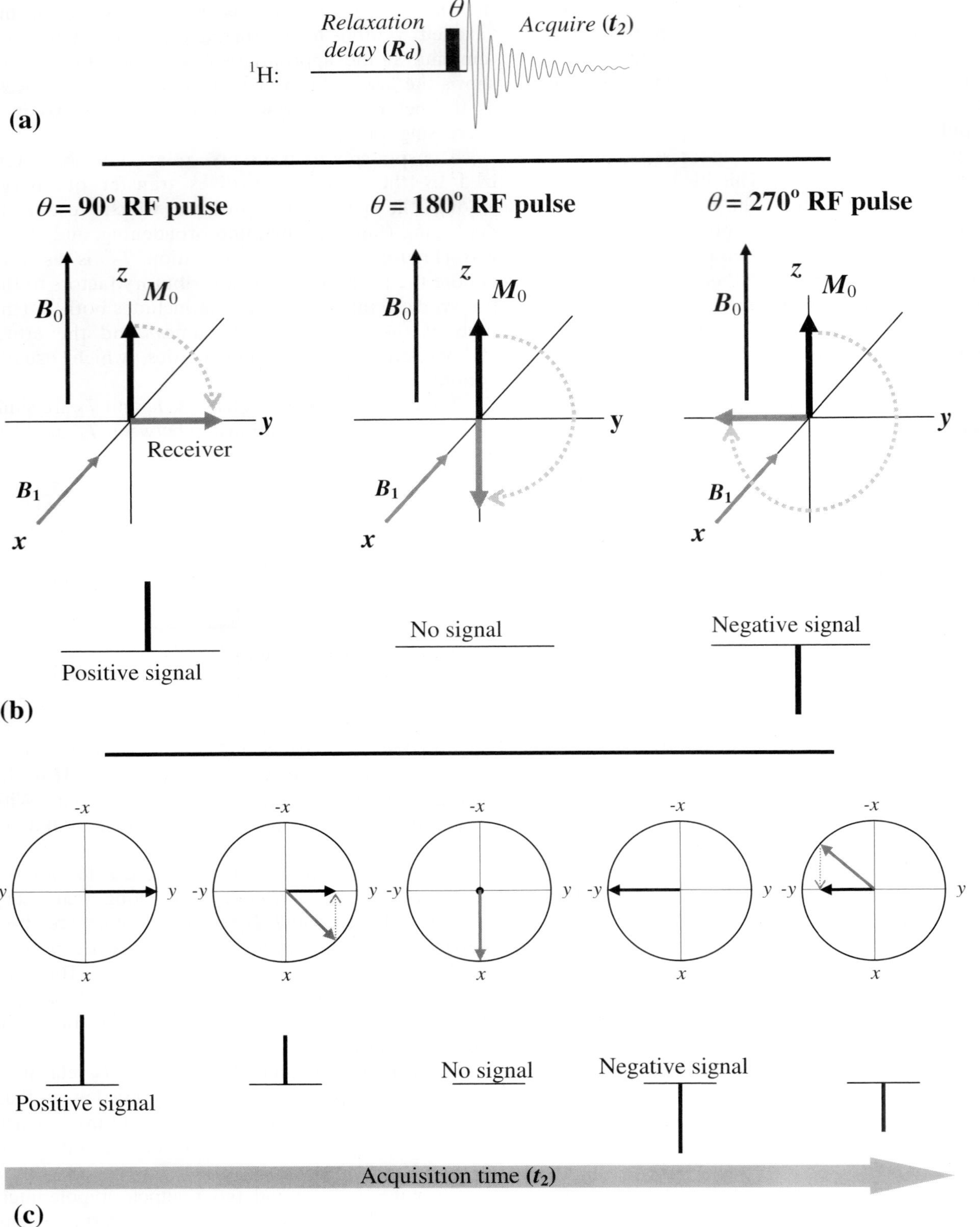

FIGURE 3.9 (A) Pulse sequence for a standard ^{1}H experiment (θ) is a variable radio frequency pulse. (B) shows the effect of various pulse angles (θ) and expected signal. (C) shows the rotation of the signal in the xy plane after the 90° pulse.

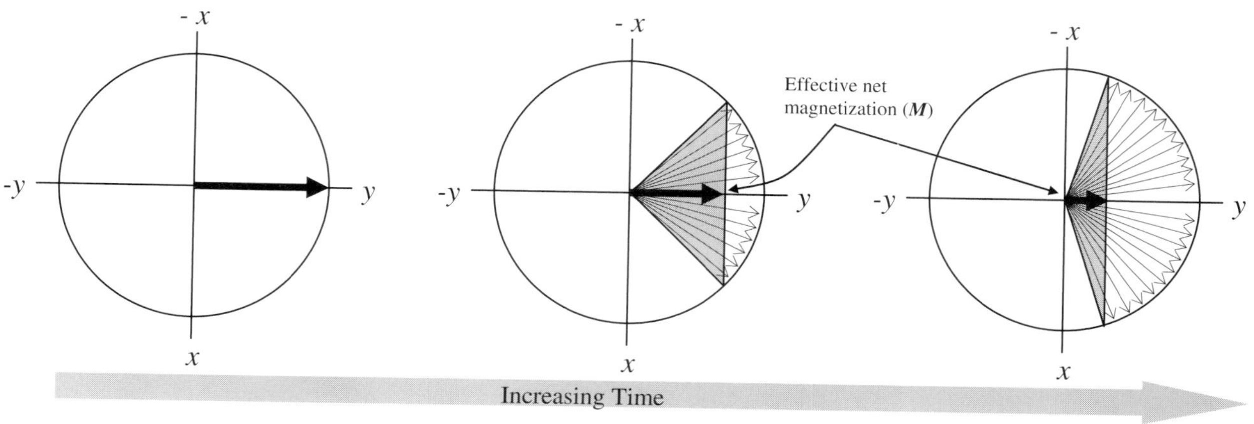

FIGURE 3.10 One spin, rotating at the same frequency as the rotating frame of reference with T_2 relaxation in the *xy* plane. T_2 relaxation will dephase or spread the magnetization in the *xy* plane, which causes line broadening and decay of the signals. T_2 effects are partially due to inhomogeneous magnetic fields.

In contrast, T_2 is involved with peak width in accordance with the Heisenberg uncertainty principle, which states that the product of the uncertainty of the frequency range and the uncertainty of the time interval is a constant:

$$\Delta\nu \cdot \Delta t \geq 1$$

If Δt (i.e., T_2^*) is small, then $\Delta\nu$ is large, and the peak is therefore broad. T_2^*, whose major component is field inhomogeneity, is the principal determining factor for peak width since T_2^* is always less than T_1, or T_2. T_2 can be measured by looking at the width of a peak at 1/2 height (Figure 3.11).

In Figure 3.16, the effect of T_2 line broadening is illustrated by adding a small amount of iron to the sample. It is quite apparent what this addition does to the linewidth of peaks in the frequency domain spectrum, what is less appreciated is what this addition does to the time domain data (FID). As mentioned above, T_2 determines the amount of time that $\boldsymbol{M}$ remains in the *xy* plane; therefore, if T_2 relaxation is short (broad signal), the FID decays to zero much faster than if T_2 is long (sharp signal).

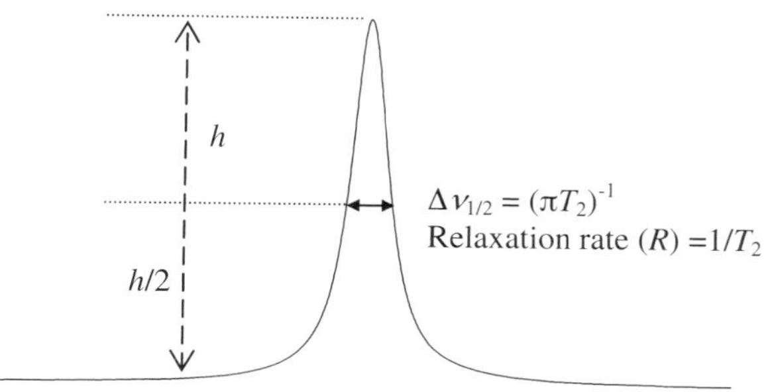

FIGURE 3.11 Peak width at half height.

Assume that the total net magnetization $\boldsymbol{M}_0$ (a broad vector in Figure 3.12) aligned with the *z* axis consists of three individual net magnetizations representing three different kinds of protons in a chemical compound. With a 90° pulse, each net magnetization is in resonance with a different frequency in the pulse, and all are rotated simultaneously onto the *y* axis. Each of the component net magnetizations (narrow vectors) now begins to precess, each at its own Larmor frequency, while relaxing by the T_1 and T_2 mechanisms. There are two practical problems: First, it is difficult to measure accurately absolute frequencies that differ over the range of, say, 5000 Hz around a pulsed central frequency of, say, 300,000,000 Hz (ν_1), also called a carrier frequency. For example, our three different kinds of protons comprising the net magnetization have Larmor frequencies (ν_L) of 300,002,000 Hz, 300,000,800 Hz, and 299,999,000 Hz. This problem is solved by measuring the difference between each Larmor frequency and the carrier frequency, which is applied in the middle of the spectral window. The frequency differences are 2000 Hz, 800 Hz, and −1000 Hz; that is, two frequencies are higher than the carrier frequency and one frequency is lower.

To simplify visualization of the problem, the static Cartesian frame is replaced by a frame rotating around the *z* axis at the carrier frequency (ν_1). The situation is similar to that of an observer who is rotating with the frame as though riding a carousel.

As mentioned above, these frequency differences comprise the FID as the signal intensities from the precessing, relaxing vectors decrease.

FID and pulsed FT spectra were described in Section 3.2.2. At this point, it will be instructive to look at both "spectra" even though vital topics such "chemical shifts" and "spin coupling" have not yet been discussed. Both the FID and FT are indeed spectra although the

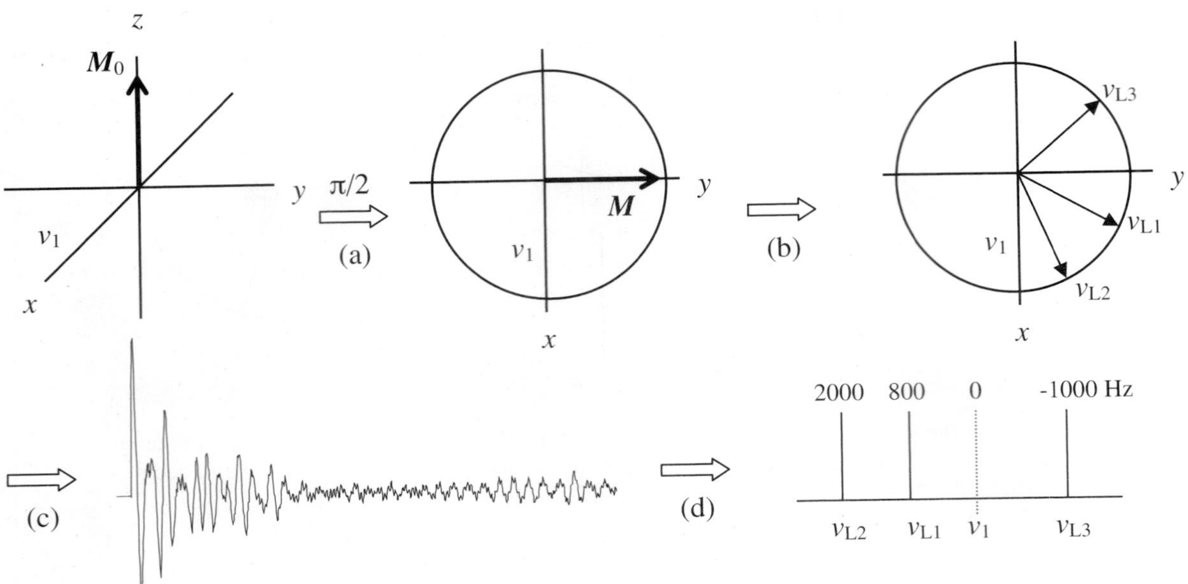

FIGURE 3.12 A rotating frame of reference. The net magnetization ***M*** (following a 90° pulse) has three components with Larmor frequencies ν_{L1}, ν_{L2}, and ν_{L3} (i.e., three different protons). The frame is rotating at ν_1, (the applied pulse). Immediately following the pulse, the components are precessing relative to ν_1, ν_{L1}, and ν_{L2} have higher frequencies than the frequency of the applied pulse ν_1, but the frequency of ν_{L3} is lower than that of ν_1.

FID, a time-domain spectrum, must be made accessible through the mathematical process of Fourier transform to a frequency-domain spectrum.

The protons in acetone, CH_3—(C=O)—CH_3, are chemically equivalent. As they relax following the pulse, the signal detected is a single, exponentially decaying sine wave (time-domain FID) with a frequency equal to the difference between the applied frequency (ν_1) and the Larmor frequency (ν_L) for the protons. The computerized Fourier transform to a frequency domain spectrum results in a single peak (Figure 3.13a) at a chemical shift of δ 2.05 relative to the arbitrary reference at δ 0.00

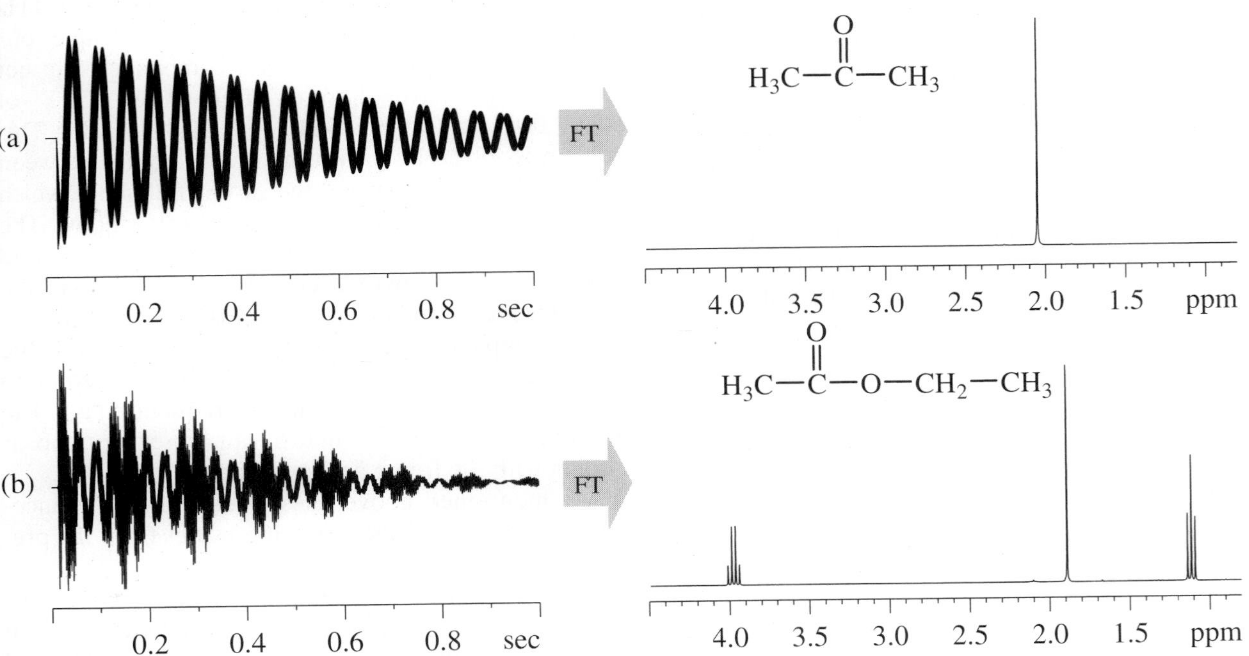

FIGURE 3.13 (A) FID and FT proton spectra of acetone. (B) FID and FT proton spectra of ethyl acetate. Both in $CDCl_3$ at 300 MHz.

(see Section 3.4). The single peak represents the conventional spectrum for the acetone protons.

In contrast with the equivalent methyl groups of acetone, the CH_3 groups and the CH_2 group of ethyl acetate (CH_3—(C=O)—OCH_2CH_3) are all chemically different from one another. Thus, the FID (Figure 3.13b) consists of three different superposed sine waves and a "beat" (interference) pattern. The FT spectrum consists of three separate clusters of peaks. From left to right (Figure 3.13b), we see a quartet, a singlet, and a triplet resulting from "chemical shifts" and "coupling" (see Sections 3.4 and 3.5).

3.3 INSTRUMENTATION AND SAMPLE HANDLING

3.3.1 Instrumentation

Beginning in 1953 with the first commercial NMR spectrometer, the early instruments used permanent magnets or electromagnets with fields of 1.41, 1.87, 2.20, or 2.35 T corresponding to 60, 80, 90, or 100 MHz, respectively, for proton resonance (the usual way of describing an instrument).

The "horsepower race," driven by the need for higher resolution and sensitivity, has resulted in wide use of 300–600 MHz instruments. All of the instruments above 100 MHz are based on helium-cooled superconducting magnets (solenoids) and operate in the pulsed FT mode. The other basic requirements besides high field are frequency-field stability, field homogeneity, and a computer interface.

The sample (routinely a solution in a deuterated solvent in a 5-mm o.d. glass tube) is placed in the probe, which contains the transmitter and receiver coils and a spinner to spin the tube about its vertical axis in order to average out field inhomogeneities.

A CW NMR spectrometer (Figure 3.14) consists of a control console, magnet (typically electromagnets or permanent magnets are used), and two orthogonal coils (transmitter and sweep coils) of wire that serve as antennas for rf radiation. One coil is attached to an rf transmitter and the other coil is the rf amplifier (pick-up) coil and is attached to the detection electronics.

Figure 3.15 shows the arrangement for a superconducting magnet. Note that, in the electromagnet, the tube spins at right angles to the z axis, which is horizontal, whereas in the superconducting magnet, the tube fits in the bore of the solenoid and spins about the z axis, which is vertical. The transmitter and receiver are coupled through the sample nuclei (protons in this chapter).

The proton spectrum obtained either by CW scan or pulse FT at constant magnetic field is shown as a series of peaks whose areas are proportional to the number of protons they represent. Peak areas are measured by an

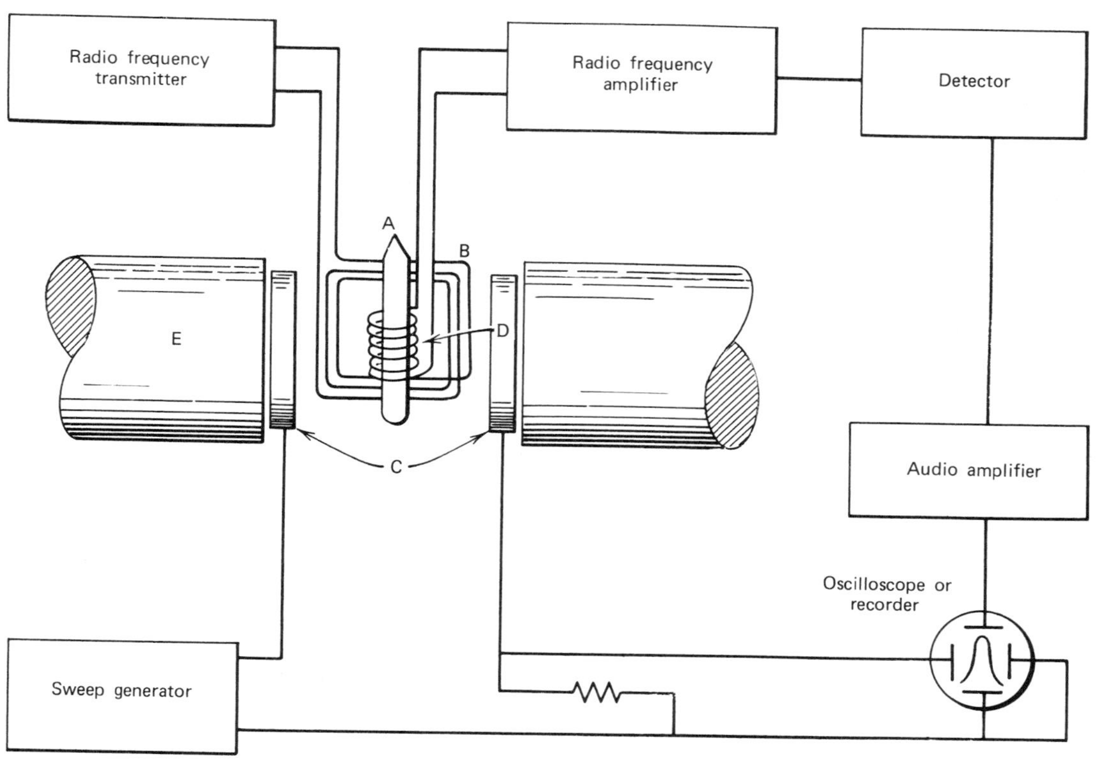

FIGURE 3.14 Schematic diagram of a CW NMR spectrometer. The tube is perpendicular to the z axis of the magnet: A, sample tube; B, transmitter coil; C, sweep coils; D, receiver coil; E, magnet

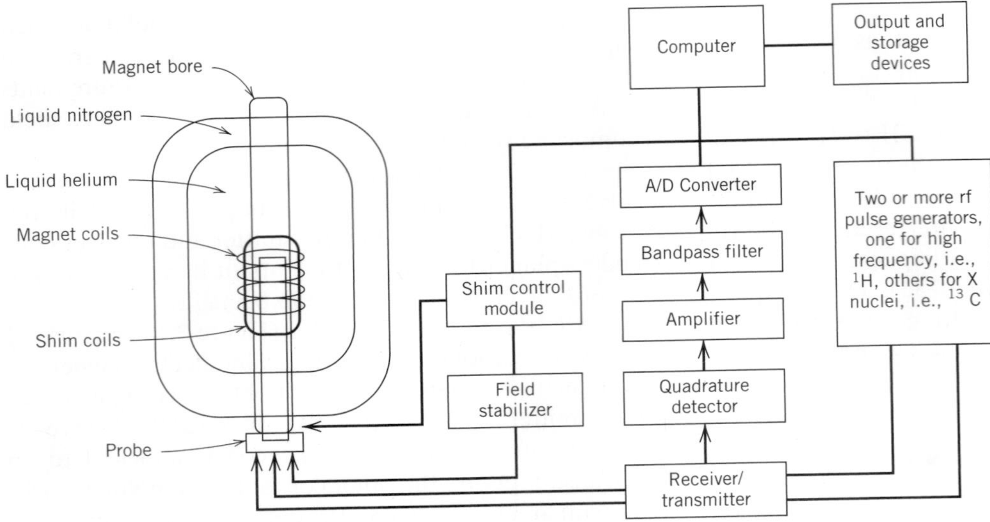

FIGURE 3.15 Schematic diagram of a Fourier transform NMR spectrometer with a superconducting magnet. The probe is parallel with the *z* axis of the magnet, which is cooled with liquid helium surrounded by liquid nitrogen in a large Dewar flask.

electronic integrator that traces a series of steps with heights proportional to the peak areas (see Figure 3.26).* A proton count from the integration is useful to determine or confirm molecular formulas, detect hidden peaks, determine sample purity, and do quantitative analysis. Peak positions (chemical shifts, Section 3.4) are measured in frequency units from a reference peak.

3.3.2 Sensitivity of NMR Experiments

The sensitivity of pulsed FT NMR experiments is given by the signal to noise ratio:

$$S/N = NT_2\gamma_{exc}\left(\frac{\sqrt[3]{B_o\gamma_{det}}\sqrt{ns}}{T}\right)$$

S/N = signal to noise ratio

N = number of spins in the system (sample concentration).

T_2 = transverse relaxation time (determines the line width)

γ_{exc} = magnetogyric ratio of the excited nucleus.

γ_{det} = magnetogyric ratio of the detected nucleus.

ns = number of scans

B_0 = external magnetic field

T = sample temperature

(Comment: for a simple ^{1}H experiment, γ_{exc} and γ_{det} are the same nucleus. It might be useful to improve S/N for less sensitive nuclei to excite one kind of nucleus and detect another kind with a better magnetogyric ratio in the same experiment. This is done in inverse experiments which are described in Chapters 5 and 6).

A routine sample for proton NMR on a 300 MHz instrument consists of about 10 mg of the compound in about 0.5 ml of solvent in a 5-mm o.d. glass tube. Microprobes that accept a 1.0 mm, 2.5 mm, or 3 mm o.d. tube are available and provide higher sensitivity. Under favorable conditions, it is possible to obtain a spectrum on 100 ng of a compound of modest molecular weight in a 1.0 mm microtube (volume 5 μl) on a 600 MHz instrument.

The development of cryogenically cooled probes has significantly decreased sample amount requirements for ^{1}H NMR. Thermal noise in the probe and the first-stage receiver electronics dominate noise in NMR experiments. These new probes have built in first-stage receivers and rf coils that are cryogenically cooled (~20°K), and have S/N improvements of ~4 × standard probes. It is obvious to users that the highest field instrument available provides the best sensitivity. For fixed concentration (N), we would need 2.8 times as much material with a 300 MHz as on a 600 MHz system to obtain spectra with identical S/N:

$$S/N = \frac{N_{600}}{N_{300}} = \sqrt[3]{\frac{600}{300}} \approx 2.8$$

So for very limited sample amounts, it usually is best to go to the highest field available, with the smallest diameter cryo-probe possible. A cryo-capillary flow microprobe that accepts a few nanograms of material in approximately 1 μl of solvent yields the highest sensitivity for mass limited samples.

* "Chemically different protons" absorb rf energy at very slightly different frequencies—differences up to around 5000 hertz at a frequency of 300 MHz (see Section 4.7). The utility of NMR spectrometry for the organic chemist dates from the experiment at Varian Associates that obtained three peaks from the chemically different protons in CH3CH2OH; the peak areas were in the ratio 3 : 2 : 1. [J.T. Arnold, S.S. Dharmatti, and M.E. Packard, *J. Chem. Phys.* 19, 507 (1951).]

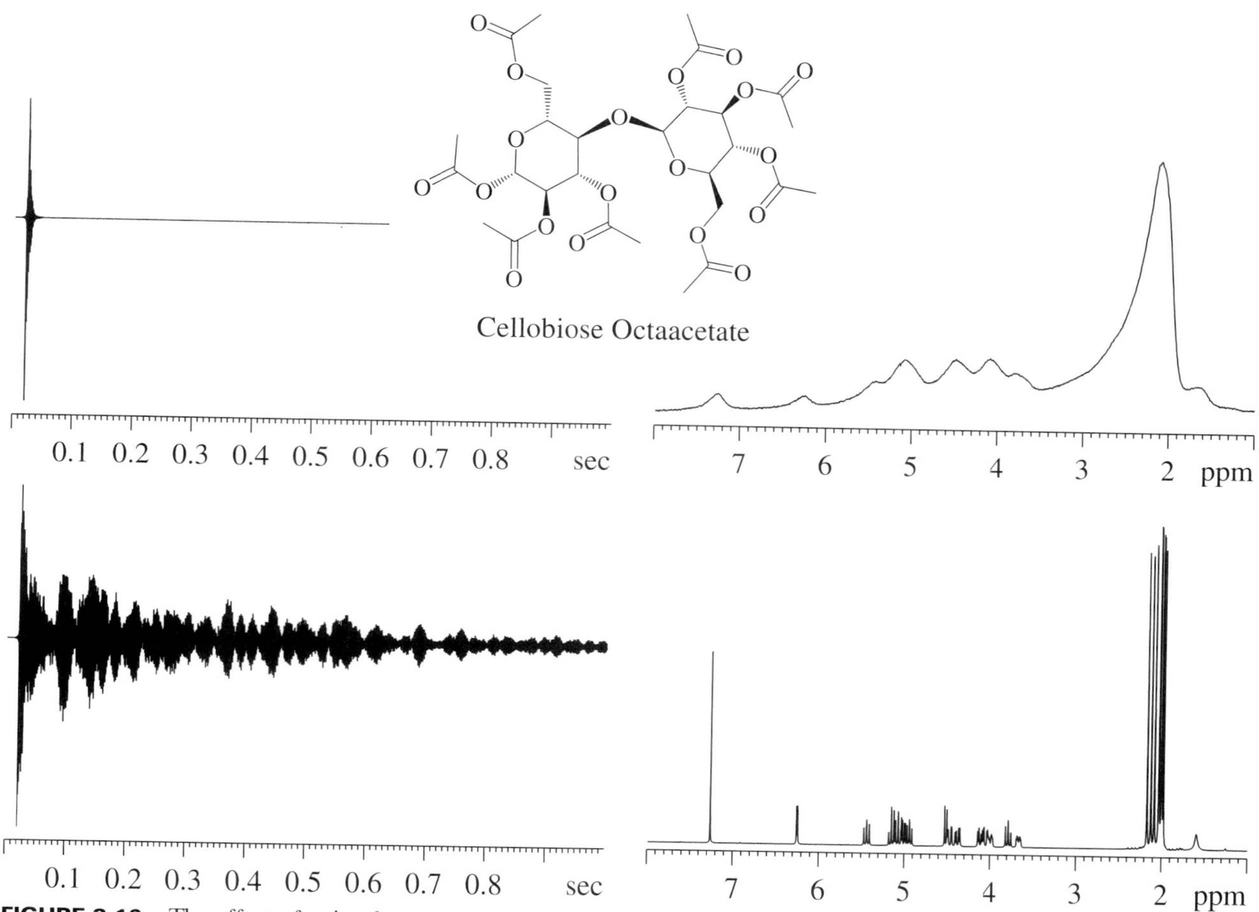

FIGURE 3.16 The effect of a tiny ferromagnetic particle on the proton resonance spectrum of cellobiose octaacetate. The top spectra are run with the particles present; the bottom curves are the spectra with the particle removed.

3.3.3 Solvent Selection

The ideal solvent should contain no protons and be inert, low boiling, and inexpensive. Deuterated solvents are necessary for modern instruments because they depend on a deuterium signal to lock or stabilize the $\boldsymbol{B}_0$ field of the magnet. Instruments have a deuterium "channel" that constantly monitors and adjusts (locks) the $\boldsymbol{B}_0$ field to the frequency of the deuterated solvent. Typically, ^{1}H NMR signals are in the order of 0.1 to several Hz wide out of 300,000,000 Hz (for a 300 MHz system), so the $\boldsymbol{B}_0$ field needs to be very stable and homogeneous.

The deuterium signal is also used to "shim" the $\boldsymbol{B}_0$ field. Instruments use small electromagnets (called shims) to bend the main magnetic field ($\boldsymbol{B}_0$) so that the homogeneity of the field is precise at the center of the sample. Most modern instruments have approximately 20–30 electromagnetic shims; they are computer controlled, and can be adjusted in an automated manner.

Deuterated chloroform ($CDCl_3$) is used whenever circumstances permit—in fact most of the time. The small sharp proton peak at δ 7.26 from the $CHCl_3$ impurity present rarely interferes seriously. For very dilute samples, $CDCl_3$ can be obtained in "100% isotope purity." A list of common, commercially available solvents with the positions of proton "impurities" (i.e. $CHCl_3$ in $CDCl_3$) is given in Appendix G.

Traces of ferromagnetic impurities cause severe broadening of absorption peaks because of reduction of T_2 relaxation times. Common sources are rust particles from tapwater, steel wool, Raney nickel, and particles from metal spatulas or fittings (Figure 3.16). These impurities can be removed by filtration.

Traces of common laboratory solvents can be annoying. See Appendix H for an extensive list of common solvent impurities. Other offenders are greases and plasticizers (phthalates in particular). NMR solvents should be kept in a desiccator.

3.4 CHEMICAL SHIFT

Only a single proton peak should be expected from the interaction of rf energy and a strong magnetic field on all of the protons in accordance with the basic NMR equation (Section 3.2.2):

$$\nu_1 = (\gamma/2\pi)\,\boldsymbol{B}_0$$

where ν_1 is the applied frequency, $\boldsymbol{B}_0$ is the flux density of the stationary magnetic field, and $\gamma/2\pi$ is a constant.

Fortunately, the situation is not so simple. A proton in a molecule is *shielded* to a very small extent by its electron cloud, the density of which varies with the chemical environment. This variation gives rise to differences in *chemical shift positions;* this ability to discriminate among the individual absorptions describes high-resolution NMR spectrometry.

The basic NMR equation for all protons is now modified for an assemblage of equivalent protons in the molecule:

$$\nu_{\text{eff}} = (\gamma/2\pi)\, \boldsymbol{B}_0\, (1 - \sigma)$$

The symbol σ is the "shielding constant" whose value is proportional to the degree of shielding by its electron cloud. At a given value of $\boldsymbol{B}_0$, the *effective* frequency at resonance is less than the applied frequency ν_1. Electrons under the influence of a magnetic field circulate and, in circulating, generate their own magnetic field opposing the applied field; hence, the "shielding" effect (Figure 3.17). This effect accounts for the diamagnetism exhibited by all organic materials. In the case of materials with an unpaired electron, the paramagnetism associated with the net electron spin far overrides the diamagnetism of the circulating, paired electrons.

The degree of shielding depends on the density of the circulating electrons, and, as a first, very rough approximation, the degree of shielding of a proton on a carbon atom will depend on the inductive effect of other groups attached to the carbon atom. The difference in the absorption position of a particular proton from the absorption position of a *reference* proton is called the *chemical shift* of the particular proton.

We now have the concept that protons in "different" chemical environments have different chemical shifts. Conversely, protons in the "same" chemical environment have the same chemical shift. But what do we mean by "different" and "same"? It is intuitively obvious that the chemically different methylene groups of $ClCH_2CH_2OH$ have different chemical shifts and that the protons in either one of the methylene groups have the same chemical shift. But it may not be so obvious, for example, that the individual protons of the methylene group of $C_6H_5CH_2CHBrCl$ do not have the same chemical shift. For the present, we shall deal with obvious cases and postpone a more rigorous treatment of chemical shift equivalence to Section 3.8.

The most generally useful reference compound is tetramethylsilane (TMS).

$$H_3C-\overset{\displaystyle CH_3}{\underset{\displaystyle CH_3}{Si}}-CH_3$$

This material has several advantages: it is chemically inert, symmetrical, volatile (bp 27°C), and soluble in most organic solvents; it gives a single, intense, sharp, absorption peak, and its protons are more "shielded" than almost all organic protons. When water or deuterium oxide is the solvent, TMS can be used as an "external reference" in a concentric capillary or the methyl protons of the water-soluble sodium 2, 2-dimethyl-2-silapentane-5-sulfonate (DSS), $(CH_3)_3SiCH_2CH_2CH_2SO_3Na$, are used as an internal reference (0.015 ppm).

Historically and now by convention, the TMS reference peak is placed at the right-hand edge of the spectrum and designated zero on the either Hz or δ scale (defined below). Positive Hz or δ numbers increase to the left of TMS, negative numbers increase to the right.* The term "shielded" means toward the right; "deshielded" means toward the left. It follows that the strongly deshielded protons of dimethyl ether, for example, are more exposed than those of TMS to the applied field; hence, resonance occurs at higher frequency—i.e., to the left—relative to the TMS proton peak. Thus, both the Hz and the δ scales reflect the increase in applied frequency, *at constant field,* toward the left of the TMS resonance frequency, and the decrease in applied frequency toward the right.

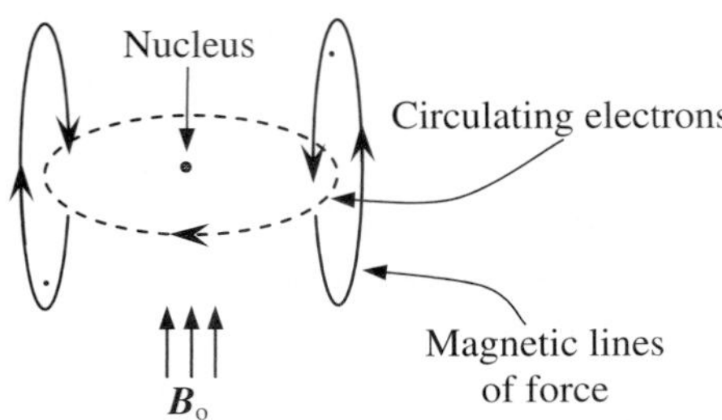

FIGURE 3.17 Diamagnetic shielding of nucleus by circulating electrons. The arrows, ↑↑↑, represent the direction of the stationary magnetic field of magnitude $\boldsymbol{B}_0$. The circulating electrons comprise the electrical current, but the current direction is shown conventionally as flow of positive charge.

* This convention, however, posed a conflict with the practice, in the earlier days of NMR, of increasing the magnetic field strength (field scan) from left to right at constant frequency and reporting the *magnetic field strength in terms of the resonance frequency units.* This "upfield scan" meant that the resonance frequency units increased from left to right, at odds with the conventional δ scale.

To resolve this conflict, the τ scale was introduced ($\tau = 10 - \delta$) and found its way into the literature from the late 1950's through the 1970's—the British literature in particular. Chemical shifts published during this period should be carefully checked for the convention used. For example, see Bovey (1967).

The τ scale disappeared with the development of frequency-scan instruments and of the pulsed FT mode, which is essentially an instantaneous frequency "scan." The terms "upfield" and "downfield" are now obsolete and have been replaced, respectively, by shielded (lower δ, or to the right) and deshielded (higher δ, or to the left).

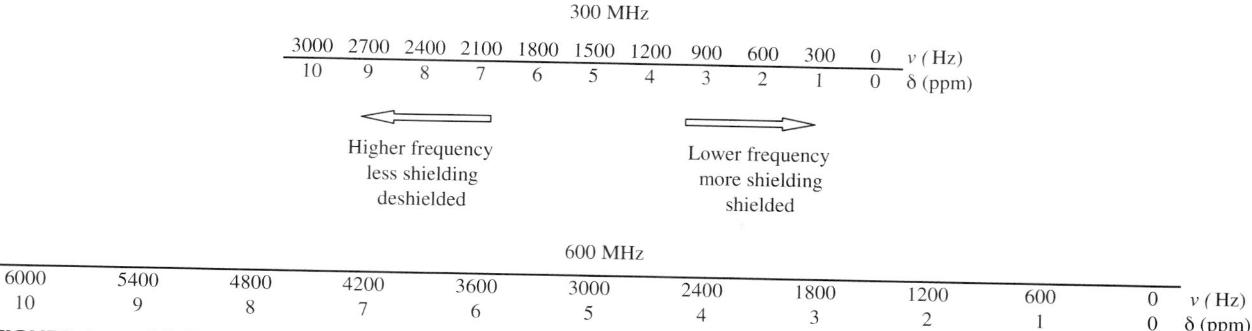

FIGURE 3.18 NMR scale at 300 MHz and 600 MHz. Relatively few organic compounds show absorption peaks to the right of the TMS peak. These lower frequency signals are designated by negative numbers to the right (not shown in the Figure).

Let us look at the NMR scales in Figure 3.18 and conventionally set the TMS peak at zero at the right-hand edge. The chemical shifts may be reported in dimensionless units (labeled δ or ppm), starting from the zero chemical shift of TMS, and are independent of the designated frequency of the spectrometer in use. This is done by dividing the chemical shift of a peak in Hz by the designated frequency of the spectrometer in MHz. In Figure 3.18, for example, we may place a proton peak at, say, 1200 Hz on the 300 MHz scale:

$$\frac{1200\ \text{Hz}}{300\ \text{MHz}} = \delta\, 4 \text{ or } 4\ \text{ppm}$$

Or we may change to a 600 MHz spectrometer. The same proton peak will now be at 2400 Hz on the 600 MHz scale, but the δ (or ppm) unit would remain the same.

$$\frac{2400\ \text{Hz}}{600\ \text{MHz}} = \delta\, 4 \text{ or } 4\ \text{ppm}$$

The strongest magnetic field necessary and available should be used to spread out the chemical shifts. This is made clear in Figure 3.18 and in Figure 3.19 in which increased applied magnetic field in the NMR spectrum of acrylonitrile means increased separation of signals.

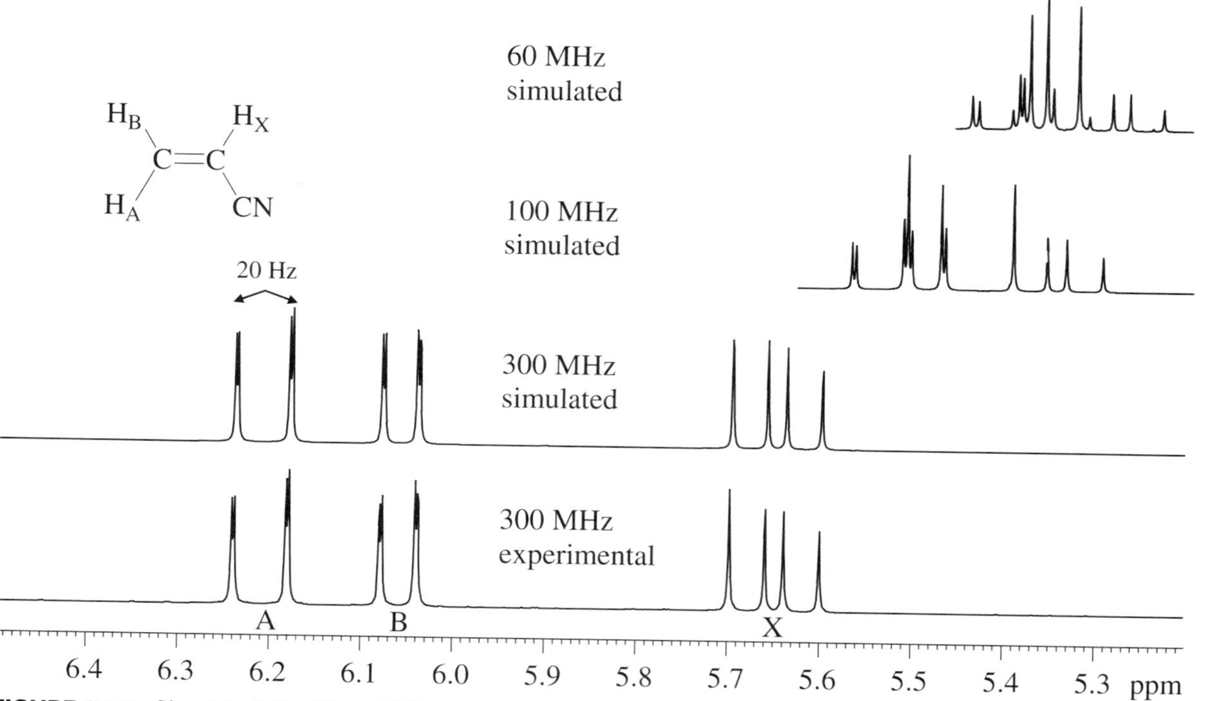

FIGURE 3.19 Simulated 60, 100, and 300 MHz spectra of acrylonitrile; 300 MHz experimental spectrum (in $CDCl_3$) for comparison.

The concept of electronegativity (see Table 3.2) of substituents near the proton in question is a dependable guide, up to a point, to chemical shifts. It tells us that the electron density around the protons of TMS is high (silicon is electropositive relative to carbon), and these protons will therefore be highly shielded (see Table 3.3). Since C is more electronegative than H, the sequence of proton absorptions in the alkyl series CH_4, RCH_3, R_2CH_2, and R_3CH is from right to left in the spectrum (Appendix A, Chart A.1).

TABLE 3.2 Electronegativity of selected elements according to Pauling.

H (2.1)						
Li (1.0)	Be (1.5)	B (2.0)	C (2.5)	N (3.0)	O (3.5)	F (4.0)
Na (0.9)	Mg (1.2)	Al (1.5)	Si (1.8)	P (2.1)	S (2.5)	Cl (3.0)
						Br (2.8)
						I (2.5)

TABLE 3.3 Chemical shift trends guided by electronegativity.

Compound	δ	Compound	δ
$(CH_3)_4Si$	0.00	CH_3F	4.30
$(CH_3)_2O$	3.27	RCO_2H	~10.80

We could make a number of good estimates as to chemical shifts, using concepts of electronegativity and proton acidity. For example, the values found in Table 3.3 are reasonable solely on the basis of electronegativity.

However, finding the protons of acetylene at δ 1.80, that is, more shielded than ethylene protons (δ 5.25), is unsettling, and finding the aldehydic proton of acetaldehyde at δ 9.97 definitely calls for some augmentation of the electronegativity concept. We shall use diamagnetic anisotropy to explain these and other apparent anomalies, such as the unexpectedly large deshielding effect of the benzene ring (benzene protons δ 7.27).

Let us begin with acetylene. The molecule is linear, and the triple bond is symmetrical about the axis. If this axis is aligned with the applied magnetic field, the π electrons of the bond can circulate at right angles to the applied field, thus inducing their own magnetic field opposing the applied field. Since the protons lie along the magnetic axis, the magnetic lines of force induced by the circulating electrons act to shield the protons (Figure 3.20), and the NMR peak is found further to the right than electronegativity would predict. Of course, only a small number of the rapidly tumbling molecules are aligned with the magnetic field, but the overall average shift is affected by the aligned molecules.

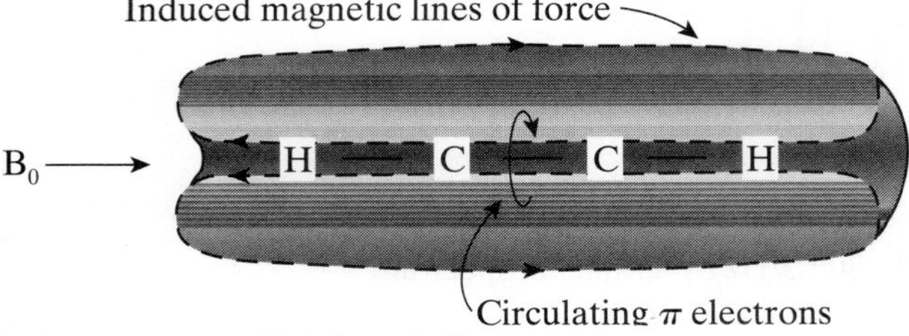

FIGURE 3.20 Shielding of alkyne protons.

This effect depends on diamagnetic anisotropy, which means that shielding and deshielding depend on the orientation of the molecule with respect to the applied magnetic field. Similar arguments can be adduced to rationalize the unexpected deshielded position of the aldehydic proton. In this case, the effect of the applied magnetic field is greatest along the transverse axis of the C═O bond (i.e., in the plane of the page in Figure 3.21). The geometry is such that the aldehydic proton, which lies in front of the page, is in the deshielding portion of the induced magnetic field. The same argument can be used to account for at least part of the rather large deshielding of alkene protons.

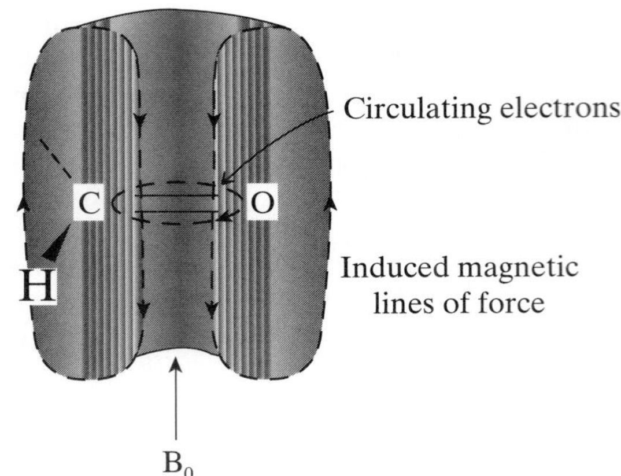

FIGURE 3.21 Deshielding of aldehydic protons.

The so-called "ring-current effect" is another example of diamagnetic anisotropy and accounts for the large deshielding of benzene ring protons. Figure 3.22 shows this effect. It also indicates that a proton held directly above or below the aromatic ring should be shielded. This has been found to be the case for some of the methylene protons in 1,4-polymethylenebenzenes.

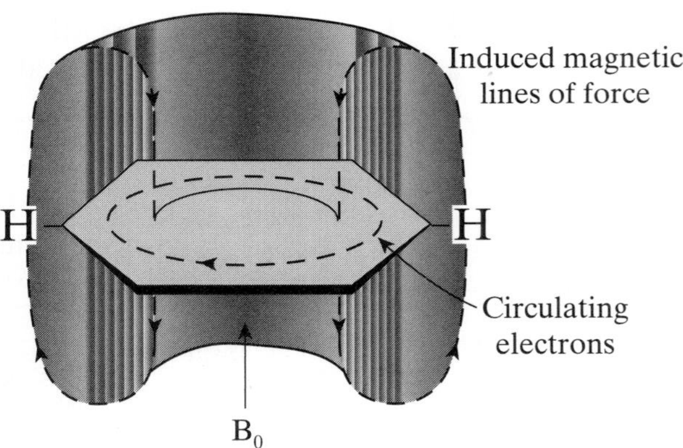

FIGURE 3.22 Ring current effects in benzene.

All the ring protons of acetophenone are deshielded because of the ring current effect. Moreover, the *ortho* protons are further deshielded (*meta, para* δ ~7.40; *ortho* δ ~7.85) because of the additional deshielding effect of the carbonyl group. In Figure 3.23, the carbonyl bond and the benzene ring are coplanar. If the molecule is oriented so that the applied magnetic field $\boldsymbol{B}_0$ is perpendicular to the plane of the molecule, the circulating π electrons of the C=O bond shield the conical zones above and below them and deshield the lateral zones in which the *ortho* protons are located. Both *ortho* protons are equally deshielded since another, equally populated, conformation can be written in which the "left-hand" *ortho* proton is deshielded by the anisotropy cone. Nitrobenzene shows a stronger effect.

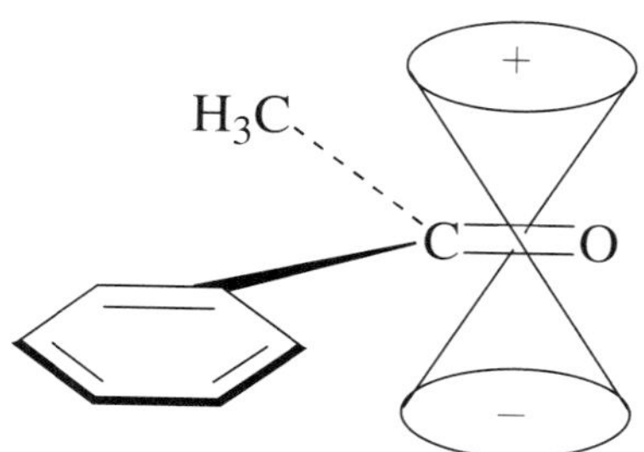

FIGURE 3.23 Shielding (+) and deshielding (−) zones of acetophenone.

A spectacular example of shielding and deshielding by ring currents is furnished by some of the annulenes. At about −60°C, the protons outside the ring of [18]-annulene are strongly deshielded (δ 9.3) and those inside are strongly shielded (δ − 3.0, i.e., more shielded than TMS).

[18]-Annulene

Demonstration of such a ring current is good evidence for planarity and aromaticity, at least at low temperature. As the temperature is raised, the signals broaden because of slow interchanges in ring conformations. At about 110°C, a single averaged peak appears at approximately δ 5.3 because of rapid interchanges in ring conformations to give an averaged chemical shift.

In contrast with the striking anisotropic effects of circulating π electrons, the σ electrons of a C—C bond produce a small effect. For example, the axis of the C—C bond in cyclohexane is the axis of the deshielding cone (Figure 3.24). The observation that an equatorial proton is consistently found further to the left by 0.1–0.7 ppm than the axial proton on the same carbon atom in a rigid six-membered ring can thus be rationalized. The axial and equatorial protons on C_1 are oriented similarly with respect to C_1—C_2 and C_1—C_6, but the equatorial proton is within the deshielding cone of the C_2—C_3 bond (and C_5—C_6).

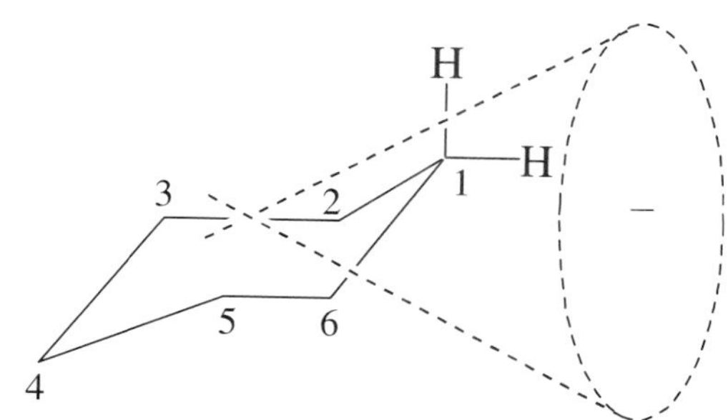

FIGURE 3.24 Deshielding of equatorial proton of a rigid six-membered ring.

Extensive tables and charts of chemical shifts in the Appendices give the useful impression that chemical shifts of protons in organic compounds fall roughly into eight regions as shown in Figure 3.25.

To demonstrate the use of some of the material in the Appendices, we predict the chemical shifts of the protons in benzyl acetate. In Appendix A, Chart A.1, we see that the chemical shift of the CH_3 group is ~δ 2.0. From Table B.1, we find that the CH_2 group is at ~δ 5.07. In Appendix D, Chart D.1, the aromatic protons are at ~δ 7.2. In the spectrum of benzyl acetate (Figure 3.26), we see three sharp peaks from right to left at δ 1.96, δ 5.00, and δ 7.22; the "integration steps" are in the ratios 3:2:5, corresponding to CH_3, CH_2, and five ring-protons.* The peaks are all singlets. This means that the CH_3 and CH_2 groups are insulated, that is, there are no protons on the adjacent carbon atoms for coupling (see Section 3.5). However, there is a problem with the apparent singlet representing the ring protons, which are not chemical shift equivalent (Section 3.8.1) and do couple with one another. At higher resolution, we would see a multiplet rather than an apparent singlet. The expanded inset shows partially resolved peaks.

* The "integration step"—i.e., the vertical distance between the horizontal lines of the integration trace—is proportional to the number of protons represented by the particular absorption peak or multiplet of peaks. These steps give ratios, not absolute numbers of protons. The ratios actually represent areas under the peaks.

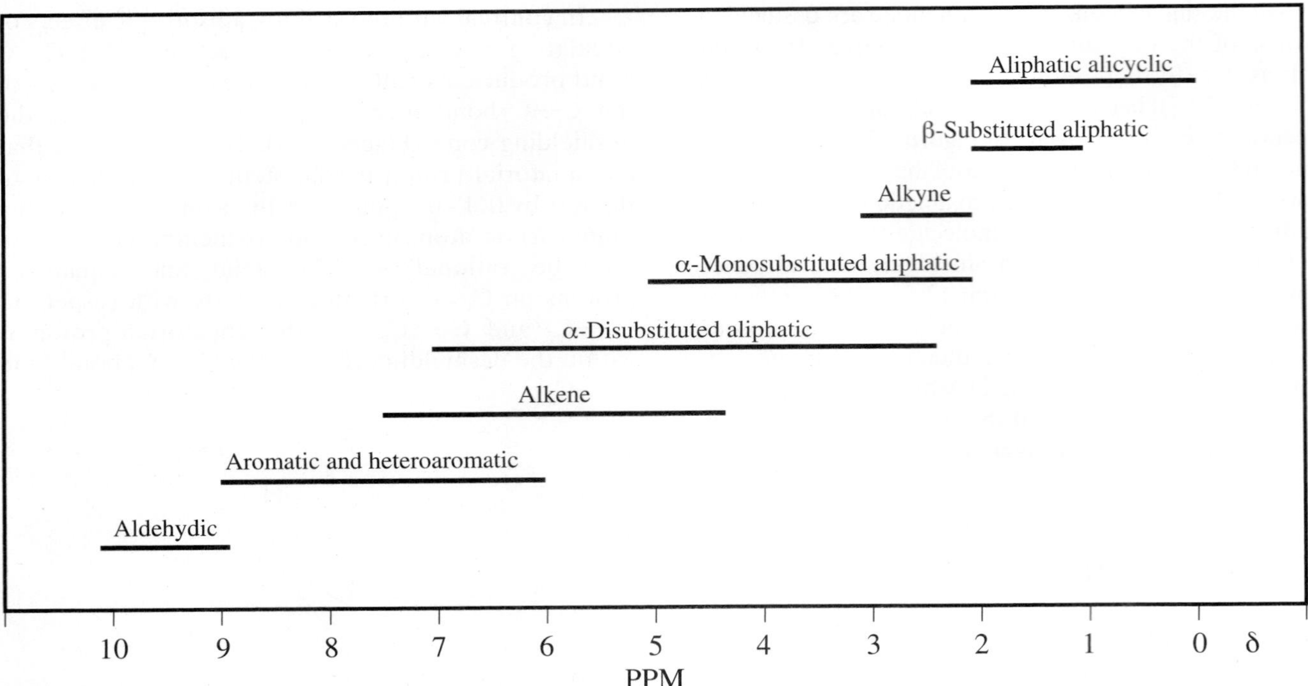

FIGURE 3.25 General regions of chemical shifts. Several aldehydes, several enols, and most carboxylic acids absorb at higher frequencies than δ 10.

We point out again that an appreciation of the concepts of electronegativity (inductive effects) and of electron delocalization—combined with an understanding of diamagnetic anisotropy—permits both rationalization and prediction of approximate chemical shift. Several examples make the point:

1. In an α,β-unsaturated ketone, resonance deshields the β-proton:

H_3C—C(H)$^{\beta}$=C(H)$^{\alpha}$—C(=O)—CH_3 ⟷ H_3C—C^+(H)—C(H)=C(O^-)—CH_3

α-proton $\delta \sim 6.2$
β-proton $\delta \sim 6.8$

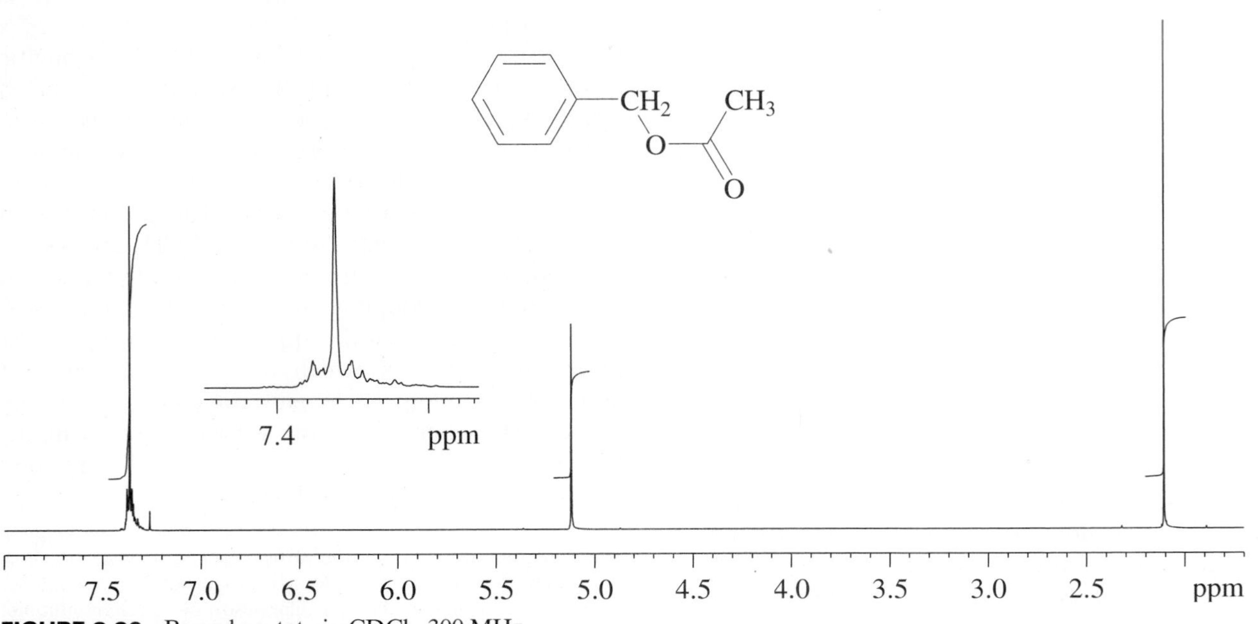

FIGURE 3.26 Benzyl acetate in $CDCl_3$, 300 MHz.

2. In a substituted vinyl ether, the oxygen atom deshields the α-proton by an inductive effect and shields the β-proton by resonance.

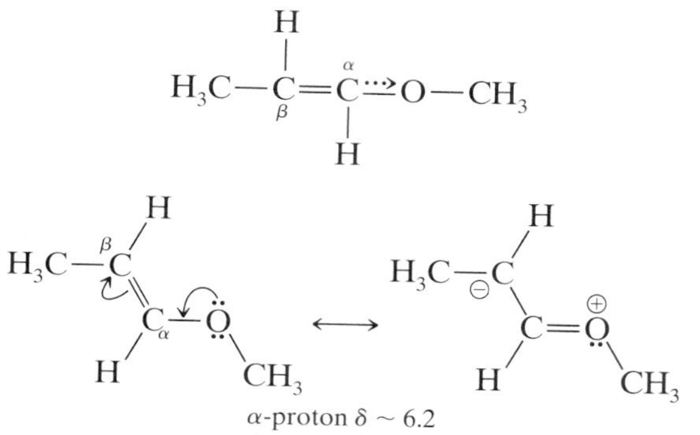

$H_3C-\overset{\beta}{C}(H)=\overset{\alpha}{C}(H)\cdots O-CH_3$

α-proton $\delta \sim 6.2$
β-proton $\delta \sim 4.6$

The above approximate values were calculated from Appendix D. In comparison the olefinic protons of *trans*-3-hexene are at δ 5.40.

3. The shifts of protons *ortho, meta,* or *para* to a substituent on an aromatic ring are correlated with electron densities and with the effects of electrophilic reagents (Appendix Chart D.1). For example, the *ortho* and *para* protons of phenol are shielded because of the higher electron density that also accounts for the predominance of *ortho* and *para* substitution by electrophilic reagents. Conversely, the *ortho* and *para* protons of nitrobenzene are deshielded, the *ortho* protons more so (see Figure 3.23).

Since chemical shift increments are approximately additive, it is possible to calculate the ring proton shifts in polysubstituted benzene rings from the monosubstituted values in Appendix Chart D.1. The chemical shift increments for the ring protons of *m*-diacetylbenzene:

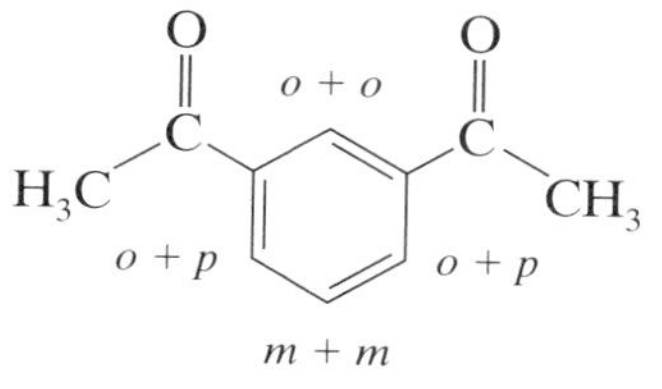

$o + o$ $o + p$ $o + p$ $m + m$

for example, are calculated as follows.

Chemical shift increments are the shifts from that of the protons of benzene (δ 7.27). Thus for a $CH_3C{=}O$ substituent (line 26, Appendix Chart D.1), the *ortho* increment is +0.63, and the *meta* and *para* increments are both +0.28 (+ being at higher frequency than δ 7.27). The C-2 proton has two *ortho* substituents; the C-4 and C-6 protons are equivalent and have *ortho* and *para* substituents; the C-5 proton has two *meta* substituents. Thus, the calculated increment for C-2 is +1.26, for C-4 and C-6 is +0.91, and for C-5 is +0.56. The spectrum shows increments of +1.13, +0.81, and +0.20, respectively. This agreement is adequate.*

Obviously, proton NMR spectrometry is a powerful tool for elucidating aromatic substitution patterns—as is ^{13}C NMR (see Chapter 4). Two-dimensional NMR spectrometry offers another powerful tool (see Chapter 5).

3.5 SPIN COUPLING, MULTIPLETS, SPIN SYSTEMS

3.5.1 Simple and Complex First Order Multiplets

We have obtained a series of absorption peaks representing protons in different chemical environments, each absorption area (from integration) being proportional to the number of protons it represents. We now consider one further phenomenon, *spin coupling.* This can be described as the coupling of proton spins through the intervening bonding electrons. According to the Pauli principle, the bonding electrons between two nuclei are paired so that the spins are antiparallel. In a magnetic field, there is some tendency for each nucleus to pair its spin with one of the bonding electrons so that most are antiparallel, this being the stable state. Coupling is ordinarily not important beyond three bonds unless there is ring strain as in small rings or bridged systems, delocalization as in aromatic or unsaturated systems, or four connecting bonds in a "W" configuration (Section 3.14). Two-bond coupling is termed *geminal;* three-bond coupling, *vicinal:*

H—C—H	H—C—C—H
Geminal coupling 2 bonds (^2J)	Vicinal coupling 3 bonds (^3J)

Suppose that two vicinal protons are in very different chemical environments from one another. Each proton will give rise to an absorption, and the absorptions will be quite widely separated, but the spin of each proton is affected slightly by the two orientations of the other proton through the intervening electrons, so that each

* Calculations for *ortho*-disubstituted compounds are less satisfactory because of steric or other interactions between the *ortho* substituents.

absorption appears as a doublet (Figure 3.27). The frequency difference in Hz between the component peaks of a doublet is proportional to the effectiveness of the coupling and is denoted by a coupling constant, J, which is independent of the applied magnetic field $\boldsymbol{B}_0$.* Whereas chemical shifts usually range over about 3600 Hz at 300 MHz, coupling constants between protons rarely exceed 20 Hz (see Appendix F).

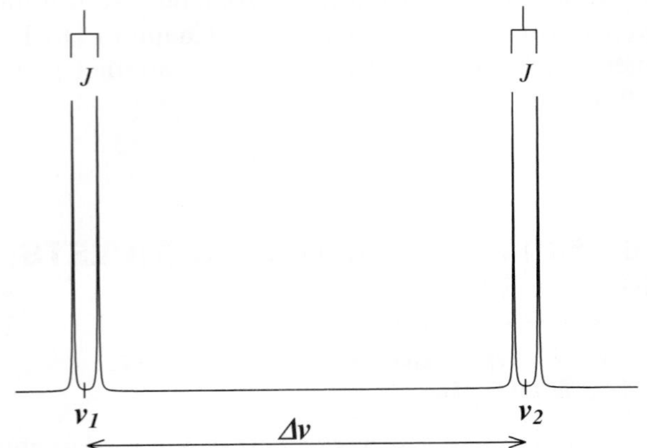

FIGURE 3.27 Spin coupling between two protons with very different chemical shifts.

So long as the chemical shift difference in hertz ($\Delta\nu$) is much larger than the coupling constant (arbitrarily $\Delta\nu/J$ is greater than about 8), the simple pattern of two doublets appear. As $\Delta\nu/J$ becomes smaller, the doublets approach one another, the inner two peaks increase in intensity, and the outer two peaks decrease (Figure 3.28). The shift position of each proton is no longer midway between its two peaks as in Figure 3.27 but is at the "center of gravity" (Figure 3.29); it can be estimated with fair accuracy by inspection.

The (d) spectrum in Figure 3.28 (and reproduced in Figure 3.29), which consists of two distorted doublets, can readily be mistaken for a quartet (see Figure 3.34). Increasing the applied frequency would not pull the true quartet apart into two doublets. As the intensities of the outer peaks of spectrum (e) continue to decrease, failure to notice the small outer peaks may lead to mistaking the inner peaks for a doublet. Eventually the outer peaks disappear and the inner peaks merge, producing a two-proton, apparent singlet.

At this stage, we can predict the spectrum, given the structure of a simple compound. Consider the methylene and methine groups in the compound:

$$\mathrm{RO{-}\underset{H}{\overset{OR}{C}}{-}CH_2{-}Ph}$$

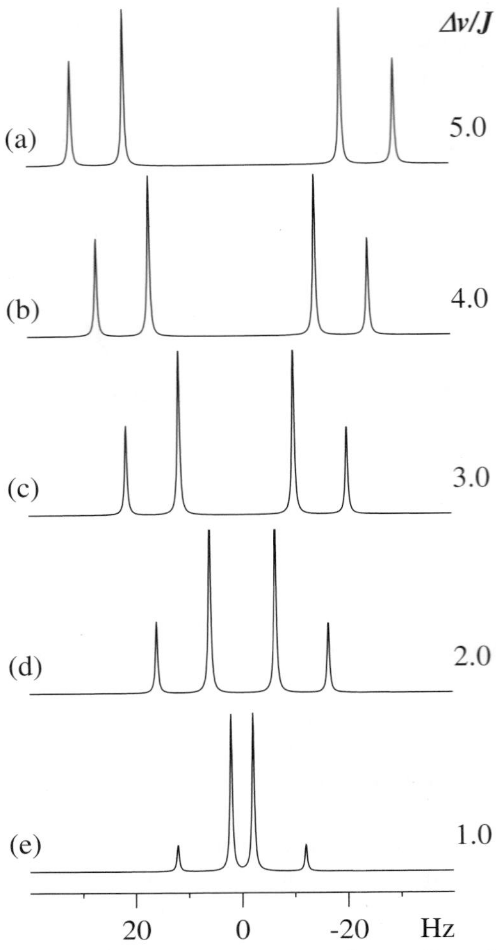

FIGURE 3.28 A two-proton system, spin coupling with a decreasing difference in chemical shifts and a large J value (10 Hz); the difference between AB and AX notation is explained in the text.

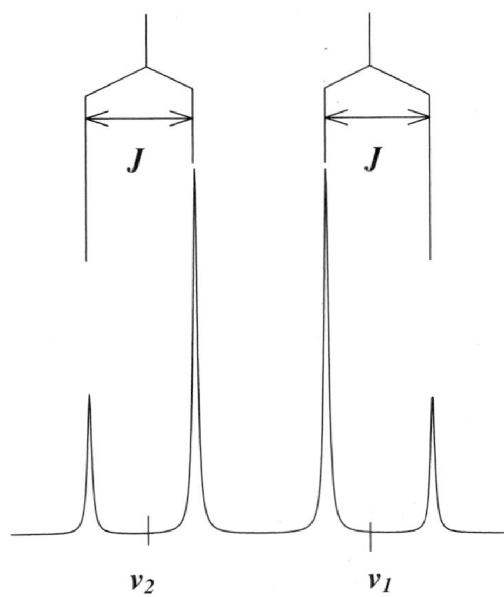

FIGURE 3.29 "Center of gravity," instead of linear midpoints, for shift location (in the case of "low" $\Delta\nu/J$ ratio).

* The number of bonds between coupled nuclei (protons in this chapter) is designated by J and a left superscript. For example, H—C—H is ^2J, H—C—C—H is ^3J, H—C=C—C—H is ^4J. Double or triple bonds are counted as single bonds.

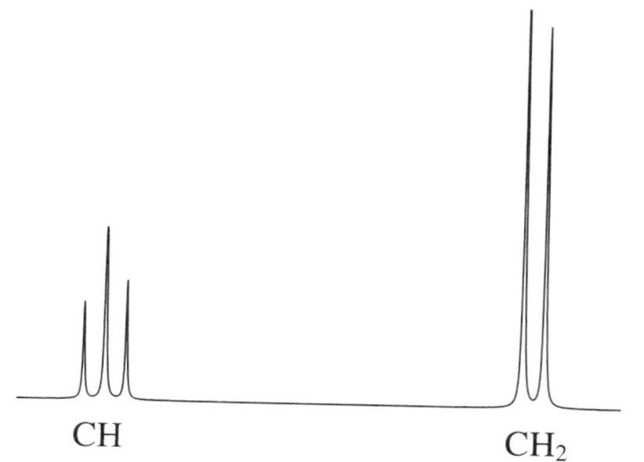

FIGURE 3.30 Spin coupling between CH and CH_2 with different chemical shifts.

in which the single methine proton is in a very different chemical environment from the two methylene protons. Both CH_2 protons couple equally with the methine proton. In Figure 3.30, we see a triplet and a doublet widely separated with an integration ratio of 1:2 respectively. The doublet is a result of splitting of the CH_2 absorption by the CH proton. The triplet can be shown as a result of consecutive splitting of the CH absorption by the CH_2 protons; the peaks overlap since the coupling constants are identical (see Figure 3.31).

In a "simple, first-order multiplet," the number of peaks is determined by the number of coupled, neighboring protons with the same (or very slightly different) coupling constants. Neighboring protons are geminal or vicinal i.e., involving two or three

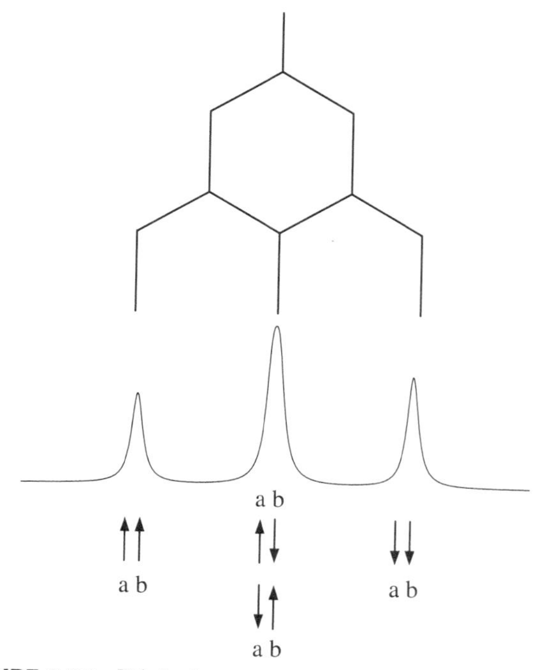

FIGURE 3.31 Triplet by consecutive splitting.

bonds; long-range coupling constants are usually much smaller (see Section 3.14). As we have seen, one neighboring proton induces a doublet, and two equally coupled neighboring protons induce a triplet. The mulplicity then is $n + 1$, n being the number of neighboring, equally coupled protons. The general formula that covers all nuclei is $2nI + 1$, I being the quantum spin number (see Section 3.2.1). The relative intensities of the peaks of a simple, first-order multiplet also depend on n. We have seen that doublet peaks ($n = 1$) are in the ratio 1:1; triplets ($n = 2$) are 1:2:1; quartets ($n = 3$) are 1:3:3:1; and so forth.

The requirements for a simple, first-order multiplet can be summarized:

- The ratio $\Delta\nu/J$ must be larger than about 8; $\Delta\nu$ is the distance in Hz between the midpoints of the coupled multiplets. J is the coupling constant.
- The number of peaks in the multiplet is $n + 1$, where n is the number of neighboring protons with the same coupling constant.
- The distance in Hz between the individual peaks of a simple, first-order multiplet represents the coupling constant.
- The simple, first-order multiplet is centrosymmetric, with the most intense peak(s) central (see the Pascal triangle, Figure 3.32).

A complex, first-order multiplet differs from a simple, first-order multiplet in that several different coupling constants are involved in the complex multiplet. The requirement that $\Delta\nu/J$ be greater than about 8 still holds, but Pascal's triangle does not hold for the complex multiplet. An example is presented later in Figure 3.37 where it can be seen in the expanded splitting pattern that the multiplet consists of a quartet of doublets; the "stick" pattern is shown in the text and consists of a sequence of two simple multiplets. Some dexterity with stick diagrams will be required throughout. Section 3.5.5 provides a more sophisticated treatment.

It should become obvious, with some experience, that the ratio $\Delta\nu/J = 8$ is not rigorous, but as the ratio decreases the interpretation gets rougher. The "horsepower race," mentioned in Section 3.3 was (and is) immensely successful in increasing this ratio. The result has been to simplify spectra and thereby permit identification of more difficult molecules. It will be useful to get some feeling for this effect at 60 MHz, 300MHz, and 600 MHz in Figure 3.33. The compound is

$$Cl—CH_2—CH_2—O—CH_2—CH_2—Cl$$

At 600 MHz, the spectrum consists of two simple first-order triplets. At 300 MHz, the spectrum is

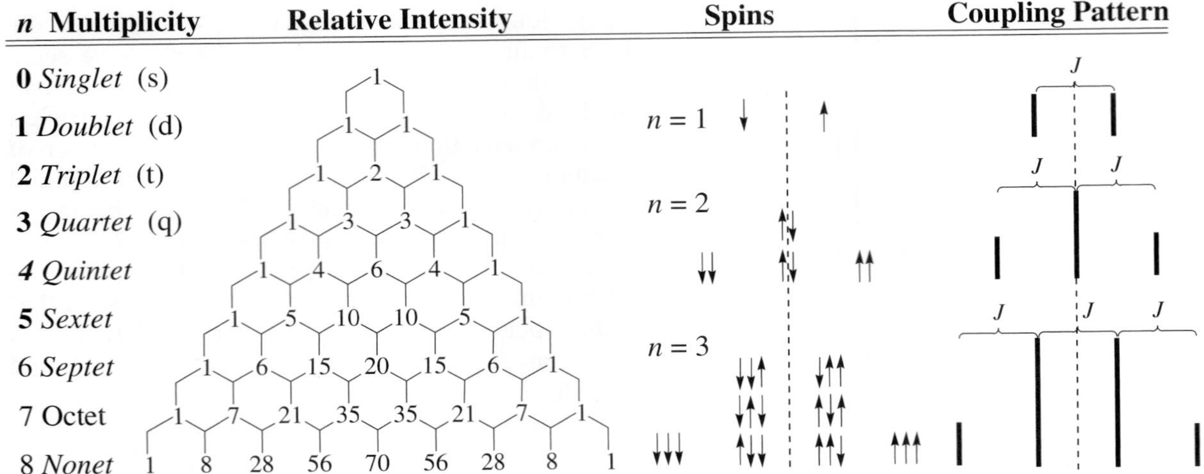

FIGURE 3.32 Pascal's triangle. Relative intensities of first-order multiplets; n = number of equivalent coupling nuclei of spin 1/2 (e.g., protons).

slightly distorted, and there are a few, smaller, extraneous peaks. At 60 MHz, there is obvious overlap and extraneous peaks. (The expanded multiplets are the result of using the same δ scale for all the spectra). The $\Delta\nu/J$ ratios are 12 at 600 MHz (certainly first-order), and 6 at 300 MHz (still easily recognized if the additional splitting is ignored). The 60 MHz overlapping multiplets with extraneous peaks are described as "higher order." A glance at Appendix A shows that substituents Cl and OR are almost equally deshielding and suggests that a 60 MHz spectrum would be risky. Return to Figure 3.19 for a more sophisticated example.

3.5.2 First Order Spin Systems

A spin system consists of multiplets that couple to one another but do not couple outside the spin system. It is not necessary for every multiplet within the spin system to be coupled to every other multiplet. A spin system is "insulated" from other spin systems by heteroatoms and by quaternary carbon (atoms without attached hydrogen atoms). The multiplets are coupled as simple first-order, complex first-order, or higher order. In a spin system there maybe long-range coupling through the insulating atom, but this usually involves small coupling constants, resulting in the appearance of peak broadening.

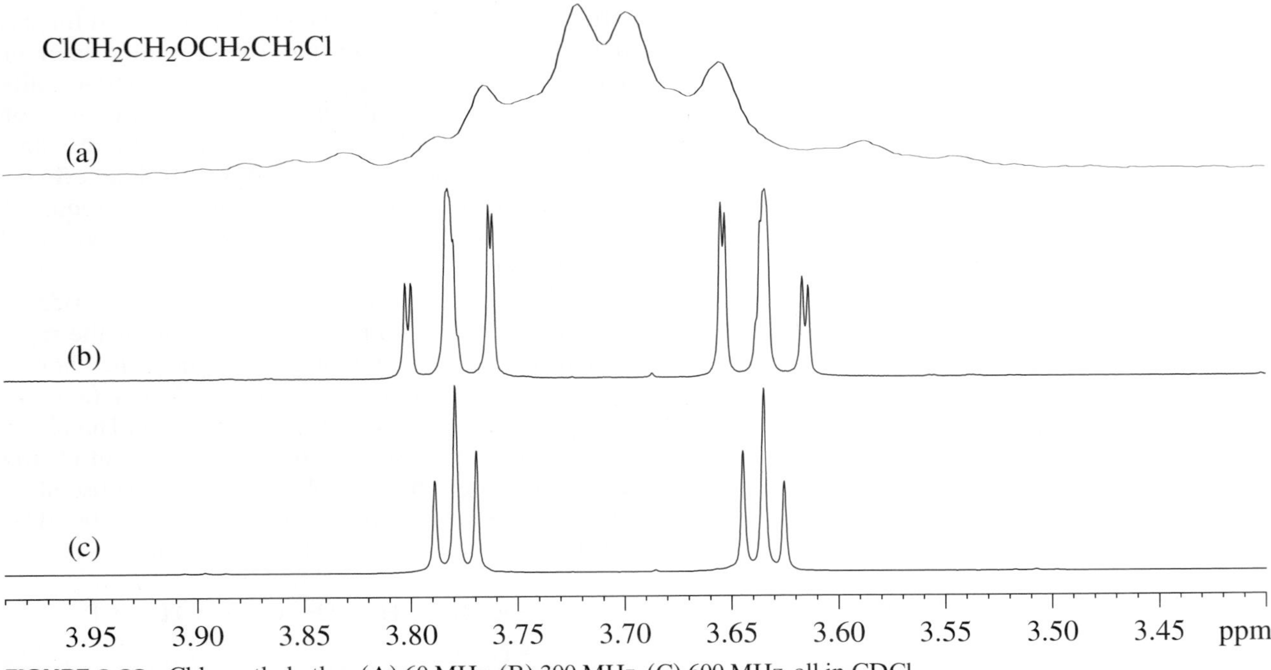

FIGURE 3.33 Chloroethyl ether (A) 60 MHz, (B) 300 MHz, (C) 600 MHz, all in $CDCl_3$.

The 600 MHz spectrum in Figure 3.33 consists of a simple first-order spin system, as does the 300 MHz spectrum but with minor distortions. The 60 MHz spectrum consists of a higher order spin system. A spin system may consist of one or more higher-order multiplets, in which case it is difficult to determine all of the δ value, coupling constants, and multiplicities by inspection. Figure 3.50 offers an example even at 600 MHz.*

It is notable that chemical shifts are usually reported to two decimal places. But are these numbers always valid? Certainly they are for the almost perfect, first-order triplets at 600 MHz in Figure 3.33. At 60 MHz, however, chemical shifts of the overlapping multiplets cannot be measured accurately by inspection. Somewhere between 300 MHz and 60 MHz, attempts to measure the chemical shifts of similar compounds would only yield approximations, but not values accurate to two decimal places.

3.5.3 Pople Notation

The Pople notation provides a means for describing multiplets, spin and spectra.† Multiplets are now termed "sets" and each is designated by a capital letter. If the $\Delta\nu/J$ ratio is larger than about 8, the coupled sets are considered to be well separated (i.e., weakly coupled), and the resulting spin systems are designated by well separated letters of the alphabet such AX. If the ratio is less than 8, such letters as AB are used. If there are three weakly coupled sets, they are designated AMX. If the first two sets are strongly coupled and the last two weakly, ABX is used. The number of protons in a set is designated by a subscript number; if there is only one proton in a set, no subscript is used.

Any such collection of sets insulated from all other sets, is a spin system. The following are examples of first-order spin system: AX (two doublets), A_2X (doublet, triplet), A_2X_2 (two triplets), A_3X_2 (triplet, quartet). The reasoning is obvious: In A_2X for example, the A_2 set is split into a doublet by the $n + 1$ neighboring protons; in this case there is one proton in the X set; the two protons in the A_2 set account for the triplet.

With two or more sets, there is the complication of several coupling constants. Thus nitropropane can be described as follows since the NO_2 group spreads out the three sets:

$$\underset{A_3}{CH_3}—\underset{M_2}{CH_2}—\underset{X_2}{CH_2}—NO_2$$

The A and X groups couple with the M group but the A group does not couple with the X group. The analysis is presented in Section 3.8.1.1 after discussion of some relevant material. At this point, we note only that the spectrum consists of two triplets with slightly different coupling constants, and a sextet with slightly broadened peaks.

In the styrene molecule,

H_X H_A H_M

all of the vinyl protons are weakly coupled with each other at 600 MHz (see Section 3.10). As mentioned above, the letters close together in the alphabet are used to describe spin systems that are not first order: AB, A_2B, ABC, $A_3B_2C_2$, etc. These spin systems are termed "higher order" and cannot be interpreted by inspection.

Beyond these spin systems are those containing "magnetically nonequivalent" protons, which are quite common and have an unpleasant aspect: They cannot become first-order systems by increasing the magnetic field. These spin systems are discussed in Section 3.12.

There are collections of calculated spectra that can be used to match complex, first-order, splitting patterns (see references to Wiberg 1962 and Bovey 1988). Alternatively, these spin systems can be simulated on the computer of a modern NMR spectrometer. For example see the NMRSIM computer program available from Bruker BioSpin.

3.5.4 Further Examples of Simple, First-Order Spin Systems

With an understanding of simple first-order spin systems and of the Pople notations, we can consider the following example.

The spectrum in Figure 3.34 consists of two spin systems insulated from each other by a carbon atom that has no attached proton. The $CH_3—CH_2$ spin system consists of a well separated triplet and quartet—that is, an A_3X_2 system as suggested by the peak multiplicities and by the 5:2:3 (left to right) ratios of integration. The favorable $\Delta\nu/J$ ratio is a result of deshielding by the benzene ring.

The ring system, which has an axis of symmetry, consists of two interchangeable *ortho,* two interchangeable *meta* protons, and one *para* proton, all in the characteristic region for ring protons. The alert student, having absorbed the concept of chemical-shift equivalence, (Section 3.8.3) and the description of the

* Coupling constants have plus and minus signs. However, these have no effect in a first-order system, in which we can measure J values by inspection but cannot determine the sign by inspection. Thus, we disregard the sign.

† Pople, J.A., Schneider, W.G., and Bernstein, H.J.(1959). *High Resolution NMR,* New York; McGraw-Hall.

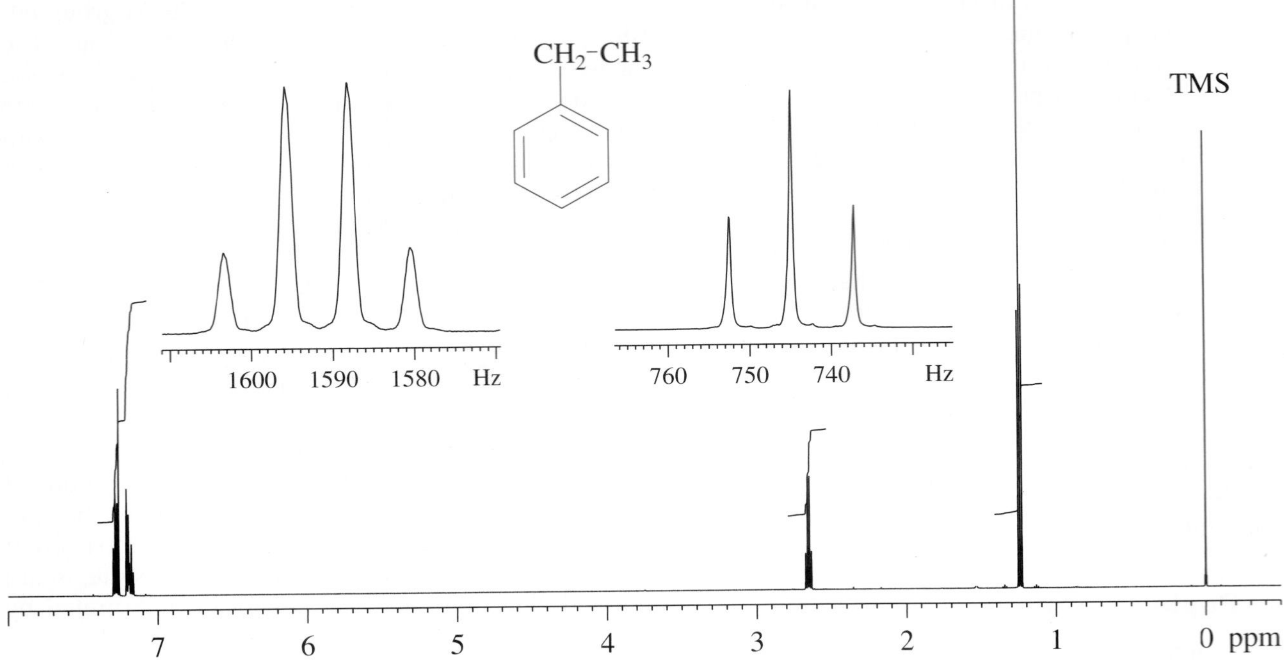

FIGURE 3.34 Ethylbenzene in $CDCl_3$ at 600 MHz. The ethyl moiety is recognized by the CH_3 triplet and the CH_2 quartet.

Pople notations (Section 3.5.3), would probably write A_2B_2C. The student might go even further and predict that a larger magnet might give a first-order ring system, A_2M_2X. However a larger magnet would separate the chemical shifts of the ring system, but Section 3.9 will explain that a first-order system would not be achieved. Briefly, the problem is that neither of the *ortho* protons nor the *meta* protons are "magnetically equivalent".

One minor but common feature may be pointed out. Note the slight broadening of the CH_2 quartet. This is merely the result of a slight "leakage" through the "insulating" carbon atom; a very small, long-range coupling exists and can be detected by a high resolution instrument.

The spectrum of isopropylbenzene in Figure 3.35 presents a simple, first-order spin system. The methine proton shows a septet because of coupling with six neighboring identical protons of the two CH_3 groups, all with the same coupling constant. The two chemical-equivalent CH_3 groups (because of free rotation) show a six-proton doublet because of coupling with the methine proton. The ring protons present the same difficulties as those of the ring protons of ethylbenzene (Figure 3.34).

3.5.5 Analysis of First-Order Patterns

The usual "stick diagram" conveniently explains the pattern after the fact. However, a first-order pattern resulting, for example, from coupling of a single proton with a CH_3 group ($J = 7$ Hz) and two different protons ($J = 4$ Hz, $J = 12$ Hz) (Figure 3.36) would be difficult to reduce to a stick diagram without prior coupling information. B.E. Mann has proposed a routine procedure for constructing such a stick diagram.* The following is a brief, slightly modified description of his procedure for analyzing the pattern resulting from the above couplings of neighboring protons with a single proton. The horizontal scale is 10 mm = 5 Hz.

1. Assign an integer intensity number centered directly under each peak in the pattern, starting with intensity 1 for the outermost peaks. A first-order pattern must have a centrosymmetric distribution of intensities (level a). If it does not, the pattern is higher order.
2. The sum of the assigned peak intensities in level a must be 2^n, n being the number of coupling protons. In this example, $n = 5$, which corresponds to coupling of three equivalent neighboring protons and two nonequivalent protons as specified above.
3. Draw a vertical line (stick) centered directly under each peak of the pattern, i.e., beneath each intensity number. All lines are drawn to equal height (level b).

* Mann, B.E. (1995). *J. Chem. Educ.*, **72,** 614.

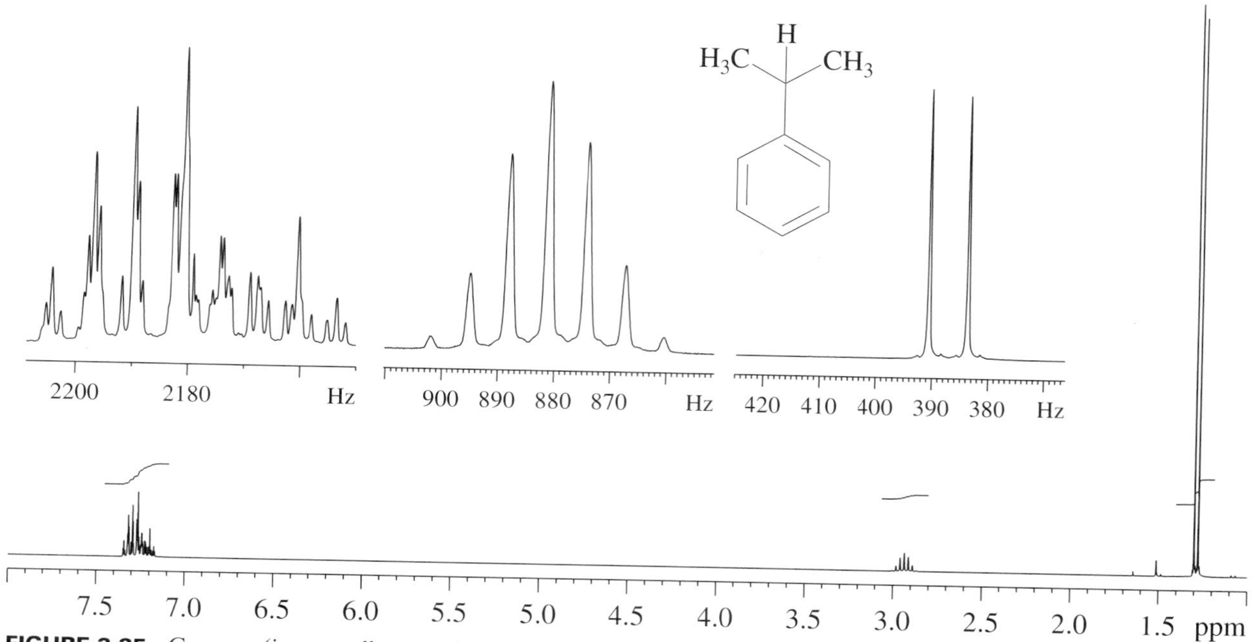

FIGURE 3.35 Cumene (isopropylbenzene) in $CDCl_3$ at 300 MHz. The isopropyl moiety is recognized by the characteristic six-proton doublet and the one-proton septet.

4. The intensity numbers of the two outer lines at either end determine the multiplicity at level b. Thus, in this example, an intensity of 1:1 ordains a series of doublets. An intensity of 1:2 would ordain a series of 1:2:1 triplets. An intensity of 1:3 would ordain a series of 1:3:3:1 quartets, and so forth through the Pascal triangle (Figure 3.32). Note that some of the doublets have intensities of 3:3, but the ratio is still 1:1.
5. The distance on the Hz scale between the first two lines in level b gives the coupling constant for the doublets in level b. All doublets, of course, have the same coupling constant as that of the end pairs. Although there is overlapping of the doublets, there is no difficulty in sorting out the doublets on the basis of equal coupling and the 1:1 ratio of paired lines. The Hz scale shows that these are the 4-Hz doublets.
6. At level c, the "stem" of each pair of lines at level b is assigned an intensity number, which is the intensity number of the first line of each pair at level b. These "stems" now become the lines of the multiplets at level c—in this example, two overlapping quartets ($J = 7$ Hz) in accordance with the 1:3 ratio of the first two lines. In turn, the stems of these quartets are assigned the intensity number of the first line of each quartet. The stems now become the lines of the multiplet of level d—in this example a doublet with $J = 12$ Hz. Note that at each level the lines should be checked for centrosymmetry of intensity and coupling.
7. In summary, we confirm, starting at level d, that there is coupling to a CH proton to give a doublet ($J = 12$ Hz). Each line of this doublet is split into a quartet ($J = 7$ Hz) by vicinal CH_3 protons at level c. Each line of the quartets is split by another CH proton to give a series of doublets ($J = 4$ Hz) at level b. The entire scheme rests on the intensity numbers of the first two lines and the distance in Hz between these lines. These intensity numbers ordain the type of multiplet at each level of lines; the separation in Hz gives the coupling constant at each level.

It should be noted that accidental degeneracy of peaks-total or partial coincidence-can cause problems in assigning intensity numbers. Several examples are given in the footnote reference. *Note the distinction between overlapping and coincident peaks. In Figure 3.36, the doublets overlap but cause no difficulty. In Figure 3.37, the end peaks of three triplets are coincident. Thus the apparent intensities of level a are incorrect and are corrected in level b. The erroneous assignment is detectable because the outer two peaks on either side are in the ratio 1:2 and must be part of a 1:2:1 triplet. The middle triplet now has the intensities 2:4:2.

* Mann, B.E. (1995). *J. Chem. Educ.*, **72,** 614.

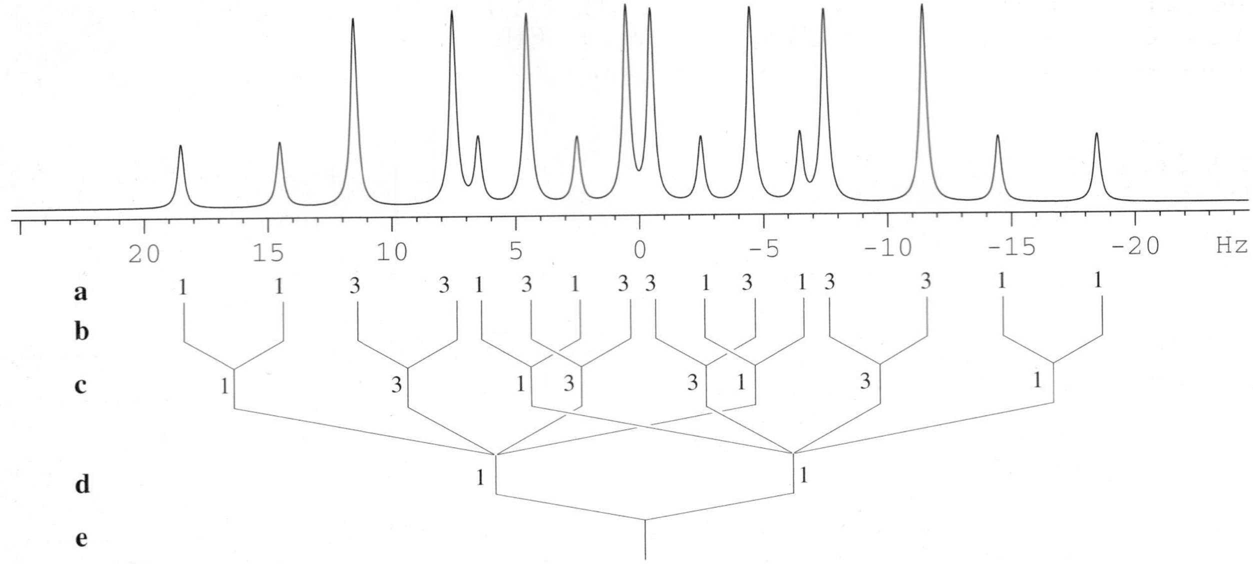

FIGURE 3.36 The coupling constants for this first-order multiplet are doublets of 4 and 12 Hz, and 2 quartets of 7 Hz. The horizontal scale is 10 mm = 5 Hz.

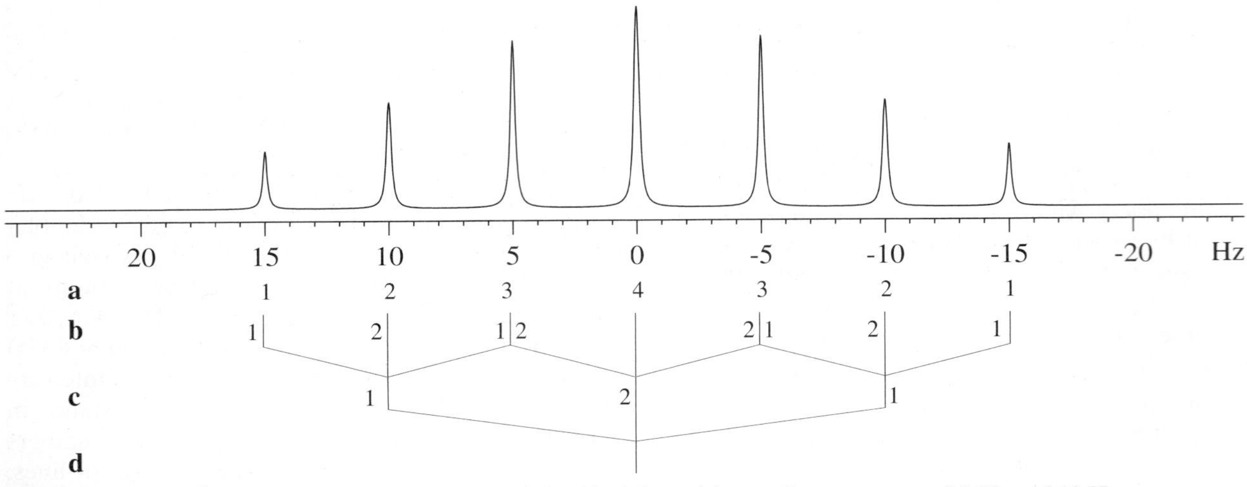

FIGURE 3.37 Peak coincidence arising from a triplet of triplets with coupling constants of 5 Hz and 10 Hz.

3.6 PROTONS ON OXYGEN, NITROGEN, AND SULFUR ATOMS. EXCHANGEABLE PROTONS

Protons directly bonded to an oxygen, nitrogen, or sulfur atom differ from protons on a carbon atom in that:

1. They are exchangeable.
2. They are subject to hydrogen bonding.
3. Those on a nitrogen (^{14}N) atom are subject to partial or complete decoupling by the electrical quadrupole moment of the ^{14}N nucleus, whose spin number is one.

Shift ranges for such protons are given in Appendix E. These variations in shift depend on concentration, temperature, and solvent effects.

3.6.1 Protons on an Oxygen Atom

3.6.1.1 Alcohols Depending on concentration, the hydroxylic peak in alcohols is found between ~δ 0.5 and ~δ 4.0. A change in temperature or solvent will also shift the peak position.

Intermolecular hydrogen bonding explains why the shift depends on concentration, temperature, and polarity of solvent. Hydrogen bonding decreases the electron density around the hydroxylic proton, thus moving the proton peak to higher frequency.

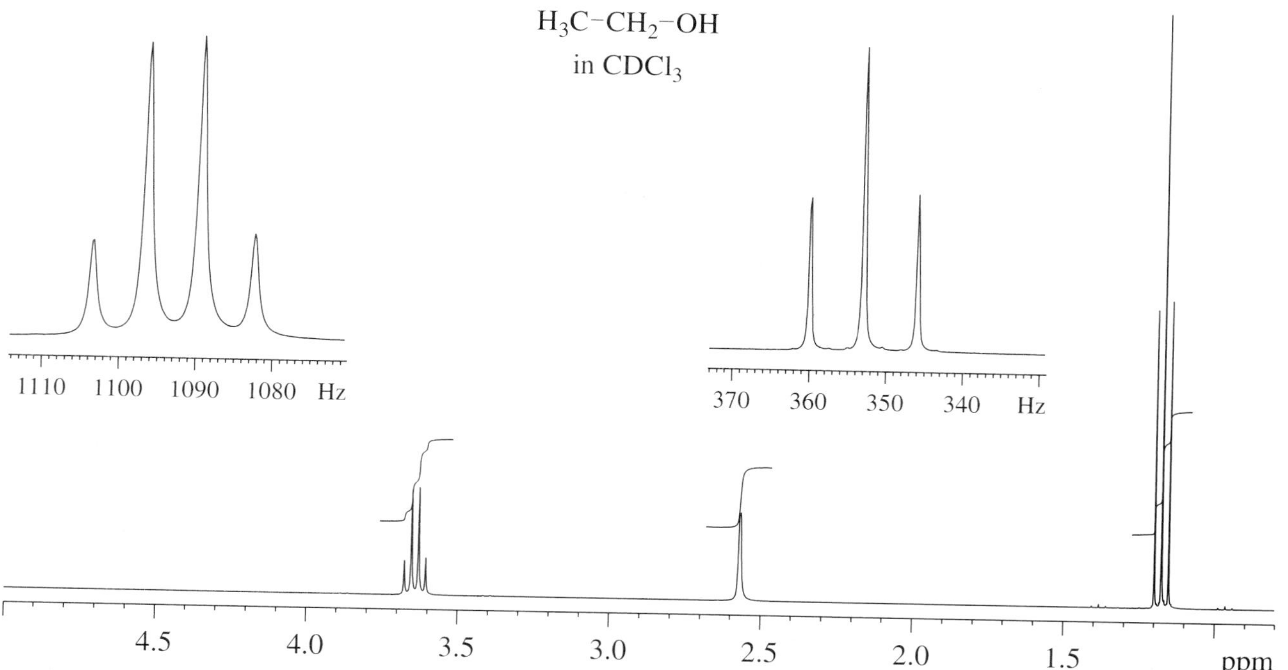

FIGURE 3.38 CH_3CH_2OH in $CDCl_3$ at 300 MHz, allowed to stand at room temperature, overnight exposed to air. The CH_2 peaks are broadened by residual coupling to OH, which also shows slight broadening. Absorbed moisture has increased the intensity of the OH signal.

Decrease in concentration in a nonpolar solvent disrupts such hydrogen bonding, and the peak appears at lower frequency—i.e., the alcohol molecules become less "polymeric." Increased temperature has a similar effect.

---O—H---O—H---O—H---
R R R

Intramolecular hydrogen bonds are less affected by their environment than are intermolecular hydrogen bonds. In fact, the enolic hydroxylic absorption of β-diketones, for example, is hardly affected by change of concentration or solvent, though it can be shifted to a lower frequency by warming. Nuclear magnetic resonance spectrometry is a powerful tool for studying hydrogen bonding.

H
O O
C C
R C R
H

Rapid exchangeability explains why the hydroxylic peak of ethanol is usually seen as a singlet (Figure 3.38). Under ordinary conditions—exposure to air, light, and water vapor—acidic impurities develop in $CDCl_3$ solution and catalyze rapid exchange of the hydroxylic proton.* The proton is not on the oxygen atom of an individual molecule long enough for it to be affected by the methylene protons; therefore there is no coupling. The OH proton shows a singlet, the CH_2 a quartet, and the CH_3 a triplet.

The rate of exchange can be decreased by lowering the temperature, by using a dilute solution, or by treating the solvent with anhydrous sodium carbonate or anhydrous alumina, then filtering through a pad of dry glass wool in a Pasteur pipette immediately before obtaining the spectrum. Under these conditions, the OH proton is coupled with the CH_2 protons, and useful information is available: An OH singlet indicates a tertiary alcohol, a doublet a secondary alcohol, and a triplet a primary alcohol.

The use of dry, deuterated dimethyl sulfoxide (DMSO-d_6) or deuterated acetone has the same effect as the above treatments. In addition, the OH proton peak is moved to higher frequency by hydrogen bonding between the solute and the solvent, thus providing

* $CDCl_3$ in small vials from Aldrich is pure enough so that a spectrum of CH_3CH_2OH taken within several hours showed the OH peak as a triplet. On standing for about 24 hours exposed to air, the sample gave a spectrum with the OH peak as a singlet (Figure. 3.38). The high dilution used with modern instruments also accounts for the persistence of the vicinal coupling of the OH proton.

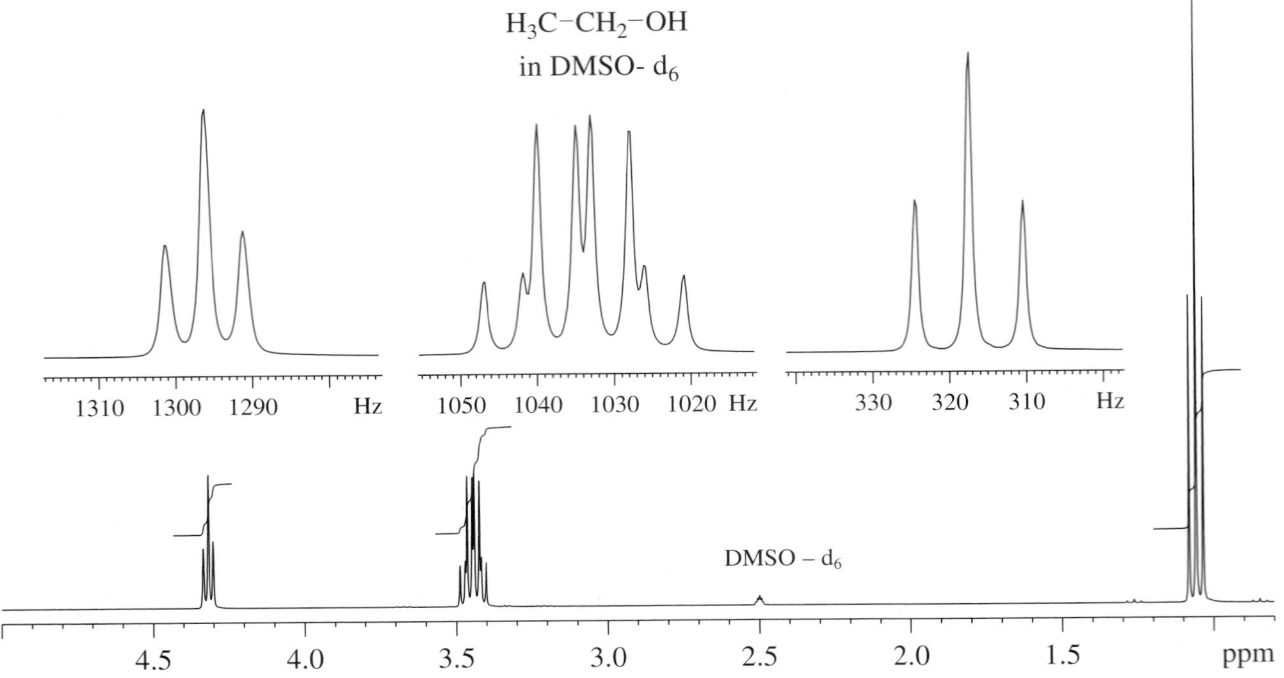

FIGURE 3.39 CH_3CH_2OH run in dry deuterated DMSO at 300 MHz. From left to right, the peaks represent OH, CH_2, CH_3. The small absorption at δ 2.5 represents the ^1H impurity in DMSO-d_6 (see appendix G).

a useful separation from overlapping peaks of other protons (see Figure 3.39).

In Figure 3.39 and in the following "stick" diagram, the CH_2 protons of ethanol are coupled to both the hydroxyl proton and the CH_3 protons. The diagram shows a somewhat overlapping quartet of doublets. The OH coupling is 5 Hz, whereas the CH_3 coupling is 7 Hz. It is usually better to start a coupling diagram with the largest coupling constant.

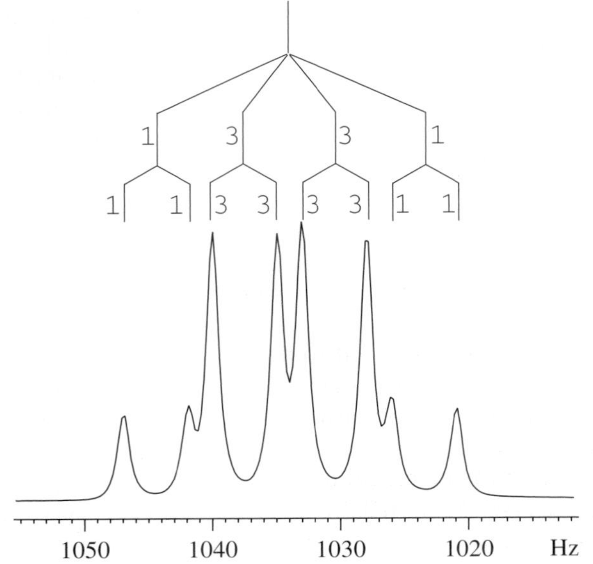

At intermediate rates of exchange, the hydroxylic multiplet merges into a broad band, which progresses to a singlet at higher exchange rates (Figure 3.38).*

A diol may show separate absorption peaks for each hydroxylic proton; in this case, the rate of exchange in hertz is much less than the difference in hertz between the separate absorptions. As the rate increases (trace of acid catalyst), the two absorption peaks broaden and then merge to form a single broad peak; at this point, the exchange rate (k) in hertz is approximately equal to twice the original signal separation in hertz. As the rate increases, the single peak becomes sharper. The relative position of each peak depends on the extent of hydrogen bonding of each hydroxylic proton; steric hindrance to hydrogen bonding moves the peak to the right.

The spectrum of a compound containing rapidly exchangeable protons can be simplified, and the exchangeable proton absorption removed, simply by shaking the solution with excess deuterium oxide or by obtaining a spectrum in deuterium oxide solution if the compound is soluble. A peak resulting from HOD will appear, generally between δ 5.0 and δ 4.5 in nonpolar solvents and near δ 3.3 in DMSO (see Appendix E). A $CDCl_3$ or CCl_4 solution in a stoppered NMR tube may be shaken vigorously for several seconds with 1 or 2 drops of D_2O, and the mixture allowed to stand

* H_2O as an impurity may exchange protons with other exchangeable protons to form a single peak at an averaged position between the proton peaks involved.

(or centrifuged) until the layers are clearly separated. The top aqueous layer does not interfere.

Acetylation or benzoylation of a hydroxyl group moves the absorption of the CH_2OH protons of a primary alcohol to the left about 0.5 ppm, and the CHOH proton of a secondary alcohol about 1.0–1.2 ppm. Such shifts provide a confirmation of the presence of a primary or secondary alcohol.

3.6.1.2 Water Aside from the problems of exchangeability, as just discussed, water is an ubiquitous impurity that faithfully obeys Murphy's law by interfering with critically important peaks. "Bulk" water as suspended droplets or wall films gives a peak at $\sim\delta$ 4.7 in $CDCl_3$ (HOD occurs in the D_2O exchange experiment mentioned in Section 3.6.1.1).

Dissolved (monomeric) water absorbs at $\sim\delta$ 1.55 in $CDCl_3$ and can be a serious interference in a critical region of the spectrum in dilute solutions.* Use of C_6D_6 (dissolved H_2O at δ 0.4) avoids this interference. A table of water peaks in the common deuterated solvents appears in Appendix H.

3.6.1.3 Phenols The behavior of a phenolic proton resembles that of an alcoholic proton. The phenolic proton peak is usually a sharp singlet (rapid exchange, no coupling), and its range, depending on concentration, solvent, and temperature, is generally to the left (δ ~7.5 to δ ~4.0) compared with the alcoholic proton. A carbonyl group in the *ortho* position shifts the phenolic proton absorption to the range of about δ 12.0–δ 10.0 because of intramolecular hydrogen bonding. Thus, *o*-hydroxyacetophenone shows a peak at about δ 12.05 almost completely invariant with concentration. The much weaker intramolecular hydrogen bonding in *o*-chlorophenol explains its shift range (δ ~6.3 at 1 *M* concentration to δ ~5.6 at infinite dilution), which is broad compared with that of *o*-hydroxyacetophenone but narrow compared with that of phenol.

3.6.1.4 Enols The familiar tautomeric equilibrium of keto and enol forms of acetylacetone is described in Section 3.8.3.1 (see Figure 3.45). The enol form predominates over the keto form under the conditions described.

Ordinarily we do not write the enol form of acetone or the keto form of phenol, although minuscule amounts do exist at equilibrium. But both forms of acetylacetone are seen in the NMR spectrum because equilibration is slow enough on the NMR scale and the enol form is stabilized by intramolecular hydrogen bonding. The enol form of acetone and the keto form of phenol are not thus stabilized; furthermore, the aromatic resonance stabilization of phenol strongly favors the enol form. Note the deshielded chemical shift of the enol proton in Figure 3.45. (See also Appendix Chart E.1).

Ordinarily, only the keto form of α-diketones such as 2,3-butanedione is seen in NMR spectra. However, if the enol form of an α-diketone is stabilized by hydrogen bonding–as in the following cyclic α-diketones—only the stabilized enol form appears in the NMR spectra.

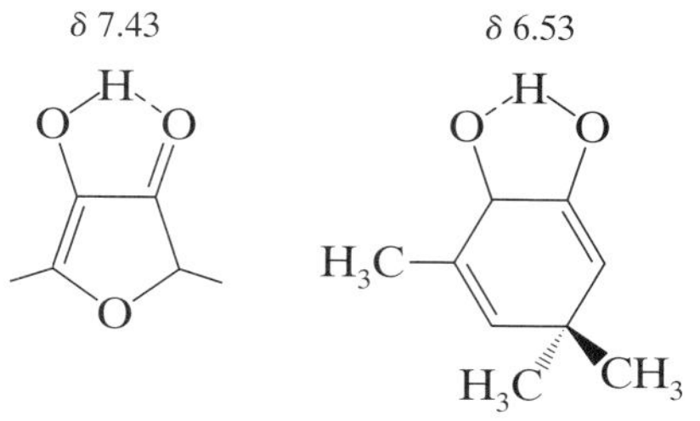

δ 7.43 δ 6.53

H O O O H O H_3C H_3C CH_3

3.6.1.5 Carboxylic Acids Carboxylic acids exist as stable hydrogen-bonded dimers in nonpolar solvents even at high dilution. The carboxylic proton therefore absorbs in a characteristic range δ ~13.2–δ ~10.0 and is affected only slightly by concentration. Polar solvents partially disrupt the dimer and shift the peak accordingly.

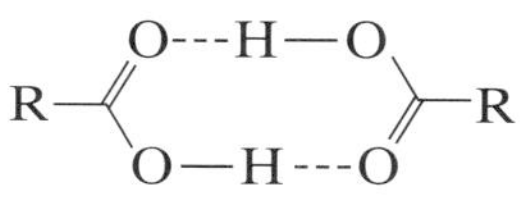

R O---H—O R O—H---O

The peak width at room temperature ranges from sharp to broad, depending on the exchange rate of the particular acid. The carboxylic proton exchanges quite rapidly with protons of water and alcohols (or hydroxyl groups of hydroxy acids) to give a single peak whose averaged position depends on concentration. Sulfhydryl or enolic protons do not exchange rapidly with carboxylic protons, and individual peaks are observed.

3.6.2 Protons on Nitrogen

The common ^{14}N nucleus* has a spin number I of 1 and, in accordance with the formula $2I + 1$, should cause a proton attached to it and a proton on an adjacent carbon atom to show three equally intense peaks. There are two factors, however, that complicate the picture: the rate of exchange of the proton on the nitrogen atom and the electrical quadrupole moment of the ^{14}N nucleus (see Section 3.2.1).

Protons on nitrogen may undergo rapid, intermediate, or slow exchange. If the exchange is rapid, the

* Webster, F.X., and Silverstein, R.M. (1985). *Aldrichimica Acta* 18 (No. 3), 58.

* ^{15}N spectra are discussed in Chapter 6.

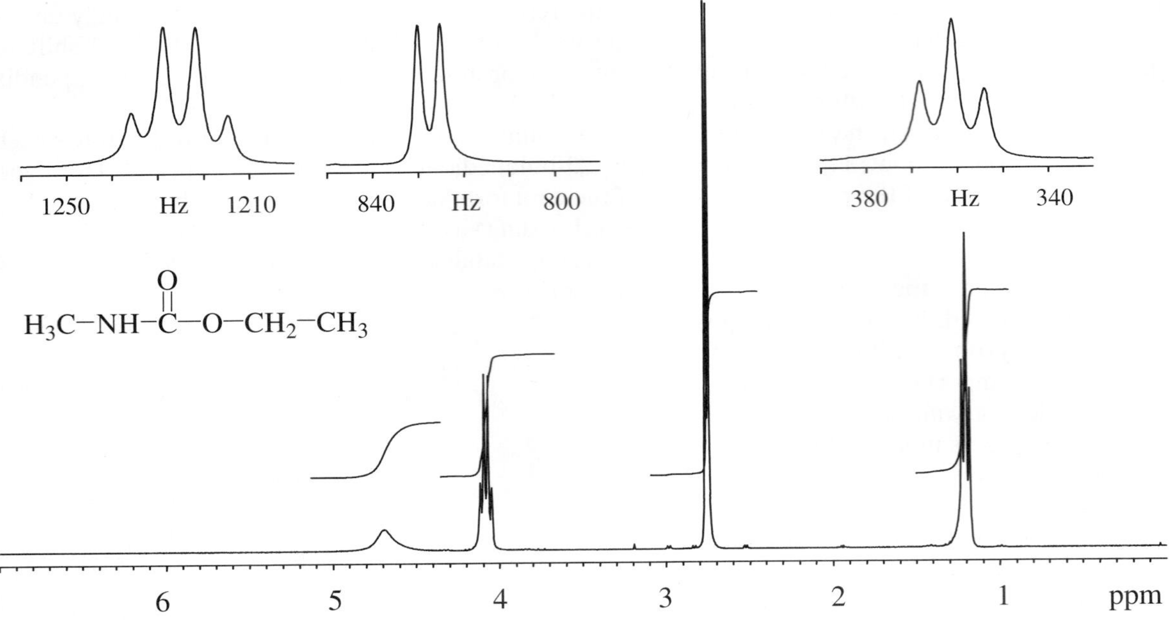

FIGURE 3.40 Ethyl *N*-methylcarbamate, at 300 MHz in $CDCl_3$.

NH proton(s) is decoupled from the nitrogen atom and from protons on adjacent carbon atoms. The NH proton peak is therefore a sharp singlet, and the adjacent CH protons are not split by NH. Such is the case for most aliphatic amines.*

At an intermediate rate of exchange, the NH proton is partially decoupled, and a broad NH peak results. The adjacent CH protons are not split by the NH proton. Such is the case for *N*-methyl-*p*-nitroaniline.

If the NH exchange rate is low, the NH peak is still broad because the electrical quadrupole moment of the nitrogen nucleus induces a moderately efficient spin relaxation and, thus, an intermediate lifetime for the spin states of the nitrogen nucleus. The proton thus sees three spin states of the nitrogen nucleus (spin number = 1), which are changing at a moderate rate, and the proton responds by giving a broad peak that may disappear in the baseline. In this case, coupling of the NH proton to the adjacent protons is observed. Such is the case for pyrroles, indoles, secondary and primary amides, and carbamates (Figure 3.40).

Note that H—N—C—H coupling takes place through the C—H, C—N, and N—H bonds, but coupling between nitrogen and protons on adjacent carbon atoms is negligible. The proton–proton coupling is observed in the signal caused by hydrogen on carbon; the N—H proton signal is severely broadened by the quadrupolar interaction.

* H—C—N—H coupling in several amines was observed following rigorous removal (with Na–K alloy) of traces of water. This effectively stops proton exchange on the NMR time scale. [K.L. Henold, *Chem. Commun.*, 1340 (1970).]

In the spectrum of ethyl *N*-methylcarbamate (Figure 3.40), the NH proton shows a broad absorption centered about δ 4.70, and the N—CH_3 absorption at δ 2.78 is split into a doublet (J ~5 Hz) by the NH proton. The ethoxy protons are represented by the triplet at δ 1.23 and the quartet at δ 4.14.

Aliphatic and cyclic amine NH protons absorb from ~δ 3.0 to 0.5; aromatic amines absorb from ~δ 5.0 to 3.0 in $CDCl_3$ (see Appendix E) because amines are subject to hydrogen bonding, the shift depends on concentration, solvent, and temperature. Amide, pyrrole, and indole NH groups absorb from ~δ 8.5 to 5.0; the effect on the absorption position of concentration, solvent, and temperature is generally smaller than in the case of amines. The nonequivalence of the protons on the nitrogen atom of a primary amide and of the methyl groups of *N*, *N*-dimethylamides is caused by slow rotation around the $\mathrm{C(=O)—N}$ bond because of the contribution of the resonance form $\mathrm{C(—O^-)=N^+}$ (Section 3.8.3.2 and Table 3.4).

Protons on the nitrogen atom of an amine salt exchange at a moderate rate; they are seen as a broad peak, (δ ~8.5 to δ –6.0), and they are coupled to protons on adjacent carbon atoms (J ~7 Hz).

The use of trifluoroacetic acid as both a protonating agent and a solvent frequently allows classification of amines as primary, secondary, or tertiary. This is illustrated in Table 3.4 in which the number of protons on nitrogen determines the multiplicity of the methylene

TABLE 3.4 Classification of Amines by NMR of their Ammonium Salts in Trifluoroacetic Acid.

Amine Precursor Class	Ammonium Salt Structure	Multiplicity of Methylene unit
Primary	$C_6H_5CH_2NH_3^+$	Quartet (Fig. 3.41)
Secondary	$C_6H_5CH_2NH_2R^+$	Triplet
Tertiary	$C_6H_5CH_2NHR_2^+$	Doublet

Source: Anderson, W.R. Jr., and Silverstein R.M. *Anal. Chem.*, **37,** 1417 (1965).

unit in the salt (Figure 3.41). Sometimes the broad ^+NH, $^+NH_2$ or $^+NH_3$ absorption can be seen to consist of three broad humps. These humps represent splitting by the nitrogen nucleus (J ~50 Hz). With good resolution, it is sometimes possible to observe splitting of each of the humps by the protons on adjacent carbons (J ~7 Hz), but it is easier to observe the splitting on the sharper α-CH signals. The behavior of the protons in the H—C—N—H sequence may be summarized as follows in Table 3.5.*

Chemical Shifts of several classes of protons attached to ^{14}N are available in Appendix E.

3.6.3 Protons on Sulfur

Sulfhydryl protons usually exchange at a low rate so that at room temperature they are coupled to protons on adjacent carbon atoms (J ~8 Hz). They do not exchange rapidly with hydroxyl, carboxylic, or enolic protons on the same or on other molecules; thus, separate peaks are seen. However, exchange is rapid enough that shaking the solution for a few minutes with deuterium oxide replaces sulfhydryl protons with deuterium. The absorption range for aliphatic sulfhydryl protons is δ 2.5 to δ 0.9; for aromatic sulfhydryl protons, δ 3.6 to δ 2.8. Concentration, solvent, and temperature affect the position within these ranges.

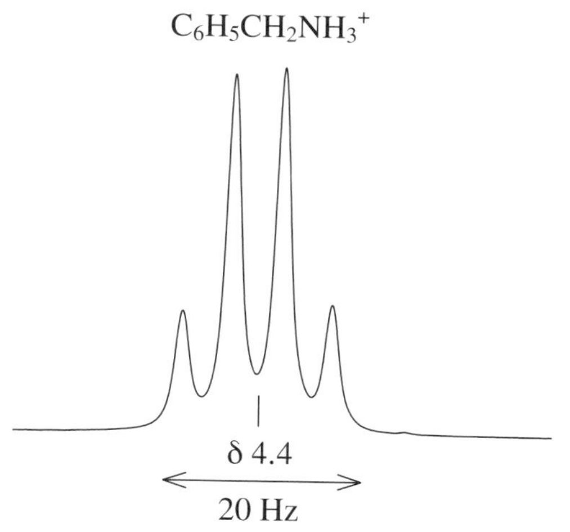

FIGURE 3.41 NMR spectrum of α-methylene unit of a primary amine at 100 MHz in CF_3CO_2H; corresponds to Table 3.4, first line.

* Courtesy of Dr. Donald C. Dittmer (Syracuse University).

TABLE 3.5 Effect of NH exchange rate on coupling.

	Rate of NH Exchange		
	Fast	Intermediate	Slow
Effect on N—H	Singlet, sharp	Singlet, broad	Singlet, broad
Effect on C—H	No coupling	No coupling	Coupling

3.6.4 Protons on or near Chlorine, Bromine, or Iodine Nuclei

Protons are not coupled to chlorine, bromine, or iodine nuclei because of the strong electrical quadrupole moments of these halogen nuclei. For example, proton–proton coupling in CH_3CH_2Cl is unaffected by the presence of a chlorine nucleus; the triplet and quartet are sharp.

3.7 COUPLING OF PROTONS TO OTHER IMPORTANT NUCLEI (^{19}F, D, ^{31}P, ^{29}SI, AND ^{13}C)

3.7.1 Coupling of Protons to ^{19}F

Since ^{19}F has a spin number of 1/2, H—F coupling and H—H coupling obey the same multiplicity rules; in general, the coupling constants for H—F cover a somewhat larger range than those for H—H (Appendix F), and there is more long-range coupling for H—F.

The spectrum of fluoroacetone CH_3—(C=O)—CH_2F, in $CDCl_3$ at 300 MHz (Figure 3.42) shows the CH_3 group as a doublet at δ 2.2 (J = 4.3 Hz) resulting from long-range coupling by the F nucleus. The doublet at δ 4.75 (J = 48 Hz) represents the protons of the CH_2 group coupled to the geminal F nucleus. The ^{19}F nucleus is about 80% as sensitive as the proton and can be readily observed at the appropriate frequency and magnetic field (see Section 6.3)

3.7.2 Coupling of Protons to D

Deuterium (D or ^2H) usually is introduced into a molecule to detect a particular group or to simplify a spectrum. Deuterium has a spin number of 1, a small

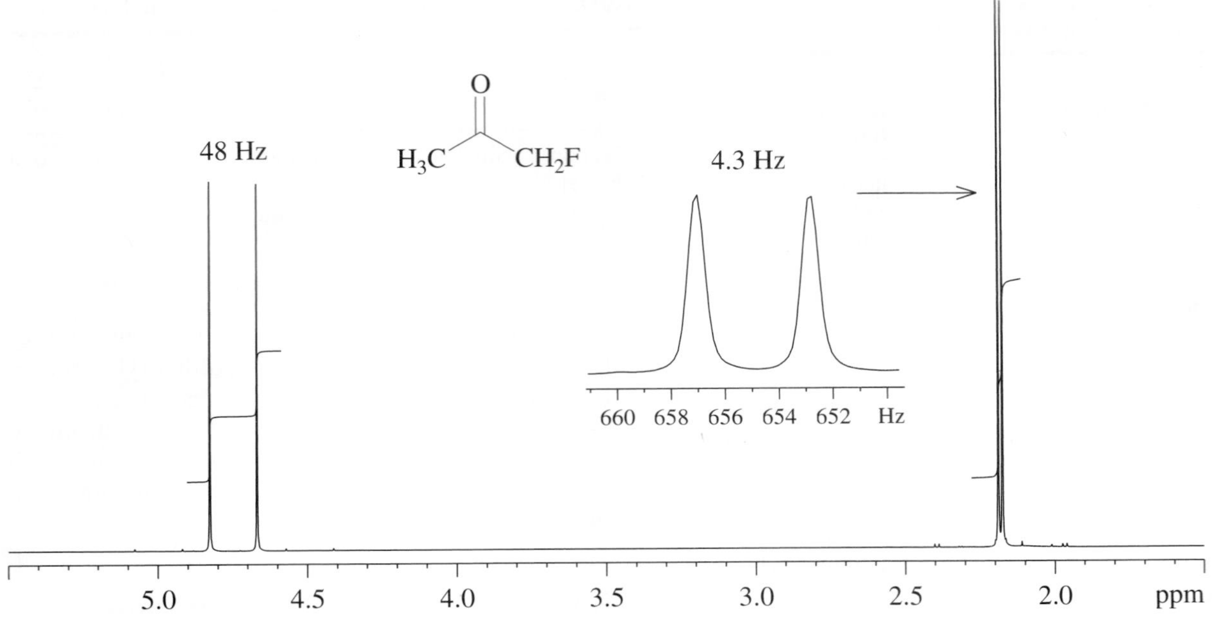

FIGURE 3.42 ^1H spectrum of fluoroacetone in $CDCl_3$ at 300 MHz.

coupling constant with protons and a small electrical quadrupole moment. The ratio of the J values for H—H to those of H—D is about 6.5.

Suppose the protons on the α-carbon atom of a ketone (X and Y contain no protons)

$$\mathrm{X{-}\overset{\gamma}{C}H_2{-}\overset{\beta}{C}H_2{-}\overset{\alpha}{C}H_2{-}\overset{O}{\overset{\|}{C}}{-}Y}$$

were replaced by deuterium to give

$$\mathrm{X{-}\overset{\gamma}{C}H_2{-}\overset{\beta}{C}H_2{-}\overset{\alpha}{C}D_2{-}\overset{O}{\overset{\|}{C}}{-}Y}$$

The spectrum of the undeuterated compound consists of a triplet for the α protons, a quintet for the β protons—assuming equal coupling for all protons—and a triplet for the γ protons. For the deuterated compound, the α-proton absorption would be absent, the β-proton absorption would appear, at modest resolution, as a slightly broadened triplet, and the γ-proton absorption would be unaffected. Actually, at very high resolution, each peak of the β-proton triplet would appear as a very closely spaced quintet ($J_{H-C-C-D}$ ~1 Hz) since $2nI + 1 = 2 \times 2 \times 1 + 1 = 5$, where n is the number of D nuclei coupled to the β protons.

Most deuterated solvents have residual proton impurities in an otherwise completely deuterated molecule; thus, deuterated dimethyl sulfoxide, $(CD_3)_2S{=}O$, contains a few molecules of $CD_2H{-}(S{=}O){-}CD_3$, which show a closely spaced quintet (J ~2 Hz, intensities 1:2:3:2:1) in accordance with $2nI + 1$ (see Appendix G).

Because of the electrical quadrupole moment of D, only broad absorption peaks can be obtained from a spectrum of deuterium nuclei.

3.7.3 Coupling of Protons to ^{31}P

The ^{31}P nucleus has a natural abundance of 100% and a spin number of 1/2 (therefore no electrical quadrupole moment). The multiplicity rules for proton–phosphorus splitting are the same as those for proton–proton splitting. The coupling constants are large (J_{H-P} ~200–700 Hz, and J_{HC-P} is 0.5–20 Hz) (Appendix F) and are observable through at least four bonds. The ^{31}P nucleus can be observed at the appropriate frequency and magnetic field (Chapter 6).

3.7.4 Coupling of Protons to ^{29}Si

The NMR active ^{29}Si isotope has a natural abundance of 4.70% (^{28}Si = 92.28%) and a spin number of 1/2. The value of J ^{29}Si—CH is about 6 Hz. The low-intensity doublet caused by the $^{29}Si{-}CH_3$ coupling can often be seen straddling (±3 Hz) an amplified peak of TMS; the low-intensity $^{13}C{-}H_3$ "satellite" doublet can also be seen at ±59 Hz (Section 3.7.5). ^{29}Si spectra can be obtained at the appropriate frequency and magnetic field (Chapter 6).

3.7.5 Coupling of Protons to ^{13}C

The isotope ^{13}C has a natural abundance relative to ^{12}C of 1.1% and a spin number of 1/2. Protons directly attached to ^{13}C are split into a doublet with a large coupling constant, about 115–270 Hz for $^{13}C—H$. The $CH_3—CH_2$ group, for example, is predominantly $^{12}CH_3—^{12}CH_2$ but contains a small amount of $^{13}CH_3—^{12}CH_2$ and of $^{12}CH_3—^{13}CH_2$. Thus, the $^{13}CH_3$ protons are split into a doublet by ^{13}C ($J \sim 120$ Hz), and each peak of the doublet is split into a triplet by the $^{12}CH_2$ protons ($J \sim 7$ Hz) as shown below. These "^{13}C satellite" peaks are small because of the small number of molecules containing the $^{13}CH_3$ group and can usually be seen disposed on both sides of a large $^{12}CH_3$ peak (e.g., the large $^{12}CH_3$ triplet shown below).

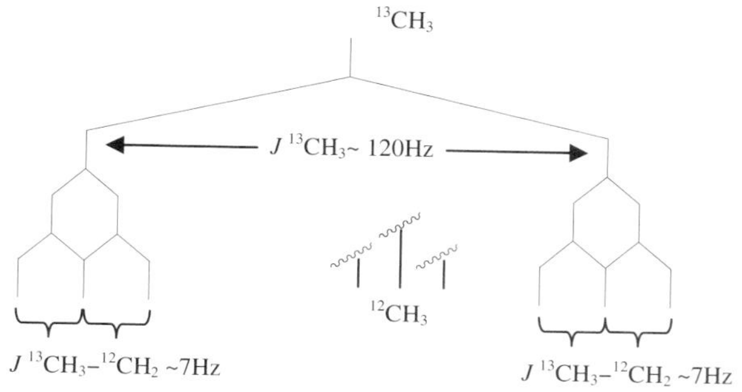

$^{13}CH_3$

$J\ ^{13}CH_3$~ 120Hz

$^{12}CH_3$

$J\ ^{13}CH_3–^{12}CH_2$ ~7Hz

$J\ ^{13}CH_3–^{12}CH_2$ ~7Hz

3.8 CHEMICAL SHIFT EQUIVALENCE

The concept of chemical shift equivalence is central to NMR spectrometry. Chemical-shift equivalent (isochronous) nuclei comprise a *set* within a *spin system* (Pople notation, Section 3.5).

The immediate question is: are selected nuclei in a molecule chemical shift equivalent, or are they not? If they are, they are placed in the *same* set. The answer can be framed as succinctly as the question: Nuclei are chemical shift *equivalent if they are interchangeable through any symmetry operation or by a rapid process.* This broad definition assumes an achiral environment (solvent or reagent) in the NMR experiment; the common solvents are achiral (Section 3.17).

We deal first with symmetry operations and later with rapid processes (Section 3.8.3).

3.8.1 Determination of Chemical Shift Equivalence by Interchange Through Symmetry Operations

There are three symmetry operations each involving a symmetry element: *rotation about a simple axis of symmetry (C_n), reflection through a plane of symmetry (σ), and inversion through a center of symmetry (i).** More rigorously, symmetry operations may be described under two headings: C_n and S_n. The latter is *rotation around an alternating axis of symmetry.* It turns out that S_1 is the same as σ, S_2 is the same as i, and higher subscripts for S_n are rare. We shall use C_n, σ, and i. The subscripts denote the number of such rotations required to make a 360° rotation. Thus, C_1 is a 360° rotation, C_2 is a 180° rotation, etc. The designation S_1 requires a 360° rotation followed by a reflection through the plane at a right angle to the axis. S_2 requires a 180° rotation followed by a reflection, and so forth.

Before we describe the three common symmetry operations to effect interchange of nuclei within a molecule, we raise another question: How do we know whether or not a symmetry operation has indeed resulted in interchange? Simply enough: Build two models of the molecule-one before the operation, one after the operation. If they are indistinguishable, interchange of the nuclei in question has occurred; or to use another term, they are *superposable.* Then graduate to "three-dimensional" drawings, or even better, three-dimensional mental images.

Of course, an "identity" operation—a 360° rotation around a symmetry axis—is not valid. The *symmetry elements* are the axis, plane, and center.

* The symmetry operation must relate to the entire molecule.

3.8.1.1 Interchange by Rotation Around a Simple Axis of Symmetry (C_n) The protons in dichloromethane (Figure 3.43, structure a) interchange by a 180° rotation around a simple axis of symmetry in the plane of the paper. This C_2 interchange shows that the protons are homotopic; they can also interchange by a σ operation (next section) through a vertical plane at right angles to the page, but the C_2 operation takes precedence. The homotopic protons are chemical-shift equivalent in both achiral and chiral environments, that is, in solvents and in contact with reagents. Drawings and models of the molecule are indistinguishable by inspection before and after the operation. During the C_n operation, the molecule is treated as a rigid object; no change in bond angles is permitted. One half of the molecule interchanges with the other.

3.8.1.2 Interchange by Reflection Through a Plane of Symmetry (σ) A plane of symmetry exists if one half of the molecule is the mirror image of the other half. The protons in chlorofluoromethane (Figure 3.43, structure b) are interchanged only by reflection at a right angle through a vertical plane of symmetry at a right angle to the page; there is no other symmetry element. The protons are mirror images (enantiotopes) of each other and are chemical-shift equivalent only in an achiral solvent or in contact with an achiral reagent. Chiral solvents or chiral reagents can distinguish between a pair of enantiotopic protons in an NMR experiment (Sections 3.17).

The methylene protons of propanoic acid (Figure 3.43, structure c) are exchangeable by reflection

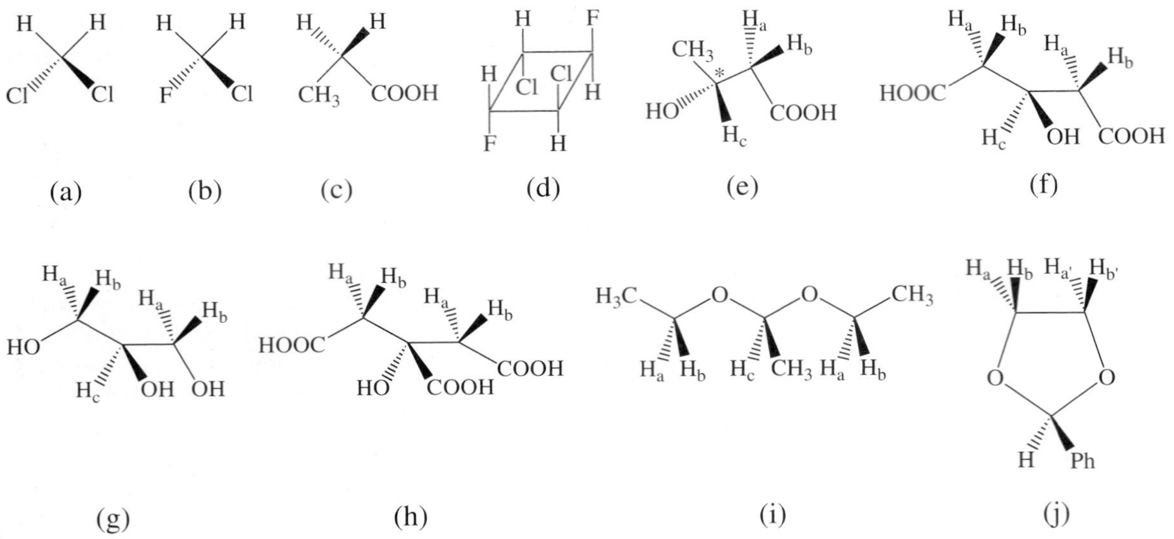

(a) (b) (c) (d) (e) (f)

(g) (h) (i) (j)

FIGURE 3.43 Examples of proton interchange by symmetry operations:
a. Homotopic protons by C_2
b, c. Enantiotopic protons by σ
d. Enantiotopic protons by i
e. Diastereotopic protons in a chiral molecule (* denotes chiral center) (see Section 3.12.1)
f, g, h , i , j Diastereotopic protons in achiral molecules (f is 3-hydroxyglutaric acid; g is glycerol; h is citric acid; i is diethyl acetal; j is a cyclic acetal of benzaldehyde, 2-phenyl-1,3-dioxolane).
The "prime" marks are explained in Section 3.9.

through the plane of symmetry in the plane of the paper. There are no other symmetry elements. The protons are enantiotopes of each other and have the same chemical shift only in an achiral environment. Drawings and models of the molecule are indistinguishable by inspection before and after the operation.

3.8.1.3 Interchange by Inversion Through a Center of Symmetry (i) A center of symmetry exists if a line drawn from each nucleus through the selected center encounters the same nucleus or group at the same distance on the other side of the center (Figure 3.43, structure d).

The inversion is equivalent to an S_2 operation, which consists of a 180° rotation around an alternating axis of symmetry followed by reflection through a symmetry plane at a right angle to the axis. Therefore, the inversion interchange involves a reflection, so that, strictly speaking, the interchangeable nuclei or groups must be *mirror images* of one other. This distinction is not necessary for nuclei or for achiral groups, but it is necessary for interchange of chiral groups. Thus, an R group and an S group are interchangeable, but two R groups (or two S groups) are not. Of course, the same restrictions hold for interchange through a plane of symmetry (Section 3.8.1.2). In both operations, the interchanged nuclei or groups are enantiotopic to each other and are chemical shift equivalent only in an achiral environment.

3.8.1.4 No Interchangeability by a Symmetry Operation If geminal protons (CH_2) in a molecule cannot be interchanged through a symmetry element, those protons are *diastereotopic* to one another; each has a different chemical shift-except for coincidental overlap. The diastereotopic geminal protons couple with each other (through two bonds). In principle, each geminal proton should show different coupling constants with other neighboring nuclei, although the difference may not always be detectable. A CH_2 group consisting of a pair of diastereotopic protons is shown in Figure 3.43, structure e; the chiral center is shown by an asterisk, but a chiral center is not necessary for the occurrence of diastereotopic protons (see Figure 3.43, structures f–j). The achiral molecule, 3-hydroxyglutaric acid, has a plane of symmetry, perpendicular to the page through the middle carbon atom, through which the two H_a protons interchange and the two H_b protons interchange, as enantiotopes. Since there is no plane of symmetry passing between the protons of each CH_2 group, protons a and b of each CH_2 group are diastereotopic. An idealized first-order spectrum for the diastereotopic protons of 3-hydroxyglutaric acid is diagrammed as follows, but in practice most of these types of compounds even at high resolution, would show partially resolved peaks, because $\Delta\nu/J$ ratios are usually small.

H_a H_b

J_{ab} J_{ba}

J_{ac} J_{bc}

The following similar achiral molecules contain diastereotopic methylene protons: 3-hydroxyglutaric acid, glycerol, citric acid, diethyl acetal, and a cyclic acetal (respectively, Figure 3.43, structures f, g, h, i, and j); structure j involves the additional concept of magnetic equivalence (Section 3.9).

From the above discussion, diastereotopic protons cannot be placed in the same set since they are not chemical-shift equivalent. However it is not uncommon for diastereotopic protons to *appear* to be chemical-shift equivalent in a given magnetic field in a particular solvent. Such *accidental* chemical shift equivalence can usually be detected by using an instrument with a higher magnetic field or by changing solvents. For an example of diastereotopic methyl groups, see Figure 3.55.

Diastereotopic protons (or other ligands) must be in constitutionally equivalent locations; that is, they cannot differ in *connectivity*. For example in structure e of Figure 3.43, the geminal protons have the same connectivity but differ in the sense that they are not interchangeable; thus they are diastereotopic. On the other hand, the proton on C-3 has a different connectivity from those on C-2, and the term, "diastereotopic" does not apply.

Students are familiar with the terms applied to relationships between stereoisomeric molecules: *homomeric* molecules (superposable molecules), *enantiomeric* molecules (nonsuperposable mirror images), and *diastereomeric* molecules (stereoisomers that are not mirror images of one another). These familiar terms are parallel to the terms that we have introduced above: homotopic, enantiotopic, and diastereotopic, which are applied to nuclei or groups within the molecule.

3.8.2 Determination of Chemical Shift Equivalence by Tagging (or Substitution)

As noted in Section 3.8.1, a clear understanding of the concepts of chemical shift equivalence and its relationship to symmetry elements and symmetry operations is essential to interpretation of NMR spectra. The question of the chemical shift equivalence of specific nuclei can also be approached by a "tagging" or substitution operation* in which two identical drawings of the same compound are made; one hydrogen atom in one of the drawings is tagged (or substituted by a different atom), and the other hydrogen atom in the second drawing is also tagged (or substituted) in the same manner. The resulting drawings (or models) are related to each other as homomers, enantiomers, or diastereomers. The H atoms are, respectively, homotopic, enantiotopic, or diastereotopic. The examples in Figure 3.44 illustrate the process.

In the first example, the models are superposable (i.e., homomers); in the second, nonsuperposable mirror images (i.e., enantiomers); in the third, not mirror images (i.e., diastereomers). Note that the tags are permanent; i.e., H and Ⓗ are different kinds of atoms. Alternatively, one proton in each structure may be replaced by Z, representing any nucleus not present in the molecule.

* Ault, A. (1974). *J. Chem. Educ.*, **51**, 729.

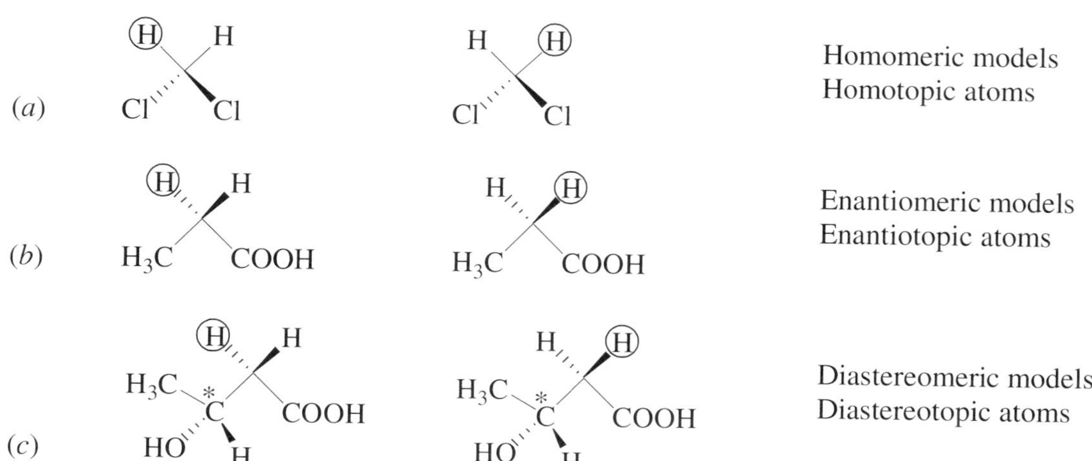

FIGURE 3.44 Tagged molecules: (A) equivalent molecules, (B) enantiomers, (C) diastereomers.

3.8.3 Chemical Shift Equivalence by Rapid Interconversion of Structures

If chemical structures can interconvert, the result depends on temperature, catalyst, solvent, and concentration. We assume a given concentration and absence of catalyst, and we treat four systems.

3.8.3.1 Keto-Enol Interconversion The tautomeric interconversion of acetylacetone (Figure 3.45) at room temperature is slow enough that the absorption peaks of both forms can be observed—i.e., there are two spectra. The equilibrium keto/enol ratio can be determined from the relative areas of the keto and enol CH_3 peaks, as shown. At higher temperatures the interconversion rate will be increased so that a single "averaged" spectrum will be obtained. Chemical shift equivalence for all of the interconverting protons has now been achieved. Note that the NMR time scale is of the same order of magnitude as the chemical shift separation of interchanging signals expressed in hertz, i.e., about $10^1 - 10^3$ Hz. Processes occurring faster than this will lead to averaged signals. Note also that the enolic OH proton peak is deshielded relative to the OH proton of alcohols because the enolic form is strongly stabilized by intramolecular hydrogen bonding.

3.8.3.2 Interconversion Around a "Partial Double Bond" (Restricted Rotation) At room temperature, a neat sample of dimethylformamide shows two CH_3 peaks because the rate of rotation around the hindered "partial double bond" is slow. At ~123°C, the rate of exchange of the two CH_3 groups is rapid enough so that the two peaks merge.

δ 2.85

δ 2.94

3.8.3.3 Interconversion Around the Single Bonds of Rings Cyclohexane at room temperature exists as rapidly interconverting, superposable chair forms.

An axial proton becomes an equatorial proton and vice versa in the interconverting structures, and the spectrum consists of a single "averaged" peak. As the temperature is lowered, the peak broadens and at a sufficiently low temperature two peaks appear-one for the axial protons, one for the equatorial protons. In other words, at room temperature, the axial and equatorial protons are chemical-shift equivalent by rapid interchange. At very low temperatures, they are not chemical-shift equivalent; in fact, in each "frozen" chair form, the protons of each CH_2 group are diastereotopic

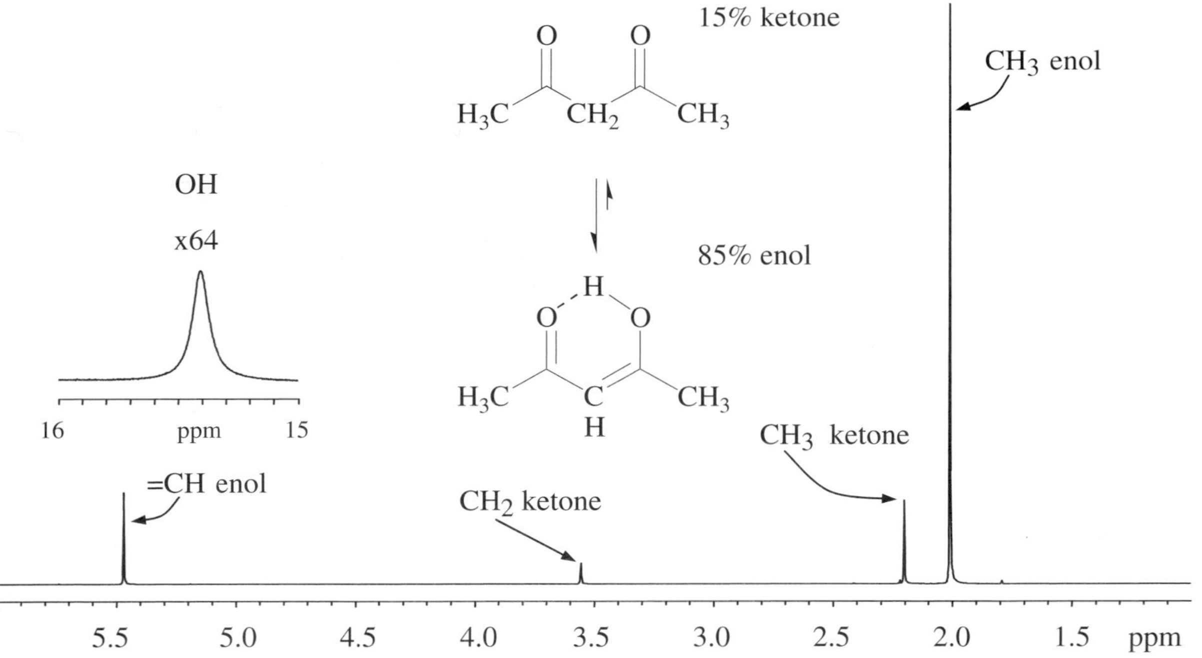

FIGURE 3.45 Acetylacetone in $CDCl_3$ at 300 MHz and 32°C. The enol-keto ratio was measured by integration of the CH_3 peaks.

pairs, but at room temperature, the rate of chair interconversion is sufficiently high to average the chemical shifts of these geminal protons.

Methylcyclohexane exists at room temperature as a rapidly interconverting mixture of axial and equatorial conformers. These conformers are not superposable, and at low temperatures a spectrum of each conformer exists.

In a fused cyclohexane ring, such as those of steroids, the rings are "frozen" at room temperature and the axial and equatorial protons of each CH_2 group are not chemical-shift equivalent.

3.8.3.4 Interconversion Around the Single Bonds of Chains Chemical shift equivalence of protons on a CH_3 group results from rapid rotation around a carbon-carbon single bond even in the absence of a symmetry element. Figure 3.46a shows Newman projections of the three staggered rotamers of a molecule containing a methyl group attached to another sp^3 carbon atom having four different substituents, that is, a chiral center. In any single rotamer, none of the CH_3 protons can be interchanged by a symmetry operation. However, the protons are rapidly changing position. The time spent in any one rotamer is short (~10^{-6} s), because the energy barrier for rotation around a C—C single bond is small. The chemical shift of the methyl group is an average of the shifts of the three protons. In other words, each proton can be interchanged with the others by a rapid rotational operation. Thus, without the labels on the protons, the rotamers are indistinguishable.

In the same way the three rotating methyl groups of a *t*-butyl group are chemical shift equivalent except for rare steric hindrance. Both the methyl group and the *t*-butyl group are described as "symmetry tops."

The staggered rotamers of 1-bromo-2-chloroethane (Figure 3.46b) are distinguishable. However, in the *anti* rotamer, H_a and H_b are chemical-shift equivalent (enantiotopic) by interchange through a plane of symmetry, as are H_c and H_d; thus, there are two sets of enantiotopic protons. In neither of the gauche rotamers is there a symmetry element, but H_a and H_b, and H_c and H_d, are chemical-shift equivalent by rapid rotational interchange between two enantiomeric rotamers. Now we have one chemical shift for H_a and H_b in the *anti* rotamer, and a different chemical shift for H_a and H_b in the *gauche* rotamers. By rapid averaging of these two chemical shifts, we obtain a single chemical shift (i.e., chemical shift equivalence)

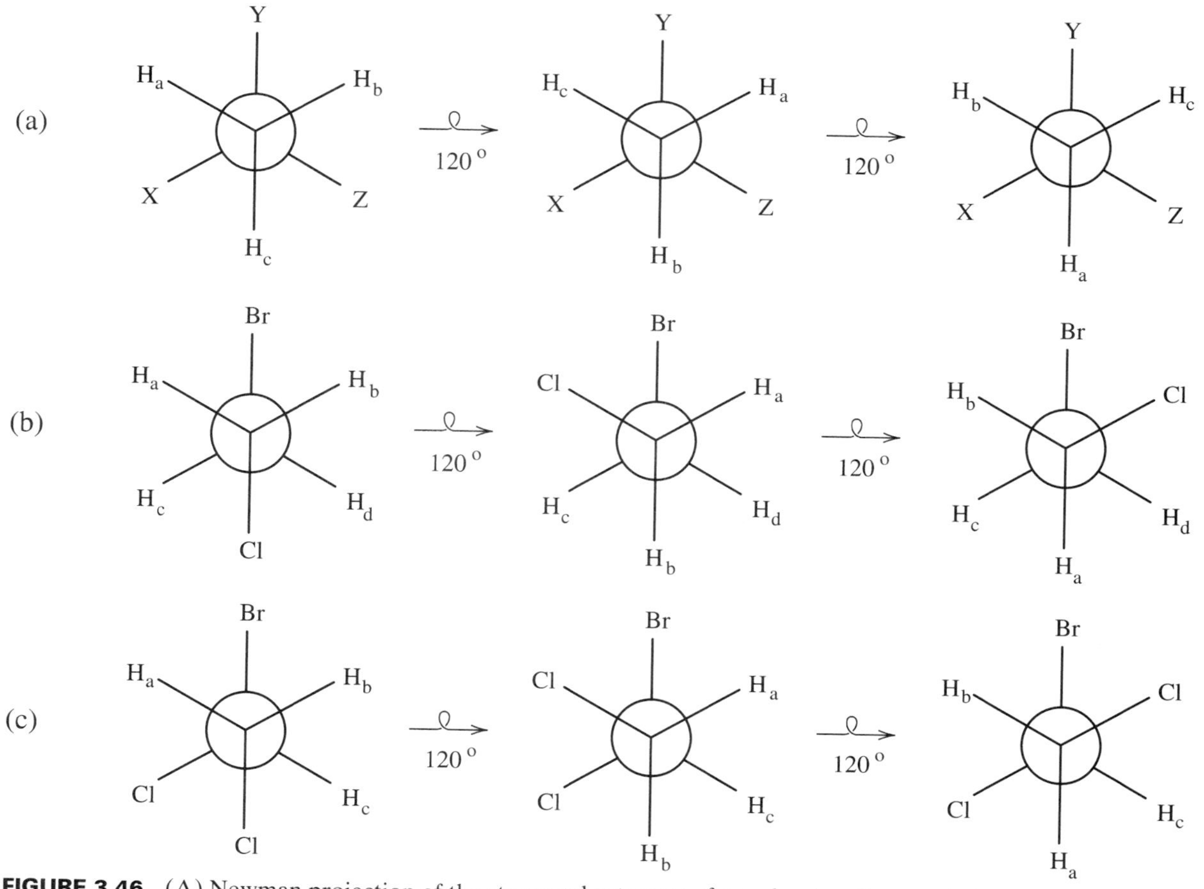

FIGURE 3.46 (A) Newman projection of the staggered rotamers of a molecule with a methyl group attached to a chiral sp^3 carbon atom. (B) 1-Bromo-1,2-dichloroethane. (C) 1-bromo-2-chloroethane.

for H_a and H_b, and of course for H_c and H_d. Strictly, the system is AA′XX′ but, as explained in Section 3.9, it is treated as A_2X_2. In general, if protons can be interchanged by a symmetry operation (through a plane of symmetry) in one of the rotamers, they are also chemical-shift equivalent (enantiotopic) by rapid rotational interchange.*

Consider a methylene group next to a chiral center, as in 1-bromo-1,2-dichloroethane (Figure 3.46c). Protons H_a and H_b are not chemical-shift equivalent since they cannot be interchanged by a symmetry operation in any conformation; the molecule has no simple axis, plane, center, or alternating axis of symmetry. Although there is a rapid rotation around the carbon–carbon single bond, the CH_2 protons are not interchangeable by a rotational operation; the averaged chemical shifts of H_a and H_b are not identical. An observer can detect the difference before and after rotating the methylene group: the protons in each rotamer are diastereotopic. The system is ABX.

3.9 MAGNETIC EQUIVALENCE (SPIN-COUPLING EQUIVALENCE)†

In addition to the above requirements for a first-order spin system, we now consider the concept of "magnetic equivalence" (or spin-coupling equivalence) by comparison with the concept of chemical-shift equivalence.

If two protons in the same set (i.e., chemical-shift-equivalent protons in the same multiplet) couple equally to every other proton in the spin system, they are also "magnetically equivalent," and the usual Pople notations apply: A_2, B_2, X_2 etc. However, if two protons in a set are not magnetically equivalent, the following notations apply: AA′, BB′, XX′, etc. To rephrase: Two chemical-shift-equivalent protons are magnetically equivalent if they are symmetrically disposed with respect to each proton in the spin system. Obviously magnetic equivalence presupposes chemical-shift equivalence. In other words, do not test for magnetic equivalence unless the two protons in question are chemical-shift equivalent.

The common occurrence of magnetic nonequivalence in aromatic rings depends on the number, kind, and distribution of the substituents.

Consider the protons in *p*-chloronitrobenzene (see Figure 3.47). There is an axis of symmetry (through the substituents) that provides two sets of chemical-shift-equivalent protons presumably A_2 and X_2. However, neither the two A protons nor the two X protons A are magnetically equivalent, and the correct labeling is AA′ and XX′. This is not a first-order spin system and is written in the Pople notation as AA′XX′. These multiplet patterns do not conform to first-order intensity patterns in the Pascal triangle (Figure 3.32) nor do the distances (in Hz) between the peaks correspond to coupling constants. Spectra such as these do not become first-order spectra regardless of the strength of the magnetic field. The choice of designating the spectrum as AA′XX′ or AA′BB′ depends quite arbitrarily on the effect of the substituents on the chemical shifts. At lower magnetic field these spectra become "deceptively simple." In the 60 MHz era, students learned to recognize two tight clusters in aromatic region as indicative of *para* disubstitution. The pattern of *o*-dichlorobenzene shows a pattern somewhat resembling that of *p*-chloronitrobenzene and for the same reasons (see Figure 3.48). Heteroaromatic rings behave similarly.

* The discussion of rotamers is taken in part from Silverstein, R.M., and LaLonde, R.T. (1980). *J. Chem. Educ.*, **57**, 343

† For excellent introductions to stereoisomeric relationships of groups in molecules, see Mislow, K., and Raban, M. (1967). In *Topics in Stereochemistry,* Vol. I. page 1; Jennings, W.B. (1975) *Chem. Rev.*, **75**, 307. These authors prefer the term "spin-coupling equivalence."

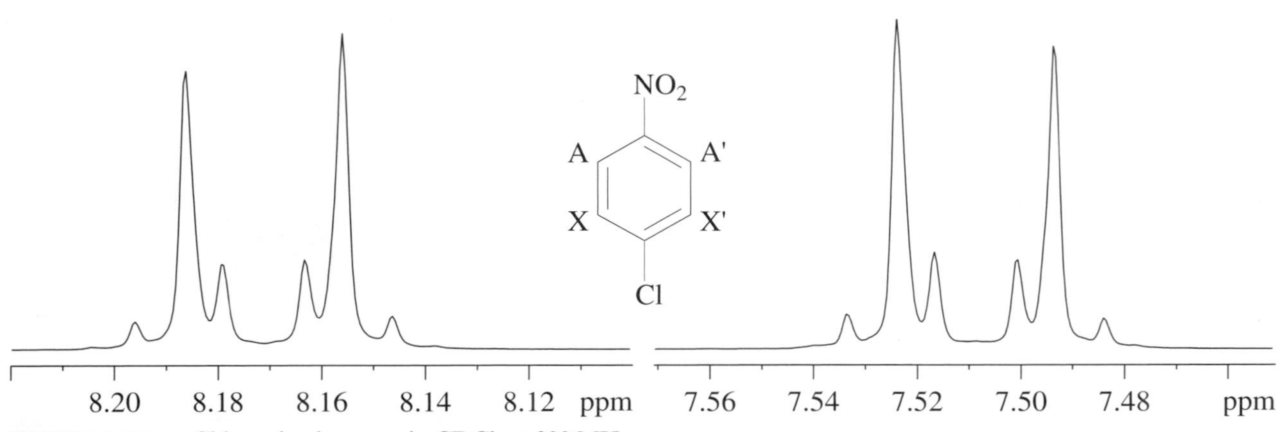

FIGURE 3.47 *p*-Chloronitrobenzene in $CDCl_3$ at 300 MHz.

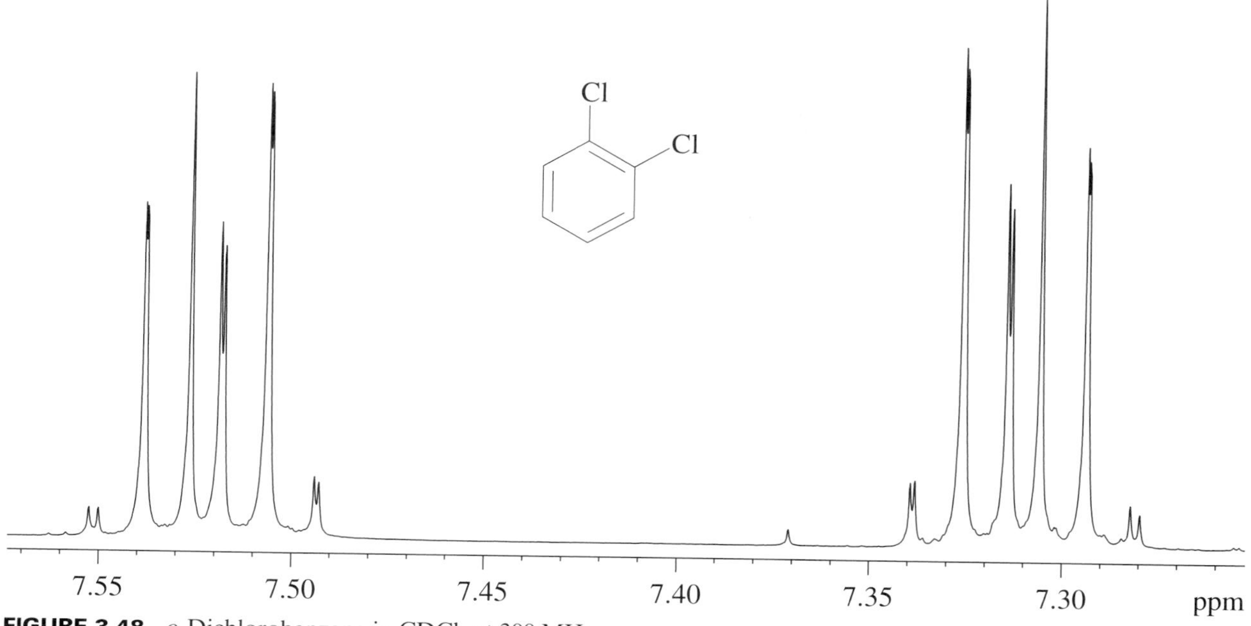

FIGURE 3.48 *o*-Dichlorobenzene in $CDCl_3$ at 300 MHz.

Three isomeric difluoroethylenes furnish additional examples of chemical-shift-equivalent nuclei that are not magnetically equivalent. The systems are AA′XX′.

$(F_{X'})(F_{X'})C{=}C(H_{A'})(H_A)$ $\quad$ $(F_X)(F_A)C{=}C(H_{X'})(H_{A'})$ $\quad$ $(F_A)(F_X)C{=}C(H_{A'})(H_{X'})$

In each system, the protons comprise a set and the fluorine nuclei comprise a set (of chemical-shift-equivalent nuclei), but since the nuclei in each set are not magnetically equivalent, the spectra are not first order. Both proton and fluorine spectra are readily available (see Chapter 6).

With some mastery of chemical-shift equivalence and magnetic equivalence, we return to structure j in Figure 3.43 whose spectrum is given in Figure 3.49. Structure

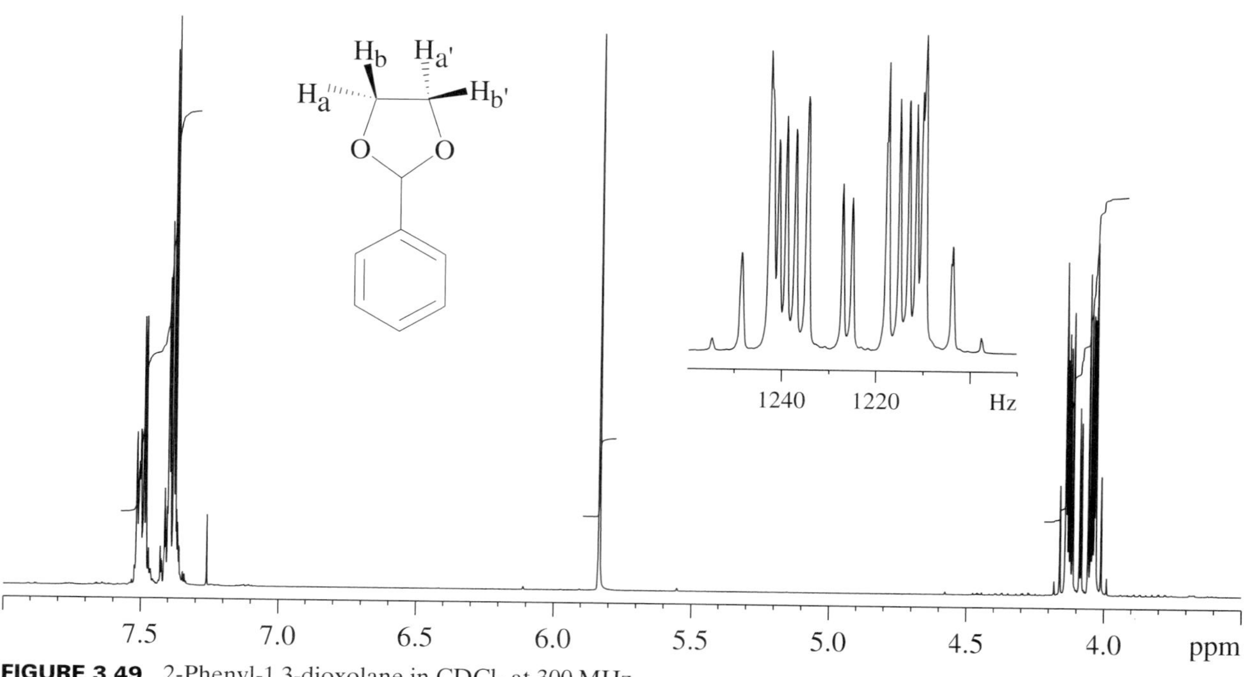

FIGURE 3.49 2-Phenyl-1,3-dioxolane in $CDCl_3$ at 300 MHz.

j resembles structures f–i in that the protons in each CH_2 group are diastereotopic since there is no plane in the plane of the paper. Structure j differs, however, in that (1) the CH_2 groups are in a ring with limited flexibility and (2) the CH_2 groups couple with one another. The question of magnetic equivalence arises: Do the protons of the "a" set which are obviously chemical-shift equivalent, couple equally with one of the protons of the "b" set:. The answer is given by the "prime" placed on one of the "a" and "b" protons. The spin system is therefore AA′BB′, and the spectrum is quite similar in complexity to the spectrum of *o*-dichlorobenzene (see Figure 3.48).

The open-chain, conformationally mobile compounds of the type:

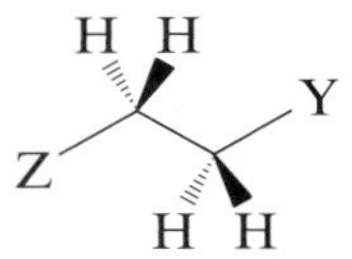

H H
Z — — Y
H H

consist of two different sets of proton coupled to each other. The groups Z and Y contain no chiral element and no protons that couple to the two sets shown; Z and Y polarities determine the difference between shifts of the sets (i.e., $\Delta\nu$). If this molecule were rigid, it would be regarded as an AA′XX′ or AA′BB′ system depending on the magnitude of $\Delta\nu/J$, and at a low enough temperature, this description would be accurate. However, for such a molecule at room temperature—barring large conformational preferences—the J values are quite similar, and, in practice, spectra resembling A_2X_2 or A_2B_2 spectra result. The "weakly coupled" A_2X_2 system would show two triplets, and a "strongly coupled" A_2B_2 system would show a complex, higher order spectrum. We shall treat these conformationally mobile systems as A_2X_2 or A_2B_2 rather than AA′XX′ or AA′BB′.

3.10 AMX, ABX, AND ABC RIGID SYSTEMS WITH THREE COUPLING CONSTANTS

In Section 3.5 we discussed the simple, first-order system AX. As the ratio $\Delta\nu/J$ decreases, the two doublets approach each other with a characteristic distortion of peak heights to give an AB system, but no additional peaks appear. However, as an A_2X system—a triplet and a doublet—develops into an A_2B system, additional peaks do appear, and the system presents a higher order spectrum; the J values no longer coincide with the measured differences between peaks [see simulated spectra in Bovey (1988) and in Wiberg and Nist (1962)].

Having considered these systems, we can now examine the systems AMX, ABX, and ABC, starting with a rigid system. Styrene, whose rigid vinylic group furnishes an AMX first-order spectrum at 600 MHz, is a good starting point (see Figure 3.50).

Correlation of the structure of a compound with its spectrum is indeed a gentle approach compared with interpreting the spectrum of an unknown compound, but it is instructive.

Since in styrene there is free rotation around the substituent bond, there are two symmetry planes through the molecule. In one conformation the vinyl group and the ring are coplanar in the plane of the page with all of the protons in the symmetry plane, hence not interchangeable. In the other conformation the vinyl group and the benzene ring are perpendicular to one another, and the symmetry plane is perpendicular to the plane of the page. Again the vinyl protons are in the symmetry plane. The following data in the vinyl group are relevant;

There are three sets in the vinyl system, AMX, each set consisting of one proton.

Chemical shifts are: X = δ 6.73, M = δ 5.75, A = δ 5.25.

$\Delta\nu_{XM}$ = 588 Hz, $\Delta\nu_{XA}$ = 888 Hz, $\Delta\nu_{AM}$ = 300 Hz.

J_{XM} = 17 Hz, J_{XA} = 11 Hz, J_{AM} = 1.0 Hz.

The $\Delta\nu/J$ ratios are: XM = 35, XA = 88, MA = 300.

Each set is a doublet of doublets (note the stick diagrams in Figure 3.50). The X proton is the most strongly deshielded by the aromatic ring, and the A proton least so. The X proton is coupled *trans* across the double bond to the M proton with the largest coupling constant, and *cis* across the double bond to the A proton with a slightly smaller coupling constant. The result is the doublet of doublets centered at δ 6.73 with two large coupling constants.

The M proton is, of course, coupled to the X proton and is coupled geminally to the A proton with a very small coupling constant. The result is a doublet of doublets centered at δ 5.75 with one large and one very small coupling constant.

The A proton is coupled *cis* across the double bond to the X proton with the slightly smaller coupling constant (as compared with the *trans* coupling). The A proton is coupled geminally to the M proton with the very small coupling constant mentioned above. The result is a doublet of doublets centered at δ 5.25 with the large coupling constant (slightly smaller than the *trans* coupling) and the very small geminal coupling constant.

In the conformation of the molecule with the perpendicular plane of symmetry, the *ortho* protons of the ring system are interchangeable with each other as are the *meta* protons. However, since the protons in neither set are magnetically equivalent, the spin system is AA′BB′C. It is a higher-order system. At 600 MHz, it is possible to assign the chemical shifts (from left to right): δ 7.42, δ 7.33, and δ 7.26.

The ratio of the integrals (2:2:1:1:1:1 from left to right) identifies the δ 7.26 proton as *para*. The *ortho* protons are the most deshielded because of the diamagnetic

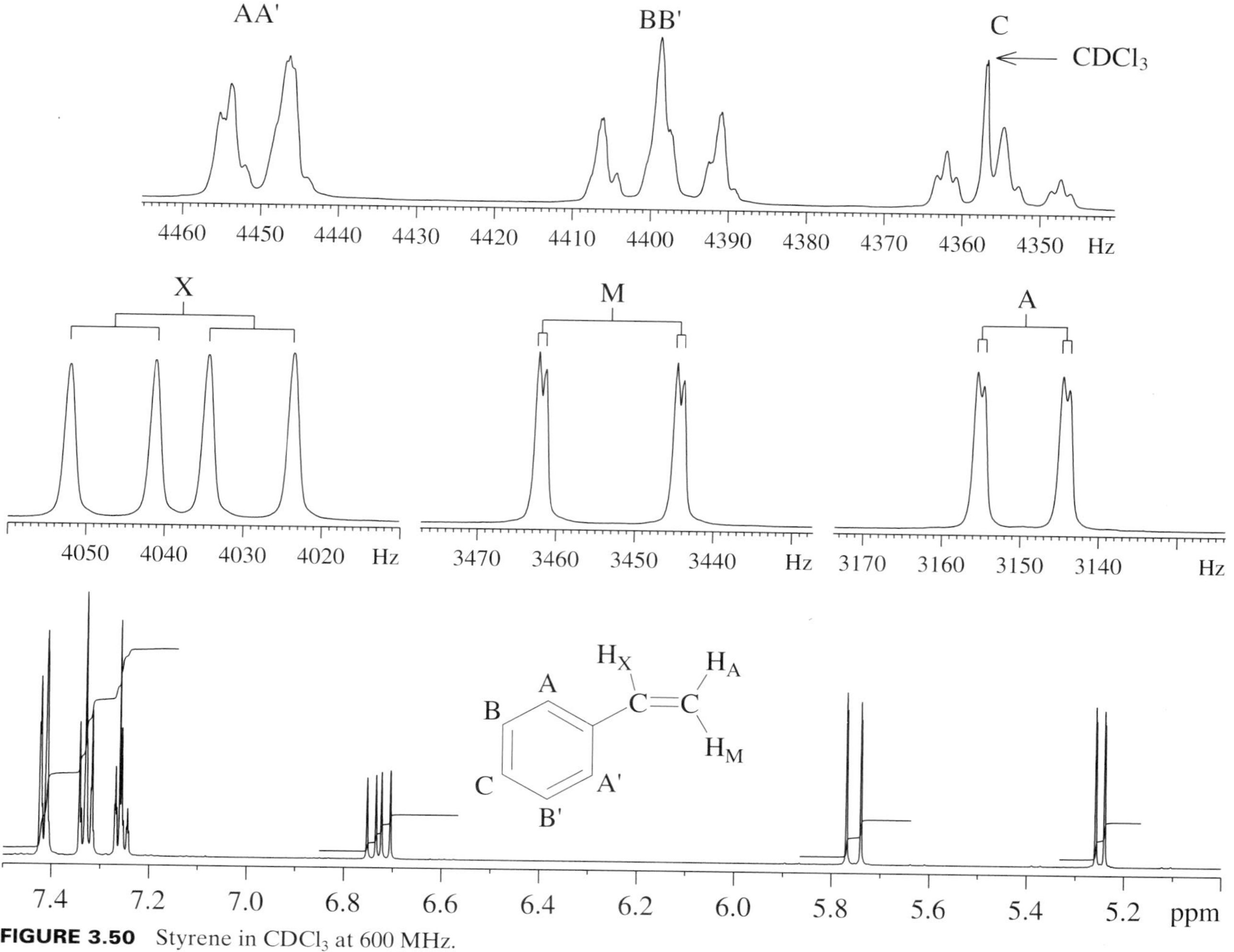

FIGURE 3.50 Styrene in $CDCl_3$ at 600 MHz.

effect of the vinyl double bond (see Figure 3.21 for the similar but stronger effect of the C=O bond).

3.11 CONFORMATIONALLY MOBILE, OPEN-CHAIN SYSTEMS. VIRTUAL COUPLING

This Section is limited to some of the more common spin systems and provides examples of pitfalls that bedevil students.

3.11. I. Unsymmetrical chains

3.11.1.1 1-Nitropropane As mentioned in Section 3.9, most open-chain compounds—barring severe steric hindrance—are conformationally mobile at room temperature; coupling constants in each set average out and become essentially chemically shift equivalent. Thus a 300 MHz, room-temperature spectrum of 1-nitropropane is described as $A_3M_2X_2$ rather than $A_3MM'XX'$, and first-order rules apply (see Figure 3.51).

The X_2 protons are strongly deshielded by the NO_2 group, the M_2 protons less so, and the A_3 protons very slightly. There are two coupling constants, J_{AM} and J_{MX} that are very similar but not exactly equal. In fact, at 300 MHz, the M_2 absorption is a deceptively simple, slightly broadened sextet ($n_A + n_X + 1 = 6$). At sufficient resolution, 12 peaks are possible: $(n_A + 1)(n_X + 1) = 12$.

The A_3 and X_2 absorptions are triplets with slightly different coupling constants. The system is described as weakly coupled, and we can justify mentally cleaving the system for analysis.

3.11.1.2 1-Hexanol In contrast, consider the 300 and 600 MHz spectra of 1-hexanol (Figure 3.52). At first glance at the 300 MHz spectrum, the three-proton triplet at δ 0.87 seems odd for the CH_3 group since the peak intensities deviate from the first-order ratios of 1:2:1. Furthermore, the "filled-in" appearance of the expanded set is obvious despite a reasonable $\Delta\nu/J$ value of 13 for the CH_2—CH_3 groups.

The problem is that the B_2, C_2, and D_2 methylene groups are strongly coupled to one another; they

1320
1300
1280 Hz
630 620 610 600 590 Hz
330 320 310 300 290 Hz
A3 M2 X2
CH3–CH2–CH2–NO2
4.0
3.5
3.0
2.5
2.0
1.5
1.0 ppm

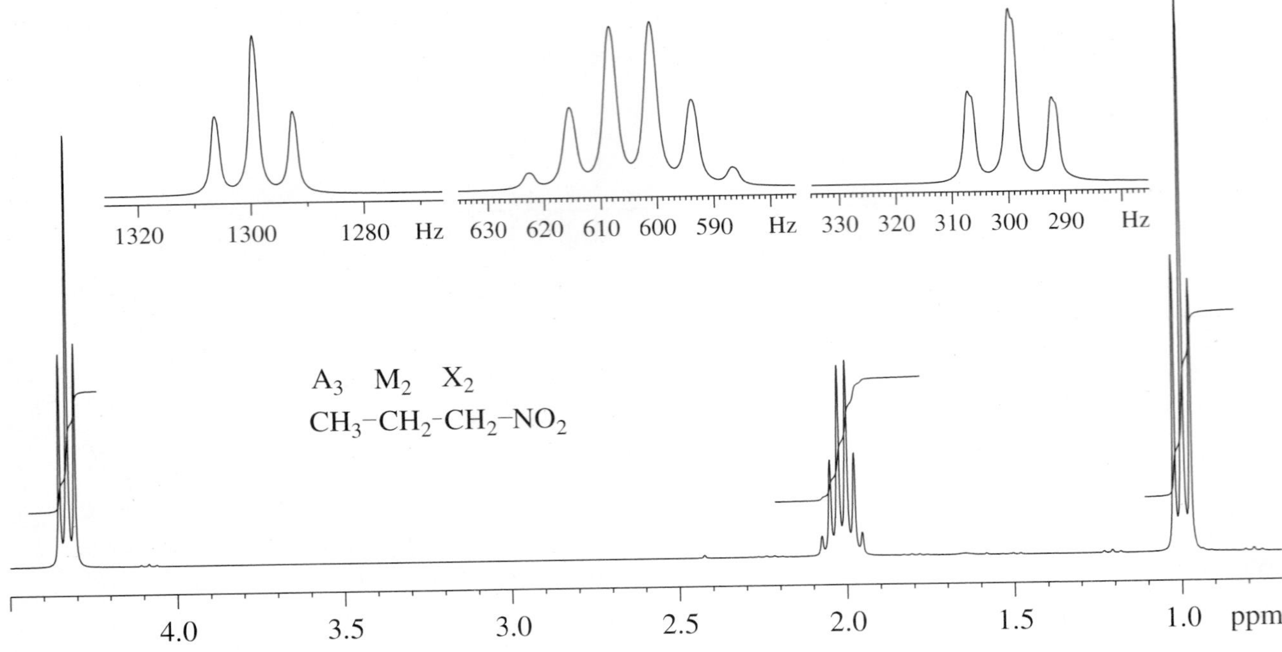

FIGURE 3.51 1-Nitropropane in $CDCl_3$ at 300 MHz.

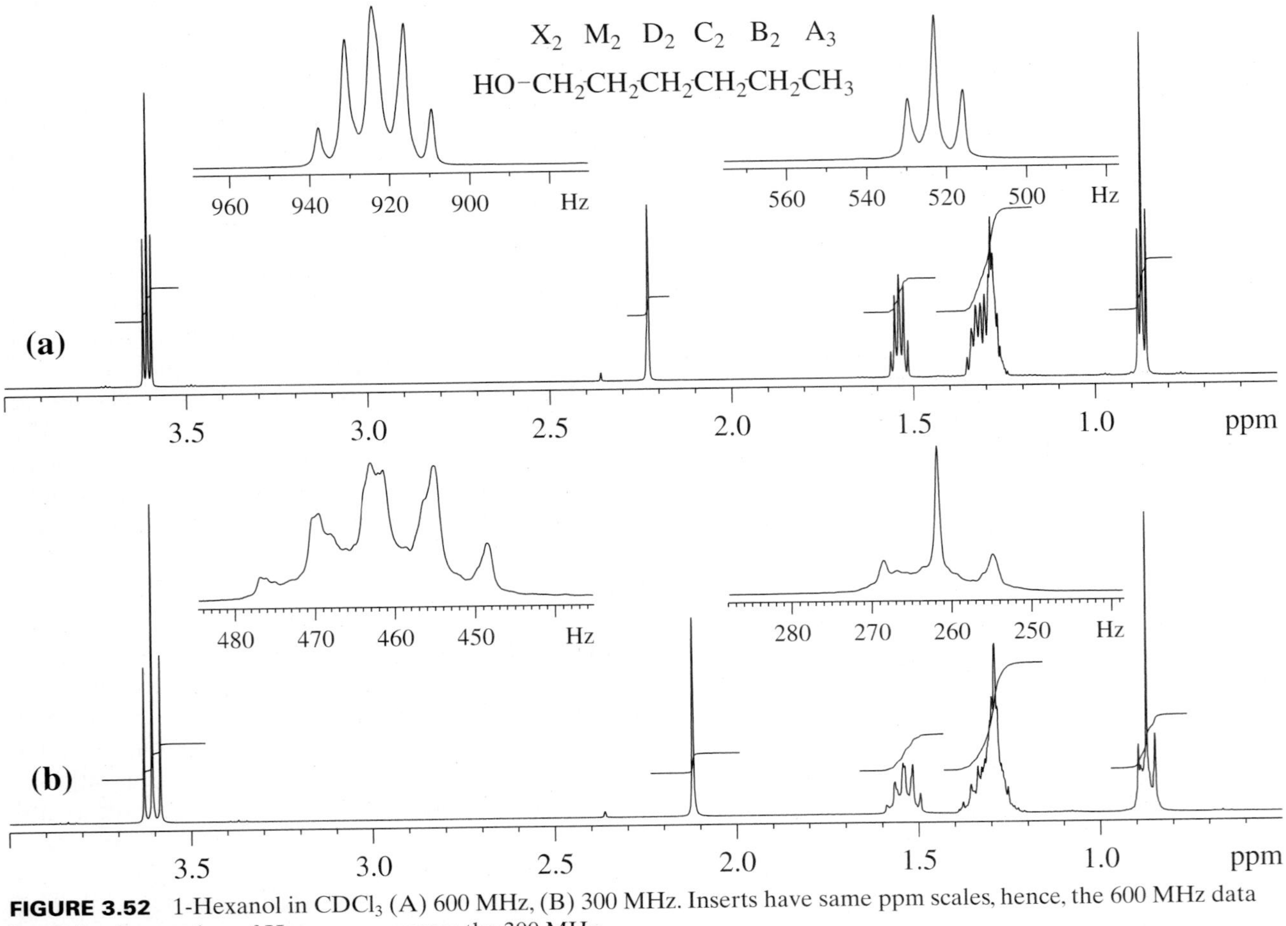

X2 M2 D2 C2 B2 A3
HO–CH2CH2CH2CH2CH2CH3
960 940 920 900 Hz
560 540 520 500 Hz
(a)
3.5 3.0 2.5 2.0 1.5 1.0 ppm
480 470 460 450 Hz
280 270 260 250 Hz
(b)
3.5 3.0 2.5 2.0 1.5 1.0 ppm

FIGURE 3.52 1-Hexanol in $CDCl_3$ (A) 600 MHz, (B) 300 MHz. Inserts have same ppm scales, hence, the 600 MHz data has twice the number of Hz per ppm verses the 300 MHz.

appear in the spectrum as a partially resolved band and act as a "conglomerate" of spins in coupling to the methyl group, which is formally coupled only to the adjoining CH_2 group. This phenomenon is termed "virtual coupling."

At 600 MHz the CH_2—CH_3 $\Delta\nu/J$ value is 26, and the CH_3 multiplet is now a first-order triplet. The B, C, and D multiplets remain as a tight conglomerate at 300 MHz but the M multiplet is deshielded enough by the OH group at 300 MHz so that a somewhat distorted quintet (splitting by the D and X protons) is apparent, as are incipient extra peaks; at 600 MHz, we clearly see a first-order quintet. In both spectra, the strongly deshielded triplet represents the X methylene protons split by M methylene protons. The one-proton singlet represents the OH proton. In order to prevent hydrogen bonding of the OH proton to the X proton, a trace of *p*-toluenesulfonic acid was added to the $CDCl_3$ solution. This was necessary because of the high purity of the $CDCl_3$ (see Section 3.6.1.1) and because of the low solute concentration used when working at high magnetic fields.

3.11.2 Symmetrical Chains

3.11.2.1 Dimethyl Succinate The symmetrical, conformationally mobile, open-chain diesters are worth examining. Dimethyl succinate:

$$\underset{}{\text{MeOOC}}-\underset{A_2}{CH_2}-\underset{A_2}{CH_2}-\text{COOMe}$$

obviously gives a four-proton singlet.

3.11.2.2 Dimethyl Glutarate Dimethyl glutarate at 300 MHz is an $X_2A_2X_2$ system, which can be written as A_2X_4 and gives a quintet and a triplet. Less electronegative substituents in place of the COOMe groups result in a complex A_2B_4 spectrum.

$$\text{MeOOC}-\underset{X_2}{CH_2}-\underset{A_2}{CH_2}-\underset{X_2}{CH_2}-\text{COOMe}$$

3.11.2.3 Dimethyl Adipate We move to dimethyl adipate and confidently cleave the molecule in the middle to produce two identical A_2X_2 systems; thus, we predict a deshielded triplet and a less deshielded triplet.

The 600 MHz spectrum comes as a shock (see Figure 3.53). Obviously, this is by no means a first-order spectrum even though $\Delta\nu/J$ for the A_2X_2 coupling is approximately 21, assuming a J value of about 7 Hz. If we redraw the structure (Figure 3.53) we see that the difficulty lies in the two central CH_2 groups (labeled 3 and 4) whose protons are all chemical-shift equivalent and are strongly coupled with each other; $\Delta\nu$ is zero, and the J value is approximately 7 Hz ($\Delta\nu/J = 0/7 = 0$, certainly not a first-order situation). To make matters worse, the protons in group 3 are not magnetically equivalent with the protons in group 4. The protons in group 3 are therefore both labeled H_A and those in group 4 are both labeled $H_{A'}$.

To test for magnetic equivalence between H_A in group 3 and $H_{A'}$, in group 4, we ask whether these protons couple equally with a "probe" proton H_X in group 2.

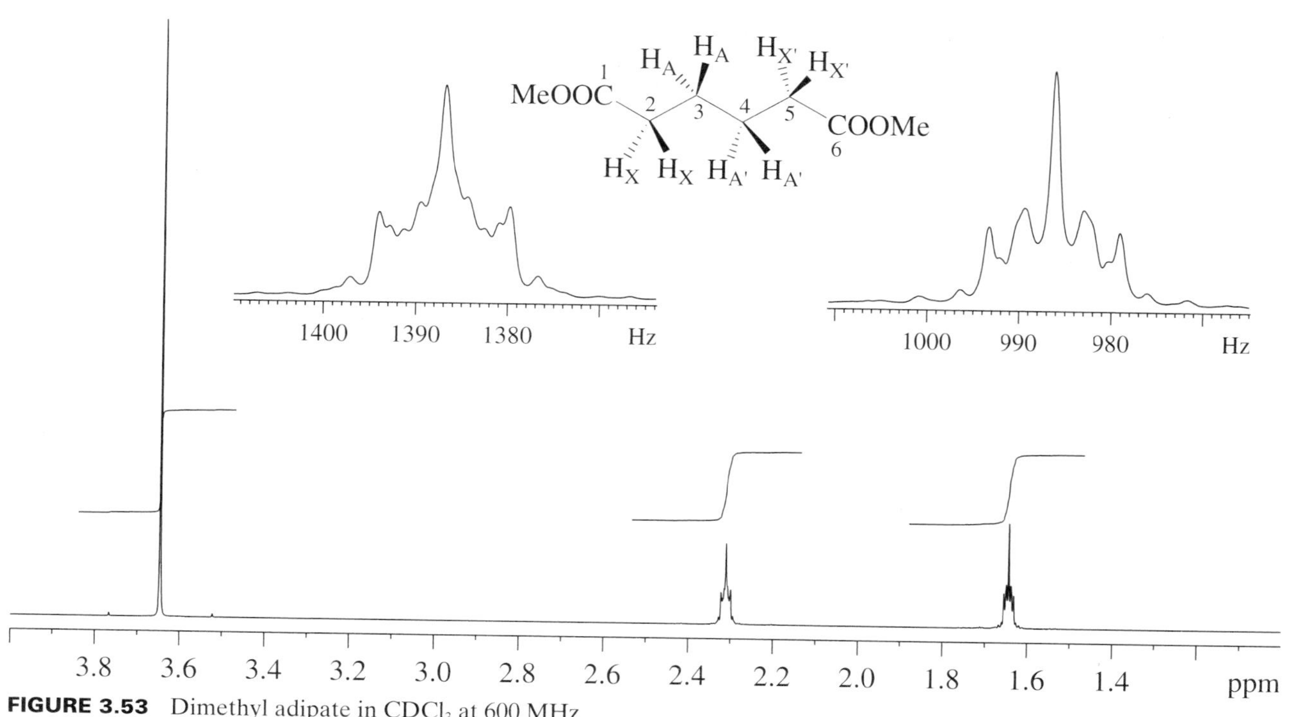

FIGURE 3.53 Dimethyl adipate in $CDCl_3$ at 600 MHz.

The answer is "no" since the H_AH_X coupling, involves three bonds, and the $H_{A'}H_X$ coupling involves four bonds; the difference in couplings is large. In the same sense, $H_{A'}$ and $H_{A'}$ do not couple equally with the "probe" proton $H_{X'}$ in group 5. The Pople notation for this spin system is XXAAA′A′X′X′, or more compactly, $X_2A_2A_2'X_2'$. This is a higher-order system and will remain so regardless of the strength of the applied magnetic field.

One possible point of confusion should be cleared up: It is often stated that chemical-shift-equivalent protons do couple with one another, but peak splitting is not observed in the spectrum. This statement is insufficient and holds only for first-order systems. But if magnetic nonequivalence is involved, the system is not first order and splitting is observed.

3.11.2.4 Dimethyl Pimelate

The next higher homologue, dimethyl pimelate, returns us to an odd number of methylene groups, hence to a first-order spectrum. (See dimethyl glutarate with three methylene groups in Section 3.11.2.2).

3.11.3 Less Symmetrical Chains

3.11.3.1 3-Methylglutaric Acid

A somewhat less symmetrical series of open-chain compounds can be obtained by placing a substituent on the center carbon atom of the chain (see Figure 3.43, structures f, g, h, and i). For example, consider 3-methylglutaric acid. Because of the 3-substituent, there is no plane of symmetry through the chain in the plane of the paper, and, as discussed in Section 3.8.1, the protons in each methylene group are not interchangeable and are thus diastereotopes.

This molecule furnishes another example of virtual coupling (see Section 3.11.1.2). Again we can be lead astray by a structure that might be expected to give a first-order spectrum-at least at 300 MHz (see Figure 3.54a).

In other words, the $\Delta\nu/J$ ratios for

$$\begin{array}{c} CH_2 \\ | \\ H_3C-CH \\ | \\ CH_2 \end{array}$$

seems adequate, and our expectations of a clean doublet for the CH_3 group seem reasonable.

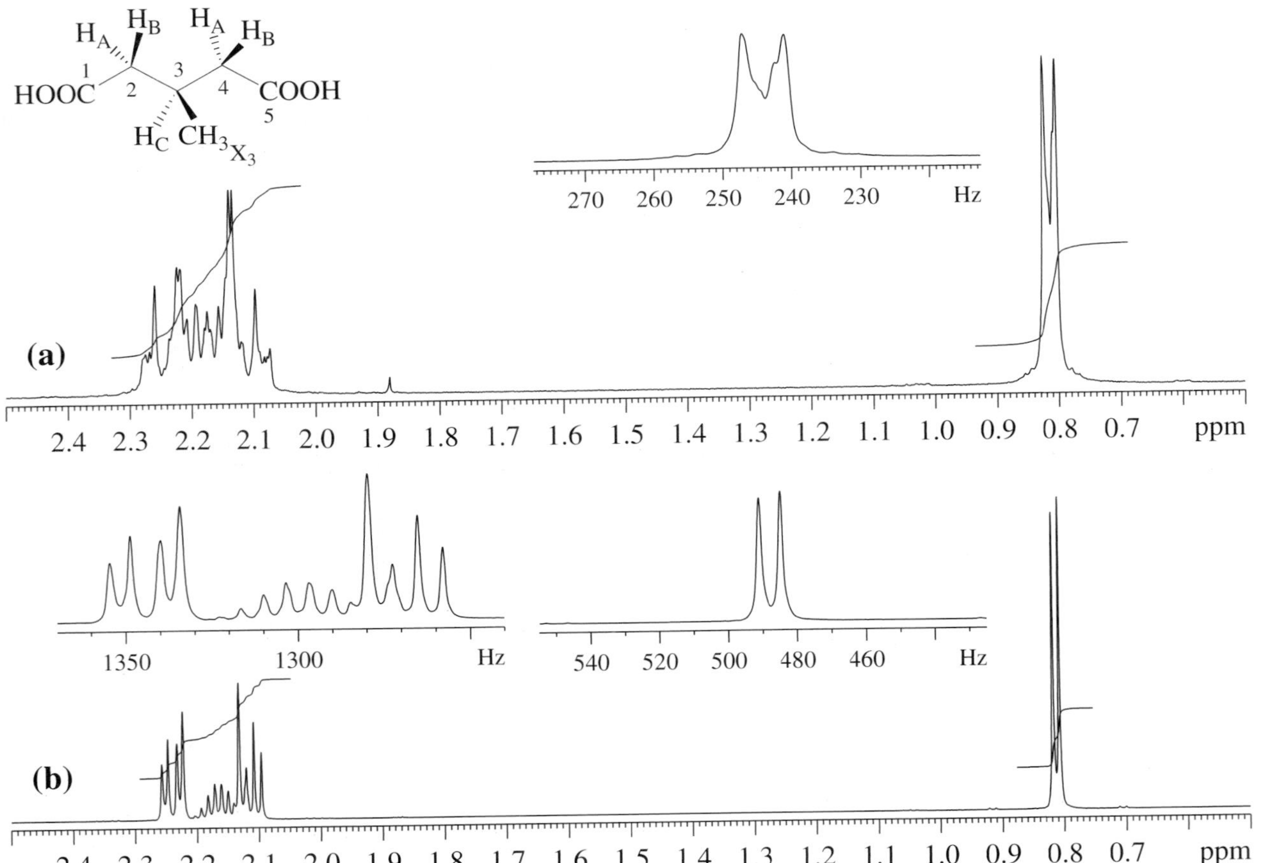

FIGURE 3.54 3-Methylglutaric acid in D_2O. (A) at 300 MHz. (B) at 600 MHz. The COOH peak exchanges with the D_2O therefore is not shown.

The difficulty is that the COOH groups deshield the CH_2 groups, thereby decreasing the $\Delta\nu/J$ ratio for CH_2—CH; the virtual coupling results in a broadened, distorted "doublet" for the CH_3 at 300 MHz. A similarly distorted "triplet" was shown in Figure 3.52. With some experience, such distortions are tolerable. But the overlapping peaks of the CH_2—CH—CH_2 moiety is still beyond interpretation by inspection at 300 MHz.

At 600 MHz (see Figure 3.54b), the CH_3 group is a clean doublet resulting from coupling with the CH proton, which is no longer superposed by the CH_2 multiplets. The CH proton is coupled, quite equally, to seven neighboring protons; this means that the CH multiplet should consist of eight peaks. And in fact it does, but the eighth peak is buried under the edge of one of the CH_2 multiplets, each of which is a doublet of doublets, somewhat distorted. As mentioned above, the protons of each CH_2 group are diastereotopes—meaning two different chemical shifts. The protons of each CH_2 group couple with each other (geminal coupling) and with the CH proton (vicinal coupling); the geminal coupling is larger—hence the two multiplets of doublets of doublets.

In summary: At 300 MHz, the spectrum is not first order since the two CH_2 groups and the CH group are severely overlapped and cannot be analyzed by inspection. At 600 MHz, they are fairly well separated and can be analyzed by inspection despite a minor overlap and some distortion. At 300 MHz, the Pople notation is $A_2B_2CX_3$, which becomes $A_2G_2MX_3$ at 600 Hz. Note that there is flexibility in choosing letters that are close together and those that are more widely separated.

3.12 CHIRALITY

The organic chemist—in particular, the natural products chemist—must always be conscious of chirality when interpreting NMR spectra. The topic was mentioned in Section 3.6. A formal definition and a brief explanation will suffice here: *Chirality expresses the necessary and sufficient condition for the existence of enantiomers.**

Impeccably rigorous but possibly a bit cryptic. The following comments may help. Enantiomers are nonsuperposable mirror images. The ultimate test for a chiral molecule is thus nonsuperposability of its mirror image. If the mirror image is superposable, the molecule is achiral. The most common feature in chiral molecules is a chiral center also called a stereogenic center. A chiral molecule possesses no element of symmetry other than possibly a simple axis or axes. For examples, see Figure 3.43, structure e, and the solved Problem 7.6 (see Chapter 7). For reassurance, consider the human hand, which has no symmetry element. The left and right hands are nonsuperposable mirror images (i.e., enantiomers). The term "chirality" translates from Greek as "handedness."

* Cahn, R.S., Ingold, C., and Prelog, V. (1966). Angew: *Chem., Int. Ed. EngL.*, **5**, 385.

3.12.1 One Chiral Center, Ipsenol

The familiar carbon chiral center has four different substituents as shown in 3-hydroxybutanoic acid (compound e in Figure 3.43). This chiral center is designated *R* in accordance with the well-known priority-sequence rules; in the enantiomeric compound, the chiral center is designated *S*. Both enantiomers give the same NMR spectrum in an achiral solvent, as does the racemate. Because of the chiral center, there is no symmetry element, and the methylene protons are diastereotopes.

Chemical shift nonequivalence of the methyl groups of an isopropyl moiety near a chiral center is frequently observed; the effect has been measured through as many as seven bonds between the chiral center and the methyl protons. The methyl groups in the terpene alcohol, 2-methyl-6-methylen-7-octen-4-ol (ipsenol) are not chemical shift equivalent (Figure 3.55). They are diastereotopic, so a strong magnetic field is usually necessary to avoid superposition.

Since the nonequivalent methyl groups are each split by the vicinal CH proton, we expect to see two separate doublets. At 300 MHz, unfortunately, the pattern appears to be a classical triplet, usually an indication of a CH_3—CH_2 moiety—impossible to reconcile with the structural formula and the integration. Higher resolution would pull apart the middle peak to show two doublets.

To remove the coincidence of the inner peaks that caused the apparent triplet, we used the very effective technique of "titration" with deuterated benzene,* which gave convincing evidence of two doublets at 20% C_6D_6/80% $CDCl_3$ and optimal results at about a 50:50 mixture (Figure 3.55). At 600 MHz, two individual doublets are seen.

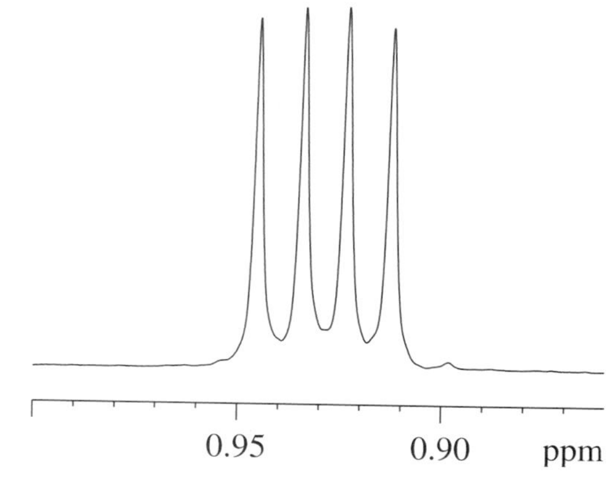

* Sanders, J.K.M., and Hunter, B.K. (1993). *Modern NMR Spectroscopy*, 2nd ed. Oxford: Oxford University Press, p. 289.

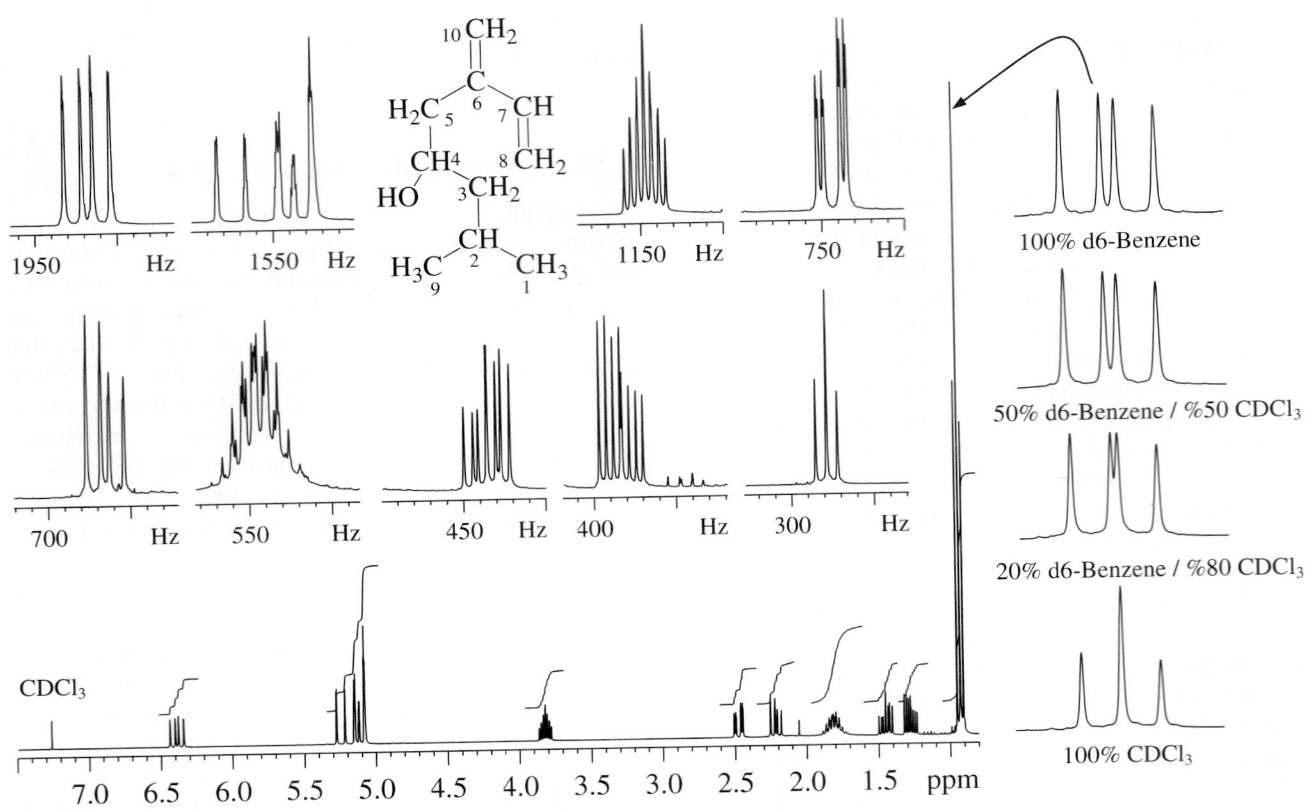

FIGURE 3.55 2-Methyl-6-methylen-7-octen-4-ol (ipsenol) in $CDCl_3$ at 300 MHz. "Titration" with C_6D_6. The sample was a gift from Phero Tech, Inc., Vancouver, BC, Canada.

Note that the chiral center accounts for the fact that the protons of each of the two aliphatic methylene groups are also diastereotopic. As a result, the rather simple structure presents a challenge in assigning the spectrum to the given structure.

The proton integration ratios from left to right are: 1:4:1:1:1:2:1:1:6 which accounts for 18 protons in accord with the molecular formula $C_{10}H_{18}O$, but there are puzzling discrepancies. The six-proton step at δ 0.92 in the integration represents the diastereotopic methyl protons described above, and the one-proton integration step at δ 3.82 with several different coupling constants is a good choice for the **CH**OH proton (see Appendix A, Chart A.1). So far, so good. But there are four CH_2 groups in the structure and apparently only one in the integration ratios and even this one is misleading.

It will help at this point to realize that the molecule consists of three spin systems with the "insulation" point at C-6 (see Section 3.5.2). The alkyl system consists of H-1, H-2, H-3, H-4, H-5, and H-9; an alkene system consists of H-7 and H-8; and another alkene system consists of H-10. The alkyl system accounts for the multiplets at the right of the spectrum, and the alkenes account for the multiplets at the left side. It will also help to reiterate that the protons of an alkyl CH_2 group will be diastereotopic in the presence of a chiral center. As will be the protons of an alkene$=CH_2$ group.

It should now be apparent that the diastereotopic protons of each alkyl group CH_2 group occur as a pair. One pair (H-3) is at δ 1.28 and δ 1.42; the other pair (H-5) is at δ 2.21 and δ 2.48. The two-proton multiplet at δ 1.80 will be discussed later.

Furthermore, the H-5 protons are at higher frequency because they are deshielded by both the OH group and the $C{=}CH_2$ group whereas the H-3 protons are deshielded only by the OH group. The H-5 protons couple geminally and each proton couples once vicinally; thus there is a doublet of doublets (first order) for each C-5 proton. The H-3 protons couple geminally and vicinally to two protons; the result for the H-3 protons (at 300 MHz) is two complex multiplets. Also in favor of the first-order status for the H-5 protons is a larger $\Delta\nu$ value (in Hz) between the two H-5 multiplets.

The mysterious multiplet at δ 1.80 may now be identified by default as the highly coupled H-2 proton superposed on the OH peak, which could be confirmed by heating, by a solvent change, or by shaking the solution with D_2O to remove it (see Section 3.6.1.1). Dilution moves it to a lower frequency. The very small peak at about δ 2.05 is an impurity.

Predictably the H-7 alkene proton is at the high frequency end of the spectrum. It couples *trans* ($J = 18$ Hz) and *cis* ($J = 10.5$ Hz) across the double bond to the H-8 protons to give a doublet of doublets (see Appendix F).

The absorptions between about 1585 Hz and 1525 Hz contain the peaks from both the H-8 and H-10 protons. At the left side, there is an 18 Hz doublet representing one of the H-8 protons coupled *trans* across the double bond to H-7. These doublet peaks are at 1585 Hz and 1567 Hz.

At 1546 Hz and 1525 Hz are the two individual peaks (not a doublet) of the H-10 protons. The very small geminal coupling results in slight broadening; the jagged edges are evidence of some long-range coupling. Note that the height of the right-hand peak is suspicious.

Having determined the *trans* coupling of one of the H-8 protons, we search for the corresponding doublet of the *cis* coupling. Unfortunately, we seem to be left with only a singlet at about 1538 Hz but quickly assume that the missing peak is buried under the suspiciously large peak at the right edge. Examination at 600 MHz of these diene spin systems clearly shows the (H-8) 10.5 Hz doublet, appropriate for the *cis* coupling, and justifies the above conclusions. Note the problems engendered by a chiral center.

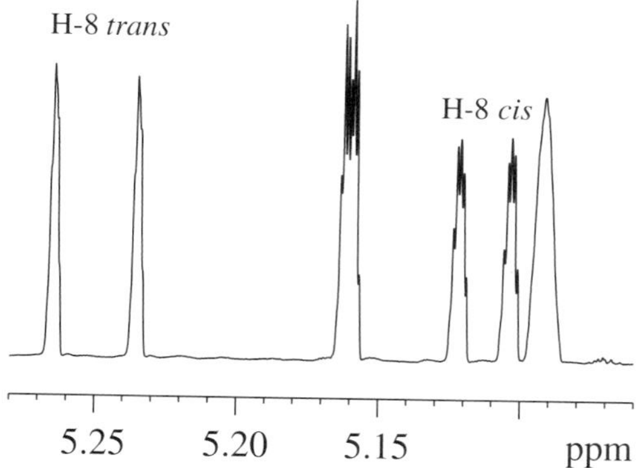

3.12.2 Two Chiral Centers

1,3-Dibromo-1,3-diphenylethane has a methylene group between two identical chiral centers (Figure 3.56). In the 1*R*,3*R* compound (one of a racemic pair), H_a and H_b are equivalent and so are H_c and H_d, because of a C_2 axis. In the 1*S*, 3*R* compound (a *meso* compound), attempted C_2 rotation gives a distinguishable structure. But H_a and H_b are enantiotopes by interchange through the plane of symmetry shown perpendicular to the plane of the page. H_c and H_d, however, cannot be interchanged since they are in the plane of symmetry; they are diastereotopes.

In the (1*R*, 3*R*) compound, H_a and H_b are not magnetic equivalent since they do not identically couple to H_c or to H_d; H_c and H_d also are not magnetic equivalent since they do not identically couple to H_a or H_b. But since the *J* values approximately average out by free rotation, the spin system is treated as A_2X_2 and the spectrum would show two triplets. In the (1*S*, 3*R*)

(1R,3R)-1,3-Dibromo-1,3-diphenylpropane

(1S,3R)-1,3-Dibromo-1,3-diphenylpropane

FIGURE 3.56 Two isomers of 1,3-dibromo-1,3-diphenylpropane. In the (1*R*,3*R*)-isomer, H_a and H_b are chemical-shift equivalent, as are H_c and H_d. In the (1*S*,3*R*)-isomer, H_a and H_b are chemical-shift equivalent, but H_c and H_d are not.

compound $J_{ad} = J_{bd}$ and $J_{ac} = J_{bc}$; thus in this molecule, H_a and H_b are magnetic equivalent. The question of magnetic equivalence of H_c and H_d is not relevant since they are not chemical-shift equivalent. The spin system is ABX_2.

In the above compound, both chiral centers have the same substituents. In a compound with two chiral centers in which such is not the case, there is no symmetry element; none of the protons would be interchangeable.

3.13 VICINAL AND GEMINAL COUPLING

Coupling between protons on vicinal carbon atoms depends primarily on the dihedral angle ϕ between the H—C—C′ and the C—C′—H′ planes. This angle can be visualized by an end-on view (Newman projection) of the bond between the vicinal carbon atoms

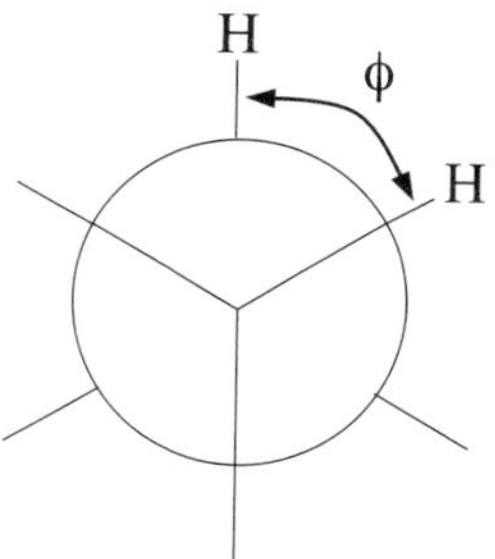

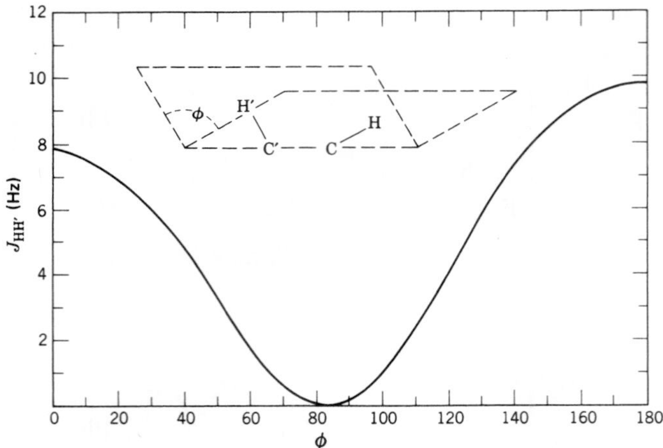

FIGURE 3.57 The vicinal Karplus correlation. Relationship between dihedral angle (ϕ) and coupling constant for vicinal protons.

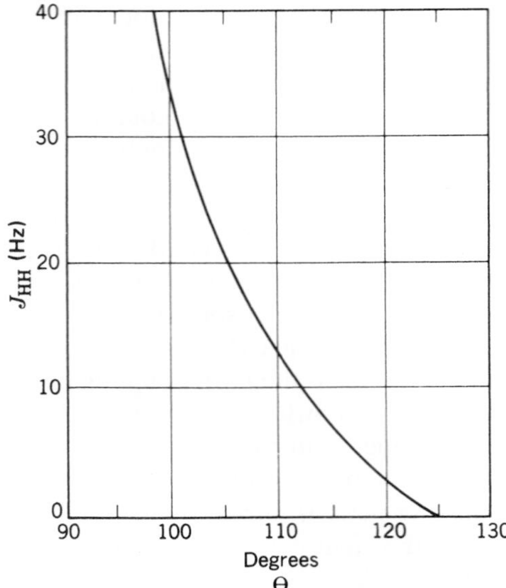

FIGURE 3.58 The geminal Karplus correlation. J_{HH} for CH_2 groups as a function of >H—C—H. Note the zero coupling at about 125°.

and by the perspective in Figure 3.57 in which the relationship between dihedral angle and vicinal coupling constant is graphed.

Karplus* emphasized that his calculations are approximations and do not take into account such factors as electronegative substituents, the bond angles θ (<H—C—C′ and the <C—C′—H′), and bond lengths. Deductions of dihedral angles from measured coupling constants are safely made only by comparison with closely related compounds. The correlation has been very useful in cyclopentanes, cyclohexanes, carbohydrates, and polycyclic systems. In cyclopentanes, the observed values of about 8 Hz for vicinal *cis* protons and about 0 Hz for vicinal *trans* protons are in accord with the corresponding angles of about 0° and about 90°, respectively. In substituted cyclohexane or pyranose rings, the chair is the preferred conformation; the following relations in Table 3.6 hold and dihedral angles of substituents follow from these ^3J proton couplings.

Note the near-zero coupling at the 90° dihedral angle. This has been a source of frustration in attempts at fitting proposed structures to the NMR spectra.

A modified Karplus correlation can be applied to vicinal coupling in alkenes. The prediction of a larger *trans* coupling (ϕ = 180°) than *cis* coupling (ϕ = 0°) is borne out. The *cis* coupling in unsaturated rings decreases with decreasing ring size (increasing bond angle) as follows: cylohexenes ^3J = 8.8—10.5, cyclopentenes ^3J = 5.1–7.0, cyclobutenes ^3J = 2.5–4.0, and cyclopropenes ^3J = 0.5–2.0.

TABLE 3.6. Calculated and observed coupling constants, J, in cyclohexanes based on bond angle.

	Dihedral Angle	Calculated J(Hz)	Observed J(Hz)
Axial-axial	180°	9	8–14 (usually 8–10)
Axial-equatorial	60°	1.8	1–7 (usually 2–3)
Equatorial-equatorial	60°	1.8	1–7 (usually 2–3)

* Karplus, M. (1959). *J. Chem. Phys.*, **30,** 11.

Two-bond geminal CH_2 coupling depends on the H—C—H bond angle θ as shown in Figure 3.58. This relationship is quite susceptible to other influences and should be used with due caution. However, it is useful for characterizing methylene groups in a fused cyclohexane ring (approximately tetrahedral, ^2J ~12–18), methylene groups of a cyclopropane ring (^2J ~5), or a terminal methylene group, i.e., = CH_2, (^2J ~0–3). Electronegative substituents reduce the geminal coupling constant, whereas sp^2 or sp hybridized carbon atoms increase it.

Geminal coupling constants are usually negative numbers, but this can be ignored except for calculations. Note that geminal couplings are seen in routine spectra only when the methylene protons are diastereotopic.

In view of the many factors other than angle dependence that influence coupling constants, it is not surprising that there have been abuses of the Karplus correlation. Direct "reading off" of the angle from the magnitude of the ^2J value is risky.

3.14 LONG-RANGE COUPLING

Proton-proton coupling beyond three bonds (^3J) is usually less than 1 Hz, but appreciable long-range coupling may occur in such rigid structures as alkenes, alkynes, aromatics, and heteroaromatics, and in strained ring systems (small or bridged). Allylic (H—C—C=C—H) couplings are typically about

1.6 Hz. The coupling through five bonds in the conjugated system of butadiene is 1.3 Hz. Coupling through conjugated polyalkyne chains may occur through as many as nine bonds. *Meta* coupling in a benzene ring is 1–3 Hz, and *para,* 0–1 Hz. In five-membered heteroaromatic rings, coupling between the 2 and 4 protons is 0–2 Hz. $^4J_{AB}$ in the bicyclo[2.1.1]hexane system is about 7 Hz.

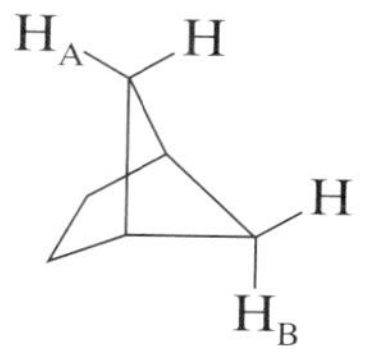

This unusually high long-range coupling constant is attributed to the "W conformation" of the four σ bonds between H_A and H_B.

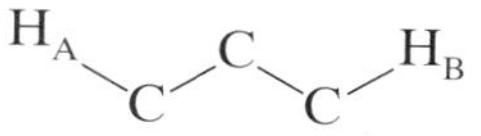

Long-range W coupling

As resolution increases, small couplings beyond three bonds often become noticeable as line broadening of the benzylic CH_2 and CH_3 peaks in Figure 3.34 and Problem 7.2, respectively (see Appendix F for coupling constants).

3.15 SELECTIVE SPIN DECOUPLING. DOUBLE RESONANCE

Intense irradiation of a proton (or equivalent protons) at its resonance frequency in a spin-coupled system removes the proton's coupling effect on the neighboring protons to which it has been coupled. Thus successive irradiation of the protons of 1-propanol, for example, yields the following results:

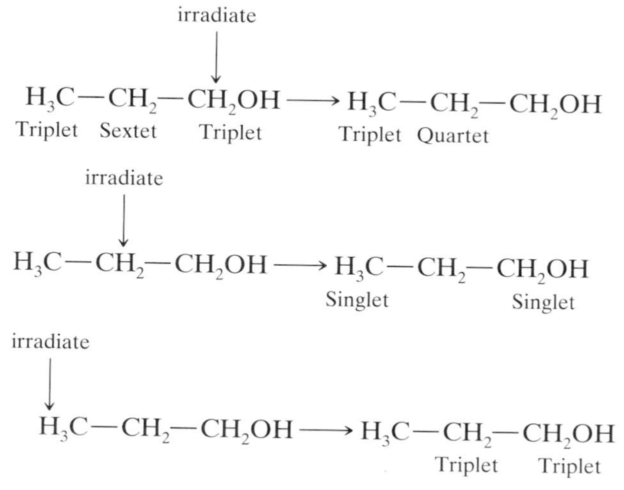

irradiate (CH_2OH)

$H_3C—CH_2—CH_2OH \longrightarrow H_3C—CH_2—CH_2OH$

Triplet Sextet Triplet → Triplet Quartet

irradiate (CH_2)

$H_3C—CH_2—CH_2OH \longrightarrow H_3C—CH_2—CH_2OH$

Singlet Singlet

irradiate (H_3C)

$H_3C—CH_2—CH_2OH \longrightarrow H_3C—CH_2—CH_2OH$

Triplet Triplet

Thus, we have a powerful tool for determining the connectivity of protons through bonds and assigning proton peaks. Furthermore, overlapping peaks can be simplified by removing one of the couplings.

Figure 3.59 displays the effect of selective spin decoupling on the neighboring protons of 3-butyne-l-ol

In spectrum 3.59a from left to right, the H-1 multiplet is a triplet ($J = 6$ Hz); the H-2 multiplet is a triplet of doublets ($J = 6$ Hz and 3 Hz); and the H-4 multiplet is a triplet ($J = 3$ Hz). The OH singlet is at δ 2.03. Again from left to right, the integration ratios of the multiplets are 2:2:1:1 (not shown).

In spectrum (b), the H-1 protons are irradiated, and the H-2 triplet of doublets is collapsed to a doublet ($J = 3$). This doublet results from long-range coupling with H-4. In spectrum (c) the H-2 triplet of doublets is irradiated, and the H-1 triplet is collapsed to a singlet, and the H-4 is collapsed to a singlet.

In spectrum (d), the H-4 triplet is irradiated, and the H-2 triplet of doublets is collapsed to a triplet ($J = 6$).

The overall information obtained from selective spin decoupling is "connectivity"—that is, progressing through the protons of the carbon chain.

It is also possible to simplify two overlapping multiplets by selectively irradiating a neighboring proton, thus disclosing multiplicity and coupling constants in the residual multiplet.

Other methods for establishing ^{1}H ^{1}H relationships are described in Chapter 5 and have certain advantages although selective spin coupling remains useful.

3.16 NUCLEAR OVERHAUSER EFFECT, DIFFERENCE SPECTROMETRY, ^{1}H ^{1}H PROXIMITY THROUGH SPACE

As described under *spin decoupling* (Section 3.15), it is possible to irradiate strongly a particular nucleus at its resonance frequency (ν_2) and detect loss of coupling (decoupling) in all the nuclei that were coupled to the irradiated nucleus. Detection of the decoupled nuclei is carried out with the usual probe; the added decoupler delivers strong, selective pulsed radiation at the resonance frequency of irradiated nucleus.

Section 3.15 describes ^{1}H ^{1}H spin decoupling through the bonds. Broad-band ^{1}H—^{13}C spin decoupling is briefly mentioned in Section 4.2.1 as is the nuclear Overhauser effect (NOE) (Section 4.2.4), which causes enhancement of the ^{13}C signal as a beneficent by-product of the intense decoupling irradiation. The NOE also occurs in ^{1}H ^{1}H spin decoupling, but the effect is small and rarely noticed.

Both spin decoupling and the NOE are examples of double resonance involving use of an additional "channel" to irradiate one nucleus at ν_2 and detect the effect on a different nucleus with a ν_1 pulse. At this point, we

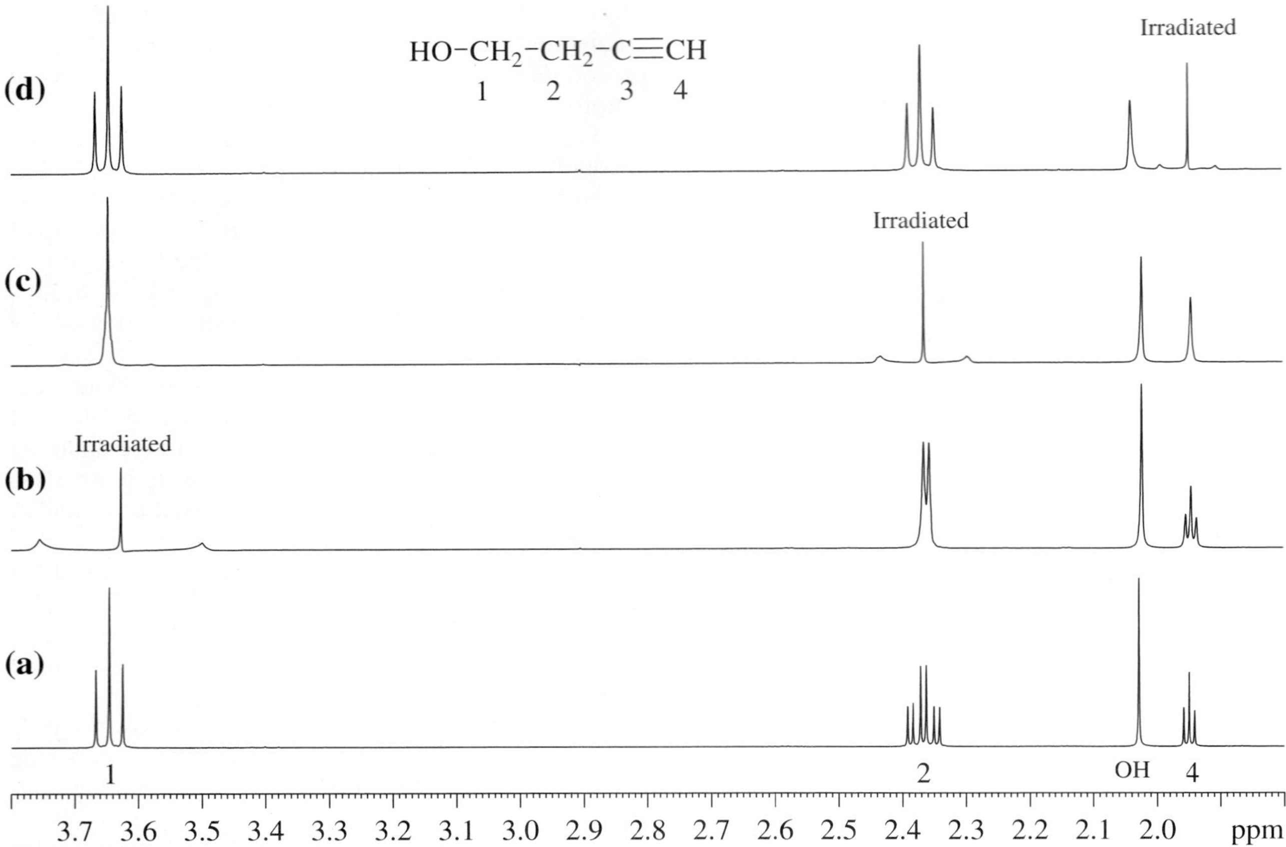

FIGURE 3.59 (a) 300 MHz spectrum of 3-butyn-l-ol in $CDCl_3$. (b) Irradiated protons on C-1. (c) Irradiated protons on C-2. (d) Irradiated protons on C-4. Irradiation may cause a detectable change in chemical shift of nearby peaks.

are interested in determining ^1H ^1H proximity within a molecule by means of the *through-space* NOE, which can be brought about by using a much weaker irradiation by ν_2 than that used for decoupling. A proton that is close in space to the irradiated proton is affected by the NOE *whether or not it is coupled to the irradiated proton;* if it is coupled, it remains at least partially coupled because the irradiation is weak in comparison with that used for a decoupling experiment. Polarization—that is, a change in population of the energy levels—by the weak ν_2 irradiation results, through space, in an increase in the population of the higher energy level in the nearby non-irradiated proton. This excess population undergoes T_1 relaxation to a lower energy level, thereby increasing the signal intensity of the nearby proton(s). In very large molecules, other factors intervene, and the results may be a decrease in signal intensity.

The through-space distances involved in ^1H ^1H proximity determinations are quite small, and the effect decreases as the inverse of the sixth power of the distance, through space, between the protons. The usual observable enhancement is less than 20%. To increase the sensitivity, we use the NOE *difference* experiment, in which a conventional ^1H spectrum* is computer subtracted from a specific proton—irradiated spectrum; this subtraction leaves only the enhanced absorptions. Under these conditions, a measurable effect can be expected between ^1H nuclei over a distance of up to about 4 Å (0.4 nm)†; for example, the distance between 1,3-diaxial protons in the cyclohexane chair form is ~2.6 Å. This procedure is a powerful tool in distinguishing among isomers, all of which should be available, if feasible, for NOE difference spectra (Figure 3.60).

A number of precautions must be taken in carrying out these ^1H ^1H proximity experiments [see Derome (1987); Sanders and Hunter (1993), or Claridge (1999).] Thus, the irradiating transmitter is "gated" off (turned off) during signal acquisition to avoid possible interference from decoupling effects. This precaution is effective since the NOE builds up slowly during irradiation and persists during acquisition, whereas decoupling effects are instantaneous. Furthermore, since relaxation rates are very sensitive to the presence of paramagnetic substances, dissolved oxygen (a diradical) should be removed by several cycles of freeze, pump, thaw, or by bubbling nitrogen or argon through the solution. The tube should then be sealed.

* Irradiation is applied to a blank region of the spectrum.

† In the familiar Dreiding models, 1 cm = 0.4 Å (0.04 nm).

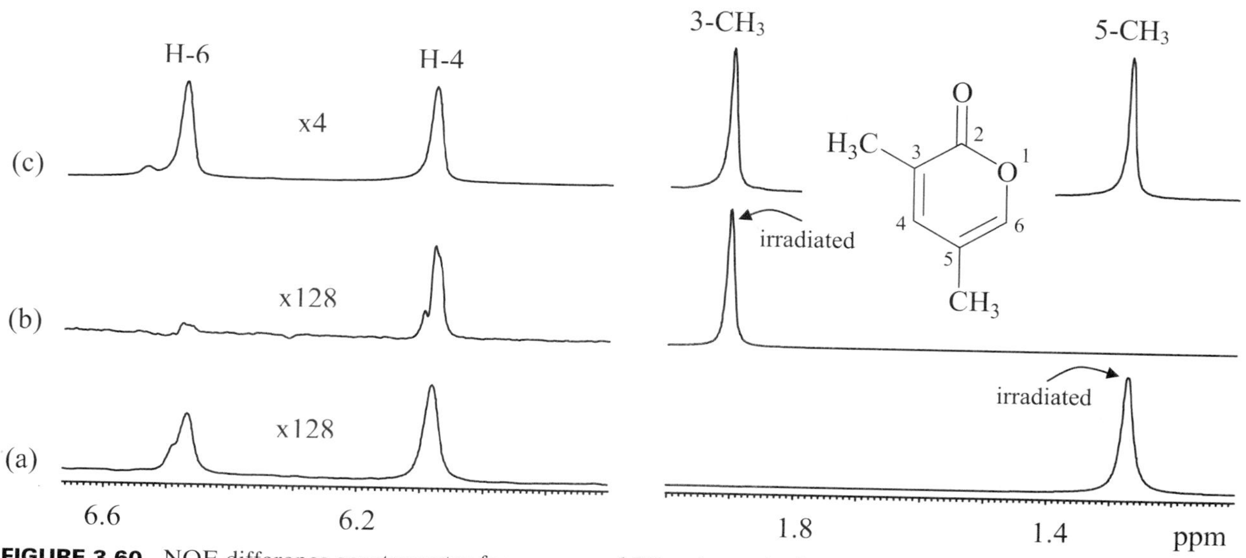

FIGURE 3.60 NOE difference spectrometry for compound III. a shows the ^{1}H NMR spectrum. b shows enhancement of the H-4 proton on irradiation of the 3-methyl group. c shows enhancement of both protons on irradiation of the 5-methyl group. 300 MHz in $CDCl_3$.

NOE difference spectrometry determined the substitution pattern of a natural product, whose structure was either I or II. The readily available dimethyl homologue III was submitted to NOE difference spectrometry (Figure 3.60).

I II III

Irradiation of the 5-methyl group resulted in enhancement of both H-4 and H-6, whereas irradiation of the 3-methyl group enhanced only H-4, since the chemical shift of the methyl absorption peak of the natural product was almost identical with that of the 3-methyl group of structure III, the natural product is structure I.

In Figure 3.60, the conventional proton spectrum is labeled a. Spectrum b shows the result of irradiation of the 3-methyl group: only H-4 is enhanced. Spectrum c shows the result of irradiation of the 5-methyl group: both H-4 and H-6 are enhanced-all predictable since enhancement is proximity driven. Some of the details may seem puzzling, but remember that this is a "difference" experiment: The conventional spectrum is subtracted from the irradiated spectrum; during the subtraction and enhancement steps, the irradiated peak returns to the "up" position; and the un-irradiated peak is not detectable because it has not been enhanced. The sensitivity numbers—×4, ×128, and ×128—refer to the vertically displayed sensitivity of the left side of each split spectrum relative to the right side, which remains ×1. These sensitivity ratios (left side: right side) are 4:1 for spectrum a, and 128: for spectra b and c. The purpose is merely to keep all peaks at convenient heights. See footnote for previous study.*

Irradiation of an OH group results in nuclear Overhauser enhancement of nearby protons. However, the usual rapid exchange rate of hydroxylic protons must be retarded either by using dry deuterated acetone or deuterated dimethyl sulfoxide as a solvent, or by simply cooling a $CDCl_3$ solution. Either process also moves the OH peak to the left; this movement can be controlled to prevent overlap of absorptions, and, of course, it serves to identify the OH peak (Section 3.6.1.1).

Distinguishing between a trisubstituted (*E*)- and a (*Z*)-double bond is not a trivial assignment. The use of NOE difference spectrometry for this purpose is illustrated in Problem 7.4.

3.17 CONCLUSION

^{1}H NMR spectrometry is the foundation upon which we will build an understanding of the magnetic resonance of other nuclei, especially ^{13}C, which leads to the important advanced correlation experiments. We began by describing the magnetic properties of nuclei, noting the special importance of spin 1/2 nuclei. For practical

* R.E. Charlton, F.X.. Webster, A. Zhang, C. Schal, D. Liang. 1. Sreng, W.L. Roelofs (1993). *Proc. Natl. Acad. Sci.*, **90**, 10202.

reasons, the historical investigations into the ^{1}H nucleus led to "high resolution" NMR and the manufacture of commercial instruments. The initial CW instruments have been largely supplanted (although not entirely) by the more powerful and versatile FT spectrometers.

Interpretation of ^{1}H NMR spectra is based on consideration of three interrelated types of information: integration of signal peaks, chemical shift, and spin-spin coupling. Integration of the signal peaks provides the ratio of hydrogen atoms in the compound. The concept of chemical shift relates the position of the signal in the spectrum with "chemical environment." A crucial concept derived from chemical shift is chemical shift equivalence. Spin-spin coupling explains the interaction of magnetic nuclei in a set of nuclei. Among other things, spin-spin coupling provides information about neighboring nuclei, giving crucial structural insights. The remaining chapters all deal with NMR either entirely (Chapters 4, 5, and 6) or in major part (Chapters 7 and 8).

REFERENCES

General

Abraham, R.J., Fisher, J., and Loftus, P. (1989). *Introduction to NMR Spectroscopy,* 2nd ed. London-New York: Wiley.

Akitt, J.W. (1992). *NMR and Chemistry: An Introduction to Modern NMR Spectroscopy,* 3rd ed. London: Chapman and Hall.

Asahi. (1997). *Handbook of Proton-NMR Spectra and Data.* New York: Academic Press.

Atta-ur-Rahman. (1986). *Nuclear Magnetic Resonance.* New York: Springer-Verlag.

Atta-ur Rahman, and Choudhary, M. (1995) *Solving Problems with NMR Spectroscopy.* New York: Academic Press.

Ault, A., and Dudek, G.O. (1976). NMR, *An Introduction to Proton Nuclear Magnetic Resonance Spectrometry.* San Francisco: Holden-Day.

Becker, E.D. (1999). *High Resolution NMR: Theory and Chemical Applications,* 3rd ed. New York: Academic Press.

Berger, S., and Braun, S. (2004). *200 and More NMR Experiments: A Practical Course.* New York: Wiley; (July 9, 2004)

Bovey, F.A. (1988). *Nuclear Magnetic Resonance Spectroscopy.* 2nd ed. New York: Academic Press.

Braun, S., Kalinowski, H., and Berger, S. (1999). *150 and More Basic NMR Experiments: A Practical Course.* 2nd ed., New York: Wiley.

Breitmaier, E. (2002). *Structure Elucidation by NMR in Organic Chemistry: A Practical Guide.* 3rd Revision ed., New York: Wiley.

Cavanagh, J., Fairbrother, W., Palmer III, A., and Skelton, N. (1995). *Protein NMR Spectroscopy: Principles and Practice* New York: Academic Press.

Claridge, T.D.W. (1999). *High-Resolution NMR Techniques in Organic Chemistry.* Elsevier Science. Oxford, UK.

Croasmun, W.R., and Carlson, R.M.K. (1994). *Two-Dimensional NMR Spectroscopy: Applications for Chemists and Biochemists,* 2nd Ed., *Fully Updated and Expanded to Include Multidimensional Work.* New York: Wiley.

Derome, A.E. (1987). *Modern NMR Techniques for Chemistry Research.* Oxford: Pergamon.

Farrar, T., and Becker, E. (1971). *Pulse and Fourier Transform NMR: Introduction to Theory and Methods.* New York: Academic Press.

Farrar, T.C. (1987). *An Introduction to Pulse NMR Spectroscopy.* Chicago: Farragut Press.

Freeman, R. (1998). *Spin Choreography: Basic Steps in High Resolution NMR.* Oxford University Press on Demand.

Friebolin, H. (1993). *Basic One- and Two-Dimensional Spectroscopy,* 2nd ed. New York: VCH.

Friebolin, H. (1998). *Basic One- and Two-Dimensional NMR Spectroscpoy,* 3rd Revised Ed., New York: Wiley.

Fukushima, E., and Roeder, S.B.W. (1986). *Experimental Pulse NMR a Nuts and Bolts Approach.* Perseus Publishing. Boulder, CO.

Grant, D.M., and Harris, R.K. (2002). *Encyclopedia of Nuclear Magnetic Resonance, Advances in NMR.* New York: Wiley.

Günther, H. (1995). *NMR Spectroscopy: Basic Principles, Concepts, and Applications in Chemistry,* 2nd Ed. New York: Wiley.

Gunther, li. (1995). *NMR Spectroscopy,* 2nd ed. New York: Wiley.

Homans, S. W. (1992). *A Dictionary of Concepts in NMR.* Clarendon Press.

Hore, P.J., Wimperis, S., and Jones, J. (2000). *NMR: The Toolkit.* Oxford University Press.

Jackman, L.M., and Sternhell. S. (1969). *Applications of NMR Spectroscopy in Organic Chemistry,* 2nd ed. New York: Pergamon Press.

Macomber, R.S. (1988). *NMR Spectroscopy—Essential Theory and Practice.* New York: Harcourt.

Macomber, R.S. (1995). *A Complete Introduction to Modern NMR Spectroscopy* New York: Wiley-Interscience.

Martin, G.E., and Zektzer, A.S. (1988). *Two-Dimensional NMR Methods for Establishing Molecular Connectivity: A Chemist's Guide to Experiment Selection, Performance, and Interpretation.* New York: Wiley.

Munowitz, M. (1988). *Coherence and NMR.* New York: Wiley-Interscience.

Nakanishi, K. (1990). *One-Dimensional and Two-Dimensional NMR Spectra by Modern Pulse Techniques.* University Science Books. Herndon, VA.

Neuhaus, D.N., and Williamson, M.P. (1992). *The Nuclear Ovcrhauser Effect in Structural and Conformational Analysis.* New York: VCH.

Paudler, W.W. (1987). *Nuclear Magnetic Resonance.* New York: Wiley.

Pihlaja, K., and Kleinpeter, E. (1994). *Carbon-13 NMR Chemical Shifts in Structural and Stereochemical Analysis.* VCH Publishing.

Roberts, J.D. (2000). *ABCs of FT-NMR.* University Science Books. Herndon, VA.

Sanders, J.K., and Hunter, B.K. (1993). *Modern NMR Spectroscopy,* 2nd ed. Oxford: Oxford University Press.

Schwartz, L.J. (1988). *A Step-by-Step Picture of Pulsed (Time-Domain) NMR. J. Chem. Ed.* 65, 752–756.

Shaw, D. (1987). *Fourier Transform NMR Spectroscopy.* Amsterdam: Elsevier.

Shoolery, J .N. (1972). *A Basic Guide to NMR.* Palo Alto, CA: Varian Associates.

Williams, K.R., and King, R.W. (1990). *J. Chem. Educ.,* **67,** A125. Sec References therein for previous papers. The Fourier Transform in Chemistry—NMR.

Yoder, C-H., and Schaeffer, C.D., Jr. (1987). *Introduction to Multinuclear NMR.* Menlo Park, CA: BenjamiñCummings.

Van de Ven, F. J. M. (1995). *Multidimensional NMR in Liquids: Basic Principles and Experimental Methods.* Wiley: New York.

Weber, U., and Thiele, H. (1998). *NMR-Spectroscopy: Modern Spectral Analysis* Wiley: New York; Book and Cassette edition.

Spectra, Data, and Special Topics

Bovey, F.A. (1967). *NMR Data Tables for Organic Compounds,* Vol. 1. New York: WileyInterscience.

Brugel, W. (1979). *Handbook of NMR Spectral Parameters,* Vols. 1–3. Philadelphia: Hoyden.

Chamberlain, N.E. (1974). *The Practice of NMR Spectroscopy with Spectra—Structure Correlation for 'H.* New York: Plenum Press.

Kiemle, D.J., and Winter, W.T. (1995) Magnetic Spin Resonance. In *Kirk-Othmer Encyclopedia of Chemical Technology,* 4th ed. Vol. 15., New York, Wiley.

Morrill, T.C., Ed. (1988) *Lanthanide Shift Reagents in Stereochemical Analysis.* New York: VCH.

Pouchert, C.J., and Behnke, J. (1993). *Aldrich Library of 13C and 1H FT-NMR Spectra, 300 MHz.* Milwaukee, WI: Aldrich Chemical Co.

Pretsch, E.. Clerc, T., Seihl, J., and Simon, W. (1989). *Spectra Data for Structure Determination of Organic Compounds,* 2nd ed. Berlin: Springer̃ Verlag.

Sadder Collection of High Resolution Spectra. Philadelphia: Sadder Research Laboratories.

Sasaki, S. (1985). *Handbook of Proton-NMR Spectra and Data,* Vols. 1̃5. New York: Academic Press. (4000 spectra)

Varian Associates. (1962, 1963). *High Resolution NMR Spectra Catalogue,* Vol. 1, Vol. 2. Palo Alto, CA: Varian.

Wiberg, K.B, and Nist, B.J. (1962). *The Interpretation of NMR Spectra.* New York: Benjamin.

STUDENT EXERCISES

3.1 For each compound given below (a–o), describe all spin systems (using Pople notation where appropriate), chemically shift equivalent protons, magnetic equivalent protons, enantiotopic protons, and diastereotopic protons.

3.2 For each compound above, predict the chemical shifts for each proton set. Give the source (Table or Appendix) that you use for your prediction.

3.3 Sketch the ^{1}H NMR spectrum for each of the compounds in question 3.1. Assume first order multiplets where possible.

3.4 Determine the structural formula for the compounds whose ^{1}H NMR spectra are given (A–W). They were all run at 300 MHz in $CDCl_3$. The mass spectra were given in Chapter 1 (Question 1.6) and the IR spectra were given in Chapter 2 (Question 2.9).

3.5 After determining the structures in exercise 3.4, calculate $\Delta\nu/J$ for appropriately coupled multiplets in the following spectra: A, E, F, G, H, I, K, L, Q, and U.

3.6 For each structure determined in exercise 3.4, describe all spin systems (using Pople notation where appropriate), chemically shift equivalent protons, magnetic equivalent protons, enantiotopic protons, and diastereotopic protons.

3.7 Sketch the following spin systems (stick diagrams are sufficient): AX, A_2X, A_3X, A_2X_2, A_3X_2.

3.8 Sketch the following spin systems (stick diagrams are sufficient): AMX, A_2MX, A_3MX, A_2MX_2, A_2MX_3, A_3MX_3. Assume that A does not couple to X, and that $J_{AM} = J_{MX}$.

3.9 Sketch the following spin systems (stick diagrams are sufficient): AMX, A_2MX, A_3MX, A_2MX_2, A_2MX_3, A_3MX_3. Assume that A does not couple to X, and that $J_{AM} = 10$ Hz and $J_{MX} = 5$ Hz.

3.10 For the three simulated spin systems (pages 185–187), draw a stick diagram for each first order multiplet in each spin system. After determining one multiplet in each spin system, the other two multiplets will be useful checks.

Exercise 3.1

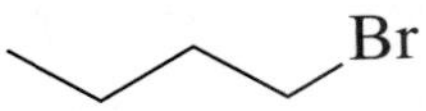

Br

a

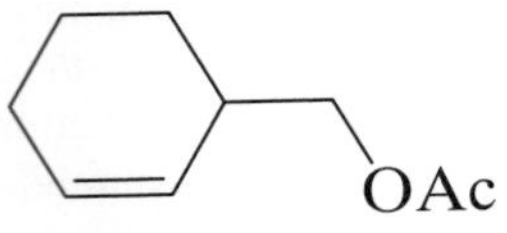

OAc

b

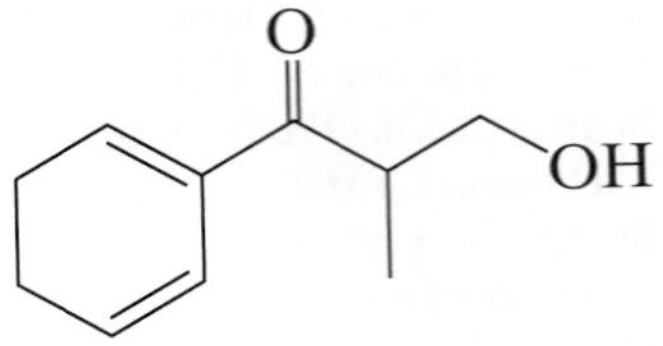

O

OH

c

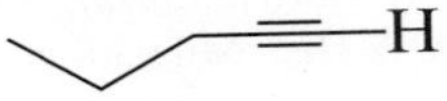

H

d

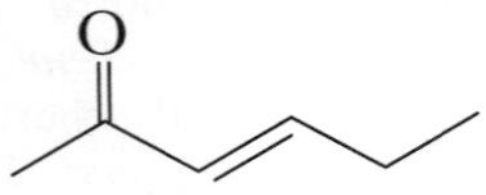

O

e

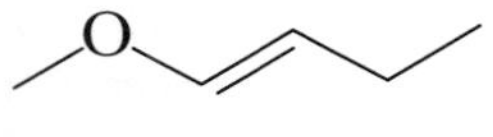

O

f

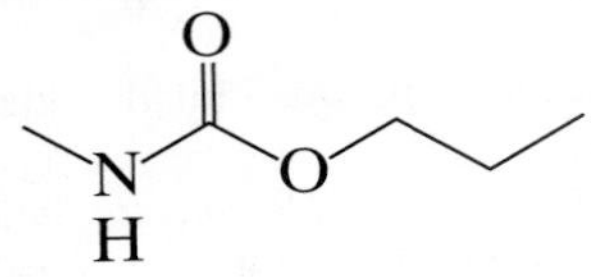

O

N

H

O

g

O O

h

OH

i

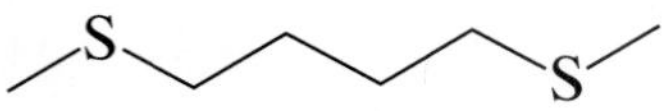

S

S

j

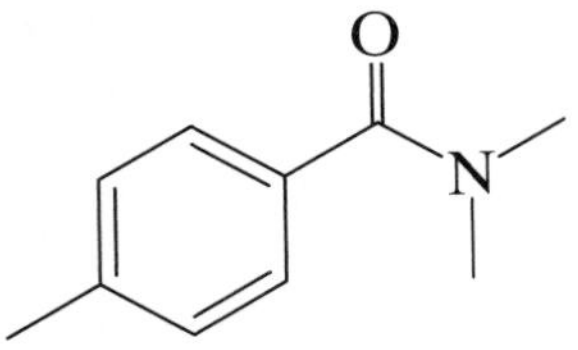

O

N

k

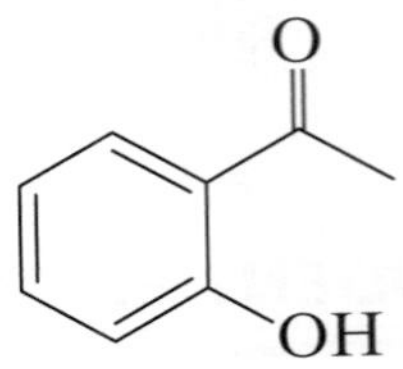

O

OH

l

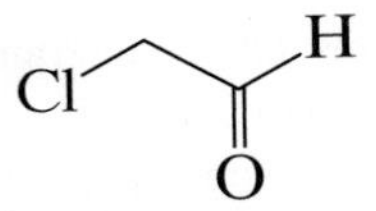

Cl

H

O

m

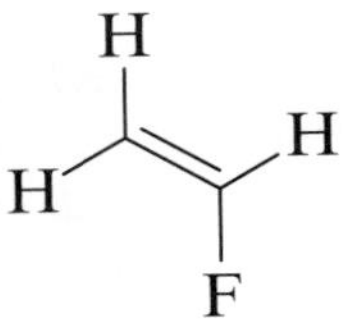

H

H

H

F

n

OAc

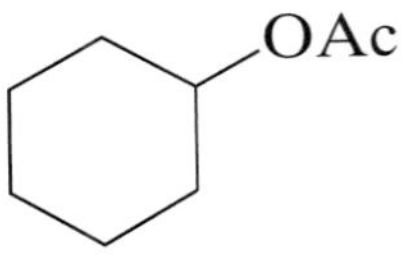

o

Exercise 3.4 (A – D)

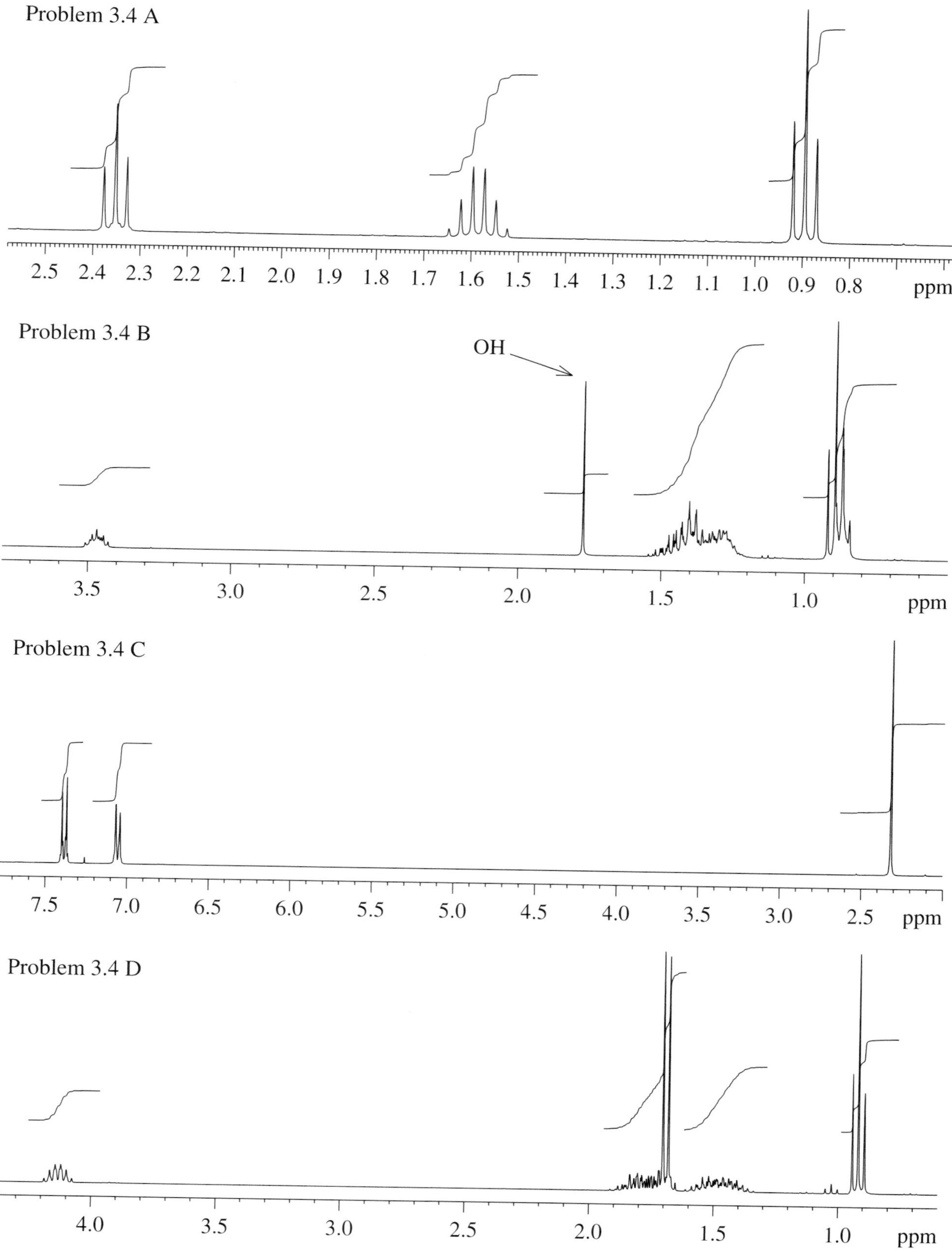

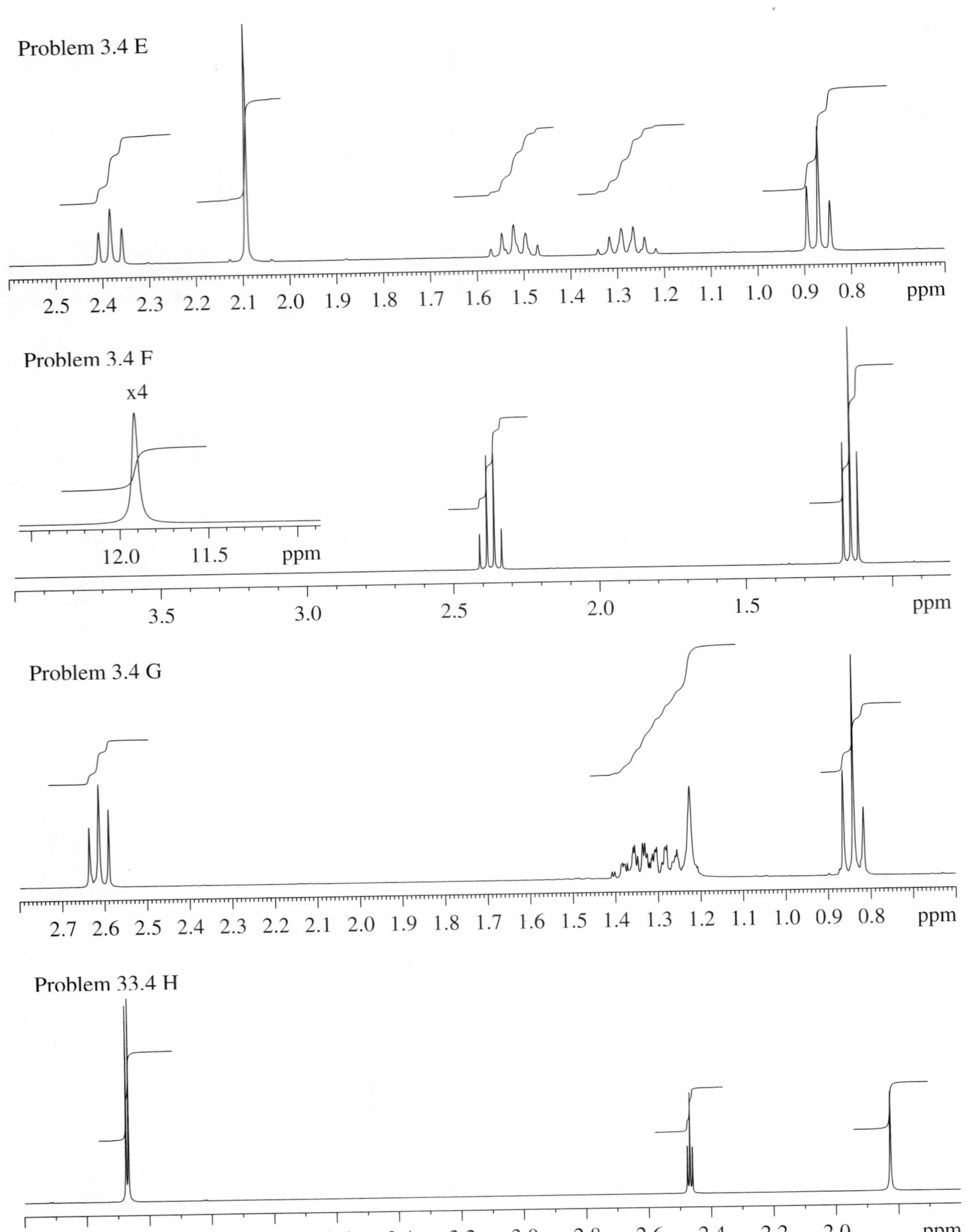

Problem 3.4 E
2.5 2.4 2.3 2.2 2.1 2.0 1.9 1.8 1.7 1.6 1.5 1.4 1.3 1.2 1.1 1.0 0.9 0.8 ppm
Problem 3.4 F
x4
12.0 11.5 ppm
3.5 3.0 2.5 2.0 1.5 ppm
Problem 3.4 G
2.7 2.6 2.5 2.4 2.3 2.2 2.1 2.0 1.9 1.8 1.7 1.6 1.5 1.4 1.3 1.2 1.1 1.0 0.9 0.8 ppm
Problem 33.4 H
4.4 4.2 4.0 3.8 3.6 3.4 3.2 3.0 2.8 2.6 2.4 2.2 2.0 ppm

Exercise 3.4 (I–L)

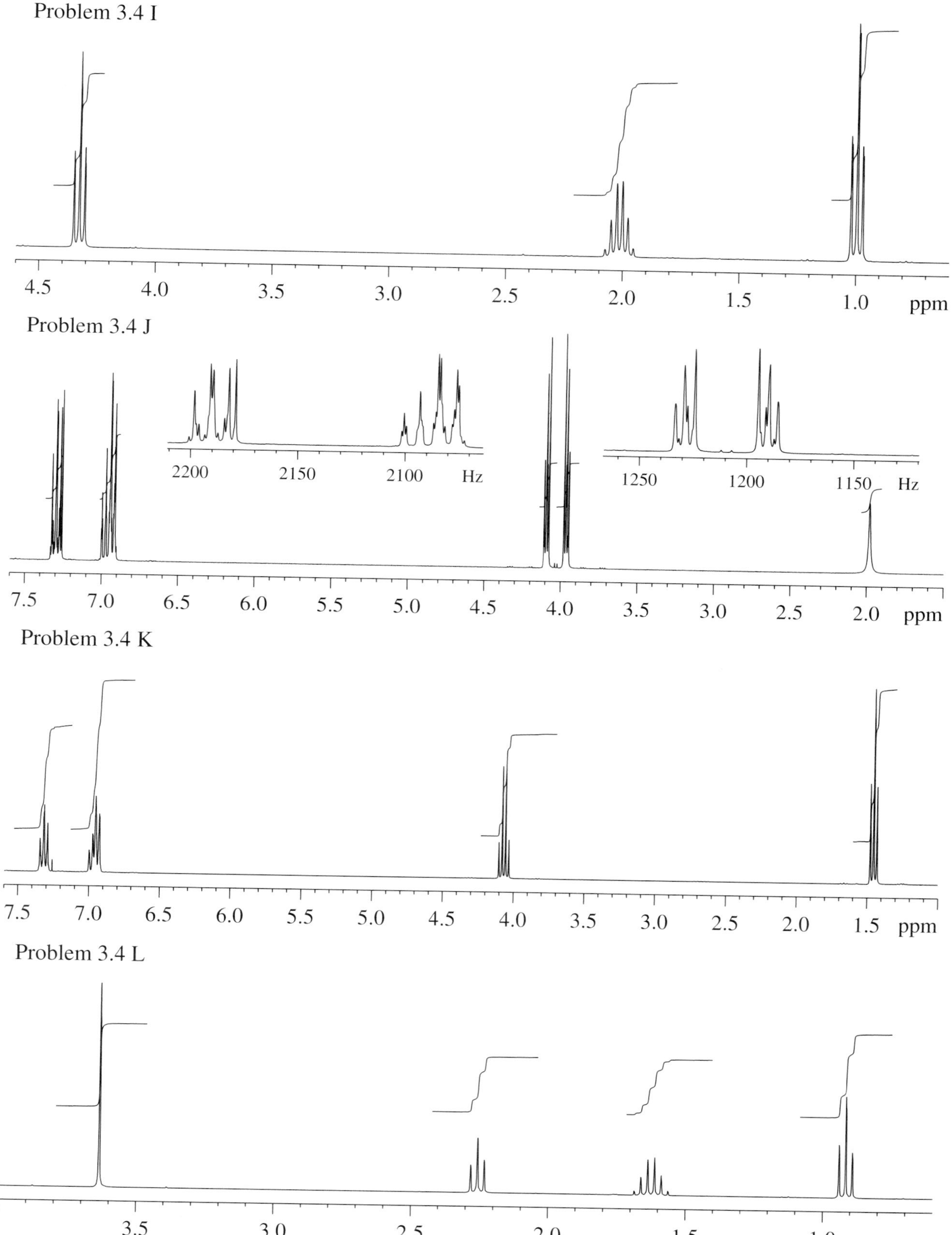

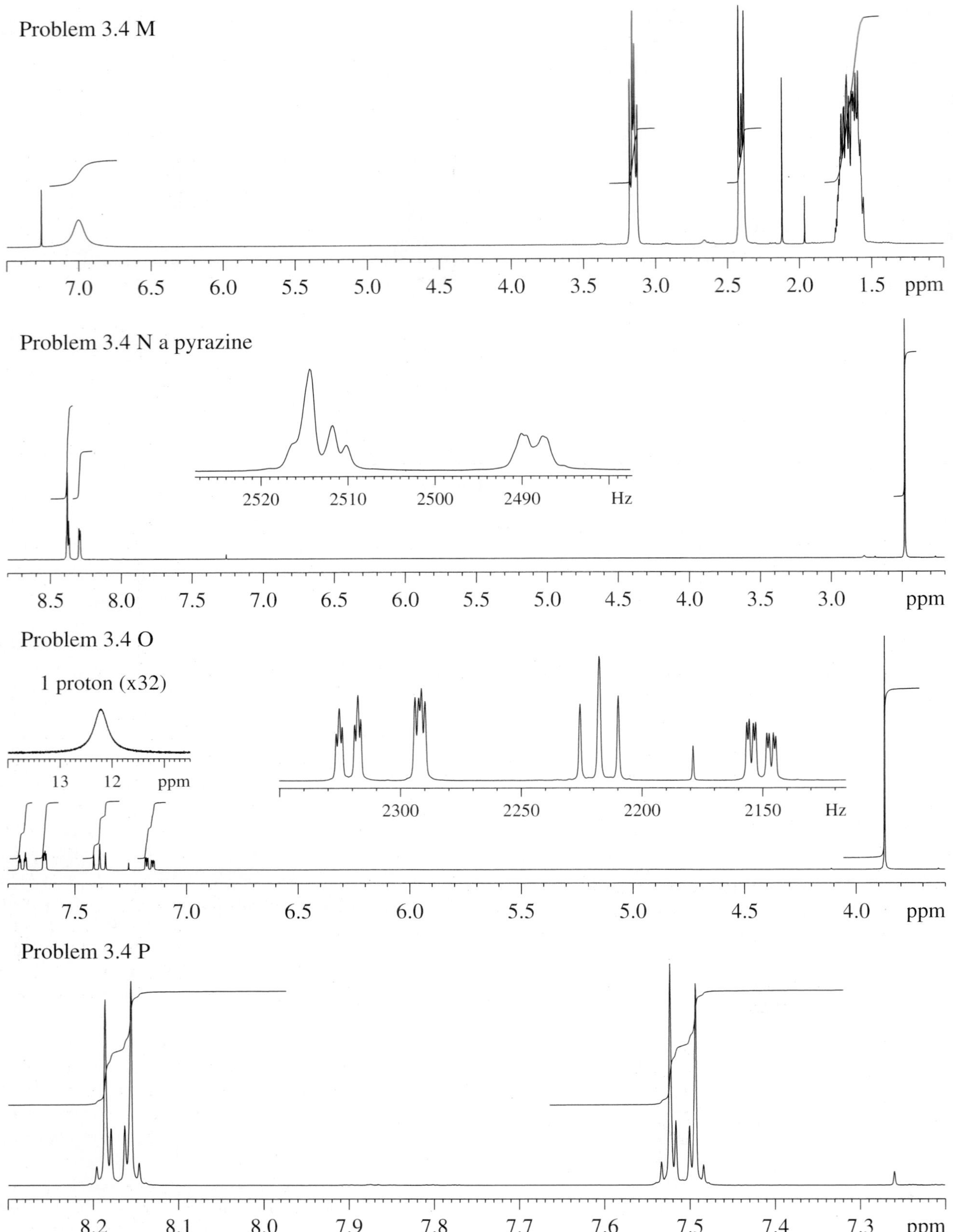
Problem 3.4 M
7.0
6.5
6.0
5.5
5.0
4.5
4.0
3.5
3.0
2.5
2.0
1.5
ppm
Problem 3.4 N a pyrazine
2520
2510
2500
2490
Hz
8.5
8.0
7.5
7.0
6.5
6.0
5.5
5.0
4.5
4.0
3.5
3.0
ppm
Problem 3.4 O
1 proton (x32)
13
12
ppm
2300
2250
2200
2150
Hz
7.5
7.0
6.5
6.0
5.5
5.0
4.5
4.0
ppm
Problem 3.4 P
8.2
8.1
8.0
7.9
7.8
7.7
7.6
7.5
7.4
7.3
ppm

Exercise 3.4 (Q – T)

Problem 3.4 Q

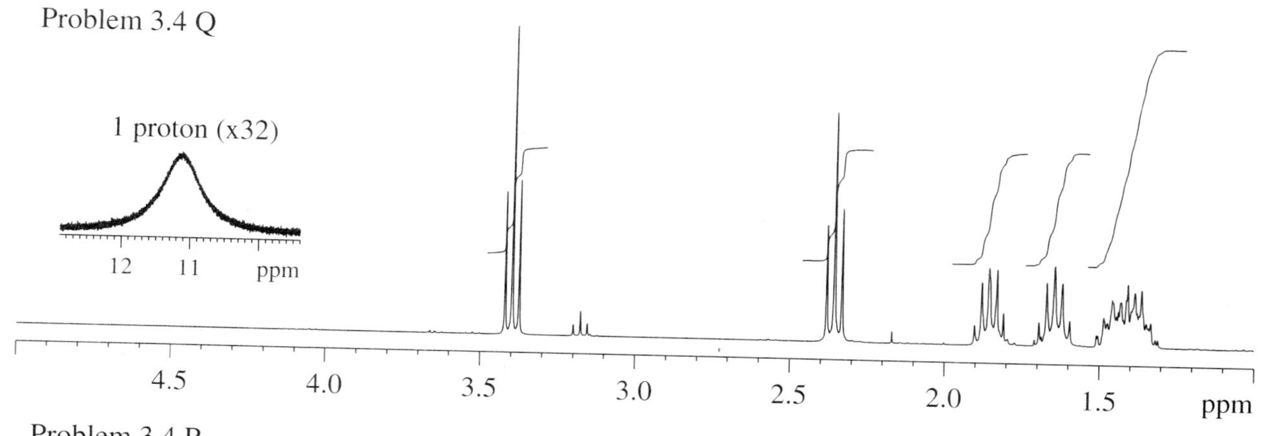

Problem 3.4 R

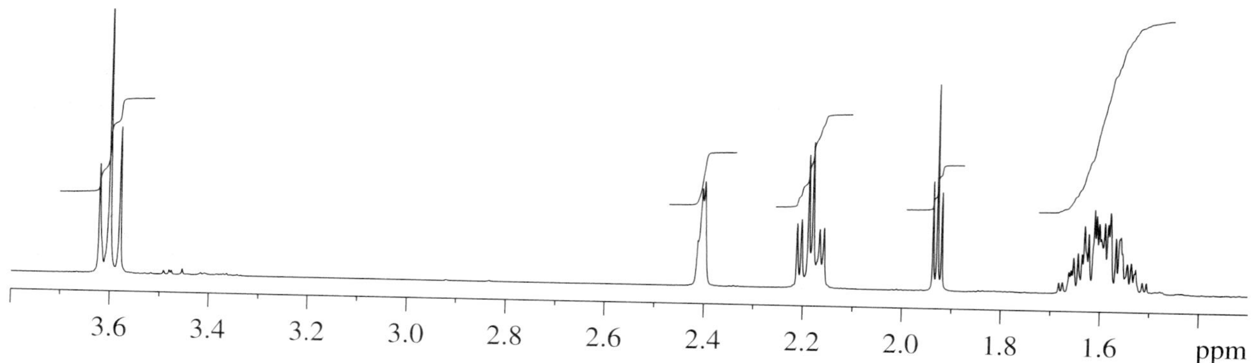

Problem 3.4 S Unsaturated ketone

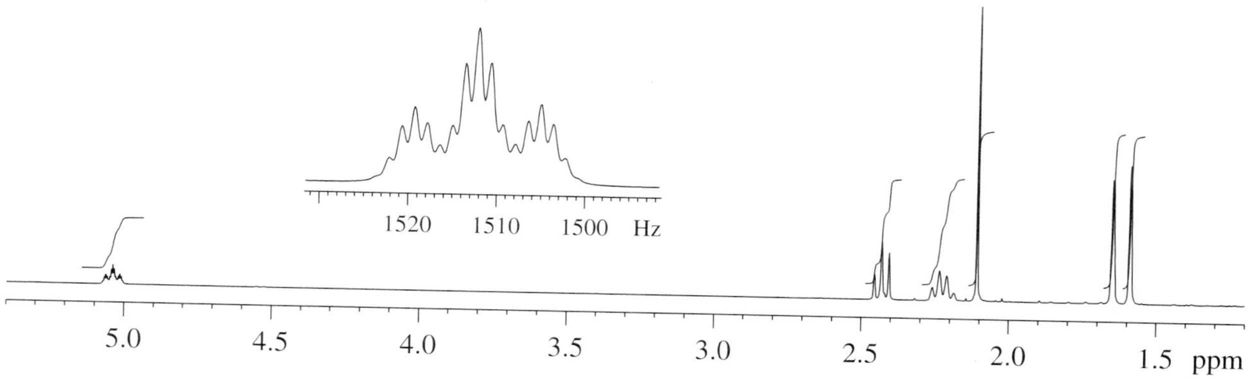

Problem 3.4 T

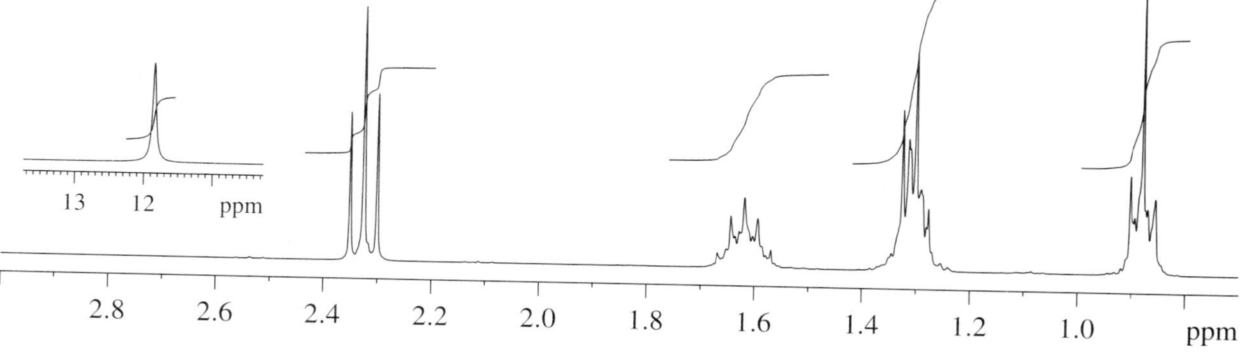

Problem 3.4 U

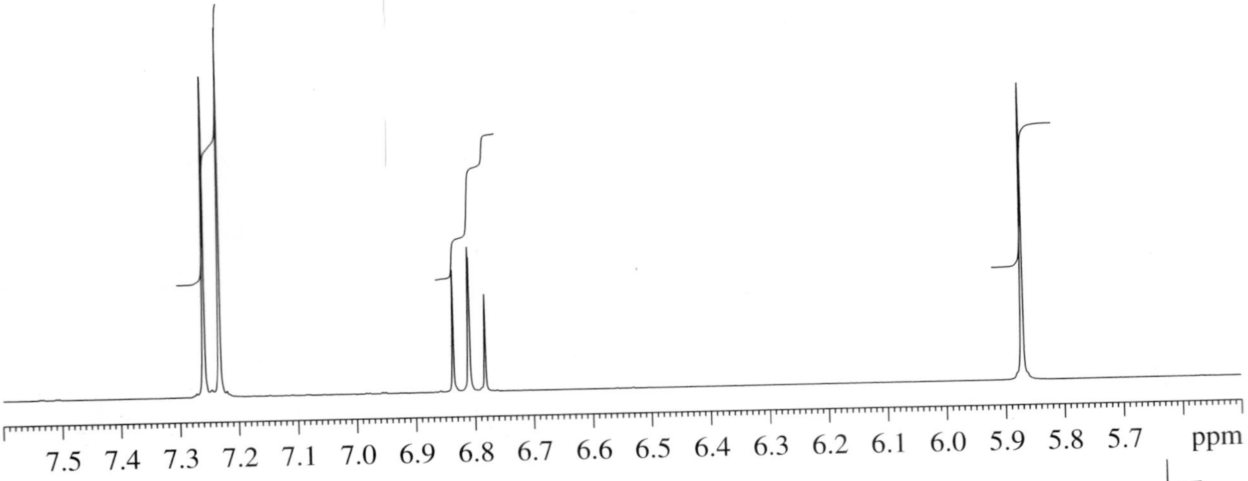

7.5 7.4 7.3 7.2 7.1 7.0 6.9 6.8 6.7 6.6 6.5 6.4 6.3 6.2 6.1 6.0 5.9 5.8 5.7
ppm

Problem 3.4 V

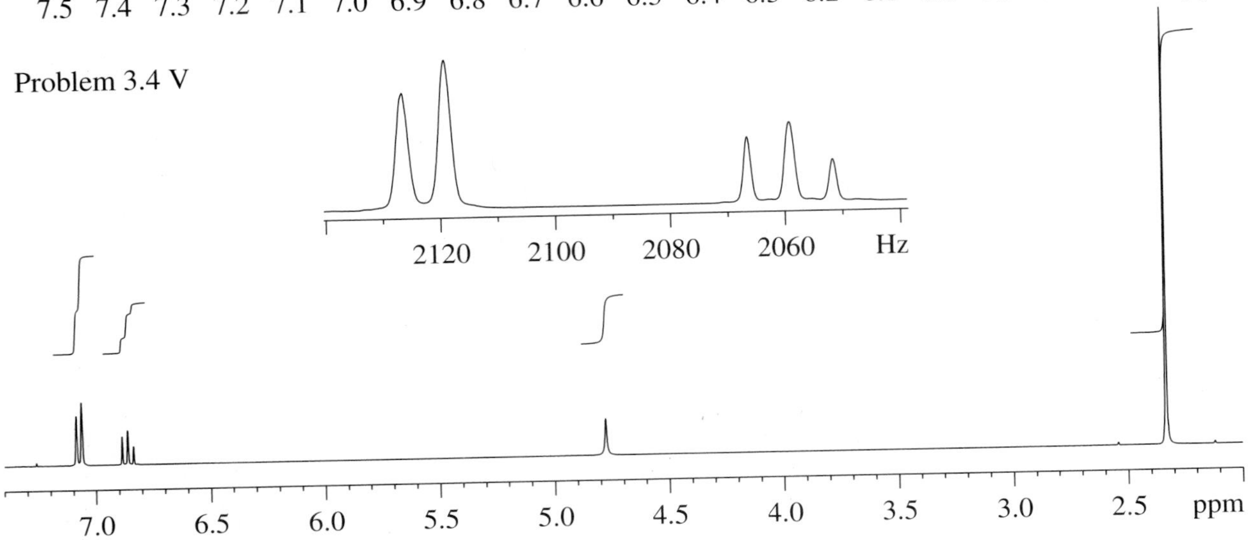

2120 2100 2080 2060
Hz
7.0 6.5 6.0 5.5 5.0 4.5 4.0 3.5 3.0 2.5
ppm

Problem 3.4 W Unsaturated ketone

7.0 6.5 6.0 5.5 5.0 4.5 4.0 3.5 3.0 2.5 2.0
ppm

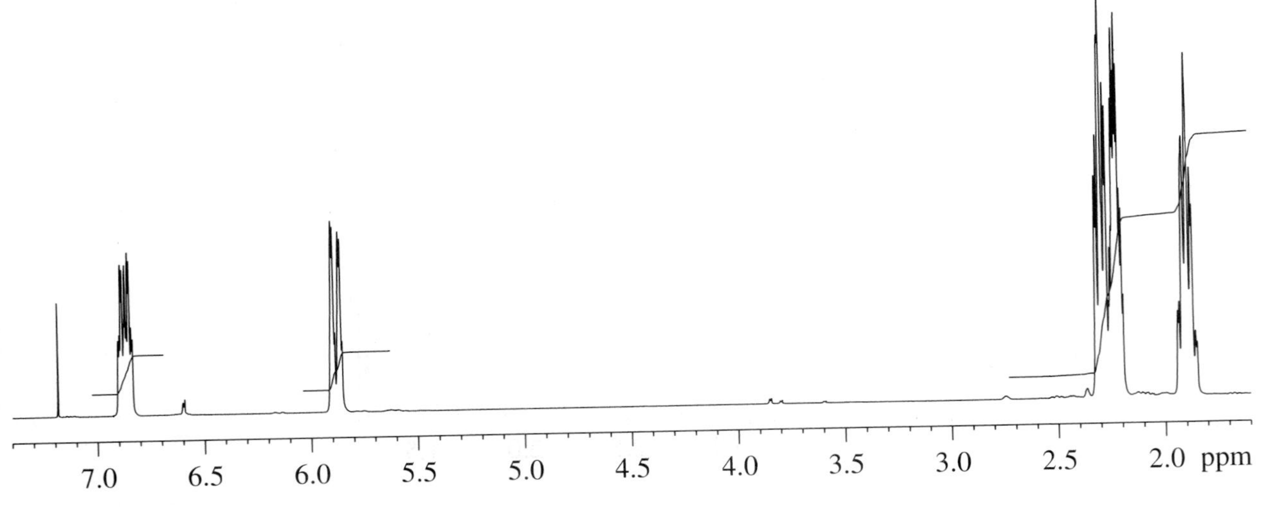

Exercise 3.10A

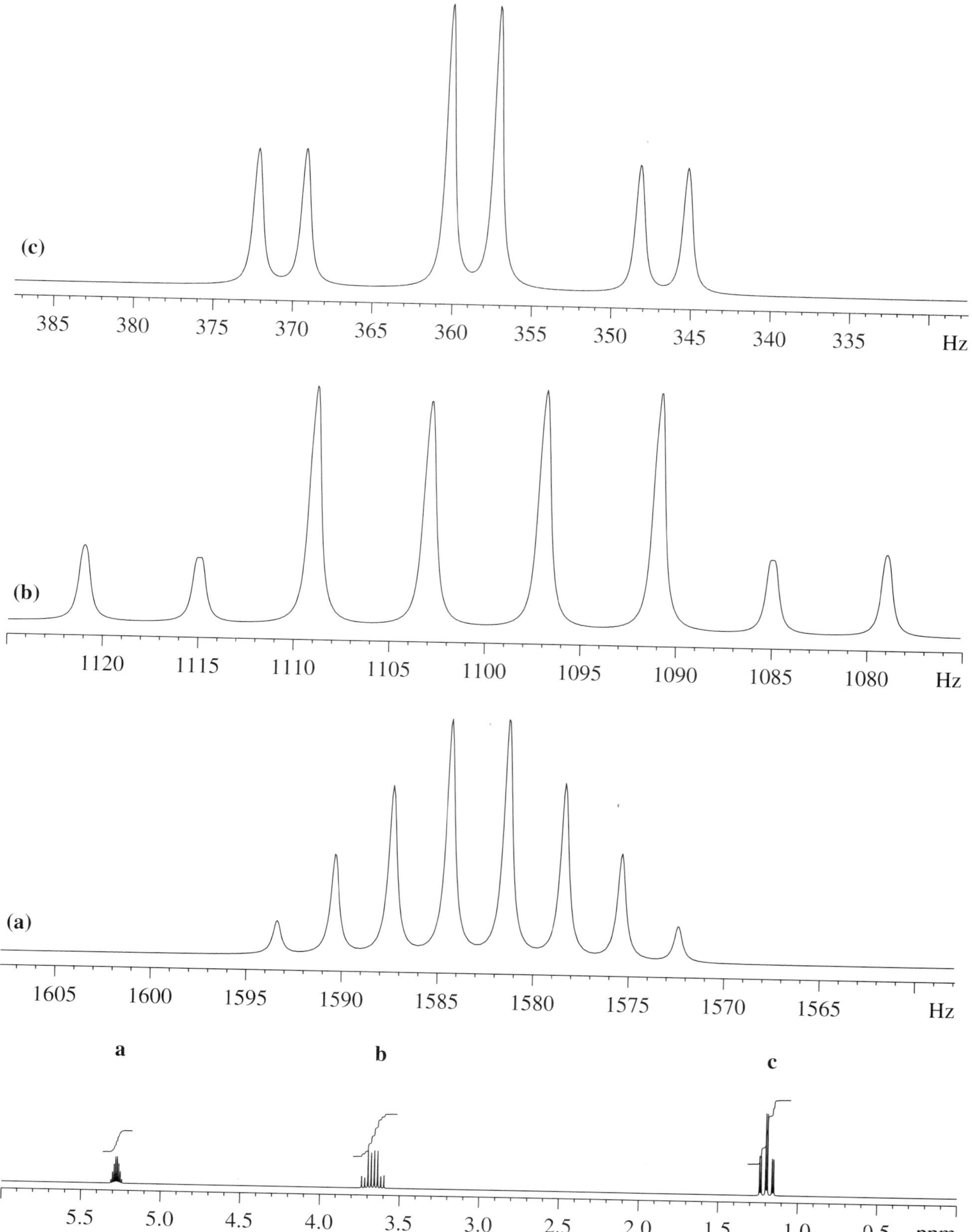

Exercise 3.10B

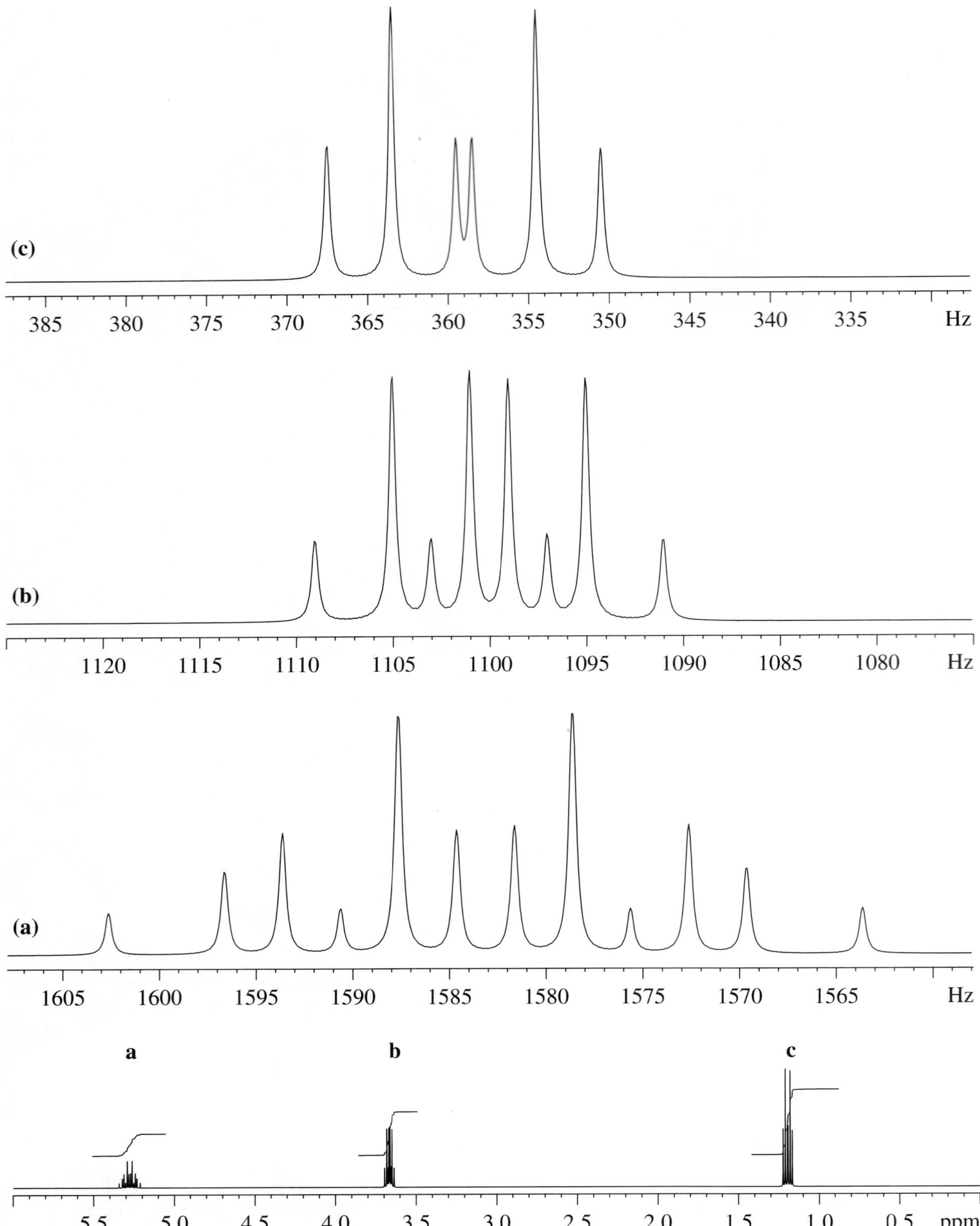

Exercise 3.10C

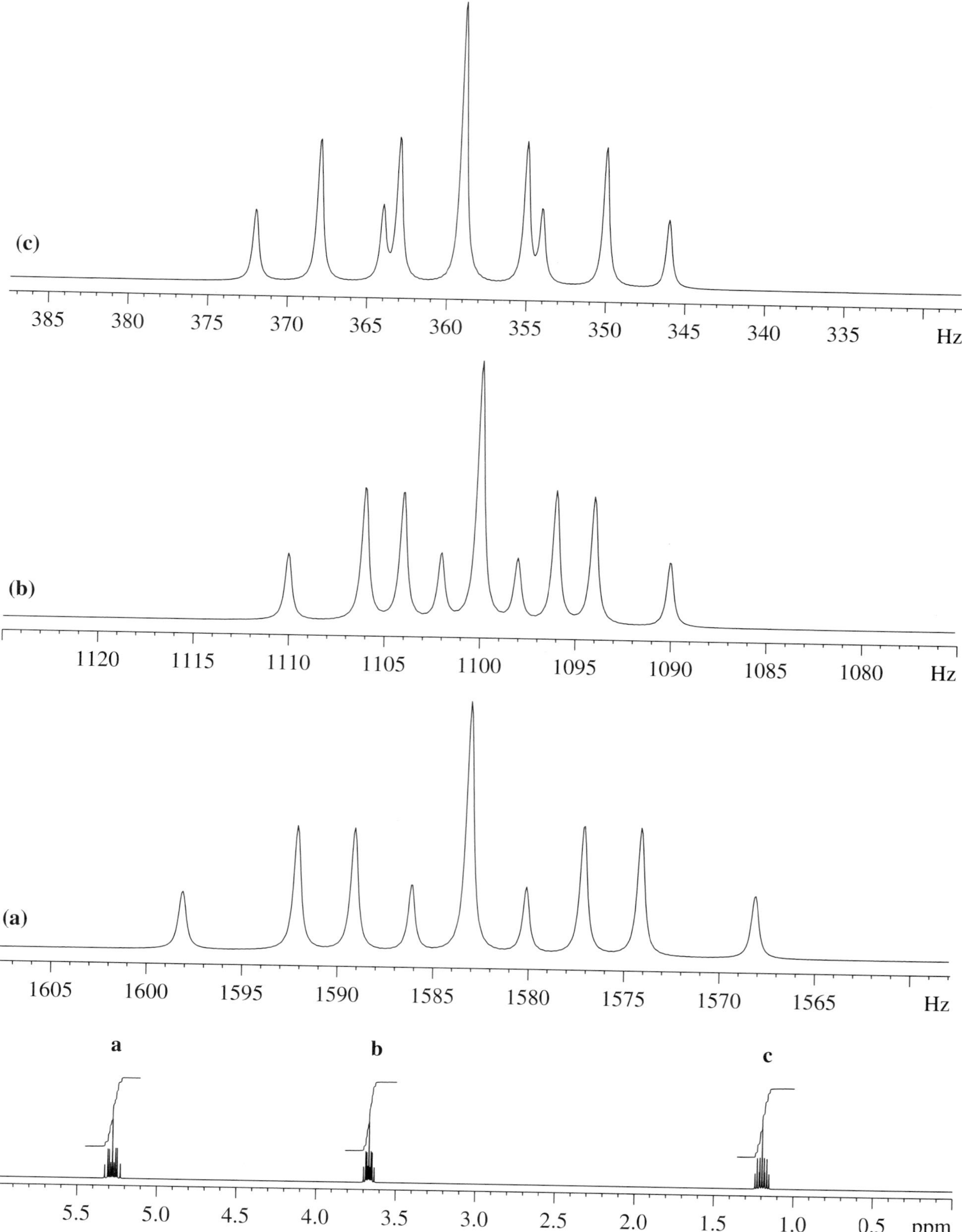

APPENDIX A CHART A.1 CHEMICAL SHIFTS OF PROTONS ON A CARBON ATOM ADJACENT (α POSITION) TO A FUNCTIONAL GROUP IN ALIPHATIC COMPOUNDS (M—Y)

I M = methyl
8 M = methylene
: M = methine

δ

.4 .2 5 .8 .6 .4 .2 4 .8 .6 .4 .2 3 .8 .6 .4 .2 2 .8 .6 .4 .2 1 .8 .6 .4 .2 0

M−CH_2R
M−C=C
M−C≡C
M—Ph
M−F
M−Cl
M−Br
M−I
M−OH
M−OR
M—OPh
M−OC(=O)R
M—OC(=O)Ph
M−OC(=O)CF_3
M−OTs*
M−C(=O)H
M−C(=O)R
M−C(=O)Ph
M−C(=O)OH
M−C(=O)OR
M−C(=O)NR_2
M−C≡N
M−NH_2
M−NR_2
M—NPhR
M−N^+R_3
M−NHC(=O)R
M−NO_2

.4 .2 5 .8 .6 .4 .2 4 .8 .6 .4 .2 3 .8 .6 .4 .2 2 .8 .6 .4 .2 1 .8 .6 .4 .2 0

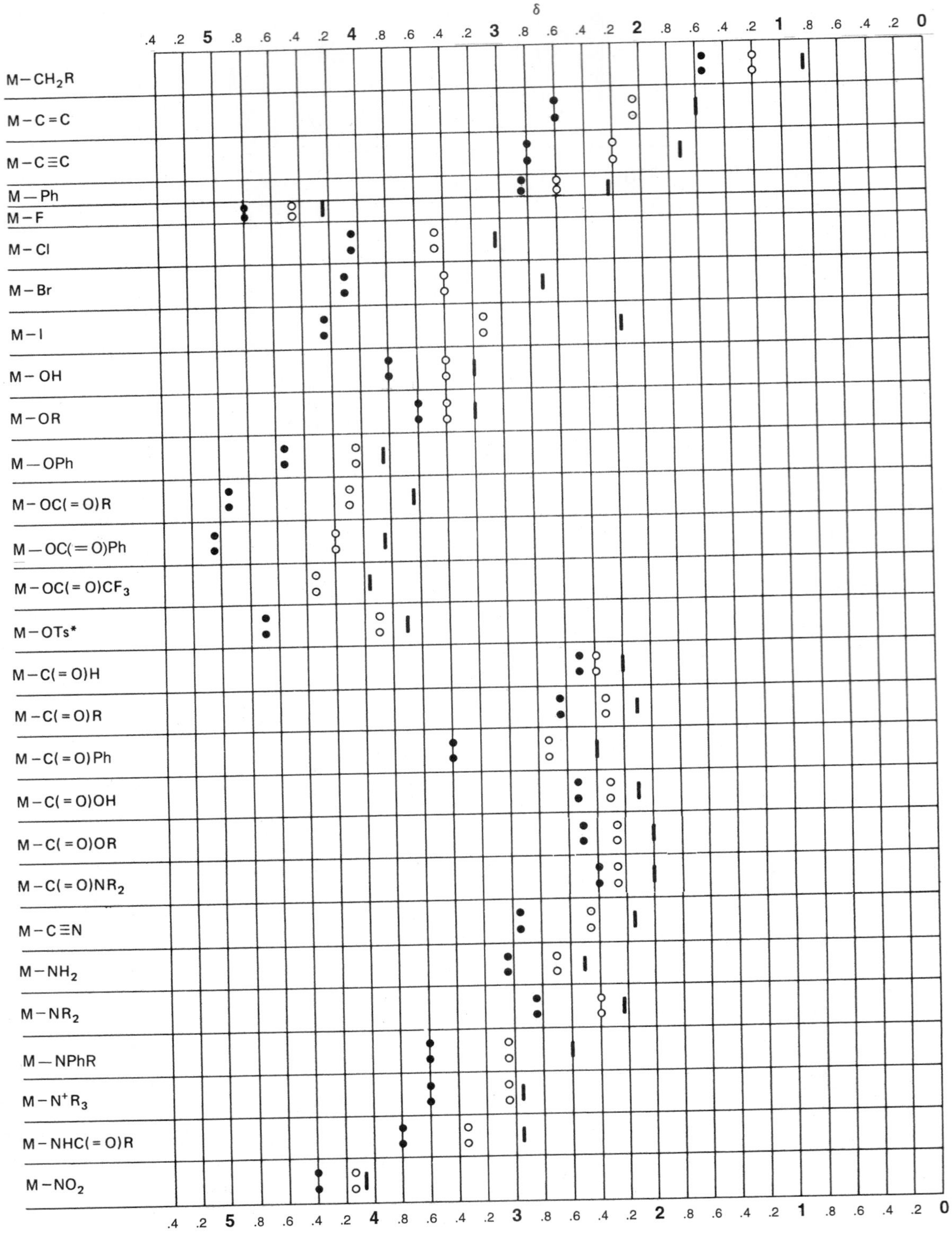

APPENDIX A *(Continued)*

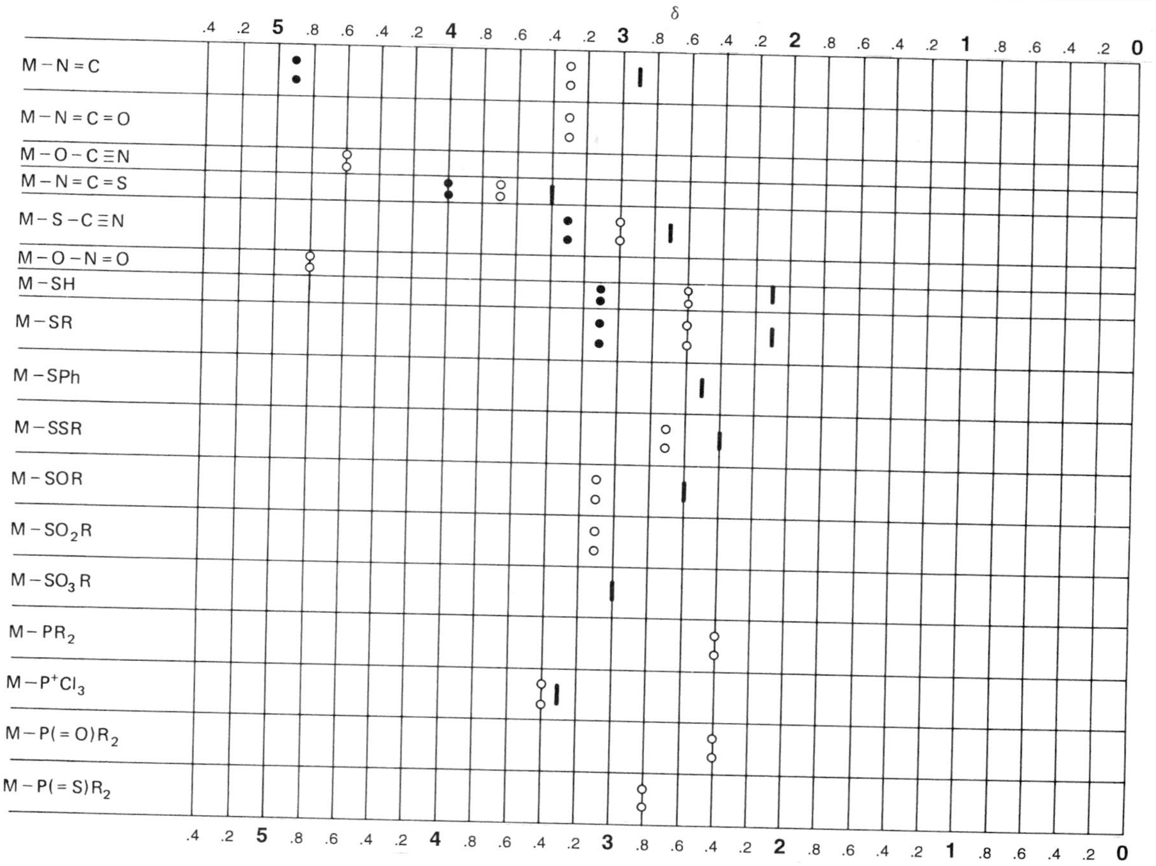

δ

.4 .2 **5** .8 .6 .4 .2 **4** .8 .6 .4 .2 **3** .8 .6 .4 .2 **2** .8 .6 .4 .2 **1** .8 .6 .4 .2 **0**

M−N=C

M−N=C=O

M−O−C≡N

M−N=C=S

M−S−C≡N

M−O−N=O

M−SH

M−SR

M−SPh

M−SSR

M−SOR

M−SO_2R

M−SO_3R

M−PR_2

M−P^+Cl_3

M−P(=O)R_2

M−P(=S)R_2

.4 .2 **5** .8 .6 .4 .2 **4** .8 .6 .4 .2 **3** .8 .6 .4 .2 **2** .8 .6 .4 .2 **1** .8 .6 .4 .2 **0**

*OTS is —O—S(=O)(=O)—C_6H_4—CH_3

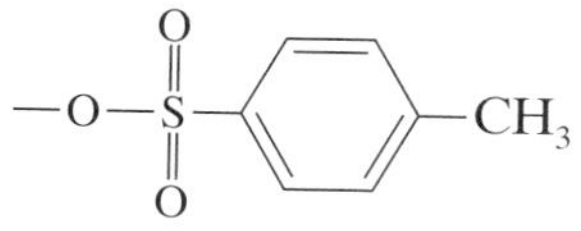

APPENDIX A CHART A.2 CHEMICAL SHIFTS OF PROTONS ON A CARBON ATOM ONCE REMOVED (β POSITION) FROM A FUNCTIONAL GROUP IN ALIPHATIC COMPOUNDS (M—C—Y)

- ▮ M = methyl
- 8 M = methylene
- ⁝ M = methine

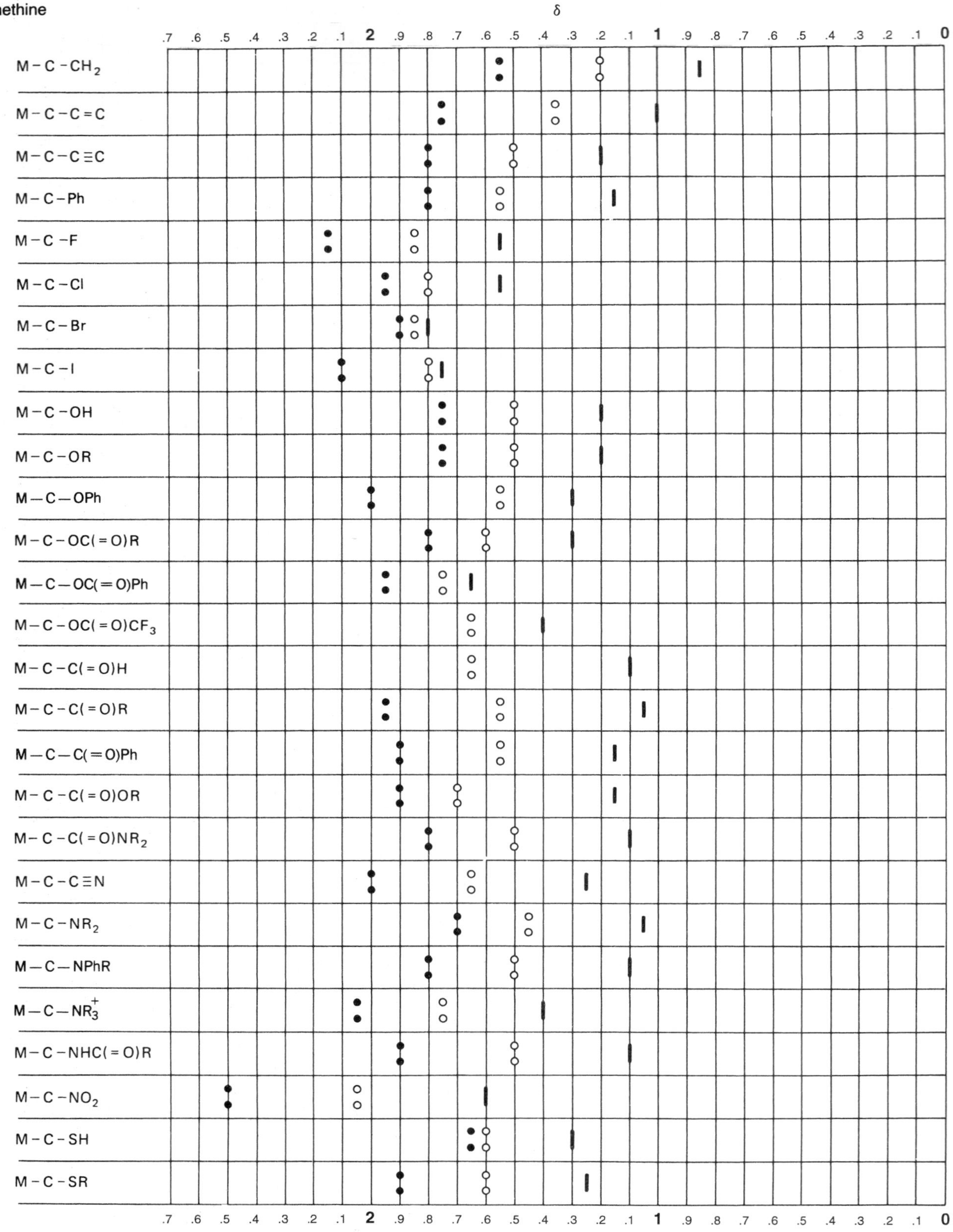

δ

.7 .6 .5 .4 .3 .2 .1 **2** .9 .8 .7 .6 .5 .4 .3 .2 .1 **1** .9 .8 .7 .6 .5 .4 .3 .2 .1 **0**

M–C–CH_2

M–C–C=C

M–C–C≡C

M–C–Ph

M–C–F

M–C–Cl

M–C–Br

M–C–I

M–C–OH

M–C–OR

M–C–OPh

M–C–OC(=O)R

M–C–OC(=O)Ph

M–C–OC(=O)CF_3

M–C–C(=O)H

M–C–C(=O)R

M–C–C(=O)Ph

M–C–C(=O)OR

M–C–C(=O)NR_2

M–C–C≡N

M–C–NR_2

M–C–NPhR

M–C–$\overset{+}{N}R_3$

M–C–NHC(=O)R

M–C–NO_2

M–C–SH

M–C–SR

.7 .6 .5 .4 .3 .2 .1 **2** .9 .8 .7 .6 .5 .4 .3 .2 .1 **1** .9 .8 .7 .6 .5 .4 .3 .2 .1 **0**

APPENDIX B EFFECT ON CHEMICAL SHIFTS BY TWO OR THREE DIRECTLY ATTACHED FUNCTIONAL GROUPS

$$Y{-}CH_2{-}Z \text{ and } Y{-}\underset{\displaystyle W}{\underset{|}{CH}}{-}Z$$

The chemical shift of a methylene group attached to two functional groups can be calculated by means of the substituent constants (σ values) in Table B.1. Shoolery's rule* states that the sum of the constants for the attached functional groups is added to δ 0.23, the chemical shift for CH_4:

$$\delta\,(Y{-}CH_2{-}Z) = 0.23 + \sigma_Y + \sigma_Z$$

The chemical shift for the methylene protons, of $C_6H_5CH_2Br$, for example, is calculated from the σ values in Table B.1.

$$\begin{aligned} & 0.23 \\ \sigma_{Ph} &= 1.85 \\ \sigma_{Br} &= 2.33 \\ \hline \delta &= 4.41 \quad \text{Found, } \delta\ 4.43 \end{aligned}$$

Shoolery's original constants have been revised and extended in Table B.1. The observed and calculated chemical shifts for 62% of the samples tested were within ±0.2 ppm, 92% within ±0.3 ppm. 96% within 0.4 ppm, and 99% within ±0.5 ppm.† Table B.1 contains substituent constants (Friedrich and Runkle, 1984) for the more common functional groups. Note that chemical shifts of methyl protons can be calculated by using the constant for H (0.34). For example H—CH_2—Br is equivalent to CH_3Br.

* Shoolery, J.N. (1959). *Varian Technical Information Bulletin*, Vol 2, No. 3. Palo Alto, CA: Varian Associates.

† Data from Friedrich, E.C., and Runkle, K.G. (1984). *J. Chem. Educ.* **61,** 830; (1986)**63,** 127.

TABLE B.1 Substituent Constants for Alkyl Methylene (and Methyl) Protons.

Y or Z	Substituent Constants (σ)	Y or Z	Substituent Constants (σ)
—H	0.34	—OC(=O)R	3.01
—CH_3	0.68	—OC(=O)Ph	3.27
—C=C	1.32	—C(=O)R	1.50
—C≡C	1.44	—C(=O)Ph	1.90
—Ph	1.83	—C(=O)OR	1.46
—CF_2	1.12	—C(=O)$NR_2(H_2)$	1.47
—CF_3	1.14	—C≡N	1.59
—F	3.30	—$NR_2(H_2)$	1.57
—Cl	2.53	—NHPh	2.04
—Br	2.33	—NHC(=O)R	2.27
—I	2.19	—N_3	1.97
—OH	2.56	—NO_2	3.36
—OR	2.36	—SR(H)	1.64
—OPh	2.94	—OSO_2R	3.13

Tables B.2a, B.2b, and B.2c: Chemical Shift Correlations for Methine Protons

Table B.2a gives the substituent constants* to be used with the formulation

$$\delta\,CHXYZ = 2.50 + \sigma_X + \sigma_Y + \sigma_Z$$

which is satisfactory if at least two of the substituents are electron-withdrawing groups. In other words, only a single substituent may be an alkyl group (R). Within these limits, the standard error of estimate is 0.20 ppm. For example, the chemical shift of the methine proton in

$$CH_3{-}\overset{\displaystyle OEt}{\overset{|}{CH}}{-}OEt$$

is calculated from Table B.2a as follows:

$$\delta = 2.50 + 1.14 + 1.14 + 0.00 = 4.78$$

The found value is 4.72.

Tables B.2b and B.2c are used jointly for methine protons that are substituted by at least two alkyl groups

* Bell, H.M., Bowles, D.B. and Senese, F. (1981). *Org. Magn. Reson.*, **16,** 285. With permission.

TABLE B.2a Substituent Constants for Methine Protons.

Group	(σ)
—F	1.59
—Cl	1.56
—Br	1.53
—NO_2	1.84
—NH_2	0.64
—NH_3^+	1.34
—NHCOR	1.80
—OH, —OR	1.14
—OAr	1.79
—OCOR	2.07
—Ar	0.99
—C=C	0.46
—C≡C	0.79
—C≡N	0.66
—COR, —COOR, —COOH	0.47
—$CONH_2$	0.60
—COAr	1.22
—SH, —SR	0.61
—SO_2R	0.94
—R	0

TABLE B.2b Observed Methine Proton Chemical Shifts of Isopropyl Derivatives.

$(CH_3)_2CHZ$		$(CH_3)_2CHZ$	
Z	**δ (ppm) obs**	**Z**	**δ (ppm) obs**
H	1.33	HO	3.94
H_3C	1.56	RO	3.55
R	1.50	C_6H_5O	4.51
XCH_2	1.85	R(H)C(=O)O	4.94
R(H)C(=O)	2.54	$C_6H_5C(=O)O$	5.22
$C_6H_5C(=O)$	3.58	$F_3CC(=O)O$	5.20
R(H)OC(=O)	2.52	$ArSO_2O$	4.70
$R_2(H_2)NC(=O)$	2.44		
C_6H_5	2.89	R(H)S	3.16
$R_2(H_2)C=CR(H)$	2.62	RSS	2.63
R(H)C≡C	2.59		
N≡C	2.67	F	4.50
		Cl	4.14
$R_2(H_2)N$	3.07	Br	4.21
R(H)C(=O)NH	4.01	I	4.24
O_2N	4.67		

TABLE B.2c Correction Factors for Methine Substituents of Low Polarity.

Open-Chain Methine Proton Systems	Δ*xy*	Cyclic Methine Proton Systems	Δ*xy*
$CH_3-C\underline{H}(Z)-CH_3$	0.00	 Z, $\underline{H}$ (cyclopropane)	−1.0
$CH_3-C\underline{H}(Z)-R$	−0.20	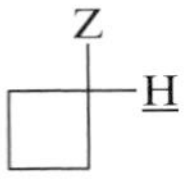 Z, $\underline{H}$ (cyclobutane)	+0.40
$R-C\underline{H}(Z)-R$	−0.40	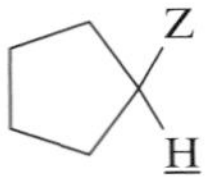 Z, $\underline{H}$ (cyclopentane)	+0.20
$CH_3-C\underline{H}(Z)-CH_2X$	+0.20	Z, $\underline{H}$ (cyclohexane)	monosub. −0.20
			axial H −0.45
$CH_3-C\underline{H}(Z)-CH=CH_2$	+0.40		equat. H +0.25
$CH_3-C\underline{H}(Z)-C_6H_5$	+1.15	 Z, $\underline{H}$ (cycloheptane)	0.00
$R-C\underline{H}(Z)-C_6H_5$	+0.90	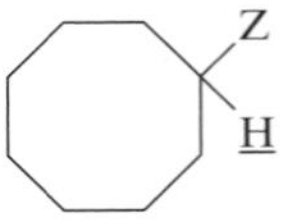 Z, $\underline{H}$ (cyclooctane)	0.00

(or other groups of low polarity). Friedrich and Runkle proposed the relationship

$$\delta_{CHXYZ} = \delta_{(CH_3)_2CHZ} = \Delta xy$$

in which the X and Y substituents are alkyl groups or other groups of low polarity. The Z susbstituent covers a range of polarities. Δ*xy* is a correction factor. The relationship states that the chemical shift of a methine proton with at least two low-polarity groups is equivalent to the chemical shift of an isopropyl methine proton plus correction factor.

The substituent constants for a Z substituent on an isopropyl methine proton are given in Table B.2b. The Δ*xy* correction factors are given in Table B.2c.

The following example illustrates the joint use of Tables B.2b and B.2c, with CH_3, $CH=CH_2$, and C_6H_5 as substituents. The most polar substituent is always designated Z.

$$\delta\ X-\overset{Z}{\overset{|}{C}}\underline{H}-Y = \delta\ CH_3-\overset{Z}{\overset{|}{C}}\underline{H}-CH=CH_2 = \delta\ CH_3-\overset{C_6H_5}{\overset{|}{C}}\underline{H}-CH_3 + \Delta xy$$

From Table B.2b, $\delta = 2.89$ for $CH_3-\overset{C_6H_5}{\overset{|}{C}}\underline{H}-CH_3$.

From Table B.2c, $\Delta xy = 0.00$ for CH_3. $\Delta xy = 0.40$ for $CH=CH_2$.

Therefore, $\delta\ CH_3-\overset{C_6H_5}{\overset{|}{C}}\underline{H}-CH=CH_2 = 2.89 + 0.00 + 0.40 = 3.29$ (Found: $\delta = 3.44$).

APPENDIX C CHEMICAL SHIFTS IN ALICYCLIC AND HETEROCYCLIC RINGS

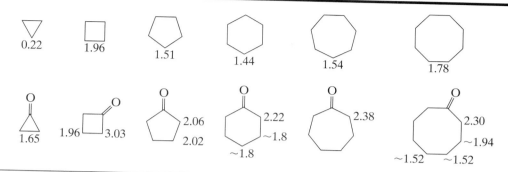

TABLE C.1 Chemicals Shifts in Alicyclic Rings.

0.22 1.96 1.51 1.44 1.54 1.78

O
O
O
O
O
O
1.65
1.96 3.03
2.06 2.02
2.22 ~1.8 ~1.8
2.38
2.30 ~1.94 ~1.52 ~1.52

TABLE C.2 Chemical Shifts in Heterocyclic Rings.

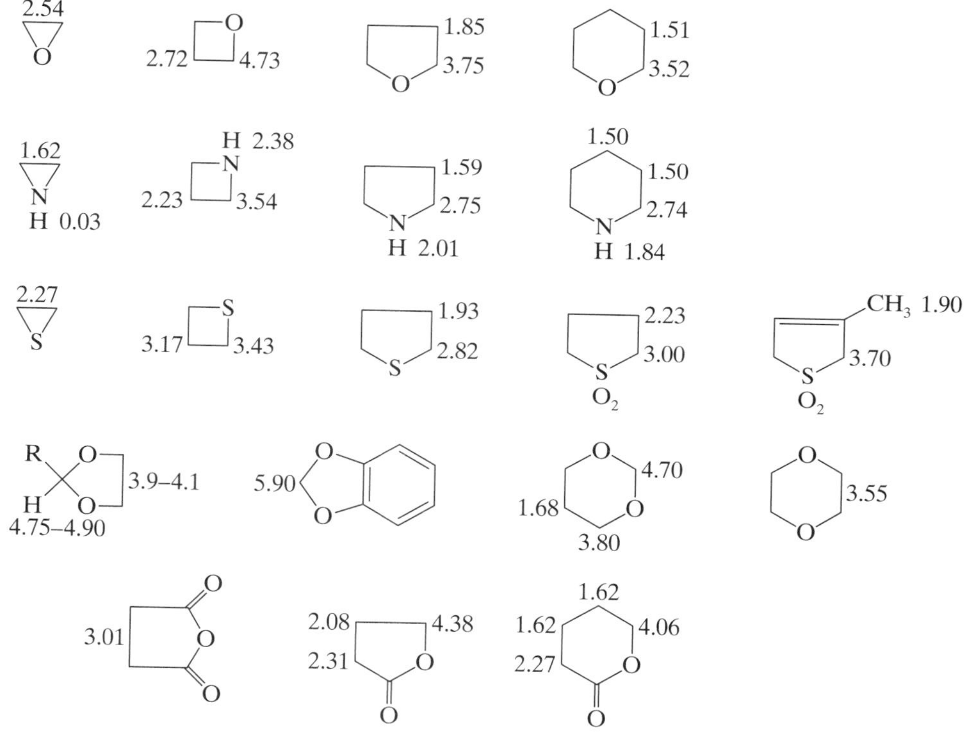

APPENDIX D CHEMICAL SHIFTS IN UNSATURATED AND AROMATIC SYSTEMS

(See Table D.1)

$$\delta_H = 5.25 + Z_{gem} + Z_{cis} + Z_{trans}$$

(R_{cis} and R_{trans} on one carbon, H and R_{gem} on the other: $R_{cis}(R_{trans})C{=}C(H)R_{gem}$)

For example, the chemical shifts of the alkene protons in

$C_6H_5(H_a)C{=}C(H_b)OC_2H_5$

are calculated:

H_a	$C_6H_5\ _{gem}$	1.35	5.25
	OR_{trans}	−1.28	0.07
		0.07	δ 5.32
H_b	OR_{gem}	1.18	5.25
	$C_6H_5\ _{trans}$	−0.10	1.08
		1.08	δ 6.33

TABLE D.1 Substituent Constants (Z) for Chemical Shifts of Substituted Ethylenes.

Substituent R	Z gem	Z cis	Z trans	Substituent R	Z gem	Z cis	Z trans
—H	0	0	0	—C(H)=O	1.03	0.97	1.21
—Alkyl	0.44	−0.26	−0.29				
—Alkyl-ring[a]	0.71	−0.33	−0.30				
$—CH_2O$, $—CH_2I$	0.67	−0.02	−0.07	—C(N)=O	1.37	0.93	0.35
$—CH_2S$	0.53	−0.15	−0.15				
$—CH_2Cl$, $—CH_2Br$	0.72	0.12	0.07	—C(Cl)=O	1.10	1.41	0.99
$—CH_2N$	0.66	−0.05	−0.23				
—C≡C	0.50	0.35	0.10				
—C≡N	0.23	0.78	0.58	—OR, R: aliph	1.18	−1.06	−1.28
—C=C	0.98	−0.04	−0.21	—OR, R: conj[b]	1.14	−0.65	−1.05
—C=C conj[b]	1.26	0.08	−0.01	—OCOR	2.09	−0.40	−0.67
—C=O	1.10	1.13	0.81	—Aromatic	1.35	0.37	−0.10
—C=O conj[b]	1.06	1.01	0.95	—Cl	1.00	0.19	0.03
—COOH	1.00	1.35	0.74	—Br	1.04	0.40	0.55
—COOH conj[b]	0.69	0.97	0.39	$—NR_2$ R:aliph	0.69	−1.19	−1.31
—COOR	0.84	1.15	0.56	$—NR_2$ R:conj[b]	2.30	−0.73	−0.81
—COOR conj[b]	0.68	1.02	0.33	—SR	1.00	−0.24	−0.04
				$—SO_2$	1.58	1.15	0.95

[a] Alkyl ring indicates that the double bond is part of the ring R(C=C).

[b] The Z factor for the conjugated substituent is used when either the substituent or the double bond is further conjugated with other groups.

Source: Pascual C., Meier, J., and Simon, W. (1966) *Helv. Chim. Acta,* **49,** 164.

TABLE D.2 Chemical Shifts of Miscellaneous Alkenes

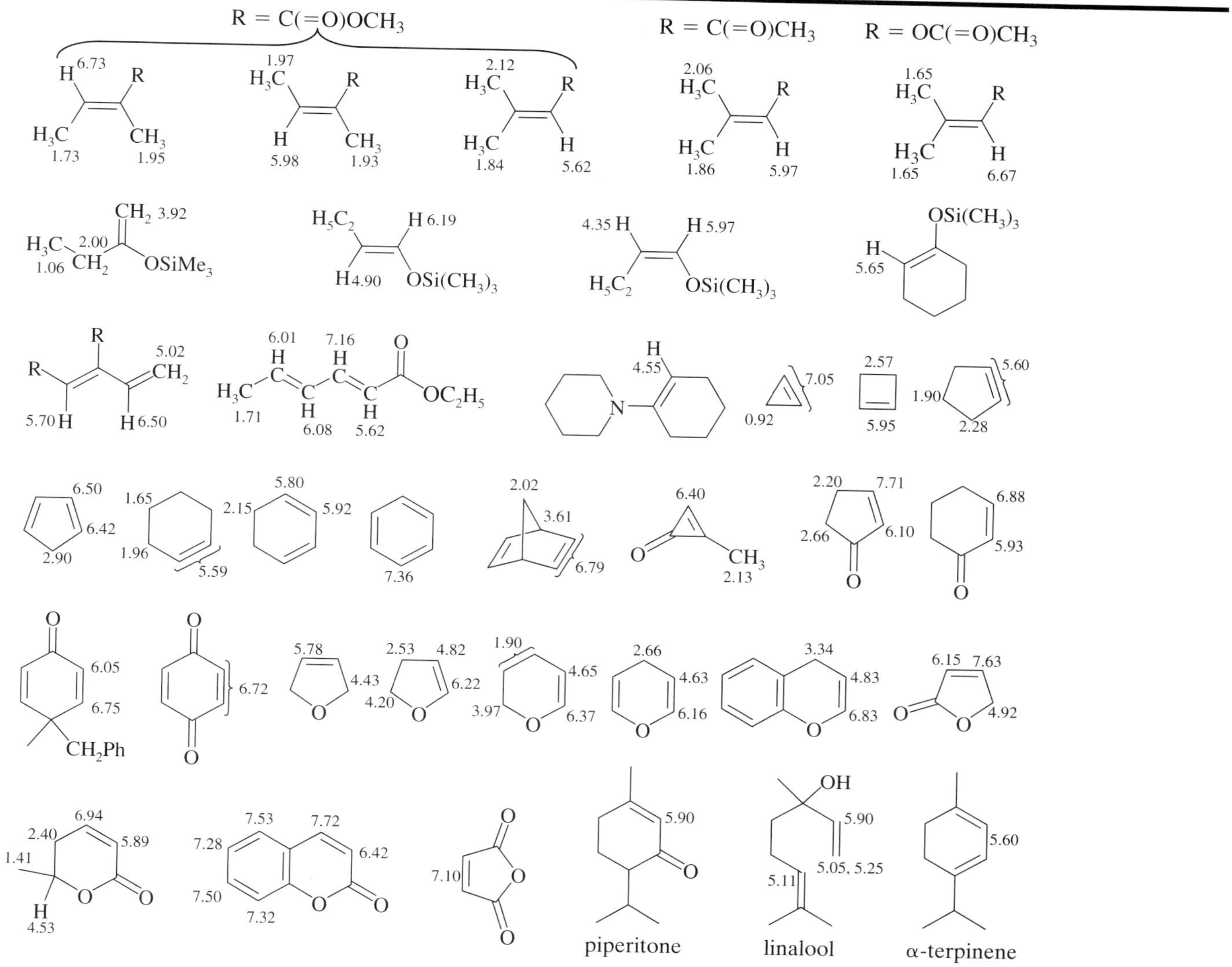

R = C(=O)OCH$_3$ R = C(=O)CH$_3$ R = OC(=O)CH$_3$

piperitone linalool α-terpinene

TABLE D.3 Chemical Shifts of Alkyne Protons

HC≡CR	1.73–1.88	HC≡C—COH	2.23
HC≡C—C≡CR	1.95	HC≡CH	1.80
HC≡C—Ph	2.71–3.37	HC≡C—CH=CR$_2$	2.60–3.10

TABLE D.4 Chemical Shifts of Protons on Fused Aromatic Rings

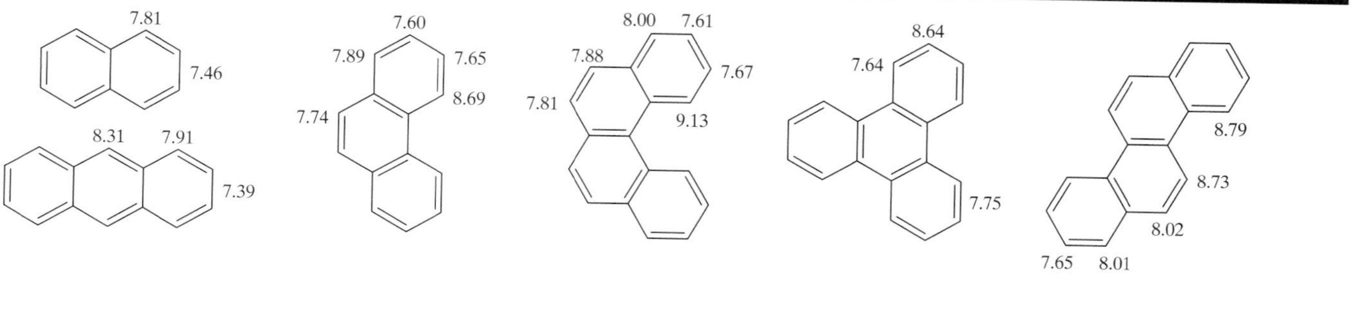

CHART D.1 CHEMICAL SHIFTS OF PROTONS ON MONOSUBSTITUTED BENZENE RINGS

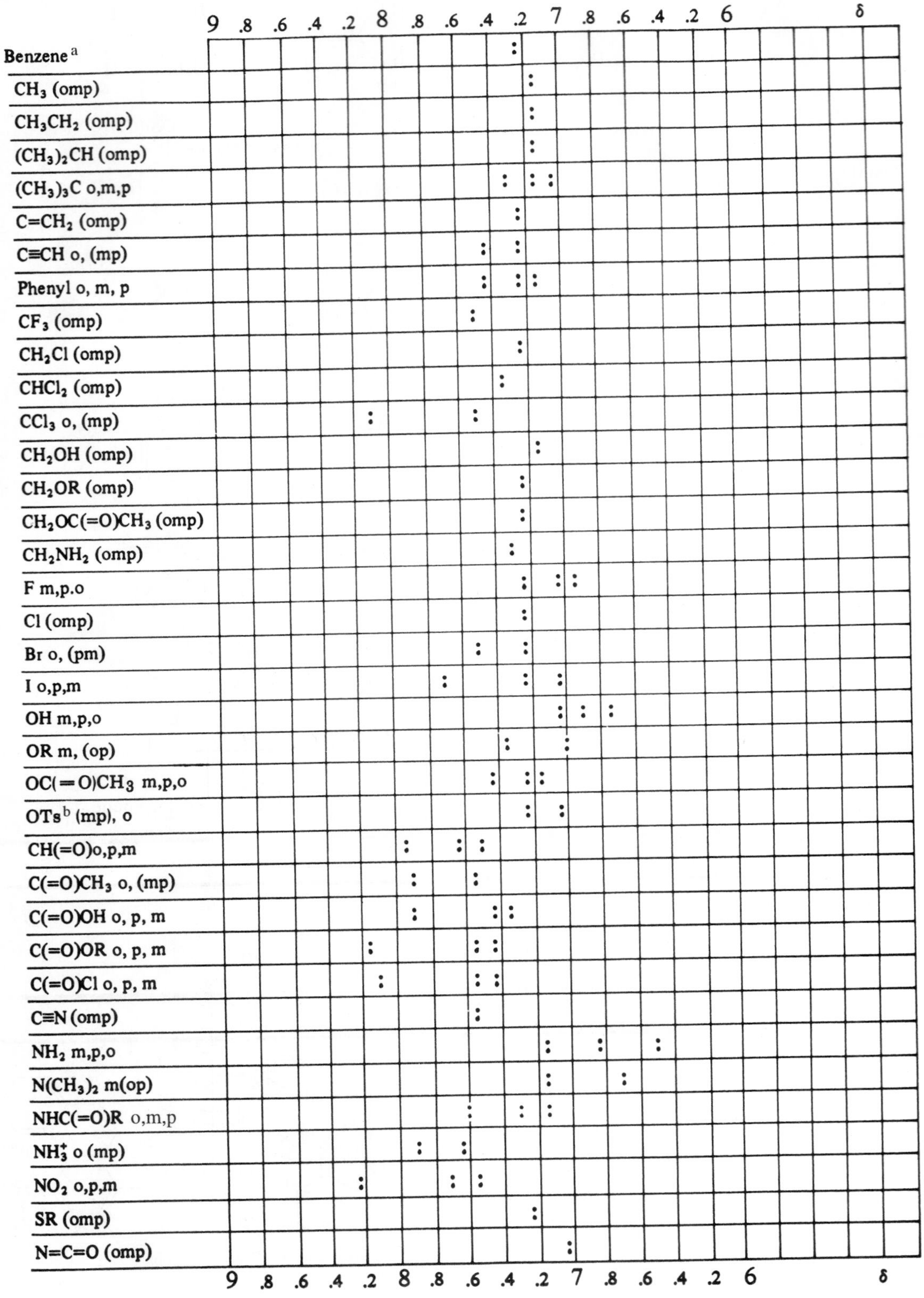

	9		.8		.6		.4		.2		8		.8		.6		.4		.2		7		.8		.6		.4		.2		6	δ
Benzene[a]																			:													
CH_3 (omp)																				:												
CH_3CH_2 (omp)																				:												
$(CH_3)_2CH$ (omp)																				:												
$(CH_3)_3C$ o,m,p																		:		:	:											
$C{=}CH_2$ (omp)																			:													
$C{\equiv}CH$ o, (mp)																	:		:													
Phenyl o, m, p																	:		:	:												
CF_3 (omp)																:																
CH_2Cl (omp)																			:													
$CHCl_2$ (omp)																		:														
CCl_3 o, (mp)										:						:																
CH_2OH (omp)																				:												
CH_2OR (omp)																			:													
$CH_2OC({=}O)CH_3$ (omp)																			:													
CH_2NH_2 (omp)																		:														
F m,p.o																			:		:	:										
Cl (omp)																			:													
Br o, (pm)																:			:													
I o,p,m														:					:		:											
OH m,p,o																					:	:		:								
OR m, (op)																		:			:											
$OC({=}O)CH_3$ m,p,o																	:		:	:												
OTs[b] (mp), o																			:		:											
$CH({=}O)$o,p,m												:			:	:																
$C({=}O)CH_3$ o, (mp)												:				:																
$C({=}O)OH$ o, p, m												:					:	:														
$C({=}O)OR$ o, p, m									:							:	:															
$C({=}O)Cl$ o, p, m										:						:	:															
$C{\equiv}N$ (omp)																:																
NH_2 m,p,o																				:			:			:						
$N(CH_3)_2$ m(op)																				:				:								
$NHC({=}O)R$ o,m,p															:			:		:												
NH_3^+ o (mp)												:			:																	
NO_2 o,p,m									:					:		:																
SR (omp)																			:													
$N{=}C{=}O$ (omp)																					:											

[a] The benzene ring proton is at δ 7.27, from which the shift increments are calculated as shown at the end of Section 3.4.

[b] OTS = *p*-toluenesulfonyloxy group.

TABLE D.5 Chemical Shifts of Protons on Heteroaromatic Rings

TABLE D.6 Chemical Shifts of HC=O, HC=N, and $HC(O)_3$ Protons

RCH=O	9.70	HC(=O)OR	8.05	RC$\underline{H}$=NOH *cis*	7.25
PhCH=O	9.98	HC(=O)NR_2	8.05	RC$\underline{H}$=NOH *trans*	6.65
RCH=CHCH=O	9.78	$HC(OR)_3$	5.00	R($\underline{H}$)C=N–NH–$C_6H_3(NO_2)_2$ (2,4-dinitrophenylhydrazone)	6.05

APPENDIX E PROTONS SUBJECT TO HYDROGEN-BONDING EFFECTS (PROTONS ON HETEROATOMS)[a]

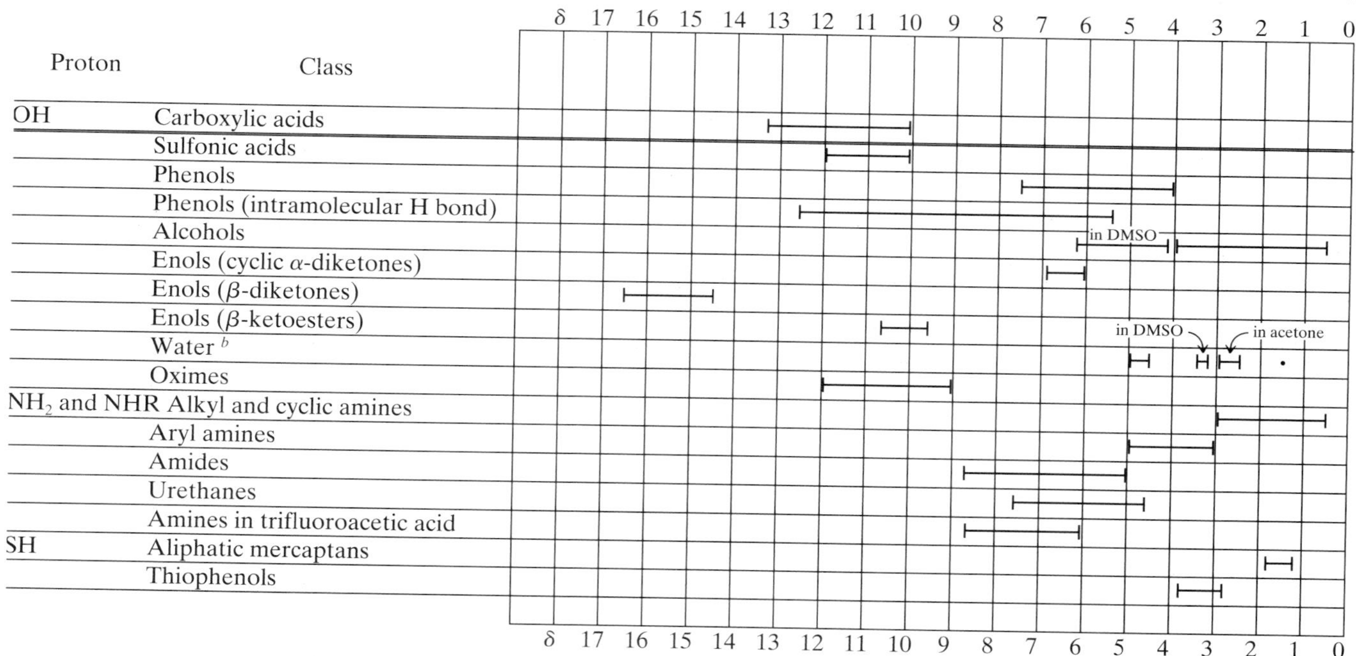

[a] Solvent $CDCl_3$. Chemical shifts within a range are a function of concentration.

[b] See Section 3.6.1.2.

APPENDIX F PROTON SPIN-COUPLING CONSTANTS

Type	J_{ab} (Hz)	J_{ab} Typical
$>C(H_a)(H_b)$ (geminal)	0–30	12–15
$CH_a—CH_b$ (free rotation)	6–8	7
$CH_a—C—CH_b$	0–1	0
Cyclohexane H_a, H_b		
ax–ax	6–14	8–10
ax–eq	0–5	2–3
eq–eq	0–5	2–3
Cyclopentane H_a, H_b (*cis* or *trans*)	*cis* 5–10 *trans* 5–10	
Cyclobutane H_a, H_b (*cis* or *trans*)	*cis* 4–12 *trans* 2–10	
Cyclopropane H_a, H_b (*cis* or *trans*)	*cis* 7–13 *trans* 4–9	
$CH_a—OH_b$ (no exchange)	4–10	5
$>CH_a—C(=O)H_b$	1–3	2–3
$C{=}CH_a—C(=O)H_b$	5–8	6
$H_a—C{=}C—H_b$ (trans)	12–18	17
$>C{=}C(H_a)(H_b)$	0–3	0–2

Type		J_{ab} (Hz)	J_{ab} Typical
$H_a—C{=}C—H_b$ (cis)		6–12	10
$CH_a—C{=}C—CH_b$ (cis)		0–3	1–2
$>C{=}C(CH_a)(H_b)$		4–10	7
$H_a—C{=}C—CH_b$ (trans)		0–3	1.5
$H_a—C{=}C—CH_b$ (cis)		0–3	2
$C{=}CH_a—CH_b{=}C$		9–13	10
$H_a—C{=}C—H_b$ (ring)		3 member	0.5–2.0
		4 member	2.5–4.0
		5 member	5.1–7.0
		6 member	8.8–11.0
		7 member	9–13
		8 member	10–13
$CH_a—C{\equiv}CH_b$		2–3	
$—CH_a—C{\equiv}C—CH_b—$		2–3	
Epoxide, H_a, H_b geminal			6
Epoxide, H_a, H_b cis			4
Epoxide, H_a, H_b trans			2.5
Benzene H_a, H_b	J (ortho)	6–10	9
	J (meta)	1–3	3
	J (para)	0–1	~0
Pyridine	J (2–3)	5–6	5
	J (3–4)	7–9	8
	J (2–4)	1–2	1.5
	J (3–5)	1–2	1.5
	J (2–5)	0–1	1
	J (2–6)	0–1	~0
Furan	J (2–3)	1.3–2.0	1.8
	J (3–4)	3.1–3.8	3.6
	J (2–4)	0–1	~0
	J (2–5)	1–2	1.5

APPENDIX F (Continued)

Type		J_{ab} (Hz)	J_{ab} Typical
Thiophene (S at 1; positions 2–5)	J (2–3)	4.9–6.2	5.4
	J (3–4)	3.4–5.0	4.0
	J (2–4)	1.2–1.7	1.5
	J (2–5)	3.2–3.7	3.4
Pyrrole (N–H at 1; positions 2–5)	J (1–3)	2–3	
	J (2–3)	2–3	
	J (3–4)	3–4	
	J (2–4)	1–2	
	J (2–5)	1.5–2.5	
Pyrimidine (positions 2, 4, 5, 6)	J (4–5)	4–6	
	J (2–5)	1–2	
	J (2–4)	0–1	
	J (4–6)	2–3	
Thiazole (positions 2, 4, 5)	J (4–5)	3–4	
	J (2–4)	~0	
	J (2–5)	1–2	

Type	J_{ab} (Hz)	J_{ab} Typical
Proton–Carbon-13 (See Tables 5.17, 5.18)		
Proton–Fluorine		
$>C(H_a)(F_b)$	44–81	
$>CH_a-CF_b<$	3–25	
$>CH_a-C-CF_b$	0–4	
H_a and F_b cis on $C=C$	1–8	
H_a and F_b trans on $C=C$	12–40	
Fluorobenzene ring–H_a	*o* 6–10 *m* 5–6 *p* 2	
$\alpha\ H_3C-C(=O)-\underset{\beta}{CH_2}F\ \gamma$	αγ 4.3 βγ 48	

Proton–Phosphorus

Type	J_{ab} (Hz)
$>P(=O)H$	630–707
$(CH_3)_3P$	2.7
$(CH_3)_3P{=}0$	13.4
$(CH_3CH_2)_3P$	0.5 (HCCP) 13.7 (HCP)
$(CH_3CH_2)_3P{=}0$	11.9 (HCCP) 16.3 (HCP)
$CH_3P(=O)(OR)_2$	10–13
$CH_3C-P(=O)(OR)_2$	15–20
$CH_3OP(OR)_2$	10.5–12
$P[N(CH_3)_2]_3$	8.8
$O{=}P[N(CH_3)_2]_3$	9.5

Source: Complied by Varian Associates. Absolute values. Reproduced with permission.

APPENDIX G CHEMICAL SHIFTS AND MULTIPLICITIES OF RESIDUAL PROTONS IN COMMERCIALLY AVAILABLE DEUTERATED SOLVENTS (MERCK & CO., INC.)

Compound[a] Molecular Weight	δ_H (multiplet)
Acetic acid-d_4 64.078	11.53 (1) 2.03 (5)
Acetone-d_6 64.117	2.04 (5)
Acetonitrile-d_3 44.071	1.93 (5)
Benzene-d_6 84.152	7.15 (br)
Chloroform-d 120.384	7.26 (1)
Cyclohexane-d_{12} 96.236	1.38 (br)
Deuterium oxide 20.028	4.63 (ref. DSS)[c] 4.67 (ref. TSP)[c]
1,2-Dichloroethane-d_4 102.985	3.72 (br)
Diethyl-d_{10} ether 84.185	3.34 (m) 1.07 (m)
Diglyme-d_{14} 148.263	3.49 (br) 3.40 (br) 3.22 (5)
N, N-Dimethylformamide-d_7 80.138	8.01 (br) 2.91 (5) 2.74 (5)
Dimethyl-d_6 sulphoxide 84.170	2.49 (5)
p-Dioxane-d_8 96.156	3.53 (m)
Ethyl alcohol-d_6 (anh) 52.106	5.19 (1) 3.55 (br) 1.11 (m)
Glyme-d_{10} 100.184	3.40 (m) 3.22 (5)
Hexafluroacetone deuterate 198.067	5.26 (1)
HMPT-d_{18} 197.314	2.53 (2 × 5)
Methyl alcohol-d_4 36.067	4.78 (1) 3.30 (5)
Methylene chloride-d_2 86.945	5.32 (3)
Nitrobenzene-d_5 128.143	8.11 (br) 7.67 (br) 7.50 (br)

Compound[a] Molecular Weight	δ_H (multiplet)
Nitromethane-d_3 64.059	4.33 (5)
Isopropyl alcohol-d_8 68.146	5.12 (1) 3.89 (br) 1.10 (br)
Pyridine-d_5 84.133	8.71 (br) 7.55 (br) 7.19 (br)
Tetrahydrofuran-d_8 80.157	3.58 (br) 1.73 (br)
Toluene-d_8 100.191	7.09 (m) 7.00 (br) 6.98 (m) 2.09 (5)
Trifluoroacetic acid-d 115.030	11.50 (1)
2,2,2-Trifluoroethyl alcohol-d_3 103.059	5.02 (1) 3.88 (4 × 3)

[a] Purity (Atom % D) up to 99.96 % ("100 %") for several solvents.

[b] The residual proton consists of one proton of each kind in an otherwise completely deuterated molecule. For example, deuterated acetic acid has two different kinds of residual protons: CD_2H—COOD and CD_3—COOH. The CD_2H proton, coupled to two D nuclei is at δ 2.03 with a multiplicity of 5 (i.e., $2nI + 1 = 2 \times 2 \times 1 + 1 = 5$). The carboxylic proton is a singlet at δ 11.53.

[c] DSS is 3-trimethylsilyl)-1-propane sulfonic acid, sodium salt. TSP is sodium-3-trimethylpropionate-2,2,3,3-d_4. Both are reference standards used in aqueous solutions.

APPENDIX H CHEMICAL SHIFTS OF COMMON LABORATORY SOLVENTS as TRACE IMPURITIES

	proton	mult	$CDCl_3$	$(CD_3)_2CO$	$(CD_3)_2SO$	C_6D_6	CD_3CN	CD_3OD	D_2O
solvent residual peak			7.26	2.05	2.50	7.16	1.94	3.31	4.79
H_2O		s	1.56	2.84[a]	3.33[a]	0.40	2.13	4.87	
acetic acid	CH_3	s	2.10	1.96	1.91	1.55	1.96	1.99	2.08
acetone	CH_3	s	2.17	2.09	2.09	1.55	2.08	2.15	2.22
acetonitrile	CH_3	s	2.10	2.05	2.07	1.55	1.96	2.03	2.06
benzene	CH	s	7.36	7.36	7.37	7.15	7.37	7.33	
tert-butyl alcohol	CH_3	s	1.28	1.18	1.11	1.05	1.16	1.40	1.24
	OH[c]	s			4.19	1.55	2.18		
tert-butyl methyl ether	CCH_3	s	1.19	1.13	1.11	1.07	1.14	1.15	1.21
	OCH_3	s	3.22	3.13	3.08	3.04	3.13	3.20	3.22
BHT[b]	ArH	s	6.98	6.96	6.87	7.05	6.97	6.92	
	OH[c]	s	5.01		6.65	4.79	5.20		
	$ArCH_3$	s	2.27	2.22	2.18	2.24	2.22	2.21	
	$ArC(CH_3)_3$	s	1.43	1.41	1.36	1.38	1.39	1.40	
chloroform	CH	s	7.26	8.02	8.32	6.15	7.58	7.90	
cyclohexane	CH_2	s	1.43	1.43	1.40	1.40	1.44	1.45	
1,2-dichloroethane	CH_2	s	3.73	3.87	3.90	2.90	3.81	3.78	
dichloromethane	CH_2	s	5.30	5.63	5.76	4.27	5.44	5.49	
diethyl ether	CH_3	t, 7	1.21	1.11	1.09	1.11	1.12	1.18	1.17
	CH_2	q, 7	3.48	3.41	3.38	3.26	3.42	3.49	3.56
diglyme	CH_2	m	3.65	3.56	3.51	3.46	3.53	3.61	3.67
	CH_2	m	3.57	3.47	3.38	3.34	3.45	3.58	3.61
	OCH_3	s	3.39	3.28	3.24	3.11	3.29	3.35	3.37
1,2-dimethoxyethane	CH_3	s	3.40	3.28	3.24	3.12	3.28	3.35	3.37
	CH_2	s	3.55	3.46	3.43	3.33	3.45	3.52	3.60
dimethylacetamide	CH_3CO	s	2.09	1.97	1.96	1.60	1.97	2.07	2.08
	NCH_3	s	3.02	3.00	2.94	2.57	2.96	3.31	3.06
	NCH_3	s	2.94	2.83	2.78	2.05	2.83	2.92	2.90
dimethylformamide	CH	s	8.02	7.96	7.95	7.63	7.92	7.97	7.92
	CH_3	s	2.96	2.94	2.89	2.36	2.89	2.99	3.01
	CH_3	s	2.88	2.78	2.73	1.86	2.77	2.86	2.85
dimethyl sulfoxide	CH_3	s	2.62	2.52	2.54	1.68	2.50	2.65	2.71
dioxane	CH_2	s	3.71	3.59	3.57	3.35	3.60	3.66	3.75
ethanol	CH_3	t, 7	1.25	1.12	1.06	0.96	1.12	1.19	1.17
	CH_2	q, 7[d]	3.72	3.57	3.44	3.34	3.54	3.60	3.65
	OH	s[cd]	1.32	3.39	4.63		2.47		
ethyl acetate	CH_3CO	s	2.05	1.97	1.99	1.65	1.97	2.01	2.07
	$\mathit{CH_2}CH_3$	q, 7	4.12	4.05	4.03	3.89	4.06	4.09	4.14
	$CH_2\mathit{CH_3}$	t, 7	1.26	1.20	1.17	0.92	1.20	1.24	1.24
ethyl methyl ketone	CH_3CO	s	2.14	2.07	2.07	1.58	2.06	2.12	2.19
	CH_2CH_3	q, 7	2.46	2.45	2.43	1.81	2.43	2.50	3.18
	$CH_2\mathit{CH_3}$	t, 7	1.06	0.96	0.91	0.85	0.96	1.01	1.26
ethylene glycol	CH	s[e]	3.76	3.28	3.34	3.41	3.51	3.59	3.65
"grease"[f]	CH_3	m	0.86	0.87		0.92	0.86	0.88	
	CH_2	br s	1.26	1.29		1.36	1.27	1.29	
n-hexane	CH_3	t	0.88	0.88	0.86	0.89	0.89	0.90	
	CH_2	m	1.26	1.28	1.25	1.24	1.28	1.29	
HMPA[g]	CH_3	d, 9.5	2.85	2.59	2.53	2.40	2.57	2.64	2.61
methanol	CH_3	s[h]	3.49	3.31	3.16	3.07	3.28	3.34	3.34
	OH	s[gh]	1.09	3.12	4.01		2.16		
nitromethane	CH_3	s	4.33	4.43	4.42	2.94	4.31	4.34	4.40
n-pentane	CH_3	t, 7	0.88	0.88	0.86	0.87	0.89	0.90	
	CH_2	m	1.27	1.27	1.27	1.23	1.29	1.29	
2-propanol	CH_3	d, 6	1.22	1.10	1.04	0.95	1.09	1.50	1.17
	CH	sep, 6	4.04	3.90	3.78	3.67	3.87	3.92	4.02

APPENDIX H (Continued)

	proton	mult	$CDCl_3$	$(CD_3)_2CO$	$(CD_3)_2SO$	C_6D_6	CD_3CN	CD_3OD	D_2O
pyridine	CH(2)	m	8.62	8.58	8.58	8.53	8.57	8.53	8.52
	CH(3)	m	7.29	7.35	7.39	6.66	7.33	7.44	7.45
	CH(4)	m	7.68	7.76	7.79	6.98	7.73	7.85	7.87
silicone grease[i]	CH_3	s	0.07	0.13		0.29	0.08	0.10	
tetrahydrofuran	CH_2	m	1.85	1.79	1.76	1.40	1.80	1.87	1.88
	CH_2O	m	3.76	3.63	3.60	3.57	3.64	3.71	3.74
toluene	CH_3	s	2.36	2.32	2.30	2.11	2.33	2.32	
	CH(*o*/*p*)	m	7.17	7.1–7.2	7.18	7.02	7.1–7.3	7.16	
	CH(*m*)	m	7.25	7.1–7.2	7.25	7.13	7.1–7.3	7.16	
triethylamine	CH_3	t, 7	1.03	0.96	0.93	0.96	0.96	1.05	0.99
	CH_2	q, 7	2.53	2.45	2.43	2.40	2.45	2.58	2.57

[a] In these solvents the intermolecular rate of exchange is slow enough that a peak due to HDO is usually also observed; it appears at 2.81 and 3.30 ppm in acetone and DMSO, respectively. In the former solvent, it is often seen as a 1:1:1 triplet, with $^2J_{H,D} = 1$ Hz.
[b] 2,6-Dimethyl-4-*tert*-butylphenol.
[c] The signals from exchangeable protons were not always identified.
[d] In some cases (see note *a*), the coupling interaction between the CH_2 and the OH protons may be observed ($J = 5$ Hz).
[e] In CD_3CN, the OH proton was seen as a multiplet at δ 2.69, and extra coupling was also apparent on the methylene peak.
[f] Long-chain, linear aliphatic hydrocarbons. Their solubility in DMSO was too low to give visible peaks.
[g] Hexamethylphosphoramide.
[h] In some cases (see notes *a, d*), the coupling interaction between the CH_3 and the OH protons may be observed ($J = 5.5$ Hz).
[i] Poly(dimethylsiloxane). Its solubility in DMSO was too low to give visible peaks.

APPENDIX I PROTON NMR CHEMICAL SHIFTS OF AMINO ACIDS IN D_2O

O O⁻ 1.46 3.76 $\overset{+}{N}H_3$
Alanine (Ala) (A)

$\overset{+}{N}H_2$ H_2N N H 3.22 1.67 1.87 3.74 O O⁻ $\overset{+}{N}H_3$
Arginine (Arg) (R)

H_2N O 2.91 3.97 O O⁻ $\overset{+}{N}H_3$
Asparagine (Asn) (N)

⁻O O 2.70 3.85 O O⁻ $\overset{+}{N}H_3$
Aspartic Acid (Asp) (D)

O O⁻ $H_3\overset{+}{N}$ 3.95 3.05 SH
Cysteine (Cys) (C)

O O⁻ 2.33 2.06 3.72 O O⁻ $\overset{+}{N}H_3$
Glutamic Acid (Glu) (E)

O H_2N 2.43 2.10 3.73 O O⁻ $\overset{+}{N}H_3$
Glutamine (Gln) (Q)

$H_3\overset{+}{N}$ 3.52 O O⁻
Glycine (Gly) (G)

N 7.84 HN 7.08 3.18 3.98 O O⁻ $\overset{+}{N}H_3$
Histidine (His) (H)

0.92 1.24, 1.45 1.95 3.64 O O⁻ 0.99 $\overset{+}{N}H_3$
Isoleucine (Ilue) (I)

0.96 1.70 1.70 3.70 O O⁻ 0.96 $\overset{+}{N}H_3$
Leucine (Leu) (L)

$H_3\overset{+}{N}$ 3.02 1.46 1.70 1.87 $H_3\overset{+}{N}$ 3.73 O O⁻
Lysine (Lys) (K)

S 2.12 2.64 2.12 3.83 O O⁻ $\overset{+}{N}H_3$
Methionine (Met) (M)

$\overset{+}{H_2}$ N 3.34 2.01 2.07, 2.32 4.11 O O⁻
Proline (Pro) (P)

3.11, 3.27 7.40 3.98 O O⁻ $H_3\overset{+}{N}$
Phenylalanine (Phe) (F)

O O⁻ $H_3\overset{+}{N}$ 3.93 3.85 OH
Serine (Ser) (S)

$\overset{+}{N}H_3$ HO 4.22 3.51 O O⁻ 1.31
Threonine (Thr) (T)

7.20 7.54 7.28 7.68 N H 7.32 3.38 4.24 O O⁻ $\overset{+}{N}H_3$
Tryptophan (Trp) (W)

HO 6.88 7.20 2.97, 3.13 $H_3\overset{+}{N}$ 4.28 O⁻ O
Tyrosine (Tyr) (Y)

0.97 1.02 2.25 3.59 $\overset{+}{N}H_3$ O⁻ O
Valine (Val) (V)

CHAPTER 4

CARBON-13 NMR SPECTROMETRY*

4.1 INTRODUCTION

Faced with a choice during the early development of nuclear magnetic resonance spectrometry, most organic chemists would certainly have selected the carbon nucleus over the hydrogen nucleus for immediate investigation. After all, the carbon skeletons of rings and chains are central to organic chemistry. The problem, of course, is that the carbon skeleton consists almost completely of the ^{12}C nucleus, which is not accessible to NMR spectrometry. The spectrometrist is left to cope with the very small amount of the ^{13}C nucleus.

There are enough differences between ^{13}C and ^{1}H NMR spectrometry to justify separate chapters on pedagogical grounds. With an understanding of the basic concepts of NMR spectrometry in Chapter 3, mastery of ^{13}C spectrometry will be rapid.

The ^{12}C nucleus is not magnetically "active" (spin number, I, is zero), but the ^{13}C nucleus, like the ^{1}H nucleus, has a spin number of 1/2. However, since the natural abundance of ^{13}C is only 1.1% that of ^{12}C and its sensitivity is only about 1.6% that of ^{1}H, the overall sensitivity of ^{13}C compared with ^{1}H is about 1/5700.†

The earlier, continuous-wave, slow-scan procedure requires a large sample and a prohibitively long time to obtain a ^{13}C spectrum, but the availability of pulsed FT instrumentation, which permits simultaneous irradiation of all ^{13}C nuclei, has resulted in an increased activity in ^{13}C spectrometry, beginning in the early 1970's, comparable to the burst of activity in ^{1}H spectrometry that began in the late 1950's.

4.2 THEORY

The theoretical background for NMR has already been presented in Chapter 3. Some of the principal aspects of ^{13}C NMR to consider that differ from ^{1}H NMR are as follows:

- In the commonly used CPD or broadband proton-decoupled ^{13}C spectrum (see Section 4.2.1), the peaks are singlets unless the molecule contains other magnetically active nuclei such as ^{2}H, ^{31}P, or ^{19}F.
- The ^{13}C peaks are distributed over a larger chemical-shift range in comparison with the proton range.
- ^{13}C peak intensities do not correlate with the number of carbon atoms in a given peak in routine spectra, due to longer T_1 values and NOE.
- The ^{13}C nuclei are much less abundant and much less sensitive than protons. Larger samples and longer times are needed.
- For a given deuterated solvent, the ^{13}C and ^{1}H solvent peaks differ in multiplicities

At first glance, some of the above summary would seem to discourage the use of ^{13}C spectra. However, the ingenious remedies for these difficulties have made ^{13}C NMR spectrometry a powerful tool, as this chapter will confirm. In fact, side-by-side interpretation of ^{13}C and ^{1}H spectra provide complementary information.

4.2.1 ^{1}H Decoupling Techniques

As mentioned in Section 3.7.5, the ^{13}C nucleus does not show coupling in ^{1}H NMR spectra (except for ^{13}C satellites) due to the low natural abundance of ^{13}C (1.1%); however, the same cannot be said about the reverse. The ^{1}H nucleus is >99% in natural abundance and effectively couples to the ^{13}C nuclei. Because of the large $^{1}J_{CH}$ values for $^{13}C—^{1}H$ (~110–320 Hz) and appreciable $^{2}J_{CH}$, $^{3}J_{CH}$ values for $^{13}C—C—^{1}H$ and $^{13}C—C—C—^{1}H$ (~0–60 Hz) couplings, proton-coupled ^{13}C spectra usually show complex overlapping multiplets that are difficult to interpret (Figure 4.1a); the proton-coupled spectrum of cholesterol is hopelessly overlapped and difficult to decipher.

To alleviate this problem, an important development was the simultaneous use of proton broadband decoupling—irradiation and saturation of the attached protons—and detection of ^{13}C signals. Irradiation of the protons over a broad frequency range by means of a broadband generator or with **C**omposite **P**ulse **D**ecoupling‡ **(CPD)** removes these couplings. In Figure 4.1b, the proton decoupled ^{13}C spectrum of cholesterol shows 27 single peaks, each representing one of the carbon atoms. Before we get too far ahead of ourselves,

* Familiarity with Chapter 3 is assumed.

† Because of the low natural abundance of ^{13}C, the occurrence of adjacent ^{13}C atoms has a low probability; thus, we are free of the complication of $^{13}C—^{13}C$ coupling.

‡ Composite pulse decoupling is described as a "continuous sequence of composite pulses." See Claridge in references for details.

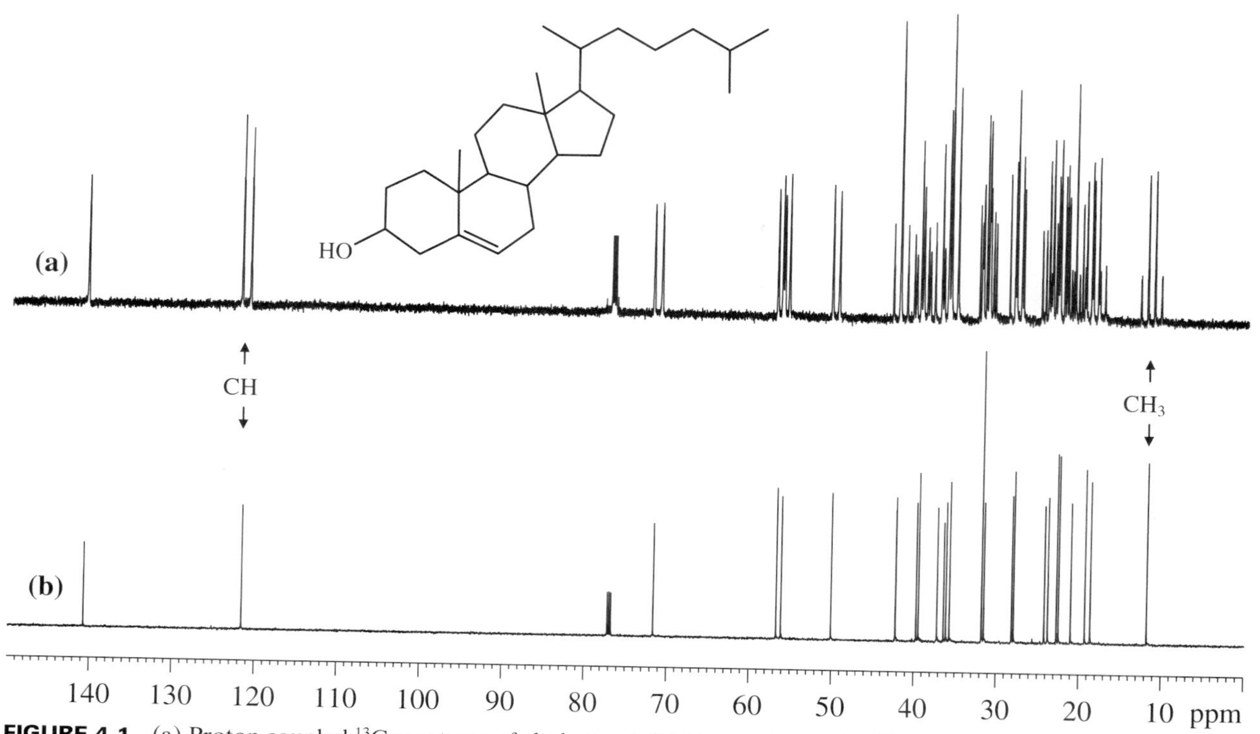

FIGURE 4.1 (a) Proton coupled ^{13}C spectrum of cholesterol. (b) Proton decoupled ^{13}C spectrum of cholesterol. Both in $CDCl_3$ at 150.9 MHz.

we defer our discussions of actual interpretation of ^{13}C spectra until Section 4.3 and 4.7.

The standardized pulse program for a proton decoupled ^{13}C spectrum is shown in Figure 4.2a. The sequence is relaxation delay (R_d) (see Section 4.2.3), rf pulse (θ), and signal acquisition (t_2). The proton "channel" has the decoupler on to remove the ^1H—^{13}C coupling, while a short, powerful rf pulse (of the order of a few microseconds) excites all the ^{13}C nuclei simultaneously. Since the carrier frequency is slightly off resonance FID (free induction decay), for all the ^{13}C frequencies, each ^{13}C nucleus shows a FID, which is an exponentially decaying sine wave.

Figure 4.2b is a presentation of the FID of the decoupled ^{13}C NMR spectrum of cholesterol. Figure 4.2c is an expanded, small section of the FID from Figure 4.2b. The complex FID is the result of a number of overlapping sine-waves and interfering (beat) patterns. A series of repetitive pulses, signal acquisitions, and relaxation delays builds the signal. Fourier transform by the computer converts the accumulated FID (a time domain spectrum) to the decoupled, frequency-domain spectrum of cholesterol (at 150.9 MHz in $CDCl_3$). See Figure 4.1b.

The result, in the absence of other nuclei such as ^2H, ^{31}P, ^{19}F, is a sharp peak for each chemically nonequivalent carbon in the compound, except for the infrequent coincidence of ^{13}C chemical shifts. See Figure 4.1b for the ^1H decoupled ^{13}C spectrum of cholesterol and compare its simplicity with the ^1H coupled spectrum in a. Note that, when ^{13}C is decoupled from ^1H, useful information on the "multiplicity" of the carbon is lost ($n + 1$ rule). For example, in Figure 4.1a, a methyl group is a quartet (3 + 1) and a methine group is a doublet (1 + 1). A more complete discussion of multiplets can be found in Sections 4.2.5 and 4.3. There are techniques, however, such as the DEPT sequence (Section 4.6) that supplies this information in a much simpler way.

4.2.2 Chemical Shift Scale and Range

As with ^1H NMR, the ^{13}C NMR frequency axis is converted to a unitless scale; this scale is in δ units (ppm). The chemical shifts in routine ^{13}C spectra range over about 220 ppm from TMS—about 20 times that of routine ^1H spectra (~10 ppm). As a result of the large range and the sharpness of the decoupled peaks, coincidences of ^{13}C chemical shifts are uncommon, and impurities are readily detected. Often, even mixtures provide useful information. (See Appendix B for an extensive list of common impurities.) For example, stereoisomers that are difficult to analyze by means of ^1H spectrometry usually show discrete ^{13}C peaks.

The fundamental NMR equation [$\nu = (\gamma/2\pi)\boldsymbol{B}_0$] is used to calculate the resonance frequency for the ^{13}C nucleus at a given magnetic field strength. For example, a 600 MHz instrument (for ^1H) is used at 150.9 MHz to produce a ^{13}C spectrum—i.e., the ratio is about 4:1. The

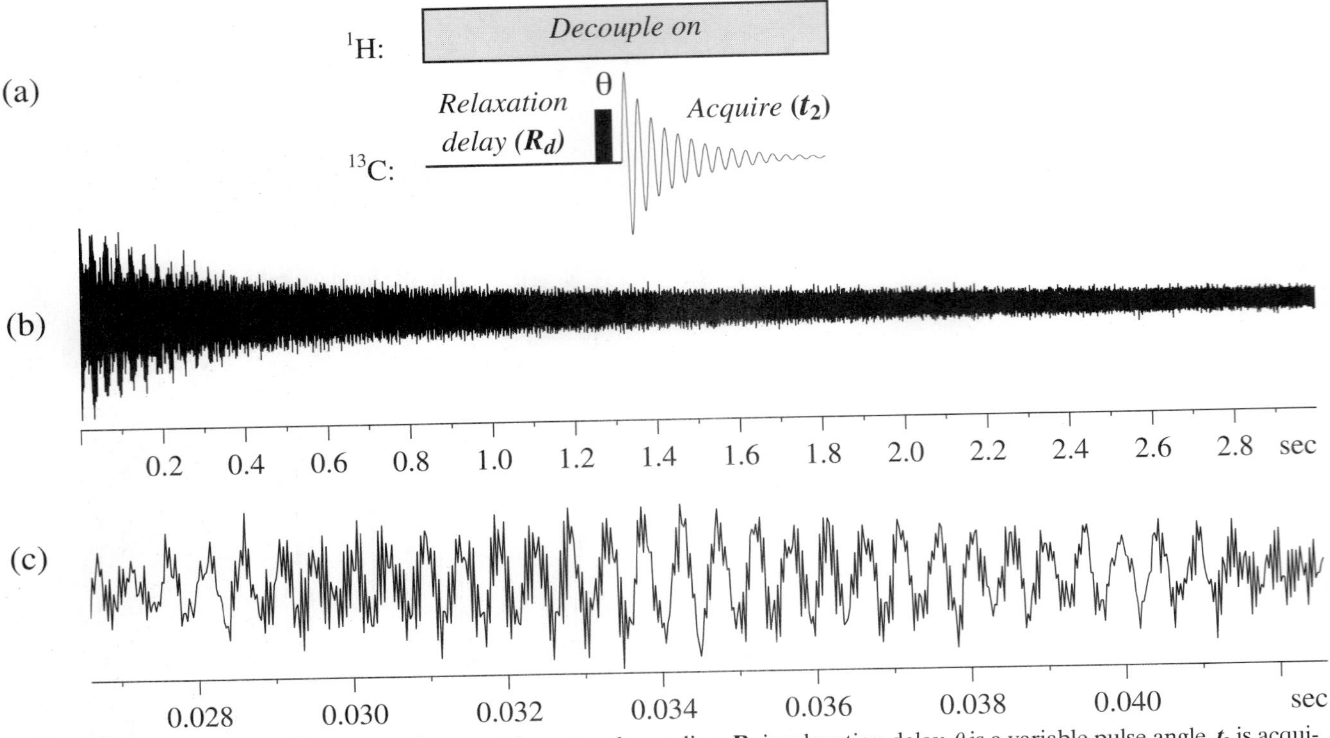

FIGURE 4.2 (a) Standard ^{13}C pulse sequence with proton decoupling. $\boldsymbol{R_d}$ is relaxation delay, θ is a variable pulse angle, $\boldsymbol{t_2}$ is acquisition time. (b) FID sinusoidal display of decoupled cholesterol at 150.9 MHz in $CDCl_3$. (c) Expanded small section of FID.

ratio of frequencies is directly related to the ratio of γ's, the magnetogyric ratios, which are (in units of 10^7 rad T^{-1} s^{-1}) 26,753 and 6,728 for ^1H and ^{13}C, respectively (see Appendix A in Chapter 6). As a quirk of history, an instrument is often referred to by its proton resonance frequency regardless of the nucleus under investigation.

Figure 4.3 gives credence to the statement that ^{13}C chemical shifts somewhat parallel those of ^1H, but we note some divergences that are not readily explainable and require development of another set of interpretive skills (see Section 4.7). In general, in comparison with ^1H spectra, it seems more difficult to correlate ^{13}C shifts with substituent electronegativity.

4.2.3 T_1 Relaxation

Unlike ^1H NMR spectra, whose integration of signals represents the ratio of protons, integration of ^{13}C peaks do not correlate with the ratio of carbon atoms in routine spectra. There are two major factors that account for the problem of peak intensities in ^{13}C spectra:

- The spin-lattice relaxation process, designated T_1 (also termed longitudinal relaxation), varies widely for different types of carbon atoms.
- The NOE accrues only for carbons with attached protons. (Section 4.2.4)

As discussed in Section 3.2.3, T_1 and T_2 relaxation times are short for protons resulting in intensities that are proportional to the number of protons involved and sharp peaks. In proton decoupled ^{13}C spectra, large T_1's values, caused by dipole-dipole interaction with directly attached protons and with nearby proton, may result in detection of only a part of the possible signal, which can be overcome by inserting a delay interval R_d between the individual pulses (see Figure 4.2a). The relaxation delay needs to be carefully considered when acquiring ^{13}C data because signals can be missed completely if the R_d is too short. For most samples, we strike a compromise between instrument time and sensitivity. The Ernst angle ($\cos\theta = e^{-T_r/T_1}$) is used to determine the repetition time (T_r) and pulse angle to optimize S/N. Typically, a 45° pulse angle (θ) and a short delay of 3–4 seconds (optimum $1.27T_1$) is used for small molecules. For larger molecules, which usually have much shorter T_1's, shorter R_d's can be used.

It is sometimes necessary to measure T_1's so that weak signals are not lost in the "noise" or in order to obtain quantitative results. The inversion-recovery method to determine T_1's is demonstrated in Figure 4.4. Generally, T_1's decrease as the number of protons directly attached to the ^{13}C nucleus increases. In other words, a quaternary ^{13}C nucleus gives a peak of lowest intensity as shown in Figure 4.4a; it also gives the slowest recovery in the inversion-recovery method. However, it is often difficult to differentiate between CH_3, CH_2, and CH on the basis of T_1's alone since other factors are involved. T_1 values cover a range of several

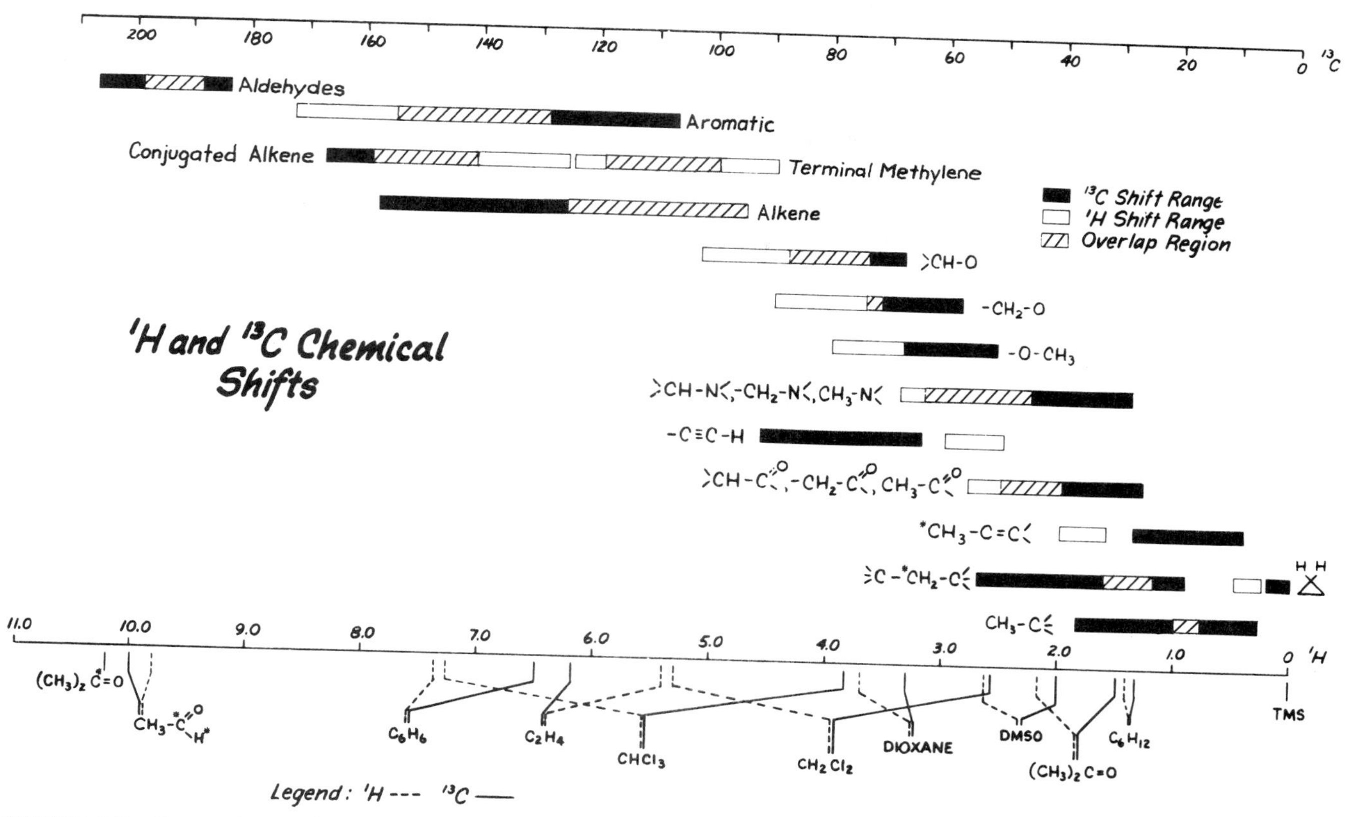

FIGURE 4.3 Comparison of ^{1}H and [13]C Chemical Shifts.

seconds for a CH_3 group to well over a minute for some quaternary ^{13}C nuclei. A delay between pulses of approximately 5 T_1's is recommended for a ^{13}C nucleus without an attached proton, and this appreciable delay must be tolerated (for quantitative purposes).

The overall pulse sequence in Figure 4.4a follows: Relaxation—180° pulse—variable time interval—90° pulse—acquire while decoupling.

In Figure 4.4b, the net magnetic moment $\boldsymbol{M}_0$ is represented by the upright, boldface vector. The first pulse inverts the vector clockwise 180° to the $-z$ axis. After a short time ($\Delta\tau$), during which the arrow shows a slight recovery toward its original, upright position, a 90° pulse rotates the vector clockwise and places it along the $-y$ axis, which is toward the left at 270°. At this point, the receiver along the y axis reads the signal as slightly diminished and negative—that is, pointed down in the spectrum. We recall that the receiver accepts the usual 90° signal as positive, hence pointing up in the phased spectrum.

In Figure 4.4c, $\Delta\tau$ has been increased so that the inverted vector was allowed to proceed further toward recovery—in fact past the null point; it is again slightly diminished and pointing up. The 90° pulse rotates it clockwise and it is recorded as a positive signal.

In Figure 4.4d, the inversion-recovery sequence is repeated eight times on diethyl phthalate (see Section 4.4 for a complete assignment of diethyl phthalate) of with increasingly longer $\Delta\tau$ time increments. It is apparent that most signals have different null point's indicating that they have different T_1 values. As implied above, the C=O signal at $\sim\delta$ 167 has the slowest recovery in the inversion-recovery experiment. Its signal inverts between the seventh and the eighth spectra, whereas the methyl signal at $\sim\delta$ 13 inverts between the fourth and fifth spectra. T_1's can be calculated with the following formula: $T_1 = t_{null}/\ln(2)$, which yields approximate values that suffices for most purposes.

4.2.4 Nuclear Overhauser Enhancement (NOE)

We described the nuclear Overhauser effect (NOE) among protons in Section 3.16; we now discuss the heteronuclear NOE, which results from broadband proton decoupling in ^{13}C NMR spectra (see Figure 4.1b). The net effect of NOE on ^{13}C spectra is the enhancement of peaks whose carbon atoms have attached protons. This enhancement is due to the reversal of spin populations from the predicted Boltzmann distribution. The total amount of enhancement depends on the theoretical maximum and the mode of relaxation. The maximum possible enhancement is equal to one-half the ratio of the nuclei's magnetogyric ratios (γ's) while the

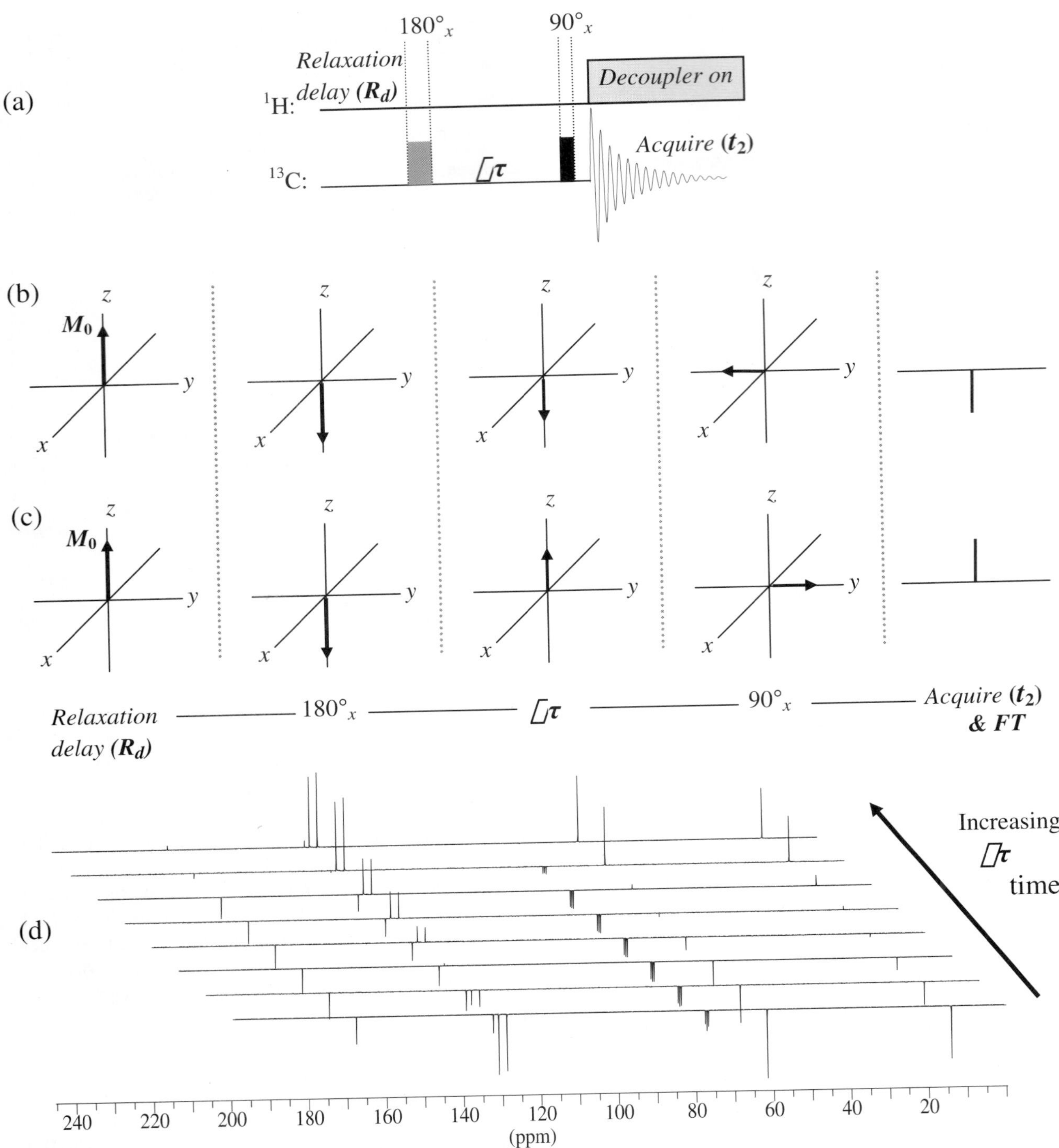

FIGURE 4.4 (a) Inversion recovery pulse sequence with inverse gated proton decoupling for T_1 measurement. (b) Short $\Delta\tau$ time, (c) long $\Delta\tau$ time. (d) Example of a ^{13}C T_1 data set of diethyl phthalate at 75.5 MHz. See Section 4.3 for complete assignment of diethyl phthalate.

actual enhancement is proportional to the extent of ^{13}C—^1H dipolar relaxation. For a proton-decoupled ^{13}C experiment, the maximum NOE enhancement is $\gamma_H/(2)\gamma_C$ or 26,753/(2)6,728 which equals 1.98. The total sensitivity increase is therefore nearly *threefold* because the NOE enhancement is added to the original intensity.

The actual enhancement for the ^{13}C—^1H system can be anywhere from 0 to 1.98 depending on the mechanism of relaxation for each individual nucleus. In practice, for carbons with no attached protons, the enhancement is essentially zero since there is practically no ^{13}C—^1H dipolar relaxation. For small to medium-sized organic molecules, $^{13}C—^{1}H$ dipolar relaxation for carbons with attached protons is very efficient yielding close to the full 200% increase in signal.

The net effect is a large reduction in the time needed to obtain a decoupled spectrum as compared with a coupled spectrum. Furthermore, this contribution to intensity increase is a nonlinear function of the number of protons directly attached to the particular ^{13}C nucleus. ^{13}C nuclei without an attached proton are characterized by peaks of distinctly low intensities (see Figure 4.1b). These erratic contributions make peak intensities unrelated to the number of ^{13}C nuclei that they should represent.

4.2.5 $^{13}C—^{1}H$ Spin Coupling (*J* values)

Spin-coupling *J* values—at least as an initial consideration—are less important in ^{13}C NMR than in ^{1}H NMR. Since routine ^{13}C spectra are usually decoupled, $^{13}C—^{1}H$ coupling values are discarded in the interest of obtaining a spectrum in a short time or on small samples—a spectrum, furthermore, free of complex, overlapping absorptions.

For the spectrum of cholesterol in Figure 4.1a the *J* couplings were retained. We saved considerable time by using the "gated proton decoupling" technique rather than the usual, continuous, broadband decoupling. The gated proton decoupling technique is explained as follows. To retain at least part of the NOE (Section 4.2.3) and still maintain C—H coupling, the "gated decoupling" technique may be employed (Figure 4.5). Briefly, the broadband proton decoupler is gated (switched) "on" during the relaxation delay period, then gated off during the brief acquisition period. Thus, the NOE (a slow process), which has built up during the lengthy delay period, decays only partially during the brief acquisition period. Coupling, a fast process, is established immediately and remains throughout the acquisition period. The result is a coupled spectrum, in which at least part of the NOE has been retained; thus time has been saved.

We did demonstrate the utility of spin coupling in Figure 4.1a in which the $^{1}J_{CH}$ coupling values are of interest. Table 4.1 gives some representative $^{1}J_{CH}$ values.

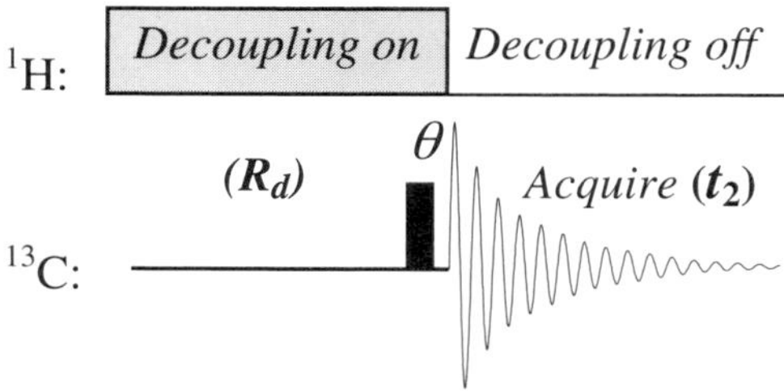

FIGURE 4.5 Gated proton decoupling pulse sequence. R_d is relaxation delay, θ is a variable pulse angle, t_2 is the acquisition time.

TABLE 4.1 Some $^{1}J_{CH}$ Values

Compound	*J*(Hz)
sp^3	
CH_4	125.0
CH_3CH_3	124.9
$CH_3\underline{C}H_2CH_3$	119.2
$(CH_3)_3CH$	114.2
—H	123.0
—H	128.0
$PhCH_3$	129.0
CH_3NH_2	133.0
H	134.0
$ROCH_3$	140.0
CH_3OH	141.0
CH_3Cl	150.0
CH_3Br	151.0
—H	161.0
$(CH_3O)_2C\underline{H}_2$	162.0
CH_2Cl_2	178.0
O —H	180.0
H	205.0
$CHCl_3$	209.0
sp^2	
$CH_3\underline{C}H{=}C(CH_3)_2$	148.4
$CH_2{=}CH_2$	156.2
C_6H_6	159.0
—H	160.0
C=C=C—H	168.0
H	170.0
$CH_3\underline{C}H{=}O$	172.4
N —H	178.0
$NH_2\underline{C}H{=}O$	188.3
=COH(OR)	195.0
$CH_3\underline{C}HX$, X=halogen	198.0
—H	238.0
sp	
CH≡CH	249.0
$C_6H_5C{\equiv}CH$	251.0
HC≡N	269.0

TABLE 4.2 Some $^2J_{CH}$ Values

Compound	J(Hz)
sp^3	
$C\underline{H}_3CH_3$	−4.5
$C\underline{H}_3\underline{C}Cl_3$	5.9
$R\underline{C}(=O)C\underline{H}_3$	6.0
$\underline{C}H_3C\underline{H}=O$	26.7
sp^2	
*C_6H_6	1.0
$C\underline{H}_2=\underline{C}H_2$	2.4
$C=C(\underline{C}H_3)\underline{H}$	5.0
$(C\underline{H}_3)_2\underline{C}=O$	5.5
$CH_2=\underline{C}HC\underline{H}=O$	26.9
sp	
$C\underline{H}\equiv\underline{C}H$	49.3
$C_6H_5O\underline{C}\equiv C\underline{H}$	61.0

* $^3J = 7.6$ Hz ($>^2J$).

One-bond $^{13}C—^1H$ coupling ($^1J_{CH}$) ranges from about 110 to 320 Hz, increasing with increased s character of the $^{13}C—^1H$ bond, with substitution on the carbon atom of electron-withdrawing groups, and with angular distortion. Appreciable $^{13}C—^1H$ coupling also extends over two or more (n) bonds ($^nJ_{CH}$). Table 4.2 gives some representative $^2J_{CH}$ values, which range from about 5 to 60 Hz.

The $^3J_{CH}$ values are roughly comparable to $^2J_{CH}$ values for sp^3 carbon atoms. In aromatic rings, however, the $^3J_{CH}$ values are characteristically larger than $^2J_{CH}$ values. In benzene itself, $^3J_{CH} = 7.6$ Hz and $^2J_{CH} = 1.0$ (see Table 4.2).

Coupling of ^{13}C to several other nuclei, the most important of which are ^{31}P, ^{19}F, and ^2D, may be observed in proton-decoupled spectra. Representative coupling constants are given in Table 4.3.

TABLE 4.3 Coupling Constants for ^{19}F, ^{31}P, D Coupled to ^{13}C

Compound	^1J(Hz)	^2J(Hz)	^3J(Hz)	^4J(Hz)
CH_3CF_3	271			
CF_2H_2	235			
CF_3CO_2H	284	43.7		
C_6H_5F	245	21.0	7.7	3.3
$(C_4H_9)_3P$	10.9	11.7	12.5	
$(CH_3CH_2)_4P^+ Br^-$	49.0	4.3		
$(C_6H_5)_3P^+ CH_3I^-$	88.0	10.9		
	^1J(Hz) of CH_3 = 52			
$C_2H_5(P=O)(OC_2H_5)_2$	143	7.1 (J_{COP})	6.9 (J_{CCOP})	
$(C_6H_5)_3P$	12.4	19.6	6.7	
$CDCl_3$	31.5			
$CD_3(C=O)CD_3$	19.5			
$(CD_3)_2SO$	22.0			
C_6D_6	25.5			

4.2.6 Sensitivity

^{13}C nuclei are much less abundant and much less sensitive than protons. Larger samples and longer times are needed to acquire sufficient signal strength. Let us turn once again to the sensitivity of pulsed FT NMR experiments in which the signal to noise ratio (S/N) is given by the equation (Section 3.3.2):

$$S/N = NT_2\gamma_{exe}\left(\frac{\sqrt[3]{B_0\gamma_{det}}\sqrt{ns}}{T}\right)$$

We notice that the S/N grows proportionately to $\sqrt{ns}$, where ns is the number of scans or repetitions of the pulse program. This relationship is not typically a problem with ^1H experiments where only a few μg to a mg of material is enough to get good S/N in a few scans. As mention previously, ^{13}C is ~6000 times less sensitive than ^1H, and therefore requires either more sample (N), higher field strengths $\boldsymbol{B}_0$ (or better probe technology, i.e., a cryo-probe), or an increase in the number of scans (ns).

In most labs, the only alternative is to increase ns, but to double the S/N you need to take four times the number of scans, which rapidly escalates into long experiment times due to 1, 4, 16, 64, 256, 1024(1k), 2k, 4k, 16k, 64k, . . . scans. Another solution is presented in Section 5.4.2, in which is described an experiment that takes advantage of the higher sensitivity of one nucleus (^1H) and transfers the energy to another less sensitive nucleus (^{13}C).

A routine ^{13}C spectrum at 75.5 MHz normally requires about 10 mg of sample in 0.5 ml of deuterated solvent in a 5-mm o.d. tube. Samples on the order of 100 μg can be handled in a probe that accepts a 2.5-mm o.d. tube, which gives higher sensitivity (see Section 3.3).

4.2.7 Solvents

Samples for ^{13}C NMR spectrometry are usually dissolved in $CDCl_3$, and the ^{13}C peak of tetramethylsilane (TMS) is used as the internal reference.* A list of the common deuterated solvents is given in Appendix A.

For a given deuterated solvent, the ^{13}C and ^1H solvent peaks differ in multiplicities. It is worth noting the difference in appearance between the solvent peaks in a proton spectrum and a ^{13}C spectrum. For example, the familiar singlet at δ 7.26 in a proton spectrum is the result of a small amount of $CHCl_3$ in the solvent $CDCl_3$. The ^1H peak is not split since ^{12}C is magnetically inactive, and the Cl nucleus has a strong electrical quadrupole moment. (See Section 3.7); the

* With modern instrumentation, TMS is usually not added; instead, the ^{13}C peak of the deuterated solvent is used as a reference. The spectrum, however, is presented with the ^{13}C peak of TMS at δ 0.00 at the right-hand edge of the scale.

small amount of ^{13}C present is insufficient to produce a visible doublet. In the case of another useful solvent, dimethyl-d_6 sulfoxide, the quintet at δ 2.49 in a proton spectrum is a result of the proton in the impurity split by D_2: $\mathbf{H}CD_2—S(=O)—CD_3$.

The proton peak is split by two deuterium nuclei with a spin number (I) of one. The multiplicity can be calculated by the familiar formula $2nI + 1$; thus $2 \times 2 \times 1 + 1 = 5$. The $—CD_3$ group does not interfere.

In a typical ^{13}C spectrum, the solvent $CDCl_3$ leaves a triplet centered at δ 77.0. But now the presence of small amount of $CHCl_3$ is irrelevant since all protons are decoupled by the usual broadband decoupling. The triplet results from splitting of the ^{13}C peak by the D nucleus. The formula $2nI + 1$ gives $2 \times 1 \times 1 + 1 = 3$. The intensity ratios are 1:1:1. In the case of $CD_3S(=O)CD_3$, the formula gives $2 \times 3 \times 1 + 1 = 7$. The multiplet at δ 39.7 is a septet with a coupling constant of 21 Hz; the intensity ratios are 1:3:6:7:6:3:1 (see Appendix A). The following diagram is the deuterium analogue of Pascal's triangle for protons (see Figure 3.32 for ^1H equivalent).

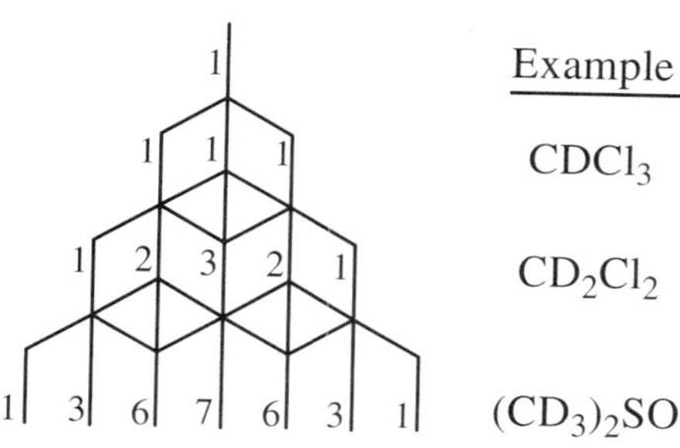

Substitution of D for H on a carbon results in a dramatic diminution of the height of the ^{13}C signal in a broadband-decoupled spectrum for the following reasons. Since deuterium has a spin number of 1 and a magnetic moment 30% that of ^1H, it will split the ^{13}C absorption into three lines (ratio 1:1:1) with a J value equal to $0.30 \times J_{CH}$. Furthermore, T_1 for $^{13}C—D$ is longer than that for $^{13}C—^1H$ because of decreased dipole-dipole relaxation. Finally, the NOE is lost, since there is no irradiation of deuterium.* A separate peak may also be seen for any residual $^{13}C—^1H$ since the isotope effect usually results in a slight shift to lower frequency of the $^{13}C—D$ absorption (~0.2 ppm per D atom). The isotope effect may also slightly shift the absorption of the carbon atoms once removed from the deuterated carbon. As an example, pure deuterated benzene, a common NMR solvent, gives a 1:1:1 triplet centered at δ 128.0 in a decoupled spectrum; J is 25 Hz. Under the same conditions, pure benzene (C_6H_6) gives a singlet at δ 128.5 (see Table 4.12 and Appendix A).

* The same explanation also accounts for the relatively weak signal shown by deuterated solvents. In addition, small solvent molecules tumble rapidly; this rapid movement makes for a longer T_1, hence for smaller peaks. Deuterated chloroform, $CDCl_3$, shows a 1:1:1 triplet, deuterated p-dioxane a 1:2:3:2:1 quintet, and deuterated DMSO $(CD_3)_2SO$, a 1:3:6:7:6:3:1 septet in accordance with the $2nI + 1$ rule (Chapter 3). The chemical shifts, coupling constants, and multiplicities of the ^{13}C atoms of common NMR solvents are given in Appendix A.

4.3 INTERPRETATION OF A SIMPLE ^{13}C SPECTRUM: DIETHYL PHTHALATE

Even though we have not discussed the chemical shift classes (Section 4.7) it is possible to discuss the peak assignments of diethyl phthalate from what we do know; that is peak intensities, $^{13}C—^1H$ coupling, and an expectation of chemical shifts relative to what we know from the proton chapter. The T_1 and NOE affect the peak intensities and therefore are not representative to the number of carbons: this, however, results in an advantage. It is usually possible by inspection of the ^{13}C spectrum to recognize the nuclei that do not bear protons by their low intensity (Figure 4.6a C=O, and the peak labeled 1). The common spin-lattice relaxation mechanism for ^{13}C results from dipole-dipole interaction with directly attached protons. Thus, nonprotonated carbon atoms have longer T_1 relaxation times, which together with little or no NOE, results in small peaks. It is therefore often possible to recognize carbonyl groups (except formyl), nitriles, nonprotonated alkene and alkyne carbon atoms, and other quaternary carbon atoms readily.

Since diethyl phthalate ($C_{12}H_{14}O_4$) has an axis of symmetry (and a plane of symmetry), the decoupled ^{13}C spectrum in Figure 4.6a consists of six peaks. Since there is no coupling, the peaks are singlets, and there is no overlap. We examine the chemical shifts and observe similarities in distribution with ^1H chemical shifts (see Figure 4.3 and Appendix C). For example, we can place the alkyl groups on the right-hand side of the ^{13}C spectrum with the deshielded CH_2 group to the left of the CH_3 group. It seems safe to designate the strongly deshielded cluster of three peaks as aromatic. The C=O group is on the far left. We also note that the ^{13}C aromatic nucleus (labeled 1) without an attached proton gives a peak that is distinctly decreased in height. The same applies to the C=O group.

The ^1J coupling in Figure 4.6b provides the multiplicities; that is the number of protons that are directly attached. These couplings confirm our chemical shifts assignments. Thus, from right to left, we see the quartet of the CH_3 group (the $n + 1$ rule) and triplet of the CH_2 group. Then there is a cluster of the two doublets of the two aromatic CH groups and the singlet of the carbon

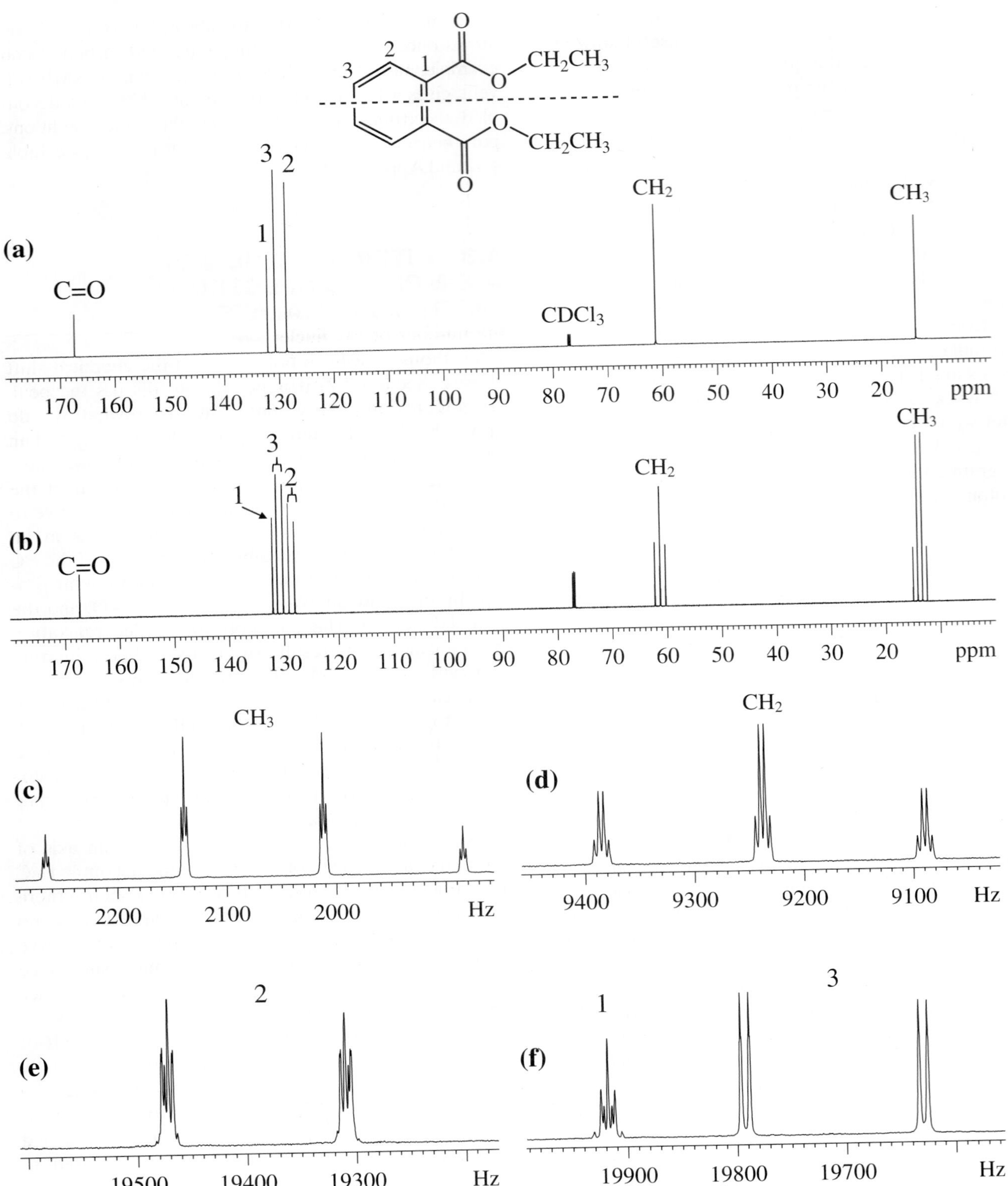

FIGURE 4.6 (a) Diethyl phthalate, decoupled ^{13}C spectrum at 150.9 MHz in $CDCl_3$, (b) coupled ^{13}C spectrum (c–f) expansions of coupled ^{13}C spectrum.

atom to which no proton is attached (labeled 1). Finally there is the C=O singlet at higher frequency.

Starting at Figure 4.6c, we see the expansion of the quartet shown in Figure 4.6b, each peak of which is now a triplet resulting from ^2J coupling of the methyl ^{13}C nucleus with the protons of the adjacent CH_2 group. These ^2J couplings are much smaller than the ^1J couplings

In d, we see the expansion of the CH_2 triplet shown in Figure 4.6b, each peak of which is now a quartet resulting from ^2J coupling with the CH_3 protons. In e, we see the expansion of the right-hand doublet of the aromatic cluster labeled 2 in Figure 4.6b. Each peak of this doublet is split by ^2J and ^3J coupling of the neighboring protons.

The expansion shown in f is the other aromatic doublet shown in Figure 4.6b labeled 3. Each peak of the doublet is split by either ^2J or ^3J coupling. Also in the same panel is the expansion of the singlet remaining in the aromatic cluster of Figure 4.6b. The singlet is not split by a large coupling since there is no attached proton. It is split only by the small ^2J and ^3J couplings and is readily assigned to the carbon nucleus labeled 1.

Carbon atom 2 is *ortho* plus *meta* to the substituents, and carbon atom 3 is *meta* plus *para*. Use of Table 4.12 (Section 4.7.4) gives δ 129.6 for peak 2 and δ 133.2 for peak 3. Measurements of peaks in Figure 4.6 give δ 128.5 for peak 2 and δ 131.2 for peak 3. Reasonable, but perhaps unexpected to some who recall from Chapter 3 that a carbonyl substituent deshields the *ortho* position of protons more so than the *para* position. Chart D.1 in Chapter 3 confirms this impression, at least for a single carbonyl substituent. And the result of comparing *ortho* plus with *meta* plus *para* gives the expected results: δ 8.0 for peak 2, and δ 7.60 for peak 3.

4.4 QUANTITATIVE ^{13}C ANALYSIS

Quantitative ^{13}C NMR is desirable in two situations. First in structure determinations, it is clearly useful to know whether a signal results from more than one shift-equivalent carbon. Second, quantitative analysis of a mixture of two or more components requires that the area of the peaks be proportional to the number of carbons atoms causing that signal.

There are two reasons that broadband-decoupled ^{13}C spectra are usually not susceptible to quantitative analysis:

- ^{13}C nuclei with long T_1 relaxation times may not return to the equilibrium Bultzmann distribution between pulses. Thus, the signals do not achieve full amplitude (see Section 4.2.3).
- The nuclear Overhauser enhancement (NOE, see Section 4.2.4) varies among ^{13}C nuclei, and the signal intensities vary accordingly.

In Section 4.2.5, we discussed the gated proton-decoupling technique. The purpose was to develop a high level of the nuclear Overhauser effect so that a coupled spectrum could be achieved in the least time. In the present Section, we deal with the inverse-gated proton decoupling technique. The proton decoupler is gated off at the beginning of the relaxation delay and gated on at the beginning of the acquisition period (see Figure 4.7). The purpose is to maintain a low, constant level of NOE. This is feasible because the NOE builds up slowly during the relatively brief time of the signal-acquisition period. The overall result is a spectrum consisting of singlets whose intensities correspond to the number of ^{13}C nuclei they represent. Lest we forget, we must still allow for the T_1 relaxation delay (R_d).

Figure 4.8 demonstrates the effects of T_1 and NOE on the peak intensities. Figure 4.8a is a standard proton decoupled ^{13}C spectrum with R_d set to $<T_1$; notice that the methyl is set to one and the rest of the integrals are labeled above the peaks. The spectrum is not quantitative due to NOE and T_1 effects. In Figure 4.4d (Section 4.2.3), the protonated aromatic carbons showed the shortest T_1's, which explains the larger integrals of those peaks. The spectrum in Figure 4.8b was obtained using the inverse-gated proton-decoupling technique. Notice that by removing the NOE effect, the intensities of the carbons with attached hydrogens are more quantitative than the ones that are quaternary. R_d is still less than T_1. Figure 4.8c illustrates that, by using long relaxation delays ($R_d > 5T_1$'s) and by minimizing the effect of NOE by using the inverse-gated proton-decoupling technique, ^{13}C data can be quantitative. The drawback is that it takes a long time to acquire the data because the R_d has to be set to at least five times the expected T_1 and the loss of part of the NOE means many more repetitions are needed to build up the signal intensity. The time required can be reduced by addition of a paramagnetic relaxation reagent. A common procedure is to add a "catalytic" amount of chromium (III) acetylacetonate (Cr(acac)$_3$), a paramagnetic substance, whose unpaired electrons efficiently stimulate transfer of spin, which effectively reduces the T_1 and T_2 (broader lines) relaxations.

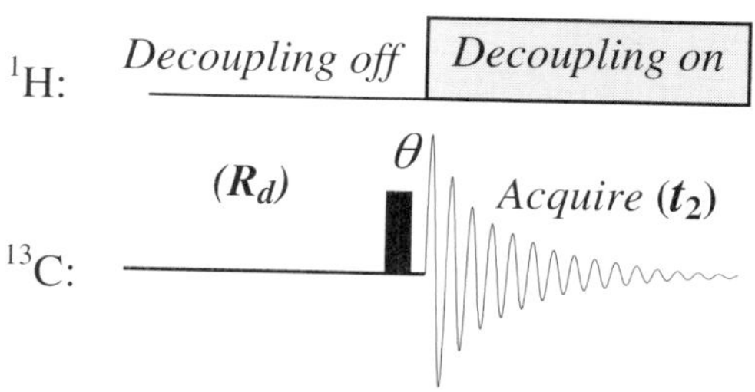

FIGURE 4.7 Inverse gated proton decoupling pulse sequence. R_d is relaxation delay, θ is a variable pulse angle, t_2 is the acquisition time.

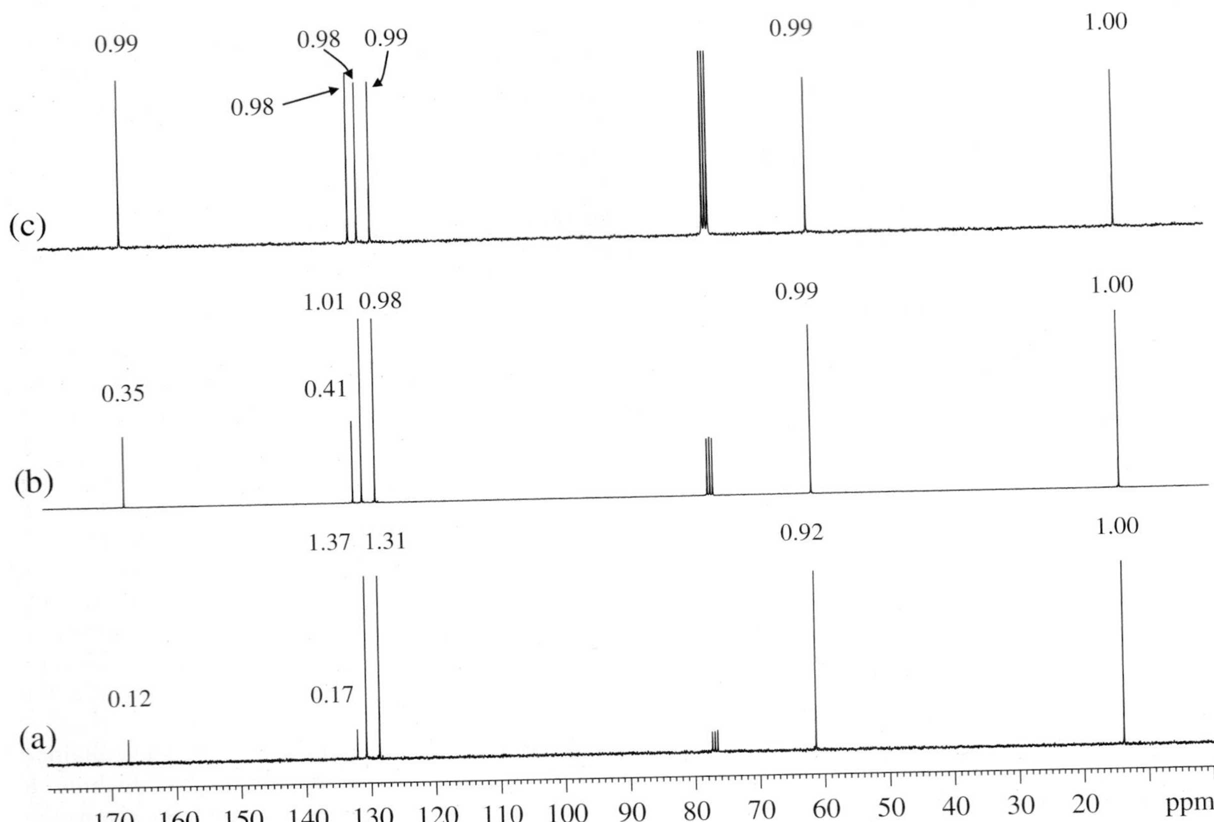

FIGURE 4.8 (a) Standard ^{1}H decoupled ^{13}C spectrum of diethyl phthalate with relaxation delay ($\boldsymbol{R}_d$) < T_1, (b) inverse-gated proton-decoupling technique with $\boldsymbol{R}_d < T_1$, (c) $\boldsymbol{R}_d > 5T_1$'s and inverse gated decoupling to compensate for NOE. The number above each peak represents its integrated value. Spectra recorded at 150.9 MHz in $CDCl_3$.

4.5 CHEMICAL SHIFT EQUIVALENCE

The definition of chemical shift equivalence given for protons also applies to carbon atoms: interchangeability by a symmetry operation or by a rapid mechanism. The presence of equivalent carbon atoms (or coincidence of shift) in a molecule results in a discrepancy between the apparent number of peaks and the actual number of carbon atoms in the molecule.

Thus ^{13}C nuclei of the methyl groups in *t*-butyl alcohol (Figure 4.9) are equivalent by rapid rotation in the same sense in which the protons of a methyl group are equivalent. The ^{13}C spectrum of *t*-butyl alcohol shows two peaks, one much larger than the other; the carbinyl carbon peak (quaternary) is much less than the

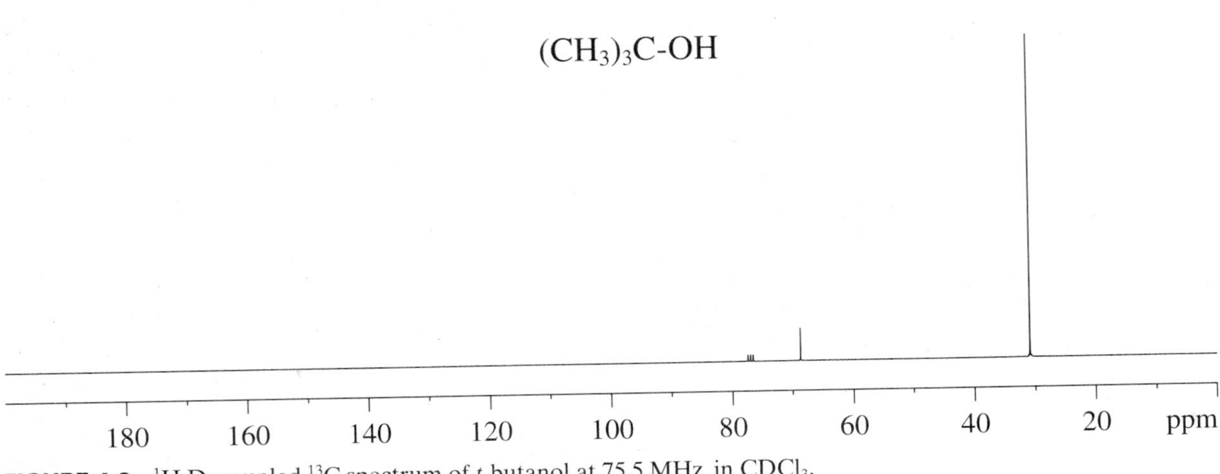

FIGURE 4.9 [1]H Decoupled ^{13}C spectrum of *t*-butanol at 75.5 MHz, in $CDCl_3$.

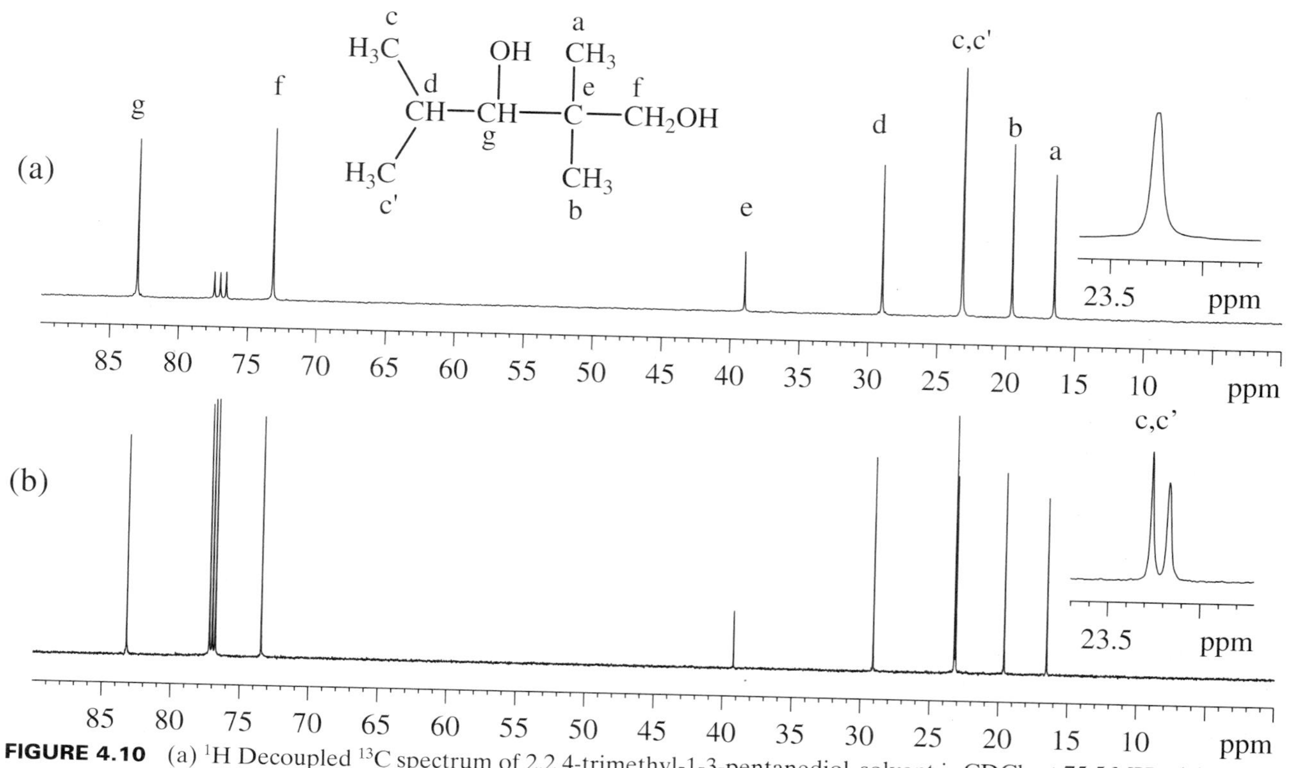

FIGURE 4.10 (a) ^{1}H Decoupled ^{13}C spectrum of 2,2,4-trimethyl-1-3-pentanediol, solvent is $CDCl_3$ at 75.5 MHz. (b) same at 150.9 MHz.

intensity of the peak representing the carbon atoms of the methyl groups.

In the chiral molecule 2,2,4-trimethyl-1,3-pentanediol (Figure 4.10) at 75.5 MHz, we note that CH_3(a) and CH_3(b) are not equivalent, and two peaks are seen. Even though the two methyl groups labeled *c* and *c′* are not equivalent, they coincidently show only one peak. Two peaks are seen at higher field strength (150.9 MHz).

In Section 3.8.3.2, we noted that the CH_3 protons of $(CH_3)_2NCH{=}O$ gave separate peaks at room temperature, but became chemical shift equivalent at about 123°. Of course, the ^{13}C peaks show similar behavior.

4.6 DEPT

The decision made early in the history of ^{13}C NMR spectrometry to completely eliminate $^{1}H—^{13}C$ coupling by irradiating the entire proton chemical shift range was always viewed as a mixed blessing. On the one hand, the simplicity of a spectrum with a single peak for each unique carbon atom in the molecule was and is extremely attractive. On the other hand, $^{1}H—^{13}C$ coupling information is valuable and clearly useful in determining structure. There have been many experiments developed over the years attempting to usefully tap this information and still maintain the simplicity of the completely decoupled spectrum. Two are worth mentioning because both were popular at one time and both are common in older literature. Also, some of the terminology that we use today grew from one of these experiments.

The first of these now obsolete experiments is called the off-resonance ^{1}H decoupled spectrum. In this experiment, the $^{1}H—^{13}C$ coupling is restricted to one bond (i.e., each carbon only shows coupling to its attached protons) by moving the broadband decoupler "off-resonance" or away from the middle of the proton chemical shift range. Thus, each carbon resonance gives a singlet (no attached protons), a doublet (one attached proton), a triplet (two attached protons), or a quartet (three attached protons). Note that these patterns are an application of first order coupling rules and the $n + 1$ rule. This experiment also gave rise to the common practice (still used today) of referring to a carbon resonance by its "multiplicity" ("s," "d," "t," or "q"). We use these descriptions of multiplicity whether or not we use this experiment.

The other experiment worth mentioning, which, by the way, is also obsolete, is the attached proton test or APT. This experiment is based on the different magnitudes of $^{1}H—^{13}C$ coupling for methine, methylene, and methyl groups. By adjusting certain delays in the pulse sequence (not given), quaternary and methylene carbons could be phased up, and methine and methyl carbons could be phased down. Since phase is arbitrary, this order could be reversed. This ability of distinguishing carbon types by the magnitude of their respective coupling constants led to the now popular DEPT.

The DEPT sequence (distortion enhancement by polarization transfer) has developed into the preferred procedure for determining the number of protons directly attached to the individual ^{13}C nucleus. The DEPT experiment can be done in a reasonable time and on small samples; in fact it is several times more sensitive than the usual ^{13}C procedure. DEPT is now routine in many laboratories and is widely used in the Student Exercises in this textbook. The novel feature in the DEPT sequence is a variable proton pulse angle θ (see Figure 4.11) that is set at 90° for one subspectrum, and 135° for the other separate experiment.

The DEPT spectrum of ipsenol is shown in Figure 4.12. It consists of the "main spectrum" (a), which is a standard ^1H-decoupled ^{13}C spectrum; the middle spectrum (b) is a DEPT 135 where the CH_3's and CH's are phased up, whereas the CH_2's are phased down. The top spectrum (c) is a DEPT 90, where only CH carbons are detected. Quaternary ^{13}C are not detected in the DEPT subspectra. We can now interpret the ^{13}C peaks in the main spectrum as CH_3, CH_2, CH, or C by examining the two subspectra peaks along with the main spectrum. The easiest way to approach the interpretation of the ^{13}C/DEPT spectra is to first look for peaks that are in the ^1H-decoupled ^{13}C spectrum and not in either of the other two DEPT spectra. These are identified as being quaternary (no protons attached), next look at the DEPT 135 and label all the CH_2; they are easily identified because they are phased down. Finally identifying CH_3 versus CH is accomplished by inspection of the DEPT 135 versus the DEPT 90. The DEPT 90 has only CH carbons, so by difference the rest that are phased up in the DEPT 135 are CH_3's.

To illustrate the simplicity of interpreting ^{13}C/DEPT spectra, we return to ipsenol in Figure 4.12, starting at

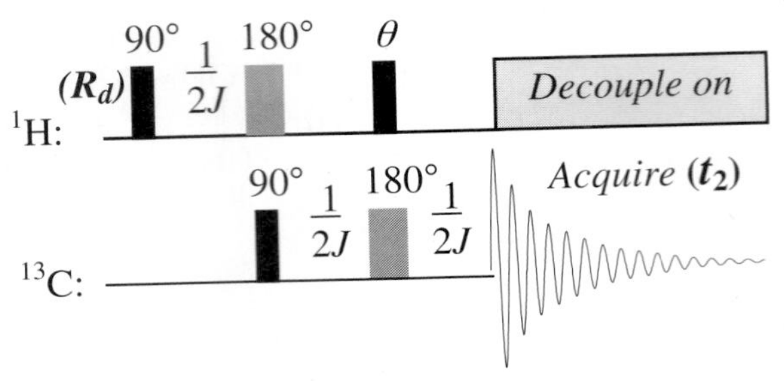

FIGURE 4.11 DEPT pulse sequence. The 1/2 J is for CH coupling constants typically 145 Hz. θ is a variable pulse angle. R_d is a relaxation delay, θ is a variable pulse angle, t_2 is the acquisition time.

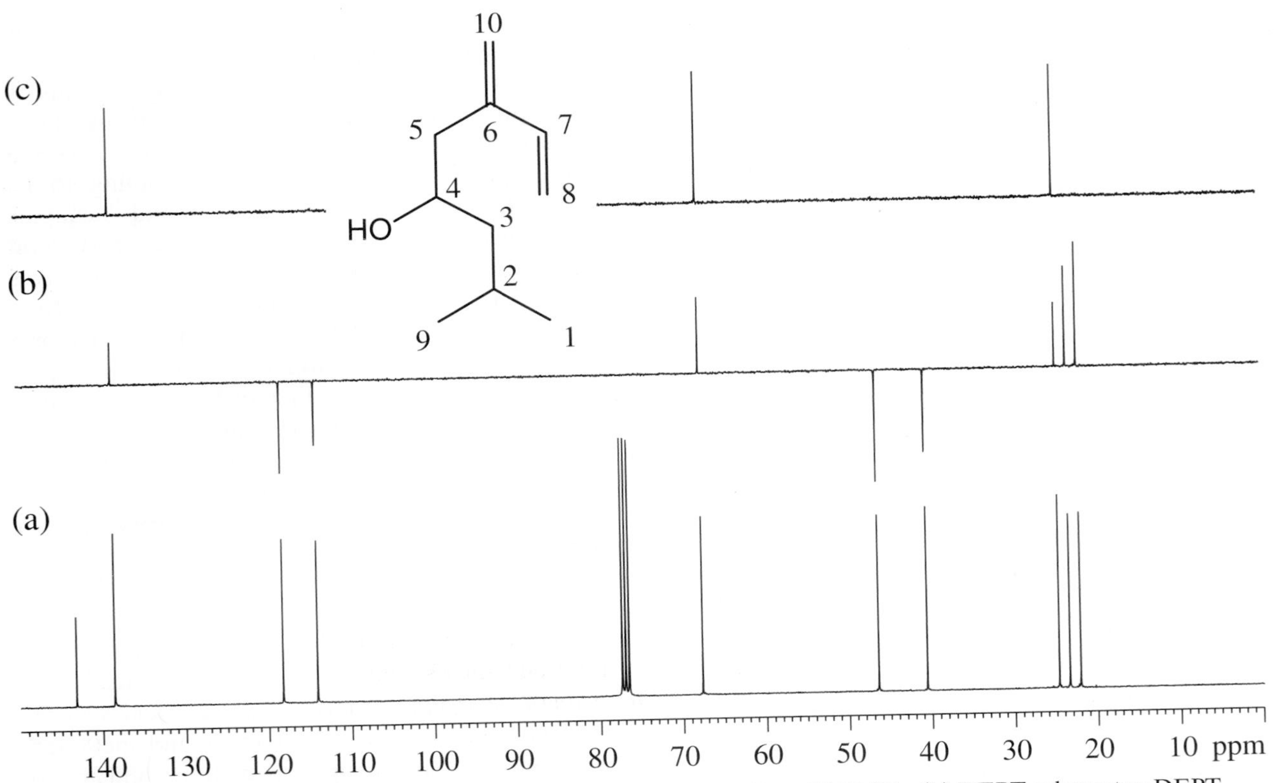

FIGURE 4.12 (a) Standard ^{13}C decoupled spectrum of ipsenol in $CDCl_3$ at 75.5 MHz. (b) DEPT subspectra: DEPT 135° CH and CH_3 up, CH_2 down. (c) DEPT 90° CH only.

the high-frequency peak at the left end of the main spectrum, we note that there are no peaks directly aligned above in the subspectra; therefore this peak in the main spectrum is a C peak—i.e., no attached proton atoms. The next peak to the right in the main spectrum is a CH peak, since the peak aligned in the lower subspectrum is pointed up, as is the aligned peak in the upper subspectrum. The next two peaks in the main spectrum are CH_2 peaks, since the aligned peaks in the lower subspectrum are pointed down. The next peak to the right of the solvent, triplet peak ($CDCl_3$) is now obviously another CH peak. The next two peaks are CH_2 peaks. The next peak is again CH. The next two are CH_3 since the aligned peaks in the lower subspectrum are pointed up, and there are no aligned peaks in the upper subspectrum. Summing up, we have (in order of frequency):

6	7	8	10		4	5	3	2	1	9
C,	CH,	CH_2,	CH_2,	solvent,	CH,	CH_2,	CH_2,	CH,	CH_3,	CH_3

Yes, interpretation of a DEPT spectrum takes a bit of practice, but the results are most instructive. Not only do we have the number of carbon and hydrogen atoms, but now we have the frequency distribution of carbon atoms with the number of hydrogen atoms attached to each carbon. However, there is a discrepancy between the proton count (see Figure 3.55 for ^{1}H spectrum for confirmation) in the proton spectrum and in the DEPT spectrum, since the OH is not recorded in the DEPT spectrum; nor are protons that are attached to such atoms as ^{15}N, ^{33}S, ^{29}Si, and ^{31}P. It is not difficult to correlate the DEPT spectrum with the ^{1}H spectrum. In fact, it is striking to observe how the olefinic and alkyl systems are widely separated in both spectra.

4.7 CHEMICAL CLASSES AND CHEMICAL SHIFTS

In this section, chemical shifts will be discussed under the headings of the common chemical classes of organic compounds. As noted earlier, the range of shifts generally encountered in routine ^{13}C studies is about 220 ppm.

As a first reassuring statement, we can say that trends in chemical shifts of ^{13}C are somewhat parallel to those of ^{1}H, so that some of the "feeling" for ^{1}H spectra may carry over to ^{13}C spectra. Furthermore, the concept of additivity of substituent effects (see Sections 4.7.1 and 4.7.6) is useful for both spectra. The ^{13}C shifts are related mainly to hybridization, substituent electronegativity, and diamagnetic anisotropy (to a lesser extent); solvent effects are important in both spectra. Chemical shifts for ^{13}C are affected by substitutions as far removed as the δ position; in the benzene ring, pronounced shifts for ^{13}C are caused by substituents at the point of attachment and at the *ortho, meta,* and *para* positions. The ^{13}C chemical shifts are also moved significantly to the right by the *γ-gauche* effect (see Section 4.7.1). Shifts to the right as much as several parts per million may occur on dilution. Hydrogen-bonding effects with polar solvents may cause shifts to the left.

As in other types of spectrometry, peak assignments are made on the basis of reference compounds. Reference material for many classes of compounds has accumulated in the literature. The starting point is a general correlation chart for chemical shift regions of ^{13}C atoms in the major chemical classes (see Figure 4.3 and Appendix C): then, minor changes within these regions are correlated with structure variations in the particular chemical class. The chemical shift values in the following tables must not be taken too literally because of the use of various solvents and concentration. For example, the C=O absorption of acetophenone in $CDCl_3$ appears at 2.4 ppm further to the left than in CCl_4; the effect on the other carbon atoms of acetophenone ranges from 0.0 to 1.1 ppm. Furthermore, much of the early work used various reference compounds, and the values were corrected to give parts per million from TMS.

A ^{13}C spectrum will often distinguish substitution patterns on an aromatic ring. If, for example, there are two identical (achiral) substituents, the symmetry elements alone will distinguish among the *para, ortho,* and *meta* isomers if the chemical shifts of the ring carbon atoms are sufficiently different. The *para* isomer has two simple axes and two planes. The *ortho* and *meta* isomers have one simple axis and one plane, but in the *meta* isomer the elements pass through two atoms. There is also a symmetry plane in the plane of the ring in each compound, which does not affect the ring carbon atoms.

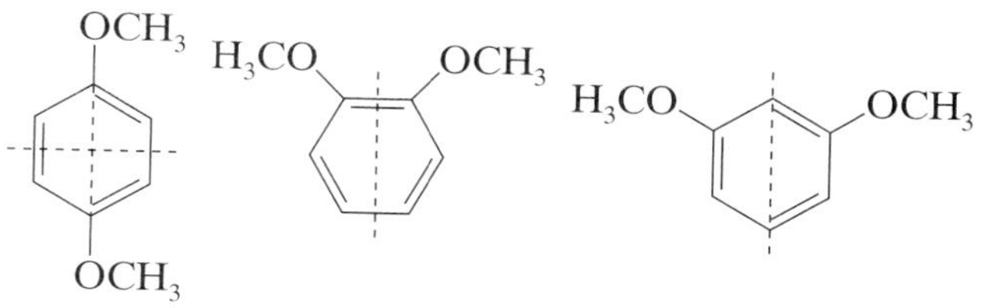

OCH_3 / OCH_3 H_3CO OCH_3 H_3CO OCH_3

The aromatic region of the ^{13}C spectrum for the *para* isomer shows two peaks: for the *ortho* isomer, three peaks: and for the *meta* isomer, four peaks. The quaternary carbon peaks are much less intense than the unsubstituted carbon peaks.

The additivity of shift increments is demonstrated in Section 4.7.6.

4.7.1 Alkanes

4.7.1.1 Linear and Branched Alkanes We know from the general correlation chart (Appendix C) that alkane groups unsubstituted by heteroatoms absorb to about 60 ppm. (Methane absorbs at −2.5 ppm). Within this range, we can predict the chemical shifts of individual ^{13}C atoms in a straight-chain or branched-chain hydrocarbon from the data in Table 4.4 and the formula given below.

This table shows the additive shift parameters (A) in hydrocarbons: the α effect of +9.1, the β effect of +9.4 ppm, the γ effect of −2.5, the δ effect of +0.3, the ϵ effect of +0.1, and the corrections for branching effects. The calculated (and observed) shifts for the carbon atoms of 3-methylpentane are:

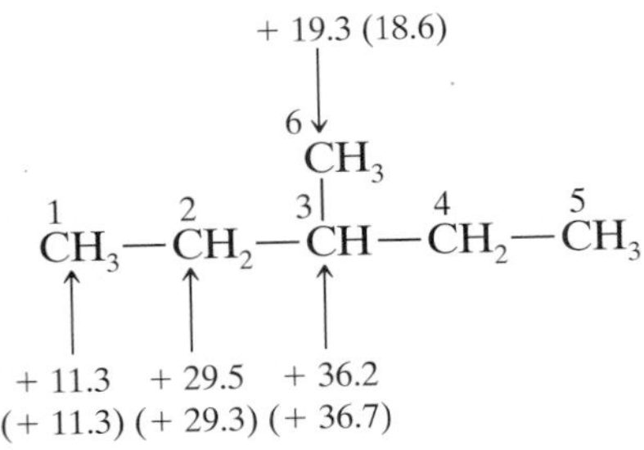

Calculations of shift are made from the formula: $\delta = -2.5 + \Sigma nA$, where δ is the predicted shift for a carbon atom; A is the additive shift parameter; and n is the number of carbon atoms for each shift parameter (−2.5 is the shift of the ^{13}C of methane). Thus, for carbon atom 1, we have 1 α-, 1 β-, 2 γ-, and 1 δ-carbon atoms.

TABLE 4.4 The ^{13}C Shift Parameters in Some Linear and Branched Hydrocarbons

^{13}C Atoms	Shift (ppm) (A)
α	9.1
β	9.4
γ	−2.5
δ	0.3
ϵ	0.1
1°(3°)[a]	−1.1
1°(4°)[a]	−3.4
2°(3°)[a]	−2.5
2°(4°)	−7.2
3°(2°)	−3.7
3°(3°)	−9.5
4°(1°)	−1.5
4°(2°)	−8.4

[a] The notations 1° (3°) and 1° (4°) denote a CH_3 group bound to a R_2CH group and to a R_3C group, respectively. The notation 2° (3°) denotes a RCH_2 group bound to a R_2CH group, and so on.

$$\delta_1 = -2.5 + (9.1 \times 1) + (9.4 \times 1) + (-2.5 \times 2) + (0.3 \times 1) = 11.3$$

Carbon atom 2 has 2 α-, 2 β-, and 1 γ-carbon atoms. Carbon atom 2 is a 2° carbon with a 3° carbon attached [2°(3°) = −2.5].

$$\delta_2 = -2.5 + (9.1 \times 2) + (9.4 \times 2) + (-2.5 \times 1) + (-2.5 \times 1) = 29.5$$

Carbon atom 3 has 3 α- and 2 β-carbon atoms, and it is a 3° atom with two 2° atoms attached [3°(2°) = − 3.7]. Thus,

$$\delta_3 = -2.5 + (9.1 \times 3) + (9.4 \times 2) + (-3.7 \times 2) = 36.2.$$

Carbon atom 6 has 1 α-, 2 β-, and 2 γ-carbon atoms, and it is a 1° atom with a 3° atom attached [1°(3°) = −1.1]. Thus,

$$\delta_6 = -2.5 + (9.1 \times 1) + (9.4 \times 2) + (-2.5 \times 2) + (-1.1 \times 1) = 19.3$$

The agreement with the determined values for such calculations is very good. Another useful calculation has been given.* The ^{13}C γ shift to lower frequency resulting from the γ carbon has been attributed to the steric compression of a γ *gauche* interaction but has no counterpart in ^1H spectra. It accounts, for example, for the shift to the right of an axial methyl substituent on a conformationally rigid cyclohexane ring, relative to an equatorial methyl, and for the shift to the right of the γ carbon atoms of the ring. Table 4.5 lists the shifts in some linear and branched alkanes.

4.7.1.2 Effect of Substituents on Alkanes Table 4.6 shows the effects of a substituent on linear and branched alkanes. The effect on the α-carbon parallels the electronegativity of the substituent except for bromine and iodine.† The effect at the β-carbon seems fairly constant for all the substituents except for the carbonyl, cyano, and nitro groups. The shift to the right at the γ carbon results (as above) from steric compression of a *gauche* interaction. For Y = N, O, and F, there is also a shift to the right with Y in the *anti* conformation, attributed to hyperconjugation.

* Lindeman. L.P. and Adams. J.Q. (1971). *Anal. Chem.*, 43,1245.
† See Section 3.4, Table 3.2 (Pauling table of electronegativity).

Y Cγ — *gauche*; Y Cγ — *anti*

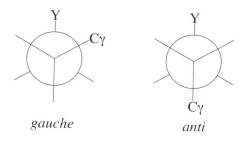

Table 4.6 provides the functional-group increments that must be added to the appropriate shift values for alkanes given in Table 4.5. For example, we can calculate the ^{13}C shifts for 3-pentanol.

$$\overset{\gamma}{CH_3}-\overset{\beta}{CH_2}-\overset{\alpha}{\underset{OH}{\underset{|}{CH}}}-\overset{\beta}{CH_2}-\overset{\gamma}{CH_3}$$

The OH substituent is attached "internally" (rather than "terminally") to the linear alkane chain of pentane; the point of attachment is labeled α, which corresponds to C-3 of pentane, for which the shift value of 34.7 is given in Table 4.6. To this value is added the increment +41, this being the increment for an OH group attached internally to the α-carbon of 3-pentanol (see line 12, 2nd column of numbers). The shift, therefore, for the point of attachment (the α-carbon) is calculated as 75.7. The β and γ shifts are calculated as follows. All of the calculated shifts are in reasonable agreement with the experimental values (Table 4.14).

	Calculated	Experimental (See Table 4.14)
C_α	34.7 + 41 = 75.7	73.8
C_β	22.8 + 8 = 30.8	29.7
C_γ	13.9 − 5 = 8.9	9.8

TABLE 4.5 The ^{13}C Shifts for Some Linear and Branched-Chain Alkanes (ppm from TMS)

Compound	C-1	C-2	C-3	C-4	C-5
Methane	−2.3				
Ethane	5.7				
Propane	15.8	16.3			
Butane	13.4	25.2			
Pentane	13.9	22.8	34.7		
Hexane	14.1	23.1	32.2		
Heptane	14.1	23.2	32.6	29.7	
Octane	14.2	23.2	32.6	29.9	
Nonane	14.2	23.3	32.6	30.0	30.3
Decane	14.2	23.2	32.6	31.1	30.5
Isobutane	24.5	25.4			
Isopentane	22.2	31.1	32.0	11.7	
Isohexane	22.7	28.0	42.0	20.9	14.3
Neopentane	31.7	28.1			
2,2-Dimethylbutane	29.1	30.6	36.9	8.9	
3-Methylpentane	11.5	29.5	36.9	(18.8, 3-CH_3)	
2,3-Dimethylbutane	19.5	34.3			
2,2,3-Trimethylbutane	27.4	33.1	38.3	16.1	
2,3-Dimethylpentane	7.0	25.3	36.3	(14.6, 3-CH_3)	

TABLE 4.6 Incremental Substituent Effects (ppm) on Replacement of H by Y in Alkanes. Y is Terminal or Internal[a]

Terminal: ...γ β α Y; Internal: ...γ β α(Y) β γ...

Y	α Terminal	α Internal	β Terminal	β Internal	γ
CH_3	9	6	10	8	−2
$CH{=}CH_2$	20		6		−0.5
$C{\equiv}CH$	4.5		5.5		−3.5
COOH	21	16	3	2	−2
COO^-	25	20	5	3	−2
COOR	20	17	3	2	−2
COCl	33	28		2	
$CONH_2$	22		2.5		−0.5
COR	30	24	1	1	−2
CHO	31				−2
Phenyl	23	17	9	7	−2
OH	48	41	10	8	−5
OR	58	51	8	5	−4
OCOR	51	45	6	5	−3
NH_2	29	24	11	10	
NH_3^+	26	24	8	6	−5
NHR	37	31	8	6	−4
NR_2	42		6		−3
NR_3^+	31		5		−7
NO_2	63	57	4	4	
CN	4	1	3	3	−3
SH	11	11	12	11	−4
SR	20		7		−3
F	68	63	9	6	−4
Cl	31	32	11	10	−4
Br	20	25	11	10	−3
I	−6	4	11	12	−1

[a] Add these increments to the shift values of the appropriate carbon atom in Table 4.5 or to the shift value calculated from Table 4.4.

Source: Wehrli. F.W., Marchand, A.P., and Wehrli, S. (1983). *Interpretation of Carbon-13 NMR Spectra.* 2nd ed. London: Heyden.

1-Pentanol would be treated similarly but as a "terminal" alcohol, and the carbon atom to which the OH group is attached would again be labeled α.

4.7.1.3 Cycloalkanes and Saturated Heterocyclics

4.7.1.3 Cycloalkanes and Saturated Heterocyclics The chemical shifts of the CH_2 groups in monocyclic alkanes are given in Table 4.7. The striking feature here is the strong shift to the right of cyclopropane, analogous to the shift of its proton absorptions.

Each ring skeleton has its own set of shift parameters, but a detailed listing of these is beyond the scope of this text. Rough estimates for substituted rings can be made with the substitution increments in Table 4.6. One of the striking effects in rigid cyclohexane rings is the shift to the right caused by the *γ-gauche* steric compression. Thus an axial methyl group at C-1 causes a shift to the right of several parts per million at C-3 and C-5.

Table 4.8 presents chemical shifts for several saturated heterocyclics.

4.7.2 Alkenes

The sp^2 carbon atoms of alkenes substituted only by alkyl groups absorb in the range of about 110–150 ppm. The double bond has a rather small effect on the shifts of the sp^3 carbons in the molecule as the following comparisons demonstrate.

$H_3C-C(CH_3)_2-CH_2-C(CH_3)=CH_2$: H_3C 30.4; C-CH_3 31.6; CH_2 52.2; =C 143.7; $=CH_2$ 114.4; CH_3 on C= 25.4

$H_3C-C(CH_3)_2-CH_2-CH(CH_3)-CH_3$: H_3C 29.9; C-CH_3 30.4; CH_2 53.5; CH 25.3; CH_3 on CH 24.7

The methyl signal of propene is at 18.7 ppm, and of propane at 15.8 ppm. In (*Z*)-2-butene, the methyl signals are at 12.1 ppm, compared with 17.6 ppm in (*E*)-2-butene, because of the γ effect. (For comparison, the methyl signals of butane are at 13.4 ppm). Note the γ effect on one of the geminal methyl groups in 2-methyl-2-butene (Table 4.9).

In general, the terminal $=CH_2$ group absorbs to the right relative to an internal $=CH-$ group, and (*Z*)$-CH=CH-$ signals are to the right from those of corresponding *E* groups. Calculations of approximate shifts can be made from the following parameters where α, β, and γ represent substitutents on the same end of the double bond as the alkene carbon of interest, and α′, β′, and γ′ represent substitutents on the far side.

α	10.6
β	7.2
γ	−1.5
α′	−7.9
β′	−1.8
γ′	−1.5
Z(*cis*) correction	−1.1

These parameters are added to 123.3 ppm, the shift for ethylene. We can calculate the values of C-3 and C-2 for (*Z*)-3-methyl-2-pentene as follows:

TABLE 4.7 Chemical Shifts of Cycloalkanes (ppm from TMS)

C_3H_6	−2.9	C_7H_{14}	28.4
C_4H_8	22.4	C_8H_{16}	26.9
C_5H_{10}	25.6	C_9H_{18}	26.1
C_6H_{12}	26.9	$C_{10}H_{20}$	25.3

TABLE 4.8 Chemical Shifts for Saturated Heterocyclics (ppm from TMS, neat)

Unsubstituted

Ring	O	S	NH
3-membered	40.6	18.1	18.2
4-membered	72.6 (α), 22.9 (β)	26.1 (α), 28.1 (β)	48.1 (α), 19.0 (β)
5-membered	67.9 (α), 25.8 (β)	31.7 (α), 31.2 (β)	47.1 (α), 25.7 (β)
6-membered	69.5 (α), 27.7 (β), 24.9 (γ)	29.1 (α), 27.8 (β), 26.6 (γ)	47.9 (α), 27.8 (β), 25.9 (γ)

Substituted

Methyloxirane: 47.6 (CH), 47.3 (CH_2), 18.1 (CH_3)

Methylthietane: 36.5 (CH_2S), 32.4 (CH), 21.8 (CH_3)

N-Methylpyrrolidine: 56.7 (α), 24.4 (β), 48.0 (NCH_3)

TABLE 4.9 Alkene and Cycloalkene Chemical Shift (ppm from TMS)

$H_2C{=}CH_2$ 123.2 | 18.7 136.2 115.9 | 27.4 113.3 13.4 140.2 | 12.1 124.6 | 126.0 17.6

13.7 36.1 114.3 22.3 138.5 | 132.7 123.2 14.0 20.5 12.3 | 14.0 133.3 17.9 25.8 123.7 | 14.0 31.4 138.7 22.4 33.7 114.5 | 13.7 29.4 12.6 22.6 137.2 124.0

13.7 35.3 125.1 23.2 131.7 17.7 | 131.2 14.5 20.6 | 14.4 131.3 20.5 | $CH_2{=}C{=}CH_2$ 74.8 213.5

117.5 137.2 | 115.9 38.0 137.3 | 114.4 129.5 17.2 137.8 133.2 | 116.5 130.9 126.4 132.5 12.8 | 18.0 130.2 13.0 128.3 127.4 123.1

109.3 30.8 149.3 22.5 12.4 | 16.9 12.4 131.4 118.7 25.3 | 109.8 41.8 144.5 23.6 20.5 14.5 | 21.2 112.9 11.7 42.1 144.9 31.2 | 19.6 19.7 14.6 131.6 126.6 25.6

30.2 137.2 | 130.8 32.6 22.1 | 127.3 24.5 22.1 | 107.1 149.7 36.2 28.9 26.9 | 26.0 124.5 | 126.1 124.6 22.3 | 126.4 137.1 128.7 135.2 113.3 128.0 126.4 128.7

$$\underset{5}{\overset{\beta}{H_3C}}-\underset{4}{\overset{\alpha}{CH_2}}-\underset{3}{C}(\overset{\alpha}{CH_3})=\underset{2}{C}(H)-\underset{1}{\overset{\alpha'}{CH_3}} \qquad \underset{5}{\overset{\beta'}{H_3C}}-\underset{4}{\overset{\alpha'}{CH_2}}-\underset{3}{C}(\overset{\alpha'}{CH_3})=\underset{2}{C}(H)-\underset{1}{\overset{\alpha}{CH_3}}$$

$$\delta_3 = 123.3 + (2 \times 10.6) + (1 \times 7.2) + (1 \times -7.9) - 1.1 = 142.7 \text{ ppm}$$

$$\delta_2 = 123.3 + (1 \times 10.6) + (2 \times -7.9) + (1 \times 1.8) - 1.1 = 115.2 \text{ ppm}$$

The measured values are C-3 = 137.2 and C-2 = 116.8. The agreement is fair.

Carbon atoms directly attached to a (Z) C=C group are more shielded than those attached to the stereoisomeric (E) group by 4–6 ppm (Table 4.9). Alkene carbon atoms in polyenes are treated as though they were alkane carbon substituents on one of the double bonds. Thus, in calculating the shift of C-2 in 1,4-pentadiene, C-4 is treated like a β-sp^3 carbon atom.

Representative alkenes are presented in Table 4.9. There are no simple rules to handle polar substituents on an alkene carbon. The shifts for vinyl ethers can be rationalized on the basis of electron density of the contributor structures

$$H_2C{=}C(H){-}\ddot{O}{-}CH_3 \longleftrightarrow H_2\bar{C}{-}C(H){=}\overset{+}{O}{-}CH_3$$

84.2 153.2

as can the shifts for α-, β-unsaturated ketones.

150.7 129.3

The same rationalization applies to the proton shifts in these compounds. Shifts for several substituted alkenes are presented in Table 4.10.

The central carbon atom (=C=) of alkyl-substituted allenes absorbs in the range of about 200–215 ppm, whereas the terminal atoms (C=C=C) absorb in the range of about 75–97 ppm.

4.7.3 Alkynes

The sp carbon atoms of alkynes substituted only by alkyl groups absorb in the range of approximately 65–90 ppm

TABLE 4.10 Chemical Shift of Substituted Alkenes (ppm from TMS)

126.1 Cl 117.4 · 122.0 Br 115.0 · 84.8 I 129.9 · 176.0 OH 88.8 · 153.2 O 55.1 84.2 · 136.4 O 136.0 192.1 · 107.7 CN 137.8 117.5

133.7 Cl 117.5 45.1 · 137.5 OH 114.9 63.4 · 128.0 COOH 131.9 173.2 · 128.7 51.6 $COOCH_3$ 129.9 166.7 · 138.5 196.9 C— 129.3 26.3 O

Br H 104.7 132.7 H CH_3 18.2 · Br 15.3 108.9 129.4 H H · 17.9 H 144.1 122.3 H $COOCH_3$ 167.0 51.3 · 12.6 166.5 $COOCH_3$ 52.0 142.9 124.4 H H · 20.2 141.7 C 167.6 O O 96.4

O 207.1 27.0 · O 208.0 9.3 46.6 · O 210.4 34.0 134.4 29.0 165.2 · O 199.7 38.1 129.8 25.7 150.9 22.8 · 128.6 152.5 134.1 193.4 129.1 128.5 131.2 O

(Table 4.11). The triple bond shifts the sp^3 carbon atoms, directly attached, about 5–15 ppm to the right relative to the corresponding alkane. The terminal ≡CH absorbs further to the right than the internal ≡CR. Alkyne carbon atoms with a polar group directly attached absorb from about 20–95 ppm.

TABLE 4.11 Alkyne Chemical Shifts (ppm)

Compound	C-1	C-2	C-3	C-4	C-5	C-6
1-Butyne	67.0	84.7				
2-Butyne		73.6				
1-Hexyne	68.1	84.5	18.1	30.7	21.9	13.5
2-Hexyne	2.7	73.7	76.9	19.6	21.6	12.1
3-Hexyne	15.4	13.0	80.9			

Polar resonance structures explain these shifts for alkynyl ethers, which are analogous to the shifts for vinyl ethers (Section 3.34).

23.2 89.4 HC≡C—O—H_2C—CH_3 28.0 88.4 H_3C—C≡C—O—CH_3

4.7.4 Aromatic Compounds

Benzene carbon atoms absorb at 128.5 ppm, neat or as a solution in $CDCl_3$. Substituents shift the attached aromatic carbon atom as much as ±35 ppm. Fused-ring absorptions are as follows:

Naphthalene: C-1, 128.1; C-2, 125.9; C-4a, 133.7.

Anthracene: C-1, 130.1; C-2, 125.4; C-4a, 132.2; C-9, 132.6.

Phenanthrene: C-1, 128.3; C-2, 126.3: C-3. 126.3; C-4, 122.2; C-4a, 131.9*; C-9, 126.6; C-10a, 130.1.*

Shifts of the aromatic carbon atom directly attached to the substituent have been correlated with substituent electronegativity after correcting for magnetic anisotropy effects; shifts at the *para* aromatic carbon have been correlated with the Hammett σ constant. *Ortho* shifts are not readily predictable and range over about 15 ppm. *Meta* shifts are generally small-up to several parts per million for a single substituent.

The substituted aromatic carbon atoms can be distinguished from the unsubstituted aromatic carbon atom by its decreased peak height; that is, it lacks a proton and thus suffers from a longer T_1 and a diminished NOE.

Incremental shifts from the carbon atoms of benzene for the aromatic carbon atoms of representative monosubstituted benzene rings (and shifts from TMS

CN Cl (I) ≡ CN (II) + Cl (III)

I C #	I Cal	I Obs	II C #	II Obs	III C #	III Obs
1	-18.0	-16.6	1	-16.0	4	-2.0
2	4.6	5.1	2	3.6	3	1.0
3	0.8	1.3	3	0.6	2	0.2
4	10.7	10.8	4	4.3	1	6.4

* Assignment uncertain.

of carbon-containing substituents) are given in Table 4.12. Shifts from benzene for polysubstituted benzene ring carbon atoms can be approximated by applying the principle of increment additivity. For example, the shift from benzene for C-2 of the disubstituted compound 4-chlorobenzonitrile is calculated by adding the effect for an *ortho* CN group (+3.6) to that for a *meta* Cl group (+1.0): 128.5 + 3.6 + 1 = 133.1 ppm.

4.7.5 Heteroaromatic Compounds

Complex rationalizations have been offered for the shifts of carbon atoms in heteroaromatic compounds. As a general rule, C-2 of oxygen- and nitrogen-containing rings is further to the left than C-3. Large solvent and pH effects have been recorded. Table 4.13 gives values for neat samples of several five- and six-membered heterocyclic compounds.

4.7.6 Alcohols

Substitution of H in an alkane by an OH group moves the signal to the left by 35–52 ppm for C-1, 5–12 ppm for C-2, and to the right by about 0–6 ppm for C-3. Shifts for several acyclic and alicyclic alcohols are given

TABLE 4.12 Incremental Shifts of the Aromatic Carbon Atoms of Monosubstituted Benzenes (ppm from Benzene at 128.5 ppm). Carbon Atom of Substituents in parts per million from TMS[a]

Substituent	C-1 (Attachment)	C-2	C-3	C-4	C of Substituent (ppm from TMS)
H	0.0	0.0	0.0	0.0	
CH_3	9.3	0.7	−0.1	−2.9	21.3
CH_2CH_3	15.6	−0.5	0.0	−2.6	29.2 (CH_2), 15.8 (CH_3)
$CH(CH_3)_2$	20.1	−2.0	0.0	−2.5	34.4 (CH), 24.1 (CH_3)
$C(CH_3)_3$	22.2	−3.4	−0.4	−3.1	34.5 (C), 31.4 (CH_3)
$CH{=}CH_2$	9.1	−2.4	0.2	−0.5	137.1 (CH), 113.3 (CH_2)
$C{\equiv}CH$	−5.8	6.9	0.1	0.4	84.0 (C), 77.8 (CH)
C_6H_5	12.1	−1.8	−0.1	−1.6	
CH_2OH	13.3	−0.8	−0.6	−0.4	64.5
$CH_2O(C{=}O)CH_3$	7.7	~0.0	~0.0	~0.0	20.7 (CH_3), 66.1 (CH_2), 170.5 (C=O)
OH	26.6	−12.7	1.6	−7.3	
OCH_3	31.4	−14.4	1.0	−7.7	54.1
OC_6H_5	29.0	−9.4	1.6	−5.3	
$O(C{=}O)CH_3$	22.4	−7.1	−0.4	−3.2	23.9 (CH_3), 169.7 (C=O)
(C=O)H	8.2	1.2	0.6	5.8	192
$(C{=}O)CH_3$	7.8	−0.4	−0.4	2.8	24.6 (CH_3), 195.7 (C=O)
$(C{=}O)C_6H_5$	9.1	1.5	−0.2	3.8	196.4 (C=O)
$(C{=}O)F_3$	−5.6	1.8	0.7	6.7	
(C=O)OH	2.9	1.3	0.4	4.3	168
$(C{=}O)OCH_3$	2.0	1.2	−0.1	4.8	51.0 (CH_3), 166.8 (C=O)
(C=O)Cl	4.6	2.9	0.6	7.0	168.5
$(C{=}O)NH_2$	5.0	−1.2	0.0	3.4	
$C{\equiv}N$	−16	3.6	0.6	4.3	119.5
NH_2	19.2	−12.4	1.3	−9.5	
$N(CH_3)_2$	22.4	−15.7	0.8	−11.8	40.3
$NH(C{=}O)CH_3$	11.1	−9.9	0.2	−5.6	
NO_2	19.6	−5.3	0.9	6.0	
N=C=O	5.7	−3.6	1.2	−2.8	129.5
F	35.1	−14.3	0.9	−4.5	
Cl	6.4	0.2	1.0	−2.0	
Br	−5.4	3.4	2.2	−1.0	
I	−32.2	9.9	2.6	−7.3	
CF_3	2.6	−3.1	0.4	3.4	
SH	2.3	0.6	0.2	−3.3	
SCH_3	10.2	−1.8	0.4	−3.6	15.9
SO_2NH_2	15.3	−2.9	0.4	3.3	
$Si(CH_3)_3$	13.4	4.4	−1.1	−1.1	

[a] See Ewing, D.E., (1979). *Org. Magn. Reson.*, **12**, 499, for 709 chemical shifts of monosubstituted benzenes.

TABLE 4.13 Shifts for Carbon Atoms of Heteroaromatics (ppm from neat TMS)

Compound	C-2	C-3	C-4	C-5	C-6	Substituent
Furan	142.7	109.6				
2-Methylfuran	152.2	106.2	110.9	141.2		13.4
Furan-2-carboxaldehyde	153.3	121.7	112.9	148.5		178.2
Methyl 2-furoate	144.8	117.9	111.9	146.4		159.1 (C=O), 51.8 (CH_3)
Pyrrole	118.4	108.0				
2-Methylpyrrole	127.2	105.9	108.1	116.7		12.4
Pyrrole-2-carboxaldehyde	134.0	123.0	112.0	129.0		178.9
Thiophrene	124.4	126.2				
2-Methylthiophene	139.0	124.7	126.4	122.6		14.8
Thiophene-2-carboxaldehyde	143.3	136.4	128.1	134.6		182.8
Thiazole	152.2		142.4	118.5		
Imidazole	136.2		122.3	122.3		
Pyridine	150.2	123.9	135.9			
Pyrimidine	159.5		157.4	122.1	157.4	
Pyrazine	145.6					
2-Methylpyrazine	154.0	141.8[a]	143.8[a]	144.7[a]		21.6

[a] Assignment not certain

in Table 4.14. Acetylation provides a useful diagnostic test for an alcohol: The C-1 absorption moves to the left by about 2.5–4.5 ppm, and the C-2 absorption moves to the right by a similar amount; a 1,3-diaxial interaction may cause a slight (~1 ppm) shift to the left of C-3. Table 4.14 may be used to calculate shifts for alcohols as described earlier.

TABLE 4.14 Chemical Shift of Alcohols (ppm from TMS)

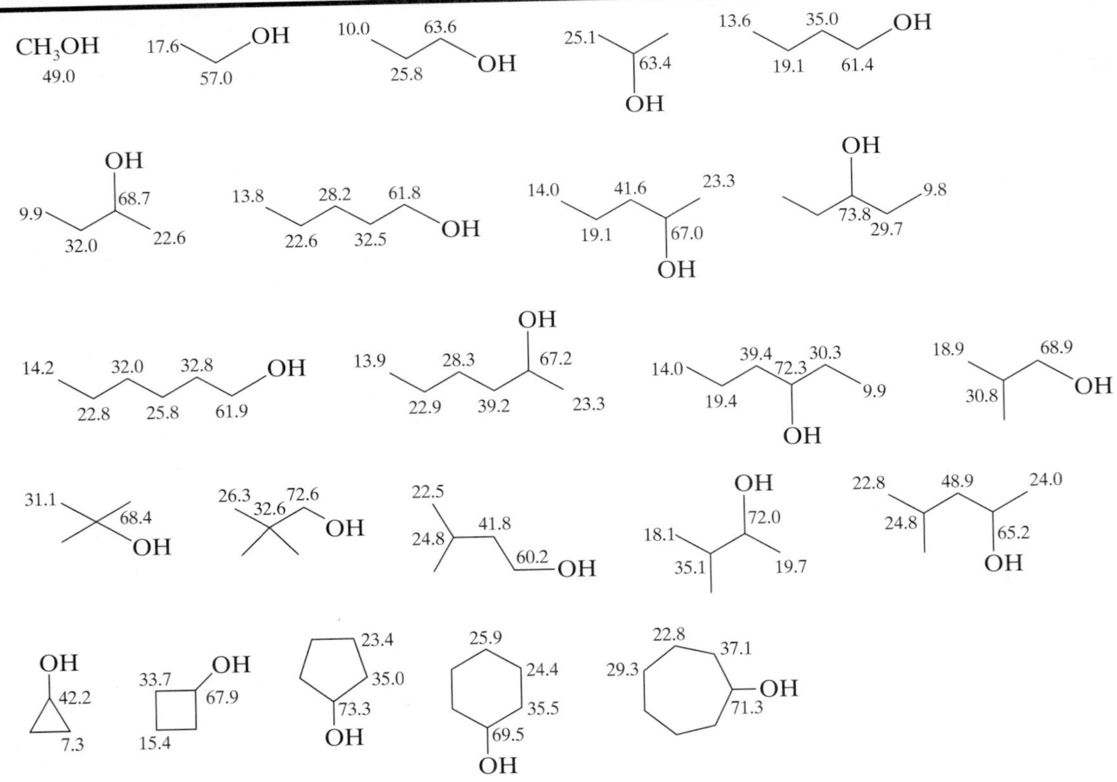

CH_3OH 49.0

17.6 57.0 OH

10.0 63.6 25.8 OH

25.1 63.4 OH

13.6 35.0 19.1 61.4 OH

OH 9.9 68.7 32.0 22.6

13.8 28.2 61.8 22.6 32.5 OH

14.0 41.6 23.3 19.1 67.0 OH

OH 9.8 73.8 29.7

14.2 32.0 32.8 OH 22.8 25.8 61.9

OH 13.9 28.3 67.2 22.9 39.2 23.3

14.0 39.4 72.3 30.3 9.9 19.4 OH

18.9 68.9 30.8 OH

31.1 68.4 OH

26.3 32.6 72.6 OH

22.5 24.8 41.8 60.2 OH

OH 18.1 72.0 35.1 19.7

22.8 48.9 24.0 24.8 65.2 OH

OH 42.2 7.3

33.7 OH 67.9 15.4

23.4 35.0 73.3 OH

25.9 24.4 35.5 69.5 OH

22.8 37.1 29.3 OH 71.3

TABLE 4.15 Chemical Shift of Ethers, Acetals, and Epoxides (ppm from TMS)

59.7 14.7 67.9 57.6 67.4 17.1 73.2 11.1 24.0 52.5 153.2 84.2

33.1 14.6 71.2 20.3 40.6 22.9 72.6 26.5 68.4 24.9 27.7 69.5

95.0 64.5 66.5 92.8 65.9 26.2 109.9 53.7 19.9 99.6 60.7 15.4

4.7.7 Ethers, Acetals, and Epoxides

An alkoxy substituent causes a somewhat larger shift to the left at C-1 (~11 ppm larger) than that of a hydroxy substituent. This is attributed to the C-1′ of the alkoxy group having the same effect as a β-C relative to C-1. The O atom is regarded here as an "α-C" to C-1.

$$\underset{17.6}{CH_3}-\underset{57.0}{CH_2}-OH \qquad \overset{2}{\underset{14.7}{CH_3}}-\overset{1}{\underset{67.9}{CH_2}}-O-\overset{1'}{\underset{57.6}{CH_3}}$$

Note also that the "γ effect" (shift to the right) on C-2 is explainable by similar reasoning. Conversely, the ethoxy group affects the OCH_3 group (compare CH_3OH). Table 4.15 gives shifts of several ethers.

The dioxygenated carbon of acetals absorbs in the range of about 88–112 ppm. Oxirane (an epoxide) absorbs at 40.6 ppm.

The alkyl carbon atoms of arylalkyl ethers have shifts similar to those of dialkyl ethers. Note the large shift to the right of the ring *ortho* carbon resulting from electron delocalization as in the vinyl ethers.

129.5 159.9 120.8 OCH_3 54.1 114.1

4.7.8 Halides

The effect of halide substitution is complex. A single fluorine atom (in CH_3F) causes a large shift to the left from CH_4 as electronegativity considerations would suggest. Successive geminal substitution by Cl (CH_3Cl, CH_2Cl_2, $CHCl_3$, CCl_4) results in increasing shifts to the left-again expected on the basis of electronegativity. But with Br and I the "heavy atom effect" supervenes. The carbon shifts of CH_3Br and CH_2Br_2 move progressively to the right. A strong progression to the right for I commences with CH_3I, which is to the right of CH_4. There is a progressive shift to the left at C-2 in the order I > Br > F. Cl and Br show γ-*gauche* shielding at C-3, but I does not, presumably because of the low population of the hindered gauche rotamer. Table 4.16 shows these trends.

I, H, CH_3, H, H, H — *gauche* ⇌ I, H, H, H, H, CH_3 — *anti*

TABLE 4.16 Shift Position of Alkyl Halides (neat, ppm from TMS)

Compound	C-1	C-2	C-3
CH_4	−2.3		
CH_3F	75.4		
CH_3Cl	24.9		
CH_2Cl_2	54.0		
$CHCl_3$	77.5		
CCl_4	96.5		
CH_3Br	10.0		
CH_2Br_2	21.4		
$CHBr_3$	12.1		
CBr_4	−28.5		
CH_3I	−20.7		
CH_2I_2	−54.0		
CHI_3	−139.9		
CI_4	−292.5		
CH_3CH_2F	79.3	14.6	
CH_3CH_2Cl	39.9	18.7	
CH_3CH_2Br	28.3	20.3	
CH_3CH_2I	−0.2	21.6	
$CH_3CH_2CH_2Cl$	46.7	26.5	11.5
$CH_3CH_2CH_2Br$	35.7	26.8	13.2
$CH_3CH_2CH_2I$	10.0	27.6	16.2

Halides may show large solvent effects; for example, C-1 for iodoethane is at −6.6 in cyclohexane, and at −0.4 in DMF.

4.7.9 Amines

A terminal NH_2 group attached to an alkyl chain causes a shift to the left of about 30 ppm at C-1, a shift to the left of about 11 ppm at C-2, and a shift to the right of about 4.0 ppm at C-3. The NH_3^+ group shows a somewhat smaller effect. *N*-alkylation increases the shift to the left of the NH_2 group at C-1. Shift positions for selected acyclic and alicyclic amines are given in Table 4.17 (see Table 4.8 for heterocyclic amines).

4.7.10 Thiols, Sulfides, and Disulfides

Since the electronegativity of sulfur is considerably less than that of oxygen, sulfur causes a correspondingly smaller chemical shift. Examples of thiols, sulfides, and disulfides are given in Table 4.18.

4.7.11 Functional Groups Containing Carbon

Carbon-13 NMR spectrometry permits direct observation of carbon-containing functional groups; the shift ranges for these are given in Appendix C. With the exception of CH=O, the presence of these groups could not be directly ascertained by ^{1}H NMR.

TABLE 4.17 Chemical Shift of Acyclic and Alicyclic Amines (in $CDCl_3$, ppm from TMS)

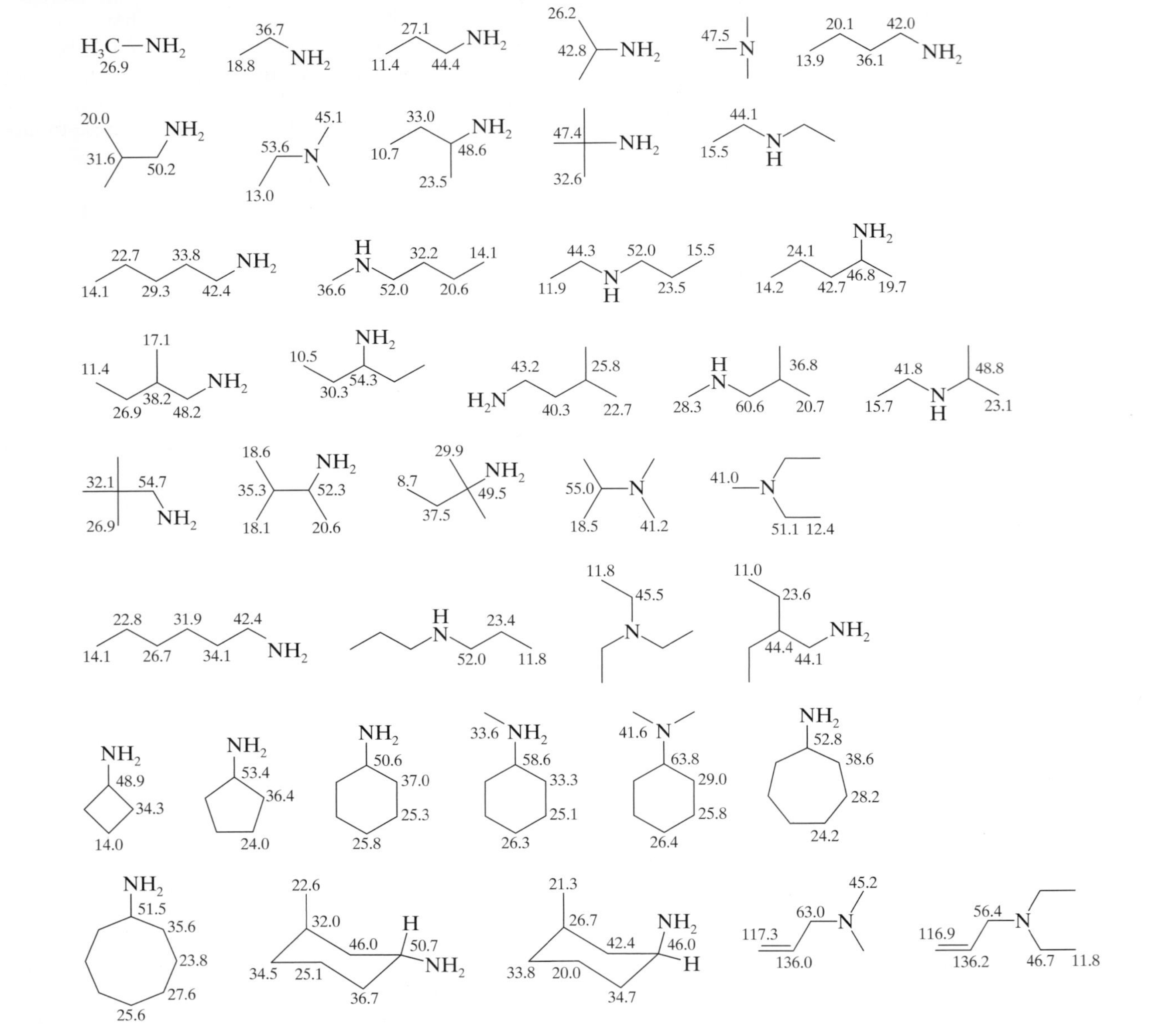

TABLE 4.18 Shift Position of Thiols, Sulfides, and Disulfides (ppm from TMS)

Compound	C-1	C-2	C-3
CH_3SH	6.5		
CH_3CH_2SH	19.8	17.3	
$CH_3CH_2CH_2SH$	26.4	27.6	12.6
$CH_3CH_2CH_2CH_2SH$	23.7	35.7	21.0
$(CH_3)_2S$	19.3		
$(CH_3CH_2)_2S$	25.5	14.8	
$(CH_3CH_2CH_2)_2S$	34.3	23.2	13.7
$(CH_3CH_2CH_2CH_2)_2S$	34.1	31.4	22.0
CH_3SSCH_3	22.0		
$CH_3CH_2SSCH_2CH_3$	32.8	14.5	

4.7.11.1 Ketones and Aldehydes The $R_2C{=}O$ and the $RCH{=}O$ carbon atoms absorb in a characteristic region. Acetone absorbs at 203.3 ppm, and acetaldehyde at 199.3 ppm. Alkyl substitution on the α-carbon causes a shift to the left of the C=O absorption of 2–3 ppm until steric effects supervene. Replacement of the CH_3 of acetone or acetaldehyde by a phenyl group causes a shift to the right of the C=O absorption (acetophenone, 195.7 ppm; benzaldehyde. 190.7 ppm); similarly, α,β-unsaturation causes shifts to the right (acrolein, 192.1 ppm, compared with propionaldehyde, 201.5 ppm). Presumably, charge delocalization by the benzene ring or the double bond makes the carbonyl carbon less electron deficient.

Of the cycloalkanones, cyclopentanone has a pronounced shift to the left. Table 4.19 presents chemical shifts of the C=O group of some ketones and aldehydes. Because of rather large solvent effects, there are differences of several parts per million from different literature sources. Replacement of CH_2 of alkanes by C=O causes a shift to the left at the α-carbon (~10–14 ppm) and a shift to the right at the β-carbon (several ppm in acyclic compounds). In a coupled spectrum, the aldehyde CH=O is a doublet.

4.7.11.2 Carboxylic Acids, Esters, Chlorides, Anhydrides, Amides, and Nitriles The C=O groups of carboxylic acids and derivatives are in the range of 150–185 ppm. Dilution and solvent effects are marked for carboxylic acids; anions appear further to the left. The effects of substituents and electron delocalization are generally similar to those for ketones. Nitriles absorb in the range of 115–125 ppm. Alkyl substituents on the nitrogen of amides cause a small (up to several ppm) shift to the right of the C=O group (see Table 4.20).

4.7.11.3 Oximes The quaternary carbon atom of simple oximes absorb in the range of 145–165 ppm. It is possible to distinguish between *E* and *Z* isomers since the C=O shift is to the right in the sterically more compressed form, and the shift of the more hindered substituent (*syn* to the OH) is farther to the right than the less hindered.

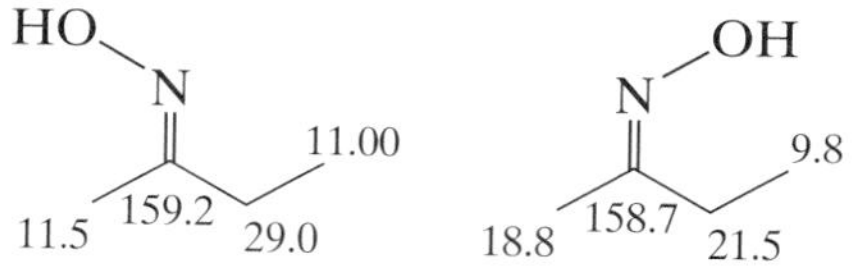

TABLE 4.19 Shift Position of the C=O Group and Other Carbon Atoms of Ketones and Aldehydes (ppm from TMS)

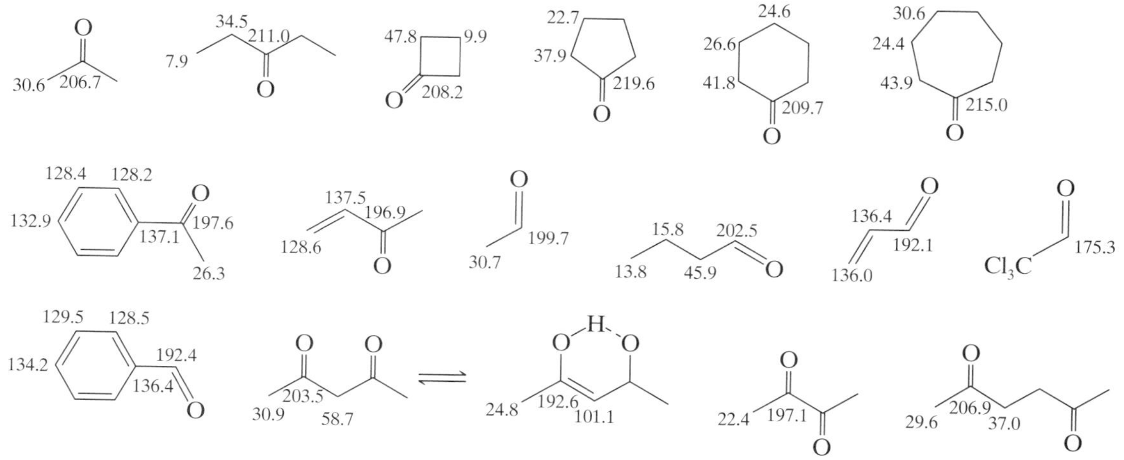

TABLE 4.20 Shift Position of the C=O Group and other Carbon Atoms of Carboxylic Acids. Esters, Lactones, Chlorides Anhydrides, Carbamates, and Nitriles (ppm from TMS)

178.1 $H_3C—COOH$ 20.6

168.0 $Cl_3C—COOH$ 89.1

O 169.5 C Cl 33.8

34.1 COOH 184.8 18.8

128.0 131.9 COOH 173.2

163.0 $F_3C—COOH$ 115.0

181.5 $H_3C—COO\ Na^+$ in D_2O

NH_2 17.2 51.5 176.5 COOH

O 60.0 20.0 170.3 O 13.8

27.2 173.3 O 9.2 50.8 O

23.4 25.5 O 51.1 14.3 32.2 33.9 172.1 O

O 66.8 21.4 170.0 O 20.4

129.9 164.5 O 128.7 52.0 O

O 142.4 20.8 168.0 O 97.4

O 61.0 F_3C 158.1 O 14.1 114.1

167.2 $COOCH_3$ 132.8 52.3 135.0 39.5 132.8 133.0

166.0 $COOCH_3$ 130.2 51.5 129.9 128.7 133.1

169.4 COOH 130.3 130.3 128.7 134.0

O 167.9 Cl 133.3 131.4 129.1 135.4

O O O 167.3 20.2

O O 8.4 27.1 170.3 O

O 28.4 171.7 O O

O 136.7 164.3 O O

125.3 O 131.1 136.1 165.5 O O

O 27.7 177.9 O 22.2 68.6

O 29.8 171.2 O 19.1 69.4 22.7

O NH_2 166.3

O —C—NH_2 25.5 174.3

31.1 O N—C 162.4 36.2 H

18.3 NH_2 141.3 171.0 124.0 O

127.2 O 135.2 37.6 128.6 169.5 N 129.8 127.2 37.6 128.6

O 60.9 H_2N 157.8 O 14.5

—C≡N 1.3 117.7

10.8 CN 10.6 120.8

18.8 H 149.3 101.2 H CN 117.6

H H 150.2 100.9 17.3 CN 116.0

118.7 CN 112.3 132.0 132.7 129.1

REFERENCES

General

Abraham, R.J., Fisher, J., and Loftus, P. (1992). *Introduction to NMR Spectroscopy.* New York: Wiley.

Beckmann, N. (1995). *Carbon-13 NMR Spectroscopy of Biological Systems.* Academic Press, New York.

Breitmaier, E. (2002). *Structure Elucidation by NMR in Organic Chemistry: A Practical Guide.* John Wiley & Sons; 3rd Revision edition. New York: Wiley.

Breitmaier, E., and Voelter, W. (1987). *Carbon-13 NMR Spectroscopy.* 3rd ed, New York: VCH Publishers.

Claridge, T.D.W. (1999). *High-Resolution NMR Techniques in Organic Chemistry.* Elsevier Science.

Farrar, T.C. (1987). *An Introduction to Pulse NMR Spectroscopy.* Chicago: Farragut Press.

Kalinowski, H.O., Berger, S., and Braun, S. (1988). *Carbon-13 NMR Spectroscopy.* New York: Wiley.

Levy, G.C., Lichter, R.L., and Nelson, G.L. (1980). *Carbon-13 Nuclear Magnetic Resonance for Organic Chemists,* 2nd ed. New York: Wiley.

Muller. K., and Pregosin, P.S. (1976). *Fourier Transform NM/EA Practical Approach.* New York: Academic Press.

Paudler, W.W. (1987). *Nuclear Magnetic Resonance.* New York, Wiley.

Pihlaja, K., and Kleinpeter, E. (1994). *Carbon-13 NMR Chemical Shifts in Structural and Stereochemical Analysis.*

Sanders, J.K. and Hunter, B.K. 1993. *Modern NMR Spectroscopy,* 2nd ed. Oxford: Wiley.

Shaw, D. (1984). *Fourier Transform NMR Spectroscopy,* 2nd ed. Amsterdam: Elsevier.

Wehrli, F.W., Marchand, A.P., and Wehrli, S. (1988). *Interpretation of Carbon-13 NMR Spectra,* 2nd ed. New York: Wiley.

Spectra, Data, and Workbooks

Bates, R.B., and Beavers, W.A. (1981). *C-13 NMR Problems.* Clifton, NJ: Humana Press.

Breitmaier, E., Haas, G., and Voelter, W. (1979). *Atlas of C-13 NMR Data,* Vols. 1–3. Philadelphia: Heyden (3017 compounds).

Bremser, W., Ernst, L., Franke, B., Gerhards, R., and Hardt, A. (1987). *Carbon-13 NMR Spectral Data* (microfiche), 4th ed. New York: VCH Publishers (58,108 spectra of 48.357 compounds, tabular).

Fuchs, P.L., and Bunnell, C.A. (1979). *Carbon-13 NMR-Based Organic Spectral Problems.* New York: Wiley.

Johnson, L. F., and Jankowski, W. C. (1972). *Carbon-13 NMR Spectra, a Collection of Assigned, Coded, and Indexed Spectra.* New York: Wiley.

Pouched, C.J., and Behnke, J. (1993). *Aldrich Library of ^{13}C and ^{1}H FT-NMR Spectra, 300 MHz.* Milwaukee, WI: Aldrich Chemical Co.

Pretsch, E., Clem, T., Seihl, J., and Simon, W. (1981). *Spectra Data for Structure Determination of Organic Compounds.* Berlin: Springer-Verlag.

Sadtler Research Lab. *^{13}C NMR Spectra.* Philadelphia: Sadtler Research Laboratories.

STUDENT EXERCISES

4.1 For each compound given below (a–o), identify all chemically shift equivalent carbons.

4.2 For each compound below, predict the chemical shifts for each carbon. Give the source (Table or Appendix) that you use for your prediction.

4.3 Sketch the proton decoupled ^{13}C NMR spectrum and DEPT spectra for each of the compounds in question 4.1.

4.4 Confirm the structure and assign all the ^{13}C resonances in spectra A–W for the compounds whose structures were determined in Problem 3.4 ^{1}H NMR. They were all run at 75.5 MHz in $CDCl_3$. The mass spectra were given in Chapter 1 (Question 1.6) and the IR spectra were given in Chapter 2 (Question 2.9).

4.5 Predict the number of lines in ^{13}C spectra for the following compounds:

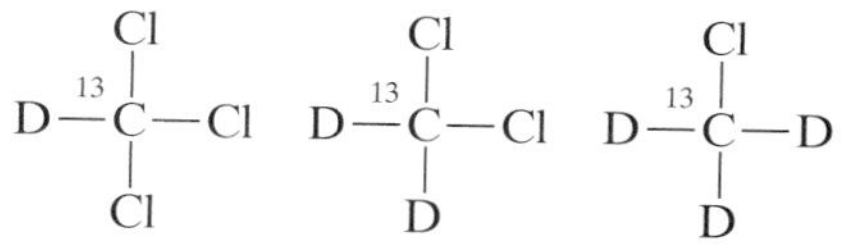

4.6 Interpret the following ^{13}C/DEPT spectra (4.6A to 4.6F). Confirm the structure and assign all the ^{13}C resonances. Give the source (Table or Appendix) that you use for your prediction.

4.7 What are the symmetry elements in *ortho, meta, para*-diethyl phthalates? How many nonequivalent carbon atoms and hydrogen atoms are there for each compound? Draw the proton decoupled ^{13}C spectrum and DEPT spectra for each compound.

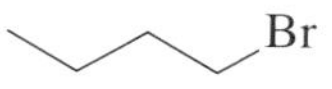
a

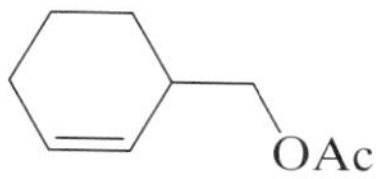
b

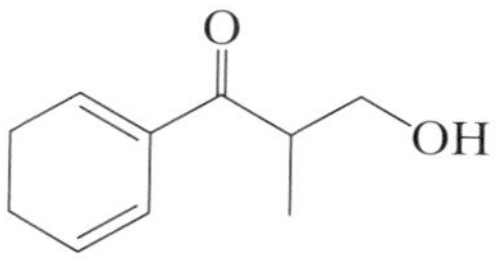
c

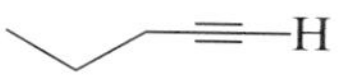
d

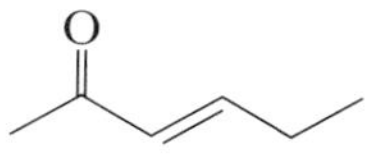
e

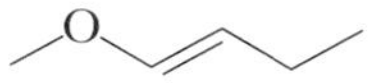
f

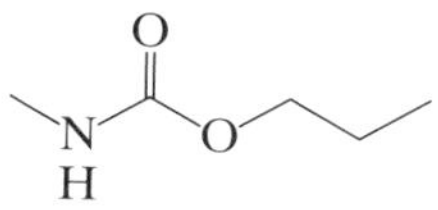
g

h

i

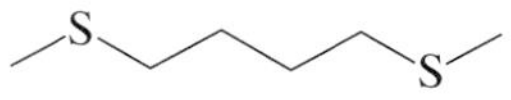
j

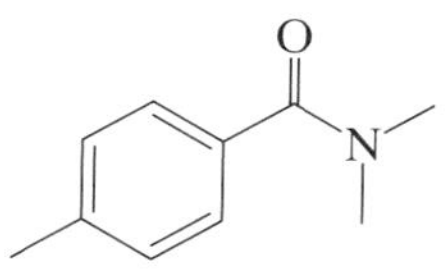
k

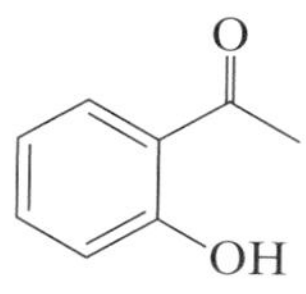
l

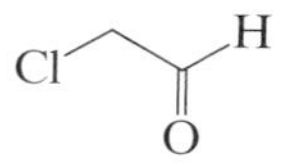
m

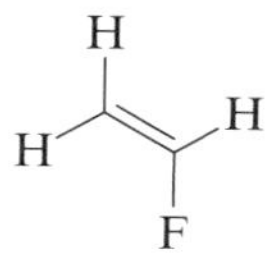
n

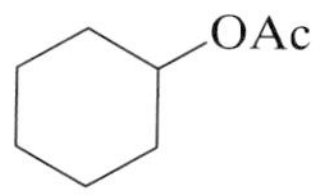
o

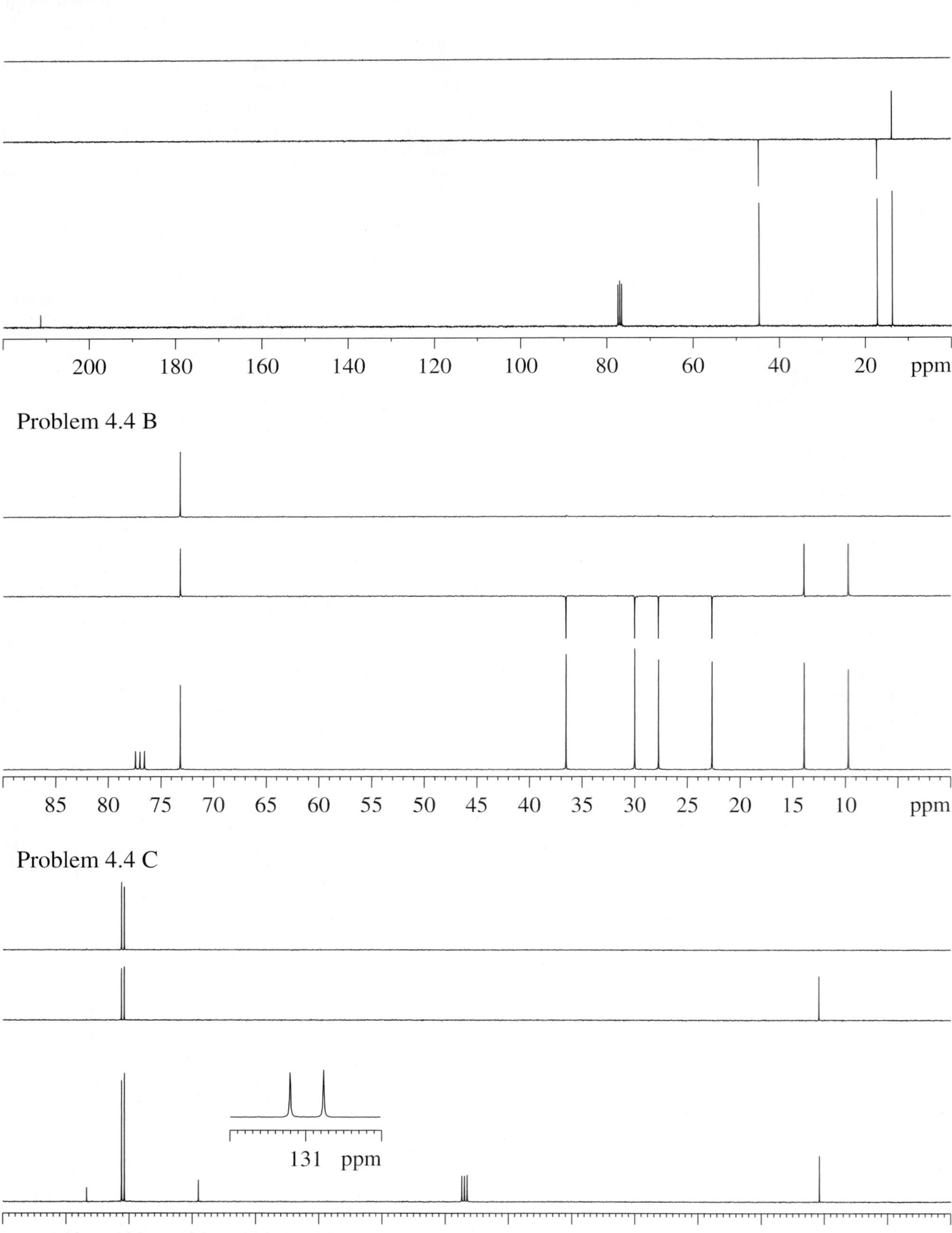
Problem 4.4 A
200
180
160
140
120
100
80
60
40
20
ppm
Problem 4.4 B
85
80
75
70
65
60
55
50
45
40
35
30
25
20
15
10
ppm
Problem 4.4 C
131 ppm
140
130
120
110
100
90
80
70
60
50
40
30
20
10 ppm

Exercise 4.4 (D–F)

Problem 4.4 D

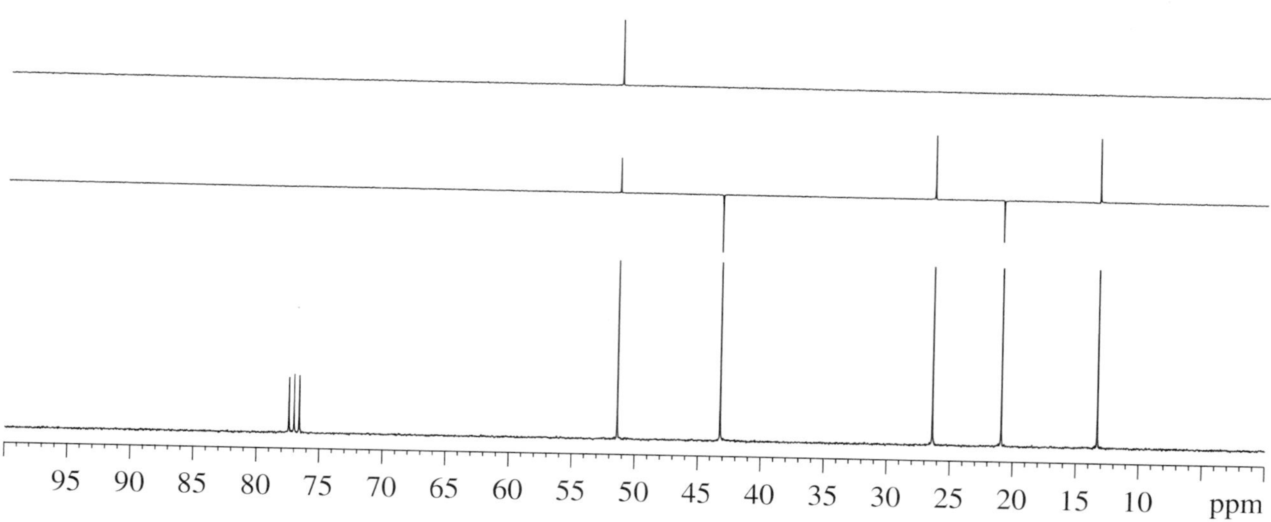

Problem 4.4 E

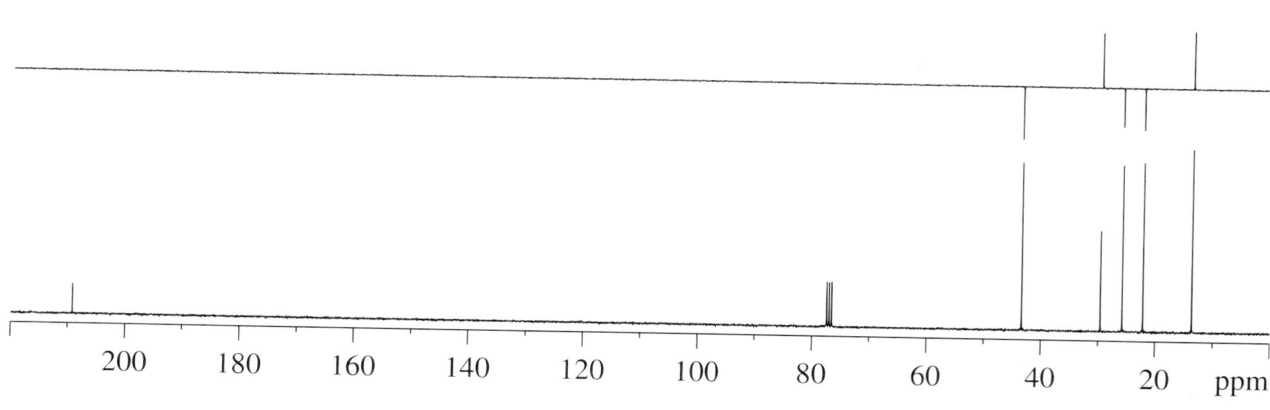

Problem 4.4 F

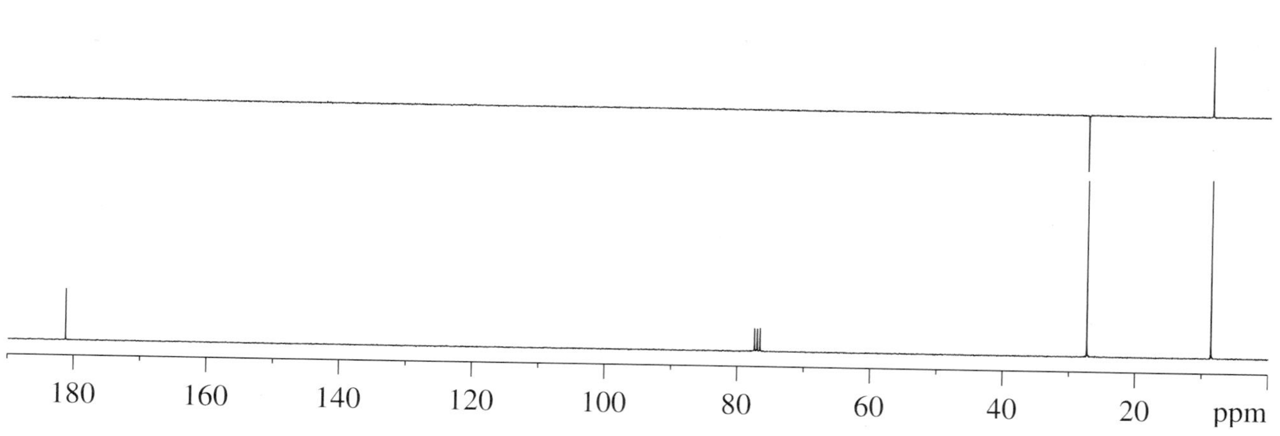

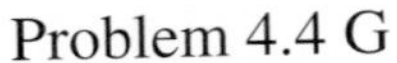

Problem 4.4 G

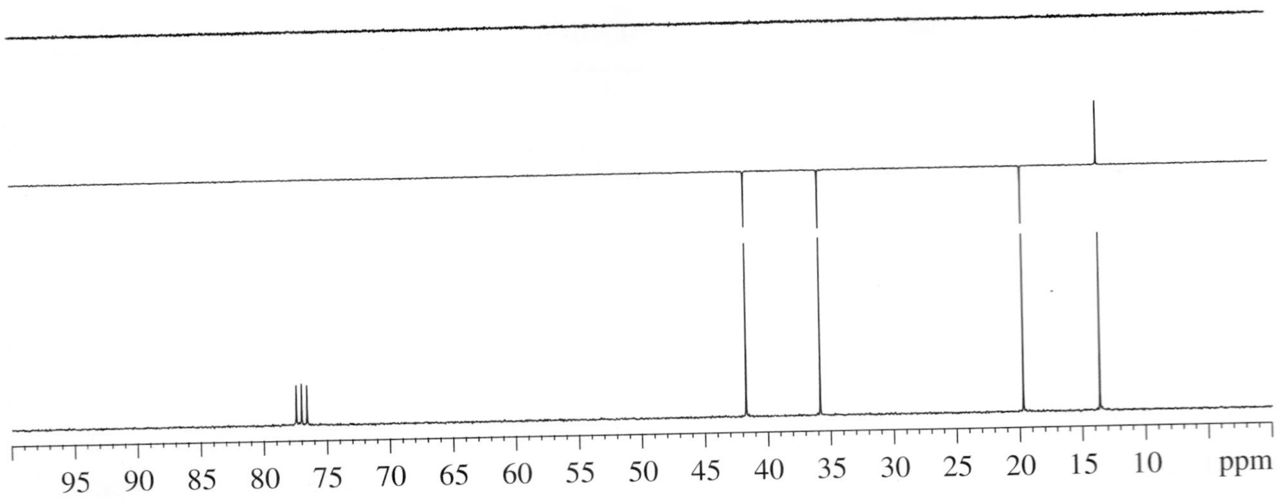

Problem 4.4 H

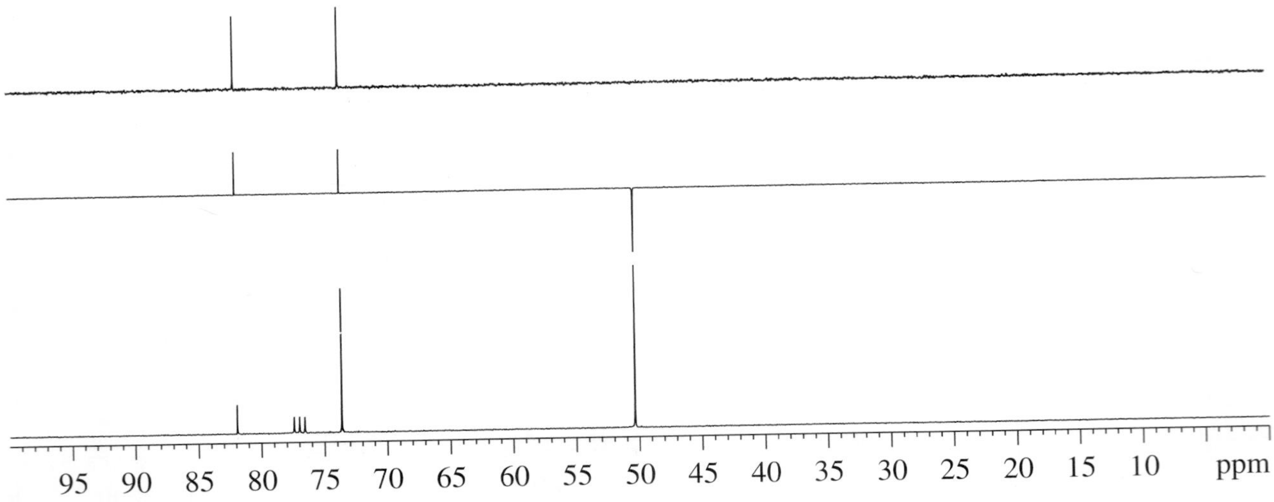

Problem 4.4 I

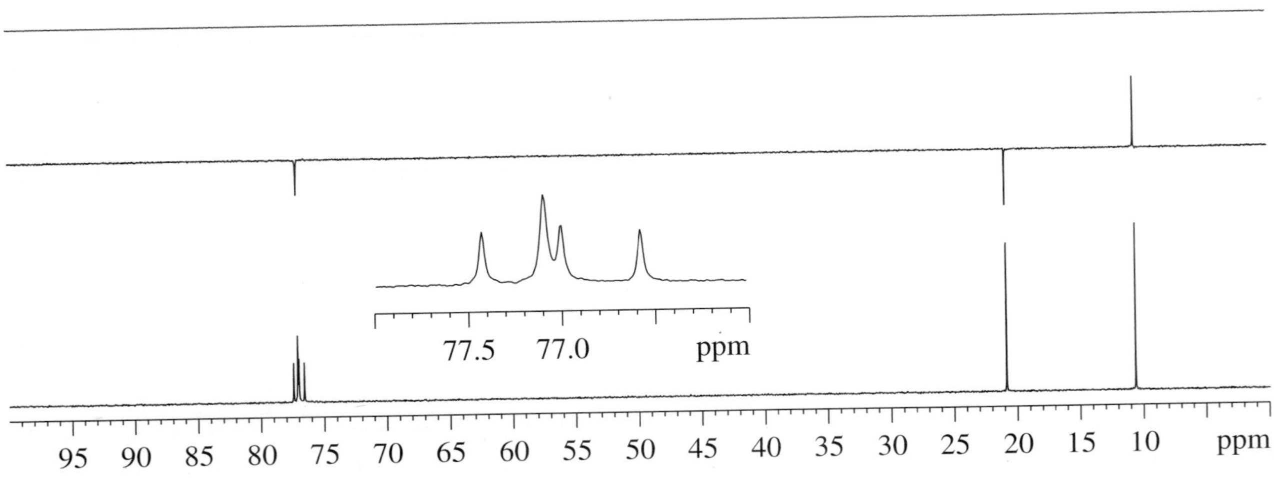

Exercise 4.4 (J – L)

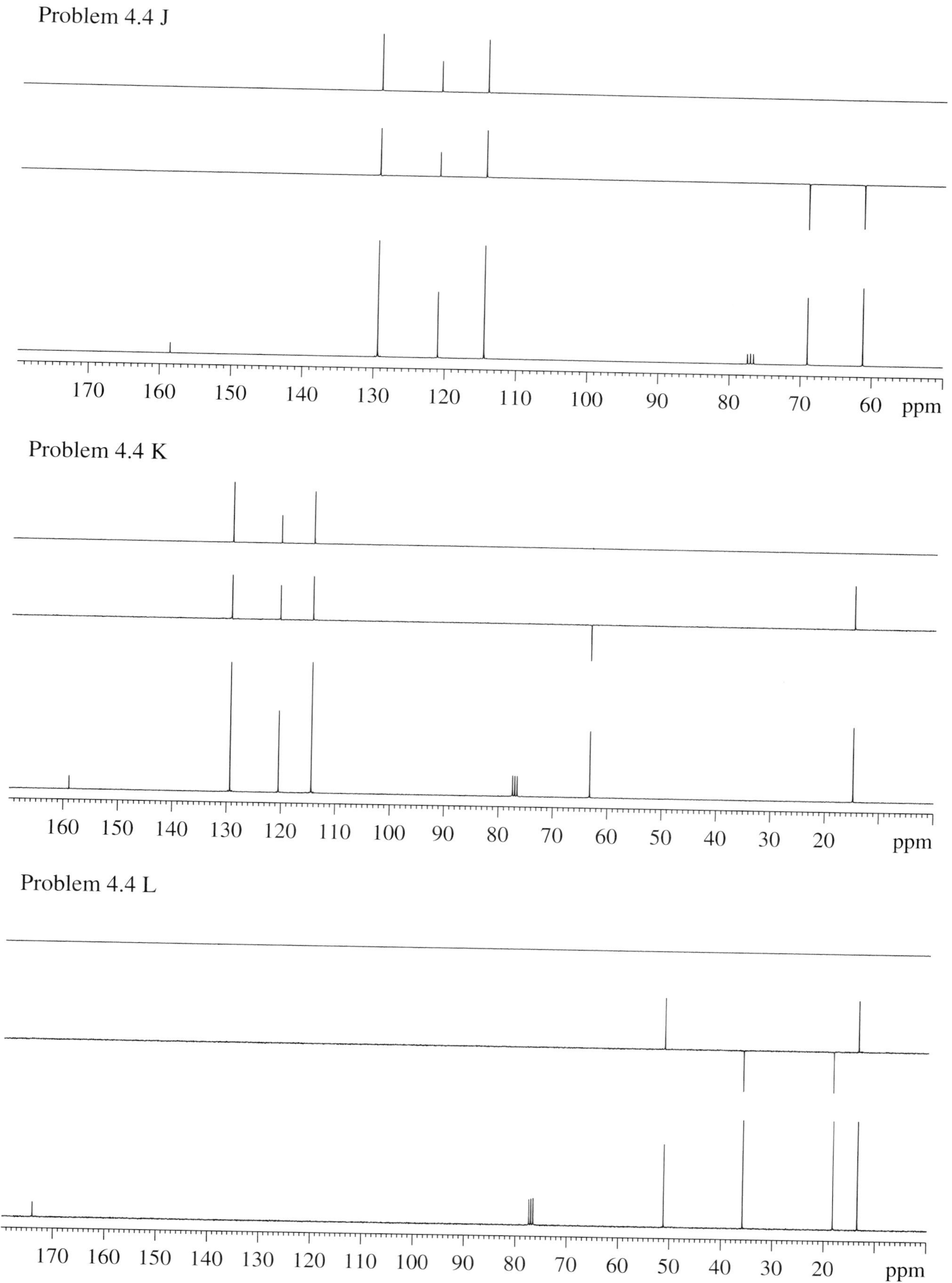

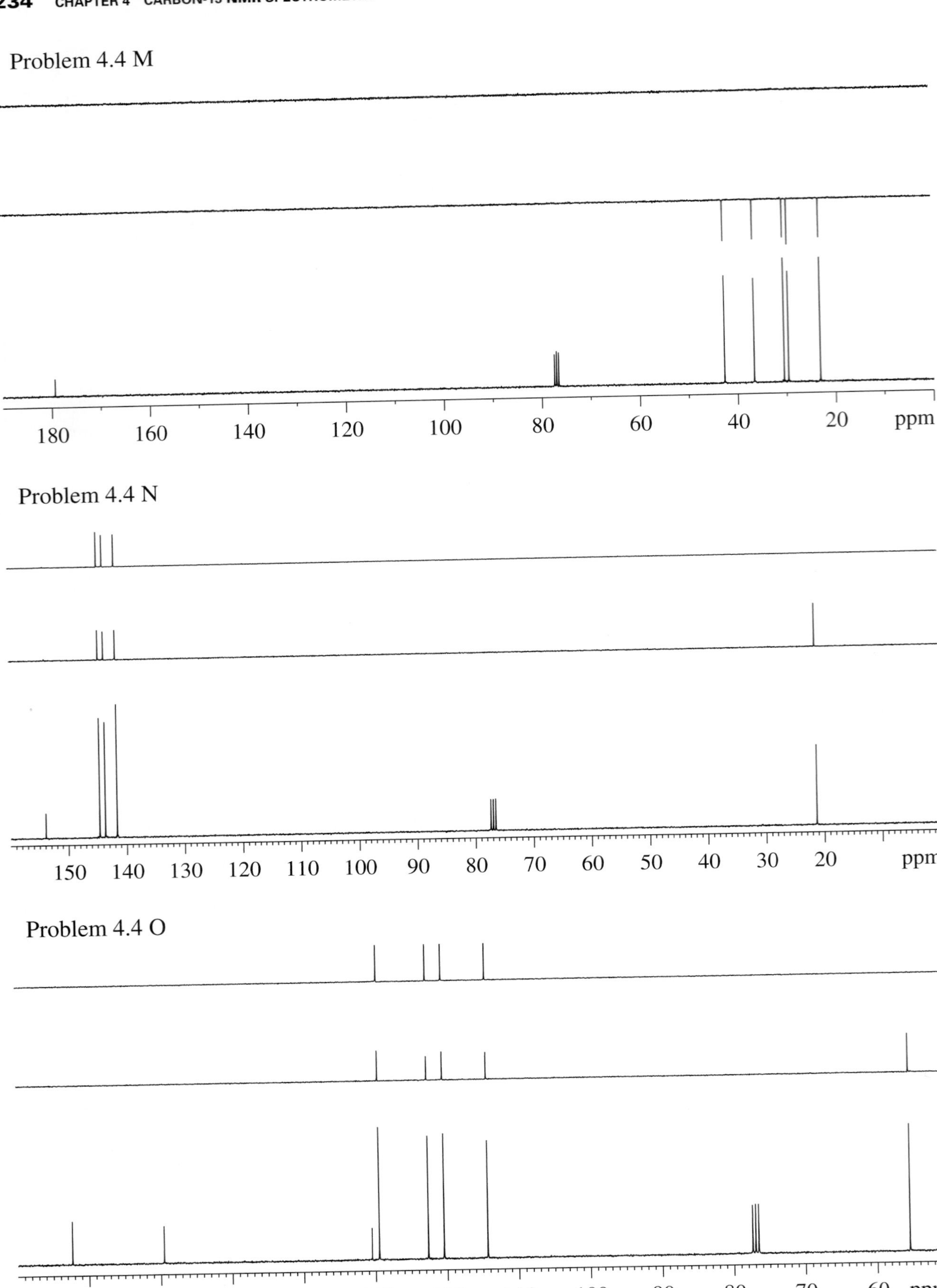

Problem 4.4 M
180
160
140
120
100
80
60
40
20
ppm
Problem 4.4 N
150
140
130
120
110
100
90
80
70
60
50
40
30
20
ppm
Problem 4.4 O
170
160
150
140
130
120
110
100
90
80
70
60
ppm

Exercise 4.4 (P – R)

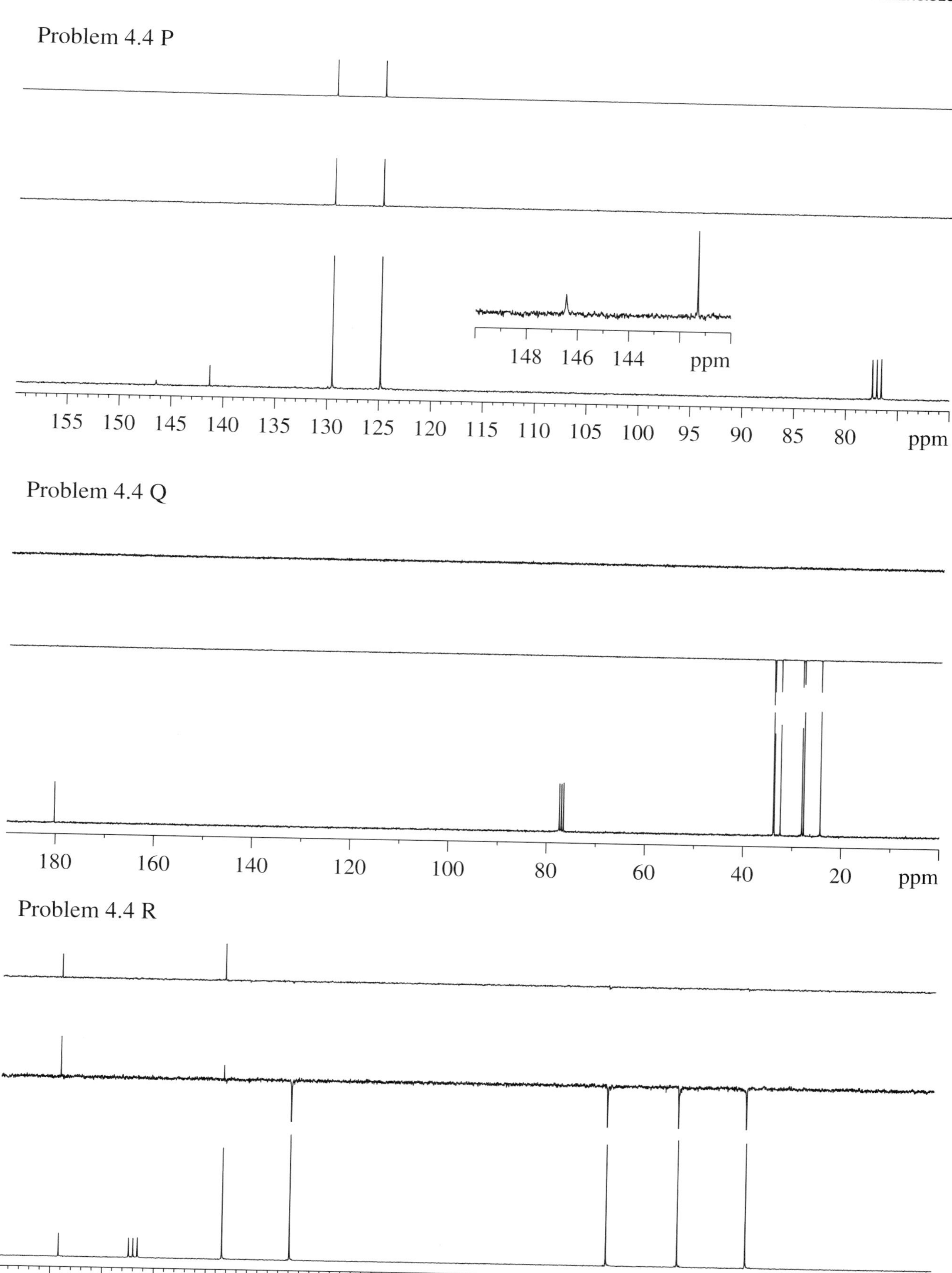

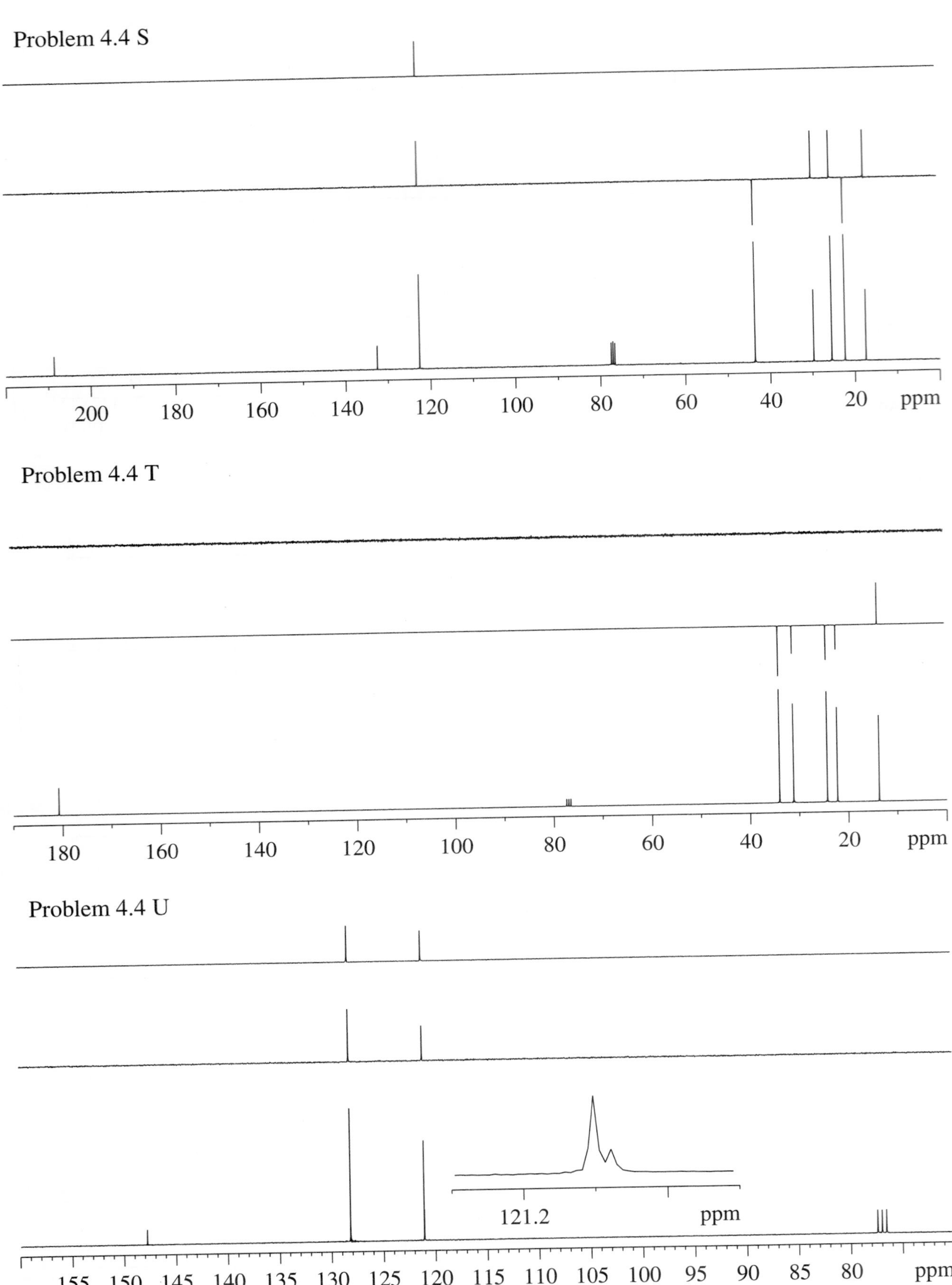
Problem 4.4 S
200
180
160
140
120
100
80
60
40
20
ppm
Problem 4.4 T
180
160
140
120
100
80
60
40
20
ppm
Problem 4.4 U
121.2
ppm
155
150
145
140
135
130
125
120
115
110
105
100
95
90
85
80
ppm

Exercise 4.4 (V – W)

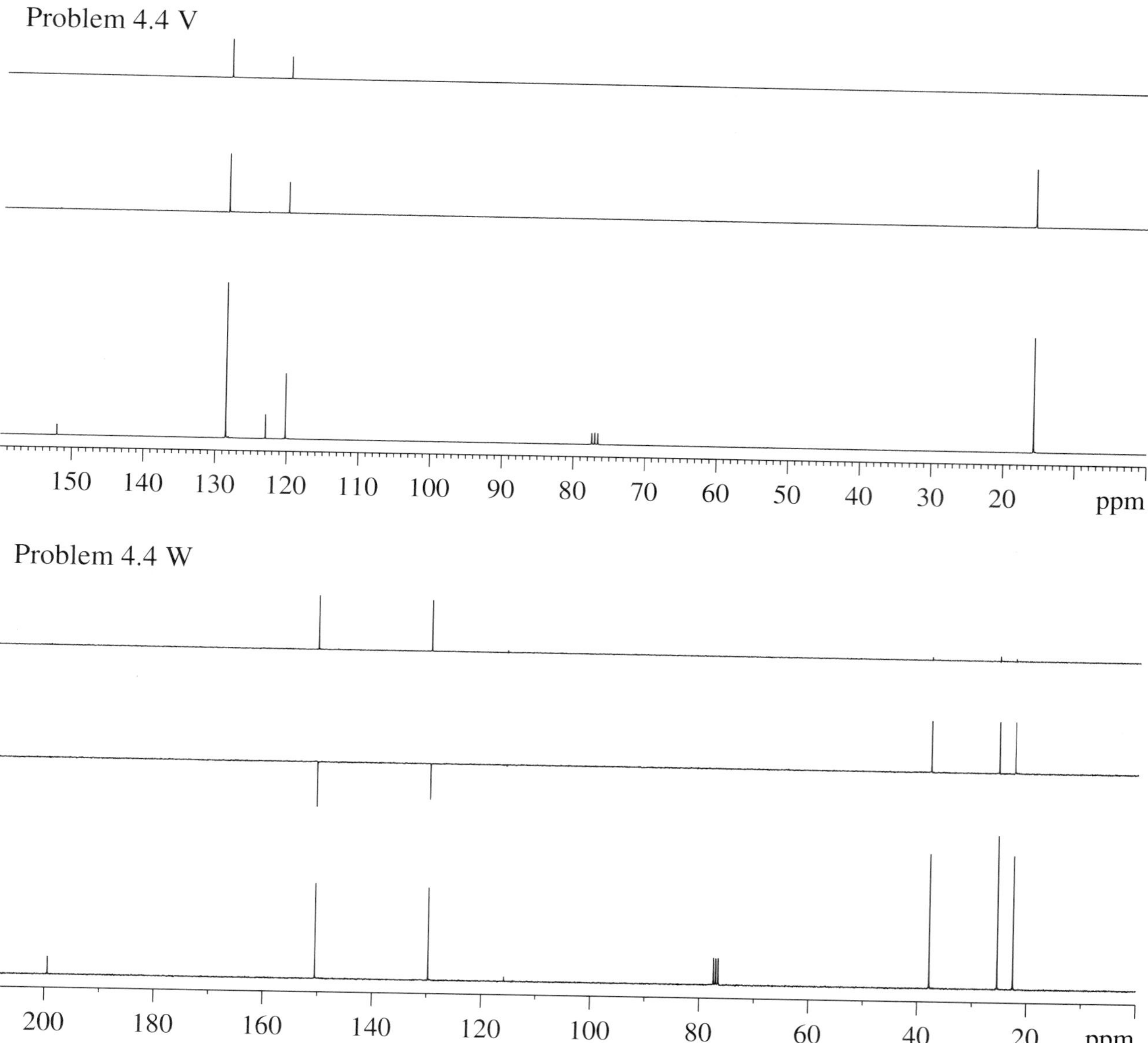

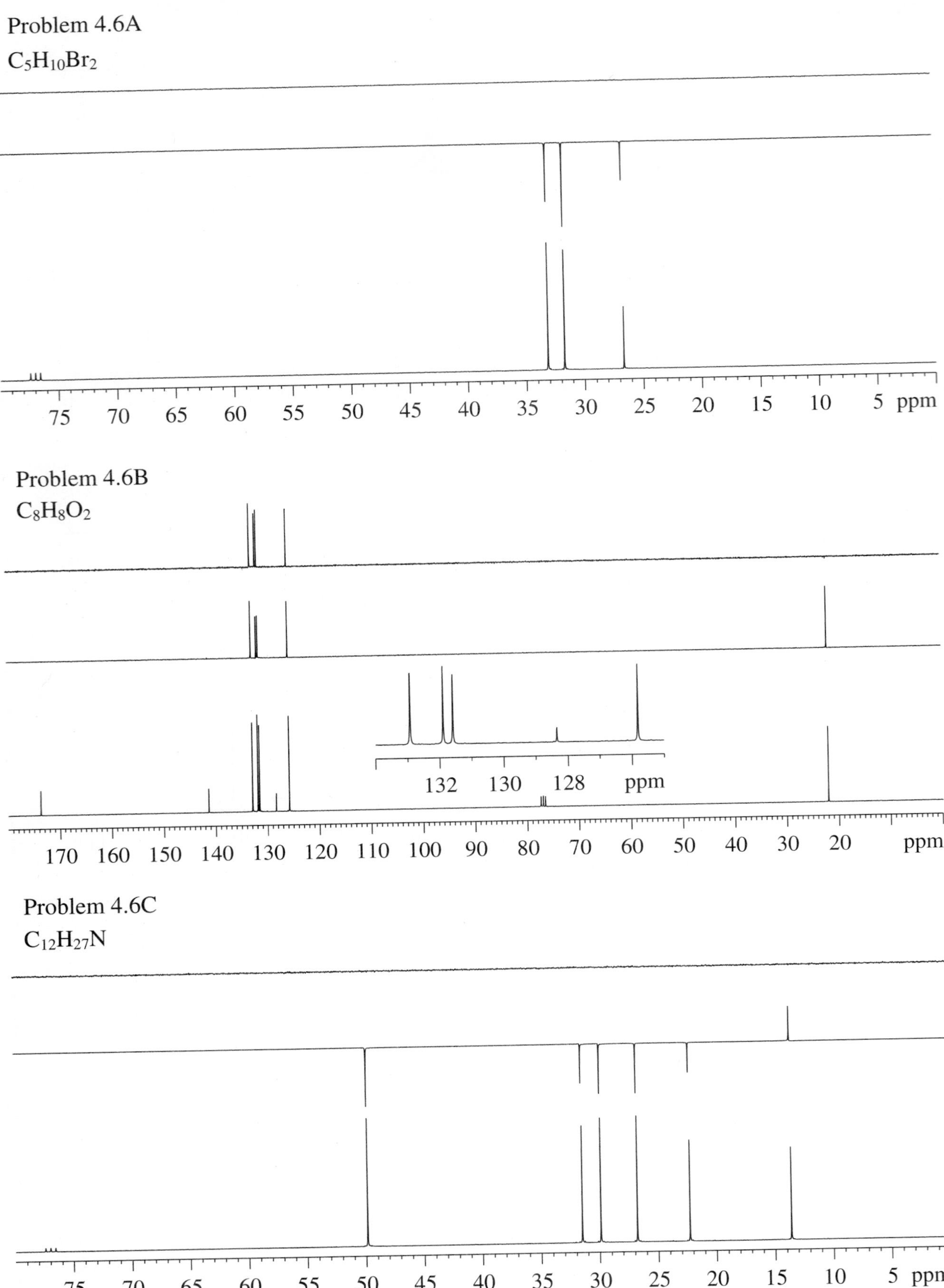
Problem 4.6A
$C_5H_{10}Br_2$
75 70 65 60 55 50 45 40 35 30 25 20 15 10 5 ppm
Problem 4.6B
$C_8H_8O_2$
132 130 128 ppm
170 160 150 140 130 120 110 100 90 80 70 60 50 40 30 20 ppm
Problem 4.6C
$C_{12}H_{27}N$
75 70 65 60 55 50 45 40 35 30 25 20 15 10 5 ppm

Exercise 4.6 (D – F)

Problem 4.6D

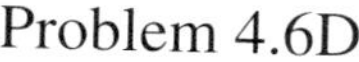

$C_{16}H_{22}O_4$

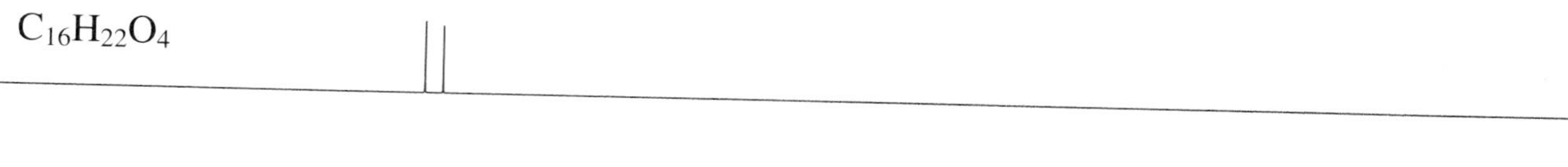

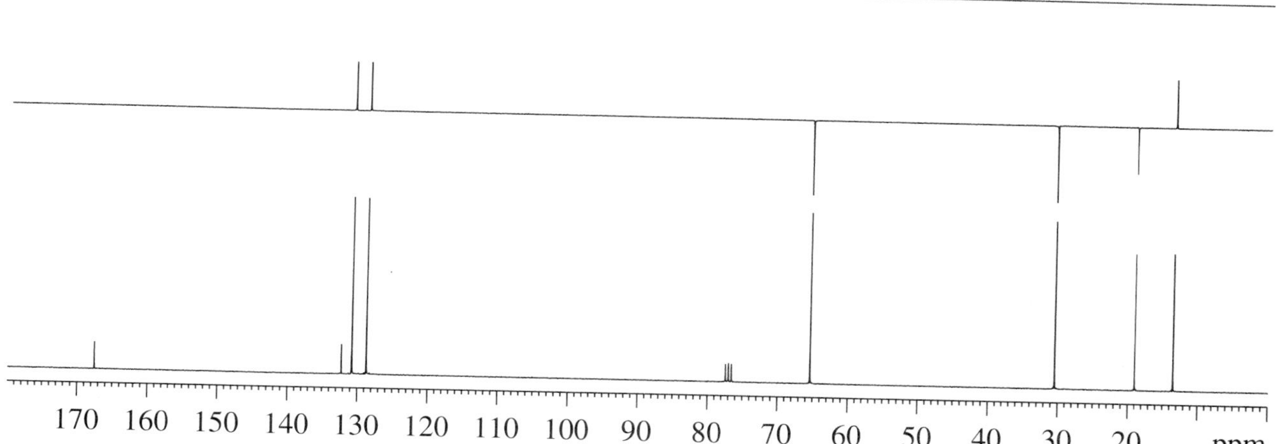

Problem 4.6E

$C_7H_{12}O$

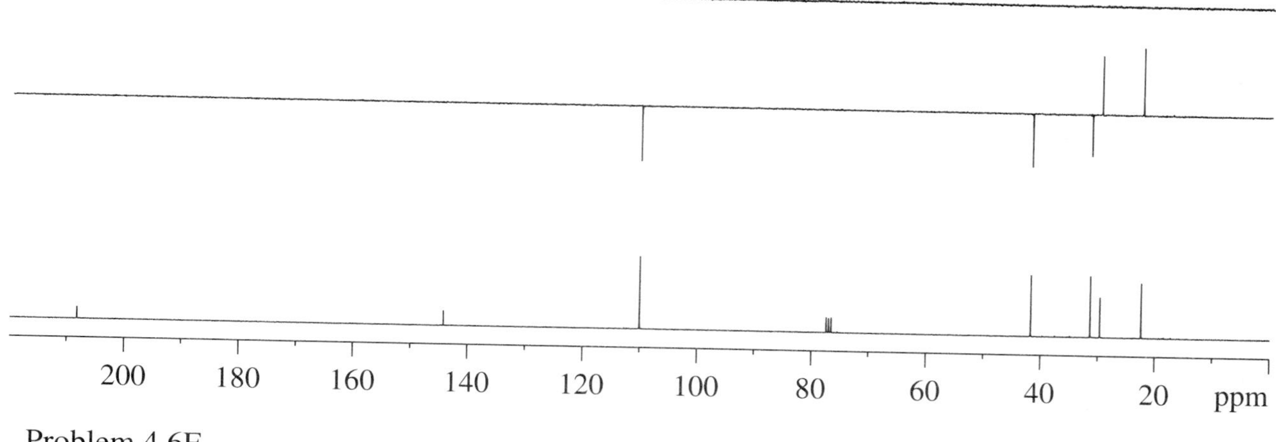

Problem 4.6F

$C_8H_{14}O$

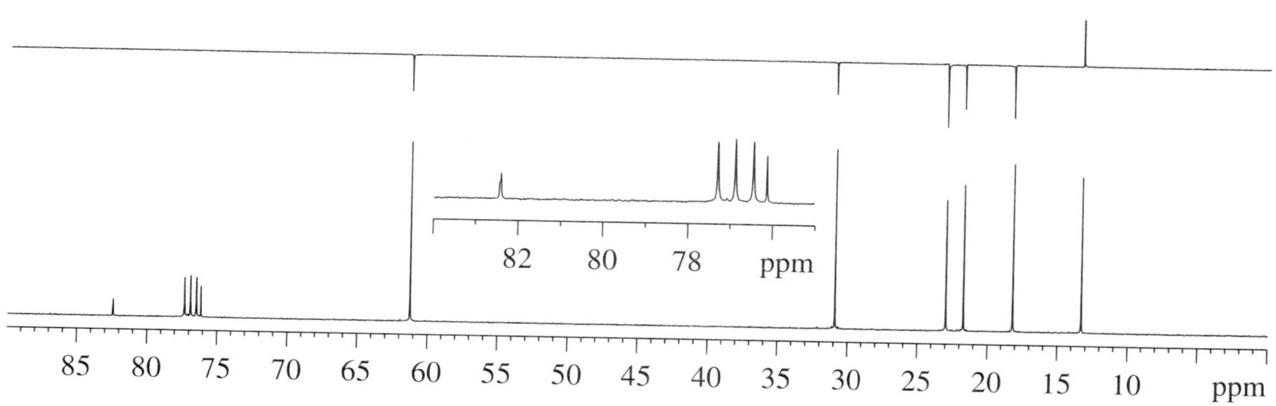

APPENDIX A THE ^{13}C CHEMICAL SHIFTS, COUPLINGS, AND MULTIPLICITIES OF COMMON NMR SOLVENTS

Structure	Name	δ(ppm)	J_{C-D}(Hz)	Multiplicity[a]
$CDCl_3$	Chloroform-d_1	77.0	32	Triplet
CD_3OD	Methanol-d_4	49.0	21.5	Septet
CD_3SOCD_3	DMSO-d_6	39.7	21	Septet
$DC(=O)N(CD_3)_2$	DMF-d_7	30.1	21	Septet
		35.2	21	Septet
		167.7	30	Triplet
C_6D_6	Benzene-d_6	128.0	24	Triplet
D_2C—CD_2, D_2C, CD_2, O (ring)	THF-d_8	25.2	20.5	Quintet
		67.4	22	Quintet
O, D_2C, CD_2, D_2C, CD_2, O (ring)	Dioxane-d_8	66.5	22	Quintet
D, D, D, D, D, N (ring)	Pyridine-d_5	123.5 (C-3,5)	25	Triplet
		135.5 (C-4)	24.5	Triplet
		149.2 (C-2,6)	27.5	Triplet
$CD_3C(=O)CD_3$	Acetone-d_6	29.8 (methyl)	20	Septet
		206.5 (carbonyl)	<1	Septet[b]
CD_3CN	Acetonitrile-d_3	1.3 (methyl)	32	Septet
		118.2 (CN)	<1	Septet[b]
CD_3NO_2	Nitromethane-d_3	60.5	23.5	Septet
CD_3CD_2OD	Ethanol-d_6	15.8 (C-2)	19.5	Septet
		55.4 (C-1)	22	Quintet
$(CD_3CD_2)_2O$	Ether-d_{10}	13.4 (C-2)	19	Septet
		64.3 (C-1)	21	Quintet
$[(CD_3)_2N]_3P{=}O$	HMPA-d_{18}	35.8	21	Septet
CD_3CO_2D	Acetic acid-d_4	20.2 (C-2)	20	Septet
		178.4 (C-1)	<1	Septet[b]
CD_2Cl_2	Dichloromethane-d_2 (Methylene chloride-d_2)	53.1	29	Quintet

[a] Triplet intensities = 1:1:1, quintet = 1:2:3:2:1, septet = 1:3:6:7:6:3:1.

[b] Unresolved, long-range coupling.

Source: Breitmaier, E., and Voelter, W. (1987). *Carbon-13 NMR Spectroscopy,* 3rd ed. New York: VCH, p. 109; with permission. Also Merck & Co., Inc.

APPENDIX B ^{13}C CHEMICAL SHIFTS of COMMON LABORATORY SOLVENTS as TRACE IMPURITIES

		$CDCl_3$	$(CD_3)_2CO$	$(CD_3)_2SO$	C_6D_6	CD_3CN	CD_3OD	D_2O
solvent signals		77.16 ± 0.06	29.84 ± 0.01 206.26 ± 0.13	39.52 ± 0.06	128.06 ± 0.02	1.32 ± 0.02 118.26 ± 0.02	49.00 ± 0.01	
acetic acid	CO	175.99	172.31	171.93	175.82	173.21	175.11	177.21
	CH_3	20.81	20.51	20.95	20.37	20.73	20.56	21.03
acetone	CO	207.07	205.87	206.31	204.43	207.43	209.67	215.94
	CH_3	30.92	30.60	30.56	30.14	30.91	30.67	30.89
acetonitrile	CN	116.43	117.60	117.91	116.02	118.26	118.06	119.68
	CH_3	1.89	1.12	1.03	0.20	1.79	0.85	1.47
benzene	CH	128.37	129.15	128.30	128.62	129.32	129.34	
tert-butyl alcohol	C	69.15	68.13	66.88	68.19	68.74	69.40	70.36
	CH_3	31.25	30.72	30.38	30.47	30.68	30.91	30.29
tert-butyl methyl ether	OCH_3	49.45	49.35	48.70	49.19	49.52	49.66	49.37
	C	72.87	72.81	72.04	72.40	73.17	74.32	75.62
	CCH_3	26.99	27.24	26.79	27.09	27.28	27.22	26.60
BHT	C(1)	151.55	152.51	151.47	152.05	152.42	152.85	
	C(2)	135.87	138.19	139.12	136.08	138.13	139.09	
	CH(3)	125.55	129.05	127.97	128.52	129.61	129.49	
	C(4)	128.27	126.03	124.85	125.83	126.38	126.11	
	CH_3Ar	21.20	21.31	20.97	21.40	21.23	21.38	
	CH_3C	30.33	31.61	31.25	31.34	31.50	31.15	
	C	34.25	35.00	34.33	34.35	35.05	35.36	
chloroform	CH	77.36	79.19	79.16	77.79	79.17	79.44	
cyclohexane	CH_2	26.94	27.51	26.33	27.23	27.63	27.96	
1,2-dichloroethane	CH_2	43.50	45.25	45.02	43.59	45.54	45.11	
dichloromethane	CH_2	53.52	54.95	54.84	53.46	55.32	54.78	
diethyl ether	CH_3	15.20	15.78	15.12	15.46	15.63	15.46	14.77
	CH_2	65.91	66.12	62.05	65.94	66.32	66.88	66.42
diglyme	CH_3	59.01	58.77	57.98	58.66	58.90	59.06	58.67
	CH_2	70.51	71.03	69.54	70.87	70.99	71.33	70.05
	CH_2	71.90	72.63	71.25	72.35	72.63	72.92	71.63
1,2-dimethoxyethane	CH_3	59.08	58.45	58.01	58.68	58.89	59.06	58.67
	CH_2	71.84	72.47	17.07	72.21	72.47	72.72	71.49
dimethylacetamide	CH_3	21.53	21.51	21.29	21.16	21.76	21.32	21.09
	CO	171.07	170.61	169.54	169.95	171.31	173.32	174.57
	NCH_3	35.28	34.89	37.38	34.67	35.17	35.50	35.03
	NCH_3	38.13	37.92	34.42	37.03	38.26	38.43	38.76
dimethylformamide	CH	162.62	162.79	162.29	162.13	163.31	164.73	165.53
	CH_3	36.50	36.15	35.73	35.25	36.57	36.89	37.54
	CH_3	31.45	31.03	30.73	30.72	31.32	31.61	32.03
dimethyl sulfoxide	CH_3	40.76	41.23	40.45	40.03	41.31	40.45	39.39
dioxane	CH_2	67.14	67.60	66.36	67.16	67.72	68.11	67.19
ethanol	CH_3	18.41	18.89	18.51	18.72	18.80	18.40	17.47
	CH_2	58.28	57.72	56.07	57.86	57.96	58.26	58.05
ethyl acetate	CH_3CO	21.04	20.83	20.68	20.56	21.16	20.88	21.15
	CO	171.36	170.96	170.31	170.44	171.68	172.89	175.26
	CH_2	60.49	60.56	59.74	60.21	60.98	61.50	62.32
	CH_3	14.19	14.50	14.40	14.19	14.54	14.49	13.92
ethyl methyl ketone	CH_3CO	29.49	29.30	29.26	28.56	29.60	29.39	29.49
	CO	209.56	208.30	208.72	206.55	209.88	212.16	218.43
	CH_2CH_3	36.89	36.75	35.83	36.36	37.09	37.34	37.27
	CH_2CH_3	7.86	8.03	7.61	7.91	8.14	8.09	7.87
ethylene glycol	CH_2	63.79	64.26	62.76	64.34	64.22	64.30	63.17
"grease"	CH_2	29.76	30.73	29.20	30.21	30.86	31.29	

APPENDIX B *(Continued)*

n-hexane	CH_3	14.14	14.34	13.88	14.32	14.43	14.45	
	$CH_2(2)$	22.70	23.28	22.05	23.04	23.40	23.68	
	$CH_2(3)$	31.64	32.30	30.95	31.96	32.36	32.73	
HMPA	CH_3	36.87	37.04	36.42	36.88	37.10	37.00	36.46
methanol	CH_3	50.41	49.77	48.59	49.97	49.90	49.86	49.50
nitromethane	CH_3	62.50	63.21	63.28	61.16	63.66	63.08	63.22
n-pentane	CH_3	14.08	14.29	13.28	14.25	14.37	14.39	
	$CH_2(2)$	22.38	22.98	21.70	22.72	23.08	23.38	
	$CH_2(3)$	34.16	34.83	33.48	34.45	34.89	35.30	
2-propanol	CH_3	25.14	25.67	25.43	25.18	25.55	25.27	24.38
	CH	64.50	63.85	64.92	64.23	64.30	64.71	64.88
pyridine	CH(2)	149.90	150.67	149.58	150.27	150.76	150.07	149.18
	CH(3)	123.75	124.57	123.84	123.58	127.76	125.53	125.12
	CH(4)	135.96	136.56	136.05	135.28	136.89	138.35	138.27
silicone grease	CH_3	1.04	1.40		1.38		2.10	
tetrahydrofuran	CH_2	25.62	26.15	25.14	25.72	26.27	26.48	25.67
	CH_2O	67.97	68.07	67.03	67.80	68.33	68.83	68.68
toluene	CH_3	21.46	21.46	20.99	21.10	21.50	21.50	
	C(*i*)	137.89	138.48	137.35	137.91	138.90	138.85	
	CH(*o*)	129.07	129.76	128.88	129.33	129.94	129.91	
	CH(*m*)	128.26	129.03	128.18	128.56	129.23	129.20	
	CH(*p*)	125.33	126.12	125.29	125.68	126.28	126.29	
triethylamine	CH_3	11.61	12.49	11.74	12.35	12.38	11.09	9.07
	CH_2	46.25	47.07	45.74	46.77	47.10	46.96	47.19

APPENDIX C THE ^{13}C CORRELATION CHART FOR CHEMICAL CLASSES

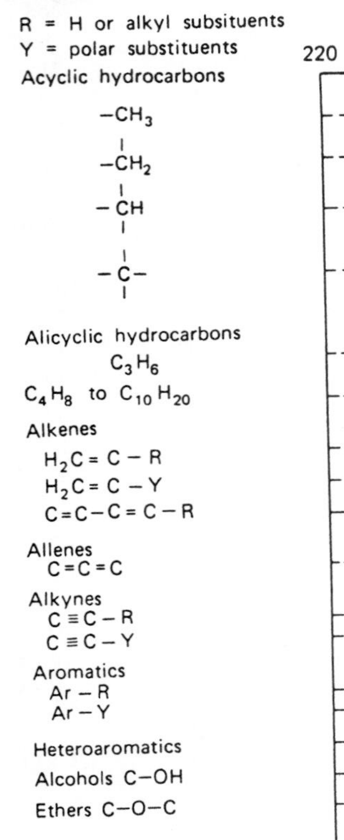

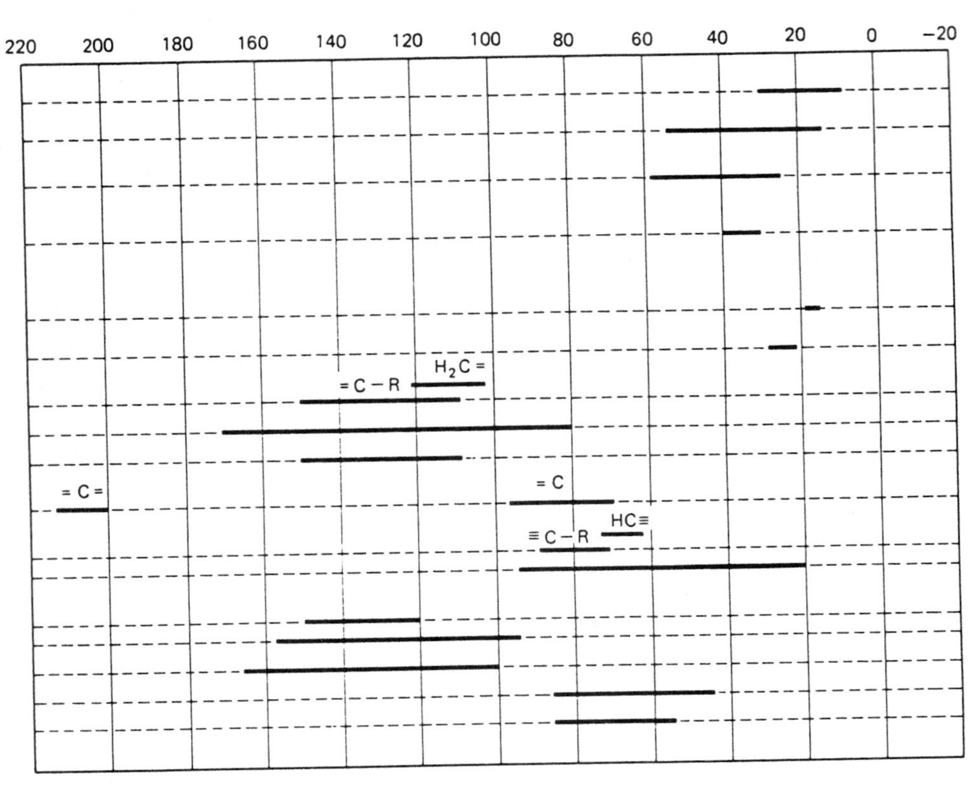

APPENDIX C *(Continued)*

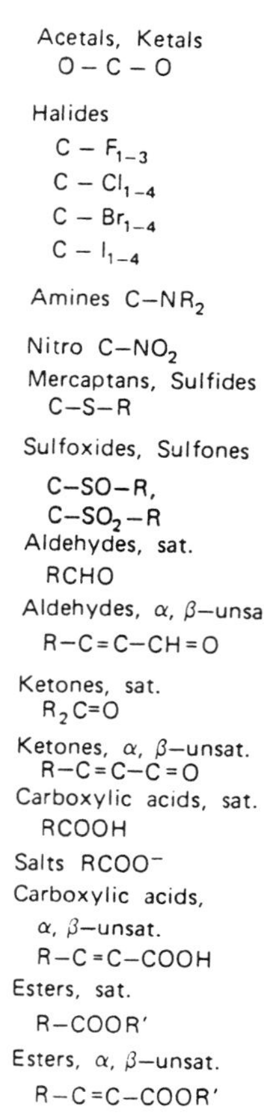
Acetals, Ketals
O – C – O
Halides
$C - F_{1-3}$
$C - Cl_{1-4}$
$C - Br_{1-4}$
$C - I_{1-4}$
Amines $C{-}NR_2$
Nitro $C{-}NO_2$
Mercaptans, Sulfides
C–S–R
Sulfoxides, Sulfones
C–SO–R,
$C{-}SO_2{-}R$
Aldehydes, sat.
RCHO
Aldehydes, α, β–unsat.
R–C=C–CH=O
Ketones, sat.
$R_2C{=}O$
Ketones, α, β–unsat.
R–C=C–C=O
Carboxylic acids, sat.
RCOOH
Salts $RCOO^-$
Carboxylic acids,
α, β–unsat.
R–C=C–COOH
Esters, sat.
R–COOR′
Esters, α, β–unsat.
R–C=C–COOR′

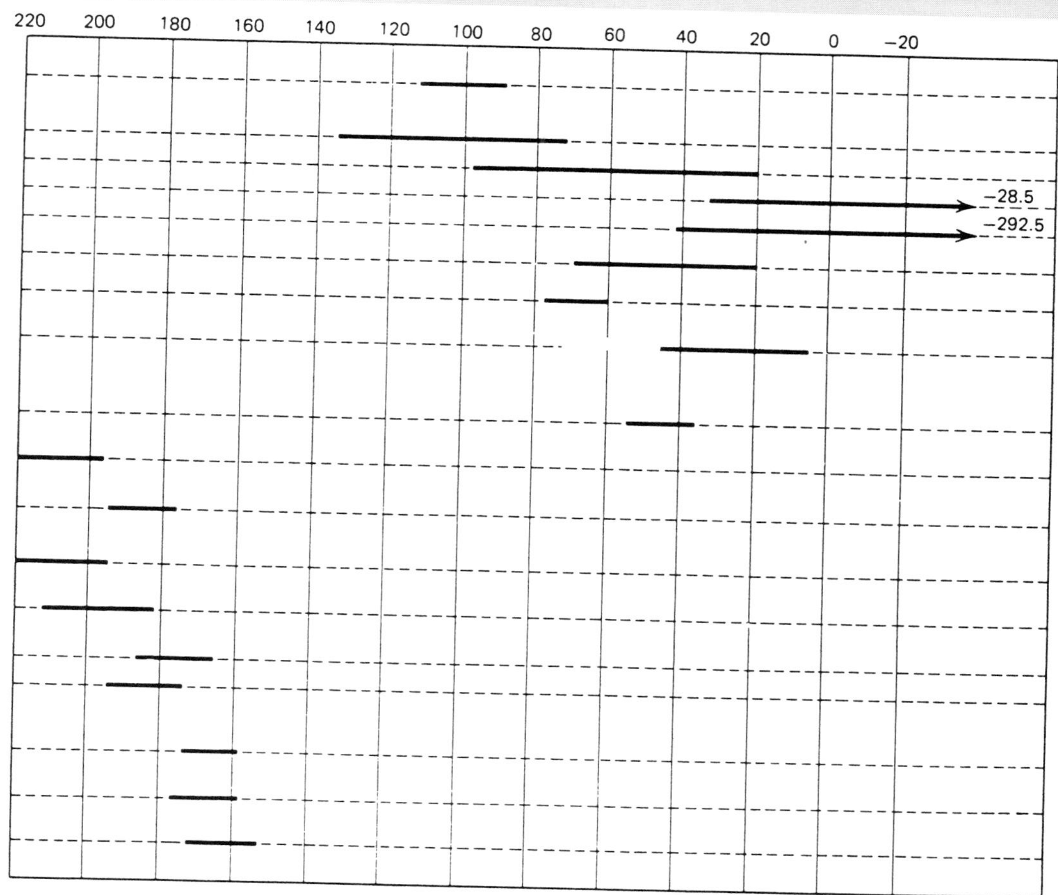
220
200
180
160
140
120
100
80
60
40
20
0
–20
–28.5
–292.5

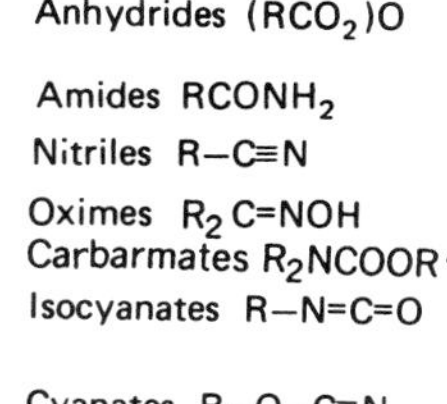
Anhydrides $(RCO_2)O$
Amides $RCONH_2$
Nitriles R–C≡N
Oximes $R_2C{=}NOH$
Carbarmates R_2NCOOR'
Isocyanates R–N=C=O
Cyanates R–O–C≡N
Isothiocyanates R–N=C=S
Thiocyanates R–S–C≡N

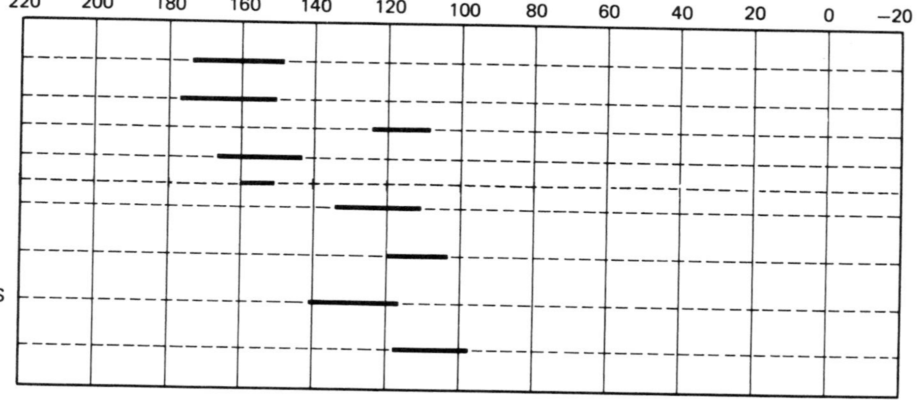
220
200
180
160
140
120
100
80
60
40
20
0
–20

APPENDIX D ^{13}C NMR DATA FOR SEVERAL NATURAL PRODUCTS (δ)

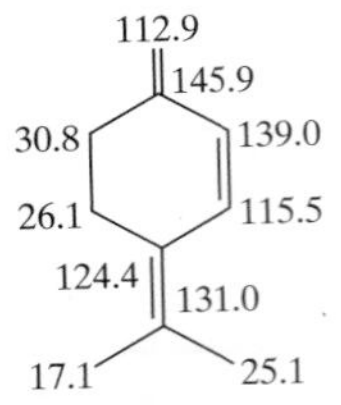

112.9, 145.9, 30.8, 139.0, 26.1, 115.5, 124.4, 131.0, 17.1, 25.1

Myrcene

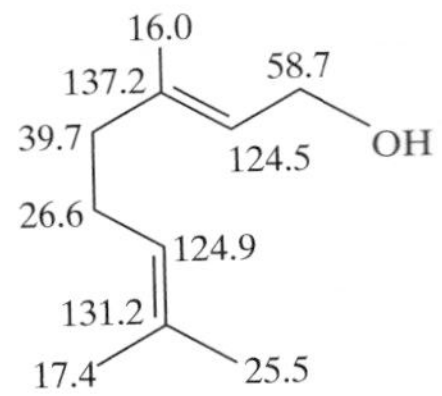

16.0, 137.2, 58.7, 39.7, 124.5, OH, 26.6, 124.9, 131.2, 17.4, 25.5

Geraniol

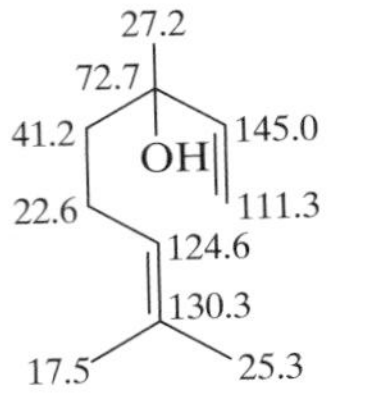

27.2, 72.7, 41.2, OH, 145.0, 22.6, 111.3, 124.6, 130.3, 17.5, 25.3

Linalool

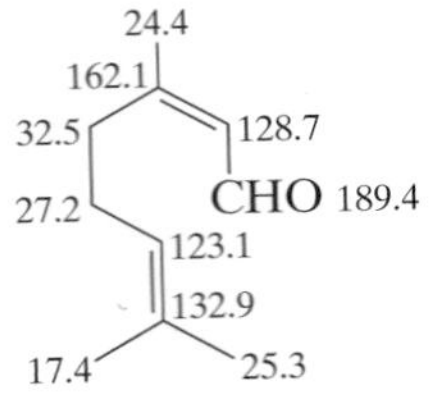

24.4, 162.1, 32.5, 128.7, 27.2, CHO 189.4, 123.1, 132.9, 17.4, 25.3

cis-Citral
(Neral, Citral b)

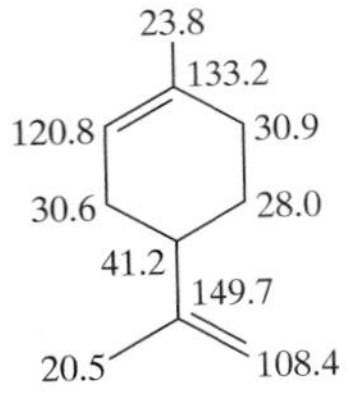

23.8, 133.2, 120.8, 30.9, 30.6, 28.0, 41.2, 149.7, 20.5, 108.4

Limonene

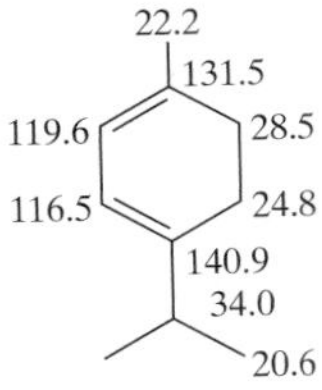

22.2, 131.5, 119.6, 28.5, 116.5, 24.8, 140.9, 34.0, 20.6

α-Terpinene

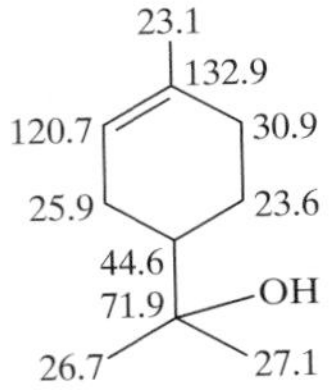

23.1, 132.9, 120.7, 30.9, 25.9, 23.6, 44.6, 71.9, OH, 26.7, 27.1

α-Terpineol

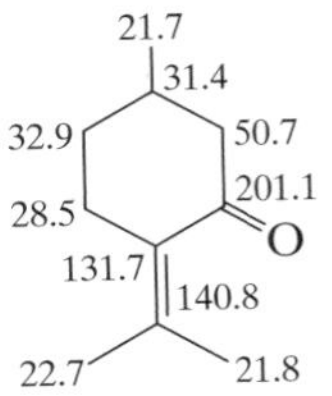

21.7, 31.4, 32.9, 50.7, 201.1, 28.5, O, 131.7, 140.8, 22.7, 21.8

Pulegone

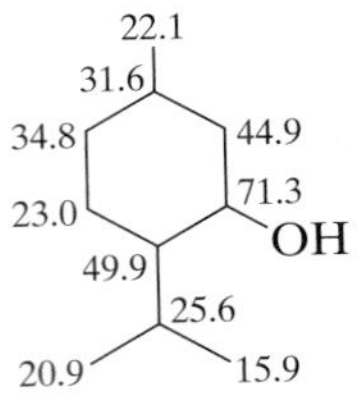

22.1, 31.6, 34.8, 44.9, 71.3, 23.0, OH, 49.9, 25.6, 20.9, 15.9

Menthol

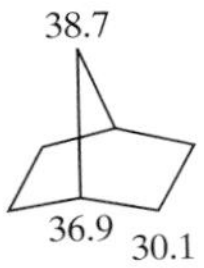

38.7, 36.9, 30.1

Norbornane

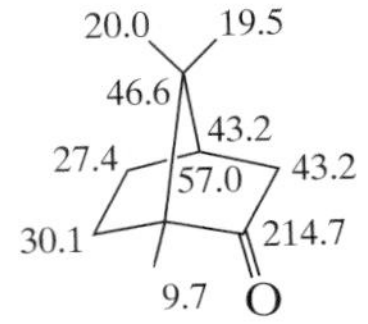

20.0, 19.5, 46.6, 43.2, 27.4, 57.0, 43.2, 30.1, 214.7, 9.7, O

Camphor

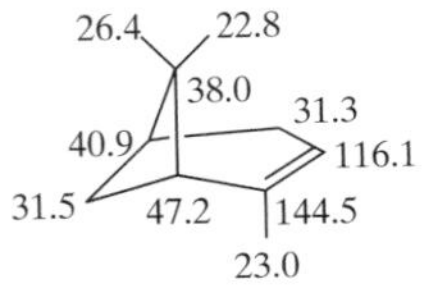

26.4, 22.8, 38.0, 31.3, 40.9, 116.1, 31.5, 47.2, 144.5, 23.0

α-Pinene

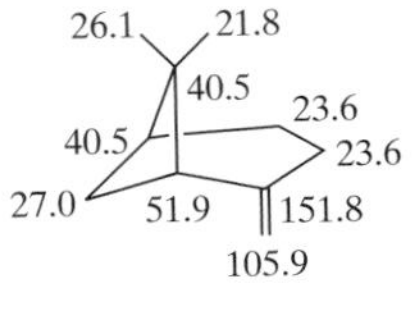

26.1, 21.8, 40.5, 23.6, 40.5, 23.6, 27.0, 51.9, 151.8, 105.9

β-Pinene

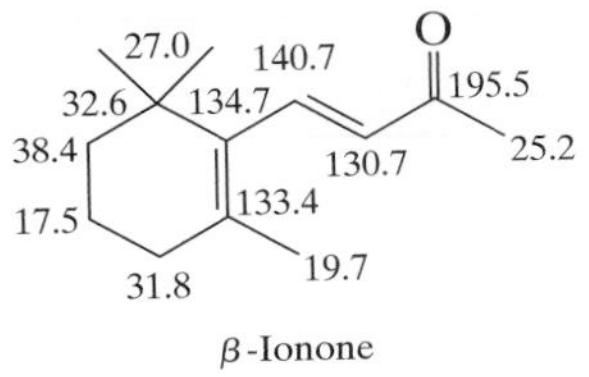

27.0, 140.7, O, 32.6, 134.7, 195.5, 38.4, 130.7, 25.2, 17.5, 133.4, 31.8, 19.7

β-Ionone

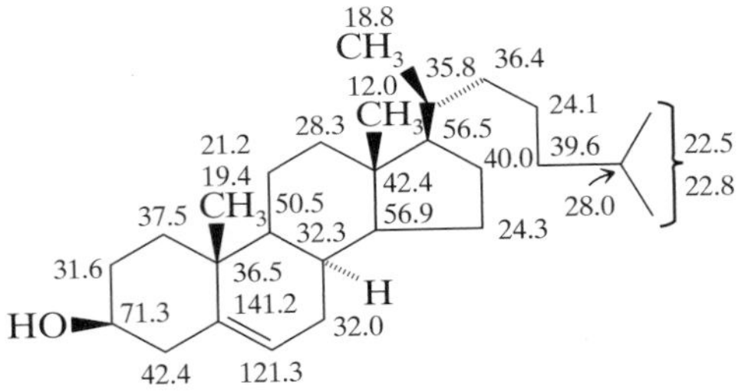

18.8, CH_3, 35.8, 36.4, 12.0, 24.1, 28.3, CH_3, 56.5, 21.2, 40.0, 39.6, 22.5, 19.4, 42.4, 22.8, 37.5, CH_3, 50.5, 56.9, 28.0, 32.3, 24.3, 31.6, 36.5, H, 141.2, HO, 71.3, 32.0, 42.4, 121.3

Cholesterol

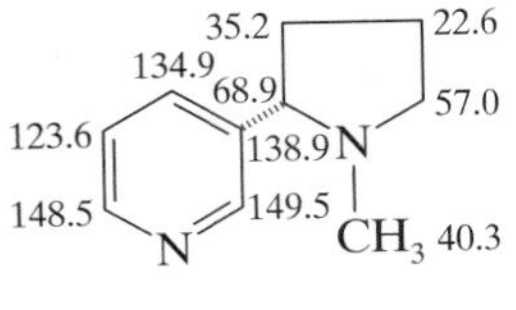

35.2, 22.6, 134.9, 68.9, 57.0, 123.6, 138.9, N, 148.5, 149.5, N, CH_3 40.3

Nicotine

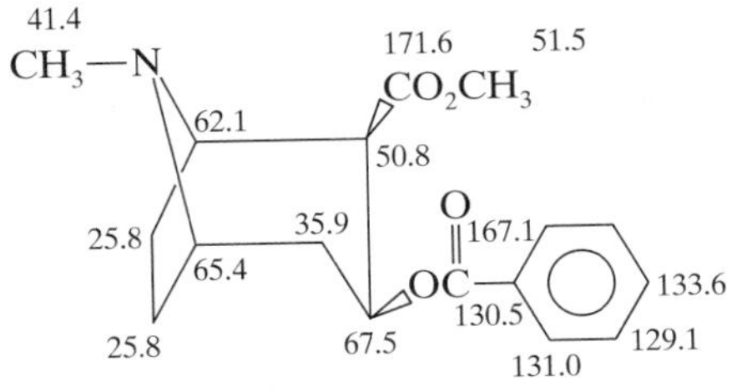

41.4, CH_3—N, 171.6, 51.5, CO_2CH_3, 62.1, 50.8, 35.9, O, 167.1, 25.8, 65.4, OC, 133.6, 130.5, 25.8, 67.5, 129.1, 131.0

Cocaine

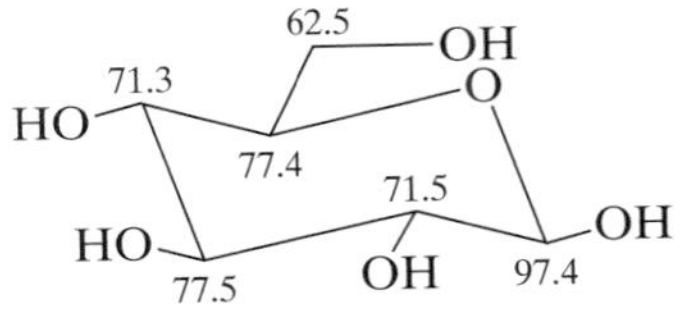

62.5, OH, 71.3, O, HO, 77.4, 71.5, HO, OH, OH, 77.5, 97.4

β-D-Glucose

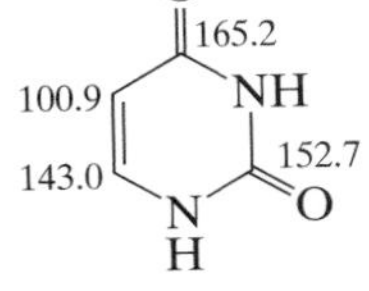

O, 165.2, 100.9, NH, 152.7, 143.0, N, O, H

Uracil

CHAPTER 5

CORRELATION NMR SPECTROMETRY; 2-D NMR

5.1 INTRODUCTION

Chapters 3 and 4 (familiarity with which is assumed) provide us with powerful techniques and methods to elucidate the structures of organic compounds especially when combined with information derived from IR and mass spectrometry. These NMR methods are collectively referred to as "one-dimensional techniques." To extend our capabilities, we turn once more to NMR. We will use four compounds as examples: ipsenol (see Chapter 3), caryophyllene oxide (a sesquiterpene epoxide), lactose (a β-linked disaccharide), and a small peptide (valine-glycine-serine-glutamate, VGSE). The structures of these compounds are shown in Figure 5.1.

These compounds provide a rich variety of structural types and features. The two terpenoids possess the typical branched carbon skeletons of isoprenoids; both compounds have diastereotopic methylene and methyl groups. Lactose is a β-1,4-linked disaccharide of galactose and glucose; glucose is the reducing residue, which at equilibrium in aqueous solution shows both anomers. The tetra-peptide, VGSE, contains four different amino acid residues and is a workable mimic for such biopolymers as proteins and nucleic acids. The NMR signals associated with these compounds can be difficult to interpret using simple, one-dimensional ^{1}H and ^{13}C NMR spectra.

In this chapter, using these four compounds as examples, we turn our attention to correlation NMR spectrometry; most (but not all) of the useful experiments fall into the category of "two-dimensional NMR." Our approach in this chapter is to present the spectra for each compound independently as a logical set. Most of the general aspects of each experiment are given in the discussion of ipsenol; others are only introduced with the more complicated compounds. The material for ipsenol should be thoroughly covered first. The other compounds can be covered independently or not at all.

Correlation in NMR is not a new concept to us. For instance, the ^{1}H NMR spectrum of ethylbenzene (Figure 3.34) shows a clean triplet and quartet for the methyl and methylene groups, respectively. These two groups are "correlated" to each other because the

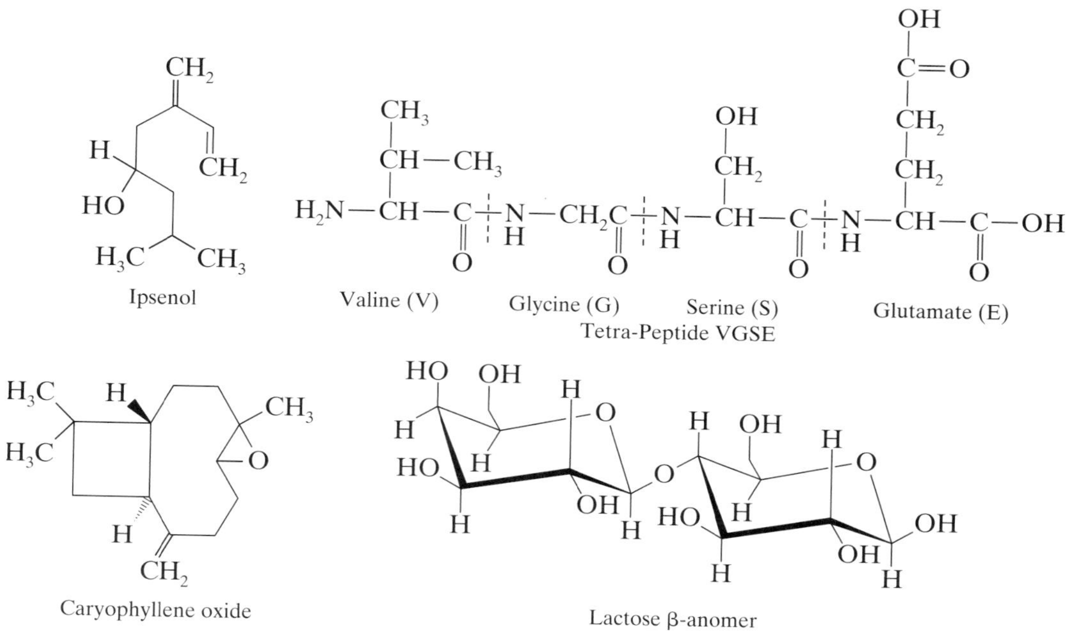

Ipsenol

Valine (V) Glycine (G) Serine (S) Glutamate (E)

Tetra-Peptide VGSE

Caryophyllene oxide

Lactose β-anomer

FIGURE 5.1 Structures of the four compounds used as examples in this chapter.

individual spins within each group are coupled (scalar coupling). First-order rules helped us to interpret these interactions among neighboring nuclei. Coupling among protons is only one type of correlation that we will be considering.

We did a credible job of interpretation of ipsenol using ^{1}H NMR in Chapter 3, but we can do a better job using correlation methods, quicker and with less ambiguity. Caryophyllene oxide, lactose, and VGSE are, however, too complex to fully analyze using one-dimensional ^{1}H and ^{13}C NMR alone. Before turning our attention to the description of specific experiments and their interpretation, we will first take a closer look at pulse sequences and Fourier transformation.

5.2 THEORY

We recall that in order to obtain a routine ^{1}H or ^{13}C NMR spectrum in a pulsed experiment, the "pulse sequence" (Figure 5.2) involves an equilibration in the magnetic field, an rf pulse, and signal acquisition. This sequence is repeated until a satisfactory signal/noise ratio is obtained; Fourier transformation of the FID results in the familiar frequency-domain spectrum.

Figure 5.2 reveals a number of interesting features. We note that there is a separate line for the ^{1}H "channel" and one for the ^{13}C "channel." These "channels" represent the hardware associated with the irradiation and signal acquisition of each relevant nucleus in our experiments. Following equilibration, the pulse sequence used to obtain a one-dimensional (1-D) proton spectrum consists of a $\pi/2_x$ pulse (θ), delay, and signal acquisition of the order of seconds (Figure 5.2a). We also notice that the ^{13}C channel (not shown) is inactive during a simple proton experiment. Normally, we will not show a given channel unless there is some activity in that channel.

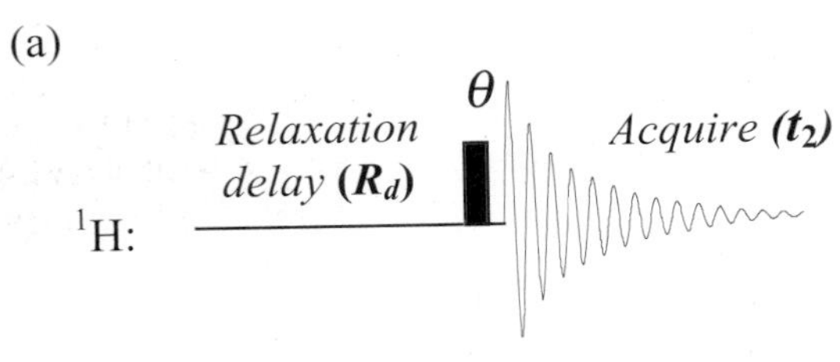

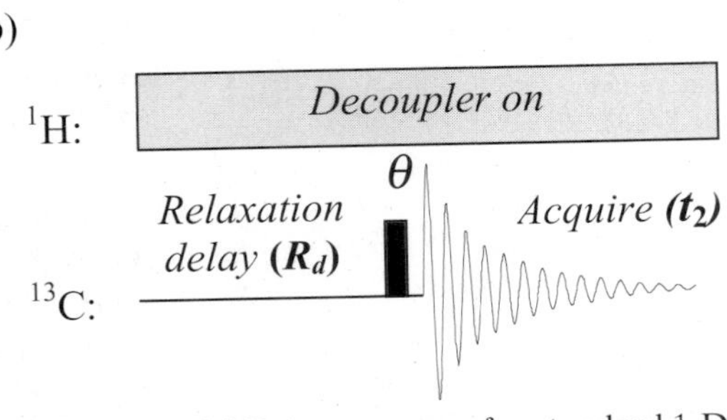

FIGURE 5.2 (a) Pulse sequence for standard 1-D ^{1}H spectrum. (b) Pulse sequence for a standard (decoupled) ^{13}C spectrum. θ is normally a $(\pi/2)_x$ or a 90° pulse along the x axis. R_d is an equilibration period in the magnetic field before the pulse.

Figure 5.2b is the pulse sequence for a ^{13}C experiment. The sequence in the ^{13}C channel is exactly the same as the sequence in the ^{1}H channel in Figure 5.2a. The protons are decoupled from the ^{13}C nuclei by irradiating the protons during the experiment; that is, the proton decoupler is turned on during the entire experiment. In other experiments, the decoupler for a given nucleus can be turned on and off to coincide with pulses and delays in another channel (i.e., for another nucleus). This process is termed gated decoupling. (See Sections 4.2.5 and 4.4)

It is worthwhile to review here (see Chapters 3 and 4) what is happening to the net magnetization vector, $\mathbf{M_0}$, for a single spin during this pulse sequence when viewed in a rotating frame of reference. In a frame of reference rotating at the Larmor frequency, $\mathbf{M_0}$ is stationary on the z-axis (equilibration period in Figure 5.2). A $\pi/2$ (θ, 90°) pulse brings $\mathbf{M_0}$ onto the y-axis; when viewed in the rotating frame, the magnetization vector appears to remain stationary although the magnitude of the vector is decreasing with time (T_1 and T_2 relaxation). Returning for a moment to the static laboratory frame, we see that the net magnetization vector is actually not static; it is rotating in the xy-plane about the z-axis at the Larmor frequency. This rotating vector generates an rf signal that is detected as an FID in an NMR experiment. The net magnetization vector soon returns to the z-axis, relaxation is complete, and the sequence can be repeated. In a simple one-pulse experiment, a $\pi/2$ pulse is used because it produces the strongest signal. A pulse (θ) less than (or greater than) $\pi/2$ leaves some of the possible signal on the z (or $-z$) -axis; only the component of the vector on the y-axis generates a signal.

We now consider multiple-pulse experiments and two-dimensional (2-D) NMR. Exactly what does the term "dimension" in NMR mean? The familiar proton spectrum is a plot of frequency (in δ units) versus intensity (arbitrary units)—obviously "2-D" but called a "1-D" NMR experiment, the one-dimension referring to the frequency axis. It is important to remember that the frequency axis, with which we are comfortable, is derived from the time axis (the acquisition time) of the FID through the mathematical process of Fourier transformation. Thus, *experimentally,* the variable of the abscissa of a 1-D experiment is in time units.

The so-called 2-D NMR spectrum is actually a 3-D plot; the omitted dimension in all NMR experiments (1-D, 2-D, 3-D, etc.) is always the intensity in arbitrary units. The two dimensions referred to in a 2-D NMR

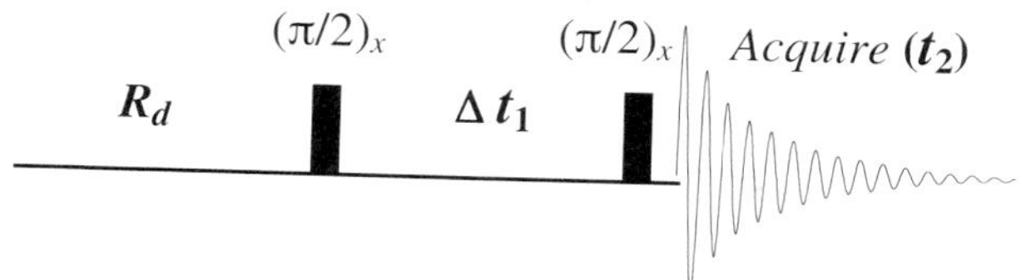

FIGURE 5.3 Prototype pulse sequence for a 2-D NMR. The incremental delay, Δt_1, and the acquisition time, t_2, are Fourier transformed into frequencies, ν_2 and ν_1, respectively. $(\pi/2)_x$ represents a 90° pulse along the x axis. The interval t_1 is of the order of microseconds; t_2 is of the order of seconds.

experiment are both frequency axes. It requires two Fourier transformations at right angles to each other on two independent time axes to arrive at two orthogonal frequency axes.

When the simple one-pulse experiment is again considered, there is only one time factor (or variable) that affects the spectrum, namely the acquisition time, t_2. We now consider a multiple-pulse sequence in which the equilibration period is followed by two pulses with an intervening time interval, the final pulse being the $\pi/2$ acquisition pulse. Thus, we have inserted an "evolution" period between the pulses. If we now vary this evolution time interval (t_1) over many different "experiments" and collect the resulting FIDs into one overall experiment, we have the basis of a 2-D experiment. Sequential Fourier transformation of these FIDs yields a set of "spectra" whose peak intensities vary sinusoidally. This first series of Fourier transformations result in the "second" frequency axis, ν_2, derived from the acquisition time, t_2, of each FID. The data are now turned by 90°, and a second Fourier transformation is carried out at right angles to the first series of transformations. This second series of Fourier transformations result in the "first" frequency axis, ν_1, a function of the evolution time, t_1, which you recall was changed (i.e., incremented) in the pulse sequence for each successive FID.

A simple prototype of a 2-D experiment should clarify some of these ideas while serving as a "template" for other more useful 2-D experiments. In this simple case, the pulse sequence (Figure 5.3) consists of a relaxation delay ($\boldsymbol{R_d}$), a $\pi/2$ pulse, a variable time interval (Δt_1, the evolution period), a second $\pi/2$ acquisition pulse, and acquisition (t_2). This pulse sequence (individual experiment) is repeated a number of times (each time resulting in a *separate* FID) with an increased t_1 interval.

We choose for this experiment a simple compound, acetone (CH_3—(C=O)—CH_3), to avoid the complication (for now) of spin-spin coupling. In Figure 5.4, we see that, after the first $\pi/2$ pulse along the x-axis $(\pi/2)_x$, the magnetization $\mathbf{M_0}$ has rotated onto the y-axis to $\mathbf{M}$. The evolution during t_1 for the spin of the equivalent protons of acetone is shown in a rotating frame. In this treatment, we ignore spin-lattice relaxation but include transverse relaxation with time constant $\boldsymbol{T_2}$. If the Larmor frequency (ν_2) is at higher frequency than that of the rotating frame, $\mathbf{M}$ precesses clockwise in the xy plane during the time interval t_1 through the angle $2\pi\nu t_1$. From trigonometry, the y component of $\mathbf{M}$ is $\mathbf{M}$ $\cos(2\pi\nu t_1)$, and the x component is $\mathbf{M}\sin(2\pi\nu t_1)$.

After time t_1, the acquisition pulse $(\pi/2)_x$ rotates the y component downward onto the $-z$-axis; this component therefore contributes no signal to the FID. The x component, on the other hand, remains unchanged (in the xy-plane) and its "signal" is recorded as the FID. When this FID is Fourier transformed, it gives a peak with frequency ν_2 and amplitude $\mathbf{M}\sin(2\pi\nu t_1)$. If we repeat this "experiment" many times (e.g., 1024 or 2^{10}), each time increasing t_1 in a regular way, we obtain 2^{10} FIDs. Successive Fourier transformation of each of these FIDs gives a series of "spectra" each with a single peak of frequency ν_2 and amplitude $\mathbf{M}\sin(2\pi\nu t_1)$. In Figure 5.5A, 22 of the 1024 spectra are plotted in a stacked plot; we see that the amplitude of the acetone peak varies sinusoidally as a function of t_1. We have now established one of the frequency axes (ν_2) for our prototype 2-D spectrum.

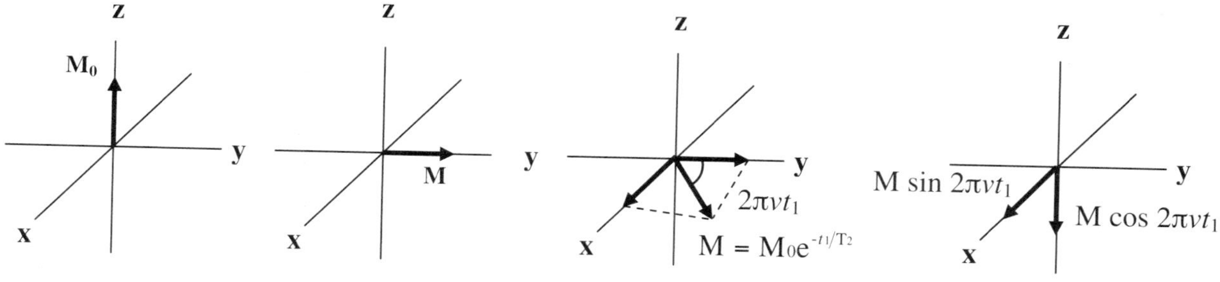

FIGURE 5.4 Evolution in a rotating frame of the acetone protons is shown during time interval t_1 following the first pulse. The second pulse and acquisition give a signal resulting only from the x component of $\mathbf{M}$; this signal amplitude varies sinusoidally with t_1. Interval t_1 is in the range of microseconds to milliseconds; t_2 is of the order of seconds. The precessional frequency of the proton is higher than that of the rotating frame.

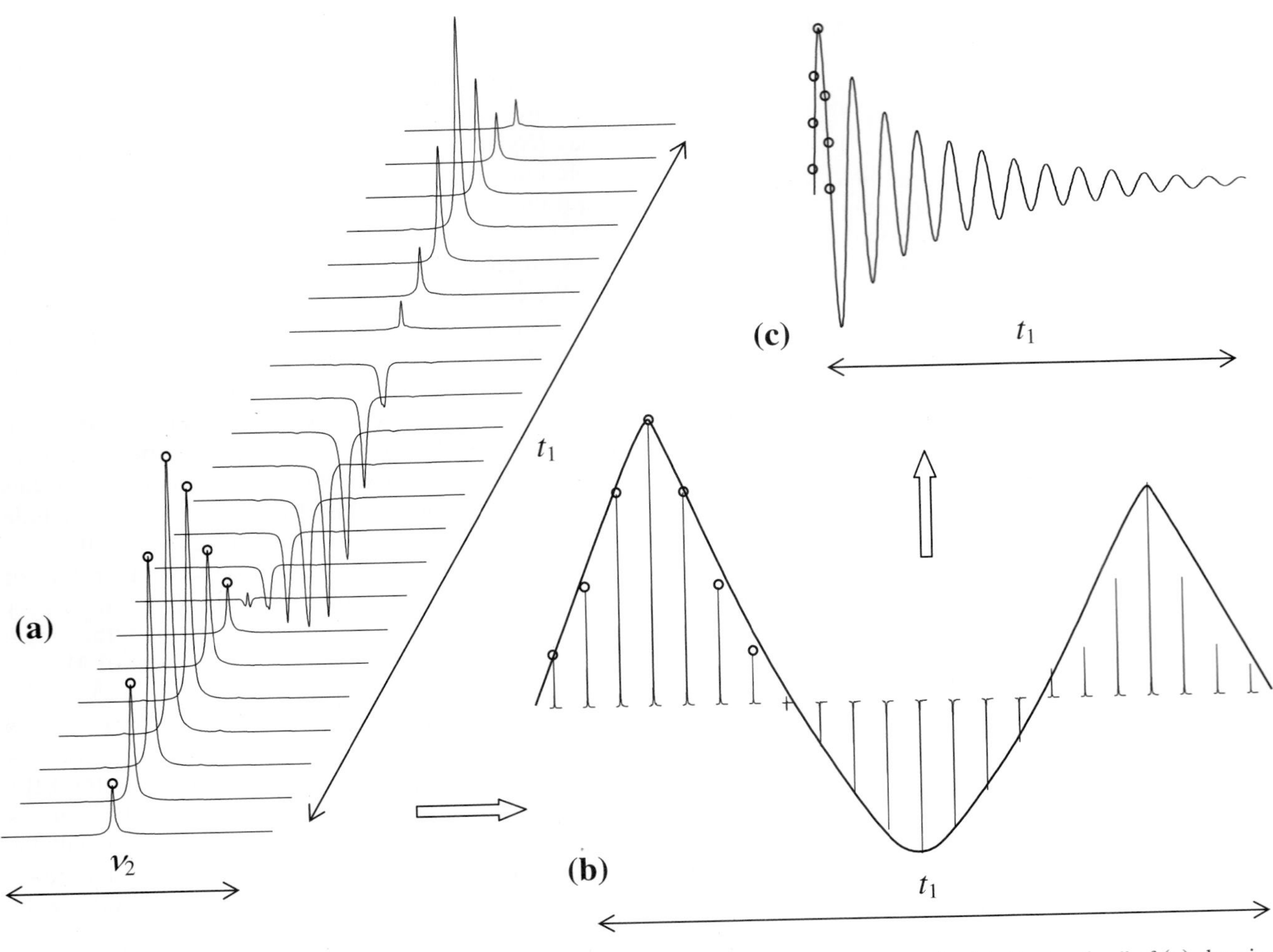

FIGURE 5.5 (a) Stacked plot of 22 "spectra" from acetone in which t_1 is varied incrementally. (b) "Projection" of (a) showing the sinusoidal behavior of the maxima and minima along t_1 for a single column. (c) Resulting interferogram representing a slice parallel with t_1 through the tops of peaks in (a) for one column.

Before we establish our second axis, let us do a little bookkeeping. Remember that each of these spectra that we now have is actually a digitized collection of points. Let us assume that each of these 1024 spectra is composed of 1024 data points. Thus, we have a square matrix of data. If we mentally rotate this collection of spectra, we can perform a second series of Fourier transformations on the data orthogonal to the first. Before we perform these transformations though, let us take a closer look at the data in Figure 5.5a.

By physically rotating the spectra of Figure 5.5a so that we are looking along the ν_2 axis, the data appear as "projections" shown in Figure 5.5b. Small circles are drawn at the maxima of the first seven rows of data of Figure 5.5a. These circles of the seven maxima are again drawn in Figure 5.5b (and Figure 5.5c) so that you can follow the rotation and projection. If we replot one column of data from Figure 5.5a (let us choose the column that corresponds to the maximum (and minimum) for the acetone peak), the data we obtain (Figure 5.5c) look like an FID or a time-domain spectrum. In fact, it is a time domain spectrum, now a function of the interval t_1 from our pulse sequence. To distinguish, we refer to data obtained in real time (a function of t_2) as an FID and data constructed point by point as a function of t_1 as an interferogram.

We now perform our second series of Fourier transformations on each of the 1024 interferograms to produce literally a transform of a transform. This result is the end product: a 2-D spectrum; we are now faced, however, with the challenge of visualizing our results. One way to plot the data is as a "stacked plot" similar to the plot that we have already seen in Figure 5.5a. This type of plot, shown in the left part of Figure 5.6, gives a sense of three-dimensions. Note that the two axes are now labeled F2 and F1, which is consistent with the rest of the text and is commonly used. For this spectrum, this

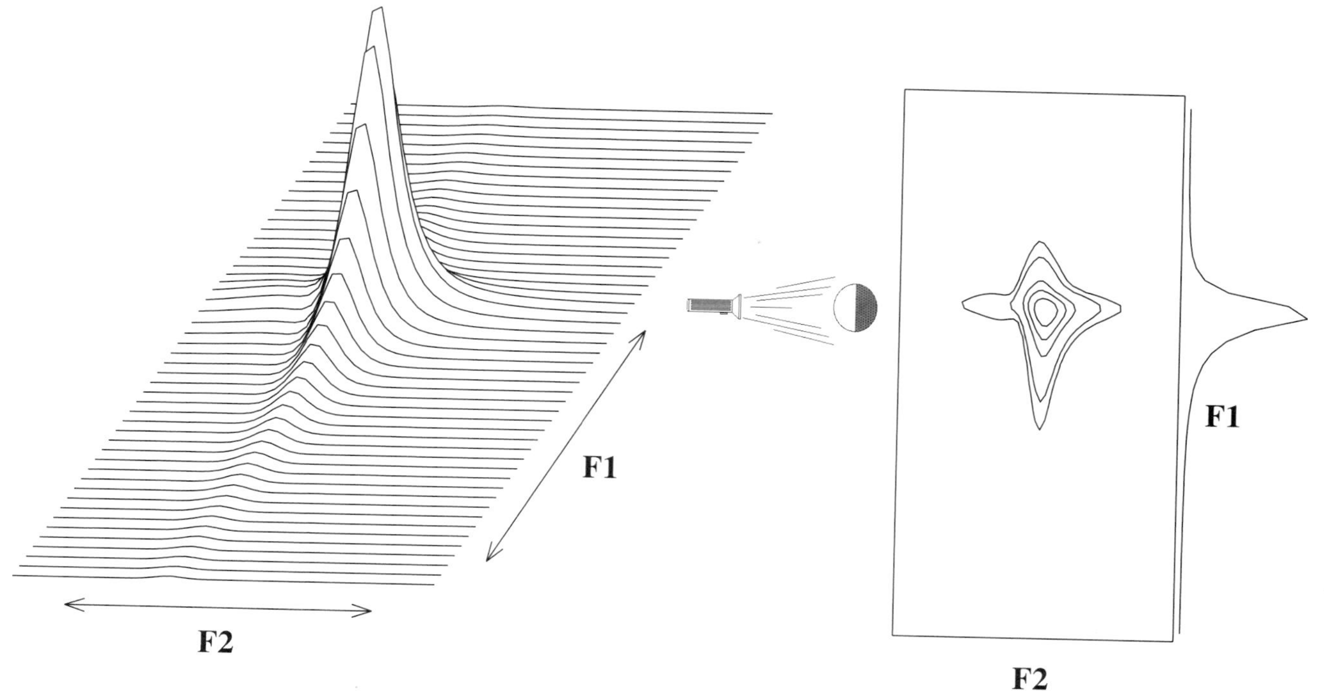

FIGURE 5.6 Fourier transformation of a series of FIDs like the ones in Figure 5.5 (C) to give the frequency-domain spectrum as both a peak and as contours. The contour plot also shows a projection parallel to **F2**.

type of plot is satisfactory because there are no peaks being blocked. Thus, one perspective is sufficient.

For more complex spectra, the data are usually presented as a series of contours just as hills and valleys are represented on a topological map. We see this representation of the data in the right part of Figure 5.6. Projections of the data are often included in 2-D spectra, which is equivalent to "shining a light" on the peak to reveal its "shadow," which is obviously 1-D. Often these projections are replaced with actual 1-D spectra that have been acquired separately. So long as there are no negative peaks (e.g., phase sensitive COSY, not covered in this book), we use this method without comment.

5.3 CORRELATION SPECTROMETRY

The observant reader will have realized by now that the above experiment, which is a type of frequency modulation, provides no additional information beyond the simple ^1H spectrum of acetone. Actually, that is the beauty of that experiment; it has all of the elements of a 2-D correlation experiment *and* we can completely follow the activity of the net magnetization vector for acetone using simple vectorial models. Let us turn this prototype pulse sequence into a general format for all 2-D experiments. If we replace the first $\pi/2$ pulse with a generalized "pulse" that contains one or more pulses and concomitant delays and replace the second acquisition $\pi/2$ pulse with a generalized "acquisition pulse" that also contains one or more pulses and their concomitant delays, we arrive at the general pulse sequence for 2-D correlation experiments, shown in Figure 5.7. We can usually describe, using simple vector models and trigonometry, the result of what is happening inside the boxes and oftentimes we will. For the time being, let us ignore "the goings on" inside the boxes and concentrate on what is happening during t_1 and t_2.

In all 2-D experiments, we detect a signal (during acquisition) as a function of t_2; this signal, however, has been ***modulated*** *as a function of t_1.* The acetone experiment above is simple because the magnetization experiences identical modulation during t_1 and t_2. In an experiment in which the magnetization is identically modulated during t_1 and t_2, the resulting peaks will be such that ν_1 equals ν_2; in the parlance of 2-D NMR, the experiment gives rise to diagonal peaks. For useful 2-D information, we are interested in experiments in which the magnetization evolves with one frequency during t_1 and a different frequency during t_2. In this case, our experiment will give rise to peaks in which ν_1 and ν_2 are different; this time we call the peaks off diagonal or cross peaks. In order to interpret a 2-D NMR spectrum, there are two things that we need to know. First, what frequencies do the axes represent? One axis (ν_2) always represents the nucleus detected during

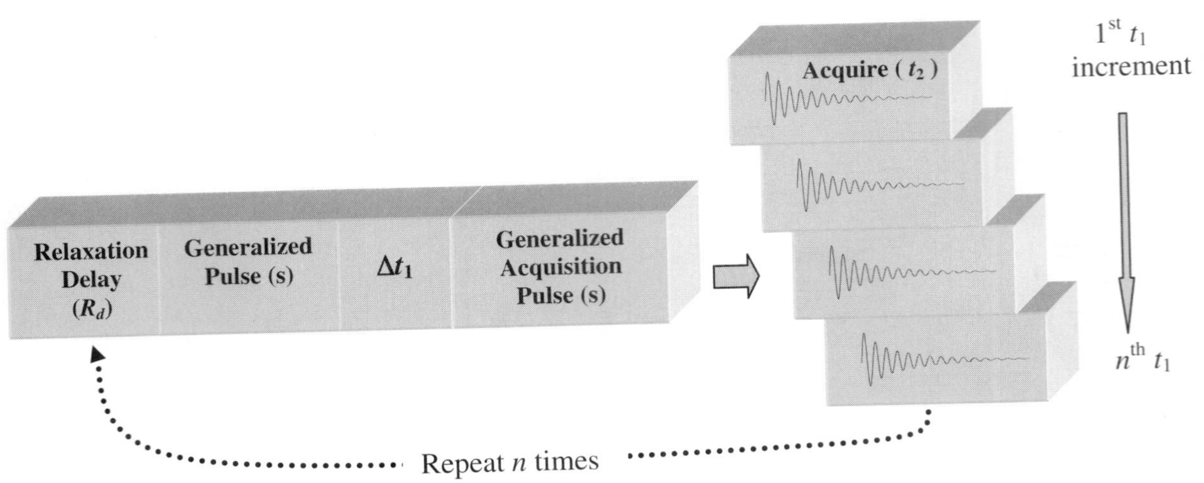

FIGURE 5.7 Generalized pulse sequence for 2-D NMR. The signal detected during acquisition, t_2, is modulated during the incremental time, t_1, thus giving rise to cross peaks in the 2-D spectrum.

acquisition (t_2). The other axis (ν_1, which obviously depends on t_1) can represent the same nucleus (e.g., $^1H—^1H$ COSY), a different nucleus (e.g., $^1H—^{13}C$ COSY also called HMQC or HETCOR), or a coupling constant (e.g., *J*-resolved spectroscopy, not covered in this chapter). Second, we need to know how the magnetizations are related during t_1 and t_2; in this way we can account for and interpret the cross peaks.

If we return to our prototype 2-D experiment and apply this pulse sequence to an AX system, we will be in a better position to appreciate the incremental time, t_1. While we can describe mathematically precisely how the spins evolve during this time, we cannot show this evolution pictorially with vector diagrams. (The mathematical description for this system requires quantum mechanics and solution of the density matrix, well beyond the scope of this text.) After the first $\pi/2$ pulse, the system can be described as a sum of two terms; each term contains the spin of only one of the two protons. During the time t_1, the spins precess (evolve) under the influences of both chemical shifts *and their mutual spin-spin coupling*. The mutual coupling has the effect of changing some of the individual spin terms into products containing magnetization components of both nuclei. Next, the second $\pi/2$ pulse causes the spins that were precessing under both chemical shift and coupling influences to redistribute magnetization among all spins (there is only one other spin in this case) with which it is associated (coupled). This redistribution of magnetization is detected in t_2; thus, a frequency detected in t_2 has its amplitude modulated as a function of other spins (again, only one here), with which it is coupled during t_1, leading to cross peaks connecting the coupled nuclei. Because the magnetization is redistributed equally in both directions (i.e., from A to X and from X to A), the cross peaks (at least for this experiment) will be symmetrically disposed about the diagonal. This description of spins precessing and mixing during t_1 and their redistribution during the acquisition pulse (and detection during t_2) is admittedly difficult to follow without pictures. The pulse sequences are not given detailed explanations because of this difficulty.

5.3.1 $^1H-^1H$ Correlation: COSY

Our simple 2-D experiment is actually a very important experiment sometimes simply called **COSY** (**CO**rrelation **S**pectroscop**Y**), and which we will call $^1H—^1H$ COSY for the time being in order to clearly indicate what is being correlated.* The pulse sequence for $^1H—^1H$ COSY is none other than the one we have already described above in Figure 5.3: two $\pi/2$ proton pulses separated by the required evolution period, t_1, which is systematically incremented, and the acquisition period, t_2.

The actual pulse sequences used by all modern spectrometers are more complicated than the idealized ones given in this text. Many spectrometers employ a technique known as "phase cycling" in which the phase of the rf pulse is changed in a regular manner (through a "cycle") for each t_1 increment. These phase cycles are extremely important *experimental factors* that help remove artifacts and other peculiarities of quadrature detection. We will ignore phase cycling in our pulse sequences and discussions because they do not affect

* Many readers will already be aware that acronyms for 2-D NMR experiments have proliferated along with available experiments. This chapter attempts neither an encyclopedic approach to describing these acronyms nor their experimental counterparts. This chapter does, however, cover enough important experiments to enable the reader to interpret nearly any 2-D experiment that one is likely to encounter. Acronyms are listed in the index.

our understanding and interpretation of these experiments. The interested reader is referred to Claridge (see References) for these and other experimental parameters important for 2-D experiments. Another factor that is ignored in our discussion is the use of "gradients." A short description of them and their purpose is given at the end of this chapter.

In the description of the 2-D experiment above for an AX spin system, we found that during t_1, spins, which are mutually coupled, precess under the influence of both nuclei's chemical shifts and thus give rise to peaks in which ν_1 does not equal ν_2. In the general case, ^{1}H—^{1}H COSY spectra are interpreted as giving rise to off diagonal or cross peaks for all protons that have spin-spin coupling; put simply, the cross peaks correlate coupled protons. In a sense, the experiment can be thought of as simultaneously doing all pertinent decoupling experiments to see which protons are coupled to which other protons. Of course, no protons are being decoupled in an ^{1}H—^{1}H COSY and this experiment should not be thought of as replacing homonuclear decoupling experiments (see Section 3.15).

5.4 IPSENOL: ^{1}H—^{1}H COSY

Let us begin our discussion of 2-D NMR by considering the ^{1}H—^{1}H COSY spectrum of ipsenol, the monoterpene alcohol considered in some detail in Sections 3.12.1 and 4.6. For reference and as a reminder, the typical 1-D NMR data at 300 MHz for ipsenol and its structure are provided in Figure 5.8.

The contour display of the simple COSY spectrum for ipsenol is shown in the top part of Figure 5.9. The presentation shown here is typical; F2 is found on the bottom (or top) with the proton scale as usual (from right to left). F1 is displayed on the right (or left) with the proton scale running from top to bottom. A proton spectrum is displayed opposite the F1 scale as a convenience instead of the poorly resolved projection; this 1-D spectrum is not part of the ^{1}H—^{1}H COSY spectrum but added later. From the upper right to the lower left runs the "diagonal," a series of absorptions in which ν_1 equals ν_2; these diagonal peaks provide nothing in the way of useful information beyond the simple 1-D ^{1}H spectrum. On either side of the diagonal and symmetrically disposed (at least theoretically) are the cross peaks. The symmetry in this type of spectrum is oftentimes imperfect.

Before undertaking detailed discussions of ^{1}H—^{1}H COSY and the structure of ipsenol, there is one further experimental refinement that decreases the "clutter" along the diagonal. Although we can interpret this spectrum without this refinement, there are instances (i.e., caryophyllene oxide) when this improvement makes a great deal of difference.

5.4.1 Ipsenol: Double Quantum Filtered ^{1}H—^{1}H COSY

By simply adding a third $\pi/2$ pulse immediately following the second $\pi/2$ pulse in our simple COSY pulse sequence and changing nothing else, we have the pulse sequence for the very popular **D**ouble **Q**uantum **F**iltered ^{1}H—^{1}H COSY (**DQF-COSY**) experiment (Figure 5.10). The purpose of the third $\pi/2$ pulse is to remove or "filter" single quantum transitions so that only double quantum or higher transitions remain. In practical terms, the double quantum filter will select only those systems with at least two spins (minimum AB or AX); thus, methyl singlets (noncoupled) will be greatly reduced. Higher quantum filtering is possible but is generally not used. For instance, in a triple quantum filtered COSY, only systems with three spins or more are selected so that AB and AX spin systems as well as noncoupled systems will be eliminated.

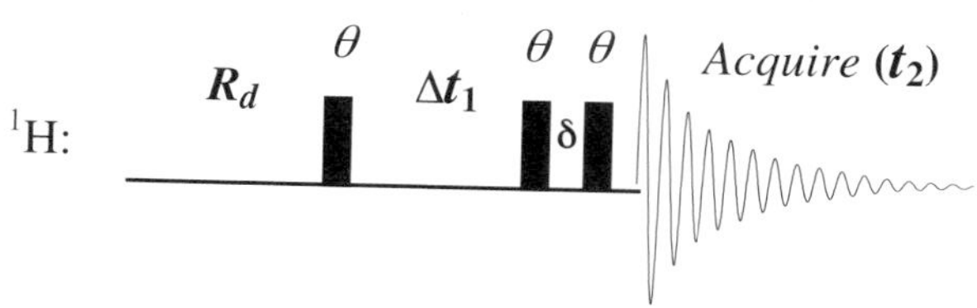

FIGURE 5.10 Pulse sequence for Double Quantum Filtered ^{1}H—^{1}H COSY (DQF -COSY). θ's are 90° pulses and δ is a fixed delay of the order of a few microseconds.

The DQF ^{1}H—^{1}H COSY spectrum of ipsenol can be found at the bottom of Figure 5.9. Note that the spectrum seems "cleaner," especially along the diagonal, making the task of interpretation significantly easier. Because of the greatly improved appearance of DQF-COSY, all COSYs in this book are double quantum filtered.

As we begin our interpretation of the DQF-COSY spectrum in Figure 5.9, let us recall that this spectrum shows correlation between coupled protons. A point of entry (i.e., a distinctive absorption) into a COSY spectrum (and other types of correlation spectra as well) is one of the keys to successfully gleaning information from it. The structure of ipsenol allows for more than one useful entry point, so let us select the carbinol methine at 3.83 ppm. If we begin at the diagonal and trace either directly to the right or directly up (we obtain the same result because the spectrum is symmetric), we intersect four off-diagonal or cross peaks. By drawing lines through these cross peaks at right angles to the one we just traced, we find the chemical shifts of the four coupled resonances. A quick check of the structure of ipsenol finds the carbinol methine adjacent to two pairs of diastereotopic methylene groups;* in other words, the proton at

* We will reserve further discussion of diastereotopic methylene groups until the next section on ^{1}H—^{13}C COSY or HMQC.

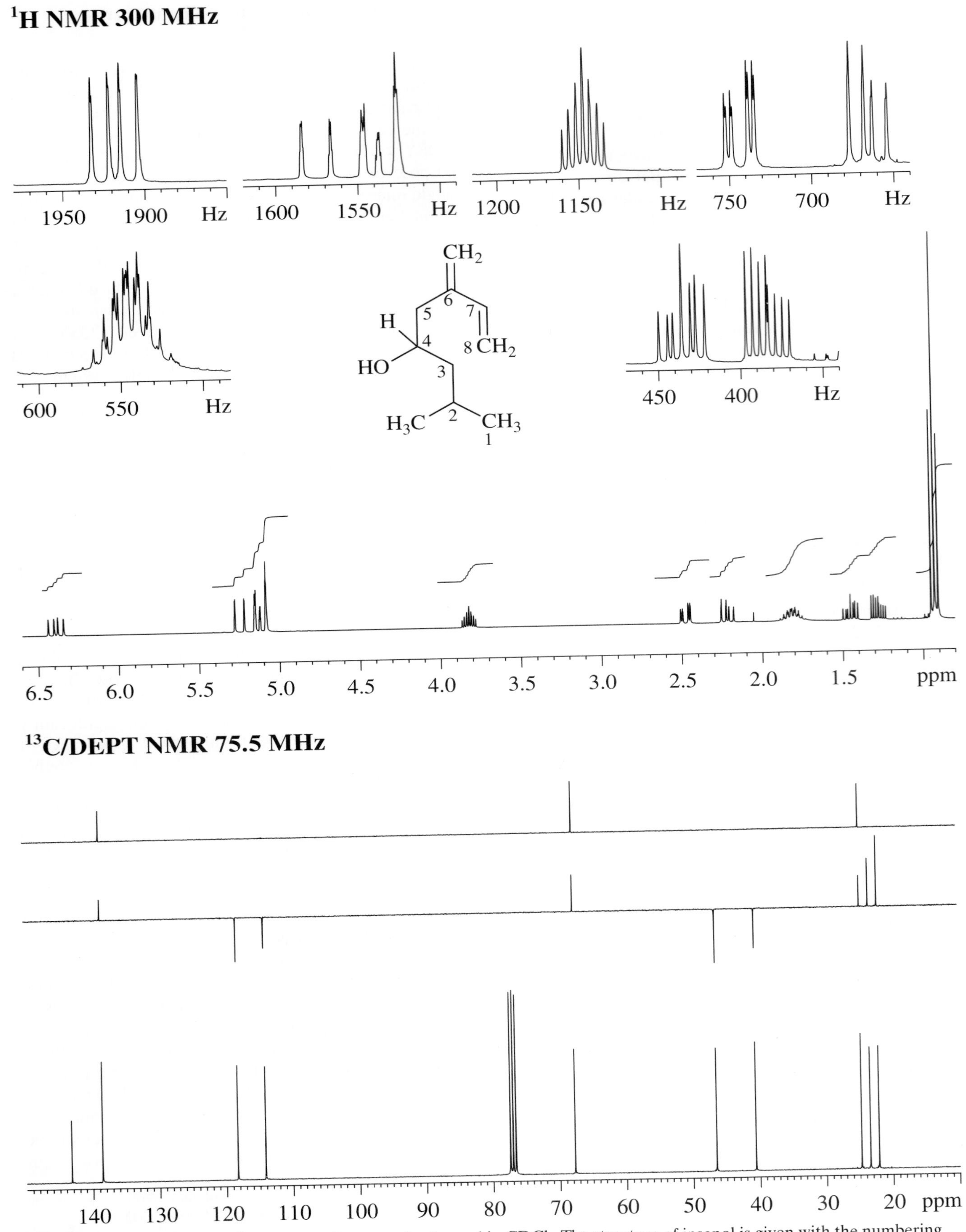

^{1}H NMR 300 MHz

^{13}C/DEPT NMR 75.5 MHz

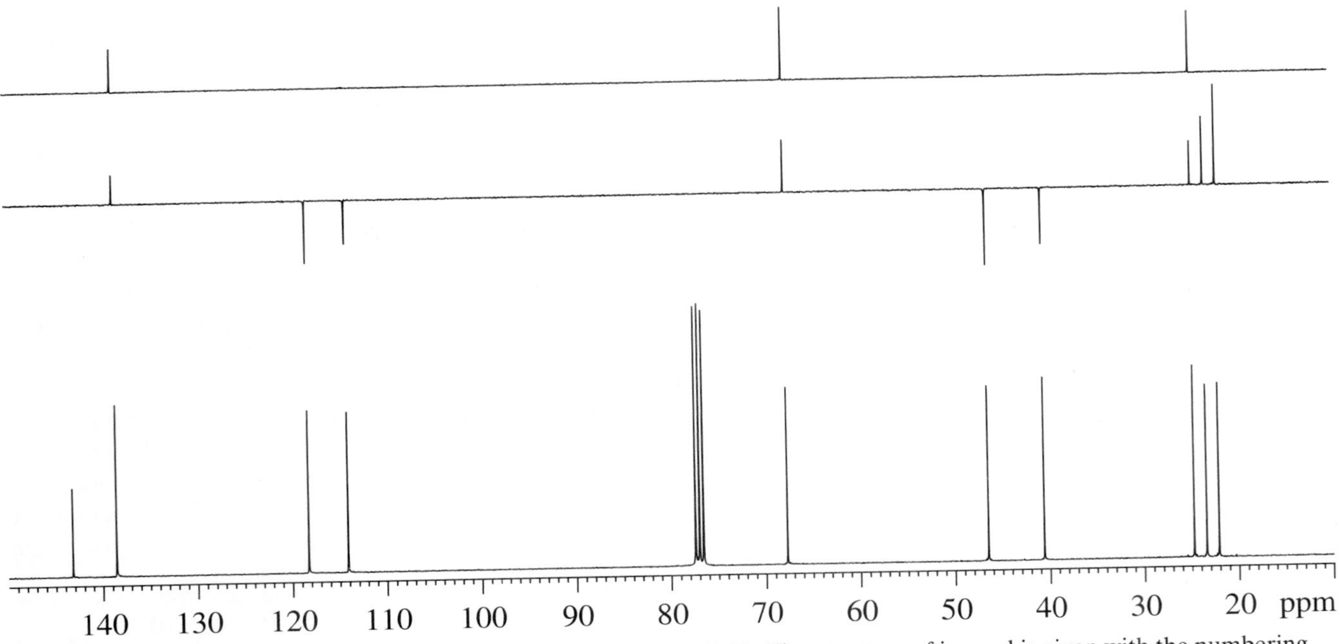

FIGURE 5.8 300 MHz ^{1}H, ^{13}C, and DEPT spectra for ipsenol in $CDCl_3$. The structure of ipsenol is given with the numbering used in the text.

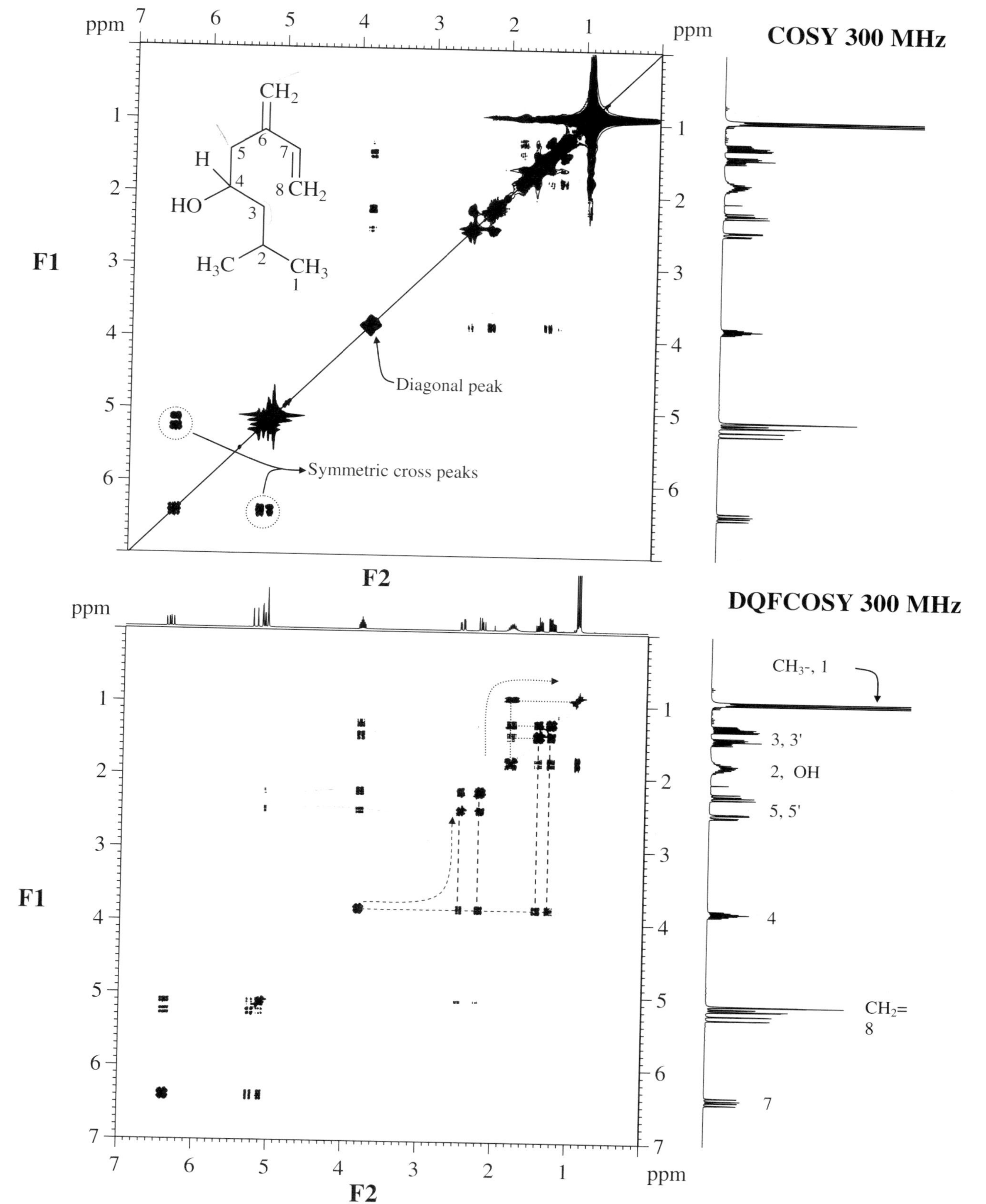

FIGURE 5.9 The top part is the 300 MHz simple COSY spectrum for ipsenol; the bottom part is the 300 MHz DQF COSY spectrum for ipsenol.

3.83 ppm is coupled to four protons, and the four protons correspond to two adjacent methylene groups.

We could continue to trace correlation paths from these four protons and the reader is invited to do so at the end of this section. Let us instead select another equally useful entry point: the isopropyl methine at 1.81 ppm. We again begin at the diagonal and this time we find that the isopropyl methine is correlated with three distinct resonances. Two of the correlations correspond to the two protons of one of the diastereotopic methylenes that also correlated with the carbinol methine above. In addition, we find a correlation to the two overlapping methyl doublets at 0.93 ppm. These correlations of course make perfect sense with the structure; in fact, by only considering these two protons (i.e., at 3.83 and 1.82 ppm) we have established correlations (also called "connectivities") through three-fifths of the molecule.

Next we consider the two protons on the C-5 methylene at 2.48 and 2.22 ppm. We have already seen that they are coupled to the carbinol methine (you can and should verify this from the methylene protons' perspective) and we see that they are also coupled to each other.* In addition, we see weaker cross peaks from both methylene protons correlating to an olefinic proton at 5.08 ppm. This correlation is due to long range coupling ($^4J_{HH}$ or four-bond coupling) of the methylene protons to the *cis*-proton of the adjacent double bond. This is a nice correlation to find because it provides H—H connectivity to the otherwise isolated diene spin system. In the absence of these long-range correlations, such isolated spin systems can be "connected" by either HMBC or INADEQUATE sequences, which are described below. At this point, the reader is invited to complete the correlations for ipsenol in this DQF-COSY spectrum. Correlations can be found for all protons except the hydroxylic proton, which is rapidly exchanging.

5.4.2 Carbon Detected ^{13}C—^{1}H COSY: HETCOR

The ^{13}C—^{1}H COSY (HETCOR) experiment correlates ^{13}C nuclei with directly attached (i.e., coupled) protons; these are one-bond ($^1J_{CH}$) couplings. The frequency domains of F1 (ν_1) and F2 (ν_2) are of different nuclei, and so there is no apparent diagonal or symmetry.

The pulse sequence for this experiment, commonly called **HETCOR** (**HET**eronuclear **COR**relation), is recorded in Figure 5.11a. During the evolution time (t_1), the large one-bond heteronuclear *J*-coupling (J_{CH}) is used for polarization transfer, and thus only ^{13}C's bonded directly to ^{1}H's are detected. To realize maximum polarization transfer, a fixed delay $\Delta_1 = 1/(2J_{CH})$ is added after t_1. The short delay (Δ_2) between the final ^{13}C pulse and the start of acquisition is a refocusing period so that the ^{13}C lines do not have opposite phase and thus do not cancel one another when ^{1}H-decoupling is applied. The optimal refocusing time (Δ_2) depends on whether the ^{13}C belongs to a CH, CH_2, or CH_3 group. Typically, a compromise value of $\Delta_2 = 1/(3J_{CH})$ is chosen.

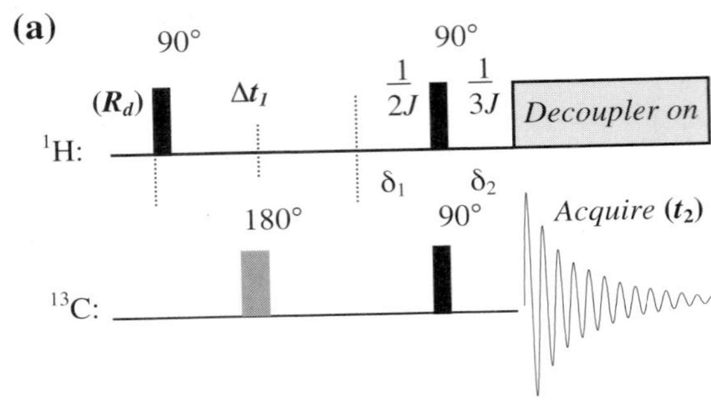

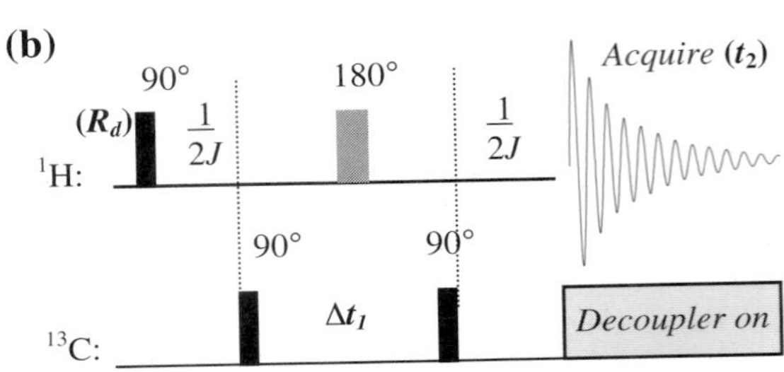

FIGURE 5.11 (a) The pulse sequence for a HETCOR. (b) The pulse sequence for a HMQC. The value for $^1J_{HC}$ is typically 145 Hz.

The F1 axis (ν_1), which is derived from the incremental delay, t_1, is the proton axis. The F2 axis (ν_2), obtained during t_2, is the carbon axis. Thus, our $\pi/2$ read pulse is in the ^{13}C channel, and the FID acquired during t_2 represents the ^{13}C nucleus. Lastly, in an ideal experiment, the carbons resonances should be singlets rather than multiplets,† composite pulse (broadband) decoupling (CPD) is applied in the proton channel during acquisition so that the carbon signals obtained from each FID are singlets. Remember, in a 2-D experiment, correlation occurs during t_1 and, hence, the proton decoupler is not turned on during this period.

5.4.3 Proton Detected ^{1}H—^{13}C COSY: HMQC

Historically, the **HMQC** (**H**eteronuclear **M**ultiple **Q**uantum **C**orrelation) experiment was preceded by

* Geminal methylene protons (sp^3 hybridized) are always coupled to each other and their coupling constant ($^2J_{HH}$) is always rather large (see Appendices, Chapter 3)

† In the presence of one-bond couplings, a methyl carbon will appear as a quartet, a methylene as a triplet, etc.

the HETCOR experiment. Although experimentally there are many differences, the essential difference is that, while the HETCOR experiment is carbon detected, the HMQC experiment is proton detected. Since there are great discrepancies between proton and carbon in their relative abundances and sensitivities, the HMQC is greatly preferred today. The advantage of inverse experiments over direct detection experiments is that with inverse experiments the nucleus with the highest γ (usually ^{1}H) is detected yielding the highest sensitivity.

As mentioned in Chapter 3, the absolute sensitivity of an NMR experiment is given by the following formula: $S/N = (NT_2\gamma_{exc}(\gamma_{det}B_0)^{3/2}ns^{1/2})/T$, where γ_{exc} and γ_{det} are both ^{13}C for the directly detected experiment whereas they are both ^{1}H for the inverse detected experiment. Replacement of the magnetogyric ratios gives a sensitivity increase of $(\gamma^1{}_H/\gamma^{13}{}_C)^{5/2}$, which is about 30-fold for ^{1}H—^{13}C HMQC compared to the directly detected method, HETCOR. The challenge of an inverse-detected chemical shift correlation experiment, however, is that the large signals from ^{1}H not coupled directly to a ^{13}C nucleus must be suppressed, which poses a dynamic range problem, and usually requires additional phase cycling. The introduction of pulsed field gradients in high-resolution NMR greatly improved the problem of suppressing signals from ^{1}H bonded to ^{12}C: The suppression is almost perfect without additional phase cycling, which significantly improves the quality of the spectrum and requires much less time.

The pulse sequence for the HMQC experiment is recorded in Figure 5.11b. Three details in this pulse sequence are worth discussing. The F1 axis (ν_1), which is derived from the incremental delay, t_1, is the carbon axis. The F2 axis (ν_2), obtained during t_2, is the proton axis. Thus, our $\pi/2$ read pulse is in the ^{1}H channel, and the FID acquired during t_2 represents the ^{1}H nucleus. Lastly, GARP decoupling—which is a low power, CPD routine—is used to decouple nuclei with large chemical shift ranges. This **GARP** (**G**lobally optimized, **A**lternating phase, **R**ectangular **P**ulses) CPD technique is applied on the carbon channel during acquisition so that the proton signals obtained are not split into doublets by the ^{13}C nuclei. Remember that the HMQC is looking at the attached ^{1}H—^{13}C coupling, which appears as ^{13}C satellite signals in the ^{1}H spectrum (1.1% natural abundance, see Section 3.7.5).

5.4.4 Ipsenol: HETCOR and HMQC

The upper portion in Figure 5.12 is the HETCOR spectrum of ipsenol, and the lower portion is a HMQC spectrum of ipsenol. The presentation of the two spectra is the same except that the axes have been switched. Thus, in the HETCOR spectrum, the F2 axis on the bottom has the carbon scale and the F1 axis on the side has the proton scale, whereas in the HMQC spectrum these axes are reversed. In theory, the information content of the two spectra is identical; in fact, we find exactly the same correlations for ipsenol. The only real practical consideration when comparing the data of the two experiments is the difference in the digital resolution of the carbon axis.

Let us familiarize ourselves with HMQC spectra by considering the HMQC spectrum of ipsenol (Figure 5.12. bottom part). Immediately obvious is the fact that there is no diagonal and no symmetry; this will be true whenever F1 and F2 represent different nuclei. In this presentation, the F1 axis (carbon) is along the left side and the F2 axis (proton) is along the bottom. Opposite these axes we find the corresponding 1-D spectra, which are given as a convenience and are not part of the actual 2-D spectrum. Interpretation of this spectrum is straightforward. We begin with any carbon atom and mentally draw a line horizontally until a cross peak is encountered.* Another line is mentally drawn perpendicular to the first to find the proton or protons with which it correlates.

There are only three cases possible for each carbon atom. If a line drawn encounters no cross peaks, then the carbon has no attached hydrogens. If the drawn line encounters only one cross peak, then the carbon may have either 1, 2, or 3 protons attached; if 2 protons are attached, then they are either chemical shift equivalent or they fortuitously overlap. If the horizontal line encounters two cross peaks, then we have the special case of diastereotopic protons attached to a methylene group. Much of this information will already be available to us from DEPT spectra (see Section 4.6); indeed, the HMQC spectrum should, whenever possible, be considered along with the DEPT.

In ipsenol, there are four methylene groups all of which possess diastereotopic pairs of protons. Resonances for two of these methylene groups occur in the carbon spectrum at 41 and 47 ppm. Note with which protons these carbon atoms are correlated and compare these results with what we found with COSY. As we expect, the results here confirm our assignments from COSY and help build an ever-strengthening basis for our assignments. The other two methylene carbon atoms are found at higher frequency in the olefinic region, and the HMQC cross peaks for these carbon resonances help clarify the overlapping proton resonances that we find in the proton spectrum.

* We could just as well start on the proton axis and in this case we would obtain exactly the same result. In cases of overlap in the proton spectrum, we will not always be able to find all of the proper starting points. Overlap is usually not a problem on the carbon axis.

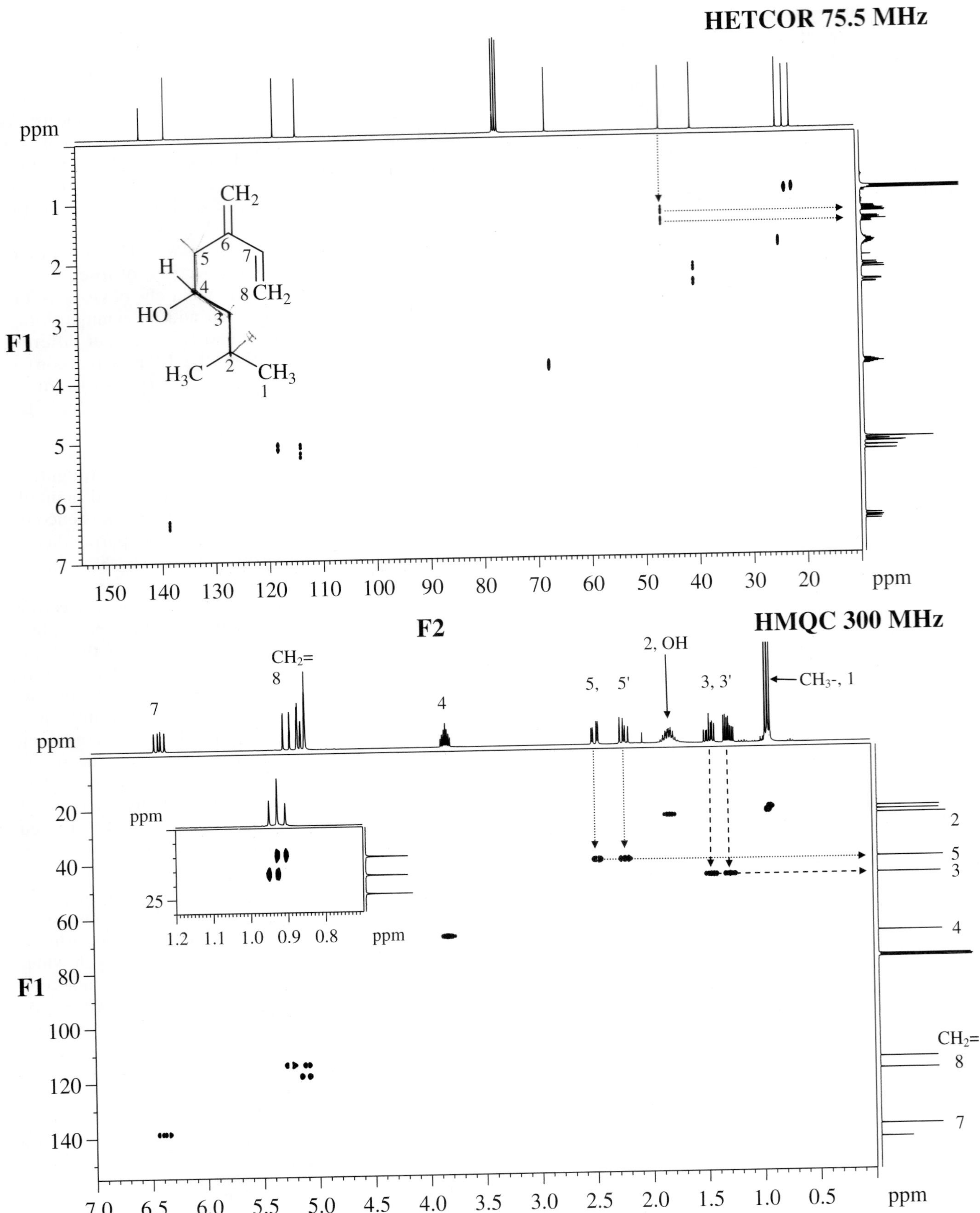

FIGURE 5.12 The bottom part of the figure is the HMQC spectrum for ipsenol while the top part of the figure is the HETCOR spectrum of ipsenol.

Ipsenol has two methyl groups that appear as a "triplet" at 0.93 ppm in the proton spectrum. A closer look, however, at the inset reveals that it is a pair of coincidental doublets. Before moving on to the next section, consider the following question: can the olefinic methylene carbon resonances be assigned on the basis of combined information from COSY and HMQC?

5.4.5 Ipsenol: Proton-Detected, Long Range $^1H—^{13}C$ Heteronuclear Correlation: HMBC

For the HMQC described above, we wanted an experiment that eliminated long-range (i.e., two and three bond) proton-carbon couplings while preserving the directly attached (i.e., one-bond) couplings, which we correlated in a 2-D experiment. The **HMBC** (**H**eteronuclear **M**ultiple **B**ond **C**oherence) experiment, on the other hand, which is also proton detected,* capitalizes on these two- and-three-bond couplings providing us with an extremely powerful (although sometimes cluttered) spectrum. In essence, we indirectly obtain carbon-carbon (although not $^{13}C—^{13}C$) correlations, and, in addition, we are able to "see" or correlate quaternary carbons with nearby protons. Since both $^2J_{CH}$ and $^3J_{CH}$ couplings are present, interpretation can be tedious; we must be methodical in our approach and keep in mind the HMQC correlations.

Interpretation of HMBC's requires a degree of flexibility because we do not always find what we expect to find. In particular, depending on the hybridization of carbon and other factors, some of the two-bond correlations ($^2J_{CH}$) or three-bond ($^3J_{CH}$) correlations are occasionally absent. To add to the confusion, infrequently we find four-bond ($^4J_{CH}$) correlations! The variations in correlations that we find are due to the variations in the magnitude of $^2J_{CH}$, $^3J_{CH}$, and $^4J_{CH}$ coupling constants.

The pulse sequence for the HMBC is given in Figure 5.13 for interested readers. The time delay, $1/(2J)$, can be optimized for different coupling constants. A typical value for J assumes an average long-range coupling constant of 8 Hz.

The HMBC for ipsenol (Figure 5.14) looks like the HMQC for ipsenol with two obvious differences: there are considerably more correlations and the one bond correlations (HMQC) are gone. (The spectrum is broken into five sections so that there is sufficient resolution to see all of the correlations.) Interpretation for ipsenol is straightforward. But first, let us note a common artifact: ^{13}C satellites of intense proton peaks,

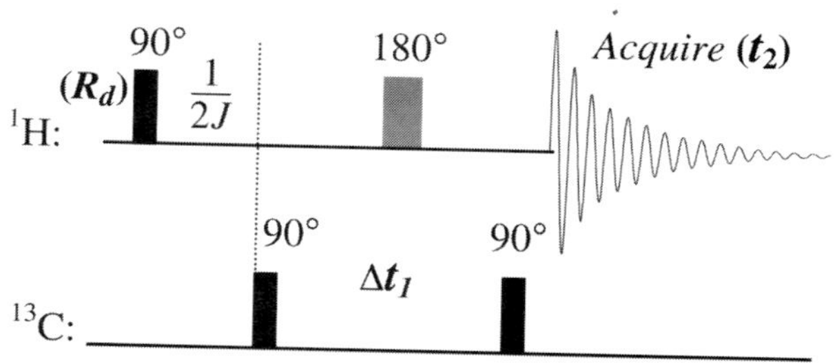

FIGURE 5.13 Pulse sequence for HMBC. J is chosen for the long-range CH coupling ($^2J_{HC}$ and $^3J_{HC}$), typically 8 Hz.

especially methyl groups. If we trace parallel to the proton axis (F2) at about 23 ppm on the carbon axis (F1), we find a cross peak at about 0.93 ppm (proton), which is real. On either side, we find two "cross peaks" that do not line up (correlate) with any protons in F2. These are ^{13}C satellites and should be ignored. The other methyl carbon resonance shows the same phenomenon; the satellite peaks are marked with arrows in the figure.

We can begin with either a carbon or a proton resonance and obtain equivalent results. We will use the carbon axis as our starting point because we usually have less overlap there. For example, a line drawn parallel to the proton axis at about 68 ppm on the carbon axis (the carbinol carbon) intersects five cross peaks; none of the five correlations corresponds to the attached proton ($^1J_{CH}$) at 3.8 ppm. Four of the cross peaks correspond to the two pairs of diastereotopic methylene groups (2.48, 2.22, 1.45, and 1.28 ppm) and these represent, $^2J_{CH}$, or two-bond couplings. The fifth interaction ($^3J_{CH}$) correlates this carbon atom (68 ppm) to the isopropyl methine proton (1.82 ppm), which is bonded to a carbon atom in the β-position. The other carbon atom in a β-position has no attached protons so we do not have a correlation to it from the carbinol carbon atom. Thus, we have indirect carbon connectivities to two α-carbons and to one of two β-carbons.

Another useful example can be found by drawing a line from the carbon resonance at 41 ppm. This carbon is the C-5 methylene and we first note that correlations to the attached protons at 2.48 and 2.22 ppm are absent. There is only one α-carbon that has one or more attached protons; its corresponding correlation is found to the C-4 carbinol methine proton at 3.83 ppm.* There are three β-carbons and they all have attached protons. The C-3 methylene carbon shows indirect correlation through both of its diastereotopic protons at 1.45 and 1.28 ppm. The C-7 olefinic methine proton gives a cross peak at 6.39 ppm as do the protons of the olefinic methylene group attached to C-6 at 5.16 and

* There is a carbon detected analogue of the HMBC experiment called **COLOC** (**CO**rrelated spectroscopy for **L**ong range **C**ouplings) that predated the experiment treated here. The COLOC is not used much any more and we will not give any examples.

* The other α-carbon at C-6, which was a β-carbon in our first example, also shows no correlation in the HMQC. The reader should show that there are useful correlations to this carbon atom in the current (Figure 5.14) figure.

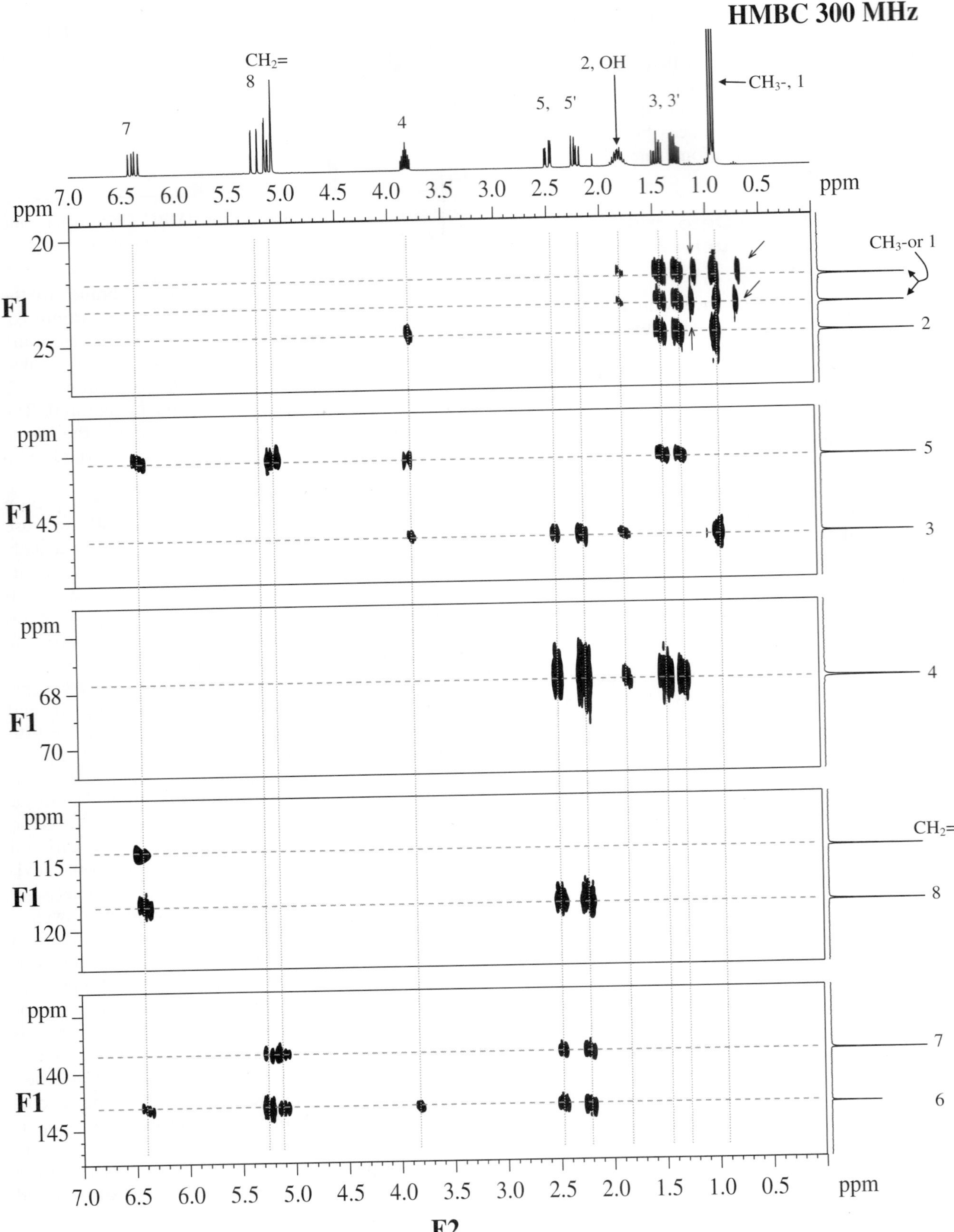

FIGURE 5.14 The HMBC spectrum for ipsenol. The spectrum is split into five sections for clarity. The arrows in the top section point to ^{13}C satellites. Lines are drawn to show correlations.

5.09 ppm. Since C-6 is a quaternary carbon, the HMBC experiment enables us to "see through" these normally insulating points in a molecule. Other assignments are left to the reader as an exercise; the quaternary carbon at about 143 ppm (C-6) is a good place to start.

5.5 CARYOPHYLLENE OXIDE

The structure of caryophyllene oxide is significantly more complicated, and is a worthy challenge for the methods that we introduced with ipsenol. For use here and for future reference, the ^{1}H, ^{13}C, and DEPT spectra are given in Figure 5.15. As an aid in discussing the DQF-COSY spectrum of caryophyllene oxide, note the following description of the proton spectrum: there are three methyl singlets, 0.98, 1.01, 1.19 ppm, two olefinic "doublets" (small geminal olefinic coupling), 4.86 and 4.97 ppm, and resonances from 13 other protons giving multiplets between 0.9 and 3.0 ppm. Even though we know the structure, it is impossible to assign any of these protons unless we make one or more unreasonable assumptions.*

5.5.1 Caryophyllene Oxide: DQF-COSY

The DQF-COSY spectrum of caryophyllene oxide can be found in Figure 5.16. The problem here is that *there is no good entry point*. The previous statement is not trivial. Without an entry point, it is impossible to relate the many obvious correlations (drawn in for convenience) that we see to a structural formula. Our approach therefore will be to record some of the correlations that we do see and wait until we have other information (i.e., HMQC) before we try to translate these correlations into a structure.

The exocyclic olefinic methylene protons show obvious COSY correlations to one another. In addition, we note weak cross peaks between the olefinic protons at 4.86 and 4.97 ppm and an apparent diastereotopic methylene group (2.11 and 2.37 ppm) and a quartet at 2.60 ppm, respectively. These interactions are reminiscent of the long-range allylic coupling that we saw in ipsenol; we could assign these correlations to the diastereotopic methylene C-7 and the methine at C-9. For now, we will be cautious and conservative, and return to this point later in the chapter.

A look at the extreme low-frequency portion of this COSY spectrum reveals an unexpected interaction. It seems that either one or both of the methyl singlets shows coupling to resonances at 1.65 and at 2.09 ppm.

* If pressed, we might assume that the allylic bridgehead methine would be the furthest downfield and assign the doublet of doublets at 2.86 ppm to this proton (wrong). The methods in this chapter will allow us to make these assignments without making unsubstantiated assumptions.

This apparent conflict can be resolved by a close examination of the methyl singlets at about 0.98 ppm. There is an unusually low-frequency multiplet, partially buried by the methyl singlets, which we had initially overlooked. This type of unexpected dividend is common in correlation spectra; both partially and completely obscured resonances usually reveal themselves in 2-D spectra (see HMQC below). Before continuing our discussion of caryophyllene oxide, let us consider ^{1}H—^{13}C correlations and how ^{1}H—^{1}H correlations interplay with ^{1}H—^{13}C correlations.

5.5.2 Caryophyllene Oxide: HMQC

The COSY spectrum for caryophyllene oxide can be understood more clearly when interpreted in conjunction with the information from an HMQC spectrum (Figure 5.17). From the DEPT spectrum (see Figure 5.15), we already know that caryophyllene oxide has three methyl carbon resonances (16.4, 22.6, and 29.3 ppm), six methylene carbon resonances (26.6, 29.2, 29.5, 38.4, 39.1, and 112.0 ppm), three methine carbon resonances (48.1, 50.1, and 63.0 ppm) and three quaternary carbon resonances (33.3, 59.1, and 151.0 ppm)

The olefinic methylene group (protons and carbon) and the three methyl groups (protons and carbons) are trivial assignments, and they correspond with our previous discussion. Of more interest and of greater utility, we assign the three methine protons: the doublet of doublets at 2.86 ppm (correlates with the carbon resonance at 63.0 ppm), the apparent quartet at 2.60 ppm (correlates with the carbon resonance at 48.0 ppm), and an apparent triplet at 1.76 ppm (correlates with the carbon resonance at 50.1 ppm). From the COSY and from the known structure, we now are now able to assign all three methine resonances and "feed" this information back into the COSY to establish other correlations.

From the long-range, allylic correlation that we found in the COSY, we now confirm our cautious assignments that we made earlier. The doublet of doublets at 2.86 ppm is assigned to the methine proton of the epoxide ring, and its chemical shift is rationalized on the basis of the deshielding effect of the epoxide oxygen. The other bridgehead methine (adjacent to the *gem*-dimethyl group) is assigned to the multiplet at 1.76 ppm. With these assignments in hand, we could "jump right back" into the COSY spectrum, but instead we will restrain our enthusiasm for now and assign the methylene protons first. Knowing these assignments first will help speed our way through the COSY.

Beginning from the low-frequency end of the ^{13}C spectrum, the following assignments can be made: the methylene carbon at 26.6 ppm correlates with proton resonances at 1.45 and 1.63 ppm, the methylene carbon at 29.2 ppm correlates with proton resonances at 2.11

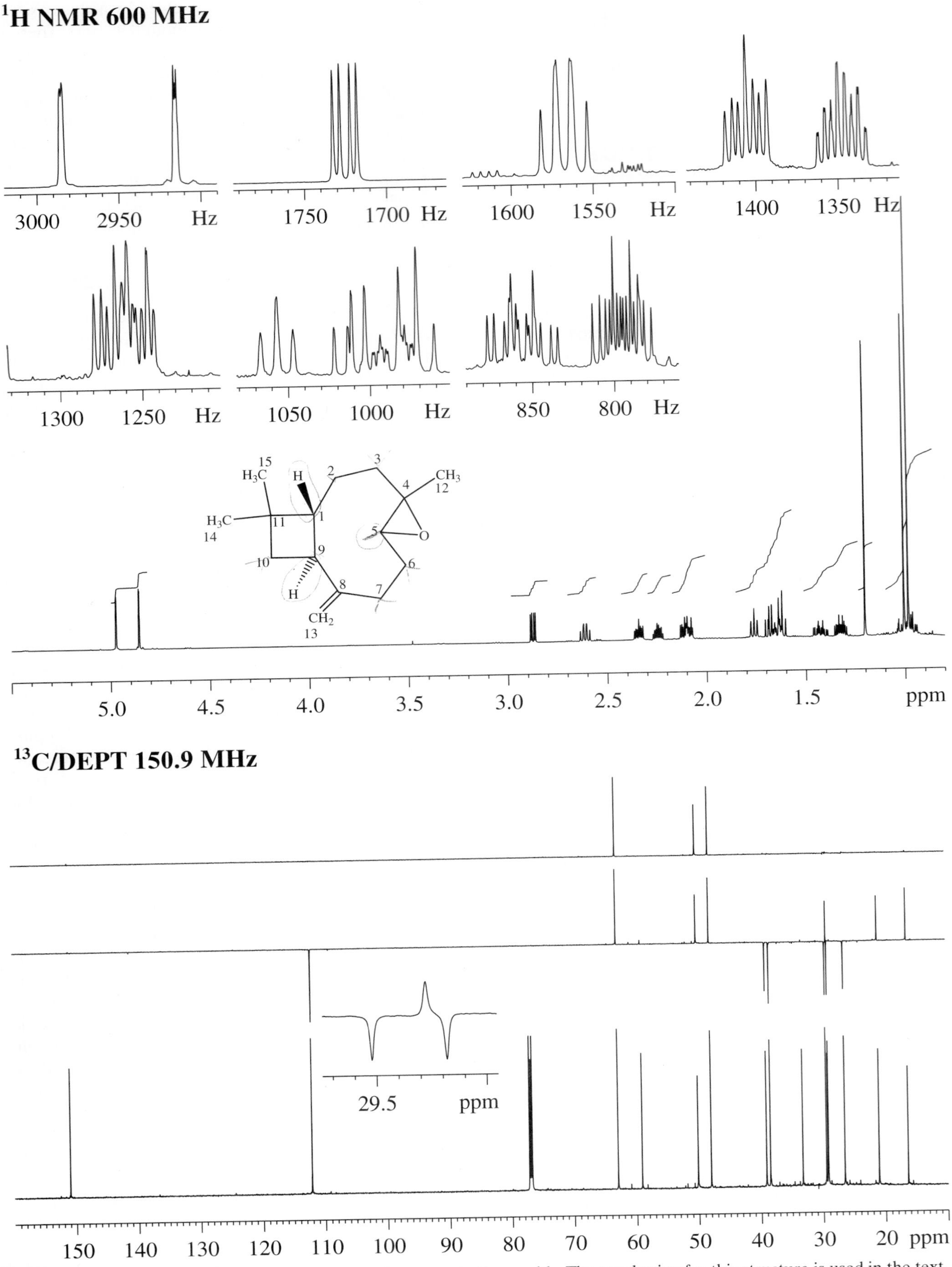

^{1}H NMR 600 MHz
3000
2950
Hz
1750
1700 Hz
1600
1550
Hz
1400
1350 Hz
1300
1250
Hz
1050
1000
Hz
850
800
Hz
15
H_3C
H
2
3
4
CH_3
12
H_3C
14
11
1
5
O
9
10
6
8
7
H
CH_2
13
5.0
4.5
4.0
3.5
3.0
2.5
2.0
1.5
ppm
^{13}C/DEPT 150.9 MHz
29.5
ppm
150
140
130
120
110
100
90
80
70
60
50
40
30
20 ppm

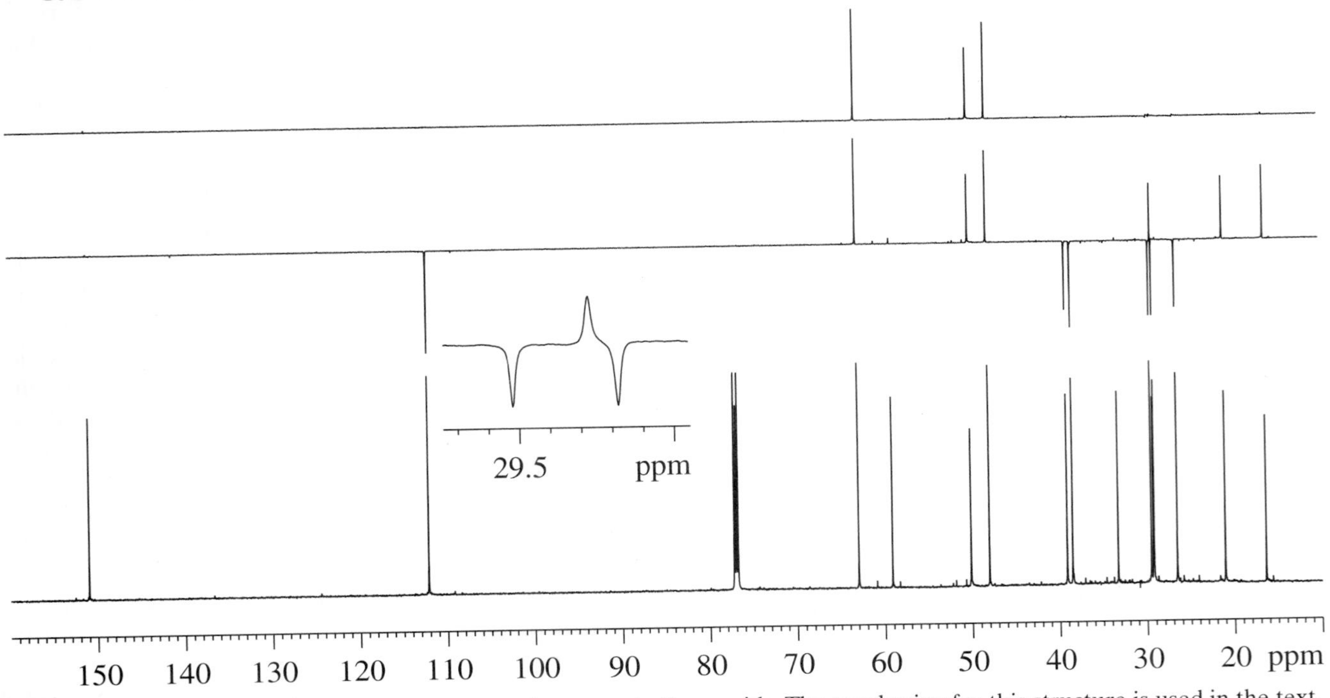

FIGURE 5.15 The ^{1}H, ^{13}C, and DEPT spectra for caryophyllene oxide. The numbering for this structure is used in the text.

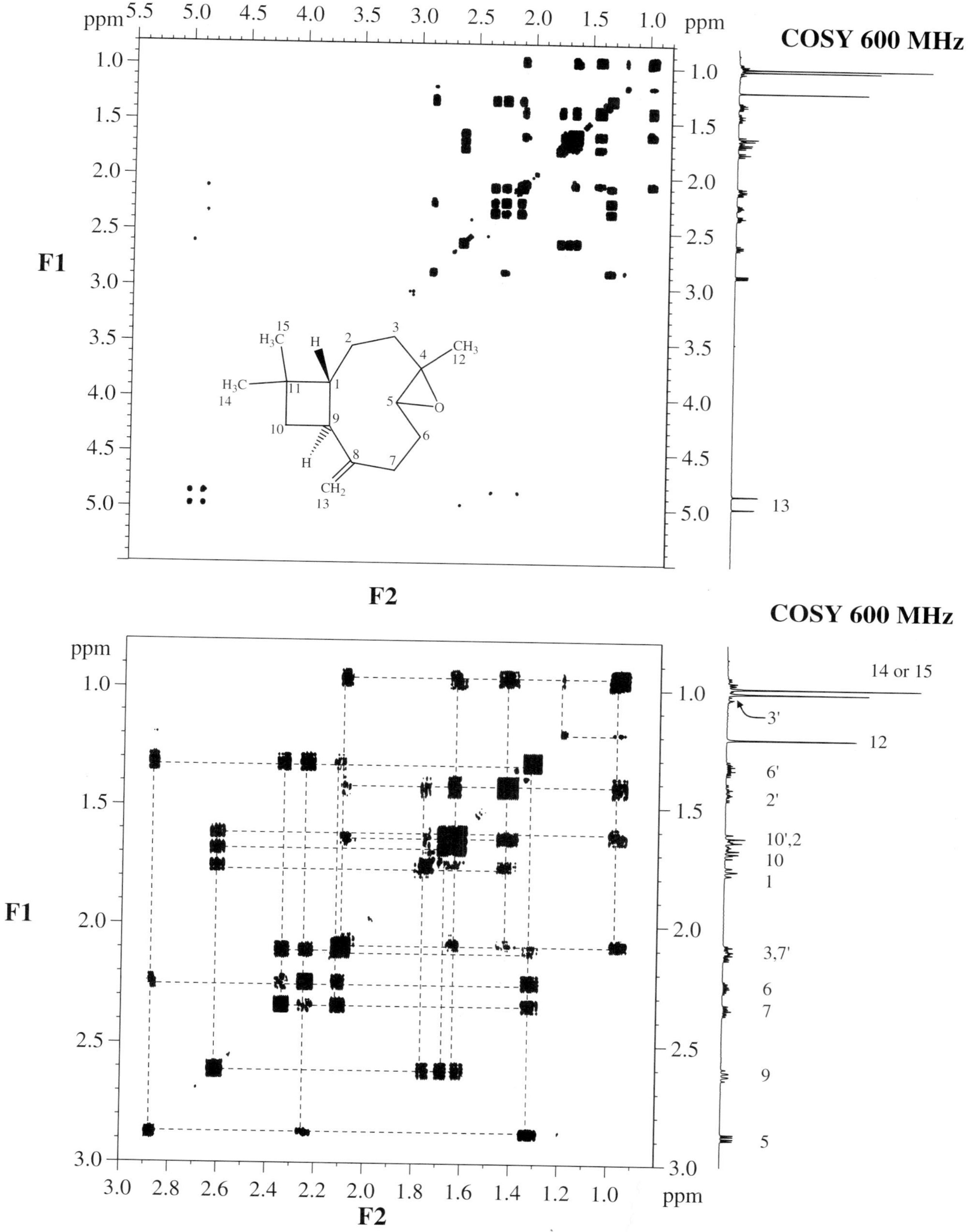

FIGURE 5.16 The DQF COSY spectrum for caryophyllene oxide. The lower portion is an expanded view with correlation lines drawn in and assignments are given as an aid.

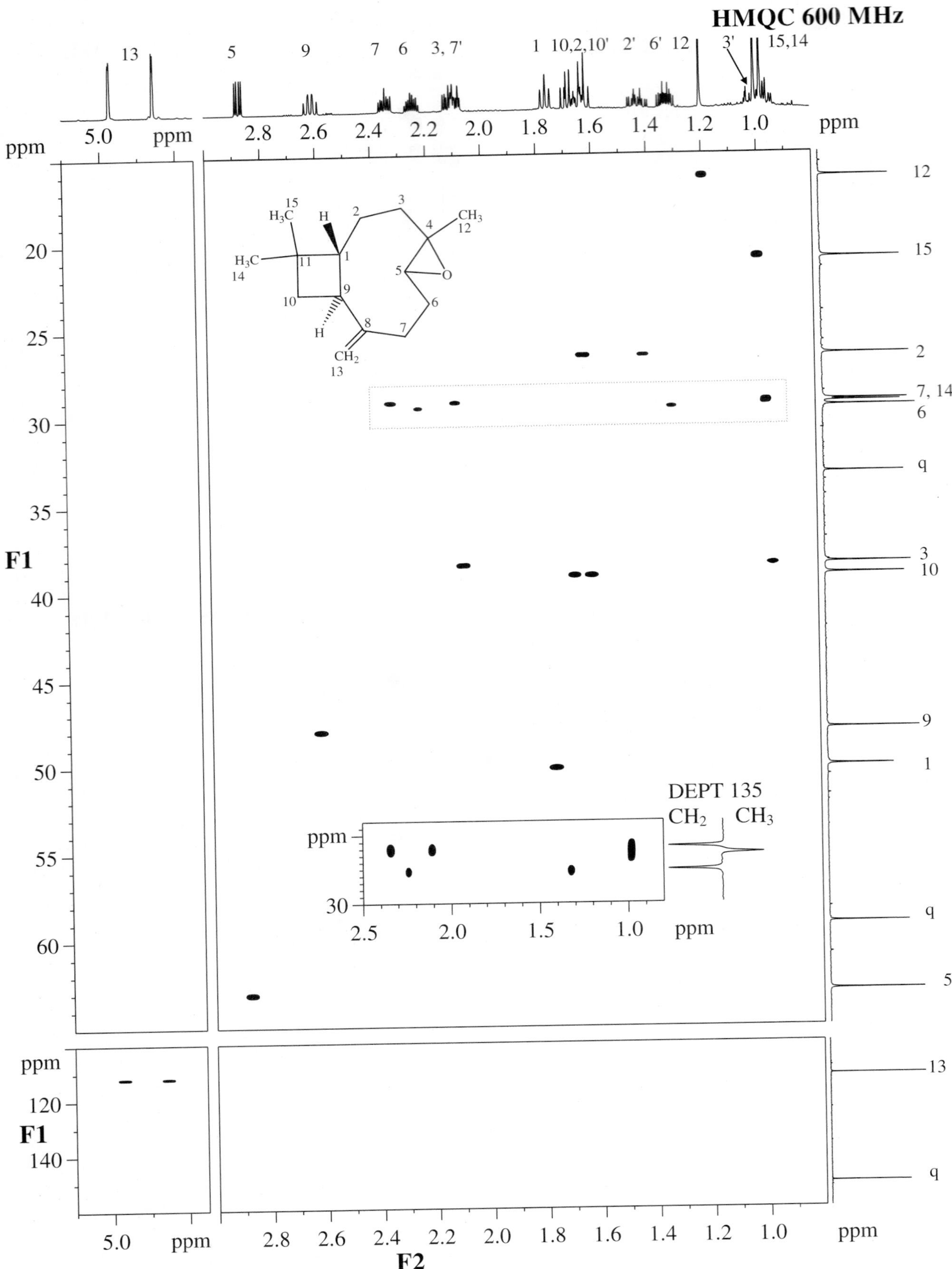

FIGURE 5.17 The HMQC spectrum for caryophyllene oxide. In place of the usual ^{13}C spectrum, the inset uses the DEPT 135 for better clarity.

and 2.37 ppm, the methylene carbon at 29.5 ppm correlates with proton resonances at 1.33 and 2.23 ppm, the methylene carbon at 38.4 ppm correlates with proton resonances at 0.96 and 2.09 ppm, the methylene carbon at 39.1 ppm correlates with proton resonances at 1.62 and 1.68 ppm, and we have already assigned the olefinic methylene group above. Thus, with little effort we have assigned a chemical shift for all of the protons in caryophyllene oxide and correlated them with a resonance from the ^{13}C spectrum; we have grouped the diastereotopic protons together for each of the methylene groups; and we have obtained three separate entry points for the COSY spectrum when before we had none. We are now ready to return to the analysis of the COSY spectrum of caryophyllene oxide and assign the correlations in light of the structure.

An expanded section from 0.8 to 3.0 ppm of the DQF-COSY of caryophyllene oxide can be found in the bottom part of Figure 5.16. Included in this portion of the figure are lines connecting proton-proton correlations to aid our discussion. The COSY "connectivities" allow us to construct structure fragments, or in this case, confirm structural segments. To correlate C-5, C-6, and C-7, we start with H-5 at 2.87 ppm. This proton shows cross peaks with two resonances at 1.32 and 2.24 ppm. From the HMQC, we know that these are diastereotopic and assign them as H-6 and H-6′. The protons attached to C-6 give correlations with protons at 2.11 and 2.37 ppm; we assign these protons, which also are diastereotopic, to C-7 at 29.2 ppm. The C-7 protons are coupled to each other, as certainly are the C-6 protons.

Other correlations are also straightforward. The C-5, C-6, C-7 spin system is isolated so we must select another entry point. We can start again with the allylic bridgehead methine (H-9) at 2.60 ppm. We have noted already the long-range allylic interaction. In addition, we find three other interactions that the HMQC helps us to assign. One of the correlations is to a methine proton at 1.76 ppm, which we assign to H-1. The other two correlations find two diastereotopic protons (again, from HMQC) at 1.62 and 1.68 ppm; we assign them as H-10 and H-10′. The C-10 protons are a dead end, and we find no other correlations to them.

H-1 shows coupling to both C-2 protons at 1.45 and 1.63 ppm. The interaction is weak between H-1 and H-2′ at 1.63 ppm. Both C-2 protons are coupled to both C-3 protons at 0.95 and 2.06 ppm and the appropriate cross peaks can be found. Thus, we have shown indirect connectivities from C-10 through C-9, C-1, and C-2 all the way to C-3. The HMQC has been invaluable in our interpretation. However, many questions still remain. We neither have correlations to the three quaternary carbons nor to the three methyl groups. The HMQC and the COSY together *support* the structure for caryophyllene oxide, but they do not preclude other possible structures.

5.5.3 Caryophyllene Oxide: HMBC

The HMBC for caryophyllene oxide (Figure 5.18) allows us to completely confirm the structure of caryophyllene oxide by giving us the required indirect carbon-carbon connectivities. An analysis of the structure of caryophyllene oxide reveals that there should be 87 cross peaks; this number is derived from considering each of the 15 carbon atoms and counting the number of chemical-shift-distinct protons at the α-positions and the number of chemical-shift-distinct protons at the β-positions. In order to keep track of all of those interactions, one must be methodical indeed.

One way to keep track of these data is to construct a table listing the carbon resonances in one direction and the proton resonances in the other. In Table 5.1, the carbons are given across the top and protons along the side. The numbering for caryophyllene oxide is the same in Table 5.1 as in all the figures.

Our approach for this spectrum is no different from any other spectrum. In this case, it is easier to start on the proton axis and look for the required cross peaks to the carbons as listed in Table 5.1. If we wished to start on the carbon axis, we would, of course, obtain the same results. If we begin at the top left of the table with H-1 at 1.76 ppm, we see first that H-1 is bonded to C-1 at 50.1, a result that we have already determined with HMQC. Going across the row, we find a total of eight interactions are expected. In the table, each interaction is labeled either α or β depending upon whether it is due to a two-bond coupling ($^2J_{CH}$) or a three-bond coupling ($^3J_{CH}$). Of course, in the spectrum itself, there is no differentiation of the two types of interactions; we label them that way for our own bookkeeping efforts. Each of the interactions for H-1 designated in the table is found in the spectrum.

There are two protons on C-2, which are labeled H-2 and H-2′; these protons have different chemical shifts, yet we expect them to act much the same way in the HMBC. Thus, we have a useful independent check of our HMBC assignments for each pair of diastereotopic protons in caryophyllene oxide. For H-2 at 1.45 ppm, we have the same five correlations that we have for H-2′ at 1.63 ppm. As we study the spectrum and the table more closely, we find that we have exquisitely detailed structural information that can be deciphered with a methodical approach.

An important point about quaternary carbons requires comment. Until now, we have had no direct correlations for carbons without protons, nor have we been able "to see through" heteroatoms such as oxygen, nitrogen, sulfur, etc. Both the two- and three-bond coupling correlations of HMBC provide us with both types of critical information. For example, C-4 of caryophyllene oxide at 59.1 ppm has no attached protons, and so far it has only appeared in the ^{13}C spectrum of the compound, and we know that it is quaternary

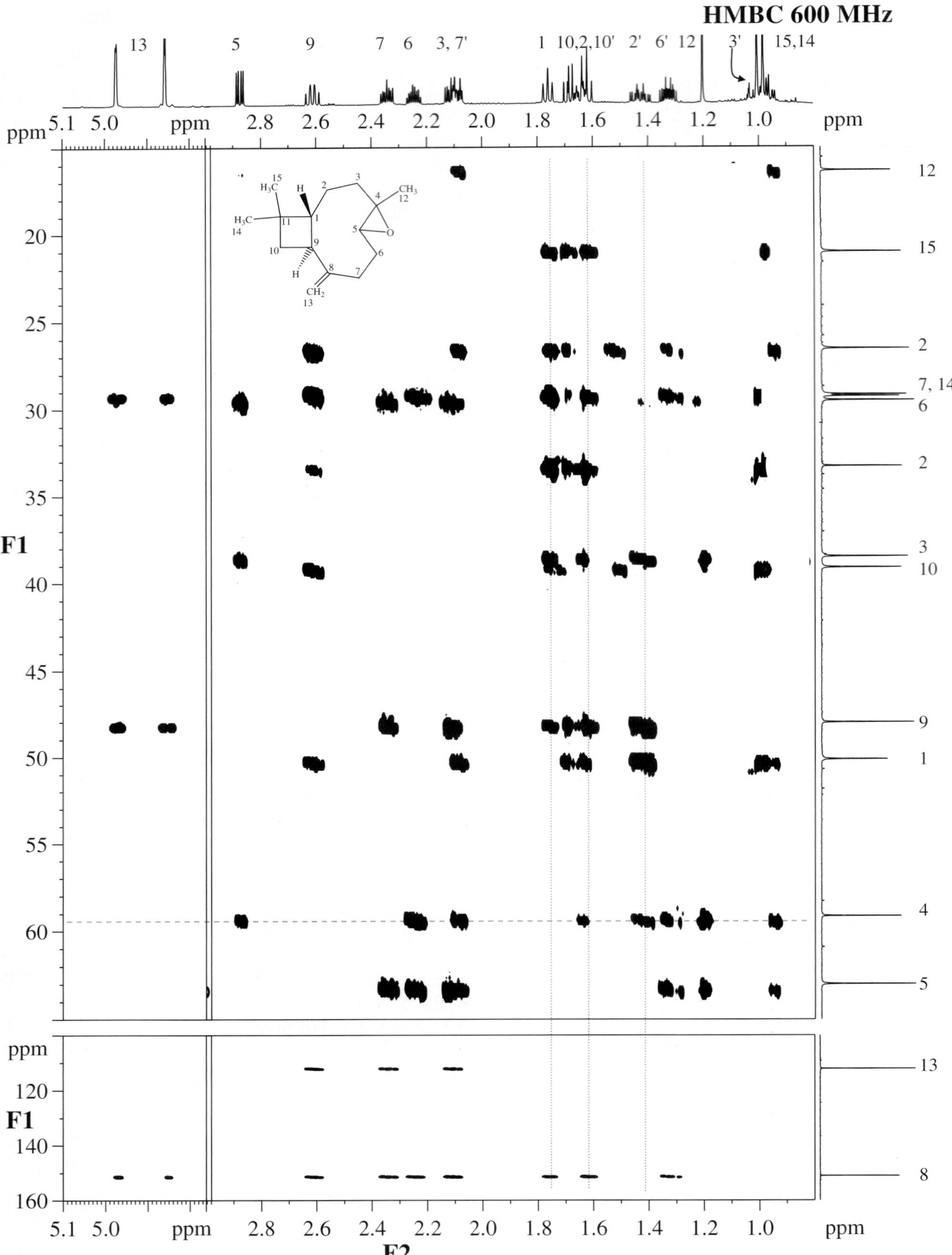

FIGURE 5.18 The HMBC spectrum for caryophyllene oxide.

TABLE 5.1 HMBC Correlations for Caryophyllene Oxide

	Carbon	**C-1**	**C-2**	**C-3**	**C-4**	**C-12**	**C-5**	**C-6**	**C-7**	**C-8**	**C-13**	**C-9**	**C-10**	**C-11**	**C-15**	**C-14**
Proton	**PPM**	**50.1**	**26.6**	**38.4**	**59.1**	**16.4**	**63.0**	**29.5**	**29.2**	**151**	**112**	**48.0**	**39.1**	**33.3**	**22.6**	**29.3**
H-1	1.76	DB*	α	β						β		α	β	α	β	β
H-2	1.45	α	DB	α	β	β						β		β		
H-2′	1.63	α	DB	α	β	β						β		β		
H-3	0.95	β	α	DB	α	β	β									
H-3′	2.06	β	α	DB	α	β	β									
H-12	1.19			β	α	DB	β									
H-5	2.86			β	α	β	DB	α	β							
H-6	1.28				β		α	DB	α	β						
H-6′	2.23				β		α	DB	α	β						
H-7	2.11						β	α	DB	α	β	β				
H-7′	2.37						β	α	DB	α	β	β				
H-13	4.86								β	α	DB	β				
H-13′	4.97								β	α	DB	β				
H-9	2.6	α	β						β	α	β	DB	α	β		
H-10	1.43	β								β		α	DB	α	β	β
H-10′	1.47	β								β		α	DB	α	β	β
H-15	0.98	β											β	α	DB	β
H-14	1.01	β											β	α	β	DB

*Directly Bonded proton-carbon, which are not seen in the HMBC

from DEPT. If we look in Table 5.1 at the C-4 column, we find four two-bond correlations and four three-bond correlations. The HMBC spectrum bears out these expectations fairly well and gives us direct evidence of the C-4 position in the molecule.

5.6 $^{13}C—^{13}C$ CORRELATIONS: INADEQUATE

The HMBC experiment allows us to trace the skeleton of organic compounds by way of indirect carbon-carbon connectivities, but the process is tedious because we do not know whether the correlations are due to two- or to three-bond couplings. The 2-D experiment, **INADEQUATE** (**I**ncredible **N**atural **A**bundance **D**oubl**E** **QUA**ntum **T**ransfer **E**xperiment), completes our set of "basic" through-bond correlations; we have COSY for proton-proton coupling, HMQC (HETCOR) for one-bond and HMBC for two- and three-bond proton-carbon coupling, and now INADEQUATE for directly attached (one-bond) carbon-carbon couplings. For elucidation of the structure of organic compounds, this experiment is, without question and without exception, the most powerful and the least ambiguous available, and, to top it off, the experiment is easy to interpret. After reading that last statement, the naturally pessimistic among us inevitably will ask: what's the "catch?" Indeed, the "catch" is plain and simple: sensitivity. Recall from Chapter 4 that the probability of any one carbon atom being a ^{13}C atom is about 0.01. Thus, the probability that any two adjacent carbon atoms will both be ^{13}C atoms (independent events) is 0.01×0.01 or 0.0001; in rounded whole numbers, that is about 1 molecule in 10,000!

This seemingly impossible task is accomplished with the aid of double quantum filtering. We recall from our DQF-COSY experiment that double quantum filtering removes all single spin transitions, which in this case corresponds to isolated ^{13}C atoms; only those transitions from systems with two spins (AB and AX systems) and higher* are detected during acquisition. The main problem facing us experimentally is sample size, assuming that the compound has the required solubility in an appropriate lock solvent. For low molecular weight compounds (atomic weight <500) run on a modern high field spectrometer, about 50–100 mg dissolved in 0.5 ml of a deuterated solvent is appropriate.

One way to imagine this experiment is as a carbon analogue of DQF-COSY in which both F1 and F2 would be carbon axes, and theoretically this experiment is possible. For practical considerations related to obtaining complete double quantum filtration, the INADEQUATE experiment is run slightly differently. In the display of the INADEQUATE spectrum of caryophyllene oxide (Figure 5.19), we find that the F2 axis is the familiar carbon axis, which we can, of course, relate to t_2 acquisition. The F1 axis looks unfamiliar and requires further explanation.

During t_1, the frequencies that evolve are not the chemical shifts of the coupled nuclei as they are in a typical DQF-COSY. Instead, it is the *sum* of their offsets

* Following the same reasoning as above, the probability of a three-spin system in an unenriched sample is 1 in 1,000,000.

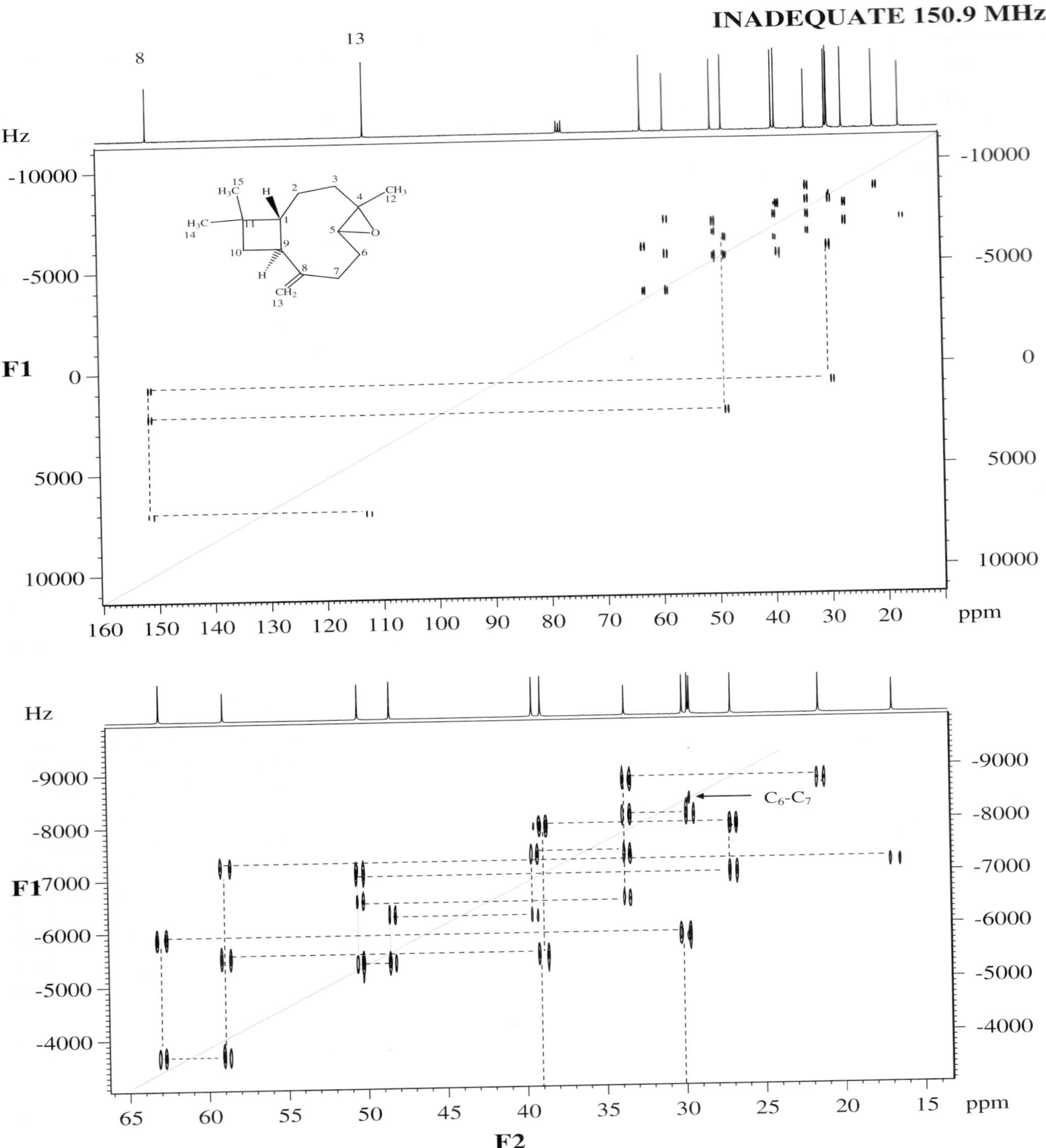

FIGURE 5.19 The INADEQUATE spectrum of caryophyllene oxide. Correlation lines are drawn as an aid.

from the transmitter frequency of the coupled nuclei that evolve during t_1, and, because they are double quantum filtered, it is only the two-spin AB and AX systems that contribute significantly to the intensity of cross peaks in the INADEQUATE spectrum. Proper selection of experimental parameters in the pulse sequence allows us to select the larger one-bond couplings ($^1J_{CC}$) thus ensuring that we are only looking at directly bonded carbon-carbon correlations. The F1 axis is usually given in Hz and it is two times the range in F2, whose units are ppm.

5.6.1 INADEQUATE: Caryophyllene Oxide

The 2-D INADEQUATE spectrum of caryophyllene oxide is presented in Figure 5.19. Cross peaks or correlations are found at $(\nu_A + \nu_X, \nu_A)$ and at $(\nu_A + \nu_X, \nu_X)$ in the (F1, F2) coordinate system for a given AX system. The actual cross peaks themselves are doublets (see the expanded portion of the spectrum, bottom of Figure 5.19) with a spacing equal to the $(^1J_{CC})$ coupling constant. The midpoint of the line connecting the two sets of doublets is $(\nu_A + \nu_X)/2$, $(\nu_A + \nu_X)$; thus, the collection of midpoints for all of the pairs of doublets lie on a line running along the diagonal. This is an important observation because it can be used to distinguish genuine cross peaks from spurious peaks and other artifacts.

With a better understanding of the F1 axis and the "diagonal," we can proceed with interpretation of the spectrum. Table 5.1 lists carbon chemical shifts and carbon numbers based on the structure given earlier; we refer to these numbers in the present discussion. From Figure 5.19, we can make the high-frequency connections quite easily. The carbon at highest frequency is C-8 at 151.0 ppm; by tracing vertically down from this peak on the F2 axis, we intersect three cross peak doublets. These cross peaks "connect" horizontally with C-7 at 29.2 ppm, C-9 at 48.0 ppm, and C-13 at 112.0 ppm. Toward lower frequencies, C-13 at 112.0 ppm comes next, and it has only one cross peak, namely the reciprocal connection to C-8 at 151.0 ppm.

In order to present the low-frequency section more clearly, the lower portion of Figure 5.19 shows an expanded view of that area. The higher resolution of this figure enables us to see the doublet fine structure more readily. Let us trace one carbon's connectivities from this expanded view. C-11, at 33.3 ppm, is a quaternary carbon, and it accordingly shows four cross peaks. We have connectivities from C-11 to C-15, at 22.6 ppm, C-14 at 29.3 ppm, C-10 at 39.1 ppm, and C-1 at 50.1 ppm.

Before we conclude our discussion, we note that the INADEQUATE spectrum of caryophyllene oxide contains an uncommon phenomenon worth exploring. Carbons 6 and 7 of caryophyllene oxide nearly overlap in the ^{13}C spectrum with each other and with the C-15 methyl; we list their chemical shifts from Table 5.1 as 29.5 and 29.2 ppm. Because they are bonded to one another in caryophyllene oxide, they should show correlation in the INADEQUATE spectrum, but, instead of an AX system, we have an AB system with $\Delta\nu/J$ being much less than ten. For this special case, we no longer expect two doublets whose midpoint lies on the diagonal; instead, we predict that an AB multiplet (see Chapter 3) should fall on the diagonal line itself. This prediction is borne out in Figure 5.19 where we find a cross peak directly below C-6 and C-7, and this cross peak intersects the diagonal line.

The other connectivities found in the expanded spectrum are left to the reader as an exercise. We summarize this section with two points:

- 2-D INADEQUATE provides direct carbon connectivities enabling us to sketch the carbon skeleton unambiguously.
- 2-D INADEQUATE has very limited applicability due to its extremely low sensitivity.

5.7 LACTOSE

The structure for the β-anomer of lactose is given in Figure 5.1. The challenges that lactose presents in the interpretation of its ^{1}H and ^{13}C NMR spectra (Figure 5.20) are obvious, but the opportunities for correlation NMR are irresistible. In solution, lactose is an equilibrium mixture of α- and β-anomers. The two diastereomers are epimeric at only one stereocenter out of ten. In addition, the protons of the two sugar residues in each diastereomer are "insulated" from each other by the glycosidic oxygen atom forming isolated spin systems. This situation is common to all oligo- and polysaccharides.

We are not going to spend too much time discussing the 1-D spectra except to note some of the obvious features. The anomeric proton resonances can be found at 4.45, 4.67, and 5.23 ppm and the anomeric carbon resonances at 91.7, 95.6, and 102.8 ppm. The reason for three anomeric protons and carbons is that the α- and β-anomers of glucose give two sets of resonances while the galactose residue, which exists only in the β-form, gives a single set of resonances in both the proton and carbon spectra. The other portions of both spectra, especially the proton, are quite complicated and show considerable overlap.

5.7.1 DQF-COSY: Lactose

The DQF-COSY spectrum for lactose (Figure 5.21) is rich with correlations, and entry points are easy to find. This figure and others of lactose use a simplified notation in which galactose resonances are labeled G*n*, where *n* is the proton or carbon position, and either αn or βn for positions in the α-glucose residue and the β-glucose residue, respectively. Each of the anomeric protons, which are the protons attached to C-1 for each sugar residue, show one and only one correlation to their respective C-2 protons. For instance, the anomeric proton at 4.67 ppm shows a correlation to a proton (obviously attached to C-2) at 3.29 ppm. This C-2 proton at 3.29 ppm shows a correlation to a C-3 proton at 3.64 ppm.

Continuation of this process quickly becomes dreadfully complicated because of the severe overlap of signals. Many of the correlations have been drawn in with different types of lines for each of the different residues in the expanded view. The reader is invited to trace some of these correlations but cautioned to be wary of frustration.

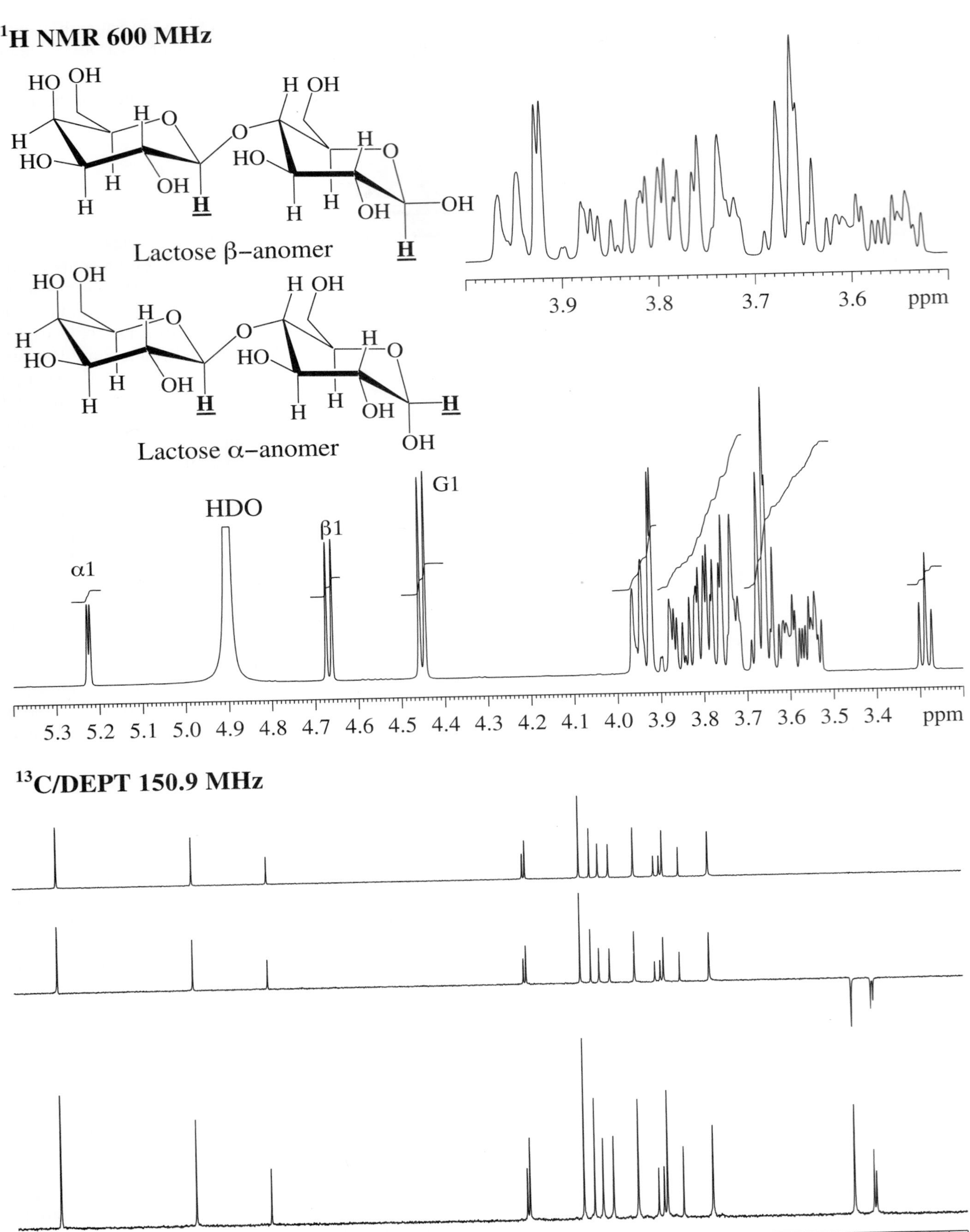

FIGURE 5.20 The ^{1}H, ^{13}C, and DEPT spectra of lactose. In solution, lactose is an equilibrium mixture of anomers.

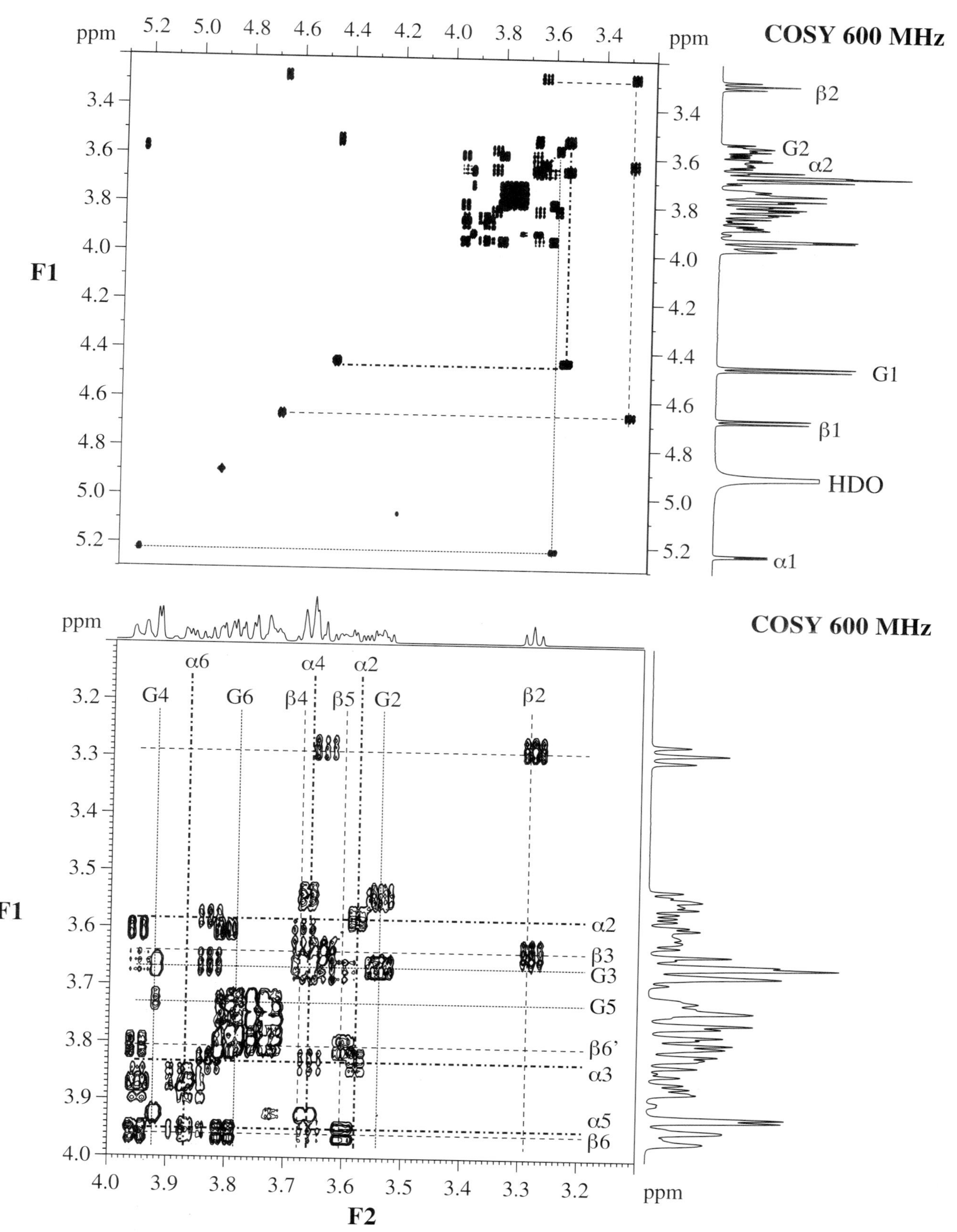

FIGURE 5.21 The DQF COSY spectrum for lactose. See the text for an explanation of the shorthand notation for proton resonances. Correlation lines are drawn in and assignments are given as an aid.

Below, lactose will again be analyzed in light of other experiments that will literally remove the overlap.

5.7.2 HMQC: Lactose

The HMQC spectrum of lactose is shown in Figure 5.23 in three sections to minimize overlap and to show good resolution on the carbon axis. As with the COSY spectrum, this figure is labeled with assignments. These assignments were made with information beyond what we have seen so far. (By the end of our discussion of lactose, it will be clear that these assignments are correct, and it should be obvious how these assignments were made.) Overlap is less severe so that some of the assignments can be made quite easily. This spectrum is useful in that it allows us to find the chemical shift of most of the protons, many of which are overlapping in the 1-D spectrum.

5.7.3 HMBC: Lactose

Considering the complexity of lactose, a complex and overlapping HMBC spectrum for lactose is expected. The spectrum found in Figure 5.24 measures up to this expectation. Students are encouraged to analyze this spectrum, which is facilitated with the labeling.

An important correlation in this spectrum is highlighted in the expanded inset that shows the correlation between C-4 of glucose (from both α- and β-anomers) and the C-1 proton of galactose. This three-bond coupling is extremely important because it shows that the glycosidic linkage is indeed from galactose C-1 to glucose C-4. Because there is no overlap in this part of the spectrum, our conclusion is unambiguous. The "reciprocal" correlation between the C-1 carbon of galactose and the C-4 protons of glucose (both anomers) is most likely there, but, because there is overlap among G3 (which also has a correlation to C-1 of galactose), α4, and β4 protons, our conclusion is ambiguous.

5.8 RELAYED COHERENCE TRANSFER: TOCSY

The common theme so far in our correlation experiments has been to allow spins to evolve during t_1 under the influence of directly coupled nuclear spins. We have seen the power of COSY, HMQC, HMBC, and INADEQUATE to provide us with detailed structural information for ipsenol, caryophyllene oxide, and lactose. In this section, we will develop another method for showing correlations and apply it to molecules with distinct, isolated proton spin systems such as carbohydrates, peptides, and nucleic acids.

Our goal is to "relay" or to transfer magnetization beyond directly coupled spins thus enabling us to see correlations among nuclei that are not directly coupled but within the same spin system. The experiment is called **TOCSY** (**TO**tally **C**orrelated **S**pectroscop**Y**) and we will consider both the 2-D and 1-D versions. The pulse sequence for a 2-D TOCSY resembles our prototype 2-D experiment, but, instead of a second $\pi/2$ pulse, we insert a "mixing period" during which the magnetization is "spin locked" on the y-axis (Figure 5.22). To understand the outcome of the experiment, we can ignore the particulars of spin locking and concentrate on the consequences of the mixing period. During this mixing period, magnetization is relayed from one spin to its neighbor and then to its next neighbor and so on. The longer the mixing period, the further the transfer of magnetization can propagate traveling, in theory, throughout an entire spin system.

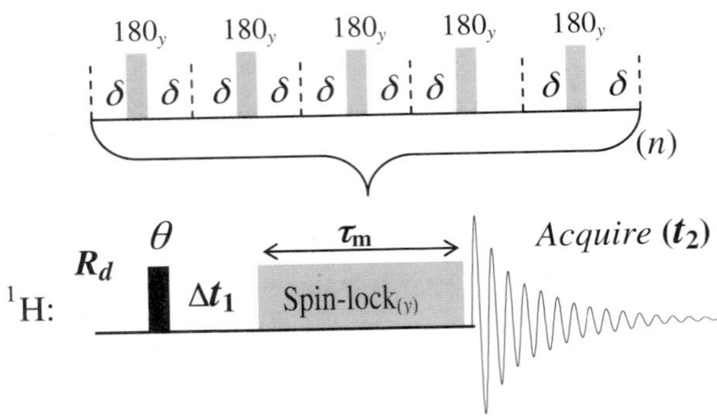

FIGURE 5.22 The pulse sequence for 2-D TOCSY.

The appearance of a 2-D TOCSY experiment resembles in all aspects a COSY. The F1 and F2 axes are for proton; the diagonal contains 1-D information; and even the cross peaks have the same appearance. The difference here is that the cross peaks in a COSY are due to coupled spins while the cross peaks in the TOCSY spectrum arise from relayed coherence transfer. For long mixing times in a TOCSY spectrum, all spins within a spin system appear to be coupled. To appreciate the advantages of TOCSY, we continue with the disaccharide lactose, which has three distinct (i.e., separate) spin systems.

5.8.1 2-D TOCSY: Lactose

The 2-D TOCSY spectrum of lactose is given in Figure 5.25. The mixing time for this 2-D spectrum was sufficiently long that magnetic coherence has been transferred more or less throughout each sugar residue's spin system. Compare this figure to the COSY spectrum for lactose in Figure 5.21 and note the similarities and differences.

As an example, we can find all of the proton resonances (and determine their chemical shifts) for the α-anomer of glucose by starting at its anomeric proton resonance at 5.23 ppm. These correlations have been

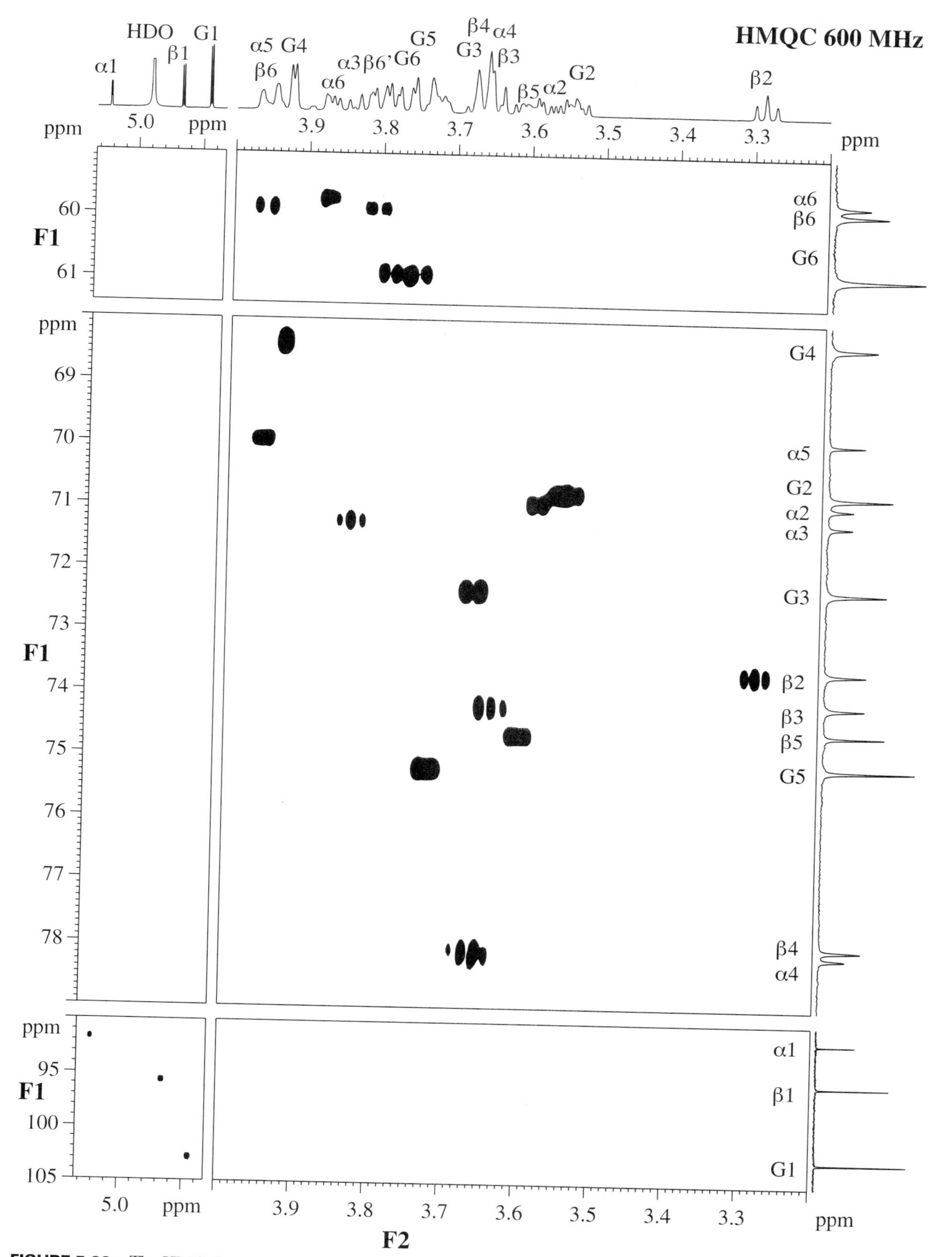

FIGURE 5.23 The HMQC spectrum for lactose. See text for an explanation of the notation. Assignments are given as an aid.

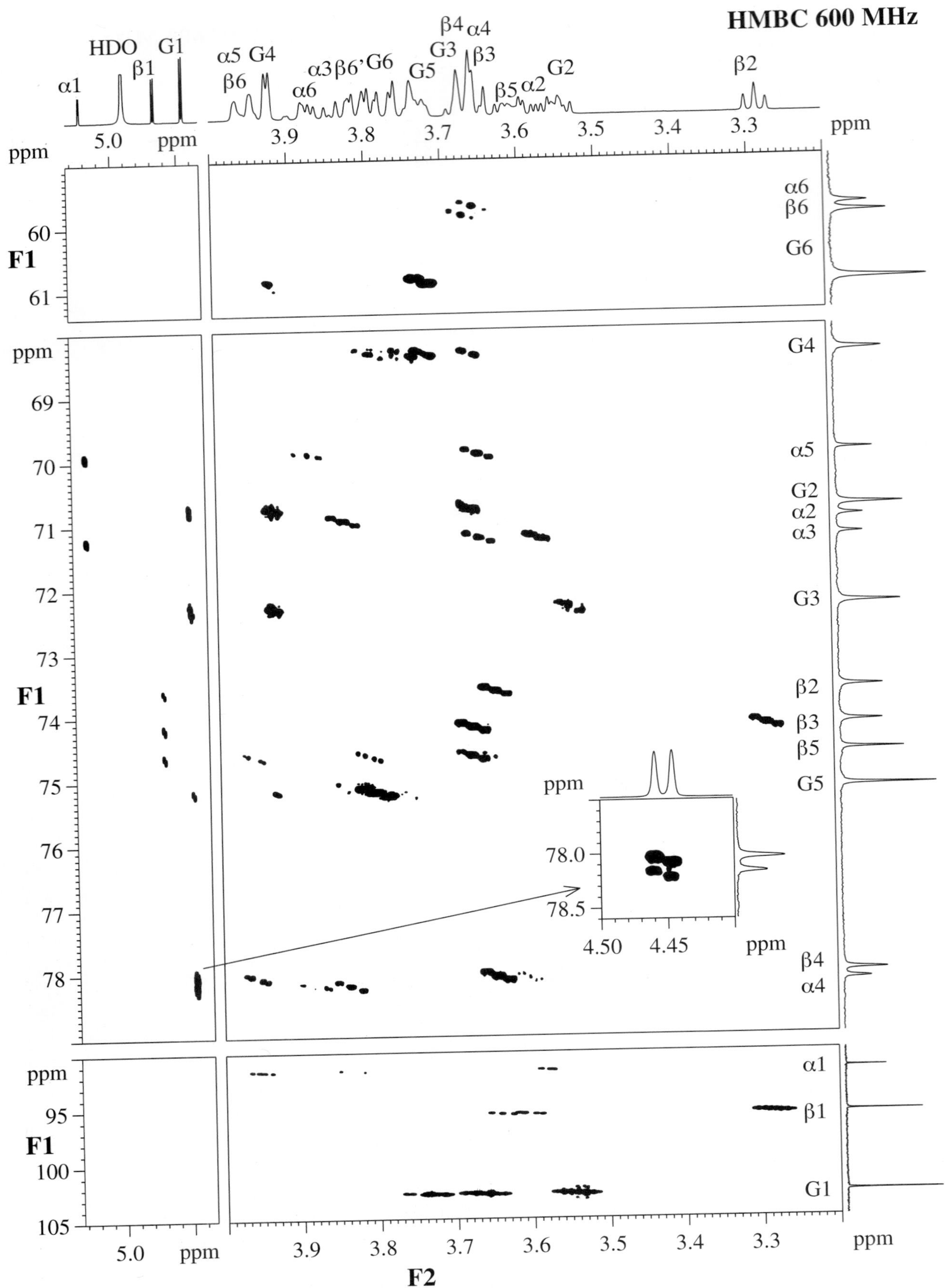

FIGURE 5.24 The HMBC spectrum for lactose. See text for an explanation of the inset.

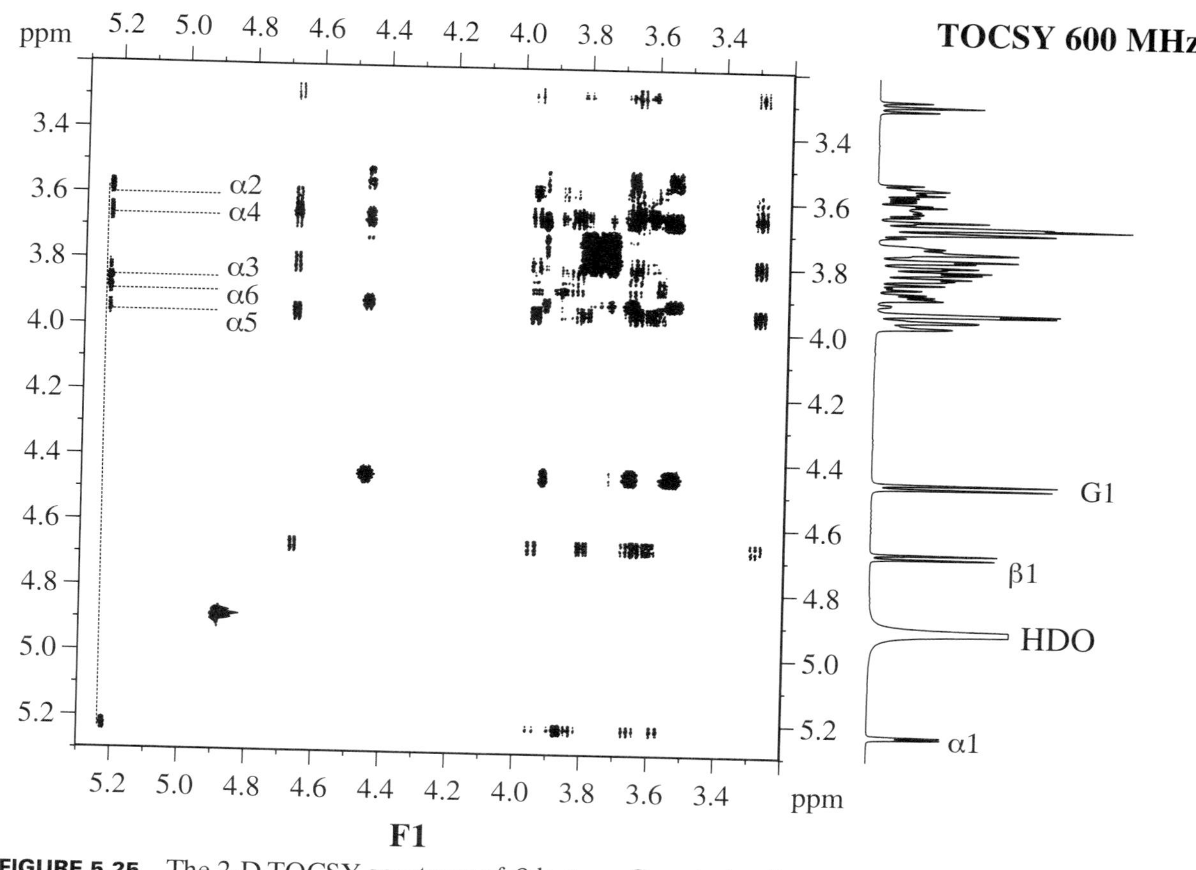

FIGURE 5.25 The 2-D TOCSY spectrum of β-lactose. Correlation lines are drawn in and some assignments are given as an aid.

marked in the figure. Assignment of the proton resonances (which are shown in the figure) cannot be made on the basis of this spectrum alone, but could be made in conjunction with the COSY spectrum. The same exercise can be carried out for the other two anomeric resonances, but these are left for the student.

5.8.2 1-D TOCSY: Lactose

Every 2-D experiment has a 1-D analogue and we tend to think that these 1-D experiments are less efficient, which they usually are. If we think again about our COSY experiment, we have said that homonuclear decoupling would give us the same type of information. We select a proton resonance, irradiate it, and compare the result with the original 1-D proton spectrum. In similar fashion for our 1-D TOCSY, often called HOHAHA (Homonuclear Hartmann-Hann), we select a proton resonance and irradiate it; we allow for an appropriate mixing time for the magnetization to be relayed during which we apply spin locking, and we acquire the 1-D spectrum. The only signals that will be recorded in this spectrum are those to which magnetization has been transferred. Put another way, all other signals that are outside the spin system do not appear.

An even better scenario is to run a series of 1-D TOCSY experiments in which the mixing time is systematically increased while the proton being irradiated is kept constant. To illustrate these experiments, we irradiated the anomeric proton from the β-anomer of the glucose ring in lactose at 4.67 ppm and run a series of experiments with mixing times ranging from 20 ms to 400 ms. The results of these experiments are shown in a series of stacked plots in Figure 5.26.

At a mixing time of 20 ms, we find only the β2 resonance, which is seen clearly as an apparent triplet at 3.29 ppm. After 40 ms of mixing time, transfer to β3 is readily discernable (another apparent triplet) and the β4 proton is just barely visible. A plot of the experiment with a mixing time of 80 ms reveals the β4 resonance a little better, while, after 120 ms, the signal from β5 has sprouted from the baseline. After 400 ms, transfer throughout the entire spin system is evident; the H-5 signal is robust as is the diastereotopic β6 methylene group. This part of the figure shows the β6 and β6′ resonances with clearly different coupling constants to β5. One negative

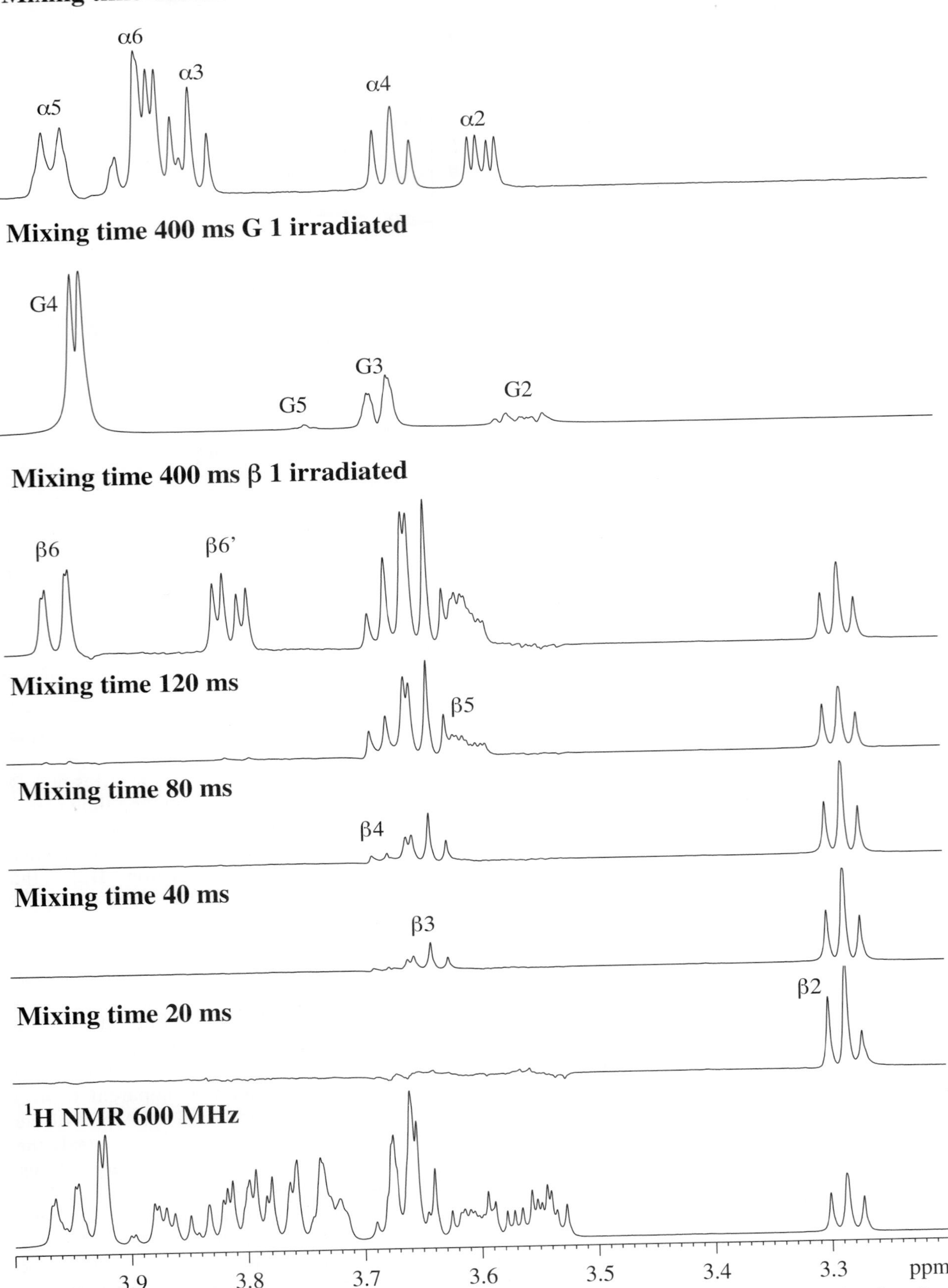

FIGURE 5.26 Stacked plots of a series of 1-D TOCSY experiments on β-lactose with increasing "mixing times." See text for an explanation. A portion of the ^{1}H NMR spectrum is reproduced for reference in the bottom plot.

aspect of long mixing times is that both resolution and signal are lost. We can usually offset signal loss by acquiring and summing more FIDs. Also shown in this figure are the 400 ms (longest mixing time) experiments for the galactose anomeric proton and the α-anomeric proton for glucose.

Both the 1-D and 2-D versions of TOCSY find wide application in deciphering overlapping signals that originate from different spin systems. The 1-D version is particularly exciting as it enables us to "walk" through a spin system as we systematically increase the mixing time.

5.9 HMQC–TOCSY

There are various "hybrid" 2-D correlation experiments that combine features of two simpler 2-D experiments. A popular and useful example is the HMQC–TOCSY spectrum that correlates one-bond $^{1}H—^{13}C$ couplings (HMQC) but shows these correlations throughout an entire spin system (TOCSY). This experiment simplifies complex carbohydrate and peptide systems and allows ready assignments of systems of protons and carbons.

5.9.1 HMQC–TOCSY: Lactose

The HMQC–TOCSY spectrum for lactose is given in Figure 5.28 with all of the proton and carbon resonances labeled. The overall appearance of this spectrum is reminiscent of an HMBC but the correlations are quite different. It is equally interesting and useful to start on the proton axis (F2) or the carbon axis (F1). If we start on the proton axis at 5.23 ppm, the anomeric proton for the α-anomer of glucose (α1), and proceed downward vertically, we find six correlations to the six carbons of this glucose residue. If we refer back to the "simple" HMQC spectrum for lactose, we find only one correlation for this proton. Likewise, the anomeric proton of the β-anomer of glucose at 4.67 ppm also shows six correlations to the carbons of its respective glucose residue.

Correlations to the anomeric proton of galactose (4.46 ppm), however, only show four interactions along the carbon axis. This result is consistent with the 1-D TOCSY of the galactose anomeric proton shown in Figure 5.26, where we find that coherence transfer does not travel beyond H-4 (G4). All six correlations are found if we start at H-4 (G4, 3.93 ppm) instead. As an exercise, try a similar process by starting on the carbon axis and tracing horizontally to the left to find HMQC–TOCSY correlations to protons. The anomeric carbon resonances are the easiest to try, but it is worthwhile to try others as well.

5.10 ROESY

In Chapter 3 (Section 3.16), there is a description of the nuclear Overhauser effect difference experiment, an experiment that provides information about $^{1}H—^{1}H$ through-space proximity. Review of this section is helpful before proceeding here. The ROESY experiment, rotating-frame Overhauser effect spectroscopy, is a useful 2-D analogue of the nuclear Overhauser effect difference experiment. This experiment is useful for molecules of all sizes whereas the related experiment, NOESY (nuclear Overhauser effect spectroscopy), is not very useful with small molecules. NOESY is used primarily with biological macromolecules. Both NOESY and ROESY experiments correlate protons that are close to each other in space, typically 4.5 Å or less.

Because ROESY correlates proton-proton (through space) interactions, its appearance and presentation resembles COSY. In fact, COSY peaks (spin–spin coupling) are present in ROESY spectra; these COSY peaks are superfluous and should be ignored. Occasionally, another complication arises from TOCSY-like transfer of magnetic coherence among J-coupled spins. The pulse sequence for a 2-D NOESY and a 2-D ROESY is given in Figure 5.27. The only difference between the two experiments is that the ROESY uses a spin lock during the mixing time τ_m.

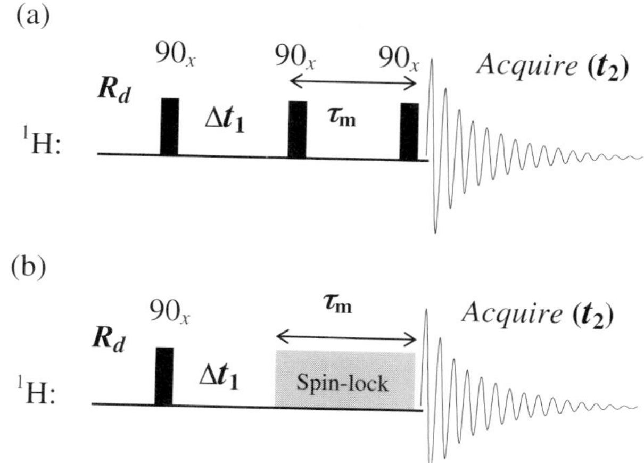

FIGURE 5.27 Pulse sequence for (a) 2-D NOESY and (b) 2-D ROESY.

5.10.1 ROESY: Lactose

The ROESY spectrum for lactose is given in Figure 5.29. Note the overall appearance in the upper part of the figure. In the lower part, the anomeric region is shown as expanded insets. The two glucose

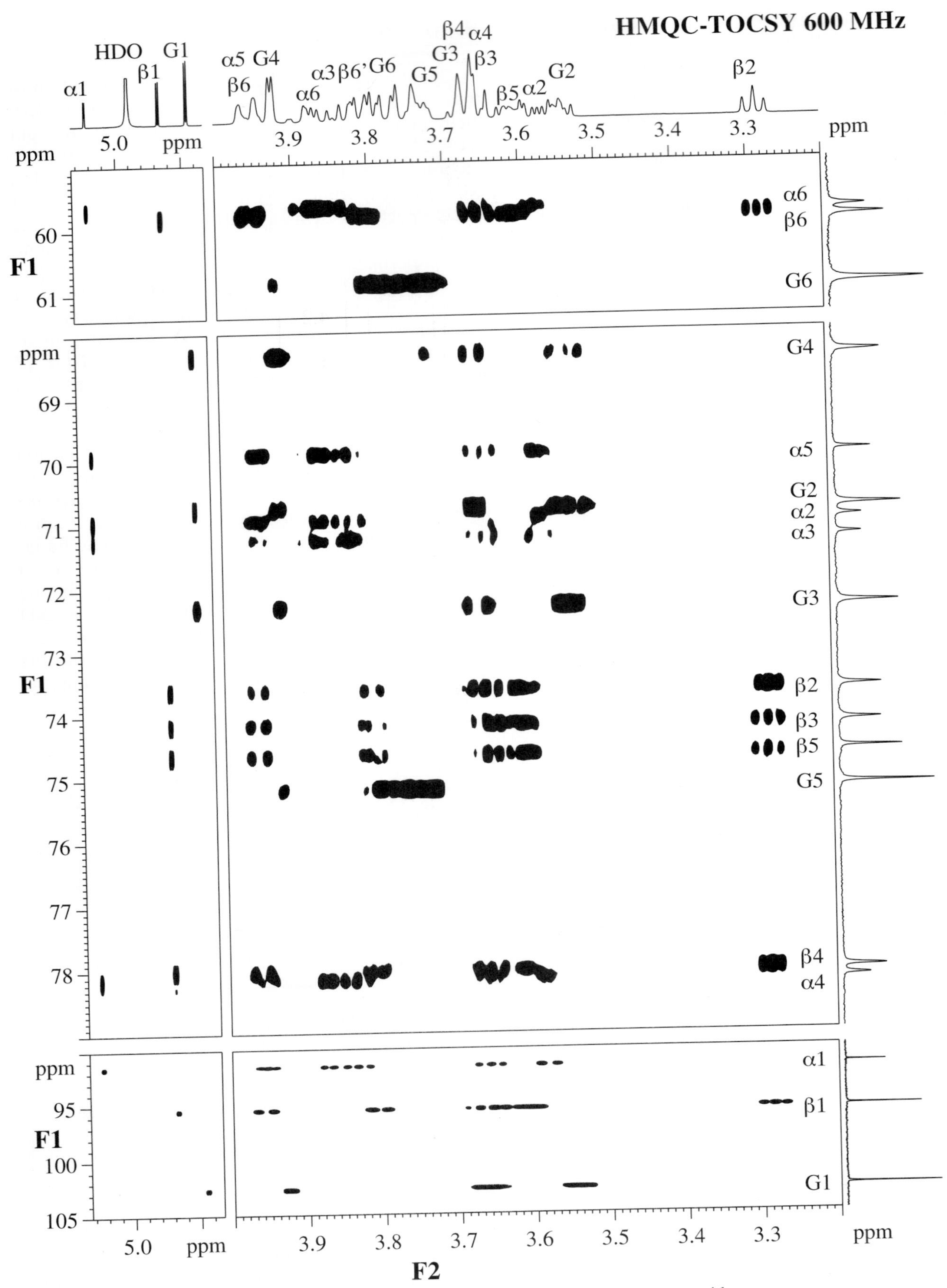

FIGURE 5.28 The HMQC–TOCSY spectrum for lactose. Assignments are given as an aid.

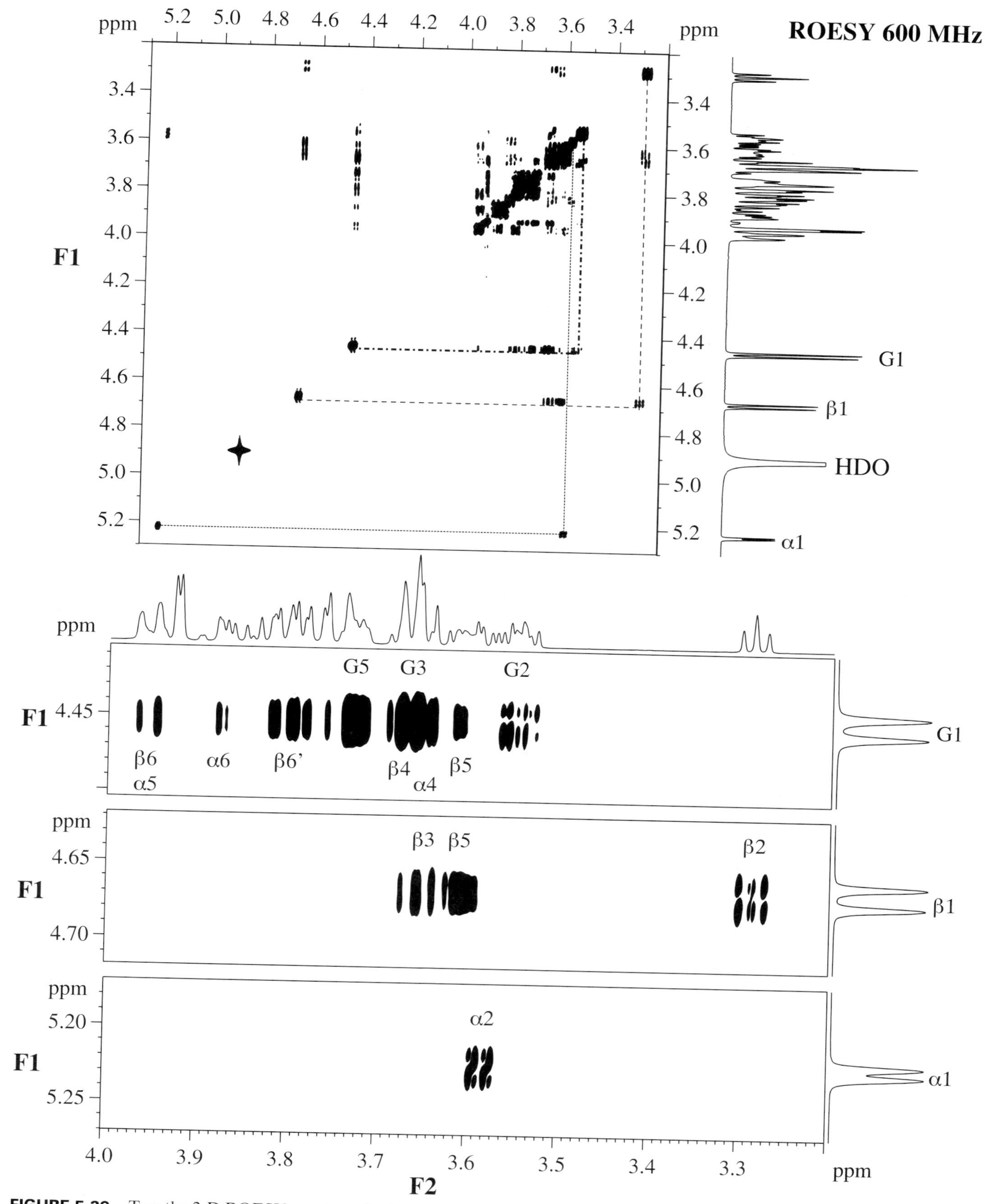

FIGURE 5.29 Top: the 2-D ROESY spectrum for lactose. Bottom: expansions of the three anomeric correlations.

residues are straightforward to interpret. The α-anomeric proton shows only one correlation and this is the expected COSY interaction. Recall that by definition, the α-anomer has its C-1 hydroxyl in the axial position and its C-1 proton ($\alpha 1$) in the equatorial position. In the equatorial position, there are no through space interactions likely and none are seen.

The other anomer of glucose, in contrast, has its anomeric proton ($\beta 1$) in the axial position. In the preferred chair conformation of glucose, protons occupy all of the other axial positions leading to presumed diaxial interactions between H-1 ($\beta 1$) and H-3 ($\beta 3$) and between H-1 ($\beta 1$) and H-5 ($\beta 5$). The ROESY spectrum reveals three interactions with the anomeric proton, $\beta 1$, at 4.67 ppm. The H-2 COSY interaction is, of course, present, and the two-diaxial NOE interactions to H-3 ($\beta 3$) and H-5 ($\beta 5$) are quite evident.

5.11 VGSE

Proteins or polypeptides are polymers or oligomers made from a limited number (although a rather large limited number >20) of (mostly) α-amino acids linked by amide or peptide bonds. A polypeptide example would be too difficult an undertaking at this point, but a tetra-peptide, on the other hand, is quite manageable while still illustrating most of the important features of a true polypeptide. Although many small, oligopeptides are found in nature, a plethora of small peptides are now manmade utilizing automated synthesis equipment.

The structure of the small peptide being used as an example in this chapter is given in Figure 5.1. Starting with the N-terminus, the peptide contains the amino acids valine, glycine, serine, and glutamic acid (VGSE) linked in the usual way. The 1-D NMR spectra for this compound (see Figure 5.30) reveal that the amino acids making up this compound are relatively simple allowing their individual components to be readily found. The positions in each amino acid have been labeled and assignments of the protons and carbons have been made. The assignments have been given in this figure to facilitate discussion; they cannot be made from these data alone.

One aspect of the experimental conditions under which these spectra were obtained is important to understand so that the spectra can be rationally interpreted. For solubility and stability purposes, peptides are generally dissolved in buffered water. Recall from Chapter 3 that compounds prepared for NMR experiments are almost always dissolved in deuterated solvents. The need for deuterated solvents is so that the spectrometer can remain "stable" for the duration of the experiment by way of the field/frequency lock. The "lock signal" comes from the deuterium NMR signal of the solvent. However, if the peptide sample were dissolved in buffered D_2O, all of the exchangeable protons, including the amide protons, would be replaced with deuterons and not seen. A compromise is necessary; the sample was dissolved in 95% buffered H_2O containing 5% D_2O. Note the presence of the three amide protons between 8.0 and 9.0 ppm. These amide resonances, as we shall see, are extremely important and they would be absent if the sample were dissolved in pure D_2O.

5.11.1 COSY: VGSE

The DQF-COSY spectrum of VGSE can be found in Figure 5.31. As with lactose, there are several good entry points, especially the three amide protons. Three of the four amino acid residues can be traced using these as starting points. The fourth amino acid can be traced starting with the methyl groups (the only ones in the molecule) at 1.0 ppm. Verify the correlations that have been drawn in Figure 5.31. Watch out for confusion between valine and serine.

5.11.2 TOCSY: VGSE

The 2-D TOCSY spectrum for VGSE is given in Figure 5.32. Like lactose, VGSE is composed of several isolated spin systems of protons; in this case there are four moieties. As an example of "total correlation," the correlations among the glutamic acid protons are drawn in the figure. As with the COSY spectrum, the amide protons are expedient points to initiate the process. Can you find an appropriate entry point for the valine, which has no amide proton?

In some ways, the information from the COSY spectrum of VGSE is complementary to its TOCSY spectrum, and in other ways the information is redundant. Both spectra allow us to individually assign all of the proton signals of VGSE in different ways. In this case, which spectrum furnishes the information more easily or more clearly?

5.11.3 HMQC: VGSE

Compared to caryophyllene oxide and lactose, the HMQC spectrum for the tetra-peptide appears relatively simple (Figure 5.33). Indeed, VGSE has only 10 carbon atoms with attached protons and the spectrum shows correlations to nine carbons. Actually, there are 10 correlations as can be seen in the inset of the shielded methyl portion of the spectrum. Let us summarize the complementary information up to and including the HMQC spectrum.

Let us start our analysis with the carbon resonance at 43 ppm in the HMQC spectrum, which is a —CH_2—

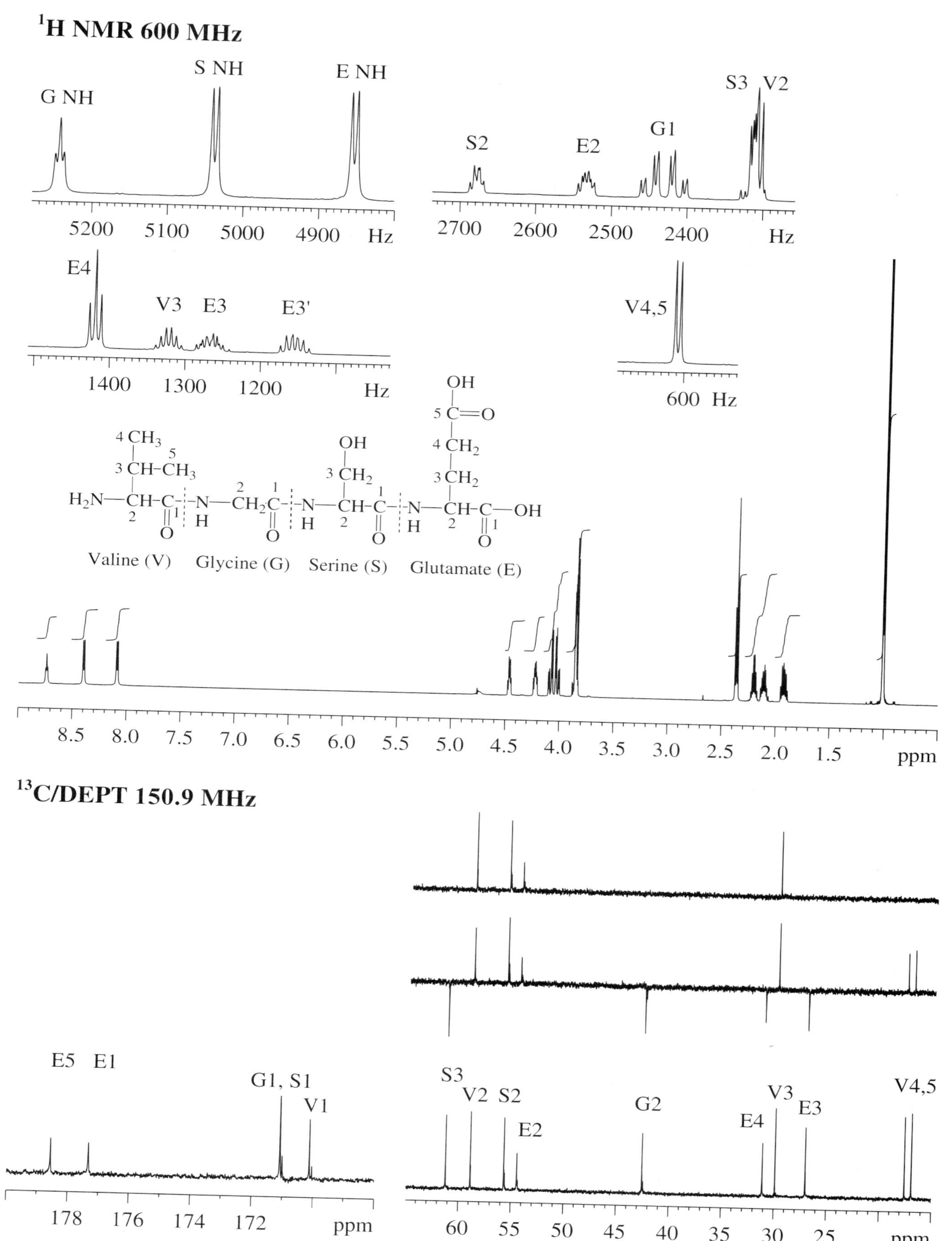

FIGURE 5.30 The ^{1}H, ^{13}C, and DEPT spectra for the tetra-peptide, VGSE, in 95% H_2O and 5% D_2O.

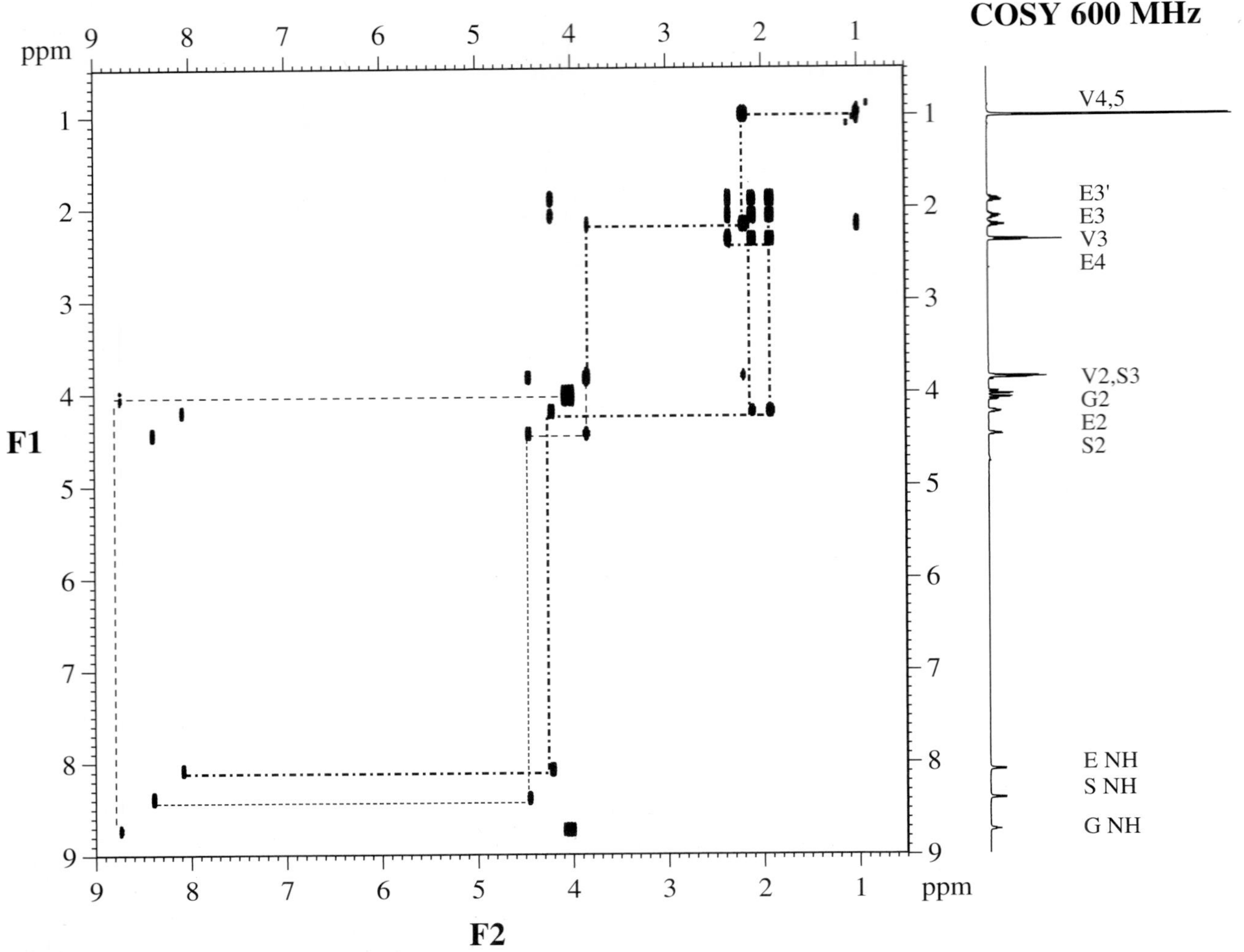

FIGURE 5.31 The DQF COSY spectrum of VGSE. Correlation lines are drawn in and some assignments are given as an aid.

group from the DEPT spectra, and note that it correlates with a diastereotopic methylene group centered at 4.04 ppm. From the DQF-COSY spectrum (see Figure 5.31), the multiplet at 4.04 ppm shows only one correlation to an amide proton at 8.75 ppm. Thus, the methylene group must belong to the glycine residue, and furthermore, the glycine residue cannot be the N-terminus, since the N-terminal amino acid must have a free amino group.

The serine residue can be accounted for by starting with the carbon resonance at 62 ppm. DEPT indicates another methylene group, and the HMQC shows that it correlates with coincident protons resonating at 3.85 ppm. These protons overlap with another proton. Careful line drawing in the DQF-COSY (or more easily in the TOCSY spectrum) suggests correlation with a proton at 4.48 ppm. This proton shows a correlation to a carbon atom at 56 ppm in the HMQC spectrum. That proton also shows a correlation to an amide proton at 8.40 ppm in the COSY. Like glycine, the serine residue cannot be the N-terminus.

Since valine is the only amino acid in the tetrapeptide that has methyl groups, and since the DEPT spectra show two methyl groups at about 17 and 18 ppm, we can safely start with those in the HMQC. Both methyl carbons correlate with the same doublet at 1.01 ppm clearly indicating that they show apparent "chemical shift equivalence" even though they are obviously diastereotopic. (Note that the integration of this doublet in the proton spectrum [Figure 5.30] corresponds to six protons.) The methyl groups are coupled to a methine proton at 2.20 ppm (COSY). In turn, the HMQC reveals a correlation from the methine proton to a carbon resonance at 31.5 ppm. The isopropyl methine (2.20 ppm) shows a further correlation in the COSY to a methine in the overlapping multiplet at 3.85 ppm. This proton shows a correlation in the HMQC to the carbon at 59 ppm. Although it is difficult to see, this methine does

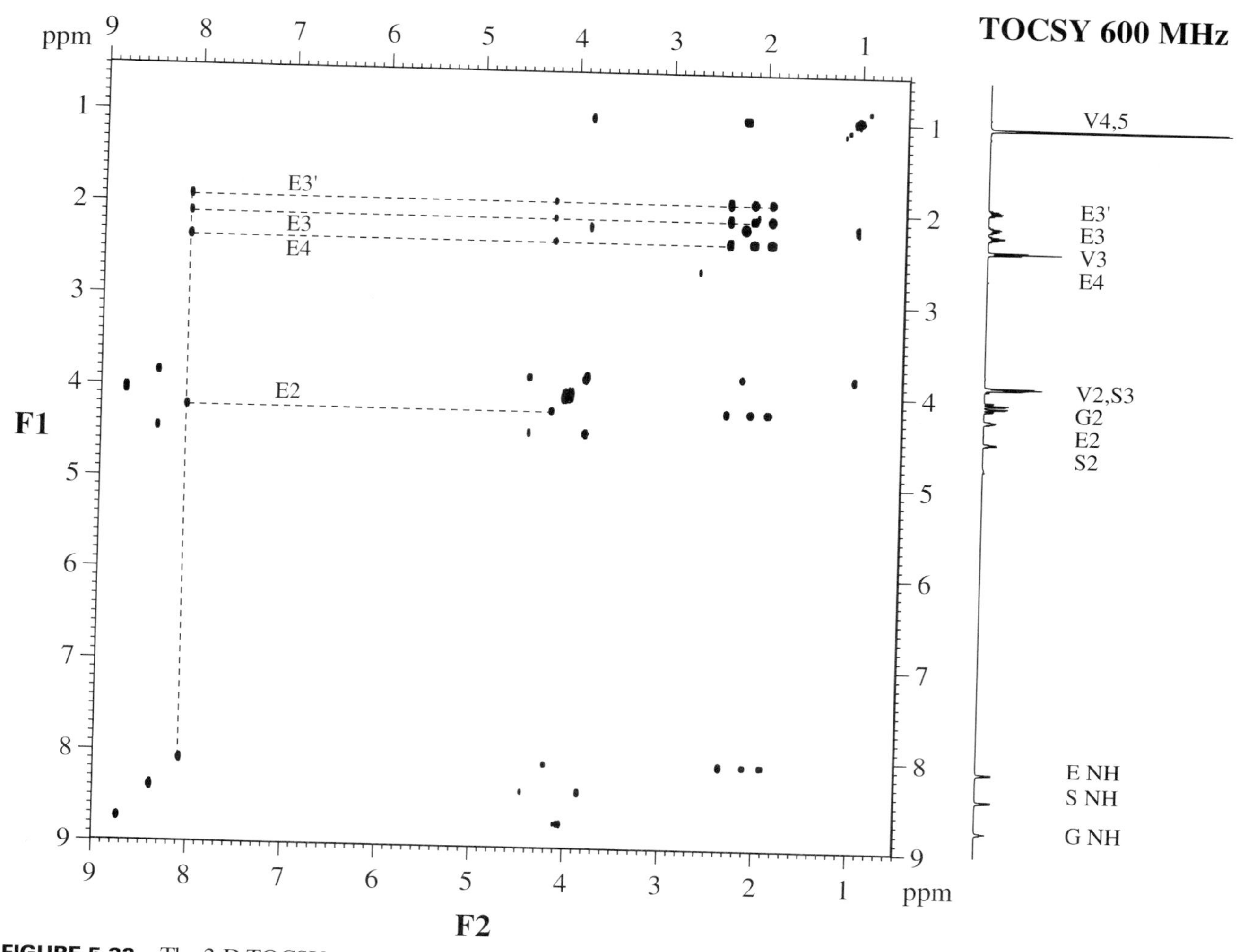

FIGURE 5.32 The 2-D TOCSY spectrum of VGSE. Some correlation lines are drawn in and assignments are given as an aid.

not correlate any further (i.e., it does not couple with any of the amide protons) making valine the N-terminus.

Assignments for the final amino acid (glutamic acid) logically can begin with the unassigned carbon at 54 ppm. The HMQC provides us with its proton partner, a multiplet at 4.21 ppm. Correlations in the COSY spectrum are found for proton resonances at 1.93 and 2.11 ppm. These two protons are readily seen as being diastereotopic because they correlate to the same carbon atom (27 ppm) in the HMQC spectrum. Both of the diastereotopic protons correlate (COSY) with a triplet at 2.37 ppm, which represents the protons of the second methylene group of glutamic acid. The carbon resonance for this methylene group is revealed in the HMQC at 31 ppm. It is worth noting from the COSY spectrum that the methine proton at 4.21 ppm is coupled to the amide proton at 8.10 ppm.

For VGSE, the HMQC is the perfect accompaniment to the DQF-COSY and TOCSY for assigning all of the protons and nearly all of the carbons (except the carbonyls). This collection of spectra, however, does not allow us to assign or confirm the order of amino acids in the peptide, an immensely important task.

5.11.4 HMBC: VGSE

The HMBC for VGSE is given in Figure 5.34; to improve the resolution of the cross peaks, the spectrum has been split into four parts. In order to assign the carbonyl carbon resonances and to completely sequence the amino acids, we can start at either end of the molecule. We have previously stated that valine is the N-terminal amino acid because its α-amino group is not part of a peptide bond. A logical place to start within the valine residue is to assign its carbonyl. This assignment can be made to the peak at

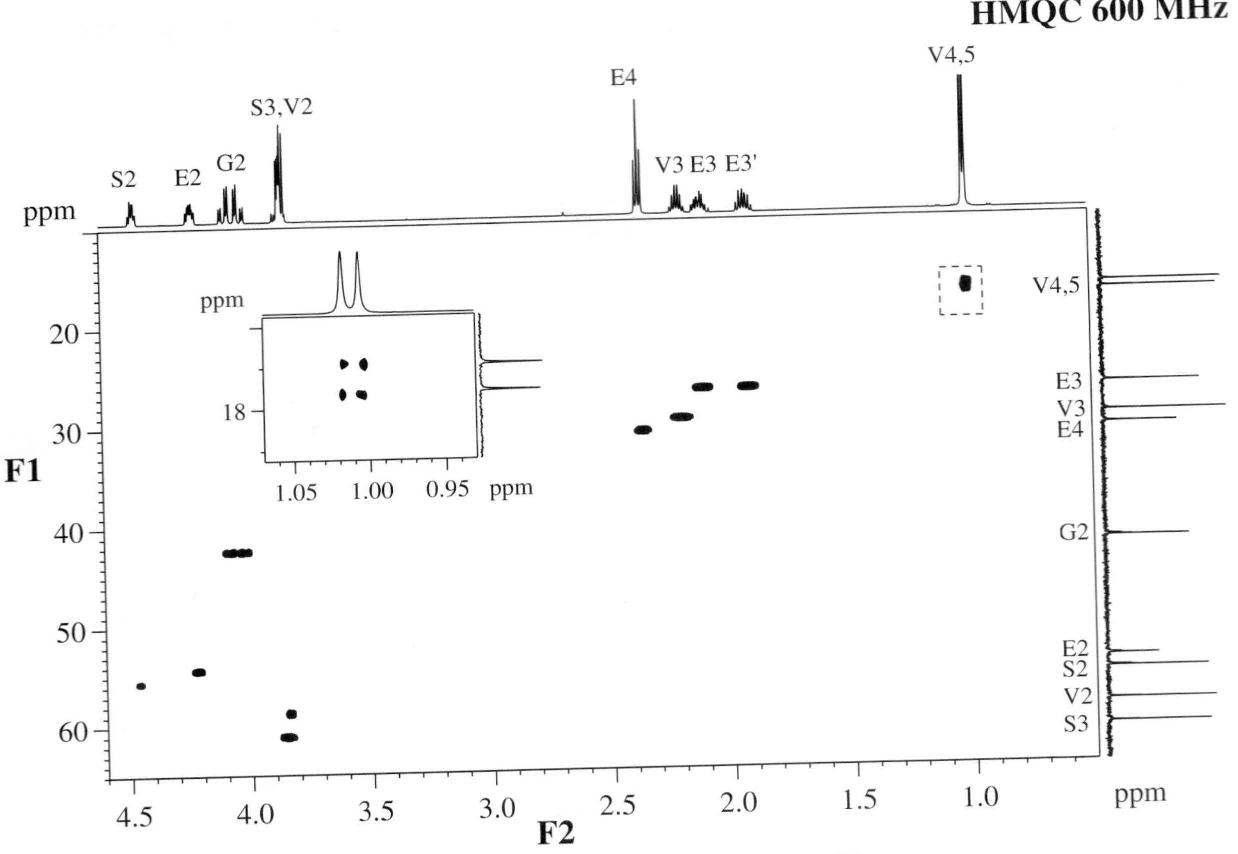

FIGURE 5.33 The HMQC spectrum for VGSE. Assignments are given as an aid.

170.1 ppm, as there are obvious correlations to the valine C-2 proton (V2) and the valine C-3 proton (V3). In addition, there are correlations to the glycine methylene group (G2) and to the glycine amide proton (G NH).

These two correlations (and generally, ones like them) are extremely important in the process of sequencing the amino acids. Until now, all of the correlations have been intra-residue, but the HMBC allows generally for inter-residue correlations. Develop a line of reasoning to sequence the remaining two amino acids (serine and glutamic acid). Note that the carbonyl carbons for glycine and serine overlap in the carbon spectrum.

5.11.5 ROESY: VGSE

We end our discussion of VGSE by comparing the ROESY correlations of the amide protons with the corresponding interactions in COSY and TOCSY. Figure 5.35 shows comparable sections of each spectrum. We have the seen the COSY and TOCSY portions earlier and showed how these spectra, along with the HMQC, can be used for intra-residue assignments. The ROESY correlations, on the other hand, reinforce the HMBC information and help to confirm the sequence of amino acids with inter-residue correlations.

The boxed cross peaks illustrate the through space interaction of the amide proton of one amino acid with the adjacent amino acid's C-2 proton. Thus, the amide proton of glycine (G NH) correlates with the C-2 proton of valine (V2), the amide proton of serine (S NH) correlates with the C-2 protons of glycine (G2), the amide proton of glutamic acid (E NH) correlates with the C-2 proton (and C-3 protons as well) of serine (S2 and S3). The manner in which the data are presented in Figure 5.35 greatly simplifies our interpretation and makes comparisons more meaningful.

5.12 GRADIENT FIELD NMR

A well-established area in the field of NMR is the use of "**P**ulsed **F**ield **G**radients," or **PFG** NMR. It is ironic to consider that so much effort has been expended over so many years to provide a homogenous or stable magnetic field. Today, most modern high field NMR spectrometers are routinely equipped with hardware (coils)

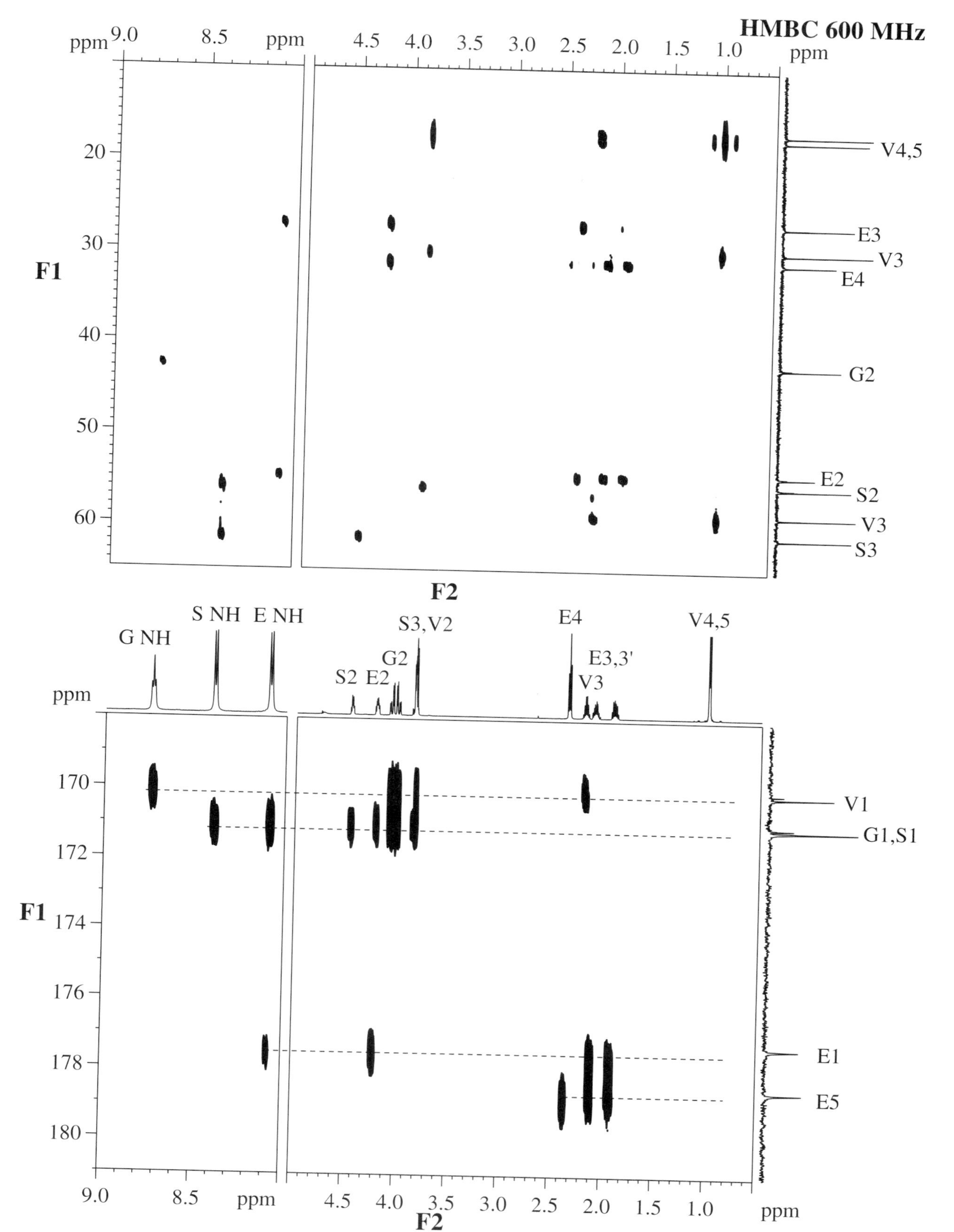

FIGURE 5.34 The HMBC spectrum for VGSE. Important correlation lines are drawn in and assignments are given as an aid.

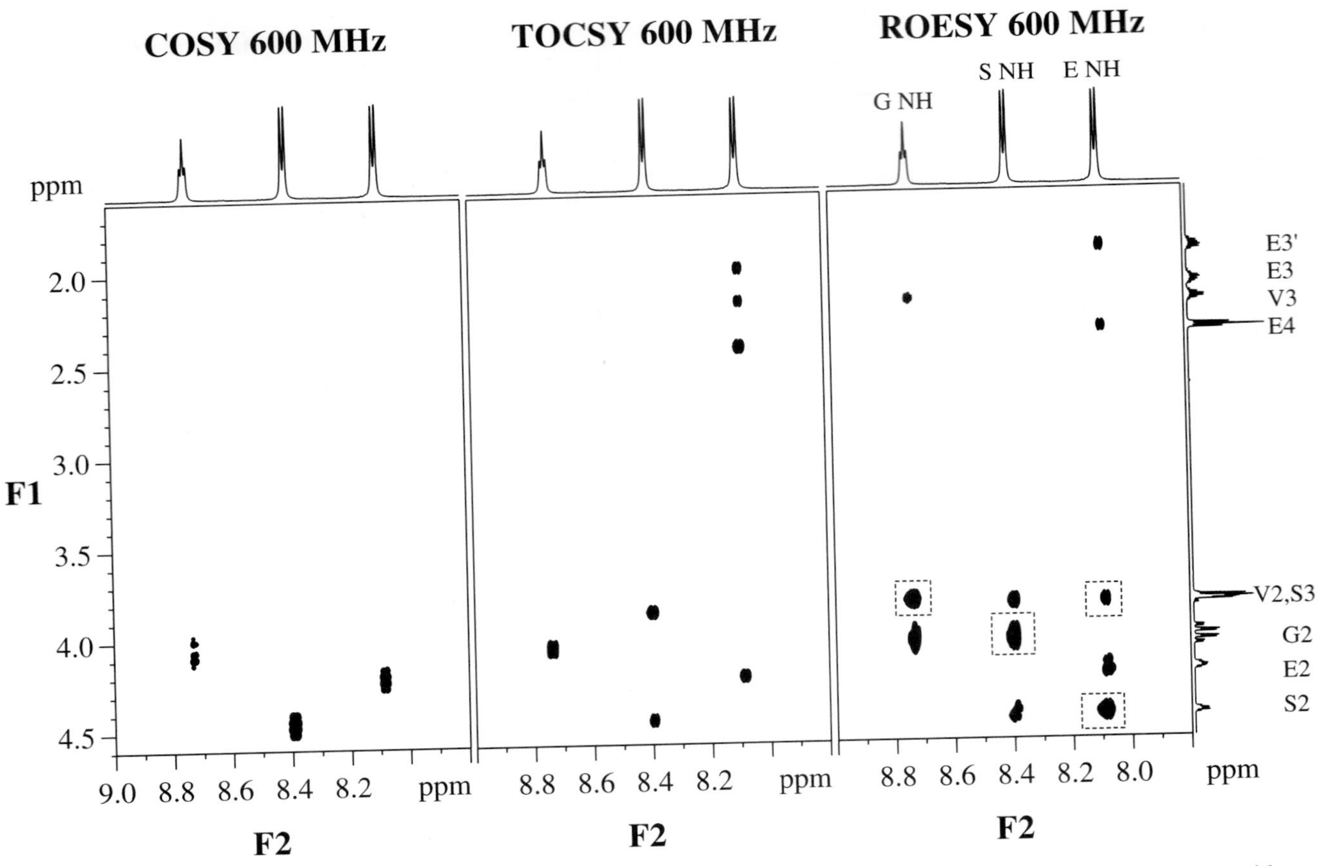

FIGURE 5.35 A comparison of the DQF COSY, 2-D TOCSY, and ROESY spectra showing the interactions of the amide protons.

that rapidly ramps the magnetic field along one of three mutually orthogonal axes. These magnetic field gradients are incorporated into the pulse sequence in a large range of applications. The 600 MHz spectrometer used to obtain spectra for this edition is equipped triple-axis gradients.

We include a brief discussion of gradients here because there are many applications to correlation experiments. We treat gradients in a general manner; for a more technical treatment and for a myriad of other applications, consult the review by Price (1996).

In Section 5.3.1, we mention phase cycling as an important part of any pulse sequence. The details of phase cycling are still beyond our treatment (see Claridge, 1999) but we can easily appreciate one of their negative aspects: time. In correlation experiments, anywhere from 4 to 64 phase cycles must be summed in order to produce one FID. If signal-to-noise is poor, then the identical cycle is repeated until sufficient signal is acquired. These phase cycles are wasteful of spectrometer time and account for at least one reason why 2-D experiments take so long.

In a PFG experiment, the pulse sequence can be rewritten so that phase cycling can be greatly reduced or eliminated altogether. Thus, if signal to noise is sufficient, each pulse can be saved as one FID in a 2-D experiment. The saving in instrument time is enormous; experiments that previously took several hours to an entire day can now be run in a matter of minutes. One reassuring point for us to realize is that, even though the experiment is run differently, the results, and hence, the interpretation remain the same.

While our initial interest in gradient NMR stems from the fact that familiar experiments can be run more quickly and more efficiently, the most exciting uses of gradients are the experiments that cannot be run without them. We find that gradients allow for improved magnetic resonance imaging (MRI), improved magnetic resonance microscopy, better solvent suppression (especially water), and entirely new areas of inquiry such as diffusion measurements.

REFERENCES

Atta-ur, Rahman, and Choudhary, M. (1995). *Solving Problems with NMR Spectroscopy.* New York: Academic Press.

Becker, E. (1999). *High Resolution NMR: Theory and Chemical Applications,* 3rd ed. New York: Academic Press.

Berger, S., and Braun, S. (2004). *200 and More NMR Experiments: A Practical Course.* Weinheim, Germany: VCH Verlagsgesellschaft MBH.

Braun, S., Kalinowski, H., and Berger, S. (1999). *150 and More Basic NMR Experiments: A Practical Course.* 2nd ed. New York: Wiley.

Brey, W., (Ed.). (1988). *Pulse Methods in 1D and 2D Liquid-Phase NMR.* New York: Academic Press.

Cavanagh, J., Fairbrother, W., Palmer III, A., and Skelton, N. (1995). *Protein NMR Spectroscopy: Principles and Practice.* New York: Academic Press.

Chandrakumar, N., and Subramanian S. (1987). *Modern Techniques in High Resolution FT NMR.* New York: Springer-Verlag.

Claridge, T.D.W. (1999). *High-Resolution NMR Techniques in Organic Chemistry.* New York: Elsevier Science.

Croasmum, W. R., and Carlson, R. M. K. (1987). *Two-Dimesional NMR Spectroscopy.* 2nd ed. New York: VCH.

Croasmun, W.R., and Carlson, R.M.K. (1994). *TwoDimesional NMR Spectroscopy: Applications for Chemists and Biochemists, Fully Updated and Expanded to Include Multidimensional Work.* 2nd ed. New York: Wiley.

Derome, A. (1987). *Modern NMR Techniques for Chemistry Research.* Oxford: Pergamon Press.

Ernst. R. R., Bodenhausen, G., and Wokaum A. (1987). *Principles of Nuclear Magnetic Resonance in One and Two Dimensions.* Oxford: Clarendon Press.

Farrar, T.C. (1987). *An Introduction to NMR Spectroscopy.* Chicago: Farragut.

Farrar, T., and Becker, E. (1971). *Pulse and Fourier Transform NMR: Introduction to Theory and Methods.* New York: Academic Press.

Freeman, R. (1998). *Spin Choreography: Basic Steps in High Resolution NMR.* Oxford: University Press on Demand.

Friebolin, H. (1998). *Basic One- and Two-Dimensional NMR Spectroscpoy. 3rd revised ed.* New York: Wiley.

Fukushima, E., and Roeder, S.B.W. (1986). *Experimental Pulse NMR a Nuts and Bolts Approach.* New York: Perseus Publishing.

Günther, H. (1995). *NMR Spectroscopy: Basic Principles, Concepts, and Applications in Chemistry. 2nd ed.* New York: Wiley.

Hoch, J.C., and Stern, A. (1996). *NMR Data Processing.* New York: Wiley.

Hore, P.J. (1995). *Nuclear Magnetic Resonance (Oxford Chemistry Primers, 32).* Oxford: Oxford University Press.

Hore, P.J., Wimperis, S., and Jones, J. (2000). *NMR: The Toolkit.* Oxford: Oxford University Press.

Kessler, H.. Gehrke, M., and Griesinger, C. (1988). *Two-Dimensional NMR Spectroscopy.* Angew. Chem, Int Ed. Engl. 27, 490–536.

Macomber, R.S. (1995). *A Complete Introduction to Modern NMR Spectroscopy.* New York: Wiley-Interscience.

Martin, G.E., and Zektzer, A.S. (1988). *Two-Dimensional NMR Methods for Establishing Molecular Connectivity: A Chemist's Guide to Experiment Selection, Performance, and Interpretation.* New York: Wiley.

Mitchell, T.N., and Costisella, B. (2004). *NMR-From Spectra to Structures: An Experimental Approach.* New York: Springer Verlag.

Munowitz, M. (1988). *Coherence and NMR.* New York: Wiley-Interscience.

Nakanishi, K. (1990). *One-Dimensional and Two-Dimensional NMR Spectra by Modern Pulse Techniques.* New York: University Science Books.

Pihlaja, K., and Kleinpeter, E. (1994). *Carbon-13 NMR Chemical Shifts in Structural and Stereochemical Analysis.* Weinheim, Germany: VCH Publishing.

Price, W.S. (1996). Gradient NMR. In G.A. Webb (Ed.). *Annual Reports on NMR Spectroscopy.* Vol. 32. London: Academic Press.

Sanders, J.K.M., and Hunter, B.K. (1993). *Modem NMR Spectroscopy.* 2nd ed. Oxford: Oxford University Press.

Schraml, J., and Bellama, J.M. (1988). *Two-Dimensional NMR Spectrometry.* New York: Wiley.

Van de Ven, F.J.M. (1995). *Multidimensional NMR in Liquids: Basic Principles and Experimental Methods.* Weinheim, Germany: Wiley-VCH.

Williams, K.R., and King, R.W. (1990). *The Fourier trans-form in chemistry*—NMR, J. Chem. Ed., 67, 125–138.

STUDENT EXERCISES

5.1 For the compounds of Problem 3.3 a-o, draw the following spectra: COSY, HMQC, HMBC, and INADEQUATE. Be sure to label the F1 and F2 axes. Assume experimental conditions are the same as in Problem 3.3.

5.2 Assign all of the correlations for ipsenol in the DQF COSY found in Figure 5.9. Indicate each type of coupling as geminal, vicinal, or long range.

5.3 Complete the assignments for ipsenol for the HMBC found in Figure 5.14. To aid in bookkeeping, you may want to construct a table similar to Table 5.1.

5.4 Identify the compound $C_6H_{10}O$ from its ^1H, ^{13}C/DEPT, COSY, and HMQC spectra and show all correlations.

5.5 Identify the compound $C_{10}H_{10}O$ from its ^1H, ^{13}C/DEPT, COSY, and HMQC spectra and show all correlations.

5.6 Identify the compound $C_8H_9NO_2$ from its ^1H, ^{13}C/DEPT, COSY, HMQC, and INADEQUATE spectra and show all correlations.

5.7 Assign all of the carbon connectivities for caryophyllene oxide using the INADEQUATE spectrum found in Figure 5.19.

5.8 Identify compound $C_{10}H_{18}O$ from its ^1H, ^{13}C/DEPT, HMQC, HMBC, and INADEQUATE spectra.

5.9 Make as many correlations as possible for lactose using the 2-D TOCSY found in Figure 5.25. Compare your results for the glucose residue to the results that were found in the 1-D HOHAHA in Figure 5.26.

5.10 Given are the structure and ^1H, ^{13}C/DEPT, COSY, 1-D TOCSY, 2-D TOCSY, HMQC, HMQC–TOCSY, HMBC, and ROESY spectra for raffinose. Confirm the structure, assign all protons and carbons, and show as many correlations as possible.

5.11 Given are the structure and ^1H, ^{13}C/DEPT, COSY, HMQC, and HMBC, spectra for stigmasterol. Confirm the structure, assign all protons and carbons, and show as many correlations as possible.

5.12 The ^1H, ^{13}C/DEPT, COSY, 2-D TOCSY, HMQC, HMBC, and ROESY for a tri-peptide containing the amino acids lysine, serine, and threonine. Sequence the peptide and assign all protons and carbons. Show as many correlations as possible.

Exercise 5.4

^{1}H NMR 600 MHz

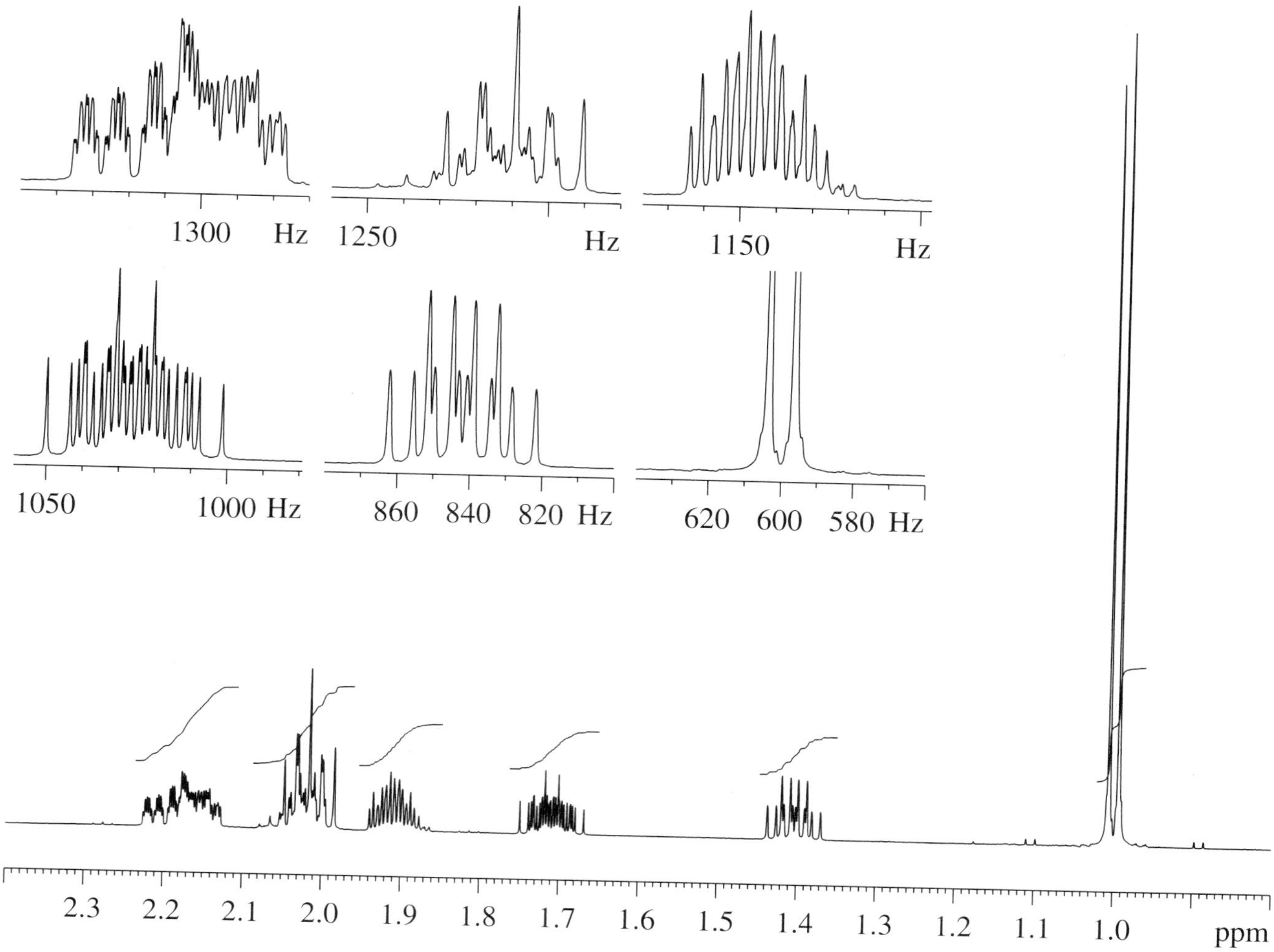

1300 Hz
1250 Hz
1150 Hz
1050 1000 Hz
860 840 820 Hz
620 600 580 Hz
2.3 2.2 2.1 2.0 1.9 1.8 1.7 1.6 1.5 1.4 1.3 1.2 1.1 1.0 ppm

^{13}C/DEPT 150.9 MHz

220 200 180 160 140 120 100 80 60 40 20 ppm

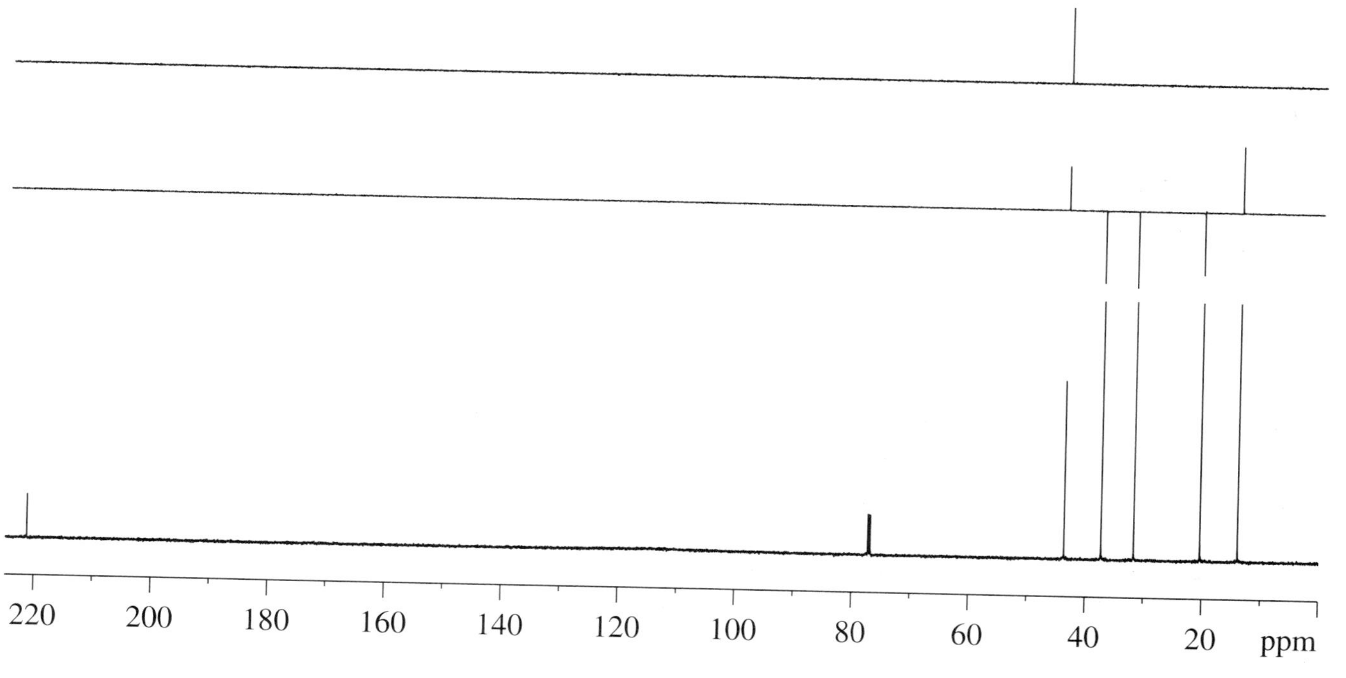

Exercise 5.4

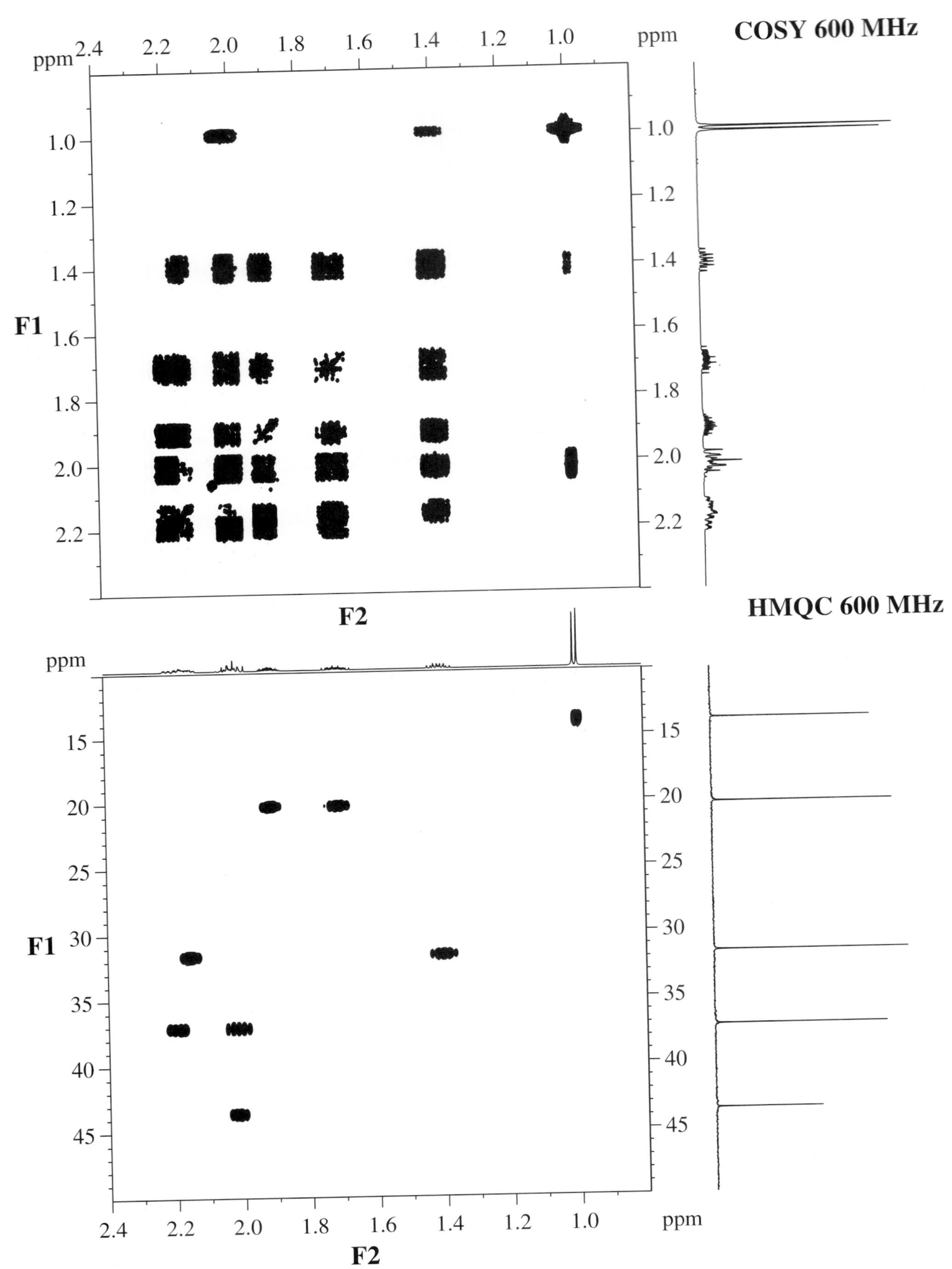

COSY 600 MHz
HMQC 600 MHz
ppm
F1
F2
2.4
2.2
2.0
1.8
1.6
1.4
1.2
1.0
15
20
25
30
35
40
45

^{1}H NMR 600 MHz

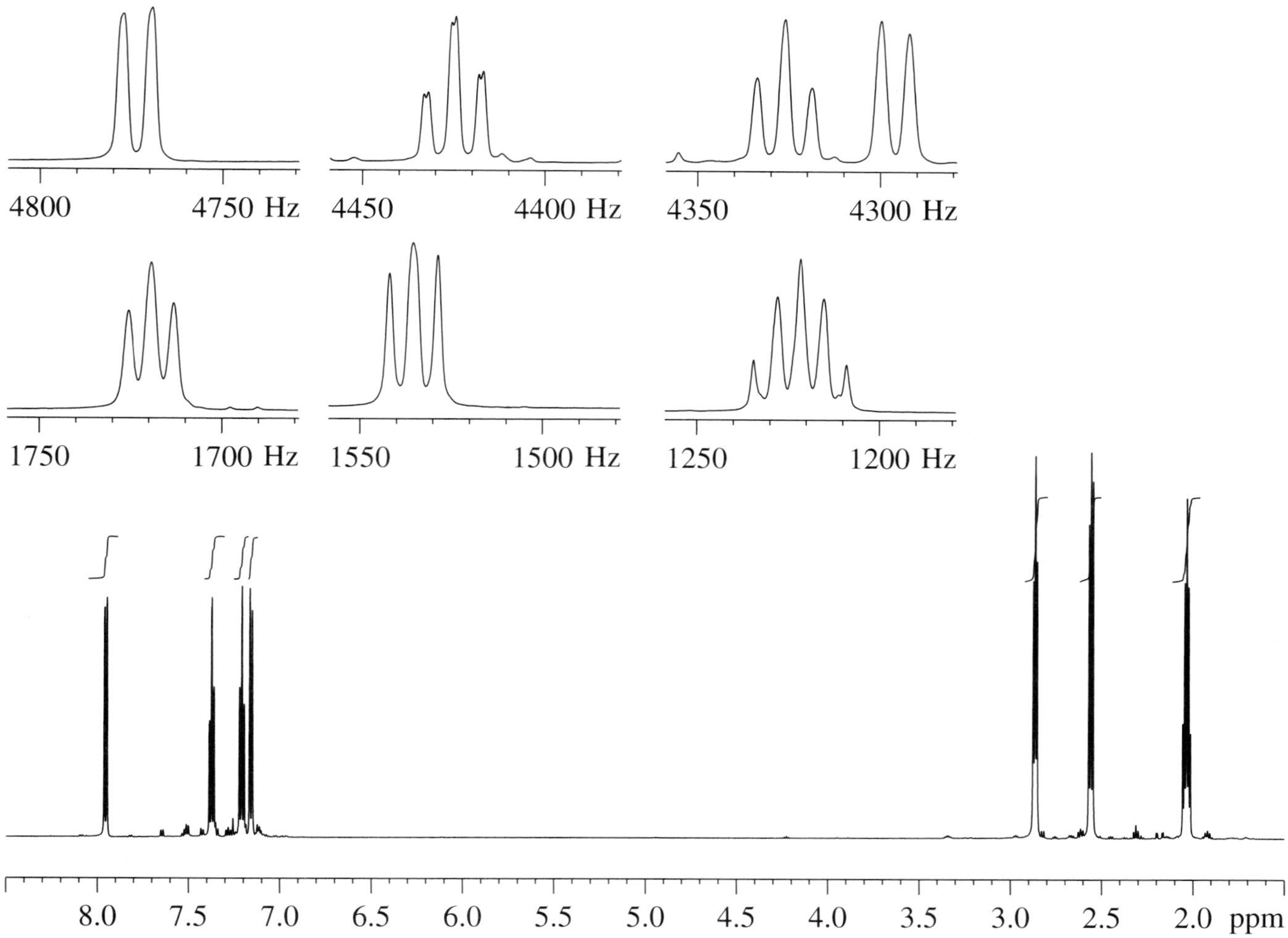

^{13}C/DEPT 150.9 MHz

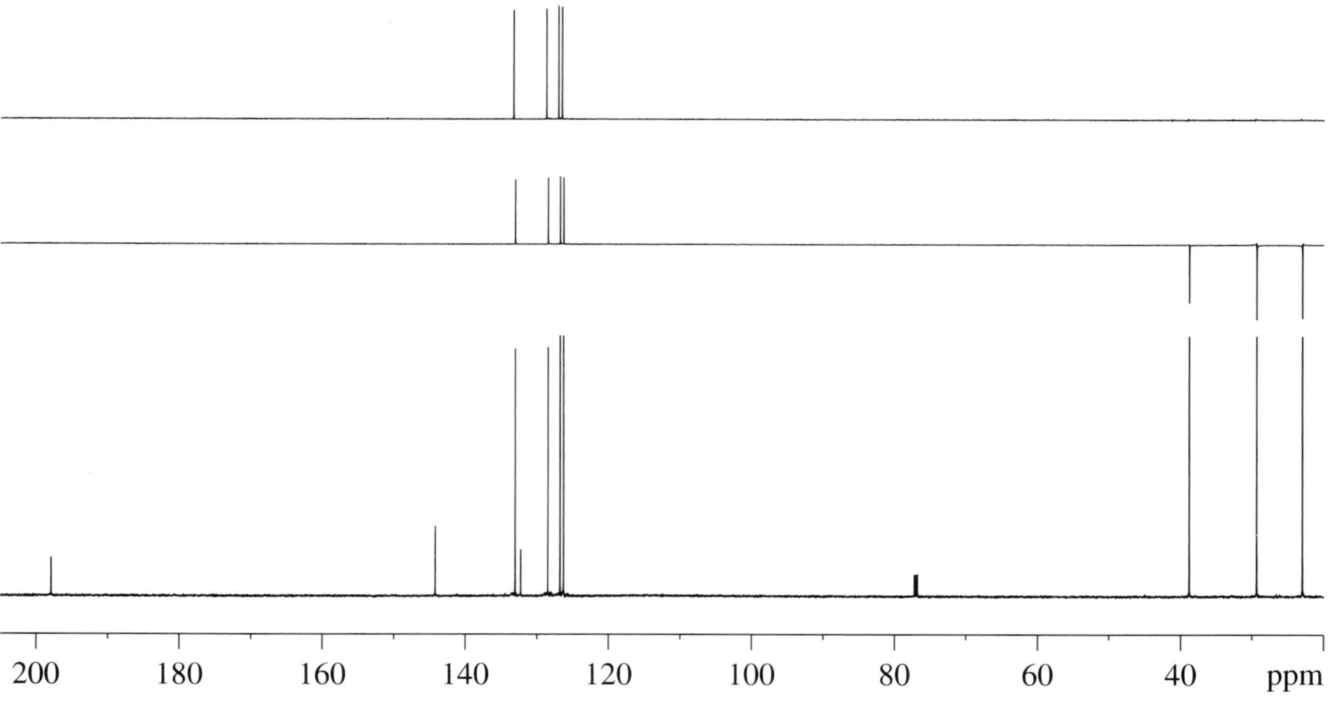

Exercise 5.5

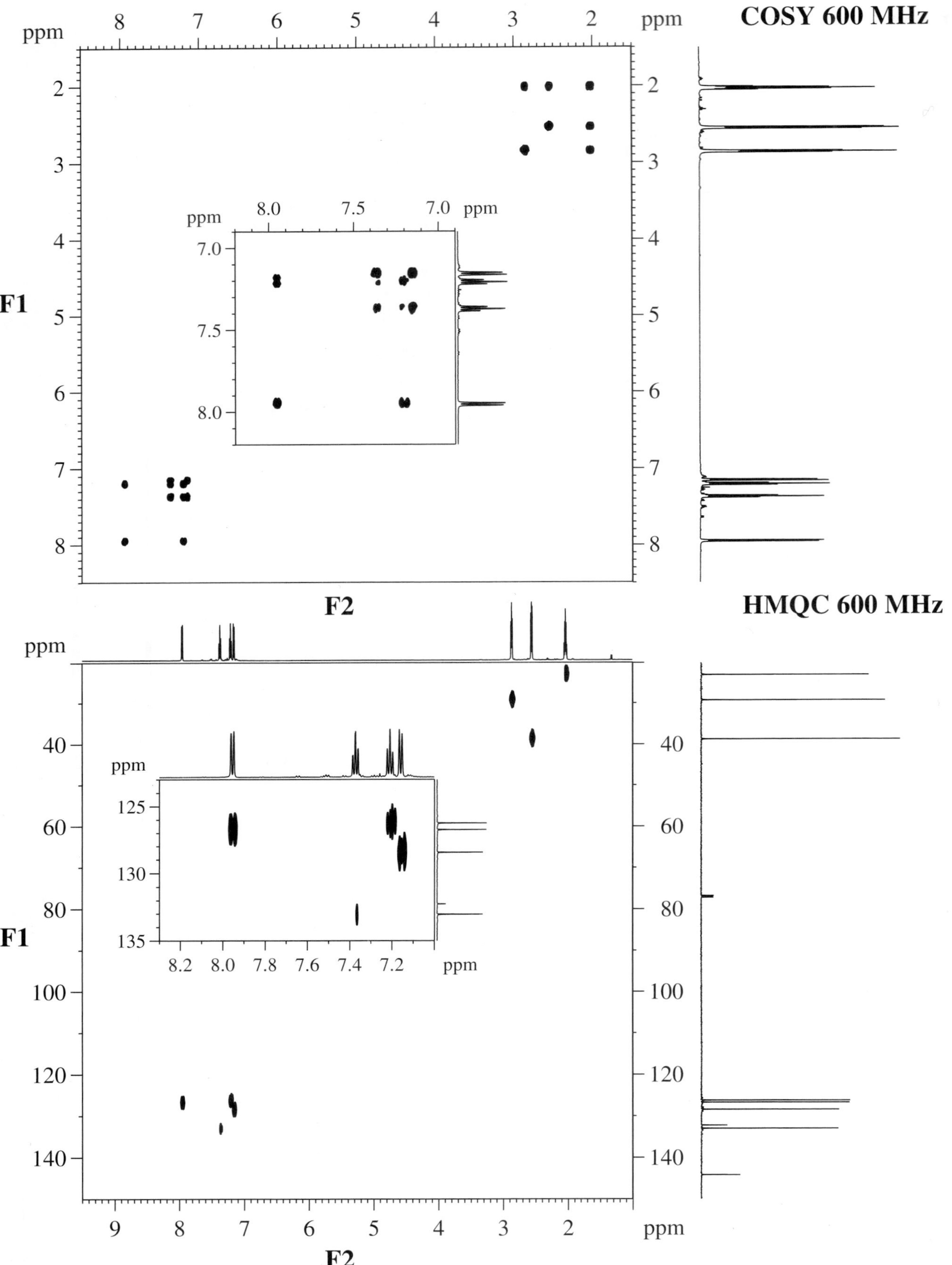

^{1}H NMR 600 MHz

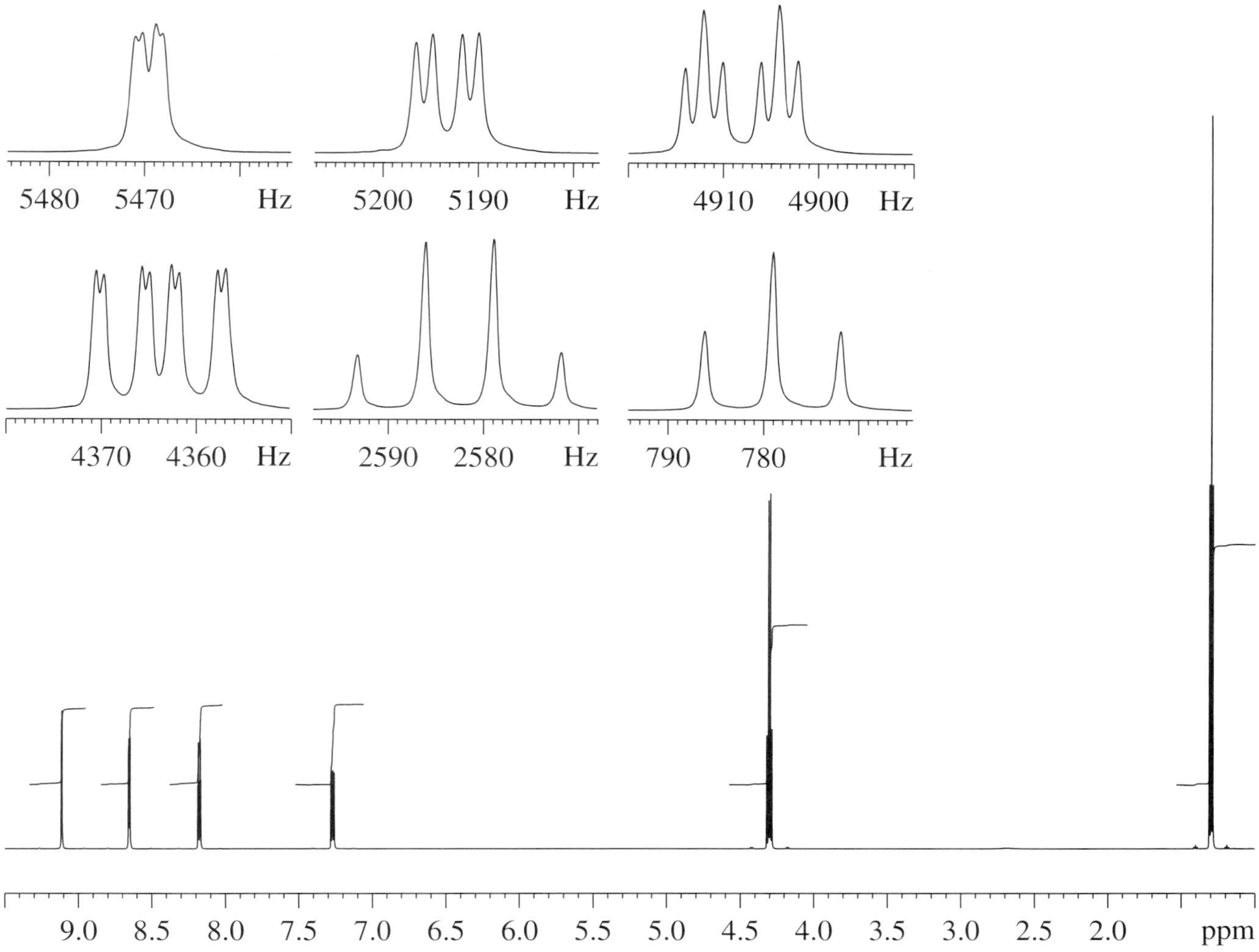

5480
5470
Hz
5200
5190
Hz
4910
4900
Hz
4370
4360
Hz
2590
2580
Hz
790
780
Hz
9.0 8.5 8.0 7.5 7.0 6.5 6.0 5.5 5.0 4.5 4.0 3.5 3.0 2.5 2.0 ppm

[13]C/DEPT 150.9 MHz

160 150 140 130 120 110 100 90 80 70 60 50 40 30 ppm

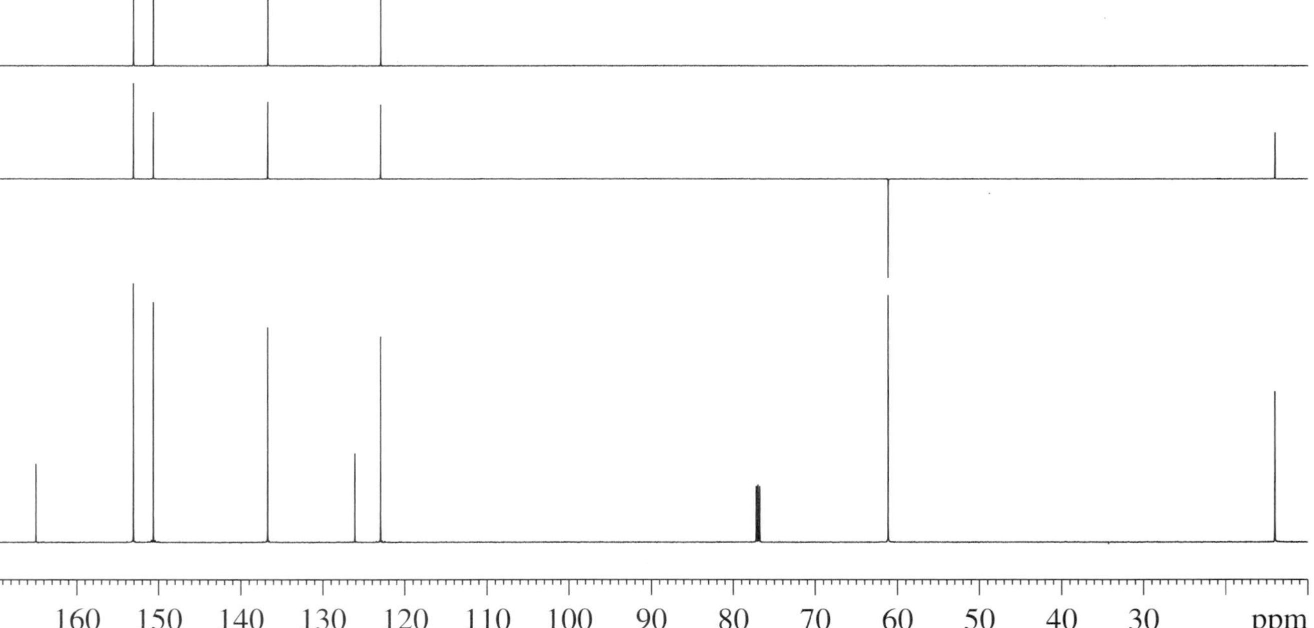

Exercise 5.6

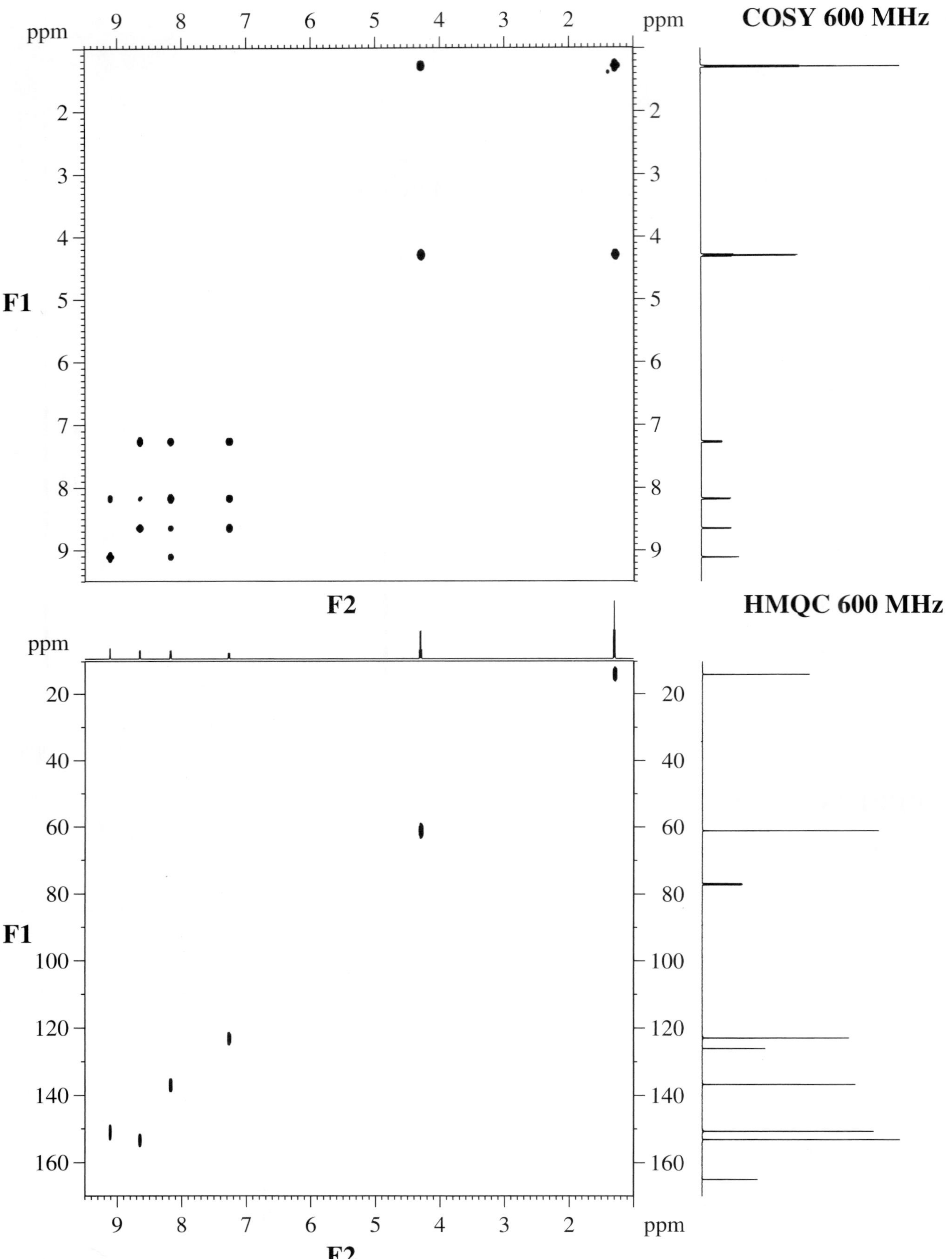
COSY 600 MHz
HMQC 600 MHz
ppm
F1
F2

Exercise 5.6

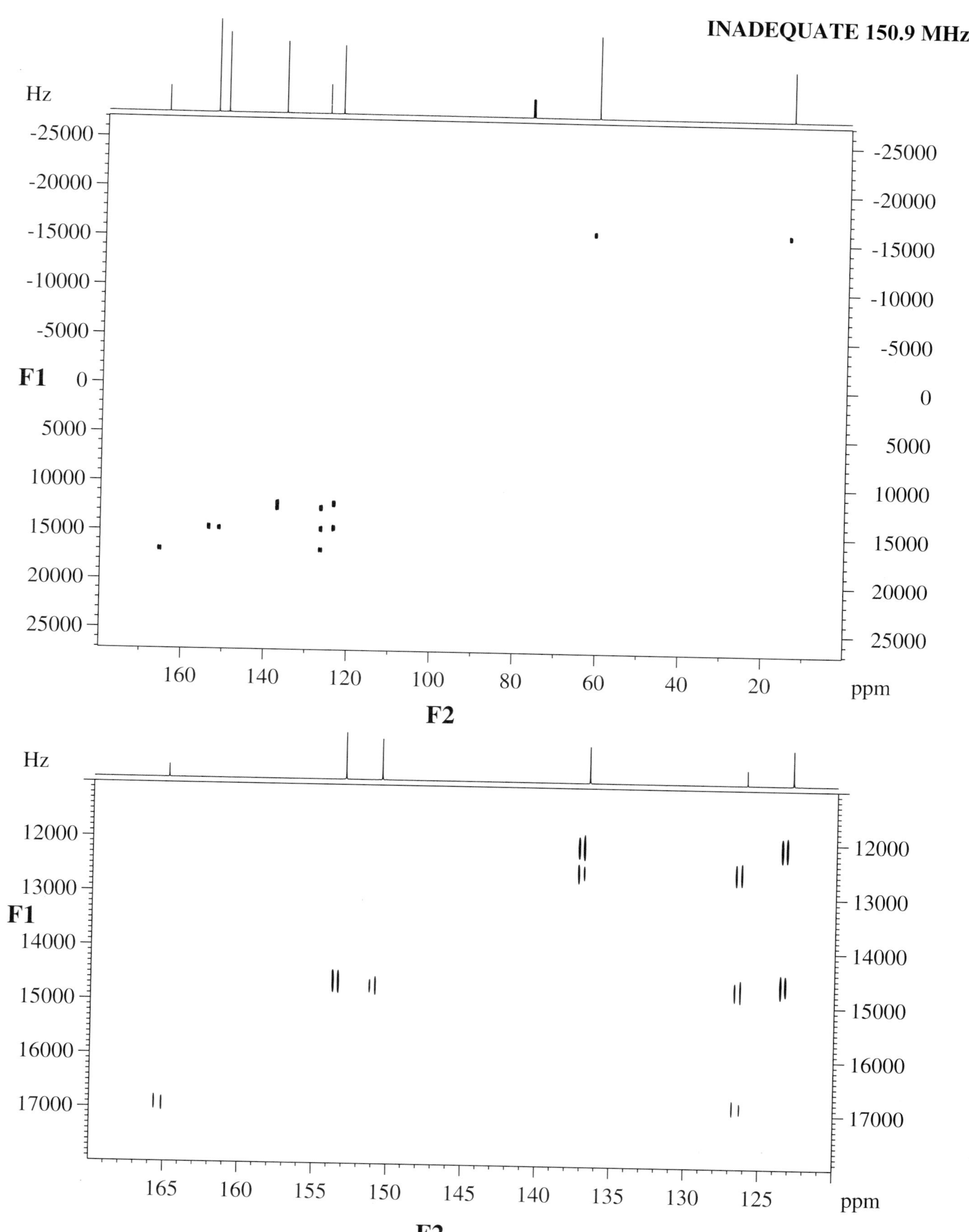

Exercise 5.8

^{1}H NMR 600 MHz

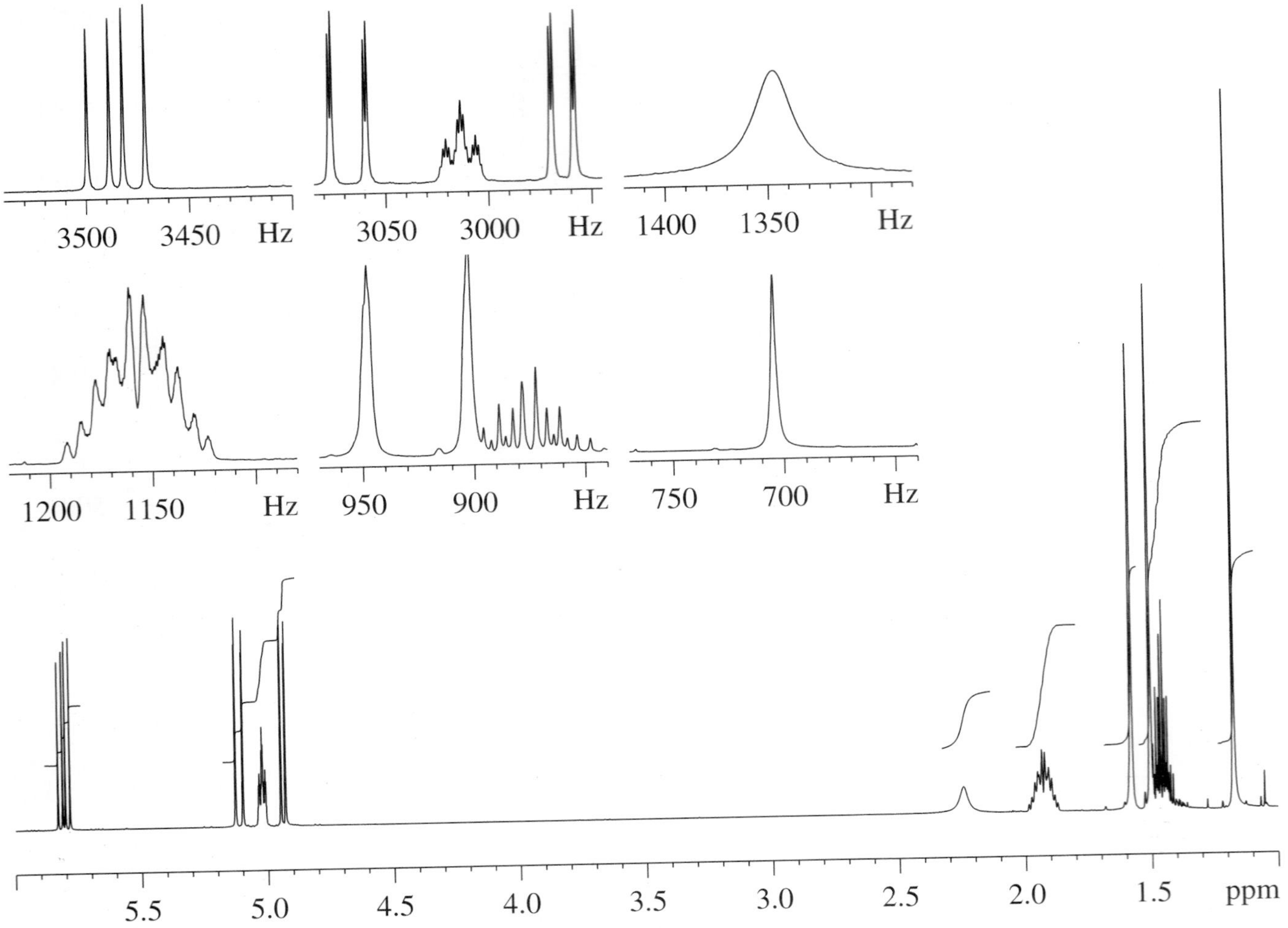

3500
3450
Hz
3050
3000
Hz
1400
1350
Hz
1200
1150
Hz
950
900
Hz
750
700
Hz
5.5
5.0
4.5
4.0
3.5
3.0
2.5
2.0
1.5
ppm

^{13}C/DEPT 150.9 MHz

150
140
130
120
110
100
90
80
70
60
50
40
30
20
ppm

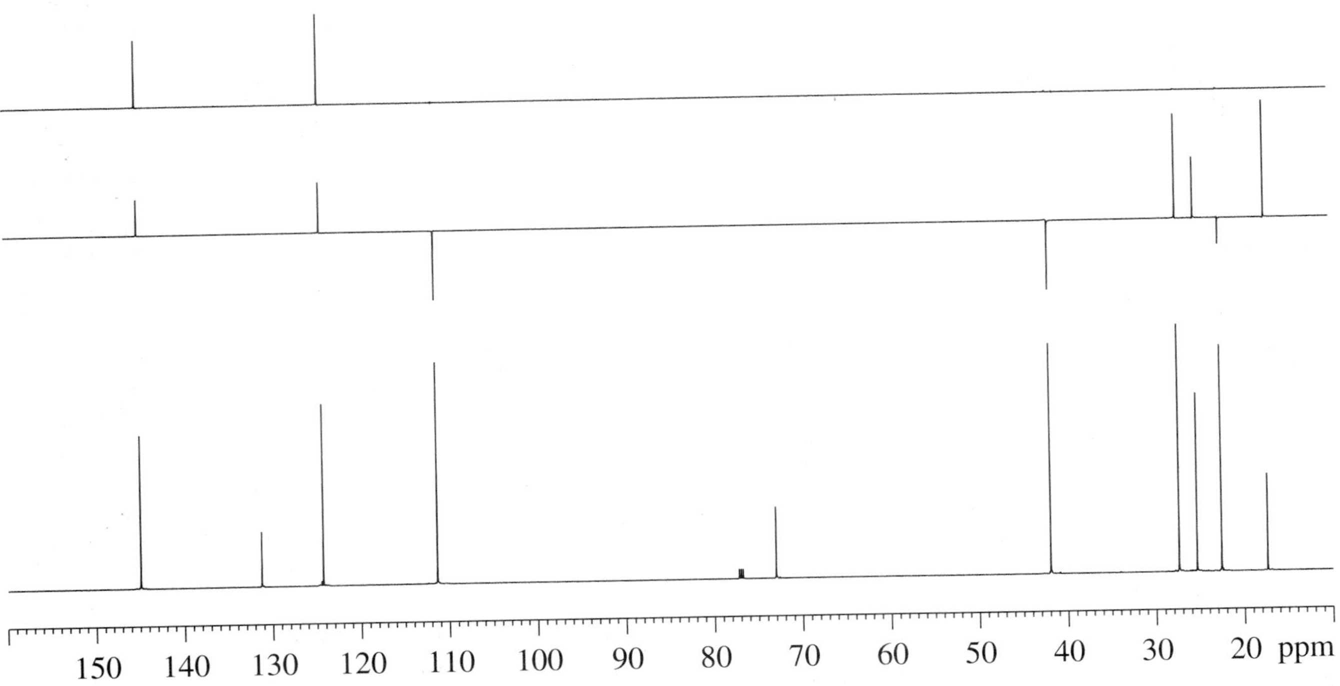

Exercise 5.8

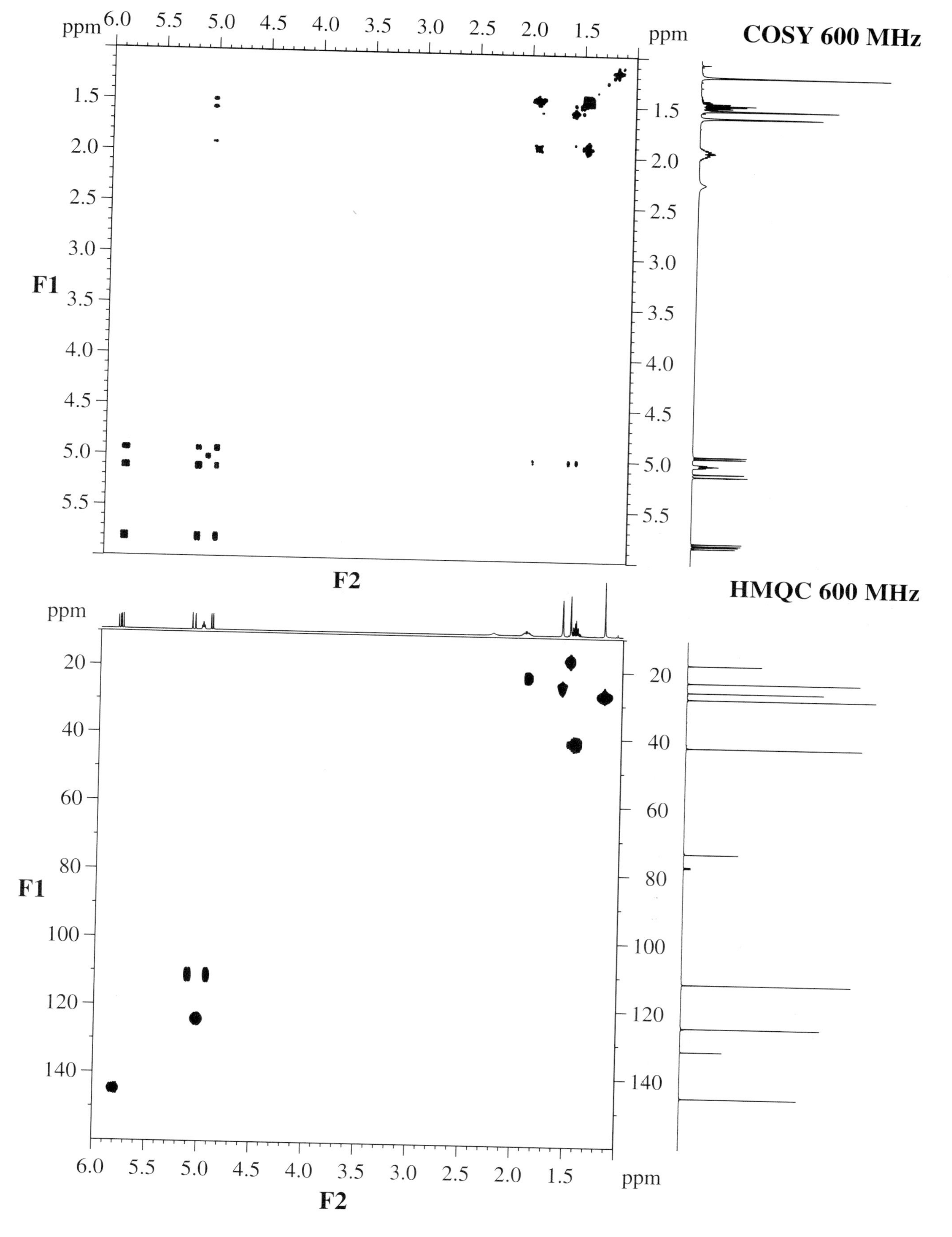

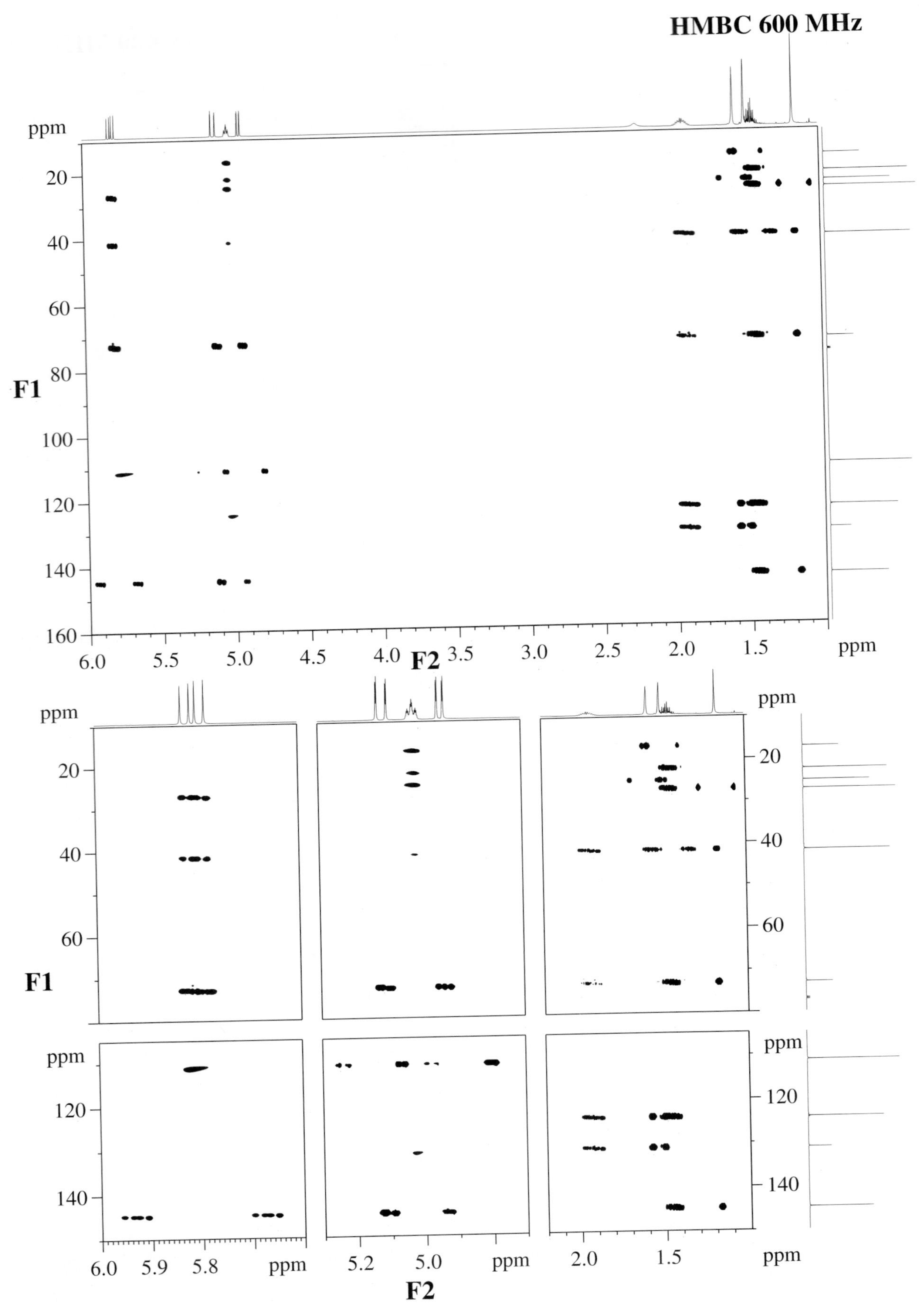
HMBC 600 MHz
ppm
F1
F2
ppm
20
40
60
80
100
120
140
160
6.0
5.5
5.0
4.5
4.0
3.5
3.0
2.5
2.0
1.5
5.9
5.8
5.2
5.0

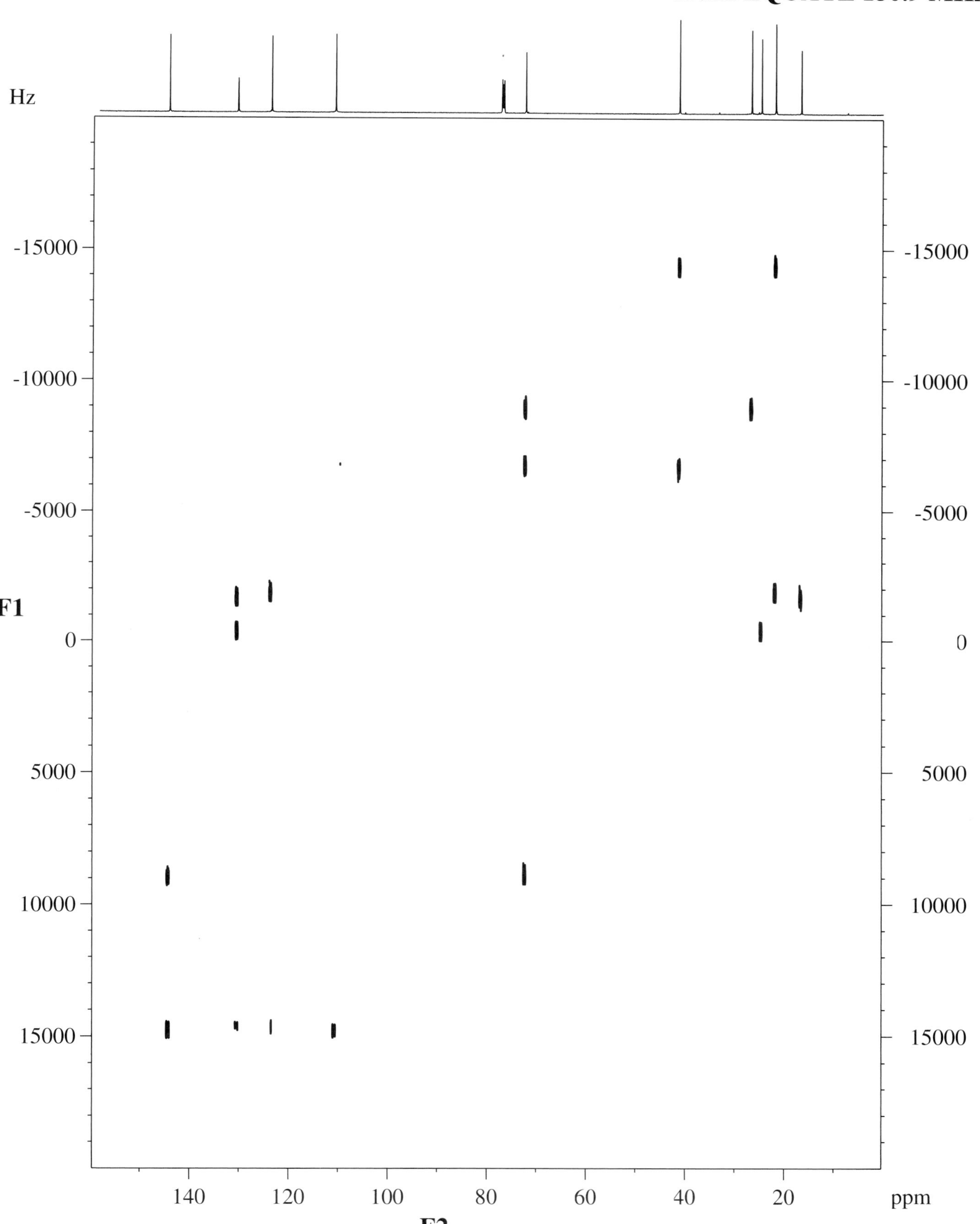
INADEQUATE 150.9 MHz
Hz
F1
-15000
-10000
-5000
0
5000
10000
15000
140
120
100
80
60
40
20
ppm
F2

Exercise 5.10

^{1}H NMR 600 MHz

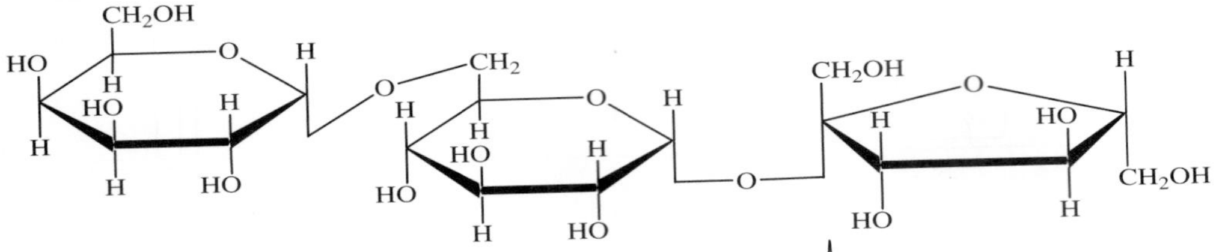

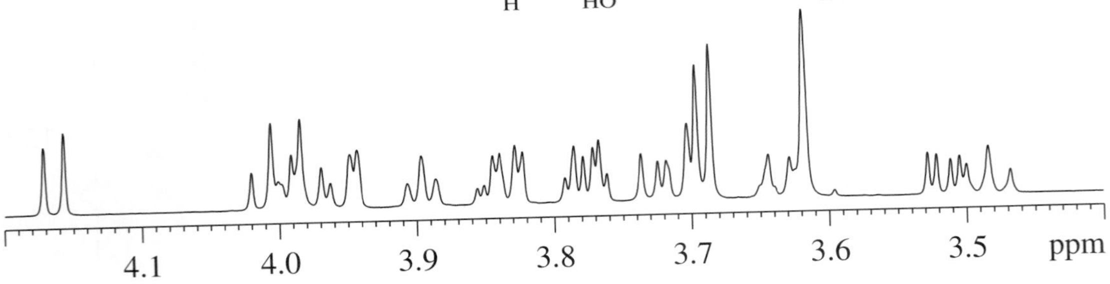

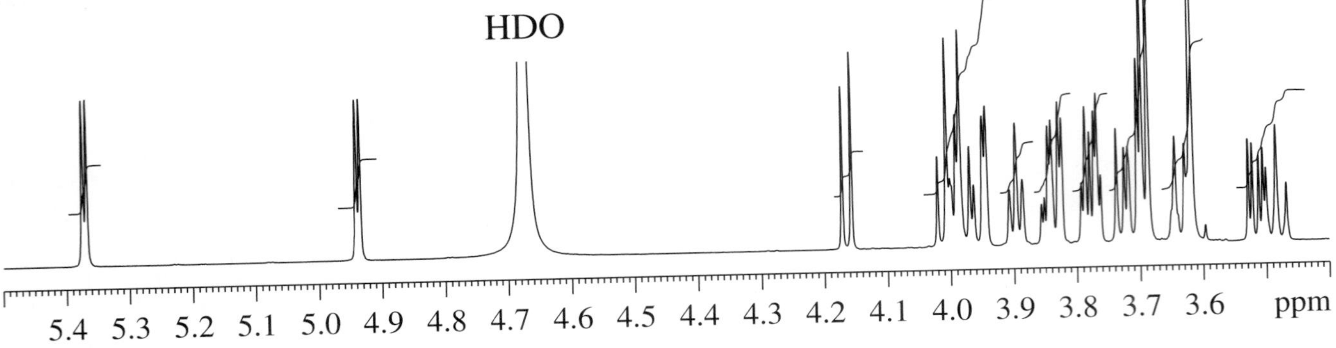

^{13}C/DEPT 150.9 MHz

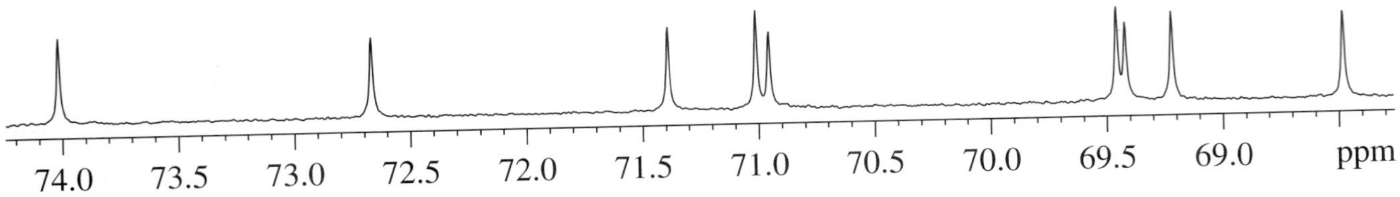

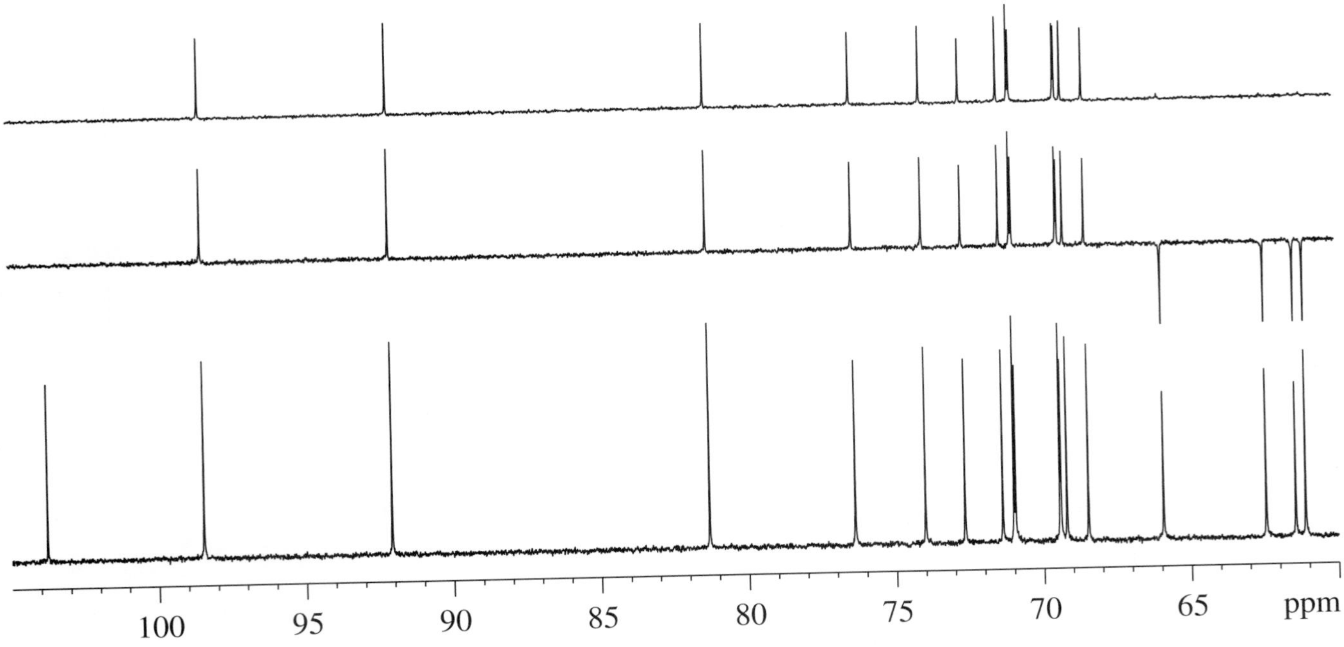

Exercise 5.10

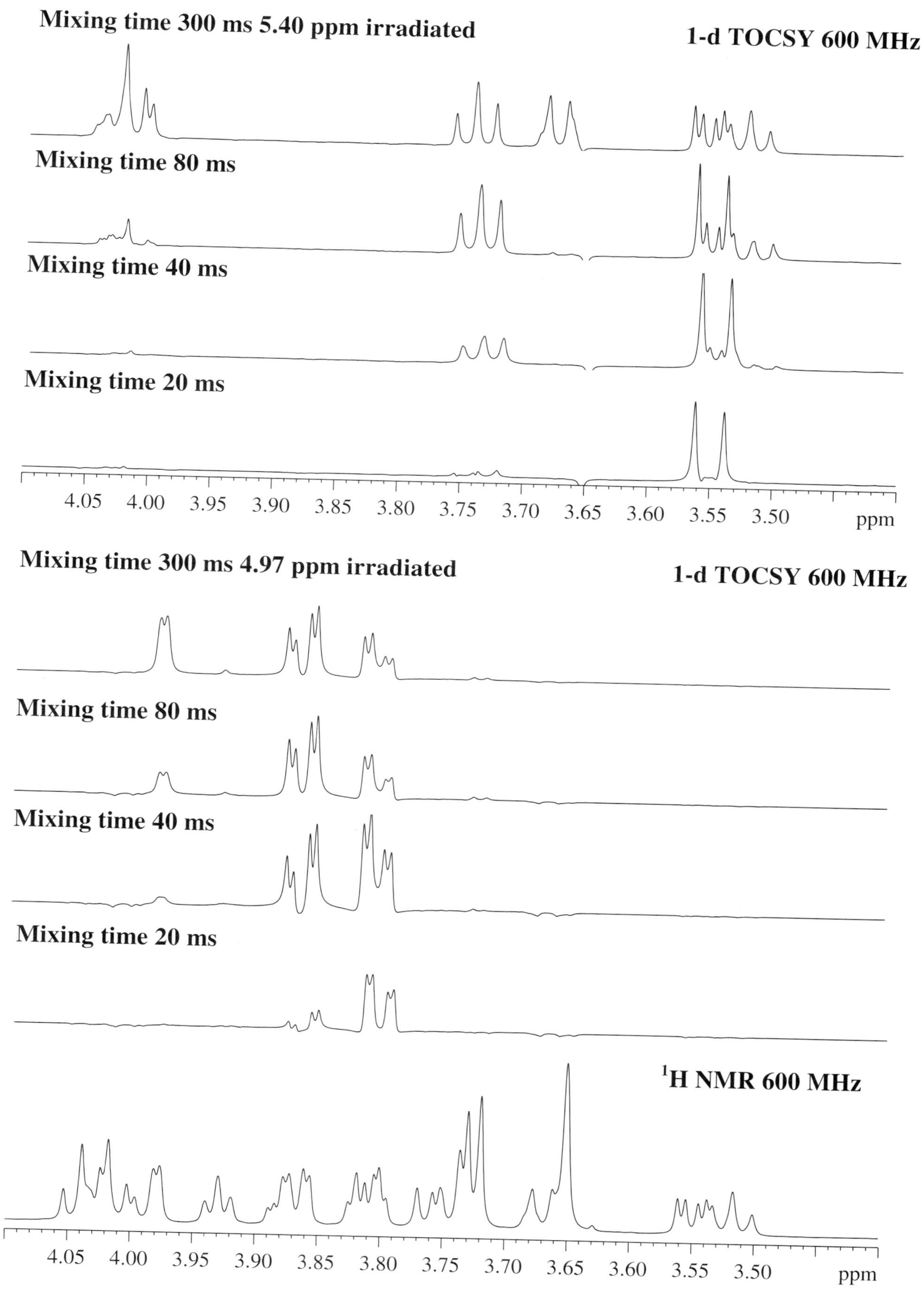

Mixing time 300 ms 5.40 ppm irradiated
1-d TOCSY 600 MHz
Mixing time 80 ms
Mixing time 40 ms
Mixing time 20 ms
4.05 4.00 3.95 3.90 3.85 3.80 3.75 3.70 3.65 3.60 3.55 3.50 ppm
Mixing time 300 ms 4.97 ppm irradiated
1-d TOCSY 600 MHz
Mixing time 80 ms
Mixing time 40 ms
Mixing time 20 ms
^{1}H NMR 600 MHz
4.05 4.00 3.95 3.90 3.85 3.80 3.75 3.70 3.65 3.60 3.55 3.50 ppm

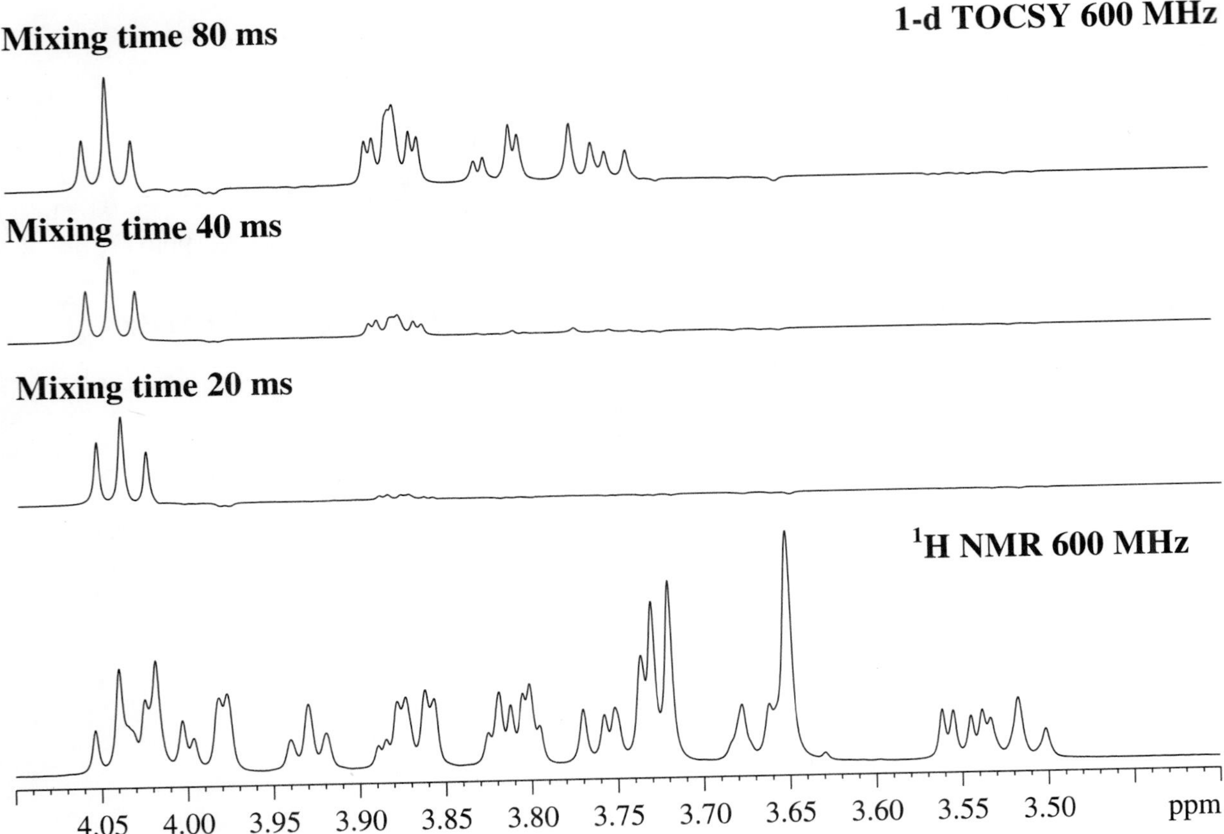
1-d TOCSY 600 MHz
Mixing time 80 ms
Mixing time 40 ms
Mixing time 20 ms
1H NMR 600 MHz
4.05
4.00
3.95
3.90
3.85
3.80
3.75
3.70
3.65
3.60
3.55
3.50
ppm

Exercise 5.10

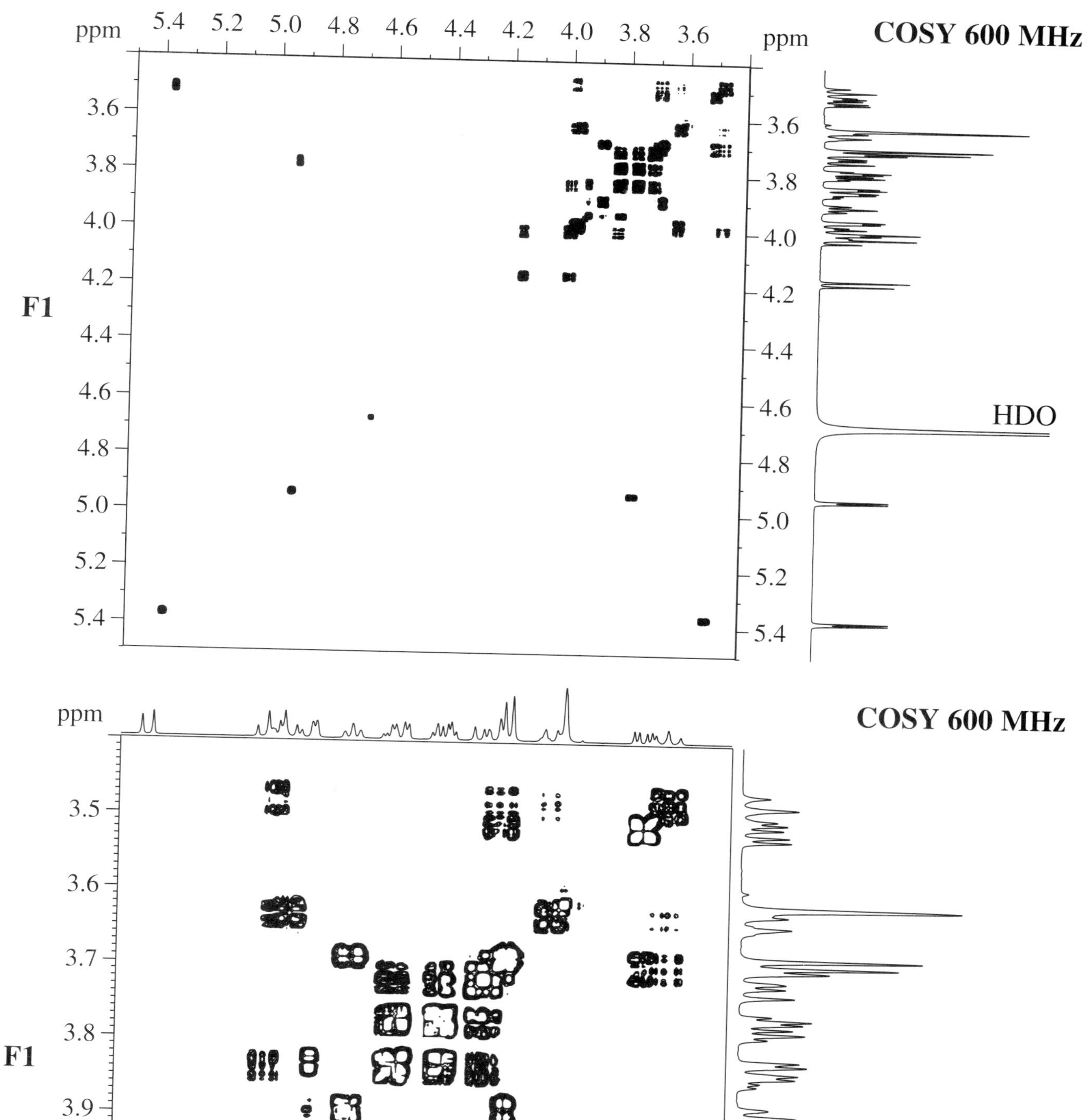

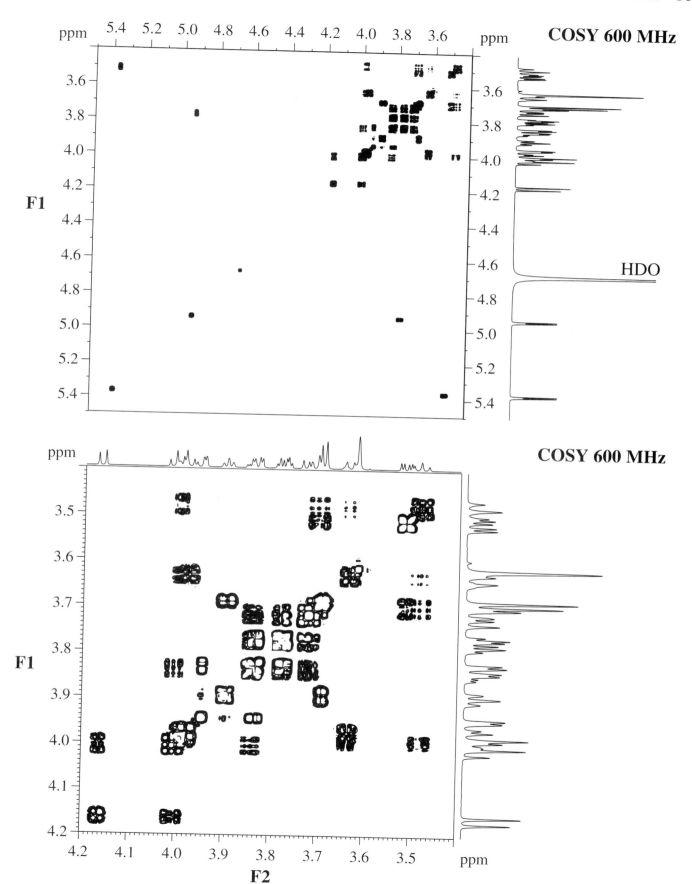

Exercise 5.10

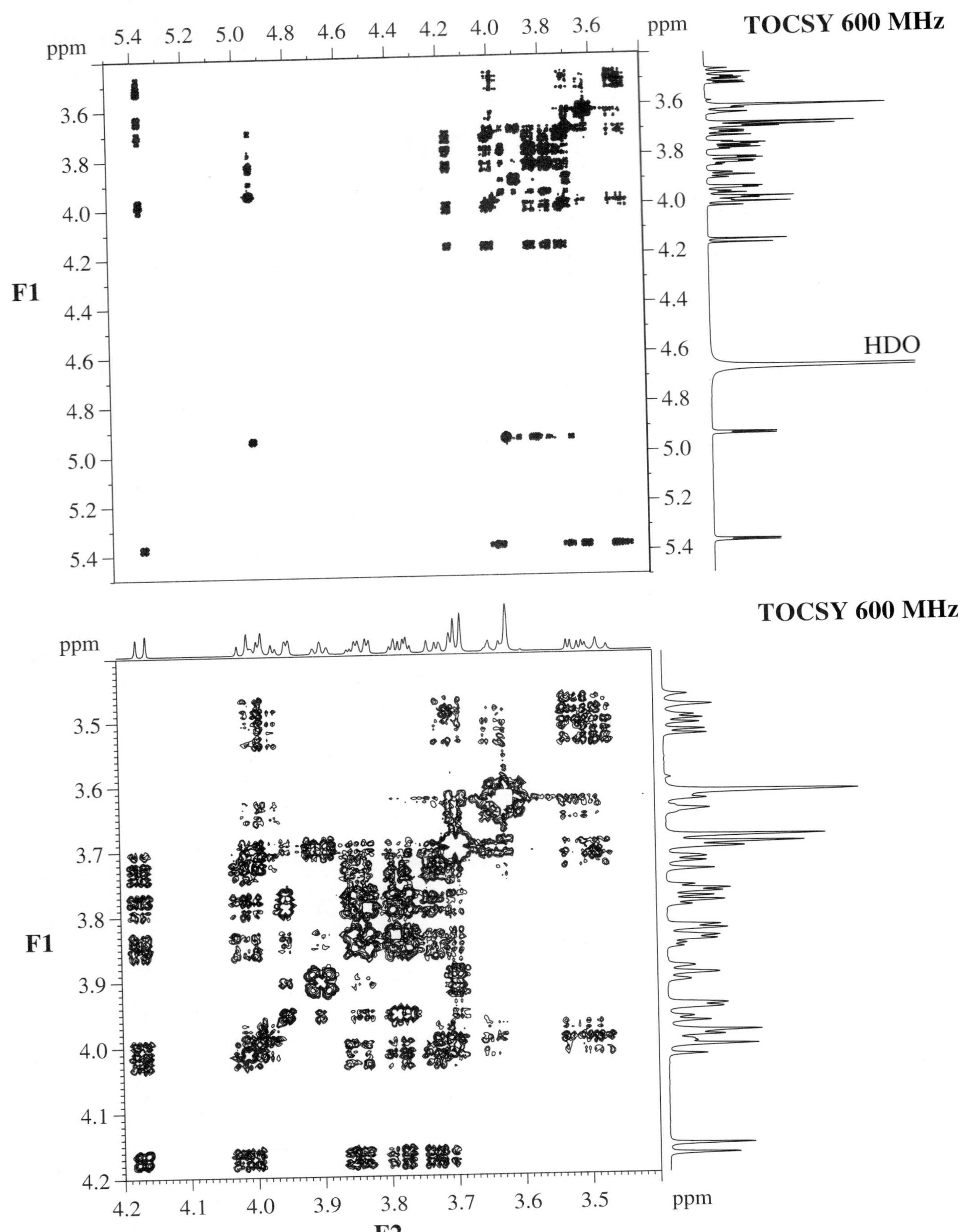

Exercise 5.10

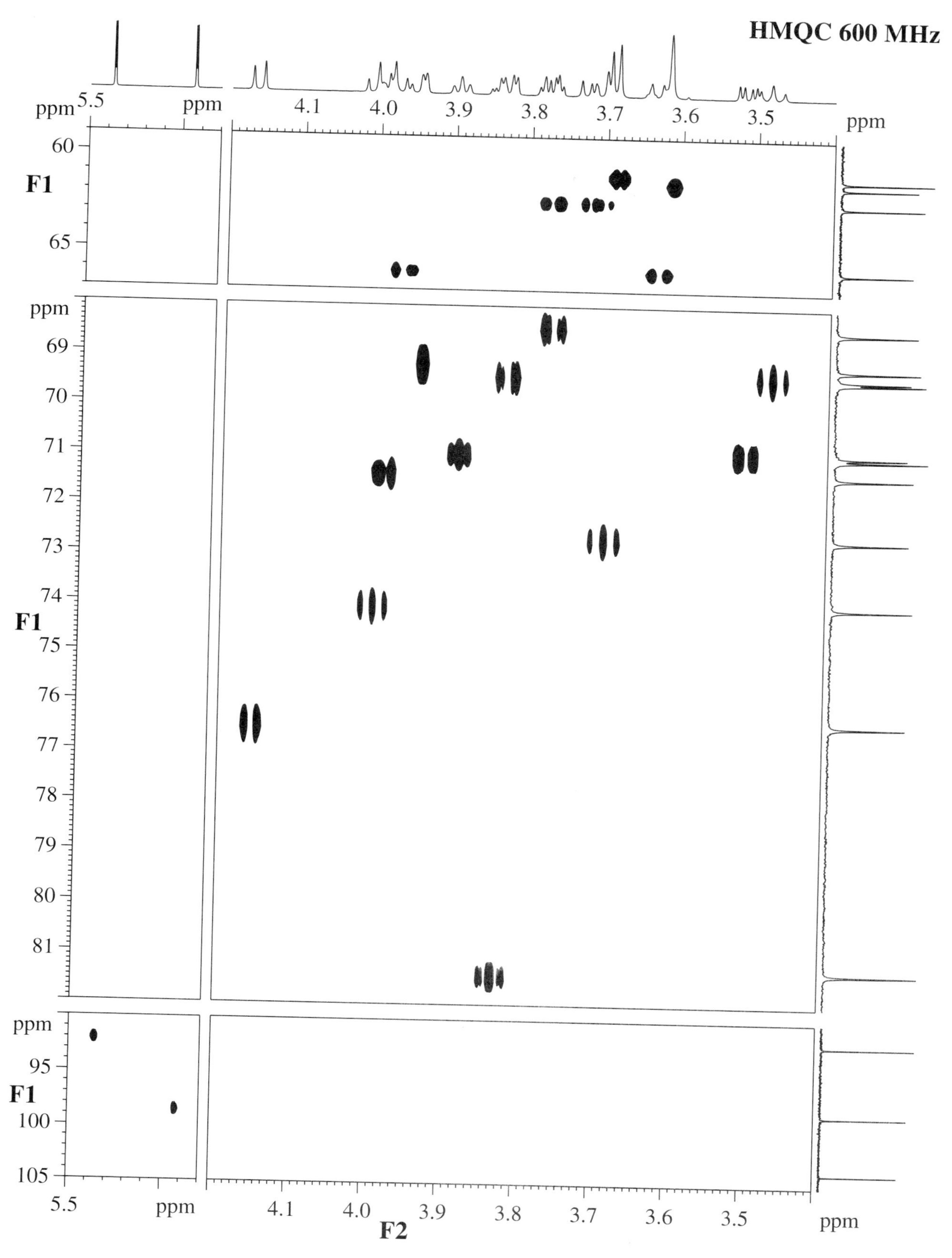

Exercise 5.10

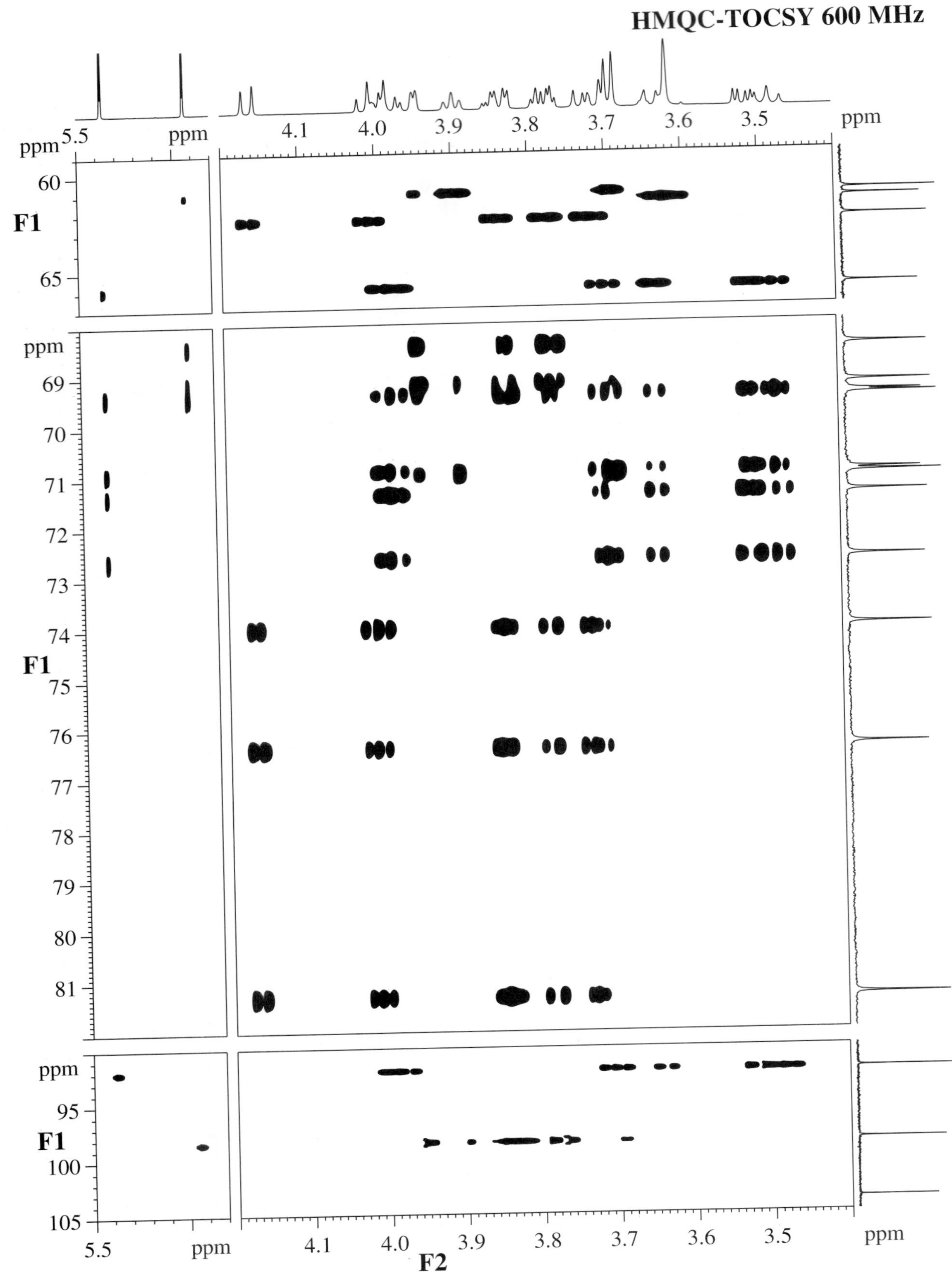

Exercise 5.10

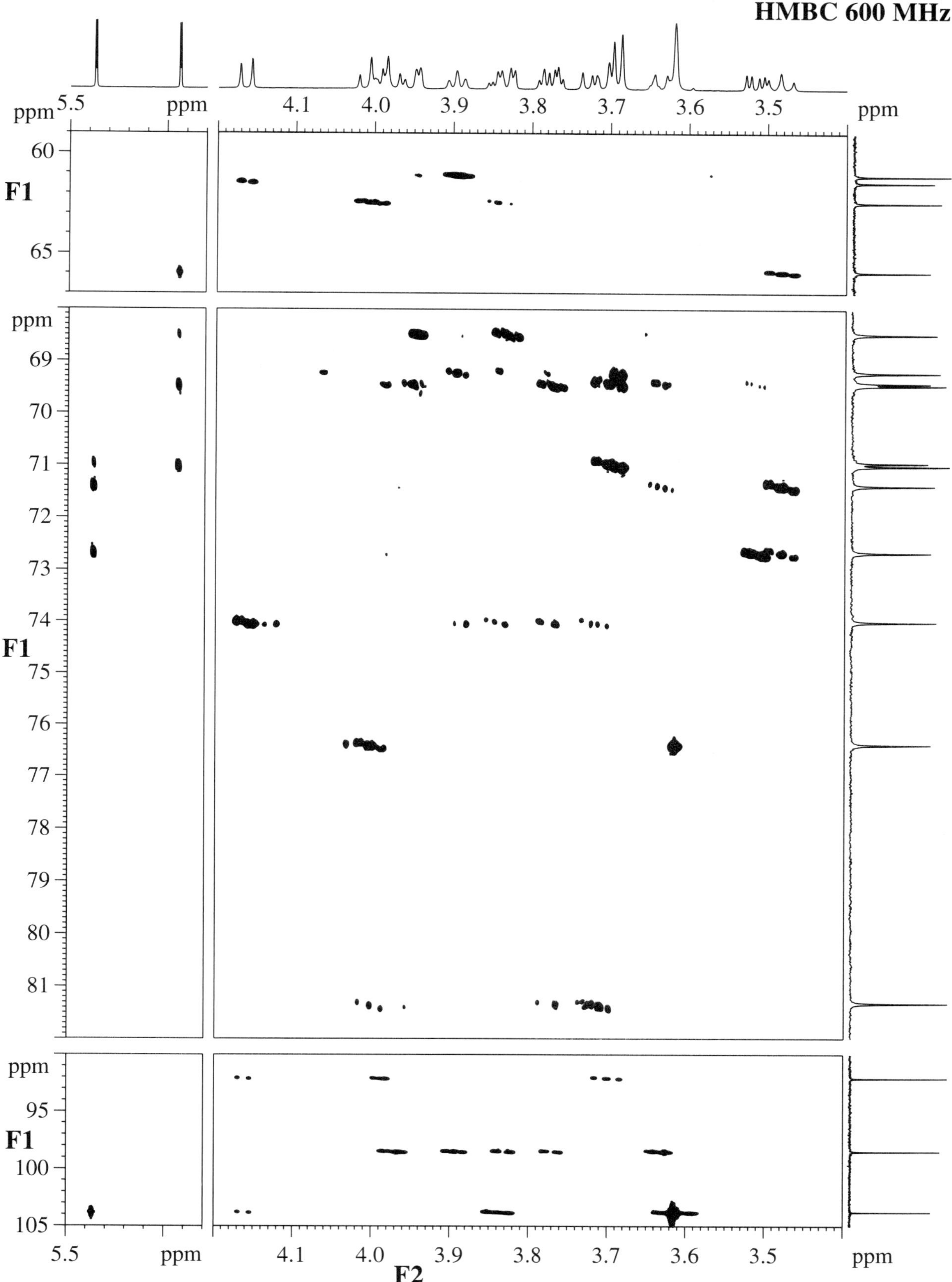

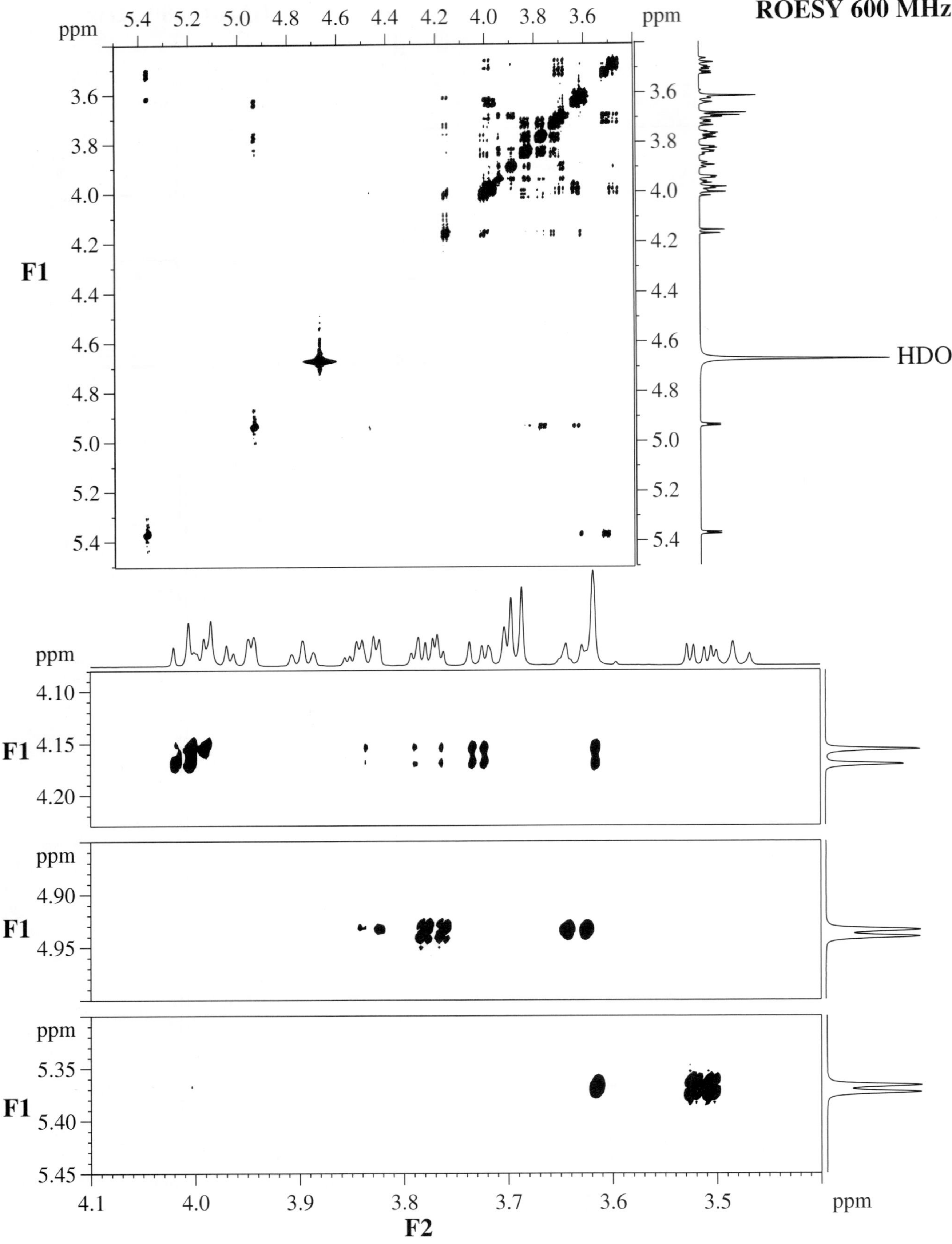
ROESY 600 MHz
ppm
5.4
5.2
5.0
4.8
4.6
4.4
4.2
4.0
3.8
3.6
F1
HDO
F2
4.10
4.15
4.20
4.90
4.95
5.35
5.40
5.45
4.1
3.9
3.7
3.5

Exercise 5.11

^{1}H NMR 600 MHz

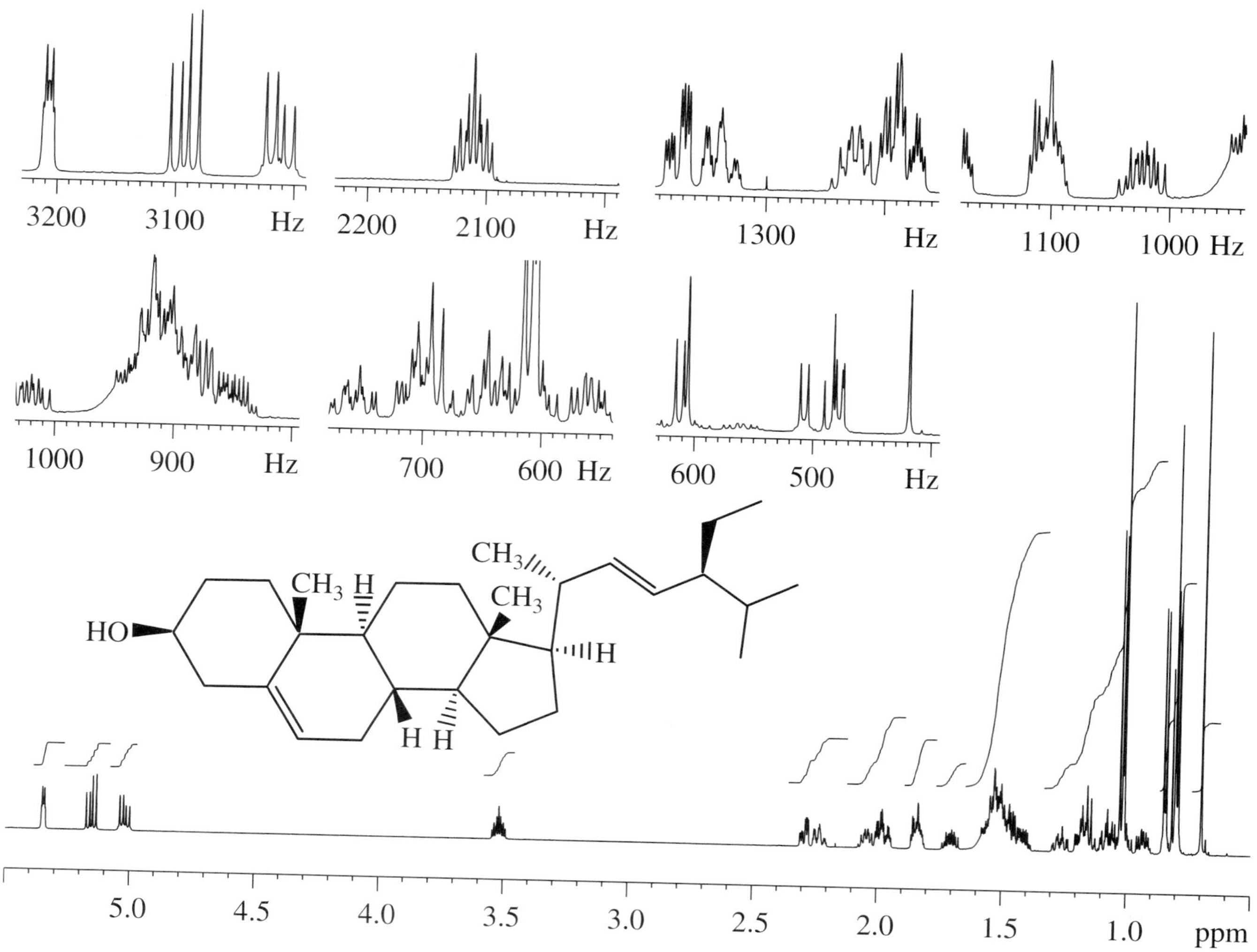

^{13}C/DEPT 600 MHz

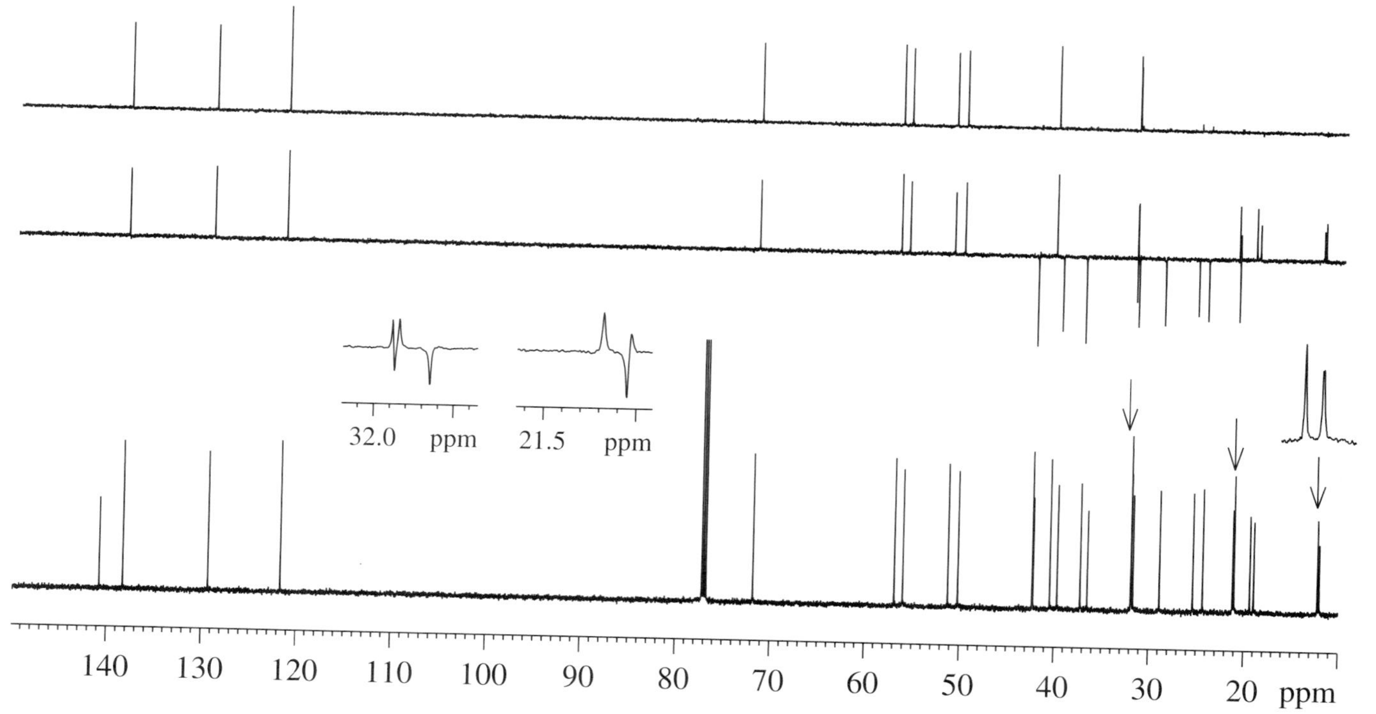

Exercise 5.11

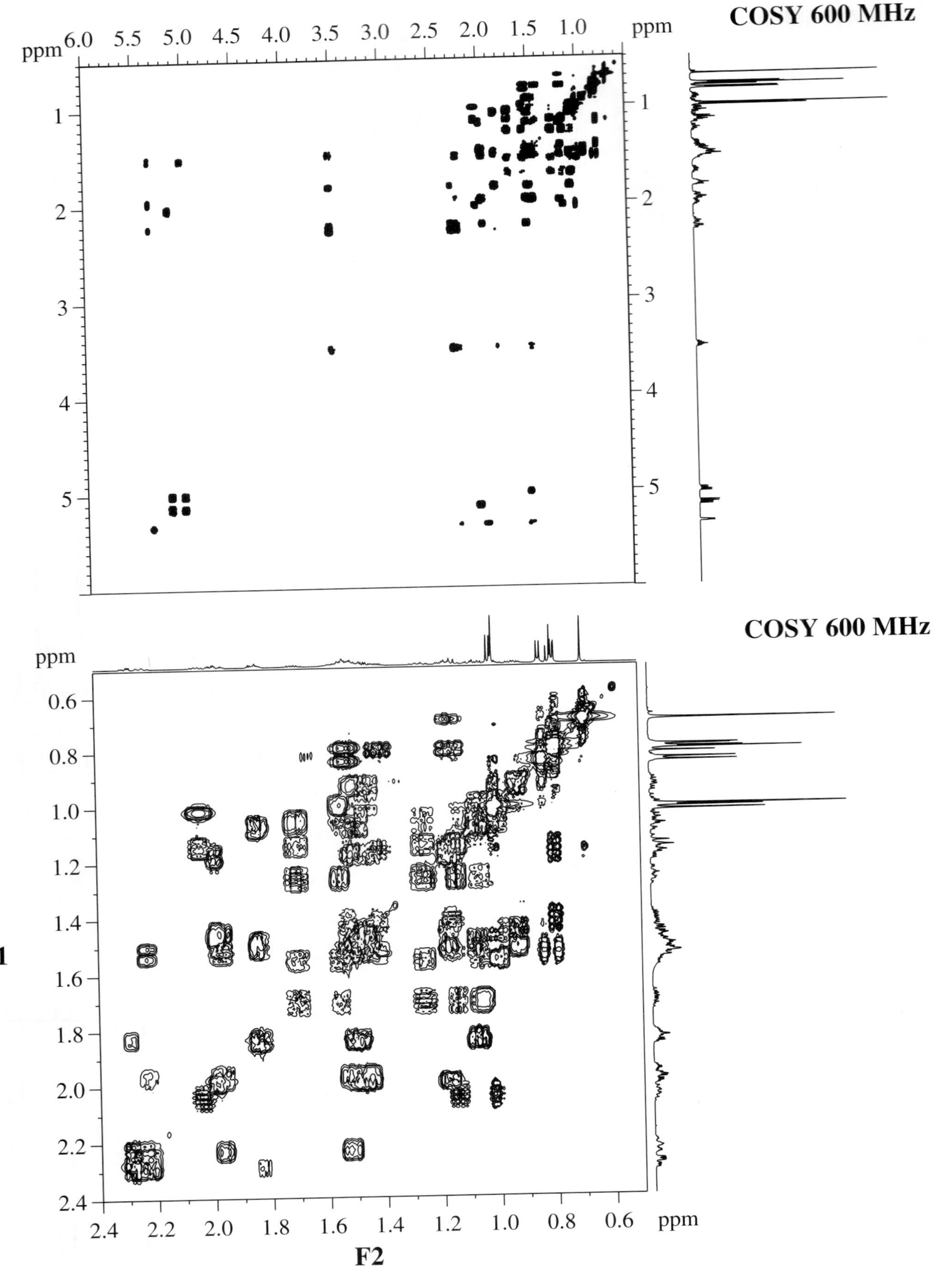

Exercise 5.11

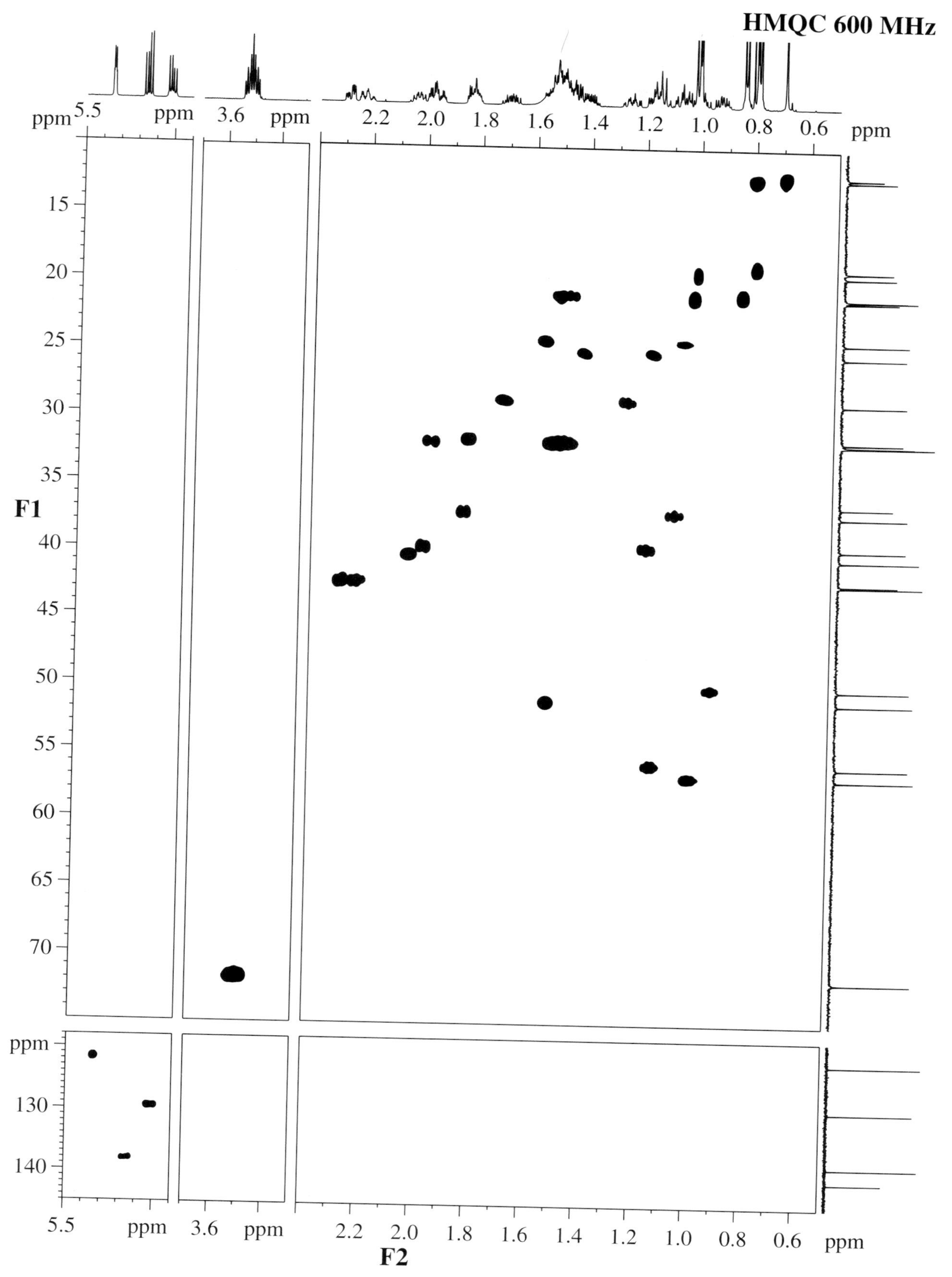

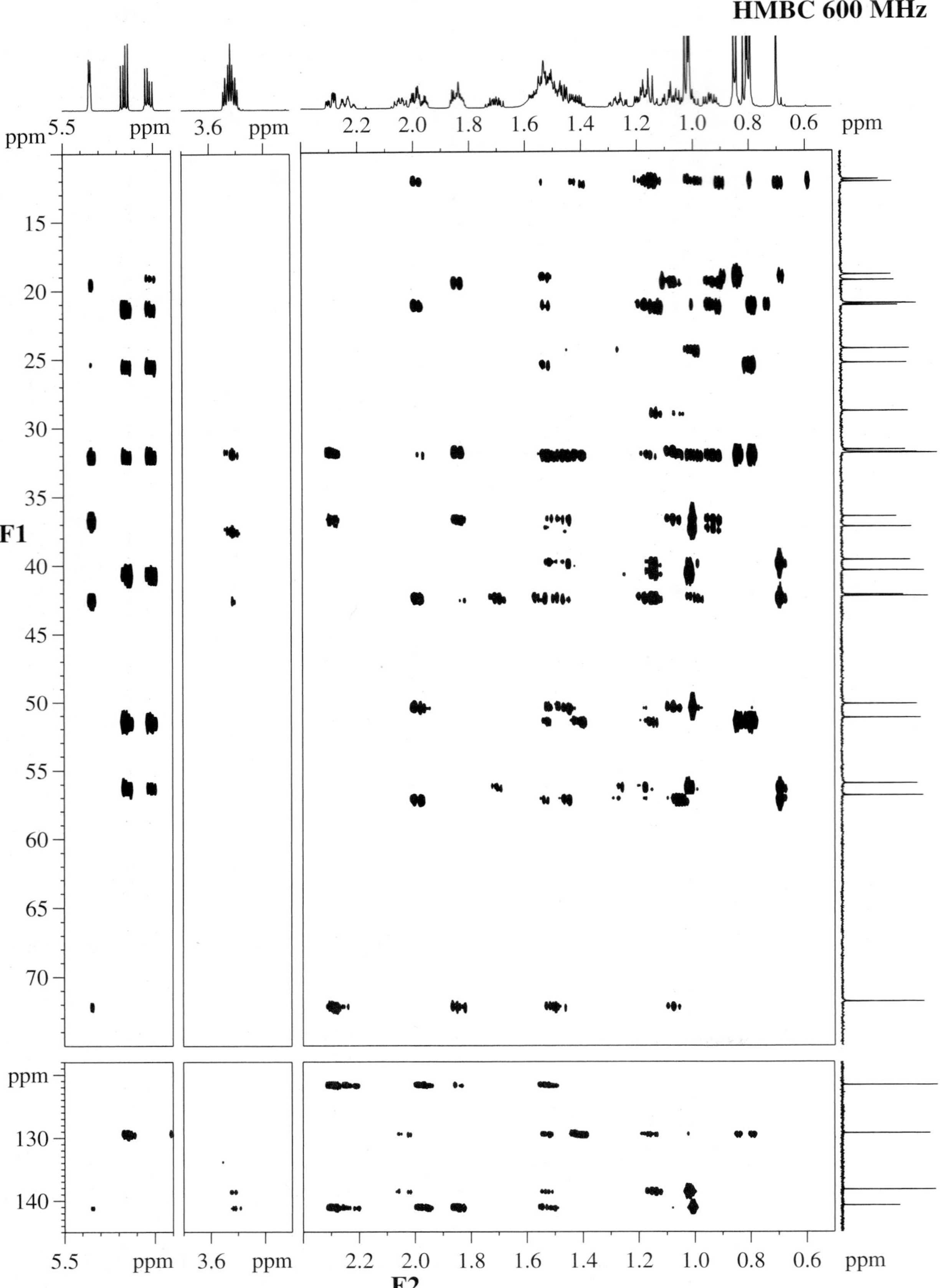
HMBC 600 MHz
ppm
5.5
ppm
3.6
ppm
2.2
2.0
1.8
1.6
1.4
1.2
1.0
0.8
0.6
ppm
15
20
25
30
35
F1
40
45
50
55
60
65
70
ppm
130
140
F2

Exercise 5.12

^{1}H NMR 600 MHz
0° C in 5%/95% D_2O/H_2O

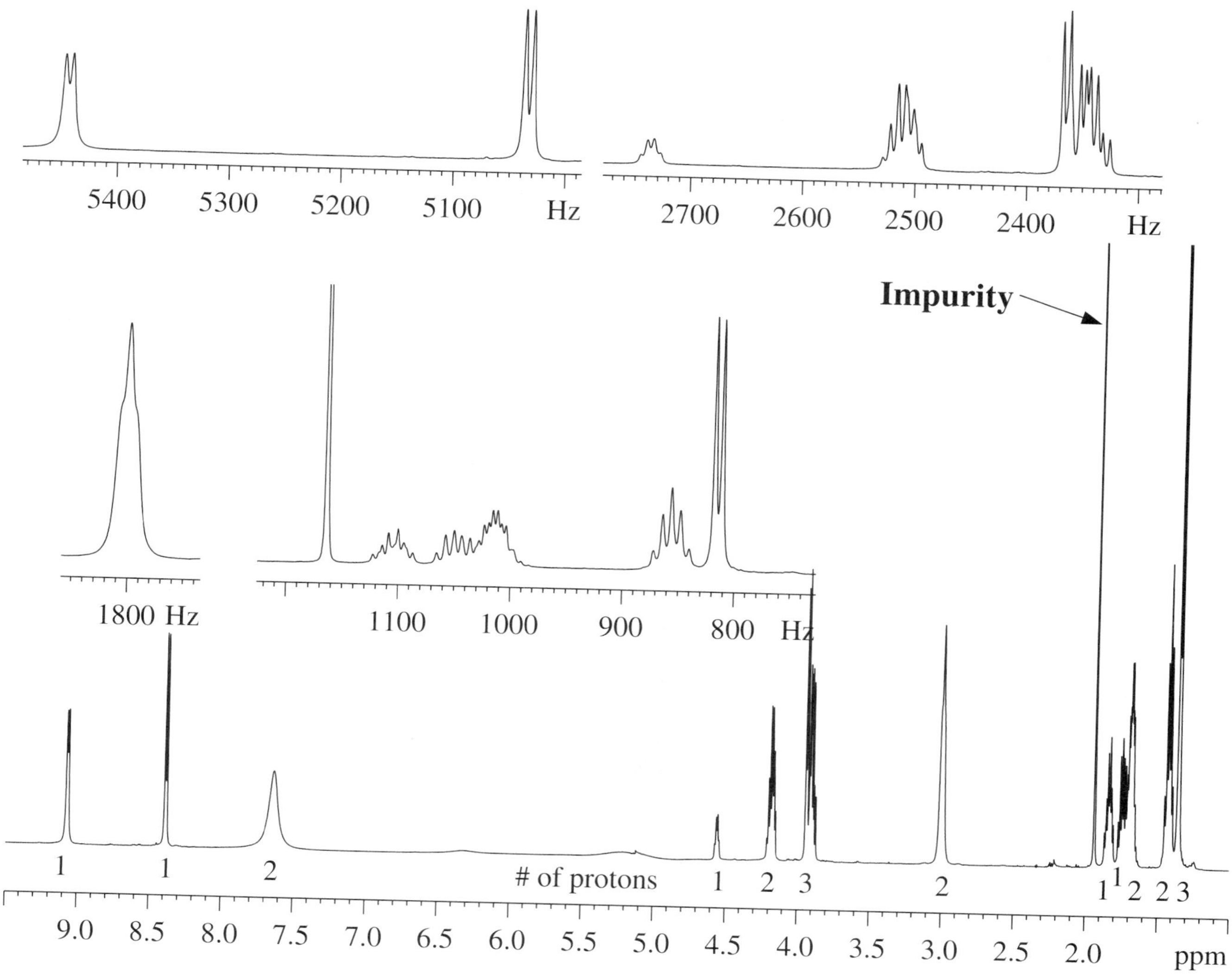

5400 5300 5200 5100 Hz

2700 2600 2500 2400 Hz

Impurity

1800 Hz

1100 1000 900 800 Hz

1 1 2 # of protons 1 2 3 2 1 1 2 2 3

9.0 8.5 8.0 7.5 7.0 6.5 6.0 5.5 5.0 4.5 4.0 3.5 3.0 2.5 2.0 ppm

^{13}C/DEPT 150.9 MHz

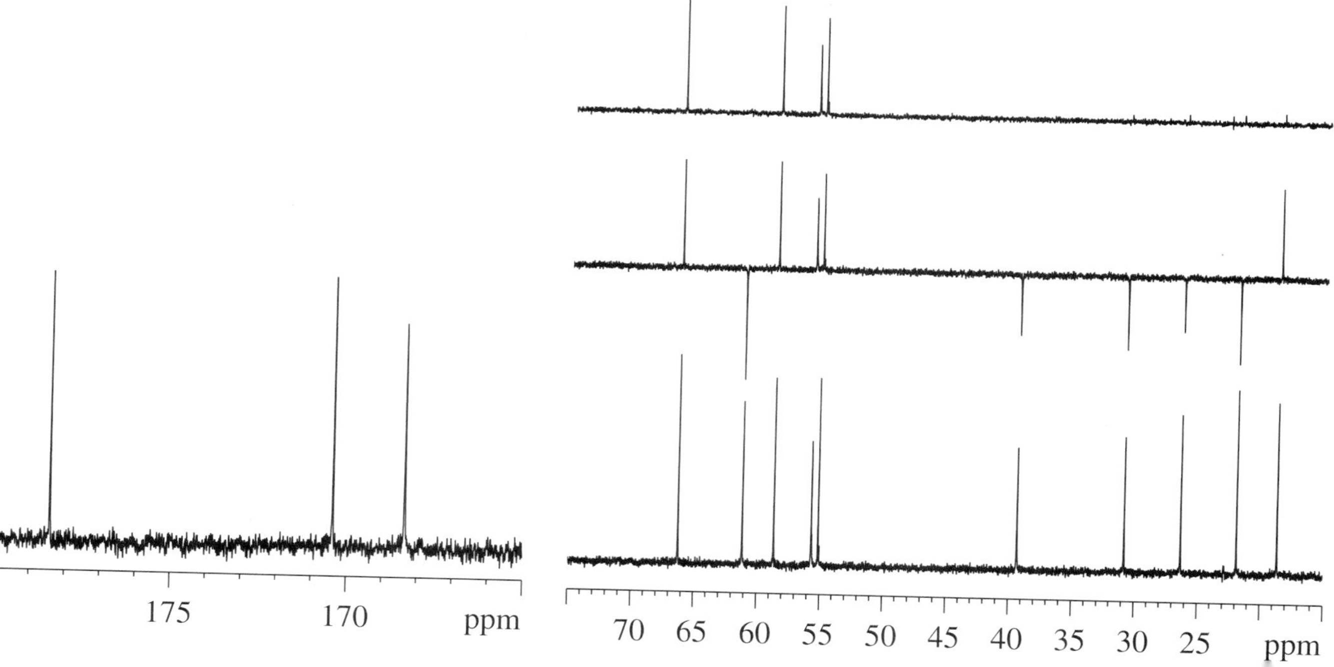

Exercise 5.12

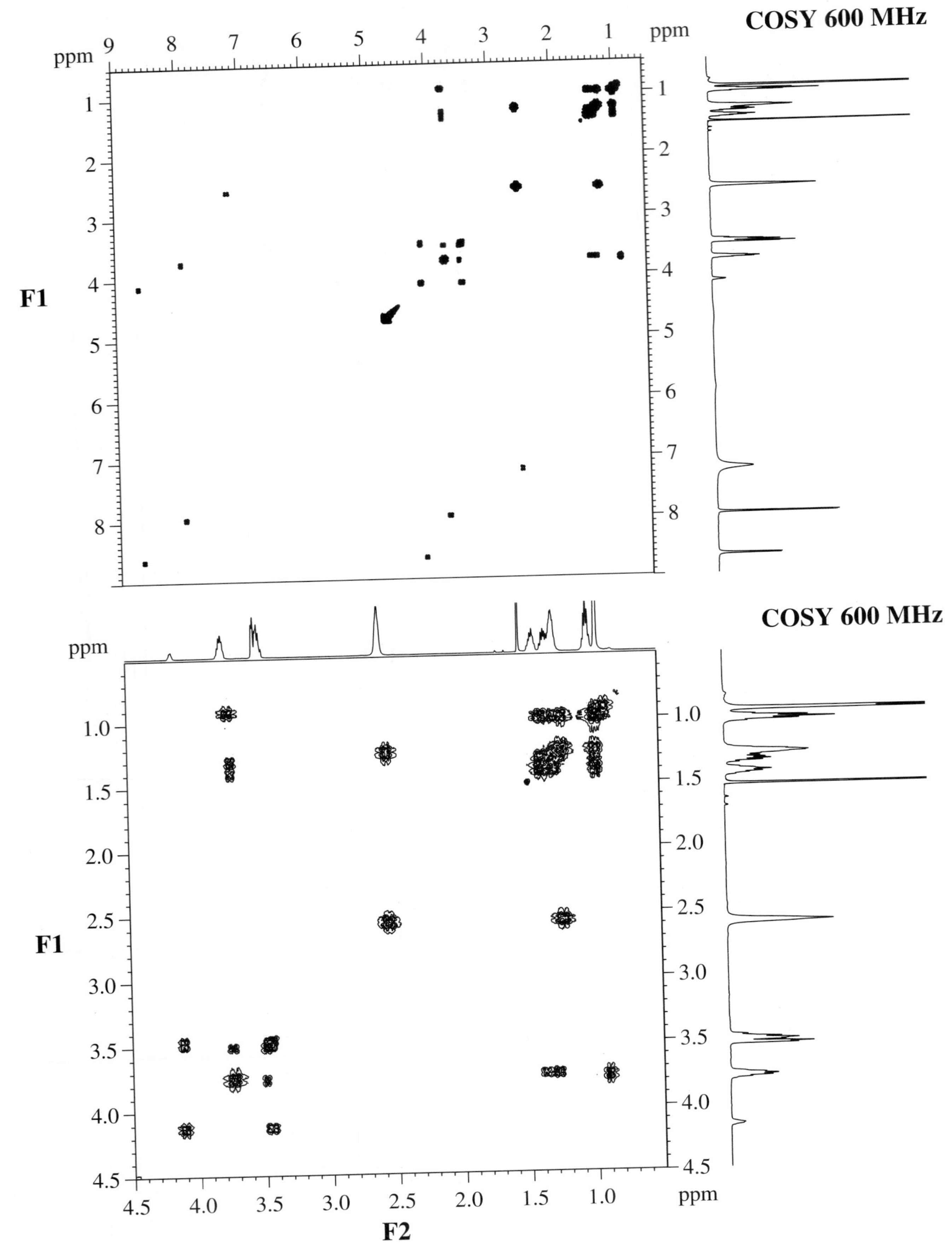

Exercise 5.12

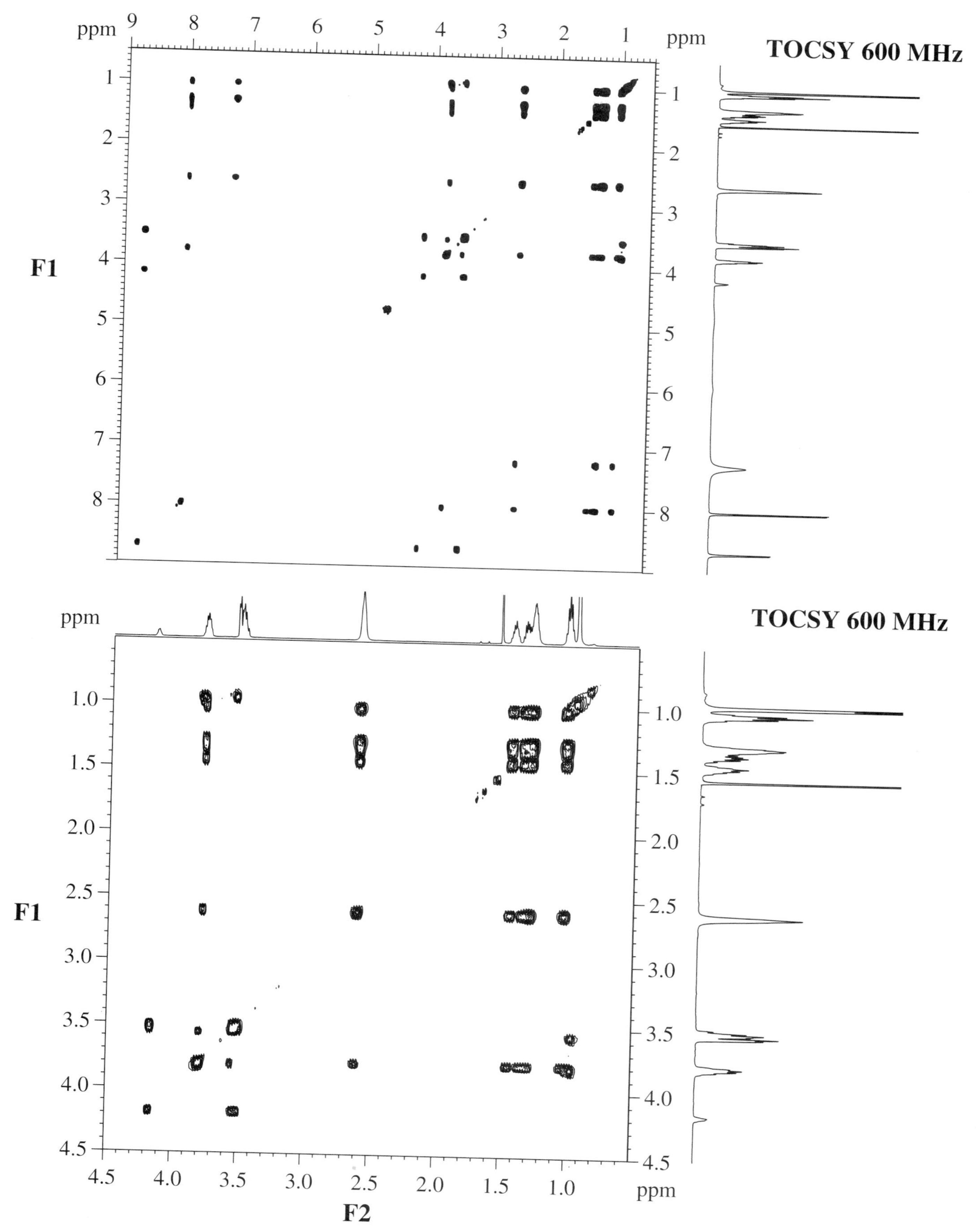

Exercise 5.12

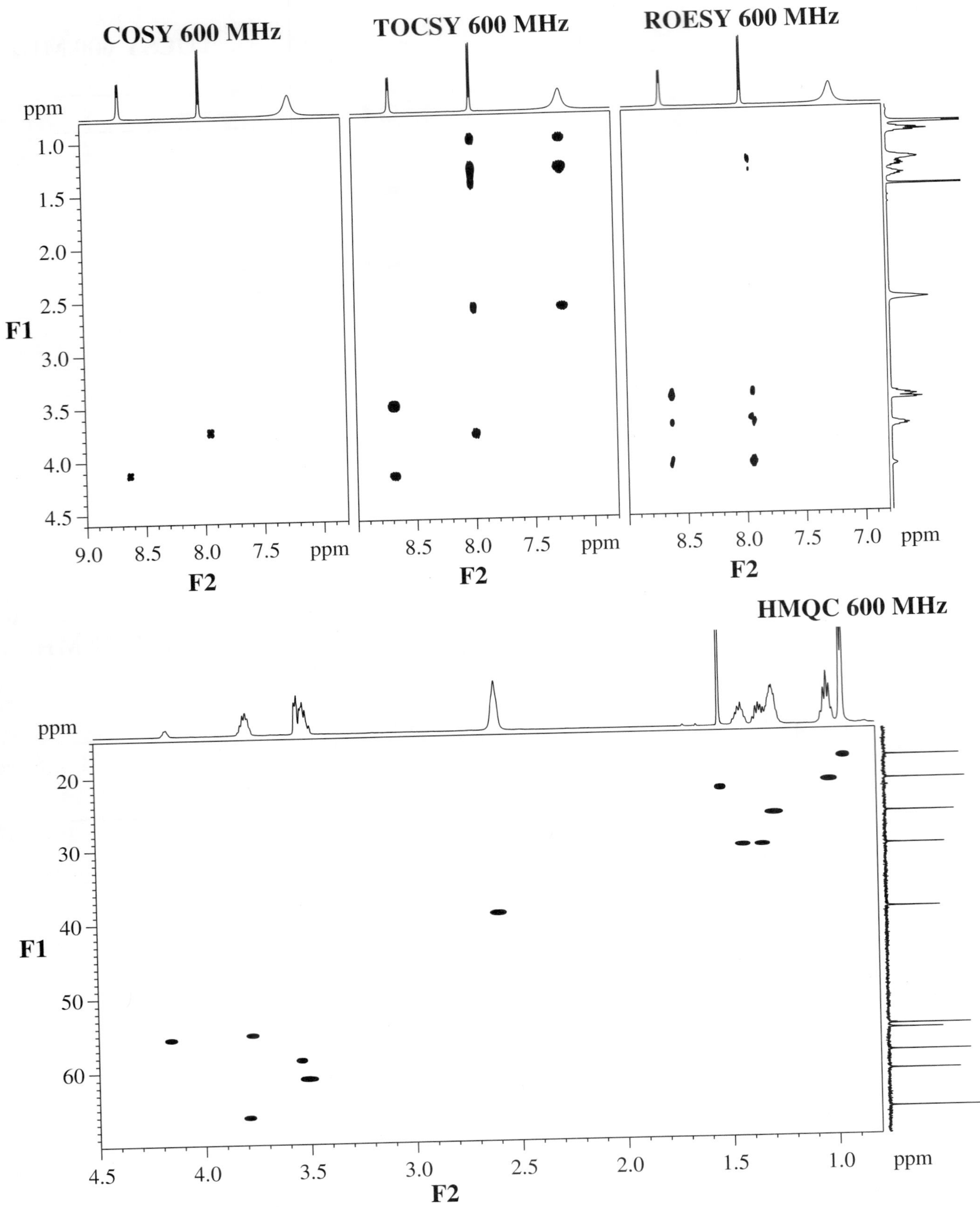

Exercise 5.12

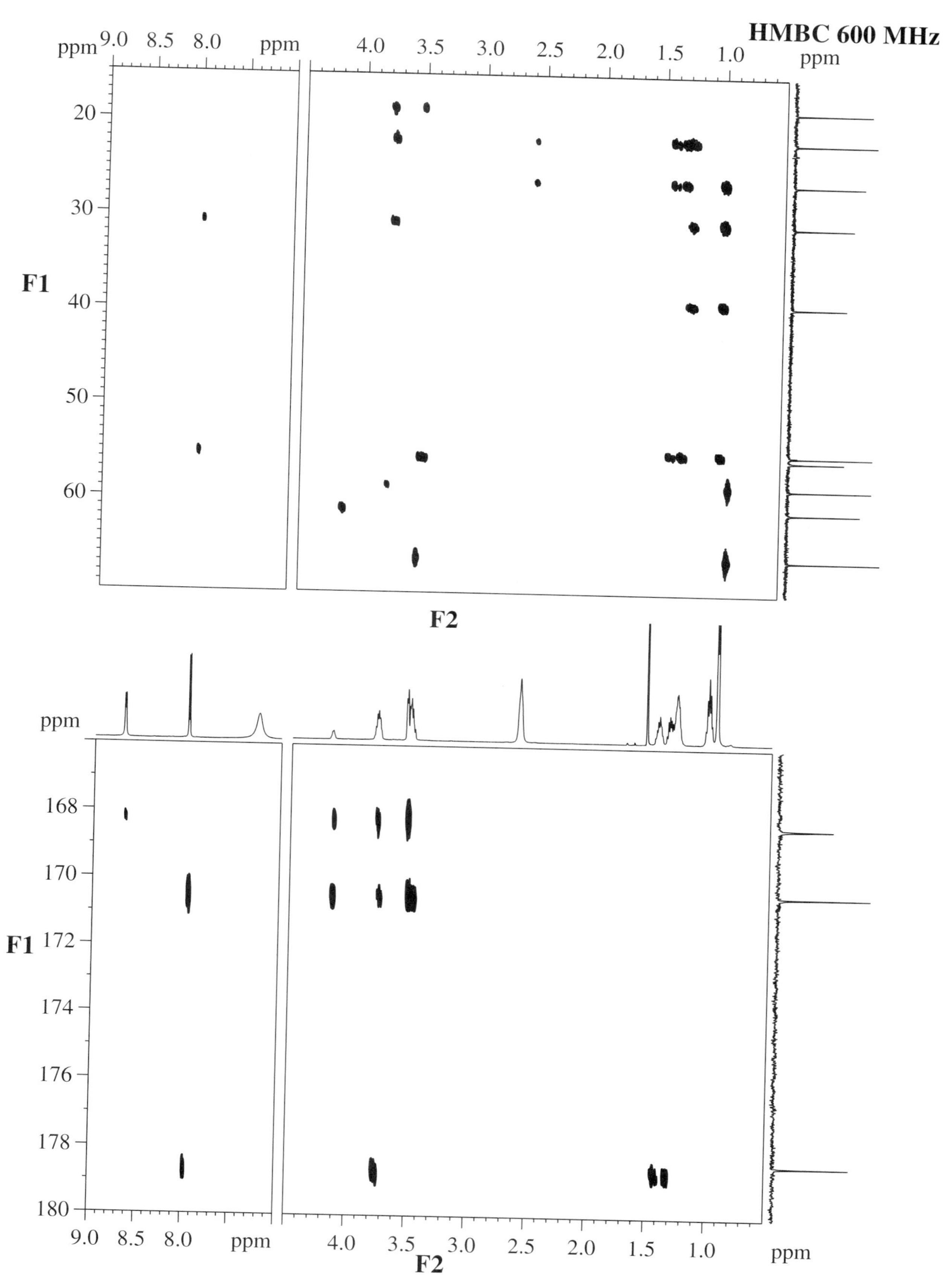

CHAPTER 6

NMR SPECTROMETRY OF OTHER IMPORTANT SPIN 1/2 NUCLEI

6.1 INTRODUCTION

The previous three chapters have shown that nuclear magnetic resonance experiments with ^{1}H and ^{13}C nuclei are enormously useful to the chemist working with organic compounds. There is no need, however, to limit ourselves to these two important nuclei. Indeed, there are 120 different nuclei whose spin number, I, is greater than zero and, therefore, theoretically observable in an NMR experiment. Of these 120 nuclei, 31 of them are dipolar, which means that their spin number is one-half ($I = \frac{1}{2}$).

Appendix A lists all magnetically active nuclei along with some associated useful data. It is worthwhile to spend some time exploring Appendix A and to compare some of the nuclei listed there with ^{1}H and ^{13}C (also listed). First, we find that many elements have more than one magnetically active isotope. A small portion of Appendix A is reproduced in Table 6.1 (reference to 300 MHz ^{1}H), with the addition of useful chemical shift ranges. Table 6.1 lists the nuclei referred to in this chapter. Let us consider the element hydrogen and its possibilities as an example. Hydrogen has three isotopes: ^{1}H (protium), ^{2}H (deuterium), and ^{3}H (tritium); each isotope has been used in NMR studies, and each has different advantages to offer. ^{1}H and ^{3}H both have a spin number of one-half and they both have high relative sensitivities, but ^{3}H has essentially zero natural abundance and it is radioactive. Tritium can therefore only be observed if we intentionally put it into a molecule synthetically, while protium is ubiquitous and we are already familiar with its uses.

Deuterium has a spin of one* and a very low natural abundance; it is not radioactive. It is a useful isotope for mechanistic studies in organic chemistry and biochemistry. Its low natural abundance allows us to ignore it when we observe other nuclei (such as ^{1}H and ^{13}C), yet with modern instruments we are still able to observe natural abundance ^{2}H NMR spectra.

Table 6.1 also lists the resonance frequency for each nucleus at 7.0463 T, which corresponds to a resonance frequency of 300 MHz for ^{1}H. Also listed in the table are useful reference compounds and useful ranges of detection in parts per million. Chemical shift data are only useful if properly referenced. In fact, excluding ^{1}H and ^{13}C data, we must exercise extreme caution when comparing chemical shifts from different sources because most other nuclei have seen more than one compound used as a reference.

We start by acknowledging that our goals are modest as we confront such a vast field as multinuclear NMR. In Section 3.7, we have seen the impact of other nuclei that possess a magnetic moment (especially those with spin one-half) on proton spectra. We will briefly examine the NMR spectrometry of four spin one-half nuclei, which were selected for their historic importance in organic chemistry (and related natural products and pharmaceutical fields), biochemistry, and polymer chemistry. These four nuclei, ^{15}N, ^{19}F, ^{29}Si, and ^{31}P, are presented with a few simple examples and a brief consideration of important experimental factors and limitations.

The theoretical background for these four nuclei has already been presented in Chapters 3 and 4. Our treatment of spin, coupling, NOE, Fourier transformation, etc. can be applied to these nuclei without modification. The concept of chemical shift we also use without modification, but we must avoid exercising the predictive skills that we have developed for ^{1}H and ^{13}C chemical shifts to these nuclei (with some exceptions as noted).

The use of NMR spectrometry of nuclei other than ^{1}H and ^{13}C to characterize and identify organic compounds is now commonplace. The use of other nuclei in NMR experiments ranges from such diverse areas as simply determining whether an unknown compound contains nitrogen to more complex questions of stereochemistry and reaction mechanisms. Although our discussions will be limited to four "other" nuclei, we should not limit our outlook with respect to the possibilities of other nuclei or other experiments. In fact, our intention here is to broaden our outlooks to the nearly limitless possibilities with NMR and the periodic table.

* Nuclei with spin greater than one-half are not covered in this treatment. We note that, with proper consideration of experimental details, we can record a spectrum from a deuterium sample.

TABLE 6.1 Useful Magnetic Resonance Data for Nuclei Discussed in this Chapter

Isotope	Spin	Natural Abundundance %	Sensitivity		MHz at T of 7.0463	Reference Compound	Detection Range ppm
			Relative[a]	Absolute[b]			
^{1}H	1/2	99.98	1.00	1.00	300.000	$Si(CH_3)_4$	10 to 0
^{2}H	1	1.5×10^{-2}	9.65×10^{-3}	1.45×10^{-6}	46.051	$Si(CD_3)_4$	10 to 0
^{3}H	1/2	0	1.21	0	319.990	$Si(CT_3)_4$	10 to 0
^{13}C	1/2	1.108	1.59×10^{-2}	1.76×10^{-4}	75.432	$Si(CH_3)_4$	220 to 0
^{14}N	1	99.63	1.01×10^{-3}	1.01×10^{-3}	21.671	$^{14}NH_3$ (l)[c]	900 to 0
^{15}N	1/2	0.37	1.04×10^{-3}	3.85×10^{-6}	30.398	$^{15}NH_3$ (l)[c]	900 to 0
^{17}O	5/2	3.7×10^{-2}	2.91×10^{-2}	1.08×10^{-5}	40.670	H_2O	1700 to −50
^{19}F	1/2	100	0.83	0.83	282.231	$CFCl_3$	276 to −280
^{29}Si	1/2	4.7	7.84×10^{-3}	3.69×10^{-4}	59.595	$Si(CH_3)_4$	80 to −380
^{31}P	1/2	100	6.63×10^{-2}	6.63×10^{-2}	121.442	85% H_3PO_4	270 to −480

[a] At constant field for equal number of nuclei
[b] Product of relative sensitivity and natural abundance
[c] At 25°C

6.2 ^{15}N NUCLEAR MAGNETIC RESONANCE

After carbon and hydrogen, oxygen and nitrogen are the next two most important elements in organic compounds. In the mind of the organic chemist, the presence of either of these elements represents the presence of one or more "functional groups" and the use of IR spectrometry is invoked. Without detracting in any way from IR, this line of reasoning is nonetheless too restrictive, especially with respect to nitrogen. Inspection of Table 6.1 reveals that in the case of oxygen we have but a single choice of a nucleus on which NMR is possible; ^{17}O has a spin of 5/2 and it is not used much in NMR studies while ^{16}O has a spin of 0 (not listed in Table 6.1).

Nitrogen, on the other hand, has two magnetically active isotopes, ^{14}N and ^{15}N. Because nitrogen compounds are so important* in organic chemistry and its applied fields of natural products, pharmacology, and biochemistry, both of these isotopes have been the subject of intensive NMR investigation.

If we again refer to Table 6.1, we find that neither of the two isotopes of nitrogen is ideal for NMR. The most abundant isotope of nitrogen, ^{14}N, which represents greater than 99% of nitrogen's natural abundance, possesses a spin of 1 and hence an electric quadrupole moment. This nucleus has an inherent low sensitivity and a very broad line due to quadrupolar relaxation. We shall not consider it any further.

The other isotope of nitrogen, ^{15}N, also has an inherent low sensitivity, which, when multiplied by a very low natural abundance, leads to an extremely low absolute sensitivity. Modern instrumentation has largely overcome the problem of sensitivity (for labeled samples or by indirect detection) and we focus our attention on ^{15}N largely because its spin number is one-half and its line-widths are quite narrow.

There are two important experimental factors that must be accounted for if we are to be successful in running ^{15}N experiments. The ^{15}N nucleus tends to relax very slowly; T_1's of greater than 80 seconds have been measured. Thus, either long pulse delays must be incorporated into our pulse sequence or, alternatively, we could provide another route for spin relaxation. A common procedure is to add a "catalytic" amount of chromium (III) acetylacetonate, a paramagnetic substance, whose unpaired electrons efficiently stimulate transfer of spin. In cases where T_1's are not known (and not intended to be measured), pulse delays and pulse angles must be considered carefully because the signal from one (or more) ^{15}N resonance can accrue too slowly or be missed altogether.

The other experimental factor, the NOE, which has already been discussed in both Chapters 3 and 4, we consider now in more detail. Recall that we routinely run ^{13}C experiments with irradiation of the protons (i.e., proton decoupled) which, aside from producing the desired effect of singlets for all carbon

* Aside from the various classes of nitrogen containing functional groups with which we are familiar, entire fields of study have developed based on nitrogen-containing compounds. These fields include: alkaloids, peptides and/or proteins, and nucleic acids. For purposes of study by NMR, the nucleic acids are a favorite subject because not only is nitrogen ubiquitous in these compounds but so is phosphorus (see Section 6.5).

signals, also enhances the signal for carbons with attached protons. This enhancement is due to NOE; the changes in signal intensity arise from polarization of spin populations away from the predicted Boltzmann distribution. The amount of enhancement depends on two factors: One-half the ratio of the nuclei's magnetogyric ratios (γ) is the maximum possible enhancement while the actual enhancement is proportional to the extent of $^{13}C—^{1}H$ dipolar relaxation. For a proton-decoupled ^{13}C experiment, the maximum NOE enhancement is $\gamma_H/(2)\gamma_C$ or 26,753/(2)6728, which equals 1.98. The total sensitivity increase is therefore nearly *threefold* because the NOE enhancement is added to the original intensity.

The actual enhancement for the $^{13}C—^{1}H$ system can be anywhere from 0 to 1.98 depending on the mechanism of relaxation for each individual nucleus. In practice, for carbons with no attached protons, the enhancement is essentially zero since there is practically no $^{13}C—^{1}H$ dipolar relaxation. For small to medium-sized organic molecules, $^{13}C—^{1}H$ dipolar relaxation for carbons with attached protons is very efficient yielding close to the full 200% increase in signal.

If we apply the same reasoning to the ^{15}N nucleus we arrive at a very different situation. The magnetogyric ratio for ^{15}N is small and *negative* ($\gamma = -2{,}712$). A quick calculation shows that the maximum NOE enhancement for ^{15}N is $\gamma_H/(2)\gamma_N$ (26,753/(2)* (−2712)) which is equal to −4.93. For the general case, a spin-one-half nucleus with a positive magnetogyric ratio gives positive NOE enhancement with proton decoupling, while a spin-one-half nucleus with a negative magnetogyric ratio gives negative NOE enhancement.

For the ^{15}N nucleus, the maximum enhancement is −4.93 + 1 or −3.93. In the case where $^{15}N—^{1}H$ dipolar spin-lattice relaxation overwhelms, the signal is inverted (negative) and its intensity is nearly four times what it would be in the absence of ^{1}H irradiation. However, since ^{15}N dipolar relaxation is only one of many relaxation mechanisms for ^{15}N, proton decoupling can lead to NOE's ranging from 0 to −4.93 or a signal ranging from +1 to −3.93. The experimental downside to this situation is that any NOE between 0 and −2.0 lowers the absolute intensity of the observed signal. In fact, an NOE of exactly −1.0 produces no signal at all! In general, as we saw for carbon and now for nitrogen, proton decoupling is commonly practiced for routine heteronuclear NMR experiments. In so doing, we must always bear in mind the practical outcome of NOE enhancement.

Let us turn our attention to ^{15}N spectra. As we have already mentioned, natural-abundance ^{15}N spectra can routinely be obtained on modern instruments even though ^{15}N is about an order of magnitude less sensitive than ^{13}C. Today, there is general agreement that liquid ammonia* is the standard reference compound for ^{15}N (used externally) although in the past, many compounds such as ammonium nitrate, nitric acid, nitromethane, and others have been used. When consulting the literature, reliable chemical shifts can usually be obtained after correcting for their reported standard.

We are by now familiar with the construction of a ppm scale and need not consider the details here. Nitrogen, like carbon, is a second-row element and in many ways experiences similar electronic influences. To a first approximation, the chemical shifts of nitrogen-containing organic compounds closely parallel carbon chemical shifts. The chemical-shift range for nitrogen in common organic compounds is about 500 ppm, which is about twice that for carbon chemical shifts.† Figure 6.1 shows the chemical shift ranges for many types of nitrogen containing compounds. The relatively large chemical shift range taken together with the very narrow lines for ^{15}N resonances means that the chances of fortuitous overlap in an ^{15}N spectrum is even smaller than in a ^{13}C spectrum. Our analogy with carbon can be furthered by pointing out that, within a given class of compounds, we can usually derive highly predictive substituent parameters as we did for carbon and that the magnitude of these parameters are similar to carbon.

Lest we take this analogy with carbon too far, we must remember that nitrogen is unique and has chemical shift features peculiar only to nitrogen. The two most important ones are both due largely to the unshared pair of electrons found on nitrogen.

Just as this electron pair has a large impact on the chemistry of these compounds, it also has a great influence on the chemical shifts of the nitrogen in certain environments. We find quite often that chemical shift positions are more sensitive to the solvent than are structurally similar carbon resonances. The other way that the lone pair on nitrogen influences chemical shift is through protonation. Surprisingly, we cannot say in which direction the signal will shift upon protonation; we can say, however, that both the direction and the magnitude of the shift are characteristic of the specific type of nitrogen.

The proton-decoupled ^{15}N spectrum of formamide is shown at the top of Figure 6.2. This spectrum looks remarkably like a ^{13}C spectrum with a single resonance. This ^{13}C-like appearance as opposed to an ^{1}H-like

* Nitromethane was used occasionally as an internal reference and set to zero ppm, but the resulting ^{15}N chemical shifts for nitrogen-containing organic compounds are generally negative. The use of liquid ammonia as an external reference precludes the need for negative numbers because virtually all ^{15}N atoms are deshielded by comparison, but handling liquid ammonia is awkward. The usual procedure is to add 380 ppm to the shift obtained by reference to nitromethane in order to report the shift relative to liquid ammonia.

† Both carbon and nitrogen chemical-shift ranges are larger than we are considering here. Chemical shifts outside of these ranges are unusual and not of interest in our discussion (See Figure 6.1).

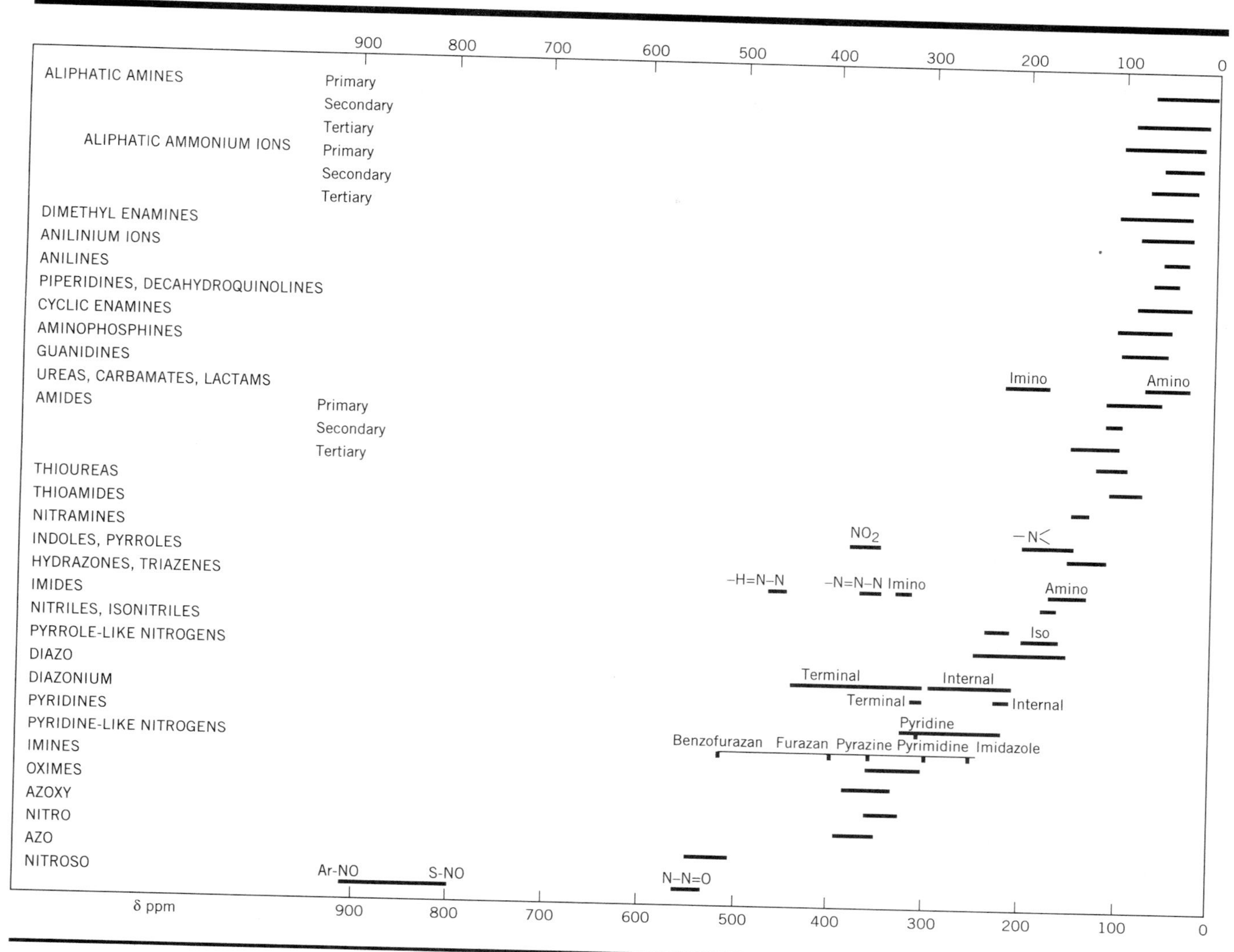

FIGURE 6.1 Chemical shift ranges for various nitrogen-containing compounds. Adapted from Levy and Lichter (1979).

appearance will be the norm throughout this chapter as long as we have proton decoupling. The other feature worth noting is the direction of the peak. With proton decoupling, formamide experiences negative NOE enhancement and should therefore be negative. In an FT experiment however, the initial phase of the FID is random and we could have phased the peak either up or down; the direction of peaks in a spectrum is arbitrary and only has meaning if both types are found in the same spectrum (see for instance DEPT). The nitrogen in formamide has partial double bond character and is shifted accordingly (deshielding effect).

The proton-coupled and proton-decoupled ^{15}N spectra of ethylenediamine are shown next in Figure 6.2. First, we note the relatively shielded position of the nitrogen here compared to formamide. The other noteworthy feature of these spectra is the inset showing proton coupling. Before we discuss this part of the figure, a few general comments are in order. One, two, and three bond ^{15}N—^{1}H couplings are common while long-range couplings usually require intervening pi bonds. The magnitude and sign of the coupling constants have been compiled but detailed consideration here is beyond our goal. We note that $^{1}J_{NH}$ varies from about 75 to 135 Hz, $^{2}J_{NH}$ between 0 and 20 Hz, and $^{3}J_{NH}$ between 0 and 10 Hz. If we again consider the inset in Figure 6.2, we note an apparent quintet with a relatively small coupling constant of about 2–3 Hz. Our interpretation is that we see no $^{1}J_{NH}$ coupling because of rapid exchange and that the two- and three-bond coupling constants are about the same. If we have equal coupling constants (careful, not always a good assumption) in a heteronuclear system (e.g. $^{n}J_{XH}$), first order rules always apply because $\Delta\nu$ is of the order of millions of Hz.

The proton decoupled ^{15}N spectrum of pyridine, which has an aromatic nitrogen, is recorded third in Figure 6.2. A comfortable pattern is now emerging and, in fact, we can safely say that there is nothing extraordinary about this or any of the spectra that we have seen thus far. Since our goal in this chapter is not to

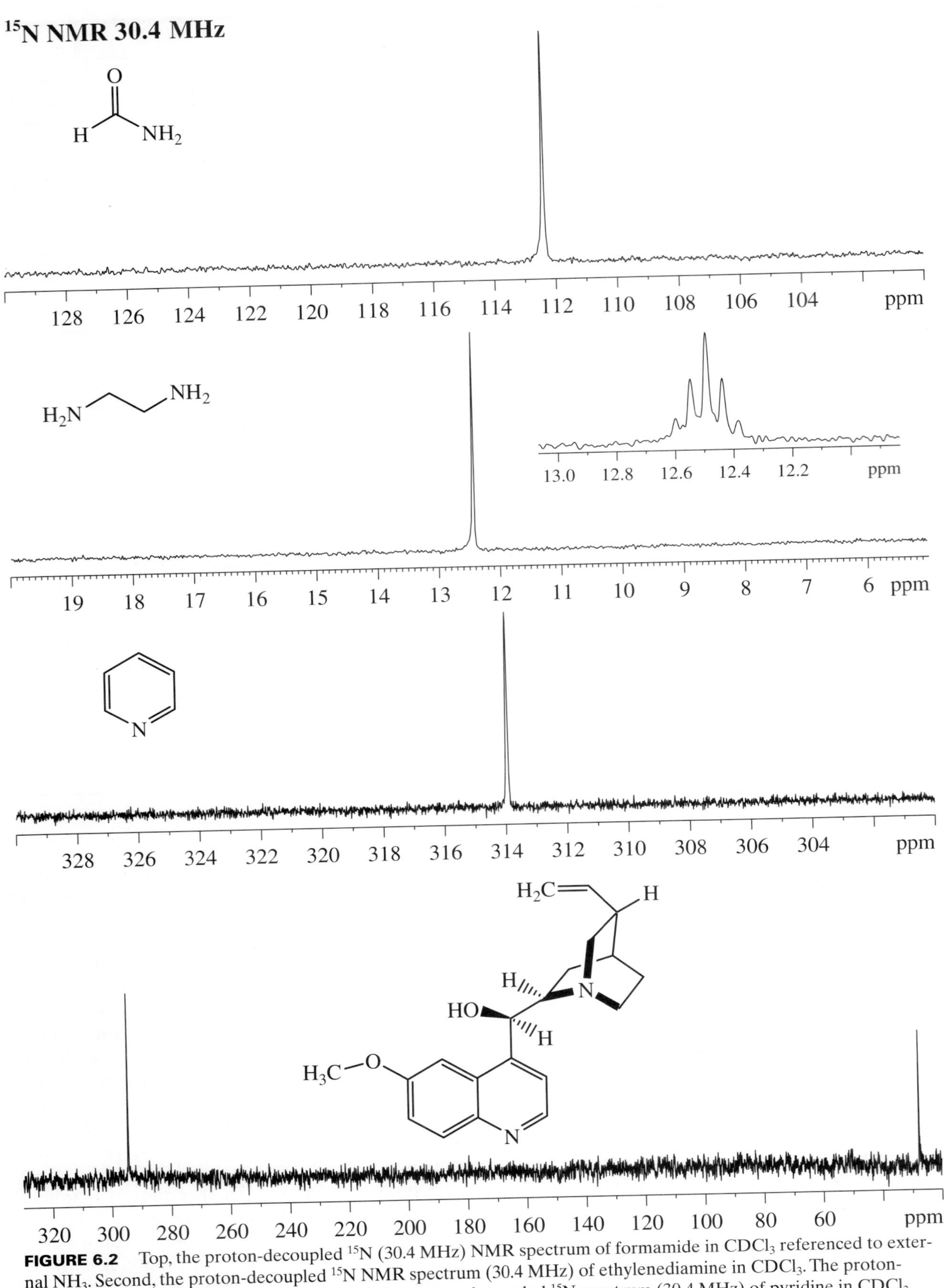

FIGURE 6.2 Top, the proton-decoupled ^{15}N (30.4 MHz) NMR spectrum of formamide in $CDCl_3$ referenced to external NH_3. Second, the proton-decoupled ^{15}N NMR spectrum (30.4 MHz) of ethylenediamine in $CDCl_3$. The proton-coupled spectrum is shown in the inset. Third, the proton-decoupled ^{15}N spectrum (30.4 MHz) of pyridine in $CDCl_3$. Fourth, the proton-decoupled ^{15}N (30.4 MHz) spectrum of quinine in $CDCl_3$.

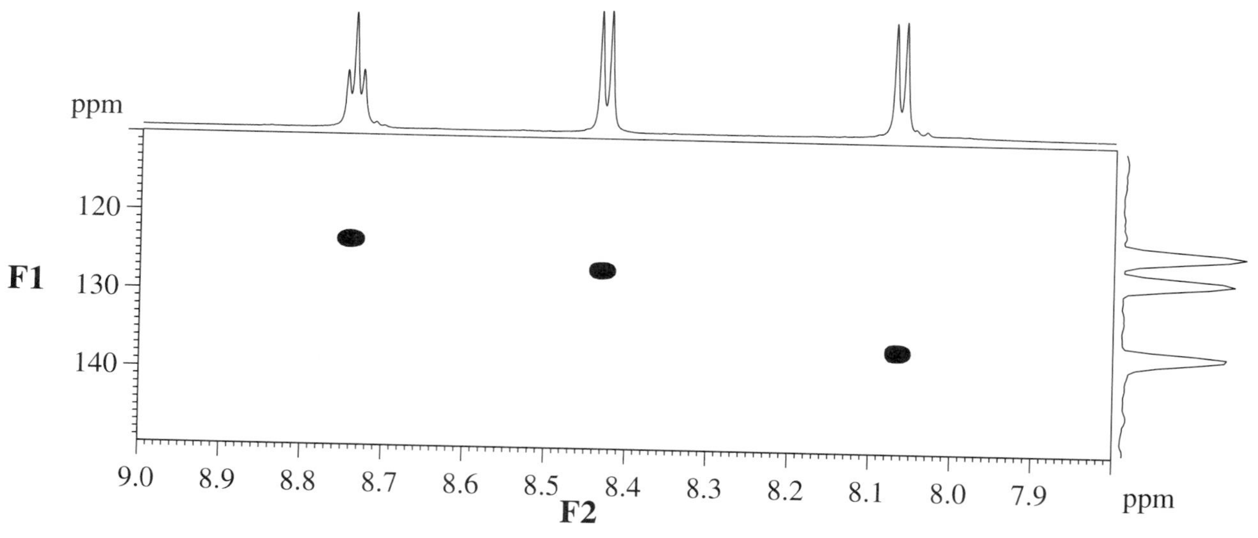

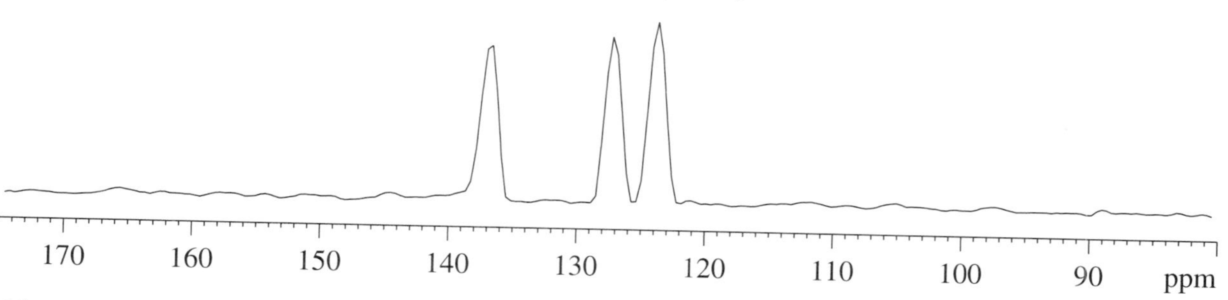

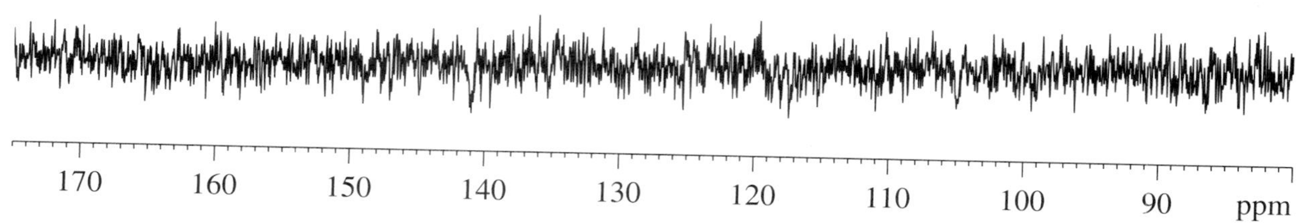

FIGURE 6.3 The ^{1}H—^{15}N HSQC spectrum and its projection of a tetra-peptide (structure shown in Figure) in dilute solution. The 2-D spectrum required 1 hour of instrument time. The bottom spectrum shows an attempt to obtain a 1-D ^{15}N spectrum for 15 hours.

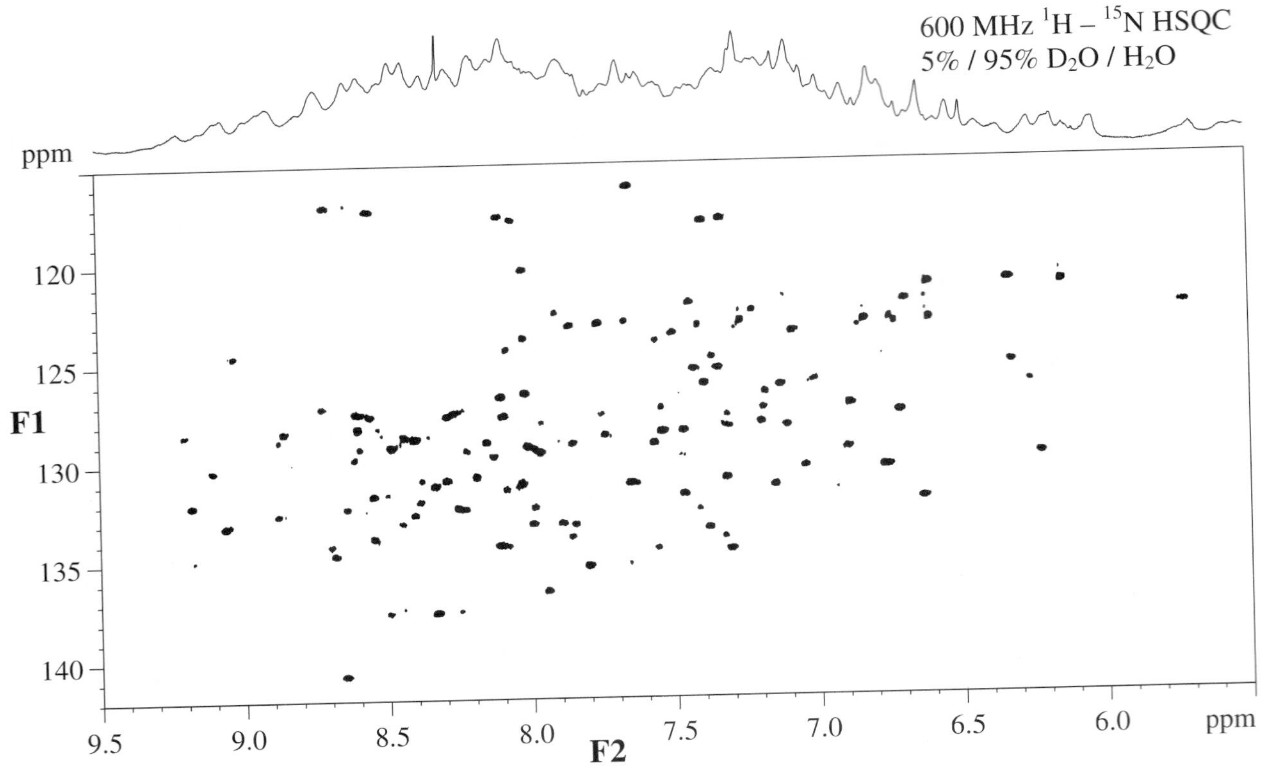

FIGURE 6.4 The $^1H—^{15}N$ HSQC spectrum of ^{15}N labeled myoglobin.

catalog the literally thousands of chemical shifts that have been reported but to "open the door" to the NMR of other nuclei besides carbon and hydrogen, let us finish the ^{15}N section of simple or one-dimensional (1-D) spectra by briefly considering the proton decoupled ^{15}N spectrum of quinine (bottom part of Figure 6.2), a well studied, naturally occurring alkaloid.

Both nitrogen atoms of quinine are evident and, without hesitation, we can make assignments. If, instead, we had isolated quinine as a new unknown, natural compound we could envision a procedure in which we extend the concept of identifying compounds from a *combination of spectra* to include heteronuclear NMR; ^{15}N NMR would assume a natural place alongside mass, infrared, and other NMR spectrometries. Furthermore, as we made the transition from simple ^1H and ^{13}C spectra to correlation spectroscopy in Chapter 5, we ask: Is there more to ^{15}N NMR than proton decoupled (and proton coupled) spectra? The question is rhetorical; it is obvious that correlation experiments are possible and indeed many are commonplace. A good example is the $^1H—^{15}N$ HSQC, which is illustrated in Figure 6.3 using the tetra-peptide example (valine-glycine-serine-glutamate) used in Chapter 5. (HSQC yields the same type of spectrum as the HMQC; the HSQC has less line broadening in the F1 dimension and therefore provides better resolution in F1.) The experiment was conducted on a dilute solution (5 mg in 0.5 ml of 95% H_2O/5% D_2O) of the compound. This spectrum gives us the chemical shifts of all three amide nitrogens indirectly. At the bottom of this figure is a comparison of the projection from the 2-D spectrum onto the ^{15}N axis, which took 1 hour of instrument time to obtain, with an attempt to obtain the ^{15}N spectrum directly, which showed no signal even after 15 hours. As discussed in Chapter 5, inverse detected experiments give a sensitivity increase of $(\gamma^1{}_H/\gamma^{15}{}_N)^{5/2}$, which is about 310-fold for $^1H—^{15}N$ HSQC compared to directly detected methods.

In the biochemical field, proteins are prepared for NMR analysis by synthesizing them with ^{15}N labeled amino acid residues. Since ^{15}N has a natural abundance of 0.368%, labeling increases the sensitivity of the experiment by a factor of about 270, which, when combined with the HSQC type experiment, yields about an 84,000-fold sensitivity increase over natural abundance directly detected methods. Thus, it is possible to obtain quality spectra on proteins up to 50 kDa.

As an illustration of the power of this technique, the 600-MHz $^1H—^{15}N$ HSQC of myoglobin is given in Figure 6.4; a complete analysis of this spectrum is beyond the scope of this text. Note however, that myoglobin, which has 153 amino acid residues, is hopelessly overlapped in the amide portion of the simple ^1H NMR spectrum. However, by separating these resonances in the ^{15}N dimension (F1), most of the $^1H—^{15}N$ correlations can be found. The amide ^{15}N region for peptides is approximately a 40-ppm window from 110–150 ppm.

6.3 ^{19}F NUCLEAR MAGNETIC RESONANCE

The NMR of ^{19}F has great historic importance. Fluorine has only one naturally occurring isotope, ^{19}F, and Table 6.1 reveals that it is an ideal nucleus for study by NMR. The sensitivity of ^{19}F is about 0.83 that of ^{1}H; this fact is the main reason why ^{19}F NMR developed contemporaneously with ^{1}H. The literature on chemical shifts and coupling constants is now old, and, again, some caution is needed when comparing older chemical shifts with newer literature because much of it is referenced to external CF_3COOH. Today, trichlorofluoromethane, $CFCl_3$, is the standard ^{19}F reference compound (0 ppm verses CF_3COOH at −78.5 ppm), which is generally inert, volatile, and gives rise to a single ^{19}F resonance.

Our approach to ^{19}F NMR is quite different than was our approach to ^{15}N NMR and this section is brief. Fluorine is monovalent and can be thought of as a substitute for hydrogen in organic compounds. Fluorine is virtually unknown in naturally occurring organic compounds; our interest in fluorine NMR involves synthetic compounds. Proton decoupling of ^{19}F NMR experiments does not involve NOE enhancement in a critical way, and there are no other experimental factors out of the ordinary. In fact, ^{19}F NMR spectra were routinely recorded using CW instruments in exactly the same way as were ^{1}H spectra. Much of the chemical shift and coupling constant data for ^{19}F were published contemporaneously with those of ^{1}H. Figure 6.5 gives chemical shift ranges for various fluorine-containing compounds.

The ^{1}H NMR spectrum of fluoroacetone was given in Chapter 3 to show the effect of fluorine on ^{1}H spectra. Recall that both the methylene group and the methyl group were split by the ^{19}F atom, each into doublets with different coupling constants. In Figure 6.6, we have an opportunity to see the reciprocal effect of the protons on the fluorine atom in fluoroacetone. This figure shows a complete set ^{19}F, ^{1}H, and ^{13}C spectra. In the proton-decoupled spectrum, of course, we see only a singlet for the fluorine atom, reminding us again of a ^{13}C spectrum. In the proton-coupled spectrum, however, we see that the fluorine is coupled to two sets of hydrogens producing a triplet with a large coupling constant; the triplet is further split into three quartets by a four-bond coupling with the methyl group. The coupling constants are listed on each spectrum for convenience. We emphasize again that the combination of ^{1}H, ^{13}C, and ^{19}F spectra is more convincing and more informative than any one spectrum is by itself.

An example of an aromatic fluorine-containing compound can be found in Figure 6.7, where we have recorded the ^{19}F spectra (both proton-coupled and decoupled) of fluorobenzene along with the ^{1}H and ^{13}C spectra. Once again we find a singlet for the fluorine atom in the proton-decoupled spectrum and a complex multiplet for the fluorine atom in the proton-coupled spectrum. The fluorine atom couples differently to the *ortho*-, *meta*-, and *para*-protons in this monosubstituted compound. Coupling constants for proton-fluorine can be found in Appendix F of Chapter 3.

It is tempting to treat the proton spectrum in a similar fashion to our treatment of fluoroacetone. What we have, however, is a higher-order system, AA′GG′MX, where X is a fluorine atom. The fluorine absorption is deceptively simple in being symmetrical; the halves are mirror images. The A and A′ protons are not magnetic equivalent since they do not couple equally with the M proton or the M′ proton. Nor are the M and M′ protons magnetic equivalent since they do not couple equally with the A proton or the A′ proton.

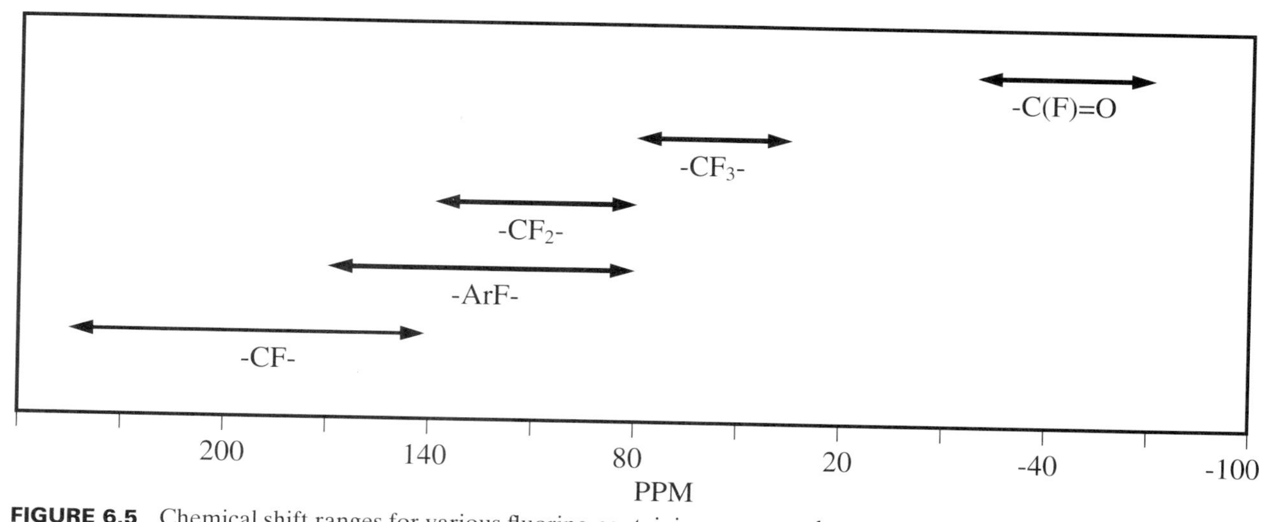

FIGURE 6.5 Chemical shift ranges for various fluorine-containing compounds.

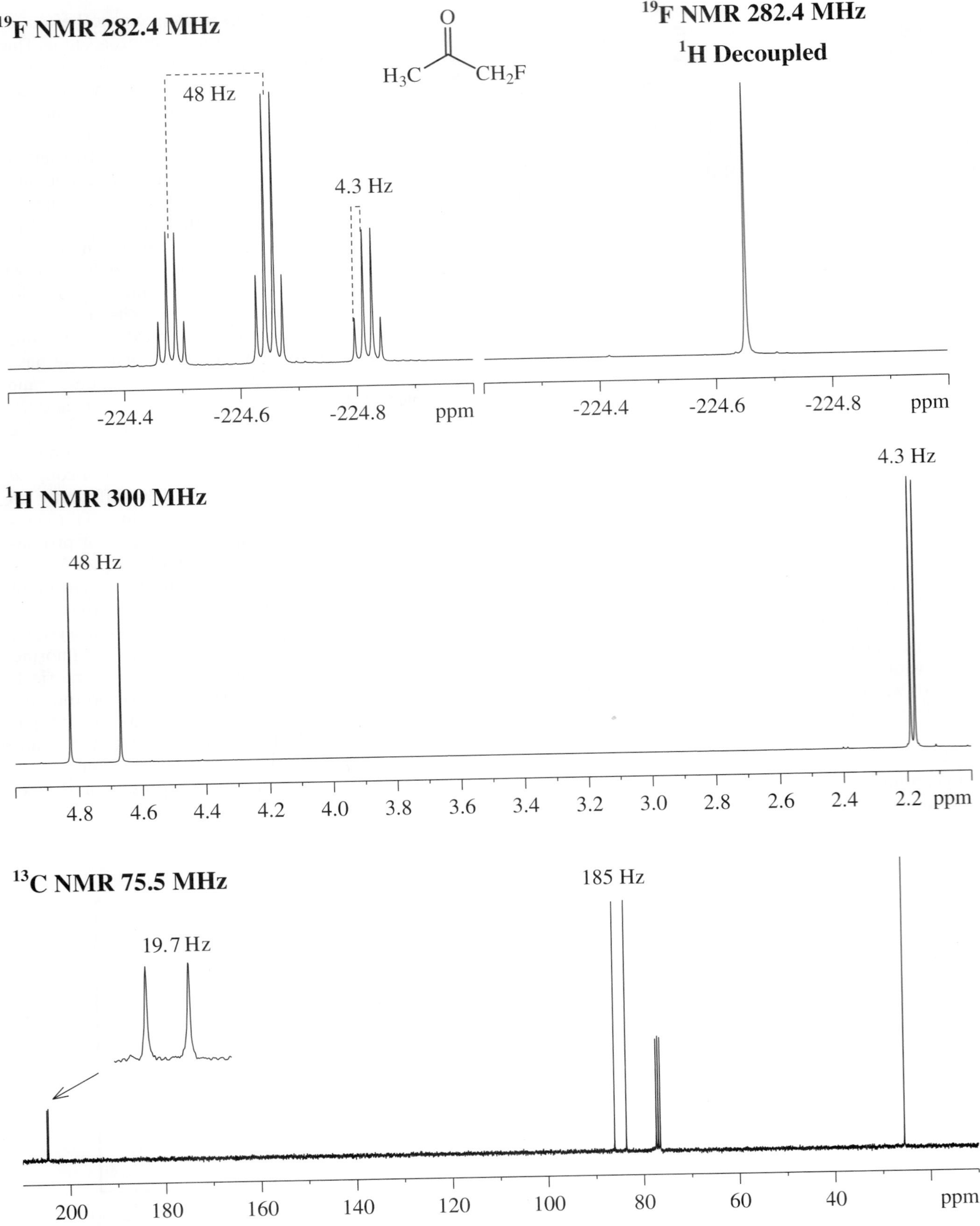

FIGURE 6.6 The proton-decoupled and proton-coupled ^{19}F NMR spectrum (282.4 MHz) of fluoroacetone in $CDCl_3$. The ^{1}H and ^{13}C spectra show reciprocal ^{19}F coupling.

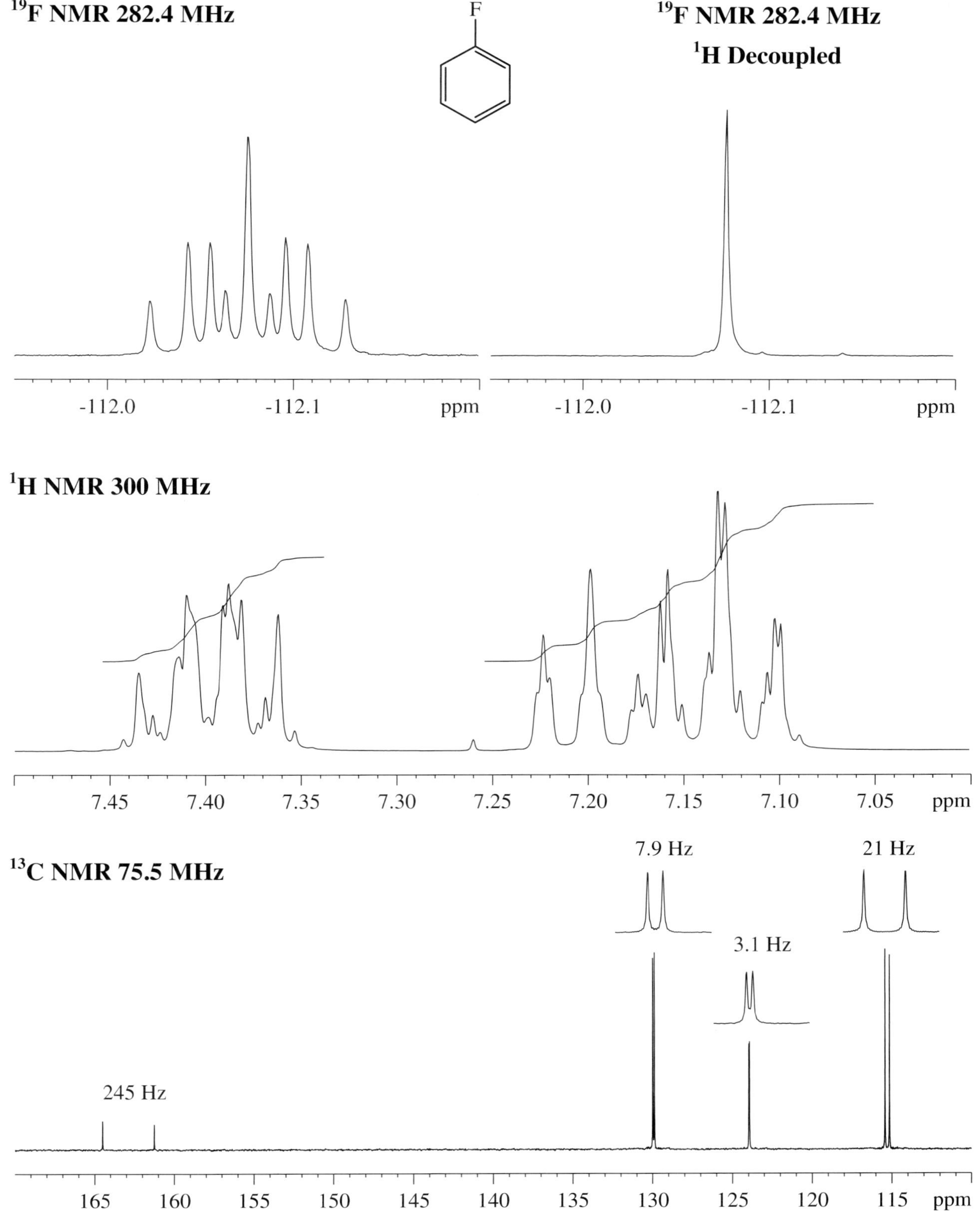

FIGURE 6.7 The proton-decoupled and proton-coupled ^{19}F NMR spectrum (282.4 MHz) of fluorobenzene in $CDCl_3$. There is long range coupling in the ^{1}H and ^{13}C spectra (see text for explanation).

Note the resemblance of the fluorine atom absorption to the symmetry of the absorptions of the magnetic nonequivalent aromatic protons described in Section 3.9. The complexity in all of these spectra is not a result of an inadequate magnetic field. In fact, a more powerful magnetic field would result in greater complexity. This is quite different from the response of a first-order system, which may not be resolved in a weak magnetic field, but would show a first-order spectrum at a sufficiently high field.

Chemical shifts for ^{19}F are difficult to predict and we will make no attempt to offer a predictive model. Table 6.2 presents an empirical compilation of ^{19}F chemical shifts for various fluorine-containing compounds. One reason that the chemical shifts for ^{19}F are difficult to predict and rationalize is that less than 1% of the shielding of the ^{19}F nucleus in organic compounds is due to diamagnetic shielding. Rather, paramagnetic shielding is the predominant factor and poorly understood.

TABLE 6.2 Chemical Shifts for Various Fluorine-containing Compounds

Compound	Chemical Shift (ppm)[a]
$CFCl_3$ Reference	0.0
CF_2Cl_2	−8.0
CF_3Cl	−28.6
$CFBr_3$	7.4
CF_2Br_2	7.0
$CFBr_3$	7.0
CFH_3	−271.9
CF_2H_2	−1436.0
CF_3H	−78.6
CF_4	−62.3
C_4F_8	−135.15
C_5F_{10}	−132.9
$(CF_3)_2CO$	−84.6
$CF_3C(O)OH$	−76.5
$CF_3C(O)OCH_3$	−74.2
CF_3COOEt	−78.7
$(CF_3)_3N$	−56.0
CH_2FCN	−251.0
$FCH{=}CH_2$	−114.0
$F_2C{=}CH_2$	−81.3
$F_2C{=}CF_2$	−135.0
C_6F_6	−164.9
C_6H_5F	−113.5
p−$C_6H_4F_2$	−106.0
$C_6H_5CFH_2$	−207
$C_6H_5C(O)OCF_3$	−73.9
$C_6H_5\,C(CF_3)_2OH$	−74.7
$C_6H_5CF_3$	−63.7
F_2 (elemental)	422.9
SF_6	57.4
SiF_4	−163.3
HF (aqueous)	−204.0
KF (aqueous F^-)	−125.3

[a] Most literature references historically reversed the sign convention (i.e., negative shifts are reported as positive).

6.4 ^{29}Si NUCLEAR MAGNETIC RESONANCE

Silicon, like fluorine, does not occur naturally in organic compounds. Silicon-containing organic compounds, however, are increasingly used by synthetic organic chemists and by polymer chemists. The ^{29}Si nucleus is the only isotope of silicon with a magnetic moment and has a natural abundance of 4.7%. We have already come across the ^{29}Si nucleus in the proton NMR spectrum of TMS; a small doublet, with a coupling constant ($^2J_{SiH}$) of about 6 Hz, with an intensity of 2–3% straddles the sharp, intense TMS singlet (Section 3.7.4). This small doublet represents the 4.7% of spin-one-half ^{29}Si, which naturally occurs in all silicon compounds.

Table 6.1 reveals that the sensitivity of the ^{29}Si nucleus is about two times that of the ^{13}C nucleus when both are recorded at natural abundance. The magnetogyric ratio for ^{29}Si (γ_{Si}) is negative (−5,319) so that for routine proton decoupled spectra we again have the possibility of negative ^{29}Si NOE enhancement depending, of course, on the relative importance of dipolar spin relaxation. In this case, the maximum NOE is −2.51. This situation is much worse than with ^{15}N NOEs because only NOEs between −2.01 and the maximum, −2.51, actually result in an "enhancement." All other values result in a net decrease in signal intensity compared to no proton decoupling. Thus, experimental conditions must be carefully controlled if we are to realize the maximum signal, especially because the ^{29}Si nucleus can have long relaxations.

The chemical shifts for ^{29}Si in common organic compounds are much smaller than for ^{13}C shifts in common compounds. This smaller shift range is probably due to the lack of multiple bonds to silicon because we find the most deshielding for ^{13}C in sp^2 hybridized atoms. Figure 6.8 gives chemical shift ranges for various silicon-containing compounds.

The proton decoupled ^{29}Si spectrum of tetramethylsilane (TMS) is shown at the top of Figure 6.9 with the proton coupled spectrum for comparison as an inset. TMS is the obvious choice for a ^{29}Si reference compound and we set it at zero ppm. The proton-coupled spectrum is quite interesting because the ^{29}Si nucleus is coupled to 12 equivalent protons in TMS. First order rules predict a multiplet with 13 peaks. There are 9 peaks clearly visible and 11 with a little imagination; we do not see the full 13 peaks because the outer ones are too weak and are lost in the noise.

The ^{29}Si NMR spectrum of triethylsilane in both proton decoupled and proton coupled modes is presented next in Figure 6.9. The proton decoupled spec-

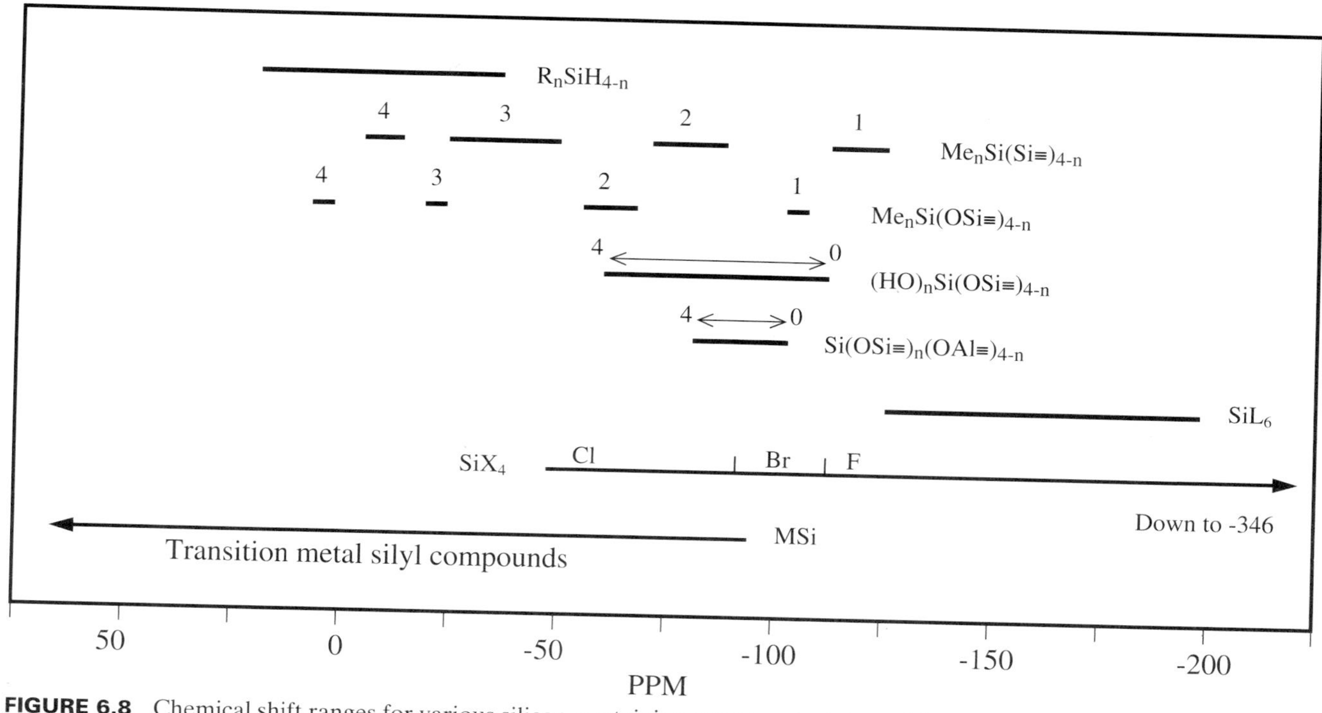

FIGURE 6.8 Chemical shift ranges for various silicon-containing compounds. Adapted from Bruker Almanac 1995.

trum gives a singlet that is only slightly shifted from TMS. In triethylsilane, there is a proton directly attached to silicon resulting in a large one-bond coupling ($^1J_{SiH}$) of about 175 Hz. There are smaller two- and three-bond couplings that lead to identical complex multiplets.

Our last example of ^{29}Si NMR is found at the bottom of Figure 6.9 where we find the proton decoupled and proton coupled ^{29}Si spectra of 1,1,3,3-tetraethyldisiloxane. This compound is commercially available and is widely used to make various silicon containing polymers. Before using the sample, a conscientious chemist might analyze it using various methods, which we now imagine might include ^{29}Si NMR. In this case, one would find that there are two types of silicon in the sample because we find a peak of about 5–10% on the shoulder of the main peak in the proton-decoupled spectrum. The chemical shift is negative, a common feature of silicon bonded to oxygen.

The proton coupled spectrum reveals a larger one-bond coupling constant ($^1J_{SiH}$) of about 215 Hz. The coupling pattern derived from the two- and three-bond coupling is complex but the pattern might serve as a starting point in the interpretation of ^{29}Si spectra of reaction products.

6.5 ^{31}P NUCLEAR MAGNETIC RESONANCE

The last of the four nuclei that we treat briefly in this chapter is ^{31}P, the only naturally occurring isotope of phosphorus. Phosphorus is of great interest to the organic chemist because reagents containing phosphorus, which range from various inorganic forms of phosphorus to the organic phosphines, phosphites, phosphonium salts, phosphorus ylides, etc., have long been used by organic chemists; the nucleus is of great interest to the biochemists primarily because of the nucleic acids that contain phosphate esters, and also smaller molecules such as ADP, ATP, etc.

NMR experiments with ^{31}P are rather straightforward; ^{31}P is a spin-one-half nucleus with a positive magnetogyric ratio (10,840). ^{31}P NOE enhancement from proton decoupling is positive with a maximum of 1.23. There has been a long history and therefore a rich literature of ^{31}P NMR. ^{31}P chemical shift data are reliable because 85% H_3PO_4 is virtually the only ^{31}P reference compound reported (external) and it remains the preferred reference today (the top spectrum in Figure 6.10). The chemical shift range for ^{31}P is rather large and generalizations are dangerous. In fact, even the different valence states of phosphorus do not fall into predictable patterns. All is not lost however, since there are many reliable published studies of ^{31}P chemical shifts. Representative proton-phosphorus coupling constants can be found in Appendix F of Chapter 3. More complete lists of ^{31}P coupling constants of phosphorus to protons and other nuclei are available.

The second spectrum from the top in Figure 6.10 displays the proton-decoupled ^{31}P spectrum of triphenylphosphine, a common reagent in organic

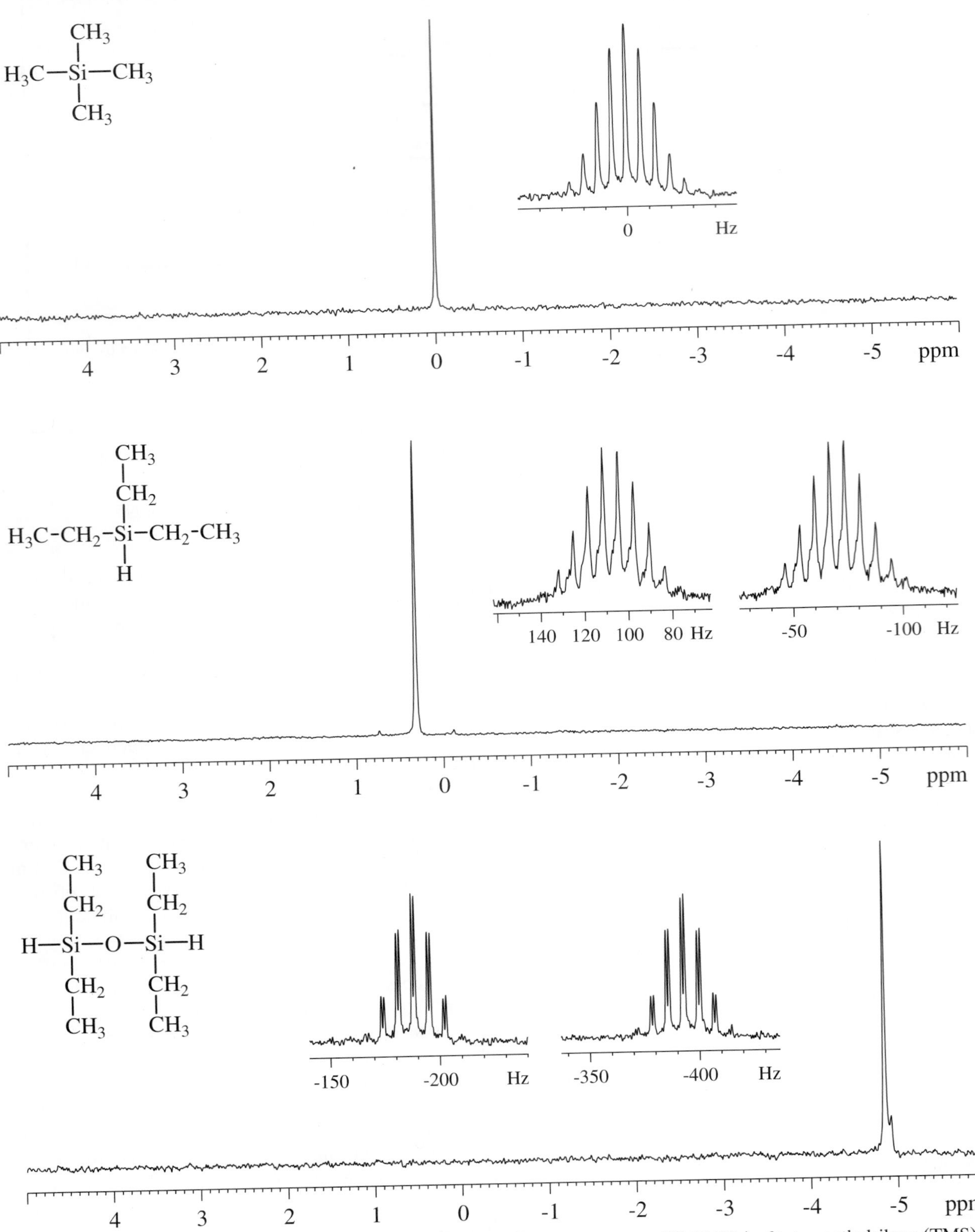

FIGURE 6.9 Top, the proton-decoupled and proton-coupled ^{29}Si NMR spectra (59.6 MHz) of tetramethylsilane (TMS) in $CDCl_3$. The outer peaks of the multiplet are not discernible because of insufficient signal/noise. Second, the proton-decoupled and proton-coupled ^{29}Si NMR spectrum (59.6 MHz) of triethylsilane in $CDCl_3$. Bottom, the proton-decoupled and proton-coupled ^{29}Si NMR spectrum (59.6 MHz) of 1,1,3,3-tetraethyldisiloxane in $CDCl_3$.

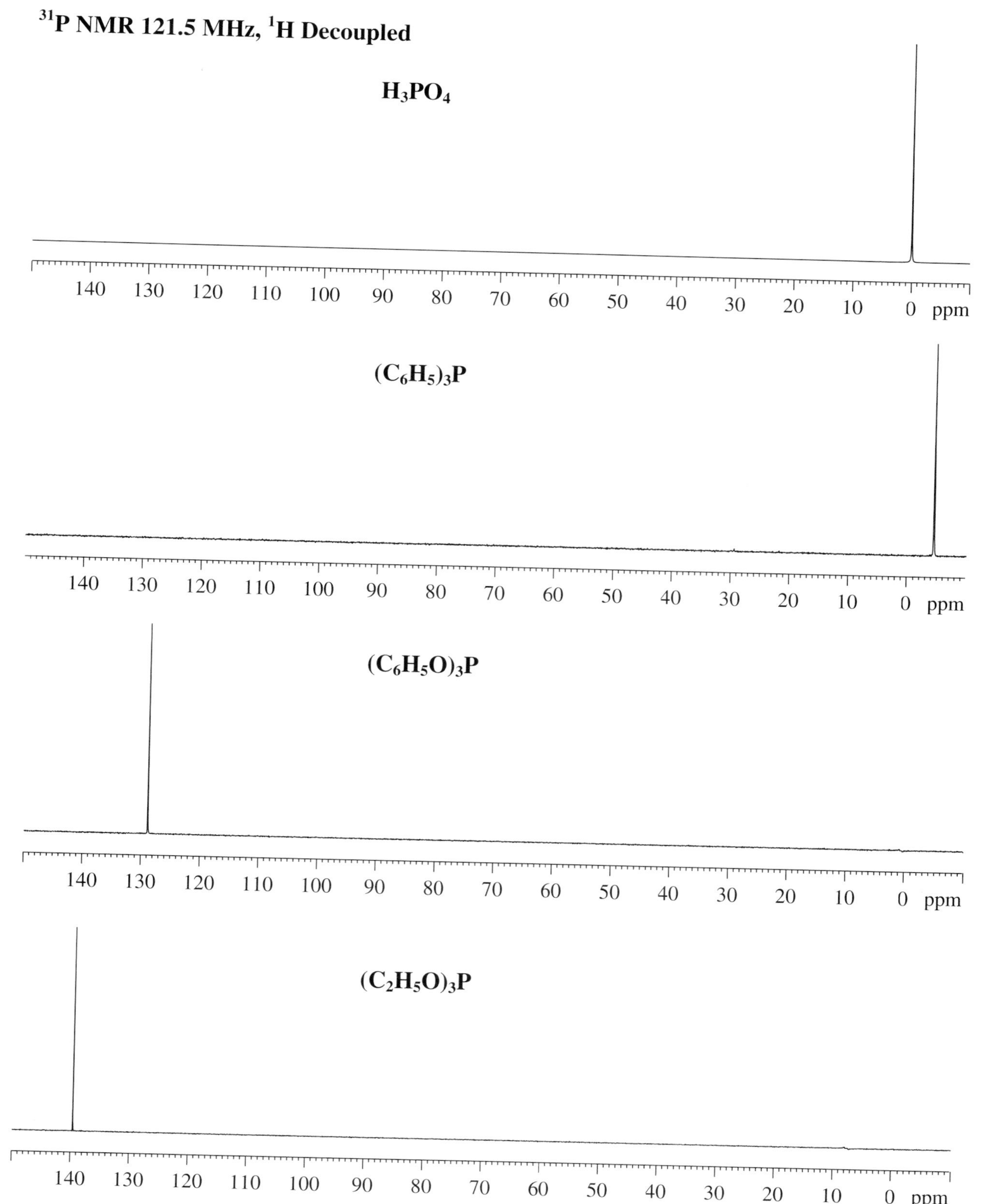

FIGURE 6.10 From top to bottom, the proton-decoupled ^{31}P NMR spectrum (121.5 MHz) of 85% phosphoric acid with a small amount of D_2O, of triphenylphosphine in $CDCl_3$, of triphenylphosphite in $CDCl_3$, and of triethylphosphite in $CDCl_3$.

TABLE 6.3 Chemical Shifts for Various Phosphorus-containing Compounds. Adapted from Bruker Almanac 1995.

Phosphorous (III) Compounds	(ppm)[a]	Phosphorous (V) Compounds	(ppm)[a]
PMe_3	−62	Me_3PO	36.2
PEt_3	−20	Et_3PO	48.3
$PPr(n)_3$	−33	$[Me_4P]^{+1}$	24.4
$PPr(i)_3$	19.4	$[PO_4]^{-3}$	6
$PBu(n)_3$	−32.5	PF_5	−80.3
$PBu(i)_3$	−45.3	PCl_5	−80
$PBu(s)_3$	7.9	$MePF_4$	−29.9
$PBu(t)_3$	63	Me_3PF_2	−158
$PMeF_2$	245	Me_3PS	59.1
$PMeH_2$	−163.5	Et_3PS	54.5
$PMeCl_2$	192	$[Et4p]^{+1}$	40.1
$PMeBr_2$	184	$[PS_4]^{-3}$	87
PMe_2F	186	$[PF_6]^{-1}$	−145
PMe_2H	−99	$[PCl_4]^{+1}$	86
PMe_2Cl	−96.5	$[PCl_6]^{-1}$	−295
PMe_2Br	−90.5	Me_2PF_3	8

[a] Reference to 85% H_3PO_4 at 0 ppm

synthesis. The spectrum itself is unremarkable in that we see a single resonance for the single phosphorus atom in triphenylphosphine. What is remarkable is the fact that there is a chemical shift range of less than 10 ppm for the three compounds (the first two in Figure 6.10 and in Figure 6.11) and yet the phosphorus atoms differ widely with respect to valence state, attached groups, and so on. We would have a very difficult job trying to rationalize these figures and it would have been foolish to have tried to predict these ^{31}P chemical shifts. Table 6.3 provides ^{31}P chemical shifts for various phosphorus-containing compounds.

The bottom two spectra in Figures 6.10 are the proton decoupled ^{31}P NMR spectra of two phosphites; they are triphenylphosphite and triethylphosphite, respectively. Both compounds give a sharp single peak as expected. The point here is to illustrate that even without a set of predictive rules for chemical shift, we can nonetheless expect useful information with little chance of overlap. By themselves, these spectra might not provide much information but when combined with other spectra, they give one more perspective on composition, structure, and stereochemistry.

The proton-decoupled ^{31}P NMR spectrum of diethyl chlorophosphate is found in Figure 6.11 along with the proton-coupled spectrum. Also included with this figure are the ^1H and ^{13}C NMR spectra for a complete set. The proton-coupled ^{31}P spectrum shows an apparent quintet, suggesting that there is no appreciable coupling of the phosphorus atom to the methyl protons of the methyl groups. Inspection of the corresponding proton spectrum reveals however, that the methyl groups are actually a triplet of doublets with the four-bond $^{31}P—^1H$ coupling constant of ca. 1 Hz. Thus, we conclude that the 1 Hz coupling is not resolved in the ^{31}P spectrum but reveals itself by broadening the lines. The four protons of the two methylene groups couple equally (approximately) to the phosphorus atom to give the observed quintet.

The methylene protons in the ^1H spectrum may at first seem too complex. The matter is resolved by noting that the methylene protons in diethyl chlorophosphate are diastereotopic (see Figure 3.42), and therefore they are not chemically shift equivalent and show strong coupling. First order analysis is impossible.

6.6 CONCLUSIONS

We have attempted in this brief chapter to introduce a few other useful nuclei and some examples of their spectra. The emphasis in this chapter is to utilize these spectra *in combination* with other spectra, especially other forms of NMR. We must concede that it is neither possible nor desirable to become "experts" on chemical shifts of these (or other) nuclei and their coupling constants. In making this concession, we can be "comfortable" with these four nuclei, and furthermore, we can easily broaden our outlook to other elements throughout the periodic table.

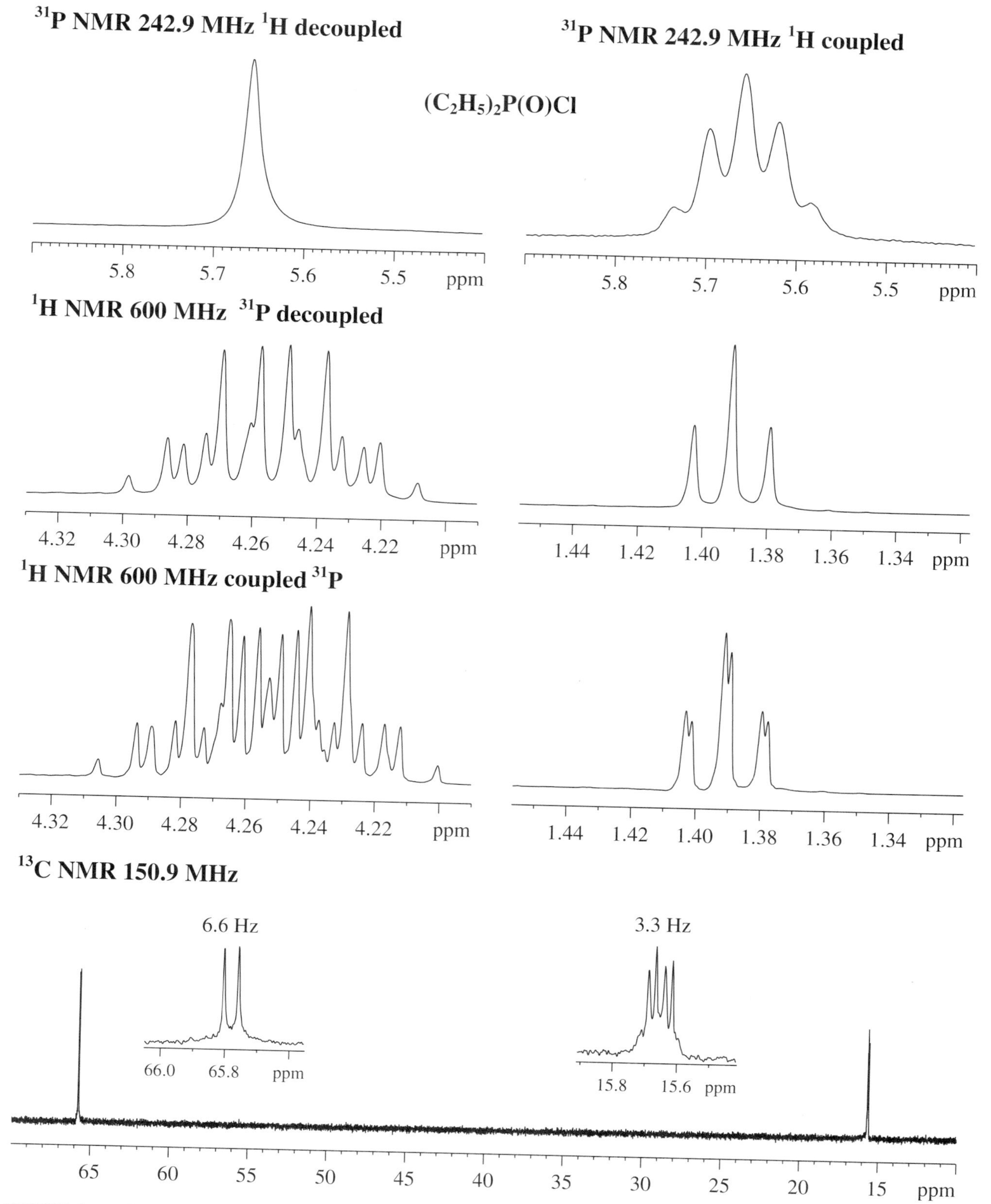

FIGURE 6.11 From top to bottom, the proton-decoupled and proton-coupled ^{31}P NMR spectra (242.9 MHz), the ^{31}P-decoupled ^{1}H NMR spectrum, the ^{31}P-coupled ^{1}H NMR spectrum, and the ^{31}P-coupled ^{13}C NMR spectrum of diethyl chlorophosphate in $CDCl_3$.

REFERENCES

Andreis, M., and Koenig, J. L. (1995). Application of nitrogen-15 NMR to polymers. *Adv. Polym. Sci.,* **124,** 191–237.

Argyropoulos, D. S. (1995). Phosphorus-31 NMR in wood chemistry: a review of recent progress. *Res. Chem. Intermediates,* **21,** 373–395.

Berger, S., Braun, S., and Kalinowski, H. O. (1992). *NMR-Spektroskopie von Nichtmetallen.* Stuttgart: Georg Thieme Verlag.

Band 1: Grundlagen, ^{17}O-, ^{33}S- and ^{129}Xe-NMR Spektroskopie.

Band 2: ^{15}N-NMR Spcktroskopie.

Band 3: ^{31}P-NMR Spektroskopie.

Band 4: ^{19}F-NMR Spektroskopie.

Brey, W., (Ed.,) (1988). *Pulse Methods in 1D and 2D Liquid-Phase NMR.* New York: Academic Press.

Buchanan, G. W. (1989). Applications of nitrogen-15 NMR spectroscopy to the study of molecular structure, stereochemistry and binding phenomena. *Tetrahedron,* **45**(3), 581–604.

Chandrakumar, N., and Subramanian, S. (1987). *Modern Techniques in High Resolution FT NMR.* New York: Springer-Verlag.

Chesnut, D. B., and Quin, L. D. (1999). The theoretical determination of phosphorus NMR chemical shielding. *Adv. Mol. Struct. Res.,* **5,** 189–222.

Coleman, B. (1983). Applications of silicon-29 NMR spectroscopy. In P. Laszlo, Ed. *NMR Newly Accessible Nuclear,* **2** 197–228. New York: Academic.

Dungan, C. Fl., and Van Wazer, J. R. (1970). *Compilation of Reported F-19 NMR Chemical Shifts.* New York: Wiley-Interscience.

Emsley, J. W., and Phillips, L. (1971). Fluorine chemical shifts. In J. W. Emsley, J. Feeney, and L. H. Sutcliffe, Eds., *Progress in Nuclear Magnetic Resonance Spectroscopy,* Vol. 7. Oxford: Pergamon Press.

Everett, T. S. (1988). The correlation of multinuclear spectral data for selectively fluorinated organic compounds. *J. Chem. Edu.,* **65,** 422–425.

Fluck, E., and Heckmann, G. (1987). Empirical methods for interpreting chemical shifts of phosphorus compounds. *Meth. Stereochemical Anal.,* **8**(Phosphorus-31 NMR Spectrosc. Stereochem. Anal.), 61–113.

Fushman, D. (2003). Determination of protein dynamics using ^{15}N relaxation measurements. *Meth. Principles Med. Chem.,* **16,** 283–308.

Gorenstein, D. G. (1983). Nonbiological aspects of phosphorus-31 NMR spectroscopy. *Prog. Nucl. Magn. Resonance Spectrosc.,* **16**(1), 1–98.

Gorenstein, D. G. (1984.) *Phosphorus-31 NMR.* London: Academic.

Gorenstein, D. G. (1992). Advances in phosphorus-31 NMR. In R. Engel, Ed. *Handbook of Organophosphorus Chemistry.* New York: Dekker.

Harris, R. K. (1983). Solution-state NMR studies of Group IV elements (other than carbon). *NATO ASI Series, Series C: Mathematical and Physical Sciences,* **103** (Multinucl. Approach NMR Spectrosc.), 343–359.

Holmes, R. R., and Prakasha, T. K. (1993). Cyclic oxyphosphoranes. Phosphorus-31 chemical shift correlations. *Phosphorus, Sulfur and Silicon and the Related Elements,* **80,** 1–22.

Horn, H. G. (1973). NMR spectra and characteristic frequencies of compounds containing nitrogen-sulfur-fluorine bonds. *Fluorine Chem. Rev.,* **6,** 135–192.

Hulst, R., Kellogg, R. M., and Feringa, B. L. (1995). New methodologies for enantiomeric excess (ee) determination based on phosphorus NMR. *Recueil des Travaux Chimiques des Pays-Bas,* **114,** 115–138.

Hutton, W. C. (1984). Two-dimensional phosphorus-31 NMR. In D. G. Gorenstein, Ed. *Phosphorus-31 [Thirty-One] NMR.* Orlando: Academic.

Karaghiosoff, K., and Schmidpeter, A. (1988). The chemical shift of two-coordinate phosphorus. II. Heterocycles. *Phosphorus and Sulfur and the Related Elements,* **36,** 217–259.

Laszlo, P., Ed. (1983). *NMR of Newly Accessible Nuclei,* Vol. 1: New York: Academic Press.

Levy, G. C., and Lichter, R. L. (1979). *Nitrogen-15 Nuclear Magnetic Resonance Spectroscopy.* New York: John Wiley & Sons.

Lichter, R. L. (1983). Nitrogen nuclear magnetic resonance spectroscopy. *NATO ASI Series, Series C: Mathematical and Physical Sciences,* **103** (Multinucl. Approach NMR Spectrosc.), 207–244.

Majoral, J. -P., Caminade, A. -M., and Igau, A. (1994). Elucidation of small ring and macrocycle structures using ^{31}P NMR. In L. D. Quin and J. G. Verkade, Eds., *Phosphorus-31 NMR Spectral Properties in Compound Characterization and Structural Analysis.* New York: VCH pp. 57–68.

Marek, R., and Lycka, A. (2002). ^{15}N NMR spectroscopy in structural analysis. *Curr. Org. Chem.,* **6**(1), 35–66.

Martin, G. J., Martin, M. L., and Gouesnard, J. P. (1981). *^{15}N-NMR Spectroscopy,* Vol. 18. Berlin: Springer-Verlag.

Mason, J. (Ed.). (1987). *Multinuclear NMR.* New York: Plenum Press.

Mason, J. (1983). Patterns and prospects in nitrogen NMR. *Chemistry in Britain,* **19**(8), 654–658.

McFarlane, W. (1987). Special experimental techniques in phosphorus NMR spectroscopy. *Methods in Stereochemical Anal.,* **8** (Phosphorus-31 NMR Spectrosc. Stereochem. Anal.), 115–150.

Mooney, E. F. (1970). *An Introduction to ^{19}F NMR Spectroscopy.* London: Hoyden.

Quin, L. D., and Verkade, J. G., (Eds.), (1994). *Phosphorus-31 NMR Spectral Properties in Compound Characterization and Structural Analysis.* New York: VCH Publishers.

Randall, E. W., and Gillies, D. G. (1971). Nitrogen nuclear magnetic resonance. In J. W. Emsley, J. Feeney, and L. H. Sutcliffe, Eds., *Progress in Nuclear Magnetic Resonance Spectroscopy,* Vol. 6. Oxford: Pergamon Press, pp. 119–174.

Rao, B. D. N. (1983). Phosphorus-31 nuclear magnetic resonance investigations of enzyme systems. *Biol. Magn. Resonance,* **5** 75–128.

Roberts, J. D. (1979). Use of nitrogen-15 nuclear magnetic resonance spectroscopy in natural-product chemistry. *Jap. J. Antibiotics,* **32**(Suppl.), S112–S121.

Schilf, W., and Stefaniak, L. (2000). Application of ^{14}N and ^{15}N NMR for determination of protonation site and hydrogen bond strength in nitrogen organic compounds. In Atta-ur-Rahman Ed., *New Adv. Anal. Chem.,* Amsterdam: Harwood Academic Publishers.

Schiller, J., and Arnold, K. (2002). Application of high resolution ^{31}P NMR spectroscopy to the characterization of the phospholipid composition of tissues and body fluids—a methodological review. *Med. Sci. Monitor,* **8**(11), 205–222.

Schraml, J., and Bellama, J. M. (1976). Silicon-29 nuclear magnetic resonance. In F. C. Nachod, J. J. Zuckerman, and E. W. Randall, Eds. *Determination Org. Struct. Phys. Methods,* **6,** 203–269. New York: Academic.

Tebby, J. C. (1987). General experimental techniques and compilation of chemical shift data. *Methods in Stereochemical Anal.,* **8** (Phosphorus-31 NMR Spectrosc. Stereochem. Anal.), 1–60.

Webb, G. A. (Ed.). (1981). *Annual Reports on NMR Spectroscopy, Vol. 11B: Nitrogen NMR Spectroscopy.* London: Academic Press.

Webb, G. A., and Witanowski, M. (1985). Nitrogen NMR and molecular interactions. *Proc. - Indian Acad. Sci. Chem. Sci.,* **94**(2), 241–290.

Witanowski, M., Stefaniak, L., and Webb, G. A. (1986). Nitrogen NMR spectroscopy. *Annu. Rep. NMR Spectrosc.,* **18,** 1–761.

Witanowski, M., and Webb, G. A. (1973). *Nitrogen NMR.* New York: Plenum.

Yeagle, P. L. (1990). Phosphorus NMR of membranes. *Biol. Magn. Resonance,* **9,** 1–54.

STUDENT EXERCISES

6.1 Deduce the structure of the compound whose molecular formula is $C_5H_{12}N_2$ from the combined information from the ^1H, ^{13}C, DEPT, and ^{15}N NMR spectra. There is no proton coupled ^{15}N spectrum provided because it provides no useful information.

6.2 The compound for this problem is a reagent commonly used in organic synthesis; its molecular formula is $C_6H_{15}SiCl$. Provided is the ^1H, ^{13}C, DEPT, and ^{29}Si (proton-decoupled) spectra.

6.3 Determine the structure of the phosphorus-containing compound whose molecular formula is $C_{19}H_{18}PBr$ from the ^1H, ^{13}C, DEPT, and ^{31}P (both proton-coupled and proton-decoupled).

6.4 Determine the structure of the fluorine-containing compound for which the mass, IR, ^1H NMR, ^{13}C/DEPT NMR, and ^{19}F NMR spectra are given.

Exercise 6.1

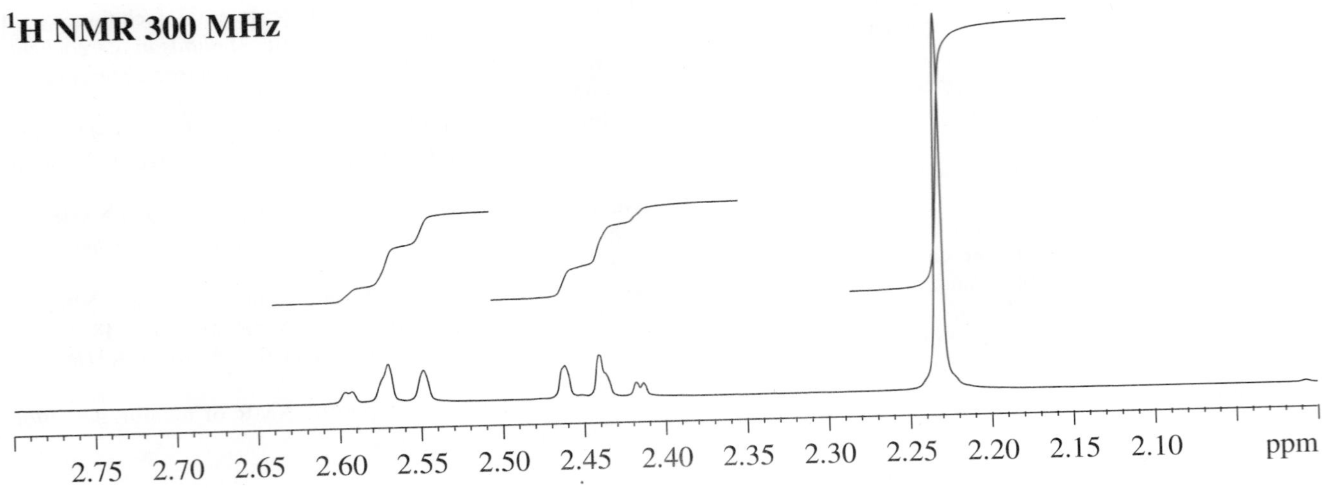
^{1}H NMR 300 MHz
2.75 2.70 2.65 2.60 2.55 2.50 2.45 2.40 2.35 2.30 2.25 2.20 2.15 2.10 ppm

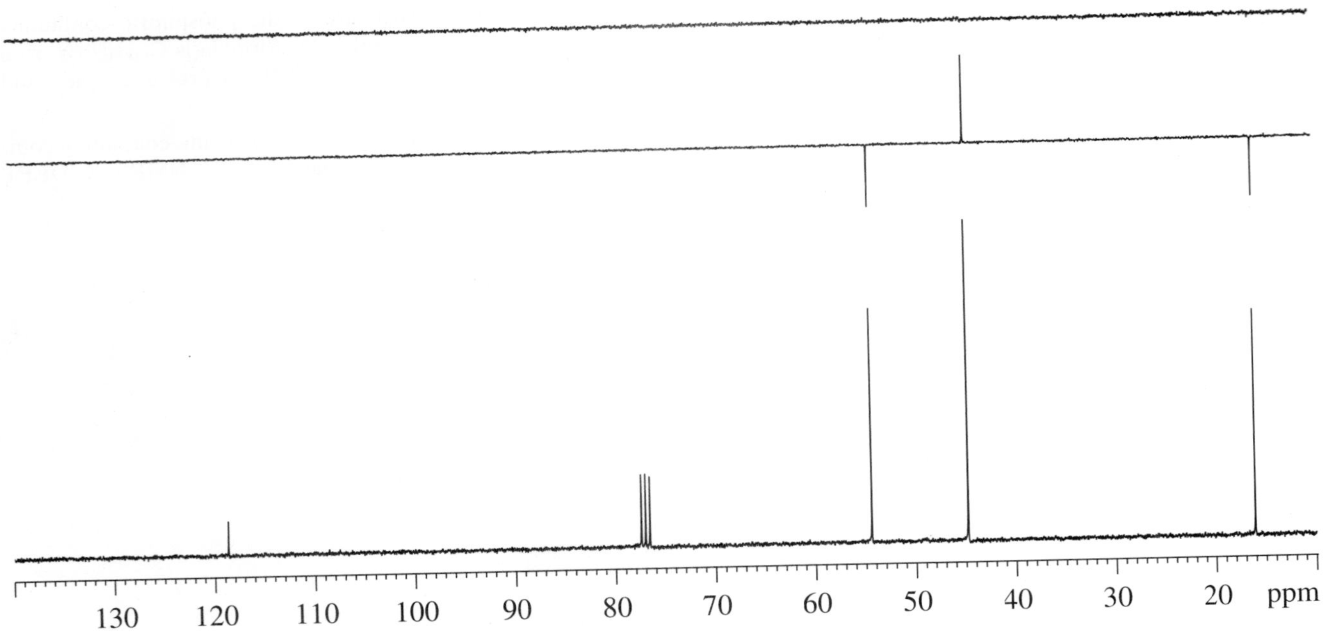
^{13}C/DEPT NMR 75.5 MHz
130 120 110 100 90 80 70 60 50 40 30 20 ppm

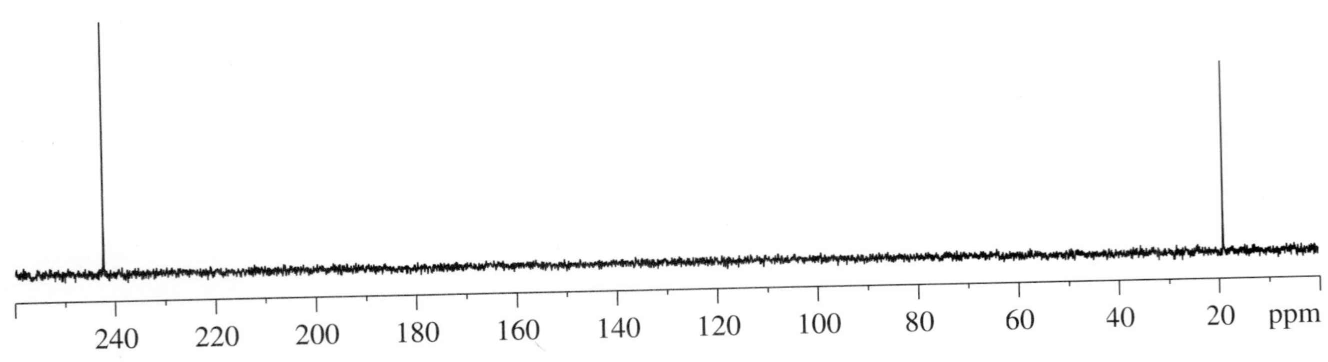
^{15}N NMR 30.4 MHz
240 220 200 180 160 140 120 100 80 60 40 20 ppm

Exercise 6.2

^{1}H NMR 300 MHz

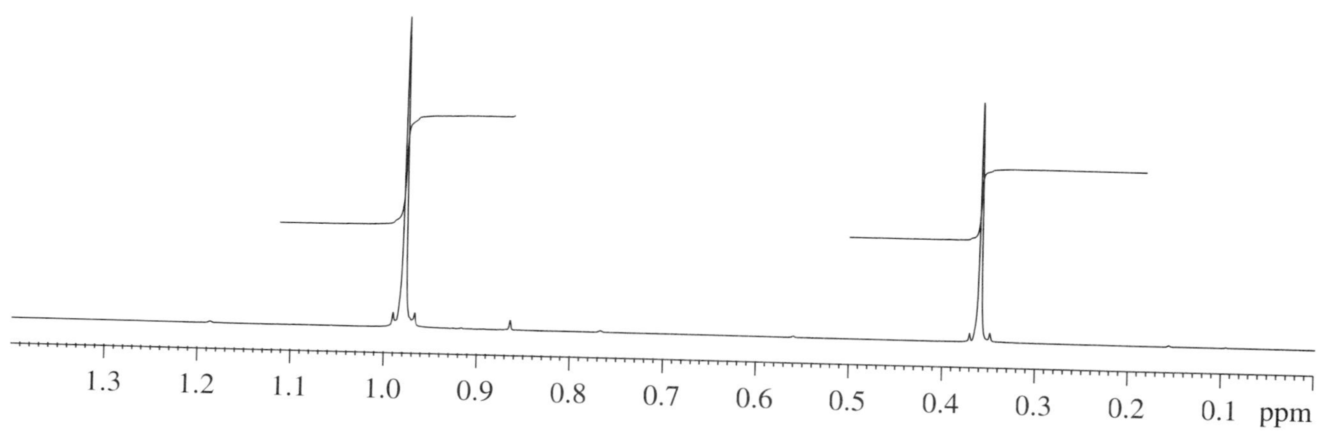

^{13}C/DEPT NMR 75.5 MHz

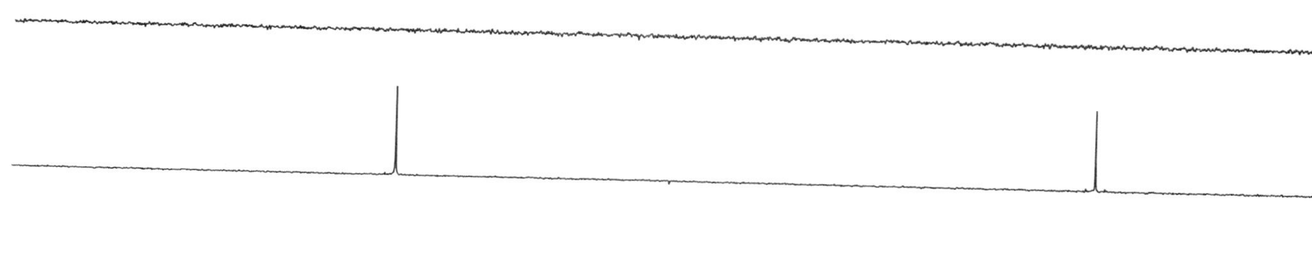

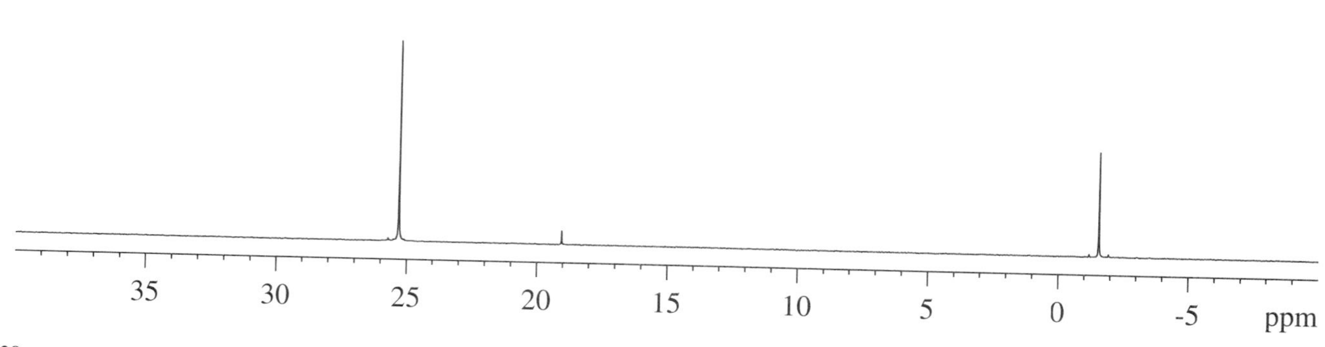

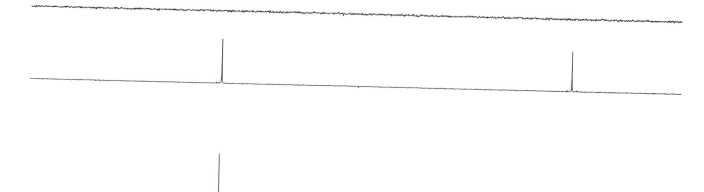

^{29}Si NMR 59.6 MHz

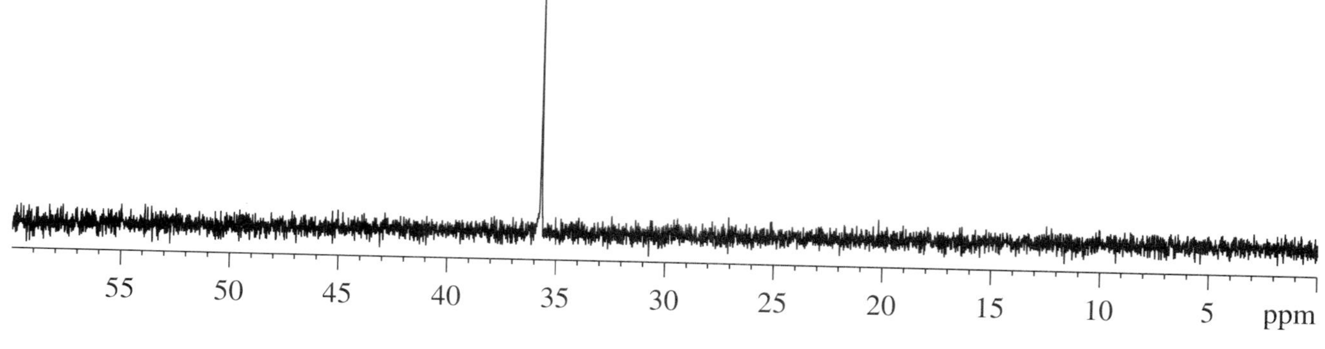

Exercise 6.3

^{1}H NMR 300 MHz

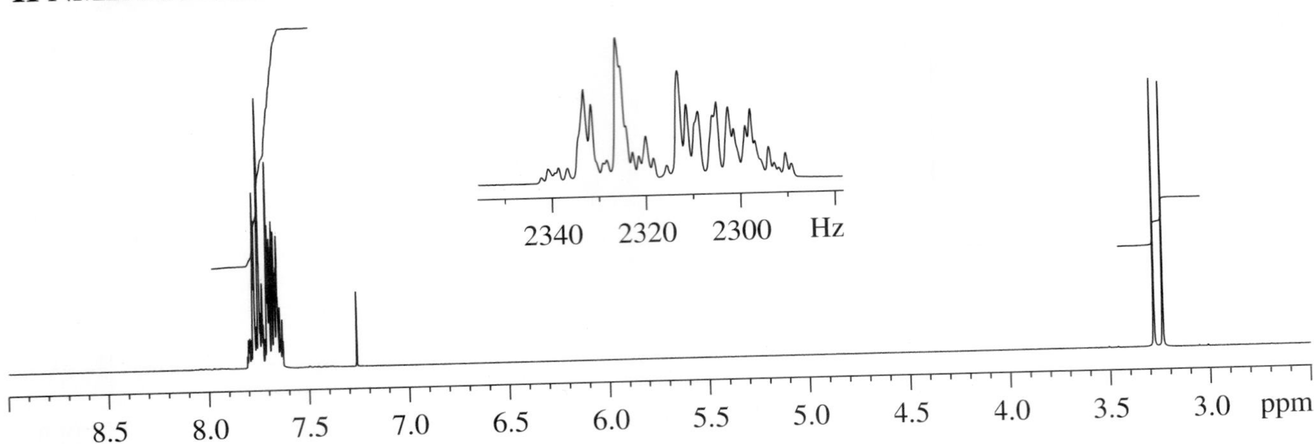

^{13}C/DEPT NMR 75.5 MHz

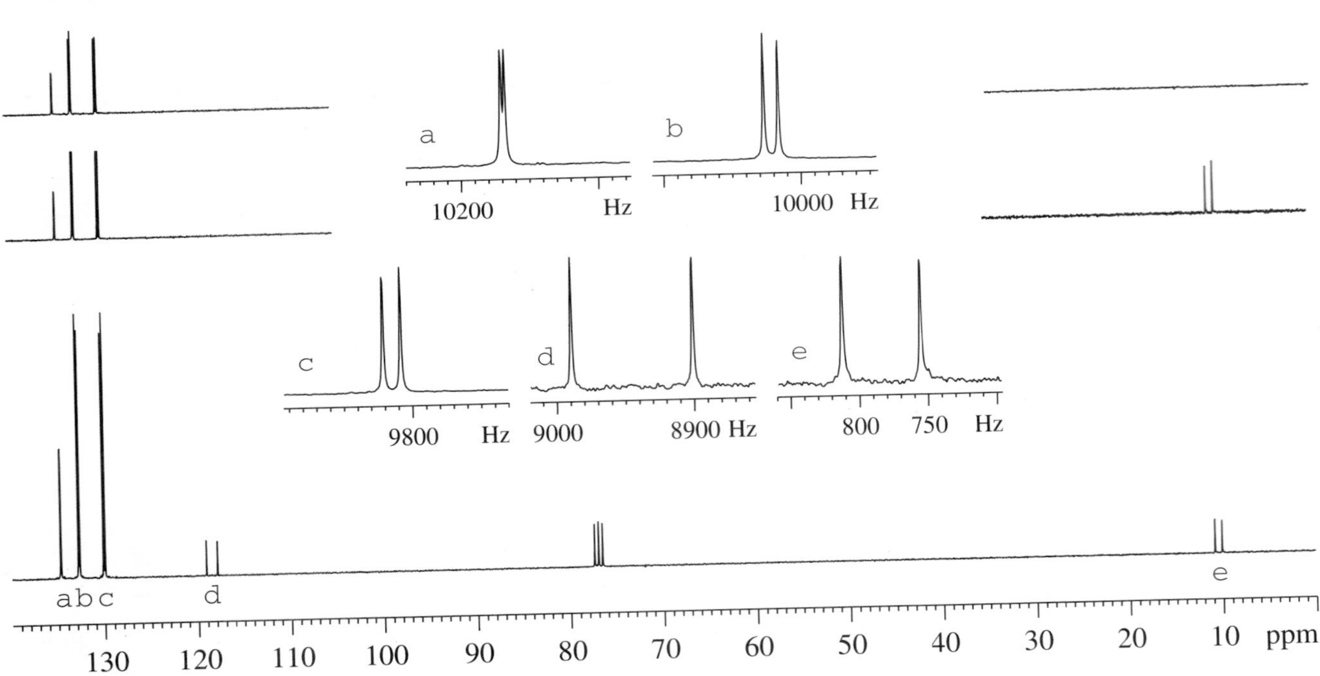

^{31}P NMR 121.5 MHz, ^{1}H Decoupled

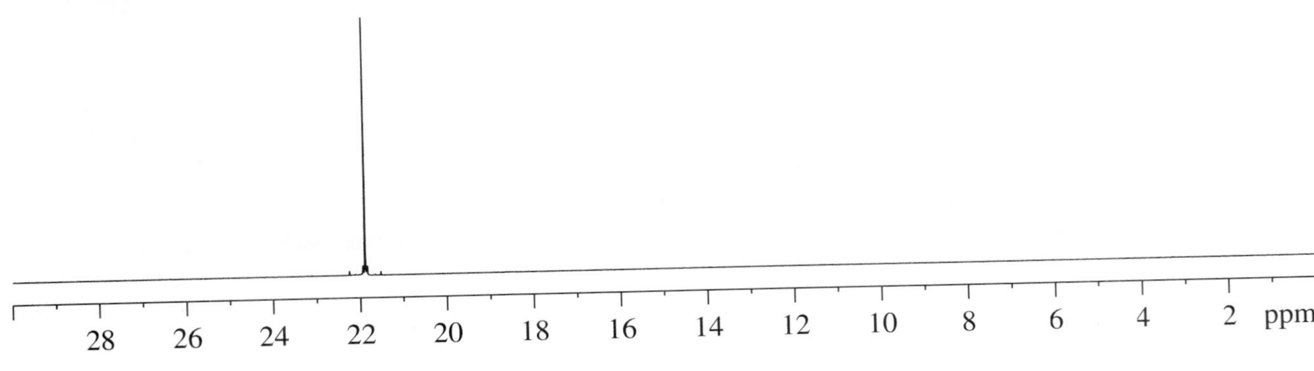

Exercise 6.4

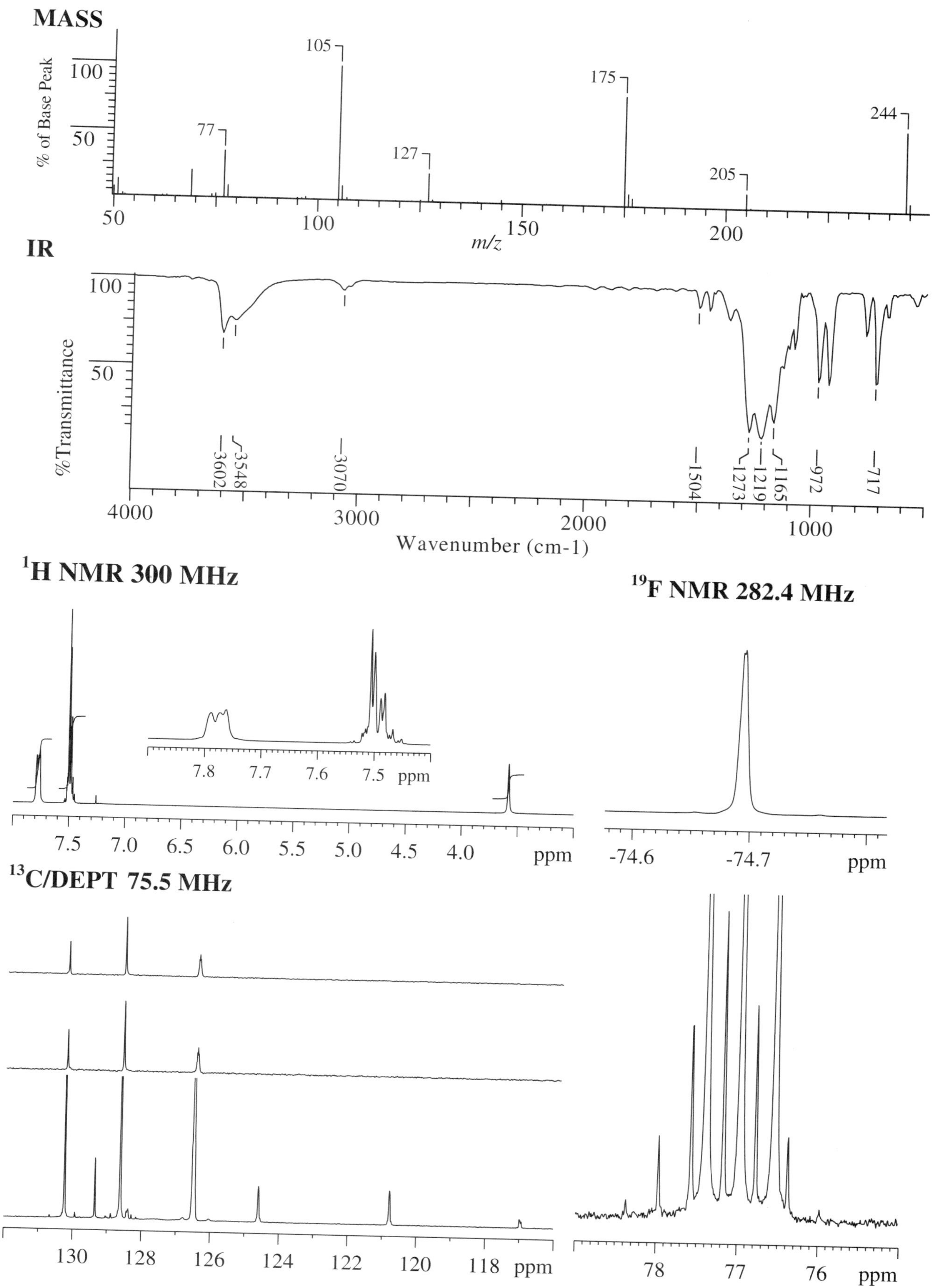

MASS
% of Base Peak
100
50
105
175
244
77
127
205
50
100
150
200
m/z
IR
%Transmittance
3602
3548
3070
1504
1273
1219
1165
972
717
4000
3000
2000
1000
Wavenumber (cm-1)
1H NMR 300 MHz
7.8
7.7
7.6
7.5 ppm
7.5
7.0
6.5
6.0
5.5
5.0
4.5
4.0
ppm
19F NMR 282.4 MHz
-74.6
-74.7
ppm
13C/DEPT 75.5 MHz
130
128
126
124
122
120
118 ppm
78
77
76 ppm

APPENDIX A PROPERTIES OF MAGNETICALLY ACTIVE NUCLEI

Isotope	Spin	Natural Abundance	Magnetic Moment μ/μ_N	Magnetogyric Ratio $\gamma/10^7$ rad s^{-1} T^{-1}	Quadrupole Moment Q/fm^2	Frequency Ratio @ 2.35 T	Reference Compound	Relative Receptivity D^{Proton}	D^{Carbon}
^{1}H	1/2	99.9885	4.837353570	26.7522128	–	100.000000	Me_4Si	1	5.87×10^{3}
^{2}H	1	0.0115	1.21260077	4.106 62791	0.286	15.350609	$(CD_3)_3Si$	1.11×10^{-6}	6.52×10^{-3}
^{3}H	1/2	–	5.159714367	28.5349779	–	106.663974	Me_4Si-t_1	–	–
^{3}He	1/2	1.37×10^{-4}	−3.685154336	−20.3801587	–	76.179437	He(gas)	6.06×10^{-7}	3.56×10^{-3}
^{6}Li	1	7.59	1.1625637	3.9371709	−0.0808	14.716086	LiCl	6.45×10^{-4}	3.79
^{7}Li	3/2	92.41	4.20407505	10.3977013	−4.01	38.863797	LiCl	0.271	1.59×10^{3}
^{9}Be	3/2	100	−1.520136	−3.759666	5.288	14.051813	$BeSO_4$	1.39×10^{-2}	81.5
^{10}B	3	19.9	2.0792055	2.8746786	8.459	10.743658	$BF_3.Et_2O$	3.95×10^{-3}	23.2
^{11}B	3/2	80.1	3.4710308	8.5847044	4.059	32.083974	$BF_3.Et_2O$	0.132	7.77×10^{2}
^{13}C	1/2	1.07	1.216613	6.728284	–	25.145020	Me_4Si	1.70×10^{-4}	1
^{14}N	1	99.632	0.57100428	1.9337792	2.044	7.226317	CH_3NO_2	1.00×10^{-3}	5.9
^{15}N	1/2	0.368	−0.49049746	−2.71261804	–	10.136767	$MeNO_2$	3.84×10^{-6}	2.25×10^{-2}
^{17}O	5/2	0.038	−2.24077	−3.62808	−2.56	13.556457	D_2O	1.11×10^{-5}	6.50×10^{-2}
^{19}F	1/2	100	4.553333	25.18148	–	94.094011	CCl_3F	0.834	4.90×10^{3}
^{21}Ne	3/2	0.27	−0.854376	−2.11308	10.155	7.894296	Ne (gas)	6.65×10^{-6}	3.91×10^{-2}
^{23}Na	3/2	100	2.8629811	7.0808493	10.40	26.451900	NaCl	9.27×10^{-2}	5.45×10^{2}
^{25}Mg	5/2	10.00	−1.01220	−1.63887	19.94	6.121635	$MgCl_2$	2.68×10^{-4}	1.58
^{27}Al	5/2	100	4.3086865	6.9762715	14.66	26.056859	$Al(NO_3)_3$	0.207	1.22×10^{3}
^{29}Si	1/2	4.6832	−0.96179	−5.3190	–	19.867187	Me_4Si	3.68×10^{-4}	2.16
^{31}P	1/2	100	1.95999	10.8394	–	40.480742	H_3PO_4	6.65×10^{-2}	3.91×10^{2}
^{33}S	3/2	0.76	0.8311696	2.055685	−6.78	7.676000	$(NH_4)_2SO_4$	1.72×10^{-5}	0.101
^{35}Cl	3/2	75.78	1.061035	2.624198	−8.165	9.797909	NaCl	3.58×10^{-3}	21
^{37}Cl	3/2	24.22	0.8831998	2.184368	−6.435	8.155725	NaCl	6.59×10^{-4}	3.87
^{39}K	3/2	93.2581	0.50543376	1.2500608	5.85	4.666373	KCl	4.76×10^{-4}	2.79
(^{40}K)	4	0.0117	−1.4513203	−1.5542854	−7.30	5.802018	KCl	6.12×10^{-7}	3.59×10^{-3}
(^{41}K)	3/2	6.7302	0.2773961	0.68606808	7.11	2.561305	KCl	5.68×10^{-6}	3.33×10^{-2}
^{43}Ca	7/2	0.135	−1.494067	−1.803069	−4.08	6.730029	$CaCl_2$	8.68×10^{-6}	5.10×10^{-2}
^{45}Sc	7/2	100	5.3933489	6.5087973	−22.0	24.291747	$Sc(NO_3)_3$	0.302	1.78×10^{3}
^{47}Ti	5/2	7.44	−0.93294	−1.5105	30.2	5.637534	$TiCl_4$	1.56×10^{-4}	0.918
^{49}Ti	7/2	5.41	−1.25201	−1.51095	24.7	5.639037	$TiCl_4$	2.05×10^{-4}	1.2
(^{50}V)	6	0.25	3.6137570	2.6706490	21.0	9.970309	$VOCl_3$	1.39×10^{-4}	0.818
^{51}V	7/2	99.75	5.8380835	7.0455117	−5.2	26.302948	$VOCl_3$	0.383	2.25×10^{3}
^{53}Cr	3/2	9.501	−0.61263	−1.5152	−15.0	5.652496	K_2CrO_4	8.63×10^{-5}	0.507
^{55}Mn	5/2	100	4.1042437	6.645 2546	33.0	24.789218	$KMnO_4$	0.179	1.05×10^{3}
^{57}Fe	1/2	2.119	0.1569636	0.8680624	–	3.237778	$Fe(CO)_5$	7.24×10^{-7}	4.25×10^{-3}
^{59}Co	7/2	100	5.247	6.332	42.0	23.727074	$K_3[Co(CN)_6]$	0.278	1.64×10^{3}
^{61}Ni	3/2	1.1399	−0.96827	−2.3948	16.2	8.936051	$Ni(CO)_4$	4.09×10^{-5}	0.24
^{63}Cu	3/2	69.17	2.8754908	7.1117890	−22.0	26.515473	$[Cu(CH_3CN)_4][ClO_4]$	6.50×10^{-2}	3.82×10^{2}

^{65}Cu	3/2	30.83	3.07465	7.60435	−20.4	28.403693	$[Cu(CH_3CN)_4][ClO_4]$	3.54×10^{-2}	2.08×10^{2}
^{67}Zn	5/2	4.10	1.035556	1.676688	15.0	6.256803	$Zn(NO_3)_2$	1.18×10^{-4}	0.692
(^{69}Ga)	3/2	60.108	2.603405	6.438855	17.1	24.001354	$Ga(NO_3)_3$	4.19×10^{-2}	2.46×10^{2}
^{71}Ga	3/2	39.892	3.307871	8.181171	10.7	30.496704	$Ga(NO_3)_3$	5.71×10^{-2}	3.35×102
^{73}Ge	9/2	7.73	−0.9722881	−0.9360303	−19.6	3.488315	$(CH_3)_4Ge$	1.09×10^{-4}	0.642
^{75}As	3/2	100	1.858354	4.596163	31.4	17.122614	$NaAsF_6$	2.54×10^{-2}	1.49×10^{2}
^{77}Se	1/2	7.63	0.92677577	5.1253857	–	19.071513	Me_2Se	5.37×10^{-4}	3.15
(^{79}Br)	3/2	50.69	2.719351	6.725616	31.3	25.053980	NaBr	4.03×10^{-2}	2.37×10^{2}
^{81}Br	3/2	49.31	2.931283	7.249776	26.2	27.006518	NaBr	4.91×10^{-2}	2.88×10^{2}
^{83}Kr	9/2	11.49	−1.07311	−1.03310	25.9	3.847600	Kr(gas)	2.18×10^{-4}	1.28
(^{85}Rb)	5/2	72.17	1.6013071	2.5927050	27.6	9.654943	RbCl	7.67×10^{-3}	45
^{87}Rb	3/2	27.83	3.552582	8.786400	13.35	32.720454	RbCl	4.93×10^{-2}	2.90×10^{2}
^{87}Sr	9/2	7.00	−1.2090236	−1.1639376	33.5	4.333822	$SrCl_2$	1.90×10^{-4}	1.12
^{89}Y	1/2	100	−0.23801049	−1.3162791	–	4.900198	$Y(NO_3)_3$	1.19×10^{-4}	0.7
^{91}Zr	5/2	11.22	−1.54246	−2.49743	−17.6	9.296298	$Zr(C_5H_5)_2Cl_2$	1.07×10^{-3}	6.26
^{93}Nb	9/2	100	6.8217	6.5674	−32.0	24.476170	$K[NbCl_6]$	0.488	2.87×10^{3}
^{95}Mo	5/2	15.92	−1.0820	−1.7510	−2.2	6.516926	Na_2MoO_4	5.21×10^{-4}	3.06
(^{97}Mo)	5/2	9.55	−1.1050	−1.7880	25.5	6.653695	Na_2MoO_4	3.33×10^{-4}	1.95
^{99}Tc	9/2	-	6.2810	6.0460	−12.9	22.508326	NH_4TcO_4	–	–
^{99}Ru	5/2	12.76	−0.7588	−1.2290	7.9	4.605151	$K_4[Ru(CN)_6]$	1.44×10^{-4}	0.848
^{101}Ru	5/2	17.06	−0.8505	−1.3770	45.7	5.161369	$K_4[Ru(CN)_6]$	2.71×10^{-4}	1.59
^{103}Rh	1/2	100	−0.1531	−0.8468	–	3.186447	$Rh(acac)_3$	3.17×10^{-5}	0.186
^{105}Pd	5/2	22.33	−0.7600	−1.2300	66.0	4.576100	K_2PdCl_6	2.53×10^{-4}	1.49
(^{107}Ag)	1/2	51.839	−0.19689893	−1.0889181	–	4.047819	$AgNO_3$	3.50×10^{-5}	0.205
^{109}Ag	1/2	48.161	−0.22636279	−1.2518634	–	4.653533	$AgNO_3$	4.94×10^{-5}	0.290
(^{111}Cd)	1/2	12.80	−1.0303729	−5.6983131	–	21.215480	Me_2Cd	1.24×10^{-3}	7.27
^{113}Cd	1/2	12.22	−1.0778568	−5.9609155	–	22.193175	Me_2Cd	1.35×10^{-3}	7.94
(^{113}In)	9/2	4.29	6.1124	5.8845	79.9	21.865755	$In(NO_3)_3$	1.51×10^{-2}	88.50
^{115}In	9/2	95.71	6.1256	5.8972	81.0	21.912629	$In(NO_3)_3$	0.338	1.98×10^{3}
(^{115}Sn)	1/2	0.34	−1.5915	−8.8013	–	32.718749	Me_4Sn	1.21×10^{-4}	0.711
(^{117}Sn)	1/2	7.68	−1.73385	−9.58879	–	35.632259	Me_4Sn	3.54×10^{-3}	20.8
^{119}Sn	1/2	8.59	−1.81394	−10.0317	–	37.290632	Me_4Sn	4.53×10^{-3}	26.6
^{121}Sb	5/2	57.21	3.9796	6.4435	−36.0	23.930577	$KSbCl_6$	9.33×10^{-2}	5.48×10^{2}
(^{123}Sb)	7/2	42.79	2.8912	3.4892	−49.0	12.959217	$KSbCl_6$	1.99×10^{-2}	1.17×10^{2}
(^{123}Te)	1/2	0.89	−1.276431	−7.059098	–	26.169742	Me_2Te	1.64×10^{-4}	0.961
^{125}Te	1/2	7.07	−1.5389360	−8.5108404	–	31.549769	Me_2Te	2.28×10^{-3}	13.40
^{127}I	5/2	100	3.328710	5.389573	−71.0	20.007486	KI	9.54×10^{-2}	5.60×10^{2}
^{129}Xe	1/2	26.44	−1.347494	−7.452103	–	27.810186	$XeOF_4$	5.72×10^{-3}	33.60
^{131}Xe	3/2	21.18	0.8931899	2.209076	−11.4	8.243921	$XeOF_4$	5.96×10^{-4}	3.50
^{133}Cs	7/2	100	2.9277407	3.5332539	−0.343	13.116142	$CsNO_3$	4.84×10^{-2}	2.84×10^{2}
(^{135}Ba)	3/2	6.592	1.08178	2.67550	16.0	9.934457	$BaCl_2$	3.30×10^{-4}	1.93
^{137}Ba	3/2	11.232	1.21013	2.99295	24.5	11.112928	$BaCl_2$	7.87×10^{-4}	4.62
^{138}La	5	0.09	4.068095	3.557239	45.0	13.19430	$LaCl_3$	8.46×10^{-5}	0.497
^{139}La	7/2	99.91	3.155677	3.8083318	20.0	14.125641	$LaCl_3$	6.05×10^{-2}	3.56×10^{2}
^{141}Pr	5/2	100	5.0587	8.1907	−5.89	30.62	–	–	–
^{143}Nd	7/2	12.2	−1.208	−1.4570	−63.0	5.45	–	–	–

(Continued)

340

APPENDIX A PROPERTIES OF MAGNETICALLY ACTIVE NUCLEI *(continued)*

Isotope	Spin	Natural Abundance	Magnetic Moment μ/μ_N	Magnetogyric Ratio $\gamma/10^7$ rad s^{-1} T^{-1}	Quadrupole Moment Q/fm^2	Frequency Ratio @ 2.35 T	Reference Compound	Relative Receptivity	
								D^{Proton}	D^{Carbon}
^{145}Nd	7/2	8.3	−0.7440	−0.8980	−33.0	3.36	–	–	–
^{147}Sm	7/2	14.99	−0.9239	−1.1150	−25.9	4.17	–	–	–
^{149}Sm	7/2	13.82	−0.7616	−0.9192	7.4	3.44	–	–	–
^{151}Eu	5/2	47.81	4.1078	6.6510	90.3	24.86	–	–	–
^{153}Eu	5/2	52.19	1.8139	2.9369	241.2	10.98	–	–	–
^{155}Gd	3/2	14.80	−0.33208	−0.82132	127.0	3.07	–	–	–
^{157}Gd	3/2	15.65	−0.4354	−1.0769	135.0	4.03	–	–	–
^{159}Tb	3/2	100	2.6000	6.4310	143.2	24.04	–	–	–
^{161}Dy	5/2	18.91	−0.5683	−0.9201	250.7	3.44	–	–	–
^{163}Dy	5/2	24.90	0.7958	1.2890	264.8	4.82	–	–	–
^{165}Ho	7/2	100	4.7320	5.7100	358.0	21.34	–	–	–
^{167}Er	7/2	22.93	−0.63935	−0.77157	356.5	2.88	–	–	–
^{169}Tm	1/2	100	−0.4011	−2.2180	–	8.29	–	–	–
^{171}Yb	1/2	14.28	0.85506	4.7288	–	17.499306	–	–	–
^{173}Yb	5/2	16.13	−0.80446	−1.3025	280.0	4.821	–	–	–
^{175}Lu	7/2	97.41	2.5316	3.0552	349.0	11.404	–	–	–
^{176}Lu	7	2.59	3.3880	2.16844	497.0	8.131	–	–	–
^{177}Hf	7/2	18.60	0.8997	1.0860	336.5	4.007	–	2.61×10^{-4}	1.54
^{179}Hf	9/2	13.62	−0.7085	−0.6821	379.3	2.517	–	7.45×10^{-5}	0.438
^{181}Ta	7/2	99.988	2.6879	3.2438	317.0	11.989600	$KTaCl_6$	3.74×10^{-2}	2.20×10^{2}
^{183}W	1/2	14.31	0.20400919	1.1282403	–	4.166387	Na_2WO_4	1.07×10^{-5}	6.31×10^{-2}
(^{185}Re)	5/2	37.4	3.7710	6.1057	218.0	22.524600	$KReO_4$	5.19×10^{-2}	3.05×10^{2}
^{187}Re	5/2	62.6	3.8096	6.1682	207.0	22.751600	$KReO_4$	8.95×10^{-2}	5.26×10^{2}
^{187}Os	1/2	1.96	0.1119804	0.6192895	–	2.282331	OsO_4	2.43×10^{-7}	1.43×10^{-3}
^{189}Os	3/2	16.15	0.851970	2.10713	85.6	7.765400	OsO_4	3.95×10^{-4}	2.32
(^{191}Ir)	3/2	37.3	0.1946	0.4812	81.6	−1.718	-	1.09×10^{-5}	6.38×10^{-2}
^{193}Ir	3/2	62.7	0.2113	0.5227	75.1	−1.871	-	2.34×10^{-5}	0.137
^{195}Pt	1/2	33.832	1.0557	5.8385	–	21.496784	Na_2PtCl_6	3.51×10^{-3}	20.7
^{199}Hg	1/2	16.87	0.87621937	4.8457916	–	17.910822	Me_2Hg	1.00×10^{-3}	5.89
^{197}Au	3/2	100	0.191271	0.473060	54.7	−1.729	-	2.77×10^{-5}	0.162
^{201}Hg	3/2	13.18	−0.7232483	−1.788769	38.6	6.611583	$(CH_3)_2Hg$	1.97×10^{-4}	1.16
(^{203}Tl)	1/2	29.524	2.80983305	15.5393338	–	57.123200	$Tl(NO_3)_3$	5.79×10^{-2}	3.40×10^{2}
^{205}Tl	1/2	70.476	2.8374709	15.6921808	–	57.683838	$Tl(NO_3)_3$	0.142	8.36×10^{2}
^{207}Pb	1/2	22.1	1.00906	5.58046	–	20.920599	Me_4Pb	2.01×10^{-3}	11.8
^{209}Bi	9/2	100	4.5444	4.3750	−51.6	16.069288	$Bi(NO_3)_2$	0.144	8.48×10^{2}
^{235}U	7/2	0.72	−0.4300	−0.5200	493.6	1.8414	–	–	–

Nuclei in parentheses are consider to be not the most favorable of the element concerned for NMR

Adapted from Harris, R.K, Becker, E.D., Cabral de Menezes, S.M., Goodfellow, R., and Granger, P. (2001). NMR nomenclature. Nuclear spin properties and conventions for chemical shifts. *Pure Appl. Chem.* **73,** 1795–1818.

CHAPTER 7

SOLVED PROBLEMS

7.1 INTRODUCTION

The perennial student question: Where do we start? The instructor will be sympathetic but not rigidly prescriptive. There are, however, guidelines that do start with the prescriptive statement: *Go for the molecular formula.* Why? Simply because it is the single most useful bit of information available to the chemist and is worth the effort sometimes necessary. It provides an overall impression of the molecule (i.e., the number and kinds of atoms), and it provides the *index of hydrogen deficiency*—in other words, the sum of the number of rings and of double and triple bonds (Section 1.5.3).

Development of the molecular formula starts with recognition of the molecular ion peak (Section 1.5). We assume the usual situation: high-resolution MS instrumentation is not readily available. Let us also assume for now that the peak of highest m/z (except for its isotope peaks) is the molecular ion peak and is intense enough so that the isotope peak intensities can be determined accurately and the presence and number of S, Br, and Cl atoms can be ascertained. Look also at the fragmentation pattern of the mass spectrum for recognizable fragments. If the molecular ion peak is an odd number, an odd number of N atoms is present.

Difficulty often starts with uncertainty in the choice of a molecular ion peak. Many laboratories use chemical ionization as a routine supplement to electron impact, and of course, access to a high-resolution instrument is desirable for more difficult problems.

A search of the infrared spectrum for the familiar characteristic groups is now in order. Note in particular C—H stretching, O—H and/or N—H, and the presence (or absence) of unsaturated functional groups.

With this information in hand, search the proton NMR spectrum for confirmation and further leads. If the spectrum allows, determine the total proton count and ratios of groups of chemical shift-equivalent protons from the integration. Look for first order coupling patterns and for characteristic chemical shifts. Look at the ^{13}C/DEPT spectra; determine the carbon and proton counts and the numbers of CH_3, CH_2, CH, and C groups. A discrepancy between the proton integration and the number of protons represented in the ^{13}C/DEPT spectra represents protons on heteroatoms.

Overlap of proton absorptions is common, but absolute coincidence of nonequivalent ^{13}C peaks is quite rare with a high-resolution instrument. Now, select the most likely molecular formula(s) from Appendix A of Chapter 1 for comparison and determine the index of hydrogen deficiency for each. In addition to difficulties caused by unresolved or overlapping peaks, discrepancies may appear between the selected molecular formula(s) and the ^{1}H and ^{13}C counts because of the presence of elements of symmetry. But this information also contributes to an understanding of the molecular structure.

Students are urged to develop their own approaches. To provide practice in the use of the newer techniques, we have sometimes presented more information than needed, but other Problems should provide compensatory frustration to simulate the real world. Remember the overall strategy: Play the spectra against one another, focusing on the more obvious features. Develop a hypothesis from one spectrum: look to the other spectra for confirmation or contradictions; modify the hypothesis if necessary. *The effect is synergistic,* the total information being greater than the sum of the individual parts.

With the high resolution now available, many NMR spectra are first order, or nearly so, and can be interpreted by inspection with the leads furnished by the mass and infrared spectra. Nevertheless, a rereading of Sections 3.8 through 3.12 may engender caution.

As an example, consider two similar compounds:

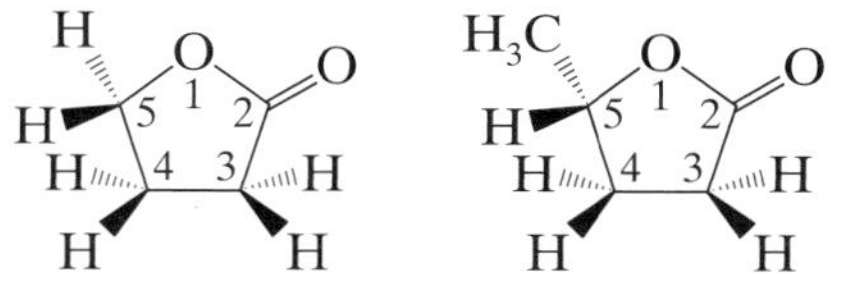

A B

Both rings exist as rapidly flexing ring conformations, but only in compound A do the protons of each CH_2 group interchange to become chemical-shift equivalent (enantiotopes). Only compound A has a plane of symmetry in the plane of the page through which the protons interchange.

From left to right in the spectrum, we predict for compound A: H-5, a two-proton triplet; H-3, a two proton triplet; H-4, a two-proton quintet (assuming nearly equal coupling constants). Given modest resolution, the spectrum is first order.

Compound B has no symmetry element in the planar conformation. C-5 is a chiral center, and the protons of each CH_2 group are diastereotopic pairs. Each proton of the pair has its own chemical shift. The H-4 proton adjacent to the chiral center is distinctly separated, but the H-3 protons are not, at 300 MHz. Each proton of a diastereotopic pair couples geminally with the other and independently (different coupling constants) with the vicinal protons to give complex multiplets.

The possibility of a chiral center should always be kept in mind; *toujours la stereochimie.*

The power of 2-D spectra will become more evident as we work through the problems in Chapters 7 and 8. It is often not necessary to examine all of the spectra in detail before proposing—tentatively—possible structures or fragments. Spectral features predicted for the postulated structures or fragments are compared with the observed spectra, and structural modifications are made to accommodate discrepancies.

These suggestions are illustrated by the following solved problems presented in increasing order of difficulty. The assigned problems of Chapter 8, again in increasing order of difficulty, will provide the essential practice.

Most students enjoy problem solving and rise to the challenge. They also begin to appreciate the elegance of chemical structure as they interpret spectra. Good sleuthing! Be wary of chirality, diastereotopes, virtual coupling, dihedral angles of about 90°, and magnetic nonequivalence.

Finally, what are the requirements for proof of structure? Ultimately, it is congruence of all available spectra with those of a pure, authentic sample obtained under the same conditions and on the same instruments. Obviously, some compromises are acceptable. Congruence with published spectra or spectral data is considered acceptable for publication, but this cannot apply to a new compound, which must then be synthesized.

Computer programs for simulation of proton NMR spectra are available.* If accurate measurements of chemical shifts and coupling constants for all of the protons can be obtained, the simulated spectrum will be congruent with the actual spectrum. In many cases, at least some of the spin systems will be first order. If not, reasonable estimates of shifts and coupling constants may be made, and the iterative computer program will adjust the values until the simulation matches the actual spectrum—assuming, of course, that the identification is valid.

Checklist for logical and pedagogical completeness, not necessarily in order:

1. Show how the molecular formula was derived.
2. Calculate the index of hydrogen deficiency.
3. ***Assign*** diagnostic bands in the IR spectrum.
4. ***Assign all*** protons in the ^{1}H NMR spectrum.
5. ***Assign all*** carbons in the ^{13}C/DEPT NMR spectra.
6. Calculate or estimate $\Delta\nu/J$ where appropriate.
7. Explain multiplicity where appropriate.
8. Assign all correlations in 2-D spectra.
9. Show how the EI mass spectrum supports the structure.
10. Consider possible isomers.

Each problem in this chapter is organized so that the molecular structure and the spectra appears first and are followed by the discussion. The molecular structure is displayed on most of the individual spectra to minimize back-and-forth page turning. The purpose of this arrangement is to encourage students to make their own tentative connections between the molecule and familiar features in the spectra. With this preparation, the subsequent discussions will be more helpful.

* Spectra can be simulated on the computer of a modern NMR spectrometer or on a PC. For example, see the NMR-SIM program, available from Bruker BIOSPIN, Billerica, MA. See also Chapter 3 (Section 3.5.3)

Problem 7.1A

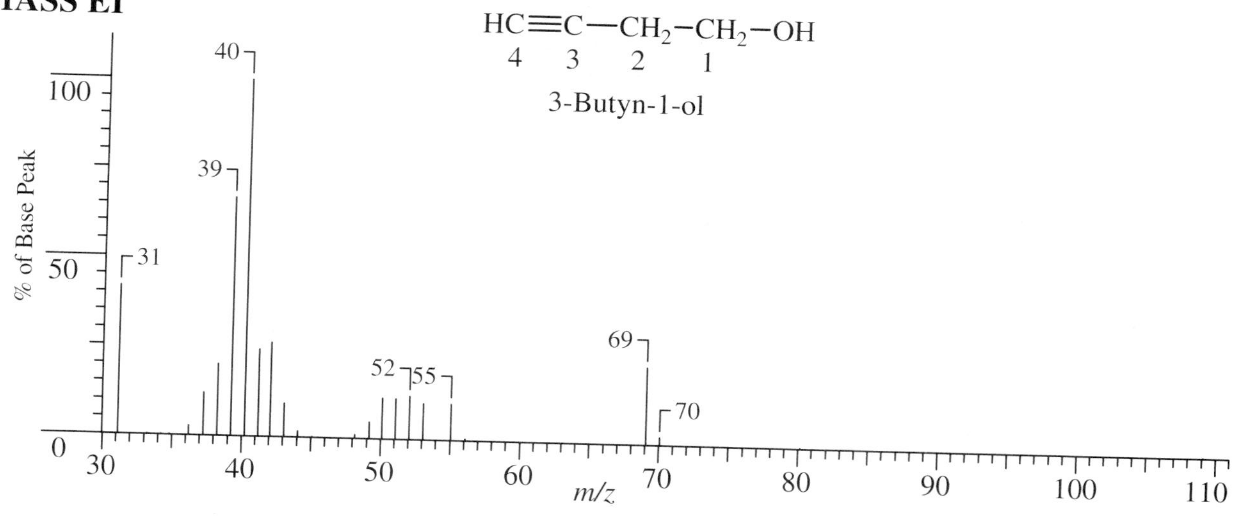
MASS EI
HC≡C—CH2—CH2—OH
4 3 2 1
3-Butyn-1-ol
% of Base Peak
100
50
0
31
39
40
52
55
69
70
30
40
50
60
m/z
70
80
90
100
110

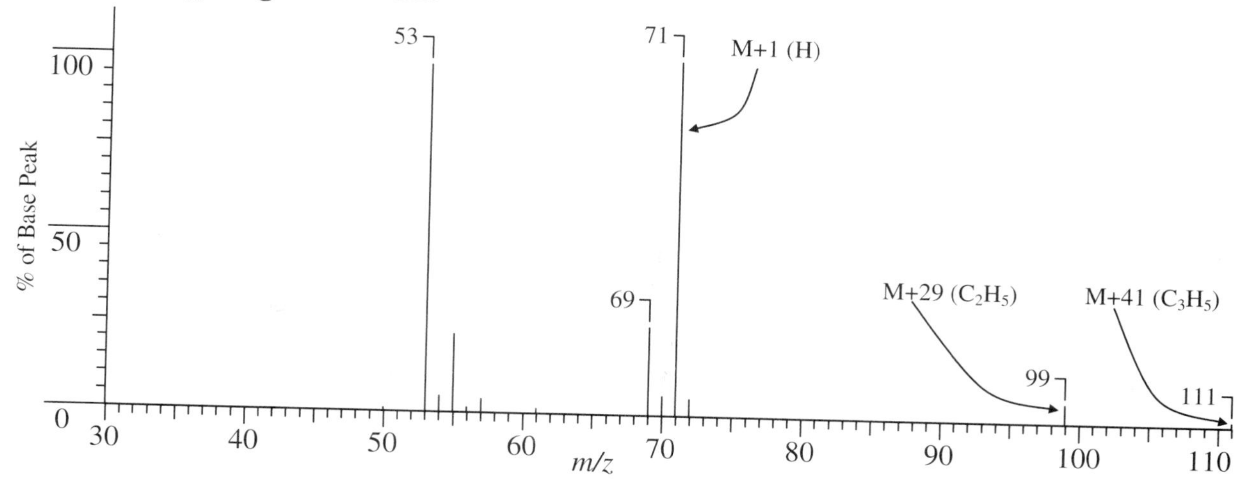
MASS CI reagent gas methane
% of Base Peak
100
50
0
53
71
M+1 (H)
69
M+29 (C2H5)
M+41 (C3H5)
99
111
30
40
50
60
m/z
70
80
90
100
110

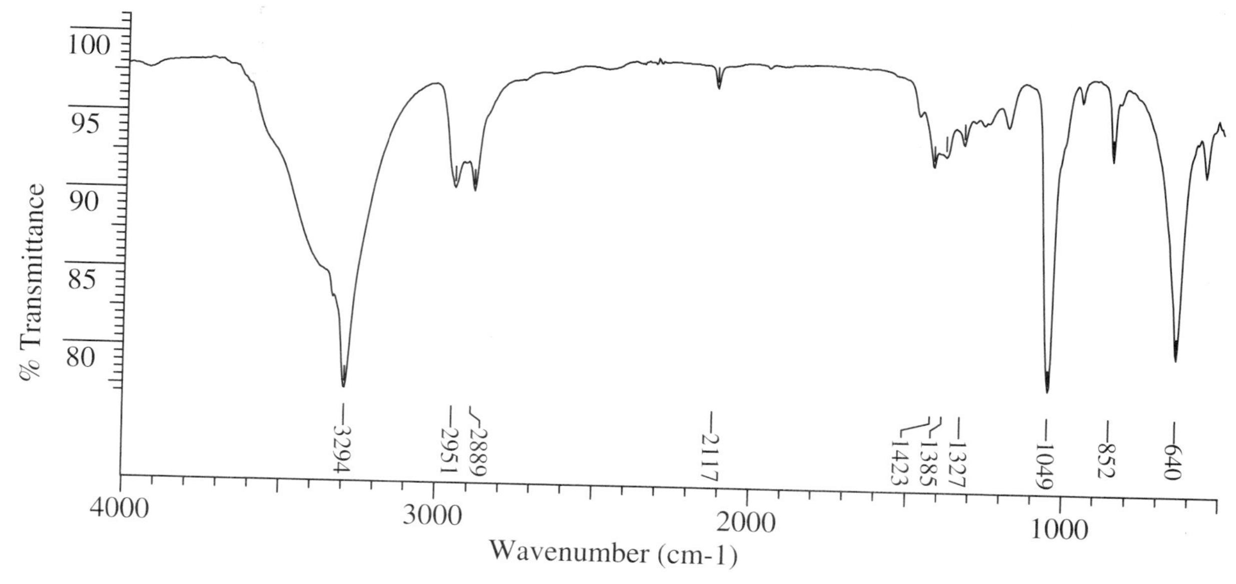
IR
% Transmittance
100
95
90
85
80
3294
2951
2889
2117
1423
1385
1327
1049
852
640
4000
3000
2000
1000
Wavenumber (cm-1)

^{1}H NMR 600 MHz

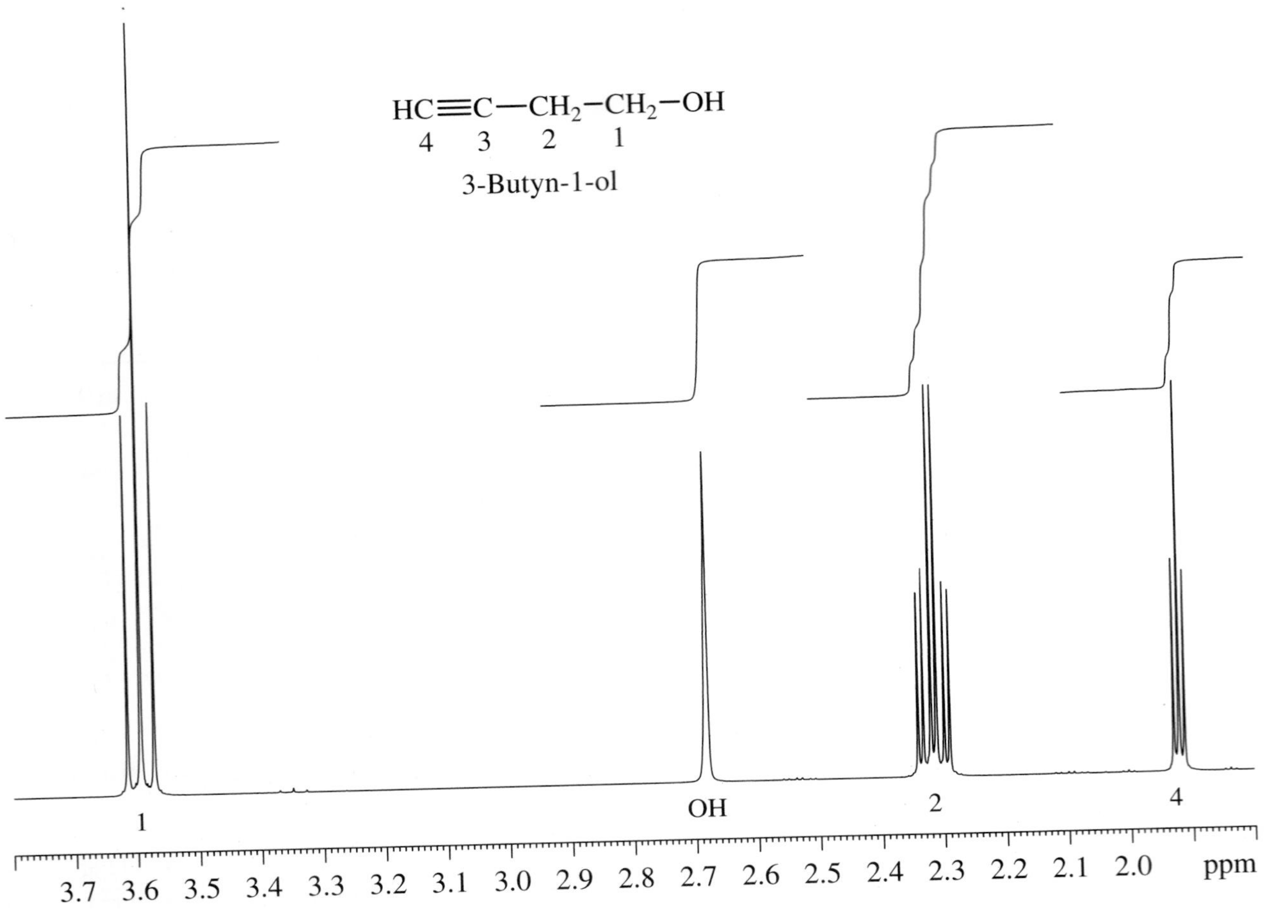

HC≡C—CH₂—CH₂—OH
4 3 2 1
3-Butyn-1-ol
1
OH
2
4
3.7 3.6 3.5 3.4 3.3 3.2 3.1 3.0 2.9 2.8 2.7 2.6 2.5 2.4 2.3 2.2 2.1 2.0 ppm

^{13}C/DEPT NMR 150.9 MHz

3
4
1
2
85 80 75 70 65 60 55 50 45 40 35 30 25 20 ppm

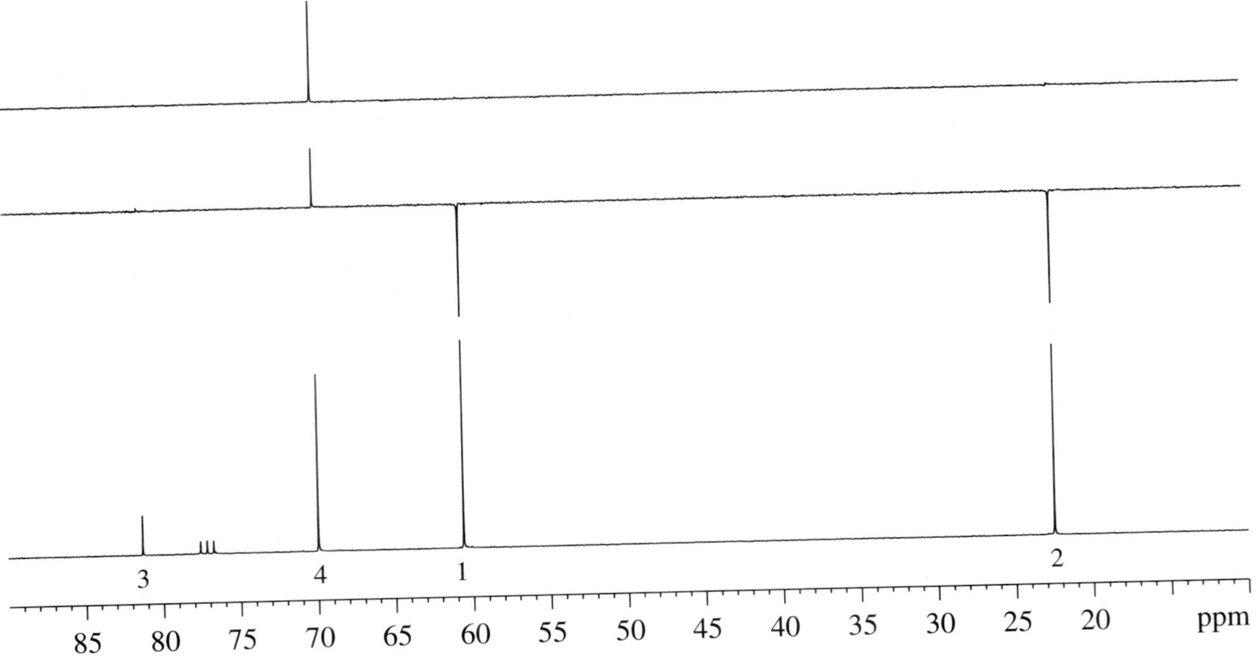

PROBLEM 7.1 DISCUSSION

Everything points to a small molecule. There appear to be no further peaks in the mass spectrum beyond *m/z* 69, but it is rejected as the molecular ion because the next peak is found at *m/z* 55, a putative loss of 14 mass units. The CI mass spectrum possesses a base peak of *m/z* 71, which represents an M + 1 pseudo-molecular ion. The molecular weight of this compound is thus taken as 70 amu. The IR spectrum suggests an alcohol with a broad O—H stretching band at about 3350 cm^{-1} and a strong C—O stretching at 1049 cm^{-1}.

The proton spectrum consists of classical first-order multiplets. From left to right, the multiplicities and integrations are: triplet (2), singlet (1), doublets of triplet (2), triplet (1), which yields six hydrogen atoms. The ^{13}C/DEPT spectra provide four carbon atoms that read from left to right: C, CH, CH_2, CH_2. This discrepancy implies that one of the protons is bonded to a heteroatom. The OH proton at 2.68 ppm in the ^1H spectrum accounts for the difference in proton count between the ^1H spectrum and the ^{13}C/DEPT spectrum.

The assumption of *m/z* 70 as the molecular ion is now quite valid. The molecular formula is now assumed to be C_4H_6O with an index of hydrogen deficiency of two. The options are: two double bonds, one double bond and a ring, two rings, or a triple bond. We can consider these options seriatim.

Consider two double bonds. Do any of the proton or carbon peaks fall in the usual ranges for alkenes? Perusal of Chapters 3 and 4 eliminates the possibility. This leaves us with rings or a triple bond.

Rings are often difficult to rule out on the basis of chemical shifts alone, but the spin couplings would be difficult to explain. Let us consider a triple bond.

Yes, a triple bound would qualify on the basis of chemical shifts for both protons and carbons. The first question is whether the triple bond is terminal or internal; in other words, is there an alkyne proton?

$$H{-}C{\equiv}C{-}R \quad \text{or} \quad R{-}C{\equiv}C{-}R'$$

The ^{13}C spectrum is unequivocal. It shows two peaks in the range for alkyne carbons. The peak at 70 ppm is about the same height as the two CH_2 peaks, but the peak at about 81.2 ppm is distinctly less intense, indicating that it has no attached proton. Furthermore, the ^{13}C/DEPT subspectra show that the peak at about 70 ppm represents a CH group. We can now write two fragments or substructures:

$$H{-}C{\equiv}C{-} \quad \text{and} \quad {-}CH_2{-}OH$$

Insertion of the missing CH_2 group gives a complete molecule:

$$H{-}C{\equiv}C{-}CH_2{-}CH_2{-}OH$$

This structure is completely in accord with the ^1H and the ^{13}C/DEPT spectra. The ^1H spectrum provides a nice demonstration of long-range coupling through the triple bond (from H-4 to H-2) splitting the triplet further into doublets.

Returning with hindsight to the infrared spectrum, we may note the strong H—C≡ stretching band at 3294 cm^{-1} superposed on the O—H band. There is also a strong ≡C—H band at 640 cm^{-1}. Furthermore, there is a weak but distinctive C≡C stretching band at 2117 cm^{-1}.

Several of the major peaks in the mass spectrum are difficult to assign since there are two closely spaced functional groups. Although trivial, verification of the assignments of the protons and their multiplicities are left as an exercise for the student. Likewise, verification of the assignments of the resonances in the ^{13}C/DEPT spectra are left for the student.

Problem 7.2A

H
H
O
6
4
2
8
CH2
H3C
5
3
1
O
7
CH3
H
H

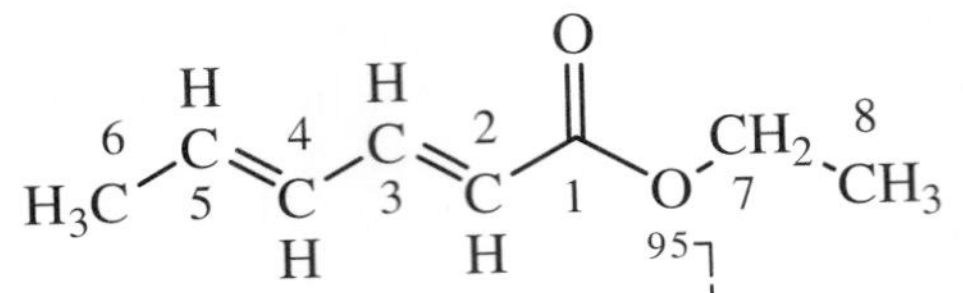

Ethyl sorbate

MASS

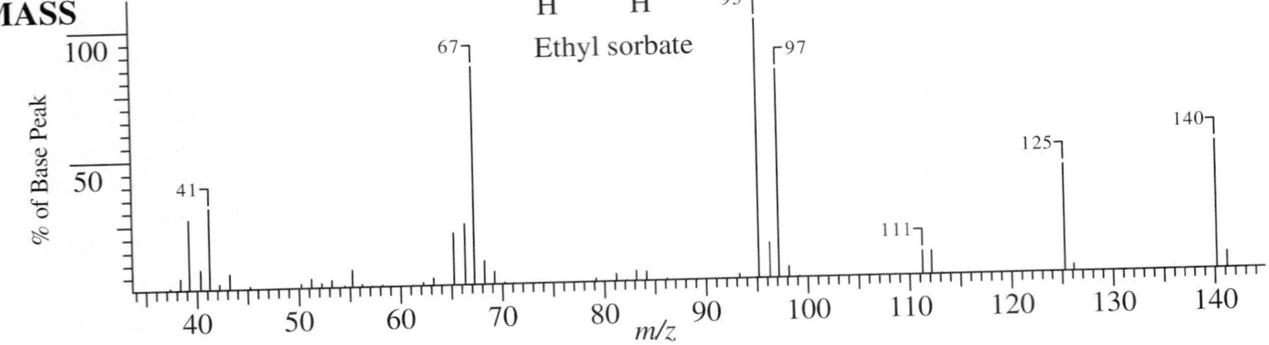

100
50
% of Base Peak
41
67
95
97
111
125
140
40
50
60
70
80
90
100
110
120
130
140
m/z

IR

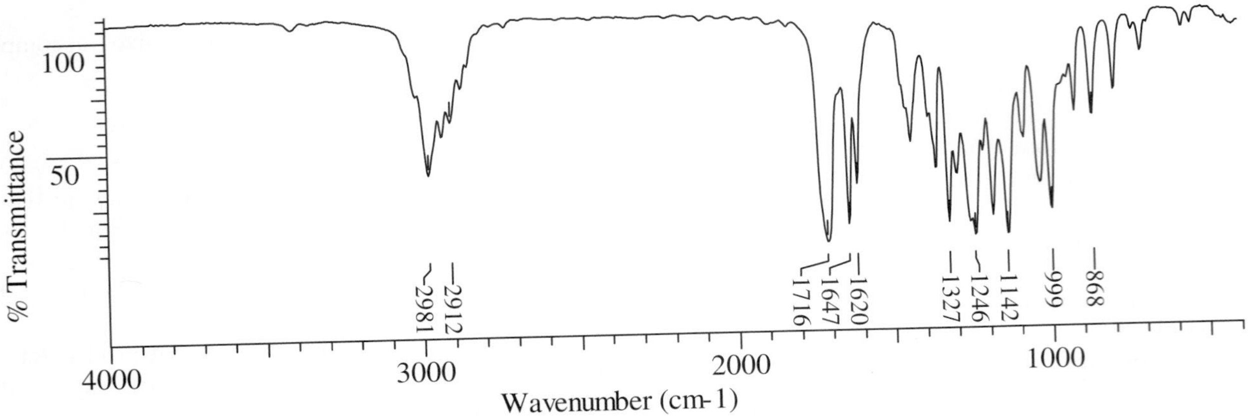

100
50
% Transmittance
2981
2912
1716
1647
1620
1327
1246
1142
999
868
4000
3000
2000
1000
Wavenumber (cm-1)

^{1}H NMR 600 MHz

4350
4300
Hz
3700
3650
Hz
3450
Hz
2500
Hz
1100
Hz
750
Hz
3
4,5
2
7
6
8
7.0
6.5
6.0
5.5
5.0
4.5
4.0
3.5
3.0
2.5
2.0
1.5
ppm

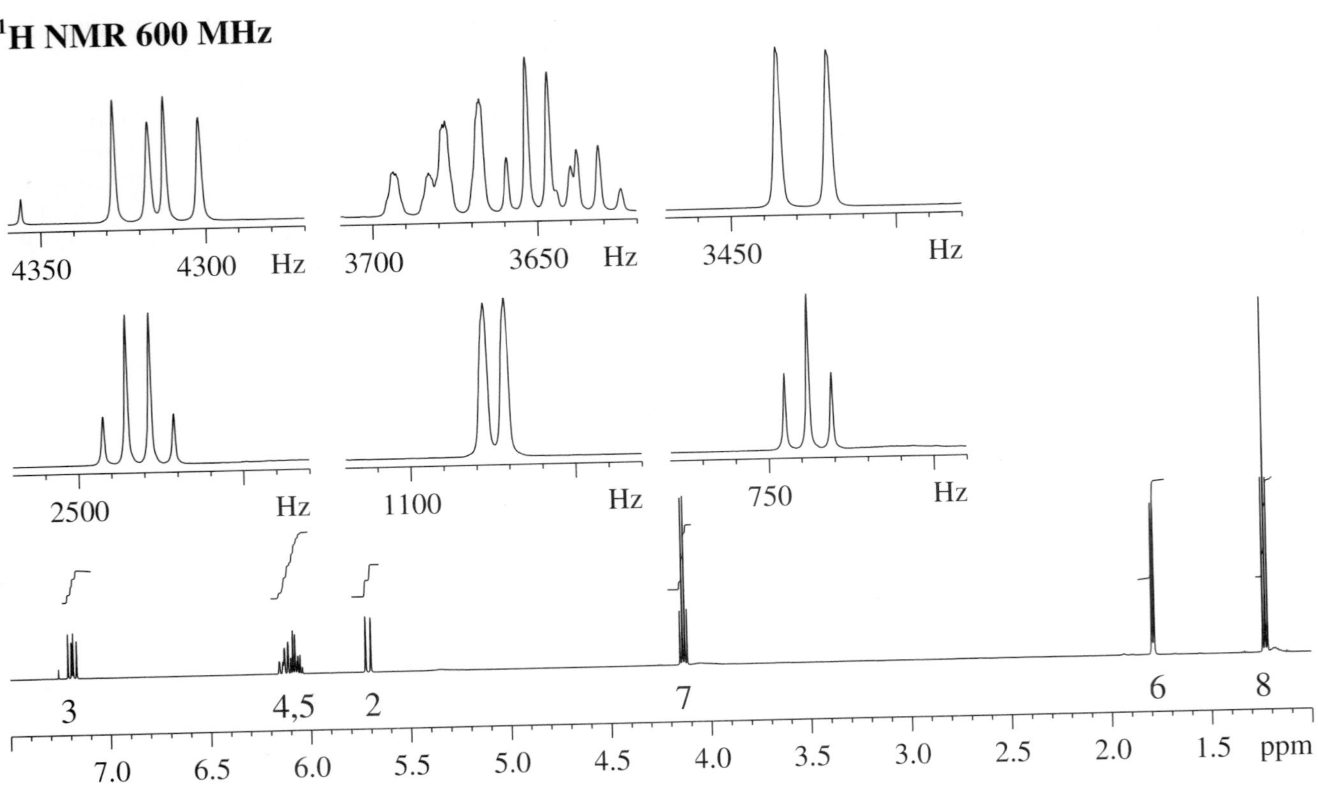

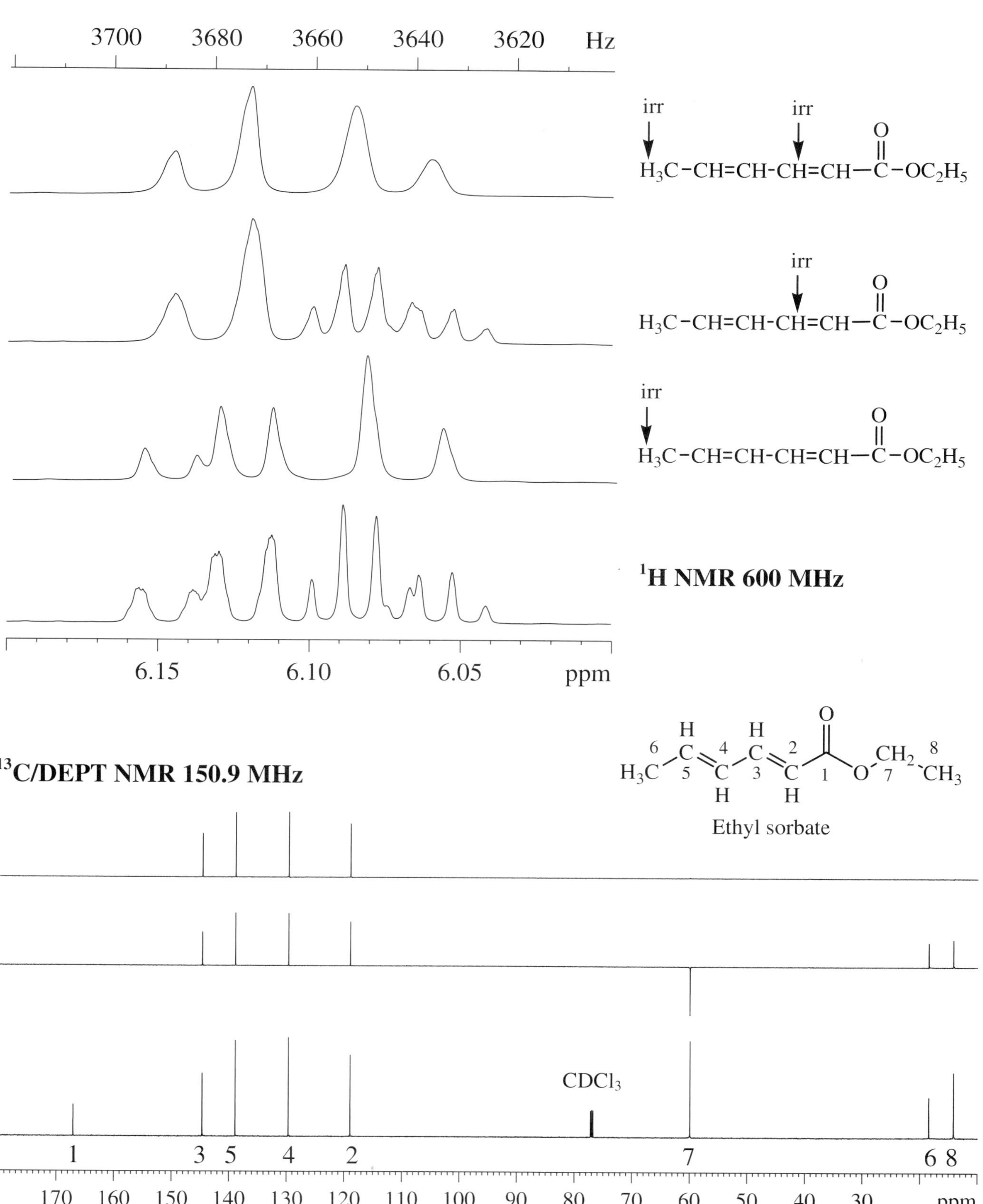

¹H Homodecoupled 600 MHz
3700
3680
3660
3640
3620
Hz
irr
irr
$H_3C-CH=CH-CH=CH-C(=O)-OC_2H_5$
¹H NMR 600 MHz
6.15
6.10
6.05
ppm
¹³C/DEPT NMR 150.9 MHz
Ethyl sorbate
CDCl₃
1
3 5
4
2
7
6 8
170 160 150 140 130 120 110 100 90 80 70 60 50 40 30
ppm

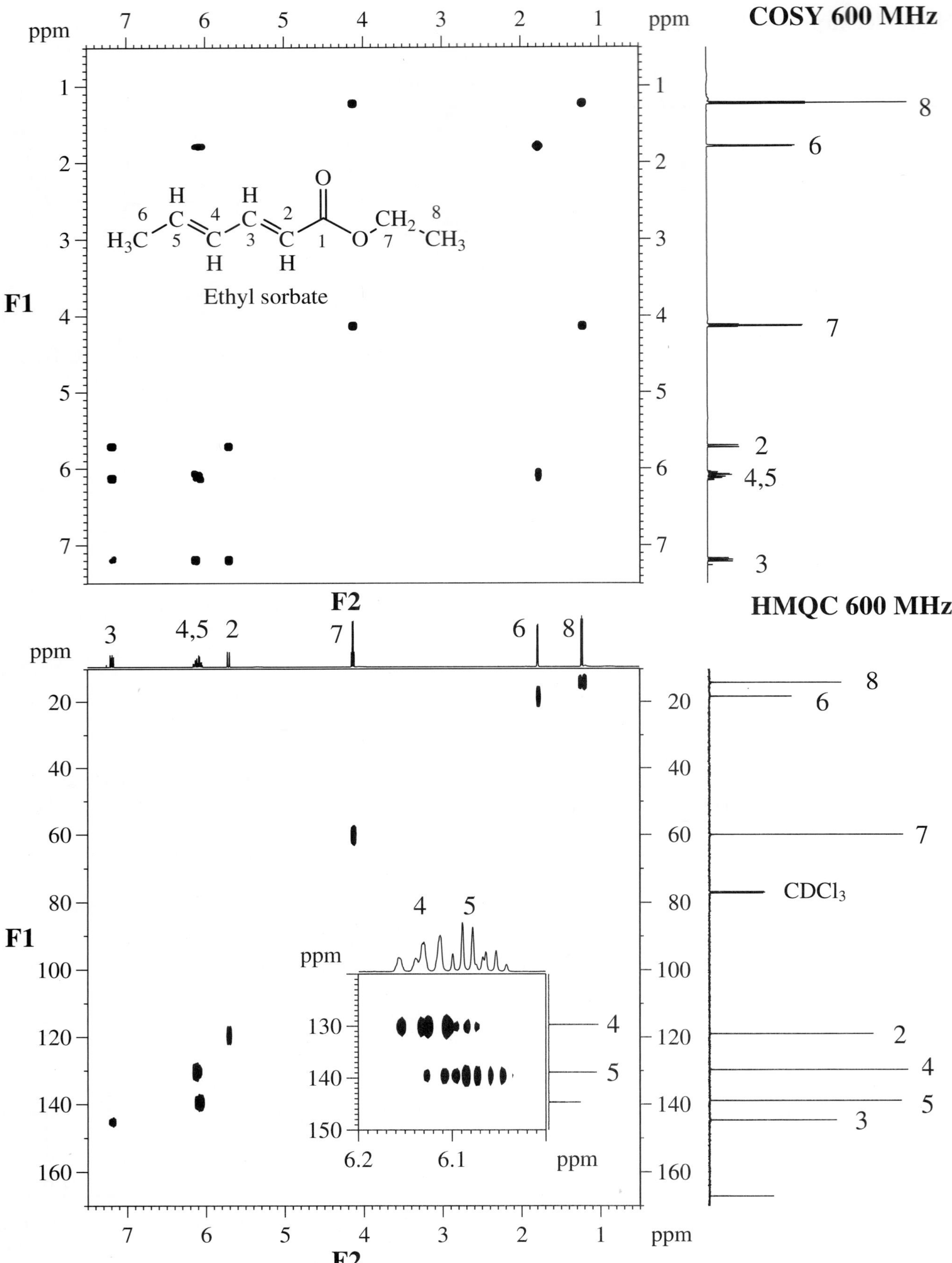

COSY 600 MHz
ppm
F1
F2
Ethyl sorbate
H
C
O
CH2
CH3
H3C
8
6
7
2
4,5
3
HMQC 600 MHz
CDCl3
4
5
ppm

PROBLEM 7.2 DISCUSSION

The relatively strong peak at *m/z* 140 in the mass spectrum is a reasonable choice for the molecular ion peak, since there are no further peaks, and the fragment at *m/z* 125 represents loss of CH_3. Since 140 is an even number, there are 0, 2, 4 . . . N atoms, and we assume 0 as a starting point. The very small M + 1 and M + 2 peaks preclude S, Cl, and Br.

The strong IR band at 1716 cm^{-1} indicates a carbonyl (C=O) group. The two sharp bands at 1647 cm^{-1} and 1620 cm^{-1} indicate one or more carbon—carbon double bonds (C=C) that may be conjugated (see Section 2.6.4.1).

There are six different kinds of protons in the ^1H spectrum in the ratios, from left to right, 1:2:1:2:3:3 with the total of 12 protons. We now count eight peaks in the ^{13}C spectrum (assuming one carbon atom per peak), and from the ^{13}C/DEPT subspectra we read (from left): (C=O) (from IR), CH, CH, CH, CH, CH_2, CH_3, CH_3. With the present information, we write $C_8H_{12}O$ with unit mass 124, which is 16 units less than then a molecular ion peak at *m/z* 140. Is there another oxygen atom in the molecular ion?

Indeed so. The chemical shift of the CH_2 group at 60 ppm suggests a —(C=O)OCH_2— sequence (see Table 4.20). Also, the chemical shift of the carbonyl carbon in the ^{13}C (168 ppm) suggests a carboxylic acid derivative such as an ester. The partial molecular formula can now be revised to $C_8H_{12}O_2$ with a hydrogen deficiency of three.

The proton NMR spectrum immediately points out that the CH_3 triplet at the extreme right is directly attached to the deshielded CH_2 group (quartet). The COSY spectrum confirms this correlation. The sequence, above, is now one end of the molecule: —(C=O)OCH_2CH_3.

In the ^{13}C/DEPT spectra, there are four CH alkene peaks between ~119 ppm and ~145 ppm. There is also the remaining CH_3 group at ~18.5 ppm, which appears in the proton spectrum at ~1.8 ppm as a doublet—obviously attached to one of the four CH groups.

It may seem presumptuous to formulate a molecular structure at this early stage, but we do have one end of the structure, four CH groups with an attached CH_3 group, no possibility for branching, and do not forget the two remaining sites of unsaturation. With some trepidation, we offer the following structure:

$$\underset{6}{CH_3}-\underset{5}{CH}=\underset{4}{CH}-\underset{3}{CH}=\underset{2}{CH}-\underset{1}{\overset{\overset{\displaystyle O}{\|}}{C}}-O-\underset{7}{CH_2}-\underset{8}{CH_3}$$

The synergism between the ^{13}C/DEPT spectra and the proton spectrum should be explored. There are two aspects to a proton spectrum: The first-order multiplets can usually be resolved, whereas the higher-order multiplets are frustrating. In the present proton spectrum, there are five first-order multiplets and two overlapping multiplets that are not first-order.

The ethyl protons are represented by the triplet at ~1.2 ppm coupled to the deshielded quartet at ~4.1 ppm. The other CH_3 group is represented by the doublet at ~1.8 ppm, coupled to one of four alkene CH protons. Rather than attempting to interpret the higher-order multiplets, we turn to the 2-D spectra.

In the COSY spectrum, one of the two overlapping CH groups, which are centered at ~6.1 ppm (labeled H-5 and H-4), couples to the CH_3 doublet at ~1.8 ppm; this coupling confirms the earlier assumption that the CH_3 group is terminal. The proton labeled H-5 also couples with the other overlapping CH group (labeled H-4), which in turn couples with the neighboring CH group at ~7.2 ppm (labeled H-3). The slightly broadened doublet at ~5.7 ppm (labeled H-2) is a result of coupling to H-3 and long-range coupling. We can summarize as follows:

$$\underset{\substack{1.8\\6}}{CH_3}-\underset{\substack{\\5}}{CH}=\underset{\substack{6.1\\4}}{CH}-\underset{\substack{7.2\\3}}{CH}=\underset{\substack{5.7\\2}}{CH}-\underset{\substack{\\1}}{\overset{\overset{\displaystyle O}{\|}}{C}}-O-\underset{\substack{4.1\\7}}{CH_2}-\underset{\substack{1.2\text{ ppm}\\8}}{CH_3}$$

With the complete proton assignments and the direct correlations between carbons and attached protons from the HMQC, we are able to assign all of the carbon resonances, except for the quaternary carbon, which is a trivial assignment in this case. An interesting example is found in the inset of the HMQC spectrum, which shows the correlations of the two overlapped protons, H-4 and H-5. Even though they are overlapped in the proton spectrum, they are well resolved in the HMQC spectrum because the carbon resonances are not overlapped.

One important question still remains: Are the double bonds *E* (*trans*) or *Z* (*cis*)? This question can be answered if the olefinic proton *J* values can be determined. One obvious starting point is the H-2 doublet, which is the result of coupling to H-3. The *J* value is about 16 Hz; this coupling constant falls within the range given for *E*-double bonds given in Appendix F, Chapter 3.

The complex, overlapping multiplets of H-4 and H-5 are not inviting. However, H-3 shows a pair of doublets as a result of the 16 Hz (*trans*) coupling to H-2 and a 10 Hz single bond coupling to H-4. Unfortunately, the coupling constant for the 4,5-double bond is not readily accessible. But spin decoupling (homodecoupling) is worth investigating (see Section 3.15). Irradiation of H-6 simplifies the overlapping H-5, H-4 complex considerably; in fact, there is a 16 Hz doublet (somewhat distorted) at the lower-frequency edge. Irradiation of H-3, individually, simplifies the complex multiplet and shows a 16 Hz doublet at the high-frequency edge. Simultaneous irradiation of H-6 and H-3 results in a pair of 16 Hz doublets. The doublet intensities are not ideal because of the small $\Delta\nu/J$ ratio. There is now no doubt that both double bonds are *E*.

Problem 7.3A

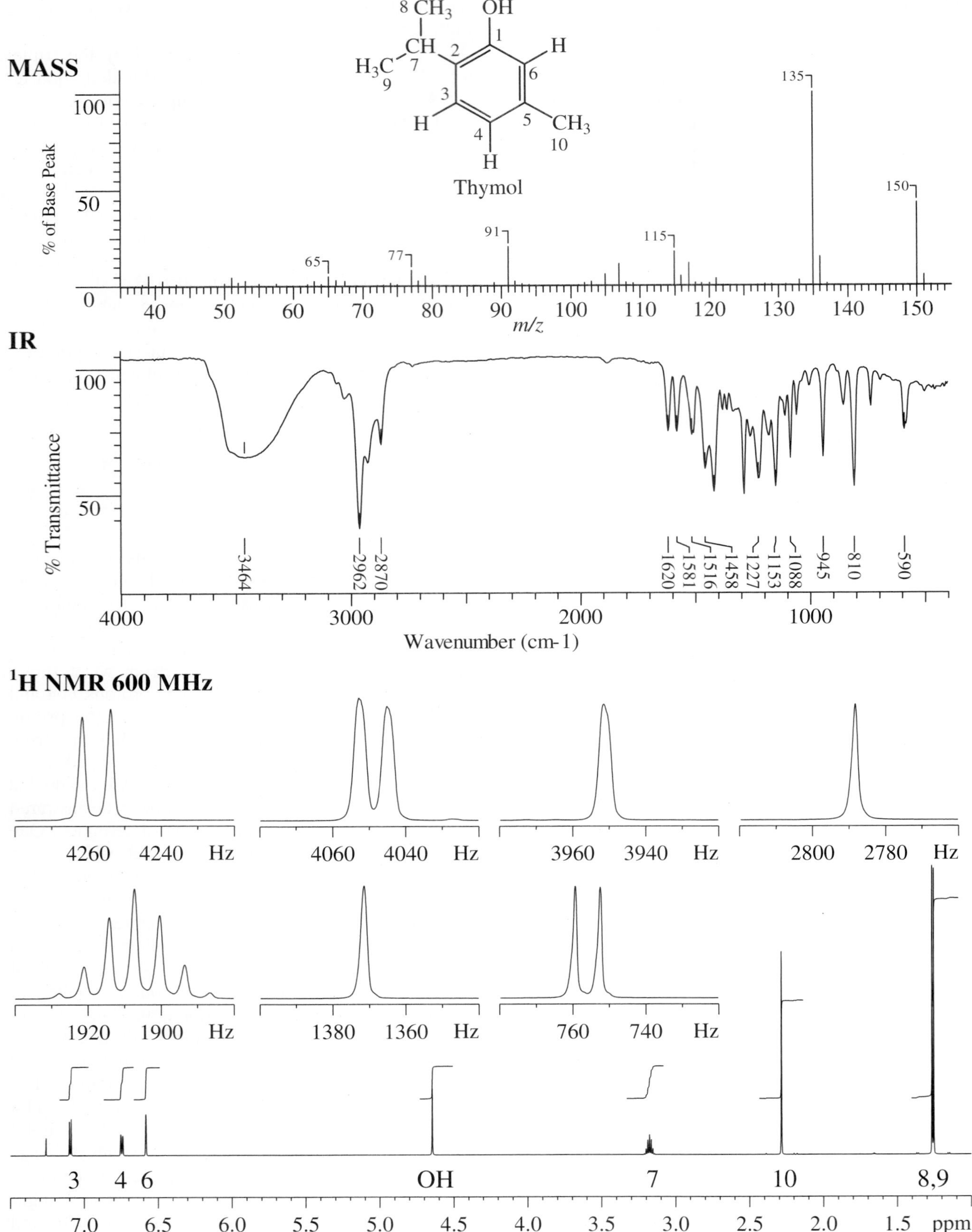

MASS
% of Base Peak
m/z
Thymol
OH
CH3
H3C
CH
H
CH3
135
150
91
115
77
65
IR
% Transmittance
Wavenumber (cm-1)
3464
2962
2870
1620
1581
1516
1458
1227
1153
1088
945
810
590
1H NMR 600 MHz
Hz
3
4
6
OH
7
10
8,9
ppm

Problem 7.3B

DQFCOSY 600 MHz

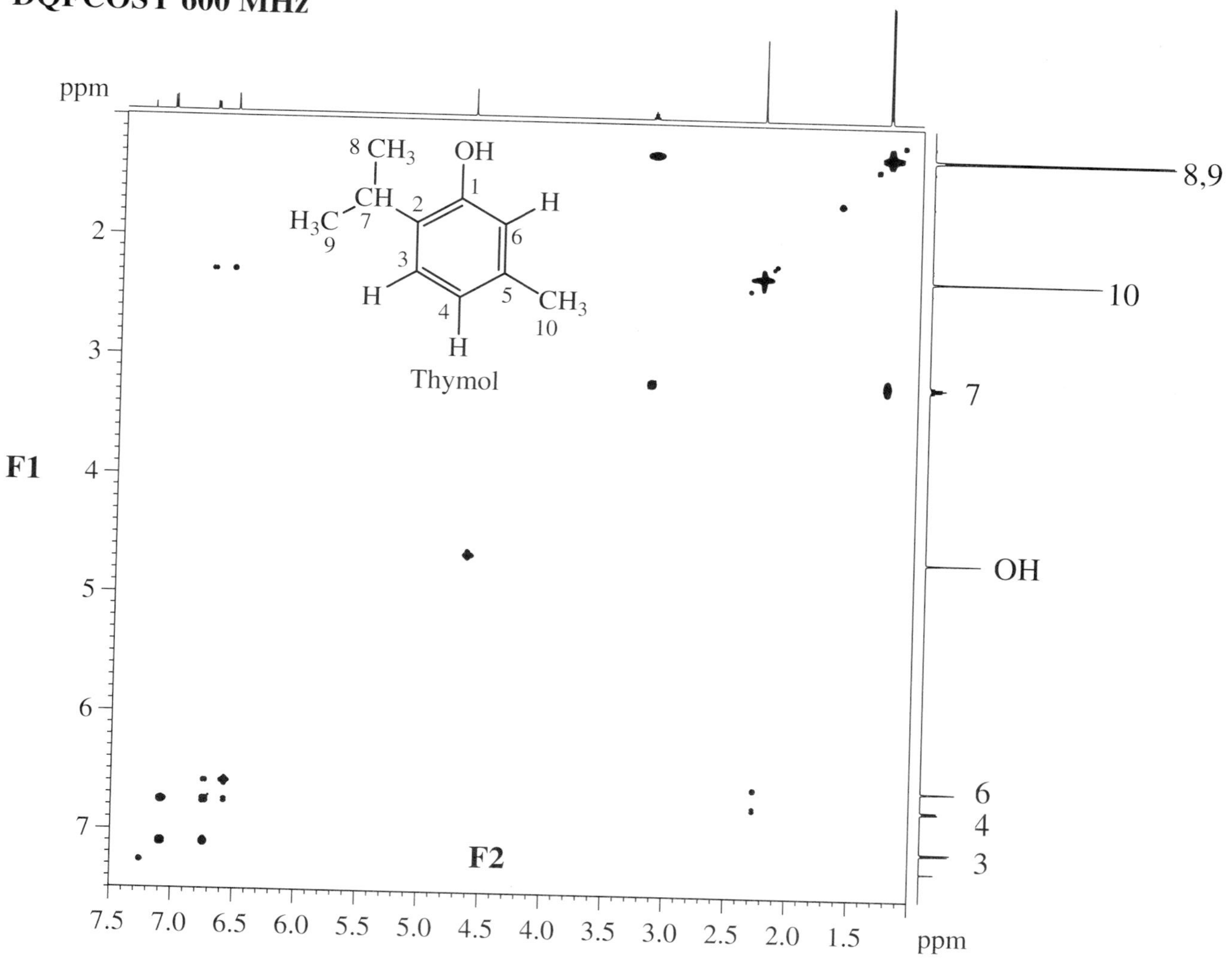

Thymol

8 CH3 OH 1 H 2 H3C 7 9 3 6 H 5 CH3 4 10 H

ppm

F1

F2

8,9
10
7
OH
6
4
3

7.5 7.0 6.5 6.0 5.5 5.0 4.5 4.0 3.5 3.0 2.5 2.0 1.5 ppm

13C/DEPT NMR 150.9 MHz

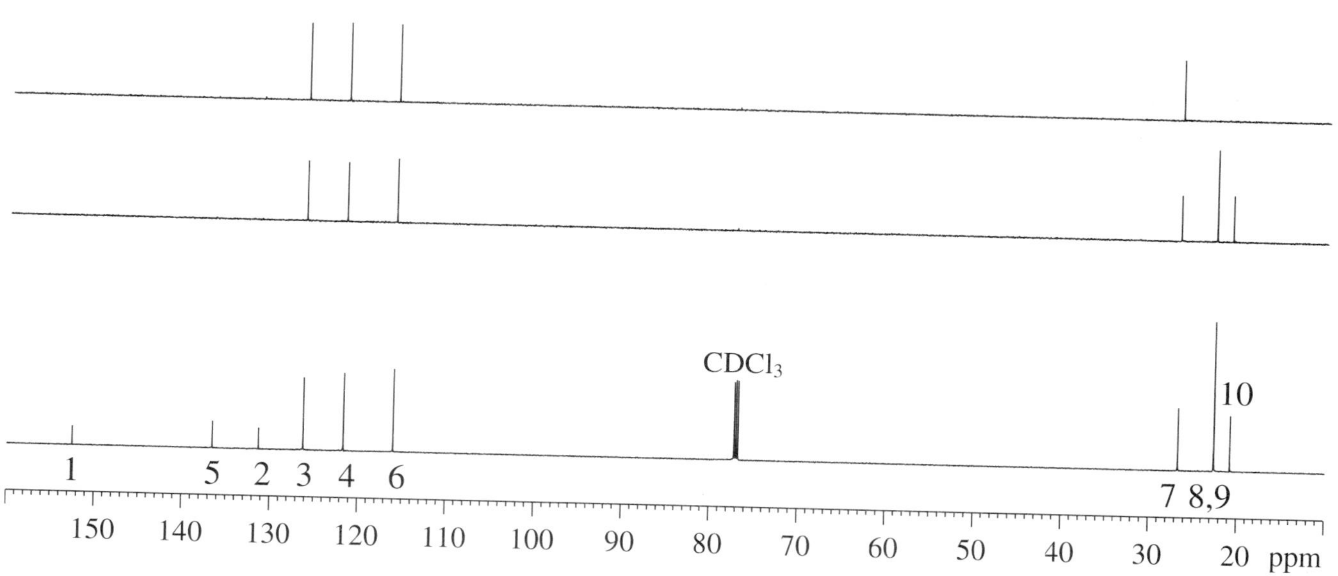

Problem 7.3C

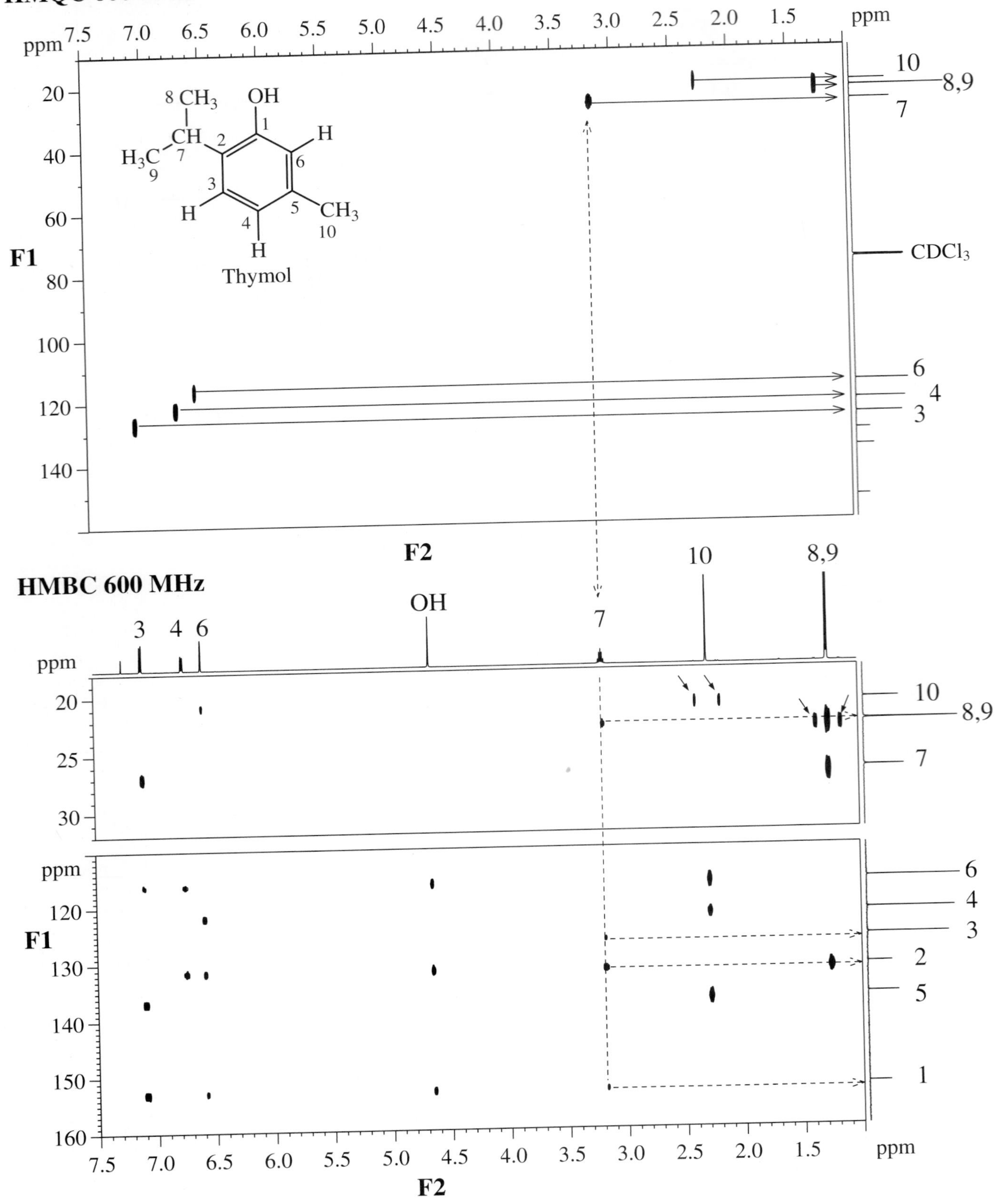

HMQC 600 MHz
Thymol
OH
H
CH3
H3C
CH
H
H
F1
F2
ppm
10
8,9
7
CDCl3
6
4
3
HMBC 600 MHz
OH
2
5
1

PROBLEM 7.3 DISCUSSION

The molecular ion is certainly the medium-intensity peak in the mass spectrum at *m/z* 150; there is a rational loss of a CH_3 group to give the base peak at *m/z* 135. The isotope peaks for the molecular ion do not permit the presence of S, Cl, or Br. Let us assume, tentatively, that the even-numbered molecular ion peak indicates the absence of N. If so, with the help of Appendix A (Chapter 1), the molecular formula can be limited to these possibilities: $C_6H_{14}O_4$, $C_8H_6O_3$, $C_9H_{10}O_2$, or $C_{10}H_{14}O$. The IR spectrum is notable for the intense OH peak at 3464 cm^{-1}. The immediate question is the presence or absence of aromaticity. If an aromatic ring is present, is it attached directly to the OH group to give a phenol? The ^1H and ^{13}C spectra provide answers with peaks in the aromatic regions. The strong IR peaks between 1600–600 wavenumbers suggest aromaticity and ions at 77 and 91 *m/z* serve to confirm our conclusion.

There are seven different kinds of protons in the ^1H spectrum in the ratios, from left to right, 1:1:1:1:1:3:6. Hence a total of 14 protons. The six-proton doublet at δ 1.25 probably represents two equivalent CH_3 groups of an isopropyl moiety; the one-proton septet at δ 3.2 is the corresponding methine group of the isopropyl group.

The ^{13}C spectrum shows nine peaks, but one of them (at 23 ppm) is suspiciously intense and since it correlates with the six-proton doublet in the HMQC, we conclude that there are two superposed CH_3 groups, which makes a total of 10 carbon atoms. The ^{13}C/DEPT spectra specify, from left to right, C, C, C, CH, CH, CH, CH, $CH_3(\times 2)$, CH_3, to which we add the OH group. Under unit mass 150, the most reasonable molecular formula is $C_{10}H_{14}O$, which has an index of hydrogen deficiency of four. This degree of unsaturation fully accounts for a benzene ring: i.e., three double bonds and one ring. Furthermore, the ^{13}C NMR spectrum consists of an aromatic region and an aliphatic region.

In the aromatic region, the three weak peaks represent three quaternary carbon atoms, and the three more intense peaks represent the carbon atoms with attached hydrogen atoms. The most deshielded, weak peak at 153 ppm represents the carbon atom to which the OH group is attached (see Table 4.12).

The substituents in the aliphatic region must be a methyl group and an isopropyl group. For confirmation, the aliphatic region in the proton spectrum shows (from left to right) a one-proton septet (i.e., CH), a three-proton singlet (i.e., CH_3), and a six-proton doublet. It is a doublet because it consists of two identical CH_3 groups coupled to the CH group—hence an isopropyl substituent.

At this point, we distribute the two alkyl substituents with reference to the OH group and do so somewhat indirectly by considering the chemical shifts and coupling constants of the three ring protons. We can assume that the proton peak at δ 6.6 is *ortho* to the OH group (see Chart D.1, Chapter 3). Since this peak is a broadened singlet, there is no adjacent hydrogen atom, but there is a hydrogen atom *meta* to it with a coupling constant too small to delineate. Furthermore, since the spectrum shows only one proton *ortho* to the OH substituent, the other *ortho* position must be attached to either the methyl or the isopropyl group.

The sharp doublet at δ 7.1 with a *J* value of about 8 Hz represents an aromatic hydrogen atom with one *ortho* coupling. Since the peaks are sharp, there is no *meta* coupling. Its chemical shift places it *meta* to the OH group, the alkyl groups having little effect on the chemical shift (see Chart D.1, Chapter 3). The broad doublet at δ 6.75 *para* to the OH group, the coupling being *ortho* and weakly *meta*. The choice is between *I* and *II*.

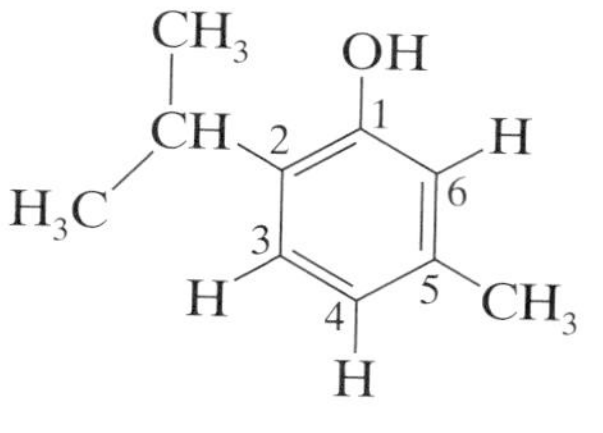

Thymol
I

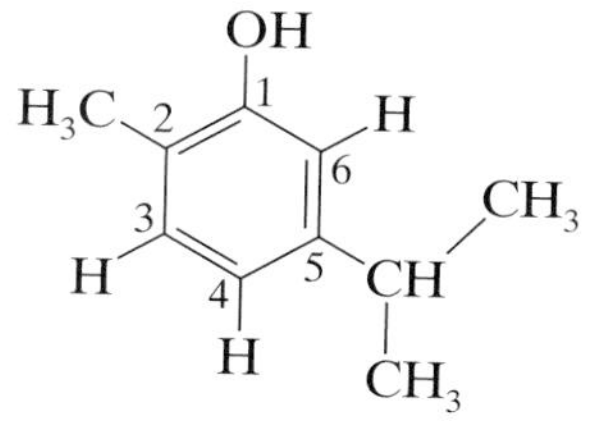

II

The COSY spectrum confirms the previous findings and shows that the protons of the methyl substituent are long-range coupled (^4J) to H-4 and H-6. Interestingly, the isopropyl CH proton does not show long-range coupling to H-3 possibly due to the high multiplicity of the CH absorption, which would produce a very diffuse (not visible) cross peak. As expected, the aromatic protons show meta coupling (^4J) between H-6 and H-4, and ortho (^3J) coupling between H-4 and H-3. Structure I (thymol) is now heavily favored. Note that the definitive long-range coupling between the CH_3 substituent and H-4 and H-6 was not resolved in the ^1H spectrum.

The HMQC shows $^1J_{CH}$ coupling. Table 4.12 in Chapter 4 allows us to arrange the aromatic unsubstituted carbon atoms as C-6, C-4, C-3 from top to bottom. The HMQC spectrum confirms the same sequence for H-6, H-4, H-3. The aromatic, unsubstituted carbon atoms can now be correlated with the firmly assigned aliphatic protons. The substituted aromatic carbon atoms cannot yet be assigned.

The HMBC spectrum permits correlation between isolated proton spin systems—i.e., bridging such "insulating" atoms as O, S, N, and quaternary carbon atoms.

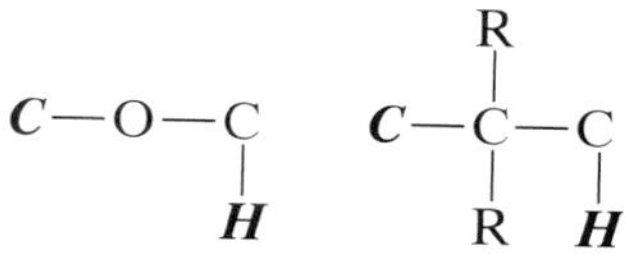

^3J $\qquad$ ^3J

Even in a molecule of modest size, the number of $^2J_{CH}$ and $^3J_{CH}$ couplings can be daunting. Where to start?

Well, simply pose an important question: How do we fully confirm the positions of alkyl substituents? The COSY spectrum did detect the long-range coupling for the methyl substituent but not for the isopropyl substituent. Confirmation can be found by looking down from the CH isopropyl septet in the HMBC spectrum and observe four cross peaks that correlate this CH proton with C-8,9 (^2J), C-2 (^2J), C-3 (^3J) and C-1 (^3J) in the thymol structure. Certainly convincing. As overkill, note that in the HMBC spectrum, the protons of the methyl substituent correlate with C-6 (^3J), C-4 (^3J) and C-5 (^2J). Further, note that the six methyl protons of the isopropyl group correlate with C-7 (^2J) and with C-2 (^3J). Interesting to note that the correlations of H-8 to C-9 (^3J) and H-9 to C-8 (^3J) exist.

The utility of HMBC in correlating quaternary carbon atoms with assigned protons can be shown by working out the correlations of C-1, C-5, and C-2. The assignment earlier of C-1 on the basis of its chemical shift is sound, but the assignment of C-5 and C-2 on the basis of chemical shift alone should be affirmed by correlations. This exercise is left to the student.

Bridging across quaternary carbon atoms has been demonstrated in the course of the above correlations. Two final points: (1) There are four contours, designated by arrows, that represent $^1J_{CH}$ couplings (large) that have not been completely suppressed. These CH doublets are obvious since they straddle the proton peaks. They can be ignored. (2) The correlations of the OH proton with C-6, C-2, and C-1 should be noted. Correlations to OH protons can be very useful, but are rarely seen in an HMBC because they are typically too broad to detect.

Problem 7.4A

MASS

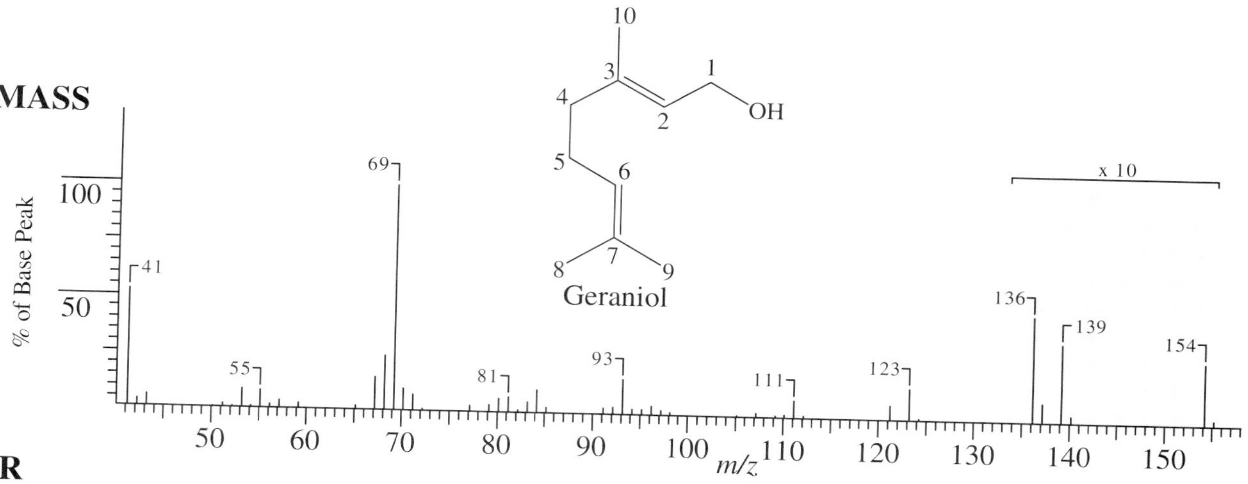

IR

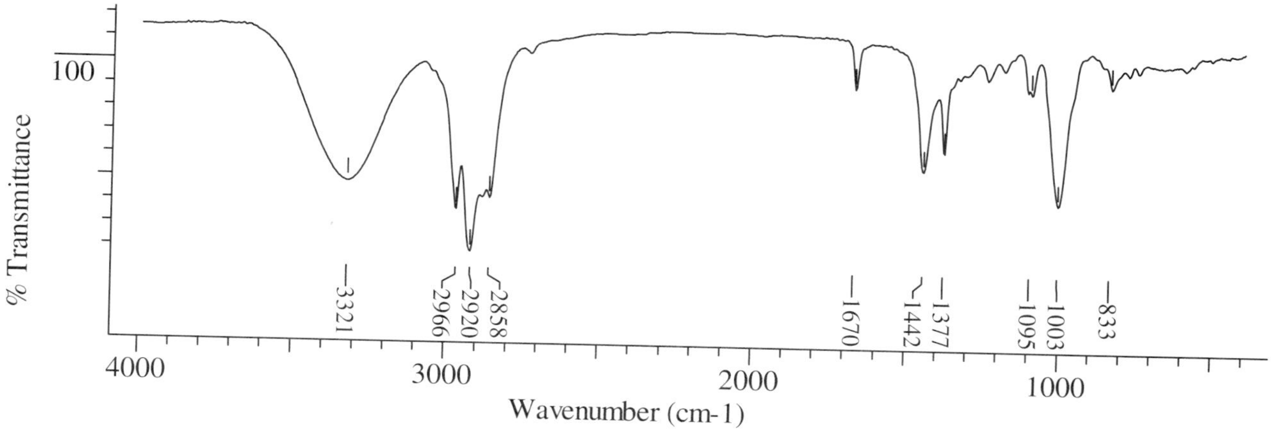

^{1}H NMR 600 MHz

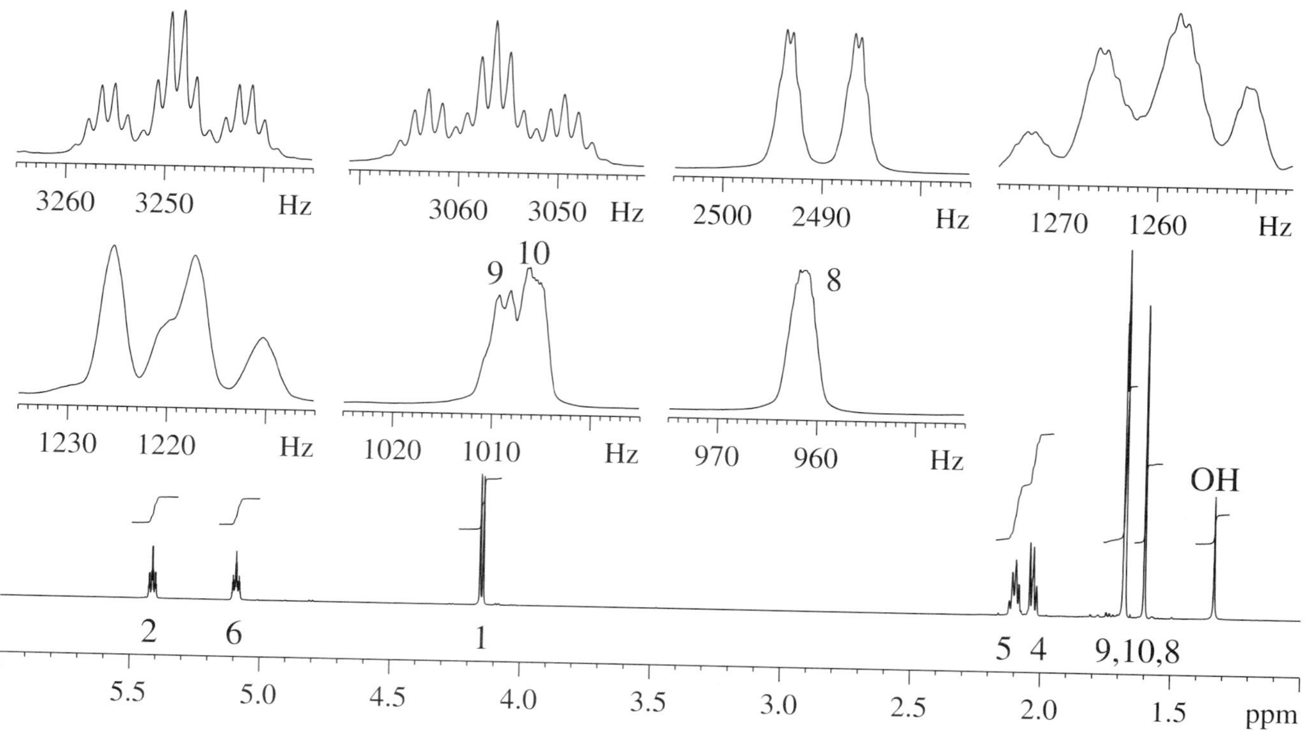

DQFCOSY 600 MHz

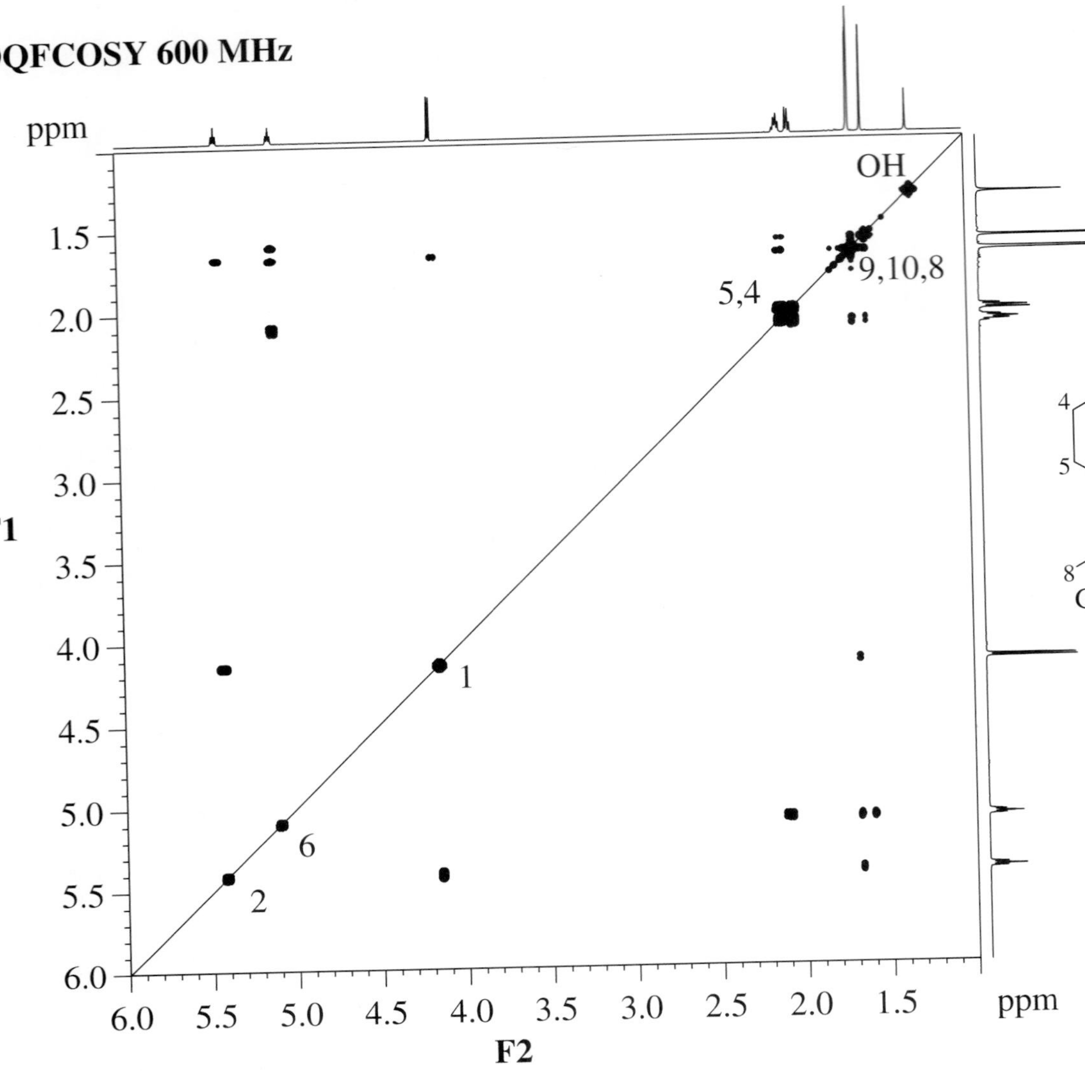

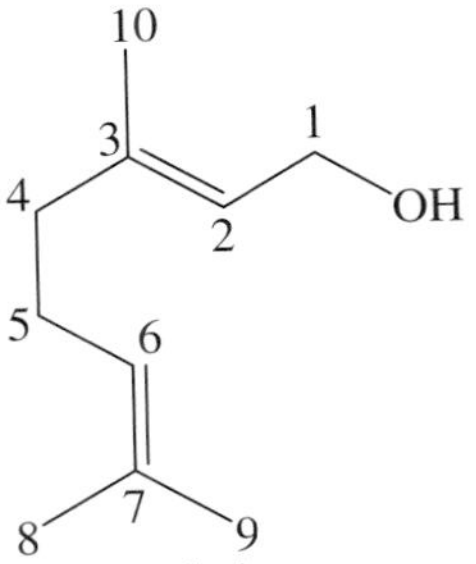

ppm

F1

F2

OH

9,10,8

5,4

1

6

2

10

1

3

4

2

OH

5

6

7

8

9

Geraniol

¹³C/DEPT NMR 150.9 MHz

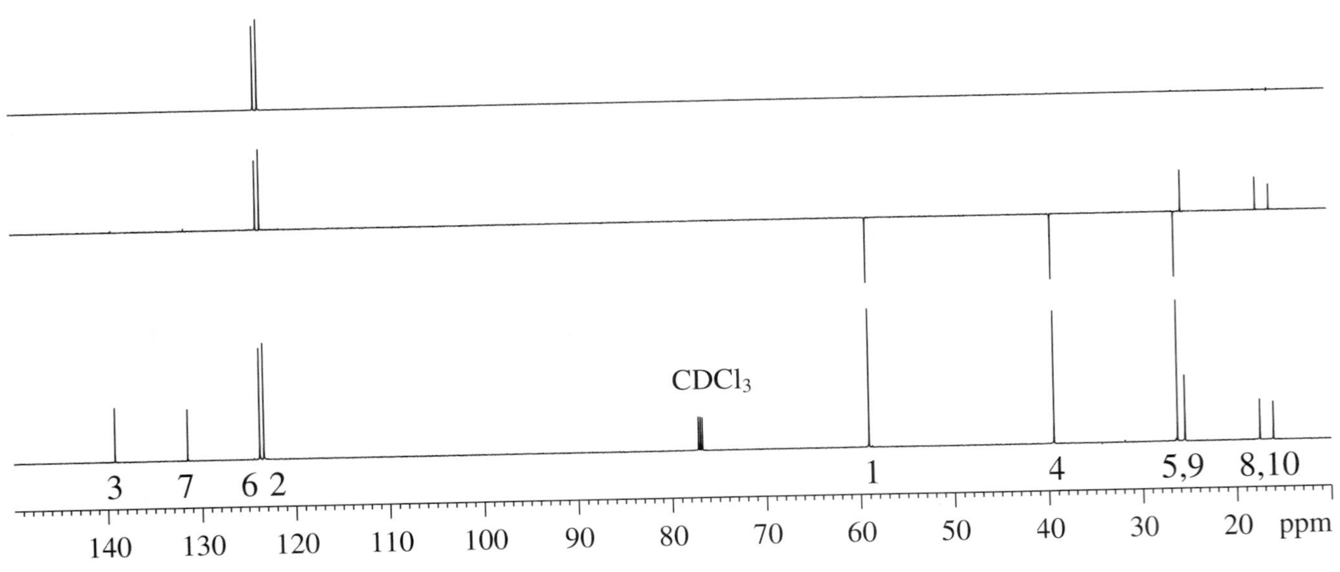

Problem 7.4C

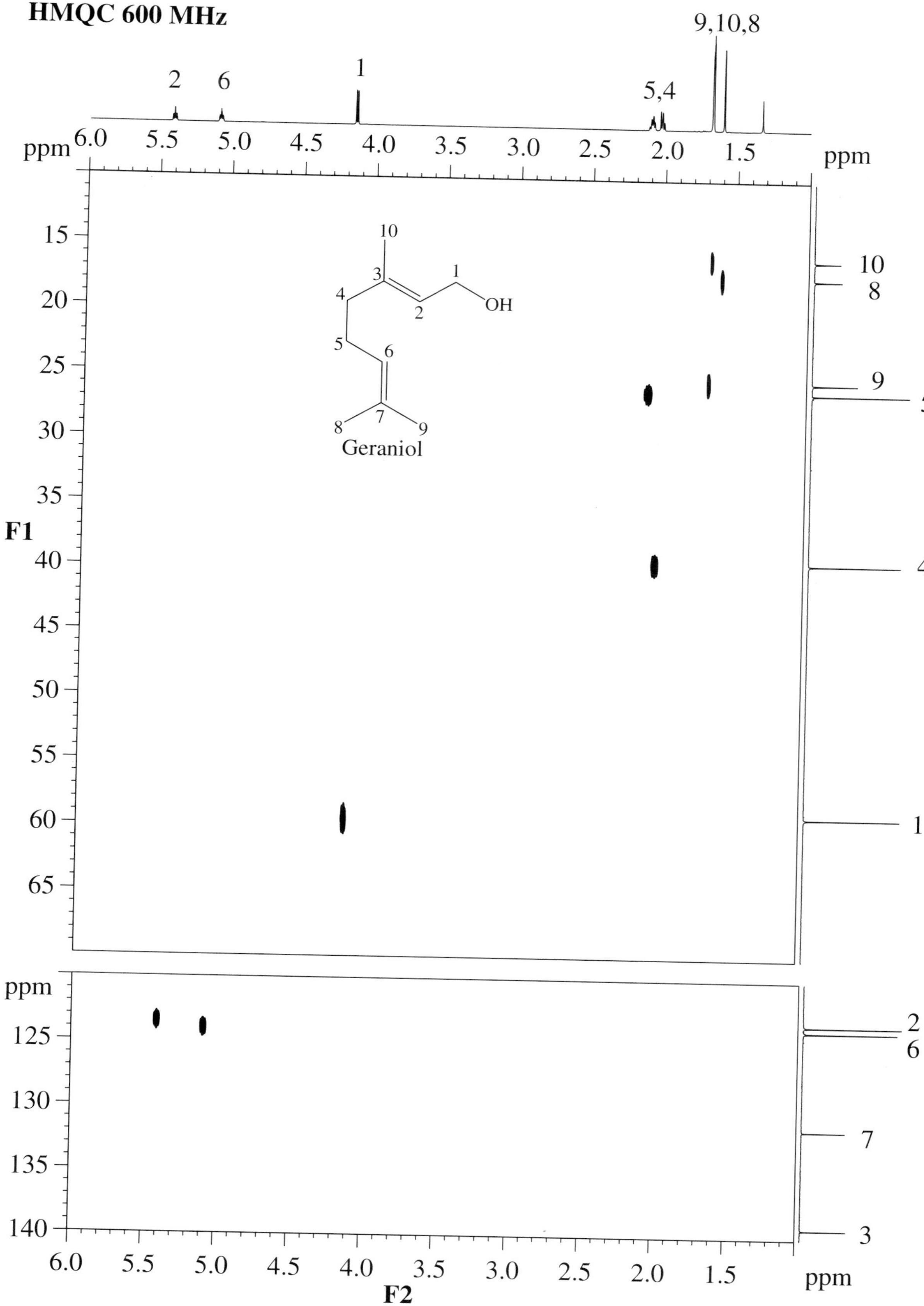
HMQC 600 MHz
9,10,8
2
6
1
5,4
ppm
6.0 5.5 5.0 4.5 4.0 3.5 3.0 2.5 2.0 1.5
10
8
9
5
4
1
2
6
7
3
15 20 25 30 35 40 45 50 55 60 65
F1
125 130 135 140
F2
Geraniol
OH

Problem 7.4D

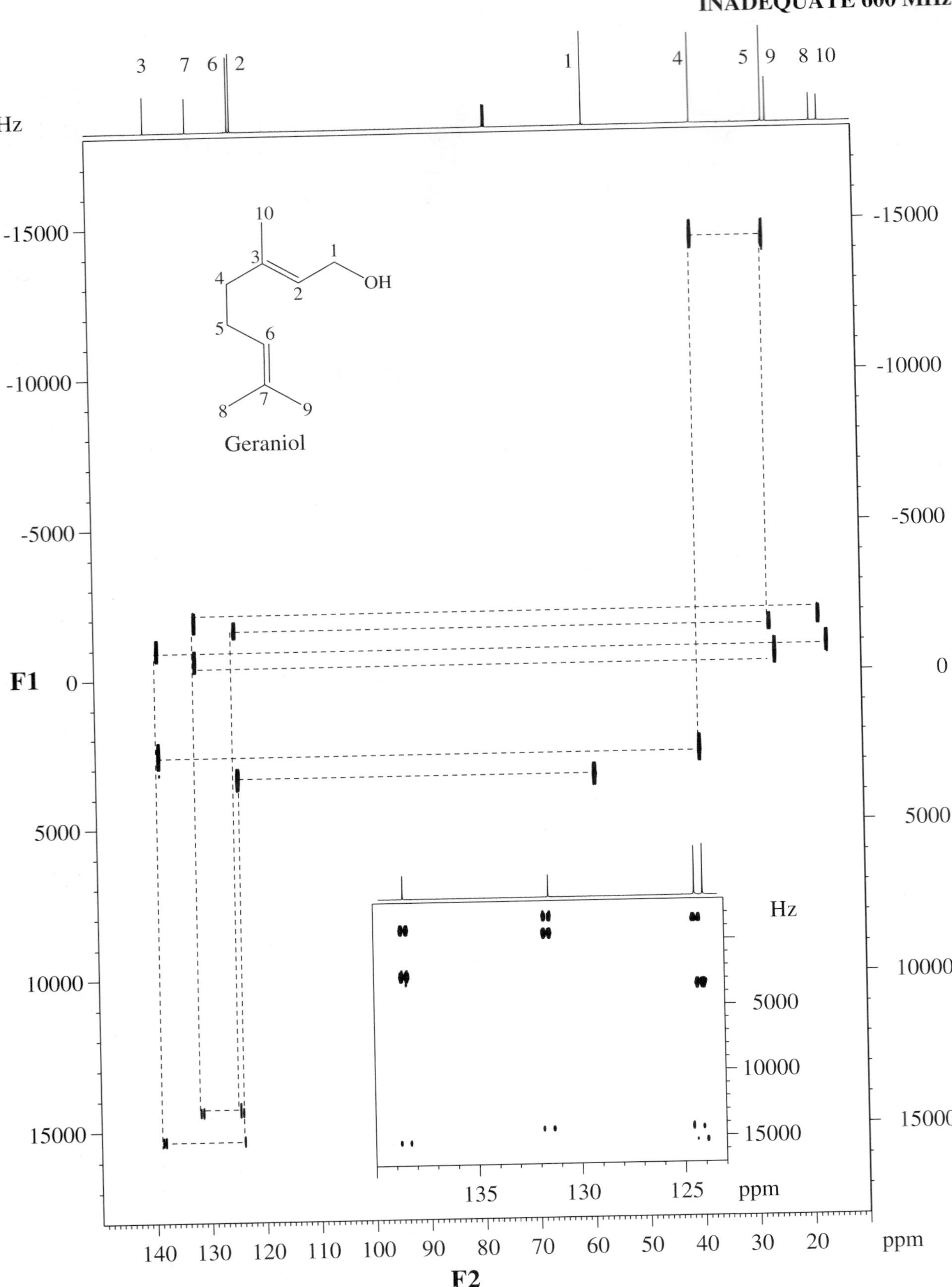
INADEQUATE 600 MHz
3 7 6 2 1 4 5 9 8 10
Hz
-15000
-10000
-5000
0
5000
10000
15000
F1
Geraniol
10
3
1
4
2
OH
5
6
7
8
9
135 130 125 ppm
5000
10000
15000
140 130 120 110 100 90 80 70 60 50 40 30 20 ppm
F2

Problem 7.4E

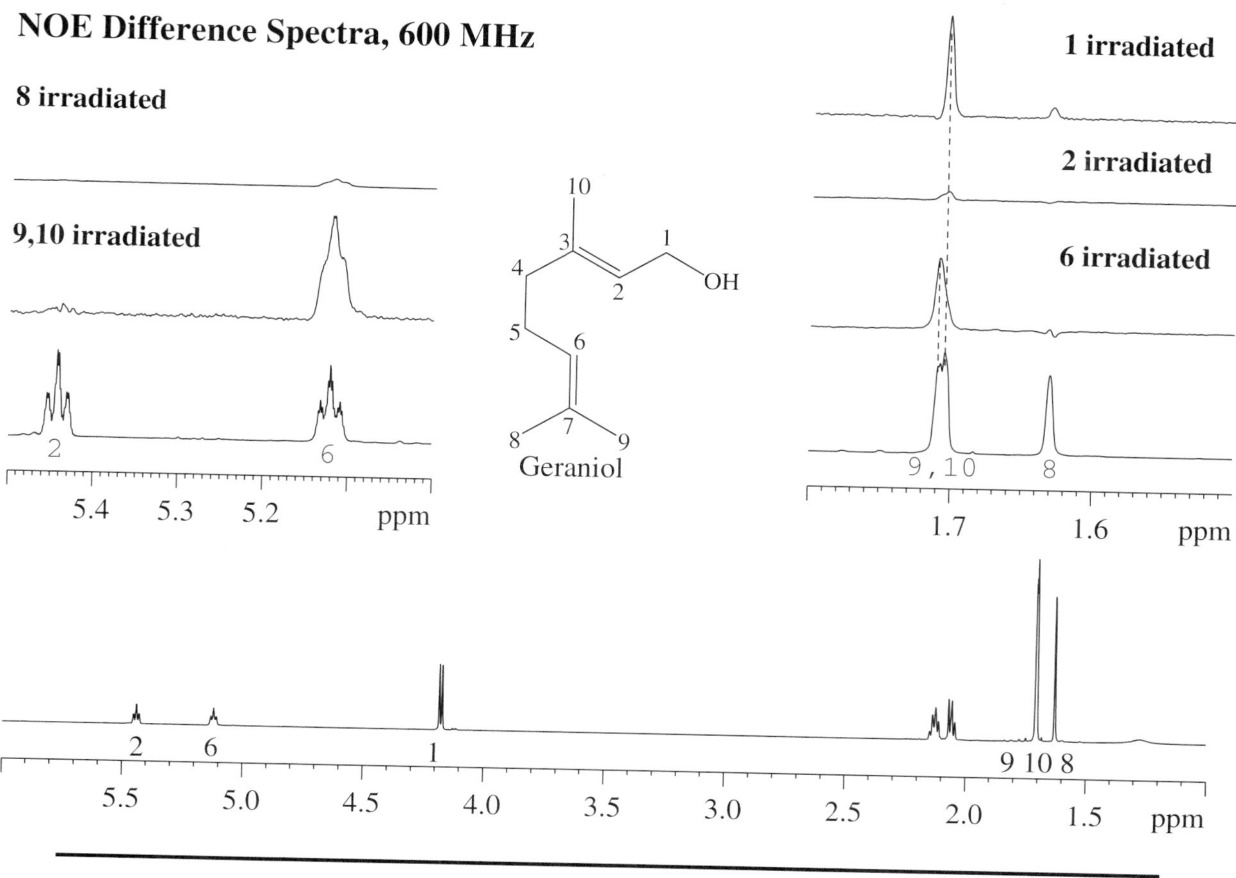
NOE Difference Spectra, 600 MHz
8 irradiated
9,10 irradiated
2
6
5.4
5.3
5.2
ppm
10
3
1
4
2
OH
5
6
8
7
9
Geraniol
1 irradiated
2 irradiated
6 irradiated
9,10
8
1.7
1.6
ppm
2
6
1
9 10 8
5.5
5.0
4.5
4.0
3.5
3.0
2.5
2.0
1.5
ppm

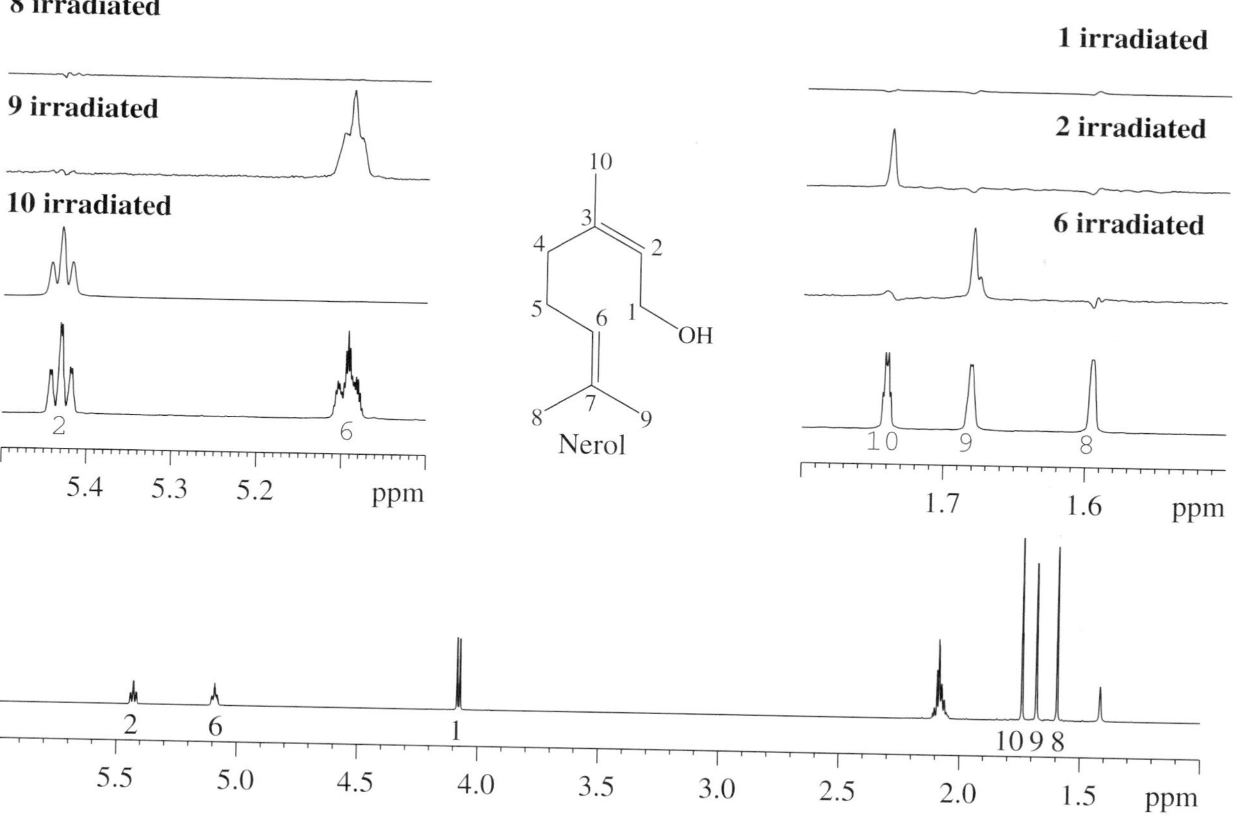
8 irradiated
9 irradiated
10 irradiated
2
6
5.4
5.3
5.2
ppm
10
3
4
2
5
6
1
OH
8
7
9
Nerol
1 irradiated
2 irradiated
6 irradiated
10
9
8
1.7
1.6
ppm
2
6
1
10 9 8
5.5
5.0
4.5
4.0
3.5
3.0
2.5
2.0
1.5
ppm

PROBLEM 7.4 DISCUSSION

It is quite likely that the m/z 154 peak, though small (the gray area is multiplied by ten), is the molecular ion peak. The m/z 139 peak, also small, results from rational loss of a methyl group. The alert interpreter also notes the M-18 peak at m/z 136 and promptly finds the intense, broad OH peak in the "neat" IR spectrum at 3321 cm^{-1} for confirmation; the intense band at 1003 cm^{-1} is probably C—O stretching. Again, as in Problem 7.3 we ask: alcohol or phenol; aromatic or not?

The very weak molecular ion peak in the present problem, together with loss of H_2O, suggests, but does not prove, an alcohol rather than a phenol. It may be worthwhile at this point to entertain the possibility that the base peak (m/z 69) represents the fragment $C_5H_9^+$ and results directly from the molecular ion peak by a strongly favored mechanism. If so, the intact molecule probably contains at least one double bond.

The ^{13}C/DEPT spectra provide ten distinct carbons and seventeen hydrogen atoms arranged thus from left to right: C, C, CH, CH, CH_2, CH_2, CH_2, CH_3, CH_3, CH_3. The first four are very likely olefinic. If the hydroxylic hydrogen atom is added, the tentative molecular formula is $C_{10}H_{18}O$, in accord with the molecular ion, m/z 154. The index of hydrogen deficiency is two, which would allow two double bonds, supported by the four olefinic carbon atoms.

At this point, it is possible to solve the overall structure by using the wealth of information in the 1-D NMR spectra. This (traditional) approach will be explored first, followed by modern use of 2-D NMR spectra. Let us note at this time that the stereochemistry of this molecule cannot be ***proved*** using simple ^{1}H and ^{13}C NMR spectra.

Beginning with the ^{1}H NMR spectrum, the integration from left to right reads: 1:1:2:2:2:6:3:1, in conformity with the 18 hydrogen atoms in the molecular formula. It can also be read: (CH, CH olefinic), (CH_2, deshielded by OH), CH_2, CH_2, CH_3, CH_3, (almost superposed), CH_3, OH. Recall from the ^{13}C/DEPT spectra that there are two carbon atoms that have no attached hydrogen atoms. Recall also that the ^{13}C/DEPT spectra showed three distinct CH_3 groups, whereas the ^{1}H NMR spectrum showed H-9 and H-10 peaks apparently superposed even at 600 MHz. However, they are not completely superposed when expanded; they are partially overlapped with some long-range coupling.

The carbinyl carbon is a methylene group (from the ^{13}C/DEPT) and it is a doublet (with some long-range coupling from H-10) at δ 4.15 in the ^{1}H NMR spectrum. Since the only methine groups in the structure are olefinic (also from the ^{13}C/DEPT), the compound must be an allylic alcohol. The three methyl groups are relatively deshielded and show no vicinal coupling forcing us to place them on olefinic carbon atoms. We consider two possible resulting allylic alcohol structures:

$$(H_3C)_2C{=}CH{-}CH_2{-}OH \quad \text{or} \quad {-}C(CH_3){=}CH{-}CH_2{-}OH$$

The structure on the left is in fact a complete molecule with no open valences; hence, it is rejected as the alcohol "fragment." The fragment on the right however seems plausible.

$$H_3C{-}C(CH_3){=}CH{-} \quad \text{and} \quad {-}C(CH_3){=}CH{-}CH_2{-}OH$$

Another "fragment" can be constructed by considering that we have another double bond with two methyl groups that have no vicinal coupling (i.e., they are geminal) and an olefinic methine, shown at left above. If we consider the two fragments that we now have and realize that the two remaining pieces that have not been used are methylene groups, it is a simple matter of inserting them between the two fragments to arrive at the structure below:

$$H_3C{-}C(CH_3){=}CH{-}CH_2{-}CH_2{-}C(CH_3){=}CH{-}CH_2{-}OH$$

This is a doubly unsaturated terpene alcohol. The stereochemistry is more accessible (and more obvious) with the structure of a conventional terpene:

10, 1, OH, 4, 3, 2, 5, 6, 7, 8, 9

The structure has no chiral center. There is a plane of symmetry in the plane of the page; thus, the protons of each methylene group are interchangeable (enantiotopic). The H-4 protons show a distorted triplet by coupling to the H-5 protons, which show a distorted quartet by coupling to the H-4 protons and to the H-6 proton with a slightly different coupling constant. The small $\Delta\nu/J$ ratio for H-4, H-5 also contributes to the distortion. The methyl groups, H-8 and H-9, are in the symmetry plane, thus not interchangeable.

Certainly the story is convincing, but the evidence is based only on the chemical shifts and on coupling patterns. It is unwarranted to base an analysis on chemical shifts and coupling patterns when a detailed analysis can be done unambiguously with 2-D experiments.

A better approach for solving structures relies less on the 1-D spectra and taps the wealth of information in the 2-D spectra. We obtain the molecular formula as we did above, noting also the presence of the alcohol function from the IR with confirmation in the ^{13}C/DEPT and ^{1}H NMR spectra. Next, we turn to the 2-D data. Evidence of diastereotopic protons is quickly ascertained in the HMQC by noting if there are two protons with different chemical shifts that are correlate to the same ^{13}C peak. No such diastereotopic correlations are seen.

The connectivity data of the COSY spectrum are most reassuring and a good place to start. The peaks along the diagonal are numbered for convenience. A good entry point for the COSY data is the carbinyl methylene at δ 4.15 (H-1). (If you need convincing that this peak is the carbinyl methylene, confirmation can be found in the HMQC and ^{13}C/DEPT.) Correlation by way of vicinal coupling is found to the olefinic methine (H-2) at δ 5.41 and correlation to a methyl group at δ 1.68 (H-10) by way of long range coupling is also evident. How do we know that the proton multiplet at δ 5.41 is a methine? By using the natural interplay of spectra, the proton multiplet at δ 5.41 correlates with a carbon resonance at 123 ppm in the HMQC; this information is fed back into the ^{13}C/DEPT spectra and we find that the carbon resonance at 123 ppm is a methine. Likewise, the proton absorption at δ 1.68 is correlated to a methyl carbon atom in the HMQC.

We can continue the connectivity pattern with the COSY to H-4 because there is a weak long range coupling from methyl H-10 to methylene H-4 at δ 2.11. (The student is encouraged to confirm that the multiplet at δ 2.11 is a methylene group by switching from the COSY to the HMQC to the ^{13}C/DEPT.) The only other correlation to H-4 is to H-5; this correlation is difficult to discern because the cross peaks are nearly on the diagonal. The H-5 methylene group shows a correlation to H-6 (the other olefinic methine) at δ 2.03. H-6 shows two other correlations, both long range, to the methyl groups H-8 and H-9. There are no other correlations in this COSY spectrum. The OH proton, of course, shows no cross peak because of rapid exchange.

At this point, we have assigned all of the protons but still cannot differentiate between the methyl groups at H-8 and H-9. Since we know all the ^{1}H assignments, it is a trivial task to transfer assignments to the ^{13}C signals through the HMQC spectrum. The quaternary carbons C-3 and C-7 have no attached protons and cannot be correlated in the HMQC spectrum. An HMBC spectrum could be used to correlate the quaternary carbon atoms, but for this problem we use carbon connectivities instead.

The INADEQUATE spectrum delineates the connectivities between adjacent ^{13}C atoms. It is a most powerful tool; after all, organic chemistry consists mainly of chains and rings of carbon atoms. Lines showing connectivities between and among correlated carbons have been added. As a starting point, consider the three methyl group carbons C-8, C-9, and C-10. We note that C-10 is connected to C-3 whereas C-8 and C-9 are both connected to an olefinic carbon (C-7). These connectivities confirm our assignment of C-10; however, we are unable to distinguish between C-8 and C-9. These assignments are made in the next section. If we continue from C-3, we see two more connectivities, one to an olefinic carbon (C-2) (see inset) and the other to an aliphatic carbon (C-4). The rest of the connectivities are left as an exercise for the student to transform the correlations into a carbon skeleton.

There are still two remaining tasks: assignment of stereochemistry of the C-2, C-3-double bond and assignment of the C-8 and C-9 methyl groups. NOE difference spectrometry is described in Section 3.16. It is a 1-D experiment that reveals ^{1}H—^{1}H proximity through space because of enhancement by the nuclear Overhauser effect. The "difference" spectrum is obtained by subtracting a standard ^{1}H spectrum from the NOE spectrum; this leaves only the enhanced peak(s).

The task we face with the present molecule—distinguishing between a trisubstituted (*E*) double bond and the corresponding (*Z*) double bond—is not a trivial assignment. Nor is the task of distinguishing H-8 and H-9 methyl groups, as has been mentioned earlier. For conclusive results, we examine both the (*E*) isomer (geraniol) and the (*Z*) isomer (nerol) at the C-2 double bond.

In the top half of the NOE Difference Spectra, the ^{1}H NMR spectrum of geraniol, along with the NOE difference subspectra resulting from irradiation of key proton groups, and, in the bottom half of the page, the ^{1}H NMR spectrum of nerol (the geometric isomer of geraniol) along with the corresponding NOE difference subspectra are given. In geraniol, irradiation of olefinic methine H-2 shows no NOE enhancement of the H-10 methyl group; the reciprocal irradiation of the H-10 methyl group shows no NOE enhancement of the H-2 methine group. We conclude that these two groups are on opposite sides of the double bond and assign geraniol an *E*-double bond. This assignment is confirmed by irradiation of the H-1 allylic methylene group and the concomitant NOE enhancement of the H-10 methyl group thereby proving their disposition on the same side of the double bond. Since methyl groups H-9 and H-10 overlap in the proton spectrum, they are irradiated together and we see an NOE enhancement of olefinic methine H-6. Check the result of irradiation of methine H-6.

Quite often in "real life" problems, especially those involving natural products, the geometric isomer is not available (although, in principle, it could be synthesized). For pedagogical purposes, the results from nerol are presented. In this case, irradiation of olefinic H-2 does result in NOE enhancement of methyl group H-10 and we conclude that nerol has a Z-double bond. The assignments of methyl groups H-8 and H-9 are left to the student.

With hindsight, we can now recognize the fragment peak at *m/z* 69 (the base peak) as the result of the allylic cleavage of an olefin. Ordinarily, reliance on this cleavage for location of a double bond is dubious, but in geraniol, cleavage of the bis-allylic bond between C-4 and C-5 results in the stabilized fragment, *m/z* 69. See Section 1.6.1.2 for the analogous allylic cleavage of β-myrcene into fragments *m/z* 69 and 67.

Problem 7.5A

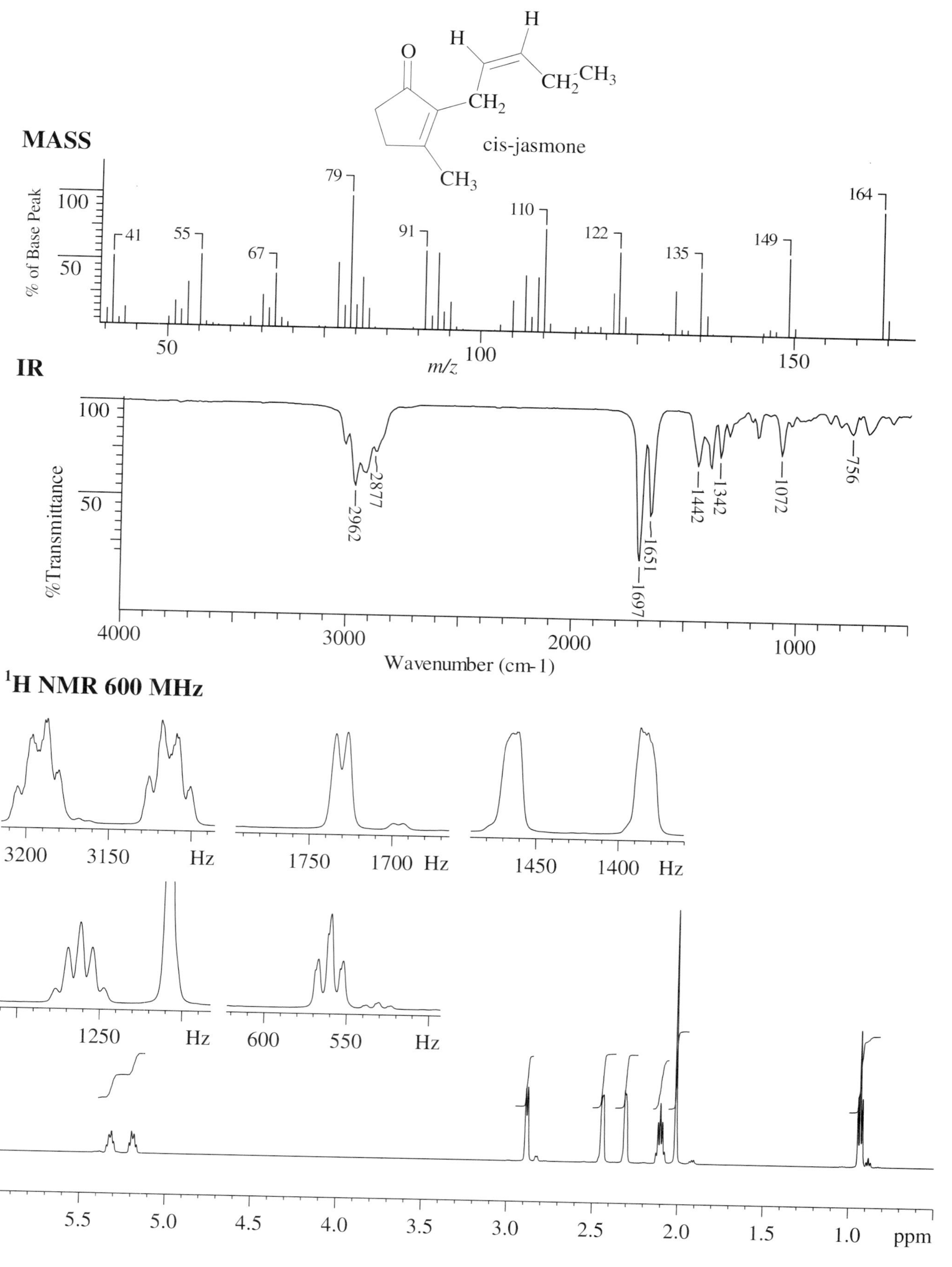

MASS
cis-jasmone
O
H
H
CH2
CH2CH3
CH3
% of Base Peak
100
50
41
55
67
79
91
110
122
135
149
164
50
100
150
m/z
IR
%Transmittance
100
50
2962
2877
1697
1651
1442
1342
1072
756
4000
3000
2000
1000
Wavenumber (cm-1)
1H NMR 600 MHz
3200
3150
Hz
1750
1700 Hz
1450
1400 Hz
1250
Hz
600
550
Hz
5.5
5.0
4.5
4.0
3.5
3.0
2.5
2.0
1.5
1.0
ppm

DQFCOSY 600 MHz

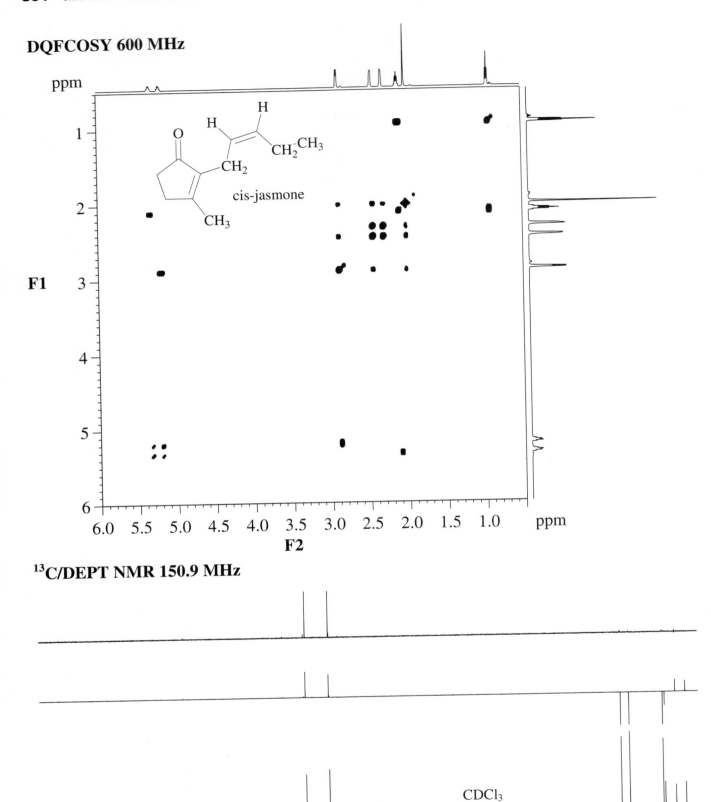

cis-jasmone

ppm

F1

F2

6.0 5.5 5.0 4.5 4.0 3.5 3.0 2.5 2.0 1.5 1.0 ppm

^{13}C/DEPT NMR 150.9 MHz

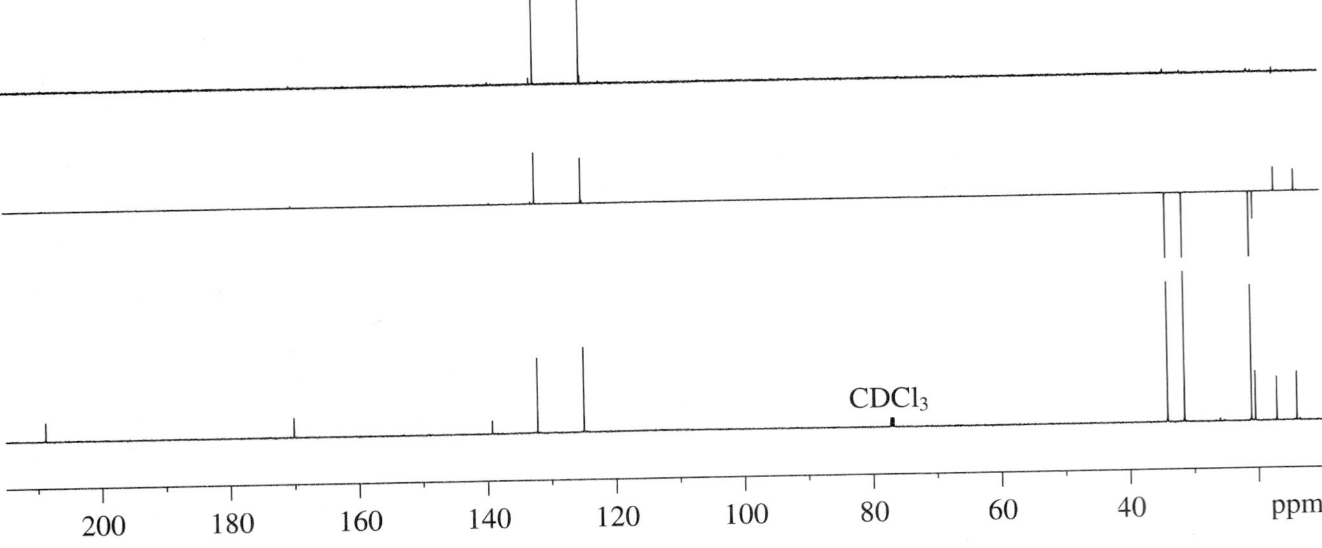

Problem 7.5C

HMQC 600 MHz

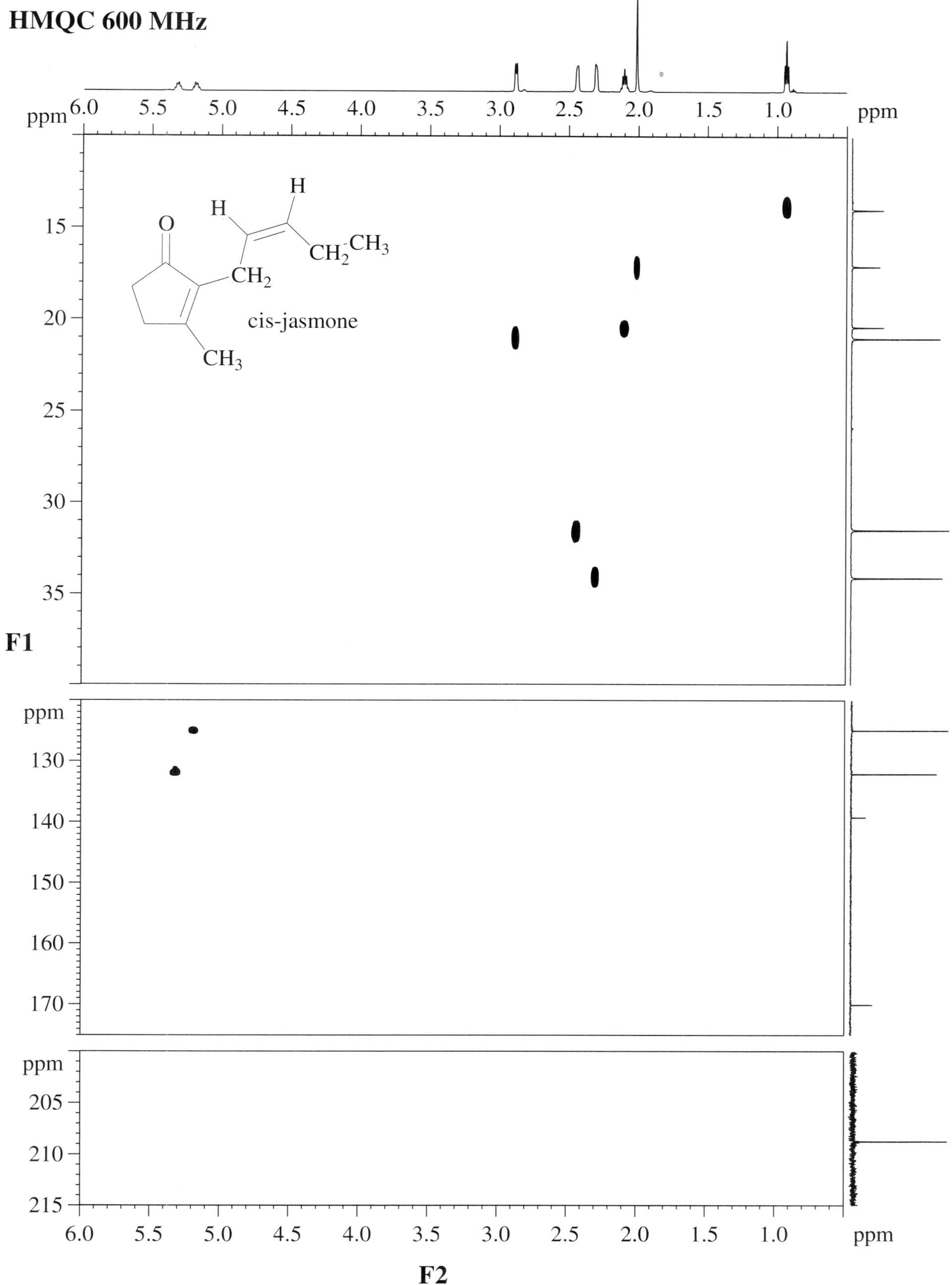

HMBC 600 MHz

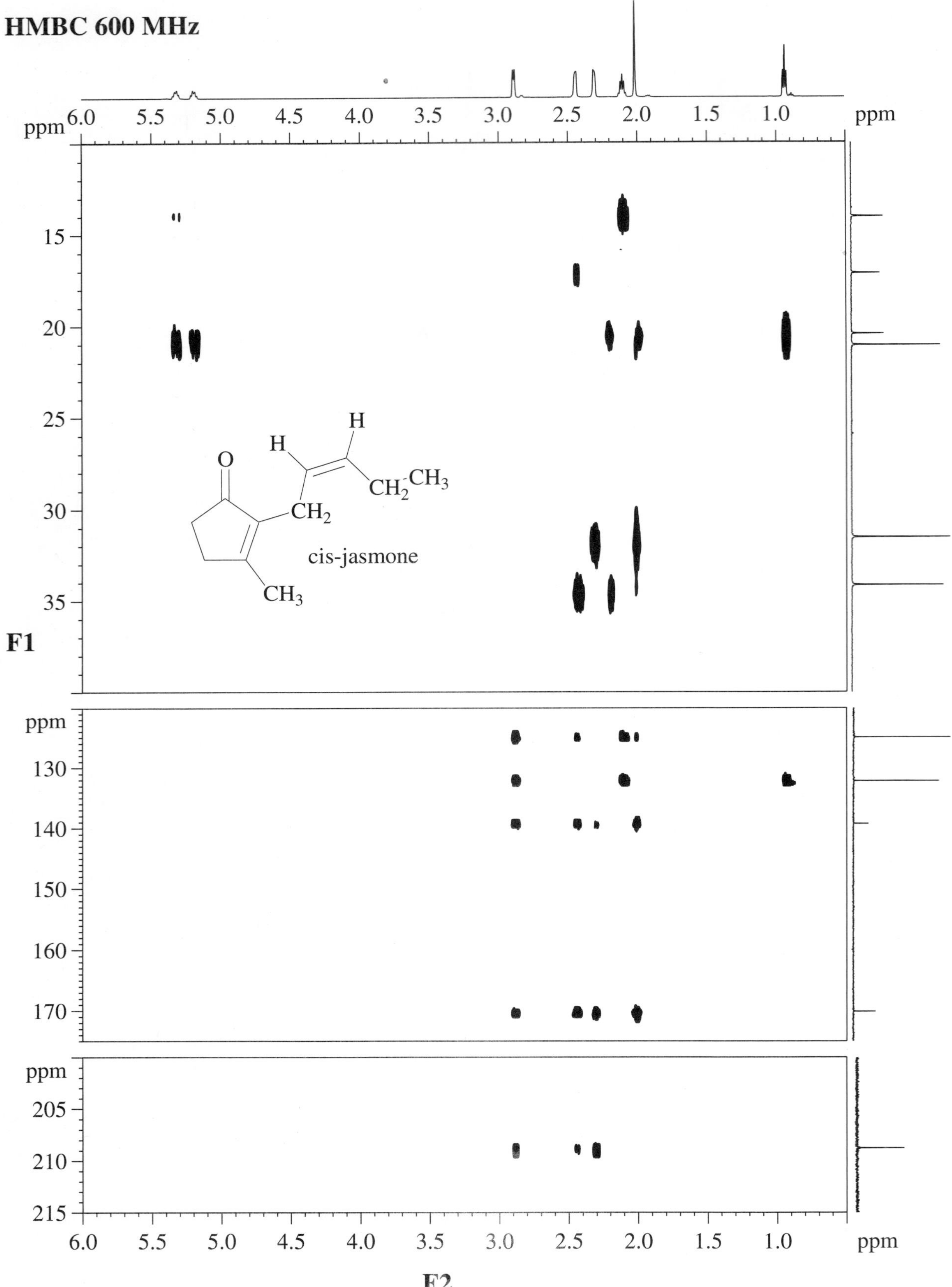

ppm
6.0
5.5
5.0
4.5
4.0
3.5
3.0
2.5
2.0
1.5
1.0
ppm
15
20
25
30
35
F1
H
H
O
CH2
CH2CH3
CH3
cis-jasmone
ppm
130
140
150
160
170
ppm
205
210
215
F2

PROBLEM 7.5 DISCUSSION

In the mass spectrum, the peak at *m/z* 149 represents a rational loss of a CH_3 group (M-15) and indicates that the *m/z* 164 peak is the molecular ion peak. The absent M + 2 indicates the absence of Cl, Br, and S. The IR spectrum shows an intense peak at 1697 cm^{-1}, suggestive of a C=O group, which is confirmed by the peak in the ^{13}C NMR spectrum at 208 ppm (in the range of ketones). In the ^{1}H NMR spectrum, the integration steps are: 1:1:2:2:2:2:3:3 (16 protons). The ^{13}C NMR spectrum indicates 11 carbon atoms. Thus, the tentative molecular formula is $C_{11}H_{16}O$, which agrees with the molecular ion peak of *m/z* 164. The index of hydrogen deficiency is four. From left to right in the ^{13}C/DEPT spectra, the number of protons attached to each carbon are: C, C, C, CH, CH, CH_2, CH_2, CH_2, CH_2, CH_3, CH_3.

Three of the four degrees of unsaturation can be dealt with directly in the ^{13}C NMR spectrum. We have already noted the carbonyl peak and because of its chemical shift we associate it with a ketone. There are four olefinic carbon resonances, which account for two more degrees of unsaturation. By inference, the fourth degree of unsaturation is attributed to a ring.

The ^{13}C/DEPT spectra have provided the molecular contents, which is a large step toward deciphering the structure. We turn to the 2-D spectra to assemble the pieces. As needed, we will jump from one spectrum to another. The more shielded of the two methyl groups at δ 0.93, which correlates with the methyl carbon resonance at 15 ppm in the HMQC, is a good place to start. This methyl group is coupled to a methylene group at δ 2.12 as evidenced by the strong cross peak in the COSY spectrum. The carbon resonance associated with this group is found at 21 ppm in the HMQC spectrum. This methylene group also correlates with one of the olefinic methines in the COSY at δ 5.33. The methylene group at δ 2.12 shows a nearly first order quintet.

The other olefinic methine at δ 5.18 shows a correlation in the COSY to the first methine (δ 5.33) strongly suggesting a disubstituted double bond. In addition, the methine at δ 5.18 is coupled to a rather deshielded methylene group at δ 2.88. This methylene is a broadened doublet and shows no other vicinal coupling. According to the COSY, this methylene group shows long-range coupling to two other groups not yet used (one of the two unused methylene groups and the other methyl group.) So far we have a 2-pentenyl group, which must be attached to one of the quaternary olefinic carbon atoms. The broadened methyl singlet at δ 2.01 must be attached to the other quaternary olefinic methine.

If we take stock of the remaining pieces from the ^{13}C/DEPT spectra (a ketone carbonyl, two methylene groups, and two quaternary olefinic carbons) and the fact that there is still the unused degree of unsaturation for a ring, we realize that we must draw an unsaturated five member-ring ketone with these pieces. Further evidence allows us to order these pieces. First, the COSY tells us that the two methylene groups (δ 2.30 and 2.45) are vicinally coupled and therefore adjacent. Second, the ^{13}C chemical shift of one of the quaternary olefinic carbons is 170 ppm; such deshielding can only be explained by conjugation with the ketone carbonyl. One obvious structure is shown below. The HMBC (see the correlations to the carbonyl carbon) confirms the ring structure and the lack of an asymmetric carbon atom (there is a plane symmetry in the plane of the page) explains the absence of diastereomeric methylene groups.

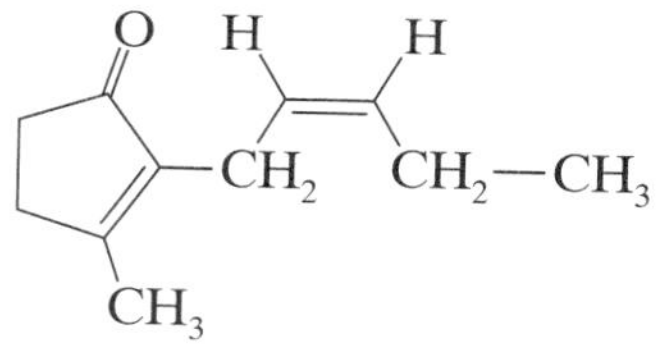

O H H

CH_2 CH_2—CH_3

CH_3

Before we finish, we ask ourselves if there are any constitutional isomers that we need to consider. Yes, if we switch the two substituents (i.e., the 2-pentenyl group and the methyl group), the resulting structure also fits the data so far. Again, the complex HMBC spectrum resolves the issue. For instance, the ring substituted methyl resonance at 17 ppm shows only one correlation to the methylene group at δ 2.45. This methylene group is in the β-position to the carbonyl, thereby confirming the structure given above. Are there other correlations in the HMBC that can confirm the structure? The student can finish the assignments.

LCMS ES

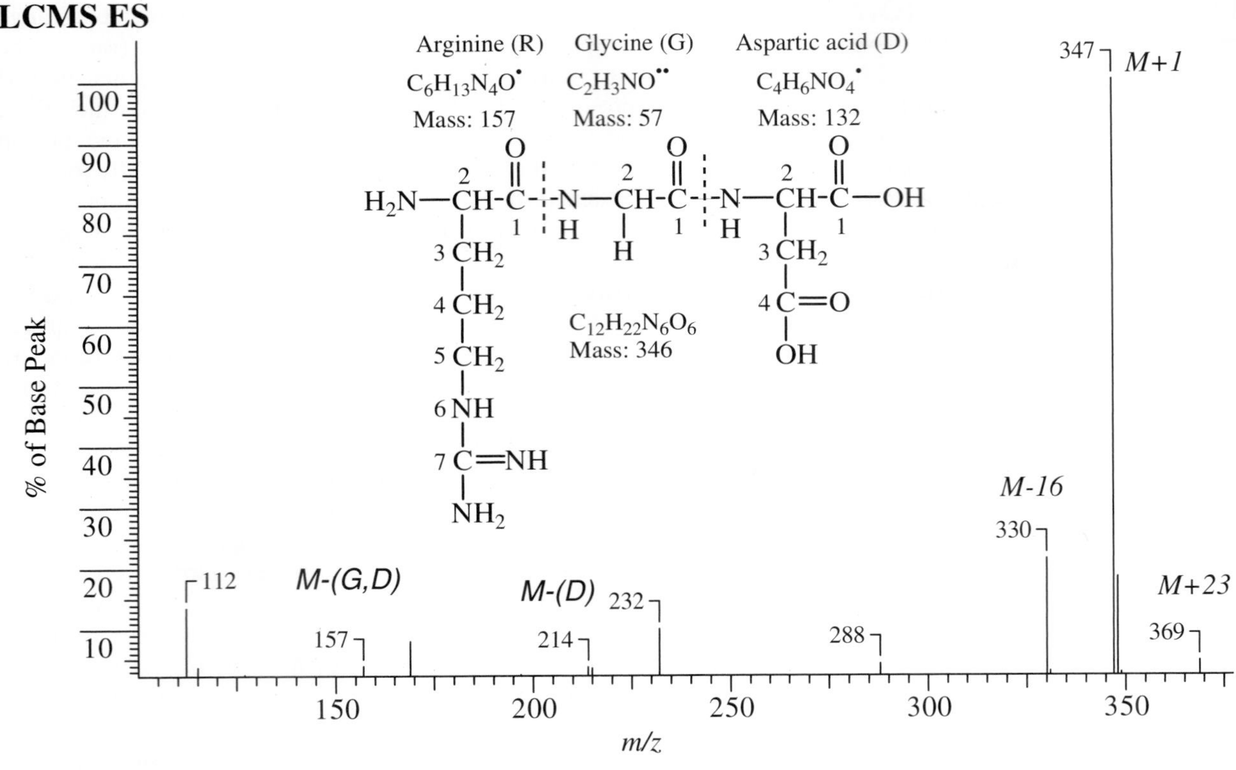

HMQC 600 MHz

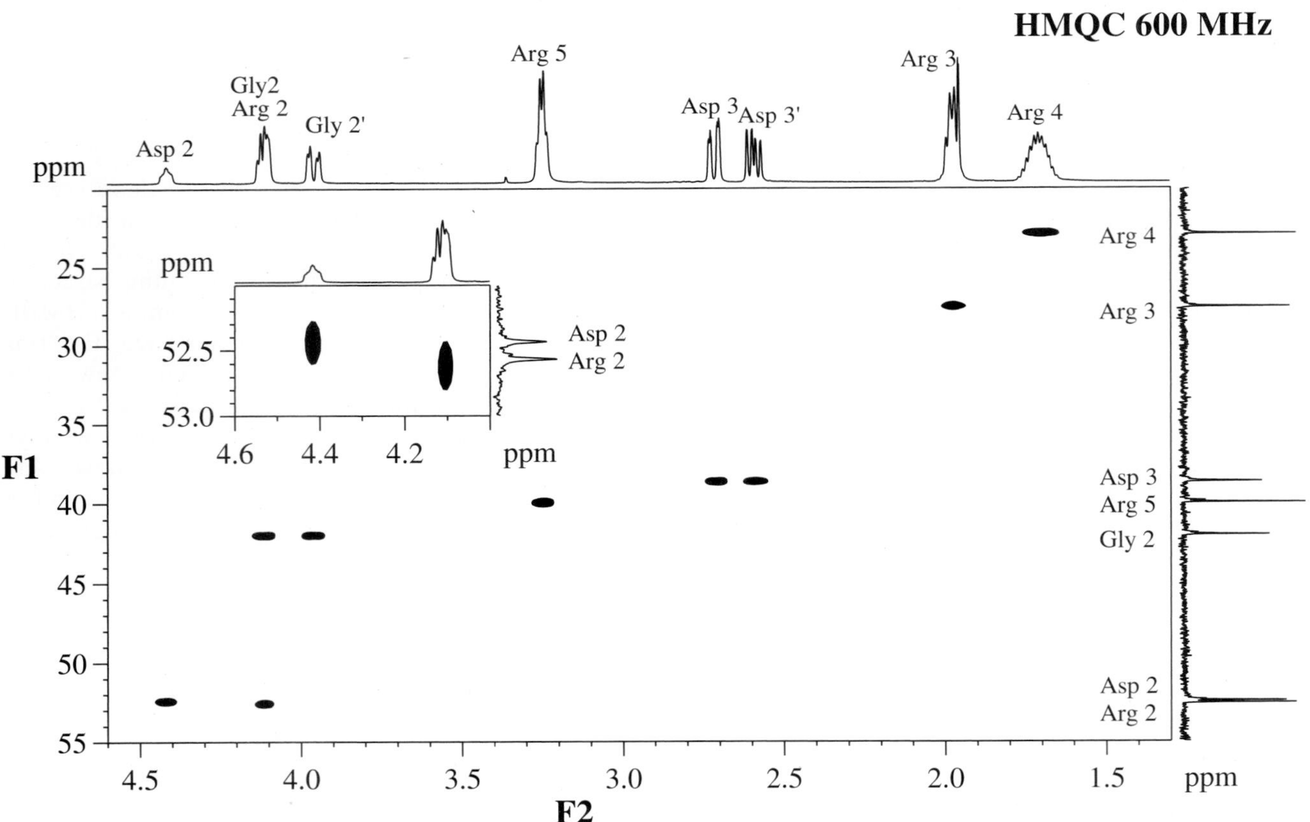

Problem 7.6B

^{1}H NMR 600 MHz

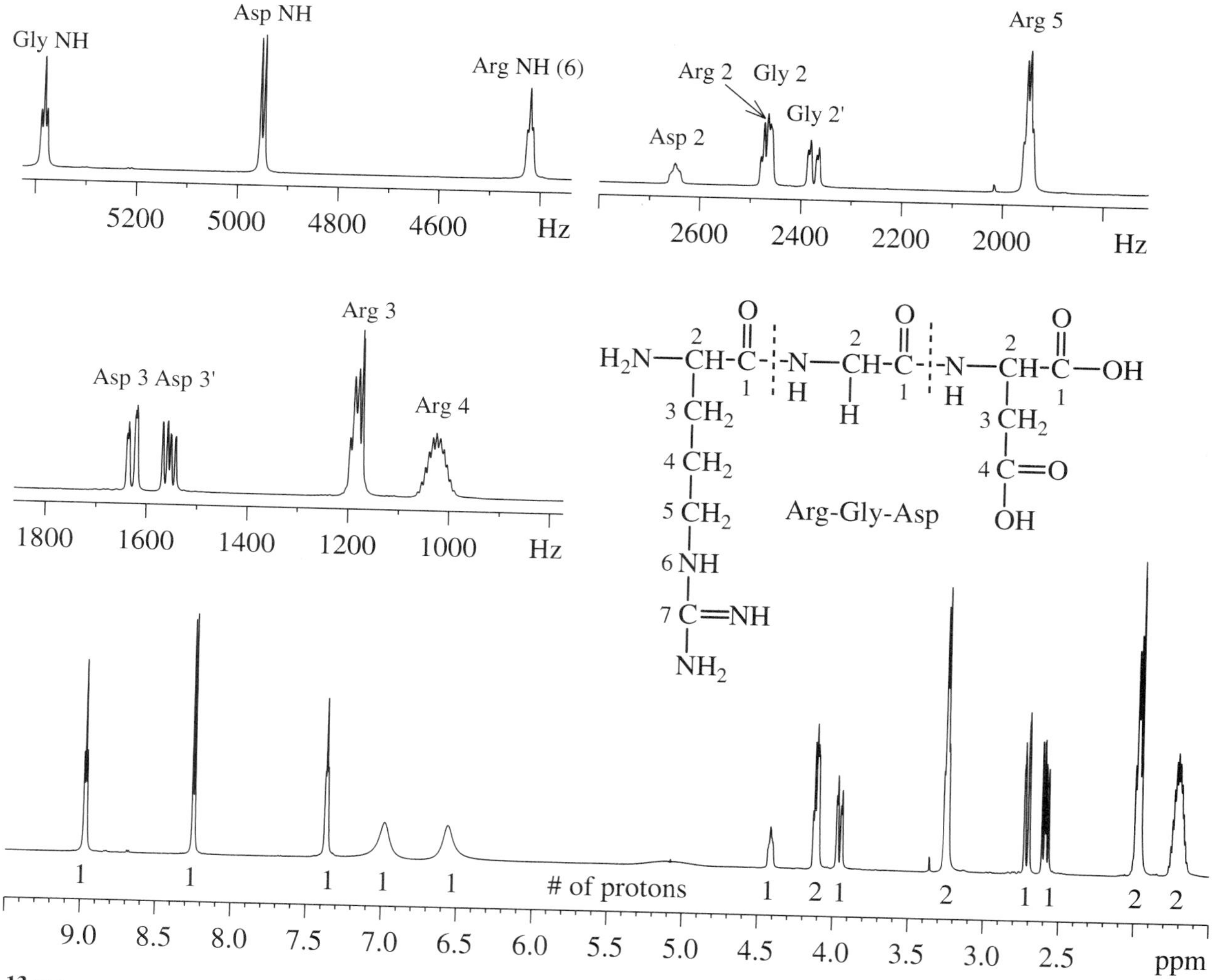

Gly NH
Asp NH
Arg NH (6)
5200
5000
4800
4600
Hz
Arg 5
Arg 2
Gly 2
Gly 2'
Asp 2
2600
2400
2200
2000
Hz
Arg 3
Asp 3 Asp 3'
Arg 4
1800
1600
1400
1200
1000
Hz
Arg-Gly-Asp
# of protons
1 1 1 1 1 1 2 1 2 1 1 2 2
9.0 8.5 8.0 7.5 7.0 6.5 6.0 5.5 5.0 4.5 4.0 3.5 3.0 2.5
ppm

^{13}C/DEPT NMR 150.9 MHz

Asp 1
Asp 4
Arg 1
Gly 1
Arg 7
175
170
165
160
ppm
Arg 2
Asp 2
Arg 5
Gly 2
Asp 3
Arg 3
Arg 4
50
45
40
35
30
25
ppm

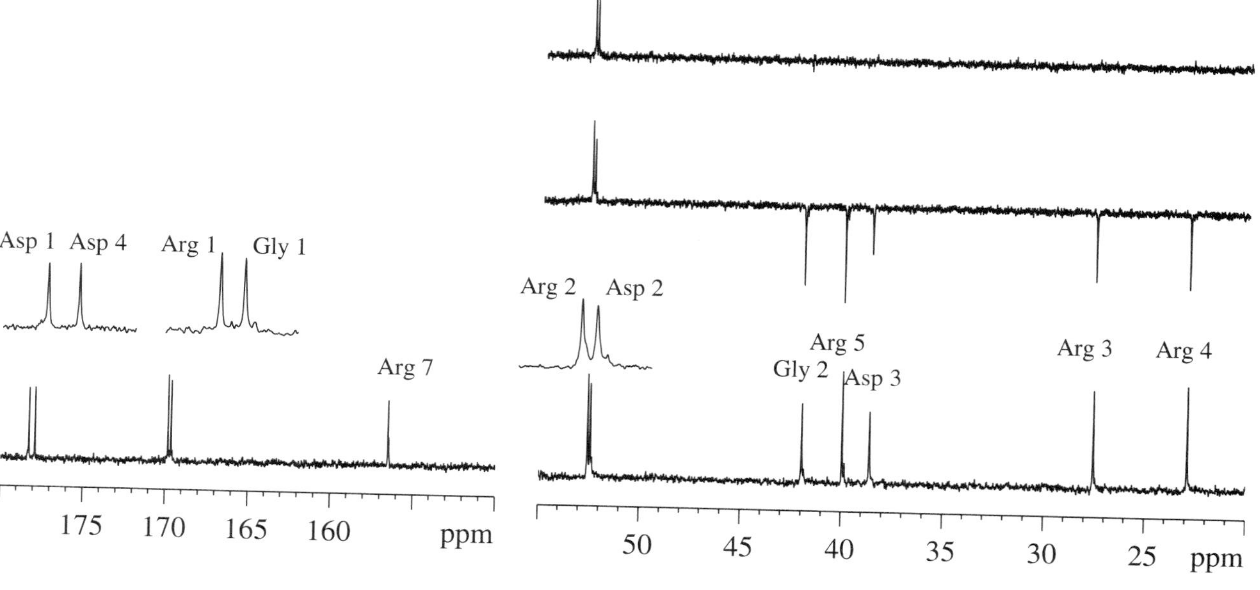

Problem 7.6C

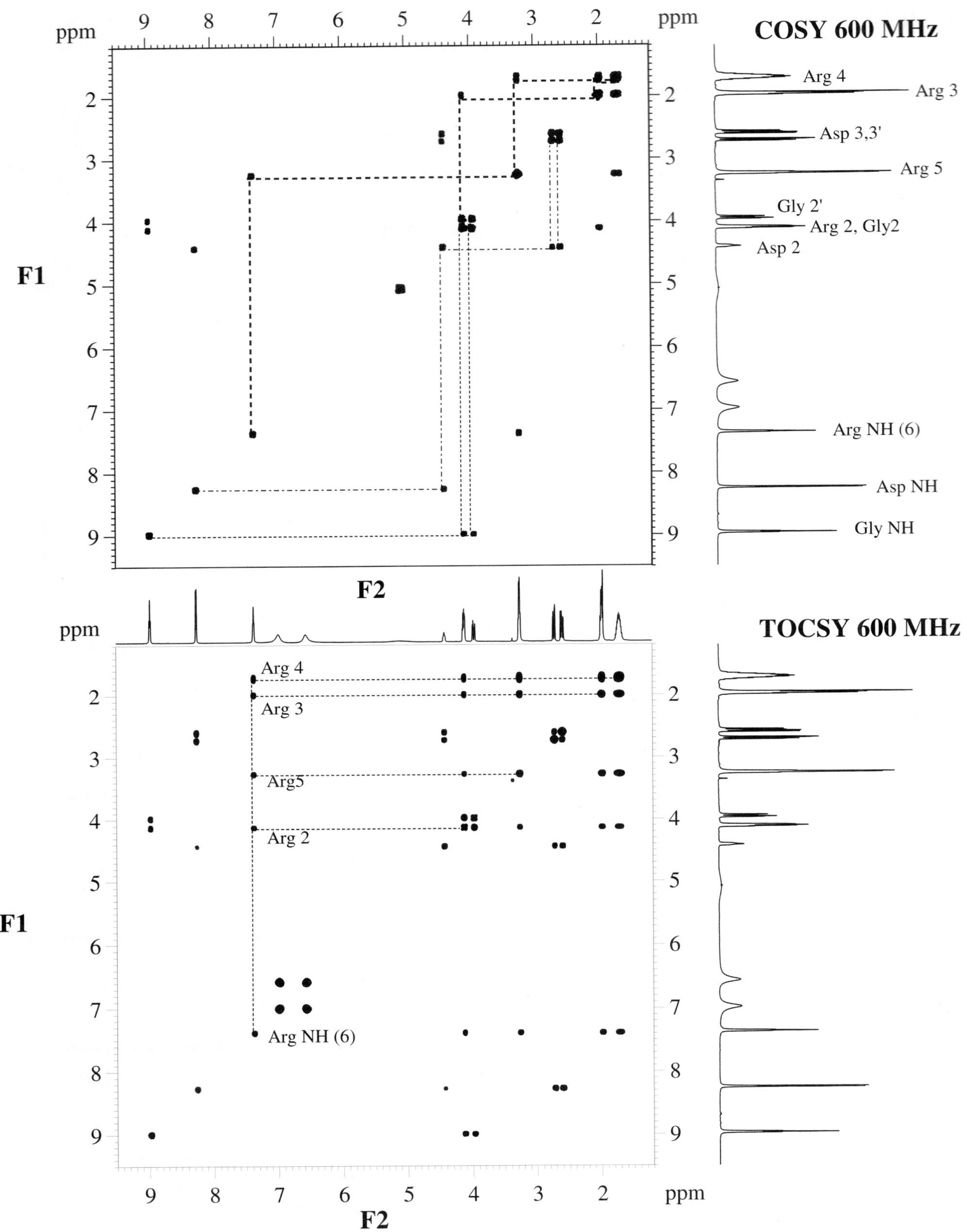

COSY 600 MHz
Arg 4
Arg 3
Asp 3,3'
Arg 5
Gly 2'
Arg 2, Gly2
Asp 2
Arg NH (6)
Asp NH
Gly NH
F1
F2
ppm
TOCSY 600 MHz
Arg 4
Arg 3
Arg5
Arg 2
Arg NH (6)

Problem 7.6D

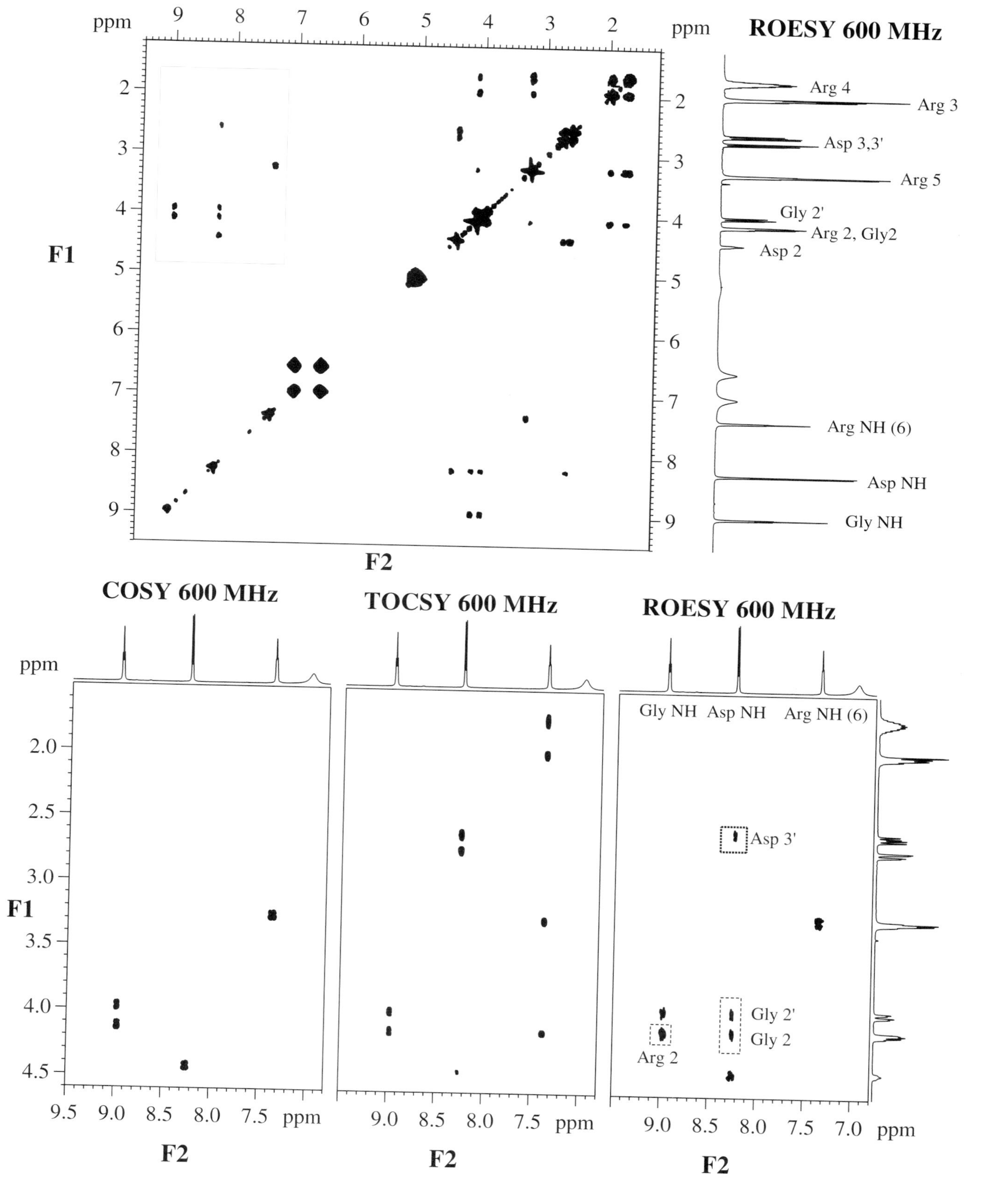
ROESY 600 MHz
ppm
F1
F2
Arg 4
Arg 3
Asp 3,3'
Arg 5
Gly 2'
Arg 2, Gly2
Asp 2
Arg NH (6)
Asp NH
Gly NH
COSY 600 MHz
TOCSY 600 MHz
ROESY 600 MHz
Gly NH
Asp NH
Arg NH (6)
Asp 3'
Gly 2'
Gly 2
Arg 2

Problem 7.6E

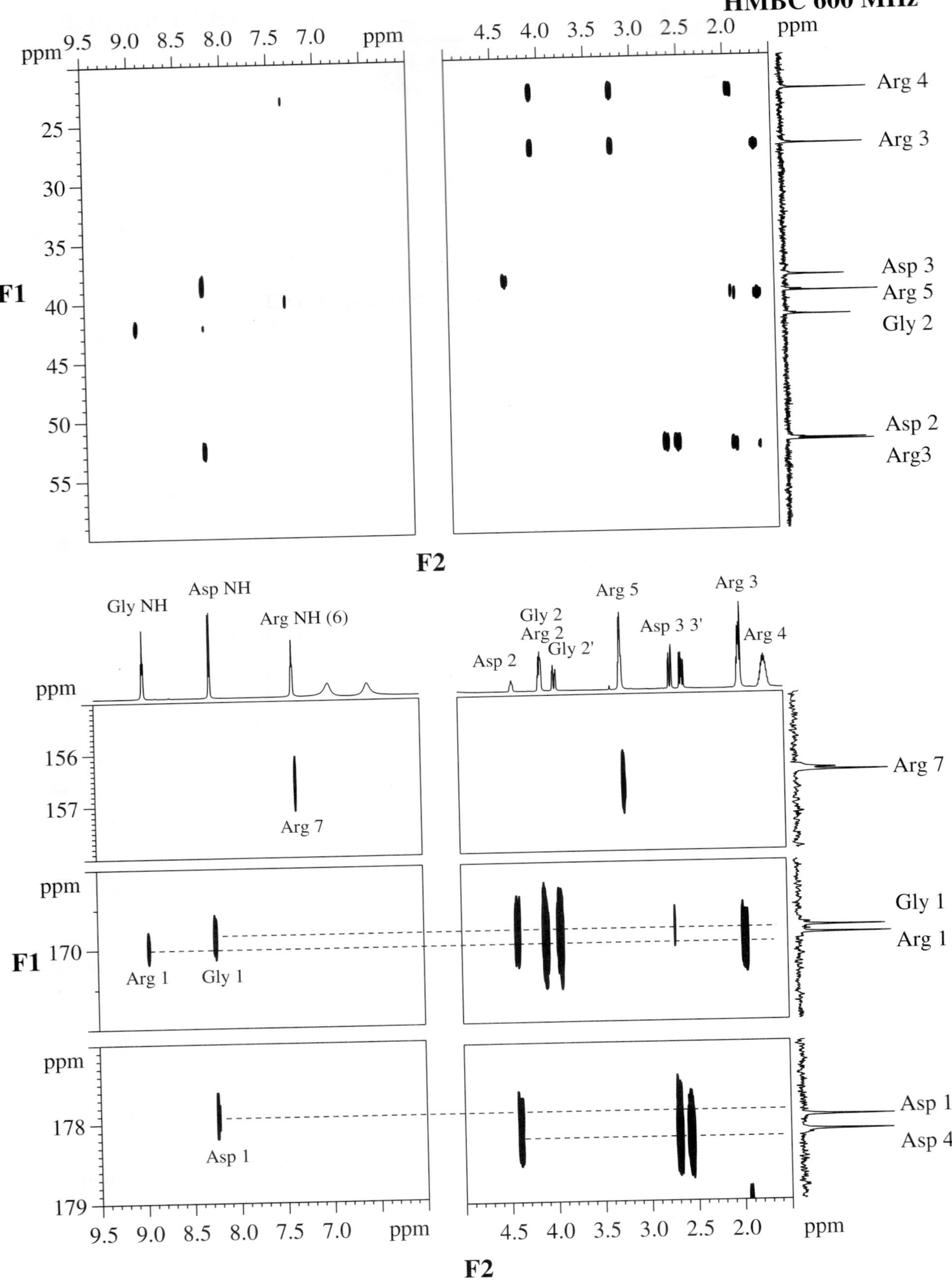
HMBC 600 MHz
ppm
F1
F2
Arg 4
Arg 3
Asp 3
Arg 5
Gly 2
Asp 2
Arg3
Gly NH
Asp NH
Arg NH (6)
Gly 2
Arg 2
Gly 2'
Arg 5
Asp 3 3'
Arg 3
Arg 4
Asp 2
Arg 7
Arg 7
Arg 1
Gly 1
Gly 1
Arg 1
Asp 1
Asp 1
Asp 4

PROBLEM 7.6 DISCUSSION

The last of the solved problems in this chapter is quite different from the other problems presented above, and our approach takes a different tack as well. The compound is a tripeptide, and structure elucidation of peptides requires two distinct "solutions." First, the individual amino acids (and their number) are determined and second, the amino acid units are put in order (sequenced). Neither of these exercises is trivial since there are more than 20 common amino acids, and the nature of the peptide bond means that they can be arranged in any order.

Before discussing the actual data, some discussion of sample handling is worthwhile. The mass spectrum of the tripeptide was obtained using an electrospray LCMS (see Chapter 1). Electrospray (ES) is a "soft" method of ionization (a type of chemical ionization), which suppresses or limits fragmentation and enhances pseudomolecular ions (depends on the number of charges on the ion, z). The NMR experiments were obtained in 95% H_2O and 5% D_2O at 0°C. The reasoning for using these solvents and the details are given in Section 5.11.

The LCMS ES gives an M + 1 of m/z 347, which corresponds to a molecular formula of $C_{12}H_{22}N_6O_6$. Derivation of this formula is not considered in detail since Appendix A in Chapter 1 only goes up to 250 amu. The small peak at m/z 369 (M + 23) is due to the presence of sodium ions, which are ubiquitous in aqueous solutions. Although there is only limited fragmentation, the fragments that do appear can be quite useful to an experienced interpreter, and some of the cleavages are shown in the Problem 7.6A.

To ascertain the three amino acids, we use information from the ^{1}H NMR spectrum, the ^{13}C/DEPT spectra, the COSY spectrum, the TOCSY spectrum, and the HMQC spectrum. Chemical shifts for protons of amino acids are given in Appendix I of Chapter 3. Peptides are chiral molecules, and all methylene groups are diastereotopic, even those of glycine. (The methylene group of free glycine is enantiotopic.) A starting point for peptides (and other compounds made of distinct units such as oligo- and polysaccharides) is the TOCSY. The 2-D TOCSY shows correlation among all spins in a spin system, but no correlations to spins outside the system. For a tripeptide, there are three distinct spin systems, and they are easy to find in the TOCSY spectrum (Problem 7.6C). The N—H resonance at 8.95 ppm reveals one spin system showing correlations to two resonances at 3.96 and 4.13 ppm. If we feed this information into the HMQC spectrum, we find that these two proton-resonances correlate to the same carbon atom at 41.9 ppm (i.e., a diastereotopic methylene group). This amino acid is identified as a glycine residue, and since there is correlation to an amide N—H group, we conclude that the glycine is not the N-terminus. (This point will be confirmed using other methods.)

Another spin system is evident with a convenient starting point with the N—H resonance at 8.26 ppm. There are three correlations to this resonance for a total of four moieties in this spin system. If we again take this information directly to the HMQC spectrum, we can find the corresponding carbon resonances. Of course, there is no correlation to the N—H resonance. The proton resonance at 4.41 ppm correlates to a carbon resonance at 52.4 ppm, and the ^{13}C/DEPT spectra confirm that this is a methine group. The remaining spins in the system are proton resonances at 2.58 and 2.72 ppm, which correlate to a single carbon resonance at 38.6 ppm. The ^{13}C/DEPT confirms that this is a methylene group. This residue is identified as aspartic acid, and again we conclude that it is not the N-terminus.

A starting point for the final spin system is the N—H resonance at 7.35 ppm. There are four other proton resonances in this spin system; the HMQC spectrum and the ^{13}C/DEPT spectra indicate that these resonances represent one methine group and three methylene groups. Thus, this amino acid residue is arginine.

The COSY spectrum could be used to confirm our analysis thus far but it has not been necessary. However, the COSY and TOCSY are not redundant, and, in fact, they are complementary in at least one important aspect. While the TOCSY shows all of the spins in a spin system, it does not reveal which spins are actually coupled to one another. For instance, the N—H proton at 8.26 ppm (from aspartic acid) shows only a correlation to a methine group in the COSY; we can safely conclude that this N—H group is involved in a peptide linkage and that the methine group is the α- or asymmetric carbon of the amino acid. (Glycine is a trivial case and not considered here.) The N—H proton at 7.35 ppm, which we assigned to an aspartic acid residue, correlates with all of the spins in the TOCSY, but only shows coupling to a methylene group at 2.24 ppm in the COSY. This information allows us to draw two conclusions that were not available from the TOCSY. First, the N—H resonance is not coupled to the α-carbon of arginine and therefore must represent the N—H from the guanadino group and the not α-amino group. Second, the arginine residue must be the N-terminal residue because there is no correlation from the methine proton at 4.13 ppm and an N—H proton in either the COSY or the TOCSY. In the discussion of the sequence of the amino acids, we will confirm this second point.

The combined information from the ^{13}C/DEPT, COSY, TOCSY, and HMQC enables us to assign all protons in the ^{1}H spectrum except those that are exchanging rapidly (i.e., the carboxyl and free amino protons) and all of the non-quaternary carbon resonances in the ^{13}C spectrum. There is no need to assign the rapidly exchanging protons, and the non-quaternary carbons will be assigned during the sequencing discussion.

The second main objective is to "sequence" the peptide or place the amino acids in their proper order. Two powerful tools from our NMR experimental repertoire are HMBC and ROESY (or NOESY). Recall that the HMBC shows long range $^1H—^{13}C$ coupling (generally 2-bond, $^2J_{CH}$, and 3-bond, $^3J_{CH}$ couplings). For sequencing purposes, this experiment shows correlation between adjacent amino acids, as it were, "seeing through" the amide (peptide) carbonyl to the amide N—H. This exercise will also enable us to assign the carbonyls.

The ROESY experiment facilitates sequencing utilizing the inevitable through space correlations between adjoining amino acids. We expect to find through space connectivities from one amino acid's N-H to the adjoining amino acid's α- or C-2 proton(s). The data from either experiment should be sufficient; together they provide strong confirmatory evidence.

The full ROESY spectrum is shown in Problem 7.6D (top part). ROESY cross peaks show both COSY correlations and NOE correlations. For easy comparison, therefore, the area of interest for sequencing (the boxed area) is shown along with the corresponding COSY and TOCSY (bottom part of Problem 7.6D). The glycine N—H, which correlates with the adjacent methylene group in both the COSY and TOCSY spectra, gives an additional correlation in the ROESY spectrum to H-2 of arginine. This correlation shows a linkage between glycine and arginine. Some might consider this correlation inconclusive because of the overlap between H-2 of arginine and one of the H-2's of glycine. In this case, confirmation is desirable (see below).

The other connectivity can be established by way of the NOE interaction between the aspartic acid N—H and the two glycine H-2's. There is no ambiguity or overlap in this correlation thus proving the sequence given in Problem 7.6A. An interesting aside worth noting is the NOE correlation between the aspartic acid N—H and only one of the two diastereotopic methylene protons of aspartic acid (H-3). This selective interaction suggests restricted rotation and allows steric differentiation and assignment between the diastereotopic protons.

Confirmation of this sequence and assignment of the quaternary carbons is accomplished with the HMBC, which is shown in Problem 7.6E. The bottom part of this page is pertinent. Before confirming the sequence, a simple assignment of a quaternary carbon is made. The assignment of the C-7 carbon of arginine can be made by noting the correlation between the arginine N—H (H-6) and the quaternary carbon at 156.5 ppm.

The analysis of the HMBC in the region of the carbonyl carbons is hampered by the lack of digital resolution along the F1 axis. Recall that the HMBC experiment is proton detected giving good resolution in the proton or F2 dimension. The only way to improve resolution along the F1 axis is to increase the number of FIDs in the experiment, which has serious practical limitations. The lines drawn in the insets help clarify the correlations.

The glycine N—H correlates with the arginine carbonyl (C-1) confirming the arginine–glycine linkage. The assignment of the arginine carbonyl is accomplished by noting the correlation of the carbonyl resonance at about 170 ppm with the arginine methine H-2. The other linkage is established by the correlation of the aspartic acid N—H and the glycine carbonyl (C-1); the glycine carbonyl is pinpointed from its correlations with the glycine diastereotopic methylene protons. The sequence of the tripeptide is confirmed by two independent methods.

STUDENT EXERCISES

The following exercises are given for the student to "solve." The structures and spectra for two compounds as shown on the next page. The student should "prove" the structure from the spectra and assign all protons and carbons.

Exercise 7.7A

MASS

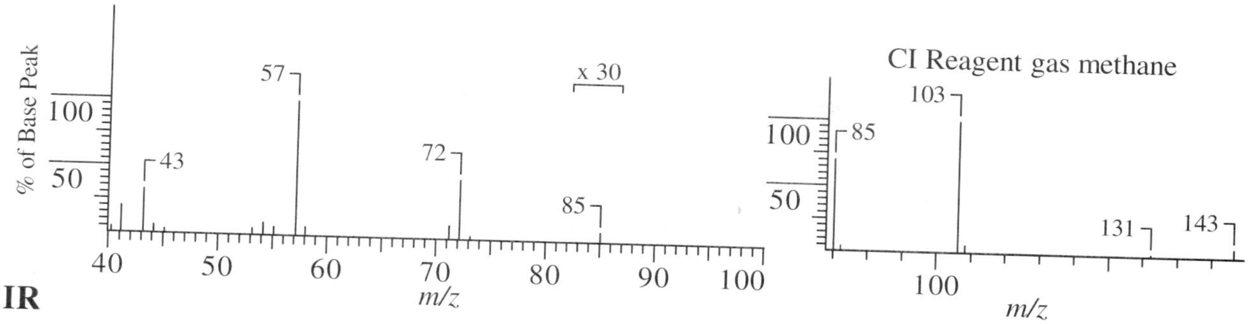

% of Base Peak
100
50
43
57
72
85
x 30
40
50
60
70
80
90
100
m/z
CI Reagent gas methane
103
85
131
143
100
50
100
m/z

IR

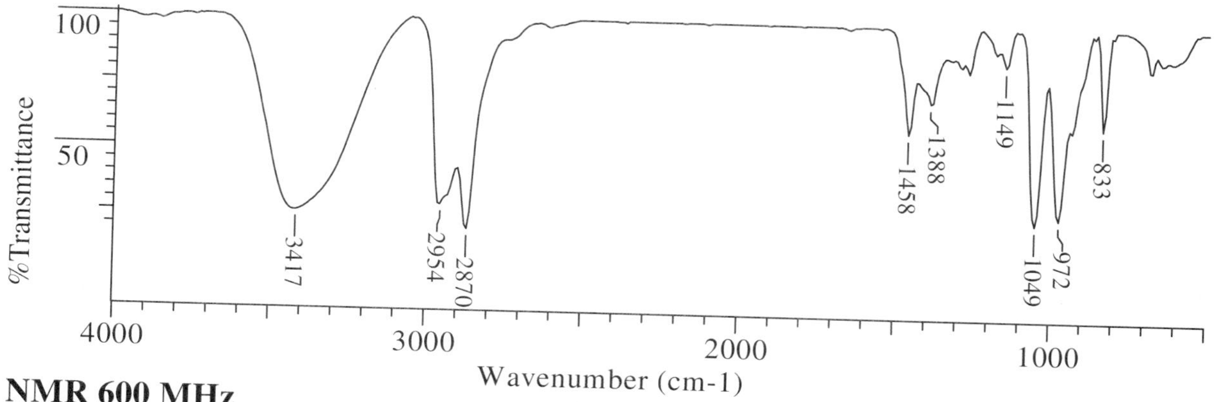

%Transmittance
100
50
3417
2954
2870
1458
1388
1149
1049
972
833
4000
3000
2000
1000
Wavenumber (cm-1)

^{1}H NMR 600 MHz

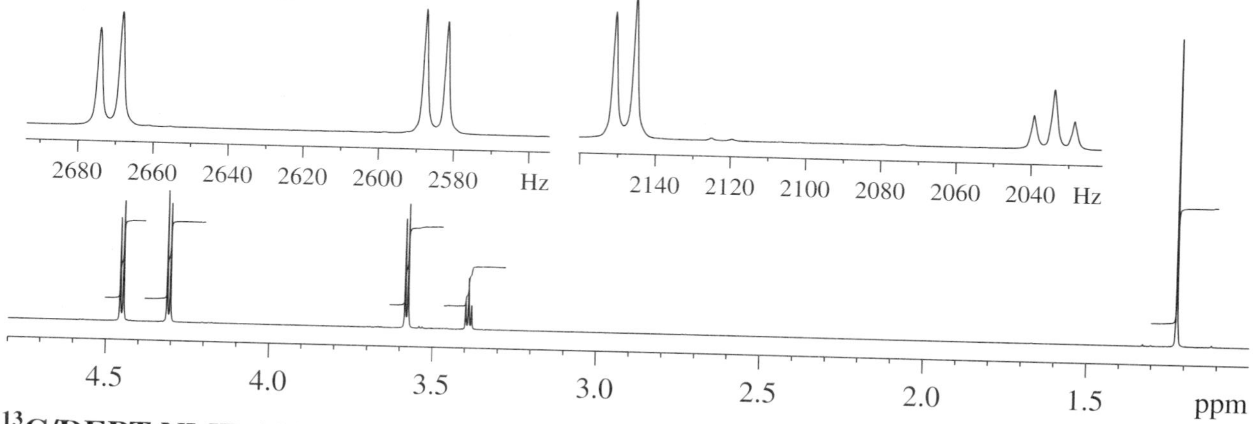

2680 2660 2640 2620 2600 2580 Hz
2140 2120 2100 2080 2060 2040 Hz
4.5
4.0
3.5
3.0
2.5
2.0
1.5
ppm

^{13}C/DEPT NMR 150.9 MHz

3-methyl-3-oxetanemethanol
CH2OH
CH3
O
80 75 70 65 60 55 50 45 40 35 30 25 20 ppm

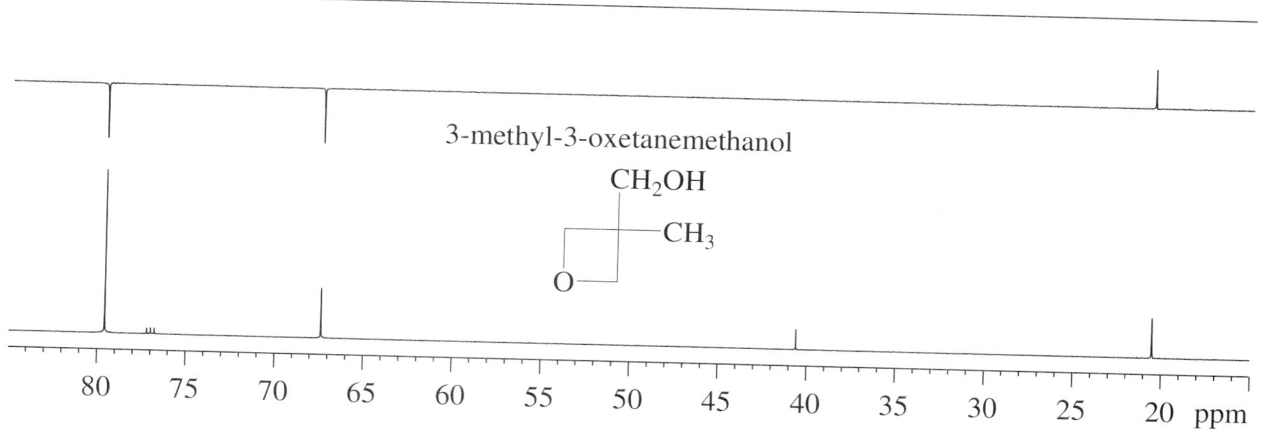

Exercise 7.7B

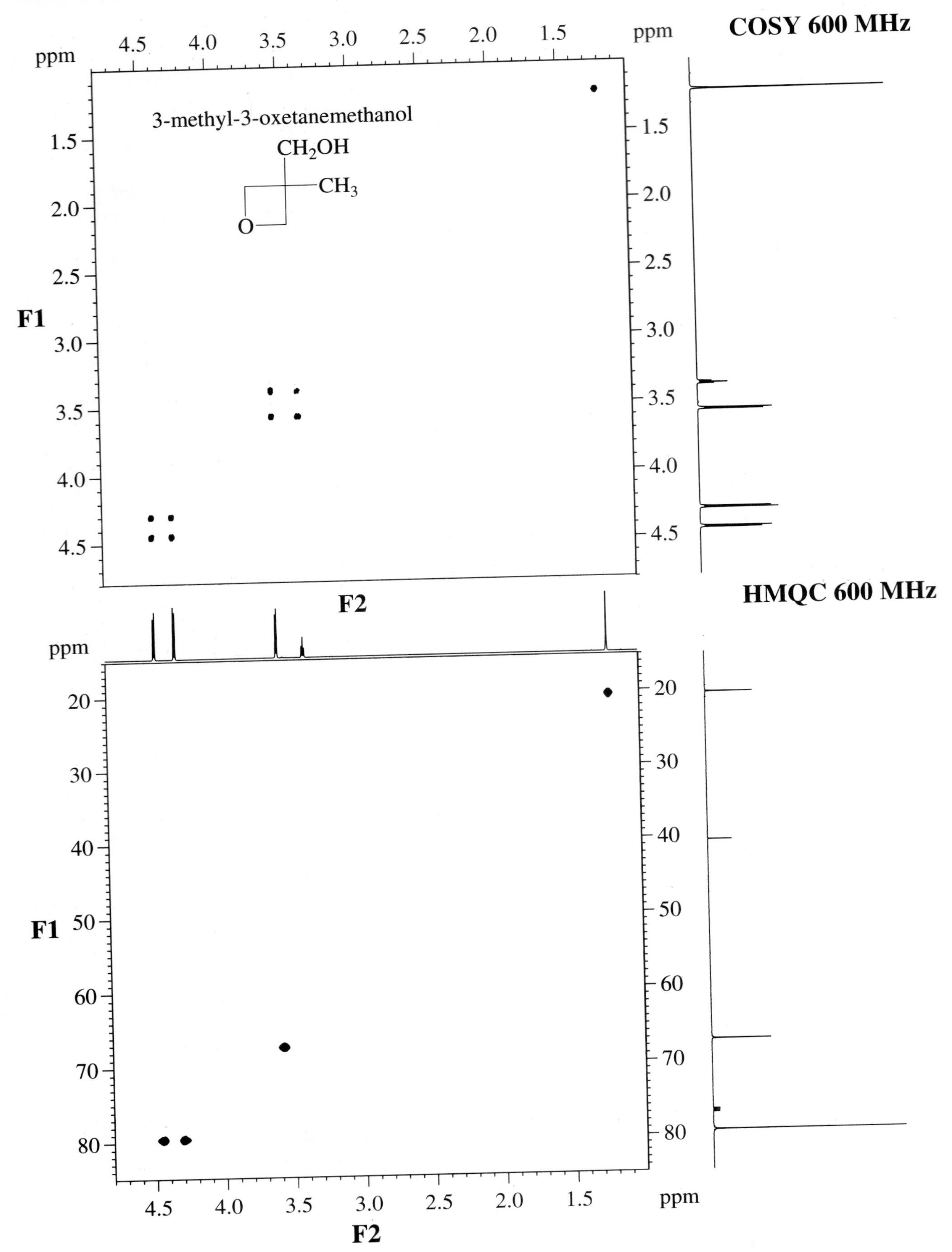

Exercise 7.8A

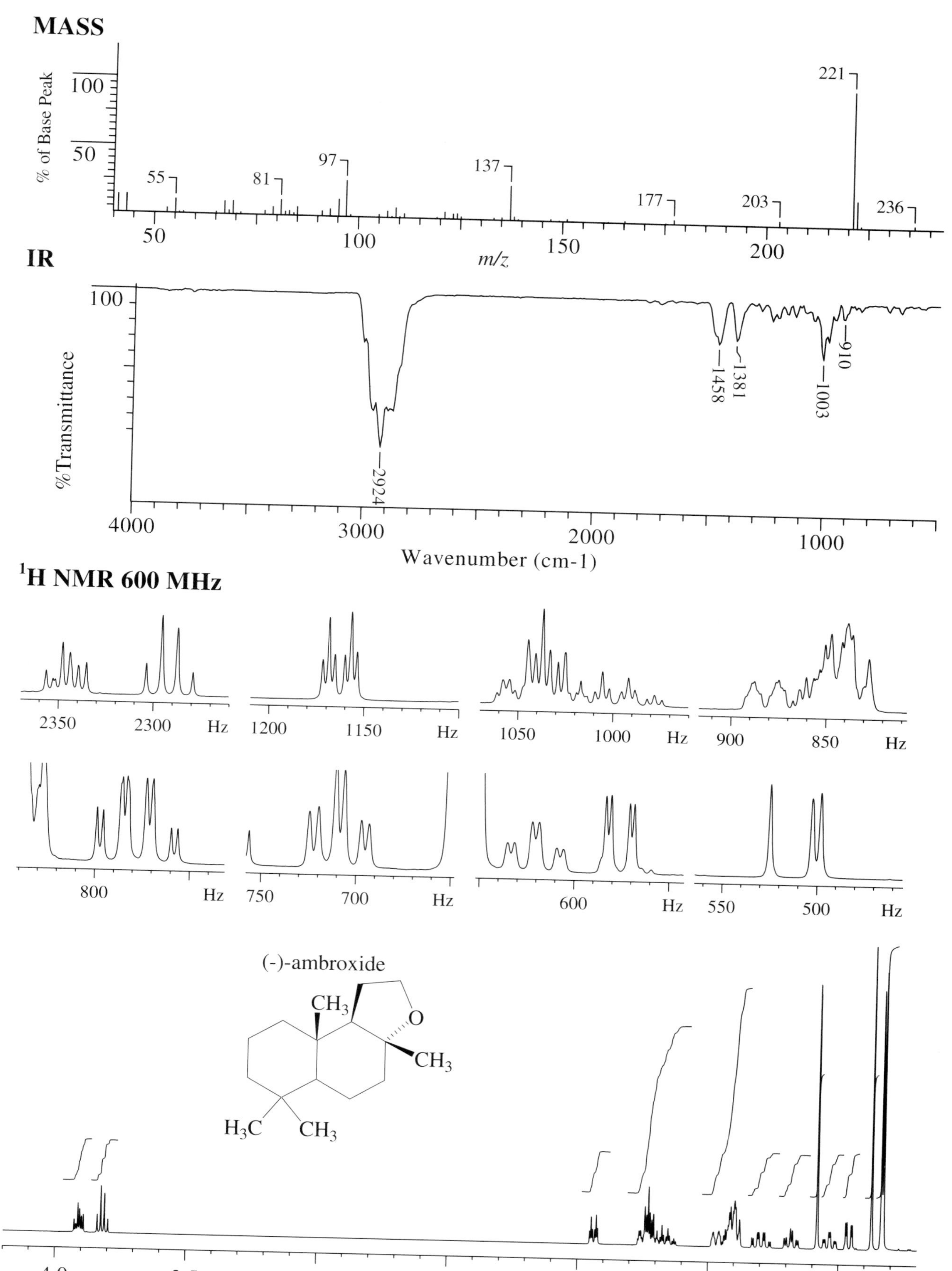

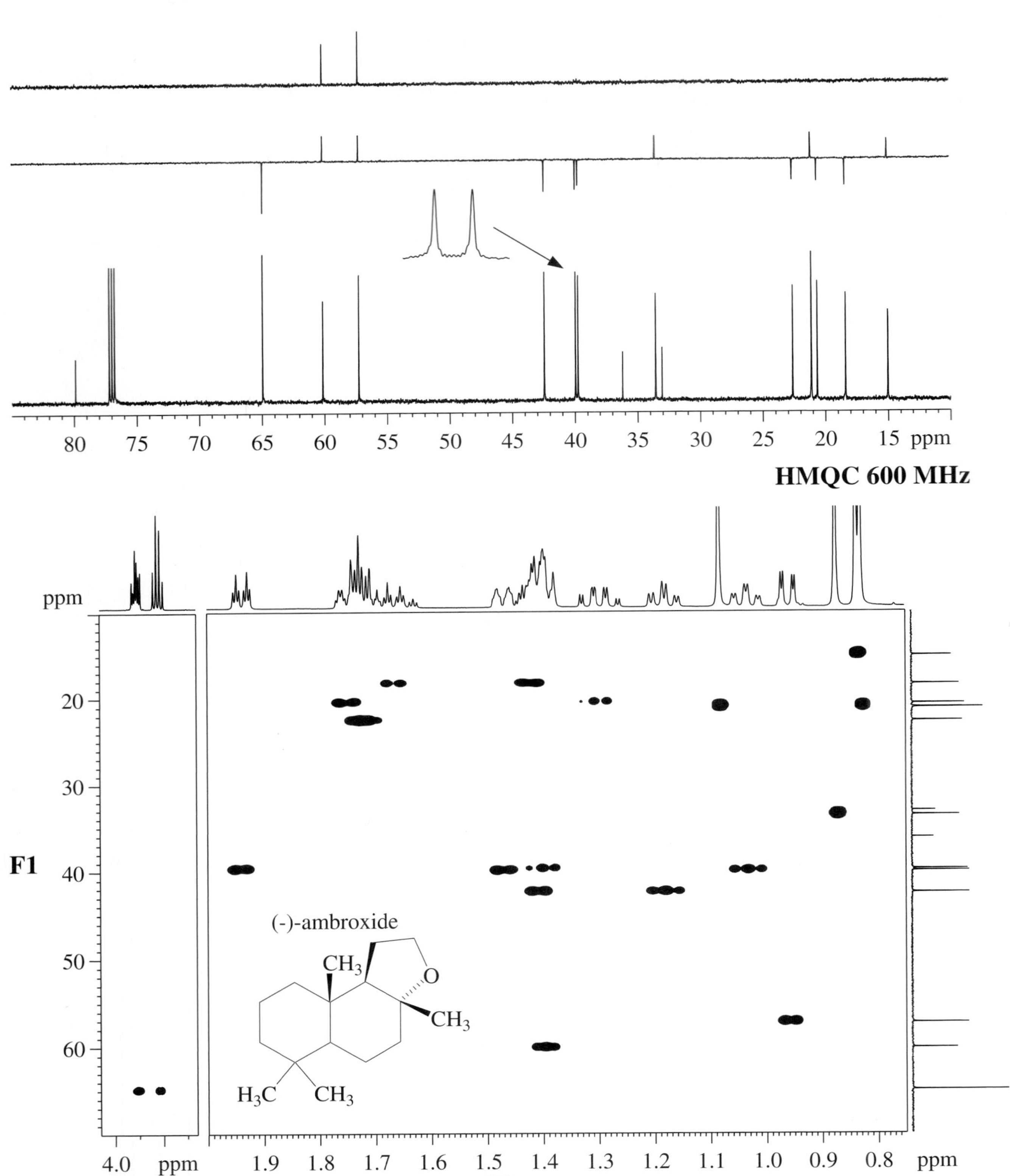

^{13}C/DEPT NMR 150.9 MHz
80
75
70
65
60
55
50
45
40
35
30
25
20
15 ppm
HMQC 600 MHz
ppm
F1
20
30
40
50
60
(-)-ambroxide
CH3
O
CH3
H3C
CH3
4.0
ppm
1.9
1.8
1.7
1.6
1.5
1.4
1.3
1.2
1.1
1.0
0.9
0.8
ppm
F2

Exercise 7.8C

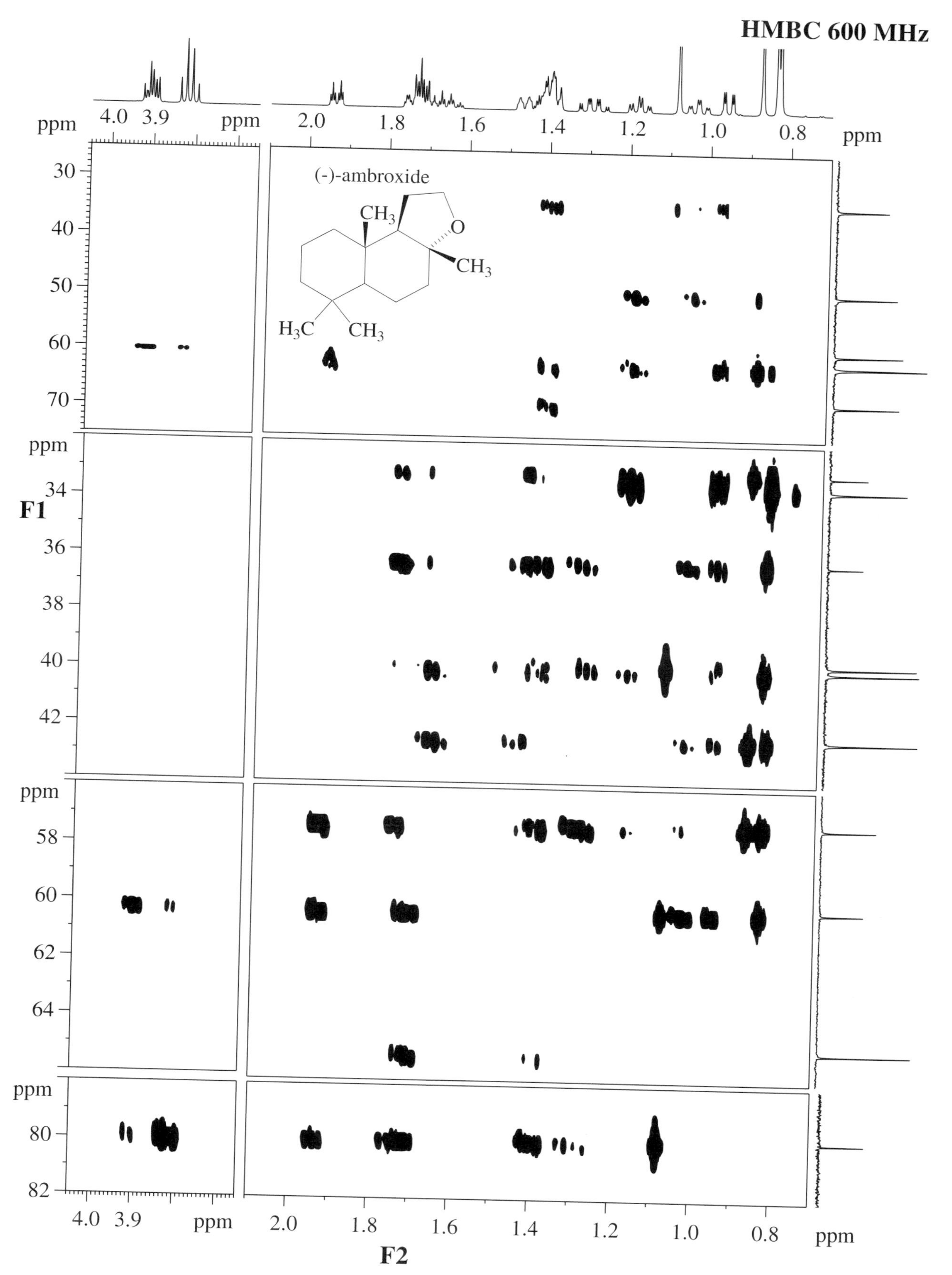

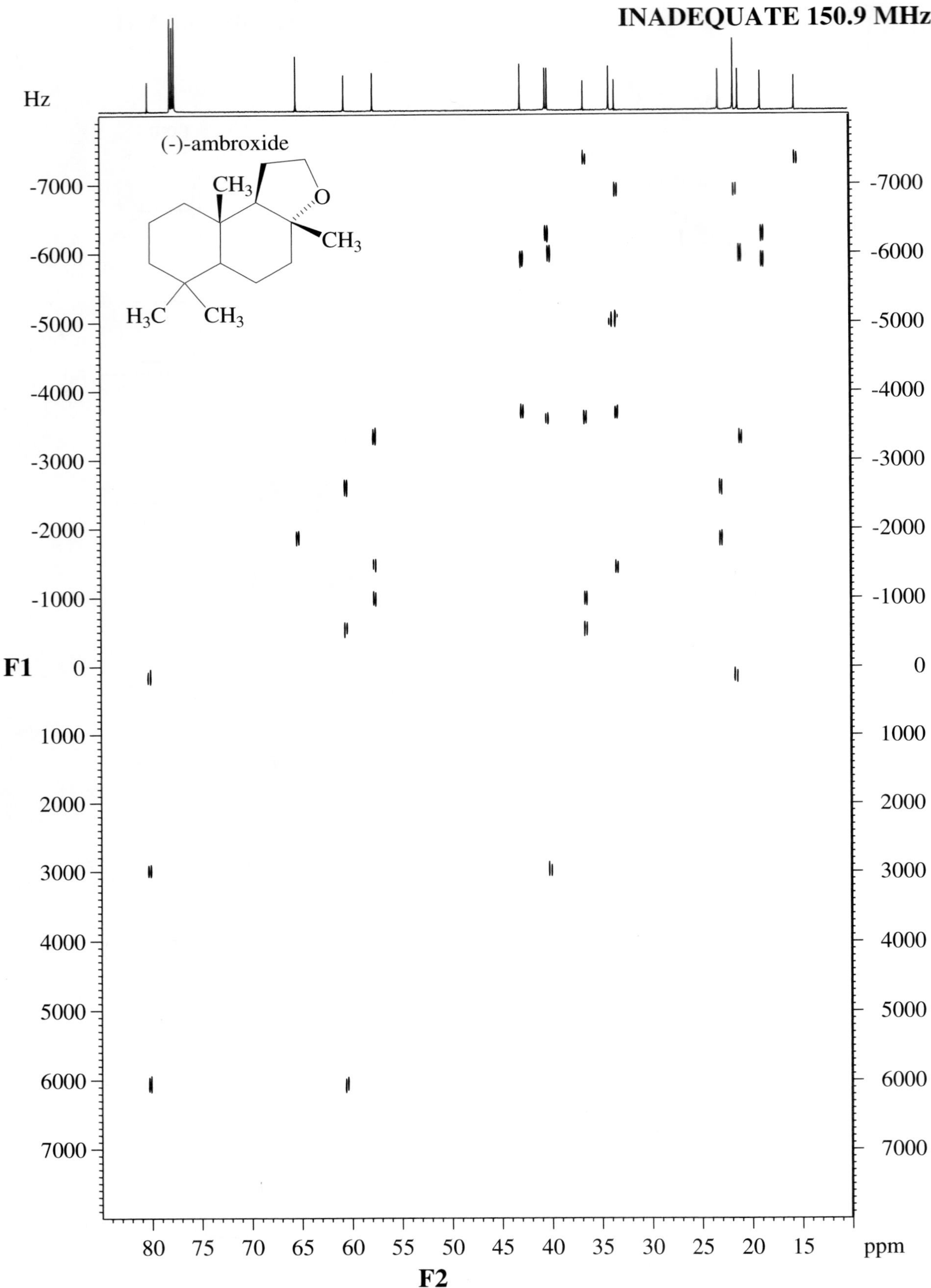
INADEQUATE 150.9 MHz
Hz
(-)-ambroxide
CH3
O
CH3
H3C
CH3
-7000
-6000
-5000
-4000
-3000
-2000
-1000
F1 0
1000
2000
3000
4000
5000
6000
7000
80 75 70 65 60 55 50 45 40 35 30 25 20 15 ppm
F2

CHAPTER 8

ASSIGNED PROBLEMS

8.1 INTRODUCTION

Chapter 7 has been propounded as though unknown problems were being resolved by the authors. Not so. It was inevitable that the authors knew the structures. The students inevitably consulted the structures as they worked through the explications. The experience, therefore, has been more rationalization of the given structures than analysis of spectra. Nonetheless, the experience was worthwhile.

Chapter 8 will be an experience in foresight rather than hindsight (except for the final group of problems), but the first few modest problems should provide encouragement, as did the problems at the end of each chapter. Admittedly, difficulties will be encountered further along, but satisfactions also.

The NMR spectra were obtained at either 300 or 600 MHz for protons and 75.5 or 150.9 MHz for carbon; $CDCl_3$ was the solvent unless otherwise labeled. At 600 MHz, all 2-D spectra were obtained with gradients except INADEQUATE. The IR spectra were obtained neat (i.e. no solvent) unless otherwise labeled. The mass spectra were obtained by EI GC-MS unless otherwise noted. Methane was used to obtain the chemical ionization mass spectra.

The problems that follow cover a wide array of compound types and range from "easy" to "moderate" to "difficult." Clearly, these terms are relative; for a beginning student, none of these problems are truly easy. With experience, many of the moderately difficult problems will seem easier especially with 2-D spectra. The first group of problems has a mass spectrum, an IR spectrum, and ^{1}H and ^{13}C/DEPT spectra. The second group of problems has, in addition to these spectra, some 2-D spectra. Many of these problems can be completed without the 2-D spectra. The third group of problems is presented with the compound's structure and generally has more 2-D spectra.

The problems in the third group are intended as an exercise so that the student can work with more complicated structures. These problems can be assigned with the objective of verifying the structure and making assignments.

For most of the problems, there is more than enough information to fully solve the problem, including stereochemistry, except, of course, absolute stereochemistry. There are, however, a few problems that lack enough information to determine stereochemistry. In these cases, the student should indicate the possible stereoisomers and suggest ways to distinguish among them. Additional Student Exercises can be found at http://www.wiley.com/college/silverstein.

At this point, we may all enjoy a more literary statement: "It is easy to be wise in retrospect, uncommonly difficult in the event."*

* Stegner, W. (1954). *Beyond the Hundredth Meridian.* Boston: Houghton Mifflin. Reprint, Penguin Books, 1992, Chapter 3.

Problem 8.1

MASS

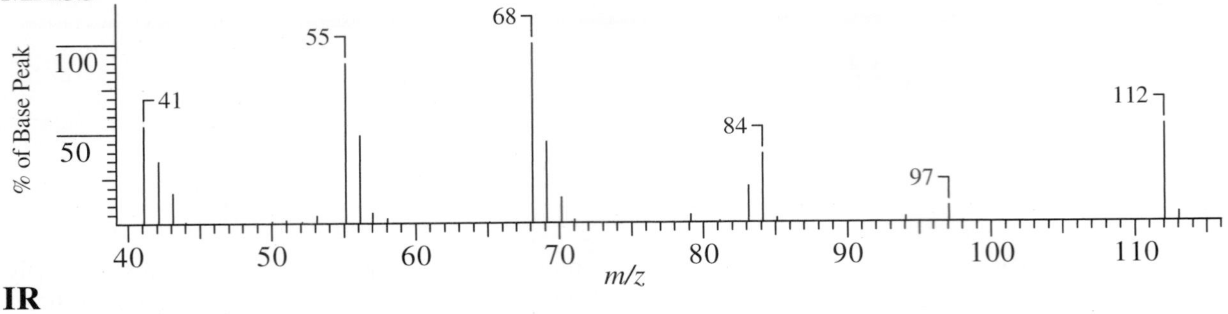

% of Base Peak
100
50
41
55
68
84
97
112
40
50
60
70
80
90
100
110
m/z

IR

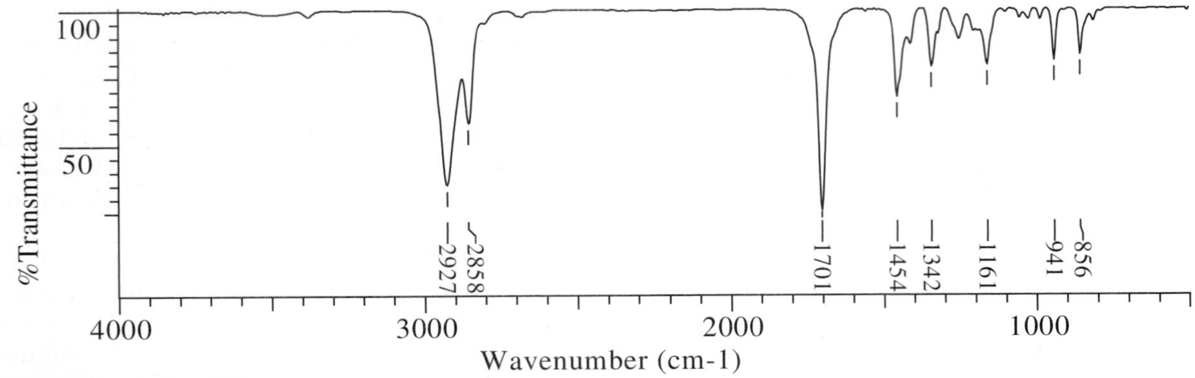

%Transmittance
100
50
2927
2858
1701
1454
1342
1161
941
856
4000
3000
2000
1000
Wavenumber (cm-1)

^{1}H NMR 300 MHz

2.7
2.6
2.5
2.4
2.3
2.2
2.1
2.0
1.9
1.8
1.7
1.6
1.5 ppm

^{13}C/DEPT NMR 75.5 MHz

200
180
160
140
120
100
80
60
40
ppm

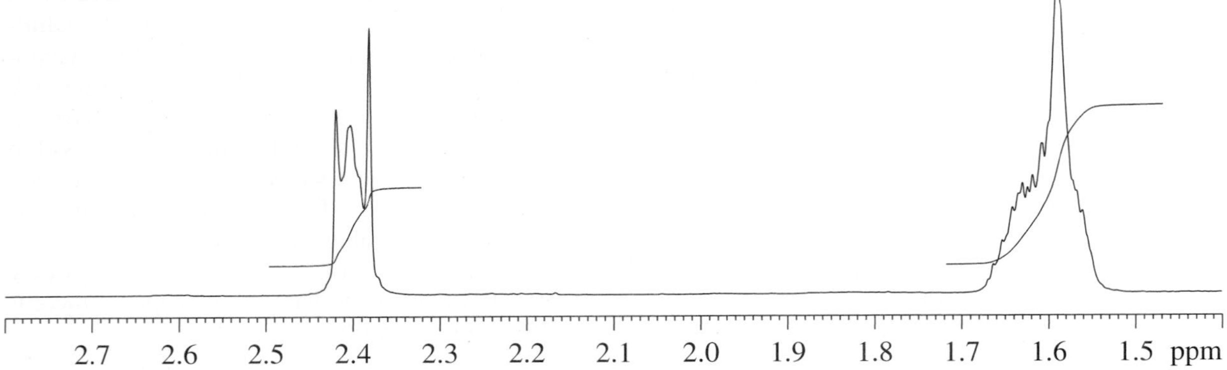

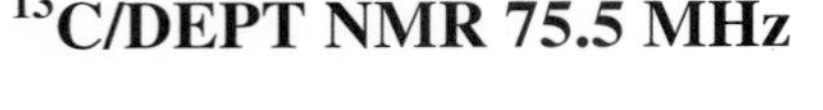

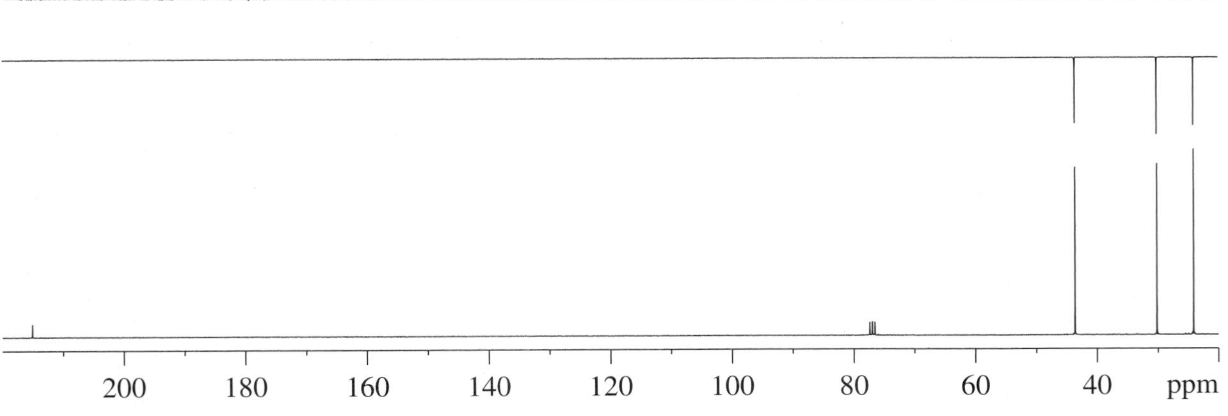

Problem 8.2

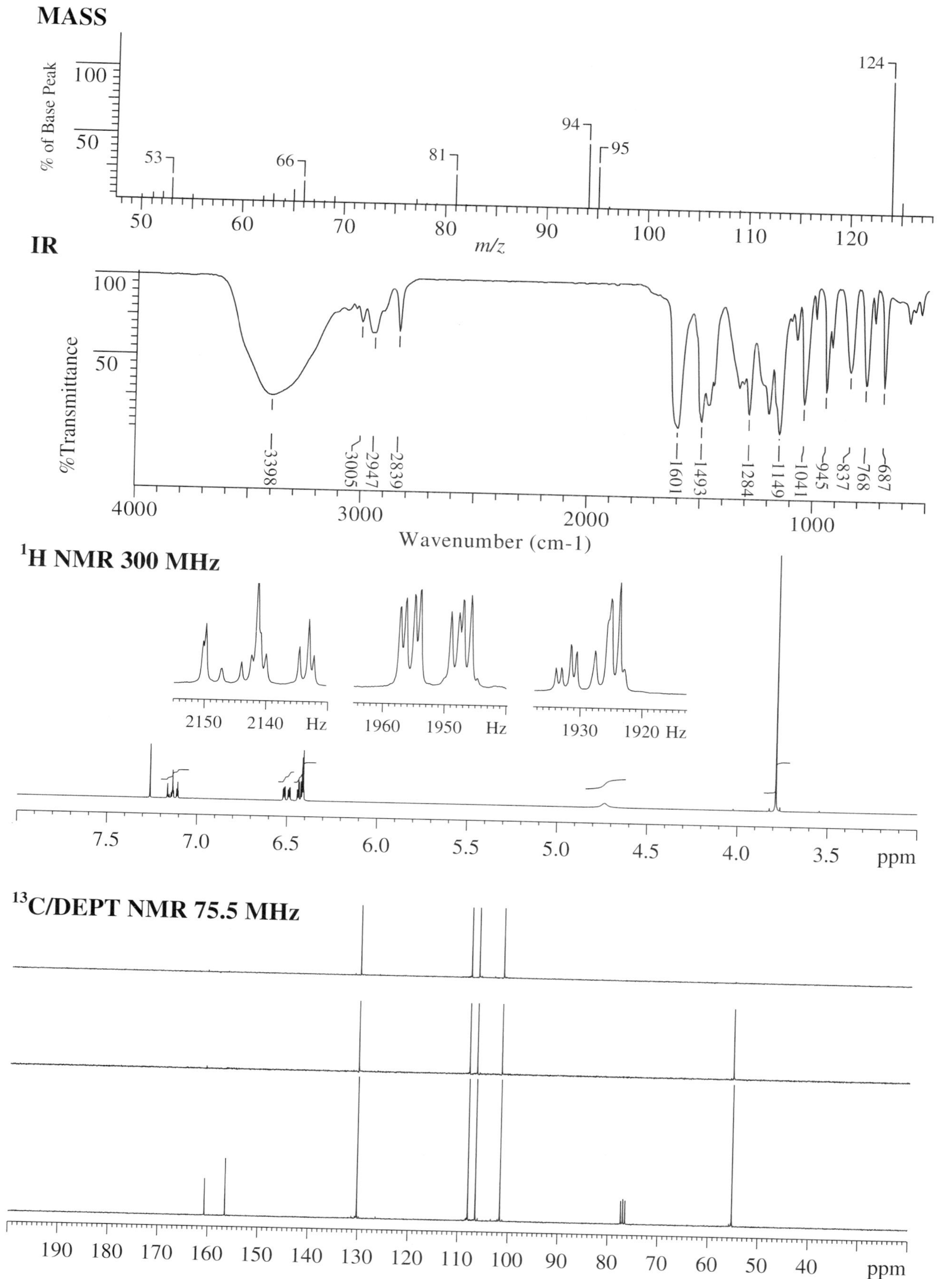
MASS
% of Base Peak
100
50
53
66
81
94
95
124
50 60 70 80 90 100 110 120
m/z
IR
% Transmittance
100
50
3398
3005
2947
2839
1601
1493
1284
1149
1041
945
837
768
687
4000 3000 2000 1000
Wavenumber (cm-1)
1H NMR 300 MHz
2150 2140 Hz
1960 1950 Hz
1930 1920 Hz
7.5 7.0 6.5 6.0 5.5 5.0 4.5 4.0 3.5 ppm
13C/DEPT NMR 75.5 MHz
190 180 170 160 150 140 130 120 110 100 90 80 70 60 50 40 ppm

MASS

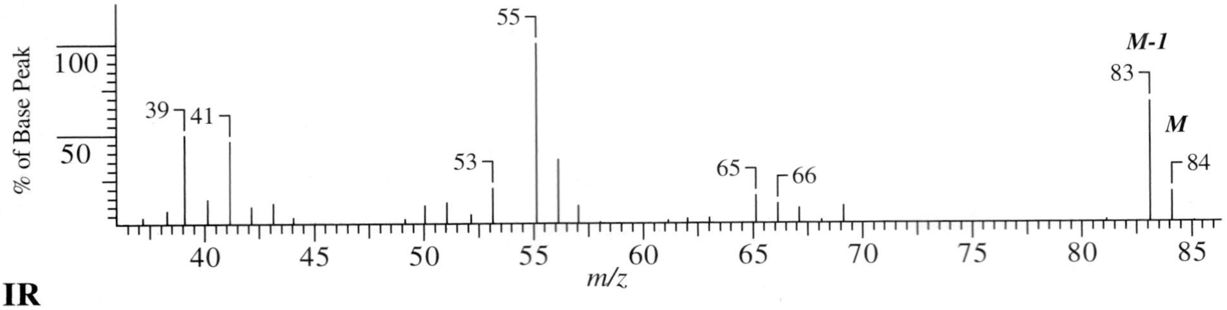

IR

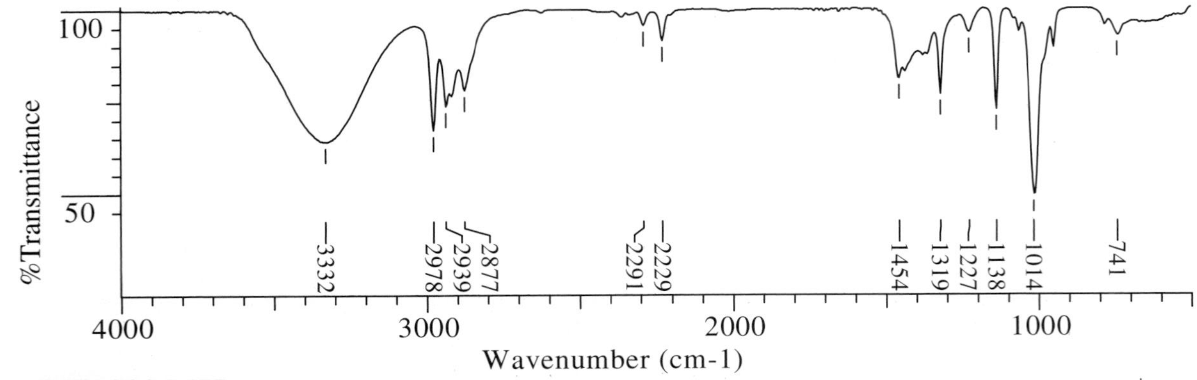

^{1}H NMR 300 MHz

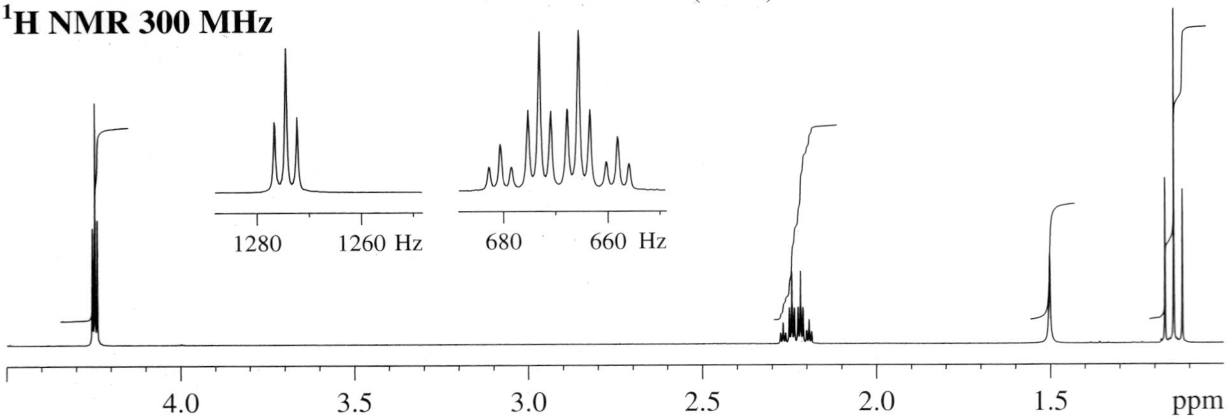

^{13}C/DEPT NMR 75.5 MHz

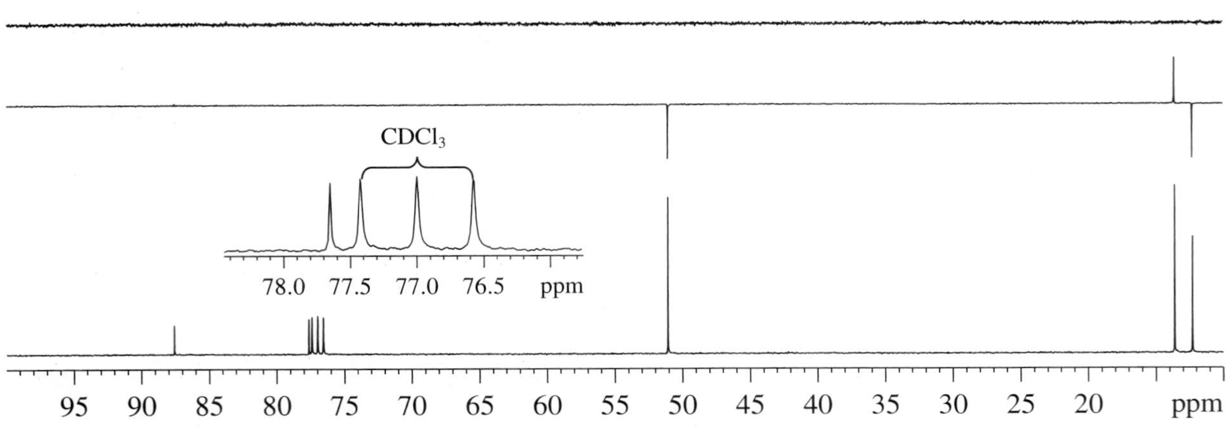

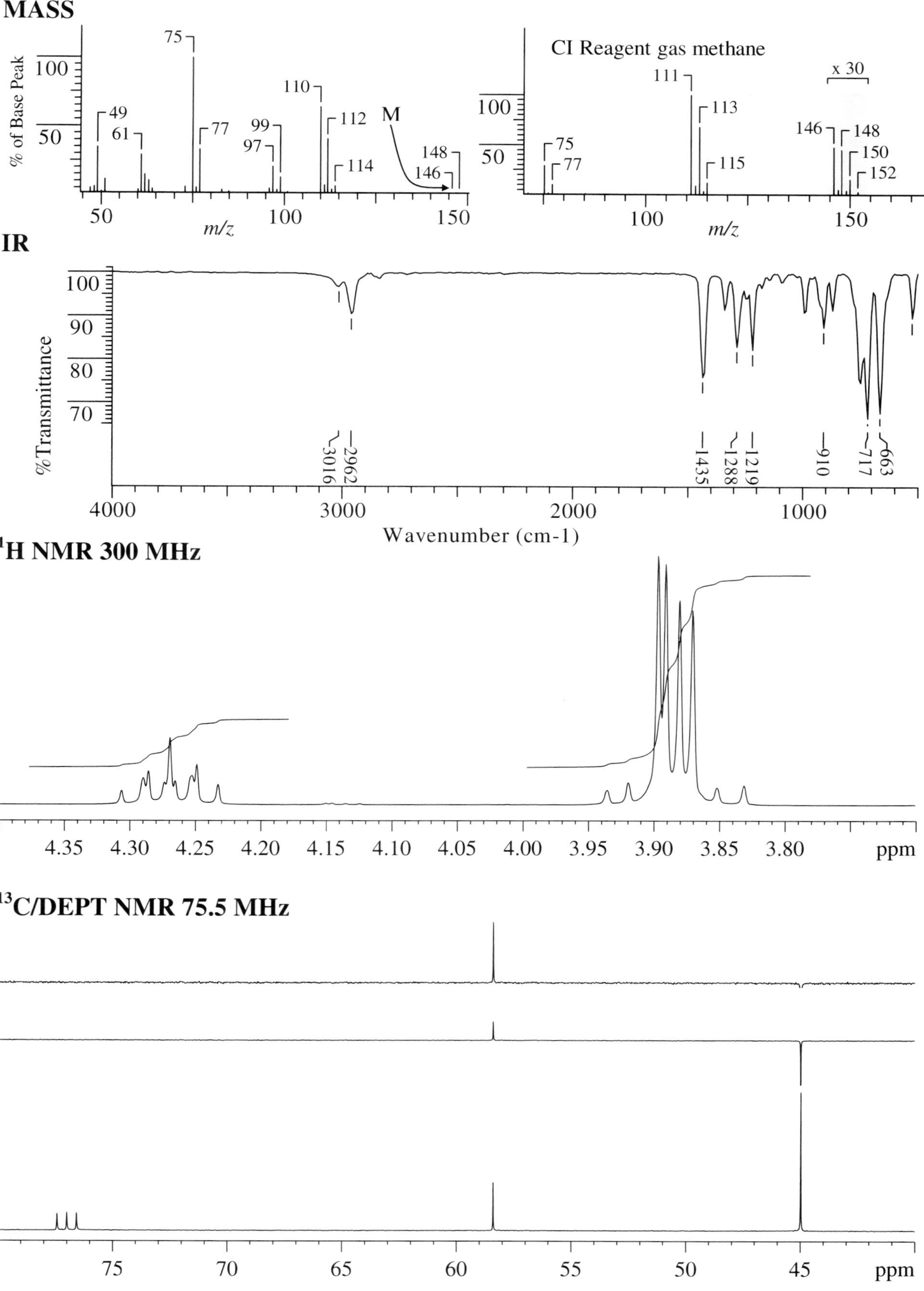
MASS
% of Base Peak
100
50
75
49
61
77
99
97
110
112
114
M
148
146
50
100
150
m/z
CI Reagent gas methane
111
113
115
x 30
146
148
150
152
75
77
100
150
m/z
IR
%Transmittance
100
90
80
70
3016
2962
1435
1288
1219
910
717
663
4000
3000
2000
1000
Wavenumber (cm-1)
¹H NMR 300 MHz
4.35 4.30 4.25 4.20 4.15 4.10 4.05 4.00 3.95 3.90 3.85 3.80 ppm
¹³C/DEPT NMR 75.5 MHz
75 70 65 60 55 50 45 ppm

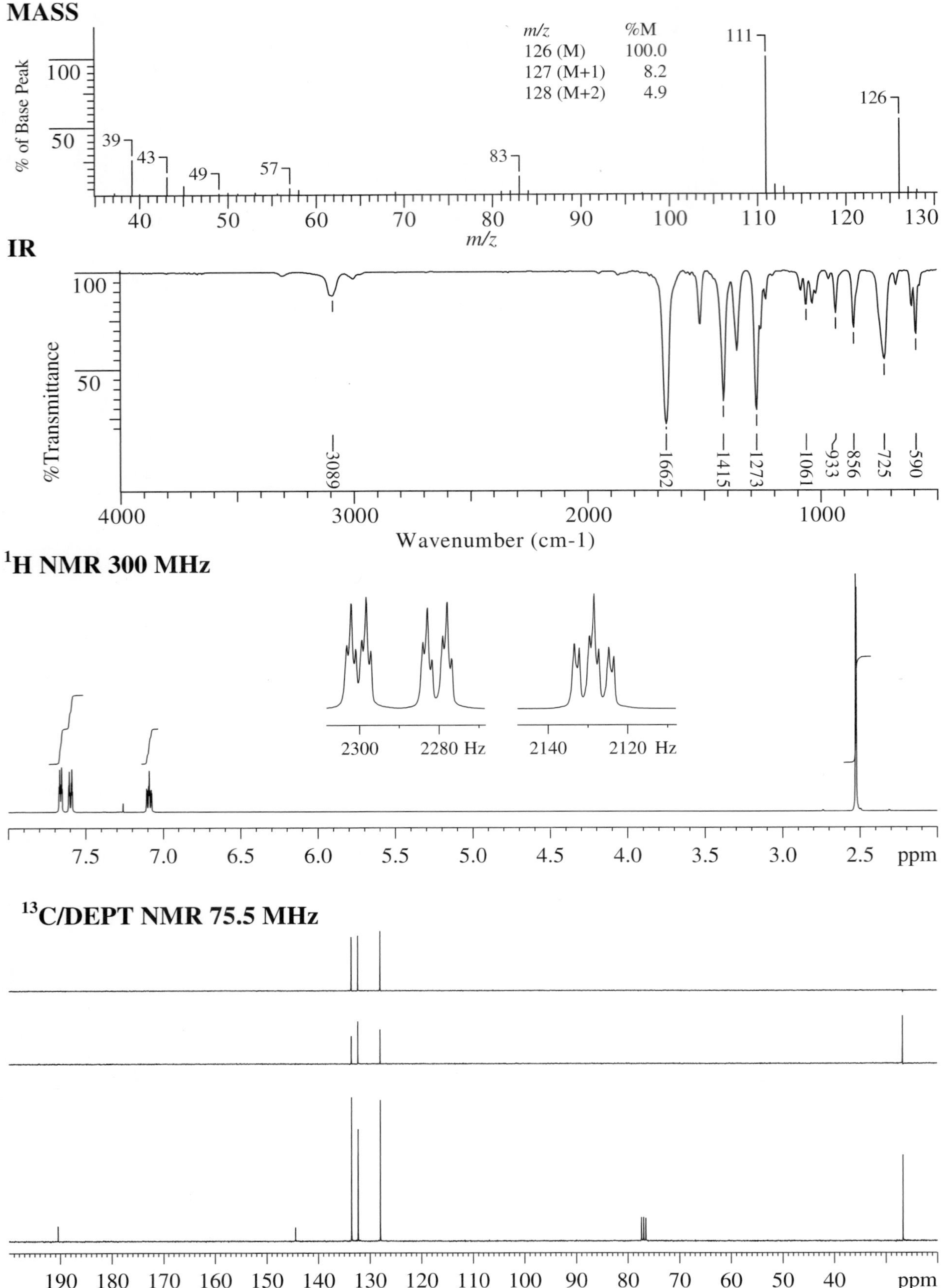

MASS
% of Base Peak
m/z
%M
126 (M) 100.0
127 (M+1) 8.2
128 (M+2) 4.9
39
43
49
57
83
111
126
IR
%Transmittance
3089
1662
1415
1273
1061
933
856
725
590
Wavenumber (cm-1)
1H NMR 300 MHz
2300
2280 Hz
2140
2120 Hz
ppm
13C/DEPT NMR 75.5 MHz
ppm

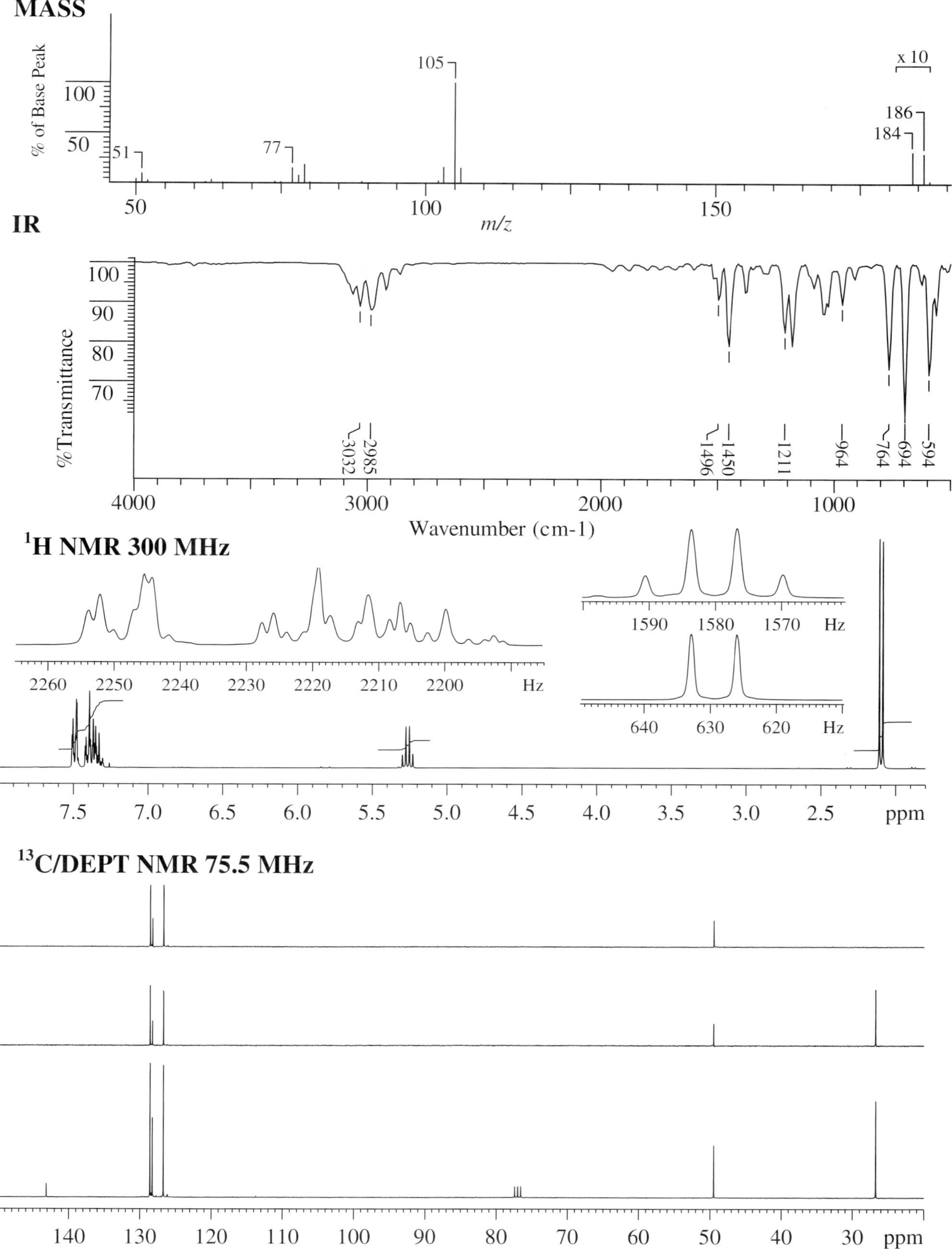
MASS
% of Base Peak
100
50
51
77
105
x 10
186
184
50
100
150
m/z
IR
% Transmittance
100
90
80
70
3032
2985
1496
1450
1211
964
764
694
594
4000
3000
2000
1000
Wavenumber (cm-1)
1H NMR 300 MHz
2260 2250 2240 2230 2220 2210 2200 Hz
1590 1580 1570 Hz
640 630 620 Hz
7.5 7.0 6.5 6.0 5.5 5.0 4.5 4.0 3.5 3.0 2.5 ppm
13C/DEPT NMR 75.5 MHz
140 130 120 110 100 90 80 70 60 50 40 30 ppm

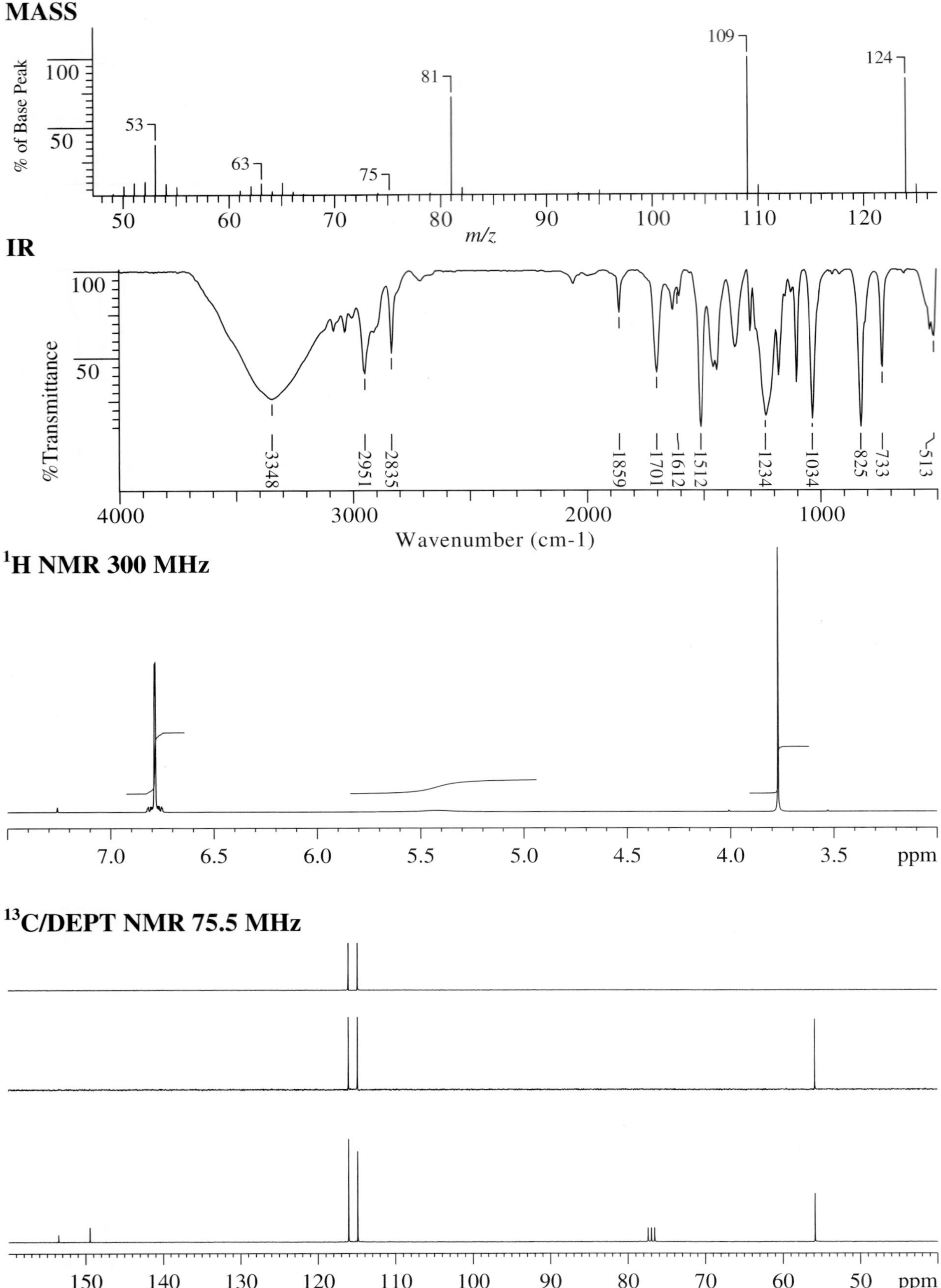
MASS
% of Base Peak
100
50
53
63
75
81
109
124
50 60 70 80 90 100 110 120
m/z
IR
%Transmittance
100
50
3348
2951
2835
1859
1701
1612
1512
1234
1034
825
733
513
4000 3000 2000 1000
Wavenumber (cm-1)
¹H NMR 300 MHz
7.0 6.5 6.0 5.5 5.0 4.5 4.0 3.5 ppm
¹³C/DEPT NMR 75.5 MHz
150 140 130 120 110 100 90 80 70 60 50 ppm

Problem 8.8

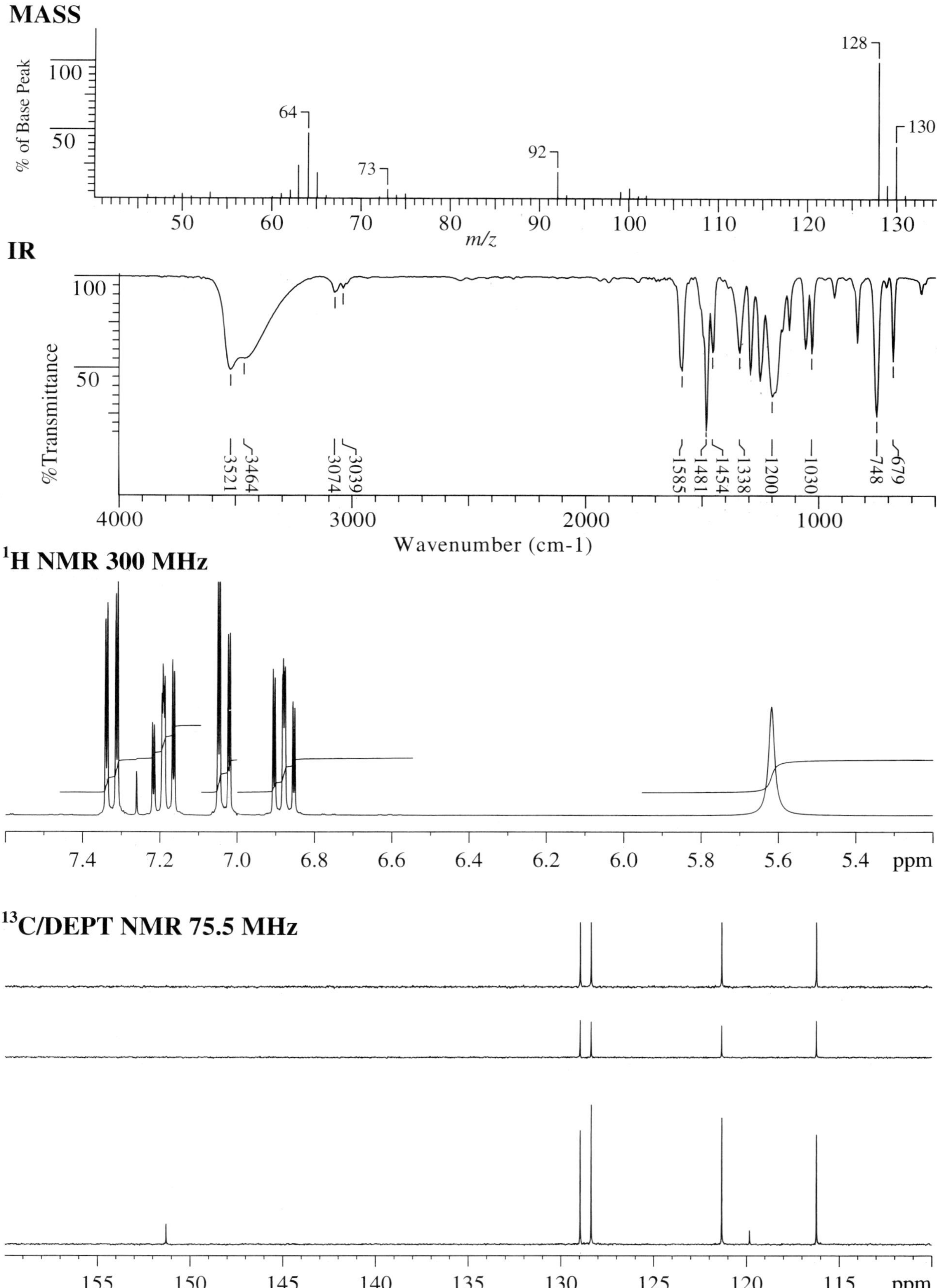
MASS
% of Base Peak
100
50
128
130
64
73
92
50
60
70
80
90
100
110
120
130
m/z
IR
% Transmittance
100
50
3521
3464
3074
3039
1585
1481
1454
1338
1200
1030
748
679
4000
3000
2000
1000
Wavenumber (cm-1)
1H NMR 300 MHz
7.4
7.2
7.0
6.8
6.6
6.4
6.2
6.0
5.8
5.6
5.4
ppm
13C/DEPT NMR 75.5 MHz
155
150
145
140
135
130
125
120
115
ppm

 Problem 8.9

MASS

% of Base Peak
100
50
0
55
73
87
115
135
137
163
165
x 20
194
196
50
100
150
m/z

IR

%Transmittance
100
50
2954
1736
1435
1365
1203
748
4000
3000
2000
1000
Wavenumber (cm-1)

^{1}H NMR 300 MHz

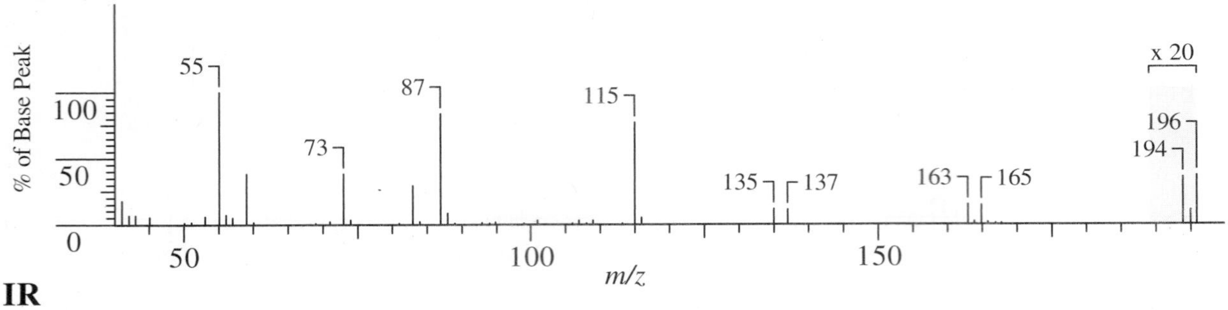

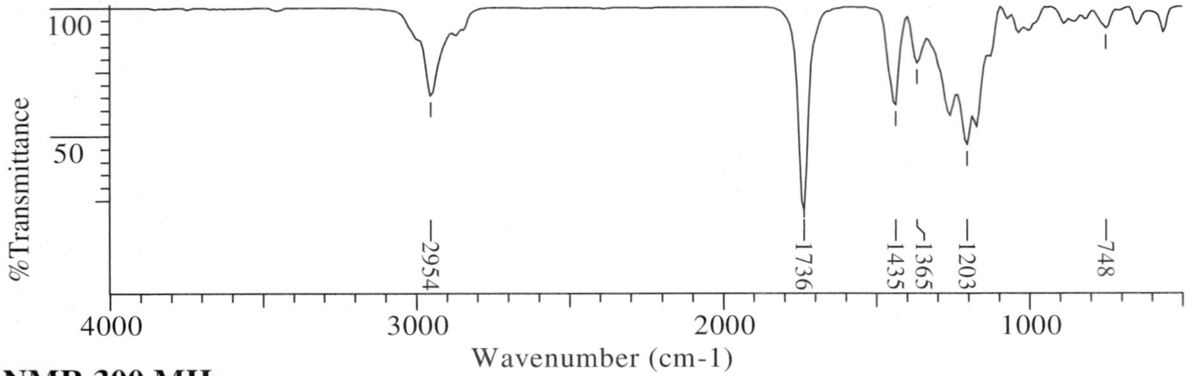

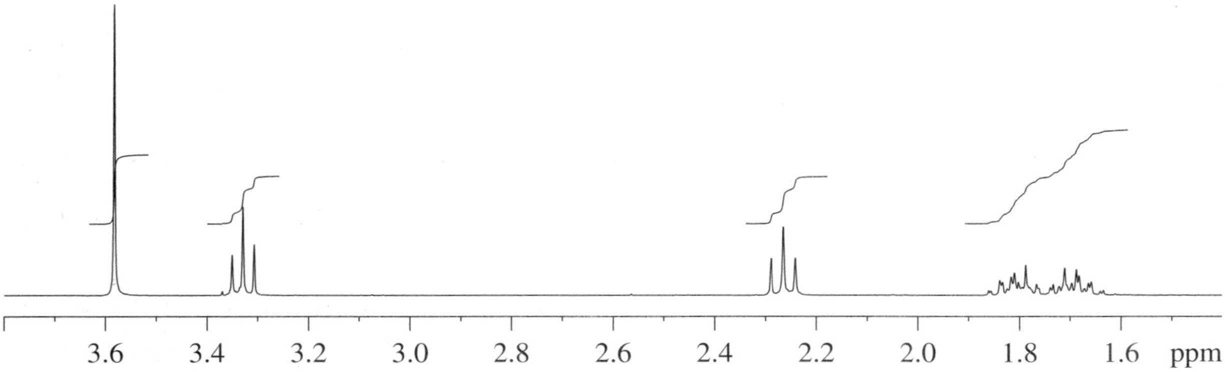

3.6
3.4
3.2
3.0
2.8
2.6
2.4
2.2
2.0
1.8
1.6
ppm

^{13}C/DEPT NMR 75.5 MHz

33.0
32.5
32.0
ppm
170
160
150
140
130
120
110
100
90
80
70
60
50
40
30
ppm

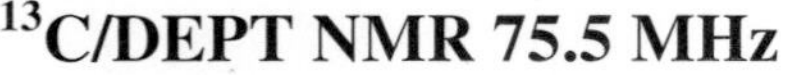

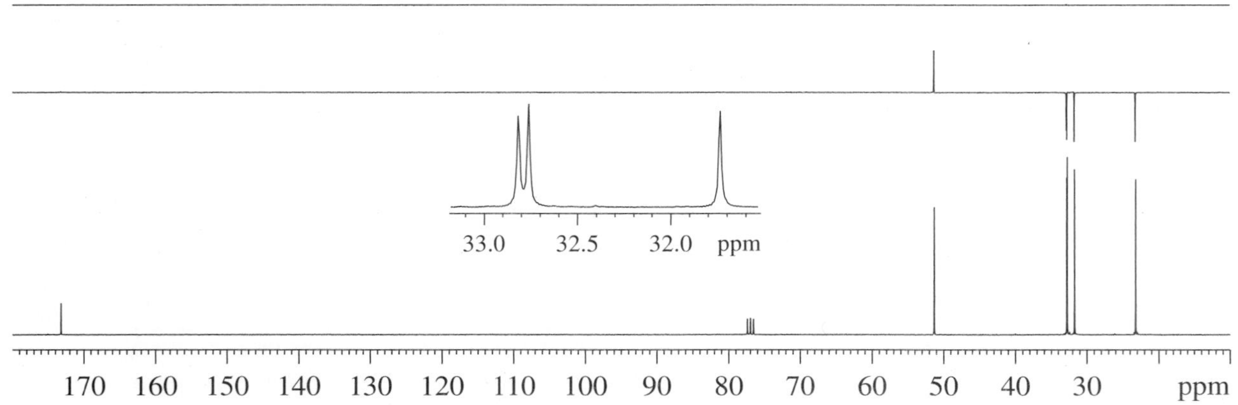

Problem 8.10

MASS

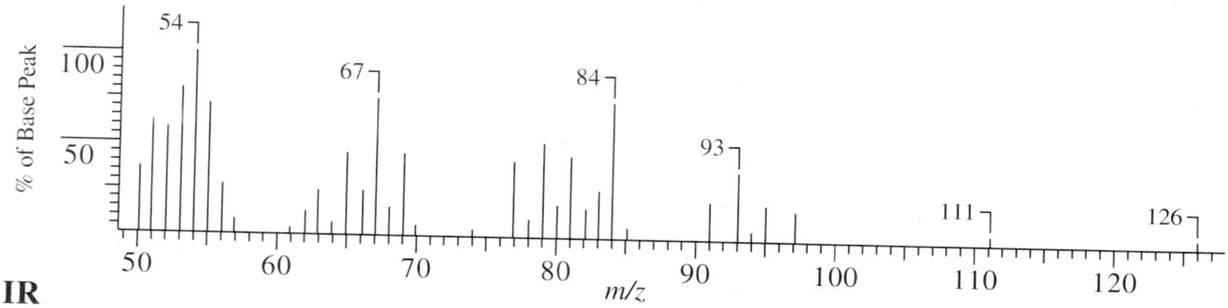

% of Base Peak
100
50
54
67
84
93
111
126
50
60
70
80
90
100
110
120
m/z

IR

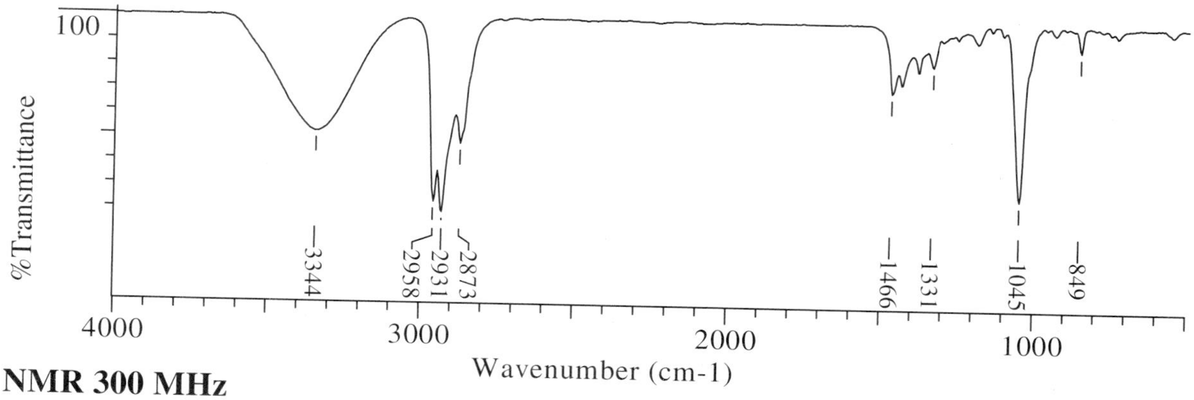

%Transmittance
100
3344
2958
2931
2873
1466
1331
1045
849
4000
3000
2000
1000
Wavenumber (cm-1)

^{1}H NMR 300 MHz

3.5
3.0
2.5
2.0
1.5
1.0
ppm

^{13}C/DEPT NMR 75.5 MHz

84
82
80
78
ppm
85
80
75
70
65
60
55
50
45
40
35
30
25
20
ppm

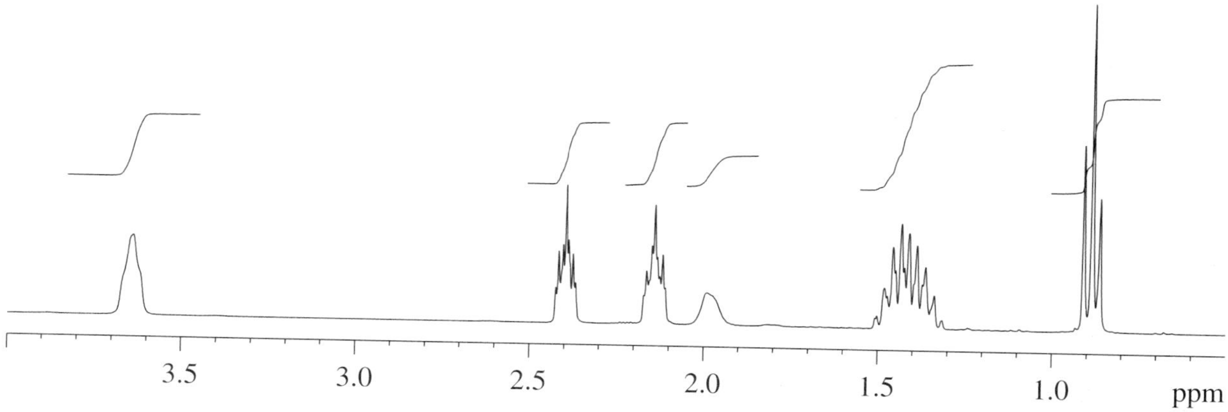

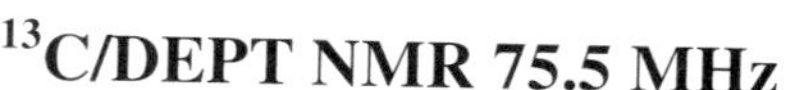

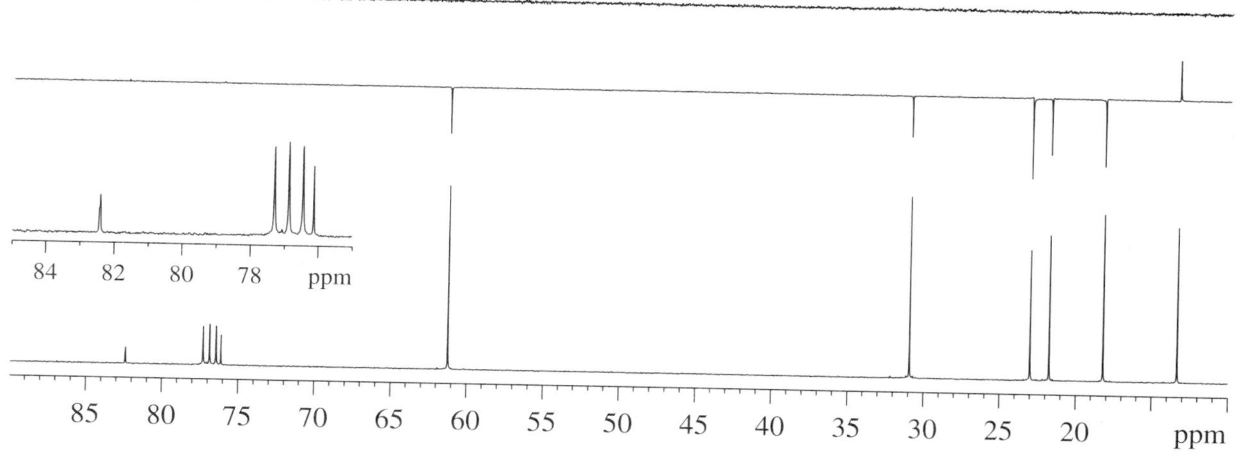

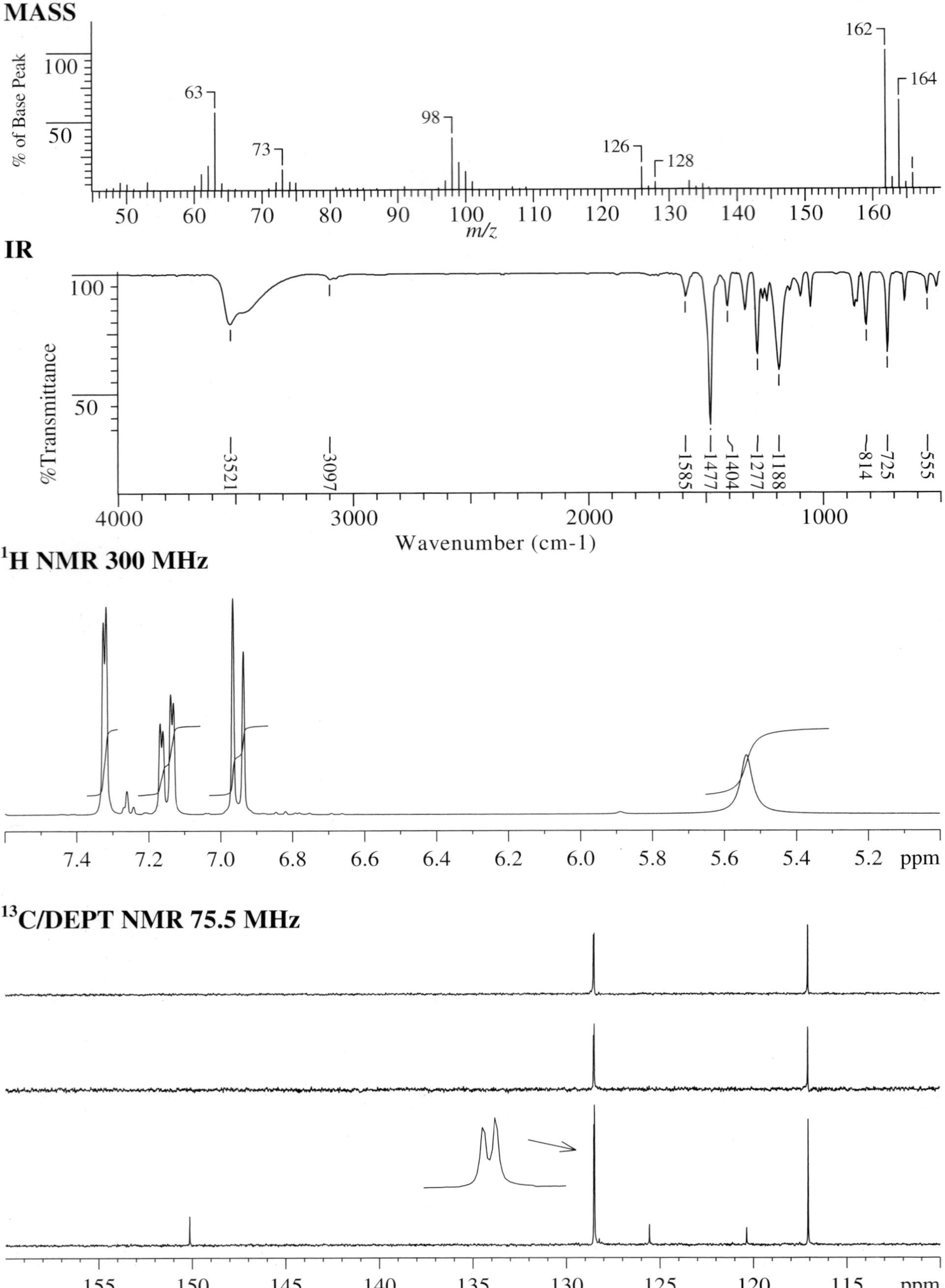
MASS
% of Base Peak
100
50
63
73
98
126
128
162
164
50 60 70 80 90 100 110 120 130 140 150 160
m/z
IR
%Transmittance
100
50
3521
3097
1585
1477
1404
1277
1188
814
725
555
4000 3000 2000 1000
Wavenumber (cm-1)
¹H NMR 300 MHz
7.4 7.2 7.0 6.8 6.6 6.4 6.2 6.0 5.8 5.6 5.4 5.2 ppm
¹³C/DEPT NMR 75.5 MHz
155 150 145 140 135 130 125 120 115 ppm

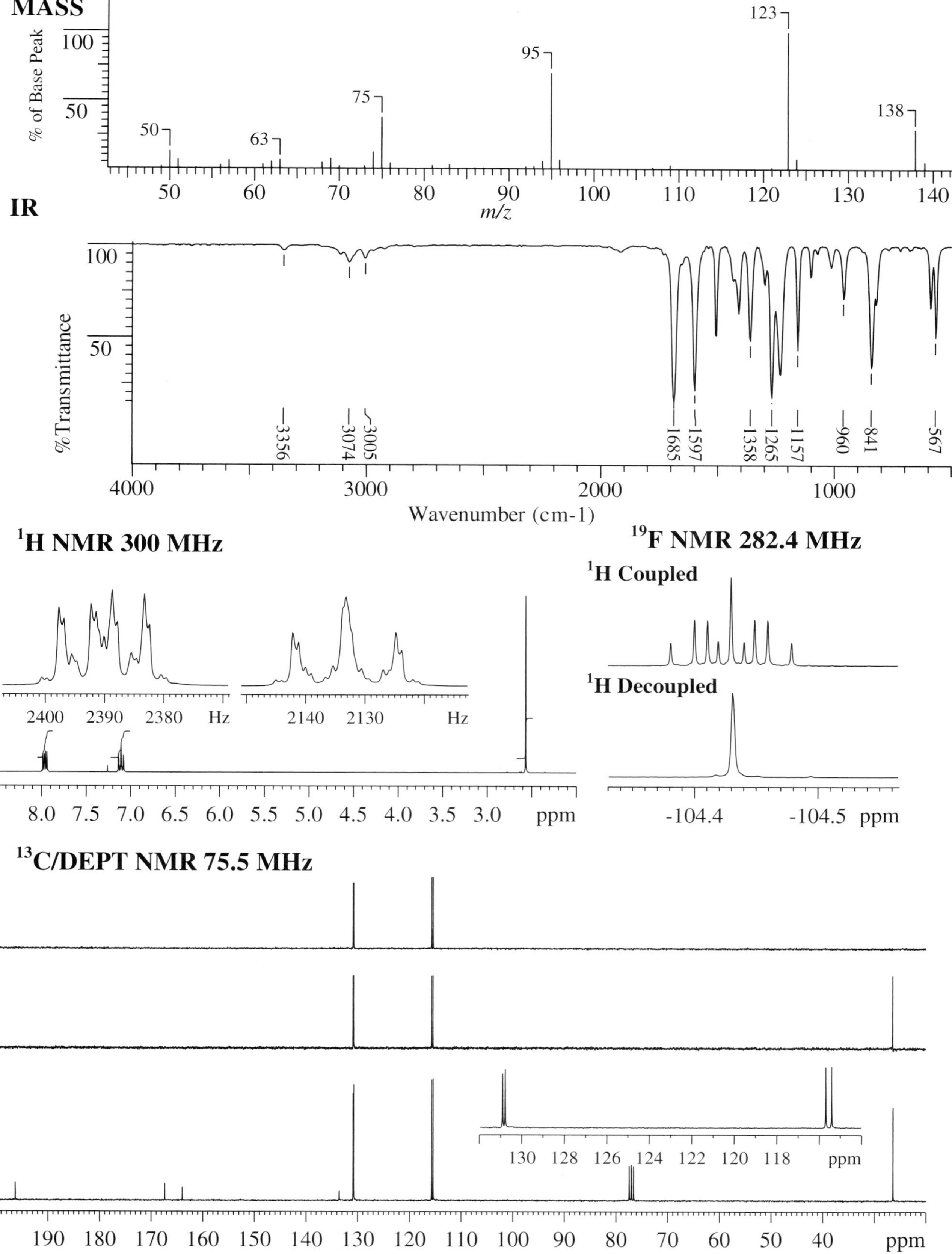
MASS
% of Base Peak
100
50
50
63
75
95
123
138
50 60 70 80 90 100 110 120 130 140
m/z
IR
% Transmittance
100
50
3356
3074
3005
1685
1597
1358
1265
1157
960
841
567
4000 3000 2000 1000
Wavenumber (cm-1)
¹H NMR 300 MHz
2400 2390 2380 Hz
2140 2130 Hz
8.0 7.5 7.0 6.5 6.0 5.5 5.0 4.5 4.0 3.5 3.0 ppm
¹⁹F NMR 282.4 MHz
¹H Coupled
¹H Decoupled
-104.4 -104.5 ppm
¹³C/DEPT NMR 75.5 MHz
130 128 126 124 122 120 118 ppm
190 180 170 160 150 140 130 120 110 100 90 80 70 60 50 40 ppm

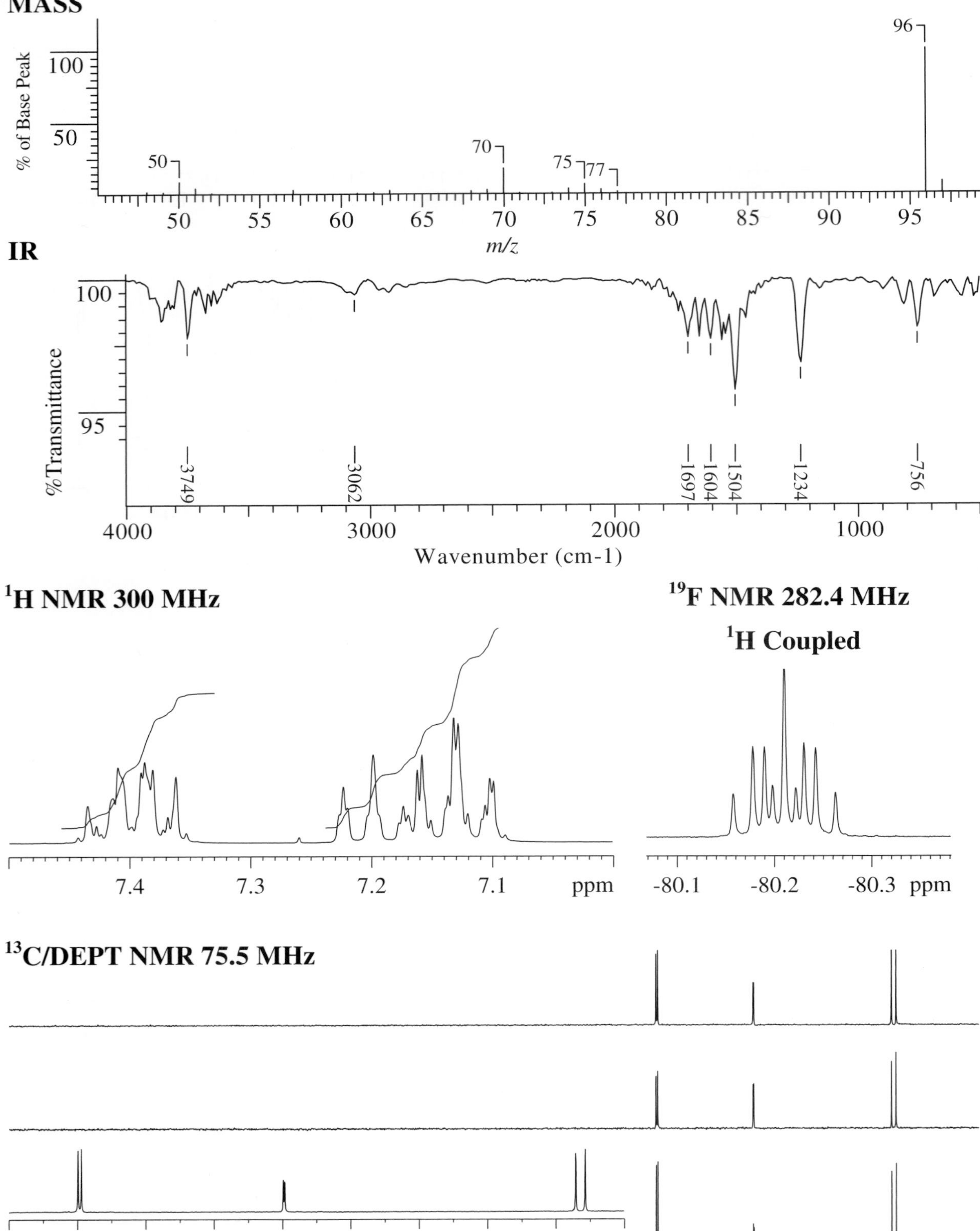

MASS
% of Base Peak
100
50
50
70
75
77
96
50 55 60 65 70 75 80 85 90 95
m/z
IR
% Transmittance
100
95
3749
3062
1697
1604
1504
1234
756
4000 3000 2000 1000
Wavenumber (cm-1)
1H NMR 300 MHz
7.4 7.3 7.2 7.1 ppm
19F NMR 282.4 MHz
1H Coupled
-80.1 -80.2 -80.3 ppm
13C/DEPT NMR 75.5 MHz
130 128 126 124 122 120 118 116 ppm
165 160 155 150 145 140 135 130 125 120 115 ppm

Problem 8.14

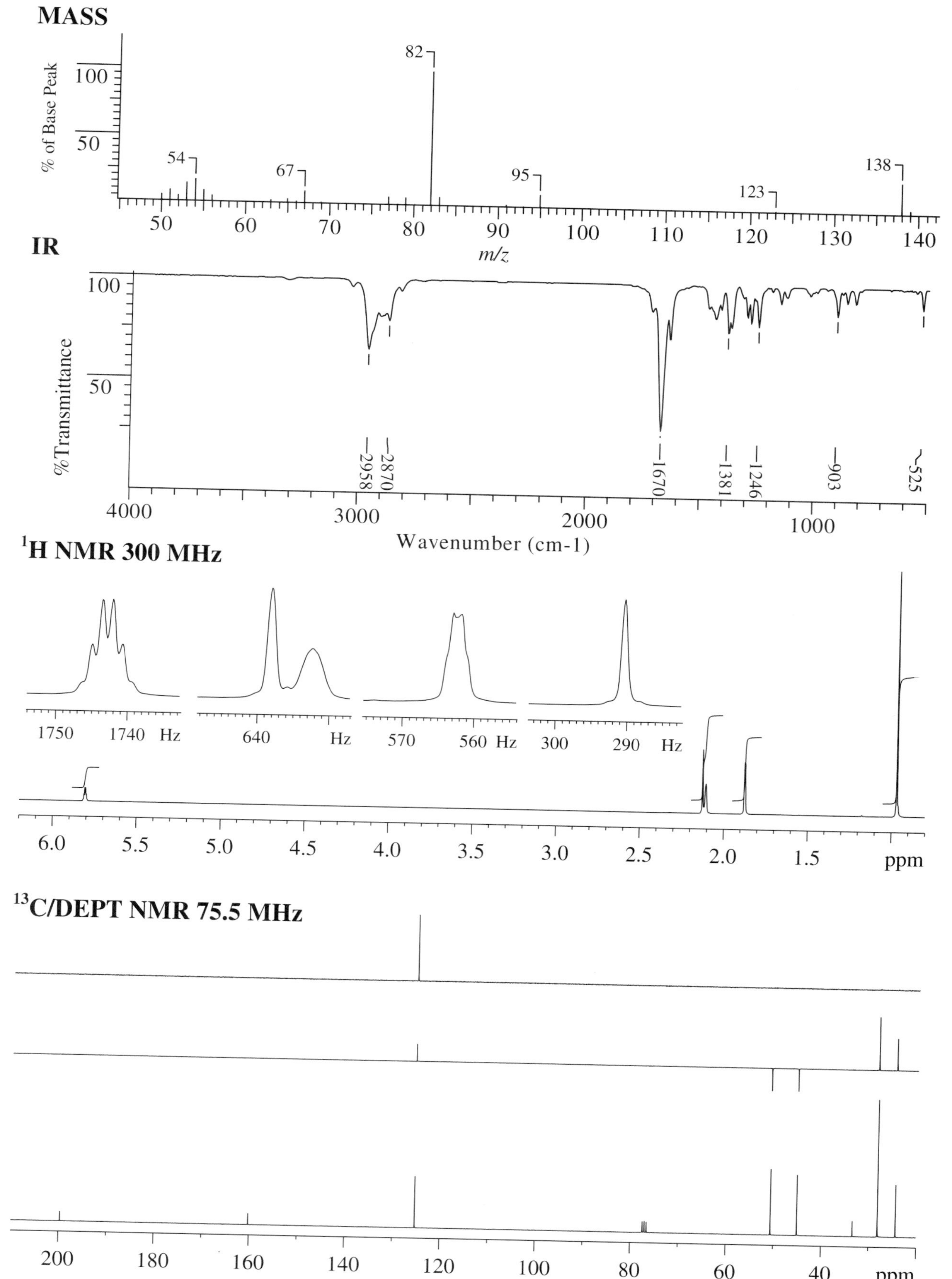
MASS
% of Base Peak
100
50
54
67
82
95
123
138
50 60 70 80 90 100 110 120 130 140
m/z
IR
%Transmittance
100
50
2958
2870
1670
1381
1246
903
525
4000 3000 2000 1000
Wavenumber (cm-1)
¹H NMR 300 MHz
1750 1740 Hz
640 Hz
570 560 Hz
300 290 Hz
6.0 5.5 5.0 4.5 4.0 3.5 3.0 2.5 2.0 1.5 ppm
¹³C/DEPT NMR 75.5 MHz
200 180 160 140 120 100 80 60 40 ppm

Problem 8.15

MASS

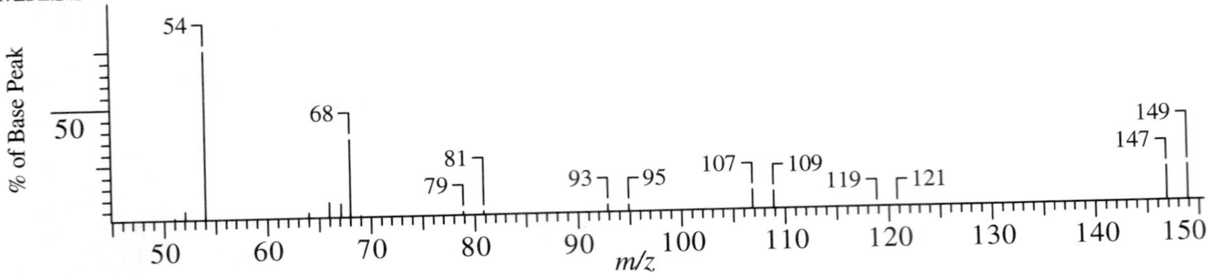

54
68
79
81
93
95
107
109
119
121
147
149
% of Base Peak
50
50 60 70 80 90 100 110 120 130 140 150
m/z

IR

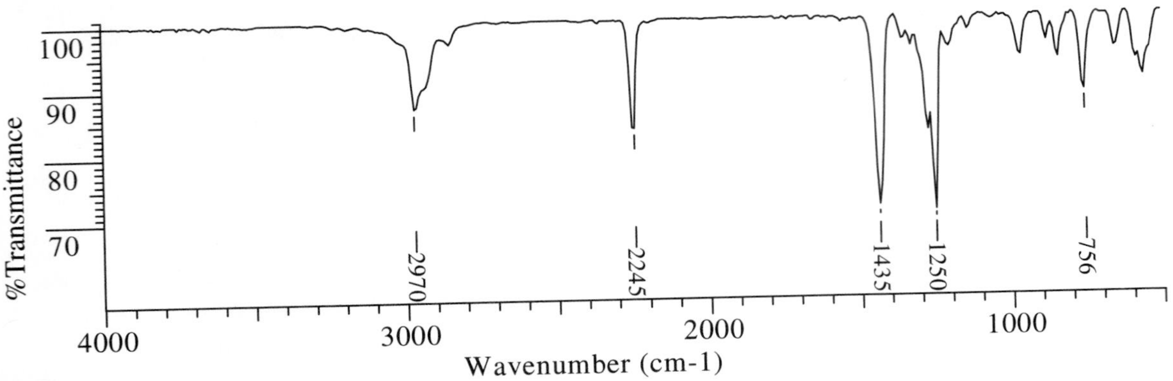

%Transmittance
100
90
80
70
2970
2245
1435
1250
756
4000 3000 2000 1000
Wavenumber (cm-1)

^{1}H NMR 300 MHz

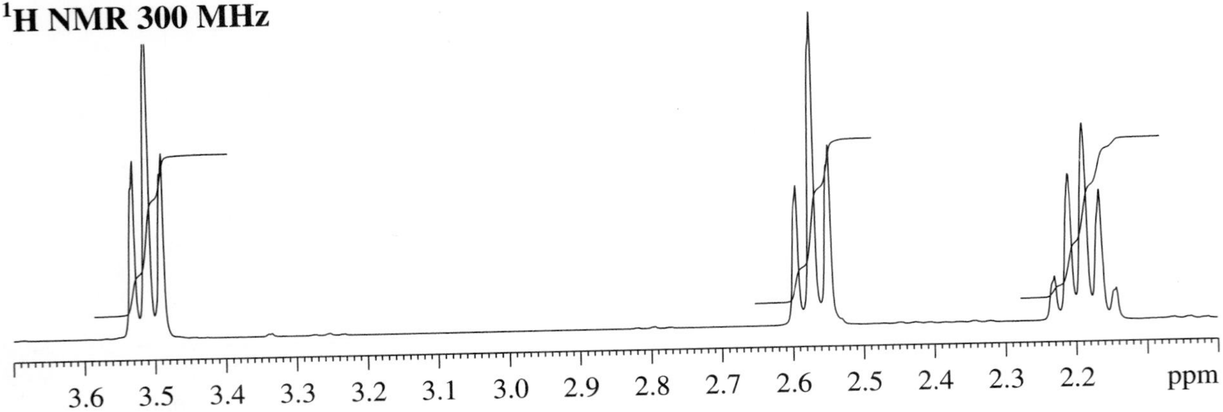

3.6 3.5 3.4 3.3 3.2 3.1 3.0 2.9 2.8 2.7 2.6 2.5 2.4 2.3 2.2 ppm

^{13}C/DEPT NMR 75.5 MHz

120 110 100 90 80 70 60 50 40 30 20 ppm

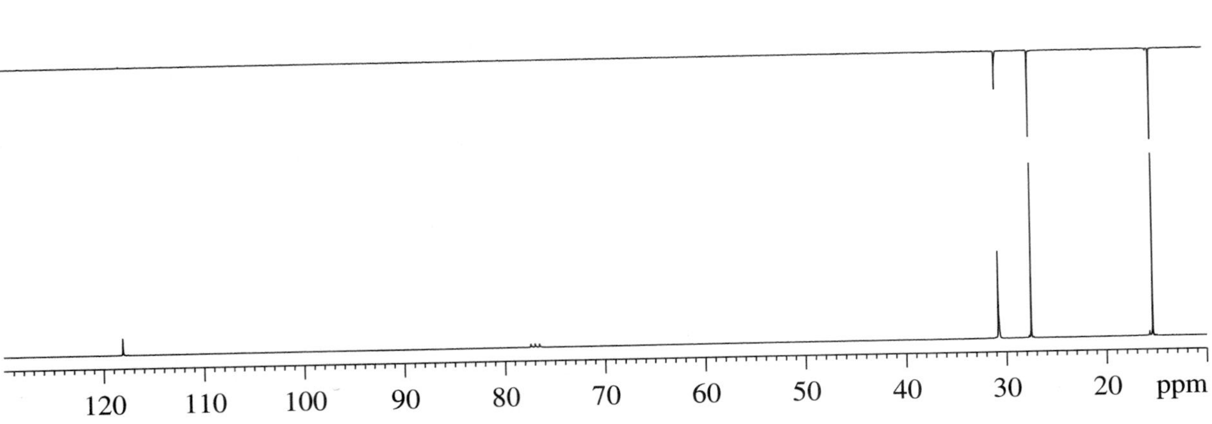

Problem 8.16

MASS

IR

^{1}H NMR 300 MHz

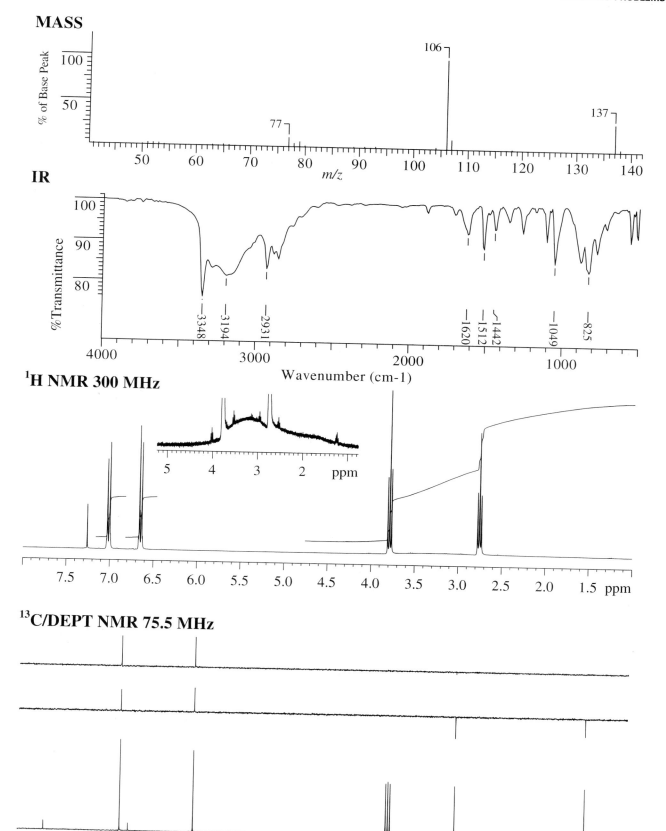

^{13}C/DEPT NMR 75.5 MHz

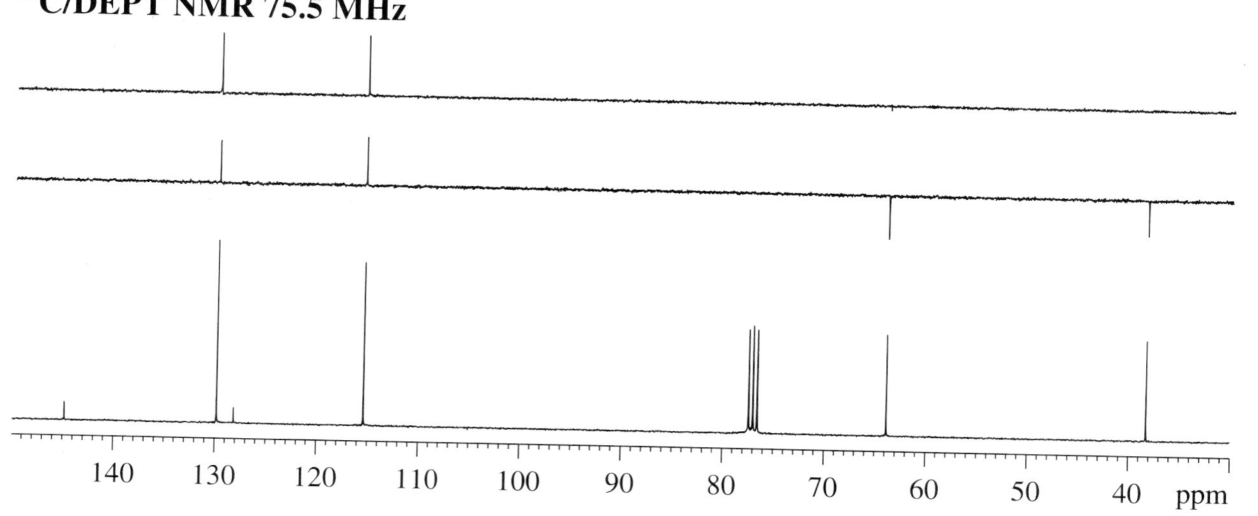

Problem 8.17

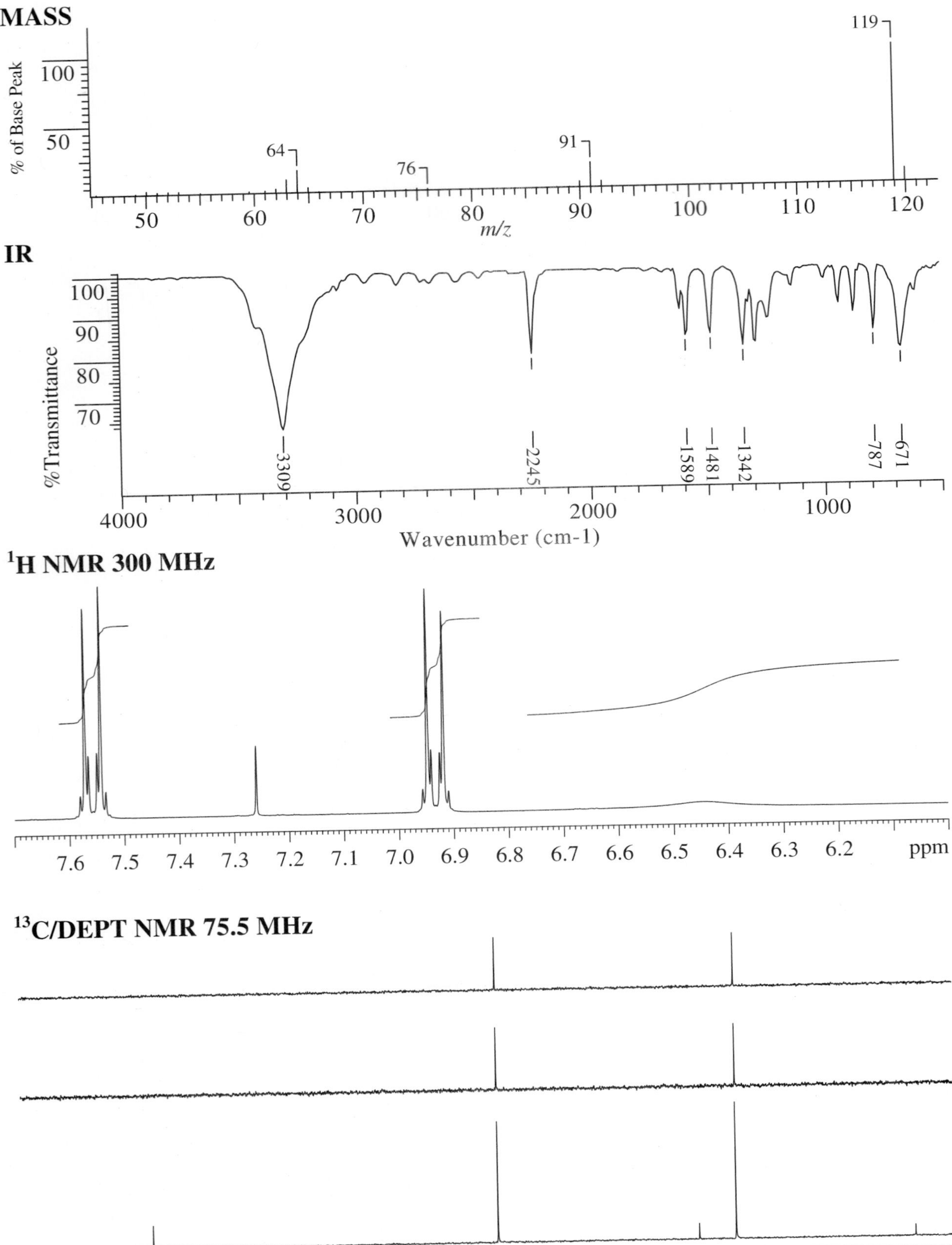

MASS
% of Base Peak
100
50
64
76
91
119
50
60
70
80
90
100
110
120
m/z
IR
%Transmittance
100
90
80
70
3309
2245
1589
1481
1342
787
671
4000
3000
2000
1000
Wavenumber (cm-1)
¹H NMR 300 MHz
7.6
7.5
7.4
7.3
7.2
7.1
7.0
6.9
6.8
6.7
6.6
6.5
6.4
6.3
6.2
ppm
¹³C/DEPT NMR 75.5 MHz
165
160
155
150
145
140
135
130
125
120
115
110
ppm

Problem 8.18

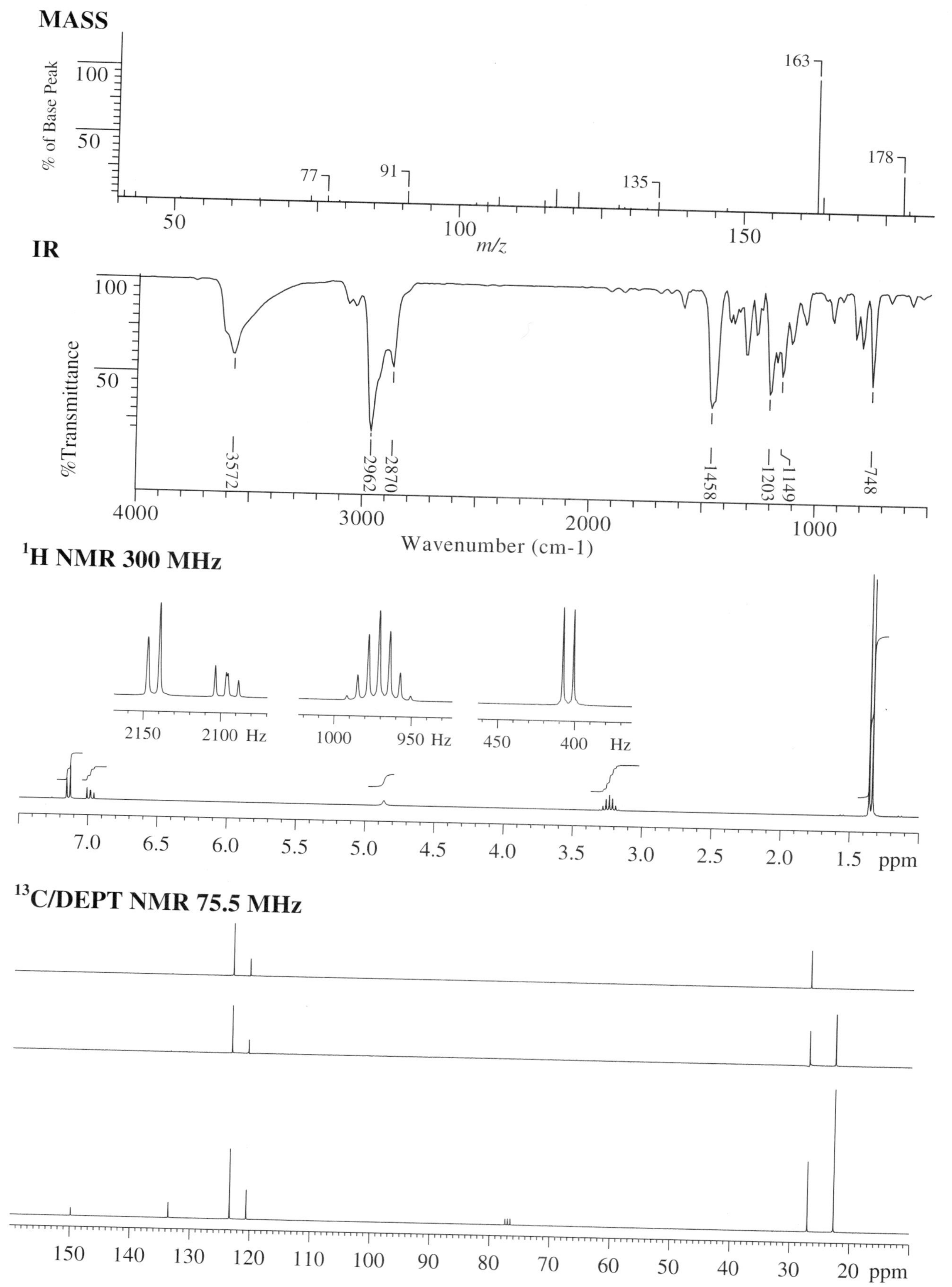
MASS
% of Base Peak
100
50
77
91
135
163
178
50
100
150
m/z
IR
%Transmittance
100
50
3572
2962
2870
1458
1203
1149
748
4000
3000
2000
1000
Wavenumber (cm-1)
1H NMR 300 MHz
2150
2100 Hz
1000
950 Hz
450
400 Hz
7.0
6.5
6.0
5.5
5.0
4.5
4.0
3.5
3.0
2.5
2.0
1.5 ppm
13C/DEPT NMR 75.5 MHz
150
140
130
120
110
100
90
80
70
60
50
40
30
20 ppm

Problem 8.19

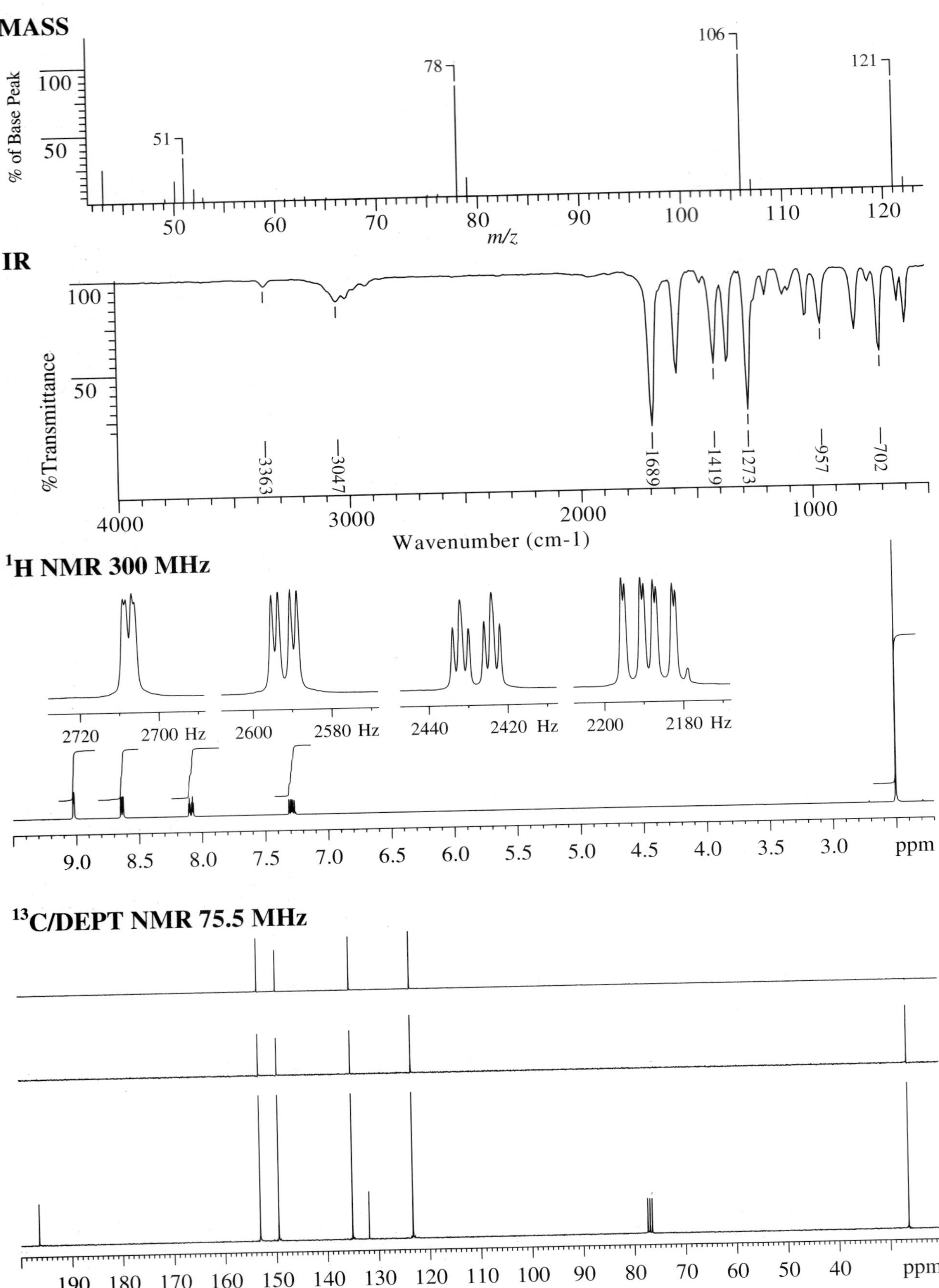
MASS
% of Base Peak
100
50
51
78
106
121
50
60
70
80
90
100
110
120
m/z
IR
% Transmittance
100
50
3363
3047
1689
1419
1273
957
702
4000
3000
2000
1000
Wavenumber (cm-1)
¹H NMR 300 MHz
2720
2700 Hz
2600
2580 Hz
2440
2420 Hz
2200
2180 Hz
9.0
8.5
8.0
7.5
7.0
6.5
6.0
5.5
5.0
4.5
4.0
3.5
3.0
ppm
¹³C/DEPT NMR 75.5 MHz
190
180
170
160
150
140
130
120
110
100
90
80
70
60
50
40
ppm

Problem 8.20

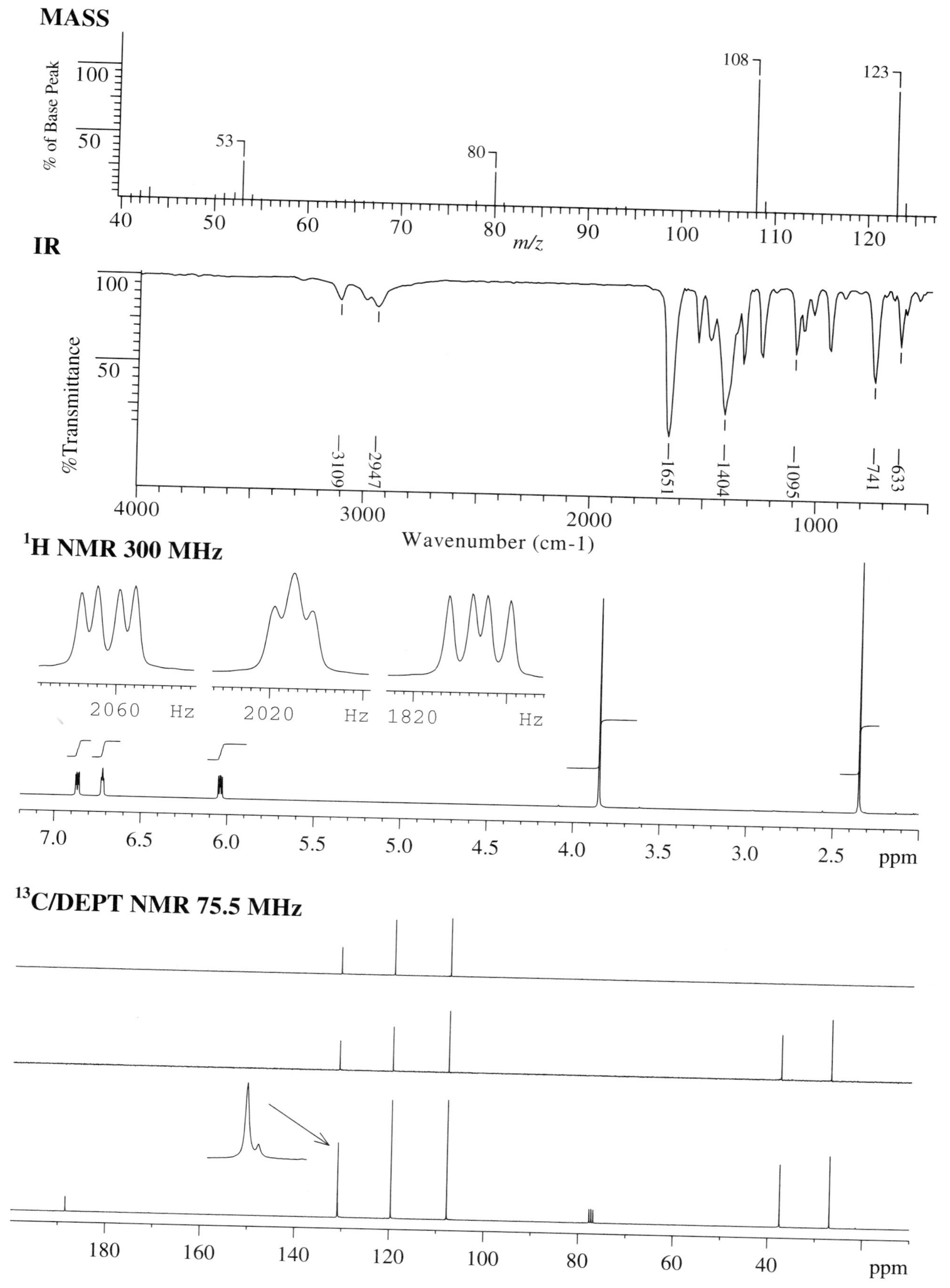

MASS
% of Base Peak
100
50
53
80
108
123
40
50
60
70
80
m/z
90
100
110
120
IR
100
50
%Transmittance
3109
2947
1651
1404
1095
741
633
4000
3000
2000
1000
Wavenumber (cm-1)
1H NMR 300 MHz
2060 Hz
2020 Hz
1820 Hz
7.0
6.5
6.0
5.5
5.0
4.5
4.0
3.5
3.0
2.5
ppm
13C/DEPT NMR 75.5 MHz
180
160
140
120
100
80
60
40
ppm

Problem 8.21

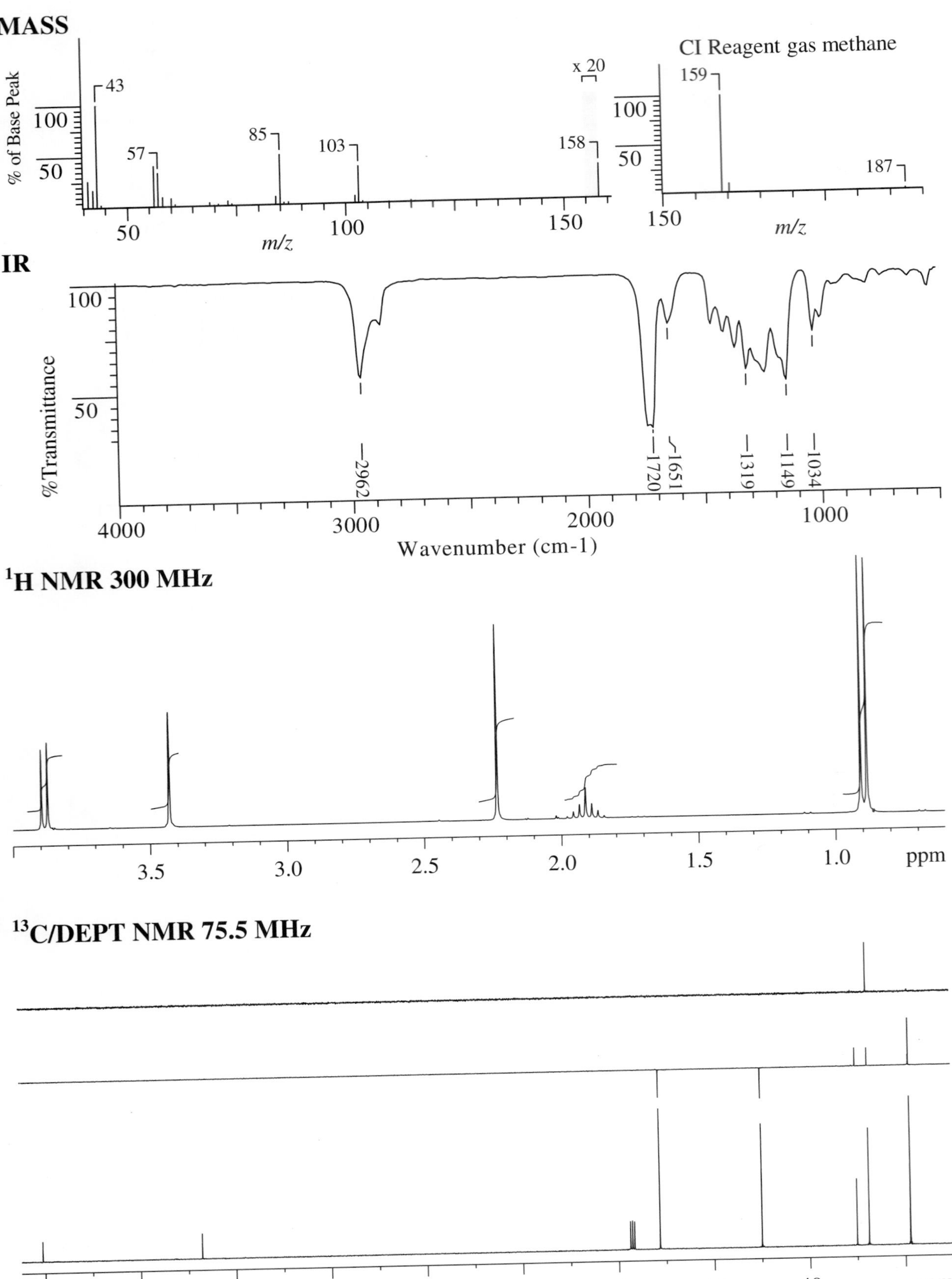

MASS
% of Base Peak
100
50
43
57
85
103
158
x 20
50
100
150
m/z
CI Reagent gas methane
159
187
150
m/z
IR
%Transmittance
100
50
2962
1720
1651
1319
1149
1034
4000
3000
2000
1000
Wavenumber (cm-1)
¹H NMR 300 MHz
3.5
3.0
2.5
2.0
1.5
1.0
ppm
¹³C/DEPT NMR 75.5 MHz
200
180
160
140
120
100
80
60
40
ppm

Problem 8.22

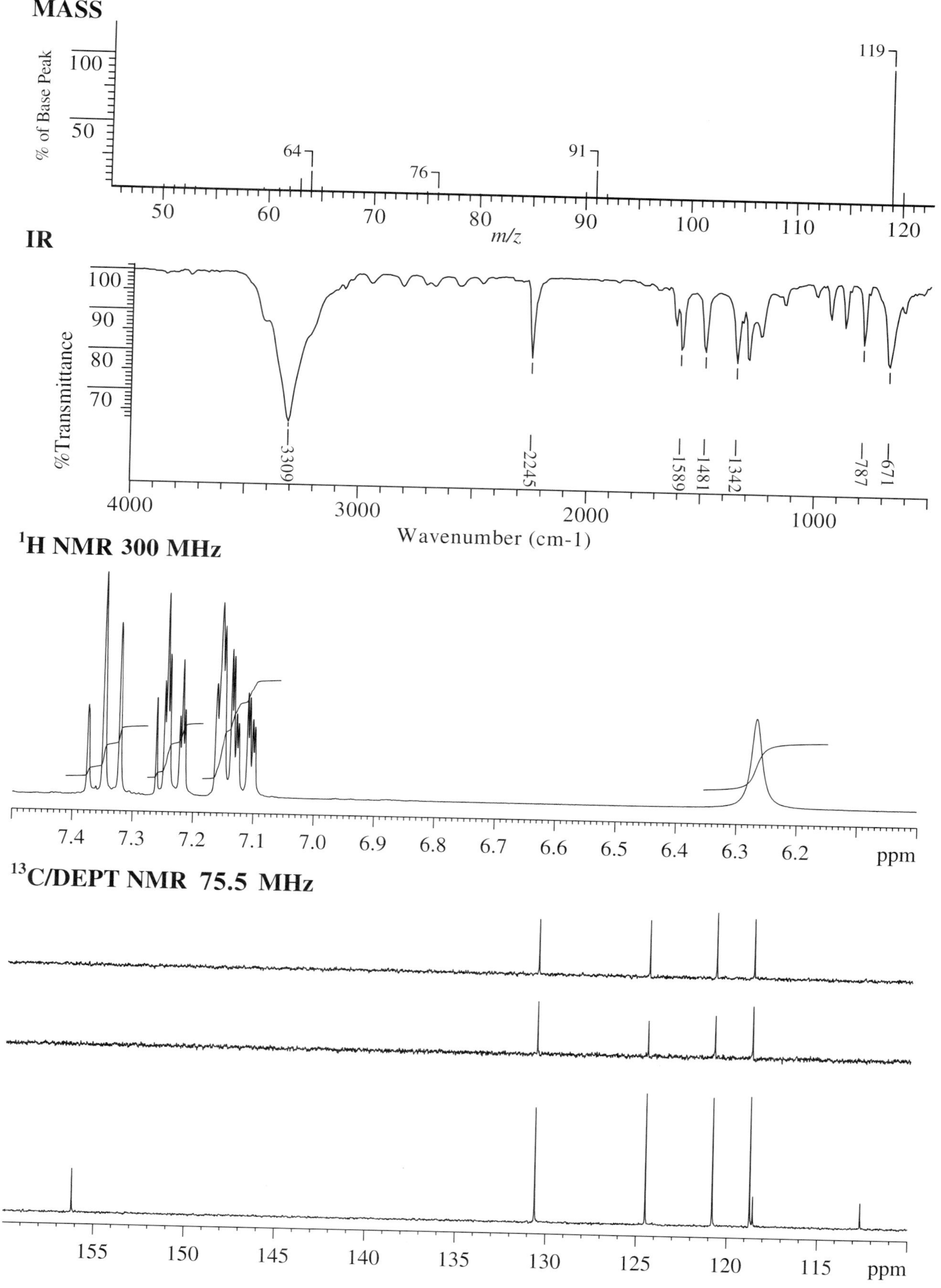
MASS
% of Base Peak
100
50
119
64
76
91
50
60
70
80
90
100
110
120
m/z
IR
%Transmittance
100
90
80
70
3309
2245
1589
1481
1342
787
671
4000
3000
2000
1000
Wavenumber (cm-1)
1H NMR 300 MHz
7.4
7.3
7.2
7.1
7.0
6.9
6.8
6.7
6.6
6.5
6.4
6.3
6.2
ppm
13C/DEPT NMR 75.5 MHz
155
150
145
140
135
130
125
120
115
ppm

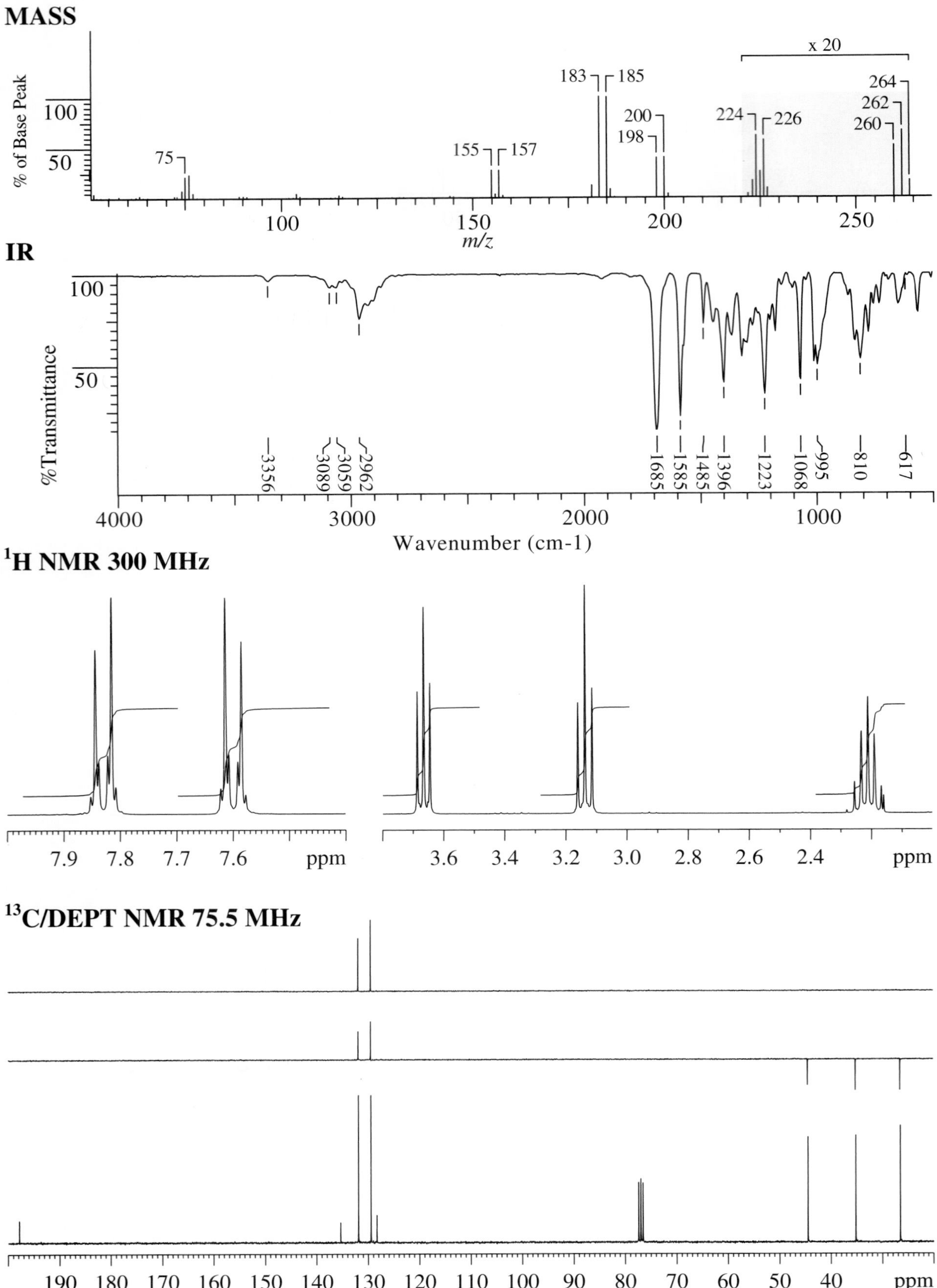
MASS
% of Base Peak
100
50
75
155
157
183
185
198
200
x 20
224
226
260
262
264
100
150
200
250
m/z
IR
%Transmittance
100
50
3356
3089
3059
2962
1685
1585
1485
1396
1223
1068
995
810
617
4000
3000
2000
1000
Wavenumber (cm-1)
1H NMR 300 MHz
7.9
7.8
7.7
7.6
ppm
3.6
3.4
3.2
3.0
2.8
2.6
2.4
ppm
13C/DEPT NMR 75.5 MHz
190
180
170
160
150
140
130
120
110
100
90
80
70
60
50
40
ppm

Problem 8.24

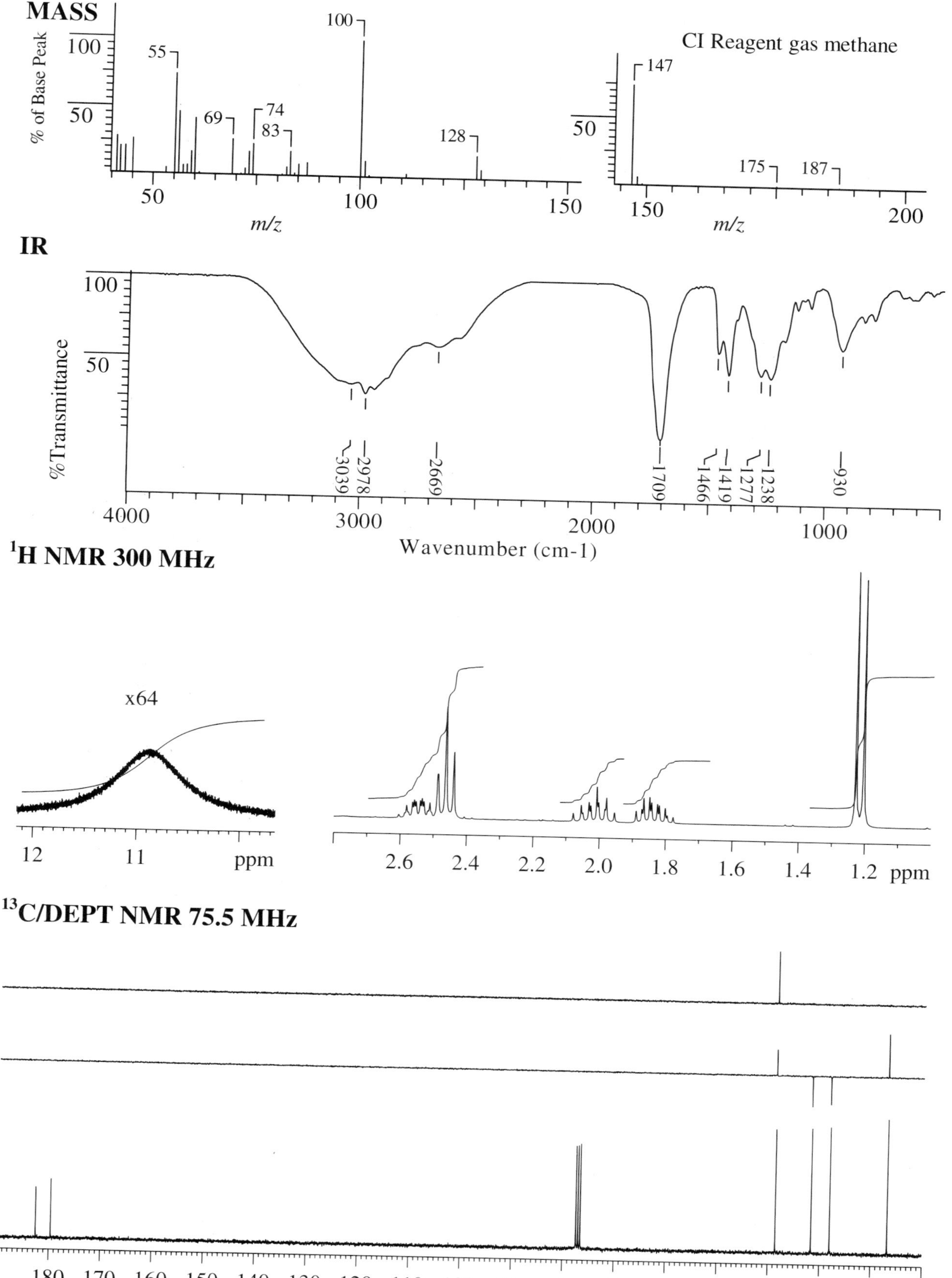
MASS
% of Base Peak
100
50
55
69
74
83
100
128
50
100
150
m/z
CI Reagent gas methane
147
50
175
187
150
m/z
200
IR
100
50
% Transmittance
3039
2978
2669
1709
1466
1419
1277
1238
930
4000
3000
2000
1000
Wavenumber (cm-1)
1H NMR 300 MHz
x64
12
11
ppm
2.6
2.4
2.2
2.0
1.8
1.6
1.4
1.2 ppm
13C/DEPT NMR 75.5 MHz
180
170
160
150
140
130
120
110
100
90
80
70
60
50
40
30
ppm

Problem 8.25

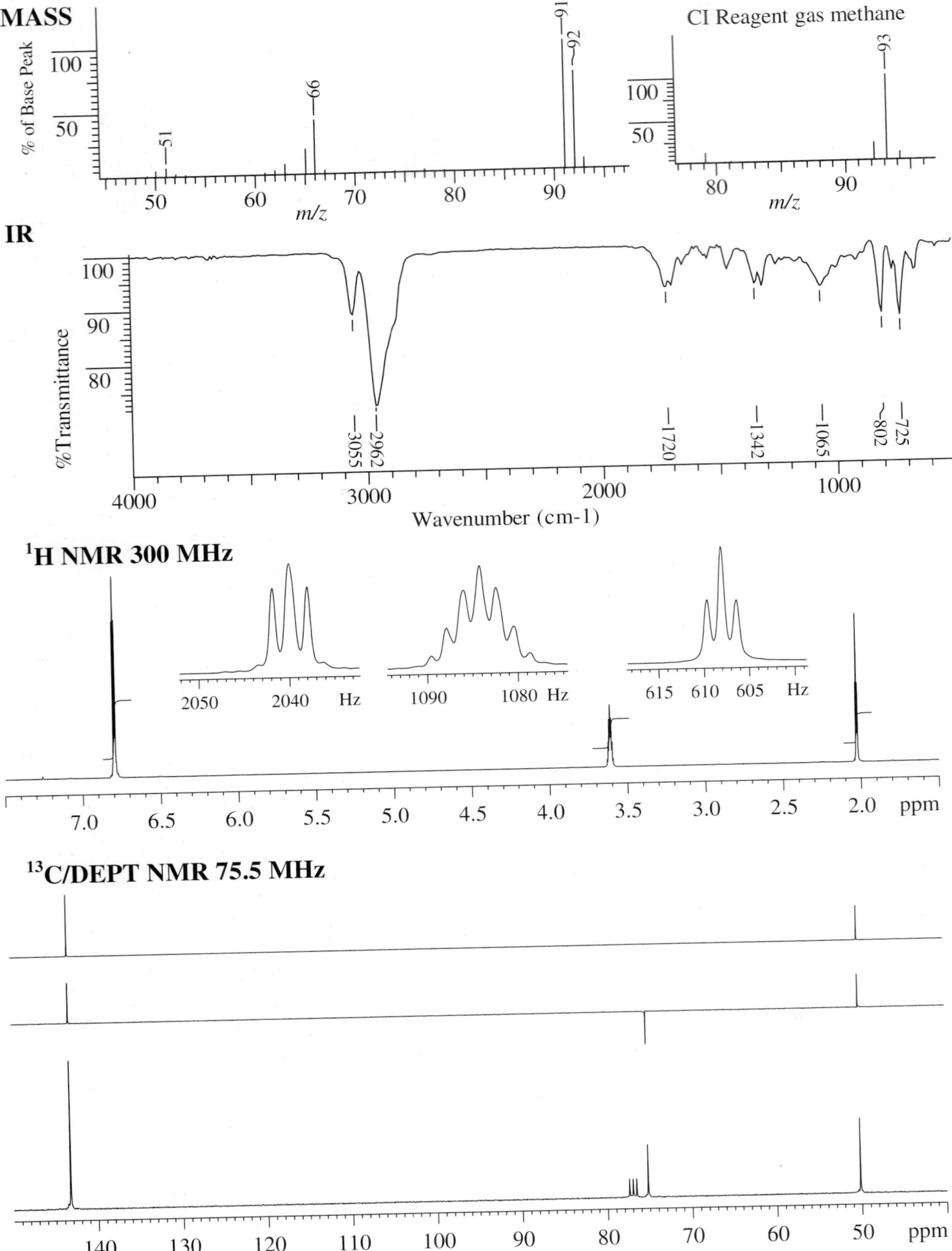

MASS
% of Base Peak
100
50
51
66
91
92
50
60
70
80
90
m/z
CI Reagent gas methane
100
50
93
80
90
m/z
IR
%Transmittance
100
90
80
3055
2962
1720
1342
1065
802
725
4000
3000
2000
1000
Wavenumber (cm-1)
¹H NMR 300 MHz
2050
2040
Hz
1090
1080
Hz
615
610
605
Hz
7.0
6.5
6.0
5.5
5.0
4.5
4.0
3.5
3.0
2.5
2.0
ppm
¹³C/DEPT NMR 75.5 MHz
140
130
120
110
100
90
80
70
60
50
ppm

Problem 8.26

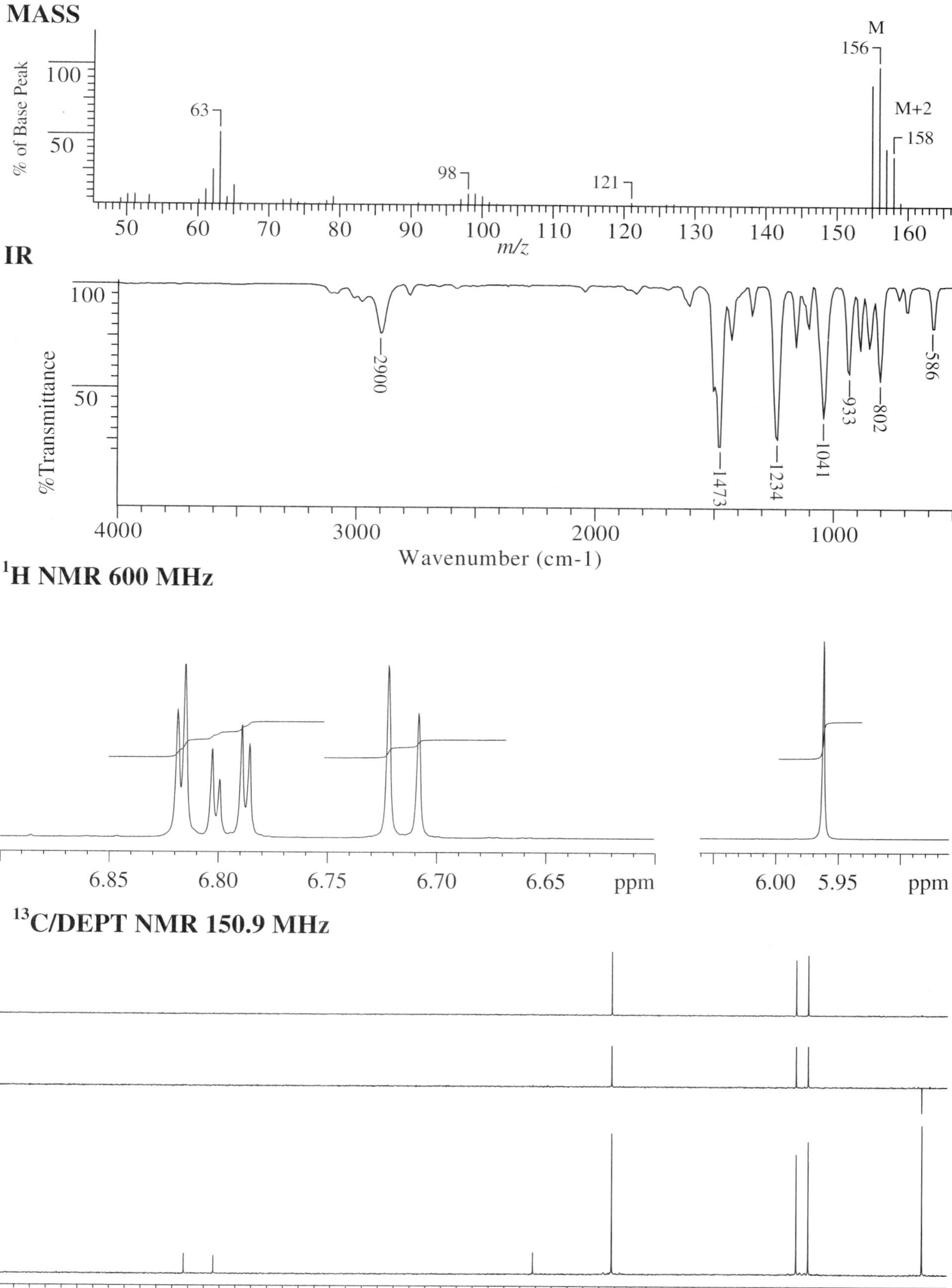
MASS
M
156
100
50
63
98
121
M+2
158
% of Base Peak
50
60
70
80
90
100
110
120
130
140
150
160
m/z
IR
100
50
%Transmittance
2900
1473
1234
1041
933
802
586
4000
3000
2000
1000
Wavenumber (cm-1)
¹H NMR 600 MHz
6.85
6.80
6.75
6.70
6.65
ppm
6.00
5.95
ppm
¹³C/DEPT NMR 150.9 MHz
155
150
145
140
135
130
125
120
115
110
105
ppm

MASS

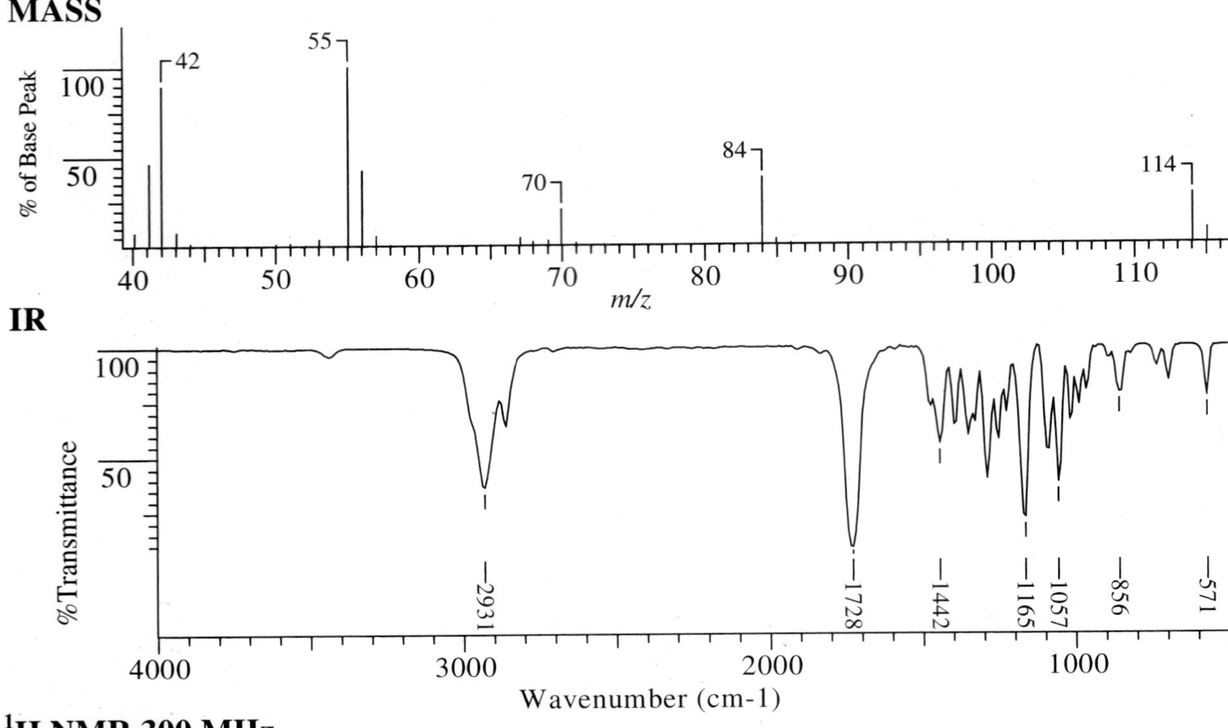
% of Base Peak
100
50
42
55
70
84
114
40
50
60
70
80
90
100
110
m/z
IR
100
50
%Transmittance
2931
1728
1442
1165
1057
856
571
4000
3000
2000
1000
Wavenumber (cm-1)

^{1}H NMR 300 MHz

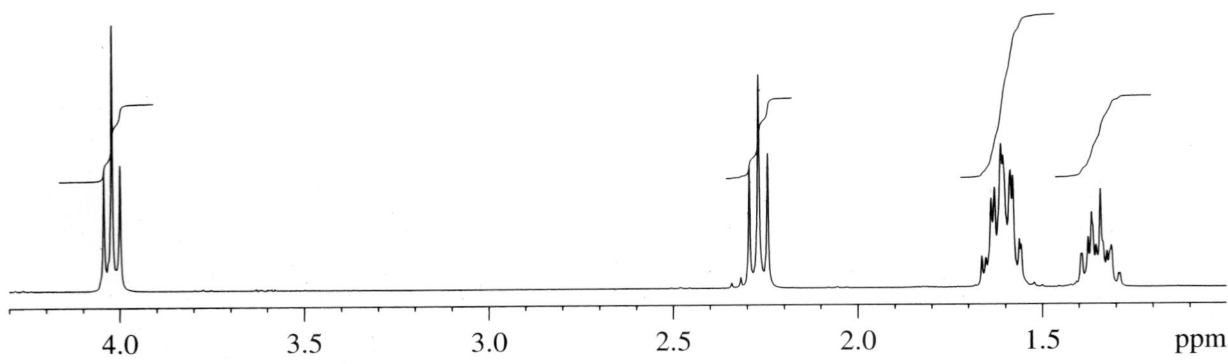
4.0
3.5
3.0
2.5
2.0
1.5
ppm

^{13}C/DEPT NMR 75.5 MHz

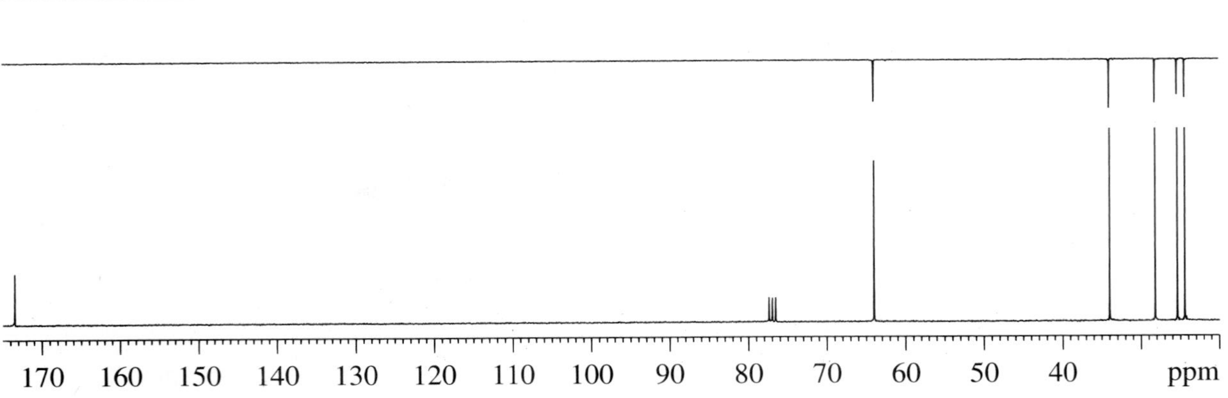
170
160
150
140
130
120
110
100
90
80
70
60
50
40
ppm

Problem 8.27B

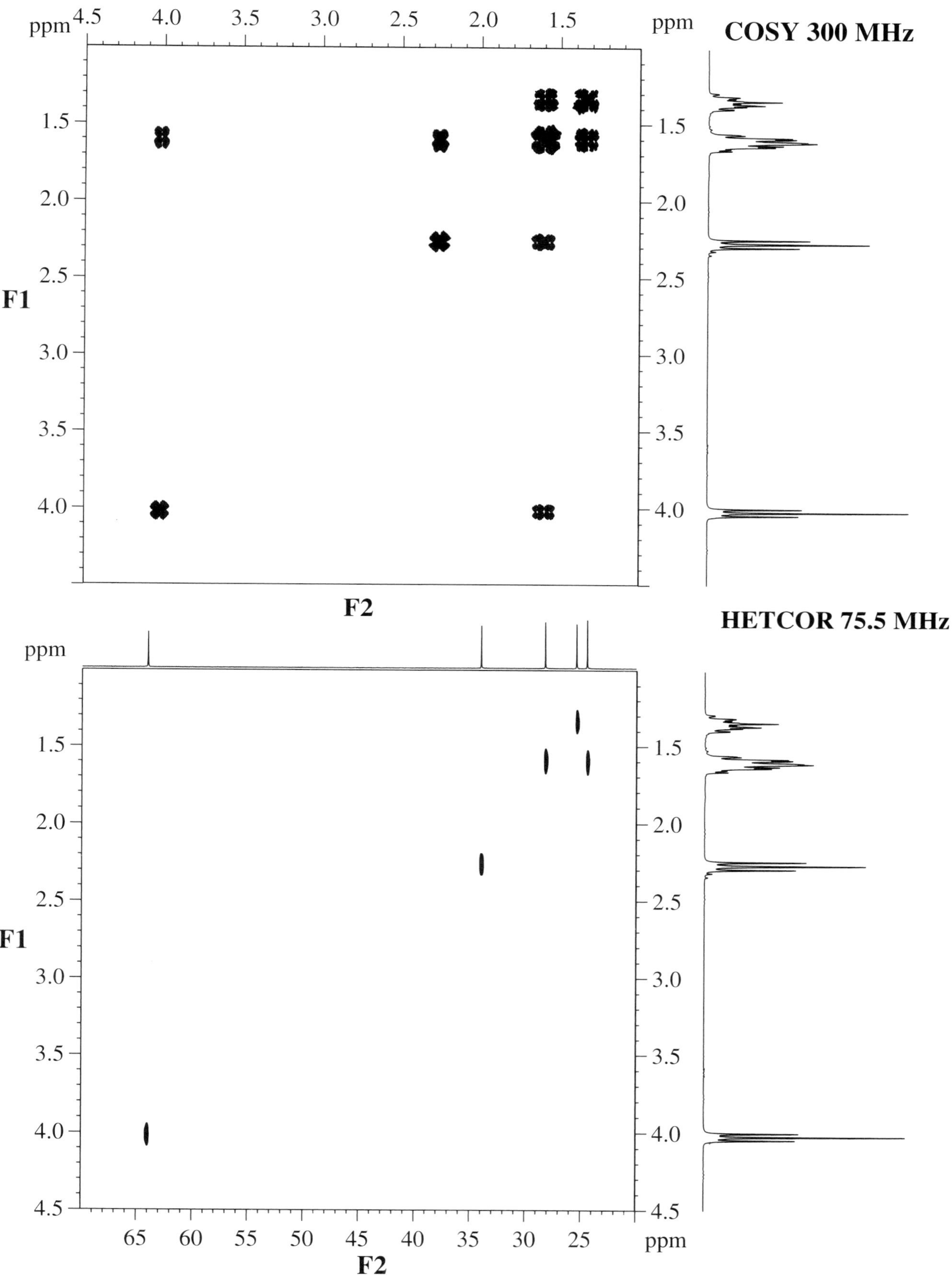

Problem 8.28A

MASS

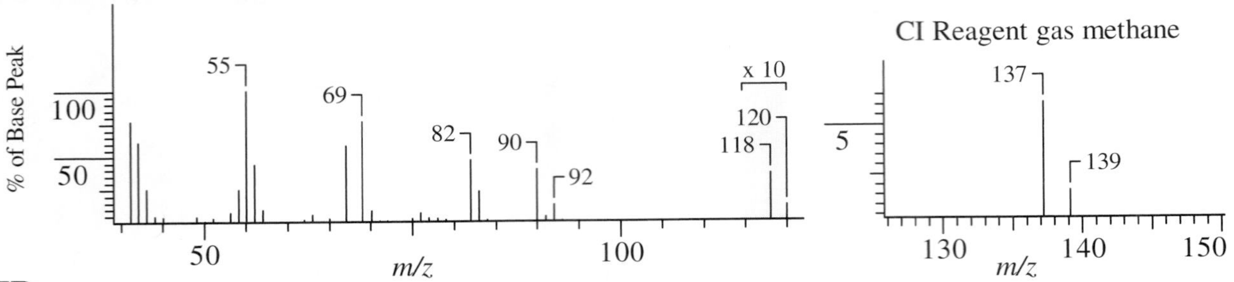

% of Base Peak
100
50
55
69
82
90
92
x 10
120
118
50
m/z
100
CI Reagent gas methane
137
139
5
130
140
150
m/z

IR

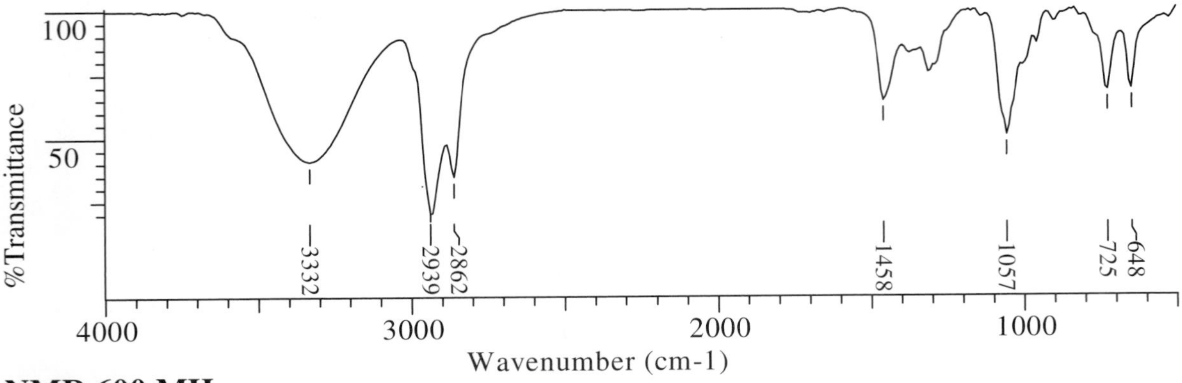

% Transmittance
100
50
3332
2939
2862
1458
1057
725
648
4000
3000
2000
1000
Wavenumber (cm-1)

^{1}H NMR 600 MHz

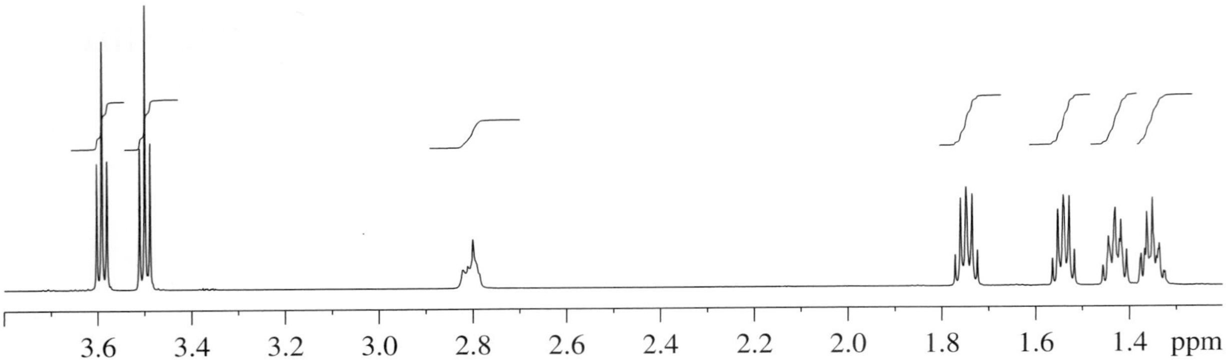

3.6
3.4
3.2
3.0
2.8
2.6
2.4
2.2
2.0
1.8
1.6
1.4
ppm

^{13}C/DEPT NMR 150.9 MHz

60
55
50
45
40
35
30
25
ppm

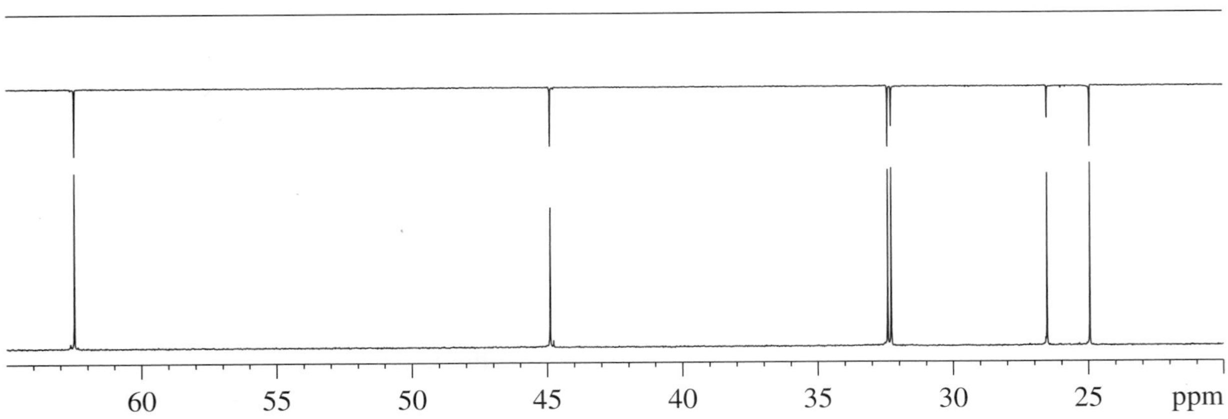

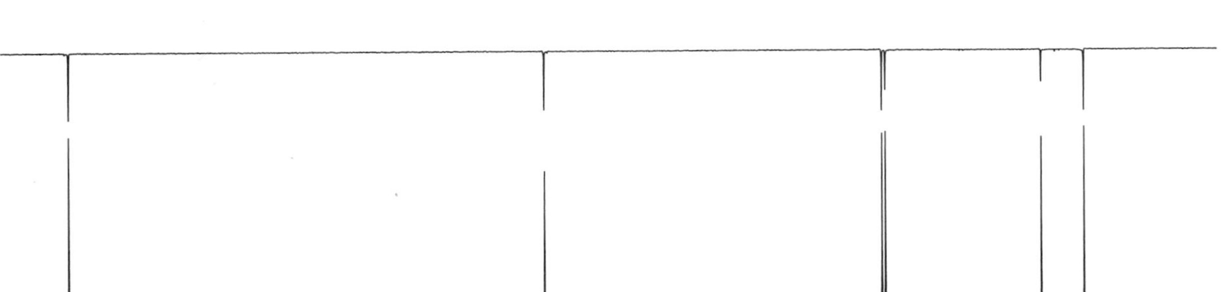

Problem 8.28B

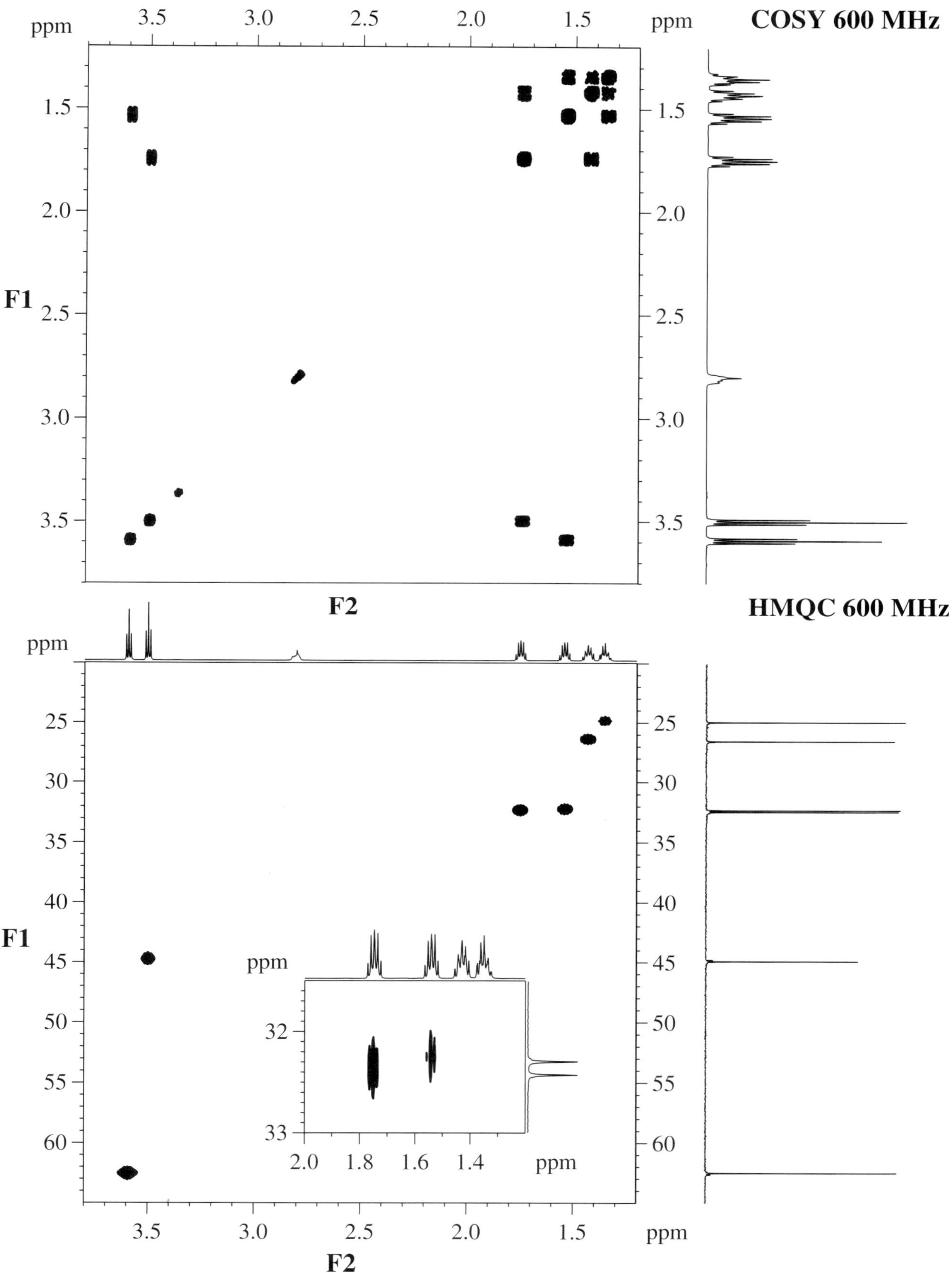

Problem 8.29A

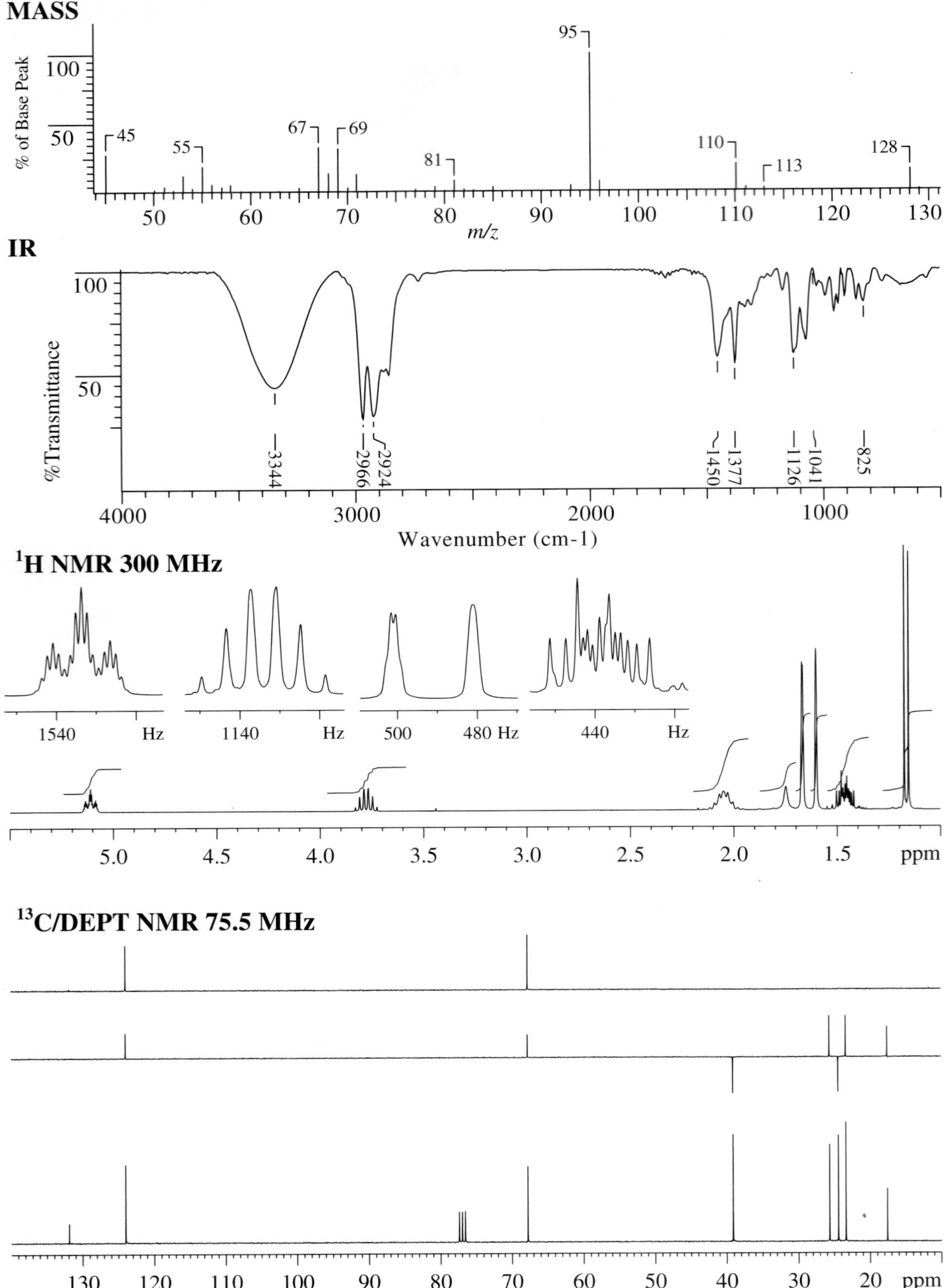

Problem 8.29B

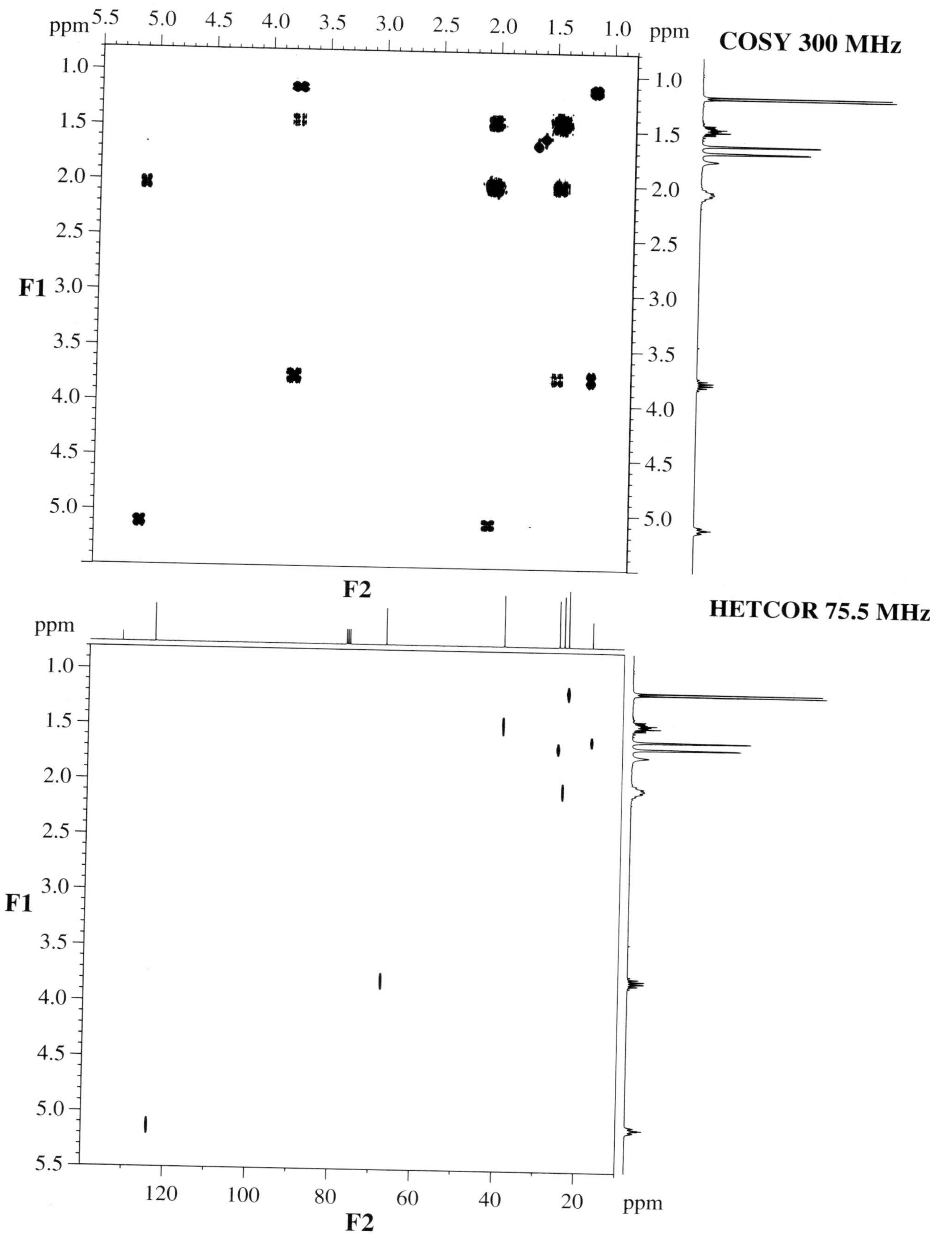

Problem 8.30A

MASS

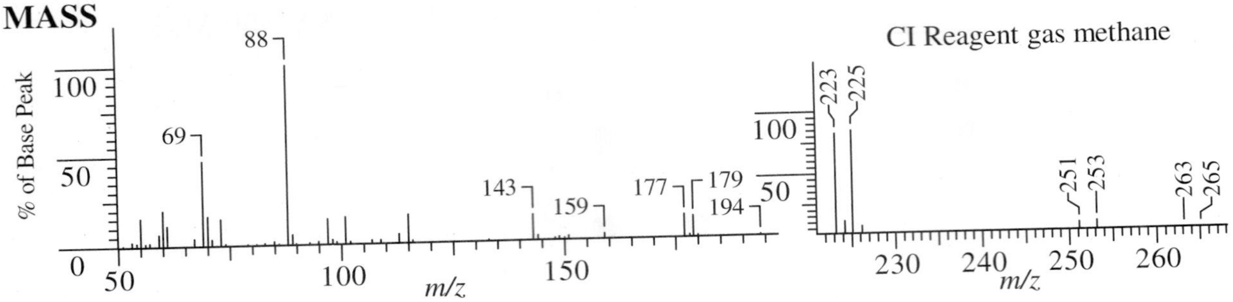

CI Reagent gas methane

% of Base Peak

0 50 100 m/z 150

88 69 143 159 177 179 194

100 50

223 225 251 253 263 265

230 240 m/z 250 260

IR

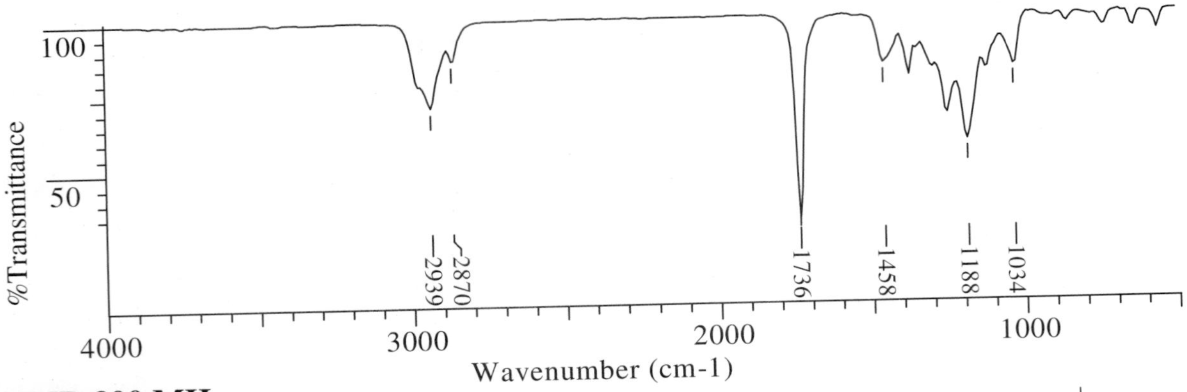

%Transmittance

100 50

2939 2870 1736 1458 1188 1034

4000 3000 2000 1000

Wavenumber (cm-1)

^{1}H NMR 300 MHz

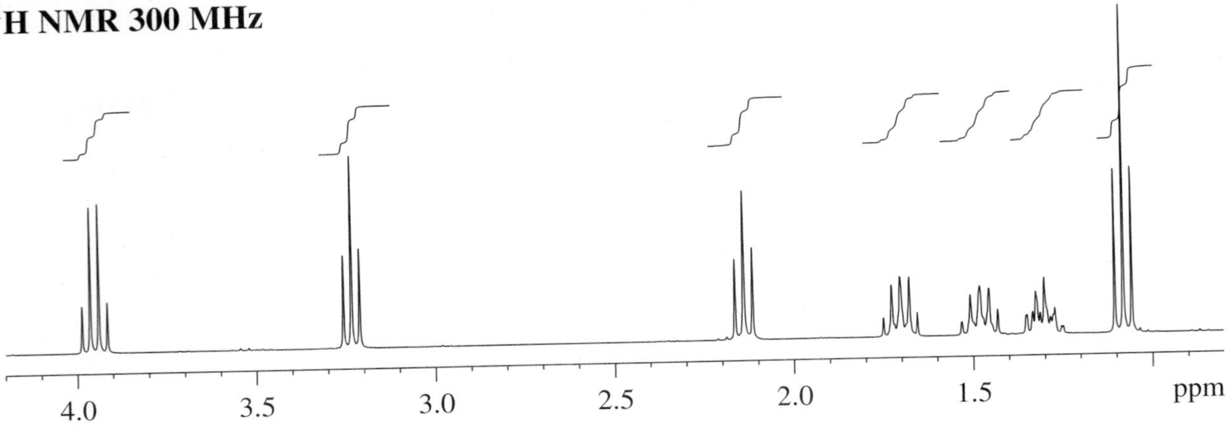

4.0 3.5 3.0 2.5 2.0 1.5 ppm

^{13}C/DEPT NMR 75.5 MHz

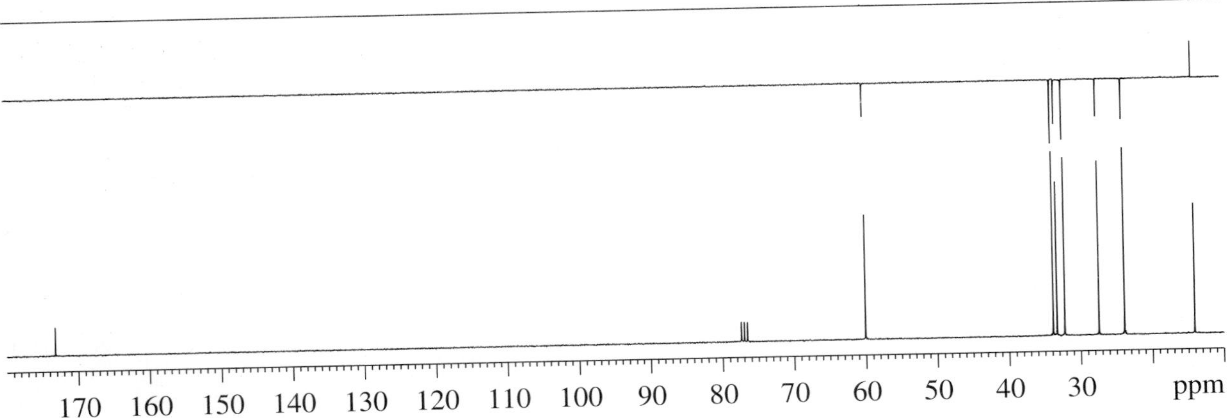

Problem 8.30B

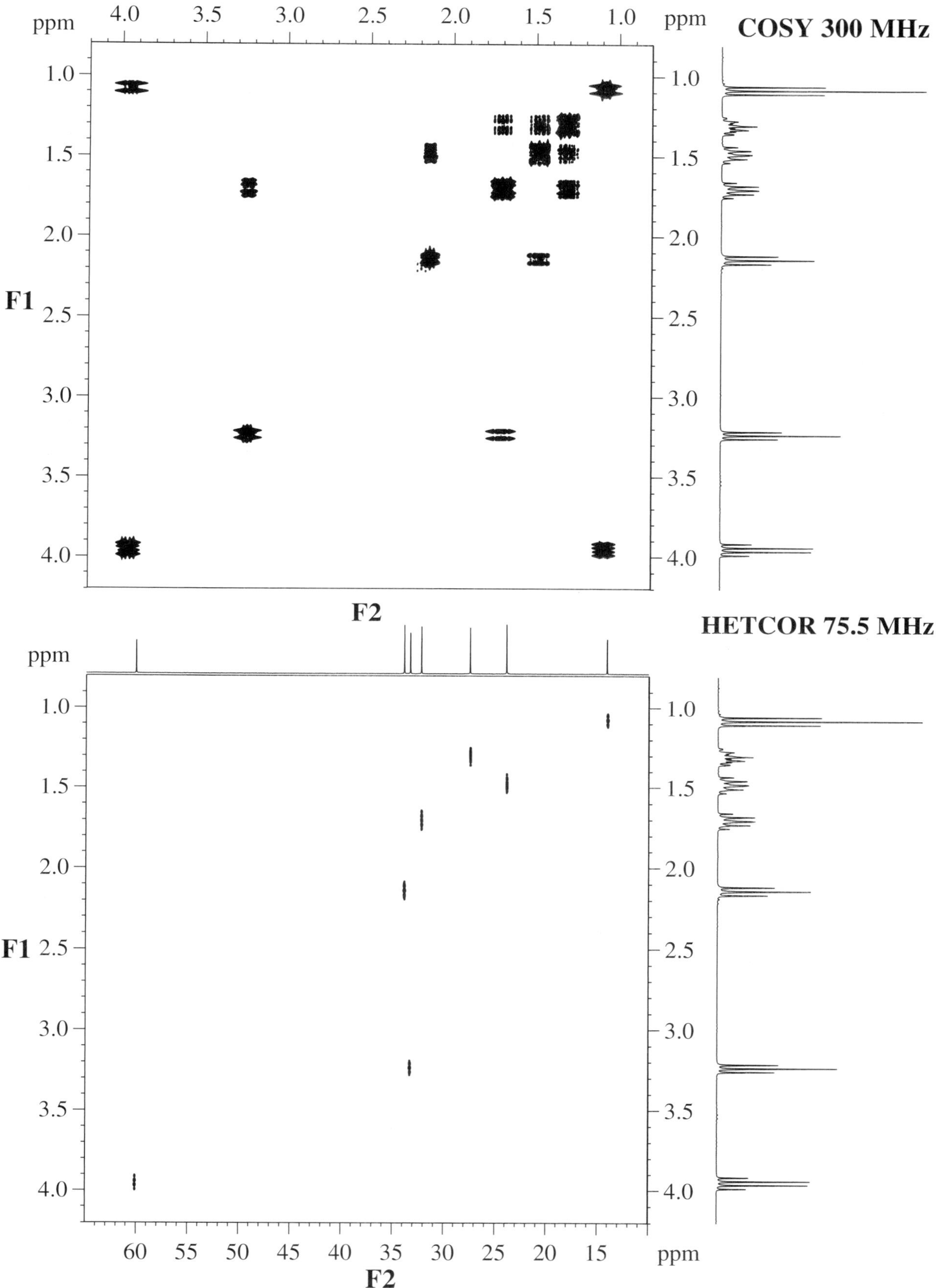

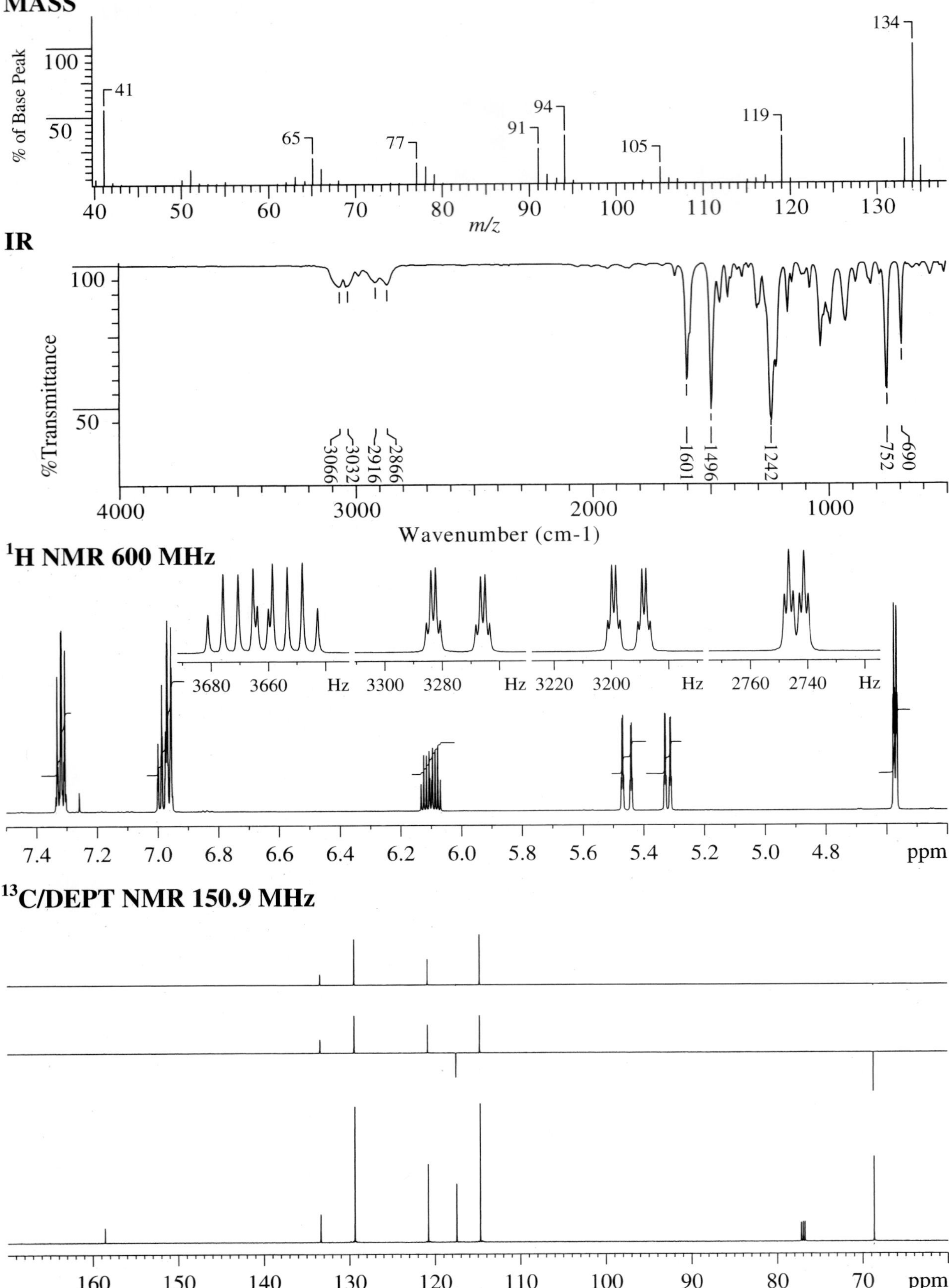
MASS
% of Base Peak
100
50
41
65
77
91
94
105
119
134
40 50 60 70 80 90 100 110 120 130
m/z
IR
%Transmittance
100
50
3066
3032
2916
2866
1601
1496
1242
752
690
4000 3000 2000 1000
Wavenumber (cm-1)
¹H NMR 600 MHz
3680 3660 Hz
3300 3280 Hz
3220 3200 Hz
2760 2740 Hz
7.4 7.2 7.0 6.8 6.6 6.4 6.2 6.0 5.8 5.6 5.4 5.2 5.0 4.8 ppm
¹³C/DEPT NMR 150.9 MHz
160 150 140 130 120 110 100 90 80 70 ppm

Problem 8.31B

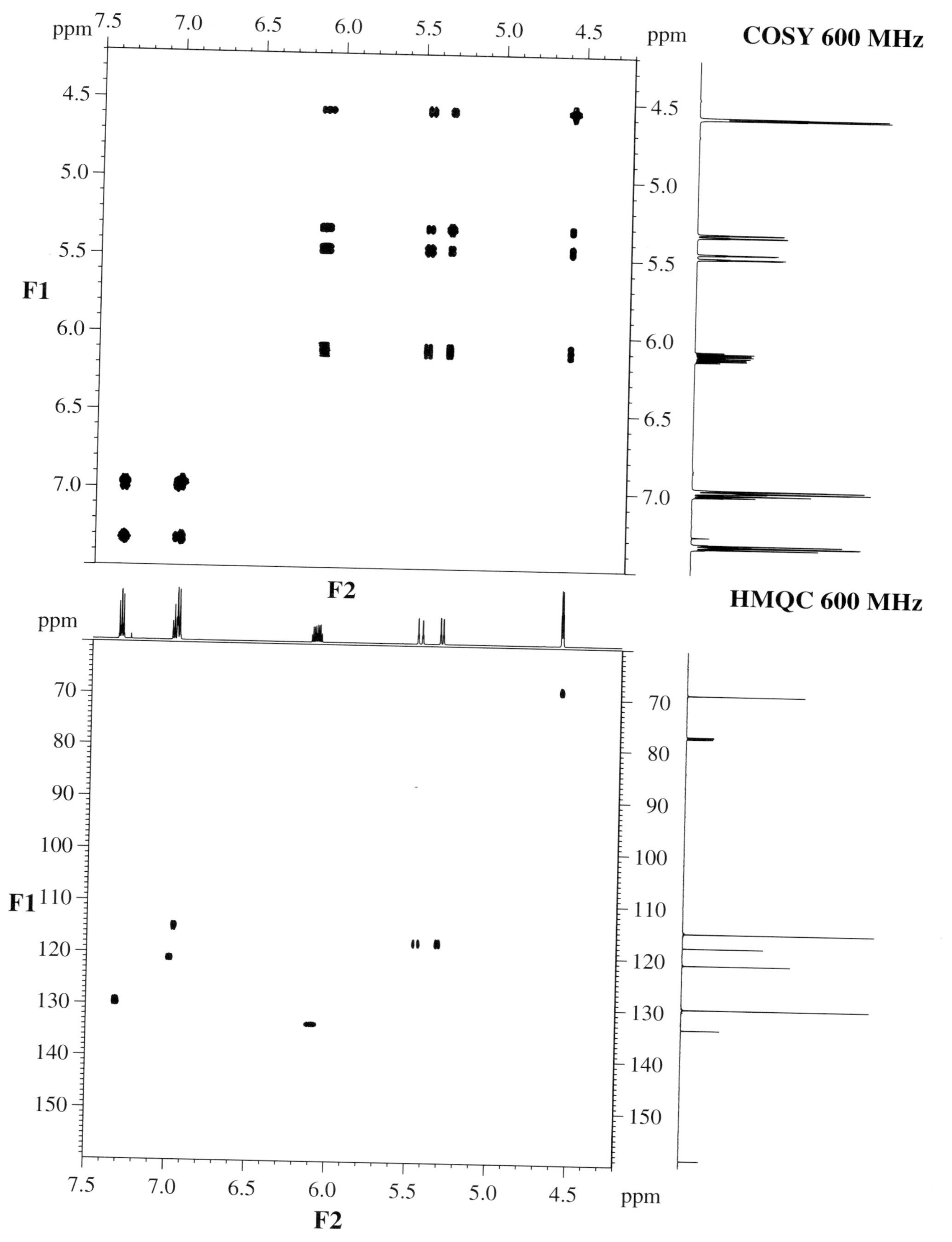

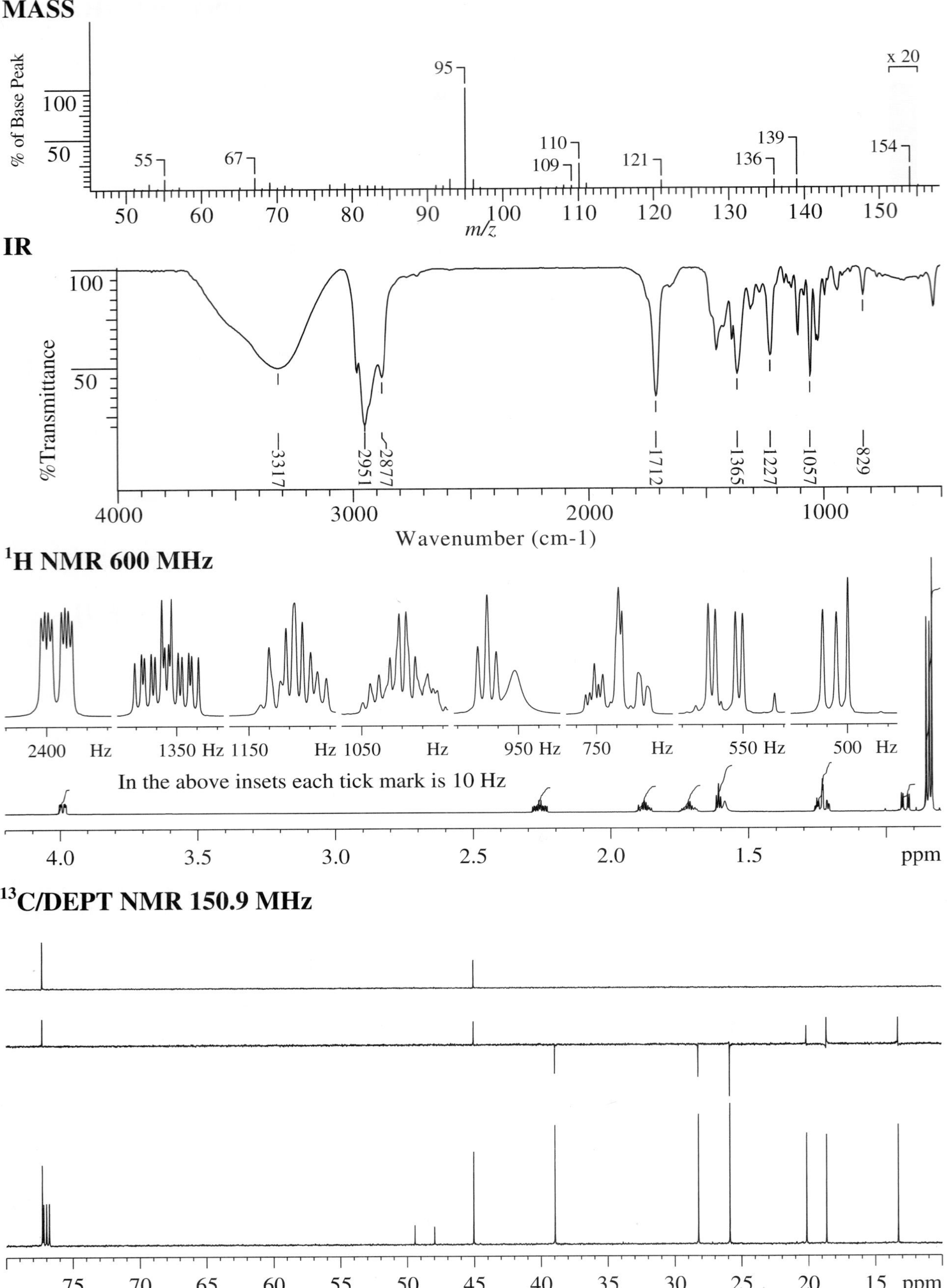
MASS
% of Base Peak
100
50
95
55
67
110
109
121
139
136
154
x 20
50
60
70
80
90
100
110
120
130
140
150
m/z
IR
%Transmittance
100
50
3317
2951
2877
1712
1365
1227
1057
829
4000
3000
2000
1000
Wavenumber (cm-1)
¹H NMR 600 MHz
2400 Hz
1350 Hz
1150 Hz
1050 Hz
950 Hz
750 Hz
550 Hz
500 Hz
In the above insets each tick mark is 10 Hz
4.0
3.5
3.0
2.5
2.0
1.5
ppm
¹³C/DEPT NMR 150.9 MHz
75
70
65
60
55
50
45
40
35
30
25
20
15 ppm

Problem 8.32B

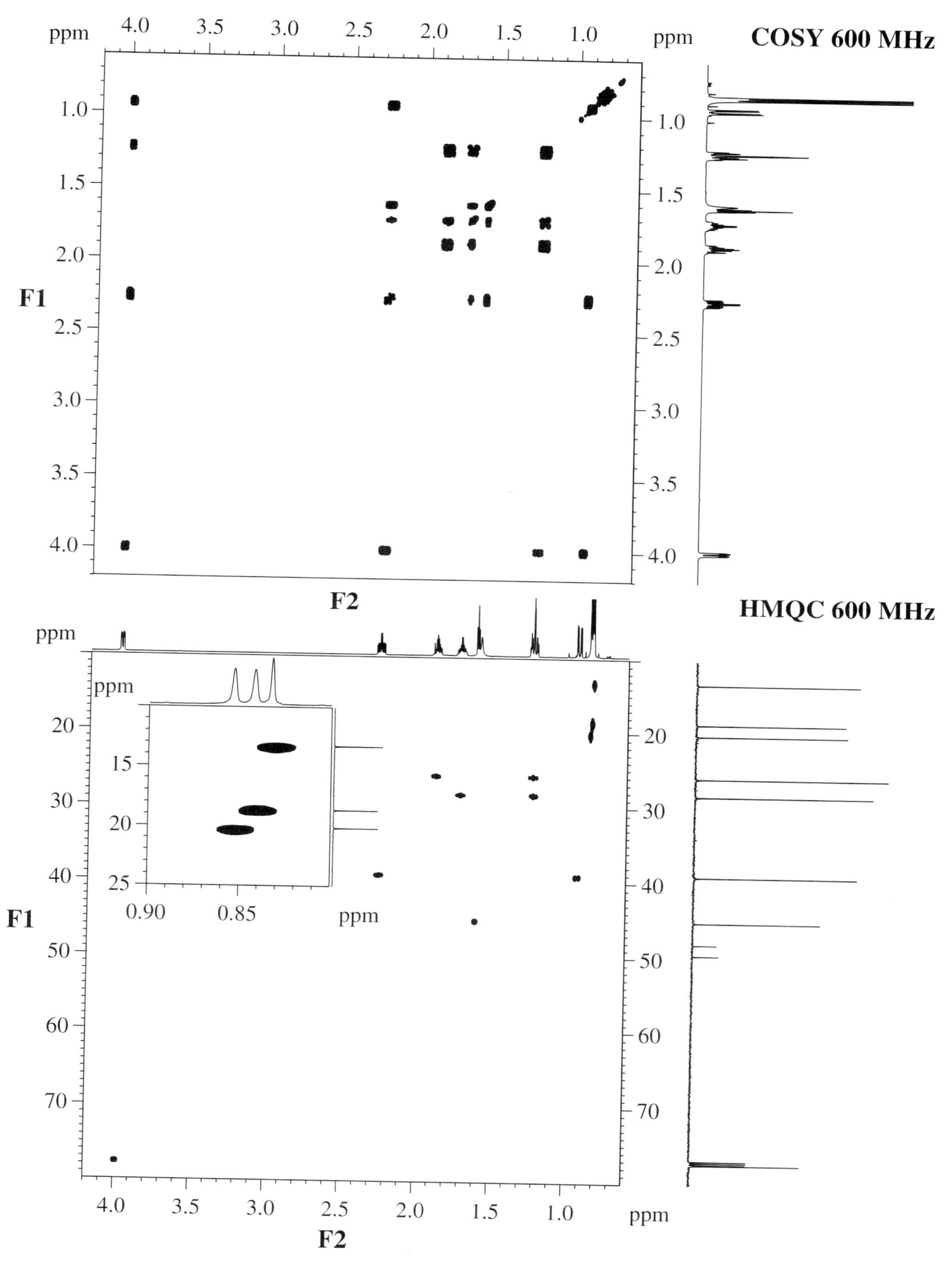

Problem 8.33A

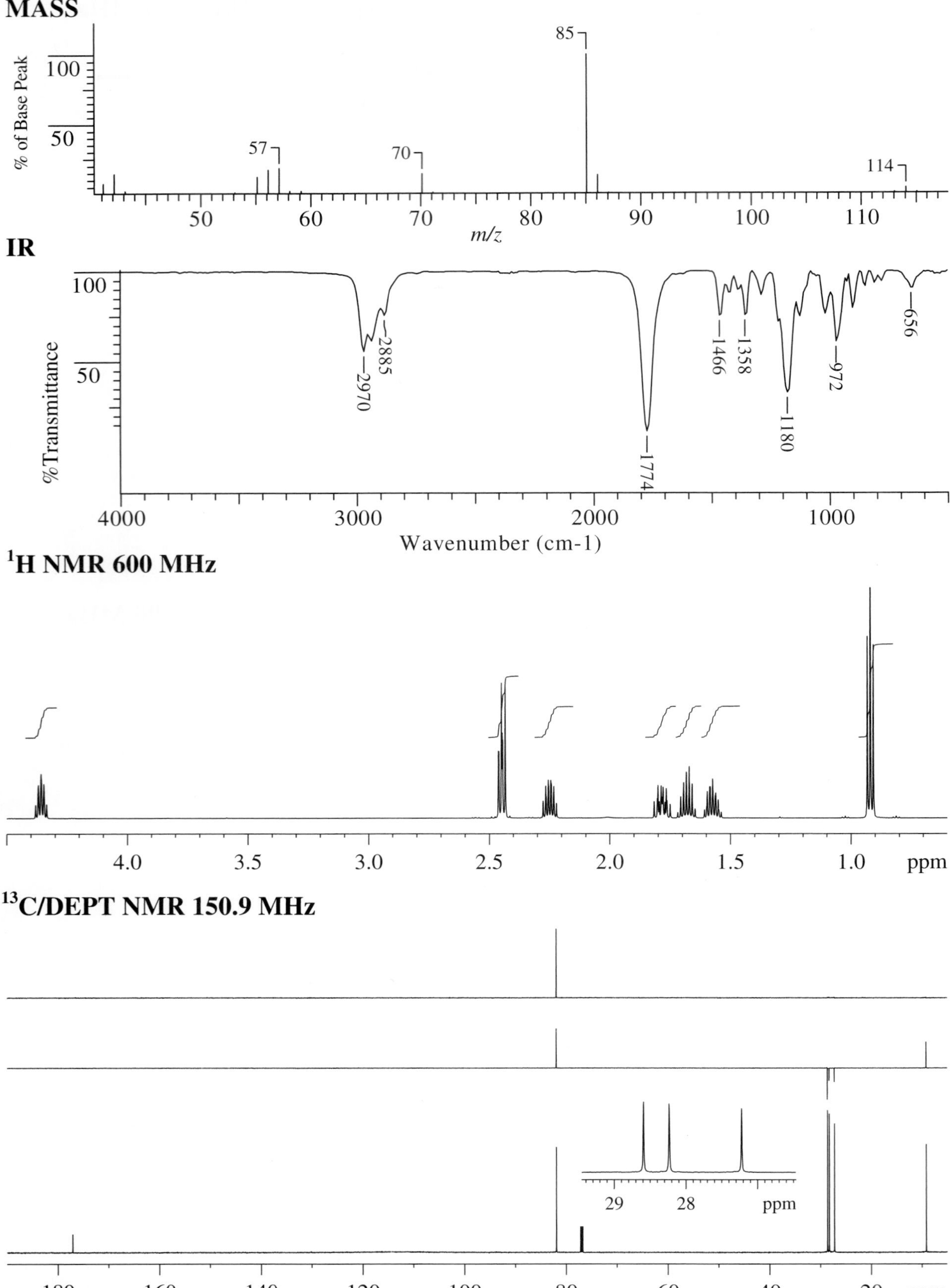

Problem 8.33B

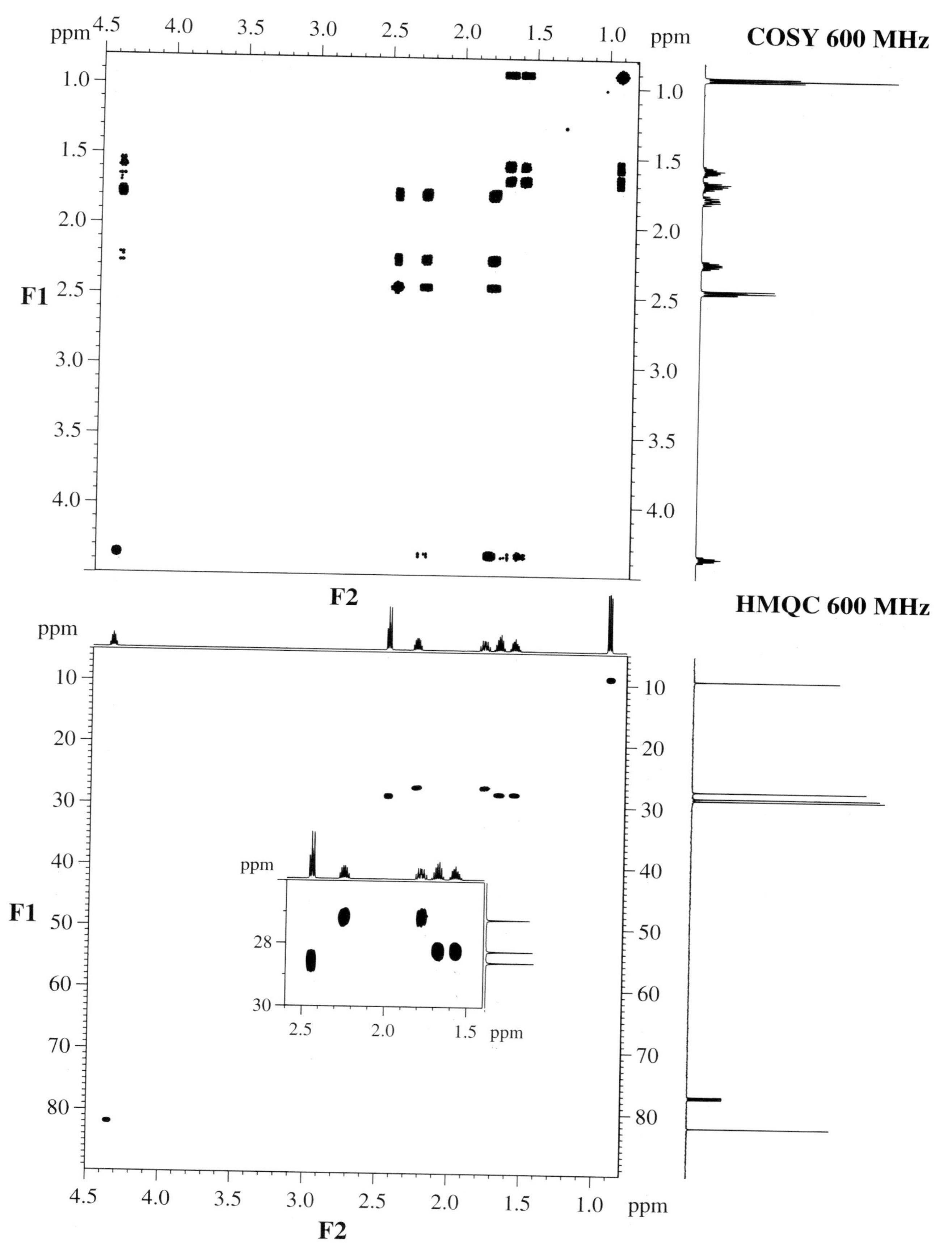

Problem 8.34A

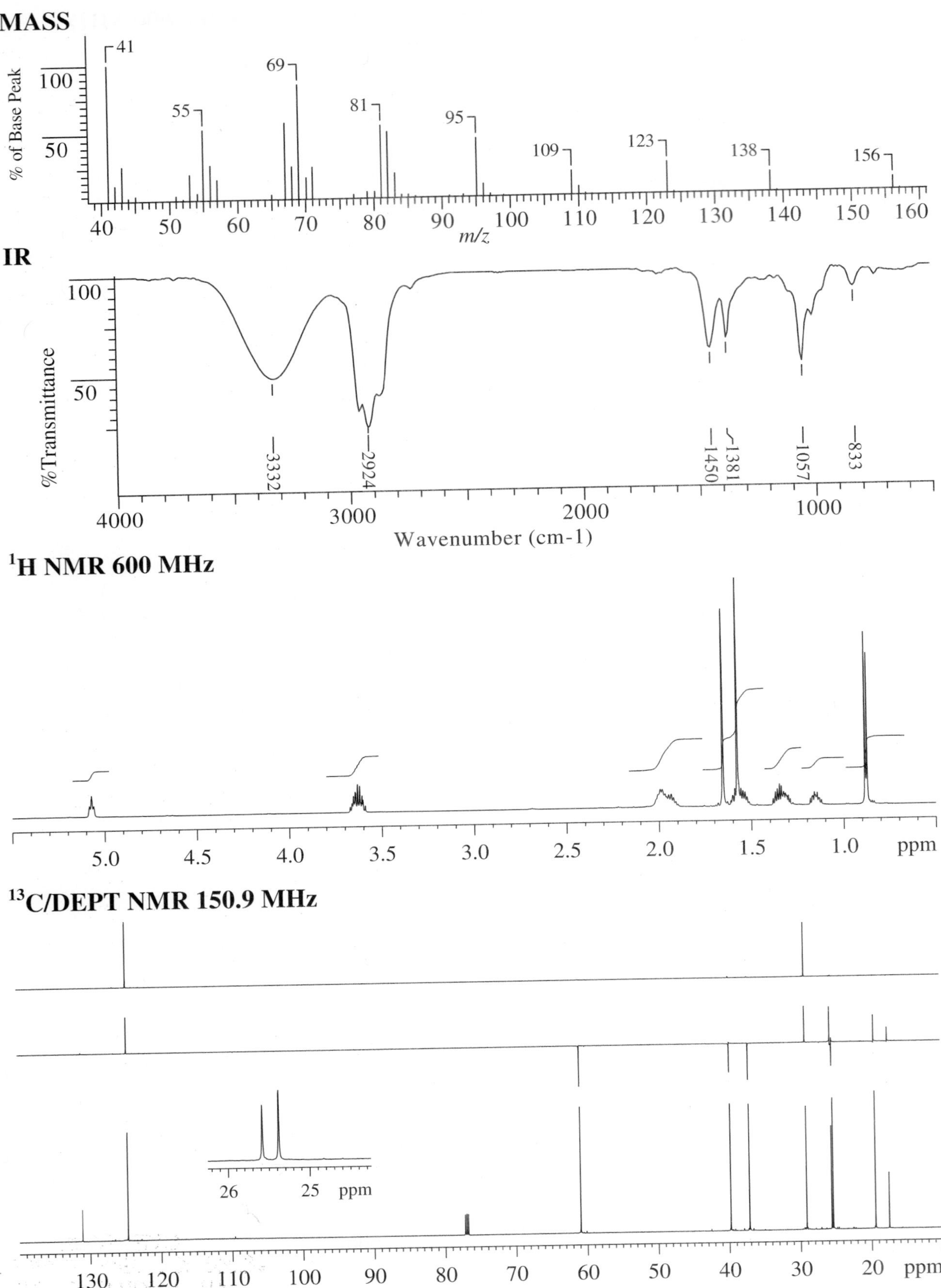

Problem 8.34B

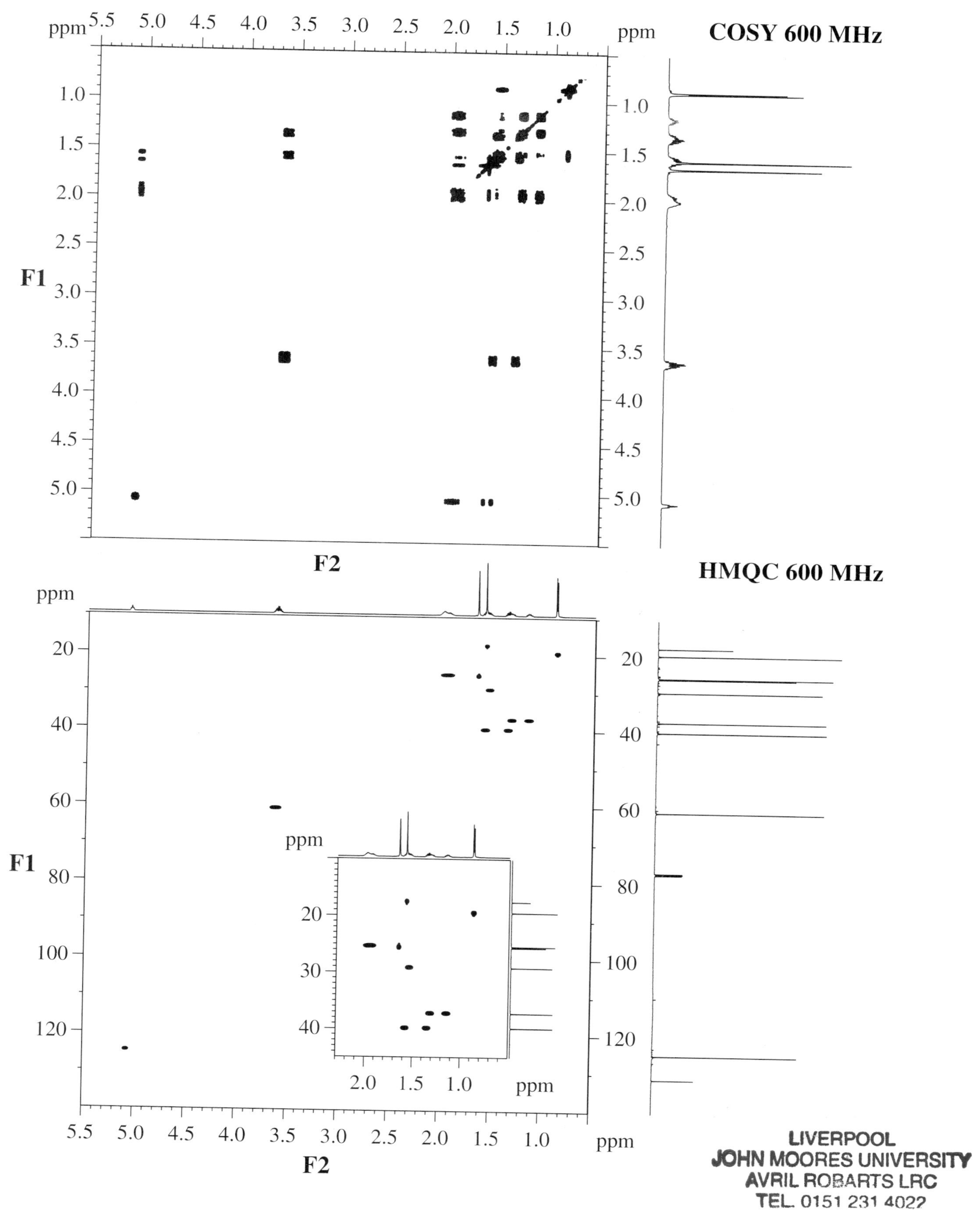

Problem 8.35A

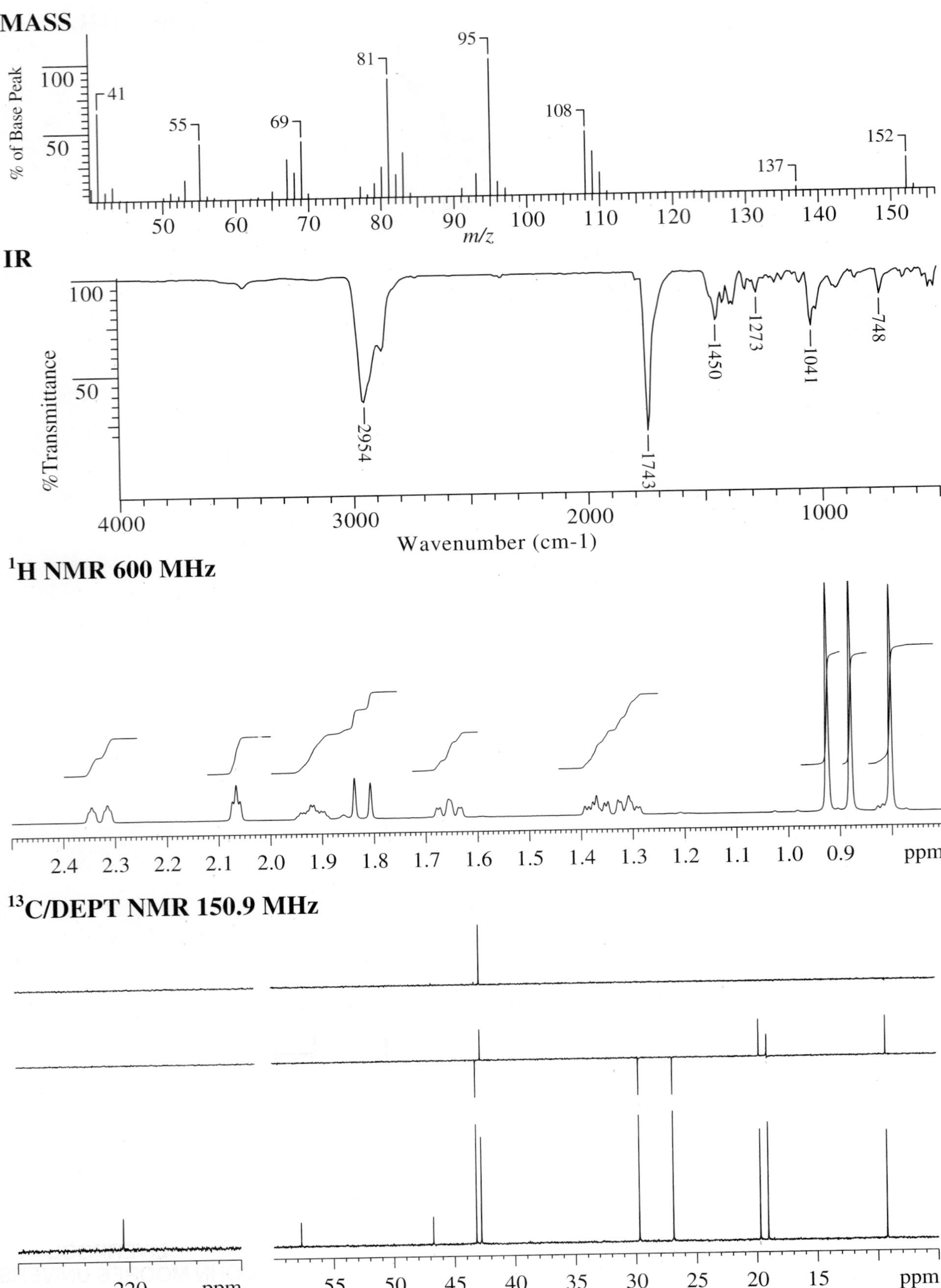

Problem 8.35B

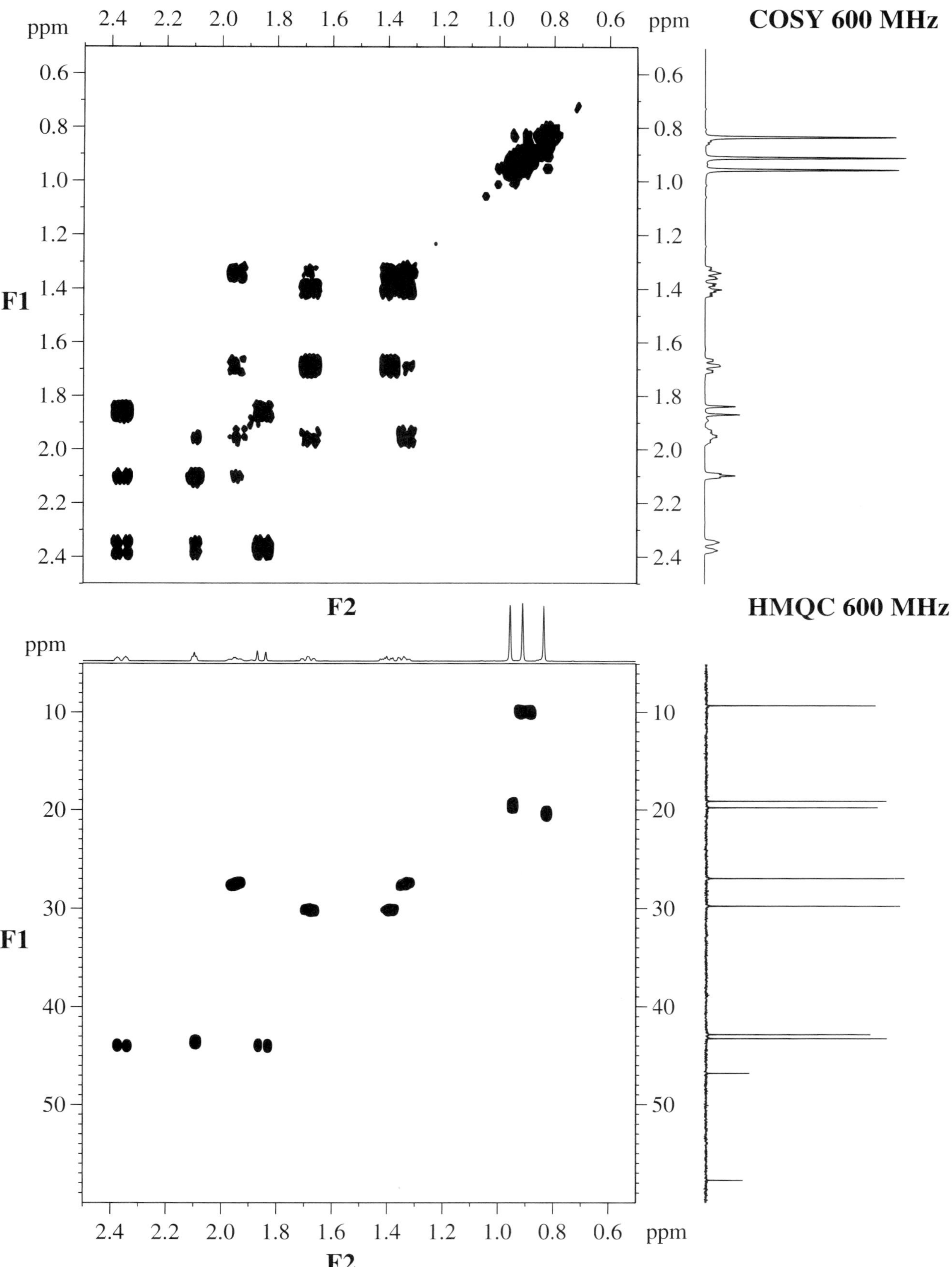

Problem 8.35C

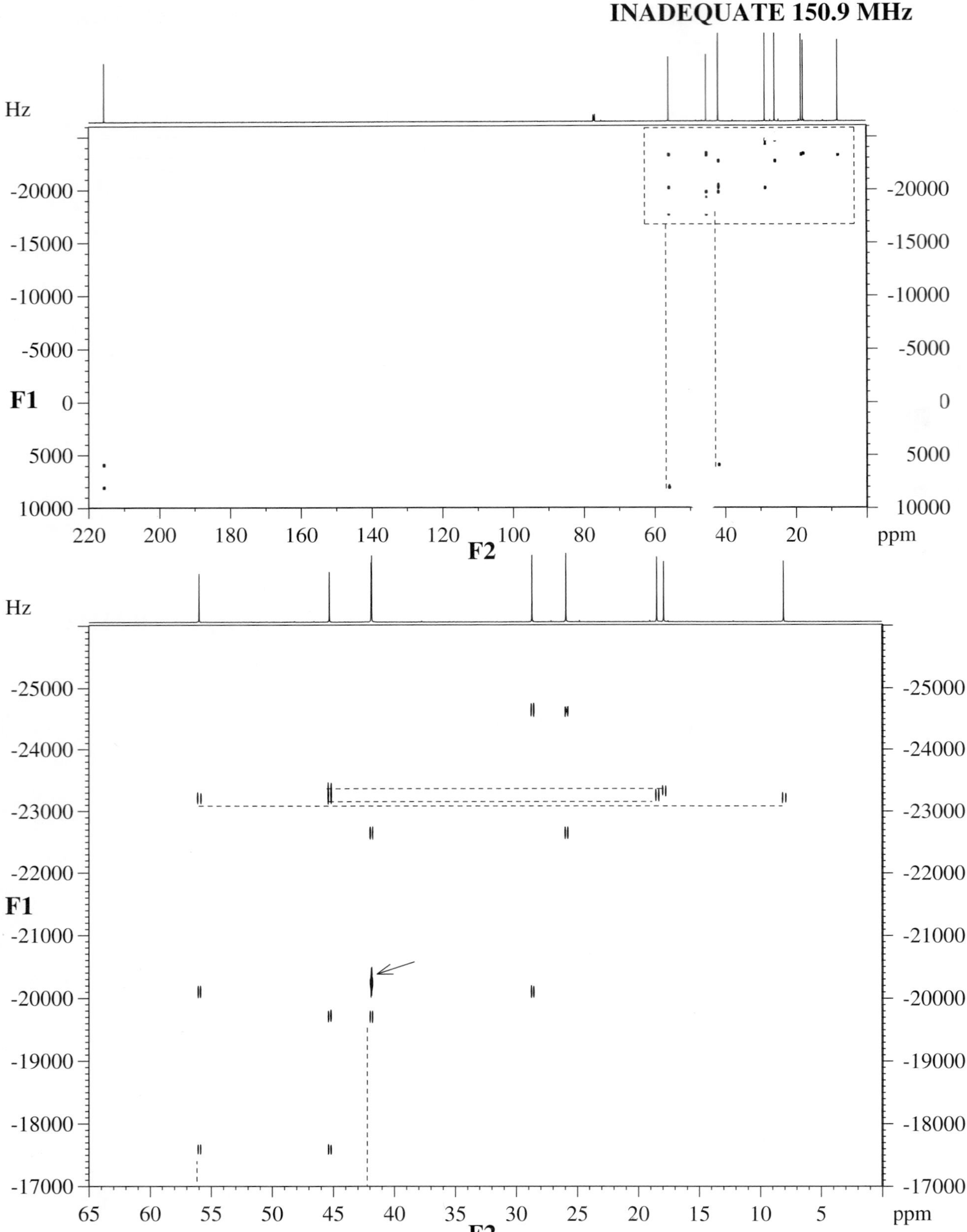

Problem 8.36A

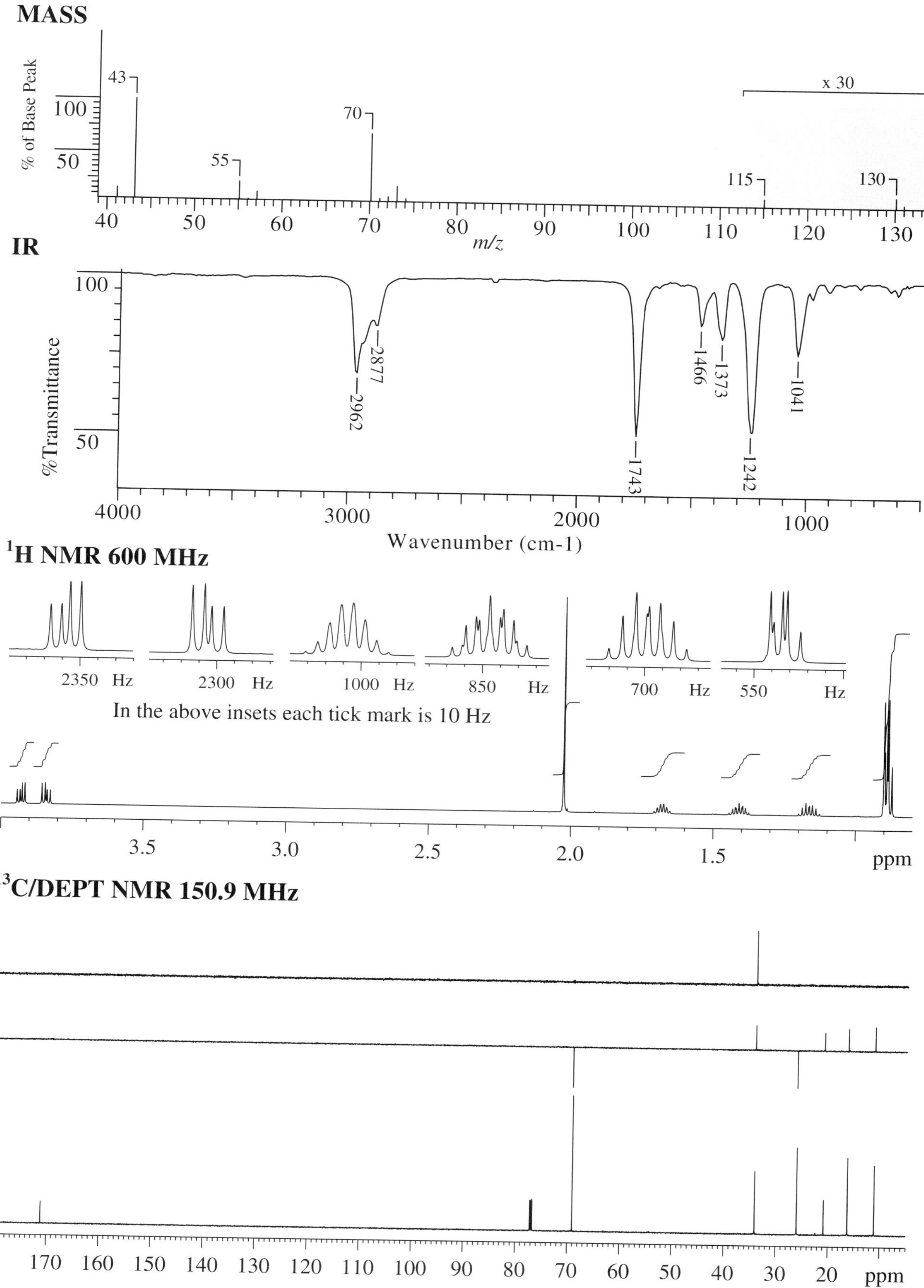

Problem 8.36B

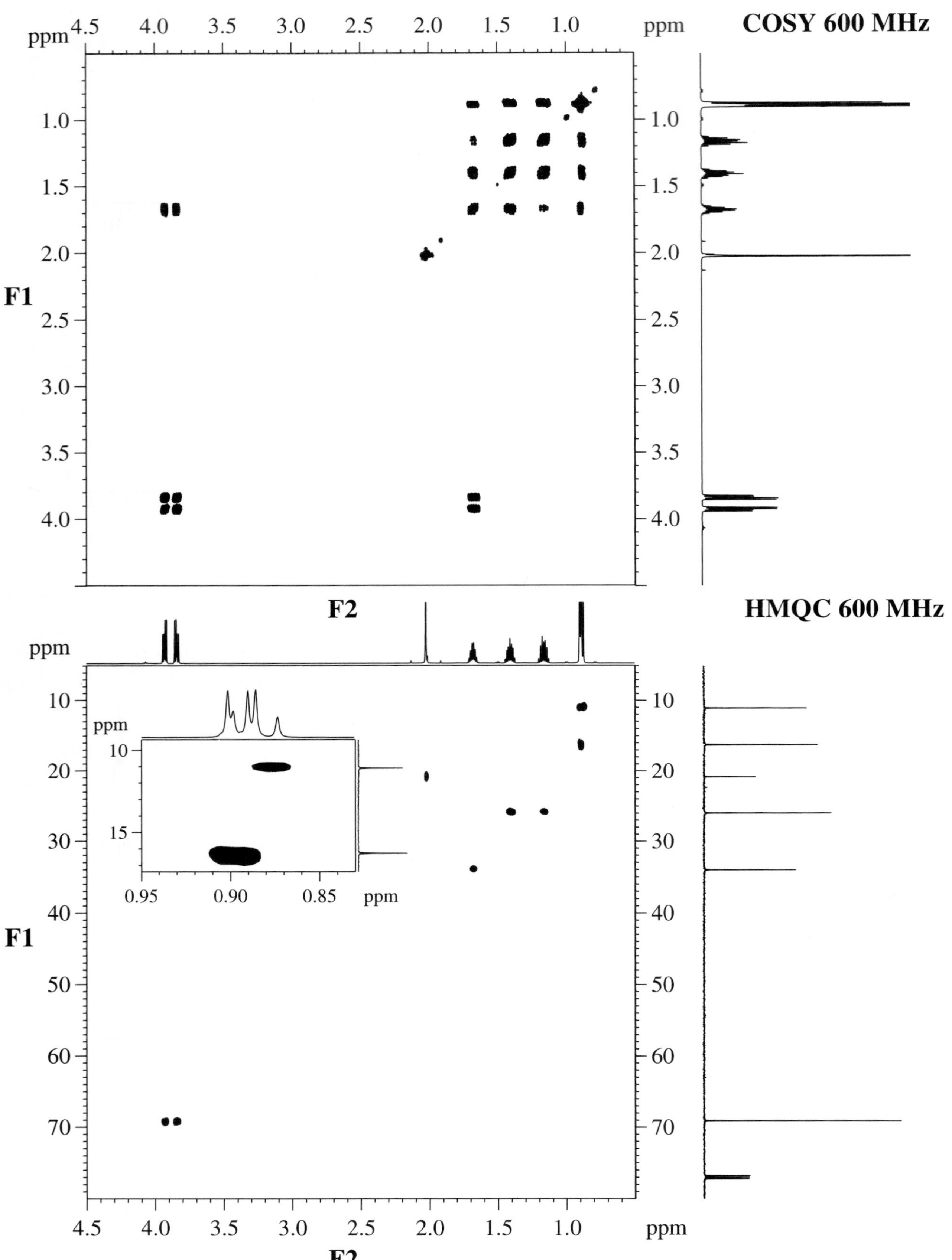

Problem 8.37A

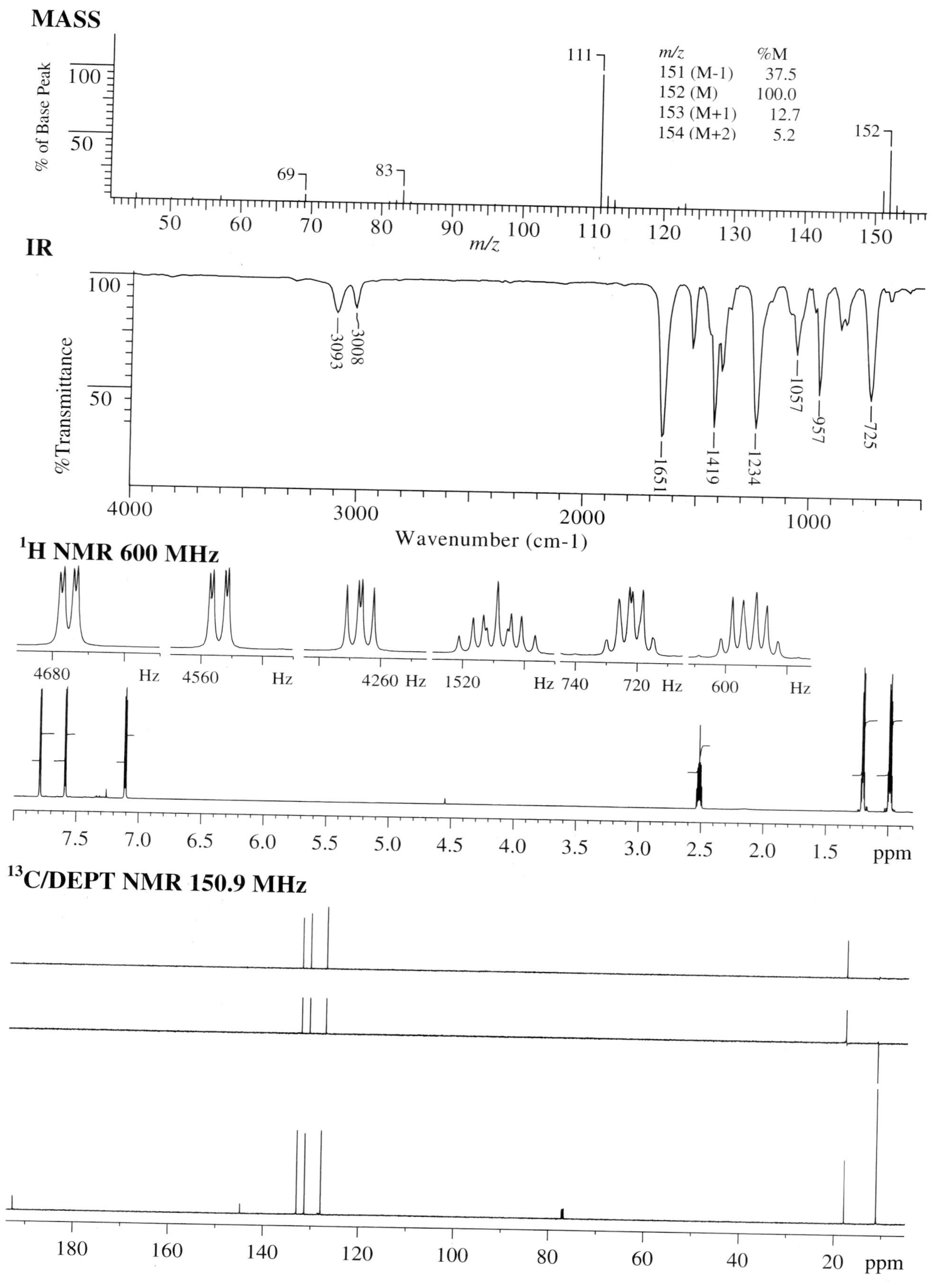

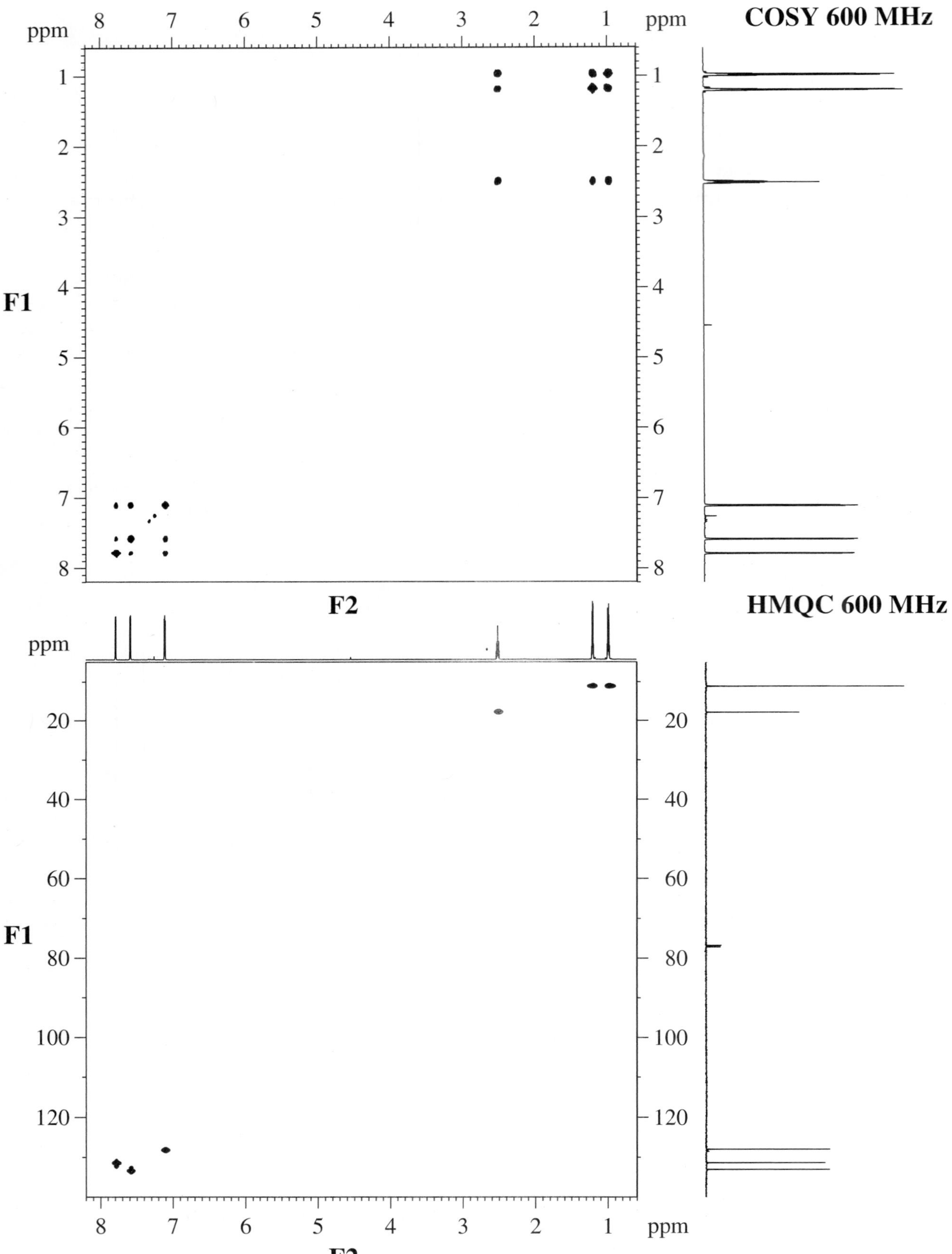

COSY 600 MHz
HMQC 600 MHz
ppm
F1
F2
8
7
6
5
4
3
2
1
20
40
60
80
100
120

Problem 8.38A

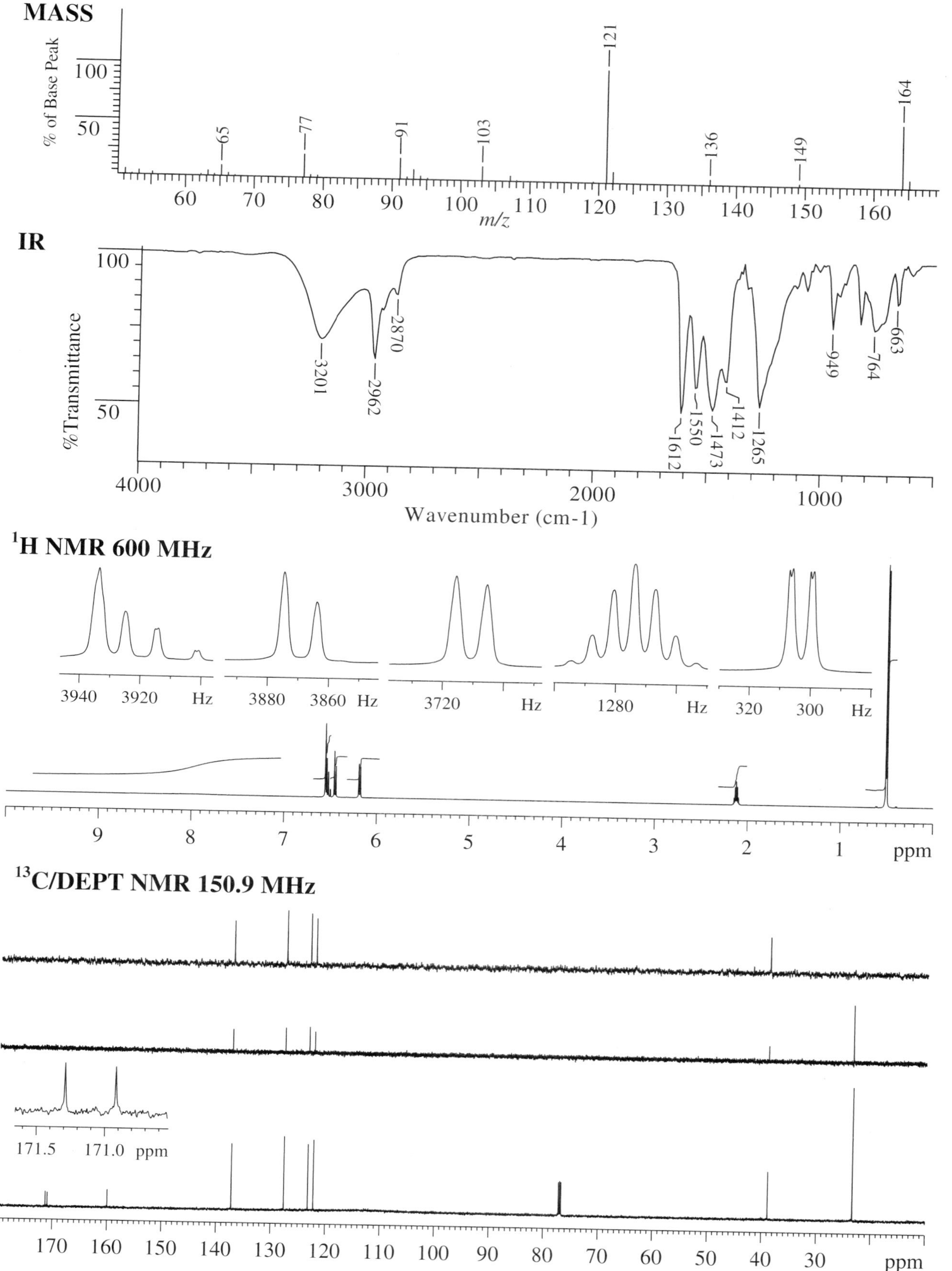

Problem 8.38B

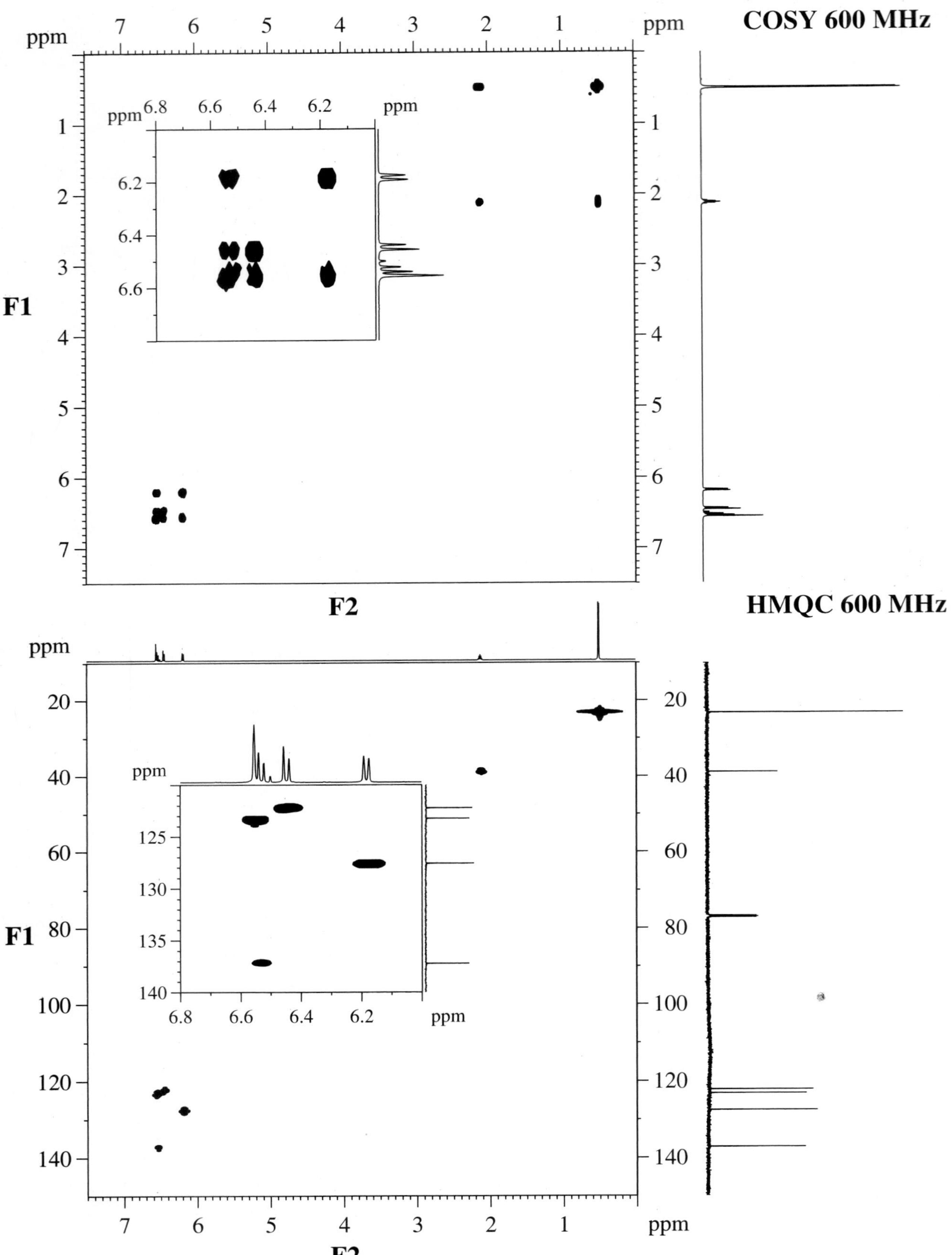

Problem 8.38C

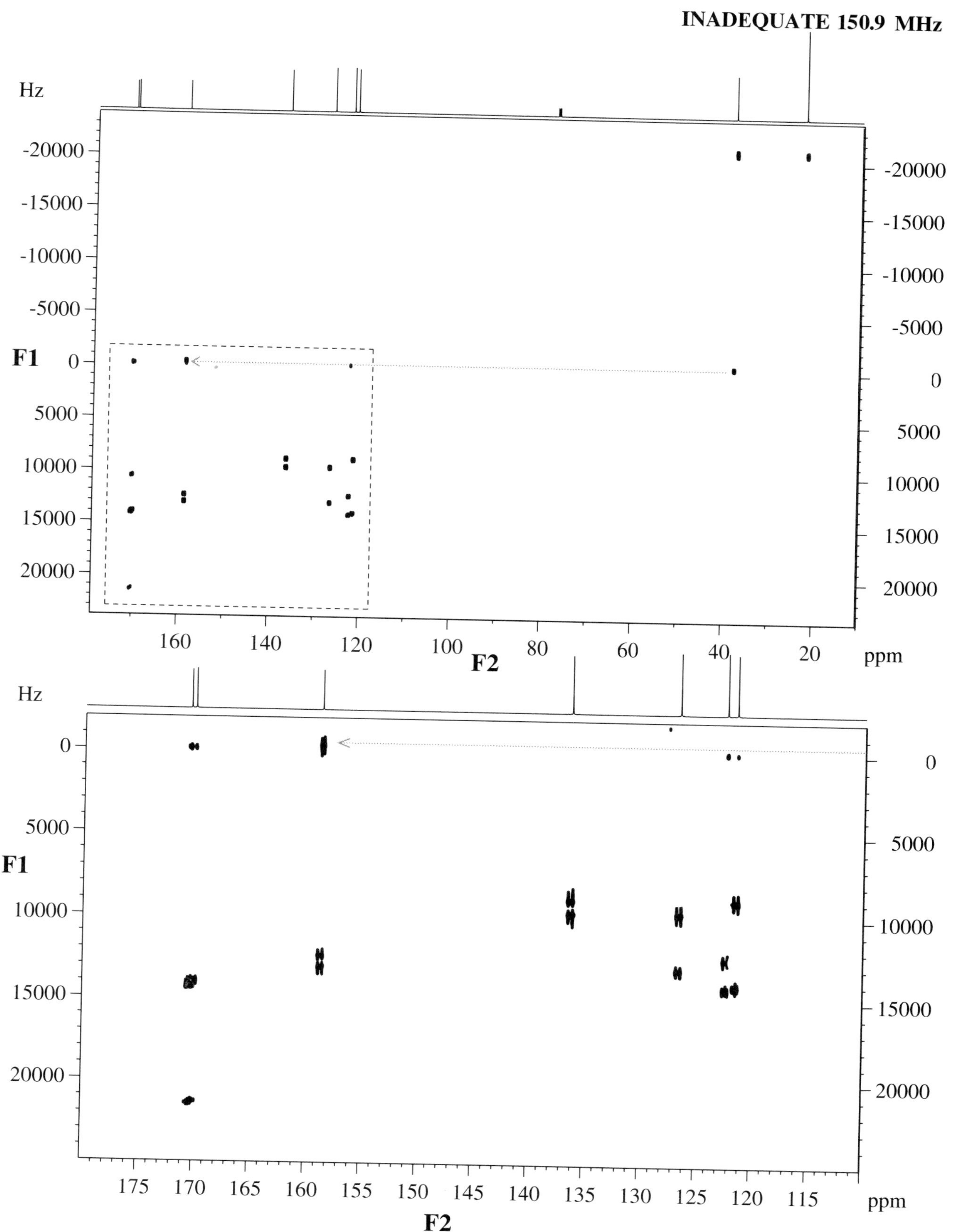

Problem 8.39A

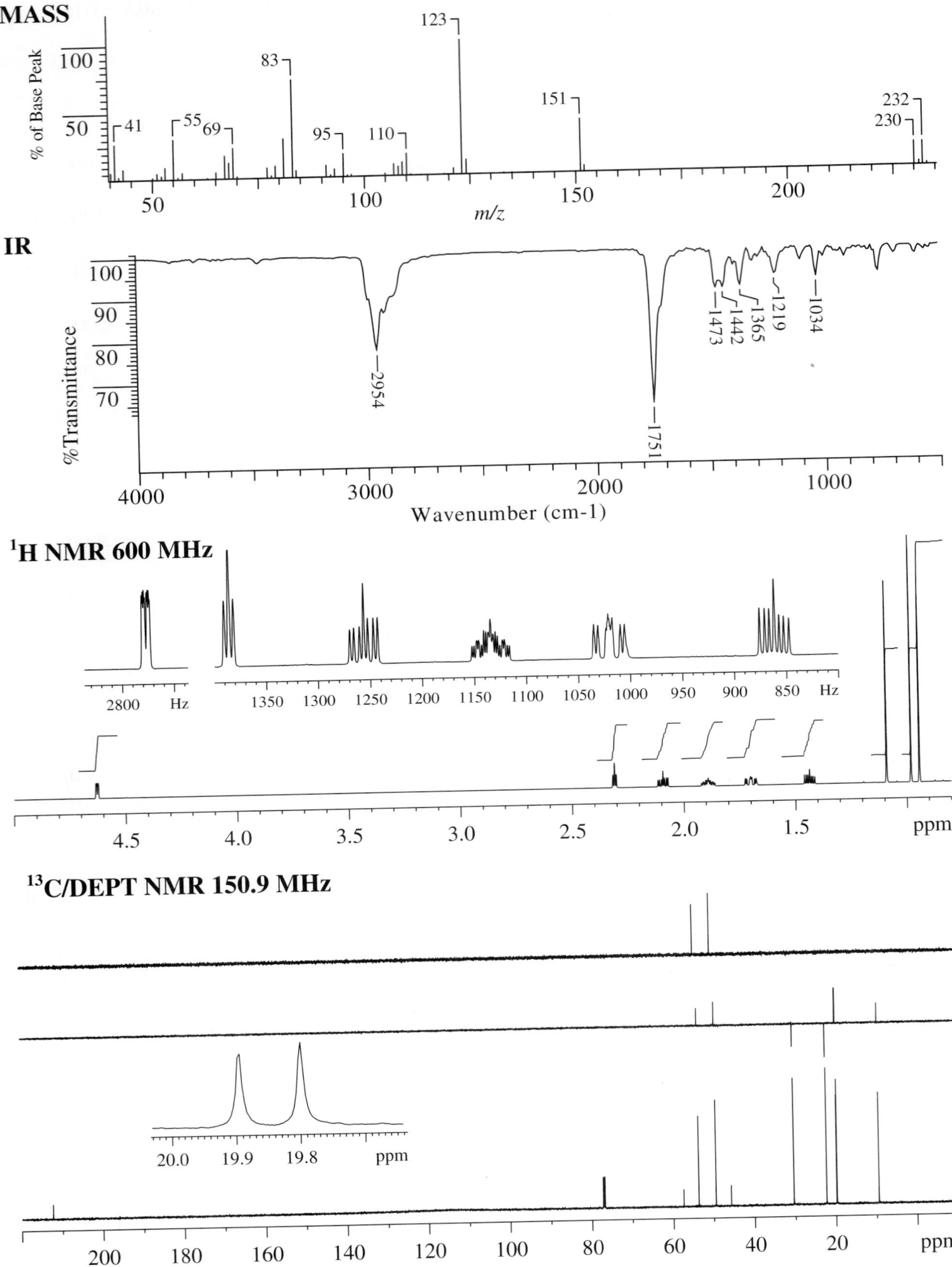

Problem 8.39B

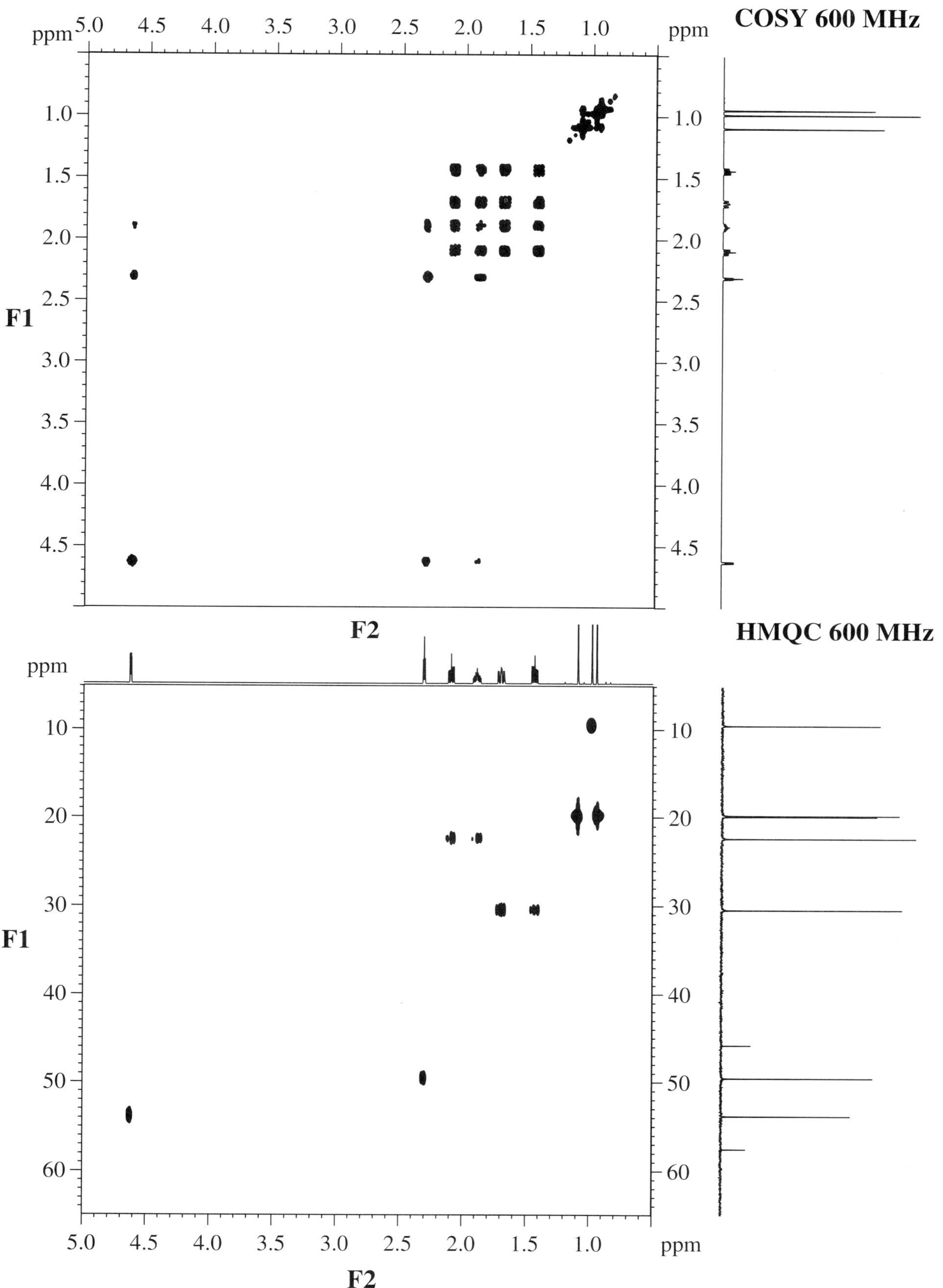

Problem 8.39C

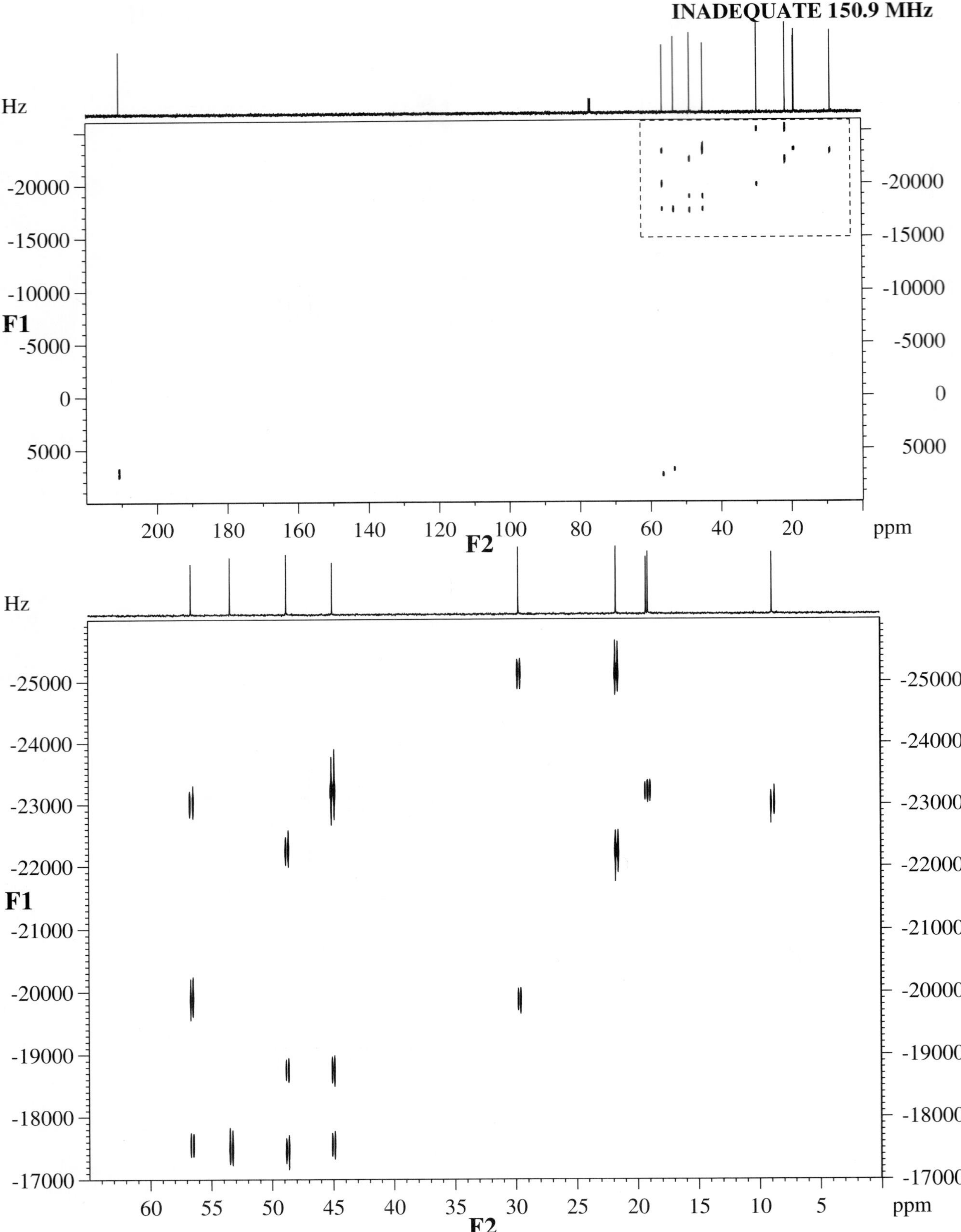

Problem 8.40A

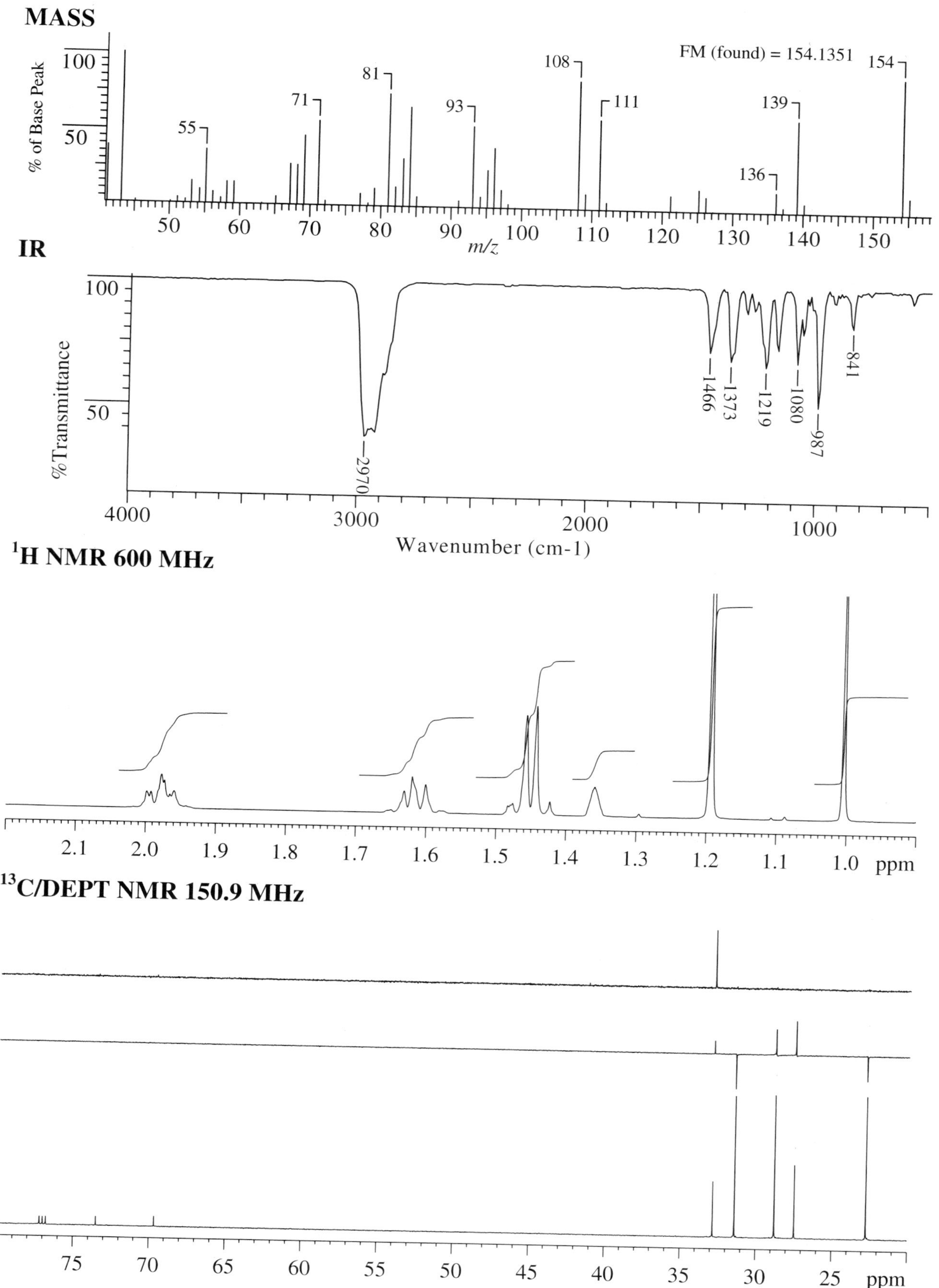

Problem 8.40B

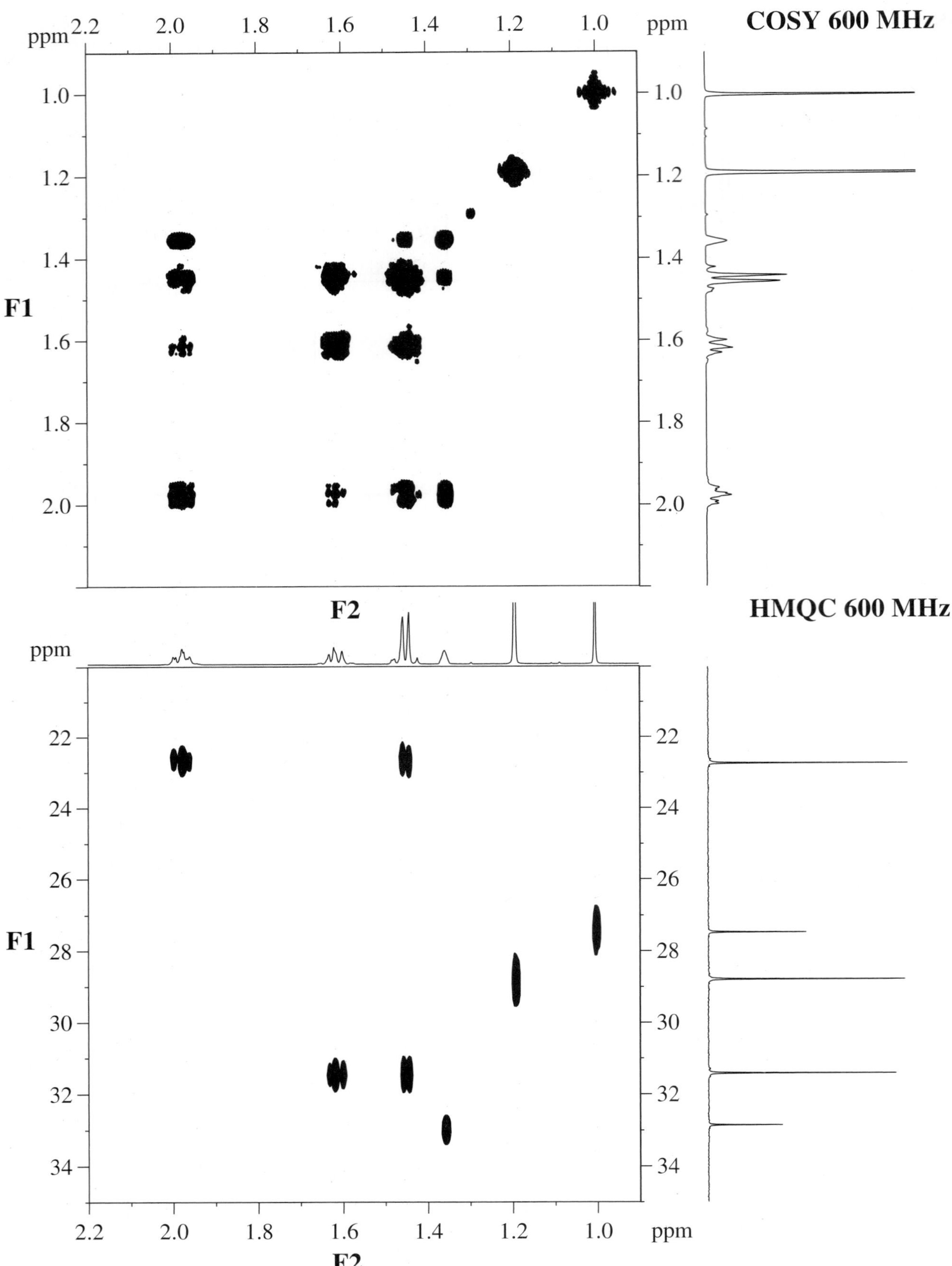

Problem 8.41A

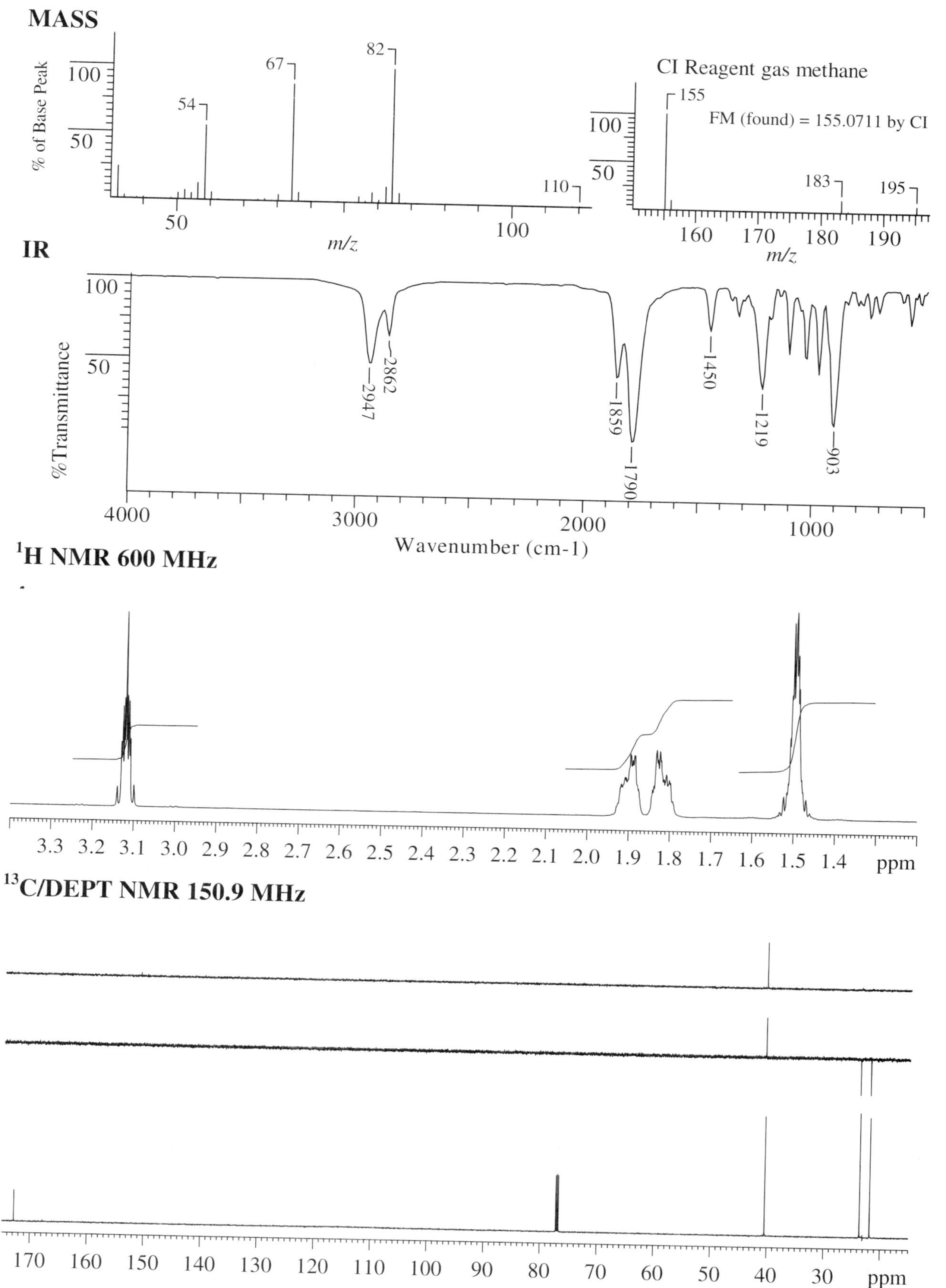

Problem 8.41B

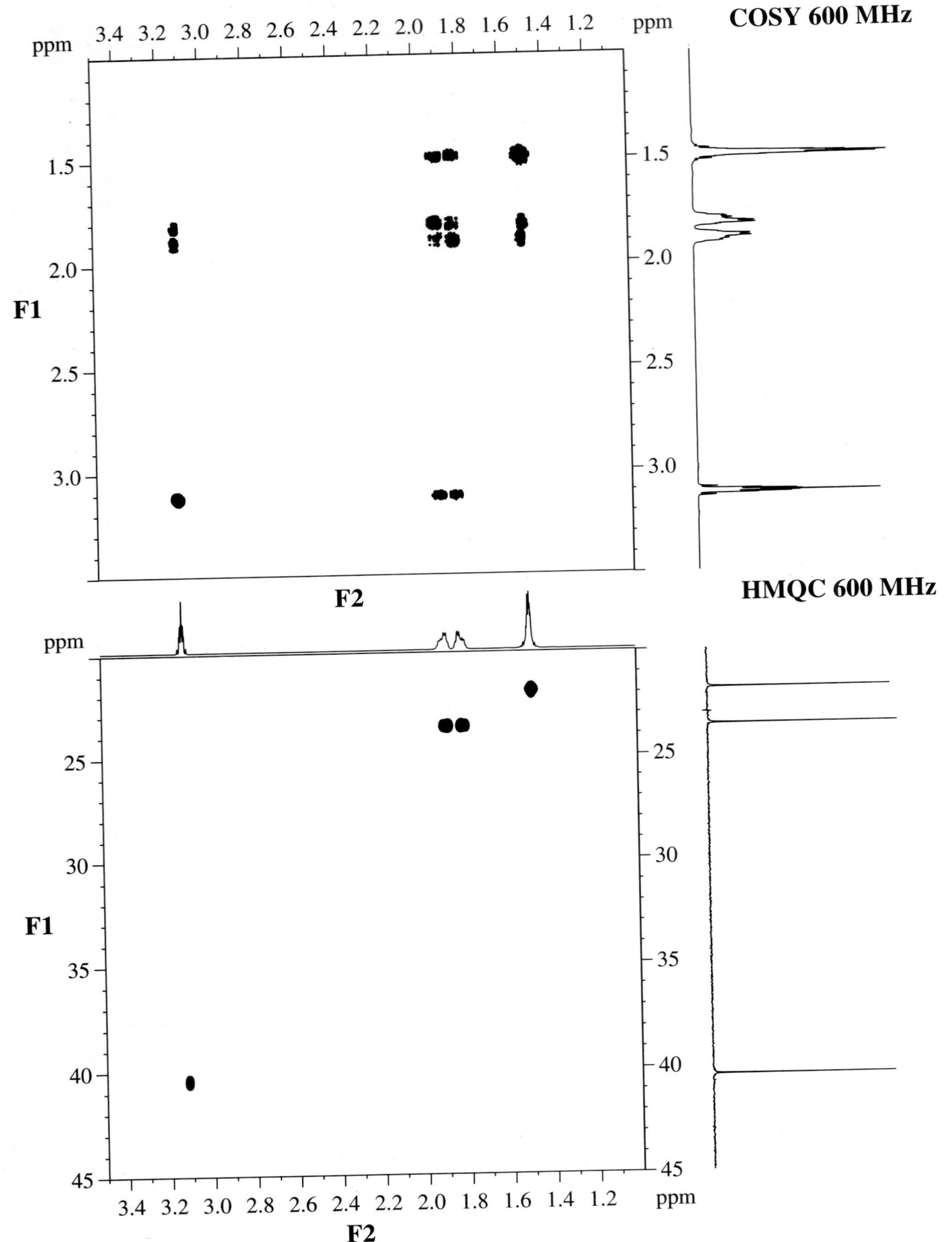

Problem 8.42A

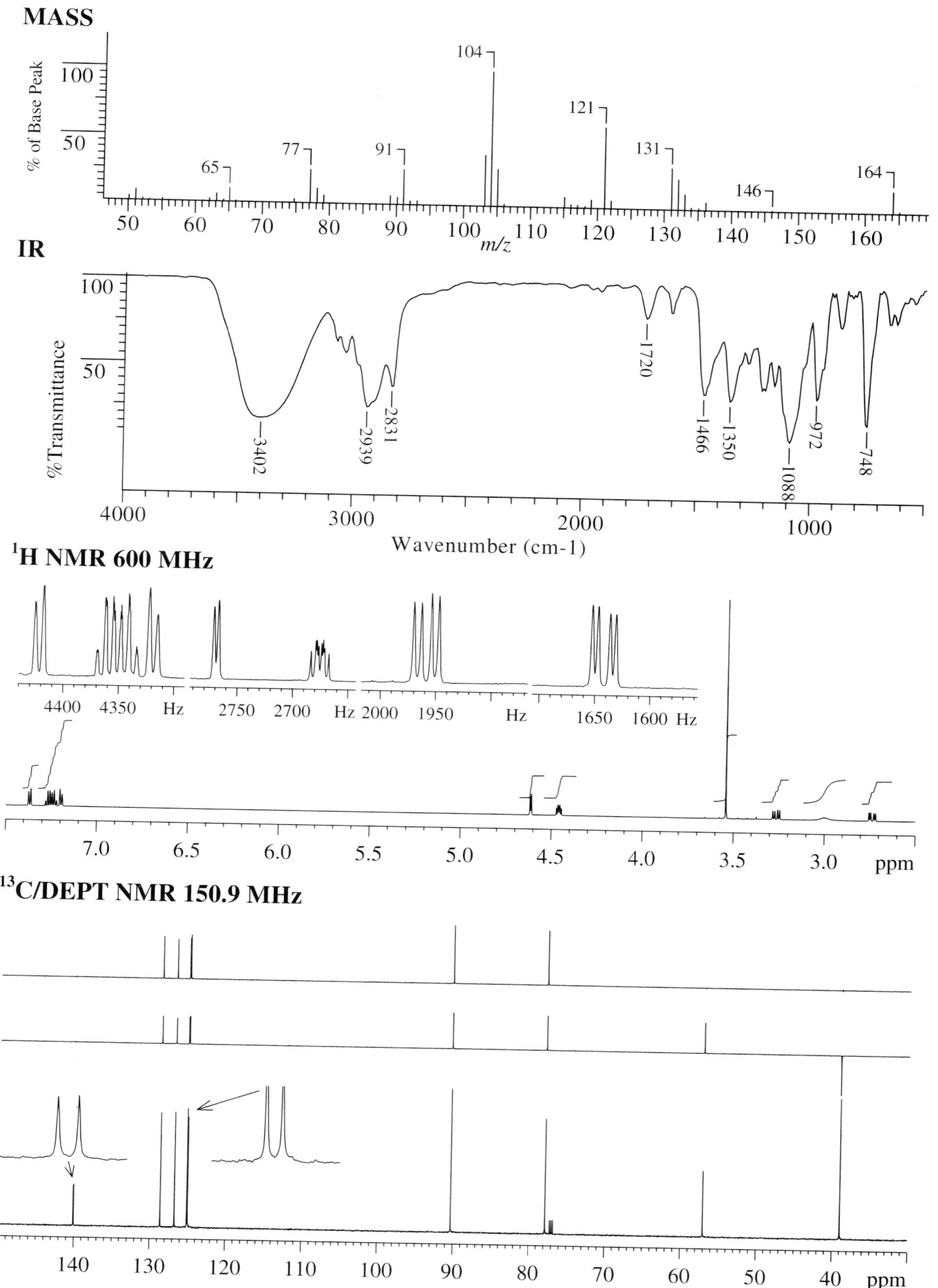

Problem 8.42B

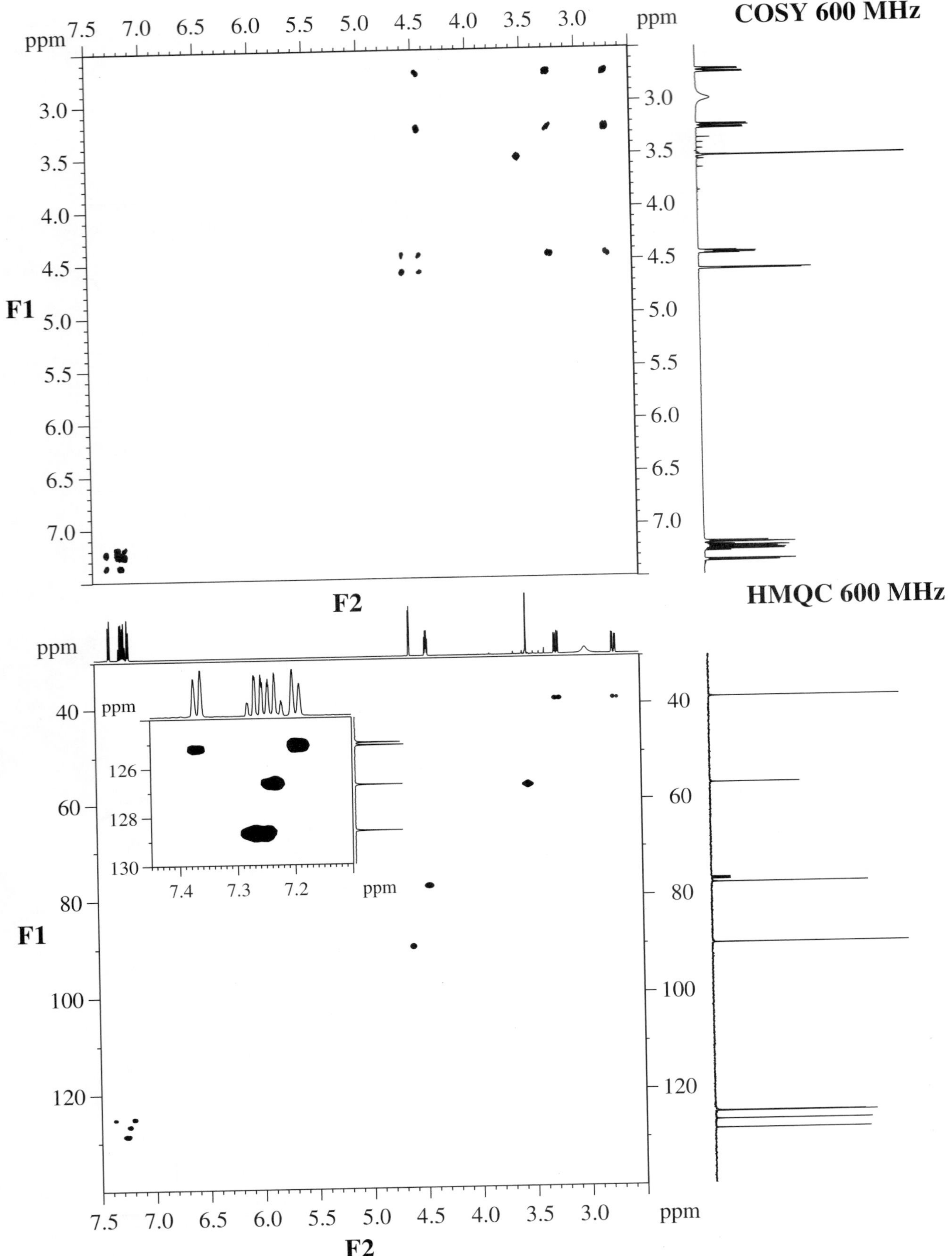

Problem 8.42C

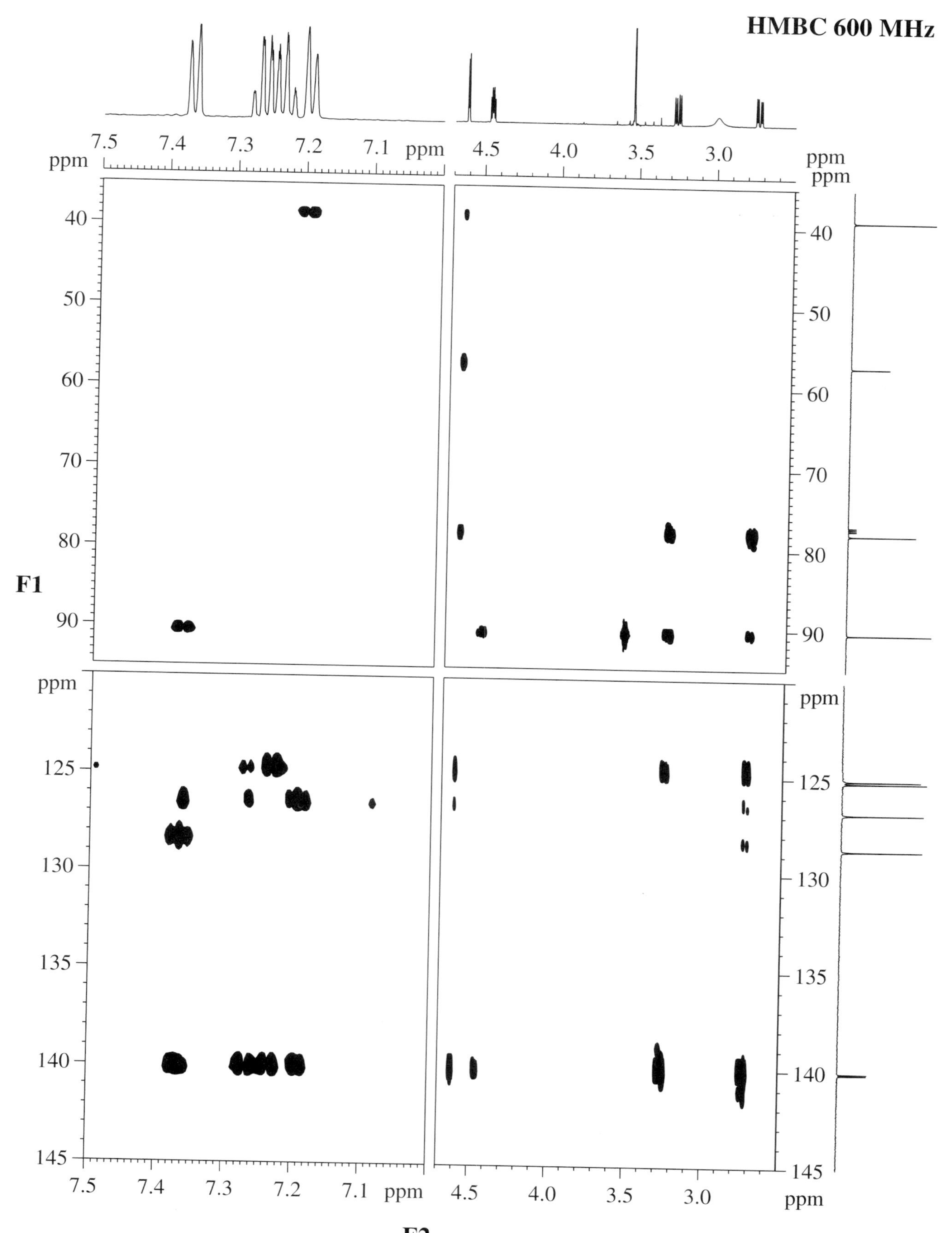

Problem 8.43A

MASS

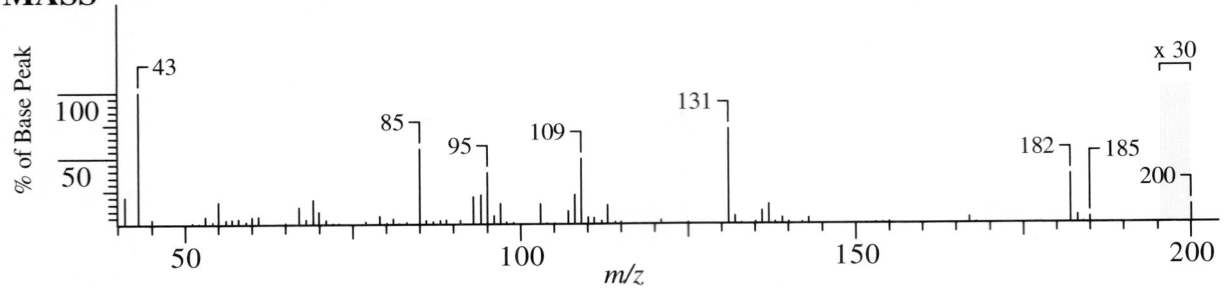

% of Base Peak: 100, 50

m/z: 50, 100, 150, 200

Peaks: 43, 85, 95, 109, 131, 182, 185, 200 (x 30)

IR

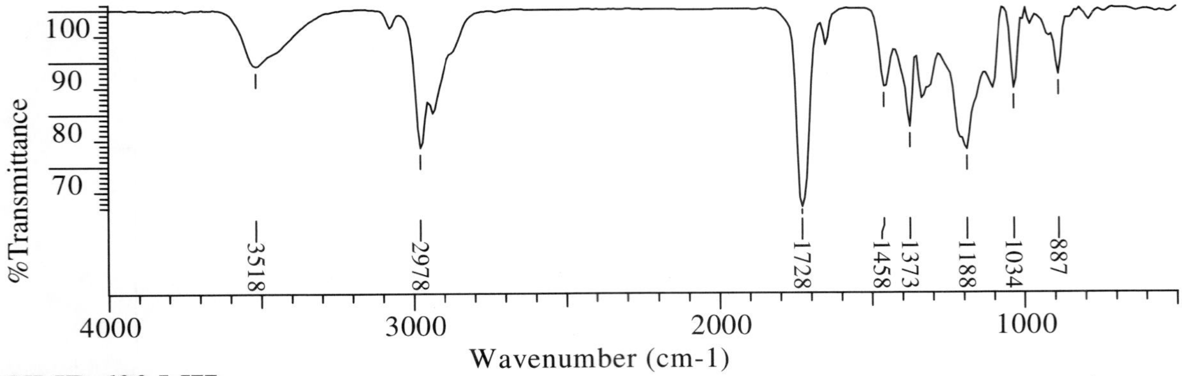

%Transmittance: 100, 90, 80, 70

Wavenumber (cm-1): 4000, 3000, 2000, 1000

Peaks: 3518, 2978, 1728, 1458, 1373, 1188, 1034, 887

^{1}H NMR 600 MHz

Insets: 1260, 1240 Hz; 780, 760, 740 Hz

ppm: 4.5, 4.0, 3.5, 3.0, 2.5, 2.0, 1.5

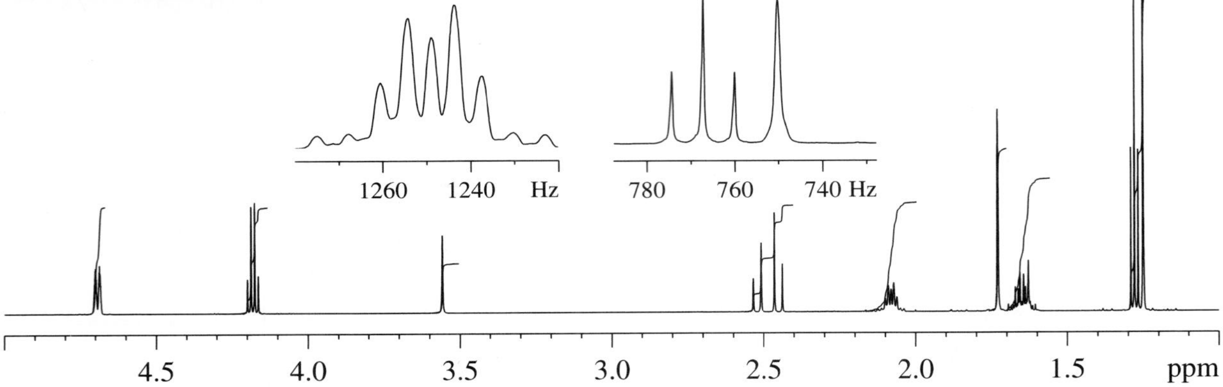

^{13}C/DEPT NMR 150.9 MHz

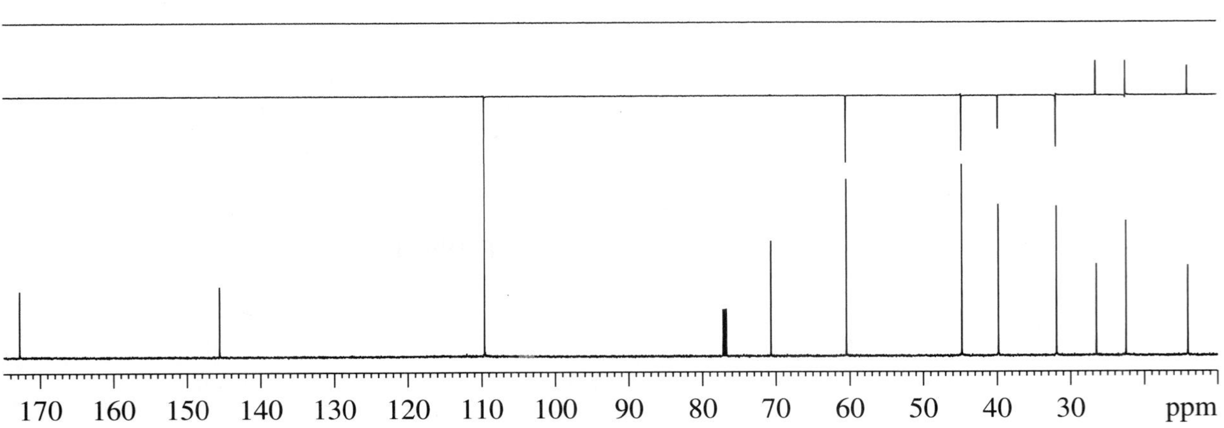

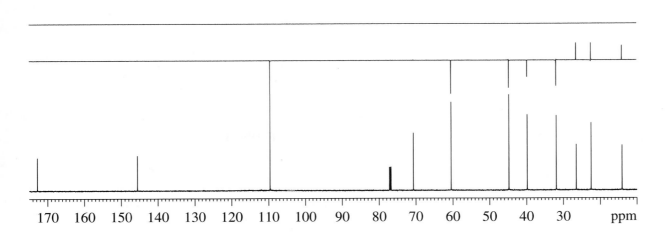

Problem 8.43B

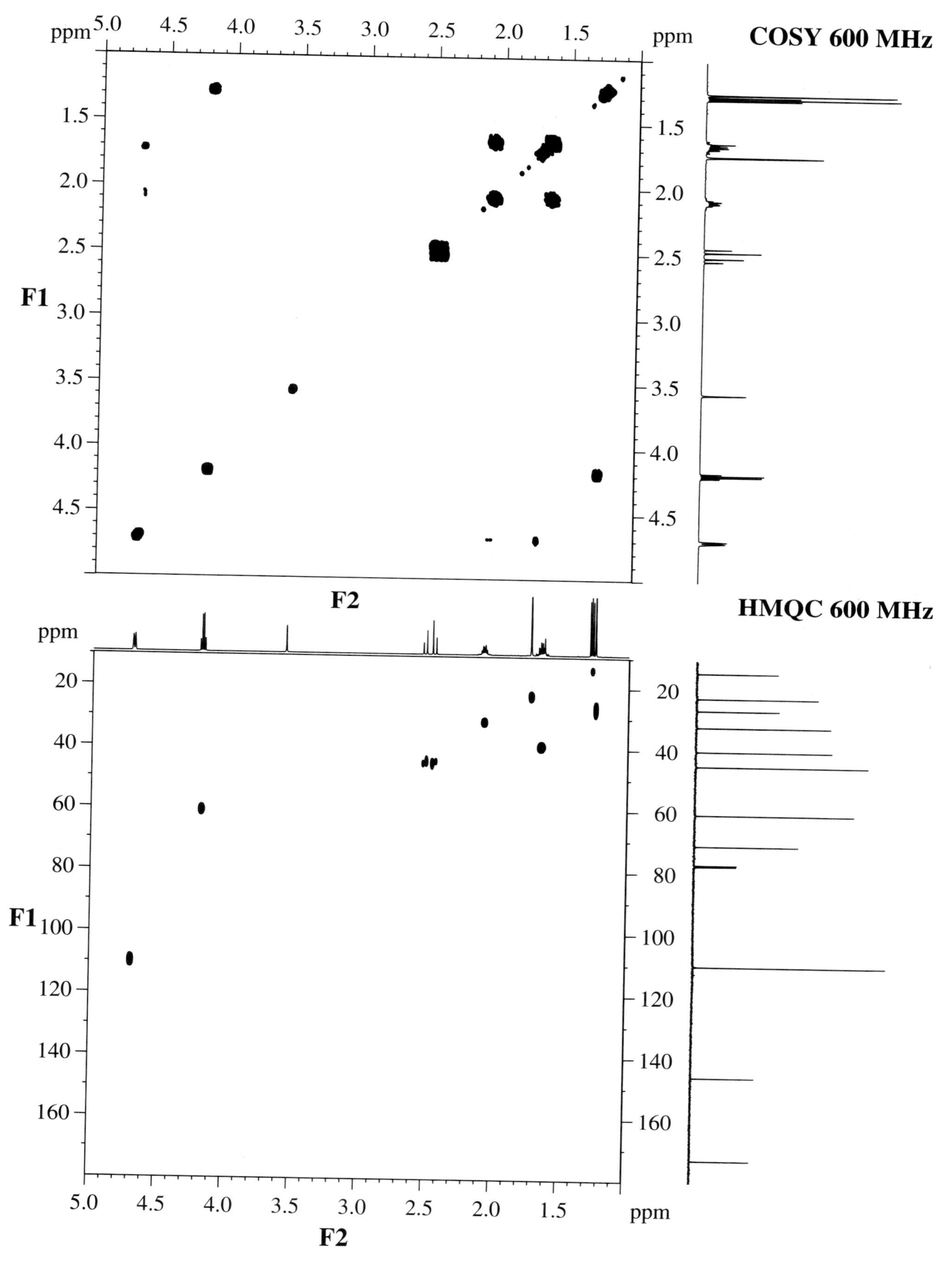

Problem 8.43C

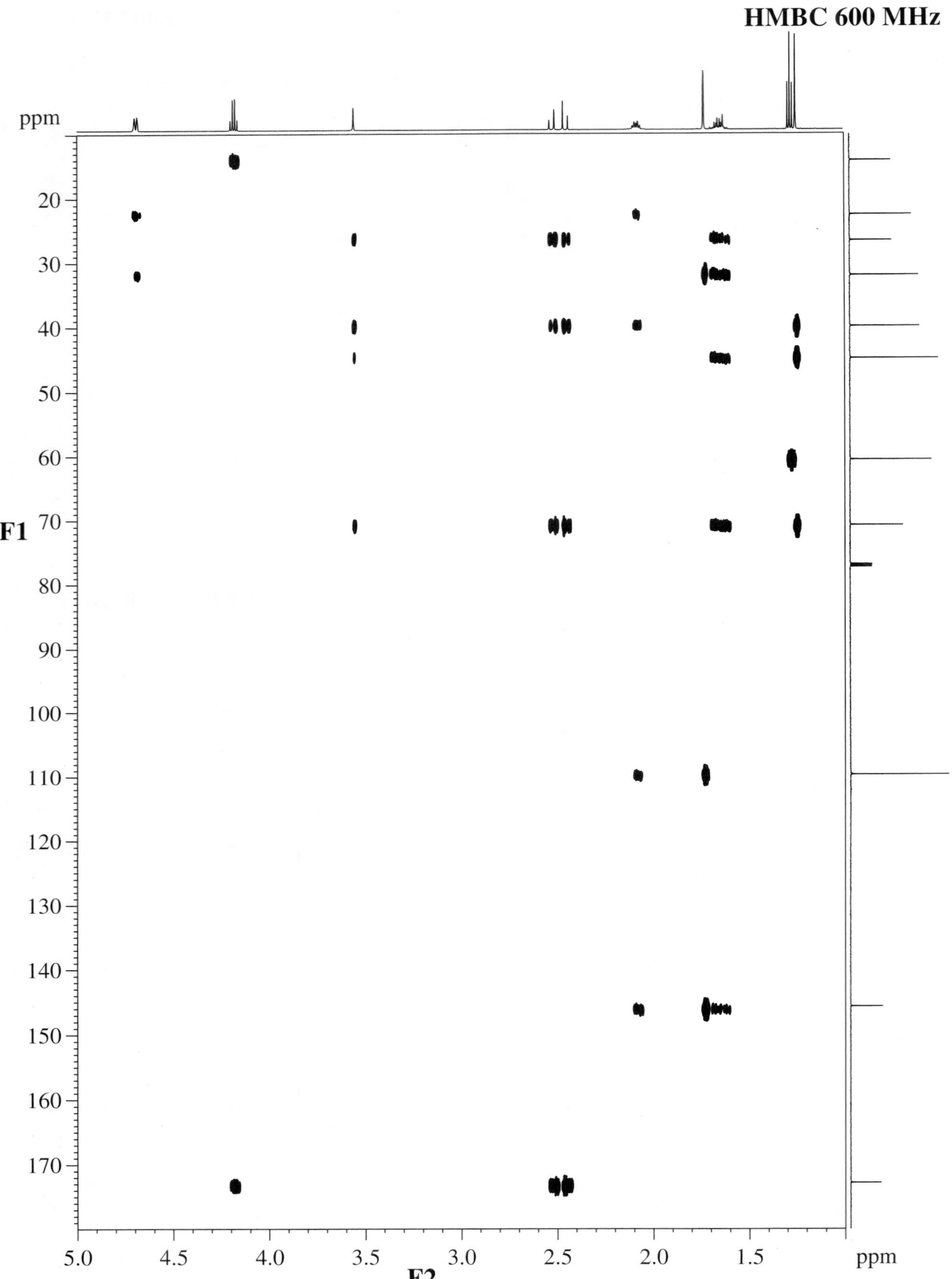

Problem 8.44A

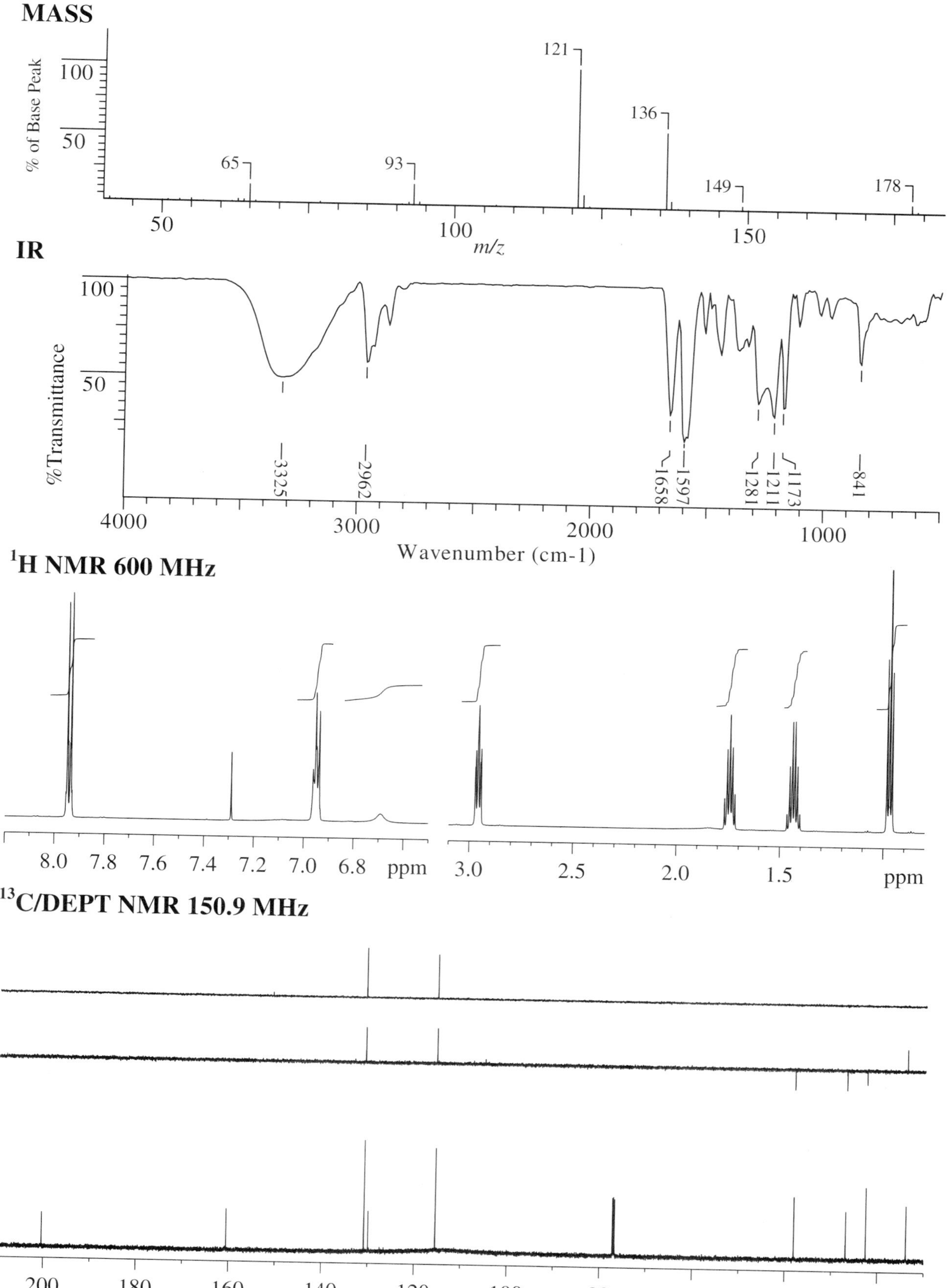

Problem 8.44B

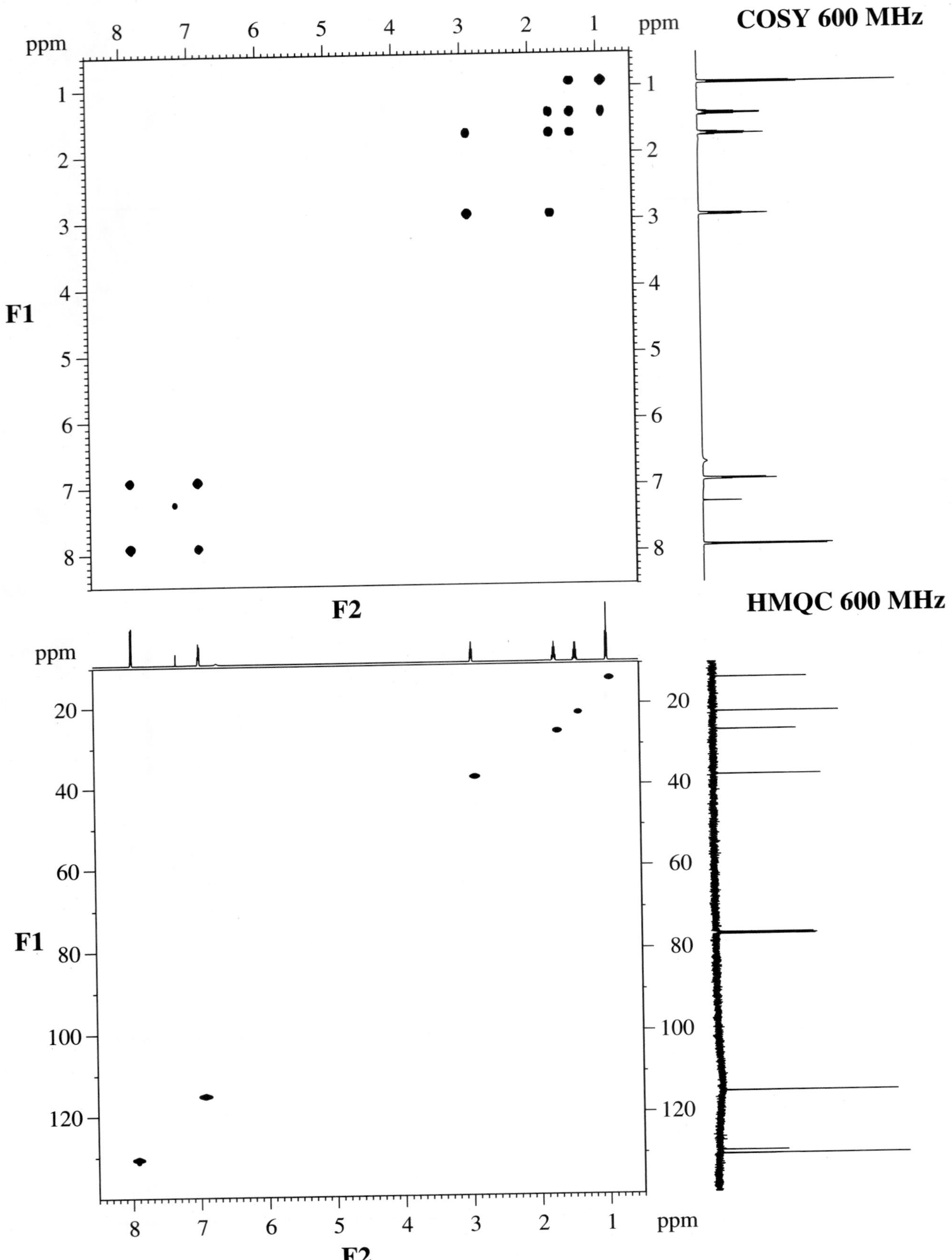

Problem 8.44C

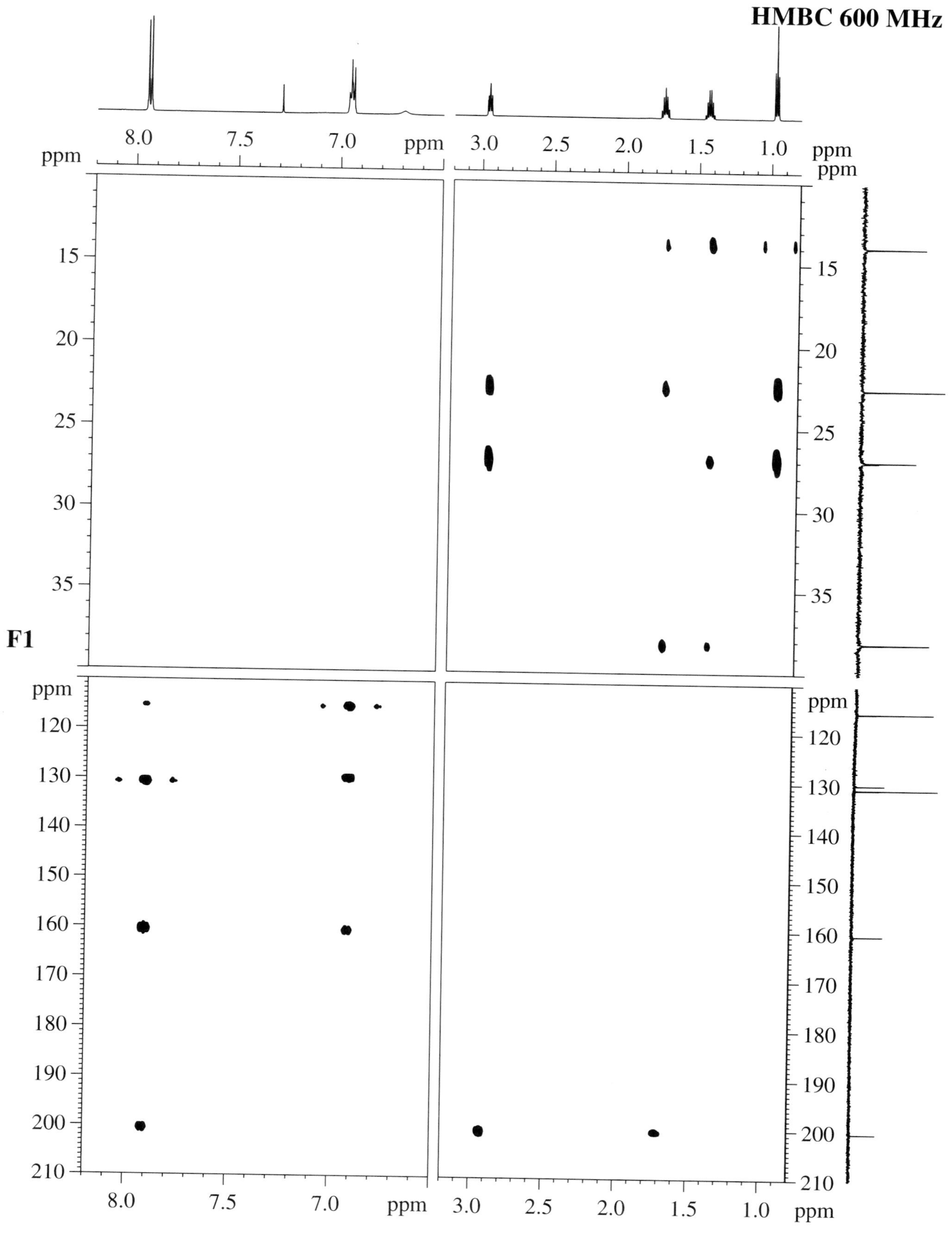

Problem 8.45A

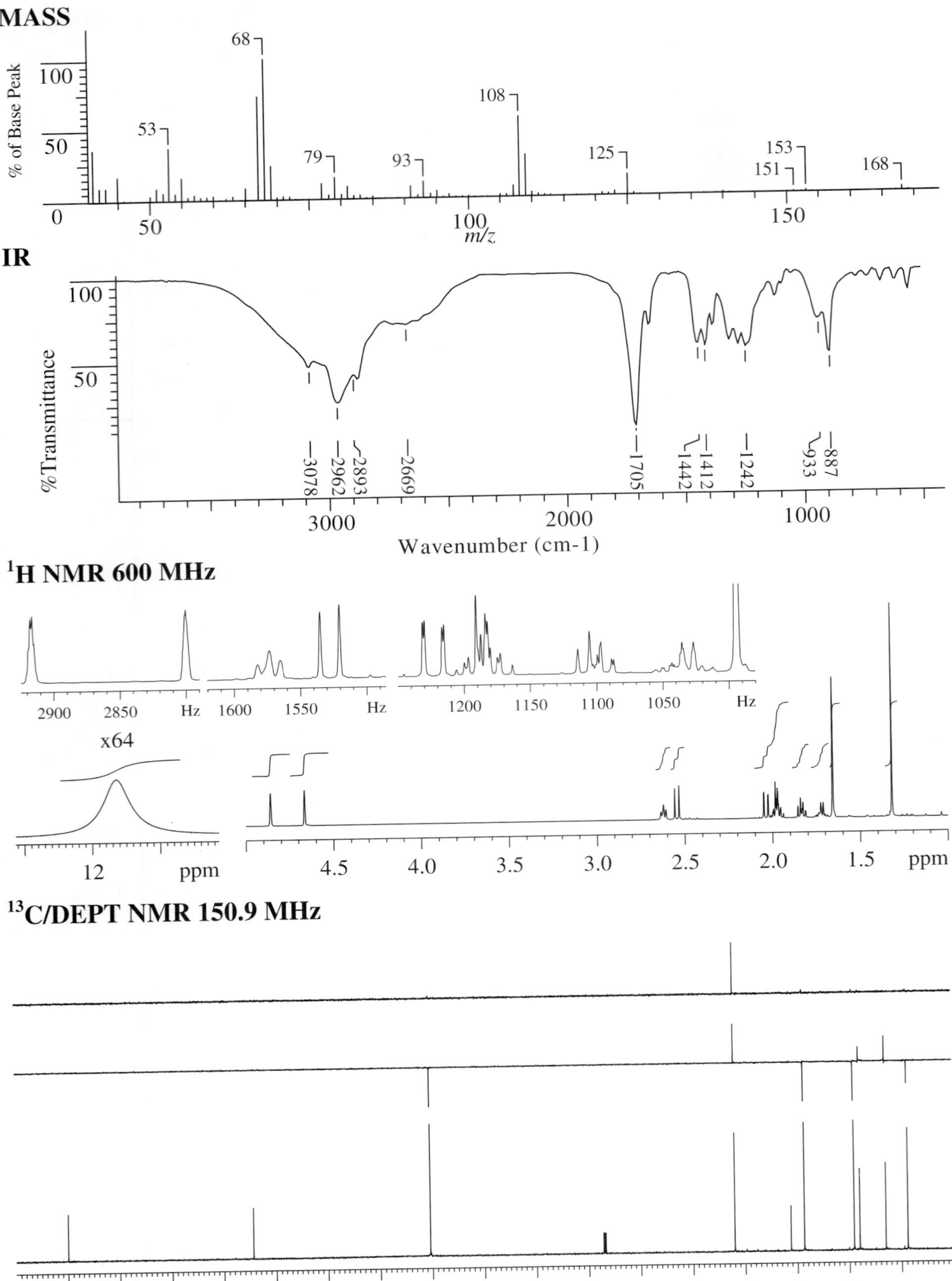

Problem 8.45B

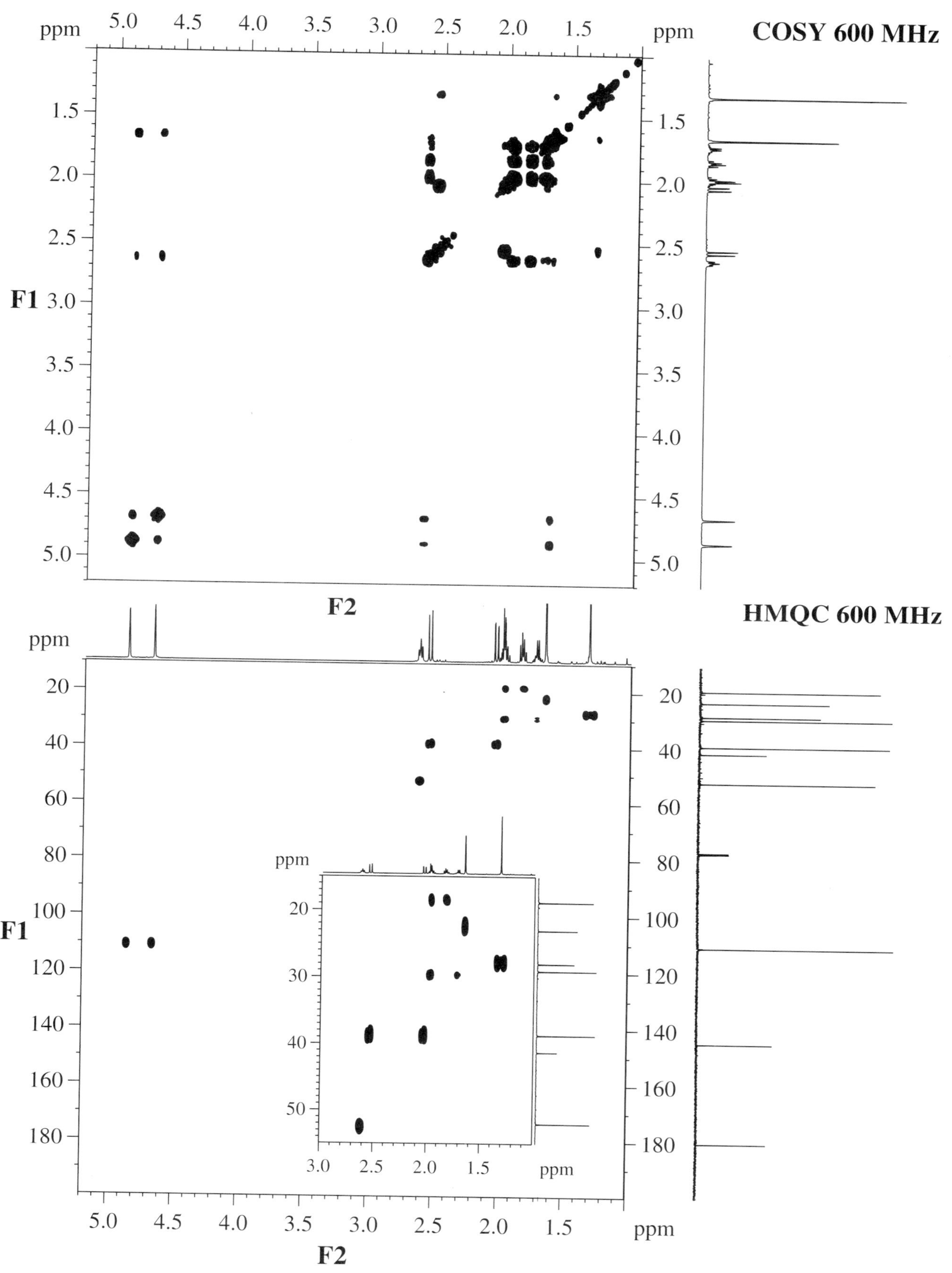

Problem 8.45C

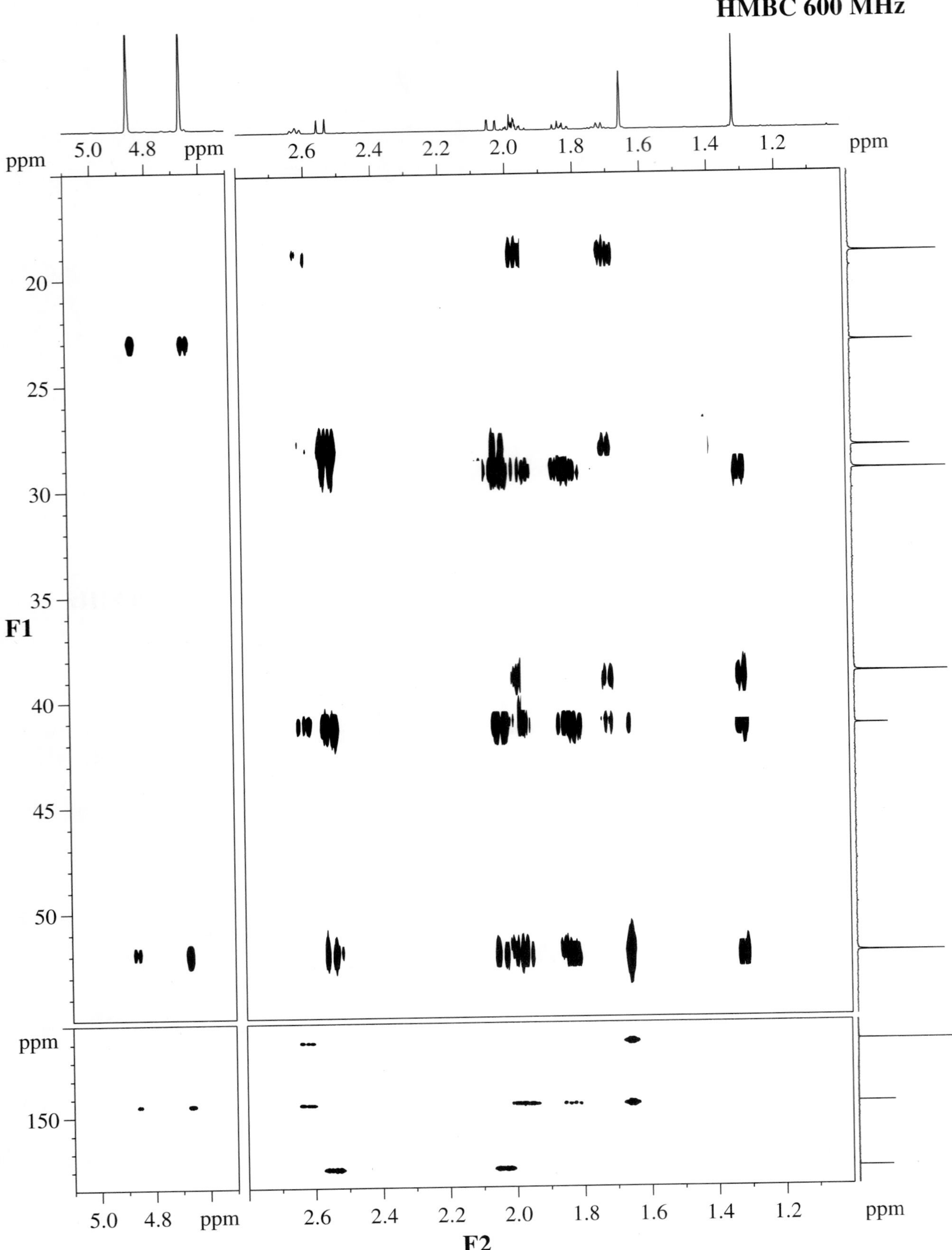

Problem 8.46A

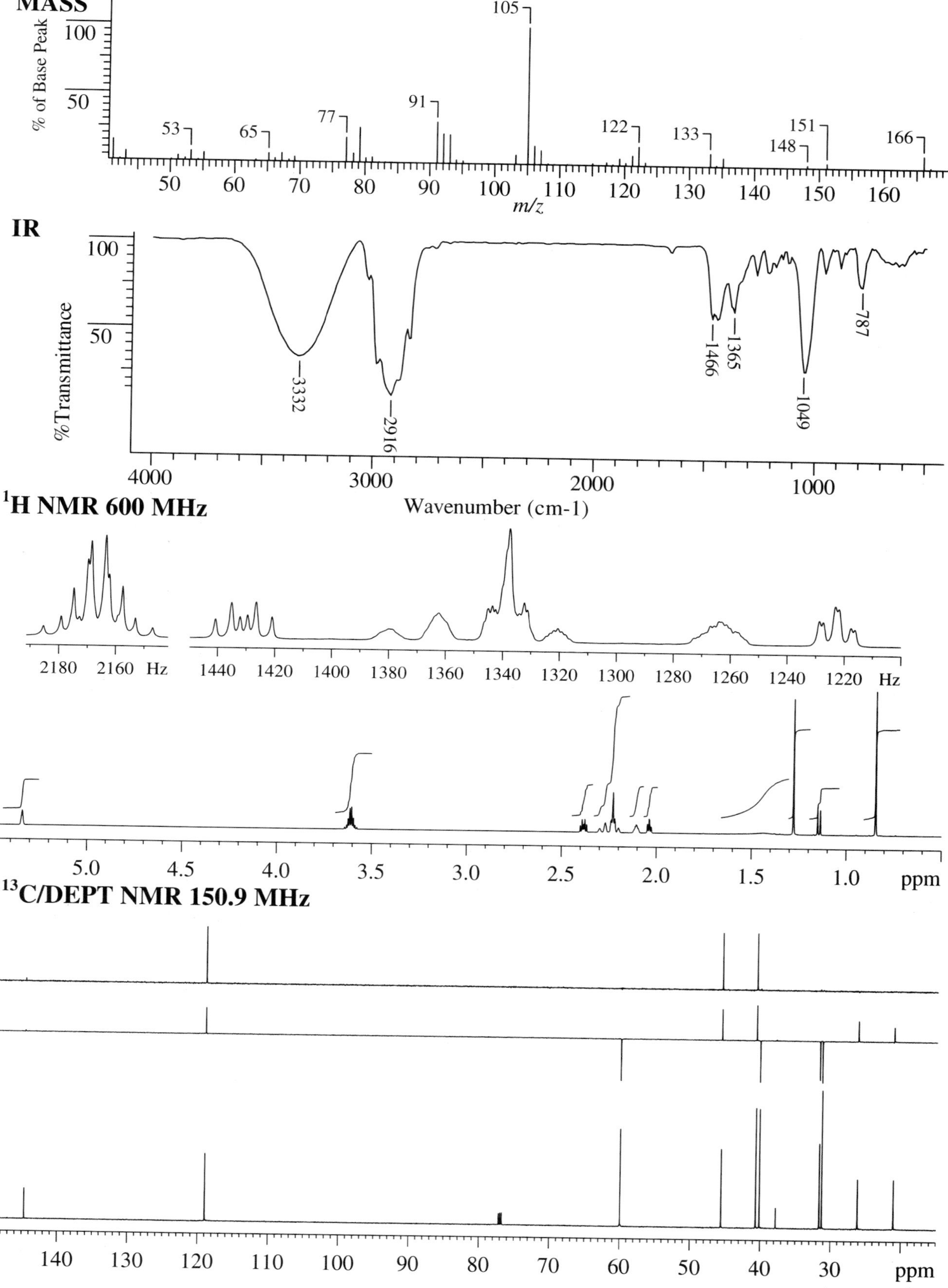

Problem 8.46B

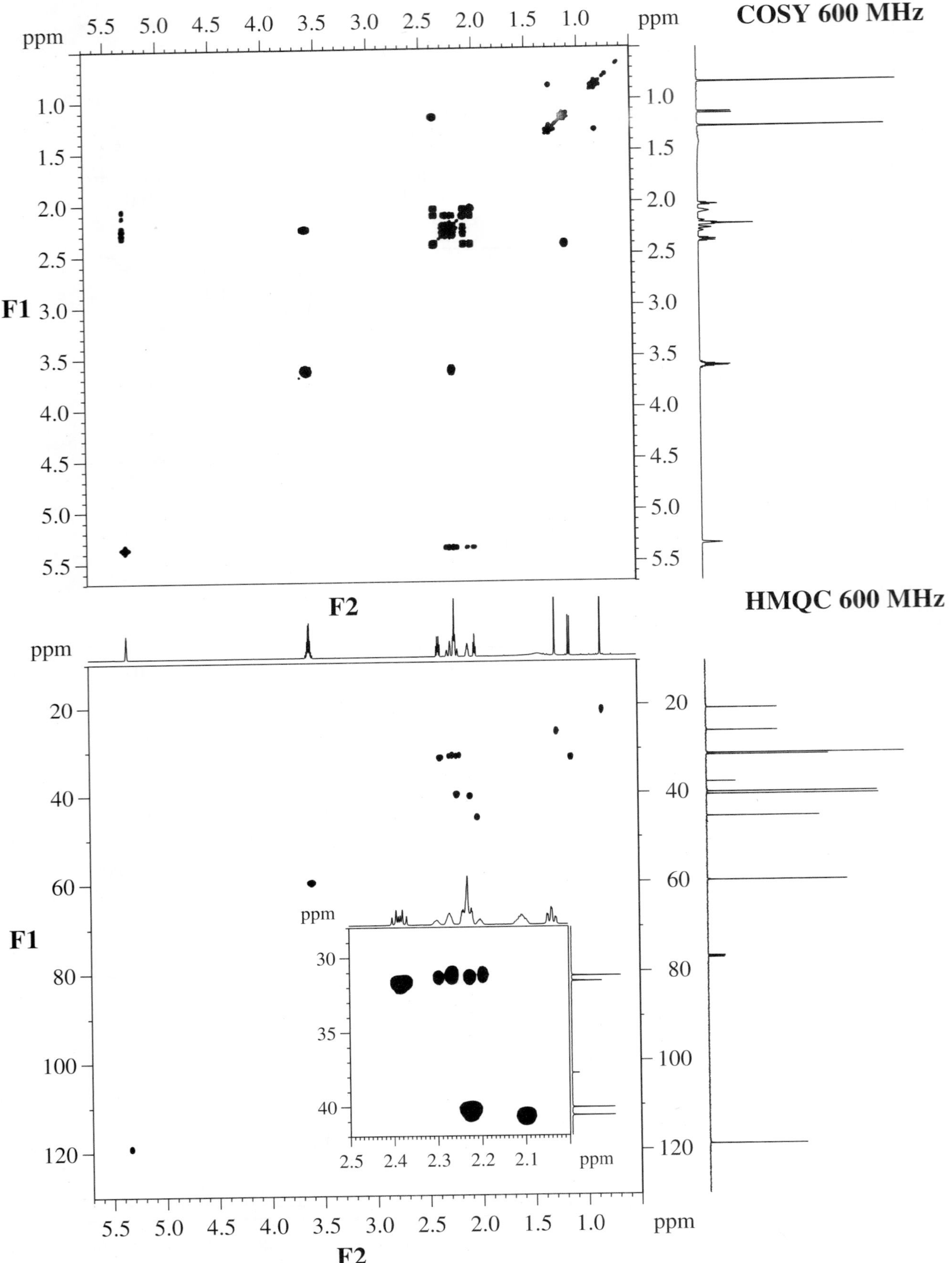

Problem 8.46C

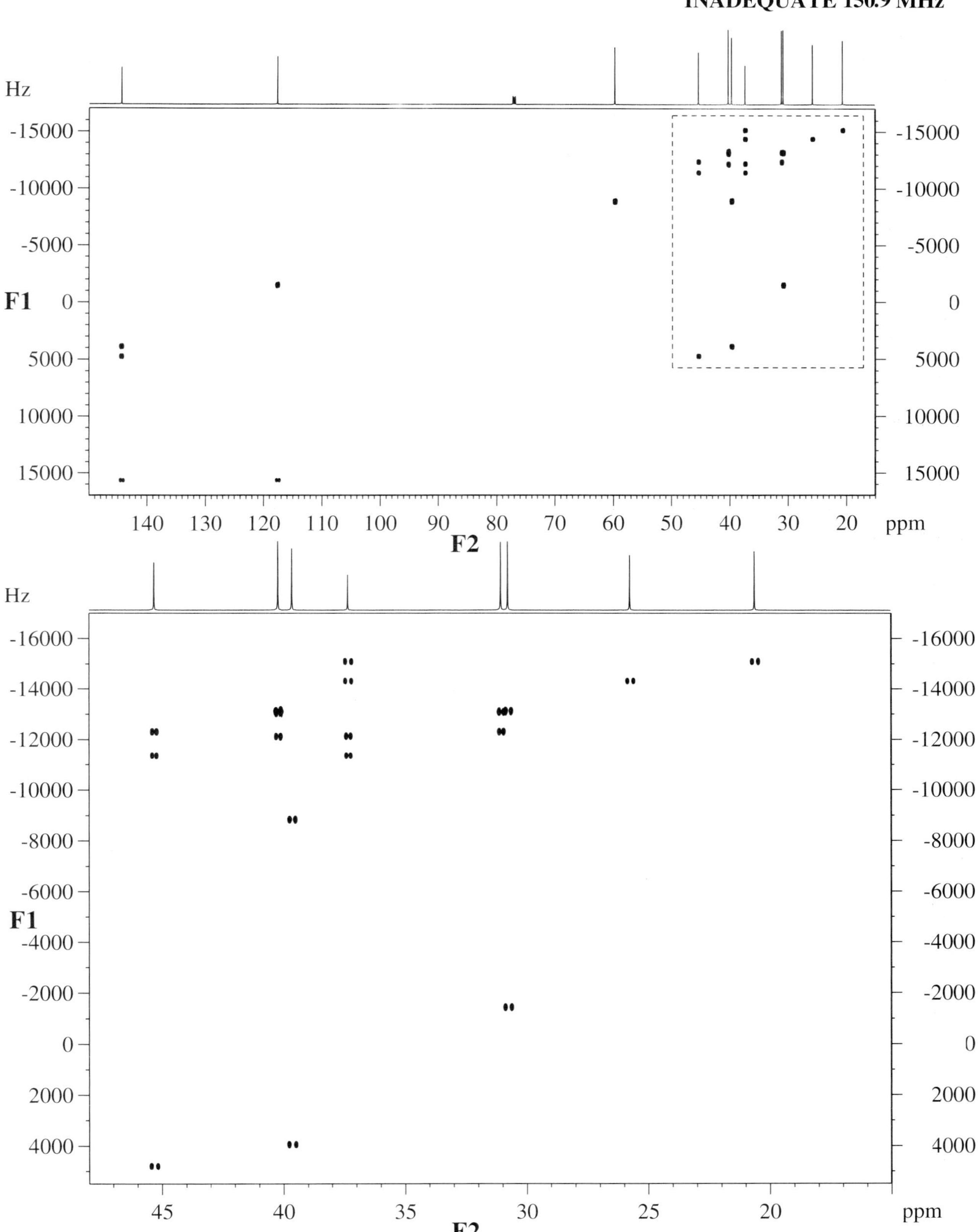

Problem 8.47A

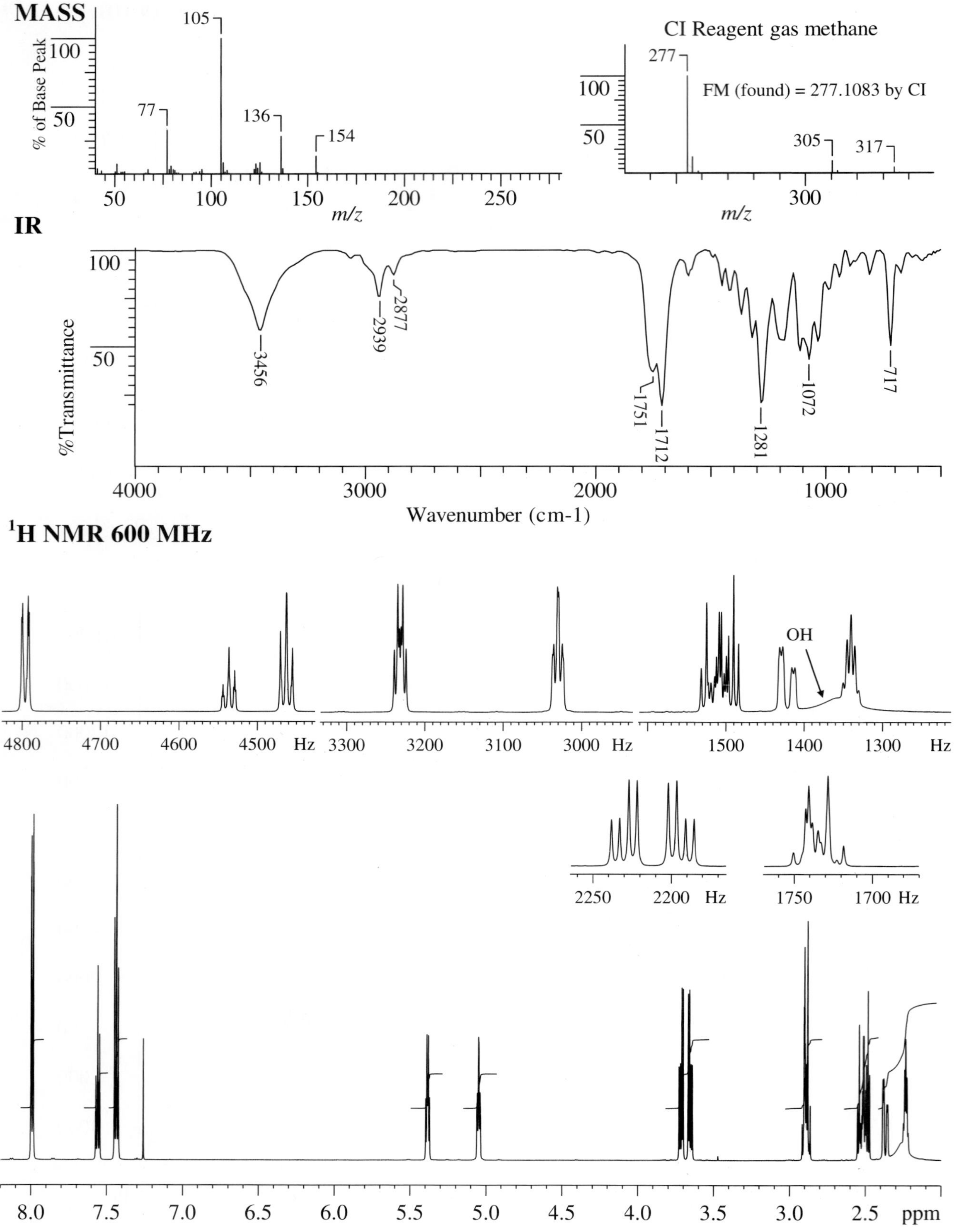

Problem 8.47B

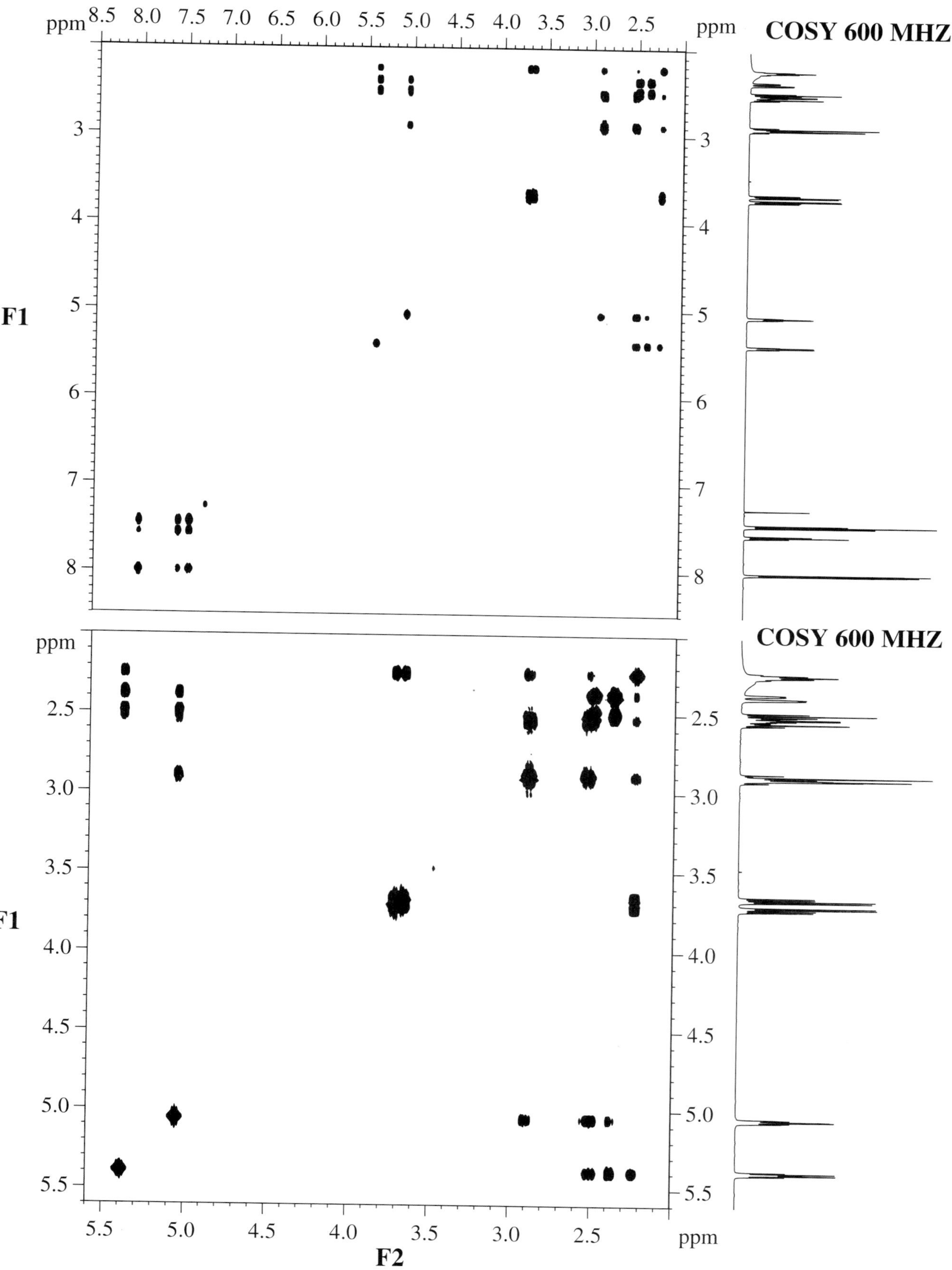

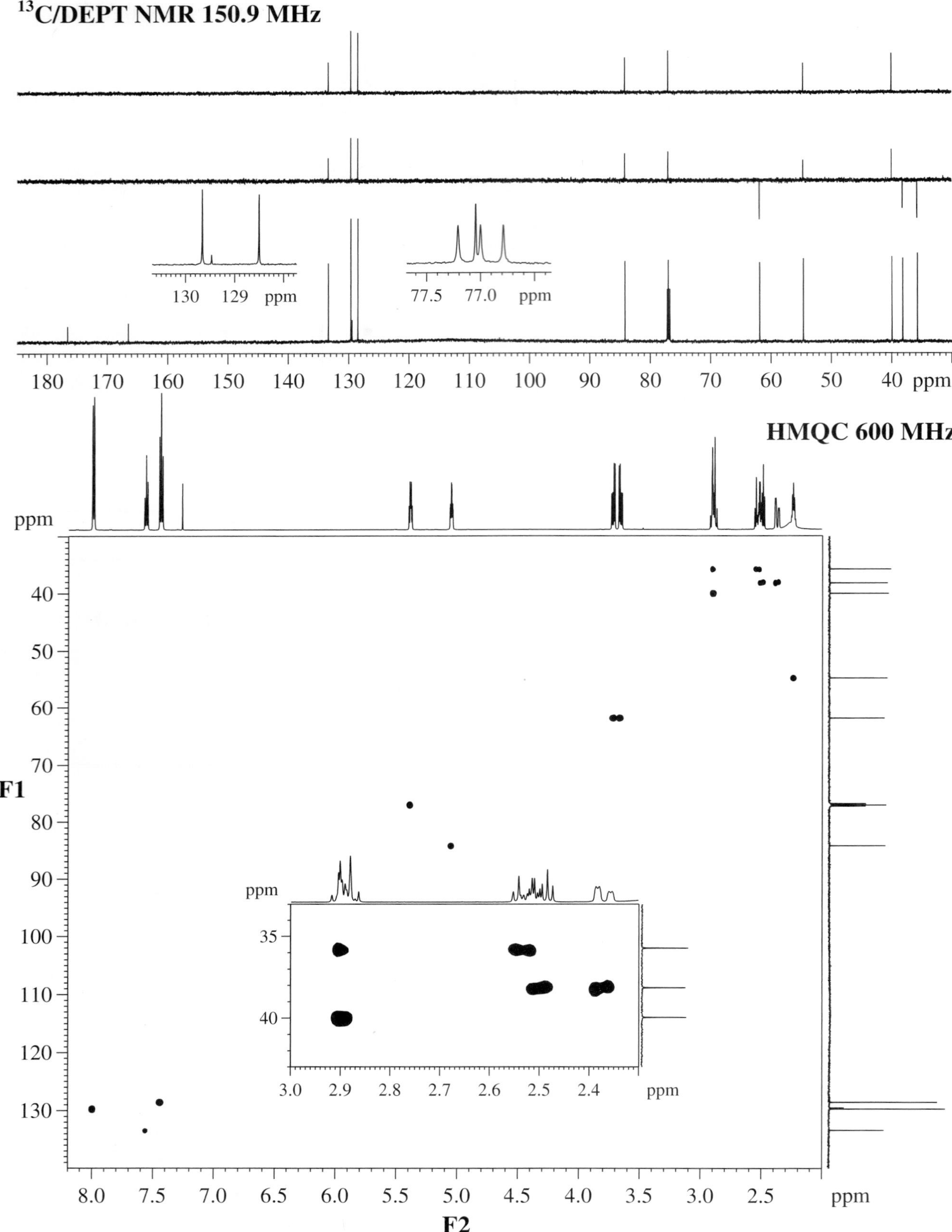

^{13}C/DEPT NMR 150.9 MHz
130 129 ppm
77.5 77.0 ppm
180 170 160 150 140 130 120 110 100 90 80 70 60 50 40 ppm
HMQC 600 MHz
ppm
40 50 60 70 80 90 100 110 120 130
F1
ppm
35 40
3.0 2.9 2.8 2.7 2.6 2.5 2.4 ppm
8.0 7.5 7.0 6.5 6.0 5.5 5.0 4.5 4.0 3.5 3.0 2.5 ppm
F2

Problem 8.47D

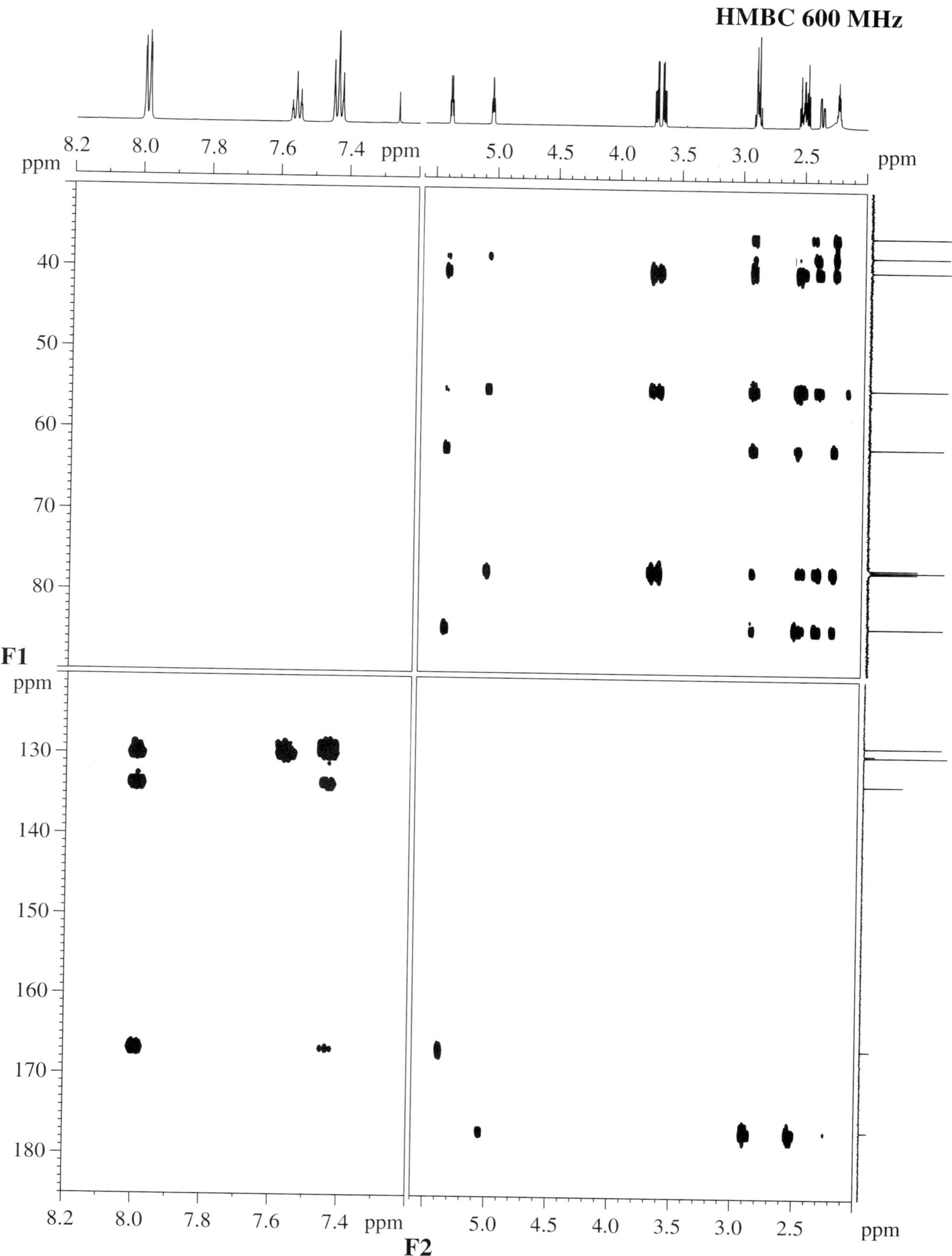

ES MASS

-O3SO
H3C
N+
NH
NH
HN
NH
OH
O
O
% of Base Peak
100
50
0
136
176
194
224
274
293
336
373
399
416
150
200
250
300
350
400
450
m/z

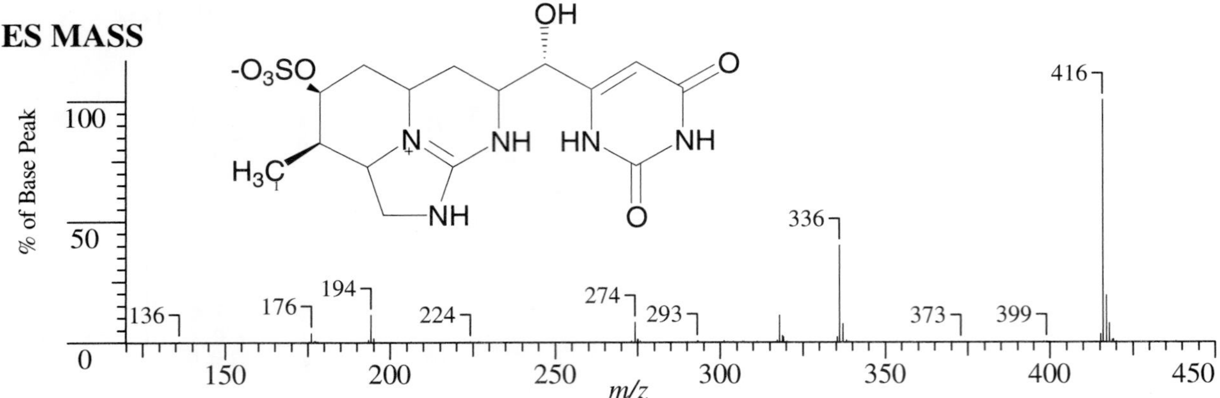

^{1}H NMR 600 MHz in D_2O

3460
3440
Hz
2800
2780
Hz
2720
Hz
2280
2260
Hz
2220
2200
Hz
2160
2140
Hz
1920
1900
Hz
1440
1420
Hz
1260
1240
Hz
1080
1060
Hz
920
900
880
860
Hz
560
550
540
530
Hz
HDO
5.5
5.0
4.5
4.0
3.5
3.0
2.5
2.0
1.5
1.0
ppm

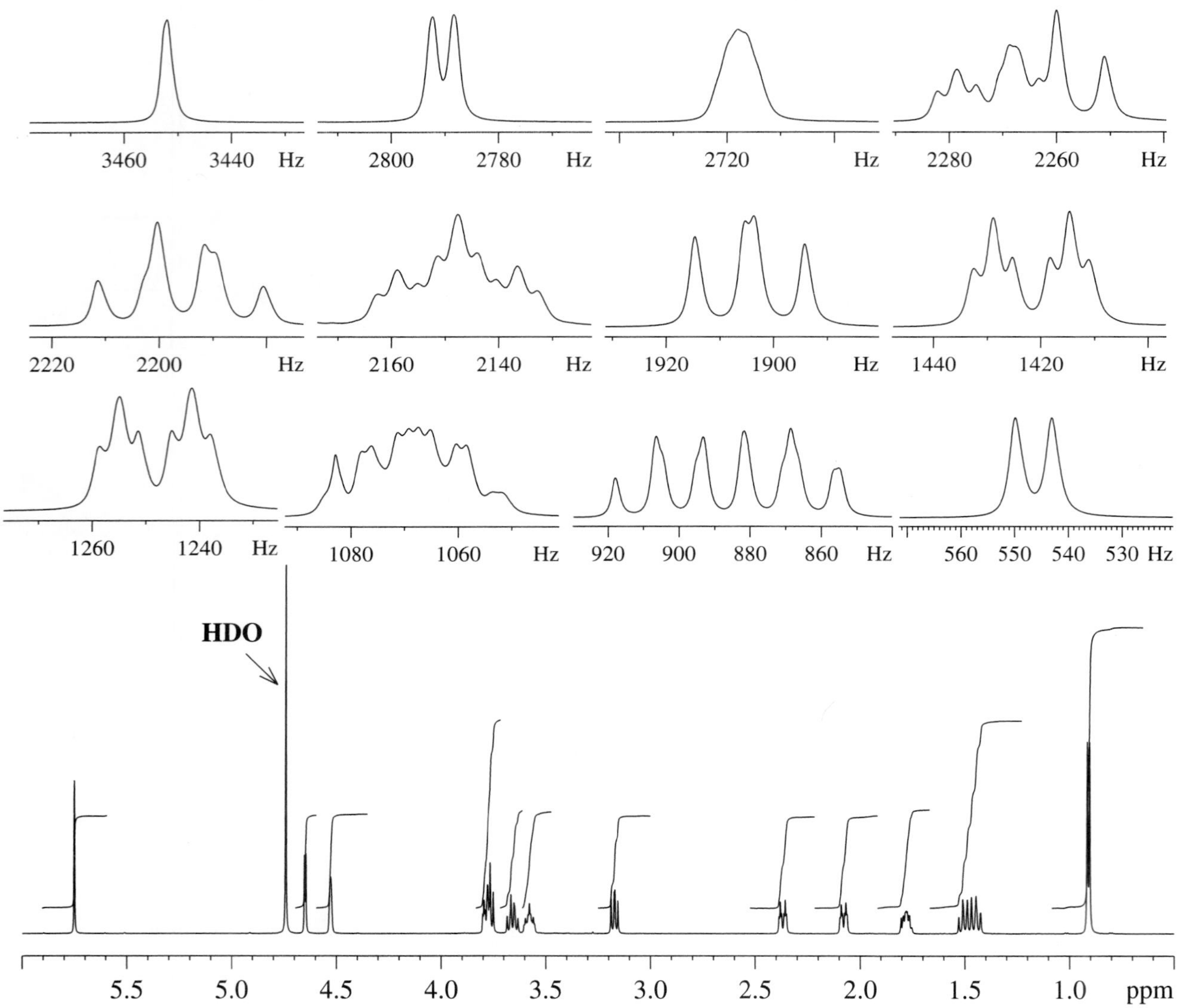

Problem 8.48B

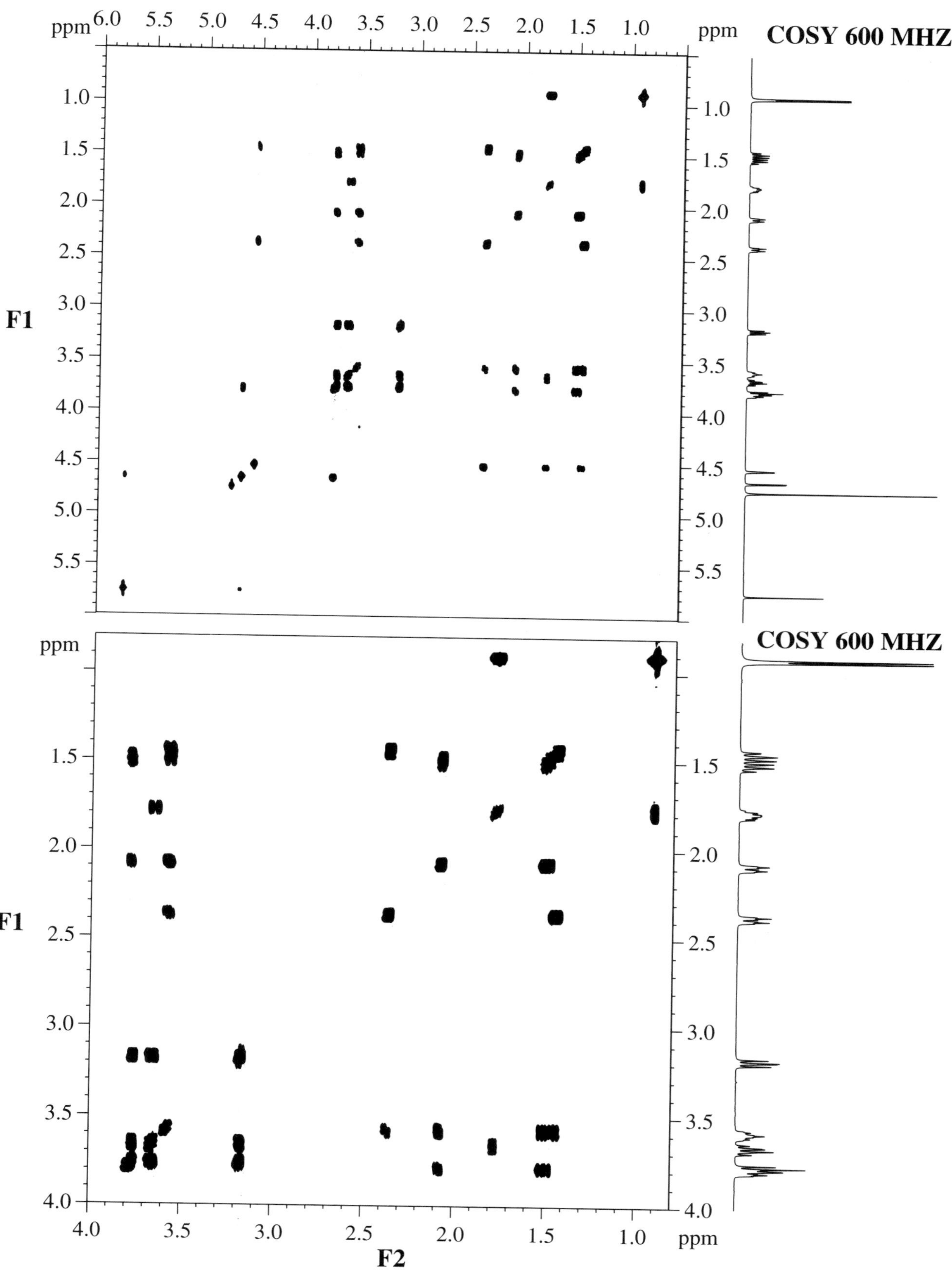

Problem 8.48C

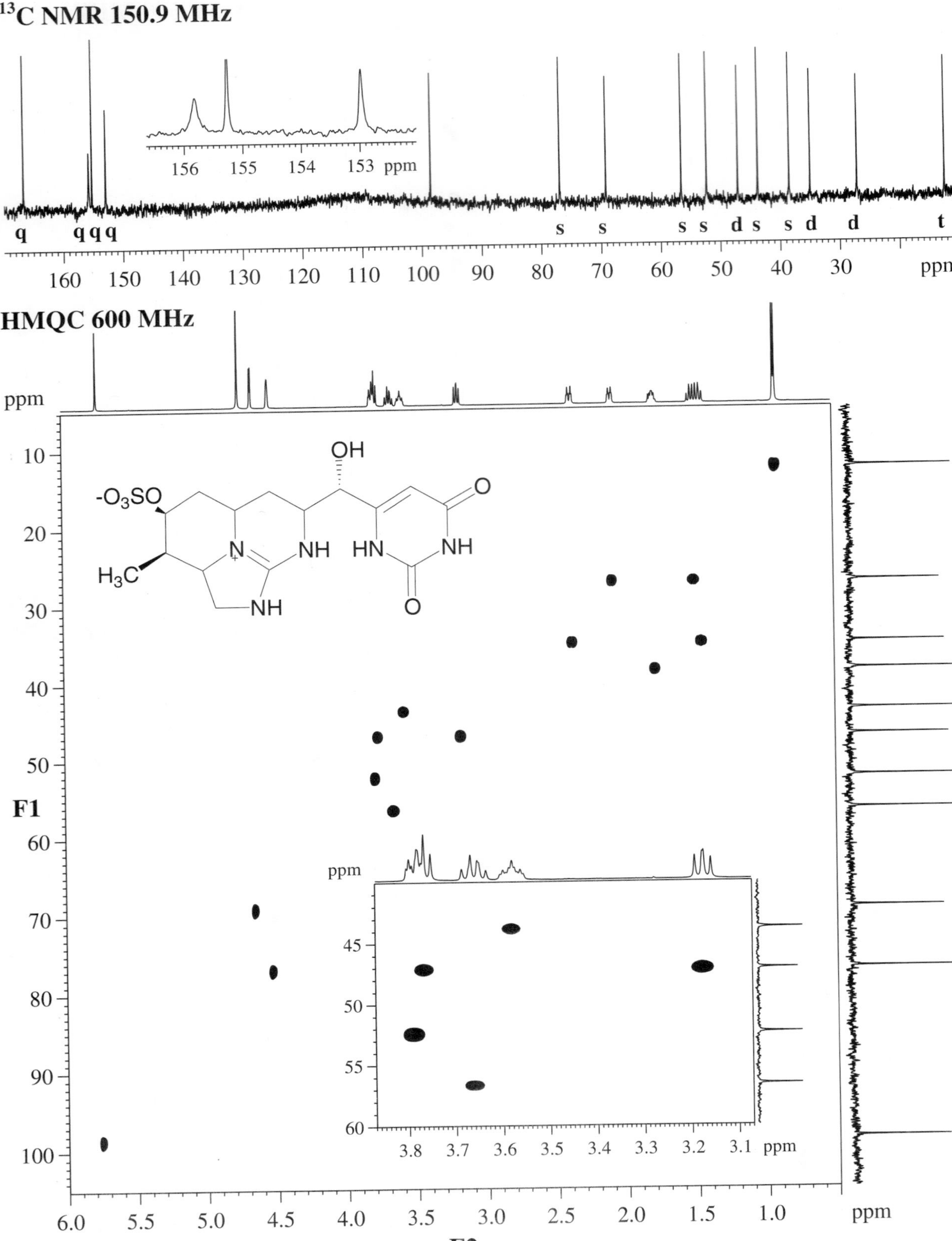

Problem 8.48D

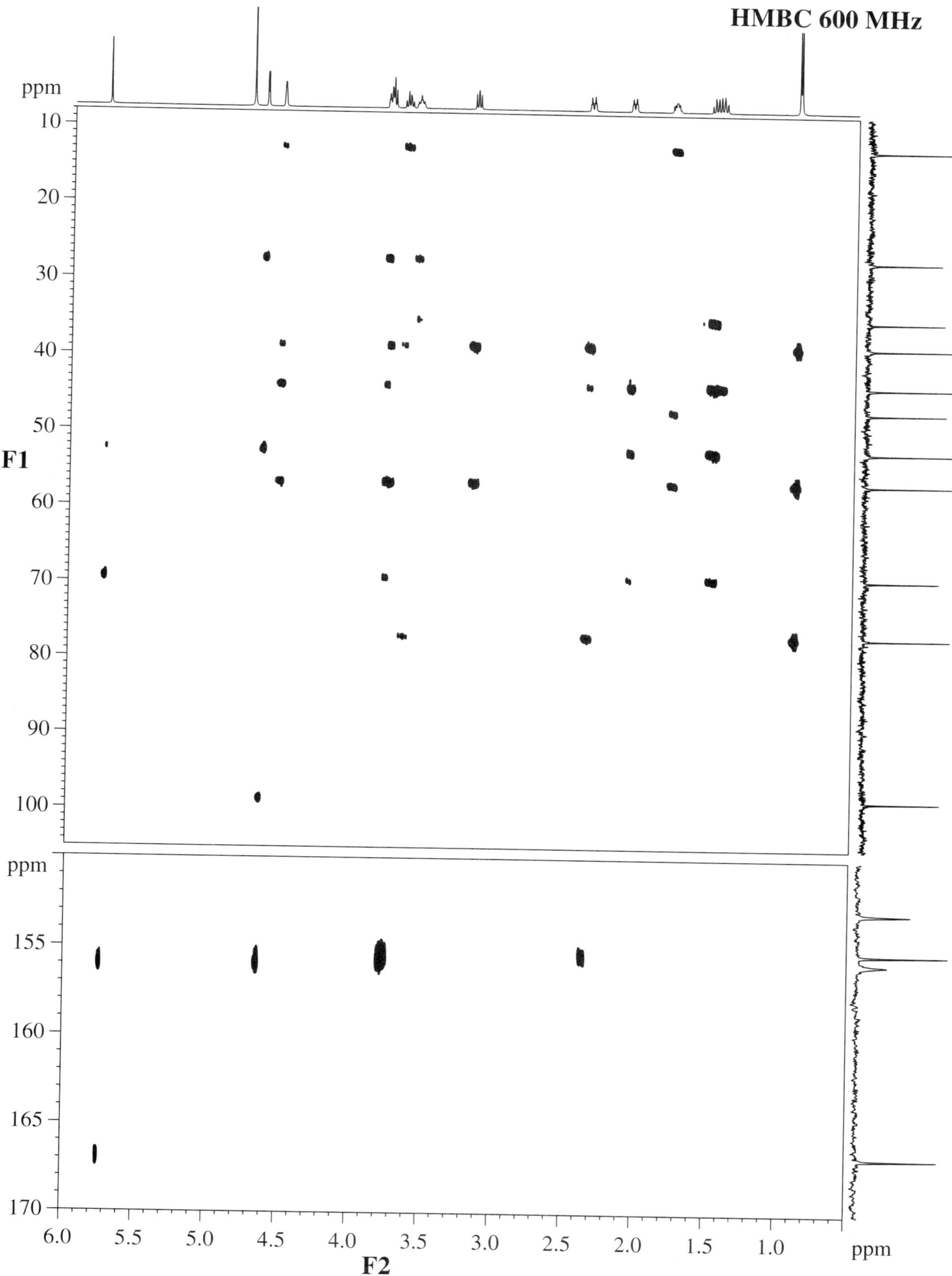

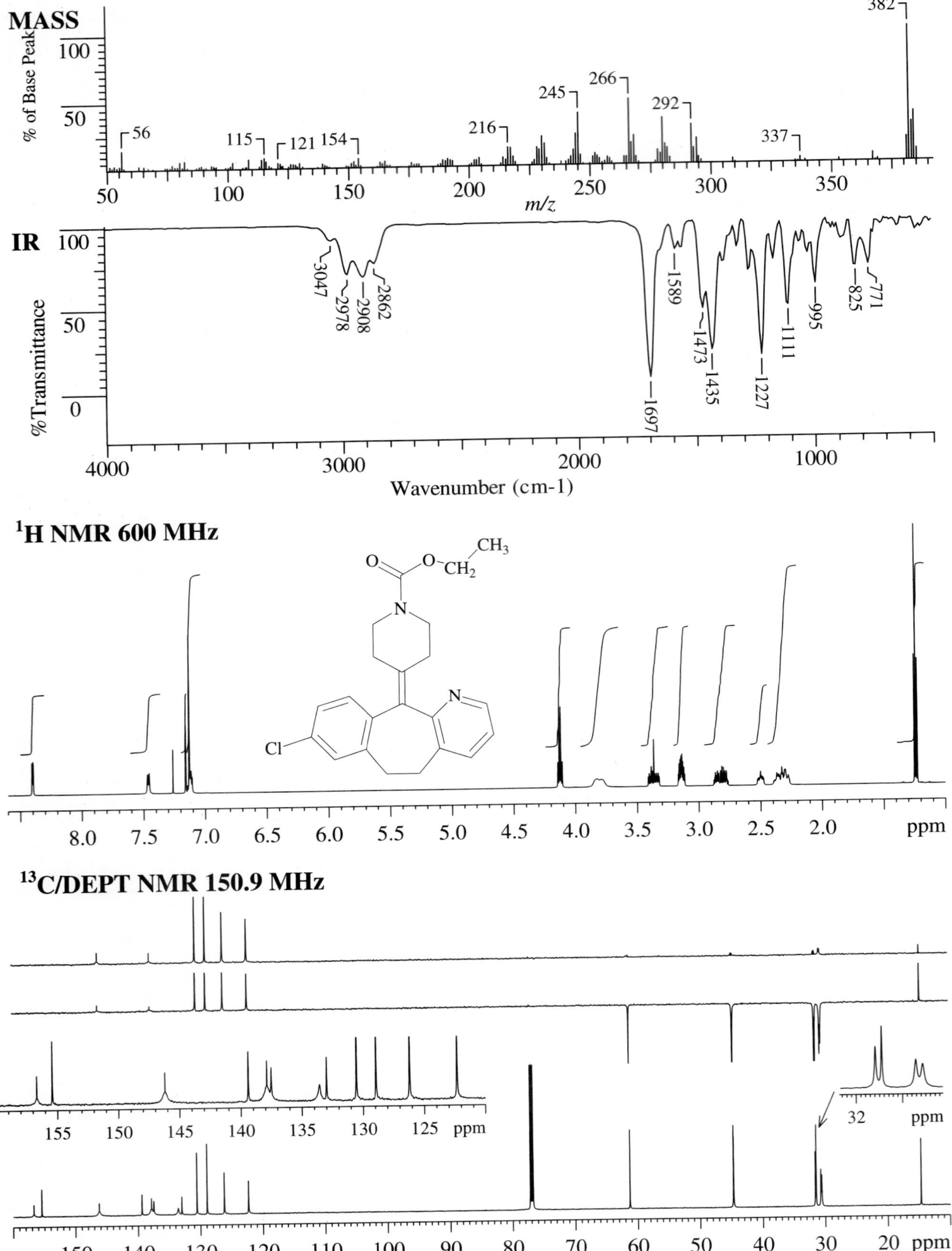
MASS
% of Base Peak
100
50
56
115
121
154
216
245
266
292
337
382
50
100
150
200
250
300
350
m/z
IR
% Transmittance
100
50
0
3047
2978
2908
2862
1697
1589
1473
1435
1227
1111
995
825
771
4000
3000
2000
1000
Wavenumber (cm-1)
¹H NMR 600 MHz
O
O
CH₂
CH₃
N
N
Cl
8.0
7.5
7.0
6.5
6.0
5.5
5.0
4.5
4.0
3.5
3.0
2.5
2.0
ppm
¹³C/DEPT NMR 150.9 MHz
155
150
145
140
135
130
125
ppm
32
ppm
150
140
130
120
110
100
90
80
70
60
50
40
30
20 ppm

Problem 8.49B

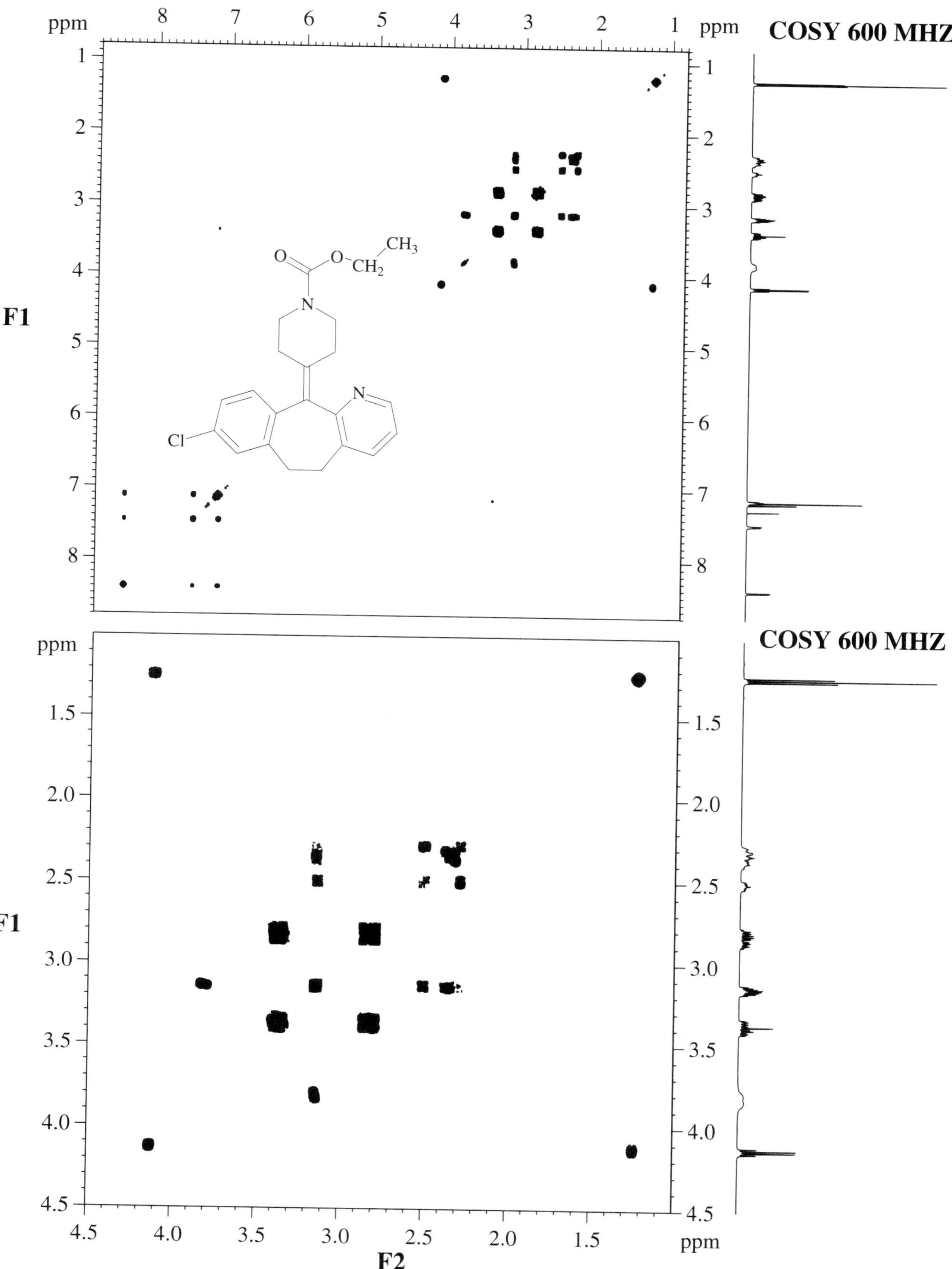

Problem 8.49C

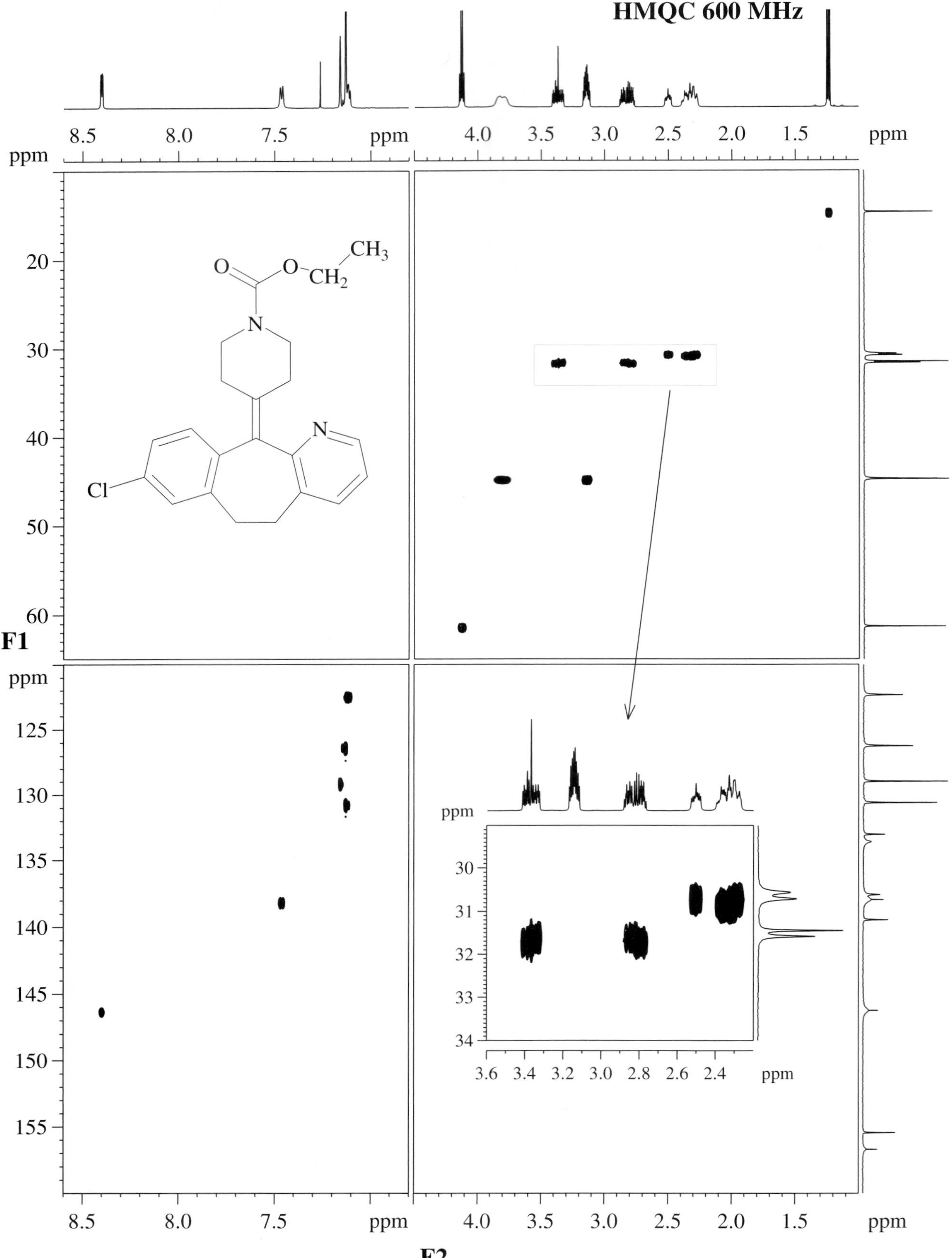

Problem 8.49D

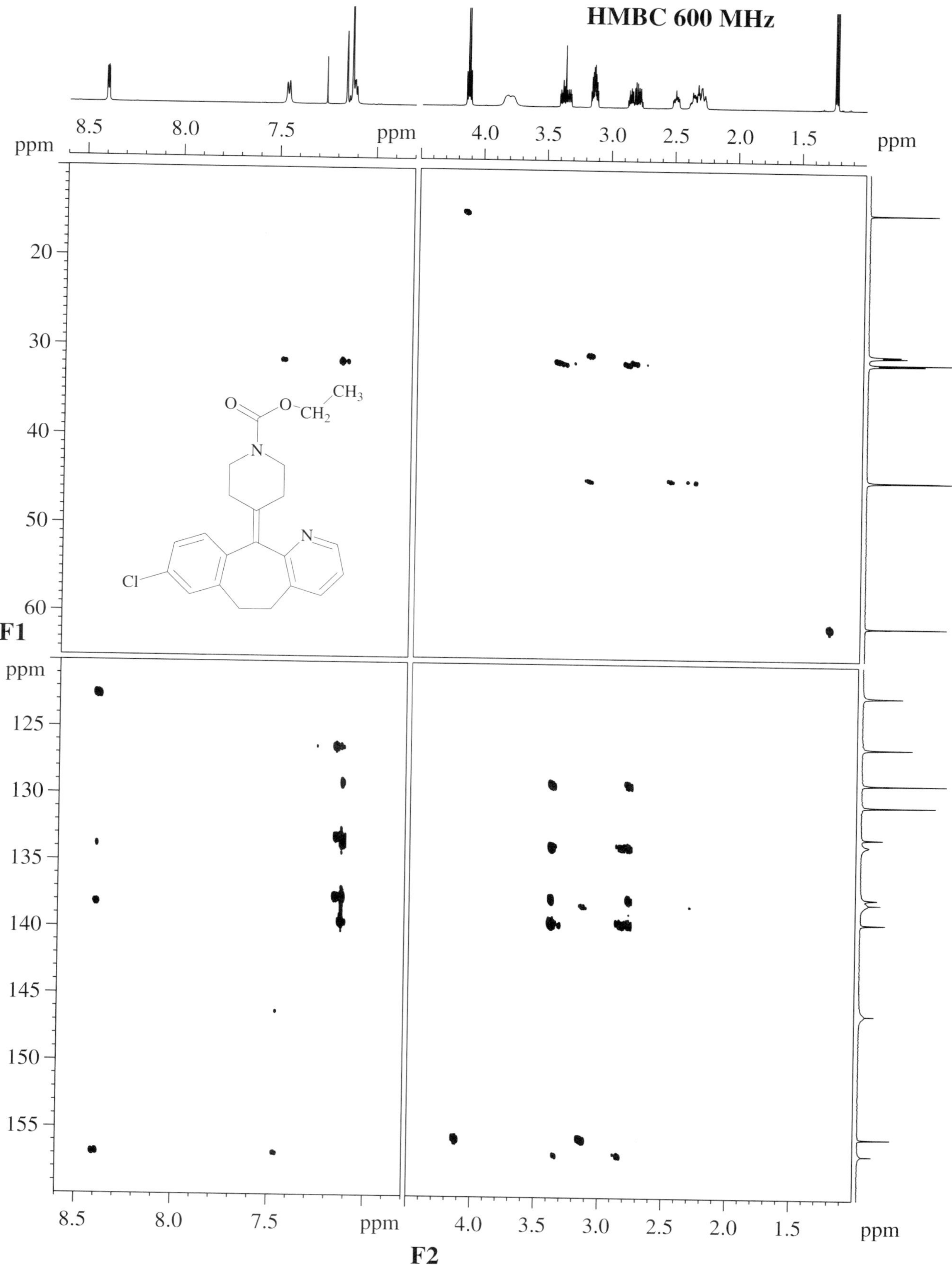

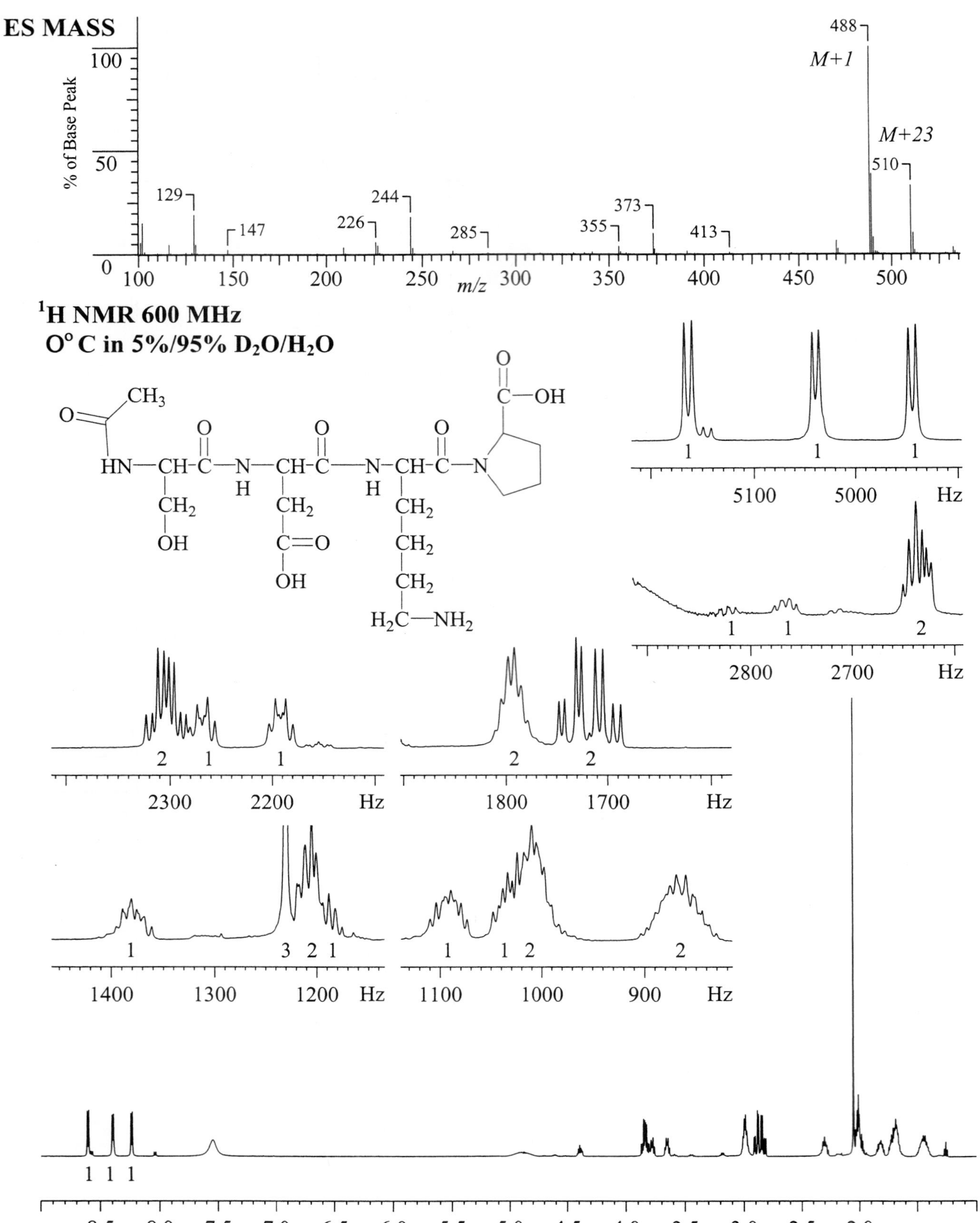

ES MASS
% of Base Peak
100
50
0
129
147
226
244
285
355
373
413
488
M+1
M+23
510
100
150
200
250
m/z
300
350
400
450
500
1H NMR 600 MHz
O° C in 5%/95% D2O/H2O
CH3
O
HN
CH
C
N
H
CH2
OH
C=O
NH2
H2C
C-OH
5100
5000
Hz
2800
2700
2300
2200
1800
1700
1400
1300
1200
1100
1000
900
8.5
8.0
7.5
7.0
6.5
6.0
5.5
5.0
4.5
4.0
3.5
3.0
2.5
2.0
ppm

Problem 8.50B

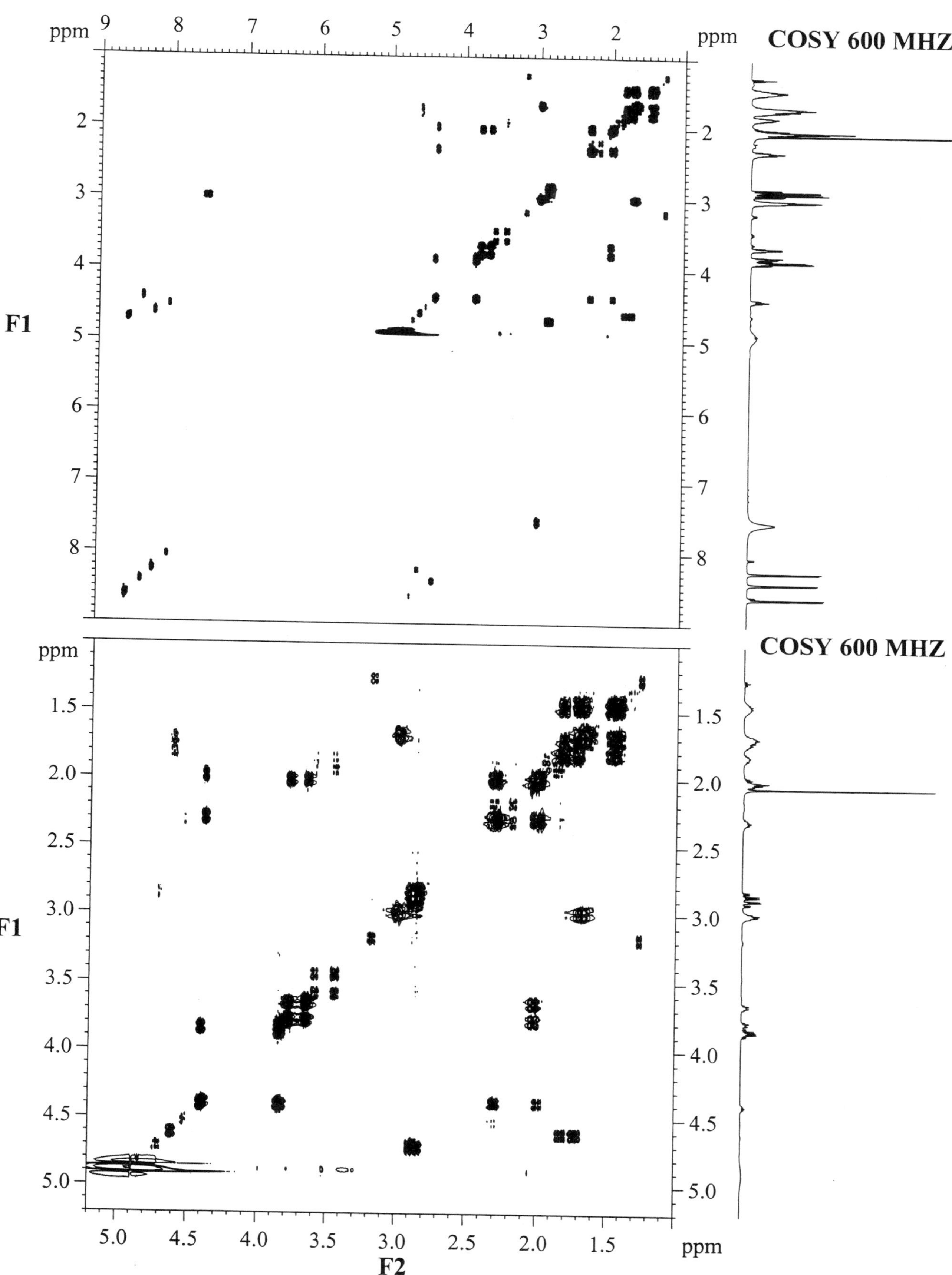

Problem 8.50C

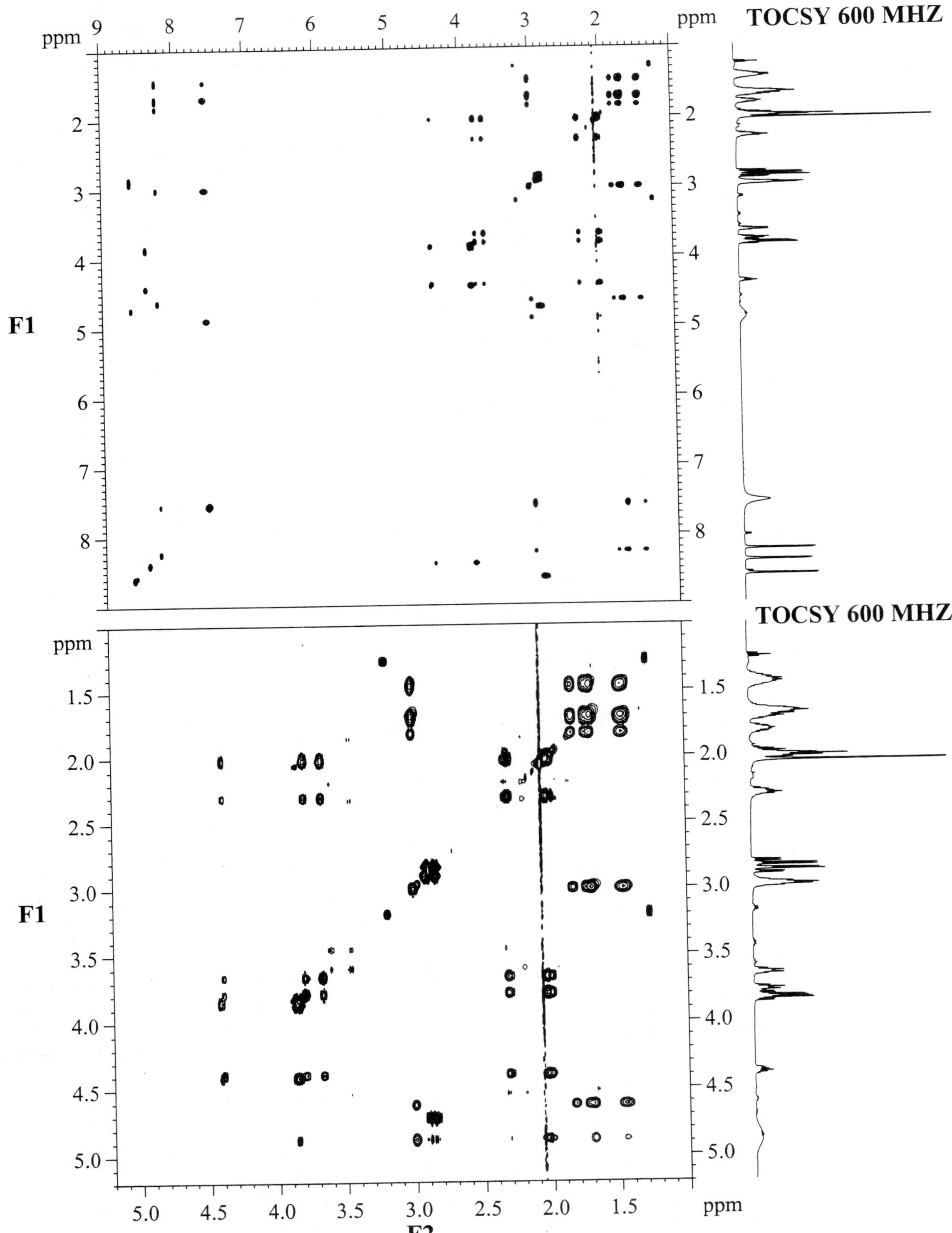

Problem 8.50D

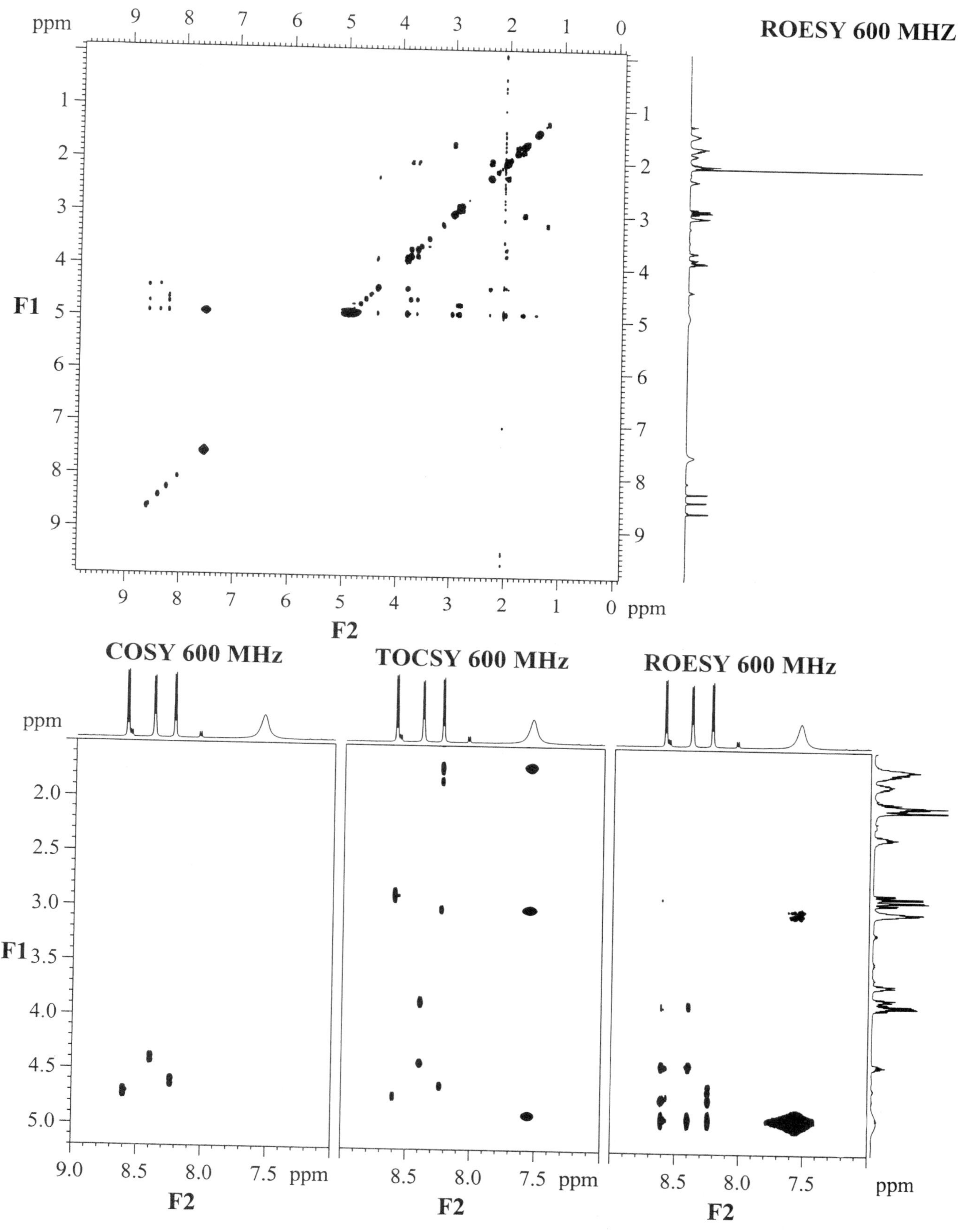

Problem 8.50E

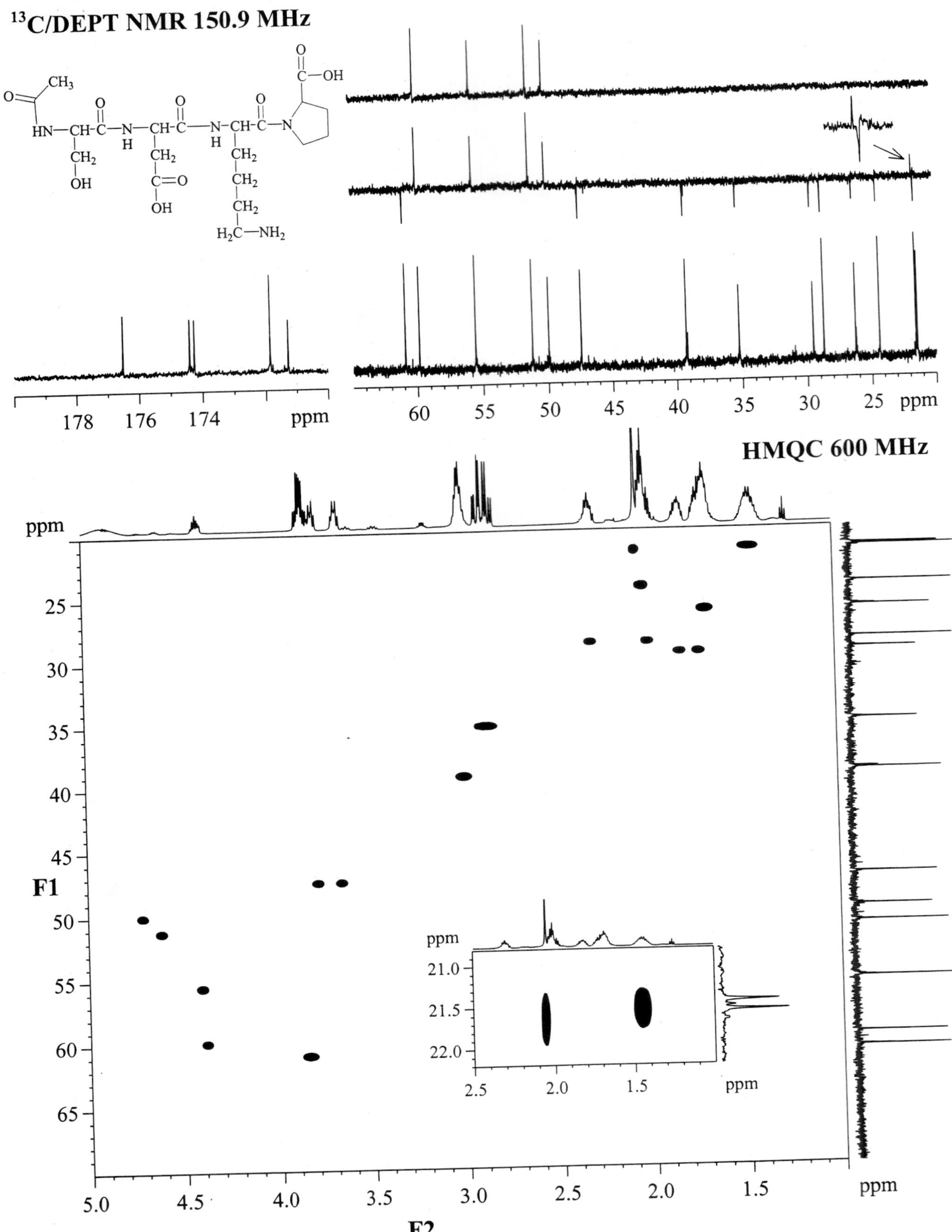

Problem 8.51A

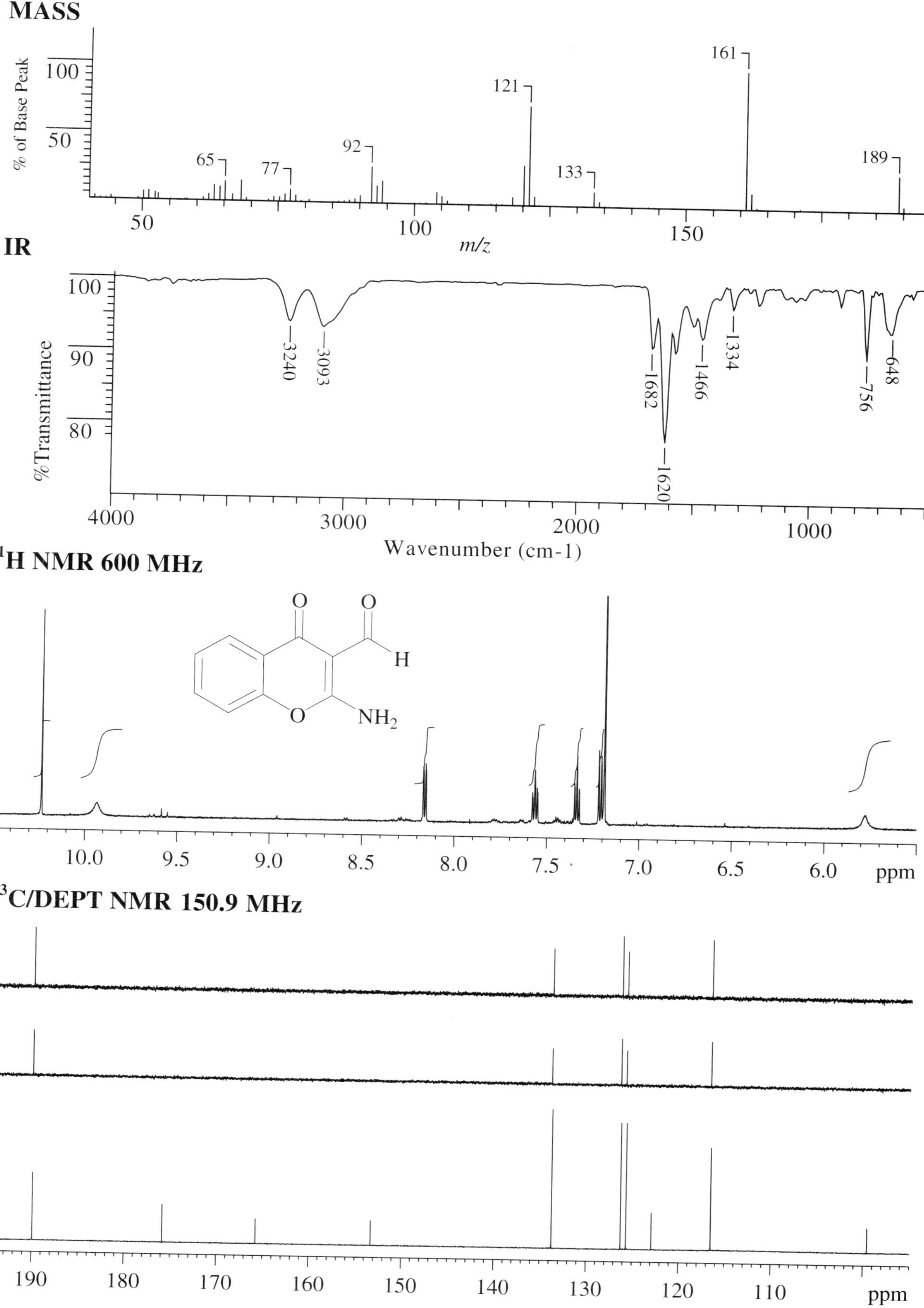

Problem 8.51B

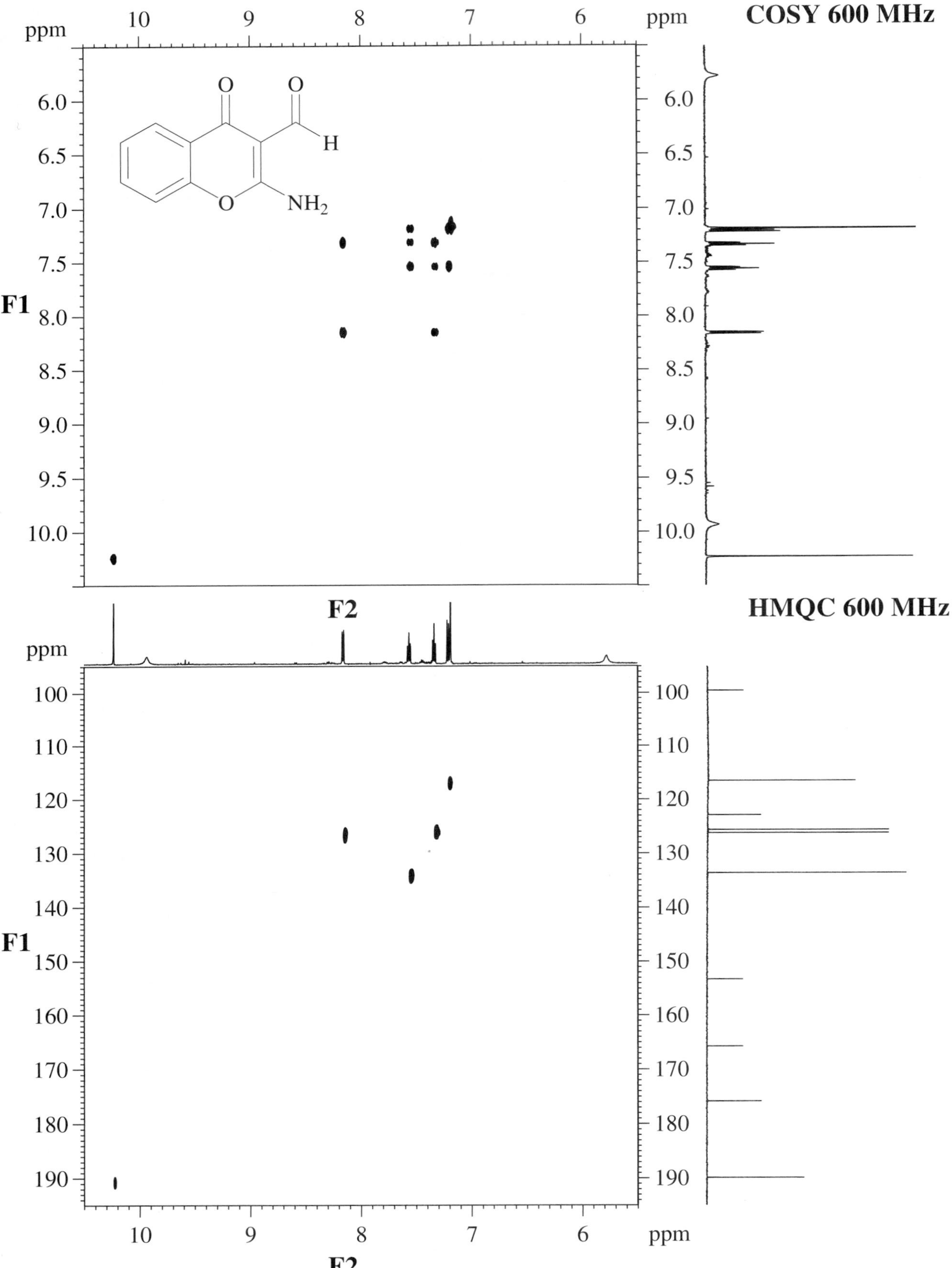

Problem 8.51C

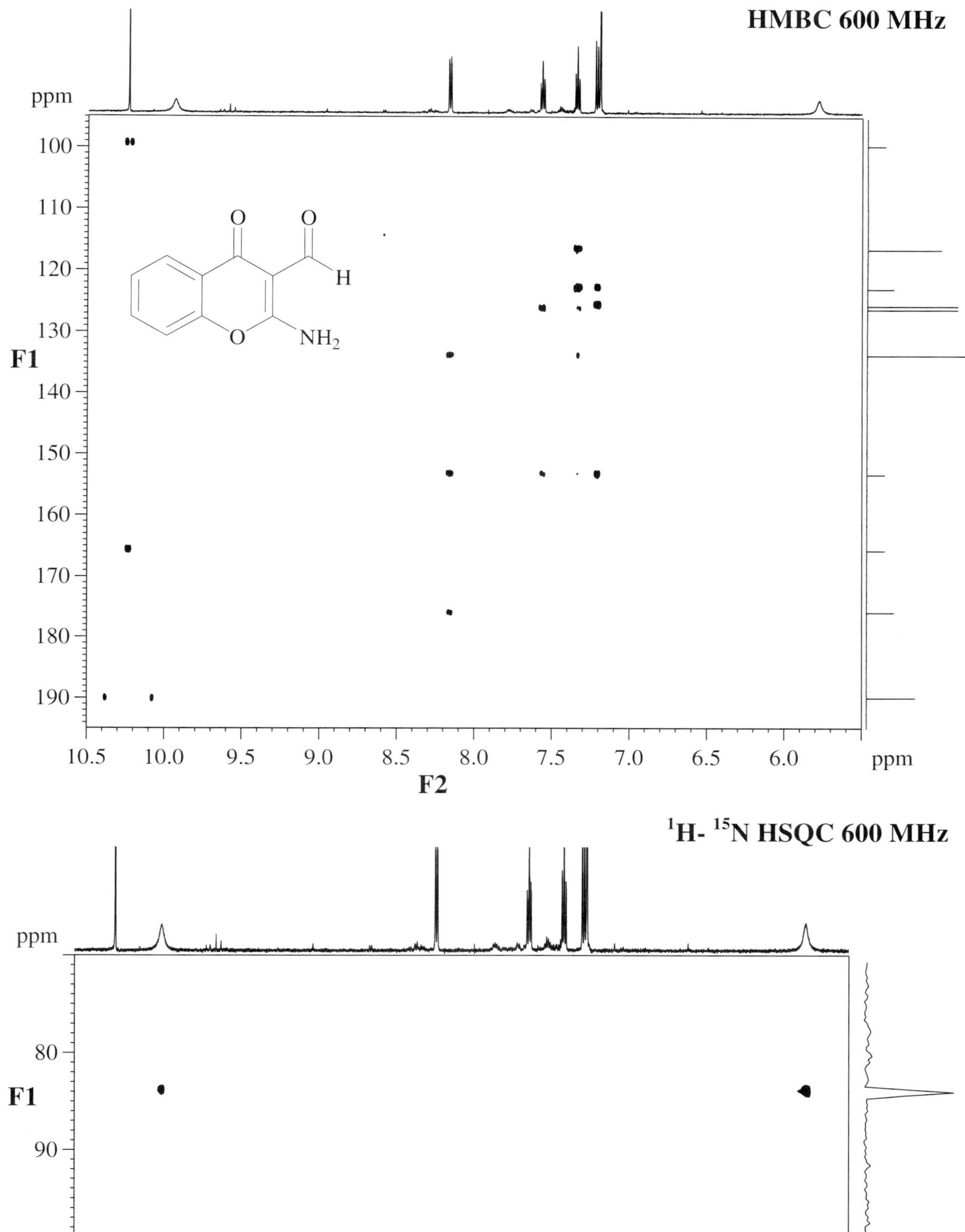

Problem 8.52A

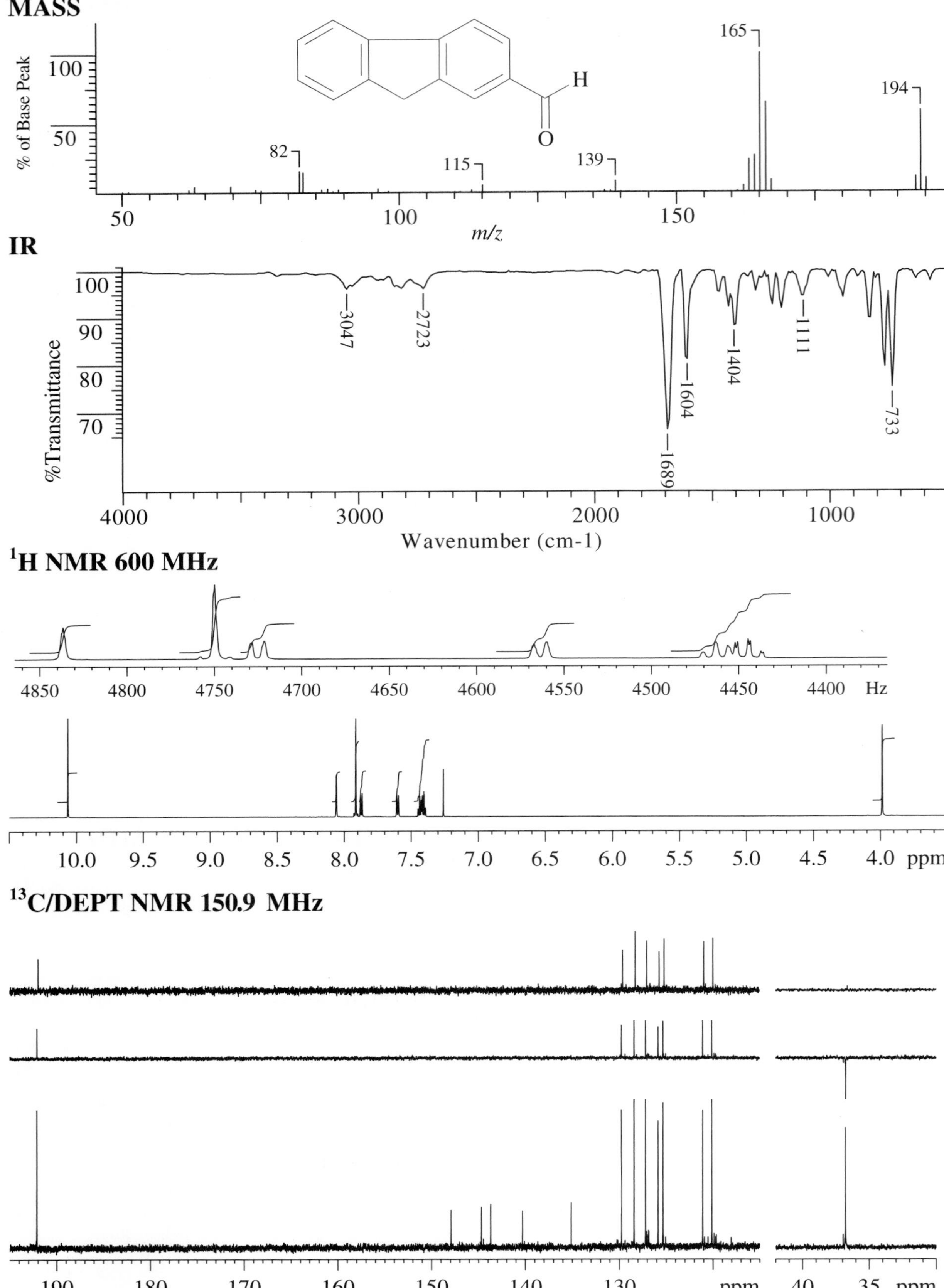

Problem 8.52B

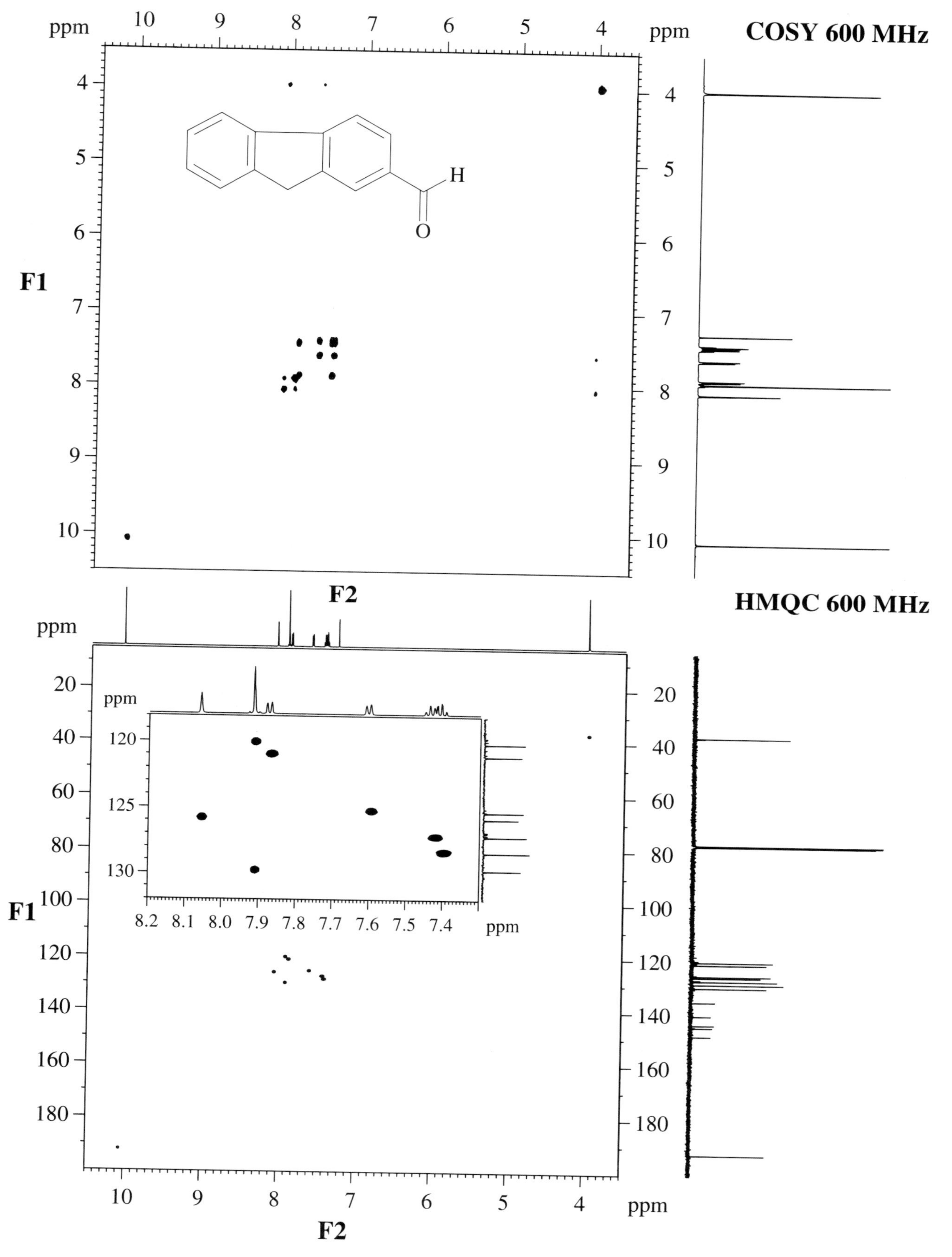

Problem 8.52C

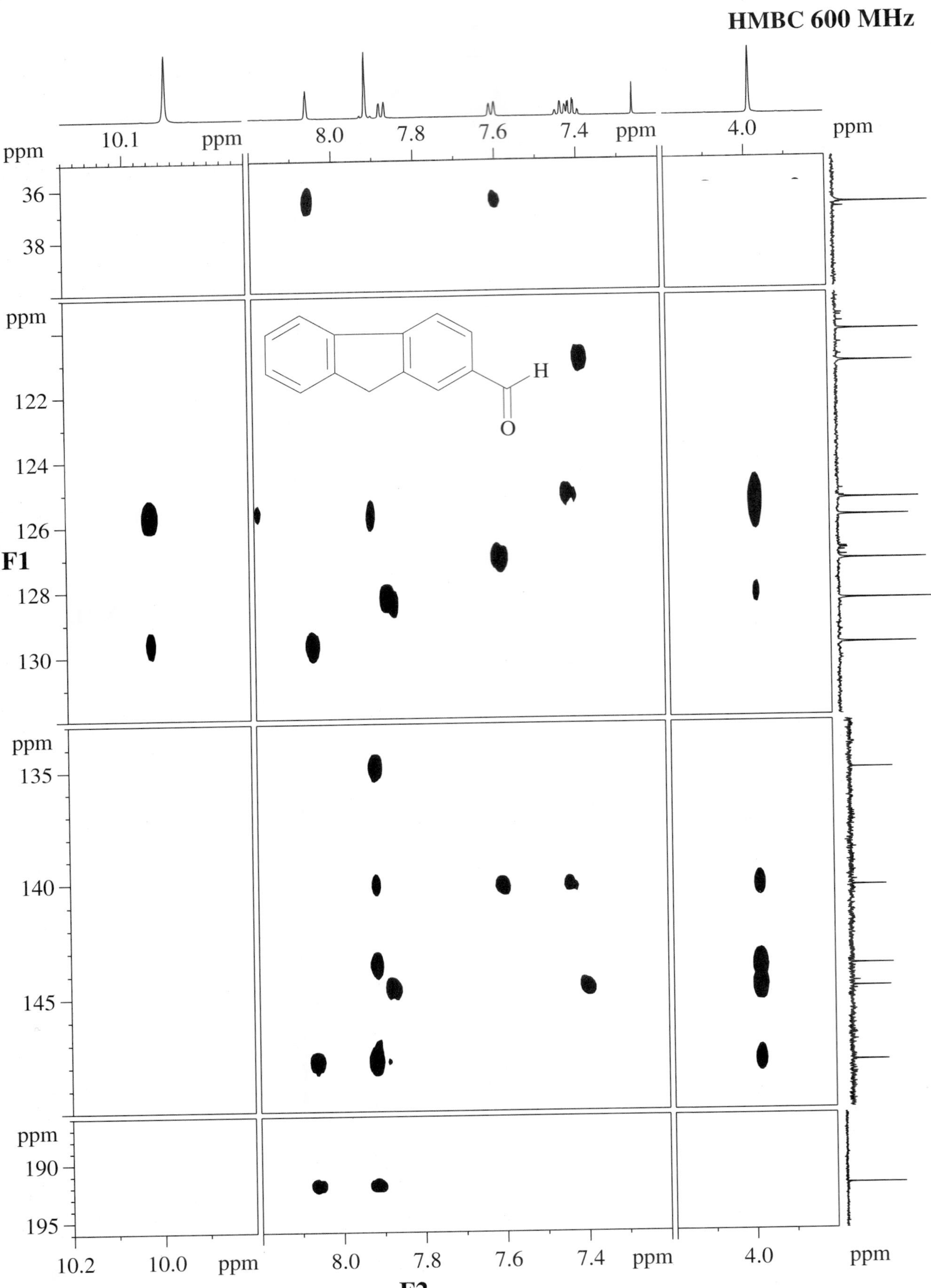

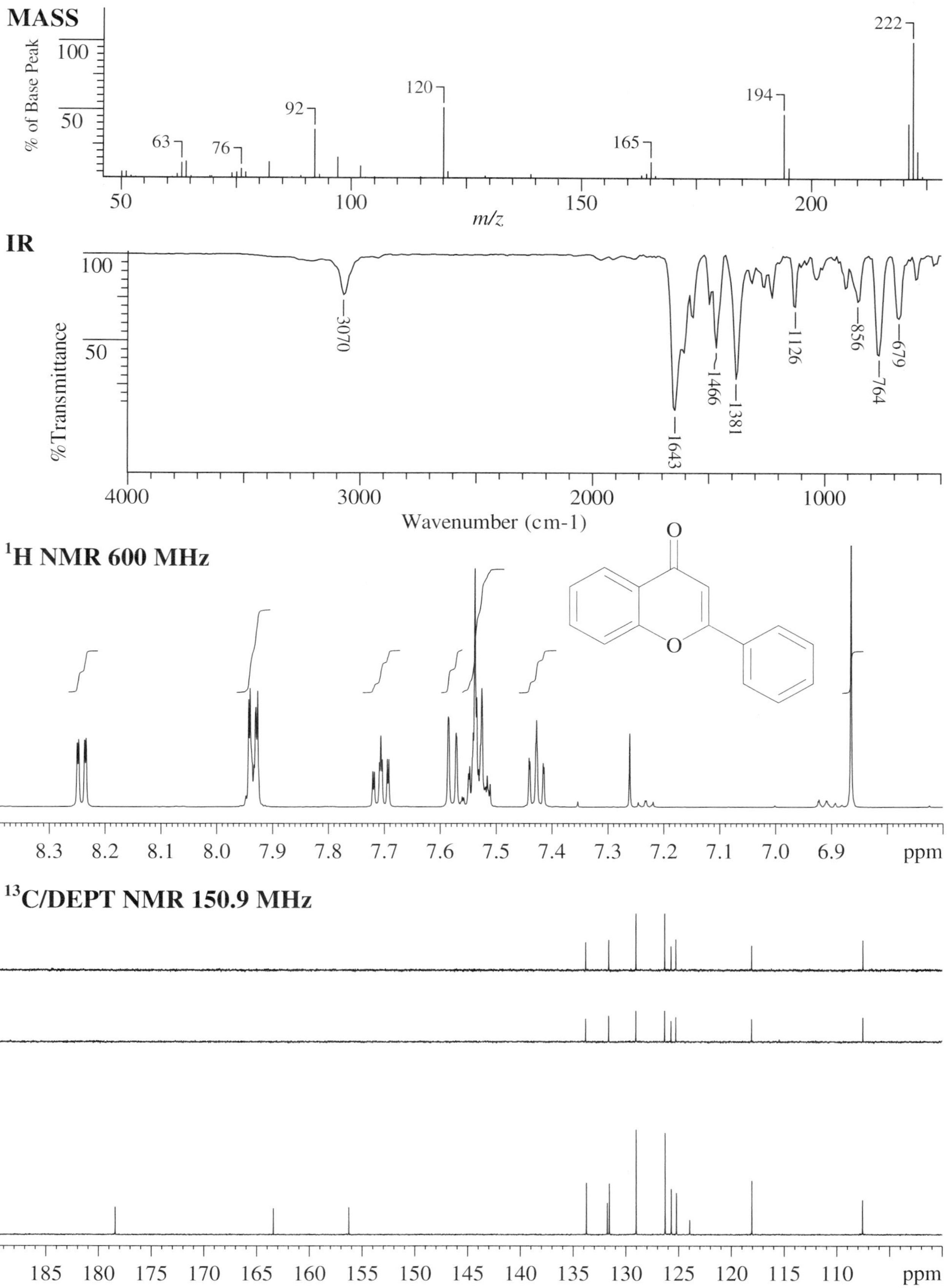
MASS
% of Base Peak
100
50
222
194
165
120
92
76
63
50
100
150
200
m/z
IR
% Transmittance
100
50
3070
1643
1466
1381
1126
856
764
679
4000
3000
2000
1000
Wavenumber (cm-1)
1H NMR 600 MHz
O
O
8.3 8.2 8.1 8.0 7.9 7.8 7.7 7.6 7.5 7.4 7.3 7.2 7.1 7.0 6.9 ppm
13C/DEPT NMR 150.9 MHz
185 180 175 170 165 160 155 150 145 140 135 130 125 120 115 110 ppm

Problem 8.53B

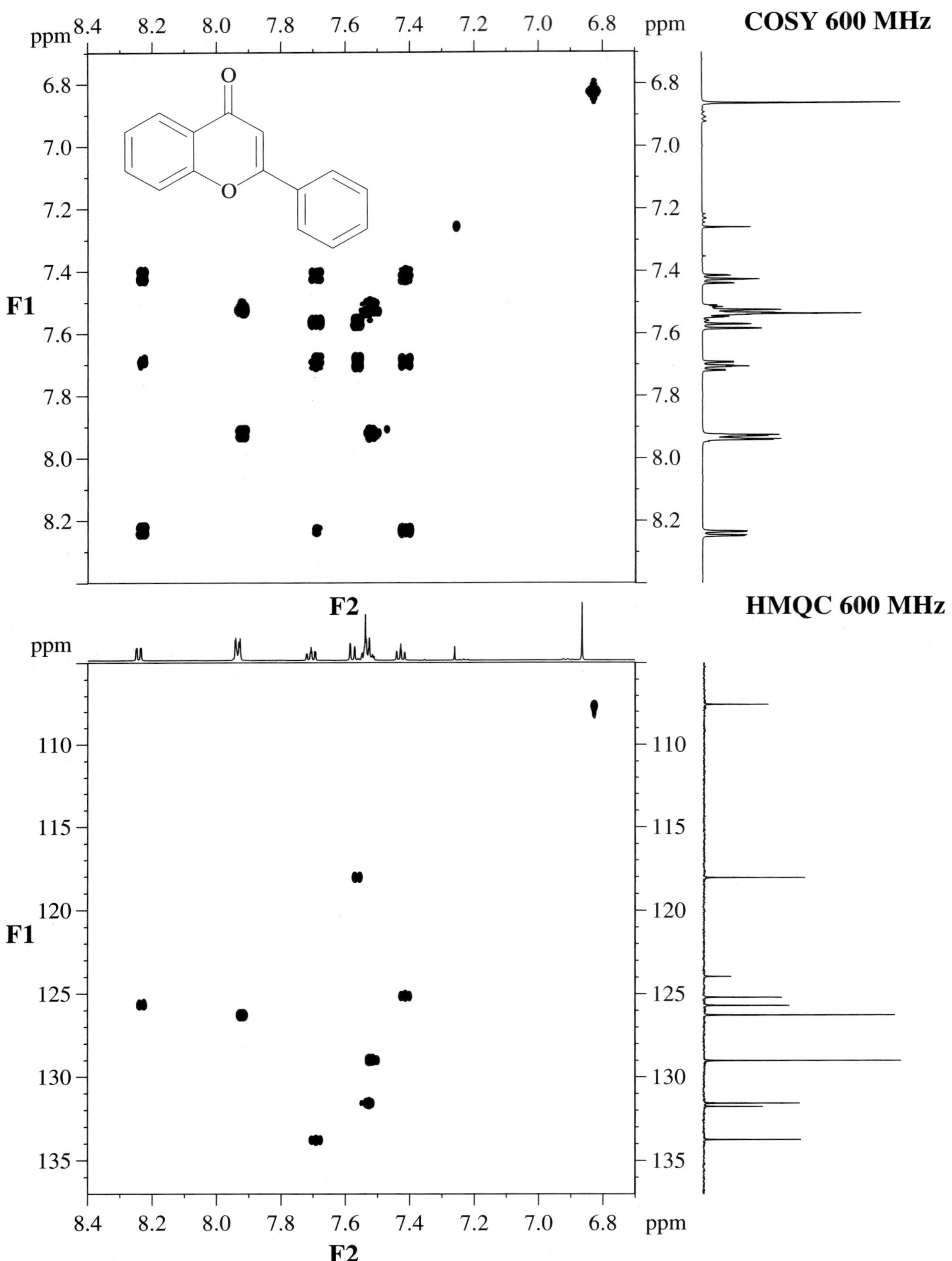

Problem 8.53C

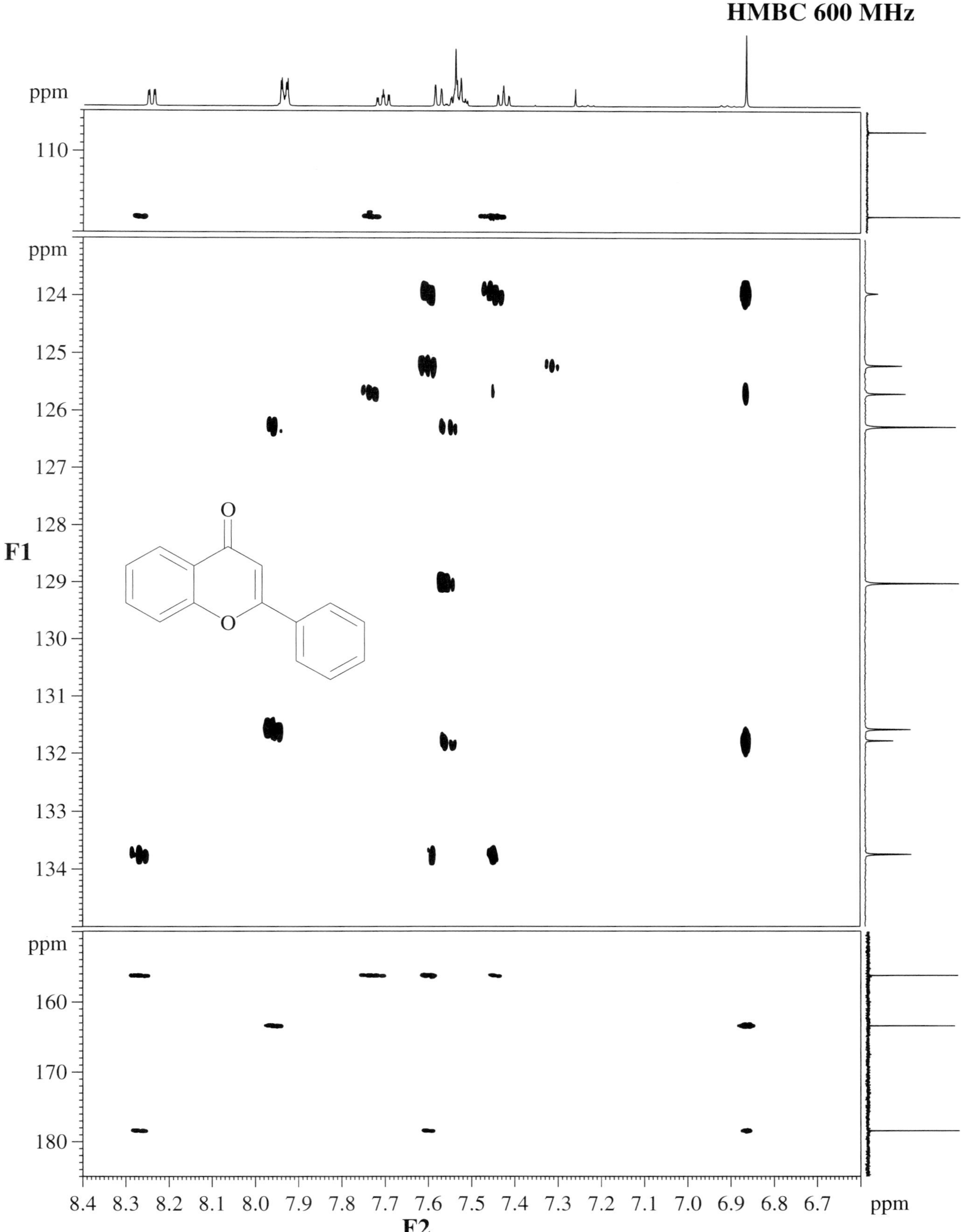

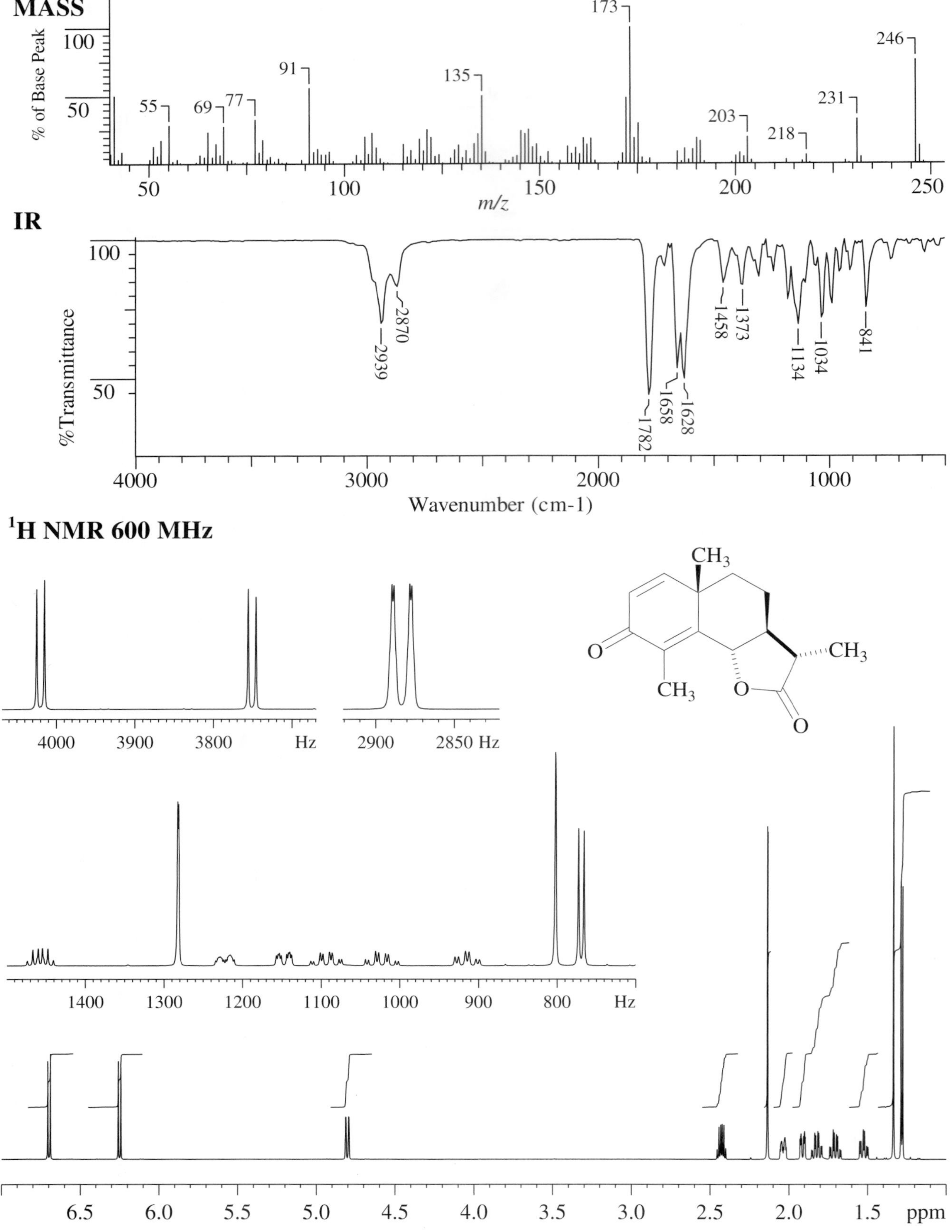
MASS
% of Base Peak
100
50
55
69
77
91
135
173
203
218
231
246
50
100
150
200
250
m/z
IR
%Transmittance
100
50
2939
2870
1782
1658
1628
1458
1373
1134
1034
841
4000
3000
2000
1000
Wavenumber (cm-1)
¹H NMR 600 MHz
4000
3900
3800
Hz
2900
2850 Hz
1400
1300
1200
1100
1000
900
800
Hz
6.5
6.0
5.5
5.0
4.5
4.0
3.5
3.0
2.5
2.0
1.5
ppm
CH3
O
CH3
O
O
CH3

Problem 8.54B

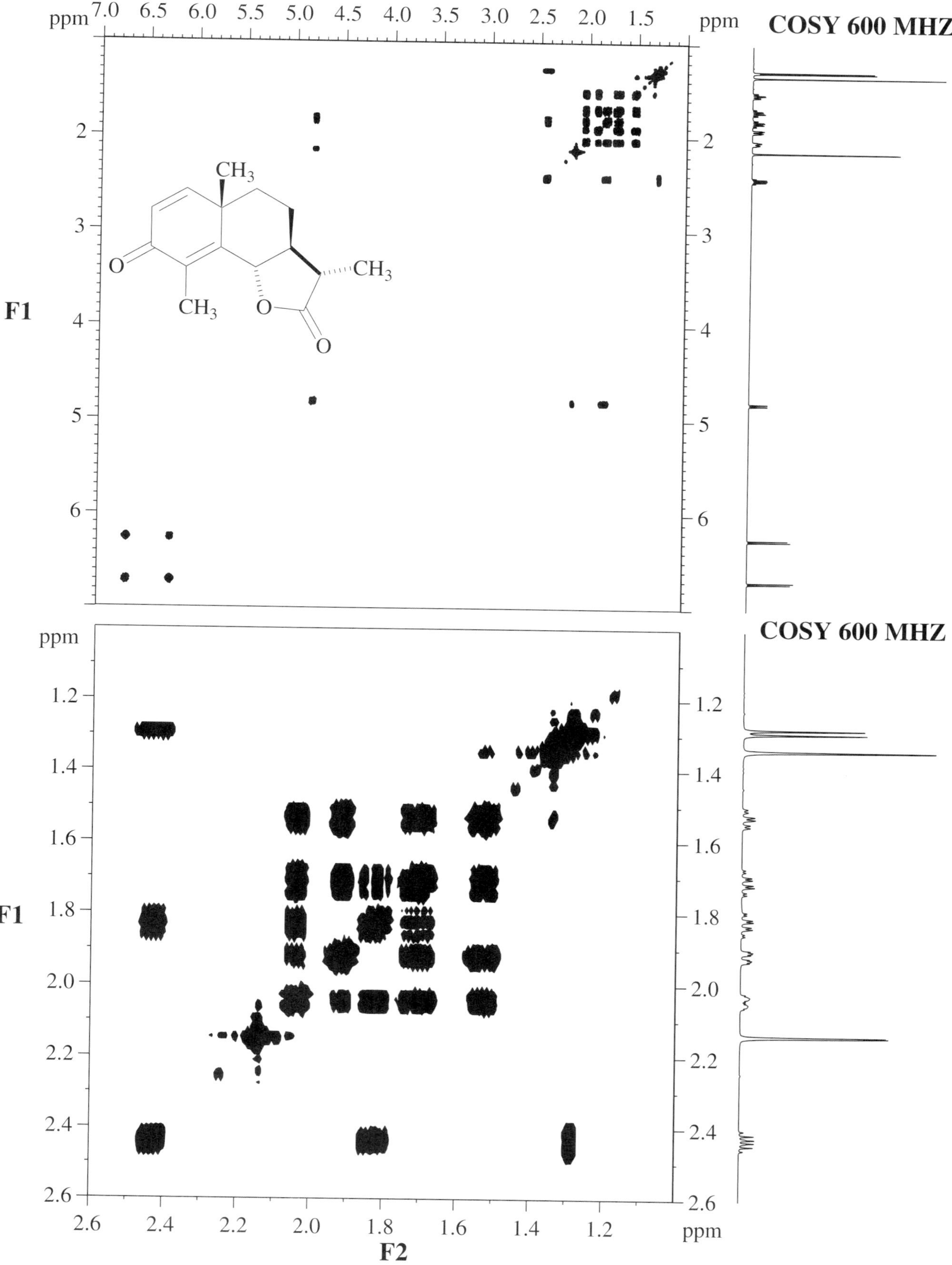

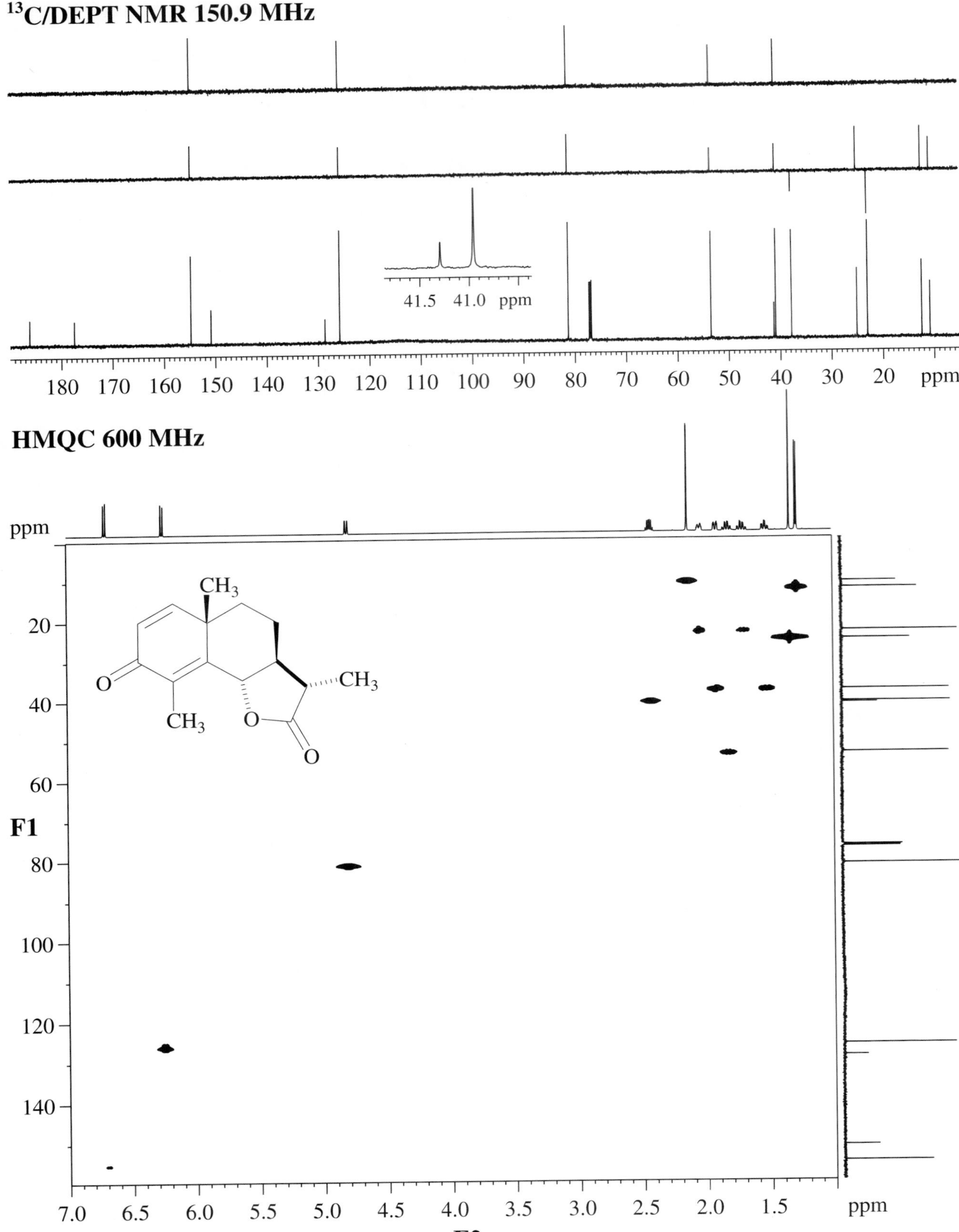

^{13}C/DEPT NMR 150.9 MHz
41.5 41.0 ppm
180 170 160 150 140 130 120 110 100 90 80 70 60 50 40 30 20 ppm
HMQC 600 MHz
ppm
20
40
60
F1
80
100
120
140
CH3
O
CH3
O
CH3
O
7.0 6.5 6.0 5.5 5.0 4.5 4.0 3.5 3.0 2.5 2.0 1.5 ppm
F2

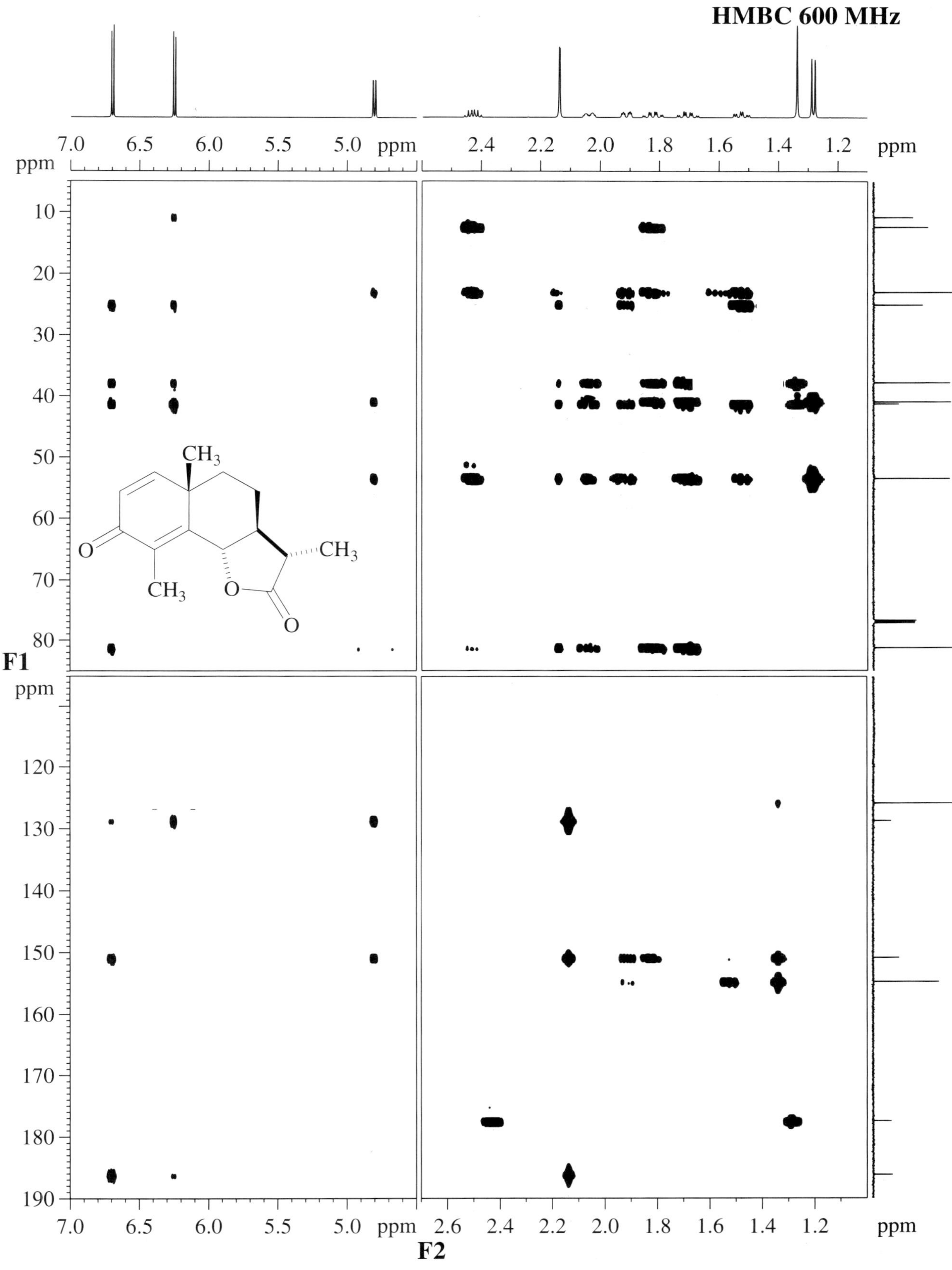
HMBC 600 MHz
ppm
F1
F2
CH3
O
CH3
O
CH3
O

Problem 8.55A

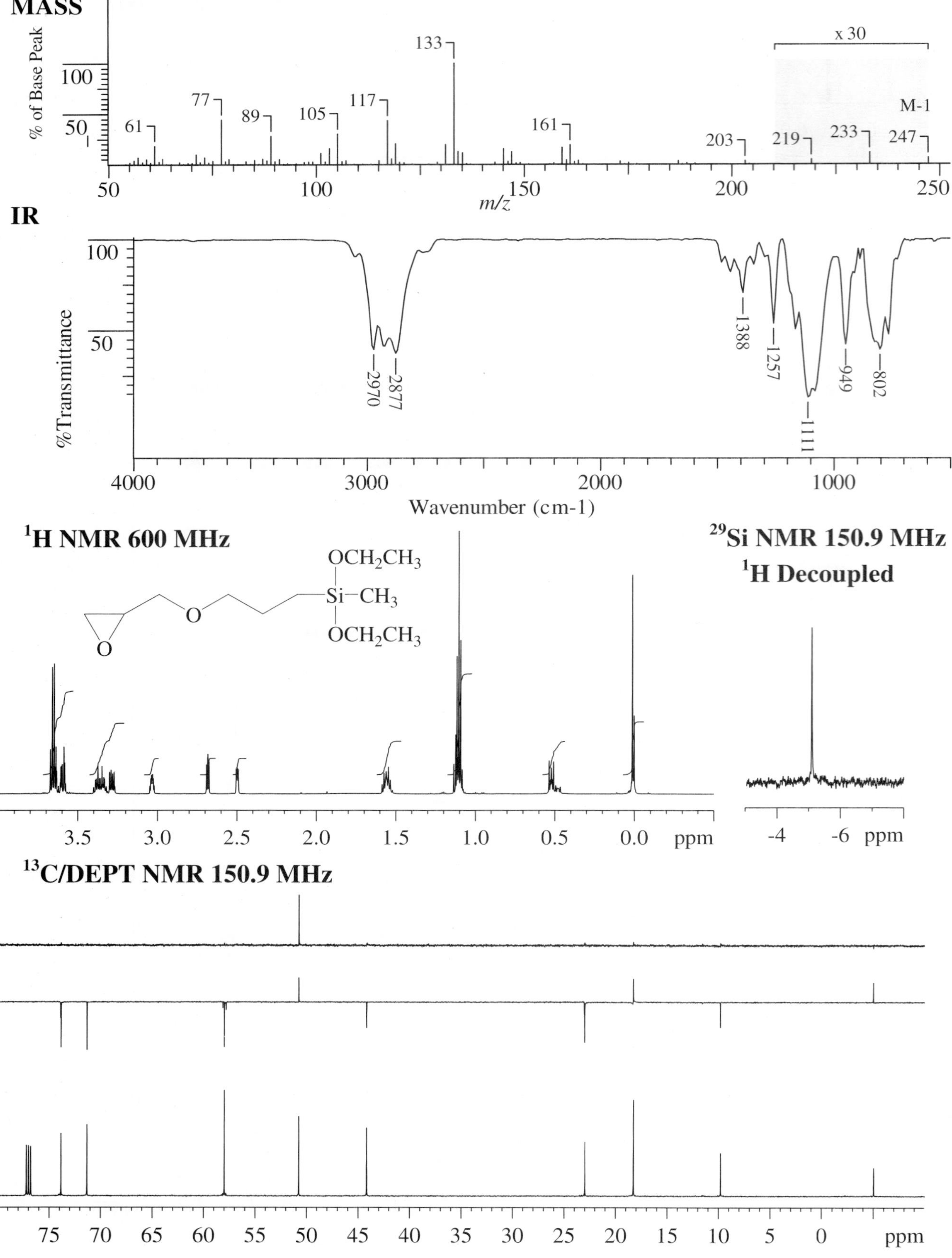

Problem 8.55B

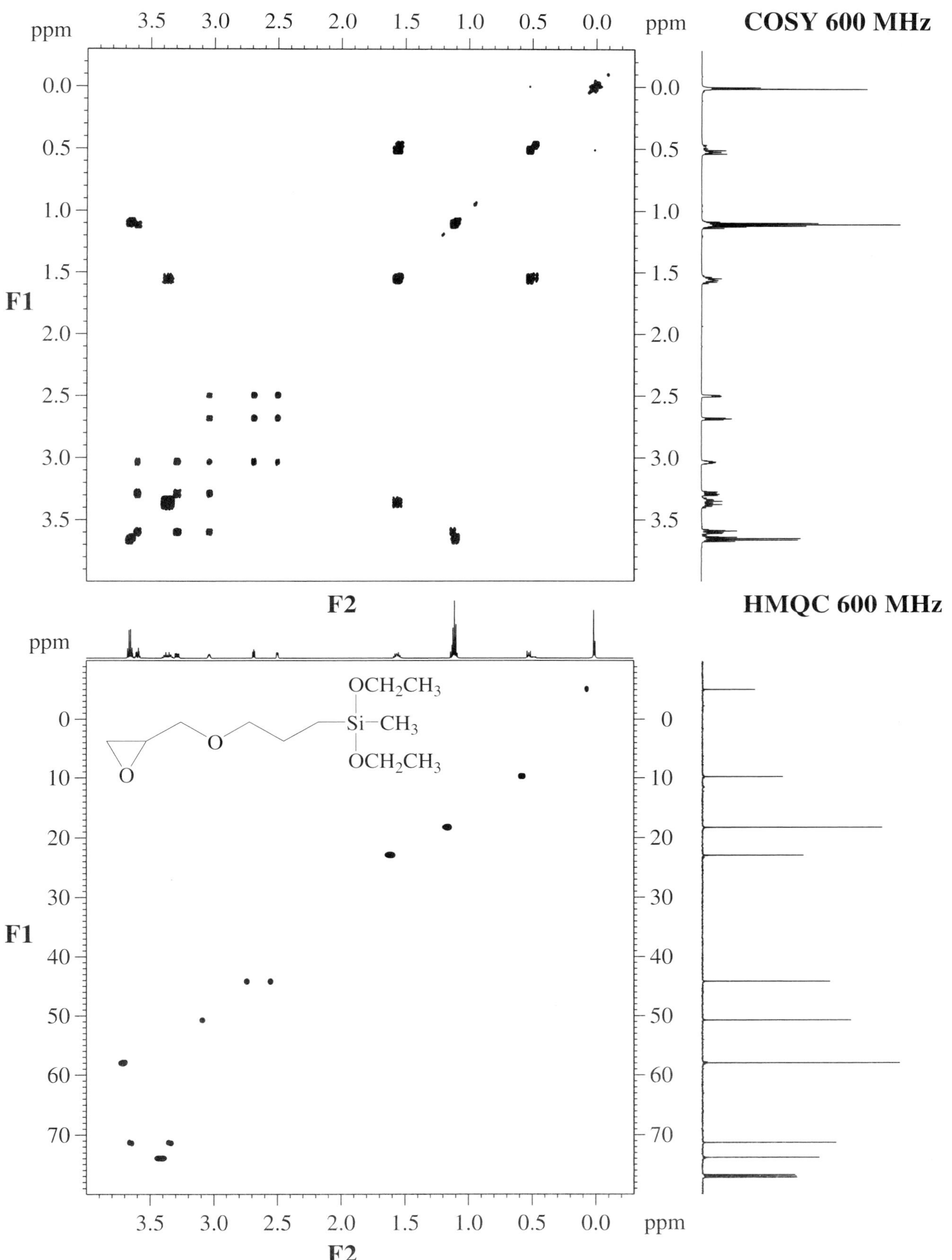

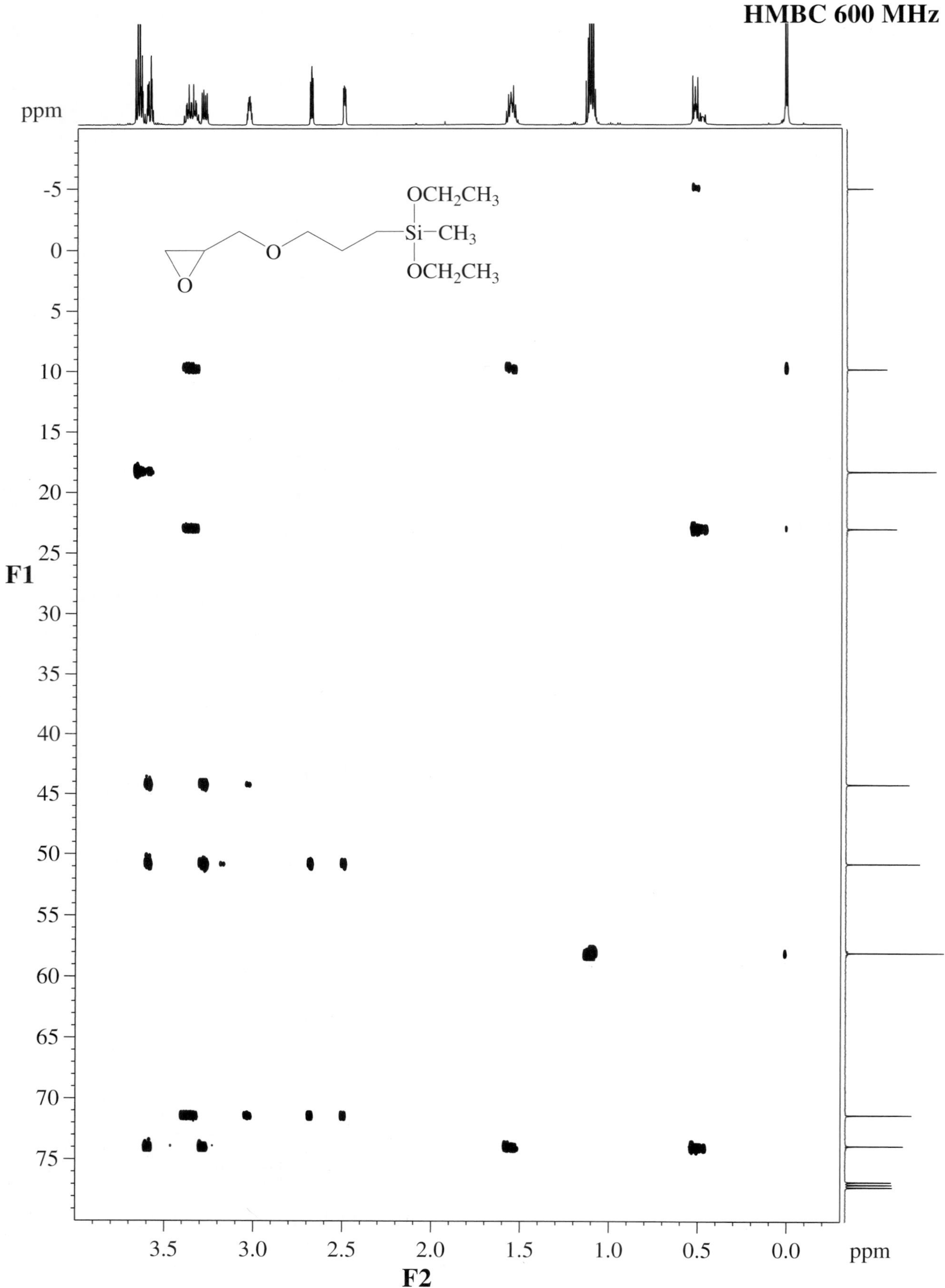
HMBC 600 MHz
ppm
F1
F2
ppm
OCH2CH3
Si−CH3
OCH2CH3
O
O

Problem 8.56A

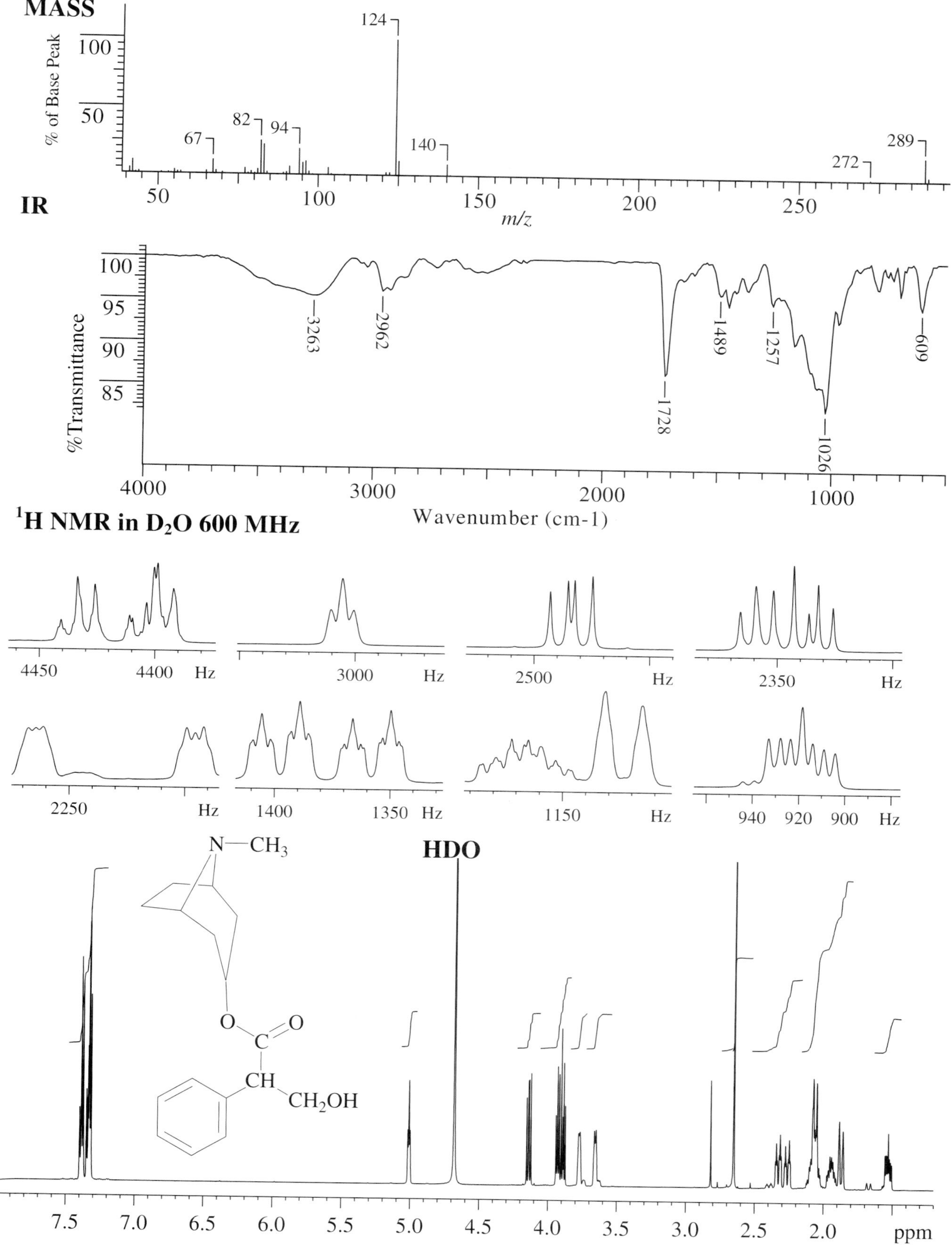

Problem 8.56B

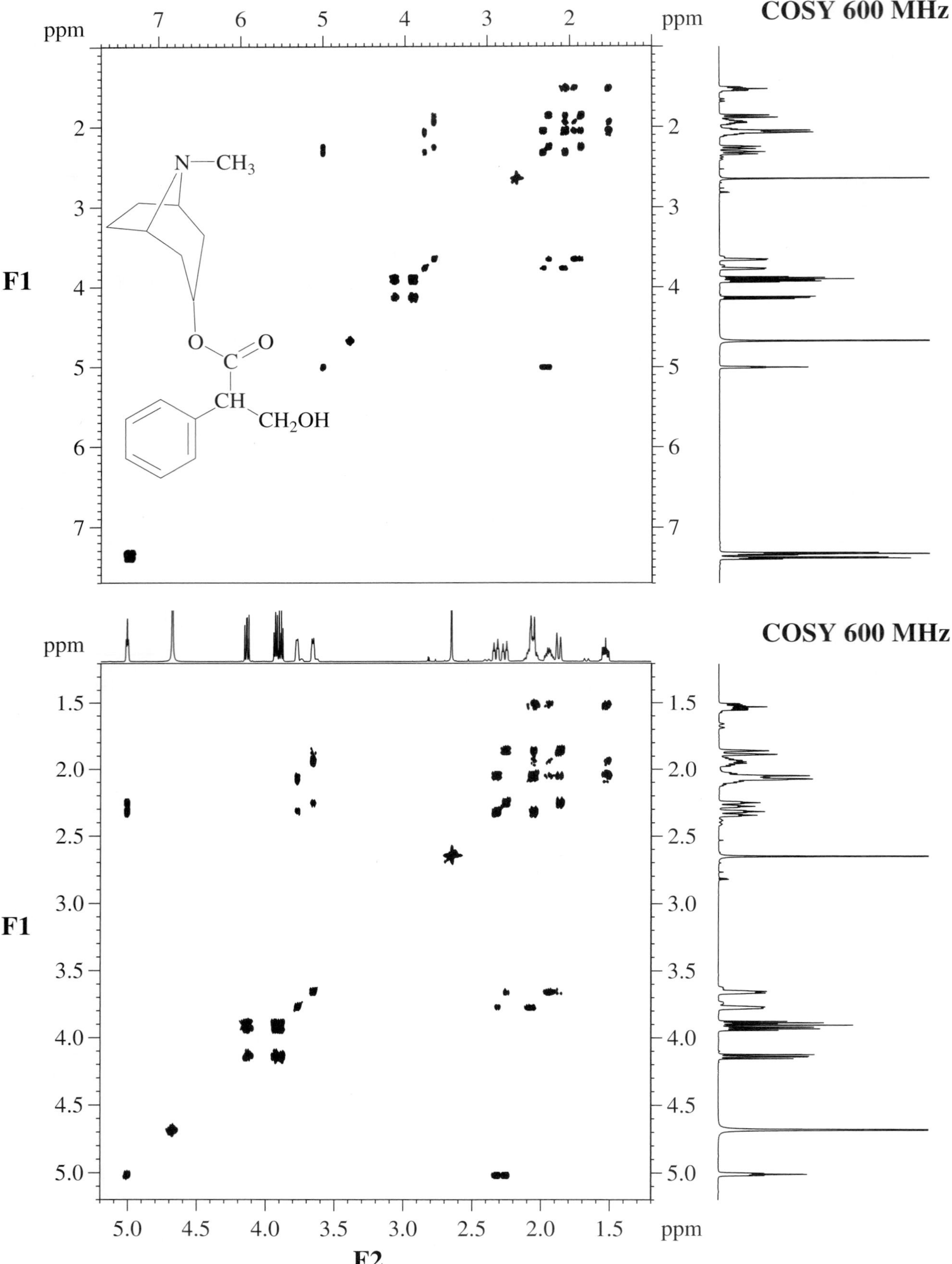

Problem 8.56C

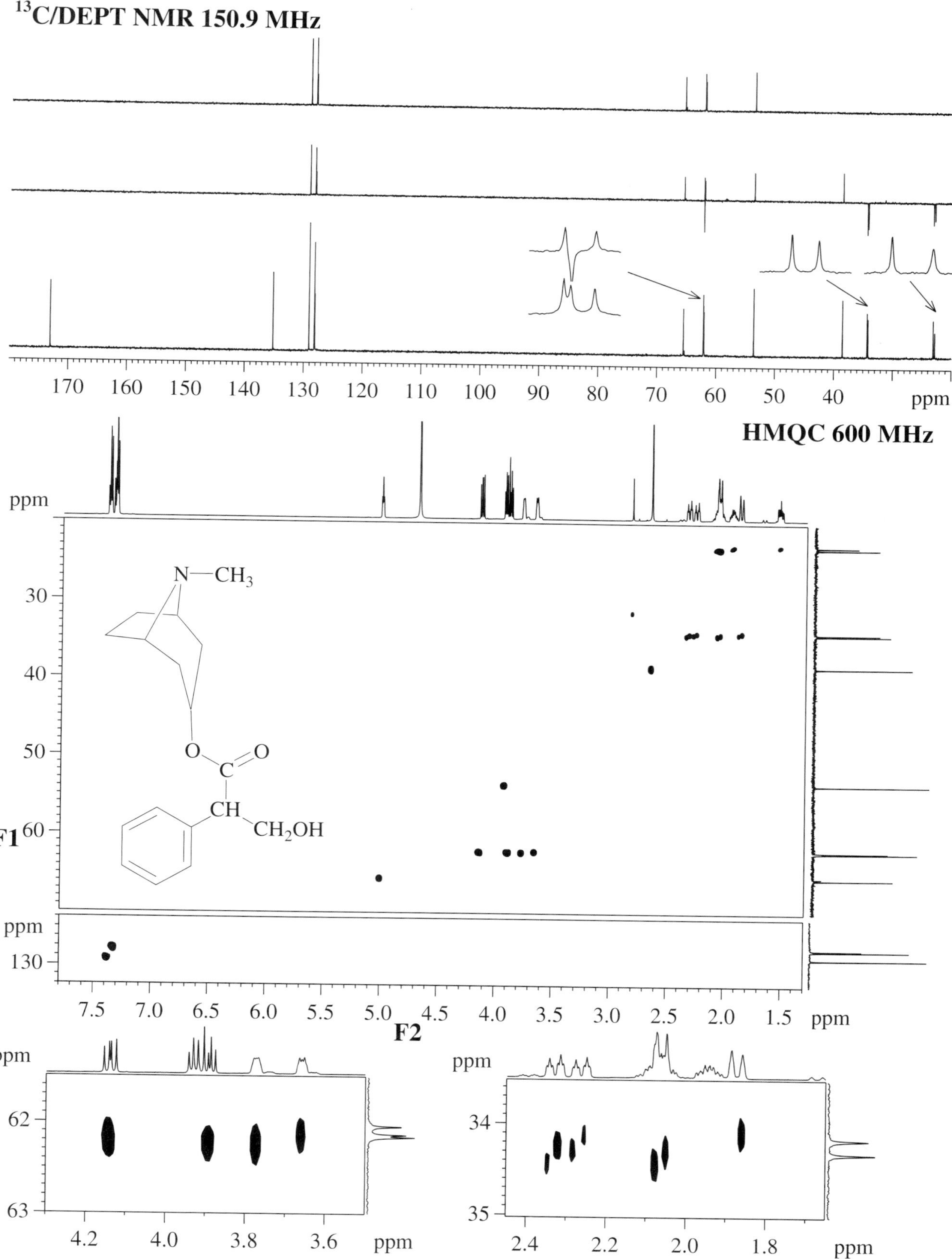

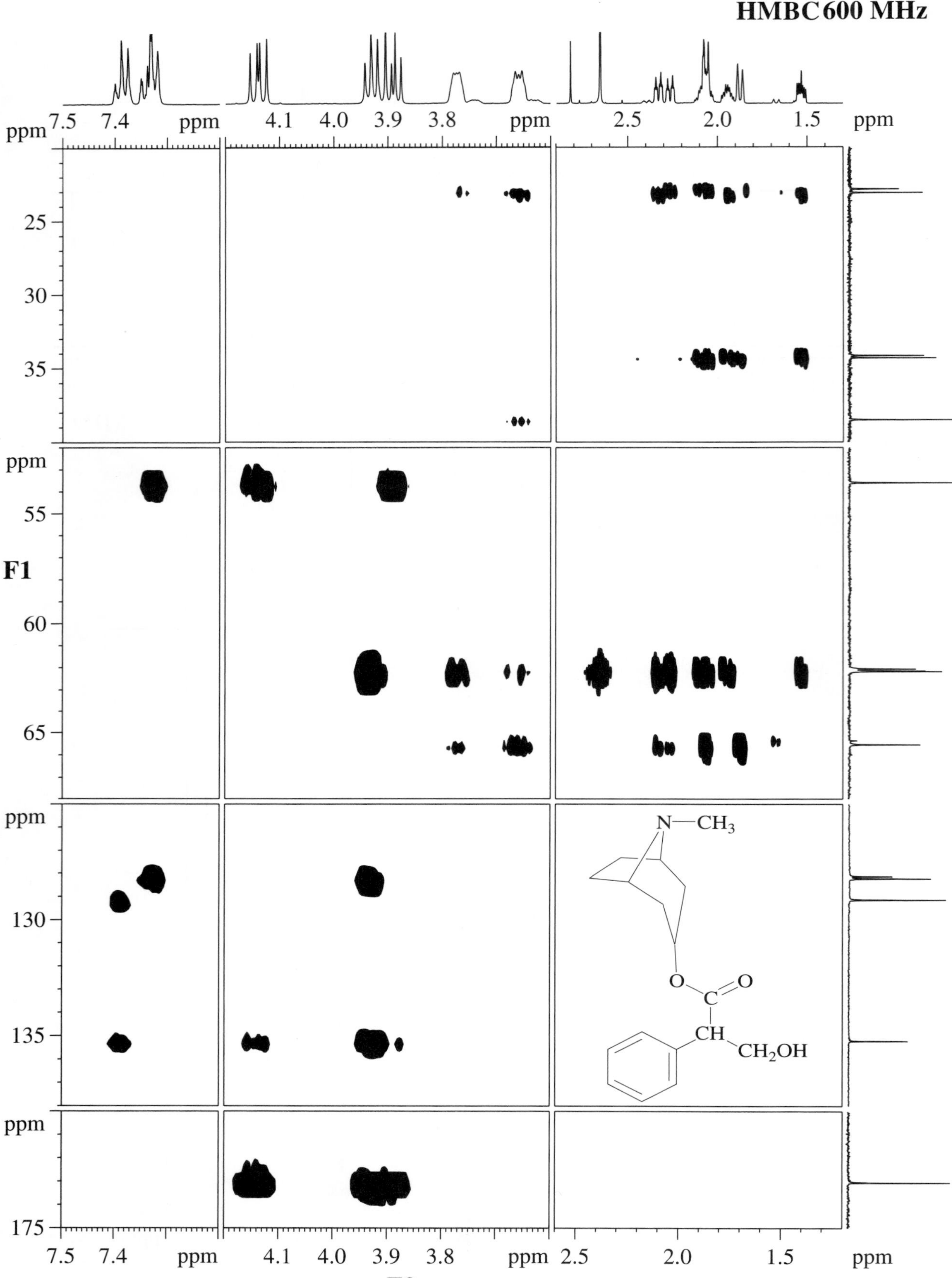
HMBC 600 MHz
ppm
7.5
7.4
ppm
4.1
4.0
3.9
3.8
ppm
2.5
2.0
1.5
ppm
25
30
35
ppm
55
F1
60
65
ppm
130
135
ppm
175
F2
N—CH3
O
C
O
CH
CH2OH

Problem 8.57A

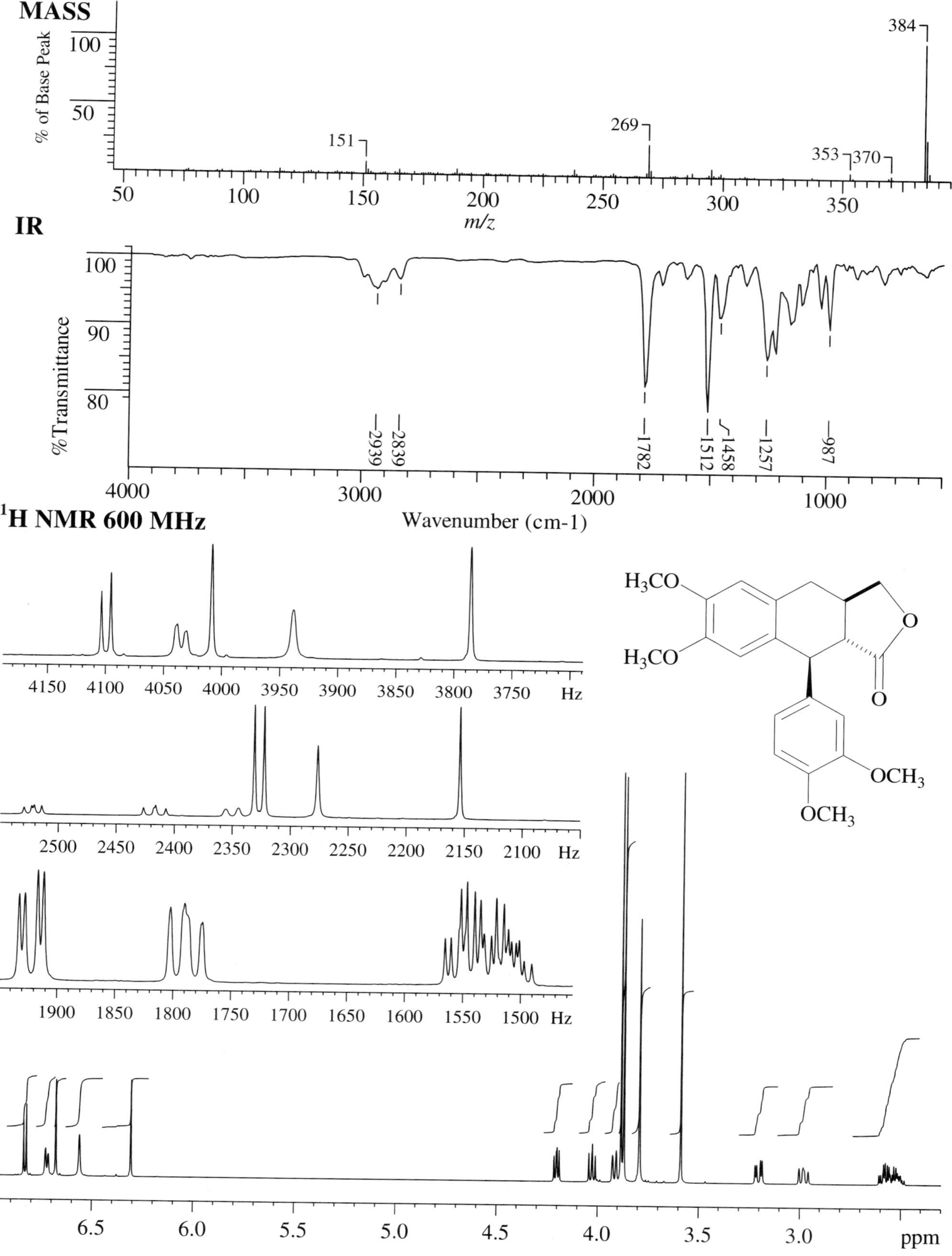

MASS
% of Base Peak
100
50
151
269
353
370
384
50
100
150
200
250
300
350
m/z
IR
%Transmittance
100
90
80
2939
2839
1782
1512
1458
1257
987
4000
3000
2000
1000
Wavenumber (cm-1)
1H NMR 600 MHz
4150 4100 4050 4000 3950 3900 3850 3800 3750 Hz
2500 2450 2400 2350 2300 2250 2200 2150 2100 Hz
1900 1850 1800 1750 1700 1650 1600 1550 1500 Hz
6.5 6.0 5.5 5.0 4.5 4.0 3.5 3.0 ppm
H3CO
H3CO
O
O
OCH3
OCH3

Problem 8.57B

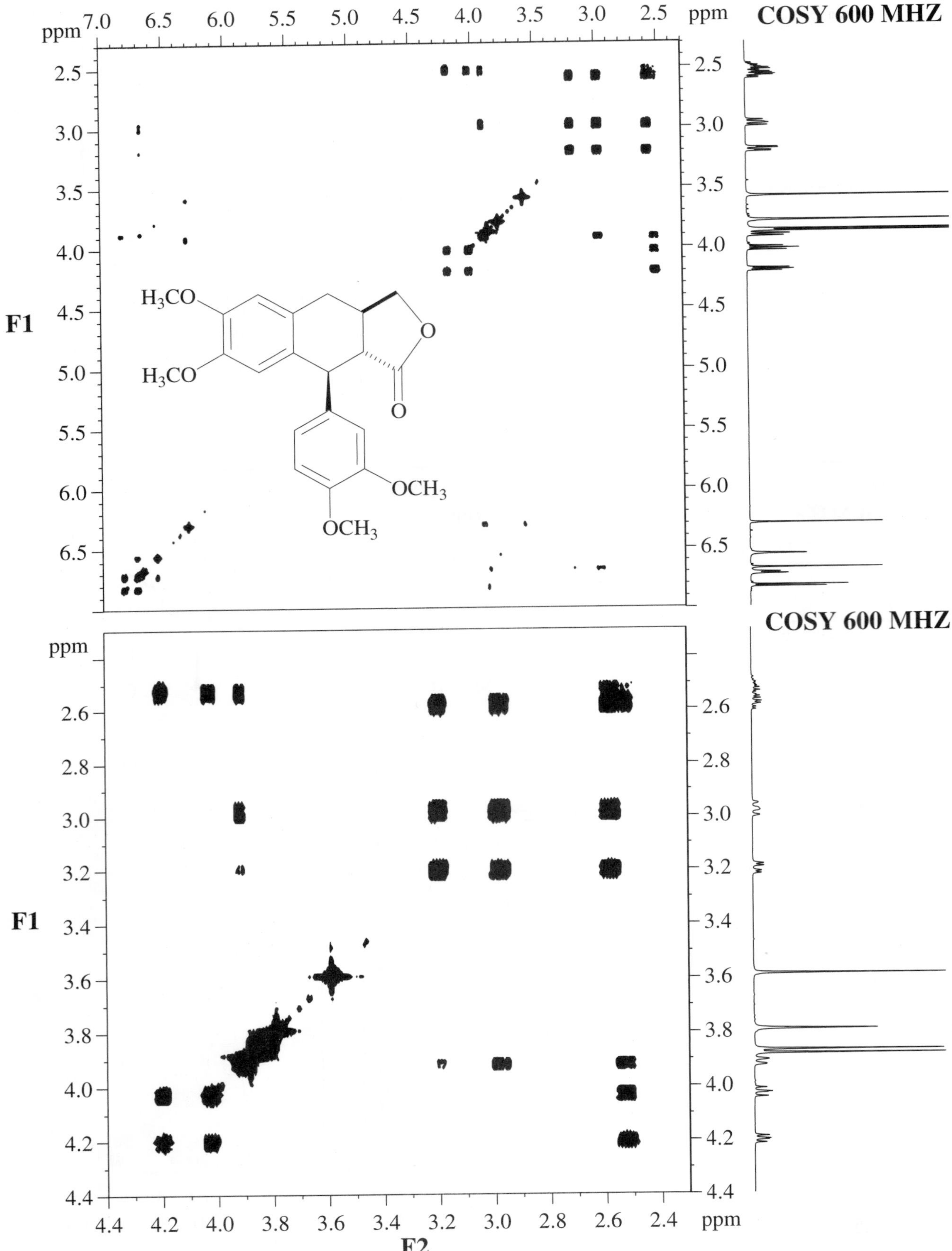

Problem 8.57C

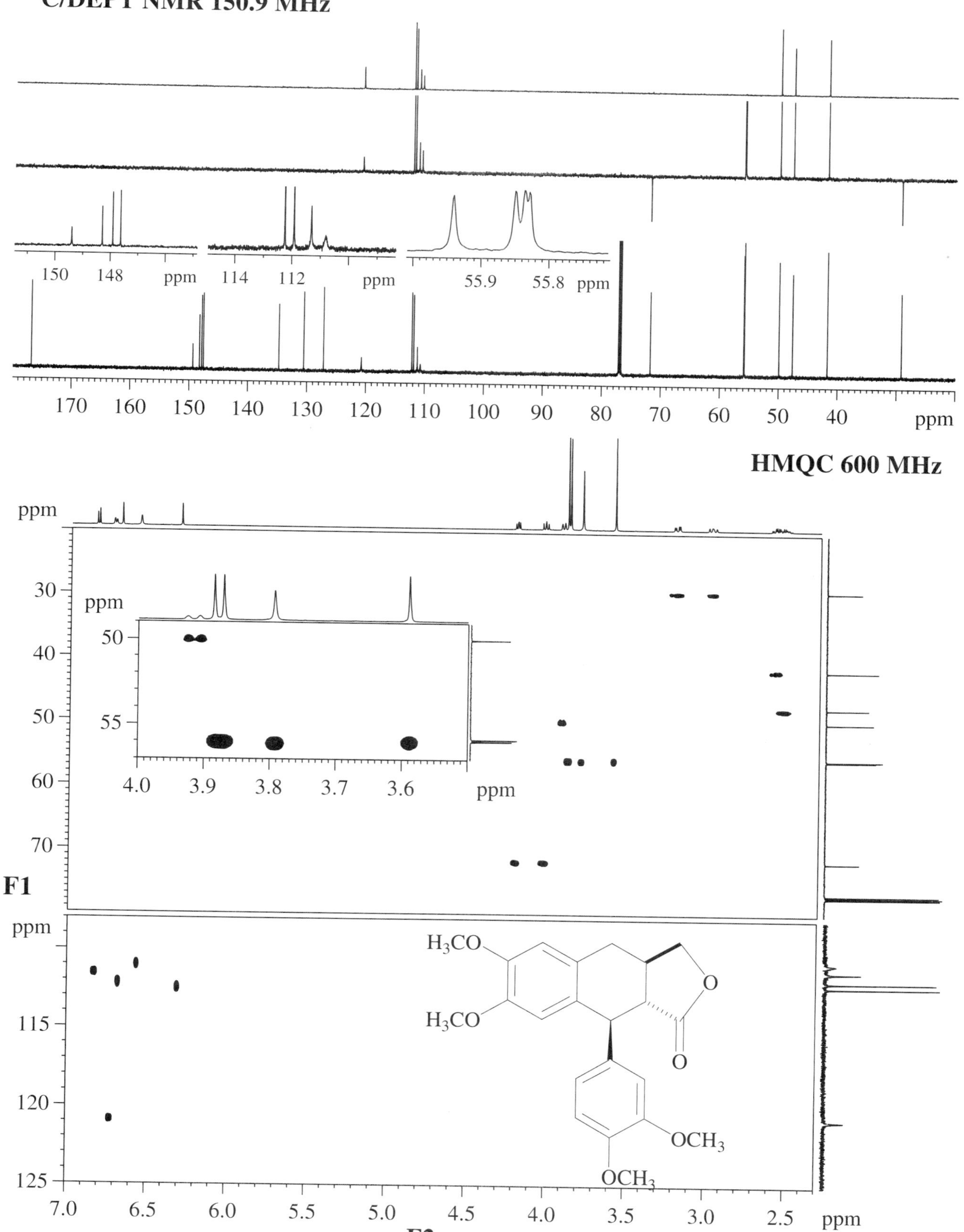

Problem 8.57D

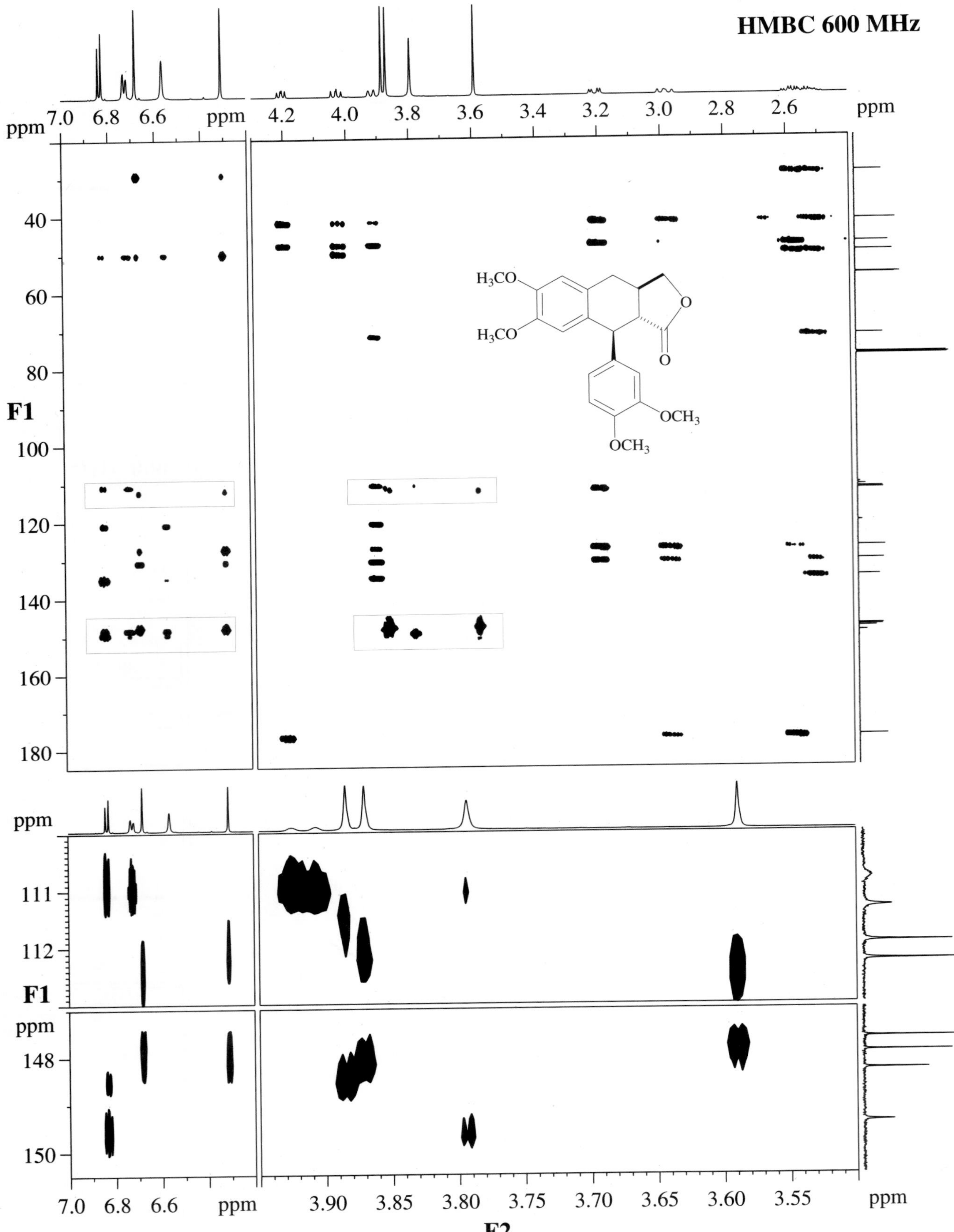

Problem 8.58A

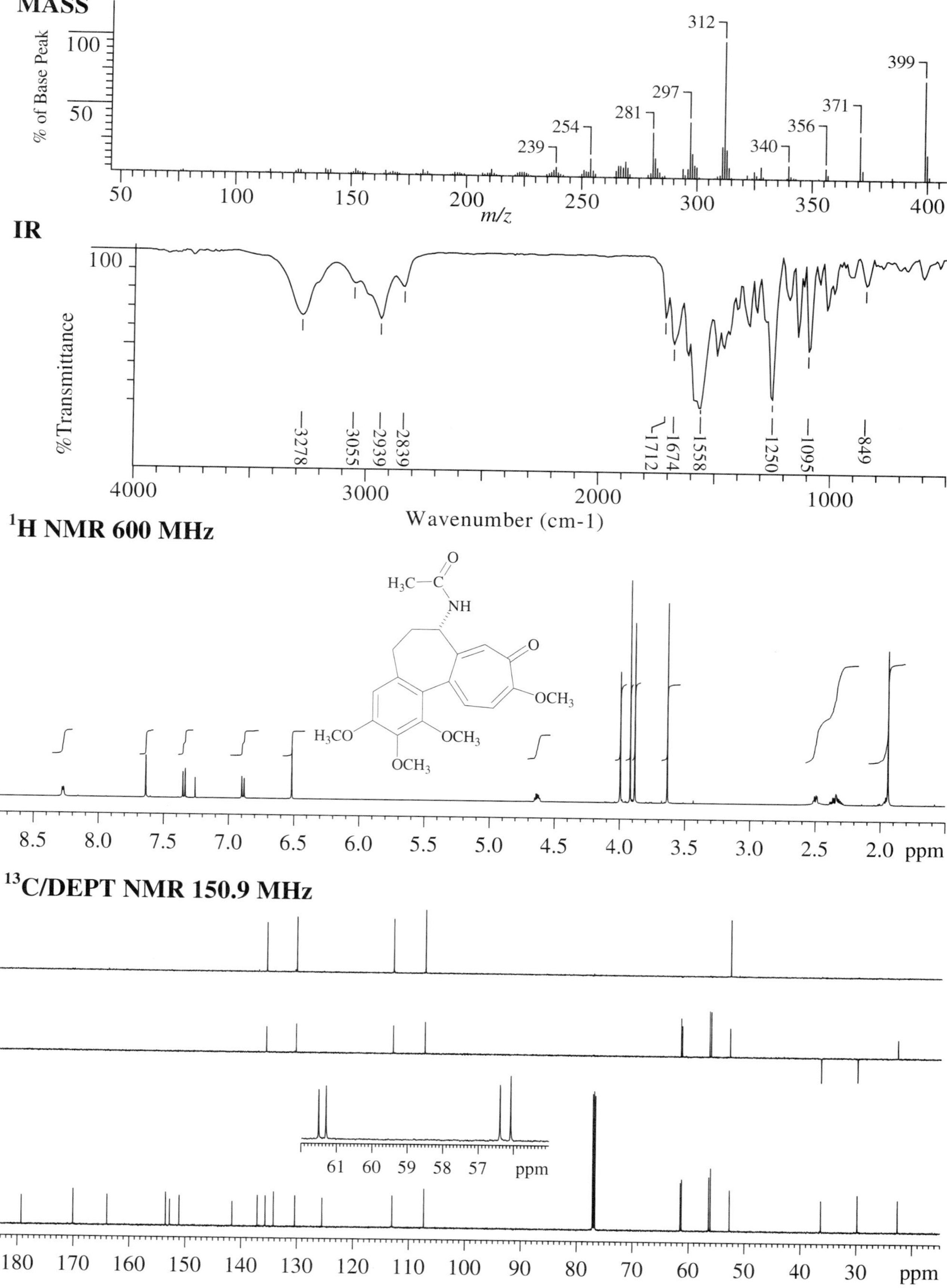

Problem 8.58B

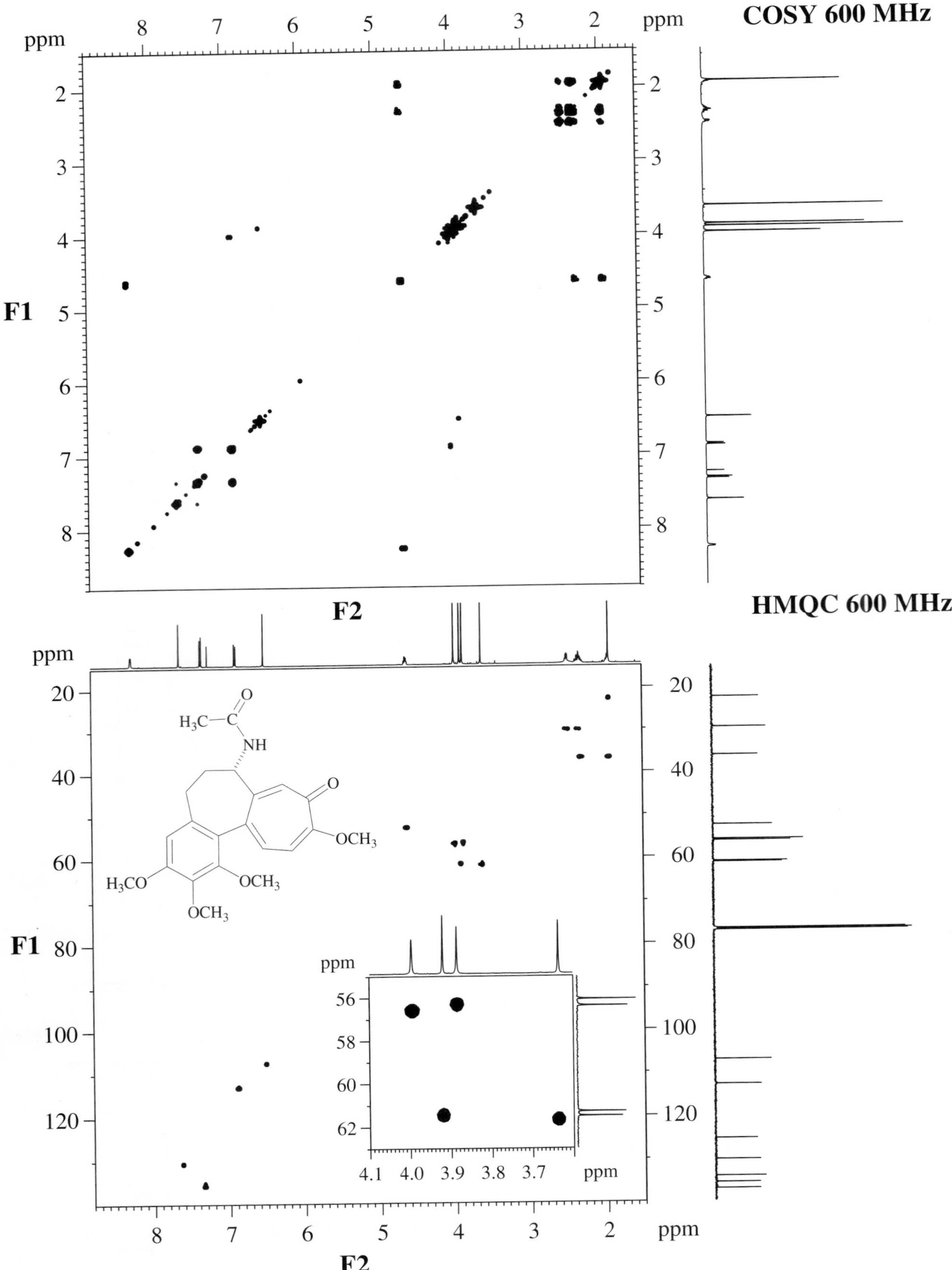

Problem 8.58C

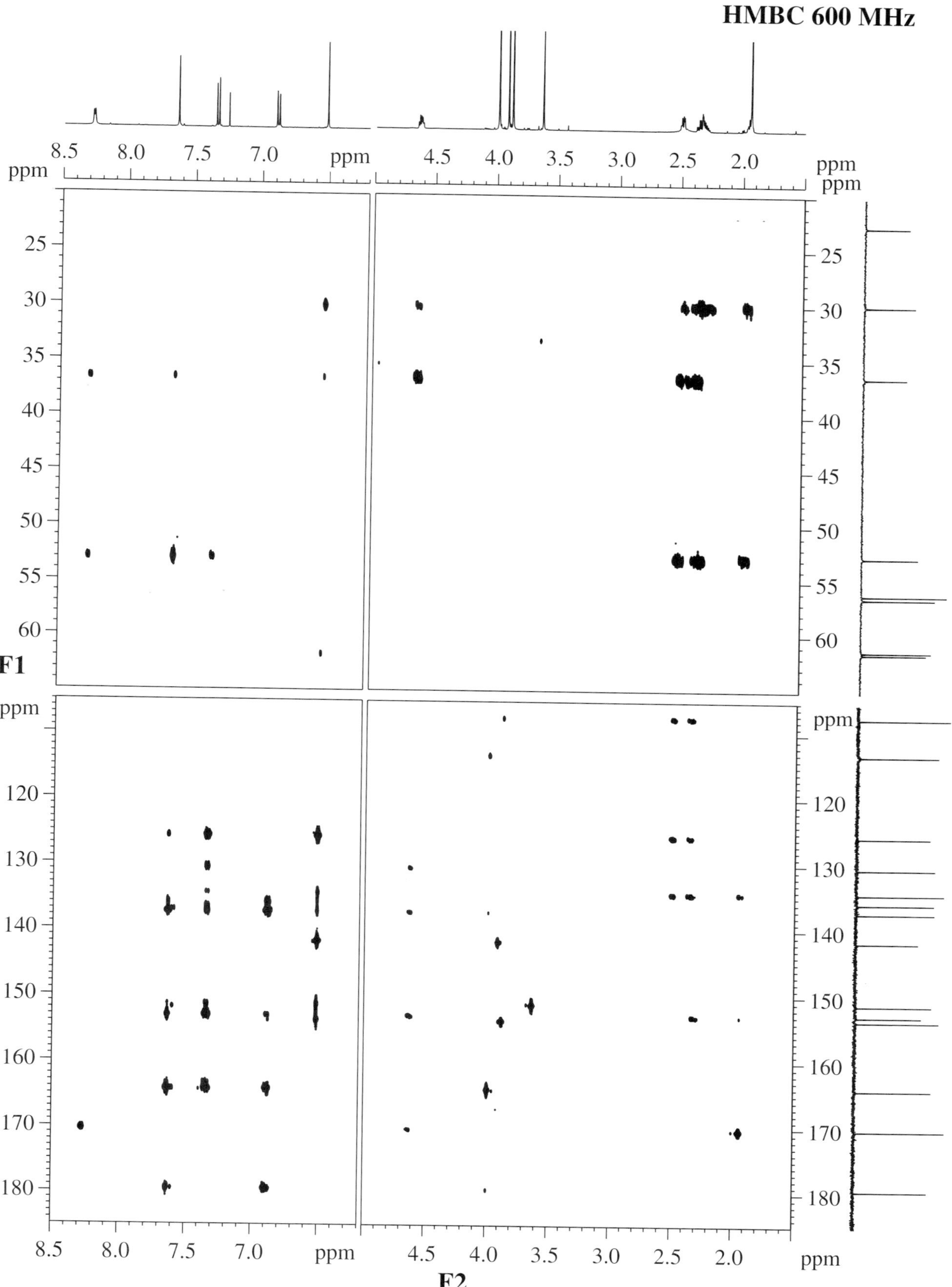

INDEX

Sidewalk Games

by Glen Vecchione
Illustrated by Blanche Sims

Sterling Publishing Co., Inc.
New York

For Briana and Nicholas Vecchione

Library of Congress Cataloging-in-Publication Data Available

2 4 6 8 10 9 7 5 3 1

Published by Sterling Publishing Co., Inc.
387 Park Avenue South, New York, N.Y. 10016
The text of this edition is based on *World's Best Street & Yard Games* by Glen Vecchione ©1989

Distributed in Canada by Sterling Publishing
℅ Canadian Manda Group, One Atlantic Avenue, Suite 105
Toronto, Ontario, Canada M6K 3E7
Distributed in Great Britain and Europe by Chris Lloyd
at Orca Book Services, Stanley House, Fleets Lane, Poole BH15 3AJ, England
Distributed in Australia by Capricorn Link (Australia) Pty. Ltd.
P.O. Box 704, Windsor, NSW 2756 Australia

Manufactured in China

Sterling ISBN 1-4027-0289-2

Contents

Pick up some chalk, a ball, and maybe a few other things you have around the house, and you have all you need to keep you and your friends entertained for all time!

The sidewalk games in this book are great to play for kids of all ages. There are tag games; circle games; games that have been played for centuries and some new surprises; roughhousing, and wacky games that will keep you laughing.

These games used to be played out on the street, when streets were really peaceful and empty most of the time. But streets and sidewalks have gotten busier and are not always easy or safe to play in as they once were.

If you can't find a really quiet sidewalk or street, you can play these games in many other places — such as schoolyards, parks and playgrounds, and backyards, and many of them can even be played on the lawn or in a grassy field. Have fun with them!

Let's Run
1

Tag can be very simple or really complicated with teams, bases, and strategies. But the idea is always the same. A player — or a group of players — is chosen to be "It." "It" chases the other players and tags them.

Sometimes a player is "captured" (put in prison or taken out of the game) or that player becomes the new "It."

The game goes on until everyone is captured — or tired out. That's all there is to it — but there are lots of great tag games. Here are some of them!

Freeze Tag

7-15 players

"It" chases the other players, trying to tag each one. When players are tagged, they "freeze" — stop right away in whatever position they were in at the very moment they were tagged — and they have to wait to be "unfrozen." The only way to be unfrozen is to be tagged by another player who hasn't been tagged yet. "It" wins if everyone gets frozen. "It" loses if some players are still running around and "It" is just too tired to play anymore.

SPECIAL: An exception to the freeze rule is called "Electricity."

When several frozen players are within touching distance of each other, one of them can call out "Electricity." This means that frozen players can move just enough to link hands. Then, if one of these players is unfrozen, everyone in the chain is free as well.

When the game seems totally lost, "Electricity" can save the day!

Statues

4-10 players

This tag game is a lot like Freeze Tag, but with less running. When players get frozen, they cannot be unfrozen by a running player. The action goes on until all the players are frozen. Then the second part of the game begins.

"It" goes to the frozen players one by one, takes them by the arm, spins them around, and then lets them go. Whatever position the players fall into, they have to hold, like some very weird statues. "It" chooses the most unusual statue to be "It" in the next game.

Sticky Tag

5-10 players

When "It" tags a player, that person becomes the new "It," but with a difference. The new "It" sticks her hand over the part of her body that was tagged. If Nancy was tagged on the shoulder, her hand is stuck to her shoulder and she can't use it to tag someone else. When she tags someone else, she is no longer "It" and she can unstick her hand.

Tip: Try tagging someone on the knee or on the foot!

Squat Tag

5-10 players

In 1559, the Flemish artist Pieter Bruegel the Elder painted children playing "Squat Tag" and other games in his famous "Children's Games" canvas.

"It" cannot tag the players if they are squatting. This makes it pretty tough to get people out. So "It" may have to be a little sneakier than usual — creeping up behind a player — or chasing one player and then turning suddenly and springing on another!

Smugglers

10-20 players (an even number)

You need two teams for this game, the "Ins" and the "Outs." The Ins have a Den, and one member of the Outs has a "jewel"(which can be any object, a key, a stone, a coin), anything small enough to hide in the palm of your hand. No one on the Ins team can know which player has the jewel.

The Ins count to 50 while the Outs move farther and farther away. When the count is over, the Ins yell, "Smugglers!" and they rush out to tag the Outs. As the members of the Out team are tagged, they must open their hands to show whether or not they have the jewel. Of course, the jewel is passed around among teammates as quickly and secretly as possible.

When the holder of the jewel is tagged, the game is over and the players change sides.

Buzzzz 10-20 players

Divide the players into two teams and draw a long line between them. Team 1 sends a player into Team 2 territory and she tags as many players as she can. While she is tagging them, she must make the sound "Buzzzz" in one long continuous breath — loud enough for everyone to hear. If she can make it back across the line to her own team without running out of breath, the players she tagged are out of the game. But if Team 2 holds her so long — by grabbing her arms or legs or pulling her back — that she runs out of breath, the players she tagged are free and she's out of the game.

Teams take turns sending a player into enemy territory. The first team to wipe out the other team wins.

Ringelevio 1-2-3

10-30 players (an even number)

Select two teams and draw a Den that is large enough to hold an entire team. One team goes out while the other, the "It" team, stands beside the Den. One member of the "It" team, the Den Guard, keeps one foot inside the Den at all times.

The "It" team counts to 100, while the Outs run off. When the count is over, the "It" team shouts, "Ready or not, here we come!" and everyone on the team except the Den Guard runs after the others.

The "It" team captures players by holding victims long enough to call out "Ringelevio 1-2-3!" three times. If a victim breaks away before the repetition is completed, he's free.

Victims are put in the Den, where they stay until tagged by a teammate, or until the Den Guard accidentally takes one leg out of the Den or puts feet both in. Players may try to pull the Guard inside or push him out. When all the Outs are captured, the game is over.

Prisoner's Base

10-30 (an even number)

A version of "Prisoner's Base" is described in Act IV of Shakespeare's *The Merry Wives of Windsor.* The game was so popular in 17th century England that it once interrupted the king's procession to Parliament!

Choose up two teams, A and B, each with its own home base. Mark out a prison for both teams. The members of each team link hands to form a chain that stretches out from their home base. The one farthest from the base of the A team breaks away and runs off. The one farthest from the base of the B team chases him. At the same time, the other A players break off from their chains and the opposite member of the B team chases him.

When a player is tagged, he goes to prison and his captor stands guard. The prisoner is released when a member of his team runs through the prison and tags him. The guard watches for this and may tag any would-be rescuer.

The game is finished when all the members of the first team are captured.

Let's Jump!

Jump Rope Games

1-10 players

Jumping rope is terrific exercise and a great test of skill and coordination. You can jump rope alone or with a partner, using a short store-bought jump rope or some clothesline you cut to about 5 feet (1.5m) long. Or you can jump inside a longer clothesline, turned by two friends.

There are two kinds of jump rope games

The first kind is fancy jumping:

Rock the Cradle. Rock the rope back and forth instead of revolving it. You can do this alone on a short rope, or on a longer rope, letting the turners rock it for you. It's a great way to warm up.

Wind the Clock. While the rope is turning, count from 1 to 12, making a quarter turn clockwise each time. You can do this alone or with two friends turning for you.

Visiting. One player jumps alone turning her own rope. Another player jumps in and faces her, "visiting" for a while, before jumping out again.

Chasing. This involves two turners and at least two jumpers. The first jumper enters, jumps over the rope once, and then rushes out as the second jumper enters, and so on.

Hopping. Two players turn as the jumper rushes in and hops, alternating legs for each turn of the rope. After 10 hops, the jumper runs out and another one jumps in for 10 hops.

Jump Rope Rhymes

Another type of rope-jumping is made up of rhymes, recited by either the turners or the jumpers. There are many of these, but most of them fall into one of the following groups:

Counting Rhymes

These rhymes end with counting to test how long a jumper can keep going. You can chant them alone or have the turners chant them for you. One popular counting rhyme goes like this:

Fire, fire, house on fire —
Mrs. Sweeny climbed up higher.
There she met the Fireman Steve —
How many kisses did she receive?
One, two, three, four, five, six. . . .

Alphabet Rhymes

The jumper or turners chant the alphabet at the end of these rhymes. The letter the jumper stumbles on means something very important — usually it's the first initial of her sweetheart's name, whether she knows it or not!

Strawberry shortcake, cream on top
Tell me the name of my sweetheart
A, B, C, D, E . . .

Switching Rhymes

These short rhymes call for the old jumper to move out and a new one to come in. They are usually chanted by the turners.

My mother and your mother
Live across the way,
Every night they have a fight
And this is what they say:
Acka baka soda cracker
Acka baka boo
Acka baka soda cracker
Out goes you!

Double Dutch

In addition to basic jump rope games, there are fancy ways to turn the rope. If you have a very long clothesline, for example, you can double it over. One turner holds both ends, while the other turner wraps the rope around her back, over her forearms, and through her hands. Now you're ready for "Double Dutch."

Here the two ropes are turned toward each other — but carefully — so that they don't hit each other. The result is a kind of eggbeater that leaves the poor jumper wiped out.

French Dutch

Here the ropes are turned away from each other — again, very carefully — with similar results. If that isn't tough enough for you, try both "Double Dutch" and "French Dutch," with a dose of "Hot Pepper." This means turning the ropes as fast as possible, until someone cries "Help!"

Chinese Jump Rope

4-8 players

A Chinese jump rope is a loop made from braided rubber bands.

Players 1 and 2 face each other with their feet apart and the rope around their ankles. They back away from each other, allowing the rope to stretch and lift off the sidewalk.

Player 3 jumps between them, placing her feet apart. At the same time, Player 1 jumps out, leaving the rope stretched between Players 3 and 2. Soon Player 4 jumps in, taking the place of Player 2.

Now Players 1 and 2 jump back into the rope as Players 3 and 4 jump out. Then Players 3 and 4 switch places with Players 1 and 2 again. Timing is very important. Jumping in and out at exactly the right moment keeps the rope stretched. One mistake and you have a snapped rope or a tripped player!

Let's Play Ball

There are lots of outdoor ball games. Some of them call for two or more players and others may be played alone. Some use pavement alone as a playing surface, while a few require a high wall. If you don't have enough people for a full-scale ball game, or even if you do, these games are great substitutes — fast and fun.

Apartments

6 players

Five players stand against the wall, about five feet (1.5m) apart, separated by chalk lines drawn up the wall. The sixth player, who stands about 30 feet (9m) away, throws the ball at any one of them. Players may twist and duck out of the way, but they may not leave their "apartments," If a player is hit, it's one count against him. Three counts and a player is out of the game. If a player catches the ball, he changes places with the thrower, who has to take on all the counts against him as well. The last remaining player wins.

Square Ball

5 players

Divide the space in front of a wall into five sections, each one about four feet (1.2m) square. If the pavement is divided into squares, let each player take a square.

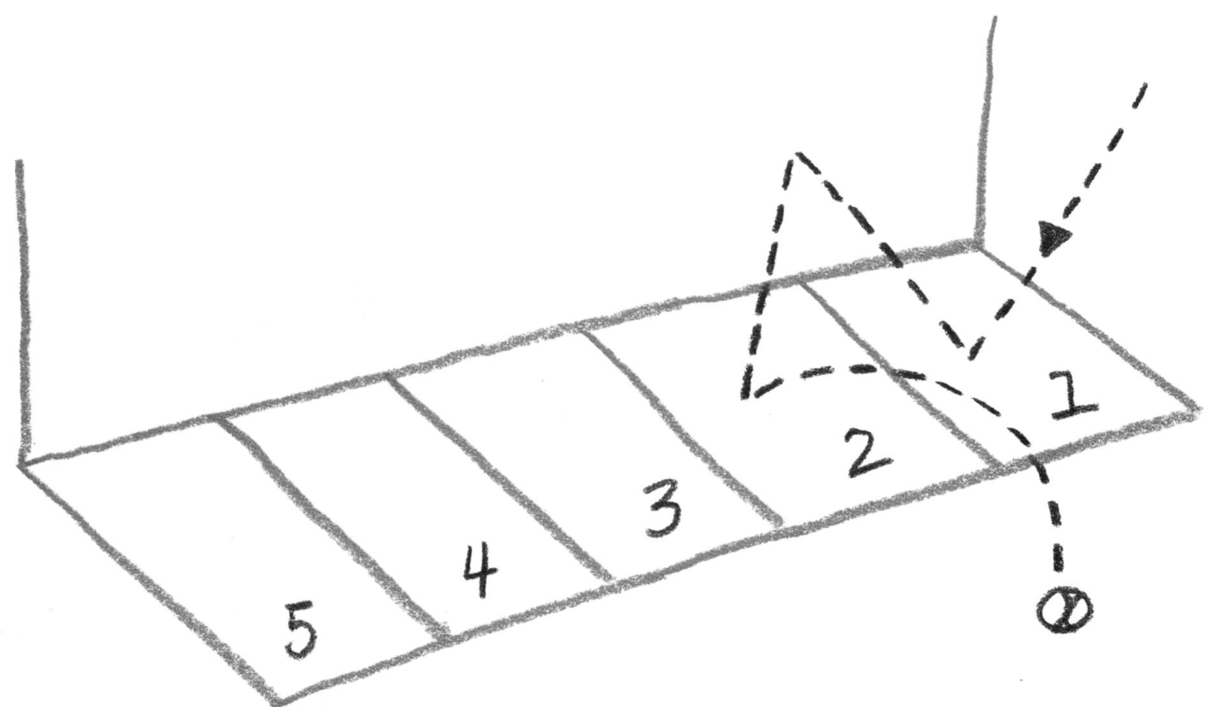

Players stand back about five feet (1.5m) from their squares. The player in Square 1 slams the ball in his own square, so that it bounces up, ricochets against the wall, and bounces in another player's square.

That player must catch the ball and toss it back the same way into someone else's square. When the game gets going there's plenty of running around.

Players who miss a catch or throw the ball without bouncing it first are out, and the remaining players move one square to the right. The last remaining player is the winner.

Kings

5 players

A variation on "Square Ball," "Kings" is played for points. Player 1 stands behind the first square and throws the ball into the square of Player 2, hitting the sidewalk first and then the wall. Player 2 catches the ball, then throws it into the square of Player 3, who throws it into the square of Player 4, and so on. When Player 5 gets the ball, she returns it to Player 4, and the ball continues back and forth.

Any player who fumbles a throw or a catch receives one count against her. Eleven counts take her out of the game and the remaining players move one square to the right. The player who eliminates all the others is the "King."

Monday, Tuesday

7 players

You need a high wall for this game. Each player takes the name of a day of the week. The first player (Sunday) throws the ball against the ground as hard as possible so that it bounces up and rebounds against the wall. At the same time, Sunday shouts out the day-name of another player — Wednesday, let's say — who must catch the ball after the first bounce and send it back in the same way, calling out the day-name of another player.

If Wednesday misses the catch, everyone scatters while Wednesday retrieves the ball and then tags one of the others. The tagged player is the next one to throw the ball against the ground, but the tag counts as a mark against him. Three tags and a player is out. The game continues until only one player is left — the winner.

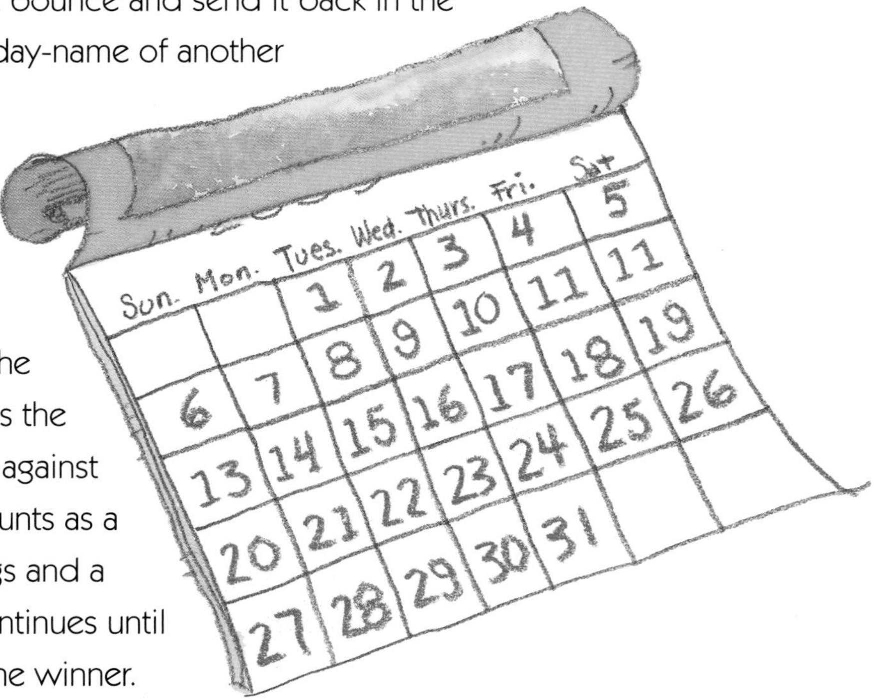

Seven Up

1 player

Draw a line about five feet (1.5m) from the base of a wall and stand behind it. Throw the ball against the wall and start with "Onesies," which means you must catch it on the fly.

Twosies: let it bounce once in front of the line before you catch it. Repeat.
Threesies: clap before you catch it. Repeat this two more times.
Foursies: catch it after the first bounce. Repeat three times.
Fivesies: clap twice behind you before catching. Repeat four times.
Sixies: do a push-up, jump up and catch the ball after the first bounce. Repeat twice.
Sevensies: clap your hands before and behind you, then catch the ball. Repeat six times.

Each catch counts as one point. If you go all the way from Onesies to Sevensies without a mistake, you've collected 25 points and win, but each miss gives the wall one point. Continue playing until you (or the wall) reaches 25.

Four Square

4 players

You'll need a soccer ball or a basketball for this game, which is lots of fun, especially when the pace heats up and the ball is flying!

With a piece of chalk, draw an area about six feet (1.8m) square and divide it into four equal compartments. A player occupies each square. One player then bounces the ball into a neighboring square. The player in that square catches the ball and bounces it into another square, and so on. You could decide ahead of time to follow a particular sequence — for example, bouncing the ball only to the player on the right, or bouncing it to the right and then diagonally across — or you can decide to let each player choose a target. If any player fails to catch a bounce in his square, the one who tossed it to him scores a point. The first player to reach a score of 21 wins.

Box Baseball

2 players

You play this game across three sidewalk squares like this:

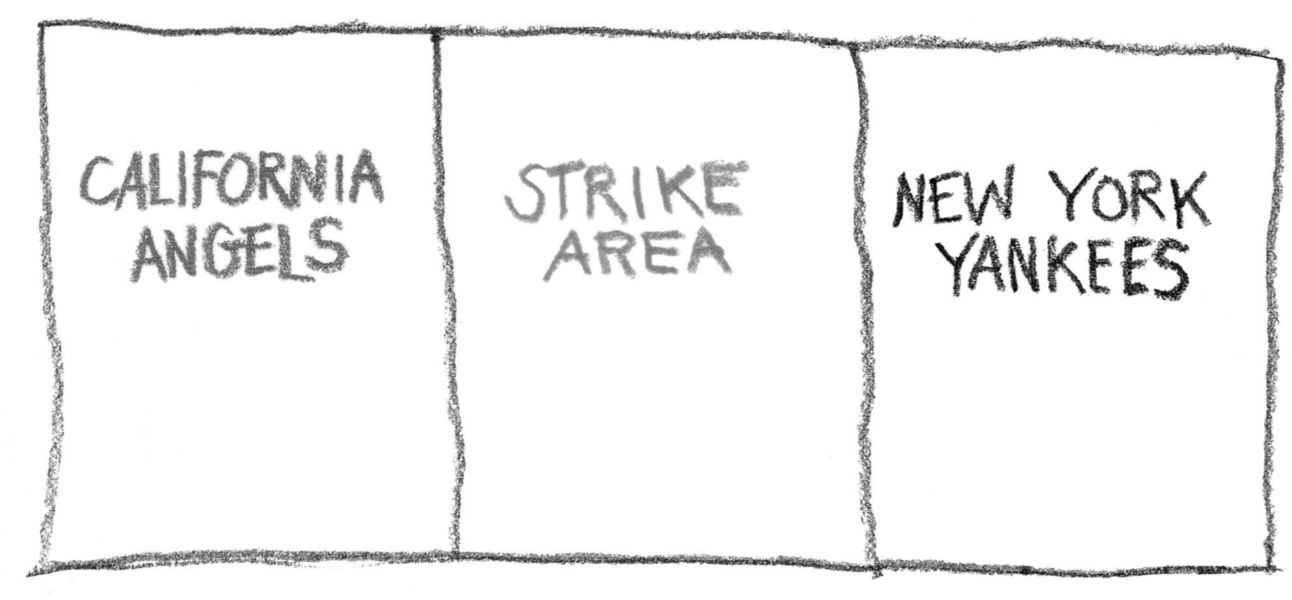

Player 1 is the California Angels and Player 2 the New York Yankees. The Angel throws the ball into the Yankee's box, passing it over the strike area. The Yankee, standing outside her box, tries to catch the ball after one bounce. If she succeeds, it counts as an out for the Angel. But if she doesn't catch the ball after one bounce, each additional bounce means one more base for the Angel. (Two bounces mean a single; three bounces a double; four bounces a triple; and five bounces a home run.) If the Angel's throw bounces in the strike area, or misses the Yankee's box, he must throw again. Three strikes make an out, and it's the Yankee's turn to throw. After nine innings, the team with the higher score wins.

Hit the Coin

2 players

Lay a coin flat on the crack between squares in the sidewalk. Or, with a piece of chalk, draw two boxes, each five feet (1.5m) square, and separate them with a straight line. One player stands in each box. The object is to bounce the ball on the coin, which scores one point. If your ball flips the coin over, you get two points. (Put the coin back on the dividing line in case it is knocked away.) If you miss the coin completely, continue to take turns bouncing the ball until someone scores a hit. The first player to reach a score of 21 wins.

Shoebox Bowling

2-5 players

You need an old shoebox, scissors, crayon, a stone, and seven marbles for this game. Turn the shoebox (without the cover) so that the open top faces the ground and cut seven triangular holes into the side. Each hole should be just wide enough for a marble to pass through. Number the holes from one to seven, but in the order shown below:

Place a stone on top of the box to weigh it down. Now stand back about five feet (1.5m) and take turns bowling marbles into the shoebox. If your marble passes through one of the holes, you score the number of points written above the hole. If you miss the shoebox, or if your marble doesn't go into a hole, you are penalized two points and lose your next turn.

The first player to reach a score of 49 wins the game. If no player reaches a winning score after all players have bowled, collect the marbles from inside the shoebox, pass them out, and play again.

Let's Compete

In these games, players either compete in teams or it's "every player for himself." The object of competition can be anything – a tin can, coin, scarf, sidewalk space, or even another player!

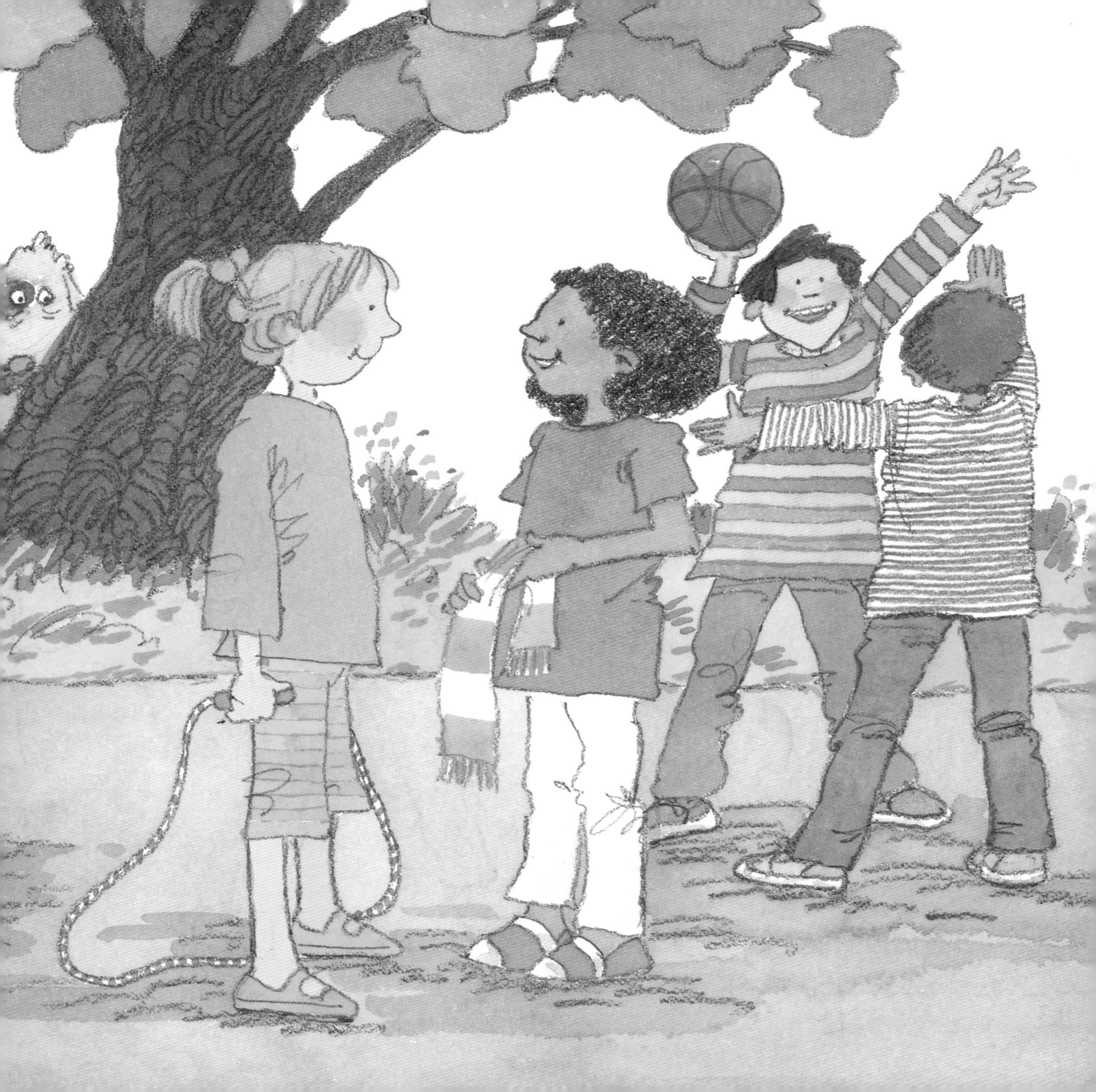

Kick the Can

4-10 players

Draw a circle about 6 feet (2m) in diameter and place an empty tin can in the center. "It"guards the can, while the others stand outside of the circle. One player at a time may rush in and try to kick the can out of the circle. If he succeeds, all players run, except for "It," who must fetch the can, carry it back to the circle, and yell "Freeze!"

"It" then takes a prisoner by calling a player's name. The prisoner enters the circle while "It" exits to touch another frozen player and take him prisoner. While "It" is out, any frozen player can kick the can, and free all prisoners. Then "It" has to fetch the can again, which gives the players time to run again.

If there are no prisoners, a player may make a dash for the circle and — if she gets inside before being tagged by "It" — shout "Home Free!" Then everyone runs into the circle. Last one in is "It" in the next game.

Sardines 5-10 players

This game is Hide and Seek backwards, with a good chase at the end. Pick a home base and one player to be "It," but in this game "It" hides, while the *others* count to 50. When the count is over, the players start searching for "It." But when one of them finds "It," instead of tagging or calling out to the others, he or she joins "It" in the hiding place. This goes on, with the players all crowding into the hiding place, like sardines. When the last player finds the hiding place, everyone jumps out and races for home base. The last player to reach home is "It" in the next game.

Rooster Romp

6-20 (an even number)

In this Mexican favorite, players tuck a handkerchief or scarf under a belt or in a pocket — in a place where it can't be snatched away easily. Players pair off, facing their partners. Then each player grasps his left shoulder with his right hand, hops on his right foot and tries to steal his partner's handkerchief. Partners must always face each other and may never run away. Pushing and bumping is allowed, but if a player drops his arm, or if his foot touches the ground, he is out of the game. If a handkerchief is stolen, its owner out of the game. Winners wait for everyone to finish and then pair off for a final challenge.

Four Corner Upset

5 players

With a piece of chalk, draw four corners, at least 25 feet apart. Then draw a small base on each corner.

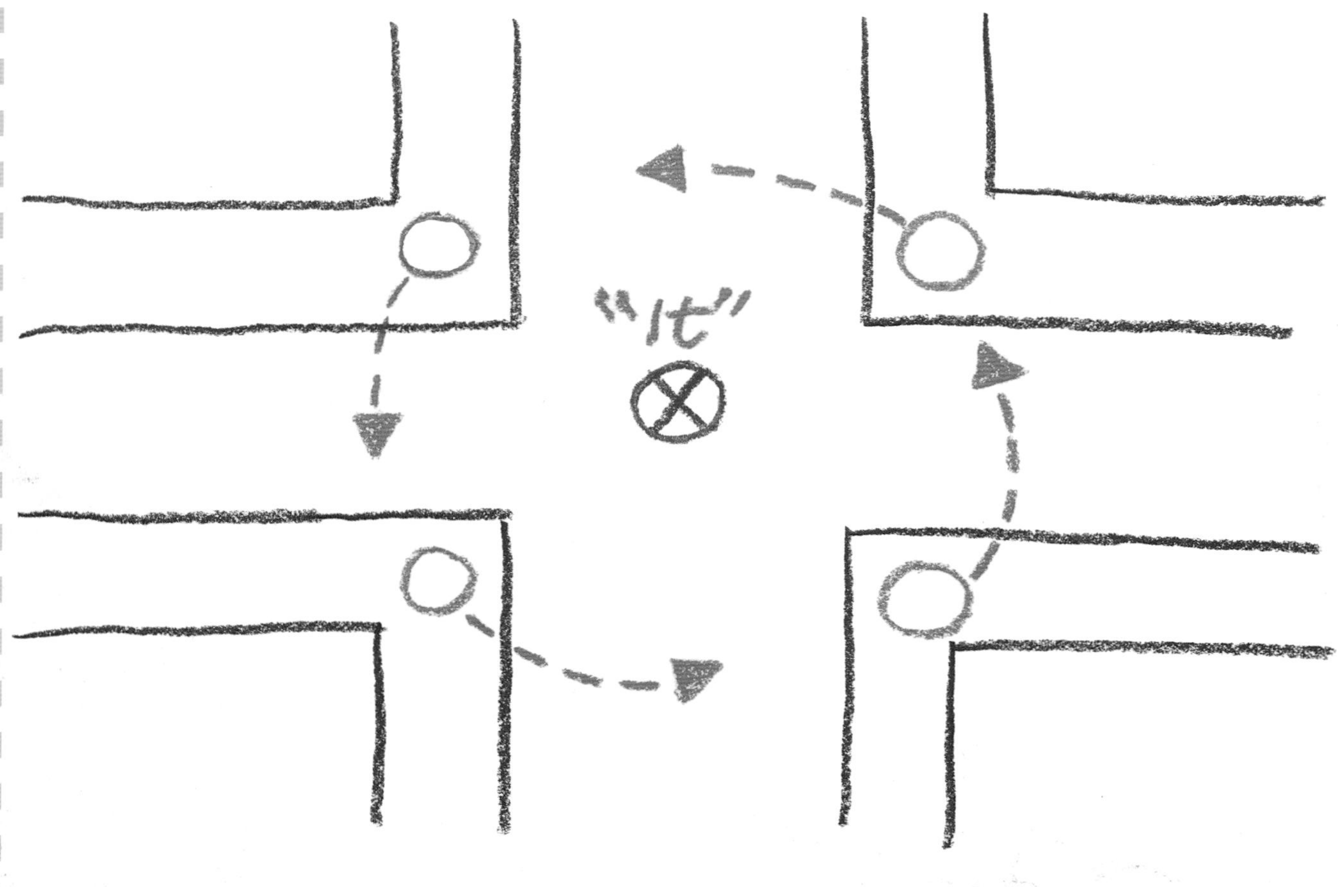

You need five players for this game. Four of them stand on the corner bases while "It" stands in the center. "It" calls out, "I want a corner. Give me *yours!*" and then points to one of the other players. If that player chooses to, he may change places with "It." Usually, though, no one wants to give up a corner, and "It" is forced to say, "I'm upset!" At that moment, each player switches corners with the player to the right. "It" tries to rush in and claim a corner for herself.

It's fun if "It" tries to fool the others into running. She can do this by calling out, "I'm up — side down!" or "I'm up — a tree!" If a player is jittery or not listening carefully, he may run away from his base and find he has no place to go!

Red Light, Green Light

5-10 players

One player — "It" — moves about 15 feet (4.5m) away from the others, who stand in a row behind a starting line.

"It" turns her back to the others and calls "Green Light," at which point the players run toward her. After a few seconds, she calls. "Red Light," which tells the running players they must freeze in position. "It" then whirls around to face the players and tries to catch someone moving. A player who gets caught must go back to the starting line.

Each time "It" turns away and calls, "Green Light," the players run closer. As they close in, "It" usually makes the green lights shorter and the red lights longer, hoping to catch the closest player and send him back to the starting line.

The player who manages to tag "It" becomes "It" in the next game.

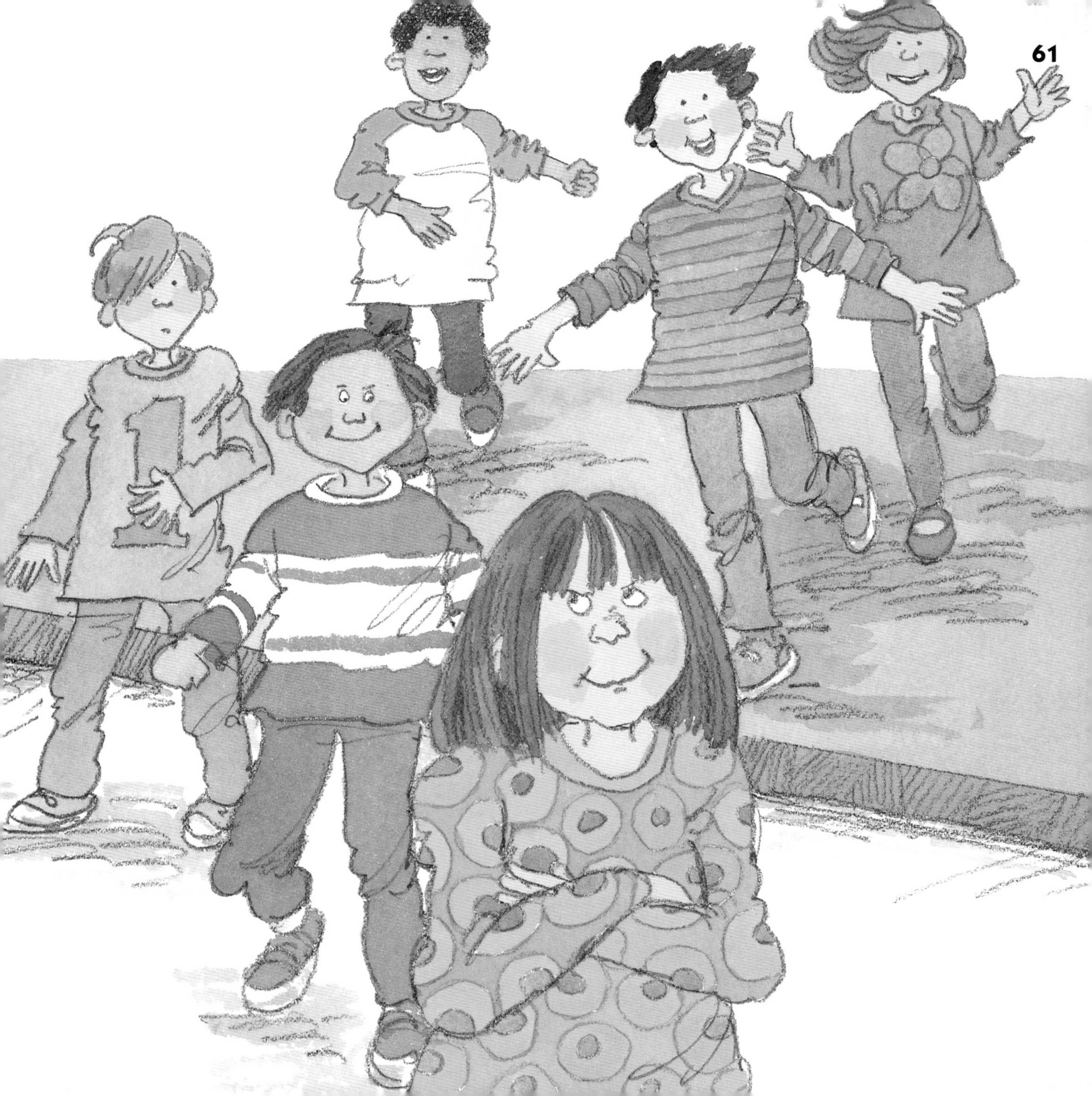

Four Coin Toss

> A popular London street game during Charles Dickens' time, "Four Coin Toss" is still great for passing an idle afternoon.

3-7 players

With a piece of chalk, draw a circle about six inches (15cm) in diameter. Then draw a starting line about five feet (1.5m) away from it. The players must stand behind this line. Each player has four coins and takes a turn pitching one coin toward the circle. The player coming closest to the center of the circle takes all the other thrown coins and is allowed to take one step closer to the circle. From this new position, he tosses all his coins towards the circle at once, keeping the ones that clearly fall inside. The coins that fall outside the circle are gathered up by the next player, who also takes a step forward before tossing all his coins. Any coins that fall outside the circle are gathered up by the third player, who also takes a step forward, and so on.

Each time a player takes a turn, he steps closer to the circle. Soon all the players surround the circle and no more steps can be taken. But the coin tossing continues until one player wins all the coins.

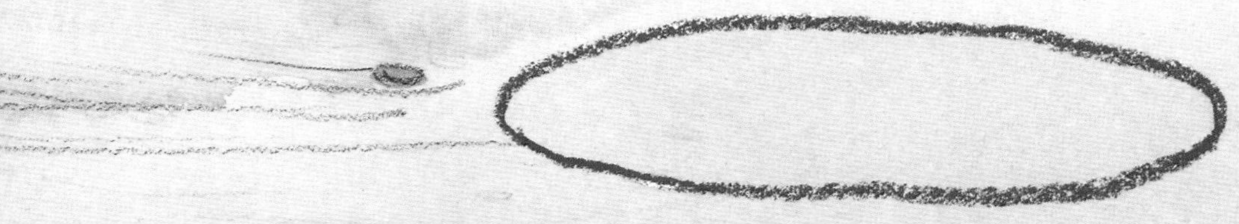

Cat and Mouse

10-20 players

One player is Mouse. The other players form a circle linking hands and holding them up high enough for Cat and Mouse to pass underneath.

Mouse walks around the outside of the circle. Then suddenly it tags one of the others and rushes away. The player tagged becomes Cat and must chase Mouse as it weaves in and out of the circle, underneath the arms of the other players. Cat must follow Mouse's moves exactly and cannot take shortcuts across the circle.

If Mouse is caught, it joins the others in the circle and Cat becomes the new Mouse.

Duck, Duck, Goose

10-20 players

This animal-name game was played on American farms about a hundred years ago. Like many other circle games, it was played in fields where the tall grass could be flattened into a playing area.

The players sit in a circle facing in. "It" walks around the outside of the circle, stopping here and there to tap a sitting player and say, "Duck." This often goes on for a while, as "It" waits for his friends to relax so he can catch them off guard.

Suddenly "It" taps a player, yells "Goose!" and rushes away. The tagged player must leap up and race around the circle in the opposite direction to get back to his place. That is — unless "It" gets there first!

The player left without a space in the circle becomes "It" in the next round.

The Wolf and the Sheep

10-20 players

Choose one player to be the "Wolf," and another to be the "Sheep." The rest of the players join hands and form a circle around Sheep, protecting it.

Wolf creeps around the outside of the circle and tries to break through while the others do their best to keep him out. Wolf might try crawling under the legs of the defending players, or he may run as fast as he can and throw himself against the line. The other players might bunch up around Sheep or fan out and run around to keep the strongest players circulating. Anything goes, as long as their hands remain linked.

When Wolf breaks through the circle, he grabs Sheep's hand and tries to break out with her. Again, the other players try to prevent this. They may close in tightly around Wolf to separate him from his catch or lift their arms to let Wolf out, and then whip them down before he can pull Sheep through.

If Wolf manages to pull Sheep through the circle, they both join the circle and choose two new players to be Wolf and Sheep.

King of the Ring

4-7 players

Called "Push-Pin" in Shakespeare's time and mentioned in Act IV of *Love's Labours Lost,* "King of the Ring" is still a challenging game of speed and strategy.

Draw a circle about five feet (1.5m) in diameter and stand in the center. As "King," you have to protect yourself against invaders who enter the circle and try to drag you out. Only one invader at a time may enter, and anything goes — pushing, shoving, tripping, lifting up and carrying — whatever! After you've tried to hang on to your kingdom for a while, you'll soon be exhausted enough to *want* out!

If two invaders enter, you may call "Foul!" and take one out of the game. And if three try to gang up on you, you may call "Double foul!" and remove two of them.

Continue playing until everyone has a chance to be King.

Limbo

6-10 players

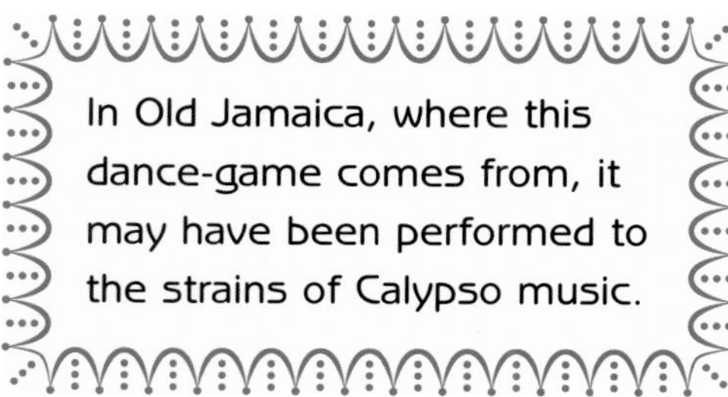

In Old Jamaica, where this dance-game comes from, it may have been performed to the strains of Calypso music.

Choose two players who will hold opposite ends of a broom or broom handle. Each end should rest on a player's upturned palm, so that the broom will fall to the ground if bumped from below. Start with the broom about chest high.

One by one, the other players walk under the broom, making themselves shorter by stretching their legs apart and bending backwards. To get in the spirit, try doing it while the others clap in rhythm. A player who stumbles or knocks the broom down while going under it, is out.

After the players go under once, the broom holders lower the broom to waist level and the players who are still in the game take turns going under it again. With each repetition the broom gets a little lower until players have to go under on their knees or practically on their backs. Eventually, all players but one are eliminated. That one is the winner.

E

Snake Eats Its Tail

10-30 players

Everyone joins hands, making one long line. Or, if you prefer, you can hold the waist or shoulders of the person in front of you. The idea is that the head of this "Snake" (the first person in line) tries to tag the last player in line, eliminating him. The players in between squirm around, trying to keep head and tail apart.

The game continues until the head of the Snake "swallows" the last morsel of the tail, or until everyone is too dizzy to go on playing.

Coin Pitching

3-6 players

Player 1 throws a coin against the curb or the lower part of a wall. Player 2 pitches one in the same way. If it lands within a handspan (tip of the thumb to the tip of the small finger) of Player 1's coin, Player 2 may claim it for his own. If Player 2's coin is further away than a handspan, Player 3 follows, trying to throw his coin so that he may claim one or both of the coins that are out there. There's no limit to the number of coins you can collect if your coin falls a handspan away.

Continue the game until one player has all the "loot."

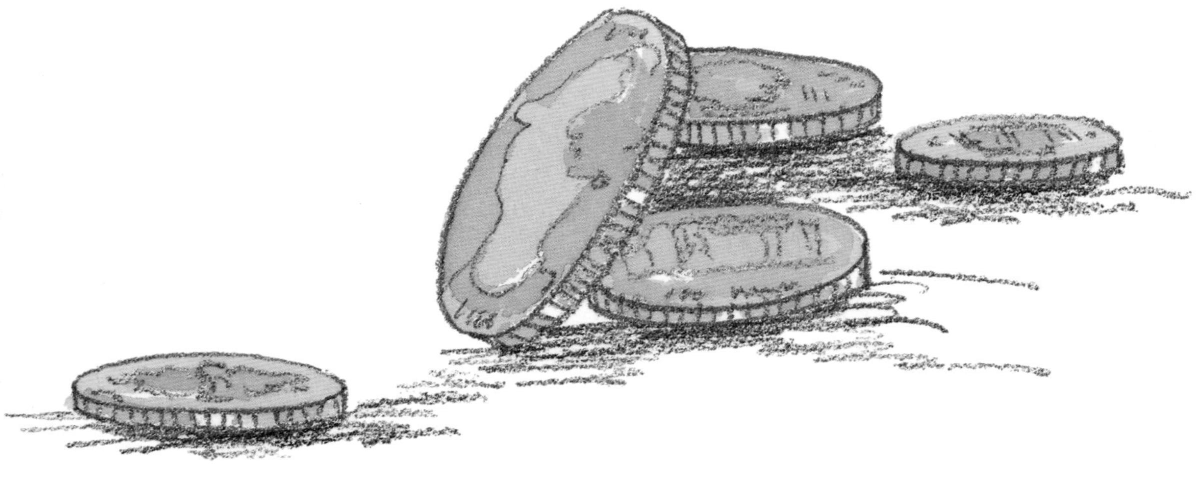

Pigs to Market

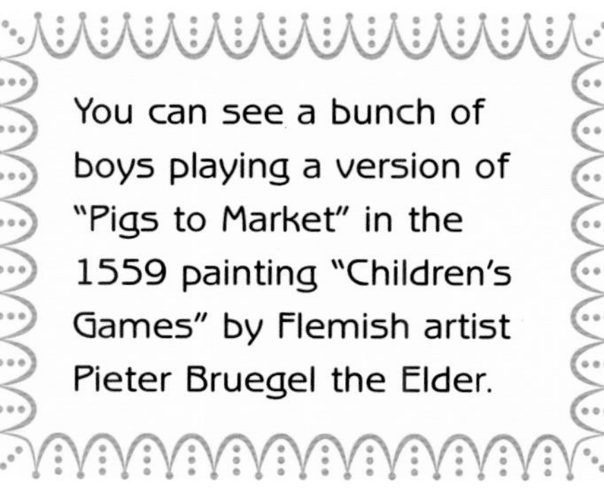
You can see a bunch of boys playing a version of "Pigs to Market" in the 1559 painting "Children's Games" by Flemish artist Pieter Bruegel the Elder.

2-10 players

This racing game is tougher than it sounds, especially when you have lots of players zigzagging in a mad dash for the finish line. Each player will need a long stick or broom handle and a plastic soda bottle filled with water – the "Pig."

Players stand beside each other holding their sticks. A pig is placed on its side in front of each player. At the starting signal, the players must push their pigs along quickly with the broomsticks, trying to race in a straight line and keep out of each other's way — easier said than done!

The first player to reach the finish line wins the game.

Crab Race

3-10 players

Players line up behind the starting rope, which is stretched along the ground. But this is no ordinary race, because each player is a "crab." To get in crab position, lie on your back and lift off the ground with your arms and legs tucked beneath you. It's peculiar — you have to look across your chest to see where you're going.

Choose a player to be the referee. He gives the "Go" signal and stands at the finish line (another rope) as the crabs race by — a hysterical sight!

Whirligig

5-10 players

A "whirligig," in this game anyway, is a long piece of rope (at least six feet or 1.8m) with an old shoe tied to the end for weight. One player holds the other end of the rope and spins around so that the rope makes a sweeping circular motion. You'll see that even though the center player is standing, the weighted end of the rope swings close to the ground. The other players jump over the rope as it sweeps past them, and they are eliminated if they stumble. The center player may spin faster, bringing the rope higher. Everyone has to keep up!

Take turns spinning the rope.

Snap the Whip

7-20 players

> Versions of this game go back hundreds of years. You can see a group of boys having fun with it in Pieter Bruegel's 1559 painting "Children's Games."

Everyone lines up, holding hands. The first player in line runs as fast as he can, dragging the others behind him. He tries to "snap the whip" by making lots of sharp turns. Any player who breaks the line is eliminated. The longer the line of players, the wilder the ride — especially for the players at the end!

Play until everyone is tired!

Index